C0-AYL-734

EXPERIMENTS, DETECTORS, AND EXPERIMENTAL AREAS FOR THE SUPERCOLLIDER

1987

Proceedings of the Workshop on

EXPERIMENTS, DETECTORS, AND EXPERIMENTAL AREAS FOR THE SUPERCOLLIDER

July 7-17, 1987

Berkeley, California

Editors

Rene Donaldson and M.G.D. Gilchriese

Sponsored jointly by the Division of Particles and Fields of the American Physical Society, National Science Foundation, SSC Central Design Group, United States Department of Energy, and Universities Research Association

Published by

World Scientific Publishing Co. Pte. Ltd.
P.O. Box 128, Farrer Road, Singapore 9128

U. S. A. office: World Scientific Publishing Co., Inc.
687 Hartwell Street, Teaneck NJ 07666, USA

Library of Congress Cataloging-in-Publication data is available.

EXPERIMENTS, DETECTORS AND EXPERIMENTAL AREAS FOR THE SUPERCOLLIDER

ISBN 9971-50-473-1

Printed by Kim Hup Lee Printing Co. Pte Ltd — Singapore

Preface

The Workshop on Experiments, Detectors, and Experimental Areas for the Supercollider was held July 7–17, 1987, at the Clark Kerr Campus of the University of California at Berkeley. Building upon the many prior studies of physics and detectors for the Supercollider, the goals of this Workshop were to examine in detail the potential experimental program at the Supercollider. The intent was to begin to address seriously the complex issue of the composition of the initial experimental program.

In order to accomplish these goals, the activities of the Workshop were divided into two broad categories. The first of these was the evaluation of interesting physics signatures and the resultant requirements for detector parameters. The second category was aimed at providing conceptual designs of first generation experiments to allow comparison among the competing designs. The results of the Workshop are contained in an impressive array of summary reports and individual contributions in these proceedings.

The success of the Workshop was primarily the result of the energy and enthusiasm provided by the many working group, subgroup, and even sub-subgroup leaders. The contents of these proceedings amply demonstrate their skill and perseverance.

As always the success of a Workshop depends on the dedication of the administrative and clerical staff. The Workshop was held with the assistance of Louise Millard and the Conference Coordination Office at LBL. Without the capable help of Chris Meyer from that department who handled registration and arrangements with Clark Kerr and Steve Merryman who became an expert at copying transparencies, the Workshop would have been in dire trouble. The SSC Central Design Group staff, in particular Nancy Talcott and Liz DiFraia, helped not only with registration but also worked much overtime during that ten-day period. Anne Cozean and Donna Matthews assisted at registration and in numerous other ways; Aletha Lundblad and Liza Goldwasser volunteered on very short notice to act as Berkeley and San Francisco guides, respectively. Dave Saucer's transportation crew moved furniture, computers, and copying equipment to Clark Kerr and back to LBL at the conclusion of the Workshop. Finally, thanks go to Ray Whitaker and his staff at the Clark Kerr Campus for the hospitality extended during the Workshop and the many hours of preplanning necessary to pull off such an event.

Financial support for the Workshop was provided by the National Science Foundation, the U.S. Department of Energy, Universities Research Association, the SSC Central Design Group, and Digital Equipment Corporation. Digital and the Physics Division of Lawrence Berkeley Laboratory provided the computing resources for the Workshop.

Rene Donaldson
M. G. D. Gilchriese
Editors

M. Gil G. Gilchriese
Editors

Table of Contents

High p_T Physics

High p_T Detector Configurations

Intermediate p_T Physics

Intermediate p_T Detector Configurations

Low p_T Physics and Detectors

Exotics

High p_T Physics

PHYSICS AT LARGE TRANSVERSE MOMENTA*

I. Hinchliffe
Lawrence Berkeley Laboratory
Berkeley, California 94720.

Abstract

The methods used in estimating the signals and backgrounds involved in searches for new particles at the SSC are reviewed. A brief summary is given of the problems involved in these searches. The required detector parameters are summarized and an indication of the open problems given.

Introduction

At this workshop various working groups were asked to devise strategies for extracting physics signals in the presence of backgrounds and to specify the detector parameters required. Since many people in these groups had considered these problems previously, the conclusions that emerged from the working groups tended to be rather realistic. Generally, groups were conservative in their choices of signal since they were influenced by what they perceived as detector limitations. For example, the heavy quark group required two leptons in their signal because they were concerned about the possibilities for triggering an event with just one. Also, the supersymmetry group made demands for particle tracking since they were skeptical about the hermiticity of calorimeters.

Work necessarily concentrated upon the more difficult signals since these place the most stringent demands upon a detector. The reader will therefore get a very distorted view if she attempts to get an overview of SSC physics from these proceedings. Such surveys[1] have been given ad nauseam, in this workshop it was time to study some hard problems.

Simulating Signals and Backgrounds.

Monte Carlo event generators[2] are used to simulate both the signals and the backgrounds in SSC studies. Two generators (ISAJET[3] and PYTHIA[4]) are used most often. While these generators are very powerful and include simulations of a very large number of processes, they do have limitations. It is important to realise what these limitations are when one is evaluating the conclusions of some physics study for the SSC. The programs should not be treated as black boxes whose output is regarded as holy script. I will make some comments here on some areas where caution should be exercised, and will refer the reader to reviews for more details.

*This work was supported by the Director, Office of Energy Research, Office of High Energy and Nuclear Physics, Division of High Energy Physics of the U.S. Department of Energy under contract DE-AC03-76SF00098.

The event generators can be thought of as having two almost distinct parts. Firstly, there is the part in which a process is be calculated using perturbative QCD, secondly, the part in which the quarks and gluons of QCD are materialized into hadrons. The first part is, in principle, fully calculable and Monte-Carlo generators are limited only by how well these calculations are implemented. The second part has to rely on parameterization of existing data.

For definiteness, consider the production of some jets at large transverse momentum (p_T) (see figure 1). The parton model divides the process into three stages; † there are the parton distribution functions ($f_i(x, Q^2)$) which are extracted from deep-inelastic scattering data at small Q^2 and then extrapolated up to the scale of order p_T appropriate to the jet process; there is the partonic process which produces outgoing partons at large transverse momenta; and finally the hadronization of these partons into the jets. At lowest order in α_s the parton scattering process is a $2 \to 2$ scattering which is order α_s^2 and one therefore expects the final state to consist dominantly of two jets.‡ The differential cross section for the production of two jets with rapidities y_1 and y_2 and transverse momentum p_t in the center-of-mass frame of the pp system is given by

$$\frac{d\sigma}{dy_1 dy_2 dp_t} = \frac{2\pi\tau}{\hat{s}} p_t \sum_{i,j} [f_i^a(x_a, M^2) f_j^b(x_b, M^2) \hat{\sigma_{ij}}(\hat{s}, \hat{t}, \hat{u}) \\ + f_j^a(x_a, M^2) f_i^b(x_b, M^2) \hat{\sigma_{ij}}(\hat{s}, \hat{u}, \hat{t})]/(1 + \delta_{i,j})$$

where $\tau = 4p_t^2 cosh^2((y_1 - y_2)/2)/s$. Here $\hat{s}, \hat{t}$ and $\hat{u}$ are the Mandelstam variables for the $2 \to 2$ partonic scattering. The inclusive jet cross-section calculated by this simple parton model formula is in excellent agreement with the data from the UA1 and UA2 collaborations. However the Monte-Carlo generators attempt to describe the complete event rather than an inclusive distribution.

The structure of a Monte-Carlo generator can be understood from an analogy with QED. Consider the case of $e^+e^- \to \mu^+\mu^-$. The total rate is given by the simple process $e^+e^- \to \mu^+\mu^-$ up to corrections of order α_{EM}^3. However, at high energy the final state actually consists of the $\mu^+\mu^-$ pair accompanied by a number of photons which are either soft or collinear with the outgoing muons or the incoming electrons. The final state of $\mu^+\mu^-$ and a hard isolated photon is suppressed by an additional power of α_{EM}.

The dominant final state can be generated as follows. The incoming e^+e^- are allowed to emit photons via a classical branching process shown in figure 2; this constitutes the initial state radiation. The branching kernel for this emission, which is proportional to α_{EM}, is such that the invariant mass of the electron before emission is as small as possible and hence, the emitted photon is either soft or collinear with the electron. After many emissions the electron (and positron) will have an (spacelike) invariant mass of order $\sqrt{s}$. The e^+e^- now annihilate with a cross-section (calculated for on-shell particles) and given by perturbative QED. The outgoing muons now have an (timelike) invariant mass of order $\sqrt{s}$, and now emit photons in the same manner as the electrons so that they eventually have invariant mass equal to

†The derivation of the model from QCD, the so-called factorization theorem, is reviewed by J. Collins and D. Soper.[5]

‡The problems of defining a jet are discussed in the article on Jets and Compositeness[6].

the rest mass when the photon emission stops. This algorithm is correct when the photon is soft or in the leading log (leading pole) approximation where $log(\sqrt{s}/M) >> 1$. This occurs when M, the invariant mass of a photon and its parent lepton, is small; *i.e.* the photons are either soft or collinear with the incoming electrons or outgoing muons. In this region many emissions are allowed since although the kernel is of order α_{EM}, the emission probability contains a factor of $log(\sqrt{s}/M)$.

This algorithm has two main deficiencies. It occasionally will generate a final state with a hard non-collinear photon. This final state is more accurately described by the fundamental process $e^+e^- \to \mu^+\mu^-\gamma$, with a shower added to each of the initial and final particles. If this process is also added, there is a problem with overcounting since both fundamental processes are capable of populating the same regions of phase space. The algorithm also neglects coherence effects[7] and so overestimates the multiplicity of final state photons. To understand this, consider the emission of a very soft (long wavelength) photon early in the final state shower. Such a photon will really see the charge of the whole final state, which is zero, rather than just that of its parent muon. This emission will therefore be suppressed and the multiplicity of photons reduced.

This basic picture also applies to the Monte-Carlo generators of QCD with some important complications. Leptons and photons are replaced by quarks and gluons which do not appear as physical states. The initial state radiation is responsible for the Q^2 dependence of the structure functions used in the parton calculation.[§] The final state evolution is stopped when the partons are off shell by some value (Λ) of order 1 GeV since $\alpha_s(\Lambda)$ is of order one and QCD perturbation theory ceases to be relevant. At this stage the parton shower is turned into the hadrons seen in the detector. Since this final stage process is non-perturbative, it cannot be calculated. The scale of this hadronization is Λ which does not increase with the jet energy. Data from jets at current energies must be used to parameterize the hadronization. Some features of the events produced by the Monte-Carlo generators are, therefore, dependent upon these parameterizations. If a simulation is sensitive to the details of hadronization, limitations from current data will limit its credibility. The increase in particle multiplicity with jet energy should be due almost entirely to the growth in the size of the parton shower and should be well predicted by the generators.

Data must also be used as input to the parts of the Monte-Carlo that generate the event structure resulting from the beam fragments. This structure consists of low transverse momentum particles. Notice that some of these fragments are due to the initial state radiation, so that the structure of this beam fragmentation event will depend on the hard scattering process. In particular, events with jets will have more low p_T particles than that in events without a hard scattering; this is borne out by data.

The Monte-Carlo generators are able to deal with the coherence effect mentioned above by modifying the shower algorithm, the effect is quite drastic as is shown in figure 3. The problem of higher order corrections is much more difficult. Three jet events will occasionally be produced by the Monte-Carlos when a gluon in one of the showers is emitted at wide angle and with high energy (analogous to the hard photon emission above). However such

---

[§] In practice the Monte-Carlo generators start with partons off-shell by Q^2 and evolve backwards[8] to generate the incoming shower.

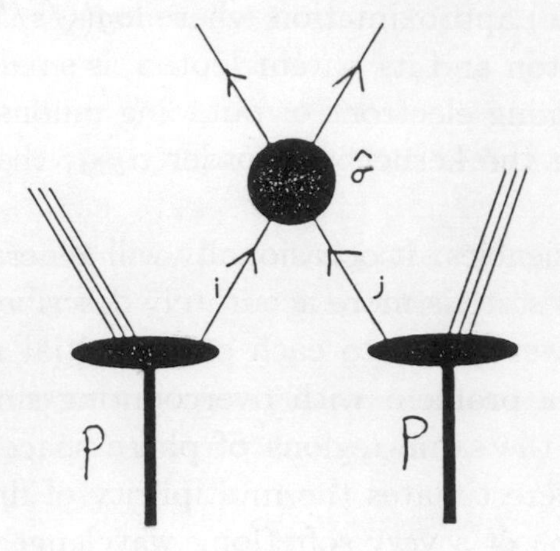

Figure 1: Diagram showing the parton model process $pp \rightarrow jets$.

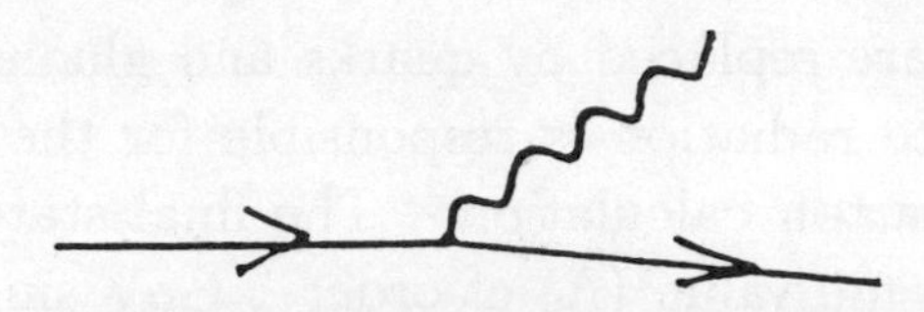

Figure 2: The branching process used to generate initial, or final, state parton showers. This kernel is iterated many times to produce the full shower.

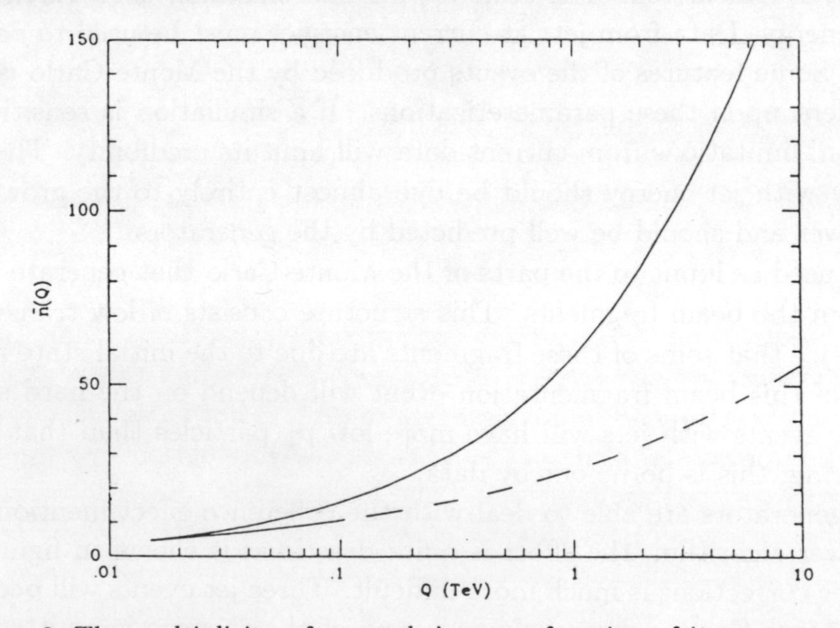

Figure 3: The multiplicity of a quark jet as a function of its transverse momentum. The solid line shows the leading log result, the dashed the result after correction for the coherence effect; figure from reference 9.

a three jet event is really due, in the partonic language, to an order α_s^3 process of the type *gluon* + *gluon* $\rightarrow$ *gluon* + *gluon* + *gluon*. In the limit where one of the outgoing gluons is either soft or collinear with another one, the event looks like a two jet event. In this region of phase space, the shower algorithm agrees with the order α_s^3 result. The accuracy of this approximation in other regions of phase space has been investigated and found to be accurate to a factor of three or so. None of the existing Monte-Carlo generators include the higher order effects properly. (In the case of e^+e^- annihilation, the problem has been solved[10] at order α_s; here the situation is simpler because there are no colored particles in the initial state.)

It is possible to do some signal/background studies using a partonic calculation that treats the final state quarks and gluons as jets and has no parton showering. In such an approach the fundamental partonic process can be used to the appropriate order in α_s. Since there is no showering and all the outgoing partons are separated, the matching problem present in the Monte-Carlos are absent. Such a simple approach can be used both to suggest areas where more detailed work is required and to dispose of lost-causes. Perhaps more importantly it can be used as a check on the Monte-Carlo's rate for the process in question.

The problem of the observation of a heavy Higgs decaying into W pairs has been analysed in this way.[11] If one W decays leptonically and the other hadronically, the final state consists of a lepton and two jets. There is a background from the production of a W in association with jets when the jets have an invariant mass close to that of the W. The lowest order partonic process which contributes to this background is $W + 2\ Jets$. The relevant partonic calculations are known so a preliminary study can be carried out. The results of such a study can be used to suggest cuts that, when applied to the di-jet invariant mass and the momentum distribution of the two jets which make up the W, which are different for the jets from a real W and those in the background, reduce the potentially enormous background to an acceptable level. A full Monte-Carlo simulation[12] incorporating detector effects can then be undertaken to determine whether these cuts can be used in a real jet environment. Unfortunately, as stated above, the Monte-Carlos use an approximation to the $W + 2\ Jets$ final state, since it is a higher order process. The parton calculation can now plays its other role; it can be used to evaluate the accuracy of the Monte-Carlo's estimate of the background.

We can summarize the main limitations of Monte-Carlos as follows.

1. Higher order QCD corrections are generally not implemented. Hence final states with more than two QCD jets are not correctly generated. The Monte-Carlos will produce events with more jets but the normalization of these rates is not to be trusted to better than a factor of five or so. In some cases, such as the production of three jets the relevant parton cross sections are known and the inclusive jet rates can be calculated in the parton model. Such a calculation can then be used to estimate the credibility of the Monte-Carlo's prediction and possibly to renormalize its event rate. Processes which are limited by this problem include; any final state with more than 2 QCD jets unless some of these jets arise from weak decays, such as the decay of a W boson; production of a W or Z boson at large p_T accompanied by more than 1 jet; events with a two gauge bosons and any jets; and production of a heavy quark-antiquark pair accompanied by more than one jet.

2. Even when some higher order processes are included, they may not be properly normalized. For example, top quarks are produced at lowest order via the processes *gluon* + *gluon* $\rightarrow t\bar{t}$ or *quark* + *antiquark* $\rightarrow t\bar{t}$. In this case the top quark and antiquark emerge with equal and opposite transverse momenta. The order α_s^3 processes such as *gluon* + *gluon* $\rightarrow t\bar{t}$ + *gluon* can result in an event where the top quark and antiquark have similar p_T and their sum is balanced by the gluon. Both of these processes are included in ISAJET and PHYTHIA. However, the relative importance of these two final states to the total rate for top production cannot be known unless the order α_s^3 corrections to *gluon* + *gluon* $\rightarrow t\bar{t}$ are included.[13] These corrections are now known, but are not included in either PYTHIA or ISAJET, hence there is some uncertainty over the interpretation of the rates given by these generators. Again a partonic calculation can be used to check the rates.

3. Some details of hadronization, such as the probability that a jet fragments into a small number of hadrons are limited by available data. For example, occasionally a QCD jet will hadronize into three fast charged particles of low invariant mass and some very soft hadrons which are indistinguishable from those of the underlying event. This "jet" may look like a tau decaying into $a_1(1270) + \nu$. Attempts to use the Monte-Carlos to estimate such "fake" taus may be very unreliable.

4. The branching processes used in Monte-Carlo generators can lead to a loss of information in the following way.[14] Consider the production of a Z pair followed by the decay of both Z's to an e^+e^- pair. There is is an azimuthal correlation of the two planes formed by the e^+e^- pairs. In a Monte-Carlo where the two Z's are produced as on shell particles and then allowed to decay, this correlation is lost. This can have serious consequences if events are generated and cuts on angular distributions made in order to reject backgrounds. This particular example can be dealt with by using the process $q + \bar{q} \rightarrow e^+e^- + e^+e^-$ as the fundamental one. However, in many cases it is not possible or practical (the exact expression can be very complicated and therefore slow to evaluate) to use the full matrix element.

A detailed check of PYTHIA and ISAJET was carried out was carried out by the "Heavy Higgs" group[15] at this meeting. They compared the rates for the production and decay of a Higgs boson and of W and Z pairs given by these Monte-Carlos with the rates predicted by the parton model. In this case the final state particles have no color and the only extra information provided by the Monte-Carlos is in the structure of the underlying event, *i.e.* the effect of the initial state radiation and of the beam remnants. This check revealed several problems with Monte-Carlo generators for these processes even in the cases where only lowest order QCD processes are relevant, most of which have now been corrected. Nevertheless this study contains an important lesson. The generators are very difficult to write and maintain; they cannot be checked too often.

Summary of Physics Signals.

This section provides a brief summary of the physics signals and the required detector parameters. A summary can be found in table 1. In this table, I have attempted to indicate the accessible mass at the SSC and at LHC. These values are a rough guide only, but the same criteria have been used at both energies. Different criteria will result in different values; this usually accounts for any discrepancy between these numbers and those given elsewhere. It is obvious that a higher energy machine has more "reach". A higher luminosity one can compensate, to some extent, for a lower energy at the cost of higher event rate and possibly larger background. Experimentation becomes increasingly difficult at luminosities greater than 10^{33} cm^{-2} sec^{-1}.

Higgs Bosons.

The production rate of a standard Higgs boson is shown in figure 4. The signature depends upon the Higgs mass. It is convenient to distinguish **four** possibilities relevant to the SSC.

First, the standard model is not complete. In this case there may be many scalar (or pseudoscalar) bosons some of which may have electric charge. At this workshop there has been a discussion of the Higgs bosons which occur in the minimal supersymmetric model.[16] The three neutral and one charged particles have production rates which are quite small and signatures which are very difficult to extract. The production rate of the neutral ones at large mass is much smaller than the rate for the standard Higgs boson since the decay width to W pairs is much smaller and the WW fusion process is much less effective. In models of this type there is always one neutral scalar of mass less than M_Z, so detection at LEPII should be possible via the process $e^+e^- \to ZH$. These models also have a large number of supersymmetric particles for which the SSC can search.

A charged Higgs boson can be produced by the process $bg \to tH^-$ or in the decay of a top quark if $M_{H^-} < m_t - m_b$. Its dominant decay is to the heaviest fermion doublet. It is is to be remarked that there is no coupling $H^+ \to W^+Z$. A H^+ of mass 400 GeV is produced in association with a t quark (100 GeV) with a cross section of 10 pb. Its dominant decay to $t\bar{b}$ is likely to be swamped by QCD background. The decay to $\tau\nu$ which has a branching ratio of order 10^{-3} is more promising and is considered in reference 16.

Second, the standard model is correct and the Higgs boson has mass less than $2M_W$. In this case the Higgs will decay dominantly into $F\overline{F}$ where F is the heaviest fermion with mass less than $M_H/2$. If the Higgs is lighter than 90 GeV or so, it will be found at LEPII via the process $e^+e^- \to HZ$. The decay of toponium to Higgs plus gamma may be able to extend this range if toponium exists with a larger mass in the range of LEPII. I shall assume a Higgs mass of 130 GeV in the following but most of my comments are not very sensitive to this choice.

The production rate for a Higgs boson in this mass range is of order 0.1 nb. I shall assume that the Higgs is too light to be able to decay into $t\bar{t}$. The decay $H \to b\bar{b}$ is likely to be almost impossible to see above the large background of $b\bar{b}$ events produced by gluon fusion. There was some hope that the production of a Higgs in association with a W would produce an observable signal.[17] Work done at this meeting has cast serious doubt over this (work by J. Brau reported in reference 18). Semi-leptonic decays of the b quark were used in the analysis which required a lepton with p_t relative to a jet of more than 1 GeV, in an attempt to reject events without a b. A signal to background ratio of order 0.05 was obtained with about 400 signal events passing the cuts. More work is needed but the mode does not

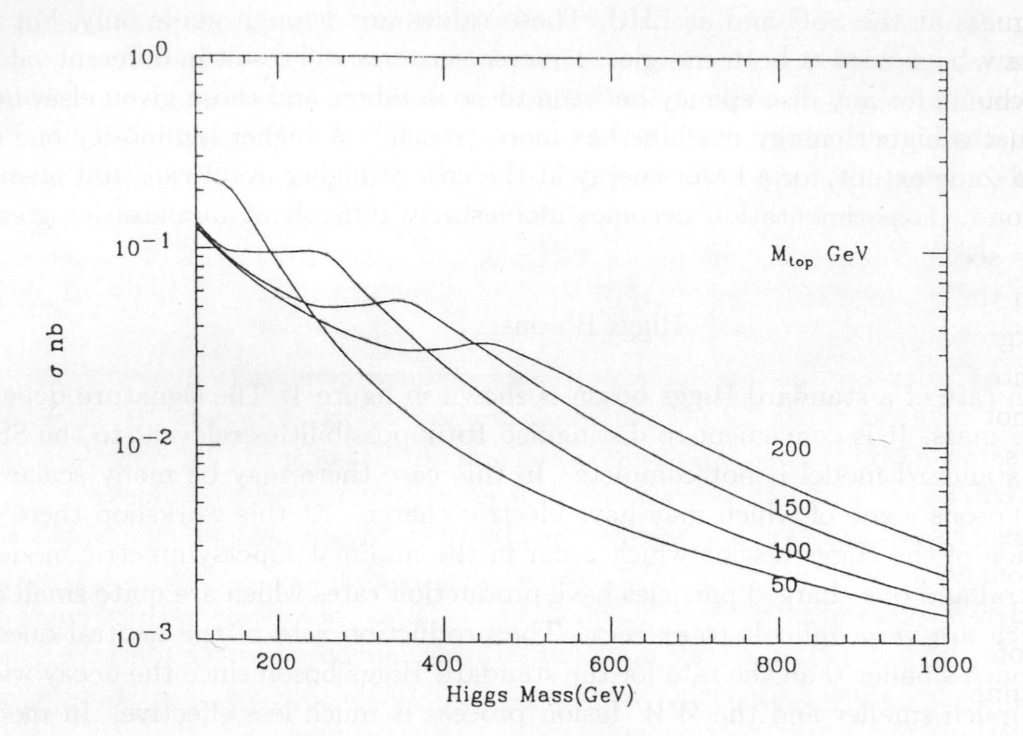

Figure 4: The total production cross-section for a standard model Higgs boson. The lines are labelled with the top quark mass assumed.

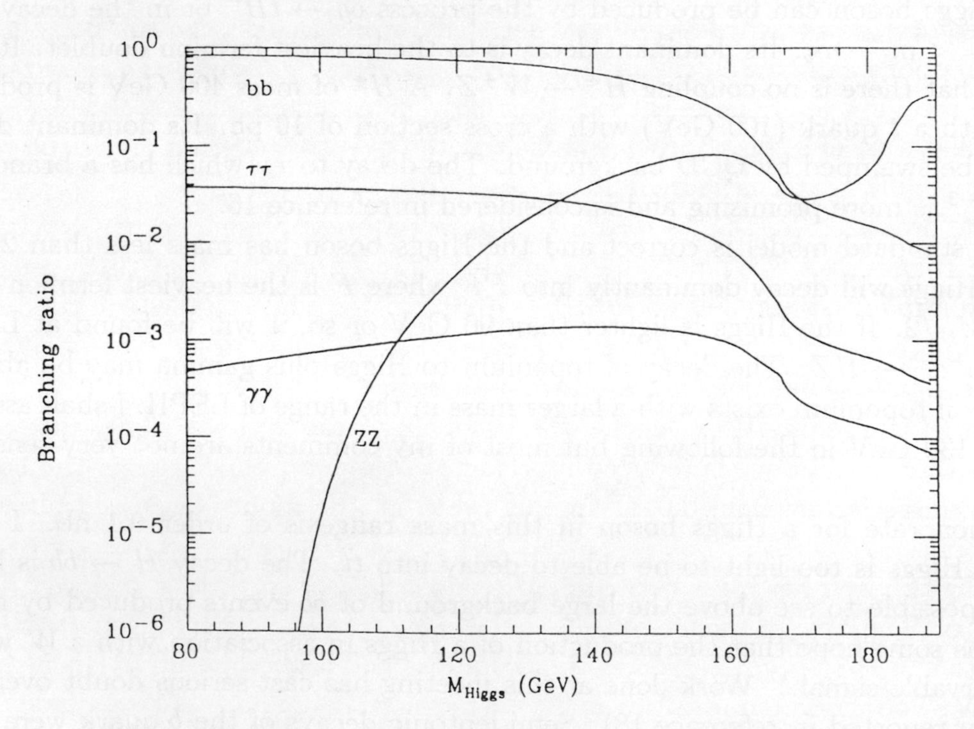

Figure 5: The branching ratios for various decays of the Higgs boson. It has been assumed that $M_H < 2m_t$.

look promising. The main requirement is the ability to detect b's with good efficiency and to reject events with lighter quarks and gluons.

Channels which have a better signal/background ratio are rare decay modes of the Higgs such as $\gamma\gamma$, $\tau\tau$(when the Higgs is produced in association with[19] or without a jet), and ZZ^* (Z^* is a virtual Z, which decays to l^+l^- or $q\bar{q}$). See figure 5 for a comparison of the relevant branching ratios. The first of these is probably the cleanest. The branching ratio is of order 10^{-3}; there are about 300 events/year on a background of about 2500 assuming a resolution of of 2 GeV in the $\gamma\gamma$ invariant mass. (Both γ's are required to have $|y| < 1.5$; this helps to reduce background.) There is also a background from isolated π^0's; attempts to estimate this from Monte-Carlos are limited by current data on jet fragmentation (see above) and are probably not reliable. Since the jet pair cross section is approximately 10^8 larger than the $\gamma\gamma$ one at same energy, a "fake γ" rejection of order 10^4 is needed. Current data on jet fragmentation can be used to estimate whether this is feasible; the problem does not appear to be too serious. Use of this channel requires the ability to trigger, at high efficiency, events with two photons with p_t or order 60 GeV. This is a formidable task.

The second rare process is $H \to \tau\tau$, which has a larger branching ratio. There is a background from the production of tau pairs via an intermediate Z or photon (Drell-Yan); the method cannot be of use if the Higgs has a mass close to the Z mass. The main problem is the identification of τ's and the reconstruction of their momenta. The presence of a neutrino in the tau decay is a serious handicap since it degrades the resolution in the tau pair invariant mass. By taking the decays of the tau to three or more hadrons the momentum lost in neutrinos is reduced at the cost of a small effective branching ratio. An alternative approach is to require the presence of an extra jet that recoils against the tau pair. In this case the taus' directions can be determined from the their decay products and their total p_t from that of the recoiling jet. The event can then be reconstructed. The problem appears to be that the jet p_t can only be of order 100 GeV if the event rate is not to be too small. Such a low p_t jet is difficult to measure and the resolution is degraded.[18]

The third rare process $H \to ZZ^* \to 4$ *charged leptons* has a branching which increases rapidly with the Higgs mass.[20] The Z's can probably only be identified in their decays to e^+e^- or $\mu\mu$ which implies a branching ratio $ZZ \to 4l$ of 0.0044. This channel will produce 30 or more events (no other cuts) for Higgs masses greater than 130 GeV or so. The background is still under investigation, but does not appear to be serious.[18]

If the $t\bar{t}$ channel is open to Higgs decay, the branching ratio and the signal rate for all the decay modes discussed above is reduced by a factor of m_b^2/m_t^2 which, using the current limit on the top mass, is less than 0.015. In this event all the modes discussed above are seriously compromised.

Third, the Higgs boson has mass between $2M_W$ and 1 TeV or so. In this region it decays dominantly to Z or W pairs. In the former channel the events can be fully reconstructed via the decay of a Z to e^+e^- or $\mu\mu$. This process is rate limited and is effective up to Higgs masses of about 600 GeV. The mode requires the ability to measure μ's and isolated electrons. The presence of two reconstructed Z's implies that the resolution in the lepton pair invariant mass is not critical. At larger masses the mode $H \to ZZ \to ll\nu\nu$ may be exploited.[21] Here the signal calls for a single reconstructed Z together with missing transverse momentum.

There is background from $Z + jet(s)$. This background can only be controlled if missing momentum can be well measured. This calls for a hermetic calorimeter with coverage out to rapidity of at least 5.5. At large values of M_H the Higgs is a very broad resonance and no clear peak will be visible. The establishment of a signal is contingent upon the ability to predict (determine) the background.[15]

The Higgs can also decay to WW. The utility of this mode is not yet clear.[12] The ability to detect $H \to WW \to l\nu + jet(s)$ depends on the precision with which the W can be observed in its hadronic decay mode. The background arises from the final states WW and $W + jets$; cuts must be imposed to reject the latter background. The problem is difficult, progress depends on improvement in the Monte-Carlos so that they correctly incorporate the $W + 2jets$ final state (see below) and upon more detailed detector simulations.

If it has large mass, the Higgs is produced in association with two jets that are at large rapidity.[22] The claim[23,24] that these jets would provide a tag to reduce the background has recently been investigated more fully at the partonic level.[25] The tag may help a little but is no panacea. More detailed work involving a full Monte-Carlo simulation is needed. The problems involved in detecting these jets at $y =$ 3-5 and $p_t \lesssim 100$ GeV have not been fully investigated. If this tagging can be utilized, it may also help to reduce the background in the channel $H \to ZZ \to ll\nu\nu$; work is also required to evaluate this.

If the top quark is able to decay into a real W then some of the backgrounds to Higgs searches in this mass range are worsened. The production of $t\bar{t}$ final states leads to a final state with a W pair. For example the total yield of W pairs from a 150 GeV top quark is about 150 times that from $q\bar{q} \to WW$).[26] This is likely to make the process $H \to WW$ very difficult to see even if the $W + jet(s)$ background can be eliminated. Chen *et al.*[27] have looked at the decay $H \to ZZ \to \mu\mu\mu\mu$ and have shown that with a typical iron muon spectrometer, the pair production of a 200 GeV top quark can give rise to a "fake Z" at a large rate. The presence of two reconstructed Z's in the mode $H \to ZZ \to 4$ *charged leptons* ensures that this mode is not compromised. However their analysis gives rise to concern that poor resolution may may compromise the $H \to ZZ \to ll\nu\nu$ mode, at least in the case where $l = \mu$. A more detailed study in needed to assess this.

Fourth, there is no elementary Higgs boson. If the Higgs mass becomes greater than 1 TeV or so, it has a very large width and cannot be thought of as a particle. In this case the couplings of W and Z bosons becomes strong. The signals[28] for such a possibility involve an excess of ZZ, ZW and WW production over that expected from processes such as $q\bar{q} \to WW$ when the gauge boson pair have invariant mass more than 1 TeV or so. Exact predictions for this scenario are not possible, but rates can be estimated fairly reliably.[28] One of the key signals is the appearance of W^+W^+ final states, for which the background is low. To detect this, one must be able to determine the sign of leptons from W decay. The transverse momenta of the leptons extend to 750 GeV. In some models (technicolor)[29] where such strong interactions occur, there are many pseudoscalar particles of mass below 1 TeV that are produced with much larger rates than the gauge boson pairs. The detection of these particles is much easier and will probably be the first indication that such strong interactions occur.

New Gauge Bosons.

These are straightforward to see via the decays $W' \to e(or\,\mu)\nu$ and $Z' \to e^+e^- or(\mu^+\mu^-)$.[30] There is essentially no background. The leptons in the signal are well isolated. It is of interest to attempt a study of the decay $W' \to \tau\nu$ both as a test of universality and as a method to determine the W' coupling via an observation of the tau polarization.[31] The decay of the tau to a single lepton may be difficult to see. The decay to one or three pions yields a very unusual jet which should be identifiable.

It appears that it will be more difficult for decay $W' \to jets$ than it was for UA2 in their attempt to find the W and Z via their hadronic decay modes at the $Sp\overline{p}S$ collider, since the relative amount of gluons and antiquarks, which control the background and the signal, is larger for the SSC search. There has been some discussion of attempts to detect $Z' \to WW \to l\nu + jets$.[32] Although a Z' is not likely to be discovered via this channel, since $\Gamma(Z' \to WW) \sim \Gamma(Z' \to e^+e^-)$, observation of this channel could be useful to pin down its couplings. As in the Higgs case discussed above, there is a background from the final state $W + jets$. Observation is likely to be challenging.

Heavy Quarks.

A new b' quark that is a member of a fourth generation doublet will decay to Wt. It can be produced in pairs at SSC. The dominant final state will therefore consist of a 6-jet final state, where two pairs of jets reconstruct to the W (assuming that the t quark does not decay into a W). There is expected to be a very large rate for 6-jet events from QCD processes. Existing Monte-Carlos will generate such final state, but they are of unknown reliability (see above). A study at the La Thuile workshop[24] employed a parton calculation that uses an approximation to the 6-jet final state which is better than the leading log one used by Monte-Carlos. They concluded that the 6-jet final state was not useful. However their calculations leave some hope that the channel may be of use at small ($\sim$ 200 GeV) b' masses. However, at these masses, the event rates are so large that other decay modes with smaller branching ratios can be used.

If the t quark can decay into a W the final state will have 4 W's, and the extra constraint may then allow the dominant decay mode to be used. A more detailed investigation of this channel is required since, if it can be exploited, it may provide a better determination of the b''s mass than the method to which I now turn.

A detailed analysis[33] of b' detection carried out at this meeting required that there be two isolated, energetic leptons in the decay of the $b'\overline{b'}$. By selecting events where both leptons come from the decay of the b', and effective trigger can be designed. The hadronic decays of the $\overline{b'}$, which have less missing energy in the form of neutrinos can then be used to determine the b' mass. The background arises dominantly from $t\bar{t}$ production and can be removed by a cut on the transverse momentum of an isolated lepton.

At the La Thuile study[24] and in a paper in these proceedings,[34] the signal and background when only one lepton is required are discussed. In this case backgrounds arise from the final states $t\bar{t}$ and $W+jets$; both can be eliminated by a cut on the leptons transverse momentum.

Signals from the decay of exotic quarks were not discussed at the meeting. In some E_6 based models, there exist charge 1/3 quarks d' which are singlets under $SU(2)_L$. Different decays are possible according to the baryon number assignment of d'. In one case the d' will mix with other down quarks and will then decay $d' \rightarrow Zd$. Pair production will then give a final state with a Z pair and jets for which there is no background. For a discussion of possible decay modes in a supersymmetric theory see reference 35. No detailed detector simulations for these modes have been undertaken, but they are unlikely to have a serious impact on detector requirements.

To summarize, the detection of a new quark will require the ability to measure jets and isolated leptons. It is useful to be able to determine the charge of leptons up to $p_t \sim 750$ GeV. Measurement of missing p_t is not important.

Heavy Leptons.

Consider a 4th generation lepton doublet (L, ν_L). An L^+L^- pair can be produced either via $q\bar{q}$ or gg annihilation. If the ν_L is much lighter than L then the production rate for $L\nu_L$ pairs dominates. L will decay to $W\nu_L$, so that if ν_L is stable, the event will consist of a W at large p_t and missing transverse momentum. The leptonic decay of the W is not observable due to the background from $q\bar{q} \rightarrow e\nu$, which occurs via annihilation through a virtual W.[36] The background to the hadronic decay of the W arises from $Z(\rightarrow \nu\nu) + jet(s)$ where the jets fake a W. The claim[24] that the background can be suppressed easily has been refuted by Barger *et al.*[36] who have conducted a detailed calculation of the signals and backgrounds associated with heavy leptons at the SSC. These calculations may be too pessimistic; cuts on the dijet system like those advocated for the similar problem of $H \rightarrow WW \rightarrow e\nu + jets$, can be applied to reduce the background.[37] Nevertheless, the detection of heavy leptons at the SSC will be a formidable task.

Supersymmetry.

Work on Supersymmetry has concentrated upon searches for the gluino ($\tilde{g}$) and squark ($\tilde{q}$). The signals depend strongly upon the decays. In previous studies it was usually assumed that the decays $\tilde{g} \rightarrow q\bar{q}\tilde{\gamma}$ and $\tilde{q} \rightarrow q\tilde{\gamma}$ are dominant. The photino ($\tilde{\gamma}$) leaves the detector without interacting and the classic supersymmetry signal of jets+missing p_t results. It was these decay modes for which the UA1 collaboration searched.[38] In this mode the background arises dominantly from the semileptonic decay of top quarks, which gives rise to neutrinos at large p_t. These background events also have isolated leptons, whose presence can be used as a veto. This decay scenario was discussed in great detail at the La Thuile meeting.[39]

The decay modes are however model dependent. As the squark and gluino masses become larger other modes are likely to dominate. From the point of view of a detector designer, this can have important consequences. For example, the decay $\tilde{g} \rightarrow q\bar{q}\tilde{W}$ ($\tilde{W}$, the Wino, is the supersymmetric partner of the W) which may be dominant, can result in a final state of $W + jets$ after the Wino decays. This final state can be identified via the leptonic decay of the W. The final state now has lepton(s) and less missing p_t. Some of these modes are assessed in detail in these proceedings.[40]

Irrespective of the mode considered the detector requirements are clear. Hermetic calorimetry and detection of isolated leptons (including taus if possible).

Jets and Compositeness.

If quarks are composite it will be revealed in deviations of the jet cross-sections from those predicted by perturbative QCD. Since jet cross-sections fall rapidly with increasing p_t, accurate measurements are contingent upon precise determinations of the jet energies and hence upon excellent calorimetry.[6] If leptons are also composite, deviations in the production cross section for dilepton pairs of large invariant mass can be expected. In this case it will be important to measure these rates accurately. Since the composite interactions may not obey universality, it will be important to measure both electrons and muons (and taus if possible).

Some of the details of jets, for example the growth of the particle multiplicity with the jet p_t are predicted by perturbative QCD. However some of these details are not predicted by QCD and must be parameterized from data (see above). It is very likely that such details will have to be measured at the SSC as a check both of QCD and, more importantly, as a check of the Monte-Carlo programs.[6] Some of the physics signals require rejection of jets at the 10^{-4} level (for example, see the Higgs discussion above); such a rejection is very sensitive to the fine structure of jets. In the measurements of jets, high statistics will be required to check the reliability of such background estimates. These measurements will require a detector which can resolve and measure all the tracks within jets. The measurements will have to be made at various values of p_t from 100 GeV to a few TeV and so cannot be done in a low luminosity interaction region where there will be insufficient events at large p_t. A special purpose detector will be needed for this task.[41] It is not clear whether this can be integrated into one of the 4π detectors.

Summary of detector requirements.

It is remarkable that many of the proposed physics signals at the SSC require similar detection techniques.

Hermetic calorimetry covering $|y| \lesssim 5.5$ is needed for missing p_t measurements required by searches for supersymmetric particles decaying to jets and missing p_t and for searches for the Higgs boson via the decay $H \to ZZ \to ll\nu\nu$. There are three main standard model sources of events with missing p_t. The production of a W or Z in association with jets followed by the decay $W \to l\nu$ or $Z \to \nu\nu$, see figure 6. In the former case there is an associated charged lepton which is usually isolated. If it is an electron or muon, the event can be tagged and vetoed if necessary. If it is a tau, some of its decays may be easy to pick up. The p_t distribution of the lepton and the neutrino are similar so that the missing p_t can be measured. Of course, the decay $Z \to e^+e^-$ can be used to normalize the missing p_t due to Z decays. Another primary source is that from the decay of top quarks. This is shown in figure 7. Again the events have an isolated lepton which can be used as a veto. (For an example of this see the Supersymmetry report.) The final source consists of neutrinos from light quarks and particles lost down the beam hole. See Figure 8. These sources are most important at small missing p_t ($\lesssim 200 GeV$). The results from PYTHIA and ISAJET for these distributions are quite different. This is perhaps not too surprising since they use different algorithms for generation of the beam fragments which populate the regions of large rapidity and small p_t.

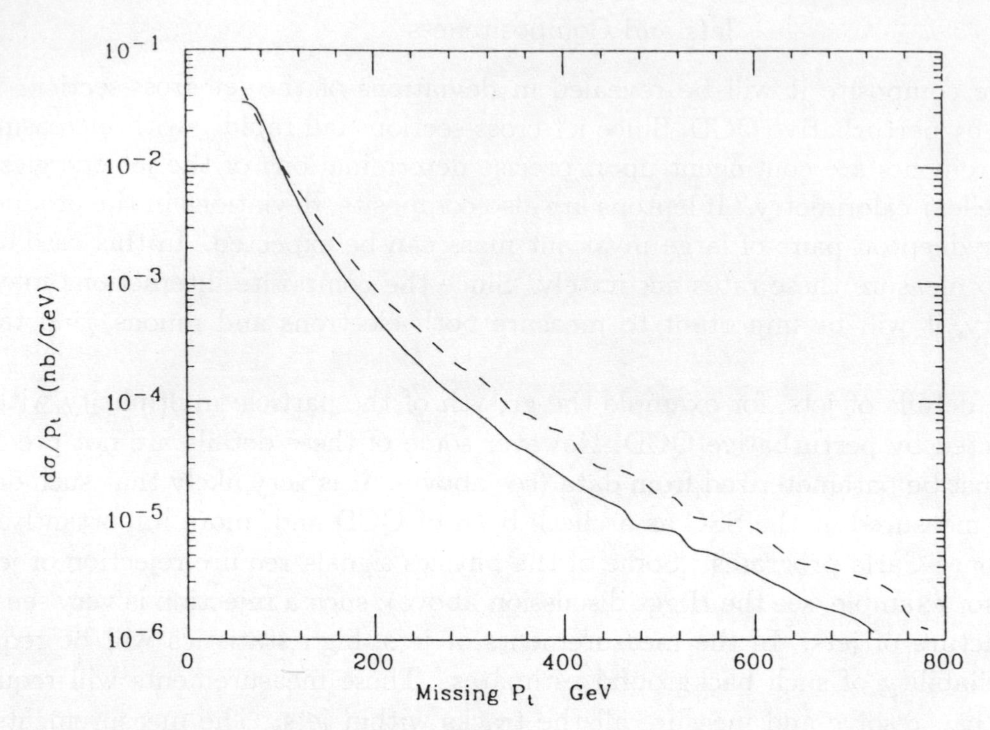

Figure 6: The missing transverse momentum distribution arising from the production and decay of a W (solid line) or Z (dashed line). The missing p_t arises only from the neutrino from $W \to \nu l$ or $Z \to \nu\nu$.

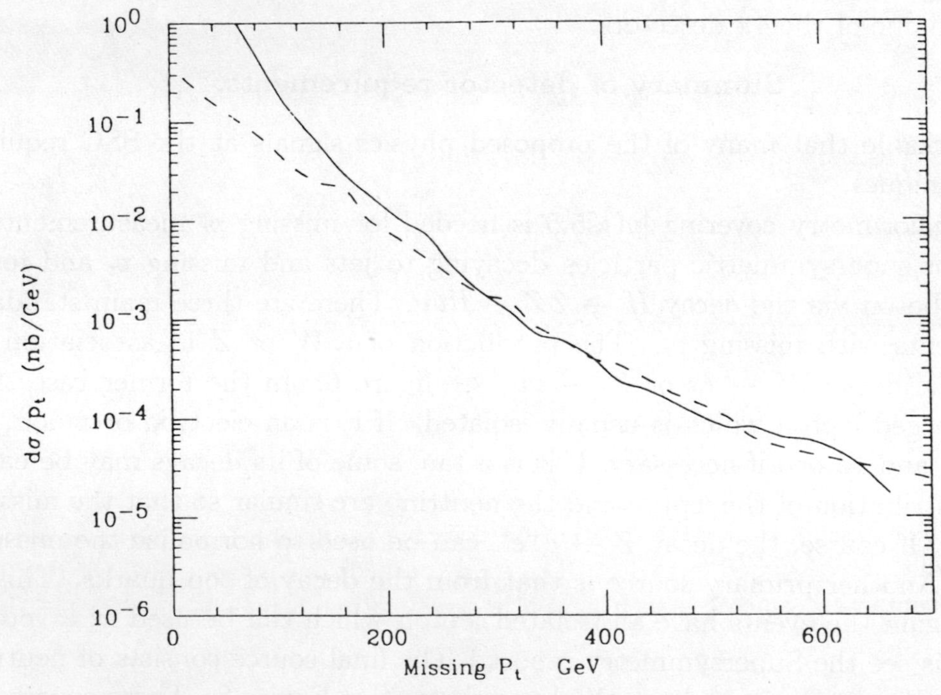

Figure 7: The missing transverse momentum distribution arising from the production and decay of a top quark pair. The missing p_t arises only from neutrinos emitted in the top quark decay chain. The solid (dashed) line corresponds to a top quark mass of 50 (150) GeV.

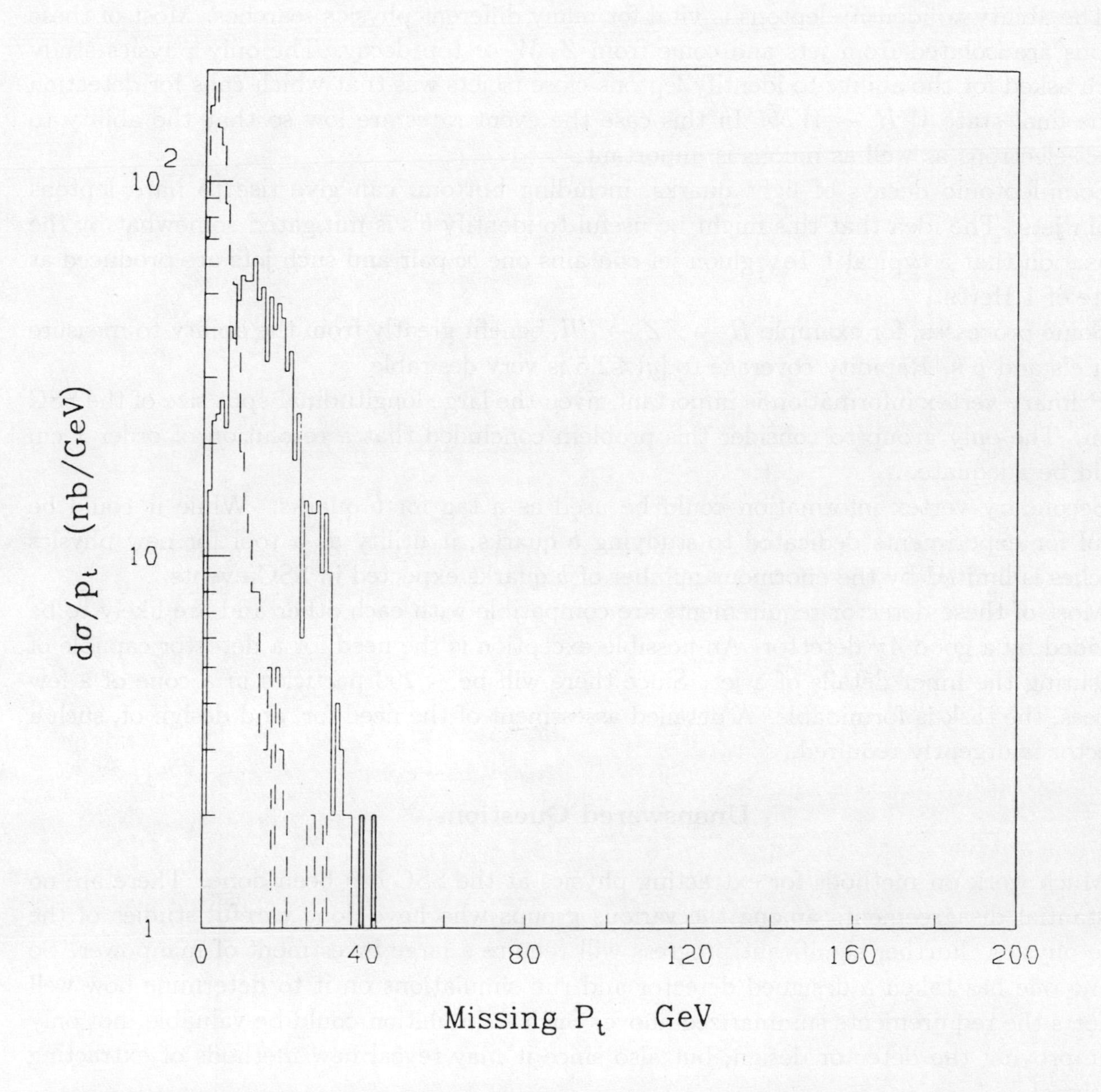

Figure 8: The missing transverse momentum distribution arising from particles with $|y| >$ 5.5 which are assumed to be lost. The solid (dashed) line is from ISAJET (PYTHIA). Jet events were generated with p_t of a jet greater than 500 GeV.

Calorimeter segmentation of order $\Delta y \Delta \eta \sim 0.05 \times 0.05$ is needed to study the details of jets and will be important if one hopes to detect the W in its hadronic modes, for example in the processes $Z' \to WW$ or $H \to WW \to l\nu q\bar{q}$.

The ability to identify leptons is vital for many different physics searches. Most of these leptons are isolated from jets and come from Z, W or top decay. The only physics study which asked for the ability to identify leptons close to jets was that which calls for detection of the final state $WH \to Wb\bar{b}$. In this case the event rates are low so that the ability to detect electrons as well as muons is important.

Semi-leptonic decays of light quarks, including bottom, can give rise to hard leptons within jets. The idea that this might be useful to identify b's is mitigated somewhat by the realization that a typical 1 TeV gluon jet contains one $b\bar{b}$ pair and such jets are produced at a rate of 1 Hertz.

Some processes, for example $H \to ZZ \to llll$, benefit greatly from the ability to measure both e's and μ's. Rapidity coverage to $|y| \lesssim 2.5$ is very desirable.

Primary vertex information is important given the large longitudinal spot size of the SSC beam. The only group to consider this problem concluded that a resolution of order 1 cm would be adequate.

Secondary vertex information could be used as a tag for b quarks. While it could be useful for experiments dedicated to studying b quarks, it utility as a tool for new physics searches is limited by the enormous number of b quarks expected in SSC events.

Most of these detector requirements are compatible with each other and are likely to be provided by a good 4π detector. An possible exception is the need for a detector capable of measuring the inner details of a jet. Since there will be $\sim$ 200 particles in a cone of a few degrees, the task is formidable. A detailed assessment of the need for, and design of, such a detector is urgently required.

Unanswered Questions.

Much work on methods for extracting physics at the SSC has been done. There are no substantial disagreements among the various groups who have done careful studies of the same physics. Further significant progress will require a large investment of manpower. So far, no one has taken a designed detector and run simulations on it to determine how well it meets the requirements summarized above. Such a simulation could be valuable, not only for improving the detector design, but also since it may reveal new methods of extracting the signal.

More work is needed on some important areas. In many cases the identification of W's via their hadronic decay modes would greatly simplify extraction of the physics. Taus can also provide a valuable tool, but we do not know how well the 1π and 3π modes can be detected above background. In some areas progress is limited by the Monte-Carlo programs. For example, a detailed evaluation of "tagging" jets at large rapidity in order to improve the $H \to WW$ signal is not possible without a reliable simulation of the $W + 4jets$ background. The appearance of isolated π^0's from the fragmentation of a jet can lead to fake photons. Fragmentation into a three pion system of low invariant mass can fake a tau. The reliability of Monte-Carlo rates for these processes can be improved once more data from the Tevatron become available.

There are some areas where only partonic calculations have been carried out and a more detailed simulation is needed. These include, heavy lepton signals and those from models of strongly interacting $W's$ and Z's.

Acknowledgment

I would like to thank my co-organizer, Pier Oddone, for his valuable insights. On behalf of the organizers of this workshop I would like to express our gratitude to the many participants at this workshop for their hard work. This work was supported in part by the Director, Office of Energy Research, Office of High Energy and Nuclear Physics, Division of High Energy Physics of the U.S. Department of Energy under Contract DE-AC03-76SF00098. Accordingly, the U.S. Government retains a nonexclusive, royalty-free license to publish or reproduce the published form of this contribution, or allow others to do so, for U.S. Government purposes.

References

1. E. Eichten *et al.*, *Rev. Mod. Phys.* **56**, 179 (1984).
2. For a recent review, see B.R. Webber *Ann. Rev. Nucl. Part. Sci.* **36**, 253 (1986).
3. F. Paige and S. Protopopescu in *Proc. of the UCLA on Observable Standard Model Physics at the SSC: Monte Carlo Simulation and Detector Studies*,Ed., H.-U. Bengtsson *et al.* World Scientific Publishing, Singapore (1986).
4. H-U. Bengtsson and T. Sjostrand in *Proc. of the UCLA on Observable Standard Model Physics at the SSC: Monte Carlo Simulation and Detector Studies*,Ed., H.-U. Bengtsson *et al.* World Scientific Publishing, Singapore (1986).
5. J. Collins and D. Soper *Ann. Rev. Nucl. Part. Sci.* (to appear).
6. R.N. Cahn *et al.*, "Jets and Compositeness", these proceedings.
7. G. Marchesini and B. Webber, *Nucl. Phys.* **B238**, 1 (1984).
8. Figure from K. Ellis in "Supercollider Physics" Ed. D. Soper World Scientific Publishing, Singapore (1986).
9. T. D. Gottschalk, Caltech preprint CALT-68-1241 (1985), T. Sjostrand *Phys. Lett.* **157B**, 321 (1985).
10. T.D. Gottschalk in *Proc. 1986 Summer Study on the Design and Utilization of the Superconducting Super Collider*, R. Donaldson and J. Marx (eds.), Fermilab, Batavia, Illinois, 1987.
11. J.F. Gunion and M. Soldate, *Phys. Rev. D* 34:826 (1986).

12. A. Savoy-Navarro, these proceedings.

13. R.K. Ellis, S. Dawson and P. Nason, in preparation.

14. J. Collins, these proceedings.

15. R.N. Cahn *et al.*"Heavy Higgs Working Group", these proceedings.

16. J.F. Gunion *et al.*, "Probing the Non-Minimal Higgs Sector at the SSC", these proceedings.

17. J.F. Gunion *et al., Phys. Rev.Lett.* 54:1226 (1985); *Phys. Rev. D* 34:101 (1986), F.J. Gilman and L. Price, *Proc. 1986 Summer Study on the Design and Utilization of the Superconducting Super Collider*, R. Donaldson and J. Marx (eds.), Fermilab, Batavia, Illinois, 1987.

18. D. Attwood *et al.*, "The Intermediate Higgs", these proceedings.

19. R.K. Ellis, *et al.*, FNAL , *Nucl. Phys*, to appear.

20. W.-Y. Keung and W. Marciano *Phys. Rev.*D30:248 (1984).

21. R.N. Cahn and M. Chanowitz, *Phys. Rev. Lett.* **56**, 1237 (1986).

22. R.N. Cahn and S. Dawson, *Phys. Lett. B* 136:196 (1986);
S. Petcov and D.R.T. Jones, *Phys. Lett. B* 84:660 (1979).

23. R.N. Cahn, S. Ellis, R Kleiss and J. Stirling, *Phys. Rev.* **D35**, 1626 (1987).

24. D. Froidevaux, "Experimental Studies in the Standard Theory Group", in the *Proc. of the Workshop on Physics at Future Colliders*, CERN-87-07.

25. R. Kleiss and W.J. Stirling, CERN-TH 4827/87.

26. G. Herten, these proceedings.

27. M. Chen *et al.*, these proceedings.

28. M. Chanowitz and M.K. Gaillard, *Nucl. Phys.*, B261:379 (1985).

29. E. Farhi and L. Susskind, *Phys. Rep.* 74:277 (1981).

30. N. Deshpande and S. Whittacker, these proceedings.

31. H. Haber in Proc. of the 1986 Snowmass Summer Study, Ed. Rene Donaldson and J. Marx, Fermilab (1987).

32. N. Deshpande *et al.*, these proceedings.

33. S. Dawson *et al.* "Heavy Quark Production at the SSC", these proceedings.

34. D. Dawson and S. Godfrey, "Single Leptons from Heavy Quark Production", these proceedings.

35. See, for example, V. Barger *et al.*, in Proc. of the 1986 Snowmass Summer Study, Ed. Rene Donaldson and J. Marx, Fermilab (1987).

36. V. Barger, T. Han and J. Ohnemus, Madison preprint MAD/PH/331 (1987).

37. G Anderson and I Hinchliffe these proceedings.

38. C. Albajar *et al.*, *Phys. Lett.* **185B**, 233 (1987), *Phys. Lett.* **185b**, 241 (1987).

39. J. Ellis and F. Pauss, "Beyond the Standard Model" in the *Proc. of the Workshop on Physics at Future Colliders*, CERN-87-07.

40. R.M. Barnett *et al.*, "Techniques for finding Supersymmetry at the SSC", these proceedings.

41. G. Van Dalen and J. Hauptmann in Proc. of the 1984 Snowmass Summer Study, Ed. Rene Donaldson and J. Morfin, Fermilab (1985).

Particle	Mass accessible at SSC, TeV	Mass accessible at LHC, TeV	Ease of Experiment	Comment
gluino, assuming $\tilde{g} \to q\bar{q}\tilde{\gamma}$	1.6	1	***	This decay mode is unlikely to dominate.
selectron, assuming $\tilde{e} \to e\tilde{\gamma}$	0.4	0.3	***	
New Z with "string inspired" coupling	7	4	****	
New d type quark	1.2	0.8	***	
New charged lepton	0.15 ?	0.1 ?	*	Serious background from $W+$ jets.
Compositeness scale in $qq \to qq$	25	14	***	Accuracy of jet rate vital.
Compositeness scale in $q\bar{q} \to ll$	20	13	****	
Higgs $\to ZZ \to llll$	$2M_Z < M < 0.6$	$2M_Z < M < 0.25$	***	
Higgs $\to ZZ \to ll\nu\nu$	$2M_Z < M < 1.0$	$2M_Z < M < 0.6$	**	No clear peak at large mass. Hermeticity vital.
Higgs $\to WW \to l\nu+$ jets	?	?	*	Background from $W+$ jets.
Higgs $\to \gamma\gamma$	$0.1 < M < 2M_W$	?	*	Luminosity vital. Large background.
No Higgs, structure in $M_{WW} > 1$ TeV	$M_{WW}, M_{ZZ} \lesssim 1.5$	–	*(*)	Very low event rate.

Table 1. The masses of various new particles which can be reached at the SSC (40 TeV) and LHC (17 TeV). Both machines are assumed to produce an integrated luminosity of $10^{40}\mathrm{cm}^{-2}$. The same criteria have been used at each machine. Numbers quoted elsewhere may differ due to the use of different criteria. The column marked "Ease of Experiment" indicates the degree of difficulty of the experiment according to the following key;
**** = straightforward experiment with no background; *** = experiment with background that can be removed by simple cuts; ** = difficult experiment that will require detailed background study and excellent detector; * = very difficult experiment whose feasibility is not fully known at this time.

Detecting the Heavy Higgs Boson at the SSC*

R.N. Cahn, M. Chanowitz, M. Golden, M.J. Herrero
I. Hinchliffe and E.M. Wang
Lawrence Berkeley Laboratory
University of California
Berkeley, CA 94720

F.E. Paige
Brookhaven National Laboratory
Upton, NY 11973

J.F. Gunion
Department of Physics
University of California
Davis, CA 95616

M.G.D. Gilchriese
SSC Central Design Group
Lawrence Berkeley Laboratory
University of California
Berkeley, CA 94720

Abstract

Detection of a heavy Higgs boson ($2M_z < M_H < 1$ TeV) is considered. The production mechanisms and backgrounds are discussed. Their implementation in the PYTHIA and ISAJET Monte Carlo programs are checked. The decay modes $H \to ZZ \to llll$ and $H \to ZZ \to ll\nu\nu$ are discussed in detail. The signal/background is evaluated and some relevant detector parameters are specified. Some remarks are also made concerning the requirements imposed on detectors by the decay mode $H \to WW \to l\nu + jets$. Experimental signatures for models in which there is no Higgs boson of mass less than 1 TeV are outlined.

* This work was supported in part by the Director, Office of Energy Research, Office of High Energy and Nuclear Physics, Division of High Energy Physics of the U.S. Department of Energy under Contracts DE-AC03-76SF00098 and DE-AC02-76CH00016.

1. Theoretical Introduction

A central motivation for the SSC is to study the mechanism of electroweak symmetry breaking which gives the W and Z their masses. Though the general framework is believed to be the Higgs mechanism, essentially nothing is known about the details. There may be a single Higgs boson as in the minimal standard model;[1] there may be several Higgs bosons as in supersymmetric theories;[2] or there may be pseudoscalar bound states as in some technicolor models.[3] The mass scale is equally uncertain and could be anywhere from a few GeV to the TeV range. Two points are assured by the general framework. First, there are new particles carrying a new force which induces the spontaneous breaking of the electroweak $SU(2)_L \times U(1)$ gauge invariance. Second, the new force causes strong scattering of longitudinally polarized W's and Z's if the mass scale of the associated new particles is of order 1 TeV or higher.

If the mass scale of symmetry breaking physics is above the WW and ZZ thresholds, then it can be studied at the SSC in events containing W^+W^-, ZZ, and in some cases $W^\pm Z$, W^+W^+, or W^-W^- pairs. In the standard model with a Higgs boson mass $m_H \lesssim 600$ GeV, the Higgs boson appears as a resonance in the W^+W^- or ZZ channels with a width of less than 100 GeV. Since the width grows like m_H^3, for larger values of m_H, the Higgs signal is an excess of W^+W^- and ZZ pairs over the WW or ZZ continuum. There is no easily recognizable peak in the mass spectrum. More generally, arguments based on unitarity show that symmetry breaking physics will be manifested in 1-2 TeV gauge boson pairs if it does not occur at a lower mass scale.[4] Detection of gauge boson pairs is therefore a critical requirement for SSC physics.

In this paper we briefly discuss the significance of the 1 TeV mass scale (Section 2), production mechanisms and decay properties of the standard heavy Higgs (Section 3), comparison of existing Monte Carlo programs with theory (Section 4), and results on the experimental detection of the heavy Higgs signal in a variety of modes (Sections 5, 6 and 7). In Section 8 we attempt to summarize the relevant desirable detector properties that are indicated by our study.

It should be noted that the studies of experimental requirements are not complete in the sense that definitive studies of signal-to-background ratios have not yet been made for all decay modes of the heavy Higgs. Nevertheless, we believe that it is possible to indicate with some confidence the necessary detector parameters required to observe the relevant signals and to reject backgrounds.

2. Significance of the TeV Scale

The 1–2 TeV scale plays a special role in the physics of electroweak symmetry breaking. First it defines the onset of strong interactions: if the typical mass scale

M_{SB} of the quanta of the symmetry breaking sector is 1 TeV or heavier, then the symmetry breaking interaction is strong. Second, as explained more precisely below, 1–2 TeV is the maximal energy scale at which the effects of the symmetry breaking must begin to emerge. In this sense it is like the 300 GeV limit of the Fermi theory of weak interactions. Unitarity requires that, at this energy scale or below, the Fermi theory would be supplanted by new physics as indeed occurred with the discovery of the W and Z.

The first of these points is easily illustrated in the minimal standard model with a single Higgs boson.[1] The lowest order relationship between the coupling constant λ, the Higgs boson mass m_H, and the vacuum expectation value v, is

$$\lambda = \frac{m_H^2}{2v^2}. \tag{1}$$

Quantum corrections are typically of order $\lambda/4\pi$ and are $O(1)$ for strong coupling, *i.e.*, if $m_H \cong \sqrt{8\pi}v \cong 1$ TeV. More precisely, unitarity fails in leading order for $m_H \cong 1$ TeV, indicating the onset of strong interactions since large quantum corrections must then arise to restore unitarity.[5]

A model-independent approach is based on symmetry principles first used in hadron physics. Just as the pion is a Goldstone boson associated with the spontaneous breaking of chiral symmetries in QCD, so the longitudinal modes W_L and Z_L are essentially (at energies large compared to M_W) Goldstone bosons arising from the breaking of gauge symmetries. The same techniques that establish pion low energy theorems,[6] such as

$$a_0(\pi^+\pi^- \to \pi^0\pi^0) = \frac{s}{16\pi F_\pi^2} \tag{2}$$

for the $J = 0$ partial wave amplitude can also be used to show that [4]

$$a_0(W_L^+ W_L^- \to Z_L Z_L) = \frac{\hat{s}}{16\pi v^2} \tag{3}$$

Eq. (2) is valid for $s \ll m_{\mathrm{hadron}}^2$ while Eq. (3) applies for $M_W^2 \ll \hat{s} \ll M_{SB}^2$ and can therefore only be relevant if $M_{SB} \gg M_W$. In that case it shows that new physics must intervene below $\sqrt{16\pi v^2} = 1.8$ TeV to preserve unitarity, $a_0 \leq 1$. Eq. (3) also shows that if $M_{SB} \geq 1$ TeV, then for $\sqrt{\hat{s}} \approx 1$ TeV, a_0 will be O(1) which is the precise indication of strong scattering (for instance, putting $\hat{s} \geq 1$ TeV^2 in Eq. (3) we find $a_0 \geq 1/3$).

The search for the mechanism of symmetry breaking is therefore not open-ended. Unlike new gauge bosons, W' or Z', which might exist at arbitrarily heavy masses or might not exist at all, the symmetry breaking sector must exist and will manifest its presence at least indirectly in strong W_L, Z_L scattering at $\hat{s} \gtrsim 1$ TeV^2 unless the mass scale M_{SB} is below 1 TeV.

3. Calculation of Higgs Production Cross Sections

Higgs production may result from gluon-gluon fusion via a heavy quark loop[7] (Fig. 3.1a) by quark-antiquark annihilation (Fig. 3.1b) and by gauge boson fusion[8] (Fig 3.1c). The first two of these depend strongly on the t quark mass — see Fig. 3.2 taken from Ref. 9 for an indication of this dependence. Note that a larger t mass generally implies a larger Higgs production cross section. For Higgs masses below about 600 GeV and above 2 M_Z, discovery of the Higgs is straightforward in the channel $H \rightarrow ZZ$, both Z decaying to e^+e^- or $\mu^+\mu^-$ as we will show in Section 5. One is therefore less sensitive to uncertainties in the theoretical calculations of the cross section, mostly due to the unknown t-quark mass which affects the gluon-gluon fusion process. Since most of our calculations have been done for a t quark mass of 40 GeV, we are somewhat conservative

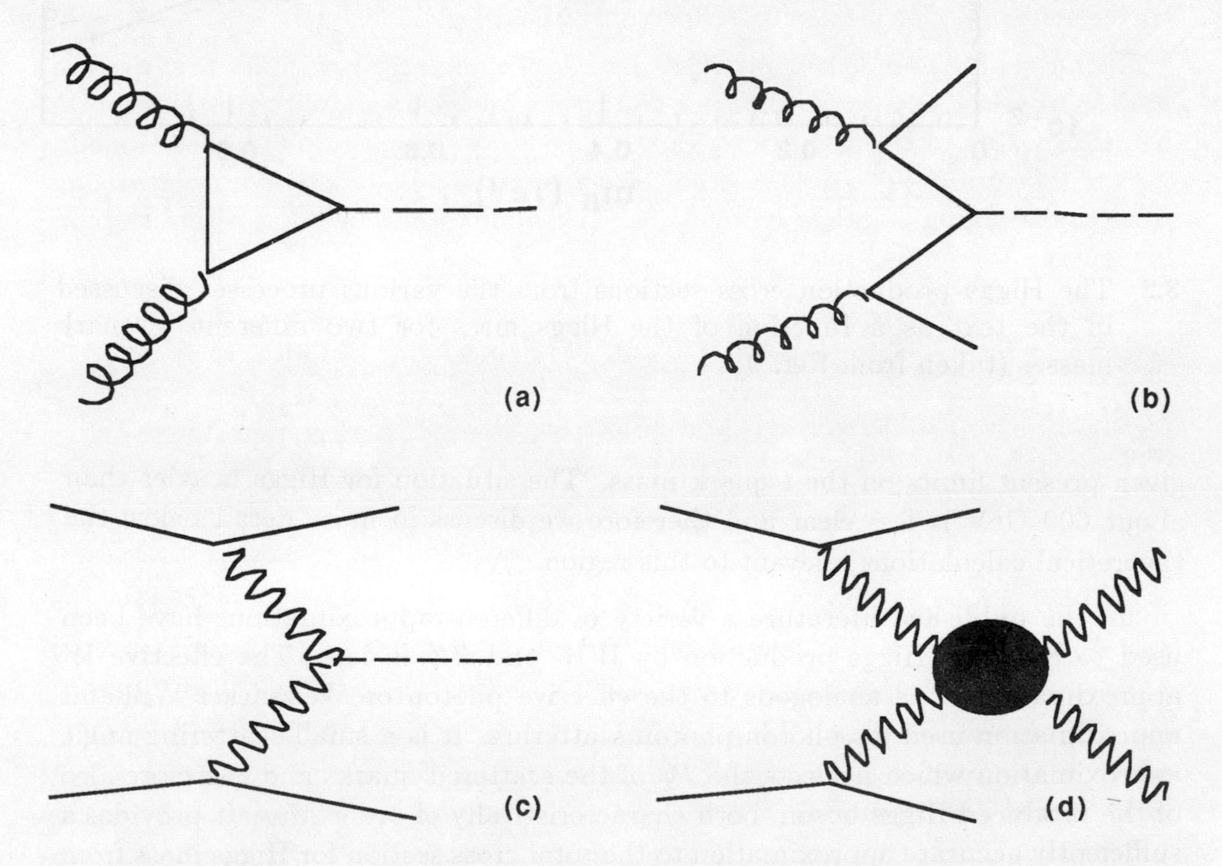

3.1 Higgs production via (a) gluon-gluon fusion; (b) quark-antiquark annihilation; (c) WW or ZZ fusion; and (d) production of gauge boson pairs in a theory with a strongly coupled gauge boson sector.

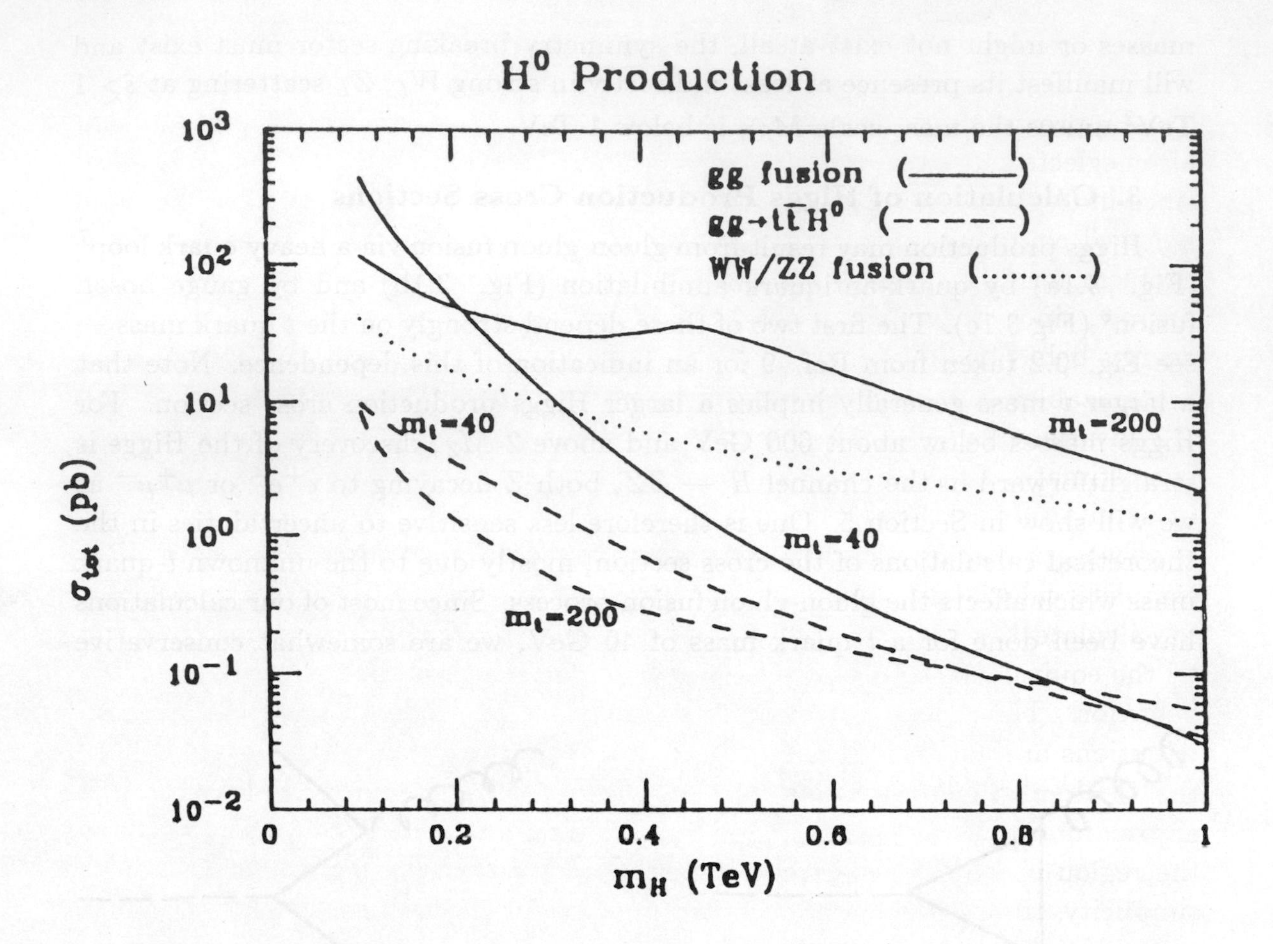

3.2 The Higgs production cross sections from the various processes discussed in the text as a function of the Higgs mass for two different t quark masses (taken from Ref. 9).

given present limits on the t quark mass. The situation for Higgs heavier than about 600 GeV is less clear and therefore we discuss in more detail below the theoretical calculations relevant to this region.

In the published literature a variety of different approximations have been used to compute Higgs production by WW and ZZ fusion. The effective W approximation[4,10] is analogous to the effective photon or Weizsacker-Williams approximation used for photon-photon scattering. It is a small scattering angle approximation which neglects the P_T of the scattered quarks and therefore also of the produced Higgs boson, both characteristically of order M_W. It provides a sufficiently accurate approximation to the total cross section for Higgs mass from 500 GeV and up, with errors of about 10%.[11]

A second common approximation is to include only the s-channel Higgs pole diagram calculated in a particular gauge (U-gauge);[8,28]

$$WW \text{ or } ZZ \to H \to WW \text{ or } ZZ$$

This approximation neglects t and u channel Higgs exchanges, if present, and also neglects gauge sector exchange diagrams. This approximation is pathological since in U-gauge good high energy behavior is obtained by cancellation of badly behaved contributions from the Higgs and gauge sectors. This is not a serious problem for Higgs masses $m_H \lesssim 600$ GeV for which the width is reasonably narrow, since the signal in the peak is not seriously affected by the bad high energy behavior. For the 1 TeV Higgs boson with $\Gamma_H \sim \frac{1}{2}$ TeV there is no recognizable peak and it is necessary to integrate over a broad range of WW and ZZ invariant masses. The s-channel U-gauge pole approximation then seriously overestimates the high mass tail and the production cross section, as discussed in detail in Section 4 below.

Two approaches can be used to overcome this problem. One is to do a complete calculation, including the gauge sector contributions.[29] The other is to calculate the Higgs sector contributions in a renormalizable gauge (justified by the equivalence theorem[4,13]) which automatically has the correct high energy behavior. This is computationally simple since the difficult gauge sector contributions are not included. In the central region (away from the singularity of the Coulomb exchange pole) they may be safely neglected.[14] Numerically the approximation is very good [15] for m_H= 1 TeV and $m_{WW} \geq$0.8 TeV, which is the region in which the signal emerges from the $\overline{q}q$ background. Because of its simplicity, the R-gauge approach can be a useful check of the "exact" U-gauge computations.

Higgs production is sometimes also computed in the zero-width approximation, as if the Higgs were a stable particle.[8,11,23] This is probably reliable for the range of Higgs masses below $\sim$ 600 GeV corresponding to "narrow" peaks but fails around $m_H = 1$ TeV. In the latter case, the signal can only be seen over the $\overline{q}q$ background on the high side of the "peak" where it suffers from falling luminosity that is not reflected in the zero-width computation.

Finally, we should remember that for $m_H = 1$ TeV the Higgs sector is strongly interacting and therefore cannot really be described by perturbation theory. Therefore even the "exact" U-gauge calculation should be regarded as heuristic in this case. There is however evidence that perturbation theory is not a completely misleading guide to Higgs production at $m_H = 1$ TeV: the Higgs sector one loop contributions only correct Γ_H by 10% at $m_H = 1$ TeV.[16]

3.1 Strongly Interacting W and Z Bosons

The WW fusion process of Fig. 3.1d is the source of the signal for strong W_L, Z_L scattering. This is a generalization of one of the Higgs production mechanisms discussed above. The strong interactions, represented by the blob in Fig. 3.1d, generate an excess of gauge boson pairs over those produced by $q\overline{q}$ and gg annihilation. The presence of gauge boson pairs with invariant mass greater than about 1 TeV is the signal for a strongly coupled symmetry breaking sector. The yield of these pairs can be estimated by modeling the gauge boson scattering amplitude as discussed in Section 2 and Ref. 4.

In addition to the neutral pairs, W^+W^- and ZZ, which are also produced by decay of a ($M_H < 1$ TeV) Higgs boson, a strongly coupled theory will produce charge 1 (ZW) and 2 ($W^\pm W^\pm$) final states.[4,17] The latter is particularly important since it has a much smaller background; $W^\pm W^\pm$ cannot be produced by $q\overline{q}$ annihilation. Many of the background issues relevant to Higgs detection are applicable to this case. Some backgrounds (e.g., the background from $Z+$jets to the $ZZ \to ee\nu\nu$ mode) may be relatively less important since the invariant mass of the boson pair is larger. Since models for a strongly coupled weak sector are not included in the ISAJET[18] and PYTHIA[19] Monte Carlo programs, we have not yet performed a detailed analysis of these signals.

3.2 $\overline{q}q$ Annihilation Background

The processes [20,21] $\overline{q}q \to WW$, ZZ are a serious background to $H \to WW$, ZZ detection in all WW, ZZ decay channels.† The gauge bosons produced by $\overline{q}q$ annihilation are predominantly transversely polarized while those from the Higgs decay are predominantly longitudinal.

For Higgs boson masses less than 0.6 TeV, for which there is a recognizable invariant mass peak, it is less critical to know the backgrounds precisely. At the SSC the Higgs boson in this mass range emerges as a peak in the $ZZ \to e^+e^-/\mu^+\mu^- + e^+e^-/\mu^+\mu^-$ channel over the smooth $\overline{q}q$ background. However for the 1 TeV Higgs boson and for the 1-2 TeV strong interaction signal discussed above, there is no peak and it is important to understand the magnitude of the $\overline{q}q$ background.

There are two ingredients in a calculation‡ of the background; perturbative QCD calculations of the relevant partonic process; and a set of structure functions. The first is the easier to deal with. There is Q^2-dependence in both $\alpha_s(Q^2)$ and $f(x,Q^2)$. For the production of W or Z pairs of invariant mass M, $Q \approx M$. The relevant processes are $q+\overline{q} \to Z+Z$[21] and $g+g \to Z+Z$.[22] Until recently only the former was calculated. The QCD corrections to this process which occur

---

† Recently, the process $gg \to ZZ$ has been computed.[22] Although this process was not included in our simulations, we shall comment on its effect when it is appropriate.

‡ Previous studies of this process were carried out by Eichten, Hinchliffe, Lane and Quigg,[23] by Owens,[24] by Tung[25], by Collins[26] and by Collins and Soper.[27]

at order α_s are not known, but the calculation is not difficult. It is very likely that they will be known by the time SSC data are available. In the absence of this calculation, there is an uncertainty of order 30% in the background from this source. The second process, gluon-gluon production, which is of order α_s^2 is also important. This process is approximately 50% of $q\bar{q}$ annihilation provided $|y_z| < 1.5$. Again it is expected that the order α_s^3 corrections will be small, although the calculation is formidable.

The uncertainties in structure functions present a more serious problem. Structure functions are measured in low Q^2 experiments and then extrapolated in Q^2 up to the appropriate scale. There are three main problems:

1) the appropriate value of Λ must be determined.
2) There is little data for $x < 0.1$ and no data at all for $x < 0.01$.
3) The gluon distribution which produces most of the antiquarks at high Q^2 is not measured directly but is inferred from the growth of the antiquarks with Q^2 and is therefore correlated with Λ.

In the case of the W or Z pair background, one is interested in M between 200 GeV and 2 TeV. For central production this corresponds to values of x between 0.005 and 0.05. To a first approximation† the structure functions depend only on Q^2/Λ^2. Consequently the uncertainty due to errors in Λ can be estimated by shifting Q^2. A look at the Q^2 dependence of the structure functions will show that the Q^2 dependence is small over the range of x that we are interested in (see EHLQ[23] Figs. 10–15). A sample is reproduced here as Fig. 3.3. Effect (3) was discussed in EHLQ where two different sets of gluon distributions at $Q^2 = 5$ GeV^2 were used (compare EHLQ Figs. 5 and 6). These different sets produced changes of less than 20% for all the SSC rates in EHLQ.

Problem (2) was also addressed by EHLQ who modified the structure functions below $x = 0.01$ (see EHLQ Eqs. 2.62 and 2.63 and Figs. 18–20). The effects wash out rapidly at Q^2 rises, and are, of course irrelevant for $x > 0.01$.

To verify experimentally these background calculations, a process that is sensitive to the same quark distributions as the background over the same range of x and Q^2 is needed. The only process available are dilepton production and the production of two large P_T photons. W or Z production is less useful since the total cross section is dominated by values of x that are much smaller than the relevant ones and Q^2 is also smaller. By looking at W's at large rapidity one becomes sensitive to the product of two distribution functions; one in the interesting x range and one at much smaller x. Jet cross sections are not useful since for total transverse energies below 6 TeV, gluon-quark and gluon-gluon scattering dominate.

† This is not quite true since there are quark mass effects, but these changes are very small unless one is sensitive to heavy (top) quark distributions.

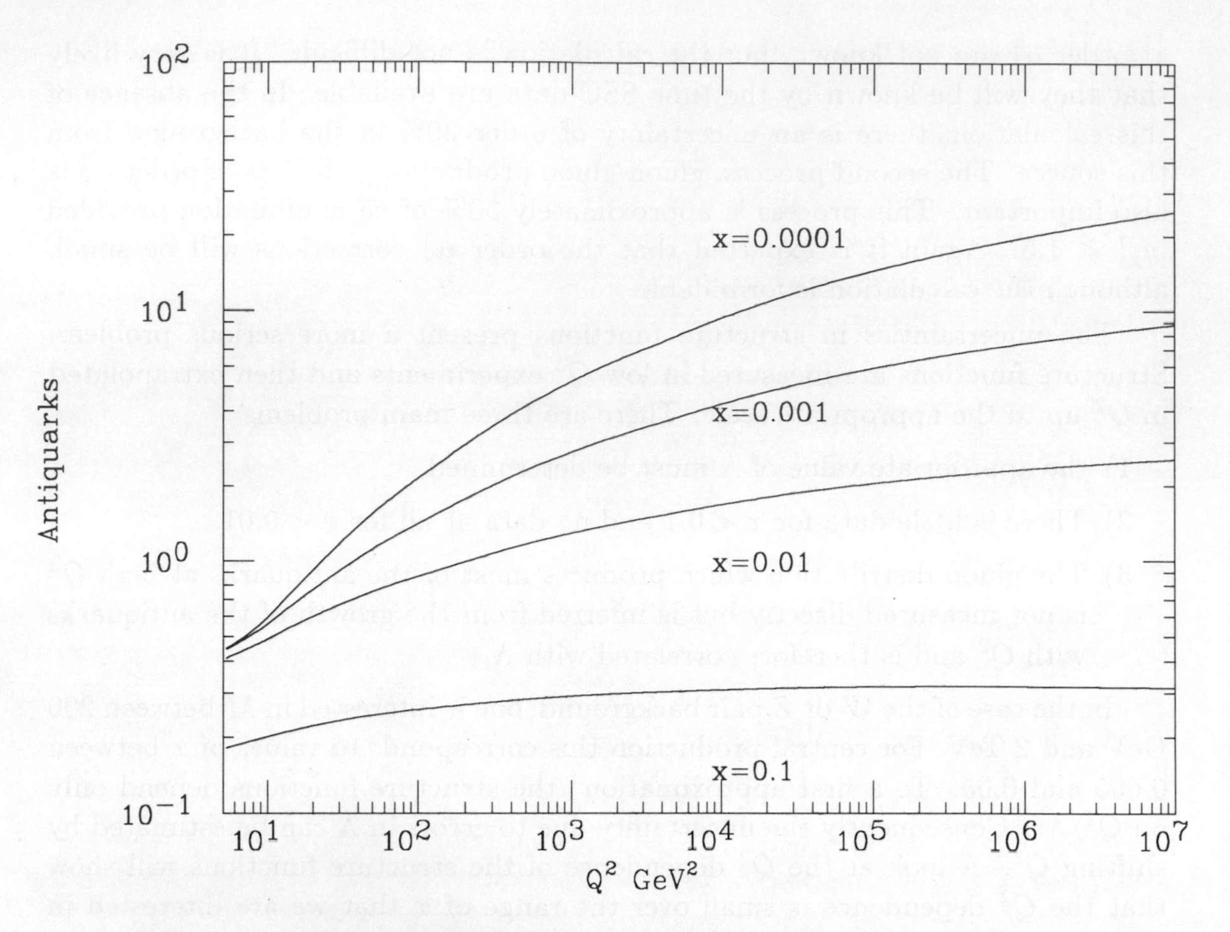

3.3 The Q^2 dependence of the antiquark distribution at fixed x.

How well can dilepton production be measured? Figure 3.4 shows the event rate which must be measured over the same range of invariant mass that will be needed for W pairs. At dilepton pair masses of 1 TeV there are enough events in a 50 GeV bin to make a 30% measurement in each bin. There will also be an error on the measurement of the dilepton mass. A 5% error on this measurement translates into a 15% error on the cross-section. This figure attempts to indicate the effect of uncertainties in structure functions; changes in Λ from 150 to 450 MeV are too small to measure.

A new set of structure functions has been invented that attempts to increase the dilepton rate (and also the W or Z pair background). The EHLQ structure functions at $Q^2 = 5$ GeV^2 have been changed as follows. Take set 2 of EHLQ and double the sea i.e., assume that the measurements of the antiquarks are a factor of 2 too small at low Q^2. Replace the gluon distribution (EHLQ 2.60) by

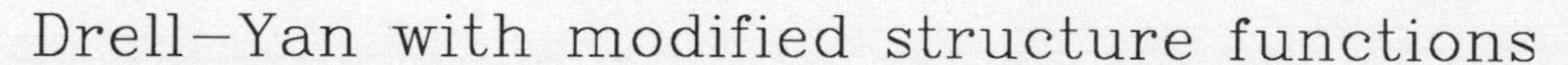

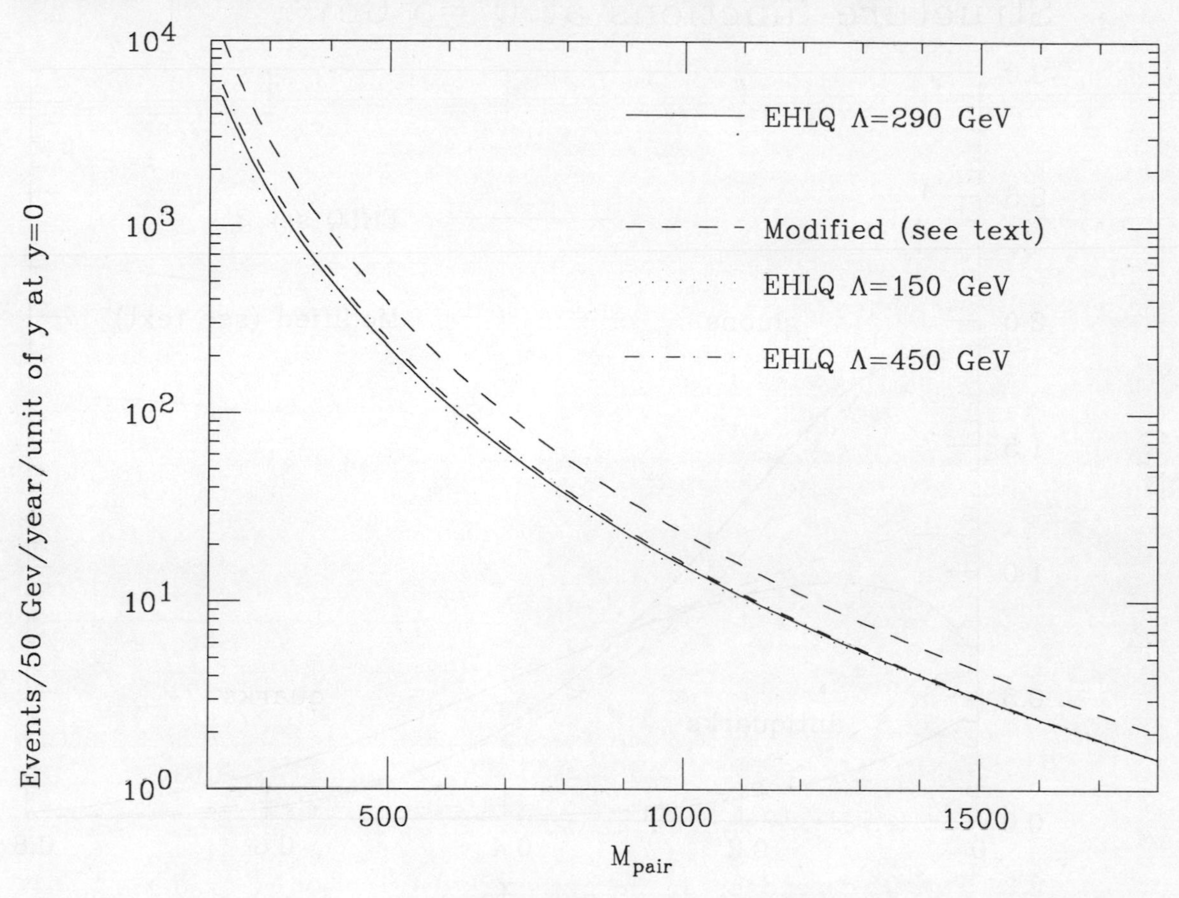

3.4 The cross section $d\sigma/dmdy$ for the production of a muon pair at $y = 0$ as a function of the muon pair invariant mass. Four curves are shown; three of these use the EHLQ set 2 structure functions with different choices for Λ_{QCD}. The fourth curve corresponds to the modified set of structure functions described in Section 3.2.

$$xg(x, 5\text{GeV}^2) = 2.6(1-x)^4 \quad ; \quad x > 0.1$$

$$xg(x, 5\text{GeV}^2) = 0.54/\sqrt{x} \quad ; \quad x < 0.1$$

This set of structure functions has a much more singular behaviour at small x than EHLQ's and is similar at larger x. These changes are unlikely to be in agreement with current data. In particular, they are likely to generate too large a value for $F_2(x, Q^2)$ at $Q^2 \sim 20$ GeV^2 for $x \sim 0.05$ where data exist. They also violate the constraint that the total momentum carried by partons must add up to one. This modified set is compared with EHLQ set 2 structure functions in Fig. 3.5. Figure 3.4 shows the effect of these structure functions on the dilepton

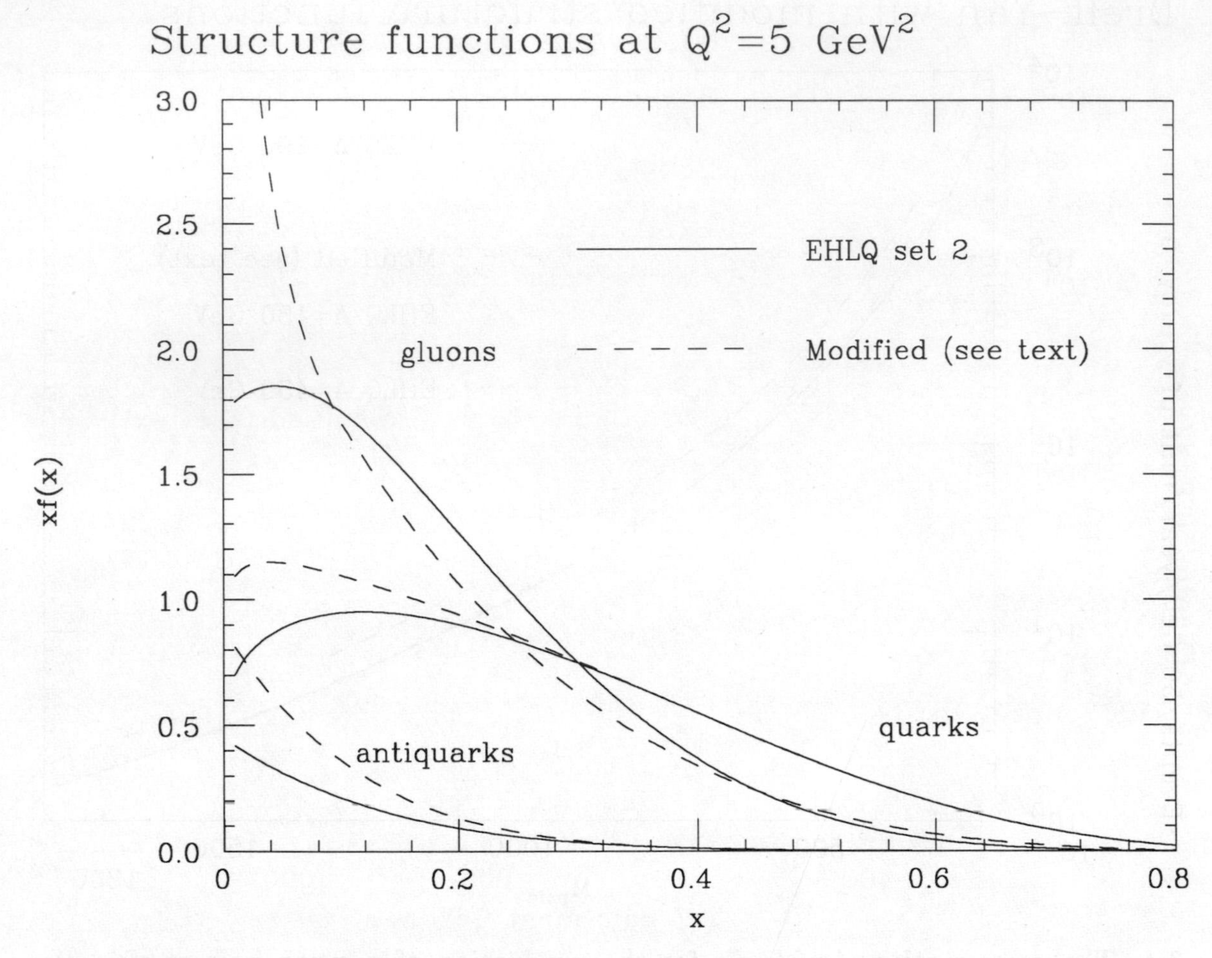

3.5 **The quark, antiquark and gluon distributions as a function of x at $Q^2 = 5$ GeV^2. Two sets are shown; those of EHLQ set 2 and those described in Section 3.2.**

rate. The effect is greatest at small values of the dilepton mass where x and Q^2 are smallest. It is likely that this process can be measured well enough to see the difference between this result and the EHLQ prediction.

Figure 3.6 shows the Z pair rate using the same structure functions as Fig. 3.5. The relative contributions of the charge 1/3 and charge 2/3 quarks to the dilepton and Z pair processes are not quite the same. Hence it is conceivable that one could be affected more than the other by changes in structure functions. Nevertheless is very difficult to double the Z pair rate while keeping the dilepton rate unchanged. In practice, a large number of processes will be measured, all of which contribute to the knowledge of structure functions. This will surely allow some of the remaining uncertainties to be eliminated. There will be enough dilepton events to provide a calibration of the background at the 60% level over

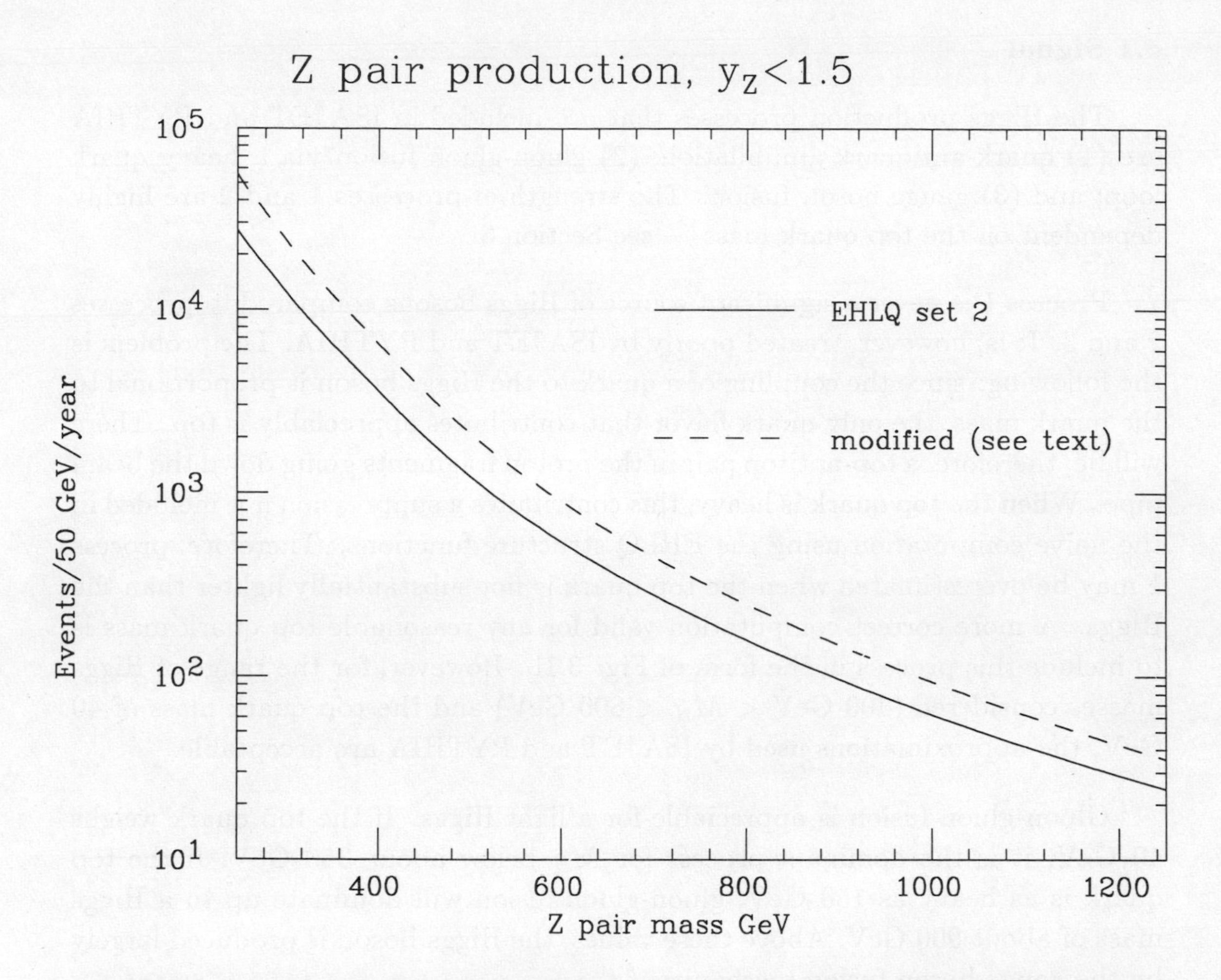

3.6 The cross section $d\sigma/dM$ for the production of a Z pair of mass M. Both Z's are required to have $|y_Z| < 1.5$. The two lines show the effect of different sets of structure functions; EHLQ set 2 and those described in Section 3.2.

the range of Z pair masses less than 1 TeV or so. At larger Z pair masses the situation is more difficult. There are only about 25 dilepton pairs with invariant mass greater than 1.5 TeV, so it becomes difficult to measure the background over the relevant range of x and Q^2. Nevertheless, it is very unlikely to be uncertain to more than a factor of two.

4. Higgs Production Processes: Comparison Between Theory and Monte Carlo Programs

In this section we test the predictions of the ISAJET[18] (v5.34) and PYTHIA[19] (v4.8) Monte Carlo programs against simple partonic calculations.

4.1 Signal

The Higgs production processes that are included in ISAJET and PYTHIA are (1) quark-antiquark annihilation; (2) gluon-gluon fusion via a heavy quark loop; and (3) gauge boson fusion. The strength of processes 1 and 2 are highly dependent on the top quark mass — see Section 3.

Process 1 is never a significant source of Higgs bosons compared to processes 2 and 3. It is, however, treated poorly by ISAJET and PYTHIA. The problem is the following: since the coupling of a quark to the Higgs boson is proportional to the quark mass, the only quark flavor that contributes appreciably is top. There will be, therefore, a top-antitop pair in the proton fragments going down the beam pipe. When the top quark is heavy, this contributes a suppression not included in the naive computation using the EHLQ structure functions. Therefore, process 1 may be overestimated when the top quark is not substantially lighter than the Higgs. A more correct computation valid for any reasonable top quark mass is to include this process in the form of Fig. 3.1b. However, for the range of Higgs masses considered (400 GeV $< M_H <$ 800 GeV) and the top quark mass of 40 GeV, the approximations used by ISAJET and PYTHIA are acceptable.

Gluon-gluon fusion is appreciable for a light Higgs. If the top quark weighs 40 GeV, it is the dominant process for M_H below about 350 GeV. If the top quark is as heavy as 150 GeV, gluon-gluon fusion will dominate up to a Higgs mass of about 900 GeV. Above these values, the Higgs boson is produced largely by the gauge boson fusion mechanism.[8,10]

In order to check that ISAJET and PYTHIA treat gluon-gluon fusion properly, their rates for this process were compared with the results of a program that computes the partonic cross section. This program was written by the authors of Ref. 23. Table 4.1 shows the results. Here and below, one SSC year corresponds to an integrated luminosity of 10^{40}cm^2. The gg fusion cross section in ISAJET v5.34 was wrong, but has since been corrected. The corrected results are given in Table 4.1.

In considering the Higgs production via gauge boson fusion it is important to note that the Higgs itself is not a final state particle, but rather a resonance that immediately decays into W or Z pairs. Therefore what one must actually calculate is the process $qq \to qqVV$, where V is W or Z. The complete gauge invariant calculation of this process[29] is quite difficult and the resulting matrix element is too unwieldy to be used in a Monte Carlo program.

Table 4.1

A comparison of the event rates for the process $gg \to H \to ZZ$ from the corrected version of ISAJET, PYTHIA and a partonic calculation *ala* EHLQ. No rapidity cuts are applied. Events per SSC year are shown.

M_H	ISAJET	PYTHIA	Partonic
400	6200	7800	7600
800	210	260	260

ISAJET and PYTHIA treat the $qq \to qqVV$ process using the effective-W approximation as described in Section 3. In this approximation, the incoming gauge bosons are treated as though they were constituents of the incoming protons. One can then compute a luminosity distribution for V pairs. This is then multiplied by the on-shell matrix element for $VV \to VV$.

ISAJET and PYTHIA take two different approaches to the computation of this matrix element. ISAJET computes the full set of $VV \to VV$ diagrams. Only longitudinally polarized incoming V's are included, though all polarizations are included on the outgoing legs. PYTHIA computes only the s-channel Higgs pole diagram and only longitudinal polarizations are included in the final state. (PYTHIA is being revised to perform the same computation as ISAJET).[30]

Each of these approximations has its pathologies. The effective-W approximation treats the incoming V's as being on mass shell while in fact they are off mass shell by approximately their mass. The net effect of this is to overestimate the total rate somewhat, especially for a Higgs near the $H \to VV$ threshold.

The approximation made by ISAJET suffers in the W^+W^- channel from having a fictitious t-channel photon exchange pole caused by putting the incoming bosons on mass shell. This leads to an infinite total cross section. One must therefore always impose some minimum P_T cut on the outgoing particles. If P_T is at least of order the W mass, then the ISAJET's result should approximate the true answer. In any case, such a cut is required to reduce the background.

PYTHIA's approximation does not have a fictitious infinity at low P_T, because the photon exchange diagram is not included. However, the s-channel Higgs exchange diagram by itself is not gauge invariant, and it violates unitarity at high energy. Examples of this are shown in Figs. 4.1a-c. These figures show the cross section for unpolarized $ZZ \to ZZ$ scattering at Higgs masses of 400, 800, and 1000 GeV. For $ZZ \to ZZ$ scattering there are three diagrams. The

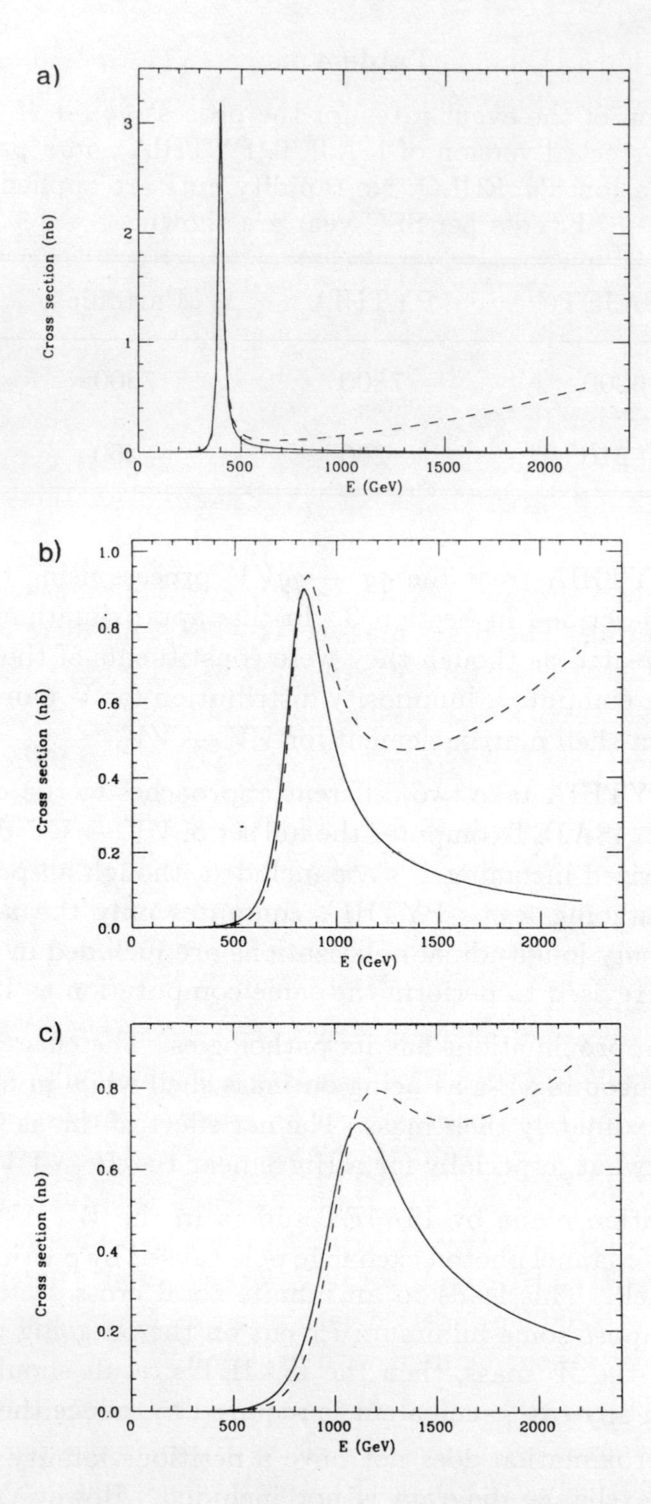

4.1 a) Higgs production via s, t and u channel diagrams (solid line) and s channel only for $M_H = 400$ GeV. b) Higgs production via s, t and u channel diagrams (solid line) and s channel only for $M_H = 800$ GeV. c) Higgs production via s, t and u channel diagrams (solid line) and s channel only for $M_H = 1000$ GeV.

Higgs may be exchanged in the s, t, and u channels. The solid line plots the correct answer, using all three diagrams. The dashed line shows the effect of leaving out the t and u channel diagrams. At large ZZ invariant mass, the dashed line grows like s^2. In the process $qq \to qqZZ \to qqH \to qqZZ$, these matrix elements will get multiplied by appropriate structure functions, which are rapidly falling functions of diboson mass. The bad high energy behavior of the non-unitary matrix element manifests itself as a long flat tail of events at high diboson mass. This approximation cannot be used for invariant masses above about 1200 GeV, and it cannot be used at all for Higgs masses of 1 TeV or above.

PYTHIA's approximation also neglects low energy scattering of the incoming V's (from the photon exchange, for example), and therefore may underestimate the cross section at ZZ invariant mass below the Higgs peak. This is unimportant in practice.

Figures 4.2a-c show the cross section vs WW invariant mass for $pp \to WW + X$ from WW scattering. The Higgs mass is 800 GeV, and there is a rapidity cut on the outgoing W's $|y_w| < 1.5$. Figure 4.2a is a partonic calculation using the effective-W approximation, including all the diagrams for $V_L V_L \to W_L^+ W_L^-$. Figure 4.2b is the result from ISAJET, and Fig. 4.2c is from PYTHIA. Note that Figs. 4.2a and 4.2b have virtually no events above about 1500 GeV, while Fig. 4.2c has a long tail extending up to high energy. This tail is fictitious, and must be removed by truncating the curve at 1200 GeV. The low energy shapes of 4.2a and 4.2b are also different from Fig. 4.2c. The agreement of Figs. 4.2a and 4.2b does not necessarily indicate that ISAJET's calculation is more correct than PYTHIA's, but only that the approximation used in the parton level program which computed Fig. 4.2a is the same as used by ISAJET.

The features of ISAJET and PYTHIA Higgs production cross-sections were checked against two different parton level programs. The first[28] computes the diagram for $qq \to qqVV \to qqH \to qqZZ$ exactly, without using the effective-W approximation. The second uses the effective-W approximation, including all diagrams for $V_L V_L \to W_L^+ W_L^-$. Of course, these two approaches suffer from the same diseases as PYTHIA and ISAJET, respectively. Therefore a cut at an invariant ZZ mass of 1200 GeV is used with program 1, and some cut away from the photon pole must always be used with program 2.

Program 1 can be used to calculate the transverse momentum of the Higgs. Figures 4.3a-c show cross-section vs. P_T of the Higgs from program 1, ISAJET and PYTHIA, respectively. The Higgs mass is 800 GeV, and a cut on the rapidity of the Higgs of $|y_H| < 1$ was imposed. (These curves are insensitive to both the Higgs mass and the rapidity cut.) These curves show that ISAJET and PYTHIA are in substantial agreement with the parton level calculation.

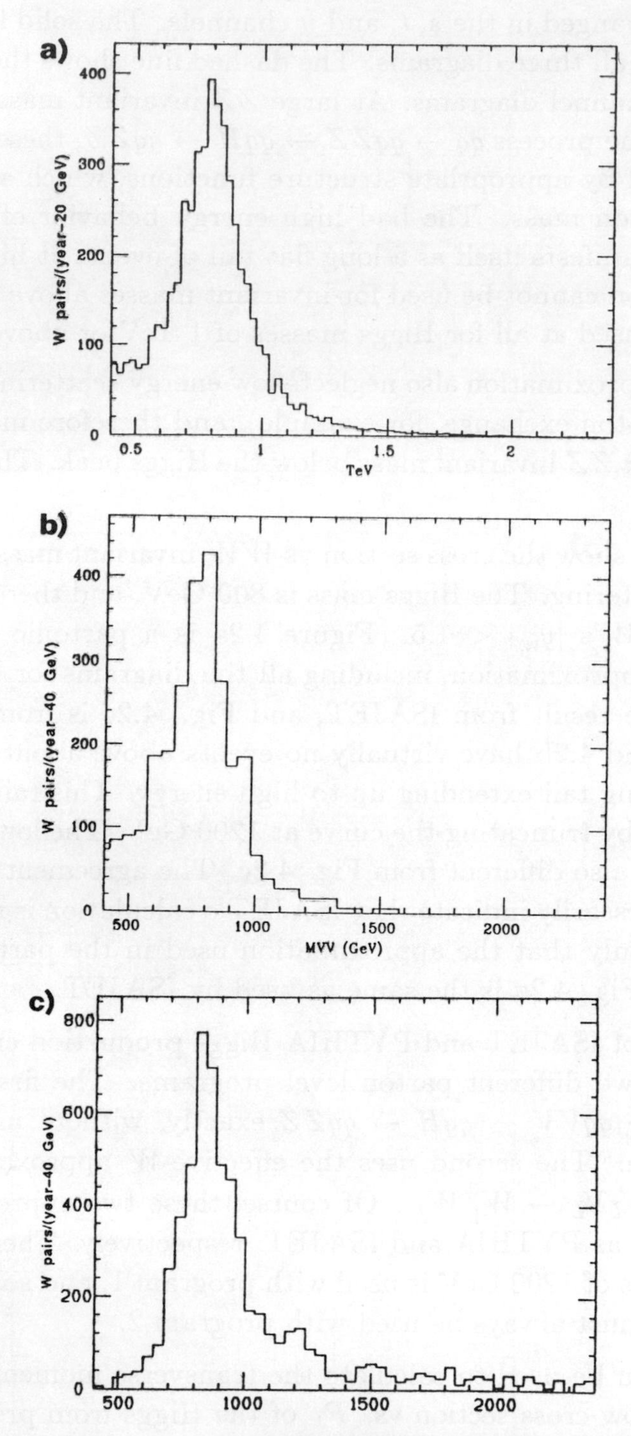

4.2 a) The WW invariant mass distribution in 40 GeV bins from an 800 GeV Higgs resulting from a partonic calculation using the effective W approximation. A cut of $|\ y_W\ | < 1.5$ is applied. b) The WW invariant mass distribution from ISAJET for $M_H = 800$ GeV and $|\ y_W\ | < 1.5$. c) The WW invariant mass distribution from PYTHIA for $M_H = 800$ GeV and $|\ y_W\ | < 1.5$.

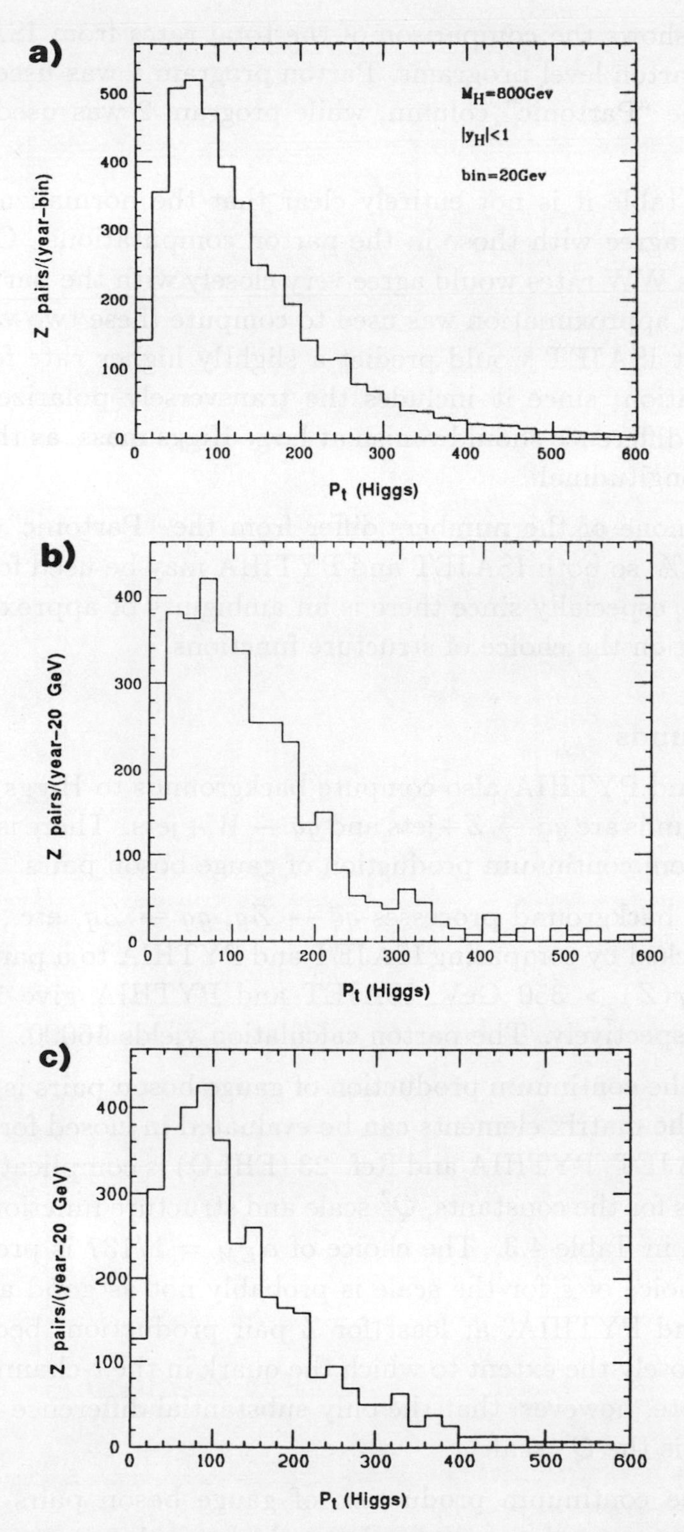

4.3 a) The P_T distribution of an 800 GeV Higgs as calculated by the parton-level program of Cahn et al. We required $| y_H | < 1$. b) The P_T distribution of an 800 GeV Higgs as calculated by ISAJET. c) The P_T distribution of an 800 GeV Higgs as calculated by PYTHIA.

Table 4.2 shows the comparison of the total rates from ISAJET, PYTHIA, and the two parton level programs. Parton program 1 was used to calculate the ZZ rates in the "Partonic" column, while program 2 was used to compute the WW rates.

From this table it is not entirely clear that the normalizations of ISAJET and PYTHIA agree with those in the parton computations. One would expect that ISAJET's WW rates would agree very closely with the parton WW column, since the same approximation was used to compute these two values. One might anticipate that ISAJET would predict a slightly higher rate for WW than the parton calculation, since it includes the transversely polarized outgoing W's. However, this difference should vanish at large Higgs mass, as the signal becomes virtually all longitudinal.

However, none of the numbers differ from the "Partonic" column by more than about 25%, so both ISAJET and PYTHIA may be used for the purposes of this workshop, especially since there is an ambiguity of approximately the same size depending on the choice of structure functions.

4.2 Backgrounds

ISAJET and PYTHIA also compute backgrounds to Higgs production. The QCD backgrounds are $q\bar{q} \to Z$+jets and $q\bar{q} \to W$+jets. There is also a significant background from continuum production of gauge boson pairs.

The QCD background processes $q\overline{q} \to Zg$, $gq \to Zq$, etc., with $Z \to e^+e^-$ only, were checked by comparing ISAJET and PYTHIA to a partonic calculation. Cutting on $P_T(Z) > 350$ GeV, ISAJET and PYTHIA give 18270 and 18700 events/year respectively. The parton calculation yields 16000.

Checking the continuum production of gauge boson pairs is made simpler by the fact that the matrix elements can be evaluated in closed form.[20,21] Comparison among ISAJET, PYTHIA and Ref. 23 (EHLQ) is complicated by their use of different values for the constants, Q^2 scale and structure functions. The situation is summarized in Table 4.3. The choice of $\alpha_{EM} = 1/137$ is probably inferior to $1/128$. The choice of $\hat{s}$ for the scale is probably not as good as the scales used by ISAJET and PYTHIA, at least for Z pair production, because it does not represent as closely the extent to which the quark in the t-channel diagrams is off mass shell. Note, however, that the only substantial difference between ISAJET and PYTHIA is the Q^2 scale.

To test the continuum production of gauge boson pairs by ISAJET and PYTHIA, a program was written that calculates these processes at the parton level. Table 4.4 shows that the results of the partonic calculations depend strongly on which set of parameters is selected from Table 4.3.

Table 4.2

Verification of the Higgs Signal from VV Fusion. Events per SSC year are shown. The numbers shown have statistical errors of approximately 5%.

	ISAJET	PYTHIA	PARTONIC	
Case I: $P_T(V) > 50$ GeV, no y_V cut.				
$M_H = 800$ GeV				
WW	13400	9000	12700	
ZZ	5600	4900	5400	
Case II: No $P_T(V)$ cut, $\mid y_V \mid < 1.5$				
$M_H = 400$ GeV				
WW	14000	15800	12400	$M_{VV} > 350$ GeV
ZZ	6000	8000	7100	
$M_H = 600$ GeV				
WW	9200	8600	7400	$M_{VV} > 400$ GeV
ZZ	3700	4100	3400	
$M_H = 800$ GeV				
WW	5900	4400	4800	$M_{VV} > 400$ GeV
ZZ	2300	2300	2300	

NB: The "Partonic" column is from two different programs; WW: from the effective W calculation; ZZ: from the program of Cahn, Ellis, Kleiss, and Stirling.[28]

Table 4.3

Constants used in $q\bar{q} \rightarrow VV$

	ISAJET	PYTHIA	EHLQ
M_W	83.38 GeV	83.0 GeV	83.0 GeV
M_Z	94.11 GeV	94.0 GeV	92.0 GeV
$\sin^2\theta_W$	0.215	0.215	0.220
$\alpha_{\rm EM}$	1/137	1/137	1/128
Struct fcns	EHLQ I	EHLQ I	EHLQ II
Q^2	$\frac{2\hat{s}\hat{t}\hat{u}}{\hat{s}^2+\hat{t}^2+\hat{u}^2}$	$\frac{1}{2}(p_{1\perp}^2 + p_{2\perp}^2 + m_1^2 + m_2^2)$	$\hat{s}$

Table 4.4

Case I: 200 GeV $< P_T <$ 600 GeV, No y_V cut

	ISAJET Parameters	PYTHIA Parameters	EHLQ Parameters
WW	30800 / 30500	20600 / 30200	38800
ZZ	6400 / 6700	4500 / 6600	8500

Case II: No P_T cut, $\mid y_V \mid\ < 1.5$ (both bosons)

	ISAJET Parameters	PYTHIA Parameters	EHLQ Parameters
WW	391000 / 306000	304000 / 344000	519000
ZZ	70000 / 59000	56000 / 65000	100000

Case III: No P_T cut, $\mid y_V \mid\ < 1.5$ (both bosons), $M_{VV} >$ 300 GeV

	ISAJET Parameters	PYTHIA Parameters	EHLQ Parameters
ZZ	16000 / 15700	14000 / 16300	22500

Note: The number of events for WW and ZZ production per SSC year. In the ISAJET and PYTHIA columns, two numbers are given. The first number is the actual value given by the programs using their default parameters. The second is the value that the parton (EHLQ) program gives using the default parameters of ISAJET and PYTHIA. The numbers shown have statistical errors of about 5%.

About half of the difference between the results with EHLQ parameters and the other two partonic calculations is due to the different Q^2. This indicates the need for the higher order calculation of these processes, since only the next order computation can resolve the differences.

Table 4.4 also shows how well ISAJET and PYTHIA agree with the parton computation. ISAJET (v5.34) was exactly a factor of 2 too large for the ZZ rates, due to a mistake in matrix element; this has been corrected in v5.35. This factor of 2 was removed in Table 4.4. Aside from this, ISAJET's numbers appear to be somewhat too large near threshold, but a cut on the invariant mass of the gauge bosons (as is likely to be made in practice, if the Higgs is substantially heavier than $2M_Z$) removes this excess of events. PYTHIA appears to be substantially below the predictions of the parton program, especially when the cut forces the gauge bosons out to large P_T. Therefore, the $q\bar{q} \to VV$ rates from PYTHIA were multiplied by a factor of 1.8 for the analyses described in subsequent sections.

The partonic cross section for the process $gg \to ZZ$ has not been given explicitly.[22] Consequently, the process cannot be incorporated into PYTHIA or ISAJET. Dicus et al.[22] give figures showing the ratio of the ZZ production at the SSC via gg and $q\bar{q}$ annihilation for various kinematic cuts. We shall use these results to estimate the effect of the $gg \to ZZ$ background process.

4.3 Conclusions

Theoretical uncertainties in the Higgs production signal are at present quite substantial, and the potential errors in ISAJET's and PYTHIA's calculations are small compared to them. We conclude that

- ISAJET and PYTHIA agree on the background $Z+$jets.
- ISAJET's $q\bar{q} \to VV$ rates may be high near threshold, but this is irrelevant for studies of a Higgs heavier than about 300 GeV.
- PYTHIA's $q\bar{q} \to VV$ rates are too low. **Rates from PYTHIA are multiplied by a factor of 1.8 for the results described in subsequent sections.**

5. Heavy Higgs Decay into All Charged Leptons

The decay of the heavy Higgs into ZZ where both $Z \to ee$ or $\mu\mu$ is the cleanest signature for a heavy Higgs at the SSC. The total Higgs production cross section depends on the t-quark mass, as previously noted in Section 3. For all $M_H > 300$ GeV, the branching ratio[31] for $H \to ZZ$ is approximately 30% but depends slightly on the t-quark mass. In most of our studies we have assumed a t-quark mass of 40 GeV.

Using PYTHIA and $m_t = 40$ GeV, we have generated Higgs events at Higgs masses of 400, 600 and 800 GeV and the continuum ZZ background (increased by the factor of 1.8 as described in Section 4) for an integrated luminosity of 10^{40} with $|y_z| < 1.5$ and perfect e and μ detection efficiency and energy resolution. The results are shown in Figs. 5.1a–c. The curves are simple polynomial and Breit-Wigner fits to the background and signal, respectively. With these assumptions, a Higgs of mass ≈ 600 GeV could be detected in this mode; above this mass the signal becomes too small and too broad to observe as a bump on top

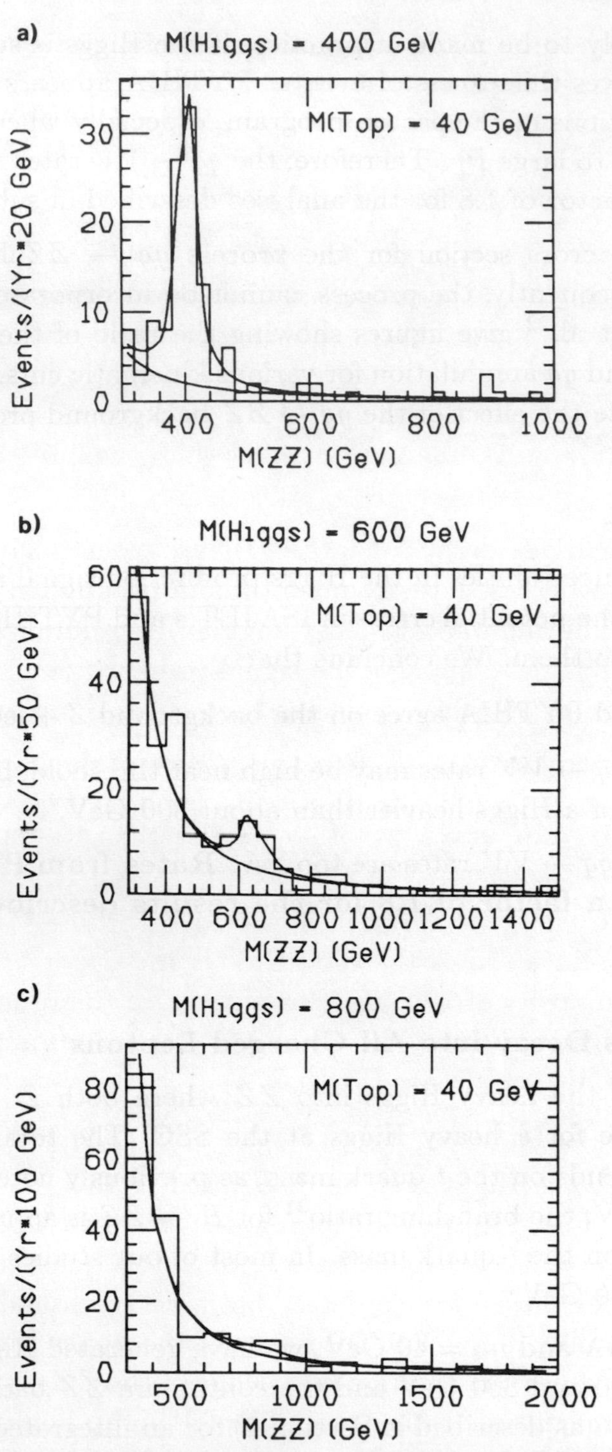

5.1 The ZZ invariant mass distribution arising from Higgs decay and from the background process $q\overline{q} \rightarrow ZZ$. The cuts are described in the text. The distribution is shown for Higgs masses of 400, 600, and 800 GeV. If the background process $gg \rightarrow ZZ$ were to be included, the background would be increased by a factor which is about 1.7 (1.6) for M_{ZZ} = 400(800) GeV.

of a smooth background. For a top quark mass of $\approx$40 GeV, it may be possible to discover the Higgs for masses somewhat above 600 GeV *if* both the shape and magnitude of the ZZ background can be determined by calculations together with other measurements which determine the structure functions in the relevant kinematic region.

If the top quark mass is larger, for example 200 GeV, the Higgs signal grows appreciably—see Fig. 5.2a,b and c—in which case discovery of a Higgs in this mode up to $\approx$800 GeV may be possible. Obviously the mass reach may also be extended by increasing the integrated luminosity. Since this signal at high mass, even with electrons in the final state, appears to be robust (see Section 8.1.4), increasing the peak luminosity rather than the integration time would be desirable.

The Z's from Higgs decay for large M_H are essentially completely longitudinally polarized whereas those from continuum production are not. The use of this polarization information, the decay angular distribution of the leptons in the Z rest frame, can slightly improve the signal to background ratio, by about one standard deviation at 600 GeV. This has been studied in some detail in Ref. 31.

We have also explored the energy dependence of the Higgs production cross section times branching ratio in this mode for a 400 GeV mass Higgs using PYTHIA (for m_t = 40 GeV) as described previously. The result of varying the center-of-mass energy from about 11 TeV ($\approx$ today's magnets in the LEP tunnel) to the SSC energy is shown in Fig. 5.3a–d. Studies[32] for the LHC (design energy $\approx$17 TeV) have shown that 300 GeV is the upper limit in this mode for an integrated luminosity of 10^{40}, in agreement with the results presented in Fig. 5.3b. Both the signal rate *and* the signal to background improve as the center-of-mass energy is increased—see Fig. 5.4.

From our studies of Higgs $\to ZZ$, both $ZZ \to ee$ or $\mu\mu$, we conclude that:

- At the SSC, the Higgs may be discovered in this mode for an integrated luminosity of 10^{40} up to a Higgs mass of $\approx$600 GeV for a top quark mass of $\approx$40 GeV and up to $\approx$800 GeV for a top quark mass of $\approx$200 GeV. This assumes the ability to detect both electrons and muons from the Z decays assuming $|y_z| < 1.5$. Discovery at the upper Higgs mass range limit will likely require quantitative knowledge of the ZZ continuum background shape and magnitude rather than simple "bump hunting".
- The use of polarization knowledge of the Z decays will help only slightly to discover the Higgs in this mode although it would provide confirmation of the nature of a resonance discovered in the ZZ channel.
- Both the Higgs cross section and ratio of Higgs production to continuum ZZ production depend strongly on the available center of mass energy.

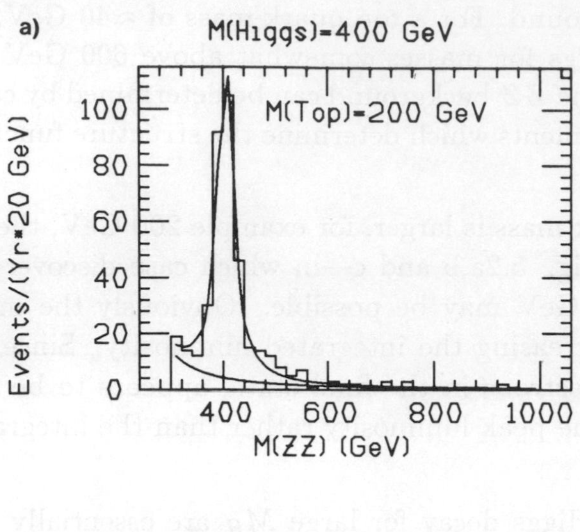

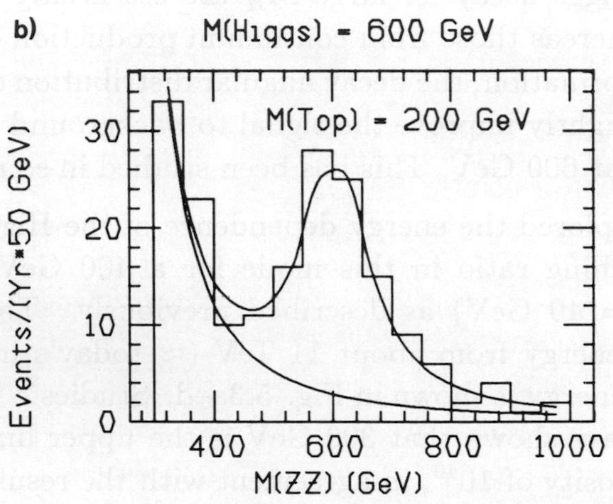

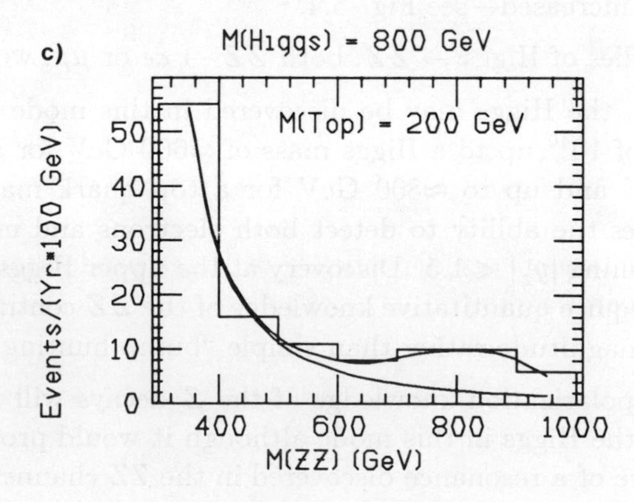

5.2 Same as Fig. 5.1 but for a top quark mass of 200 GeV.

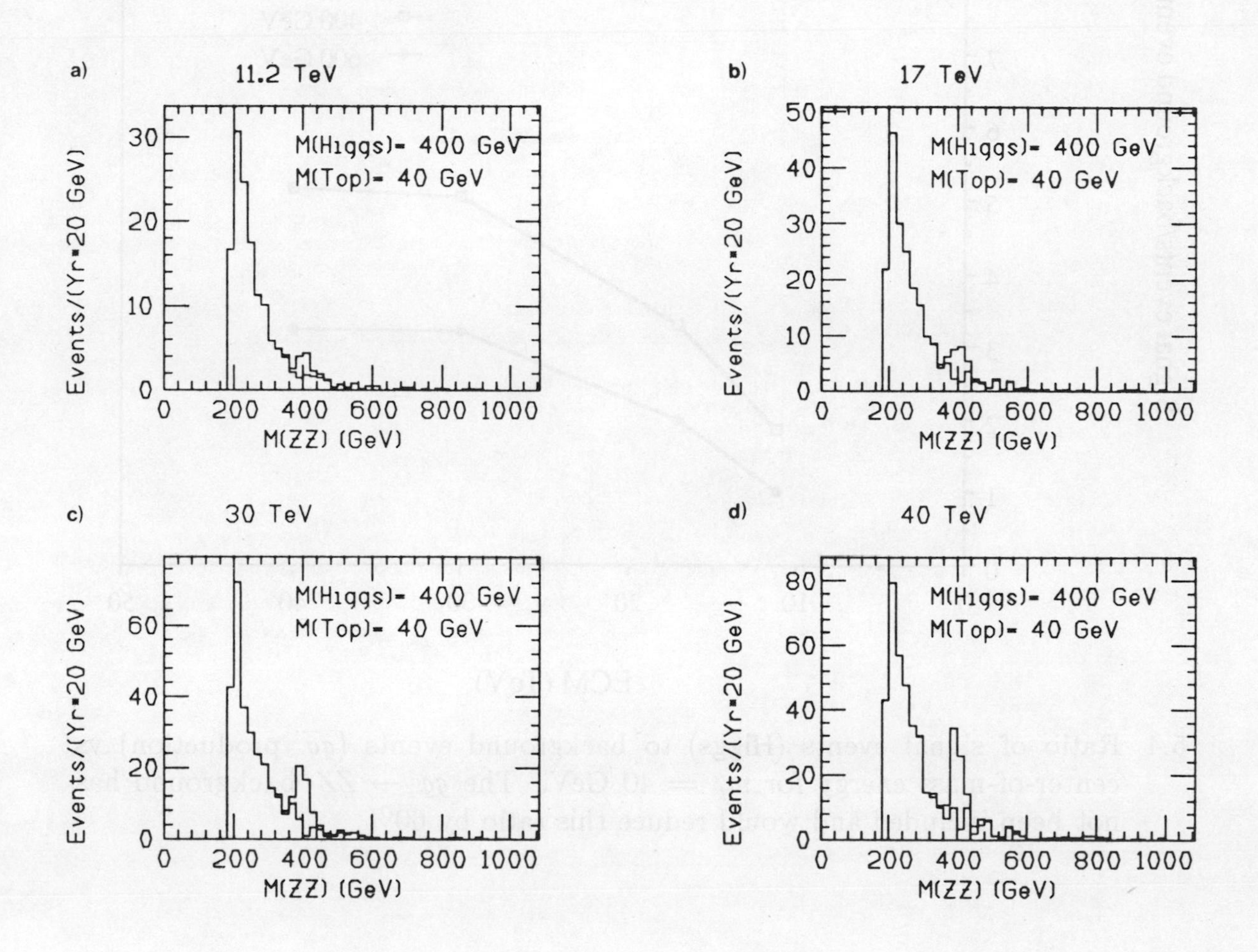

5.3 The center-of-mass energy dependence of Higgs production for $M_H = 400$ GeV for the energies shown.

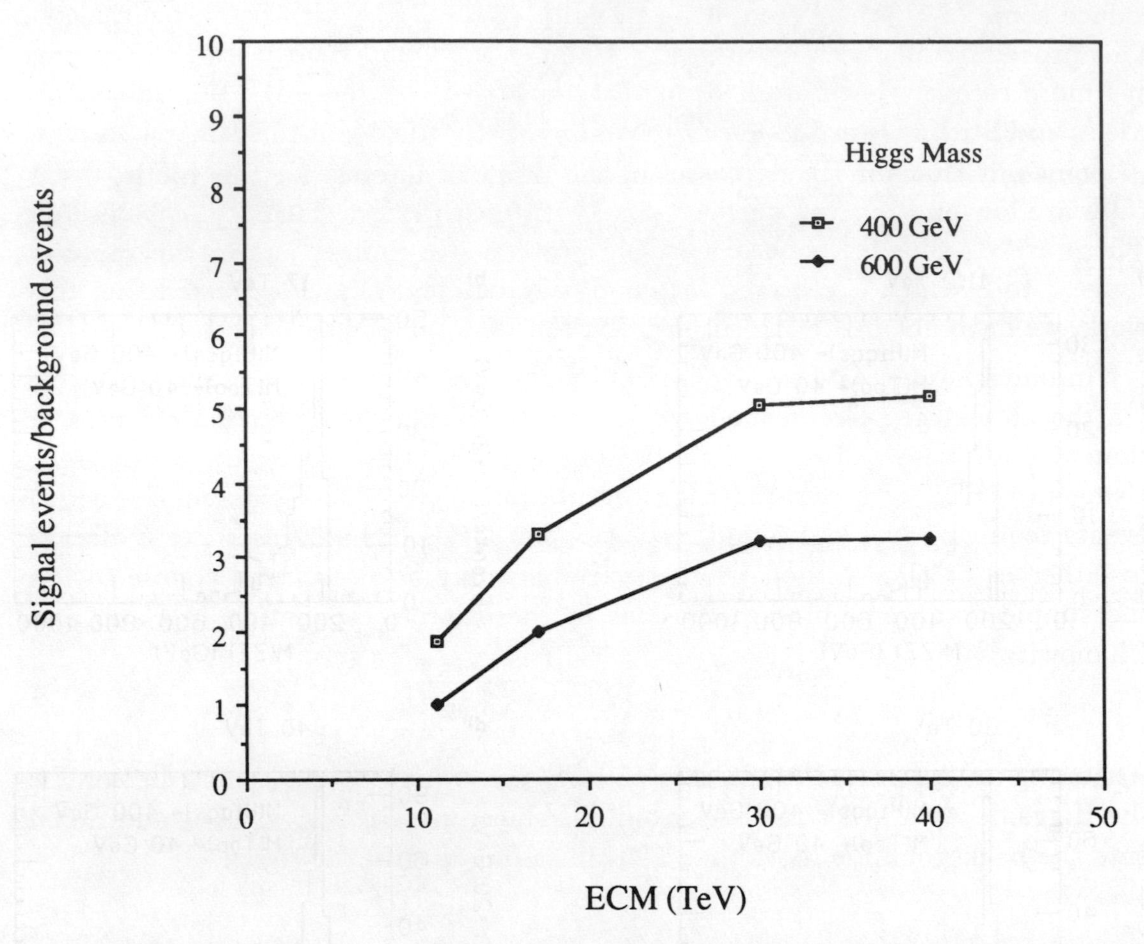

5.4 Ratio of signal events (Higgs) to background events ($q\overline{q}$ production) vs center-of-mass energy for $m_t = 40$ GeV. The $gg \to ZZ$ background has not been included and would reduce this ratio by 60%.

6. Heavy Higgs Decay into Z $\to ee$ or $\mu\mu$ and Z $\to \nu\overline{\nu}$

The advantage of this mode is that the branching ratio is a factor of six larger than the all charged lepton mode. The disadvantage is that single Z + jet(s) production becomes a serious background because its rate is much larger than the Higgs production rate. To remove this background it is important to reject events in which jet activity balances the transverse momentum of the observed $Z \to ee$ or $\mu\mu$.

It may be possible to tag the quarks recoiling against the gauge bosons that interact to form the Higgs. As shown in Fig. 4.3, the Higgs typically carries about 100 GeV of transverse momentum. This is balanced by the two quark jets, which are typically at large rapidities. By identifying these jets, it may be possible to reduce some of the backgrounds, especially $q\bar{q} \to ZZ$, which has no such jets. This procedure has been discussed by Cahn et al.,[28] and Barger et al.[33] Tagging has more recently been studied, in the context of the $H \to W^+W^-$ mode, by Kleiss and Stirling (see Section 7).[34] We have little to add on this subject, except to point out that for Higgs masses in the range of interest for this mode, event rates are low and tagging must be done with near perfect efficiency to be useful. There are also likely to be formidable problems in finding jets in the rapidity range 3 to 5 which is required for good tagging efficiency. For purposes of this work, we have not made the assumption that the tagging can be used.

In our analysis, we have focused on Higgs masses above the range accessible via the all charged lepton mode described in Section 5, i.e. for masses greater than ≈600–700 GeV. To be specific we have analyzed the case for $M_H = 800$ GeV. Using ISAJET, we generated Higgs events for $M_H = 800$ GeV and $Z +$ jet(s) events requiring $|y_z| < 1.5$ and $P_T(Z) > 350$ GeV. In both cases the Z decays to either ee or $\mu\mu$. Although other variables may be used,[33] we choose to use the simple transverse mass defined as in the original calculation of Cahn and Chanowitz[44]

$$M_T = 2\sqrt{P_T^2(Z) + M_Z^2}$$

Assuming an integrated luminosity of 10^{40}, the transverse mass distribution for the Higgs decay and for the background is shown in Fig. 6.1; without additional cuts the background is more than 100 times larger than the signal.

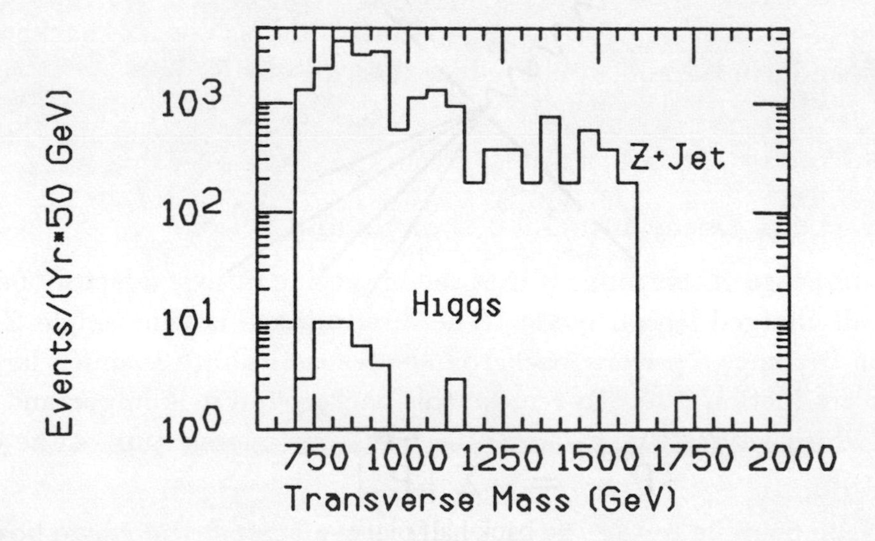

6.1 The transverse mass distribution for Higgs events and for the Z+jet(s) background without cuts on jet activity.

There are a number of ways to quantify and characterize the missing energy signal—see the contribution of A. Savoy-Navarro to these Proceedings,[35] Barger et al.[33] and previous work.[12,36] We have chosen a very simple method to explore the consequences of the lack of detector hermeticity on the background level. First, we assume a quasi-ideal detector with perfect energy resolution, no cracks and a beam hole of $|y| > 5.5$ i.e., particles with rapidity greater than $|5.5|$ are not detected. In the transverse plane of the event (see Fig. 6.2), the total scalar P_T in the half plane opposite to the direction of the Z

$$P_T^{back} \equiv \sum_{back-half\ plane} |\vec{P}_T|$$

is computed. For Higgs events this should be small since the only jet activity is from the recoil quarks and beam remnants. For the Z + jet(s) background this should be larger since there will be at least one jet in the event, roughly opposed to the Z. The distribution of P_T^{back} for signal and background is shown in Fig. 6.3a. Under these quasi-ideal assumptions, the Z + jet(s) background can be completely eliminated by cutting on P_T^{back}, at least within the statistics of our background simulation. The remaining background will then be true ZZ continuum production which has a P_T^{back} distribution similar to that of the Higgs.

Clearly as one degrades the hermeticity of the detector, this clean separation is likely to be diminished. We have explored the consequences of

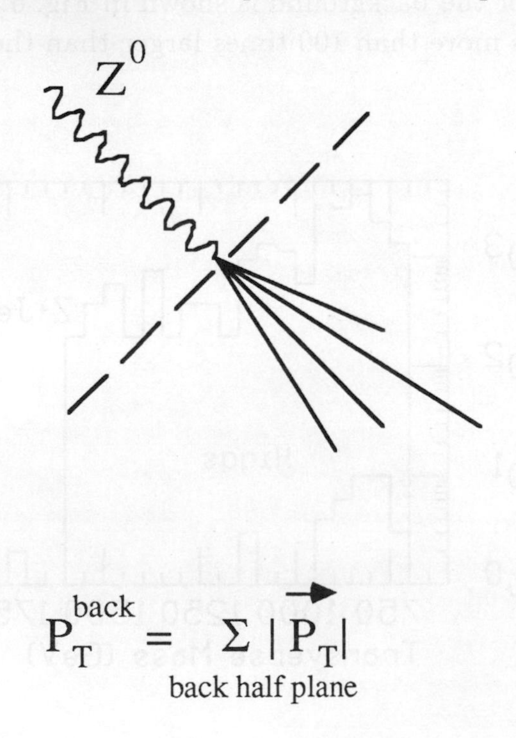

6.2 Definition of P_T^{back} which is a simple characterization of jet activity.

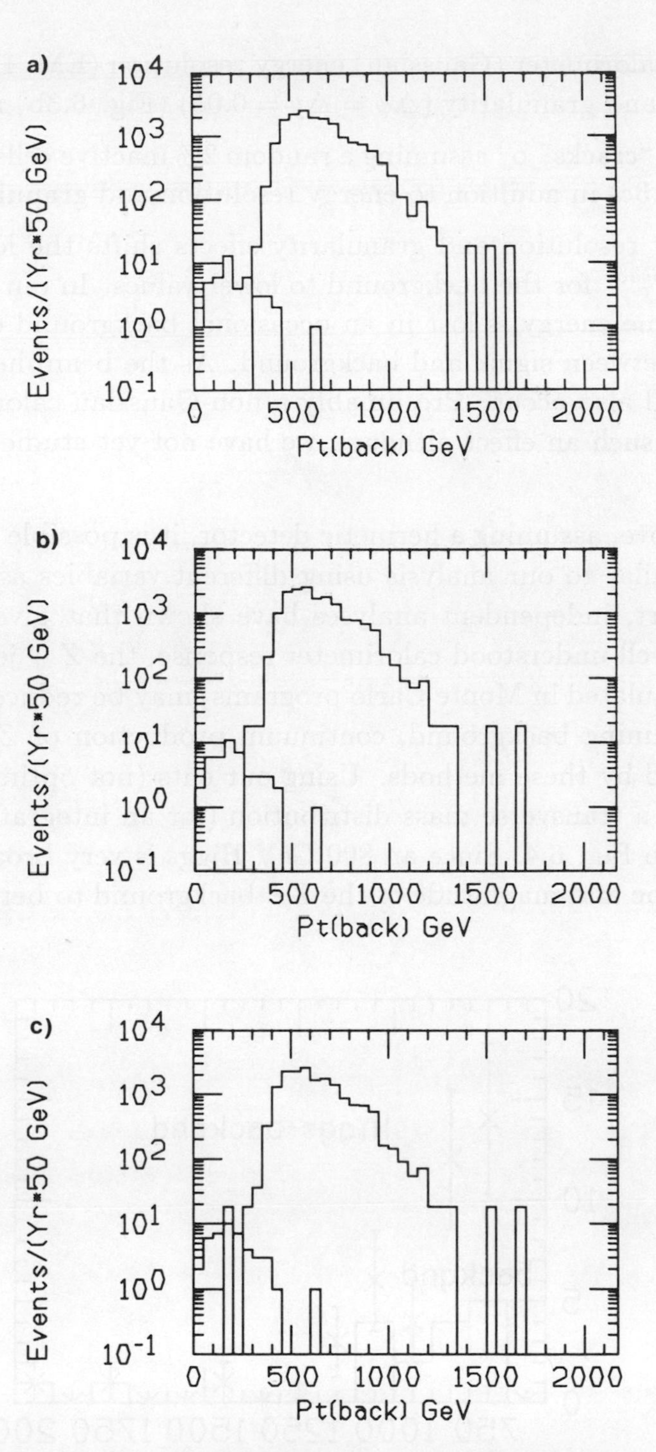

6.3 a) Distribution of $P_T{}^{back}$ for a hermetic detector (for $|y| < 5.5$) with perfect efficiency and energy resolution.
b) Same as (a) but including calorimeter energy resolution and granularity as described in the text.
c) Same as (b) but 2% of the calorimeter cells are dead, which crudely simulates cracks.

- including calorimeter (Gaussian) energy resolution (EM: 15% / $\sqrt{E}$, HAD: 40%/$\sqrt{E}$) and granularity ($\Delta\phi = \Delta y = 0.05$) (Fig. 6.3b) and
- simulating "cracks" by assuming a random 2% inactive cells in the calorimeter (Fig. 6.3c) in addition to energy resolution and granularity

Including energy resolution and granularity effects shifts the lower edge of the distribution of P_T^{back} for the background to lower values. In our simple model of cracks, again some energy is lost in an occasional background event, worsening the separation between signal and background. As the beam hole is enlarged, a similar effect will also occur. Presumably a non-Gaussian calorimetry response would also have such an effect although we have not yet studied this in quantitative detail.

As noted above, assuming a hermetic detector, it is possible to obtain results qualitatively similar to our analysis using different variables as measures of jet activity. In short, independent analyses have shown that given excellent hermeticity and a well understood calorimeter response, the $Z+$jet(s) background (as presently simulated in Monte Carlo programs) may be reduced to a negligible level. The remaining background, continuum production of ZZ pairs, cannot be easily reduced by these methods. Using our cuts (not optimized) one would therefore obtain a transverse mass distribution (for an integrated luminosity of 10^{40}) as shown in Fig. 6.4. Since an 800 GeV Higgs is very broad, it is *essential* to know the shape and magnitude of the ZZ background to better than 30%, in

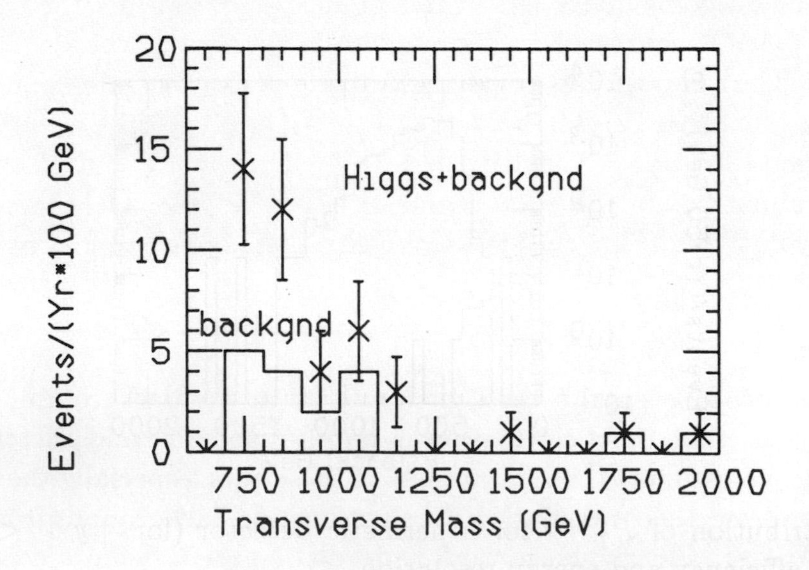

6.4 The transverse mass distribution for $M_H = 800$ GeV after applying cuts which eliminates all the $Z+$jet(s) background. The remaining background shown arises only from $q\overline{q} \rightarrow ZZ$. If the process $gg \rightarrow ZZ$ were included, the background would increase by about 60%.

order to exploit this decay mode.† A more realistic detector simulation must be done in order before a definitive statement about the utility of this mode can be made.

7. Heavy Higgs Decay into WW, $W \to e\nu$ or $\mu\nu$ and $W \to q\overline{q}$

The advantage of this mode is that it has a much larger event rate ($B(H \to llll) = 1.4 \times 10^{-3}$ and $B(H \to W^+W^- \to l\nu q\overline{q} = 0.16)$. There is a large background from $W +$ jet(s) production where the jet system has an invariant mass close to the W mass. This rate is substantially larger than the Higgs rate. To separate the Higgs signal from this background, one must be able to distinguish $W \to q\overline{q} \to$ jets from QCD jets of similar invariant mass with rejection factors of the order of 100:1 or more.[37]

Our investigations concerning the observability of this mode are incomplete, and we have very little to add to published studies[36,38] and other contributions to these Proceedings.[35] Detailed studies[12] have shown that, at the parton level, it is possible to reduce the $W +$ jet(s) background to a level comparable to the Higgs signal if the t quark mass is less than the W mass. Savoy-Navarro[35] in her contribution to these Proceedings states that a 5σ (signal to background of 1:12) effect can be found via a series of cuts, but this analysis neglects the continuum WW production and assumes that the $W +$ jet(s) rate is precisely known.

In a recent LHC study, Kleiss and Stirling[34] have examined the effectiveness of tagging the outgoing quark jets in this mode. They consider the QCD process $qq \to Wjjjj$ as a source of background to tagged gauge boson fusion Higgs production. They impose cuts requiring two jets in the central region ($|y_{jet}| < 2$) which reconstruct to a $W(|M_{jj} - M_W| < 5$ GeV), and two high energy jets in the forward region ($3 < y_{jet} < 5$, and $E_j > 1$ TeV). They find about 43 events in the signal, and about 260 in the background, at $\sqrt{s} = 17$ TeV. They then impose an "asymmetry cut" requiring that the two central jets have energies which are not too different.

$$(E_1 - E_2)/(E_1 + E_2) < 0.5$$

They find about 31 events in the signal, and 35 in the background. It must be noted, however, that their calculation is purely partonic, and therefore it may significantly overestimate the effectiveness of these cuts, especially the asymmetry cut.[35,38] Moreover, their calculation does not include backgrounds from processes like $gg \to WWjj$. Clearly, more work is needed before a definite conclusion can be reached regarding the effectiveness of tagging.

---

† If the t quark mass is very large, the production rate is enhanced and a greater uncertainty in the background could be tolerated.

If the t quark mass is larger than the W mass, then t quarks will decay into real W's and the WW background will be substantially increased, making it impossible to exploit this mode.[39]

8. Detector Requirements

Considerable work remains to be done before the optimal set of criteria for extracting a heavy Higgs signal can be determined. However, the detector parameters required both to observe the heavy Higgs signal in various decay modes and to reduce or eliminate backgrounds can now be reliably estimated. Even though the best set of experimental cuts is not well understood, detector parameters can be specified reasonably well. In the sections below we describe the motivation for the choice of detector parameters for the three decay modes of interest. We also describe the requirements for the observation of gauge boson pair production including the $W^{\pm}W^{\pm}$ and WZ channels.

8.1 Heavy Higgs $\to ZZ$, both $Z \to ee$ or $\mu\mu$

The parameters of interest include:

- the angular or rapidity acceptance for the leptons
- the transverse and total momentum distribution of the leptons
- the opening angle distribution
- the momentum or energy resolution required
- lepton sign determination
- lepton identification criteria, particularly the jet rejection needed to observe electrons, and
- vertex criteria, both primary and secondary vertices.

In the sections below, each of these items is discussed. A summary is given in Table 8.1.

8.1.1 Rapidity Acceptance

In order to reduce the background from continuum ZZ production, a cut of $|y_z| < 1.5$ is usually applied,[23] restricting the rapidity range of the leptons from the Z decay. With this cut, the four-lepton acceptance vs. the lepton rapidity coverage is shown in Fig. 8.1 for 400 and 800 GeV Higgs masses. Electron and muon coverage of $|y| < 2.5 - 3$ is adequate.

8.1.2 Transverse and Total Momentum Distribution of the Leptons

In Figs. 8.2 and 8.3 we plot the P_t distributions of leptons in Higgs events after requiring $|y_z| < 1.5$, for $M_H = 400$ and 800 GeV. The hatched bins in these

Table 8.1: Summary of experimental requirements for $H \to ZZ \to \ell^+\ell^-\ell^+\ell^-$. For details see the text.

z vertex resolution		$\approx 1cm$.
Secondary vertex		Not needed.
Charged hadron tracking	y coverage	Backup for calorimetry.
	Resolution	Not critical.
Electron identification	y coverage	$\|y\| \lesssim 2.5$-3. See Fig. 8.1.
	$\Delta E/E$	$\approx .15/\sqrt{E}$. Must reconstruct M_Z for all M_H.
	P_T range	$\lesssim M_H$. See Figs. 8.2 and 8.3.
	Jet rejection	Z mass constraint adequate for $M_H \gtrsim 500GeV$. Not known for smaller masses.
	Charge measurement	Not needed for $M_H \gtrsim 400GeV$.
Muon identification	y coverage	$\|y\| \lesssim 2.5$-3. See Fig. 8.1.
	$\Delta p/p$	$\lesssim 15\%$ if $m_t \lesssim 100GeV$. Better if m_t is larger.
Calorimetry	y coverage	Same as for $e^\pm$.
	$\Delta E/E$	See electrons. Not important for hadrons.
	Segmentation	$\Delta y = \Delta\phi \approx 0.03$-0.05.
	Hermeticity	Not important.
Luminosity		Need $\mathcal{L} \sim 10^{33} cm^{-2}s^{-1}$. Perhaps can use $\mathcal{L} \sim 10^{34} cm^{-2}s^{-1}$.
Trigger		Require leptons above $P_T \sim 10GeV$. Higher level: make loose cuts on Z mass and P_T.

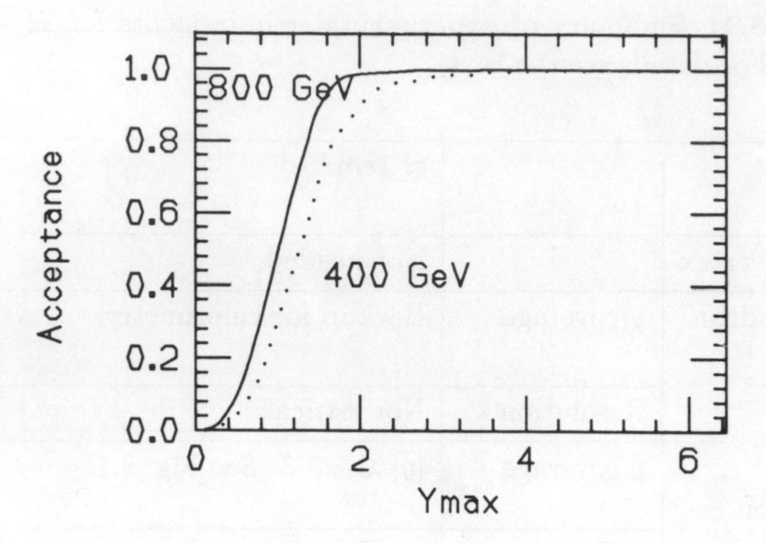

8.1 The four-lepton acceptance vs lepton rapidity coverage.

figures indicate the contributions from leptons in the same events which do not come from the Higgs decay; for example, they are from π^0 Dalitz decay. This contribution is small and limited to low P_T.

By examining the distribution of the lowest P_T lepton from the Higgs decay, we also find that identification of leptons down to $P_T \sim$ 20–50 GeV is required for good efficiency.

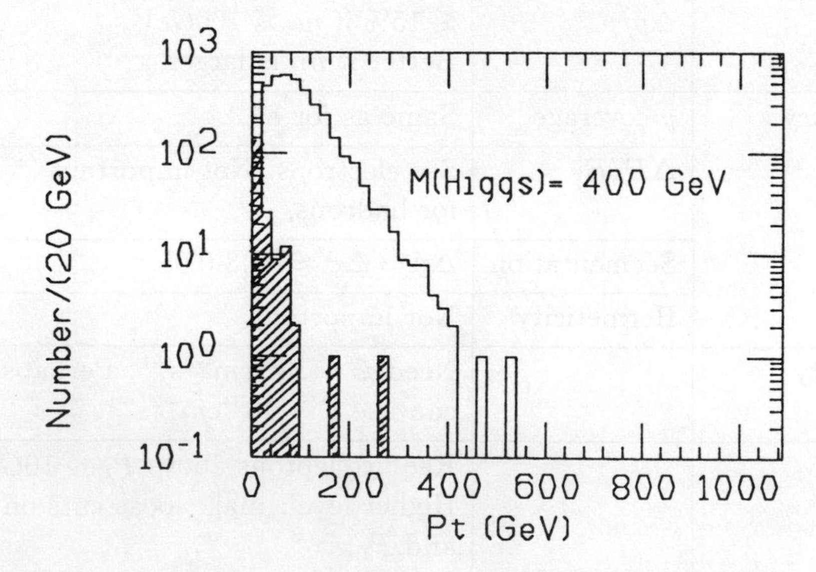

8.2 The P_T distribution of electrons in Higgs events for M_H = 400 GeV. The hatched bins indicate the contribution from non-Z sources in Higgs events.

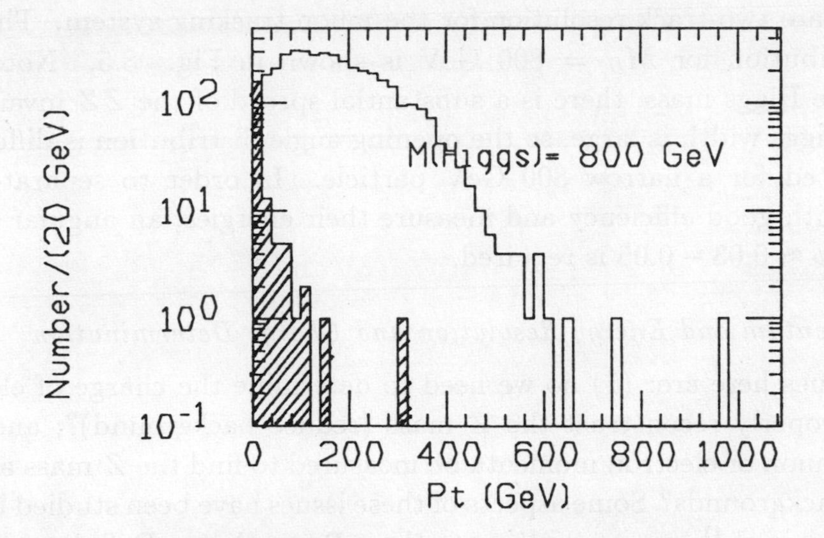

8.3 Same as 8.2 but for $M_H = 800$ GeV.

In Fig. 8.4 we show the four-lepton acceptance vs. total momentum of the most energetic lepton. The acceptance is about 50% for a total momentum of about one-half of the Higgs mass.

8.1.3 Opening Angle Distribution

The opening angle distribution between the two leptons from the Z decay is of interest because good angular granularity in a calorimeter is required to separate the two electrons and reconstruct the Z mass. It is also important to

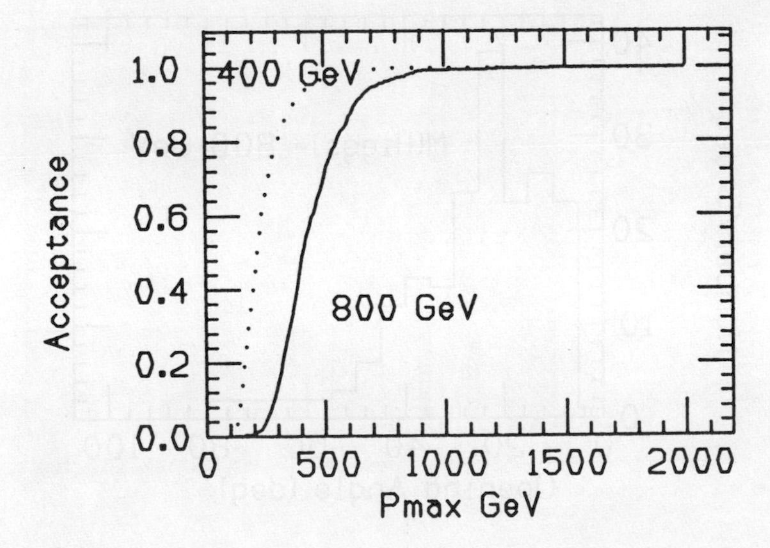

8.4 The four-lepton acceptance vs the total momentum of the most energetic lepton.

have adequate two-track resolution for the muon tracking system. The opening angle distribution for $M_H = 800$ GeV is shown in Fig. 8.5. Note that for such a large Higgs mass, there is a substantial spread of the ZZ invariant mass since the Higgs width is large, so the opening angle distribution is different from that expected for a narrow 800 GeV particle. In order to separate the two electrons with good efficiency and measure their energies, an angular resolution of $\Delta\eta = \Delta\phi \approx 0.03 - 0.05$ is required.

8.1.4 Momentum and Energy Resolution and Charge Determination

The issues here are: (1) do we need to determine the charge of electrons in order to properly reconstruct the Z mass (reduce background)?; and (2) how well must muon or electron momenta be measured to find the Z mass and reduce potential backgrounds? Some aspects of these issues have been studied by Paige[40] and by Chen et al.[41] in contributions to these Proceedings. Definite resolution of these issues is not simple since the results depend on the Higgs mass, the t quark mass and on modeling of jet rejection for identification of electrons or muons.

Paige[40] has shown that, for a Higgs mass of 800 GeV, it is possible to eliminate jet backgrounds which simulate electrons using simple calorimetric and isolation cuts. Electrons from the Z's in Higgs decay have high P_T, are isolated and can be paired to form the Z mass; electron candidates from QCD jets do not have these properties. Determination of the electron charge is not required in this analysis. Although this analysis has been done for an 800 GeV Higgs mass, it would likely apply down to Higgs masses in the 500–600 GeV range. For lower masses the jet background, relative to the signal, becomes larger and P_T cuts are less effective.

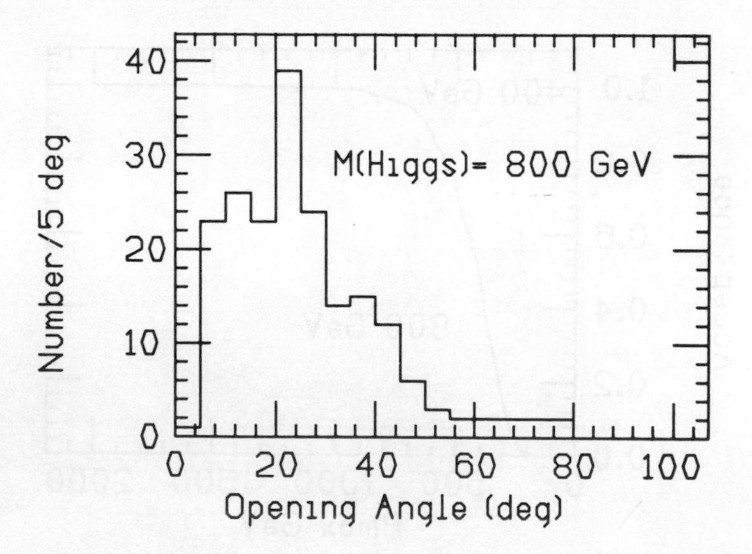

8.5 The opening angle distribution of electrons from $Z \to ee$ for 800 GeV Higgs decays.

Chen et al.[41] have emphasized the need for good momentum resolution to observe the Z mass peak in the presence of a dilepton background resulting from heavy top quark (m_t = 200 GeV) decay. If the t quark is sufficiently massive, there is a copious dilepton background from $t\bar{t} \rightarrow WW + X$ production which populates the invariant mass region in the neighborhood of the Z. In their contribution, Chen et al. show that muon momentum resolution characteristic of an iron spectrometer results in a dimuon muon invariant mass distribution in which the Z is barely visible because of the $t\bar{t}$ background.

However, for the all charged lepton mode and for Higgs masses above about 400 GeV, this background can very likely be eliminated by simply cutting on the transverse momenta of the dimuon pairs. To test this hypothesis we generated $t\bar{t}$ events for m_t = 200 GeV and required events with at least two ee or $\mu\mu$ (or the combination) pairs with $|y_{e \ or \ \mu}| < 3$. In order to increase the stastical power of our estimates, we required $W \rightarrow e\nu$ or $\mu\nu$ and b-hadron semileptonic decay. Analysis of a smaller sample of events without this restriction indicates that these decays are the dominant source of ≥ 4 lepton events. We also require the dilepton mass of both pairs, formed from different leptons, to be between 70 and 110 GeV. In Fig. 8.6 we show a scatter plot of the P_T of one lepton pair vs. the P_T of the other, corresponding approximately to 1/100 of an SSC year. The magnitude of this background is reasonably described by an exponential distribution when plotted against a minimum P_T for both pairs. For example, about 100(10) events/year would remain requiring a minimum P_T of 150(200) GeV. Hence this background should be negligible for masses of the Higgs greater

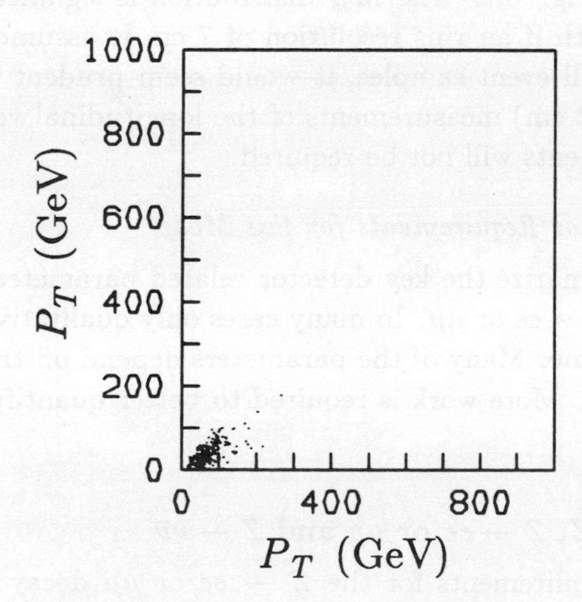

8.6 A scatter plot of P_T of one dilepton pair from $t\bar{t}$ events vs. P_T of the other pair with cuts described in the text.

than about 400 GeV. We note that Chen et al. in their analysis also applied a cut (of 300 GeV) on the P_T of the dimuon pair but some $t\bar{t}$ background remained. Requiring a second Z candidate eliminates the remaining background. Of course, this method only works for the all charged lepton mode of Higgs decay. For the mode with only one Z decay to ee or $\mu\mu$, and $Z \to \nu\bar{\nu}$ good resolution may be more important. A similar study for this mode has not yet been done.

8.1.5 Longitudinal Vertex Resolution

At the SSC, the rms longitudinal bunch length is expected to be about 7 cm and hence the protons interact in a region about 20 cm long. In a high luminosity environment it may be difficult to determine the precise longitudinal location of the event vertex. For events containing muons, previous studies have shown that it is possible to trace stiff muon trajectories through absorber back to the origin[42] with reasonable accuracy. For electrons this requires central tracking information, which may be unattainable at the highest luminosities.

To estimate the effect of the uncertainty in the longitudinal vertex position, we generated 800 GeV Higgs events decaying into four electrons for different rms beam spot length sizes. The electrons are assumed to be measured with perfect energy and position resolution at a radius of 1 meter. The four highest P_T electrons in the event are selected (these are almost always the electrons from the Higgs). The pair, M_{12}, forming an invariant mass closest to the Z is found and the invariant mass of the other pair, M_{34}, calculated. This was done for three different assumptions about the accuracy to which the longitudinal vertex position is known—see Fig. 8.7. The M_{34} distribution is significantly broader than the Z intrinsic width if an rms resolution of 7 cm is assumed. Since one will be dealing with small event samples, it would seem prudent to have some means for crude ($\approx 1 - 2$ cm) measurements of the longitudinal vertex position but millimeter measurements will not be required.

8.1.6 Summary of Detector Requirements for this Mode

In Table 8.1 we summarize the key detector related parameters determined by Higgs $\to ZZ$, both $Z \to ee$ or $\mu\mu$. In many cases only qualitative conclusions can be reached at this time. Many of the parameters depend on the Higgs mass and on the t quark mass. More work is required to better quantify some of the requirements.

8.2 Higgs Decay to ZZ, $Z \to ee$ or $\mu\mu$ and $Z \to \nu\bar{\nu}$

The experimental requirements for the $Z \to ee$ or $\mu\mu$ decay in this mode are obviously very similar to those described above. Dilepton mass resolution may be more important in this mode than in the all charged lepton mode. The $t\bar{t}$ background to this mode for large m_t is not known. Rejection of fake Z

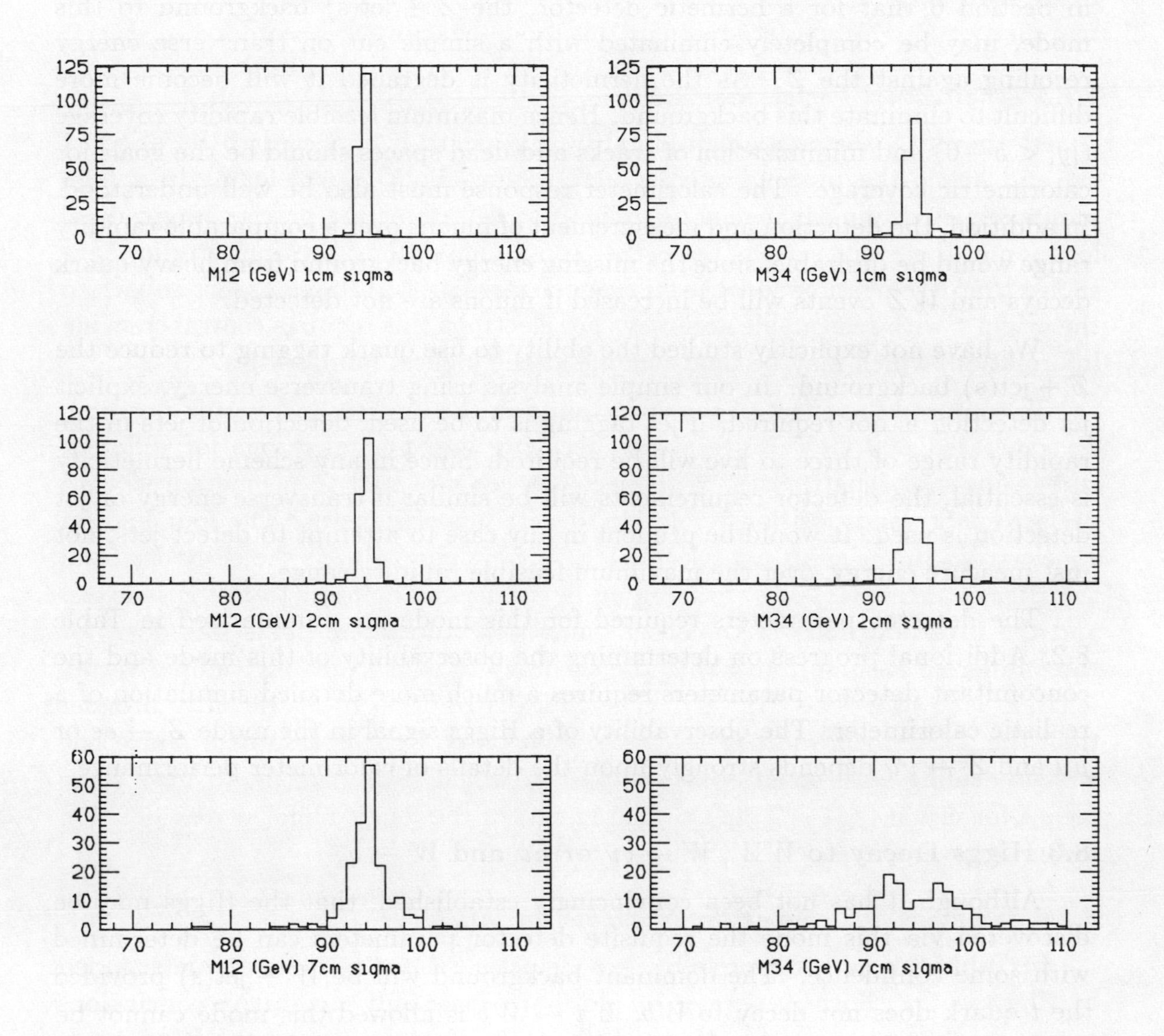

8.7 Dilepton invariant mass distributions M_{12} and M_{34} (see text) for different assumptions regarding the accuracy of determination of the longitudinal vertex of the event.

candidates requires better lepton identification since the event contains one Z and is less constrained. We have not studied this issue in sufficient detail to make a quantitative statement.

Measurement of missing energy is obviously crucial for this mode, We stated in Section 6 that for a hermetic detector, the Z + jet(s) background to this mode, may be completely eliminated with a simple cut on transverse energy recoiling against the Z. As the hermeticity is degraded it will become more difficult to eliminate this background. Hence maximum feasible rapidity coverage ($|y| < 5-6$) and minimization of cracks and dead spaces should be the goals for calorimetric coverage. The calorimeter response must also be well understood. In addition, the detection and measurement of muons over a comparable rapidity range would be desirable, since the missing energy background from heavy quark decays and WZ events will be increased if muons are not detected.

We have not explicitly studied the ability to use quark tagging to reduce the Z + jet(s) background. In our simple analysis using transverse energy, explicit jet detection is not required. If jet tagging is to be used, detection of jets in the rapidity range of three to five will be required. Since in any scheme hermeticity is essential, the detector requirements will be similar if transverse energy or jet detection is used. It would be prudent in any case to attempt to detect jets, not just measure energy, over the maximum feasible rapidity range.

The detector parameters required for this mode are summarized in Table 8.2. Additional progress on determining the observability of this mode and the concomitant detector parameters requires a much more detailed simulation of a realistic calorimeter. The observability of a Higgs signal in the mode $Z \to ee$ or $\mu\mu$ and $Z \to \nu\overline{\nu}$ depends strongly upon the details of calorimeter performance.

8.3 Higgs Decay to WW, $W \to e\nu$ or $\mu\nu$ and $W \to q\overline{q}$

Although it has not been convincingly established, that the Higgs may be discovered via this mode the requisite detector parameters can be determined with some confidence. The dominant background will be W + jet(s) provided the t quark does not decay to Wb. If $t \to Wb$ is allowed this mode cannot be utilized.

Rejection of the W+jet(s) background requires reconstruction of the $W \to q\overline{q}$ invariant mass with good resolution. This will require fine grained calorimetry with good energy resolution. This has been studied in quantitative detail by Freeman and Newman-Holmes in their contribution to these Proceedings.[43] They find that calorimetric tower sizes of ≈ 0.03 are required; tower sizes of 0.1 significantly degrade the W mass resolution. Relatively modest energy resolution of $0.15/\sqrt{E}$ for electromagnetic calorimetry and $0.5/\sqrt{E}$ for hadronic is adequate but an electron-to-hadron response near one (within ± 0.1) is very desirable. They also

Table 8.2: Summary of experimental requirements for $H \to ZZ \to \ell^+\ell^-\nu\bar{\nu}$. For details see the text.

z vertex resolution		$\approx 1cm$.
Secondary vertex		Not needed.
Charged hadron tracking	y coverage	Backup for calorimetry. Important
	Resolution	Not critical.
Electron identification	y coverage	$\lvert y\rvert \lesssim 2.5$-3. See Fig. 8.1.
	$\Delta E/E$	$\lesssim .15/\sqrt{E}$. Must reconstruct M_Z for all M_H.
	P_T range	$\lesssim M_H$. See Fig. 8.2, 8.3.
	Jet rejection	Z mass constraint adequate for $M_H \gtrsim 500GeV$. Not known for smaller masses.
	Charge measurement	Not needed for $M_H \gtrsim 400GeV$.
Muon identification	y coverage	$\lvert y\rvert \lesssim 2.5$-3. See Fig. 8.1.
	$\Delta p/p$	$\lesssim 15\%$ if $m_t \lesssim 100GeV$. Better if m_t is large.
Calorimetry	y coverage	$\lvert y\rvert \le 5.5$ or better.
	$\Delta E/E$	$0.15/\sqrt{E}$ (electrons), $0.5/\sqrt{E}$ (hadrons). $e/\pi = 1.0 \pm 0.1$.
	Segmentation	$\Delta y = \Delta\phi \approx 0.03$-$0.05$.
	Hermeticity	Crucial
Luminosity		Need $\mathcal{L} \sim 10^{33}cm^{-2}s^{-1}$. Limited by hermeticity requirement.
Trigger		Require leptons above $P_T \sim 10GeV$. Higher level: require loose Z mass and missing $P_T \gtrsim 200GeV$.

show that the W mass resolution is very sensitive to pile-up of events within the detector resolving time. Low energy particles make a significant contribution to the W mass even for high P_T W's, so the addition of low P_T particles from out-of-time events shifts the reconstructed W mass to higher values. To avoid this problem, calorimetry for W mass reconstruction must be able to time-tag energy deposition. A Monte Carlo study assuming the ability to tag energy deposition but with long integration time should be done.

Calorimetry that is sufficient to permit reconstruction of the W mass should be adequate to allow kinematic cuts on the di-jet system such as those suggested by Gunion.[12] Detection of jets at large rapidity to tag the recoil quarks would have the same requirements and difficulties as discussed in Section 8.2.

The other aspect of this mode is detection of the $W \to e\nu$ or $\mu\nu$ signal. This requires detection and measurement of single e or μ over a rapidity and P_T range comparable to that in the other Higgs decay modes. More stringent jet rejection is required since the Z mass constraint is now absent. The required rejection depends on the P_T of the lepton and has not been investigated in detail; we expect a rejection of at least 10^4 will be required when applied in conjunction with the missing energy signature. A larger rejection may be required at low P_T and a smaller one at high P_T. Measurement of missing transverse energy is also required to reconstruct the $W \to l\nu$ with a zero constraint fit so hermetic calorimetry covering $|y| < 5 - 6$ will be required.

The detector parameters for this mode are summarized in Table 8.3.

8.4 Detection of Other Gauge Boson Pairs

In Sections 1 and 3.1 we discussed the motivation for the detection of all gauge boson pairs, not just those arising from Higgs decay. Detection of $W^\pm W^\pm$ and WZ events is of great importance, since these final states are characteristic of a strongly interacting symmetry breaking sector. Many of the requirements for the detection of same sign W pairs are the same as those for opposite sign pairs, but the charge of the lepton from the W decay(s) must also be determined. Since event rates in the interesting mass region are low, determination of both electron and muon charges is essential.

As discussed in Section 3.2, models for $W^\pm W^\pm$ or WZ production predict an enhanced rate for these events if there is no Higgs boson of mass less than about 1 TeV. The P_T distribution of the lepton with the largest P_T arising from $W^+W^+ \to l^+\nu l^+\nu$ is shown in Fig. 8.8. The model of Ref. 4 is used with the requirement that $M_{WW} > 0.5$ TeV and $|y_w| < 1.5$. From this figure, it is clear that one needs to determine the sign of leptons with $P_T < 750 - 1000$ GeV.

Table 8.3: Summary of experimental requirements for $H \to W^+W^- \to \ell^\pm \nu q\bar{q}$. For details see the text.

z vertex resolution		Probably not important.
Secondary vertex		Not needed.
Charged hadron tracking	y coverage	Backup for calorimetry. If used for $W \to q\bar{q}$ identification, need tracking inside jets.
	Resolution	Not critical.
Electron identification	y coverage	$\lvert y\rvert \lesssim 2.5$-3. See Fig. 8.1.
	$\Delta E/E$	Not critical.
	P_T range	$\lesssim M_H$. See Fig. 8.2, 8.3.
	Jet rejection	Required. $e/jet \sim 10^{-4}$.
	Charge	Not needed.
Muon identification	y coverage	$\lvert y\rvert \lesssim 2.5$-3. See Fig. 8.1.
	$\Delta p/p$	$\lesssim 15\%$ for any m_t.
Calorimetry	y coverage	$\lvert y\rvert \leq 5.5$ or better to measure P_T^{miss}.
	$\Delta E/E$	$0.15\sqrt{E}$ (electrons), $0.5/\sqrt{E}$ (hadrons). $e/\pi = 1.0 \pm 0.1$.
	Segmentation	Needed for background rejection. $\Delta y = \Delta\phi \approx 0.03 - 0.05$.
	Hermeticity	Needed to see ν.
Luminosity		Need $\mathcal{L} \sim 10^{33} cm^{-2} s^{-1}$. Limited by hermeticity requirement and $W \to q\bar{q}$ reconstruction.
Trigger		Not studied in sufficient detail.

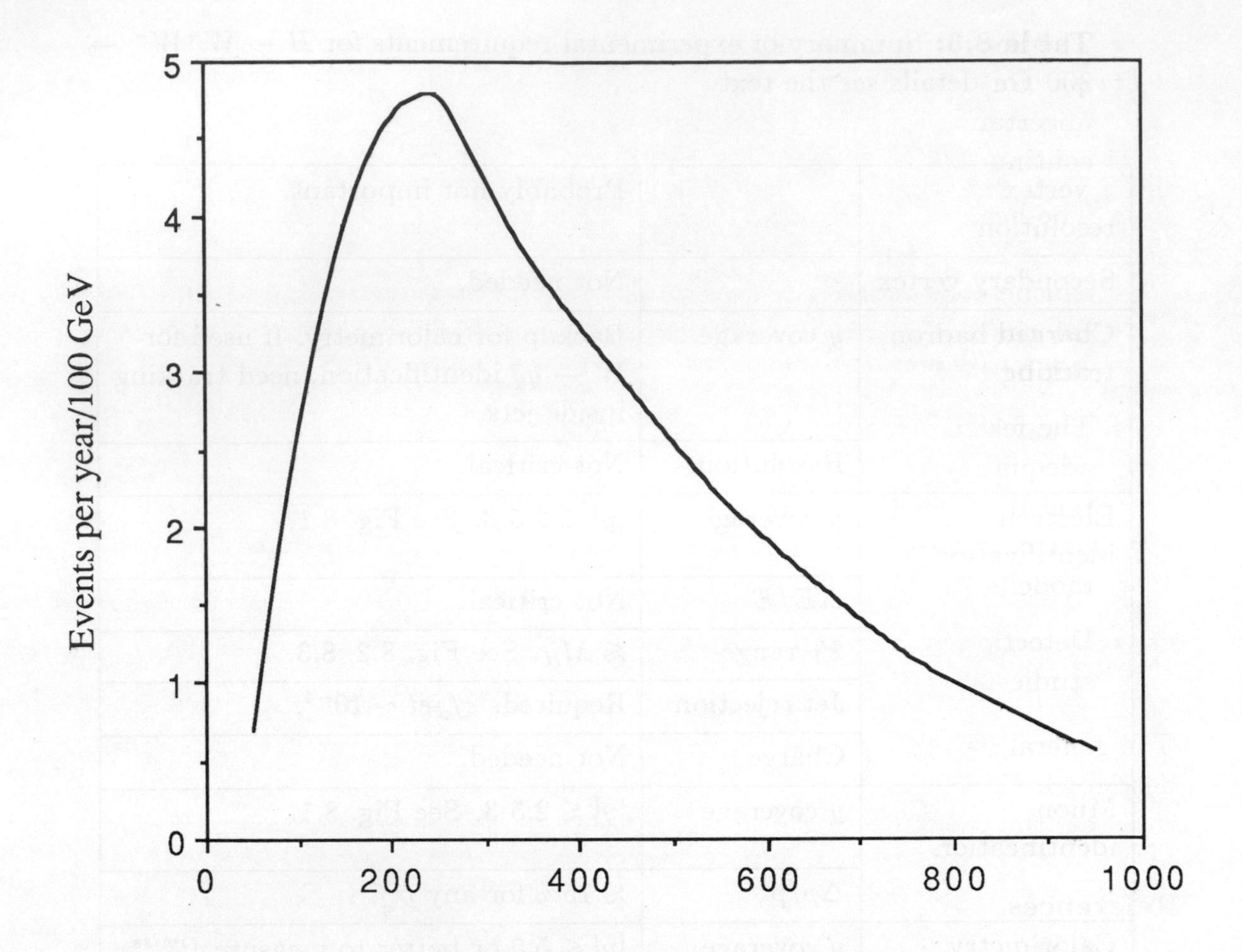

8.8 The P_T distribution of the lepton with the larger P_T from W^+W^+ events for $\mid y_W \mid < 1.5$ and $m_{WW} > 0.5$ TeV as predicted by the model of Chanowitz and Gaillard (Ref. 4).

There are no other important sources of W^+W^+ in the standard model, so that, although the event rates of Fig. 8.8 are small, detection may be possible. Backgrounds in the W^+W^+ and W^-W^- channels are likely to be equal whereas the W^+W^+ rate from stongly interacting gauge bosons is approximately three times that from W^-W^-.[17] Background from events with $W^\pm$ + jet(s) wherein the jets yield an isolated lepton has not been evaluated.

9. Conclusions

In this paper we have explored the observability of a heavy Higgs boson at the SSC. We conclude that:

- The heavy Higgs boson may be observed at the SSC up to masses of 600–800 GeV in the mode $H \to ZZ, Z \to ee$ or $\mu\mu$. The range results from uncertainties in the t quark mass and our inability to quantify the ZZ continuum background.
- The heavy Higgs boson may be observable for higher masses in the mode $H \to ZZ, Z \to ee$ or $\mu\mu$ and $Z \to \nu\overline{\nu}$ if detectors of sufficient hermeticity can be constructed and operated. A detailed and more realistic study of calorimeter hermeticity and response is required before definite conclusions can be reached.
- The feasibility of detecting the Higgs in the WW mode has not yet been adequately demonstrated here or elsewhere. If the t quark can decay into a real W, the WW background will be overwhelming. Studies going beyond partonic level calculations which include fragmentation effects and realistic modeling of calorimeter response are required.
- Detection of $W^{\pm}W^{\pm}$ pairs and WZ events is important and requires more studies at the Monte Carlo level.

The general detector parameters for detection of heavy Higgs decays have been described. Considerable effort will be needed to fine-tune these requirements and to assess the feasibility of construction of actual experiments which meet them.

References

1. S. Weinberg, Phys. Rev. Lett. 19 (1967) 264; A. Salam, Elementary Particle Physics, ed. N. Svartholm, p. 367.
2. See for example, I. Hinchliffe, Ann. Rev. Nuc. Part. Sci. 36 (1986) 505.
3. S. Weinberg, Phys. Rev. D13 (1976) 974; Phys. Rev D19 (1977) 1277; L. Susskind, Phys. Rev. D20 (1979) 2619.
4. M. Chanowitz and M. Gaillard, Nucl. Phys. B261 (1985) 379.
5. B. Lee, C. Quigg, and H. Thacker, Phys. Rev. D16 (1977) 1519.
6. S. Weinberg, Phys. Rev. Lett. 17 (1966) 616.
7. H. Georgi et al. Phys. Rev. Lett. 40 (1978) 692.
8. R. Cahn and S. Dawson, Phys. Lett. 136B (1984) 196.
9. J.F. Gunion et al., Univ. of California Davis preprint 86-15 (1986).
10. M. Chanowitz and M. Gaillard, Phys. Lett. 142B (1984) 85; S. Dawson, Nucl. Phys. B249 (1985) 42; G. Kane, W. Repko and W. Rolnick, Phys. Lett. 148B (1984) 367.
11. R. Cahn, Nucl. Phys. B255 (1985) 341.

12. J. Gunion and M. Soldate, Phys. Rev. D34 (1986) 826.

13. J. Cornwall, D. Levin and M. Tiktopolous, Phys. Rev. D10 (1974) 1145.

14. M. Bento and C.H. Llewellyn Smith, Nucl. Phys. B289 (1987) 36; G. Altarelli, B.. Mele, and F. Pitolli, Nucl. Phys. B287 (1987) 205; M. Duncan, G. Kane, and W. Repko, Nucl. Phys. B272 (1986) 517.

15. M. Bento and C.H. Llewellyn Smith, *op. cit.*.

16. J. Fleischer and F. Jegerlehner, Phys. Rev. D23 (1981) 2001.

17. M. Chanowitz and M. Golden, in preparation; M. Chanowitz, in the Proceedings of the UCLA Workshop on Observable Standard Model Physics at the SSC: Monte Carlo Simulation and Detector Capabilities, H-U. Bengtsson, C. Buchanan, T. Gottschalk and A. Soni, eds., World Scientific Publishing, Singapore (1986).

18. F. Paige and S. Protopopescu, to appear in in the Proceedings of the UCLA Workshop on Observable Standard Model Physics at the SSC: Monte Carlo Simulation and Detector Capabilities, H-U. Bengtsson, C. Buchanan, T. Gottschalk and A. Soni, eds., World Scientific Publishing, Singapore (1986).

19. H-U. Bengtsson and T. Sjöstrand, in in the Proceedings of the UCLA Workshop on Observable Standard Model Physics at the SSC: Monte Carlo Simulation and Detector Capabilities, H.-U. Bengtsson, C. Buchanan, T. Gottschalk and A. Soni, eds., World Scientific Publishing, Singapore (1986).

20. R.W. Brown and K. Mikaelian, Phys. Rev. D19 (1979) 922.

21. R.W. Brown, D. Sahdev and K. Mikaelian, Phys. Rev. D20 (1979) 1164.

22. D.A. Dicus, C. Kao, and W.W. Repko, Phys. Rev. D36 (1987) 1570.

23. E. Eichten, I. Hinchliffe, K. Lane and C. Quigg, Rev. Mod. Phys. 56 (1984) 579.

24. J. Owens, in the Proceedings of the 1984 Summer Study on the Design and Utilization of the Superconducting Super Collider, Snowmass, Colorado, 1984.

25. W.K. Tung, in the Proceedings of the 1984 Summer Study on the Design and Utilization of the Superconducting Super Collider, Snowmass, Colorado, 1984.

26. J. Collins, in these Proceedings.

27. J. Collins and D. Soper, Ann. Rev. Nucl. Part. Sci., to appear.

28. R. Cahn, S. Ellis, R. Kleiss, and W. Stirling, Phys. Rev. D35 (1987) 1626.

29. D.A. Dicus and R. Vega, Phys. Rev. Lett. 57 (1986) 1110; J.F. Gunion, J. Kalinowski, and A. Tofighi-Niaki, Phys. Rev. Lett. 57 (1987).

30. H-U. Bengtsson, private communication.
31. The branching ratio formulae are conveniently summarized in R. Thun et al., "Searching for the Higgs $\to Z^0Z^0 \to \mu^+\mu^-\mu^+\mu^-$ at SSC," in these Proceedings.
32. D. Froidevaux in "Experimental Studies" in the Proceedings of the Workshop on Physics at Future Accelerators, Vol. 1, CERN 87-07.
33. V. Barger, T. Han and R.J.N. Phillips, University of Wisconsin Preprint MAD/PH/368 (revised), September 1987 and references therein.
34. R. Kleiss and W.J. Stirling, CERN-TH-4828-87.
35. A. Savoy-Navarro, contribution to these Proceedings and D. Hedin, private communication.
36. See the reports of the Higgs and W Pair Physics Group in the Proceedings of the Summer Study on the Physics of the Superconducting Super Collider, Snowmass, 1986.
37. F.E. Paige, in the Proceedings of the 1984 Summer Study on the Design and Utilization of the Superconducting Super Collider, Snowmass, Colorado; J.F. Gunion, Z. Kunszt and M. Soldate, Phys. Lett. 163B (1985) 389; Erratum-ibid. 168B (1986) 427; W.J. Stirling, R. Kleiss and S.D. Ellis, Phys. Lett. 163B (1985) 261.
38. J.F. Gunion et al., in the Proceedings of the UCLA Workshop on Observable Standard Model Physics at the SSC: Monte Carlo Simulation and Detector Capabilities.
39. G. Herten, in these Proceedings.
40. F. Paige, "ELMUD: An Electron Muon Detector for Higgs Physics at the SSC," to appear in the Proceedings of the Workshop, from Colliders to Supercolliders, Madison, 1987.
41. M. Chen et al., "High P_T Weak Bosons as Signatures for Higgs-like Heavy Particles," in these Proceedings.
42. Report of the Particle ID Group in the Proceedings of the Workshop on Physics at Future Accelerators, Vol. 1, CERN 87-07.
43. J. Freeman and C. Newman-Holmes, "Detector Dependent Contributions to Jet Resolution," in these Proceedings.
44. R. Cahn and M. Chanowitz, Phys. Rev. Lett. 56 (1986) 1327.

EXPLORING THE HIGGS SECTOR

A. Savoy-Navarro, CEA-Saclay, 91191 Gif-sur-Yvette, France

1. Introduction

Within the framework of the Snowmass '86 Summer Study, much effort has been devoted to the search for heavy Higgs and heavy WW or ZZ pairs produced in super pp colliders, on the theoretical as well as experimental side.[1,2] Two ways to pursue this search have been explored in more detail. The first is the possibility to identify a heavy Higgs by its decay into WW pairs where the W's subsquently decay, one in the leptonic mode, and the other in the hadronic mode. The second is to search for $H^0 \to Z^0Z^0$ where the Z^0's decay into 4 charged leptons or 2 charged plus 2 neutral leptons. This work has now been further pursued, and the main purpose of this paper is to review the present status of this study carried out by H.-U. Bengtsson, J. Hauptman, S. Linn and A. Savoy-Navarro.[3,4]

The organization of this paper is as follows: Section 2 deals with the Monte Carlo, including recent improvements, and presents raw cross-sections for signals and backgrounds under consideration; Section 3 presents an investigation, using a realistic detector simulation, into the detectability of an 800 GeV Higgs via the decay chain $H^0 \to WW \to e\nu q\bar{q}'$; Section 4 reports on a similar investigation of $H^0 \to ZZ \to 4e$ or $2e2\nu$; and Section 5, finally, summarizes the results of the previous sections and contains our conclusions as to the detectability of a Higgs (or heavy WW and ZZ pairs) at a very high energy pp collider.

2. Monte Carlo and Raw Cross-Sections

The Monte Carlo that has been used in this study is PYTHIA 4.8.[5] The features of this programme will not be repeated here; however, we will make a short comment on the issue of gauge invariance and IVB fusion.

For the Higgs signal, PYTHIA version 4.8 includes the three processes $q\bar{q} \to H^0$, $gg \to H^0$ and $VV \to H^0$ (where $V = W, Z$). In the latter case, the effective W approximation is used to yield the total cross-section; however the exact matrix element is used to generate the p_T distribution of interacting V's (and consequently of the produced H^0). Unfortunately, the matrix elements for $VV \to H^0$ show bad high energy behaviour, which can be cured only by the inclusion of the full gauge invariant set of $VV \to V'V'$ graphs (the dominant decay mode of a heavy H^0 being $H^0 \to VV$). This complete set of processes has been included in a new version of PYTHIA, still under development, enabling us to make a comparison between the simple s-channel Higgs contribution and the full gauge invariant set.

Fig. 1 shows the dN/dm_{ZZ} distribution from the two programmes, for two different cases of the Higgs mass. We note that for a mass m_{ZZ} within a few widths around the nominal Higgs mass, the differences in cross-sections are clearly negligible (of the order of 10 or 20%). We may point out that our results here agree with an independent check carried out as regards both the s-channel Higgs and the full gauge invariant set of graphs.[6] A reasonable way to proceed, using the s-channel Higgs graph only, is then simply to throw away events in the fictitious tail; for a Higgs of nominal mass 300 (800) GeV, a cut at 600 GeV (1200) GeV retains 95% (84%) of the total cross-section. Although the simplified treatment of Higgs production and IVB fusion in the version 4.8 may lead to an overestimate of the Higgs cross-section by a factor 2, and to an overly clean polarization signal for the V's from the Higgs decay, we feel justified in using PYTHIA 4.8 for our simulations, partly because of the above considerations and partly because the backgrounds to be overcome are typically 4 or 5 *orders of magnitude* larger than the signal, which makes an argument about factors of 2 rather pointless.

In Tables 1 and 2, we present the raw cross-sections for the signals and backgrounds under consideration as obtained from PYTHIA 4.8. From the numbers quoted in these tables it is clearly evident that a 10 TeV collider is not adequate even for a relatively low mass heavy Higgs, whereas at LHC energies and 10^{40} cm^{-2} integrated luminosity per year it will be possible to study the lower mass range of heavy Higgs, but not the upper range. The SSC project, and an LHC with $> 10^{41}$ cm^{-2} integrated luminosity per year, should both be able to explore the heavy Higgs at around 1 TeV (at least in terms of rates). But we note that multiplying the luminosity by a factor 10 not only increases the signal but also the background by this amount, and that therefore the backgrounds rates at the LHC would be more than a factor 2 worse than at the SSC in a corresponding situation. Furthermore, the detection problems and strategies would certainly be different. From now on we will only consider the case of the SSC.

Taking all these facts into account, the main conclusion to be drawn from the tables is that tremendously small signal to background ratios (S/B) will have to be faced. We expect a total of 1800 events due to $H^0 \to WW \to e\nu_e q\bar{q}'$; only 8 events due to $ZZ \to 4e$, and 94 due to $ZZ \to e^+e^-\nu\bar{\nu}$. The low p_T contribution of the V + jets background gives a S/B equivalent to $2 \cdot 10^{-4}$, 10^{-5} and 10^{-4}, respectively, for each of the three signals mentioned above. The high p_T tail of the standard QCD backgrounds gives results worse by a few additional orders of magnitude.

3. Is It Possible to Detect High Mass W Pairs at Super pp Colliders?

We describe here an attempt to reconstruct high mass W pairs of 800 GeV, subsequently decaying into $e\nu_e q\bar{q}$ in the realistic experimental simulation of a non-magnetic 4π detector running in the environment of a super pp collider. The possible ways to filter, reconstruct and fully analyze these events are discussed in detail. The final detection efficiencies, signal to background ratio and statistical significance of the results are presented within the present status of our work.[3] The sources of true and fake WW signals that we consider in this study are the processes $pp \to H^0 \to WW$ and $pp \to W$ + jets. A detailed simulation of a non-magnetic 4π general purpose detector is used. The detector simulation is described in detail in [2,3].

The selection and reconstruction consists of three steps: a high p_T electron trigger, in conjunction with a certain amount of missing transverse energy ($E_{T,miss}$); the identification of one or two jets; and the reconstruction of the $W \to q\bar{q}'$ system. This selection strategy has already been used and described in [2,3]. Let us briefly summarize it here.

The high p_T electron trigger requires:

i) at least one cluster in the electromagnetic part of the calorimeter within $|\eta| < 3$ and with $E_T > 25$ GeV and a clustering radius of 0.5 units of ΔR $(= \sqrt{\Delta\eta^2 + \Delta\phi^2})$;

ii) at least 95% of the energy to be contained in the first 40 radiation lengths of the calorimeter;

iii) an isolation parameter, defined as the ratio of the E_T of the hit cell to the total E_T measured in the surrounding cells, to be bigger than 95%;

iv) an r.m.s. for the cluster radius less than 0.1 in ΔR units;

v) the e.m. cluster to be associated with the track in the TRD with more than 400 keV of ionization.

In addition, a minimum amount of 25 GeV of $E_{T,miss}$ is required. The 4-momentum of the electron and the two components of the missing momentum along with the W-mass constraint yield a quadratic equation in longitudinal neutrino momentum p_z. We choose the smaller value of p_z (correct about 60% of the time).

Next we look for one or two jets, each with $E_{T,jet}$ bigger than 25 GeV, in the opposite hemisphere to the reconstructed $W \to e\nu_e$. The trigger jet is combined with a second jet to give an invariant mass $m_{j_1j_2}$ as close as possible to m_W (imposing $m_{j_1j_2} < 150$ GeV); if no such second jet is found, only the trigger jet is kept if this by itself fits m_W better. Fig. 2 shows a comparison between the invariant jet-jet mass ($m_{j_1j_2}$) from the 800 GeV Higgs signal and the W + jets background which for events which pass this stage of the selection. The distribution peaks nicely around 80 GeV in the case of the Higgs signal, but is much broader in the case of the W + jets events. However, as we already know[2] it is not enough to apply even a stringent cut on $m_{j_1j_2}$ to get rid of the background. This is due to the tremendously small value of S/B before any further selection is applied.

The last step consists mainly of refining the definition of the $W \to q\bar{q}$ system so as to distinguish it from ordinary QCD jets. This can be done in two ways. To study the internal properties of a certain cluster pattern, one usually considers its representation in the (E_T, η, ϕ) reference frame (the LEGO plot). A cluster finder is used to reconstruct the cluster pattern and compute its main characteristics; this is the first way to attack our problem.

By projecting the energy of each calorimeter cell onto the thrust axis of the event in the (η, ϕ) plane of the calorimeter, another representation of the cluster pattern is obtained. This is in fact an energy spectrum corresponding to the distribution $1/EdE/dS$ (Fig. 3). Moreover, a set of variables characterizing not only the magnitude but also the shape of the energy pattern may be defined within this framework. Such a technique has been expressly developed to study the main characteristics of a $W \to q\bar{q}'$ system as compared to those of various ordinary QCD jet systems.[2,3] We apply this technique here to study the properties of the 'W-blob' localized in the hemisphere opposite to the reconstructed $W \to e\nu_e$. The technique consists of a set of combined cuts on variables which define the main features of the $1/EdE/dS$ distribution, on $m_{j_1j_2}$ of the reconstructed $q\bar{q}'$ system, and on the variable R_{min}[7] characterizing the reconstructed WW system.

In Table 3 we summarize the values of the detection efficiencies at the various stages of the analysis. At the moment we have only examined the medium and high $p_{T,W}$ contribution of the W + jets background. If we apply the detection efficiency ϵ_f obtained both for the Higgs signal and the W + jets background, correctly folding in the latter according to p_T bins and applying, for consistency, a cut on $p_{T,W}$ of 250 GeV also on the signal, we get a S/B ratio of 0.06. This corresponds to 220 remaining Higgs events and 3200 W + jets background events. The statistical significance of this result is given by $S/\sqrt{B} = 4\sigma$.

In Fig. 4 we plot the distribution m_{WW} for the reconstructed and selected 800 GeV $H^0 \to WW$ and compare it with m_{W+jets} given by the selected W + jets background events which pass all these cuts and have $p_{T,W} > 350$ GeV. We also show the $p_{T,WW}$ distribution for the 800 GeV H^0 signal passing our overall selection and the one directly given by primary partons (Fig. 5). In Fig. 6 the corresponding $p_{T,W+jets}$ distributions as given by the data and calculated from the primary partons are shown. In either case good agreement is obtained between the reconstructed and calculated curves. The results lead us to believe that a 3σ effect can be achieved in this channel.

Present improvements include a refinement of the cluster algorithm to reconstruct the $W \to q\bar{q}'$ energy pattern in the (E_T, η, ϕ) reference frame better. Much work is going on to improve and combine the cuts for the alternative energy pattern representation. The optimization of the matching between these two approaches is crucial for a better definition of the region of the hadronic W (in relation to the leptonic W), and for a better estimate of the characteristics of the $W \to q\bar{q}'$ system.

The problems still to be solved concern first the low p_T contribution of the W + jets background. To get a good estimate of the effects of our selection on this background one has to run very high statistics. This problem has been solved in the case of the ZZ signal (see Section 4); we are still facing it in the case of the WW signal. Another important problem is related to the case of a top quark heavier than the W. If this is the case, the top will predominantly decay into $W + b$ and therefore the direct production of $t\bar{t}$ in pp collisions could be a major background in the search for the $H \to WW$ signal. If the top has a mass of 120 GeV, PYTHIA 4.8 gives a cross-section of $1.2 \cdot 10^{-5}$ ($3.3 \cdot 10^{-7}$) mb for a p_T cut on the top quark of 100 (350) GeV. The corresponding numbers for a top quark mass of 200 GeV are $2.7 \cdot 10^{-6}$ ($2.4 \cdot 10^{-7}$) mb. The numbers are fairly insensitive to the structure functions used (quoted numbers are for EHLQ set 1). Thus, the expected rate of the background is of the order of 10^6 events per year even applying the high p_T cut, *i.e.* a factor 10 higher than the W + jets background. In a forthcoming study, these events will be submitted to our selection criteria.

4. Detection of Higgs Decaying Into Z Pairs

As we have seen, the detection of heavy W pairs at super pp colliders requires very highly performing detectors, in particular from the calorimetric point of view. Conversely, heavy Z pairs decaying leptonically seem, *a priori*, much easier to identify. The events are characterized by 4 high p_T charged leptons (e's and muons) or 2 charged high p_T leptons and a fair amount of missing transverse energy. No special hadronic pattern is demanded.

Therefore, at least for a first level analysis, it is not really necessary to have a detailed detector simulation as in Section 3. Much more important is to have a realistic Monte Carlo to reproduce the extra jets (mainly due to gluon bremsstrahlung) or extra leptonic activity of the underlying part of the event. For this we use PYTHIA 4.8. To emphasize the main features of the $4e$ and $2e2\nu$ signatures, we use the simplified simulation of a 4π fine grained calorimeter. The calorimeter is segmented into cells in $(\Delta\eta, \Delta\phi)$ of (0.05,0.05) within ± 6 units in pseudorapidity. A Gaussian smearing in transverse energy defined by $\Delta E_T = 0.2\sqrt{E_T}$ is applied to each cell to simulate calorimetric resolution effects. Clusters in energy are reconstructed using an algorithm which considers all cells which have a minimum transverse energy E_T of 2 GeV, starting from an initiator cell with $E_T > 5$ GeV. It sums all the corresponding cells within a cone in $\Delta R = \sqrt{\Delta\eta^2 + \Delta\phi^2} < 0.5$. The reconstructed cluster is then required to have a minimum E_T of 20 GeV in order to be considered a jet. Furthermore, perfect electron recognition is assumed within the range $|\eta| < 2.5$.

In order to investigate the detectability of such signals, we are going to define filter strategies. To do so we take into account not only the theoretical predictions concerning these types of events but also their main features as given by a realistic Monte Carlo. A detailed study of the main characteristics of the signal and the background has recently been finished.[4] One important result is that stiff jets and a fair amount of $E_{T,miss}$ are added to the core by the underlying event; this degrades the signal even in the case of purely leptonic signatures.

We have considered two different filters, based mainly on theoretical and experimental prejudices, respectively. Filter 1 is defined according to the following four steps. (Requirements for the $4e$ signature are given in brackets.)

1. Demand at least one [two] pair[s] of electrons, e^+e^-, within $|\eta^e| < 2.5$ and with $E_T^e > 30$ GeV.
2. Veto events with a jet of $E_T^{jet} > 50\,(200)$ GeV within $|\eta| < 6.0$, for the case of a 300 (800) GeV Higgs.
3. Demand [veto] events with $E_T^{miss} > 200\,(400)\,[50\,(100)]$ GeV, for the case of a 300 (800) GeV Higgs.
4. Demand one [two] Z^0['s] reconstructed from the [two] e^+e^- pair[s].

Filter 2 is mainly based on a redefinition of the missing energy. Instead of considering the total missing transverse energy, E_T^{miss}, we prefer to use the projected values of $\bar{E}_T^{miss}$ onto two axes in the plane transverse to the beam: the first along the reconstructed Z^0 momentum; the second orthogonal to this. We end up with two new variables:[8]

$$\epsilon_{L,T} = \frac{\bar{E}_T^{miss} \cdot \hat{e}_{L,T}}{E_T^{tot}},$$

where $\hat{e}_L$ is a unit vector antiparallel to the reconstructed Z^0 direction (in the plane transverse to the beam), and $\hat{e}_T$ the orthogonal unit vector in the transverse plane: $\hat{e}_L \cdot \hat{e}_T = 0$. This applies to the case of the $2e2\nu$ signature; for the $4e$ signature, $\bar{E}_{T,miss}$ is replaced by the $\bar{p}_T$ of the second reconstructed Z^0. These two variables allow a measurement of the balance in energy between the reconstructed Z^0 and E_T^{miss} (or the second reconstructed Z^0). Filter 2 is then defined according to the following steps (requirements for the $4e$ signal are given in brackets):

1a. Demand at least two [four] electrons, within $|\eta| < 2.5$ and with $E_T^e > 20$ GeV.
1b. Demand [veto] events with $E_T^{miss} > 50$ (100) GeV
1c. Demand one [two] Z^0['s] reconstructed from the electrons.
2. Demand $\epsilon_L > 0.35\,(0.50)\,[0.05\,(0.10)]$.

The results provided by these two filtering strategies are summarized in Table 4 and discussed in detail in [4] (as are the criteria for successful reconstruction of the Z^0's). Let us concentrate on the following important remarks:

i) Filter 1 is much more severe than Filter 2 as regards the low mass case.
ii) From the results given by both filters it seems very unlikely that the $4e$ signature can be detected unless the luminosity is increased by at least a factor 10. There are simply too few events to start with!
iii) The $2e2\nu$ signature, on the other hand, seems very promising. Both filters give reasonable promise for the detection of the 800 GeV mass signal. Moreover, Filter 2 should make it possible to identify even the 300 GeV case.

In Table 5 we give a summary of results for Filter 2 on the $2e2\nu$ signature. Note that the background we refer to is only the Z^0 + jets background. For a discussion of the QCD background, we refer again to [4]. We have performed a run generating 10^5 events Z^0 + jets, with a required minimum p_T of the Z^0 of 100 GeV and submitting them to the requirements for the 800 GeV Higgs $2e2\nu$ signature. As not a single event out of this sample passes this filter, we may set an upper limit on the number of Z^0 + jets events with $p_T > 100$ GeV which pass Filter 2 of < 10. This allows us to compute the results of Table 5 applying a cut of 100 GeV *only* on the p_T of the Z^0 from the H^0 signal at 800 GeV. This retains 97% of the total cross-section, whereas imposing a 350 GeV cut to get rid of the low p_T contribution of the background would leave only 50% of the H0 signal.

Let us stress the importance of this fact, namely *to relax as much as possible p_T cut to be applied on the signal.* This not only serves to keep the signal at the maximum, but also preserves its intrinsic properties. Cutting on essential parameters such as the p_T of the Z^0's, or their allowed pseudorapidity range, one may bias quite strongly the study of the general properties of the signal, once it is discovered. Furthermore, the statistical significance of the signal is also strongly affected by too severe cuts. The prize to pay for not making these stringent cuts is the necessity to find more sophisticated filtering, or even prefiltering, procedures and to run sufficiently high statistics on the background. By doing so in the case of the low p_T Z^0 + jets contribution we have succeeded in solving the problem at least for this background. The question still to be solved concerns the QCD background. This is now underway.

The main result at this stage is that it seems feasible to discover a heavy ZZ pair with the signature $ZZ \to 2e2\nu$.

Once the ZZ pair signal is filtered from the standard background, it is possible to distinguish the H^0 signal from the ZZ continuum by using the polarization properties of the respective events.[9] The differences between the respectively predominantly longitudinal and transverse Z's clearly appear in the distribution of the cosine of the decay angle (in the Z rest frame) between the two electrons from the Z (Fig. 7). For the $2e2\nu$ Higgs signature, imposing a cut on $\cos\theta^* < 0.6$ yields a signal to background ratio (where now the background is the ZZ continuum only) of 0.72 and 8.9 for the 300 and 800 GeV case, respectively. This corresponds to 47 (20) remaining H^0 events, and 67 (2) ZZ continuum events for the 300 (800) GeV mass case.

5. Some Concluding Remarks

Although the studies reported here are not yet definite we may make the following remarks concerning the search of a possible heavy Higgs signal at a super pp collider such as the SSC.

- The fact that a relatively light (*i.e.* $m_H \leq m_Z$) and apparently conventional H^0 signal could be discovered by the e^+e^- or $\bar{p}$ machines working at the W mass range, should not prevent or discourage one from looking for heavy VV pairs — quite the contrary!
- In the present situation a super pp collider such as the SSC project or the LHC (provided it runs with $> 10^{34}\,\mathrm{cm}^{-2}\mathrm{s}^{-1}$ peak luminosity and detectors can survive this) has a unique chance to scan the TeV range and in particular search for heavy IVB pairs.
- From the works reported in Sections 3 and 4, which are a continuation of the exploration started at Snowmass '86 and pursued since then by H.-U. Bengtsson, J. Hauptman, S. Linn and A. Savoy-Navarro as present members of this expedition, it seems possible to extract a clear heavy Higgs signal at least for the high mass case (around 800 GeV).
- The best way is by looking for the $2e2\nu$ signature provided by the decay chain $H^0 \to Z^0Z^0 \to e^+e^-\nu\bar{\nu}$.[10] A detector having a good (muon and) electron recognition for leptons with $E_T > 20$ GeV in the central pseudorapidity range ($|\eta| < 2.5$) and a full 4π coverage for the calorimetry is required for this purpose. The identification of the charged lepton pair reconstructing a Z^0 and a certain energy balance between this leptonic part and the reconstructed total missing transverse energy should be good enough to extract such a signal even for the low mass case
- A complementary way to extract a possible heavy Higgs signal seems to be by studying WW pairs which decay into $e\nu_e q\bar{q}'$. This would require highly performing 4π fine grained calorimetry able to measure properly refined cluster structures to distinguish $W \to q\bar{q}'$ systems from simple or multiple QCD jet(s) patterns.
- The importance of the standard QCD and V + jets backgrounds in strongly hiding the heavy Higgs signal is not only clearly emphasized for the WW case but also for the ZZ case even in the case of the so-called gold plated signature (*i.e.* the 4 charged leptons signal).
- We stress that no stringent cuts on the fundamental variables of the events should be applied to get rid of the backgrounds. Derived parameters and a careful study of the low p_T tail background are mandatory. Otherwise the final signal is not only strongly depleted from a statistics point of view, but may actually be castrated and left void of its fundamental characteristics.

References

[1] J. F. Gunion *et al.*, *Probing the W-Z-Higgs Sector of Electroweak Gauge Theories at the Superconducting Super Collider*, Theoretical Report of the W/Z/Higgs SSC Working Group, coördinated by J. F. Gunion and A. Savoy-Navarro, in *Proceedings of the 1986 Summer Study on the Physics of the Superconducting Supercollider*, eds. R. Donaldson, J. Marx (1987), p.142.

[2] G. Alverson *et al.*, *Experimental Search for W/Z Pairs and Higgs Bosons at Very High Energy Hadron-Hadron Colliders* and *Detecting W/Z Pairs and Higgs at High Energy pp Colliders: Main Experimental Issues*, Experimental Reports of the W/Z/Higgs SSC Working Group, coördinated by J. F. Gunion and A. Savoy-Navarro, in *Proceedings of the 1986 Summer Study on the Physics of the Superconducting Supercollider*, eds. R. Donaldson, J. Marx (1987), p. 93, p. 114.

[3] H.-U. Bengtsson, J. Hauptman, S. Linn, A. Savoy-Navarro, in preparation.

[4] H.-U. Bengtsson, A. Savoy-Navarro, Saclay preprint DPhPE-87-08 and UCLA preprint UCLA-87-004, submitted to Phys. Rev. **D**.

[5] H.-U. Bengtsson, T. Sjöstrand, Computer Phys. Comm. **46** (1987) 43.

[6] I. Hinchliffe, M.-J. Herrero, private communication.

[7] R_{min} was first introduced by J. F. Gunion and M. Soldate; see [1].

[8] S. Dawson, A. Savoy-Navarro, *Searching for Supersymmetry at the SSC*, Report of the Supersymmetry Working Group, in *Proceedings of the 1984 Summer Study on the Design and Utilization of the Superconducting Super Collider*, eds. R. Donaldson, G. Morfín (1984), p. 263.

[9] M. J. Duncan, G. L. Kane, W. W. Repko, Nucl. Phys. **B272** (1986) 517.

[10] M. S. Chanowitz, M. K. Gaillard, Phys. Lett. **142B** (1985) 85.

Figure Captions

Fig. 1 Comparison between the invariant mass distributions (in GeV) of final state Z pairs from $VV \to H^0 \to ZZ$ (s-channel Higgs exchange only; full line) and the complete gauge invariant set of diagrams $VV \to ZZ$ (dashed line); a) $m_H = 300$ GeV, b) $m_H = 800$ GeV.

Fig. 2 Invariant jet-jet mass distribution (in GeV) from $pp \to H^0 \to WW \to e\nu q\bar{q}'$ ($m_H = 800$ GeV; full line) and from $pp \to W\,(\to q\bar{q}')+$ jets ($p_{T,W} > 350$ GeV; dashed line)

Fig. 3 The distribution $1/EdE/dS$ for $W \to q\bar{q}'$.

Fig. 4 Invariant mass distributions (in GeV): m_{WW} for $pp \to H^0 \to WW$ ($m_H = 800$ GeV) (signal), and m_{W+jets} for $pp \to W+$ jets ($p_{T,W} > 350$ GeV) (background). Full line: background; dashed line: signal + background; insertion: signal only. All histograms properly normalized.

Fig. 5 Distribution of $p_{T,WW}$ (in GeV) for $pp \to H^0 \to WW$ ($m_H = 800$ GeV). Full line: parton calculation; dashed line: reconstruction.

Fig. 6 Distribution of $p_{T,W+jets}$ (in GeV) for $pp \to W+$ jets ($p_{T,W} > 350$ GeV). Hatched: parton calculation; full line: reconstruction.

Fig. 7 Comparison between distributions of $\cos\theta^*$ (cosine of angle between electrons from Z in c.m. of Z) in $pp \to H^0 \to ZZ$ (s-channel Higgs exchange, $m_H = 800$ GeV; full line) and $pp \to ZZ$ (ZZ continuum, $p_{T,Z} > 350$ GeV; dashed line) for $2e2\nu$ signature.

Table 1: Cross-sections (mb) and Rates per Year for WW

Process	$\sqrt{s} = 10$ TeV, $\int \mathcal{L} = 10^{39}$ cm^{-2}		$\sqrt{s} = 17$ TeV, $\int \mathcal{L} = 10^{40}$ cm^{-2}		$\sqrt{s} = 40$ TeV, $\int \mathcal{L} = 10^{40}$ cm^{-2}	
	300 GeV	800 GeV	300 GeV	800 GeV	300 GeV	800 GeV
$pp \to H^0 \to WW$	$1.5 \cdot 10^{-10}$	$4.8 \cdot 10^{-12}$	$4.4 \cdot 10^{-10}$	$2.3 \cdot 10^{-11}$	$2.0 \cdot 10^{-9}$	$1.8 \cdot 10^{-10}$
$WW \to e\nu q\bar{q}'$	150	5	4400	230	$2.0 \cdot 10^{4}$	1800
$pp \to WW$	$4.0 \cdot 10^{-10}$	$5.2 \cdot 10^{-12}$	$7.9 \cdot 10^{-10}$	$1.6 \cdot 10^{-11}$	$2.1 \cdot 10^{-9}$	$6.3 \cdot 10^{-11}$
$WW \to e\nu q\bar{q}'$	400	5	7900	160	$2.1 \cdot 10^{4}$	630
$pp \to W g,\ q$	$8.7 \cdot 10^{-8}$	$7.4 \cdot 10^{-10}$	$2.0 \cdot 10^{-7}$	$2.5 \cdot 10^{-9}$	$6.4 \cdot 10^{-7}$	$1.2 \cdot 10^{-8}$
$W \to e\nu$	$8.7 \cdot 10^{4}$	740	$2.0 \cdot 10^{6}$	$2.5 \cdot 10^{4}$	$6.4 \cdot 10^{6}$	$1.2 \cdot 10^{5}$
$pp \to qq,\ qg,\ gg$	$1.1 \cdot 10^{-6}$	$2.6 \cdot 10^{-9}$	$2.8 \cdot 10^{-6}$	$1.1 \cdot 10^{-8}$	$1.1 \cdot 10^{-5}$	$7.4 \cdot 10^{-8}$
$\times 10^{-3}$ (rad. W)	$1.1 \cdot 10^{6}$	$2.6 \cdot 10^{3}$	$2.8 \cdot 10^{7}$	$1.1 \cdot 10^{5}$	$1.1 \cdot 10^{8}$	$7.4 \cdot 10^{5}$

Table 2: Cross-sections (mb) and Rates per Year for ZZ

Process	$\sqrt{s} = 10$ TeV, $\int \mathcal{L} = 10^{39}$ cm^{-2}		$\sqrt{s} = 17$ TeV, $\int \mathcal{L} = 10^{40}$ cm^{-2}		$\sqrt{s} = 40$ TeV, $\int \mathcal{L} = 10^{40}$ cm^{-2}	
	300 GeV	800 GeV	300 GeV	800 GeV	300 GeV	800 GeV
$pp \to H^0 \to Z^0 Z^0$	$5.7 \cdot 10^{-13}$	$2.2 \cdot 10^{-14}$	$1.7 \cdot 10^{-12}$	$1.0 \cdot 10^{-13}$	$7.0 \cdot 10^{-12}$	$7.9 \cdot 10^{-13}$
$Z^0 Z^0 \to eeee$	0	0	17	1	70	8
$pp \to H^0 \to Z^0 Z^0$	$6.8 \cdot 10^{-12}$	$2.6 \cdot 10^{-13}$	$2.0 \cdot 10^{-11}$	$1.2 \cdot 10^{-12}$	$8.4 \cdot 10^{-11}$	$9.4 \cdot 10^{-12}$
$Z^0 Z^0 \to ee\nu\bar{\nu}$	7	0	200	12	840	94
$pp \to Z^0 Z^0$	$6.4 \cdot 10^{-13}$	$9.0 \cdot 10^{-15}$	$1.3 \cdot 10^{-12}$	$2.8 \cdot 10^{-14}$	$3.6 \cdot 10^{-12}$	$1.1 \cdot 10^{-13}$
$Z^0 Z^0 \to eeee$	1	0	13	0	36	1
$pp \to Z^0 Z^0$	$7.6 \cdot 10^{-12}$	$1.1 \cdot 10^{-13}$	$1.5 \cdot 10^{-11}$	$3.4 \cdot 10^{-13}$	$4.3 \cdot 10^{-11}$	$1.3 \cdot 10^{-12}$
$Z^0 Z^0 \to ee\nu\bar{\nu}$	8	0	150	3	430	13
$pp \to Z^0 g,\ q$	$1.3 \cdot 10^{-8}$	$1.2 \cdot 10^{-10}$	$3.0 \cdot 10^{-8}$	$4.1 \cdot 10^{-10}$	$9.9 \cdot 10^{-8}$	$2.0 \cdot 10^{-9}$
$Z^0 \to ee$	$1.3 \cdot 10^{4}$	120	$3.0 \cdot 10^{5}$	4100	$9.9 \cdot 10^{5}$	$2.0 \cdot 10^{4}$
$pp \to b\bar{b},\ t\bar{t}$	$1.2 \cdot 10^{-5}$	$2.6 \cdot 10^{-8}$	$3.2 \cdot 10^{-5}$	$1.1 \cdot 10^{-7}$	$1.2 \cdot 10^{-4}$	$8.1 \cdot 10^{-7}$
	$1.2 \cdot 10^{7}$	$2.6 \cdot 10^{4}$	$3.2 \cdot 10^{8}$	$1.1 \cdot 10^{6}$	$1.2 \cdot 10^{9}$	$8.1 \cdot 10^{6}$

Note: All cross-sections, except the ones for Higgs production, include a p_T cut on the parton level of 100 (350) GeV/c for the 300 (800) GeV case. If the same cut is applied also to the Higgs cross-sections, these should be multiplied by $\sim (1 - 4p_{T,min}^2/m_H^2)^{1/2} \approx 0.75\ (0.50)$. Also, the Higgs cross-sections do not include the O(10%) correction necessary because of the fictitious high mass tail.

Table 3: Detection Efficiencies for WW

Process	Efficiencies in %		
	$\epsilon_{e\nu}$	ϵ_{jj}	ϵ_f
$pp \to H \to WW;\ m_H = 800$	89 ± 4	88 ± 4	16 ± 1
$pp \to W + \text{jets};\ p_T^W \in [250, 350]$	77 ± 6	77 ± 6	0.5 ± 0.2
$pp \to W + \text{jets};\ p_T^W > 350$	84 ± 3	84 ± 3	1.3 ± 0.3

Table 4: Efficiencies and Rates of $2e2\nu$ Signature

Process	Filter 1		Filter 2	
	ϵ	Final rate	ϵ	Final rate
	(%)	(# events)	(%)	(# events)
$pp \to H^0 \to Z^0Z^0$	0.2	2	5.6	48
	(34)	(32)	(27)	(25)
$pp \to Z^0Z^0$	5.2	22	21	91
	(42)	(5)	(30)	(4)
$pp \to Z^0$+ jets	0	< 100	0.01	99
	(0)	(< 2)	(0)	(< 2)
$pp \to b\bar{b},\ t\bar{t}$	0	$< 1.2 \cdot 10^5$	0	$< 1.2 \cdot 10^5$
	(0)	(< 810)	(0)	(< 810)

Note: Numbers in parentheses refer to the case of an 800 GeV Higgs (p_T cut on background of 350 GeV/c); other numbers to the case of a 300 GeV Higgs (p_T cut on background of 100 GeV/c).

Table 5: Detectability of $2e2\nu$ Signature, Filter 2

Process	Final rate (# events)	S/B	S/$\sqrt{\text{B}}$
$pp \to H^0 \to Z^0Z^0$ $m_H = 300, p_T^Z > 100$	36	> 0.36	> 3.6
$pp \to H^0 \to Z^0Z^0$ $+pp \to Z^0Z^0$ $m_H = 300, p_T^Z > 100$	104	> 1.0	> 10
$pp \to H^0 \to Z^0Z^0$ $m_H = 800, p_T^Z > 100$	24	> 2.4	> 7.5

Fig. 1a

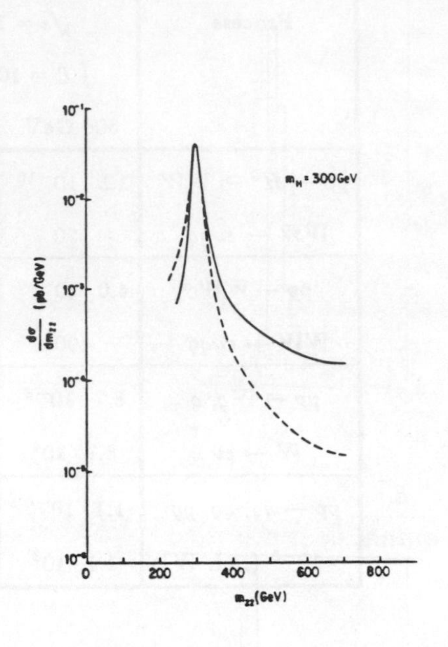

Fig. 1b

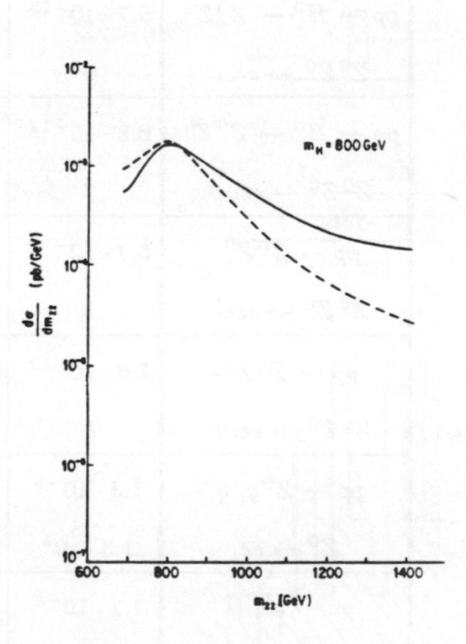

Fig. 2

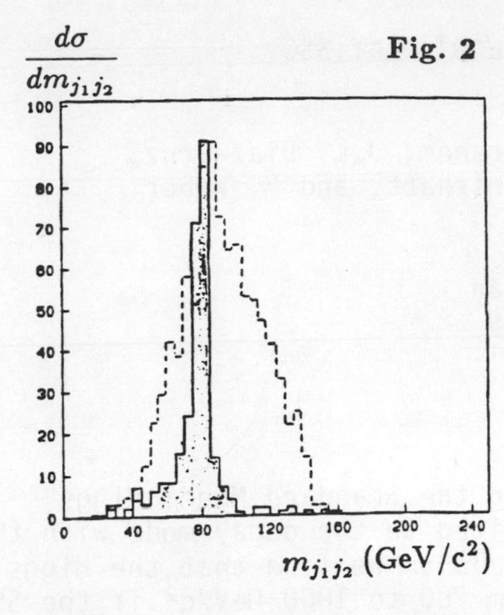

Fig. 3

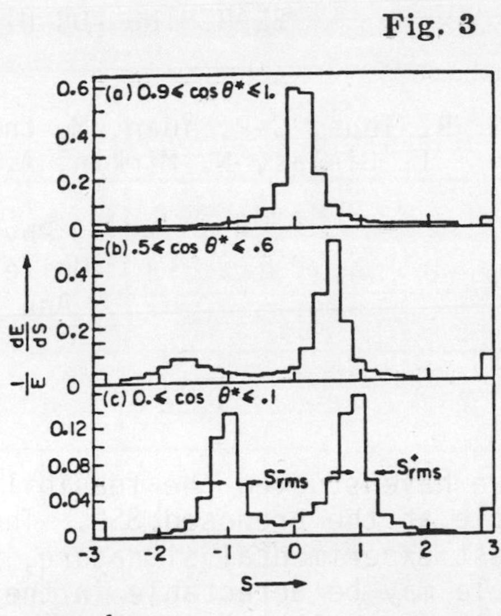

Fig. 4

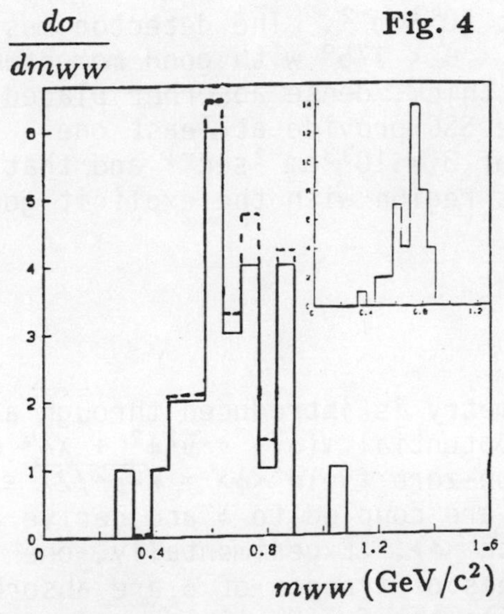

Fig. 5

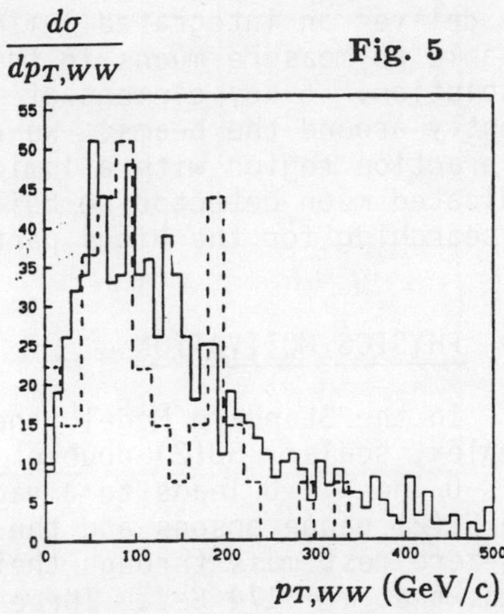

Fig. 6

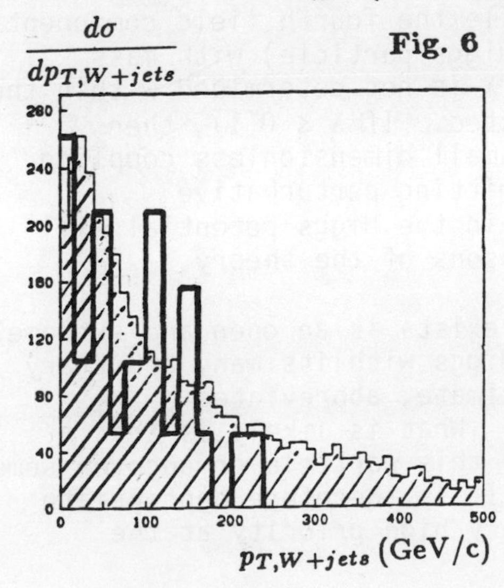

Fig. 7

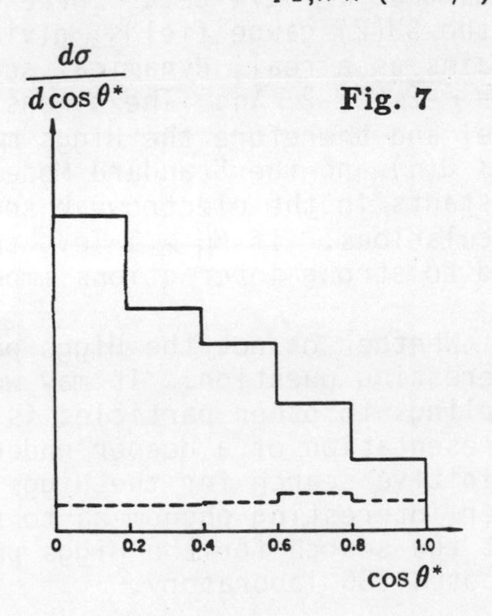

SEARCHING FOR HIGGS $\rightarrow Z^{\circ}Z^{\circ} \rightarrow \mu^{+}\mu^{-}\mu^{+}\mu^{-}$ AT SSC

R. Thun, C-P. Yuan, M. Chmeissani, T. Dershem, J.L. Diaz-Cruz,
I. Gialas, N. Mirkin, A. Nguyen, R. Tschirhart, and M. Weber

Physics Department
University of Michigan
Ann Arbor, MI 48109

ABSTRACT

We have studied the feasibility of detecting the Standard-Model Higgs particle at the proposed SSC. The study was limited to the decay mode with the cleanest experimental signature, $H \rightarrow Z^{\circ}Z^{\circ} \rightarrow \mu^{+}\mu^{-}\mu^{+}\mu^{-}$. We find that the Higgs particle may be detectable in the mass range from 200 to 1000 GeV/c^2 if the SSC can deliver an integrated luminosity of at least 10^{41}cm^{-2}. The detector must be able to measure muons in the angular range $5 < \theta < 175°$ with good momentum resolution. A key element of the detector is a thick, dense absorber placed tightly around the beams. We recommend that the SSC provide at least one interaction region with a luminosity in excess of 3×10^{33}cm^{-2}sec^{-1} and that a dedicated muon detector be built to exploit this region with the explicit goal of searching for the Higgs particle.

1.) PHYSICS MOTIVATION

In the Standard Model[1] the breaking of symmetry is introduced through a complex, scalar, SU(2) doublet ϕ for which the potential $V(\phi^2) = \mu^2\phi^2 + \lambda\phi^4$ with $\mu^2 < 0$ and $\lambda > 0$ leads to a vacuum state with non-zero field $\langle\phi\rangle = \sqrt{-\mu^2/2\lambda} \equiv \eta$. The SU(2) gauge bosons and the charged fermions are coupled to ϕ and derive a non-zero rest mass through their interaction with $\langle\phi\rangle$. Experimentally, one finds that η = 174 GeV. Three of the four degrees of freedom of ϕ are absorbed by the SU(2) gauge fields, giving them mass, while the fourth field component remains as a real, dynamical scalar field (the Higgs particle) with mass $M_H = \sqrt{-2\mu^2} = 2\sqrt{\lambda}\eta$. The dimensionless constant λ is not determined within the model and therefore the Higgs mass is not predicted. If $\lambda \lesssim O(1)$, then $M_H \lesssim O(\eta)$ and the Standard Model contains only small dimensionless coupling constants in the electroweak sector thereby permitting perturbative calculations. If $M_H \gtrsim 1$ TeV, then the $\lambda\phi^4$ term in the Higgs potential will lead to strong interactions among the massive bosons of the theory.

Whether or not the Higgs particle actually exists is an open and extremely interesting question. It may well be that the Higgs with its many arbitrary couplings to other particles is simply an approximate, abbreviated representation of a deeper underlying principle. What is likely is that a definitive search for the Higgs will either find this particle or uncover some other interesting phenomena to take its place. It is therefore appropriate that the search for the Higgs particle have a very high priority at the proposed SSC laboratory.[2]

The question arises immediately if the Standard-Model Higgs is detectable at the SSC. The following numbers give some perspective to this question. Let us assume that the Higgs mass is $M_H = 300$ GeV/c^2. Then the production cross section for Higgs particles is expected to be about 12 pb for a pp c.m. energy of 40 TeV. Running one calendar year with a machine efficiency of 30% and peak luminosity of 10^{33} cm^{-2}sec^{-1} yields an integrated luminosity of 10^{40} cm^{-2} (=10^4 pb^{-1}) and a total of 120,000 Higgs particles. The total number of inelastic pp collisions for the same SSC run is expected to be $\sim 10^{15}$. To find the Higgs one must be able to discriminate events at the level of one in 10^{10}. Instantaneous event rates are about 10^8 Hz and single-particle rates about 10^{10} Hz. The problems of particle detection at such rates and of implementing workable trigger schemes are truly awesome. In addition, most decay modes of the Higgs particle give signals with severe physics backgrounds.

Considerations such as those above, have led us to investigate the possibility of searching for the decay mode $H \to Z°Z° \to \mu^+\mu^-\mu^+\mu^-$. Muons are able to penetrate thick absorbers placed around the interaction region for the shielding of muon detector and triggering elements from the huge flux of secondary particles. Known physics backgrounds to this signal are estimated to be small. The price one pays is that the branching ratio of Higgs to four muons is also small. For a Higgs mass of 300 GeV/c^2, this ratio is 2.5×10^{-4}, yielding a sample of 30 events during one "standard" SSC year (= 10^{40} cm^{-2} integrated luminosity). While this is a small number, the events would show two reconstructed Z°'s, a powerful signature against spurious backgrounds.

2.) HIGGS PRODUCTION AT SSC

Cross sections for producing Higgs particles in high-energy proton-proton collisions have been calculated by a number of authors.[3] The calculations are based on the Standard Model and use as input measured parton distribution functions extrapolated to the higher momentum-transfer scales of such collisions.

The dominant production mechanisms for Higgs particles with mass greater than 200 GeV/c^2 are expected to be the gluon-gluon and heavy gauge-boson fusion mechanisms. The cross section from the gluon-fusion process depends on the, as yet unknown, top quark mass. The cross section increases with increasing t-quark mass. We shall use the cross sections calculated by Cahn[4] who assumed a t-quark mass of 40 GeV/c^2. Table 2.1 summarizes these cross sections for Higgs masses in the range from 200 to 1000 GeV/c^2 for two c.m. pp energies. The contribution from WW and ZZ fusion dominates that from gg (gluon) fusion for Higgs masses above 300 GeV/c^2. It is interesting to note that a factor of two decrease in the c.m. pp energy from 40 to 20 TeV reduces the Higgs production cross section by a factor of 3 to 4.

The rapidity distribution of Higgs particles produced via gluon fusion has been given by Eichten et al.[3] Production is approximately uniform in the rapidity y and limited roughly to the region $-2 < y < 2$, the range shrinking somewhat as the Higgs mass increases. Transverse momentum distributions of Higgs particles have been calculated by Cahn et al.[5] for the W-fusion process and are parameterized by:

$$\frac{dN}{dp_\perp^2} = \frac{1}{\langle x\rangle M_W^2} F(z)$$

with $\langle x\rangle = 1 - M_H/\sqrt{s}$; $z = P_\perp^2/(\langle x\rangle M_W^2)$

and $$F(z) = \frac{2z^2 - 4z}{(z^2 + 4z)^2} + \frac{16z(1+z)}{(z^2+4z)^{5/2}} \tanh^{-1} \frac{1}{\sqrt{1 + 4/z}}$$

The distribution $dN/dP_\perp$ peaks near $P_\perp = M_W$ and falls off as $P_\perp^{-3}$ at large $P_\perp$.

TABLE 2.1 HIGGS PRODUCTION CROSS SECTIONS[4] (in picobarn)
[Mass of t-quark assumed to be 40 GeV/c^2]

M_H GeV/c	$\sqrt{s}$ = 20 TeV				$\sqrt{s}$ = 40 TeV			
	WW Fusion	ZZ Fusion	gg Fusion	SUM	WW Fusion	ZZ Fusion	gg Fusion	SUM
200	2.6	0.9	8.8	12.3	6.8	2.5	22.8	32.1
300	1.6	0.6	1.9	4.1	4.5	1.7	5.8	12.0
500	0.7	0.3	0.2	1.2	2.4	0.9	0.8	4.1
700	0.4	0.15	0.03	0.58	1.5	0.6	0.2	2.3
1000	0.19	0.07	0.005	0.27	0.8	0.31	0.03	1.1

3.) HIGGS DECAY

For Higgs masses above 200 GeV/c^2 the dominant decay modes will be $H \to W^+W^+$, $H \to Z^\circ Z^\circ$ and $H \to t\bar{t}$. The partial decay widths are given by[6]:

$$\Gamma(H \to W^+W^-) = \frac{G_F M_H^3}{32\pi\sqrt{2}} (4 - 4a_W + 3a_W^2)(1 - a_W)^{1/2}$$

$$\Gamma(H \to Z^\circ Z^\circ) = \frac{G_F M_H^3}{64\pi\sqrt{2}} (4 - 4a_Z + 3a_Z^2)(1 - a_Z)^{1/2}$$

$$\Gamma(H \to t\bar{t}) = \frac{3G_F m_t^2 M_H}{4\pi\sqrt{2}} \left(1 - \frac{4m_t^2}{M_H^2}\right)^{3/2}$$

where G_F = Fermi constant = 1.166×10^{-5} GeV^{-2}

M_H = Higgs Mass

m_t = t-quark mass, assumed to be 40 GeV/c^2

$$a_W = \frac{4M_W^2}{M_H^2} \; (M_W = 81.8 \text{ GeV/c}^2) \qquad a_Z = \frac{4M_Z^2}{M_H^2} \; (M_Z = 92.6 \text{ GeV/c}^2)$$

Table 3.1 displays the total decay width and branching ratios as a function of Higgs mass.

The decay $H \to Z°Z°$ yields an isotropic angular distribution of the Z°'s in the rest frame of the Higgs. One can show by a straight-forward calculation that the fraction of Higgs decays into longitudinally polarized Z°'s is given by:

$$f_L = \left(\frac{M_H^2}{2M_Z^2} - 1\right)^2 \Big/ \left[\left(\frac{M_H^2}{2M_Z^2} - 1\right)^2 + 2\right]$$

and the fraction into transversely polarized Z°'s is:

$$f_t = 2/\left[\left(\frac{M_H^2}{2M_Z^2} - 1\right)^2 + 2\right]$$

These fractions are also displayed in Table 3.1. For Higgs masses above 300 GeV/c^2, nearly all Z°'s are longitudinally polarized.

The angular distributions of the μ^- from the decay $Z° \to \mu^+\mu^-$, in the rest frame of the Z°, are given by:

$$\frac{dN}{d\Omega} = (1 + \cos^2\theta - a\cos\theta) \text{ for } \lambda = +1$$

$$= (1 + \cos^2\theta + a\cos\theta) \qquad \lambda = -1$$

$$= \sin^2\theta \qquad \lambda = 0$$

the first two corresponding to transverse polarizations, and the last to longitudinal polarization. The angle θ is measured with respect to an axis given by the direction of flight of the Z° in the Higgs rest-frame. In the Standard Model $a = 2(1-2X)/(2X^2-2X+1)$ with $X = 2\sin^2\theta_W = 2(0.226) = 0.452$, so that $a = 0.380$.

4.) GENERAL REMARKS ON DETECTING $H \to Z°Z° \to \mu^+\mu^-\mu^+\mu^-$

Searches for the Higgs particle can be carried out with two very different techniques characterized by "open" and "shielded" detector geometries. The "open" detector consists of devices for tracking, calorimetry, and muon detection configured in such a way as to be sensitive to all proton-proton collision products. Such a detector can be used to search for $H \to Z^\circ Z^\circ \to q\bar{q}\ l^+l^-$, $l^+l^-\ \bar{\nu}\nu$, and $l^+l^-\ l^+l^-$ where $q\bar{q}$ represents a pair of quark jets and l^+l^- stands for either e^+e^- or $\mu^+\mu^-$. The branching ratios for these three decay modes are 0.0260, 0.0065 and 0.0011, respectively. The physics background to $q\bar{q}\ l^+l^-$ arising from the production of a single Z° in association with hadronic jets is potentially severe.[7] The $l^+l^-\ \bar{\nu}\nu$ decay mode has the disadvantage that the neutrino pair is undetectable although the transverse momentum distribution of l^+l^- pairs may still show structure from Higgs decays.[8] In addition to such background and kinematical problems, the "open" detector is vulnerable to the intense flux of secondary particles. It will be difficult to define discriminating and efficient triggers for the detection of Higgs particles.

TABLE 3.1 HIGGS DECAY PARAMETERS ($m_t \equiv 40$ GeV/c^2)

		BRANCHING RATIOS			Z° POLARIZATION FRACTIONS	
M_H (GeV/c^2)	Γ (GeV)	W^+W^-	$Z^\circ Z^\circ$	$t\bar{t}$	LONG.	TRANS.
200	1.8	0.548	0.187	0.264	0.470	0.530
300	9.1	0.628	0.279	0.093	0.900	0.100
400	25.2	0.650	0.303	0.047	0.972	0.028
500	53.2	0.657	0.314	0.028	0.989	0.011
600	95.9	0.661	0.320	0.019	0.995	0.005
700	156	0.662	0.324	0.014	0.997	0.003
800	238	0.664	0.326	0.010	0.998	0.002
900	343	0.664	0.328	0.008	0.999	0.001
1000	474	0.665	0.329	0.007	0.999	0.001

The great advantage of a Higgs search limited to $H \to Z^\circ Z^\circ \to \mu^+\mu^-\mu^+\mu^-$ is the ability to shield the active detector elements from most secondary particles produced in the proton-proton collisions. The rate of prompt muons and other "punch-thru" particles that will penetrate a thick absorber surrounding the interaction region is a factor of at least 10^4 lower than the total flux of produced particles.[9] The signal for $H \to Z^\circ Z^\circ \to \mu^+\mu^-\mu^+\mu^-$ is potentially very clean in that two neutral muon-pair combinations must yield effective masses consistent with the Z° mass. A disadvantage of this kind of

Higgs search with a "shielded" detector is the small branching ratio for the decay mode $H^\circ \to \mu^+\mu^-\mu^+\mu^-$. This ratio is at the 3×10^{-4} level and dictates a detector design with high muon detection efficiencies. In principle, the shielded detector should be able to operate at luminosities that are several orders of magnitude larger than those permissible for the open detector. If the SSC can deliver such higher luminosities, then this will compensate for the low $H \to \mu^+\mu^-\mu^+\mu^-$ branching ratio.

It is beyond the scope of this study to make a detailed comparison of the relative competitiveness of "open" and "shielded" detectors for Higgs searches. A general look at physics in the TeV energy regime will dictate the construction of at least one large, complex, open detector. From an experimental and financial viewpoint, the shielded muon detector is much simpler, cheaper and more focussed on certain physics goals. In particular, the elimination of high-resolution calorimetry removes a tremendous burden of complexity from the detector design and construction. It may well happen that, if the Higgs is detectable at all, it will first be observed with an optimized, dedicated muon detector.

We conclude this section, with a statement on the general requirements that a muon detector must satisfy. Three important parameters are the minimum angle with respect to the beam directions at which muons must be detectable, a transverse momentum requirement to suppress backgrounds, and the muon momentum resolution necessary for reconstructing the $H^\circ \to Z^\circ Z^\circ \to \mu^+\mu^-\mu^+\mu^-$ signal. Figure 4.1 shows the efficiency for detecting all four muons as a function of the minimum angle of detection. An angle of 5° appears feasible for a detector and yields an efficiency around 85%. Prompt muons from charm and beauty decay and hadronic punch-thru products are expected to produce "muon" signals at relatively low transverse momentum. Figure 4.2 shows the efficiency for detecting $H^\circ \to \mu^+\mu^-\mu^+\mu^-$ as a function of the minimum transverse momentum, $P_\perp$ satisfied by each muon and assuming a minimum detection angle of 5°. A requirement of $P_\perp > 10$ GeV/c and $\Theta > 5°$ is seen to yield a detection efficiency for $H \to \mu^+\mu^-\mu^+\mu^-$ around 78%.

The question of mass resolution will be discussed in detail after presenting a possible design for a muon detector. We make the following general observation. The natural width of the Higgs particle increases as the third power of the Higgs mass and will dominate the detector mass resolution for $M_H \gtrsim 500$ GeV/c^2. For masses below this value, the mass resolution dominates the observed signal width. On the other hand, the production cross section rises sharply as the Higgs mass decreases so that the ability to discriminate a signal becomes less dependent on good mass resolution at lower Higgs masses. The ability to identify Z°'s in the four-muon final state is important for eliminating spurious backgrounds. For this purpose, the natural width of the Z°, $\Gamma_Z \approx 2.8$ GeV, places a lower limit on useable mass resolution.

5.) DETECTOR DESIGN

We show the general lay-out for a muon detector in Fig. 5.1. At angles within 30° of the beam directions, muon momenta are measured with a set of magnetized iron toroids, whereas at larger angles a superconducting solenoid is used for this purpose. The basic reason for this choice of magnets is that

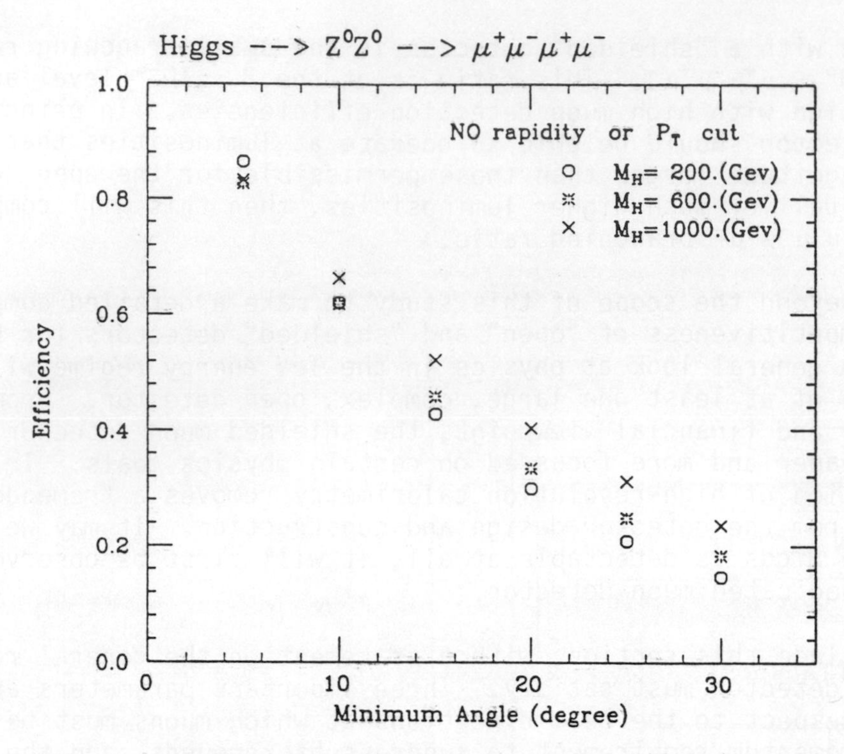

Fig. 4.1 Detection efficiency for $H \rightarrow Z^{\circ}Z^{\circ} \rightarrow 4\mu$ as a function of the minimum muon-detection angle with respect to the beams.

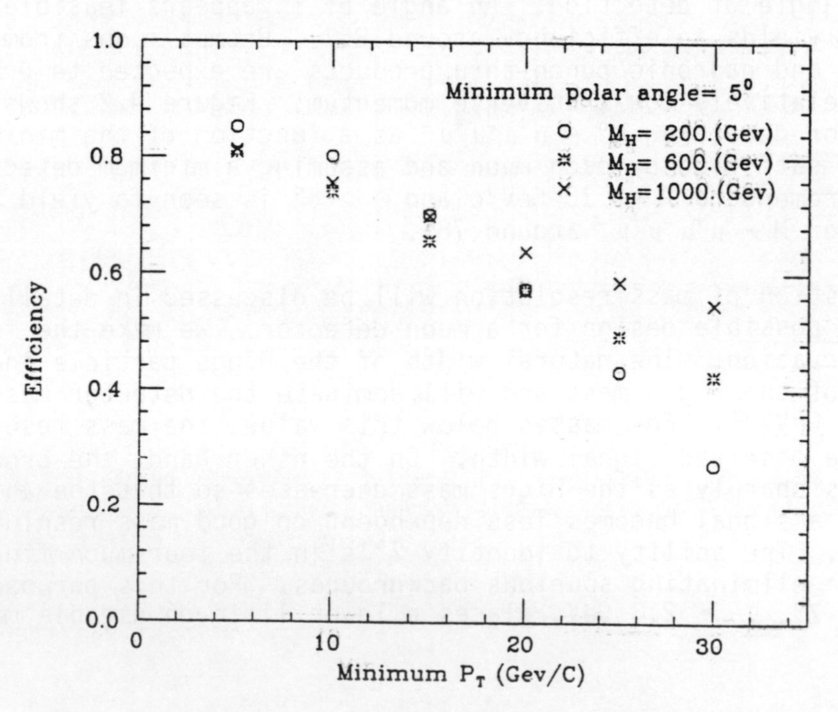

Fig. 4.2 Detection efficiency for $H \rightarrow Z^{\circ}Z^{\circ} \rightarrow 4\mu$ as a function of a minimum P_T requirement on each muon.

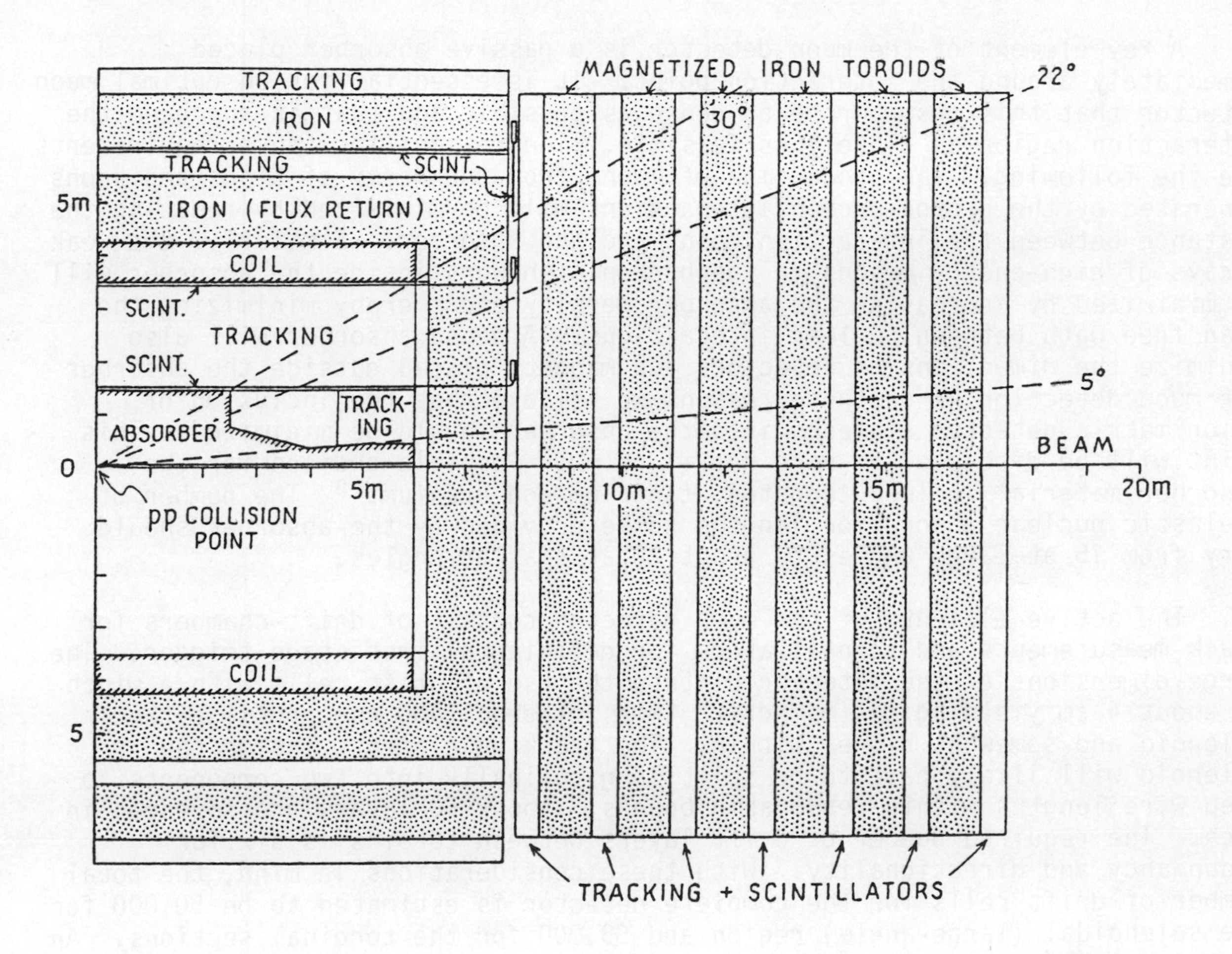

Fig. 5.1 Side view of muon detector. The detector is symmetric about the collision point and only the right half is shown.

at small angles the muon momenta can be expected to be as high as several TeV/c. It would be very difficult to design and construct open dipole magnets of the requisite dimensions to handle such momenta for muon angles in the range 5-30° with respect to the beams. At larger angles, a solenoidal magnetic geometry is appropriate and is provided most naturally by a large open superconducting magnet. The precision of muon momentum measurements with such a solenoid exceeds that which could be achieved with magnetized iron for momenta below several hundred GeV/c. The muon momenta from $H \to \mu^+\mu^-\mu^+\mu^-$ at large angles are expected to be $\lesssim$ 200 GeV/c for an observable Higgs signal.

A key element of the muon detector is a passive absorber placed immediately around the interaction point. It is essential for an optimal muon detector that this absorber be as dense as possible and that it surround the interaction region as closely as feasible. The reasons for these requirements are the following. The production of muons from the decay of pions and kaons generated by the proton-proton interactions will be minimized by reducing the distance between the interaction point and the absorber. Similarly, the weak decays of high-energy mesons in the hadronic showers inside the absorber will be minimized by increasing the absorber density and thereby minimizing the mean free path between nuclear interactions. A dense absorber will also minimize the dimensions of detectors and magnets placed outside the absorber for muon detection. For these reasons we believe that the inclusion of calorimetric detector elements into the absorber should be minimized. This point will be discussed in more detail below. Possible choices for the absorber material include tungsten, tantalum and uranium.[10] The number of inelastic nuclear absorption lengths to be provided by the absorber should vary from 15 at large angles to about 30 at smaller angles.

The active elements of the muon detector consist of drift-chambers for track measurements and scintillators for defining a first-stage trigger. The large dimensions of the detector dictate the use of drift cells with a width of about 4 cm yielding maximum drift times of about 400 nsec outside the solenoid and somewhat higher within. The tracking elements inside the solenoid will likely have to be split longitudinally into two components to keep wire lengths within reasonable bounds. About 50 layers are foreseen in each. The required number of drift layers between toroids is six for redundancy and directionality. With these considerations in mind, the total number of drift cells for the complete detector is estimated to be 50,000 for the solenoidal (large-angle) region and 63,000 for the toroidal sections. An additional small-angle tracker inside the central-magnet is estimated to require another 7,000 wires bringing the total to about 120,000.

A layer of scintillators, divided into 96 pie-shaped sections, is positioned behind each toroidal iron magnet. The signals from these layers are required to be coincident within well-defined azimuthal sectors so as to define a set of particle trajectories pointing toward the proton-proton collision point. Similarly, three concentric layers of scintillators placed within the solenoidal magnet also define particle paths pointing to the interaction region. Unlike the scintillators behind the toroids, those in the solenoid define a minimum transverse momentum by suitable coincidence requirements. Each solenoid layer contains 96 azimuthal elements, divided longitudinally at the detector mid-plane. The total number of phototubes for the entire detector is 1728. Table 5.1 summarizes the configuration of drift cell and scintillator channels.

TABLE 5.1 DETECTOR CONFIGURATION

REGION	NUMBER OF DRIFT CELLS	NUMBER OF SCINTILLATORS
SOLENOID	CENTRAL = 38000 IRON = 12000 SMALL ANGLE = 7000	3 LAYERS × 96 × 2 SCINTILLATORS = 576
TOROIDS	14 LAYERS × 4500 CELLS = 63000	12 LAYERS × 96 SCINTILLATORS = 1152
TOTAL	120000	1728

We considered briefly the question of hadronic calorimetry in the above description of the detector. In principle, it is not necessary to measure the hadronic energy flow for the reconstruction of the $H \to \mu^+\mu^-\mu^+\mu^-$ signal. However, a crude hadronic shower measurement might identify multi-muon events arising from the decay of heavy quarks. Whether or not this is necessary for background rejection has not been studied in detail. The inclusion of several layers of proportional wires (say 5, or 6) within the absorber would not seriously degrade the performance of the absorber and might give an effective albeit crude identification of hadronic jets.

6.) MASS RESOLUTION

We have calculated the mass resolution for $Z° \to \mu^+\mu^-$ and $H \to Z°Z° \to \mu^+\mu^-\mu^+\mu^-$ with a simulation program which generates Higgs production and decay according to the parameters summarized in Sections 2.) and 3.) above. For the purpose of this particular study the intrinsic widths of the Higgs and Z°'s were set to zero so that the observed mass smearing is due entirely to the assumed instrumental effects. The momentum vectors of the generated decay muons were smeared using the following gaussian parameterization of the detector resolution:

a.) Magnitude of Momentum:

(i) Toroidal magnets ($5° < \theta < 30°$, $150° < \theta < 175°$)

$$\sigma_p = 0.08\ p$$

(ii) Solenoidal magnet ($30° < \theta < 150°$)

$$\sigma_p = 4.5 \times 10^{-4}\ p^2 \quad (p \text{ in GeV})$$

b.) Polar angle:

$$\sigma_\theta = \sqrt{(\sigma_\theta^{MS})^2 + (\sigma_\theta^{TR})^2}$$

where the multiple scattering contribution from the absorber, assumed to be tungsten, is given by:

$$\sigma_\theta^{MS} = \frac{0.014}{p}\sqrt{X/X_R}$$

with X, the path length in the absorber, equal to 144 cm/sinθ, not to exceed 300 cm, and where the radiation length is X_R = 0.35 cm.

The contribution from tracking errors is given by:
σ_θ^{TR} = 0.002 radian

c.) Azimuthal angle

$$\sigma_\phi = \sigma_\theta^{MS}/\sin\theta$$

This expression neglects intrinsic tracking errors since they are dominated by multiple scattering for measurements of azimuth.

The results of this mass resolution study are summarized in Table 6.1 which gives the measurement error for the reconstructed Z° and Higgs masses in H → Z°Z° events. The measurement error dominates the intrinsic Z° width, expected to be $\Gamma_Z \sim$ 2.8 GeV, for all cases. For Higgs masses exceeding 500 GeV, the intrinsic Higgs width exceeds the measurement error.

TABLE 6.1 MEASUREMENT ERROR FOR Z° AND HIGGS MASSES IN H → Z°Z° EVENTS ($5 < \theta_\mu < 175°$, $P_{\perp\mu} > 10$ GEV/C, $|Y_Z| < 2.5$) THE ERROR IS GIVEN AS THE FULL WIDTH AT HALF MAXIMUM (FWHM) OF THE DETECTOR-SMEARED MASS DISTRIBUTIONS ASSUMING ZERO INTRINSIC WIDTH FOR THE Z° AND HIGGS

M_H (GeV)	Z° FWHM (GeV)	HIGGS FWHM (GeV)
200	5.9	11
300	7.4	23
400	8.7	36
500	10.0	53
600	11.2	75
700	12.3	98
800	13.3	124
900	14.1	157
1000	14.7	190

7.) SIGNAL AND BACKGROUND

In this section we summarize relevant cross sections for $pp \rightarrow H + \ldots \rightarrow \mu^+\mu^-\mu^+\mu^- + \ldots$ and for known backgrounds to this process. We give event rates and discuss the requirements for observing a significant Higgs signal.

An unavoidable background to the $H \rightarrow Z^\circ Z^\circ$ signal arises from $p\bar{p}$ collisions where $q\bar{q}$ parton pairs annihilate into $Z^\circ Z^\circ$ pairs.[11] Whereas the Z°'s from Higgs decay will have longitudinal polarization, those from the background are expected to be polarized mostly in the transverse direction. One can therefore improve the signal-to-background ratio by making requirements on the angular distribution of muons in the respective Z° rest frames.[12] However, such a selection will improve only modestly the statistical significance of the signal because of the loss of signal when making cuts on the muon decay angles. We define:

$$L(x) = \int_{-x}^{x} (1 - \cos^2\theta)\, d(\cos\theta) / \int_{-1}^{1} (1 - \cos^2\theta)\, d(\cos\theta)$$

$$T(x) = \int_{-x}^{x} (1 + \cos^2\theta)\, d(\cos\theta) / \int_{-1}^{1} (1 + \cos^2\theta)\, d(\cos\theta)$$

where L stands for the normalized integrated angular distribution from the decay of longitudinally polarized Z°'s and T from transversely polarized Z°'s. Since we have two Z°'s per event the signal-to-background ratio as a function of the requirement $|\cos\theta| < x$ will be proportional to L^2/T^2 whereas the statistical significance of the signal will go as $L^2/\sqrt{T^2} = L^2/T$. The ratio L^2/T reaches a maximum of 1.26 when $x = \cos\theta$ is 0.68. For this cut the signal-to-background ratio is enhanced by a factor of $L^2/T^2 = 0.746/0.348 = 2.1$.

Table 7.1 presents cross sections and corresponding event numbers for an integrated luminosity of 10^{41} cm^{-2}. The requirements placed on the four-muon events are:

a.) $5^\circ < \theta_\mu < 175^\circ$

b.) 10 GeV/c $< (P_\perp)\mu$

c.) rapidity of Z°'s, $|y_Z| < 2.5$

For this study, Higgs particles and the Z°-pair background were assumed to be produced with zero transverse momentum. To estimate the statistical significance of the Higgs signal we compare the signal and background over a mass interval equal to $\sqrt{(\Gamma_H)^2 + (\mathrm{FWHM})^2}$, which is the total measured width of the Higgs expected from the natural width Γ_H and the mass resolution (FWHM) given in Table 6.1. Assuming a Breit-Wigner shape, half of the total signal is contained in this interval. The background cross section is approximated by the simple product of the differential cross section of the background[3] given in Table 7.1 and the mass interval defined above. This yields a slight underestimate because of the exponential shape of the background mass distribution. The estimated background detection efficiency in Table 7.1 is higher than for the Higgs since the Z° rapidity cut of $|Y_Z| < 2.5$ is already implicit in the quoted background cross section. As already mentioned, the signal can be enhanced relative to background by placing a requirement of $|\cos\theta| < 0.68$ on the decay angle of muons in the Z° rest-frames. Except when the Higgs mass is near the Z°-pair threshold, this requirement enhances the statistical signficance of the signal by a factor of 1.26 as indicated in the last column of Table 7.1.

We make the following observations regarding the feasibility of observing a Higgs signal in the four-muon decay mode. The minimum integrated luminosity required for a significant signal is 10^{41} cm^{-2}. This corresponds to ten years of running with an SSC designed for a peak luminosity of 10^{33} cm^{-2} sec^{-1}. The first conclusion is therefore that the design luminosity of SSC should be raised by at least a factor of 3 (and preferably 10) if this kind of Higgs search is a goal of the SSC physics program. Even with such a higher luminosity, the success of the muon-based Higgs search is not assured. The background events estimated in Table 7.1 correspond only to the unavoidable Z°-pair background from $q\bar{q}$ annihilation. Multi-muon backgrounds arising from possible combinations of pion, kaon, charm, beauty, top, $W^{\pm}$ and Z° decays as well as absorber punch-thru are very difficult to estimate at the required level of sensitivity and are deferred to a future study. Such potential background events would contain energetic hadronic jets and might be identified with a crude calorimetric capability as mentioned at the end of Section 5.

8.) TRIGGER SCHEME AND RATES

D. Carlsmith et al.[13] have made a detailed study of muon rates and possible triggering schemes for a large general detector at SSC. Many of the results of that study apply here with one important exception. The detector considered for this Higgs search has an absorber which is thicker and much closer to the collision region. Hence, single-muon and punch-thru rates will be roughly one order of magnitude less than in the general detector.

The detector configuration described in Section 5 allows a three-level trigger. The first level is based entirely on fast signals derived from the scintillators. The time resolution of this level is limited by the lengths of the scintillator elements which are typically 5 meter. The corresponding timing gates have widths of order 30 nano-seconds. The configuration of scintillators allow the imposition of a transverse momentum requirement ($\gtrsim$ 10 GeV/c) on muons at angles greater than 30° to the beams. Approximately 86% of Higgs decays to four muons will have at least one muon at such large angles. At forward angles the scintillators cannot be reasonably configured to provide a transverse momentum requirement.

A second-level trigger which uses information from the drift tubes can be set up to define a momentum requirement on muons in this forward-angle (< 30°) region of $p \gtrsim 40$ GeV/c. The absorber and magnet iron already impose $p \gtrsim 15$ GeV/c. This leads to an angle-dependent transverse momentum requirement of $p_\perp \gtrsim 50 \sin\theta$ GeV/c. The drift tubes have maximum drift times of order 0.4 μsec which sets the time resolution of the second-level trigger.

TABLE 7.1 $H \to Z^\circ Z^\circ$ AND Z°-PAIR BACKGROUND ($\int Ldt = 10^{41}$ cm^{-2})

M_H(GeV)	Cross Section $H \to ZZ \to 4\mu$ (10^{-3} pb)	Meas. Width (GeV)	Higgs* Detection Efficiency	Higgs Events within Meas. Width	Background $ZZ \to 4\mu$ (10^{-6} pb/GeV) $\lvert Y_Z \rvert < 2.5$	Background* Detection Efficiency	Background Events within Meas. Width	Std. Dev. of Signal	Std. Dev. with Decay Angle Cut †
200	5.4	11	0.80	220	160	0.80	140	18	N/A
300	3.0	25	0.74	110	50	0.78	97	11	14
400	1.72	44	0.74	64	17	0.78	59	8	10
500	1.16	75	0.71	41	7.2	0.76	41	6	8
600	0.86	122	0.70	30	3.4	0.75	31	5	6
700	0.68	184	0.69	23	1.9	0.75	26	4	5
800	0.53	268	0.69	18	1.1	0.75	23	3.8	4.8
900	0.44	377	0.70	15	0.63	0.76	18	3.5	4.4
1000	0.34	511	0.70	12	0.40	0.77	15	3.1	3.9

* Events are required to satisfy $5 < \theta_\mu < 175^\circ$, $P_{\perp\mu} > 10$ GeV/c, $\lvert Y_Z \rvert < 2.5$.
† Muon decay angles in Z° rest-frames satisfy $\lvert \cos\theta \rvert < 0.68$.

Finally a third-level trigger can be implemented which makes a detailed evaluation of all the scintillator and drift-tube information. Such a microprocessor based trigger would yield as output a list of tracks with crudely determined momenta and would make decisions on the time scale of about 50 microseconds.

Trigger rates will depend sensitively on the number of muons demanded at each level. It is difficult to estimate the rate of four-muon triggers because of the combinatorial nature of processes yielding four muons. Moreover, the rates from the majority of muon sources depend very sensitively on the transverse momentum requirement that can be imposed by the trigger. A copious physics source of events with four muons is $b\bar{b}$ production. The $b\bar{b}$ system will decay to a final state containing 4 muons roughly at the 10^{-4} level. If we assume a cross section of 100 μb for $b\bar{b}$ production and a luminosity of 10^{34} $cm^{-2}sec^{-1}$ (!), then the rate of four-muon production from this source is about 100 per second. Demanding that all four muons have transverse momenta in excess of 5 GeV/c (corresponding to the trigger requirement at 5°) can be expected to reduce this rate several orders of magnitude. Crude estimates of four-muon trigger rates from punch-thru (assumed conservatively to be at the 10^{-4} level) yield similar numbers. Cosmic rays are not a significant source of multi-muon triggers with the absorber and scintillator configuration described above. Our best guess is that the overall trigger rate for four-muon events will be at the level of 10 Hz for a luminosity of 10^{34} $cm^{-2}sec^{-1}$. Clearly, detailed simulations are necessary to define this important number with real confidence.

The trigger must also allow a significant admixture of two-muon events for two reasons. First, Z° production and decay to muon-pairs will provide an essential source for checking the performance of the spectrometer system. Second, there is the possibility of new massive bosons which decay to muon pairs. At a luminosity of 10^{34} $cm^{-2}sec^{-1}$ the rate of di-muon triggers is expected to be well above 1000 Hz so that a large fraction of di-muon events may have to be rejected to obtain an analyzable data sample. We note that the cross section for Z° production in the rapidity interval $|y| < 1.5$ is at the 20 nb level.[3] An integrated luminosity of 10^{41} cm^{-2} will yield 6×10^7 $Z^\circ \to \mu^+\mu^-$ events in this interval. Even if the di-muon trigger is greatly suppressed, there will be no shortage of Z° particles!

9.) PERSONNEL AND COST

The detector outlined in Section 5 can be constructed and operated by a team of about 100 physicists. The design and construction phase would require about five years.

A crude cost estimate is summarized in Table 9.1. The total cost is roughly at the level of 100 million dollars.

TABLE 9.1 COST ESTIMATE

Component	Cost M$
Solenoidal Magnet	20
Toroidal Magnets	20
Absorber	10
Drift Chambers	24
Scintillators	5
Assembly	5
Computers	5
Trigger Electronics	3
Contingency	10
TOTAL	102

10.) CONCLUSION

We have made a preliminary feasibility study of an experiment that would search for the Standard-Model Higgs particle via the decay mode $H \to Z^\circ Z^\circ \to \mu^+\mu^-\mu^+\mu^-$. Our most important conclusion is that a definitive search in the mass range 200-1000 GeV/c^2 would require a minimum integrated luminosity of 10^{41} cm^{-2}. We therefore recommend strongly that the SSC provide at least one high-luminosity region ($L > 3 \times 10^{33}$ $cm^{-2} sec^{-1}$) to allow a Higgs search by this, the cleanest decay mode, on a reasonable time scale. Another conclusion is that the detector must be optimized for muon detection by utilizing a dense, close-in absorber surrounded by large magnetic spectrometers. We recommend against attempting this measurement with only a large all-purpose detector. A specialized muon detector suited to the SSC environment can be constructed for about $100 Million. Two major topics must still be studied in depth. One is an estimate of spurious backgrounds to the $Z^\circ Z^\circ \to \mu^+\mu^-\mu^+\mu^-$ signal. The other is an estimate of multi-muon trigger rates. These are very difficult topics because of the varied sources of muon signals and the large number of proton-proton collisions ($\gtrsim 10^{16}$) that can, in principle, contribute to spurious backgrounds and triggers.

We express our appreciation for useful discussions with R. Cahn, D. Green, D. Hedin, G. Kane, P. Limon, F. Paige and L. Sulak. This work was supported by the U.S. Department of Energy.

References

1. For a pedagogical review of the Standard Model see, for example, C. Quigg Gauge Theories of the Strong, Weak, and Electromagnetic Interactions, Benjamin Cummings, Menlo Park, California, 1983.

2. For a general review of SSC related physics and machine topics see: Proceedings of the 1982 DPF Summer Study on Elementary Particle Physics and Future Facilities;
Proceedings of the 1984 Summer Study on the Design and Utilization of the Superconducting Super Collider;
Superconducting Super Collider Conceptual Design, SSC-SR-2020, (1986);
Report of the Task Force on Detector R & D for the Superconducting Super Collider, SSC-SR-1021, (1986);
Cost Estimate of Initial SSC Experimental Equipment, SSC-SR-1023, (1986);
Proceedings of the 1986 SSC Summer Study (to be published).

3. E. Eichten, I. Hinchliffe, K. Lane and C. Quigg, "Supercollider Physics", Rev. Mod. Phys. 56 (1984) 579 and references therein.

4. R.N. Cahn, Nucl. Phys. B255, (1985) 341.

5. R.N. Cahn, S.D. Ellis and W.J. Stirling, Preprint LBL-21285, (1986).

6. B.W. Lee, C. Quigg and H.B. Thacker, Phys. Rev. D16, (1977) 1591.

7. J.F. Gunion, Z. Kunszt and M. Soldate, Phys. Lett. 163B, (1985) 389.

8. R.N. Cahn and M.S. Chanowitz, Phys. Rev. Lett. 56, (1986) 1327.

9. Some recent references on muon identification include:
A. Bodek, University of Rochester Report UR-911, (1985);
D. Green et al., Nucl. Inst. Meth. A244, (1986) 356;
F. Merrit et al., Nucl. Inst. Meth. A245, (1986) 27.

10. In terms of cost and handling, tungsten may be the most attractive material. Teledyne Powder Alloys Corp. of Clifton, New Jersey gives an informal quote of about $18 per pound for a large tungsten absorber. To keep the total cost within reasonable bounds, one might consider a hybrid absorber with an inner core consisting of tungsten and an outer shell made from a cheaper material such as copper.

11. R.W. Brown and K.O. Mikaelian, Phys. Rev. D19, (1979) 922.
see also reference 3 above.

12. M.J. Duncan, G.L. Kane and W.W. Repko, Nucl. Phys. B272, (1986) 517;

M.J. Duncan, University of Pennsylvania Preprint UPR-0299-T, (1986).

13. D. Carlsmith et al., "SSC Muon Detector Group Report", 1986 SSC Summer Study.

HIGH p_t WEAK BOSONS As SIGNATURES for HIGGS-LIKE HEAVY PARTICLES

M. Chen, E. Nagy, G. Herten

June 26, 1987

Abstract

We use the Higgs as an example to demonstrate the importance of reconstructing the invariant mass of high pt weak bosons with a mass resolution comparable to their natural widths in order to identify new heavy particles. Detectors with fine energy resolution and granularity for di-electrons, dimuons and di-jets are required to carry out this reconstruction.

The high luminosity and high center of mass energy which will be available at SSC give us hope that heavy particles such as the Higgs, top or fourth generation quarks or leptons with masses up to a few TeV can be produced abundantly. The main challenge has been to find a suitable signal which can stand out among huge QCD backgrounds. It is the purpose of this paper to point out that most of these heavy particles, when their masses are far above 200 GeV, decay predominantly into at least one real high p_t weak boson whose mass can be reconstructed from the final state leptons or jets to tag these high mass particles.

For example, heavy Higgs could be produced via either the quark-antiquark annihilation channel or the gluon-gluon interaction or through the virtual weak boson interactions. Since the partial decay width of a Higgs is proportional to M_H^3 for bosons and to $M_H \cdot M_F^2$ for fermions, the weak bosons will be the main decay channel independent of the heavy quark masses. These weak bosons will subsequently decay into pairs of leptons or quarks, so that we have

$$\begin{aligned} p + p &\rightarrow H^o + x, \\ &\rightarrow W^+W^- \ \textit{or} \ Z^oZ^o, \\ &\rightarrow \textit{lepton pairs and/or quark pairs} \end{aligned} \tag{1}$$

Even though the contribution of the ZZ channel is only half as large as that of the WW channel, the useful constraint that the final state di-lepton can be precisely reconstructed to reproduce Z^o mass, makes it much less contaminated by background from QCD and from heavy quark decays. Thus we limit ourselves here to the ZZ channel. Since the muons are relatively easy to trigger and to reconstruct, we discuss the final states containing dimuons

$$4\ \mu\text{'s};\ 2\mu\text{'s} + 2\ \text{e's}\ \text{or di-jets}\ \text{or}\ 2\ \nu's. \tag{1a}$$

The neutrinos can be experimentally identified using the missing pt of the event. Fig. 1 shows the Higgs production rates via the ZZ channel as functions of its mass for the inclusive dimuon, four muon and all final states respectively. Were there be no background, one would be sensitive up to a Higgs mass as high as 1 TeV, 3 TeV and 5 TeV using the 4μ, 2μ or all ZZ final states of the Higgs decay. The acceptance of dimuon events expected as function of the minimun muon detection angle for 1 TeV Higgs mass is shown in Fig. 2. Furthermore, for a muon detector covering down to $\theta = 10^o$, the acceptance for all four μ's is about 80% for 1 TeV Higgs.

One major background of reaction (1) is the single production of weak bosons via QCD processes. It has been shown that this background can be kept smaller than the Higgs signal after proper cuts[1]. Here we consider the background contribution from heavy quark decays, i.e. a t quark with mass equal to 100 and 200 GeV, produced via the following reaction:

$$\begin{aligned} p + p \to t\bar{t} &\to W + b, \\ &\to \textit{lepton pairs and/or quark pairs} \end{aligned} \tag{2}$$

To reduce this background one could reject events with an additional low p_t jet, which is expected in the t $\to$ W+b decay. But this appears only possible with high efficiency, if $m_t$is greater than 150 GeV, because only then the p_t of the b jet is sufficiently large for observation in the calorimeter. Top production is also a background to the reconstruction of Z^0's, which is needed in the study of the Higgs decay H $\to ZZ \to \mu^+\mu^-$ + jets. The tail of the invariant mass distribution of the μ^+ and μ^- produced in the decay t $\to W^+ + b \to \mu^+\nu + \mu^-\bar{\nu} + c$ could overlap with the Z^0 mass.

The classical methods of finding new particles such as the J, the Υ and even the weak bosons themselves have been the observation of an excess of events with a pair of leptons over a narrow region of their invariant mass distribution. These leptons are characterized by the fact that although the pt of the individual lepton is large (about half of the mass of the parent particle), the pt of the pair is small. The topology of these events is dominated by the configuration that one lepton lies on each side of the beam line with the sum of their transverse momenta close to zero.

Fig. 1 The Higgs event rate (assuming M_t= 80 GeV) produced via the ZZ channel as functions of its mass with an integrated luminosity of 10^{40} cm^{-2} at 40GeV for the four muon; inclusive dimuon; and all final states. Also shown is the dimuon rate of a new heavy Z' (dashed curve) as a function of its mass.

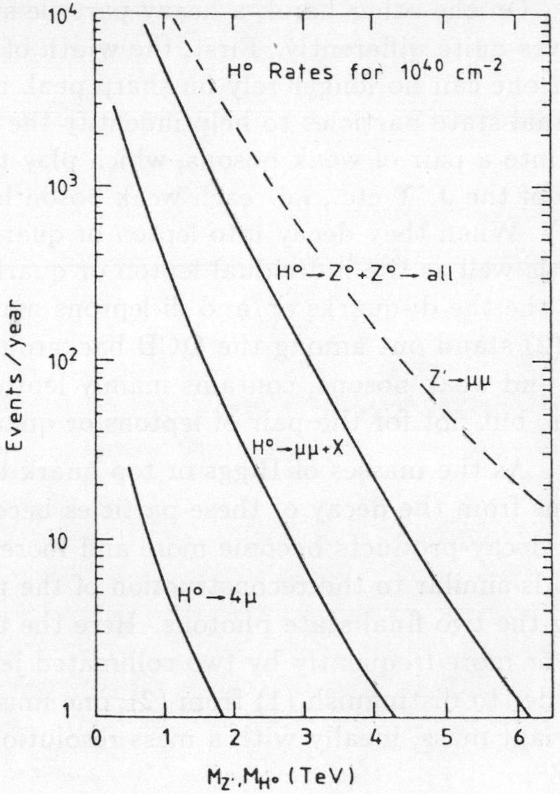

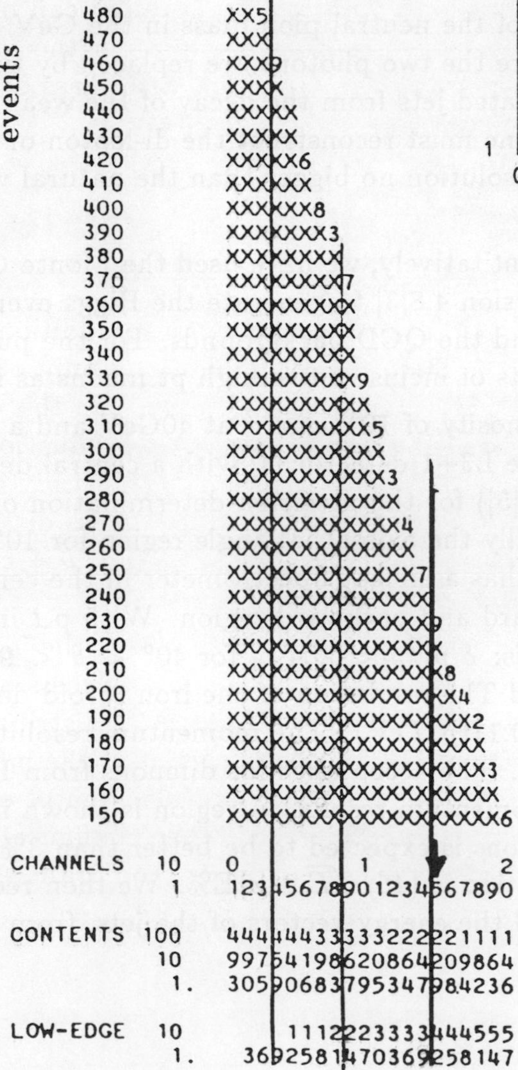

Fig. 2 The no. of dimuon events from Higgs decays as a function of θ_{min}, with both $\theta_\mu > \theta_{min}$ for 1 TeV Higgs mass. A total of 500 events were generated. The vertical lines shows the number of events in the three different regions with $\delta\ p_t/p_t \approx$ 0.04 p_t for $40° < \theta_\mu < 90°$, $\delta\ p_t/p_t \approx$ 0.16 p_t for $22° < \theta_\mu < 40°$ and $\delta p/p = 0.09 + 0.11(p/TeV)$ for $10° < \theta_\mu < 22°$.

On the other hand, a heavy particle such as the Higgs in the TeV region behaves quite differently. First, the width of the Higgs is expected to be quite large, one can no longer rely on sharp peak in the invariant mass distribution of the final state particles to help indentify the Higgs. Secondly, the Higgs first decays into a pair of weak bosons, which play the same role as the leptons in the cases of the J, Υ etc., i.e. each weak boson has large pt (about half of the Higgs mass). When they decay into lepton or quark pairs, the di-lepton or di-quark system, as well as the individual lepton or quark, also carries large pt. These high pt of the the di-quarks or/and di-leptons make these events from reactions (1) and (2) stand out among the QCD background., which, similar to the cases of the J, Υ and weak bosons, contains mainly leptons or quarks with high pt for themselves, but not for the pair of leptons or quarks.

As the masses of Higgs or top quark become larger ($>$ 0.5 TeV), the weak bosons from the decay of these particles become more and more energetic and their decay products become more and more collimated. Experimentally, the situation is similar to the reconstruction of the neutral pion mass in the GeV region, using the two final state photons. Here the two photons are replaced by two leptons or more frequently by two collimated jets from the decay of the weak boson. In order to distinguish (1) from (2), one must reconstruct the di-lepton or di-jet invariant mass, ideally with a mass resolution no bigger than the natural width of the Z^o.

To demonstrate this point quantitatively, we have used the Monte Carlo programs ISAJET [2], and Pythia version 4.8[3] to generate the Higgs events from reaction (1), top from reaction (2), and the QCD backgrounds. For the purpose of clear identification, we accept events of inclusive two high pt muons as in (1a).

We assume an integrated luminosity of $10^{40}\ cm^{-2}$ at 40GeV and a large solid angle precision detector (e.g. the L3+1 detector [4] with a central detector such as the Xenon Olive Detector[5]) for the precision determination of the muons, electrons and jets, covering fully the azimuthal angle region for $10^o < \theta < 170^o$. The L3+1 muon detector has a air gap spectrometer in the central region and an iron toroid in the forward and backward region. With p_t in TeV, the momentum resolution for muons is: $\delta\ p_t/p_t \approx 0.04\ p_t$ for $40^\circ < \theta < 90^\circ$, $\delta\ p_t/p_t \approx 0.16\ p_t$ for $22^\circ < \theta < 40^\circ$ and The resolution of the iron toroid detector for $10^\circ < \theta < 22^\circ$ is $\delta p/p = 0.09 + 0.11(p/TeV)$. The momentum resolution as a function of theta is shown in Fig. 3. The acceptance for dimuons from Higgs or a heavy weak boson Z' in the best momentum resolution region is shown in Fig. 4. The energy resolution of the electrons is expected to be better than $3\%/\sqrt{E}$. The hadron calorimeter energy resolution is $(45/\sqrt{E} + \epsilon)$E%. We then reconstruct the momenta of the muons and the energy vectors of the jets, from which

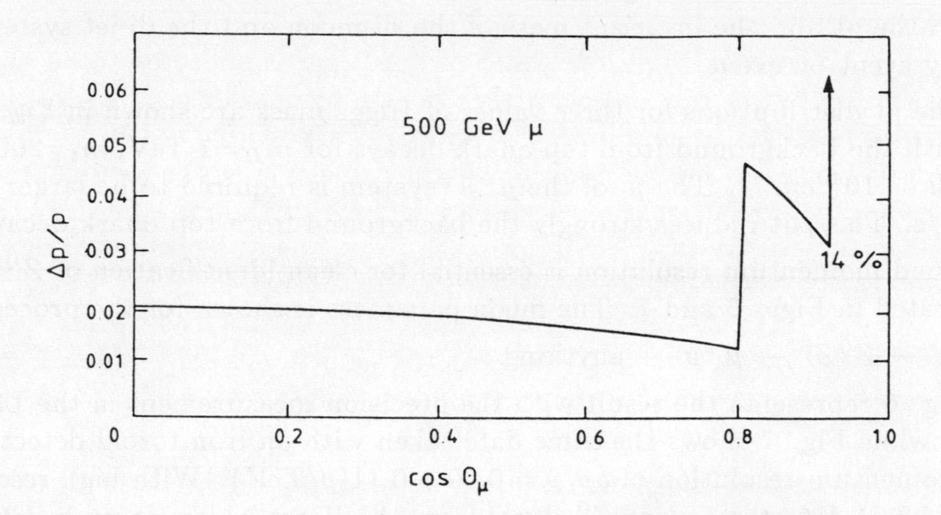

Fig. 3 The muon momentum resolution of the L3+1 detector as a function of the muon production angle.

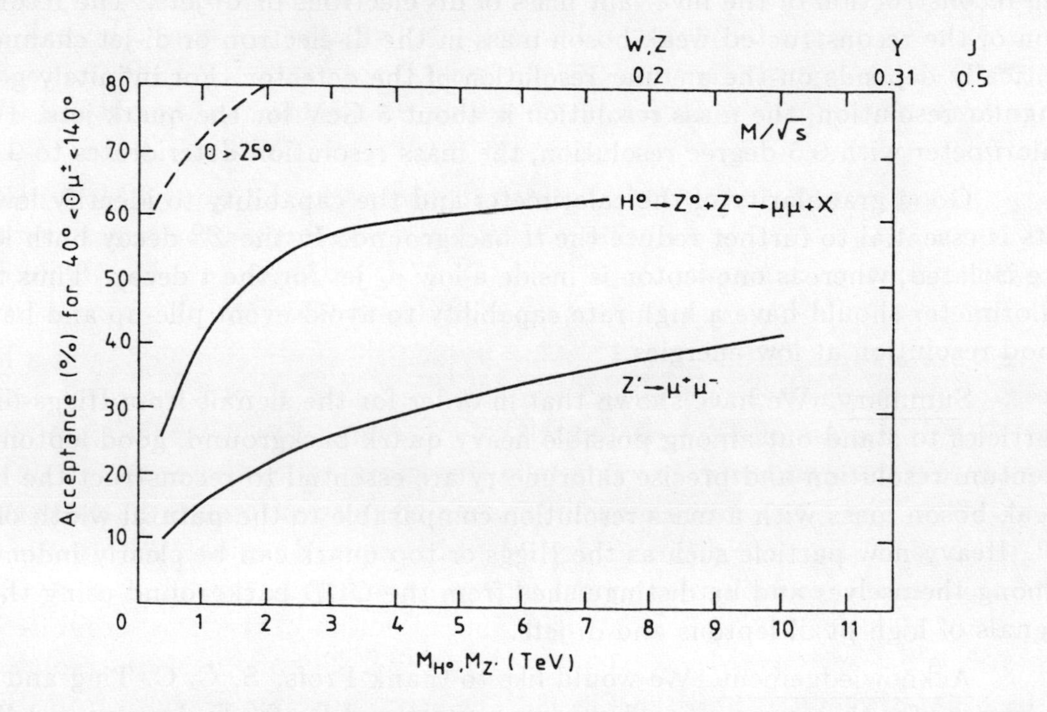

Fig. 4 The di-muon acceptance in the good momentum resolution regions, i.e. $\theta > 40^o$ (solid) and $\theta > 25^o$ (dashed) of the L3+1 detector for the Higgs and for a new heavy Z'. The top scale shows the scaling variable x = the ratio of the mass of H^o or Z' to the cm energy of the machine. The J particle, the Υ and the W and Z were originally discovered at x= 0.5, 0.31 and 0.2 respectively, which illustrates the mass region SSC should in principle be sensitive to (with good ideas, detectors and unlimited luminosity etc.).

we obtain the pt and the invariant mass of the di-muon and the di-jet systems separately event by event.

The pt distributions for three values of Higgs mass are shown in Fig. 5 together with the background from top quark decays for m_H=1 TeV, m_t=200 GeV and $\int Ldt = 10^{40} cm^{-2}$. The p_t of the $\mu^+\mu^-$system is required to be larger than 300 GeV/c. This cut reduces strongly the background from top quark decay.

Good momentum resolution is essential for clean identification of Z^0's as demonstrated in Figs. 6 and 7. The muon pair mass is shown for the process

$H^o \rightarrow Z^0 Z^0 \rightarrow \mu^+\mu^- +$ anything

Fig. 6 represents the result with the precision measurement in the L3+1 detector, while Fig. 7 shows the same date taken with an iron toroid detector with a momentum resolution of $\delta p/p = 0.09 + 0.11(p/TeV)$. With high resolution as in the L3+1 detector a clear Z^0 signal from the Higgs boson decay is detected, whereas only a shoulder is vaguely visible for an iron toroid detector.

Fine granularity and resolution of the calorimeter is most important for the reconstruction of the invariant mass of di- electrons or di-jets. The resolution of the reconstructed weak boson mass in the di-electron or di-jet channels critically depends on the angular resolution of the detector. For infinitely good angular resolution, the mass resolution is about 5 GeV for the quark jets. For a calorimeter with 0.5 degree resolution, the mass resolution deteriorates to 9 GeV.

Good granularity of the calorimeter and the capability to identify low p_t jets is essential to further reduce the $t\bar{t}$ background. In the Z^0 decay both leptons are isolated, whereas one lepton is inside a low p_t jet for the t decay. Thus the calorimeter should have a high rate capability to avoid event pile-up and have a good resolution at low energies.

Summary. We have shown that in order for the signals from Higgs-like particles to stand out among possible heavy quark background, good lepton momentum resolution and precise calorimetry are essential to reconstruct the high pt weak boson mass with a mass resolution comparable to the natural width of the Z^o. Heavy new particle such as the Higgs or top quark can be clearly indentified among themselves and be distinguished from the QCD background using these signals of high pt di-leptons and di-jets.

Acknowledgement. We would like to thank Profs. S. C. C. Ting and G. Gilchriese for their support and encouragement and Profs. D. Luckey and R. Yamamoto for many useful discussions.

References.

1. T. Gottschalk, Calt-68-1400, also Procedings of SSC Summer Study, Snow Mass(1986).

2. Isajet: F.E. Paige, S.D. Protopopescu, BNL 29777 (1981).

3. Phytia: H.U. Bengtsson, T. Sjöstrand, LU TP 87-3, UCLA-87-001(1987).

4. L3+1: U. Becker et al., Nucl. Inst. Meth. A253(1986) 15.

5. Xenon Olive Detector: M. Chen et al., MITLNS Rep. No. 162 and contribution to this proceedings

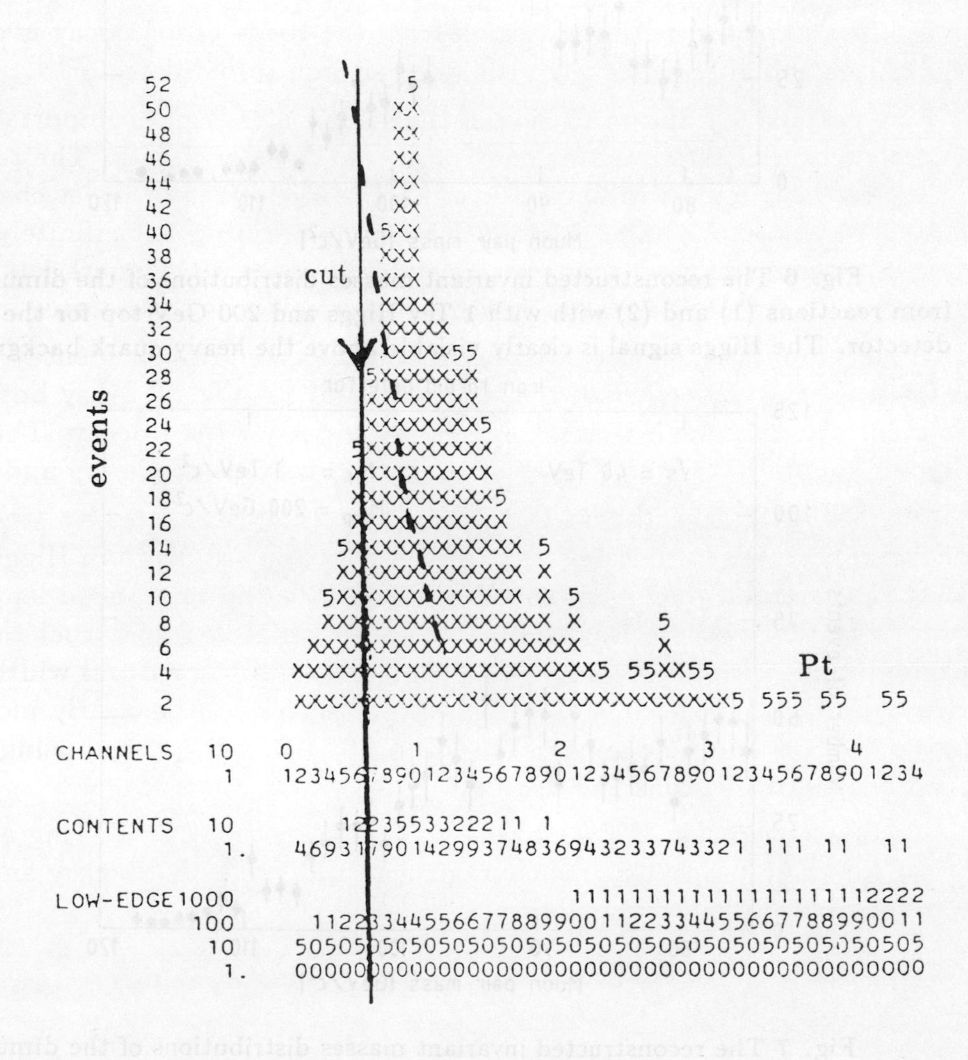

Fig. 5 The Pt distributions of the dimuons from reactions (1) and (2) with 1 TeV Higgs (histograms) and 200 GeV top (dashed curve). The cut at 300 GeV Pt greatly reduces the contribution from top relative to the Higgs.

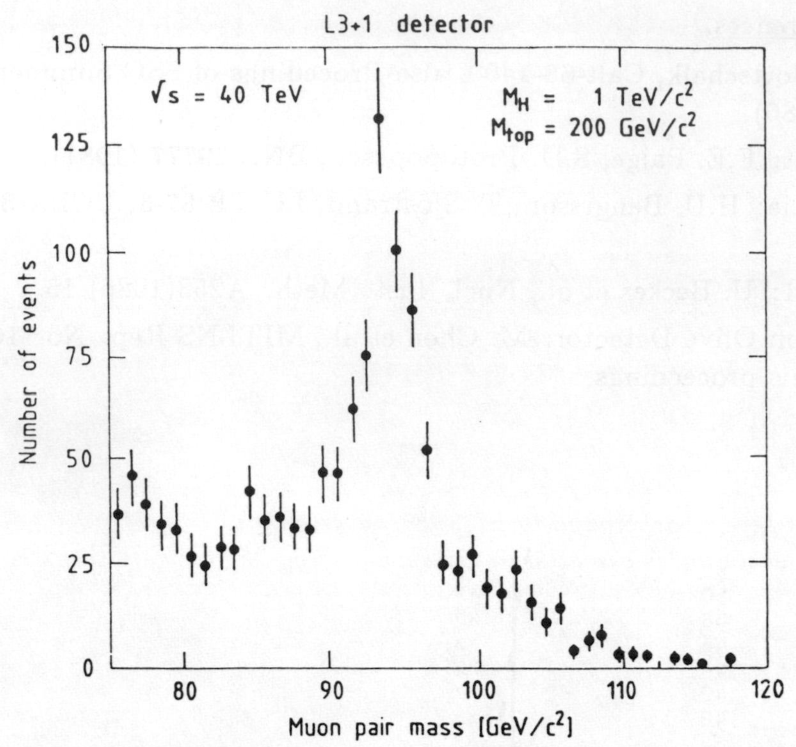

Fig. 6 The reconstructed invariant masses distributions of the dimuons from reactions (1) and (2) with with 1 Tev Higgs and 200 GeV top for the L3+1 detector. The Higgs signal is clearly visiable above the heavy quark background.

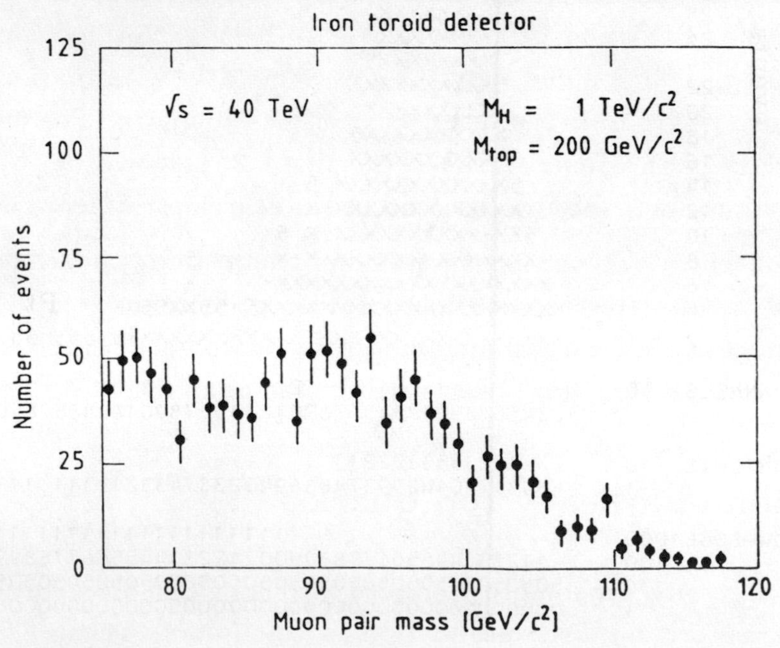

Fig. 7 The reconstructed invariant masses distributions of the dimuons from reactions (1) and (2), same as in Fig. 6, but with an iron toroid.

Heavy Top Quark Production as a Background in the Search for the Higgs Boson

Gregor Herten

Massachusetts Institute of Technology, Cambridge, MA 02139, USA

Abstract

The production of a heavy top quark at SSC with mass larger than the W mass is studied. Leptonic cascade decays of the top quark provide a good signal to study top quark decays and to determine its mass precisely. Heavy top decay into an on-shell W boson is a severe background to Higgs boson decay into W or Z pairs. Good muon momentum resolution is essential to extract the Higgs signal from the $t\bar{t}$ background.

1 Introduction

Recent limits [1] from UA1 on the mass of the top quark indicate that m_t is larger than 45 GeV. Since t quark masses up to 90 GeV are accessible at LEP200, the mass range $m_t > m_W$ is interesting at SSC for the discovery of the top quark. Studies of heavy top production at TEVATRON and LHC energies have been performed earlier [2]. In this paper we study the production and signatures of top quarks with $m_t > m_W$ at SSC and examine the lepton spectra from cascade decays of the top quark. The results have been obtained with the ISAJET Monte Carlo [3] (version 5.25). It will be shown that heavy top quark production is a severe background in the search for a heavy Higgs boson.

The studies concentrate on the measurement of muons from t quark decays assuming a high resolution muon detector, like L3+1. The momentum resolution is taken as $\delta p/p = 4 \times 10^{-5} p$ (p in GeV/c). An angular coverage of $5° < \theta < 175°$ is assumed and only muons with $p_t > 10$ GeV have been used for the analysis.

2 Production of a heavy top quark

Top quarks with $m_t > m_W$ are mainly produced through gluon gluon fusion [4]. Figure 1 shows the total cross section for $t\bar{t}$ production as function of the mass for $p_t > 10$ GeV and $p_t > 100$ GeV. One should note that the cross sections have an uncertainty of about a factor 2, because of the poor knowledge of the gluon density function. For $\int Ldt = 10^{40} cm^{-2}$ at $\sqrt{s}$=40 TeV one expects 1.8×10^8 events (2.6×10^7) for m_t =100 GeV (200 GeV) and $p_t(top) > 100$ GeV. At $p_t > 100$ GeV the cross section for b quark production is 3 (21) times larger than for a top quark with mass m_t = 100 GeV (200 GeV).

The t quark decays into a W boson and a b quark. The final state gg $\rightarrow t\bar{t} \rightarrow W^+W^- \, b\bar{b}$ consists of two W bosons and two low p_t b jets. The p_t of the b jets is especially small if the mass of the top quark is only slightly larger than the W mass. The decay of a heavy top quark represents a large source of W pairs at the SSC. Figure 2 shows the WW invariant mass from the t decay compared to the continuum production of W pairs. The WW production from t and $\bar{t}$ decays for $m_t < 200$ GeV is about two orders of magnitude larger than the continuum production. The cross section in figure 2 corresponds to $\sqrt{s}$= 40 TeV and pseudorapidity $|\eta| < 2.5$ for each W. Therefore heavy top quark production is a severe background in the search for the Higgs boson decay into W pairs, since this cross section is even smaller than the WW continuum production. This background in the search for the Higgs boson will be studied in more detail later.

A relatively clean signal of heavy t quark production is obtained if one W boson decays into a muon and the other into quarks. Then one observes an isolated muon and a low p_t b jet balanced on the other side by 2 jets from the W decay and a low p_t b jet. To further reduce QCD background from high p_t W production one can study events with two muons produced in the cascade decay of the top quark $t \rightarrow W^+ + b \rightarrow \mu^+\nu + \mu^-\bar{\nu} + c$. The background from high p_t $b\bar{b}$ and $c\bar{c}$ production can be reduced in requiring that the high p_t muon is isolated. If one requires the energy in the calorimeter to be less than 10 GeV inside a cone of $\Delta R < 0.3$ around the high p_t muon, one can reduce the light quark background sufficiently.

Figure 3 shows the mean p_t of muons from the W decay, μ(W), and the b decay, μ(B), as function of the p_t of the primary t quark. The signature of heavy top decay is an isolated high p_t muon and a second low p_t muon inside the bottom jet. The mean p_t of both muons has a strong dependence on the top mass. The ratio of the mean transverse momenta of the muons $\langle p_t(\mu_W)\rangle / \langle p_t(\mu_B)\rangle$ is about 10 for m_t = 100 GeV. This ratio can be used to measure the mass of the t quark.

Another method to determine the t quark mass is shown in figure 4. The distribution of $R = \sqrt{(\Delta\eta)^2 + (\Delta\phi)^2}$ is shown, which measures the separation in space between the muons from the W and b decay. $\Delta\phi$ is the difference in azimuthal angle with

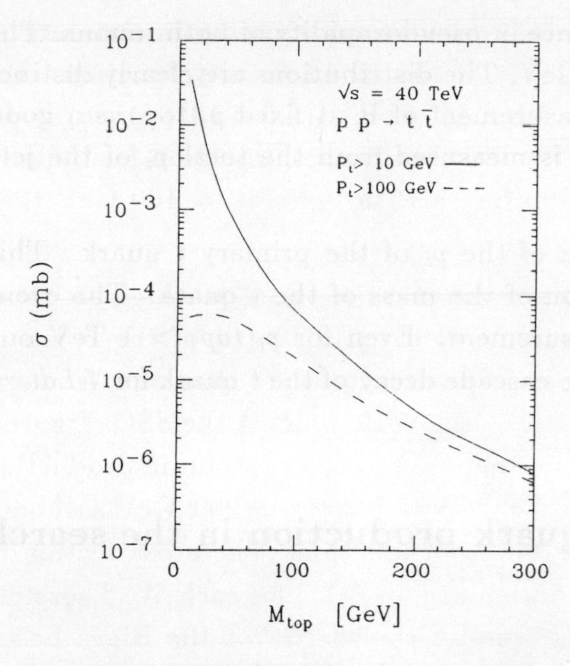

Figure 1: Cross section for $t\bar{t}$ production in pp collisions at $\sqrt{s} = 40$ TeV as function of the mass of the top quark.

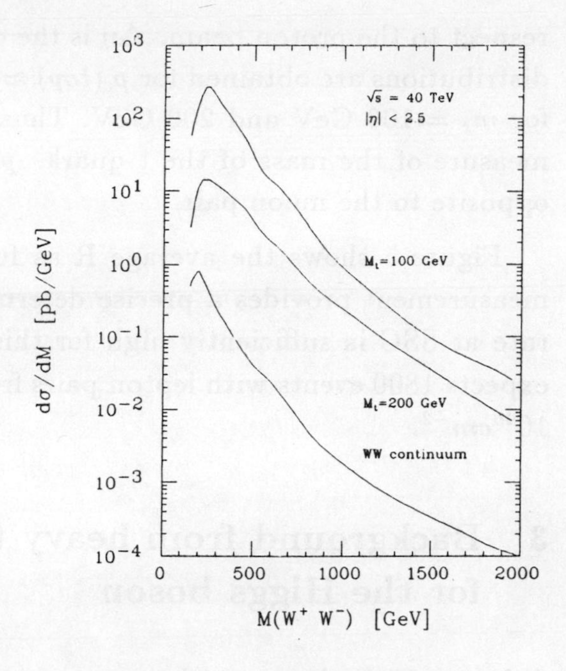

Figure 2: Invariant mass of W^+W^- pairs at $\sqrt{s} = 40$ TeV and pseudorapidity $|\eta| <$ 2.5 for each W boson from $t\bar{t}$ production.

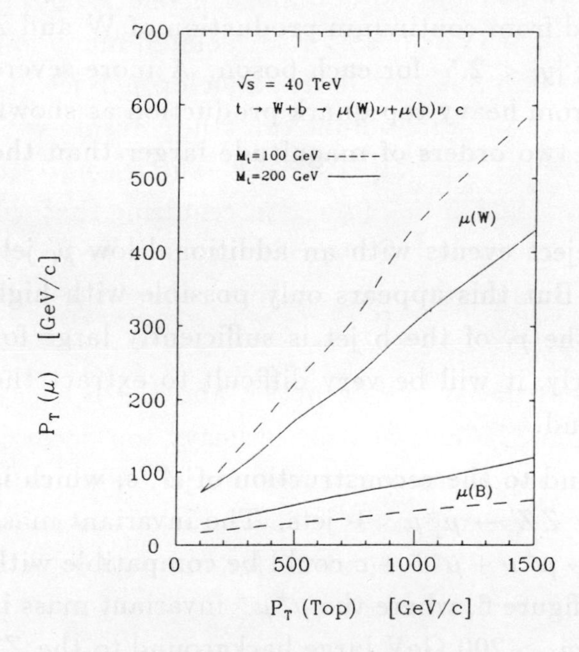

Figure 3: The average p_t of muons from the cascade decay of the $t \rightarrow W + b$ is shown for $m_t = 100$ GeV and 200 GeV.

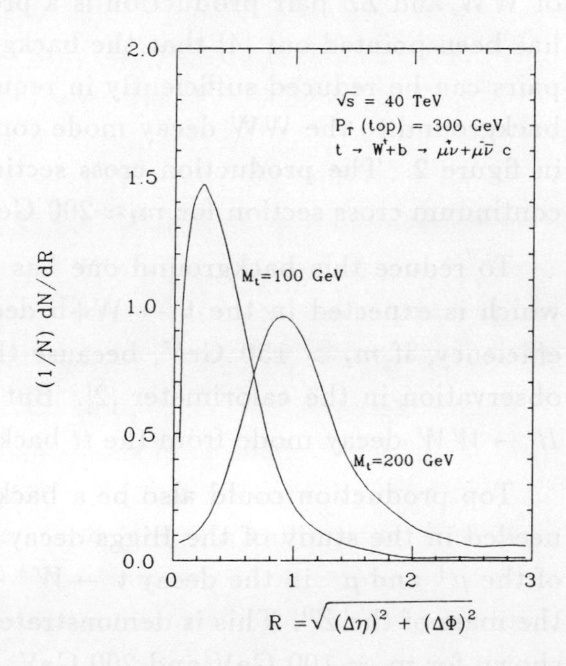

Figure 4: The distribution of R is shown at $\sqrt{s} = 40$ TeV for p_t(top)=300 GeV.

respect to the proton beam. $\Delta\eta$ is the difference in pseudorapidity of both muons. The distributions are obtained for $p_t(top) \approx$ 300 GeV. The distributions are clearly distinct for m_t = 100 GeV and 200 GeV. Thus a measurement of R at fixed p_t(top) is a good measure of the mass of the t quark. p_t(top) is measured from the total p_t of the jets opposite to the muon pair.

Figure 5 shows the average R as function of the p_t of the primary t quark. This measurement provides a precise determination of the mass of the t quark. The event rate at SSC is sufficiently high for this measurement. Even for $p_t(top) > 1$ TeV one expects 1800 events with lepton pairs from the cascade decay of the t quark for $\int Ldt = 10^{40} cm^{-2}$.

3 Background from heavy top quark production in the search for the Higgs boson

A heavy Higgs boson with mass $m_H \gg 2m_W$ decays predominantly into boson pairs, WW and ZZ, with B(H $\rightarrow$ WW) $\approx$ 2 B(H $\rightarrow$ ZZ). This decay mode dominates even for very high top quark masses, since $\Gamma_{W,Z} \propto (m_H)^3$, but $\Gamma_{f\bar{f}} \propto m_H(m_f)^2$. The study of WW and ZZ pair production is a promising method to detect the Higgs boson. It has been pointed out [4] that the background from continuum production of W and Z pairs can be reduced sufficiently in requiring $|y| < 2.5$ for each boson. A more severe background to the WW decay mode comes from heavy top quark production as shown in figure 2. The production cross section is two orders of magnitude larger than the continuum cross section for $m_t \approx$ 200 GeV.

To reduce this background one has to reject events with an additional low p_t jet, which is expected in the t $\rightarrow$ W+b decay. But this appears only possible with high efficiency, if $m_t > 150$ GeV, because then the p_t of the b jet is sufficiently large for observation in the calorimeter [2]. But clearly, it will be very difficult to extract the $H \rightarrow WW$ decay mode from the $t\bar{t}$ background.

Top production could also be a background to the reconstruction of Z^0's, which is needed in the study of the Higgs decay H $\rightarrow ZZ \rightarrow \mu^+\mu^-$ + jets. The invariant mass of the μ^+ and μ^- in the decay t $\rightarrow W^+ + b \rightarrow \mu^+\nu + \mu^-\bar{\nu} + c$ could be compatible with the mass of the Z^0. This is demonstrated in figure 6, where the $\mu^+\mu^-$ invariant mass is shown for m_t = 100 GeV and 200 GeV. For $m_t > 200$ GeV large background to the Z^0 reconstruction has to be expected. The amount of background is determined by:

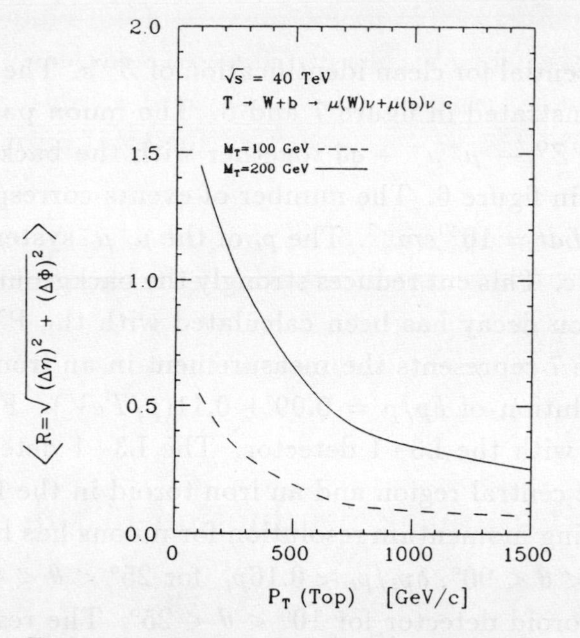

Figure 5: Average value of R as function of the transverse momentum of the primary top quark.

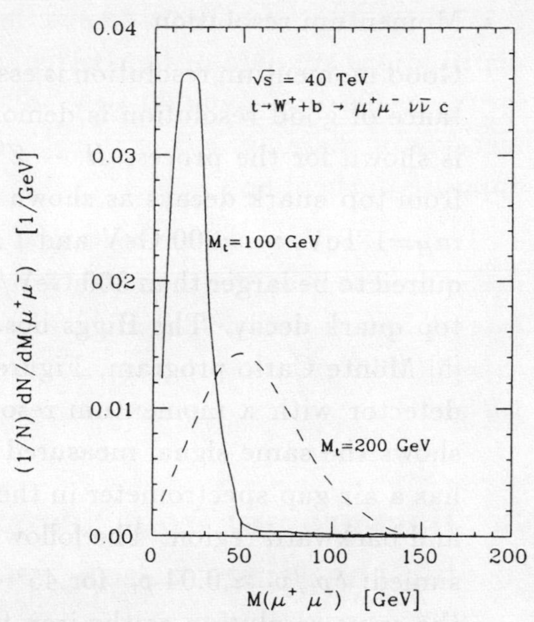

Figure 6: Distribution of the invariant mass of the μ^+ and μ^- from the cascade decay of the t quark.

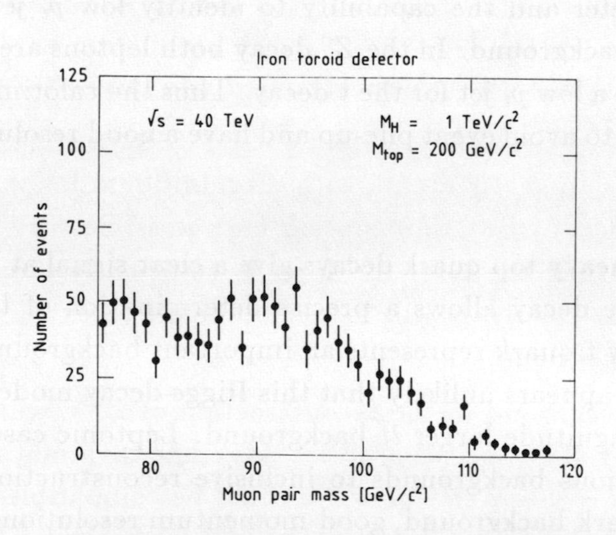

Figure 7: $M(\mu^+\mu^-)$ with an iron toroid detector for the process $H^0 \rightarrow Z^0Z^0 \rightarrow \mu^+\mu^- X$ plus the background from t decays.

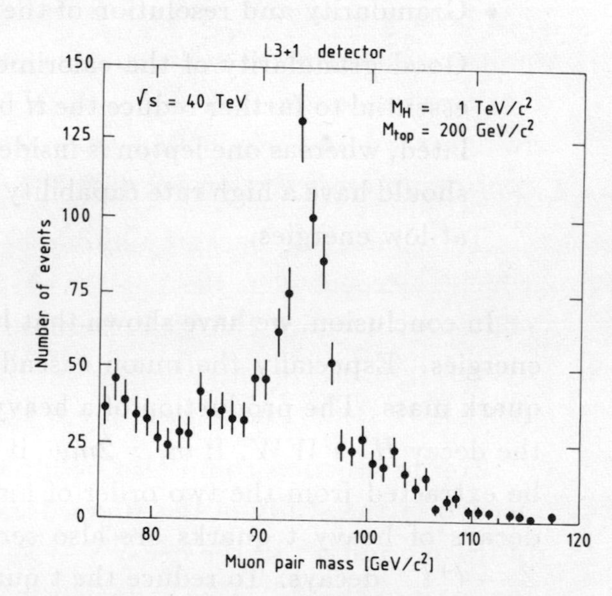

Figure 8: Same process as figure 7 with the L3+1 detector.

- Momentum resolution

  Good momentum resolution is essential for clean identification of Z^0's. The importance of good resolution is demonstrated in figure 7 and 8. The muon pair mass is shown for the process $H \rightarrow Z^0Z^0 \rightarrow \mu^+\mu^- + q\bar{q}$ together with the background from top quark decays as shown in figure 6. The number of events correspond to m_H=1 TeV, m_t=200 GeV and $\int Ldt = 10^{40} cm^{-2}$. The p_t of the $\mu^+\mu^-$system is required to be larger than 300 GeV/c. This cut reduces strongly the background from top quark decay. The Higgs boson decay has been calculated with the PYTHIA [5] Monte Carlo program. Figure 7 represents the measurement in an iron toroid detector with a momentum resolution of $\delta p/p = 0.09 + 0.11(p/TeV)$. Figure 8 shows the same signal measured with the L3+1 detector. The L3+1 detector [6] has a air gap spectrometer in the central region and an iron toroid in the forward and backward region. The following momentum resolution for muons has been assumed: $\delta p_t/p_t \approx 0.04\ p_t$ for $45° < \theta < 90°$, $\delta p_t/p_t \approx 0.16 p_t$ for $25° < \theta < 45°$ and the same resolution as the iron toroid detector for $10° < \theta < 25°$. The resolution is symmetric for the forward and backward region.

  With high resolution as in the L3+1 detector a clear Z^0 signal from the Higgs boson decay is detected, whereas only a shoulder is visible for an iron toroid detector.

- Granularity and resolution of the calorimeter

  Good granularity of the calorimeter and the capability to identify low p_t jets is essential to further reduce the $t\bar{t}$ background. In the Z^0 decay both leptons are isolated, whereas one lepton is inside a low p_t jet for the t decay. Thus the calorimeter should have a high rate capability to avoid event pile-up and have a good resolution at low energies.

In conclusion, we have shown that heavy top quark decays give a clear signal at SSC energies. Especially the muon cascade decay allows a precise determination of the t quark mass. The production of a heavy t quark represents an important background to the decay $H \rightarrow WW$. If $m_t > 2m_W$, it appears unlikely that this Higgs decay mode can be extracted from the two order of magnitude larger $t\bar{t}$ background. Leptonic cascade decays of heavy t quarks are also serious backgrounds to inclusive reconstruction of $Z \rightarrow \ell^+\ell^-$ decays. To reduce the t quark background, good momentum resolution and precise calorimetry is essential.

I thank Prof. S.C.C. Ting for supporting this work, as well as Prof. M. Chen for many fruitful discussions and for providing the calculations for the Higgs production. I thank A. Di Ciaccio for help with the ISAJET program.

References

[1] UA1 collaboration, C. Albajar et al., "Search for a new Heavy Quark at the CERN Proton-Antiproton Collider", in preparation;
P. Erhard, Proc. Int. Europhysics Conf. on High Energy Physics, Uppsala, Sweden (June 1987).

[2] P. Colas, D. Denegri, DPhPE 86-24;
D. Denegri, Proc. of the Workshop on Physics at Future Accelerators La Thuile (Val d'Aoste) and CERN, 7-13 January 1987.

[3] F.E. Paige, S.D. Protopopescu, BNL 29777 (1981).

[4] E. Eichten et al., Rev. Mod. Phys. 56 (1984), 579.

[5] H.U. Bengtsson, T. Sjöstrand, LU TP 87-3, UCLA-87-001. (1987)

[6] U. Becker et al., Nucl. Inst. Meth. A253(1986) 15;
M. Chen et al., these proceedings

Probing the Non-Minimal Higgs Sector at the SSC*

J.F. GUNION
Department of Physics
University of California, Davis, CA 95616

H.E. HABER
Santa Cruz Institute for Particle Physics
University of California, Santa Cruz, CA 95064

S. KOMAMIYA
Stanford Linear Accelerator Center
Stanford University, Stanford, CA 94305

H. YAMAMOTO
Department of Physics
University of California, Los Angeles, CA 90024

A. BARBARO-GALTIERI
Lawrence Berkeley Laboratory, Berkeley, CA 94720

Abstract

Non-minimal Higgs sectors occur in the Standard Model with more than one Higgs doublet, as well as in theories that go beyond the Standard Model. In this report, we discuss how Higgs search strategies must be altered, with respect to the Standard Model approaches, in order to probe the non-minimal Higgs sectors at the SSC.

A. Introduction

A great deal of effort has been focused on the search for the minimal Higgs of the Standard Model at the SSC. The term "minimal Higgs" implies that the $SU(2) \times U(1)$ electroweak theory consists of the minimal choice of one complex Higgs doublet. In such a theory, there is only one physical Higgs scalar in the spectrum, whose mass is a free parameter not fixed by the theory. This minimal choice is somewhat arbitrary. Given the fact that there is no experimental information concerning the Higgs sector at present, one must resort to theoretical arguments to constrain the unknown Higgs sector, even in the context of the Standard Model.

Two theoretical constraints exist. First, it is an experimental fact that $\rho = m_W^2/(m_Z^2 \cos^2\theta_W)$ is very close to 1. This almost certainly implies that the Higgs bosons are either $SU(2)$ weak doublets or singlets. (Other choices are possible, but rather ugly.) Second, there are severe limits on the existence of flavor-changing neutral currents (FCNC's). In the model with the minimal Higgs, tree-level flavor changing neutral currents are automatically absent. This continues to be true in non-minimal models in which fermions of a given electric charge couple to no more than one Higgs doublet.[1] An example of a model satisfying this requirement is the minimal supersymmetric extension of the Standard Model. This model (of which we will have more to say below) possesses two Higgs doublets of opposite hypercharge; the $Y = -1$ doublet couples only to down-type quarks and leptons, and the $Y = 1$ doublet couples only to up-type quarks and leptons. In this report, we shall concentrate on the two-Higgs doublet extension of the Standard Model. In addition, we will choose the Higgs–fermion coupling described above, which is compatible with the supersymmetric extension of the Standard Model. This framework is useful in that it adds new phenomena (*e.g.*, charged Higgs), introduces a minimal number of new parameters, and satisfies the theoretical constraints mentioned above. Two-Higgs-doublet models possess five physical Higgs bosons: a charged pair ($H^\pm$), two neutral CP-even scalars (H_1^0 and H_2^0), and a neutral CP-odd pseudoscalar (H_3^0). Here, we have made an implicit assumption that the Higgs potential is CP-invariant, so that the neutral Higgses have definite CP quantum numbers. The terms "scalar" and "pseudoscalar" refer to the way in which these neutral Higgses couple to fermion pairs. Instead of the one free parameter of the minimal model, this model has at least six free parameters: four Higgs masses, the ratio of vacuum expectation values:

$$\tan\beta = v_2/v_1, \tag{1}$$

and a Higgs mixing angle, α. The angle α arises when one diagonalizes the 2×2 neutral scalar Higgs mass matrix, whose eigenstates are H_1^0 and H_2^0. For definiteness, we will always take $m_{H_1^0} \geq m_{H_2^0}$. Note that $v_1^2 + v_2^2$ is fixed by the W mass. (Additional Higgs self-coupling parameters do not concern us here.)

There are two phenomenologically crucial types of Higgs couplings: those to fermion-antifermion pairs and those to two vector bosons. The couplings of the physical Higgses to fermion pairs are rather similar to those of the minimal model Higgs, especially if $\tan\beta$ is around 1. Let v_1 (v_2) be the vacuum expectation value of the Higgs field which couples only to down-type (up-type) fermions. Then (in 3rd generation notation), the $H_3^0 t\bar{t}$ ($H_3^0 b\bar{b}$) coupling is suppressed (enhanced) if $\tan\beta > 1$, and vice versa if $\tan\beta < 1$. Similar results hold for H_1^0 and H_2^0, although the couplings also involve the mixing angle α which can reduce the size of the couplings somewhat. For the charged Higgs we have:

$$g_{H^+t\bar{b}} = \frac{g}{2\sqrt{2}m_W}[m_t \cot\beta(1+\gamma_5) + m_b \tan\beta(1-\gamma_5)]. \quad (2)$$

Of even greater importance are the Higgs couplings to vector bosons. The H_3^0 couplings to vector boson pairs are forbidden at tree level. Of course, the $Z^0 H_i^0 H_j^0$ couplings are forbidden, when $i = j$ by Bose symmetry. When $i \neq j$, this coupling is only present when the two Higgses have opposite CP quantum numbers. Vertices involving neutral particles only and one or two photons clearly vanish at tree level, although they are generated at one-loop. (The same is true for the coupling of all neutral Higgs bosons to a pair of gluons. The radiatively generated $H_k^0 gg$ vertex is important since the two-gluon fusion is one of the major production mechanisms for neutral Higgses at a hadron collider.) Two other vertices, $H^+W^-Z^0$ and $H^+W^-\gamma$, also vanish at tree level. This turns out to be a general feature of models with only Higgs doublets and singlets.[2] Again, these vertices are radiatively generated at one-loop, and lead to interesting rare decays of the charged Higgs.[3] All other three-point tree-level vertices involving gauge and Higgs bosons are allowed. Probably the most important vertices for phenomenology are couplings of H_1^0 and H_2^0 to W^+W^- and Z^0Z^0. These couplings tend to be somewhat suppressed compared to their values in the minimal-Higgs model. However, there is a sum rule:

$$g^2_{H_1^0VV} + g^2_{H_2^0VV} = [g^{minimal}_{H^0VV}]^2 \quad (3)$$

which holds separately for $V = W$ or Z. Without further information, one cannot be certain as to how the H^0VV coupling strength is divided between H_1^0 and H_2^0. However, as discussed later in this report, in supersymmetric models the coupling of the heavier Higgs (H_1^0) to vector bosons is severely suppressed.

Having summarized the general properties of the two-Higgs doublet model, we briefly turn to the implications for the Higgs search at the SSC. Because of the fact that we now have (at least) six free parameters in the Higgs sector, there are only a few general statements one can make concerning the phenomenology of the Higgs at the SSC. First, if the scalar Higgs has couplings to WW and ZZ which are similar to their values in the Standard Model, and its mass is between about $2m_W$ and 800 GeV, then it should be possible to detect this Higgs at the SSC by observing its decay into a pair of vector bosons (followed by subsequent decay of the vector bosons into lepton pairs), as described by the Heavy Higgs Group.[4] On the other hand, for masses less than $2m_W$, we are in the regime of the "intermediate mass Higgs", in which the dominant Higgs decay is into the heaviest quark pair which is kinematically allowed. While the two-photon and ZZ^* (rare) decay modes may be useful over a large portion of the m_t-m_H parameter space, the ability to successfully observe such a Higgs at the SSC is not certain, especially if the top is not very heavy (e.g. $m_t \sim 55\ GeV$) and the Higgs has mass just above the $t\bar{t}$ decay threshold.[3][5] In supersymmetric models, detection of the heavy Higgs (H_1^0) via its Standard Model decay modes is particularly problematical. As alluded to above, the decay of H_1^0 into vector boson pairs (even when kinematically allowed) is extremely suppressed. Similarly, the vector boson fusion production mechanism is numerically unimportant, thereby reducing the production cross-section for the heaviest Higgs case. Since m_H is almost certain to be above $2m_t$ for this Higgs, even the two photon decay mode (useful for $m_H < 2m_t$) cannot be employed and detection would be extremely difficult.

Consider next the pseudoscalar Higgs. As described above, the pseudoscalar does not couple to vector boson pairs at tree level. The phenomenological implications of this fact are devastating. First, the important vector boson fusion mechanism for production of a Higgs boson is absent. Second, the dominant decay of the pseudoscalar Higgs will be into the heaviest quark pair available, independent of the Higgs mass. Thus, the search for the pseudoscalar Higgs will be very difficult once $m_{H_3^0} > 2m_t$, while for $m_{H_3^0} < 2m_t$ the 2γ decay mode may allow observation. Note, however, that if such an object could be found in the mass region above $2m_W$, then the absence of decays into vector boson pairs would be strong evidence for the pseudoscalar nature of the object. (An exception to this rule occurs in supersymmetric models, which predict Higgs *scalars* with suppressed couplings to the vector boson channels. Nevertheless, such an observation would be definitive evidence for a non-minimal Higgs sector.)

Finally, consider the charged Higgs boson. Because of the absence of tree-level coupling of the charged Higgs to vector boson pairs (WZ and $W\gamma$), the detection of the charged Higgs is likely to be at least as difficult as detection of the pseudoscalar Higgs. An exception, to be discussed later, occurs if $m_t > m_{H^\pm} + m_b$, and the decay $t \to H^+ + b$ has a large branching ratio. In contrast, for $m_t < m_{H^\pm} + m_b$ the total cross-section for the production of a single charged Higgs is smaller than that typical of a neutral Higgs, since the gluon-gluon fusion and vector-boson-fusion mechanisms are not available in this case. Instead, we must rely on the coupling to heavy quarks. This point will be discussed further below.

In summary, the detection of non-minimal Higgses is at best equivalent to the detection of the heavy minimal Higgs when the dominant decay is into vector boson pairs. Otherwise, (*e.g.* in the case of the pseudoscalar and charged Higgs) the prospect for detection is substantially worse, since it is very difficult to detect a Higgs whose primary decay products contain t and b quark initiated hadron jets. Thus, in order to have any hope for observing such Higgs bosons at the SSC, alternative decay modes must be studied. There are two basic approaches. The first approach involves the search for rare decay modes, with the hope that the decrease in background will compensate the decrease in signal due to a presumably small branching ratio. The second approach is to look for completely new final states which may constitute an important fraction of all Higgs boson decays. An example of this approach is to make use of the supersymmetric model, and investigate the branching ratio of the various Higgses into supersymmetric final states.

In this report, we will focus much of our attention on the charged Higgs boson, since its discovery would unequivocably signal the presence of a non-minimal Higgs sector. The report (a preliminary version of which appears in ref. 6) is organized as follows. In Sections B and C, we assess the feasibility of observing the charged Higgs boson at the SSC. QCD backgrounds to observing the H^+ via its $t\bar{b}$ decay are large, so we concentrate on the search for the charged Higgs boson via rarer decay modes. We briefly survey a number of possible charged Higgs decays: $H^\pm \to W^\pm\gamma$, $H^\pm \to W^\pm + quarkonium$, $H^\pm \to W^\pm H^0$, and $H^\pm \to W^\pm H^0\gamma$, and in Section C, we turn to an extensive study of the $\tau\nu$ decay. A detailed Monte Carlo analysis is presented, and various possible regions of parameter space are examined. In Section D, implications of the previous section for detector design are considered. Finally, in Section E, we briefly consider the implications of the "low-energy" supersymmetry approach for the phenomenology of Higgs bosons at the SSC. Our conclusions are summarized in Section F.

B. Search for the Charged Higgs Boson–General Considerations

First, let us make a few remarks about the production mechanism of a singly charged Higgs boson at the SSC. Which production mechanism is dominant depends upon the relative mass of the t quark and the $H^\pm$. If the top quark has a moderate mass, but $m_t > m_{H^\pm} + m_b$, then the rate for $gg \to t\bar{t}$ followed by t decay to the $H^\pm$ is very large. Relative to the t decay rate to charged W's we have:

$$\frac{\Gamma(t \to H^+ b)}{\Gamma(t \to W^+ b)} = \frac{p_{H^+}}{p_{W^+}} \frac{m_t^2(m_t^2 - m_{H^+}^2)}{(m_t^2 + 2m_W^2)(m_t^2 - m_W^2)} \cot^2\beta, \tag{4}$$

where p_{H^+} and p_{W^+} are the center-of-mass momenta of the H^+ and W^+ for the respective decays. Thus, the H^+ channel is fully competitive with the W^+ mode. If $m_t < m_{H^\pm} + m_b$, then one must turn to other production modes for the charged Higgs. Naively, one might expect that, at SSC energies, there will be a non-negligible amount of top-quarks (and bottom-quarks) inside the proton, so that one could use $t\bar{b}$ fusion to create the H^+ (and, of course, $b\bar{t}$ fusion for H^-). Although this is in some sense true, it turns out that the application of the parton model to this subprocess overestimates the charged Higgs cross-section by about a factor of 2. The reason for this is that, even at SSC energies, the top quark distribution function is not present at full strength (as compared to other massless quarks). So, effectively, the t-quark distribution function is of $\mathcal{O}(\alpha_s)$. This means that other partonic subprocesses which are $\mathcal{O}(\alpha_s)$ down from $t\bar{b} \to H^+$ are competitive with $t\bar{b}$ fusion, if they do not involve an initial t-quark.

The sub-process which turns out to be most important is $\bar{b} + gluon \to \bar{t} + H^+$. In fact, there is a subtle point involved here, since the leading logarithm of this process (for $m_{H^+} \gg m_t$) corresponds precisely to the $t\bar{b}$ fusion process. In ref. 7 a method of calculation is developed which avoids any problem of double counting, and shows how to correctly evaluate the charged Higgs production cross-section. Numerical analysis [8] shows that the correct procedure at SSC energies, for Higgs and top masses of interest, is to omit entirely the $t\bar{b}$ fusion contribution, and include only the exact $2 \to 2$ subprocess $\bar{b}g \to \bar{t}H^+$ We will make use of this result in the analysis presented below. For ease of reference, we give the cross section for $H^\pm$ production from ref. 8 in fig. 1.

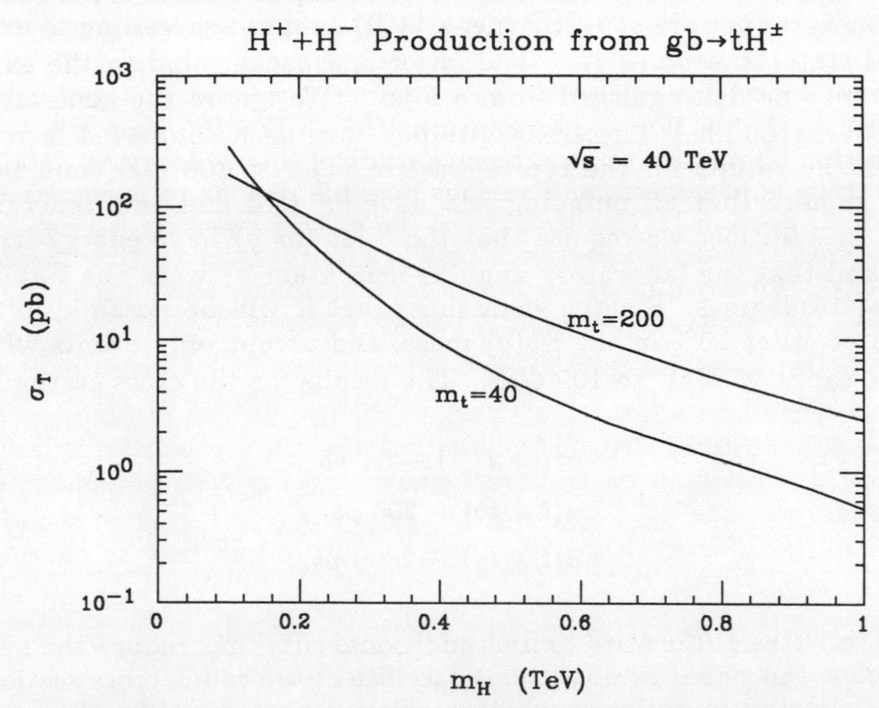

Figure 1: Charged Higgs total cross section from ref. 8. The cross section is computed for $\tan\beta = 1$, and is displayed for $m_t = 40$ and 200 GeV.

Even in the case where $m_{H^\pm} > m_t$, the raw number of charged Higgs events is substantial. However, for a given Higgs decay mode, the desired signal is generally swamped by huge backgrounds. To have any chance of seeing a signal, a trick must be employed. One trick that we shall explore is that of a 'stiff lepton trigger', first proposed in ref. 8. In the production mechanisms, $\bar{b}g \to \bar{t}H^+$ and $bg \to tH^-$, one attempts to trigger on the t or $\bar{t}$ produced in association with the charged Higgs. One approach to doing this is to note that the final state t and $\bar{t}$ quarks are typically moving nearly parallel to the original beam. Ordinarily, they would just be lost inside the beam jets. However, if the t-quark decays semi-leptonically, the electron or muon will be kicked out with sufficiently large p_T (of order $m_t/2$), so that it can be used to trigger the desired event. Even the leptons coming from decays of the secondary b quarks that arise from t decay will contribute to this trigger, so that a trigger in which a stiff lepton with $p_T^l > 10\ GeV$ is required retains $\sim 45\%$ of the $H^\pm$ events, while rejecting all but 1% to 2% of most types of background processes.[8]

The first question that one must ask is whether the possibility of such a trigger could even make the observation of the charged Higgs in its major tb decay mode feasible. For simplicity of notation let us consider production and detection of the single charge state, H^-. If we imagine for a moment that the stiff lepton trigger is 100% efficient in eliminating events without a spectator t quark and that the $\bar{t}$ quark can also be triggered upon with 100% efficiency, then the only backgrounds are $gb \to t\bar{t}b$ (i.e. a QCD subprocess leading to exactly the same final state) and $gg \to t\bar{t}g$. The latter is a background to the extent that a g jet cannot be distinguished from a b jet. (We ignore the generally smaller $gq \to t\bar{t}q$ backgrounds.) These backgrounds have been computed in ref. 9. We will sketch the results for the typical case of $m_{H^\pm} = 100\ GeV$ and $\tan\beta = 1$. First, we require that all outgoing jets have $|y| < 5$ and total energies above $10\ GeV$. In addition, we require that the b jet (or g) have energy larger than $50\ GeV$, and that the laboratory angular separation between the $\bar{t}$ and b (or g) be at least 15 degrees. Finally, we assume that it will be possible to achieve a resolution of order 10% in the $\bar{t}$-$b(g)$ mass, and accept only events with $M_{\bar{t}b(g)}$ within the range 95 GeV to 105 GeV. The results for the cross sections are:

$$\begin{aligned} \sigma(t + H^-) &\sim 64\ pb \\ \sigma(t + \bar{t}b) &\sim 200\ pb \\ \sigma(t + \bar{t}g) &\sim 2600\ pb. \end{aligned} \tag{5}$$

It is relatively straightforward to find additional cuts that reduce the $t + \bar{t}b$ background below the signal rate (without sacrifice of too much cross section). Thus the most important question is whether efficient procedures for distinguishing b jets from g jets to one part in 40 can be developed. Of course, the above discussion has so far ignored backgrounds such as $gg \to ggg$ (and similar ones involving light quarks) which enter to the extent that t jets cannot be distinguished from g and light quark jets. It was shown in ref. 9 that a set of cuts can be found that reduce the $gg \to ggg$ cross section sufficiently, so that discrimination between b, t and g jets to one part in ~ 40 would make this background comparable to the signal (which is about 1 pb after cuts). This factor of 40 might be achievable, based on the stiff lepton trigger example discussed earlier. The major problem will be whether or not this can actually be done with high efficiency. In addition, there is the question of what mass resolution in the $\bar{t}b$ channel can actually be achieved. Clearly these are questions requiring a detailed Monte Carlo study. We do not feel that one should be very optimistic about direct detection of the $H^\pm$ in the tb channel; however, the above results do suggest that further study is warranted.

Since the dominant decay mode of the charged Higgs is a very problematical mode of discovery, it is imperative to examine other possible rarer decay modes. The obvious strategy is to choose distinctive final states in order that the increased signal-to-background can more than compensate the reduced branching ratio. In this report we shall consider five interesting rare decays, involving standard model particles and/or Higgs bosons, neglecting temporarily the possibility of exotic final states (*e.g.* containing supersymmetric particles). First, among the possible final state fermion pairs, we consider the $\tau\nu$ decay mode. The branching ratio in the two-Higgs-doublet model is:

$$BR(H^{\pm} \to \tau\nu) \approx \frac{m_{\tau}^2 \tan^2\beta}{3(m_t^2 \cot^2\beta + m_b^2 \tan^2\beta)}, \tag{6}$$

where we have assumed that the dominant decay of the charged Higgs is into $t\bar{b}$ (or $b\bar{t}$). Thus, unless $\tan\beta$ is quite large, we expect a branching ratio of $BR(H^{\pm} \to \tau\nu) \lesssim 10^{-3}$, when $m_{H^+} > m_t + m_b$ and $m_t \gtrsim 55\ GeV$. Of course, if the top quark mass is larger than the charged Higgs mass, then $BR(\tau\nu)$ can be substantially bigger; at $\tan\beta = 1$ roughly 35% of the charged Higgs decays are to $\tau\nu$, and the number could be substantially higher if $\tan\beta > 1$. To be more specific requires a definite model, which also includes the W^+H^0 modes to be discussed later. As an example, if the branching ratio for $H^+ \to \tau^+\nu$ is computed in the minimal supersymmetry model,[8] for $m_{H^{\pm}} < m_t + m_b$ and all supersymmetric particle modes forbidden, a typical choice of parameters yields a $\tau\nu$ mode branching ratio ranging between 10% and 40%, and even higher for small $m_{H^{\pm}}$. To evaluate whether it is feasible to detect the charged Higgs in this mode, we must carefully evaluate the charged Higgs production and the competing backgrounds to the $H^+ \to \tau\nu$ final state. A detailed discussion of the detectibility of the charged Higgs via the $\tau\nu$ mode is presented in Section C.

What about other rare decay modes of the charged Higgs boson? Within the context of Standard Model-particle final states, the only possibilities that come to mind are $H^{\pm} \to W^{\pm}\gamma$, $H^{\pm} \to W^{\pm} + quarkonium$, $H^{\pm} \to W^{\pm}H^0$, and $H^{\pm} \to W^{\pm}H^0\gamma$ (where H^0 can, in principle, be either of the neutral scalars, H_1^0 or H_2^0). The rate for the first mode has been computed in ref. 3. In general the branching ratio is quite small, and the event rate too low to compete with the $W^{\pm}\gamma$ continuum background. An exception to this statement occurs when the charged Higgs mass is much smaller than the mass of the heavier neutral Higgs (H_1^0). However, note that in the supersymmetric models to be discussed later the $H^{\pm}$ and H_1^0 masses are always quite similar.

The $W^{\pm} + quarkonium$ mode branching ratios were considered in ref. 10. The modes $H^+ \to W^+\Upsilon$ and $H^+ \to W^+\Theta$ (where Θ is the $t\bar{t}\ {}^3S_1$ bound state) were computed; both are quite sensitive to the value of m_t which enters the loop diagram calculations and controls the phase space. The conclusions of ref. 10 are easily summarized. If $H^+ \to t\bar{b}$ is not allowed, then the branching ratio for $H^+ \to W^+\Upsilon$ is quite significant (typically $1 - 3 \times 10^{-2} BR(H^+ \to \tau^+\nu)$)

when $m_{H^\pm}$ is just below $m_t + m_b$, although it falls rapidly with increasing m_t. Together with $t\bar{t}$ production followed by $t \to H^+ b$ and $\bar{t} \to W^- \bar{b}$, one finds a significant rate for production of two b jets, two leptonically decaying W's and a leptonically decaying Υ. In contrast, since $H^+ \to t\bar{b}$ is always allowed if $H^+ \to W^+ \Theta$ is allowed, the latter decay always has a very small branching ratio (typically $\lesssim 10^{-5}$).

The $H^+ \to W^+ H_1^0$ and $H^+ \to W^+ H_2^0$ decays are potentially quite important due to the large contributions from longitudinal W polarization states. These modes have been explored in ref. 3. Defining the Feynman coupling for $H^+ W^- H^0$ as the coefficient of $-i(p+p') \cdot \epsilon_W$ (where p and p' are the four-momenta of H^+ and H^0, respectively) we have a sum rule analogous to that of eq. (3):

$$g^2_{H^+W^-H_1^0} + g^2_{H^+W^-H_2^0} = g^2/4. \tag{7}$$

Again, a specific model is required to determine both the division of the coupling strengths and the relation between the H^+, H_1^0 and H_2^0 masses. Defining the ratio $R_{WH^0} \equiv BR(H^+ \to W^+ H^0)/BR(H^+ \to t\bar{b})$, we obtain:

$$R_{WH_2^0} = \frac{2\cos^2(\beta-\alpha) p_W^3 m_{H^+}^2}{3p_{\bar{b}}[(m_t^2 \cot^2\beta + m_b^2 \tan^2\beta)(m_{H^+}^2 - m_t^2 - m_b^2) - 4m_t^2 m_b^2]} \tag{8}$$

where p_W and $p_{\bar{b}}$ are the center-of-mass momenta of the indicated final state particles, and α is the scalar Higgs mixing angle. The corresponding formula for H_1^0 is obtained by replacing $\cos(\beta-\alpha)$ with $\sin(\beta-\alpha)$. To determine just how important these modes could potentially be, we consider the case where the outgoing H_2^0 has a mass of 40 GeV and saturates the allowed coupling strength (*i.e.*, $\cos(\beta-\alpha) = 1$ in eq. (8)). At $\tan\beta = 1$ and $m_t = 55$ GeV the ratio $R_{WH^0} \equiv BR(H^+ \to W^+ H^0)/BR(H^+ \to t\bar{b})$ rises from ~ 0.17 at $m_{H^\pm} = 140$ GeV to ~ 1.2 at $m_{H^\pm} = 200$ GeV, passing 10 in the vicinity of $m_{H^\pm} = 460$ GeV. However, if the minimal supersymmetric model is employed, the importance of such modes is greatly reduced. First, the mass relations are such that $H^+ \to W^+ H_1^0$ is never allowed. Second, $g_{H^+W^-H_2^0}$ has the same severe suppression that characterizes the $H_1^0 WW$ and $H_1^0 ZZ$ couplings, as discussed earlier. At $m_t = 55$ GeV the resulting $R_{WH_2^0}$ value is 0 at $\tan\beta = 1$ (since by eqs. (13) and (17), $\cos(\beta-\alpha) = 0$ at this point), and it reaches a maximum as a function of m_{H^+} of $\sim 2\times10^{-2}$ at $\tan\beta = 1.5$ and $\sim 7\times10^{-2}$ at $\tan\beta = 3$. As m_{H^+} increases beyond the location of the maximum $R_{WH_2^0}$ falls slowly. The H_2^0 masses implied by the choices of m_{H^+} and $\tan\beta$ in the above range are of order 20 to 40 GeV. Thus H_2^0 would decay to $b\bar{b}$. In fact, the above branching ratios for $\tan\beta \geq 1.5$ imply an effective $W^+ H_2^0$ associated production cross section (~ 1 pb) that is not very different from the associated production cross section considered in searches for the intermediate mass Standard Model neutral Higgs (using $W^* \to WH^0$).[11] Of course, outside the context of the minimal supersymmetry model considerably larger cross sections are possible. The studies of the intermediate mass Higgs region[11] suggest that the $b\bar{b}$ mass resolution will be sufficient to recognize a 1 pb level W^+H^0 signal over backgrounds coming from mixed QCD/Electroweak processes such as $qq' \to W^+b\bar{b}$, when the H^0 mass is in the vicinity of 120 GeV (and the $t\bar{t}$ mode is not allowed). This would undoubtedly be much more difficult at lower $b\bar{b}$ invariant mass; problems would include decreased $b\bar{b}$ mass resolution and significantly larger backgrounds. On the other hand, charged Higgs production can be tagged using the "stiff-lepton" trigger discussed earlier. In addition charged Higgs decay would lead to Jacobian peaks in the outgoing W^+ and $b\bar{b}$-system transverse momenta that might allow for effective cuts that would further reduce backgrounds. Clearly, a detailed Monte Carlo study is required to fully assess the situation, but this mode looks relatively promising.

Finally, we summarize the results of ref. 3 for the decays $H^+ \to W^+H^0\gamma$. As in the previous case, the strength for such modes is divided between the H_1^0 and H_2^0. The useful branching ratio relative to the $t\bar{b}$ decay channel of the H^+ depends upon the minimum energy allowed for the γ. Let us adopt a requirement of $E_\gamma > 20\ GeV$ in the H^+ rest frame. For $\tan\beta = 1$, $m_t = 40\ GeV$, an H^0 mass of 55 GeV, and maximal coupling, the ratio $R_{WH^0\gamma} \equiv BR(H^+ \to W^+H^0\gamma)/BR(H^+ \to t\bar{b})$ is 0.01 at $m_{H^\pm} = 300\ GeV$ and rises steadily with increasing $m_{H^\pm}$, passing 0.1 by $m_{H^\pm} \sim 520\ GeV$. This would clearly provide a very viable signature and rate for H^+ detection, even accounting for the need to reconstruct the H^0 in a $b\bar{b}$ decay mode. However, just as in the W^+H^0 mode case, the minimal supersymmetric model predicts that $H^+ \to W^+H_1^0\gamma$ is phase-space-forbidden, while $H^+ \to W^+H_2^0\gamma$ is severely suppressed. In this particular model, $R_{WH_2^0\gamma}$ (where we have taken $E_\gamma > 20\ GeV$, $m_t = 55\ GeV$, and $\tan\beta = 1.5$, as an example) reaches a maximum of $\sim 2.6 \times 10^{-5}$ at $m_{H^\pm} \sim 400\ GeV$ and decreases (slowly) thereafter. The resulting event rate at the SSC would not be useful!

Thus, to summarize, the 'rare' decay mode of the charged Higgs that is significant in the largest class of models (including, in particular, the minimal supersymmetric model) is the $\tau\nu$ channel. Thus, we shall focus on this mode in the following section.

C. Search for the Charged Higgs Boson Via its $\tau\nu$ Decay

As discussed in Section B, there are two different scenarios to consider when discussing the detectability of the $H^\pm$ in any of its decay modes. In the first, $m_t > m_{H^\pm} + m_b$ and we will look for $H^\pm$ in the decays of the t and $\bar{t}$ quarks produced via $gg \to t\bar{t}$. In the second, $H^\pm$ must be produced inclusively via the $gb \to H^\pm t$ fusion processes.

The first case, $m_{H^\pm} + m_b < m_t$, can be expected to provide the clearest signal for the charged Higgs, because both the production cross section, $\sigma(gg \to t\bar{t})$, and the branching ratio for $H^+ \to \tau^+\nu_\tau$ are large. The main difficulty is to distinguish the charged Higgs signal, $t \to bH^+$, from the standard decay, $t \to bW^+$, with the subsequent decay of $H^\pm$ and $W^\pm$ to a lepton and neutrino. Unlike the case of W decay, charged Higgs decay yields a violation of "lepton universality" that can be used to distinguish between the two possible top decay modes. Specifically, in the case of charged Higgs decay, the branching ratio for decay to the τ lepton is much larger than that for decay to an electron or a muon. But, in the case of W boson decay, all three leptons are produced equally.

As is discussed in ref. 12, events containing τ's can be enhanced by triggering on isolated charged tracks. The isolated track can be either a lepton or a hadron. Of course, W decay to e, μ, τ can also produce an isolated charged track, so we will have to use a statistical technique to uncover a signal for charged Higgs production against the background of charged W production in top quark decays. If the source of the isolated track is the decay of the charged Higgs, the ratio of the probability that the isolated track is a hadron to the probability that the track is a lepton (hereafter referred to as $R(h/l)$) is the same as that in the decay of the τ. For simplicity, we will assume that the decay of the charged Higgs and W boson to quarks can be rejected completely. In the case of W boson decay, the ratio $R(h/l)$ is smaller than for tau decay, because an electron or muon directly produced from the W decay cannot be distinguished from a single track coming from a tau decay. Thus, $R(h/l)$ for W decay is $h(\tau)/(l(\tau)+2)$, where $h(\tau)$ and $l(\tau)$ are the branching fractions for single charged hadrons and leptons, respectively, in tau decay. Using the measured decay branching fractions of the τ, we find that if there is no charged Higgs in the t decays, the measured $R(h/l)$ will be 0.22. In contrast, if we assume for the moment that the branching ratio for $t \to W^+b$ is equal to that for $t \to H^+b$, one finds $R(h/l) = 0.59$.

We should not ignore the possibility that the top quark is light enough to have a large production cross section at existing accelerators such as the Tevatron, yet heavy enough that $m_t > m_{H^\pm} + m_b$, so that $BR(H^\pm \to \tau\nu)$ is large. In such a case, the SSC might not be required for the discovery of the $H^\pm$. Even in the supersymmetric model where $m_{H^\pm} \geq m_W$, the charged Higgs could be quite near its lower limit and the t quark could be only moderately heavier, yet still light enough to be produced at the Tevatron with a substantial rate. We have made a first exploration of this possibility. As an example we shall take $m_{H^\pm} = 85\ GeV$ and consider two possible t quarks masses: 110 GeV and 120 GeV. The signal for $H^\pm$ production that we focus on is that discussed above, namely an excess number of isolated singly charged hadrons (h) produced in t quark decays via the chain

$$t \to H^+b \qquad H^+ \to \tau^+\nu, \text{ with } \tau^+ \to e^+, \mu^+ \text{ or } h^+ + X, \tag{9}$$

where X can contain only π^0's and/or ν's and $h^+ = \pi^+$ or K^+, as compared to the normal sequence

$$t \to W^+b \qquad W^+ \to e^+, \mu^+ \text{ or } \tau^+ + \nu, \tag{10}$$

with the τ^+ decaying as above. The first process produces substantially more isolated singly charged hadrons than does the second. Assuming $\tan\beta = 1$ we find, using eq. (4), that for $m_{H^\pm} = 85\ GeV$, $BR(t \to H^+) = 31\%(27\%)$ for $m_t = 120\ GeV (110\ GeV)$. Larger masses for the $H^\pm$ result in smaller branching ratios, and, therefore, less sensitivity.

In order to see whether charged Higgs from t-quark decay are observable at the Tevatron, one of us (L. Galtieri) has performed a Monte Carlo study using ISAJET. This study does not incoporate any detector simulation. In addition, detailed QCD background studies were not done. The conclusions that we reach are probably optimistic; therefore, this procedure only sets an upper bound to the discovery limit of the charged Higgs. The branching ratios for $t \to H^+b$ and $t \to W^+b$ were computed using eq. (4) with $\tan\beta = 1$. Assuming only Standard Model decays, the branching ratio for $H^\pm \to \tau\nu$ (for $\tan\beta = 1$) is approximately 35%. We select events in which the ISAJET generated particle list shows that there is a lepton (e or μ), with laboratory transverse momentum $p_T > 20\ GeV$, coming from the decay of one of the top quarks (t_1). Experimentally there is a significant chance that such a lepton will be relatively isolated, and we shall call it an 'isolated lepton'. However, isolation criteria were not actually implemented (and events chosen accordingly) for this first study. In the present context, such a lepton is most likely to originate from the $W^\pm$ or $H^\pm$ appearing in the decay of t_1. Note that leptons from τ decay (for example, $t_1 \to W^\pm$ (or $H^\pm$) $b \to \tau\nu b \to \ell\nu\nu\nu b$) are also included if the p_T condition is satisfied. For the second top (t_2) decay we require a 'prompt' charged particle coming from the $W^\pm$ or $H^\pm$ having $p_T > 10\ GeV$ and count how often this is a hadron or a lepton; an excess of hadrons is the signature for Higgs decay of the top. The 'prompt' charged particle can be a charged lepton (e or μ) as in the case of t_1 decay, or a single charged hadron ($\pi^\pm$ or $K^\pm$) from τ decay. Thus, only decays of the τ's into one charged prong are considered. We assume that experimentally a narrow cone surrounding a 'prompt' charged hadron can be defined which contains all the energy of the neutral hadronic decay products that might be associated with this charged track. Thus, in calculating the p_T of the charged hadron coming from τ decay, associated photon (π^0) momentum (for example, π^0's from ρ or K^* decays of the τ) is added to the charged particle momentum. This is done in order to have larger detection efficiency for the given p_T cut. Finally, to reduce possible QCD background, we require the lepton or hadron from the second top to be central, *i.e.*, $|\eta| < 1.5$.

Two choices of integrated luminosity are considered: a) the Tevatron (TEV-I) to run 5 years at its maximum yearly yield of 10 pb^{-1}; and b) an upgraded Tevatron (TEV-II) to run for two years at 500 $pb^{-1}/year$. The results are shown in Table 1, where l stands for an isolated charged lepton and h for an isolated charged hadron; the notation " N_{evts}" refers to number of events. It is clear from this table that it is not possible to discover the $H^\pm$ at TEV-I, whereas the possibility of its discovery at TEV-II cannot be excluded. For example, at TEV-II the difference between the $H^\pm + W^\pm$ and the $W^\pm$ alone (for $m_t = 120\ GeV$) is $\Delta R(h/l) = (.351 \pm .032) - (.176 \pm .018) \simeq 0.18 \pm .04$, a nominally significant effect. Nonetheless, it is certainly possible that one may not be able to rule out charged Higgs with $m_{H^\pm} < m_t - m_b$ by the time the SSC is ready to turn on.

Table 1

Sensitivity of Tevatron to Charged Higgs with $m_t > m_{H^\pm} + m_b$

	$\int \mathcal{L}\, dt$ (pb^{-1})	m_t (GeV)	σ (pb)	$gg \to t\bar{t}$ N_{evts}	top decay	$t_1 \to l$ N_{evts}	$t_2 \to l$ N_{evts}	$t_2 \to h$ N_{evts}	$R(h/l)$
TEV-I	50	120	16.6	830	$W^\pm$	158	32.4	5.7	$.18 \pm .08$
					$W^\pm + H^\pm$	132	22.5	7.9	$.35 \pm .15$
TEV-I	50	110	27.0	1350	$W^\pm$	252	50.6	9.3	$.18 \pm .06$
					$W^\pm + H^\pm$	216	37.0	12.6	$.34 \pm .11$
TEV-II	1000	120	16.6	16600	$W^\pm$	3160	648	114	$.18 \pm .02$
					$W^\pm + H^\pm$	2640	450	158	$.35 \pm .03$
TEV-II	2000	110	27.0	27000	$W^\pm$	5049	1012	185	$.18 \pm .01$
					$W^\pm + H^\pm$	4320	740	252	$.34 \pm .03$

At the SSC, when $m_t > m_{H^\pm} + m_b$ one anticipates that this type of procedure will provide a clear signal for $H^\pm$ production, due to the high machine luminosity and energy. To verify this, one of us (H. Yamamoto) has performed a Monte Carlo study using PYTHIA. Here a Monte Carlo simulation was performed using experimentally defined triggering procedures developed in ref. 12. However, this simulation did not include the effects of (unknown) momentum and energy resolution.

The technique for triggering on τ leptons from charged Higgs decay relies on the event shape of the charged Higgs events and on the topology of possible background events.

(1) The charged particle with highest p_T relative to the beam direction is chosen as a candidate for the isolated charged particle emerging from τ decay. We demand that no other charged particles are within a narrow cone around this candidate particle. The narrow cone, illustrated in fig. 2, is defined by requiring that the half angle ψ satisfies $\cos\psi = 0.999$.

(2) The candidate track can be associated with γ's (π^0's). The charged particle momentum and the photon momenta in the narrow cone are summed vectorially. The resultant transverse momentum magnitude, $|\vec{p}_T(narrow\ cone)|$,

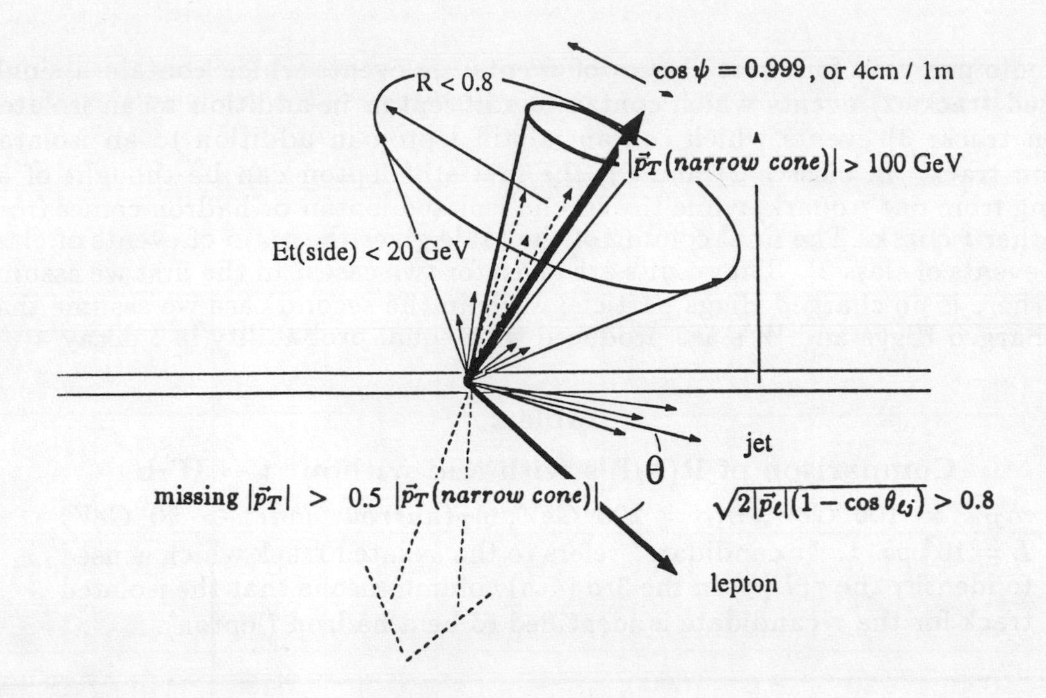

Figure 2: We illustrate the cones used to define an isolated charged track.

must exceed 100 GeV and the pseudo-rapidity of the narrow cone three-momentum must satisfy $|\eta(narrow\ cone)| < 1.5$.

(3) Around the narrow cone, a broader region is defined by requiring that $\Delta R = \sqrt{\Delta\phi^2 + \Delta\eta^2}$ satisfy $\Delta R < 0.8$. We require that the E_T sum in this region (excluding the region in the narrow cone) be smaller than 20 GeV. This defines what we mean by an *isolated* charged particle track.

(4) Next, we require that the missing $|\vec{p}_T|$ of the whole event must exceed 50 % of the $|\vec{p}_T(narrow\ cone)|$. The idea of this requirement is to select events containing a rather energetic neutrino, such as that which would emerge in charged Higgs decay.

(5) Finally, the event should have an additional 'stiff' lepton (e or μ) with $p_T >$ 5 GeV, an angle with respect to the beam that satisfies $5° < \theta_{lepton} < 175°$, and the charge of this lepton must be opposite in sign from the charge of the observed isolated high p_T charged track. Since the lepton is supposed to come from the other top quark, that is produced in association with the charged Higgs, it is required to be isolated from the nearest reconstructed jet. The isolation condition[13] is given by $\sqrt{2|\vec{p}_\ell|(1-\cos\theta_{\ell j})} > 0.8$, where $\vec{p}_\ell$ is the momentum of the lepton and $\theta_{\ell j}$ is the angle between the lepton and the nearest reconstructed jet which is found by the Lund cluster algorithm.[14]

The results of this study are summarized in Table 2. Table 2 gives the number of events per year for three classes of events: 1) events which contain a single isolated track; 2) events which contain a stiff lepton in addition to an isolated lepton track; 3) events which contain a stiff lepton in addition to an isolated hadron track. In classes 2) and 3), the first stiff lepton can be thought of as coming from one t quark, while the second isolated lepton or hadron comes from the other t quark. The final column of the table gives the ratio of events of class 3) to events of class 2). The results are given for two cases: in the first we assume that there is no charged Higgs particle; while in the second case we assume that the charged Higgs and W's are produced with equal probability in t decay.

Table 2

Comparison of R(h/l)'s with and without $t \to H^{\pm}b$

$m_{H^\pm} = 100$ GeV, $m_t = 200$ GeV, $p_T(narrow\ cone) > 50$ GeV, $L = 10^4$ pb^{-1}. "τ candidate" refers to the isolated track which is used to identify the τ; l (h) in the 3rd (4th) column means that the isolated track for the τ candidate is identified to be a hadron (lepton).

Process	τ candidate	l + stiff lepton	h + stiff lepton	$R(h/l)$
$t \to W^{\pm} + b$ (no $H^{\pm}$)	2.1×10^6	2.5×10^5	2.3×10^4	0.09
$t \to W^{\pm}$ or $H^{\pm} + b$	1.2×10^6	7.5×10^4	3.0×10^4	0.38

One important observation from this table is that $R(h/l)$ differs by a factor of 4 in these two cases. This number is the ratio of two different kinds of events, and is independent of the distribution functions or the total production cross section of the $t\bar{t}$ pair. Secondly, the difference in the number of isolated charged tracks (*i.e.* τ candidates) comes from the fact that an electron or muon directly produced from the W boson is more easily tagged than a charged particle produced from the τ decay, because the former is more energetic and is isolated in nature. But the difference between the absolute production rate for τ candidates, when the charged Higgs is present and when it is not, is useful only if one can estimate the total cross section of $t\bar{t}$ pair production.

The main problem in the overall significance of the signal for charged Higgs production obtained from $R(h/l)$ is misidentification of leptons (electrons and muons directly produced from the W boson) as isolated hadrons. However, even if we allow for a reasonable level of the misidentification probabilities of electrons and muons, the tighter cuts mentioned below should allow us to distinguish between the case where the charged Higgs is present in top decay from the case in which it is not.

As state earlier, the numbers given in Table 2 are based on the PYTHIA Monte Carlo, which carries out the hadronization of partons and decays of particles. When tracing back to the parent of the isolated tracks, it was found that sometimes a hadron which comes from the few body decay of a charmed particle produced in the b quark jets was identified as an isolated charged track. This means that the results shown in Table 2 are somehow dependent on the fragmentation models and the decay model used in the PYTHIA Monte Carlo. The PYTHIA Monte Carlo is fairly well tuned using the currently available data, but still it is desirable to find a way to make the result independent of the treatment of hadronization and fragmentation. A tighter selection of isolated charged tracks was tried in order to reject this kind of background. The efficiency dropped, but the dependence on hadronization became weaker. This analysis is still going on, and no conclusion can be drawn as to the dependence on the hadronization models, but the results shown in Table 2 seem to still be valid.

The background due to misidentification of a charged lepton as a charged hadron, and vice versa, in processes with larger cross section, *i.e.* pure QCD processes (in which only light quarks and gluons are involved) or processes in which a single gauge boson is produced in association with one or more jets, has not yet been studied. However, the background situation is far better than that which we encounter below in the case where $m_{H^\pm} + m_b > m_t$. In the present situation, the events of interest arise from $t\bar{t}$ production which is a strong QCD process with a cross section that is of order a hundred times larger than the $bg \to H^\pm t$ production process that is the dominant mechanism for charged Higgs production when $m_{H^\pm} + m_b > m_t$.

We turn now to the second case, where $m_{H^\pm} + m_b > m_t$. Existing theoretical models suggest that this is more likely to be the case than $m_t > m_{H^\pm} + m_b$. Detection of a charged Higgs that does not result from t decay is impossible at the Tevatron, so we shall focus only on the SSC. Before presenting a detailed Monte Carlo analysis of the charged Higgs search via its $\tau\nu$ mode in this case, let us make some quick estimates to determine the extent of the difficulty of separating the signal from the anticipated background. Using the cross-sections discussed in fig. 1, one can make a quick comparison between signal and likely backgrounds. It immediately becomes clear that the signal-to-noise is much smaller than 1, except for the limited mass range where $m_t - m_b \lesssim m_{H^\pm} \lesssim m_b + m_t$ where the top quark is still not heavy enough to decay to $H^\pm + b$. The problem here is that both real and virtual W bosons can also decay into $\tau\nu$ final states. Even when we employ the trick of a stiff-lepton trigger, there is an irreducible background process which is not rejected, namely $g\bar{b} \to \bar{t}W^+$ and its charge conjugate. The event topology for this reaction is identical to that of the signal, and only the lower mass of the W^+ can be used to separate this background from the signal (using rapidity cuts and the Jacobian peak in the p_T spectrum of the single charged particles from τ decay [15]). To illustrate the magnitude of the problem, we give in fig. 3 the cross section for $g\bar{b} \to \bar{t}W^+$ [6] [16] and its charge conjugate, compared to $g\bar{b} \to \bar{t}H^+$ and its charge conjugate, as a function of the t quark mass at $m_{H^\pm} = 300\ GeV$.

No branching ratios for the $W^\pm \to \tau\nu$ or $H^\pm \to \tau\nu$ decays have been incorporated. In comparing m_t dependence it should be kept in mind that $BR(W^\pm \to \tau\nu)$ is m_t independent once the tb channel is closed, while $BR(H^\pm \to \tau\nu)$ falls

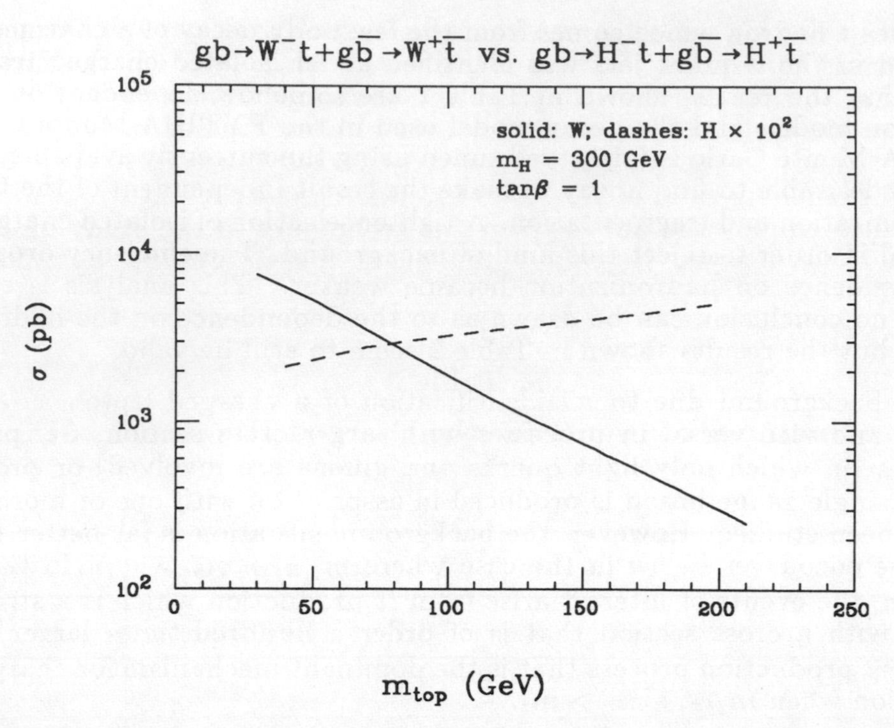

Figure 3: We give the cross section for $g\bar{b} \to \bar{t}W^+$ and its charge conjugate, compared to $g\bar{b} \to \bar{t}H^+$ and its charge conjugate, as a function of the t quark mass at $m_{H^\pm} = 300$ GeV. No branching ratios for the $W^\pm \to \tau\nu$ or $H^\pm \to \tau\nu$ decays have been incorporated. In comparing m_t dependence it should be kept in mind that $BR(W^\pm \to \tau\nu)$ is m_t independent once the tb channel is closed, while $BR(H^\pm \to \tau\nu)$ falls like $1/m_t^2$. This calculation is taken from ref. 6.

like $1/m_t^2$. The cross section for $gb \to tH^\pm$ presented in fig. 3 can be adjusted for $\tan\beta \neq 1$ by using

$$\sigma(gb \to tH^\pm) \propto m_t^2 \cot^2\beta + m_b^2 \tan^2\beta. \tag{11}$$

We choose "typical" values of $m_{H^\pm} = 300$ GeV, $\tan^2\beta \sim 3$ and $m_t = 70$ GeV. Including branching ratios for the $\tau\nu$ decays of the W and H, we have $S/B \sim 10^{-4}$ with a signal of 200 events where the small signal rate is due in part to the small $H^\pm \to \tau\nu$ branching ratio which is of order 10^{-3}. The means for discriminating between this background and the charged Higgs signal are limited and can probably never achieve better than a factor of 10 discrimination. Using such a factor it quickly becomes clear that $BR(H^\pm \to \tau\nu) \gtrsim 0.5$ is required before one could detect the charged Higgs in this manner.

Despite this pessimistic outlook, we have pursued the detection of the charged Higgs when $m_t \leq m_{H^\pm} + m_b$ using the PYTHIA Monte Carlo. We employ the techniques of ref. 12 outlined earlier for identifying the τ lepton from charged Higgs decay and for triggering on the stiff lepton from the decay of the top quark, produced in association with the charged Higgs. Even though the production mechanism is $gb \to H^{\pm}t$ (instead of $gg \to t\bar{t}$ with $t \to H^{\pm}b$) the final state is very similar to that considered earlier and the same techniques apply. The case of $m_{H^{\pm}} = 300\ GeV$ and $m_t = 40\ GeV$ was already studied in ref. 12, with the conclusion that it is very difficult to observe charged Higgs production because of the small branching ratio of $H^{+} \to \tau^{+}\nu_{\tau}$. During the present workshop we have continued to employ the above triggering techniques in the study of additional choices for the charged Higgs and top quark mass. Table 3 summarizes our results.

Table 3

Expected number of events for the charged Higgs and backgrounds
$L = 10^4\ pb^{-1}$

Process	Isolated Lepton	Stiff Lepton	$m_{H^{\pm}}(GeV)$	$m_t(GeV)$
$H^{\pm} + t$	13	1.7	300	40
$H^{\pm} + t$	10	1.5	400	40
$H^{\pm} + t$	8	1.0	500	40
$H^{\pm} + t$	1.7	0.7	300	200
$W^{\pm} + q(g)$	7×10^5	9×10^3	300	40
$W^{\pm} + q(g)$	2×10^3	1.3×10^3	300	200
W^*	7.5×10^4	< 80	300	40
$W^{\pm} + Z$	400	4	300	40

Table 3 shows the expected number of events per year for several choices of $m_{H^{\pm}}$ and m_t after selecting events with an isolated charged particle as described earlier, and after imposing the additional requirement of the existence of a stiff lepton on the opposite side of the τ jet. Also shown in Table 3 are the main sources of background. As can be seen, the main conclusion of ref. 12 is not changed; that is, the event rate is very small for the mass range of $m_{H^{\pm}} = 300 - 500\ GeV$, and the background is much larger than the signal.

Finally we have examined the special case where $m_{H^{\pm}} = m_t$, for the particular choice of $m_{H^{\pm}} = m_t = 200\ GeV$. In this case, the main source of charged Higgs production is, once again, $bg \to H^{-}t$, plus the charge conjugate process. The production rate, $\sigma(bg \to H^{\pm}t) \times BR(H^{\pm} \to \tau\nu)$ is shown in Table 4, for an integrated luminosity of $L = 10^4\ pb^{-1}$ per year.

Table 4

τ signal in $H^{\pm}$ decay and background from $W^{\pm}$ decay

$m_{H^{\pm}} = m_t = 200\ GeV,\ L = 10^4\ pb^{-1}$

Process	$\sigma \cdot BR(pb)$	Events/year
$H^{\pm} + t,\ H^{\pm} \to \tau\nu_\tau$	100	2960
$W^{\pm} + t,\ W^{\pm} \to \ell\nu_\ell$	240	9000

The expected number of events per year after the signal selection is around 3000. The main source of background is associated $W^{\pm}t$ production with the subsequent decay of the $W^{\pm}$ to a lepton and a neutrino. The Higgs particle is tagged by a single charged track from the τ decay, and it is impossible to clearly separate this Higgs decay mode from the W decay. Also, this W is associated with a top quark on the opposite side, and the stiff lepton cut does not help to reduce this background. As can be seen from Table 4, the S/B ratio is around 0.3 to 0.4, both in the production rate $\sigma \times BR$ and the observed number after signal selection. The statistical significance of the signal is reasonable, and if one can estimate the background rate, it may be possible to establish the excess over other known sources. However, because this method for the identification of the Higgs particle is not a direct reconstruction of the particle by an invariant mass, it may be difficult to prove that this excess originates from the production of a new particle, *i.e.* the charged Higgs particle.

In summary, if $m_{H^{\pm}} > m_t$, it will be very difficult to observe a signal for the charged Higgs boson at a hadron collider via the decay mode $H^{\pm} \to \tau\nu$; but the present studies indicate that it should be possible to find evidence for the charged Higgs if $m_{H^{\pm}} \leq m_t$.

D. Detector Requirements for the Charged Higgs Search

We have seen that detection of a charged Higgs boson requires triggering on high p_T isolated charged particles, in this case the τ and the relatively isolated electrons and muons from top quarks. Obviously, we need a detector with good lepton identification power and good momentum resolution for isolated particles. In addition, we have to identify taus through their specific decay modes ($\tau^{\pm} \to \pi^{\pm} + \nu(+\pi^{o}$'s) or $\pi^{\pm}\pi^{-}\pi^{+} + \nu$) and they should be distinguished from e's, μ's and QCD jets. Similar requirements emerge in tagging the leptons from top semi-leptonic decays and our techniques can be applied to the search for the intermediate mass neutral Higgs boson and for heavy quarks or lepto-quarks. We have developed a number of tricks to enhance the distinctive topologies typical of such triggers. However, their implementation places definite demands upon the detector. In this section, we give the basic detector requirements and discuss technical innovations that will be needed to implement them.

1. Tracking Devices

Vertex detector

It is not easy to select taus by using the impact parameter method at SSC, since the impact parameter is relatively small for taus, of $\mathcal{O}(100\ \mu m)$, and it is almost a Lorentz boost invariant variable. Therefore, high momentum does not help to reconstruct the secondary vertex, unless we can find the tau decays in the fiducial volume of the vertex detector. The problem of tagging would be less severe if the decay mode $H^{\pm} \rightarrow t + b$ were large enough to be observed above background, since the vertex detector *could* be used to find secondary vertices from B-hadron decays.

Central tracking device

It is essential to have a tracking device to detect charged Higgs bosons. However, reconstruction of the charged tracks is necessary only in the vicinity ($\Delta R = \sqrt{\Delta\eta^2 + \Delta\phi^2} < 1$) of the isolated large p_T energy clusters or high p_T muon candidates. The magnetic field should not be so high as to disturb the calorimetry. The requirements for the tracking devices are:

(1) Momentum resolution: $\Delta p_T / p_T \approx 0.0005 p_T\ (GeV) \oplus 0.02^*$. The muon momentum must be measured with an accuracy of 10% for $p_T = 200\ GeV$. Since electron identification will require using $p_{TR} \approx E_{EM}$ (where p_{TR} is the momentum of the track measured by the tracking device and E_{EM} is the electromagnetic energy measured by the calorimeter), good resolution will be necessary. Good resolution is also essential in order to select a charged pion from tau decay ($\tau^{\pm} \rightarrow \pi^{\pm} + \nu\ (+\pi^{o}\text{'s})$), by efficiently rejecting isolated electrons, for example, from W-boson decays.

(2) Double track resolution: The three charged pions from $\tau^{\pm} \rightarrow A_1^{\pm} + \nu \rightarrow \pi^{\pm}\pi^{-}\pi^{+} + \nu$ must be reconstructed as three separated charged tracks.

(3) Rapidity range: Not optimized yet, but we need at least $|\eta| < 2.5$.

(4) Pattern recognition: This is related to double track resolution but is a more complicated issue. We do not have to reconstruct all the tracks but charged particles from isolated taus and leptons from top quarks must be reconstructed. Even for these cases, we have to deal with a high multiplicity of charged particles in a narrow cone.

In order to have a cheap and fast tracking device with both good momentum resolution and powerful pattern recognition ability, we recommend using a combination of two devices. First, a small number of expensive layers with very good position accuracy ($\sigma \approx \mathcal{O}(10\mu m)$) separated by large distances can provide good momentum resolution. However, it is difficult to connect the hit points if the layers are far from each other. For connecting the hits between the expensive layers, a second, cheaper, device with a large number of layers may facilitate

* $a \oplus b$ means quadratic sum of a and b, *i.e.* $\sqrt{a^2 + b^2}$.

pattern recognition. A candidate for the expensive layers might be a silicon strip detector, assuming it can be made to work in such an enviornment, while the cheaper layers could be straw chambers. Possible problems are that such a tracking detector might require more material than a conventional chamber and that it could be difficult to make radiation hard.

2. Calorimetry

Calorimetry is essential to look for isolated energy clusters and for determining the missing p_T at an early stage of the on/off-line analysis. The calorimeter must be designed to have a fast high p_T isolated energy trigger with or without accompanying missing p_T.

EM-Calorimeter

(1) Energy resolution: $\Delta E/E \approx 0.15/\sqrt{E(GeV)} \oplus 0.02$ is sufficient.

(2) Segmentation: The lateral segmentation of the electromagnetic and hadronic parts should be matched. A lateral cell size of $\Delta\eta \times \Delta\phi$ which is at least 0.03×0.03 is needed to separate the isolated energy clusters. Longitudinal segmentation is necessary for e/π separation, especially to identify the charged pions associated with π^o's from electrons. For example, the decay products of a charged rho arising from $\tau^\pm \rightarrow \rho^\pm \nu \rightarrow \pi^\pm \pi^o \nu$ should be efficiently distinguished from electrons and from other hadronic jets, *e.g.* by examining the longitudinal shower pattern.

(3) Hermeticity: The electromagnetic calorimeter should be hermetic over as large an $|\eta|$ range as possible (say, $|\eta|_{max} \approx 5$), so that missing p_T of $\mathcal{O}(70-80\ GeV)$ can be measured.

Hadron-Calorimeter

(1) Energy resolution: $\Delta E/E \approx 0.5/\sqrt{E(GeV)} \oplus 0.02$ is sufficient.

(2) Segmentation: The lateral segmentation of the electromagnetic and hadronic parts of the calorimeter should be matched. Therefore, a lateral cell size of $\Delta\eta \times \Delta\phi$ of 0.03×0.03 is required for the first few layers of the hadronic calorimeter. The longitudinal segmentation must be designed to have good μ-identification.

(3) Hermeticity: The hadronic calorimeter should be hermetic over as large an $|\eta|$ range as possible (say, $|\eta|_{max} \approx 5$), in order to measure missing p_T over the same region as the electromagnetic calorimeter.

3. Lepton identification efficiencies and hadron rejection factors

We have not evaluated the necessary lepton identification efficiency and the hadron rejection factor for electrons and muons. In any case, compared with the requirements for particle searches with good signal to background ratio, we need even better identification efficiency for isolated leptons and greater hadron rejection power. Charged pions (associated with π^0's) from isolated taus must be distinguished from isolated electrons and muons.

E. The Supersymmetric Two-Higgs-Doublet Model and Beyond

In this report, we have endeavored to discuss the consequences of a non-minimal Higgs sector, in the framework of the Standard Model, without further theoretical assumptions. However, as stressed in the Introduction, the number of new parameters increases rapidly as additional Higgs doublets are added. In this section, we wish to examine briefly the consequences of "low-energy" supersymmetry for the phenomenology of the Higgs sector. The advantages of imposing such a theoretical framework are twofold. First, supersymmetry imposes strong constraints on the form of the Higgs potential, thereby reducing the number of free parameters and providing more predictive power. Second, supersymmetry may be the only consistent theory which contains weakly coupled Higgs bosons (with mass of order m_Z) and can explain the origin of the electroweak scale.

We shall briefly describe the main features of the Higgs sector in the minimal supersymmetric extension of the Standard Model. Details of the model can be found in refs. 17 and 18. [Many of our conclusions below continue to hold in non-minimal supersymmetric models; see ref. 6.] The notation for the Higgs bosons will be the same as introduced earlier. The effect of the supersymmetry is to introduce relations between the various parameters; the end result is that two parameters suffice to determine all Higgs masses and nearly all of their couplings. Here, we shall take $\tan\beta$ and $m_{H_3^0}$ as the free parameters. The minimal supersymmetric model also has the property that the Higgs potential is automatically CP invariant. Furthermore, we are free to choose the phases of the scalar fields so that the vacuum expectation values are real and positive. Hence, we will choose $0 \le \beta \le \pi/2$. The other (tree-level) masses are then given by:

$$m_{H^+}^2 = m_{H_3^0}^2 + m_W^2 \tag{12}$$

$$m_{H_1^0,H_2^0}^2 = \frac{1}{2}\left[m_{H_3^0}^2 + m_Z^2 \pm \sqrt{(m_{H_3^0}^2 + m_Z^2)^2 - 4m_Z^2 m_{H_3^0}^2 \cos^2 2\beta}\right], \tag{13}$$

where, by definition, $m_{H_2^0} \le m_{H_1^0}$. Note that eqs. (12) and (13) imply that $m_{H^+} \ge m_W$, $m_{H_1^0} \ge m_Z$, and

$$m_{H_2^0} \le m_Z \cos 2\beta \le m_Z. \tag{14}$$

The result of eq. (14) is remarkable, in that it guarantees that the theory must possess at least one light Higgs boson. Unless $\cos 2\beta$ is near its maximum of 1, this relation implies that the lightest scalar Higgs will be observable at SLC, LEP or LEP-II.

The one additional parameter which is determined is the scalar Higgs mixing angle α. Using the definition given in ref. 17, it turns out that α is constrained to lie in the range $-\pi/2 \le \alpha \le 0$. Explicitly, we have:

$$\cos 2\alpha = -\cos 2\beta \left(\frac{m^2_{H^0_3} - m^2_Z}{m^2_{H^0_1} - m^2_{H^0_2}} \right); \tag{15}$$

$$\sin 2\alpha = -\sin 2\beta \left(\frac{m^2_{H^0_1} + m^2_{H^0_2}}{m^2_{H^0_1} - m^2_{H^0_2}} \right). \tag{16}$$

Probably one of the most interesting implications of the minimal supersymmetric model is obtained by examining the coupling of the heavy Higgs scalar (H^0_1) to vector boson pairs. We already know from eq. (3) that this coupling will be suppressed compared to the minimal Higgs model coupling. In fact, in a general two-Higgs doublet model, the suppression factor turns out to be $\cos(\beta - \alpha)$. In the minimal supersymmetric model, this factor is given by:

$$|\cos(\beta - \alpha)| = \left[\frac{m^2_{H^0_2}(m^2_Z - m^2_{H^0_2})}{(m^2_{H^0_1} - m^2_{H^0_2})(m^2_{H^0_1} + m^2_{H^0_2} - m^2_Z)} \right]^{\frac{1}{2}}. \tag{17}$$

If one computes $|\cos(\beta - \alpha)|$ over the allowed range of parameters, one quickly sees that it is a very small number, except for a small range where $m_{H^0_1} \to m_Z$. In particular, for the so-called "heavy" Higgs range ($m_{H^0_1} \ge 2m_W$), we find that $|\cos(\beta - \alpha)|$ never exceeds 0.15; and for heavier masses, it goes to zero like $1/m^2_{H^0_1}$. Thus, in the supersymmetric model, the H^0_1 totally decouples from the theory in the limit that its mass gets large, and H^0_2 becomes identical to the minimal Standard Model Higgs (as is evident from eq. (3)). It is this same suppression factor that also enters the $H^+ \to W^+ H^0_2$ and $H^+ \to W^+ H^0_2 \gamma$ modes discussed earlier.

Based on the discussion above, the consequences for the search for neutral Higgs bosons at the SSC are twofold. First, a light Higgs scalar may already have been discovered before the SSC turns on. (We would argue that such a discovery would be the first experimental evidence for supersymmetry!) Second, the heavy Higgs scalar will be extremely difficult to observe, due to its suppressed decay rate into vector boson pairs. (The pseudoscalar has no tree-level couplings to vector boson pairs, and is therefore just as hard to detect.) Assuming that the dominant decay of such heavy Higgses is into $t\bar{t}$ pairs, we know of no technique for observing these Higgs at the SSC, either through their $t\bar{t}$ decays or through rare decays into non-supersymmetric particles. The decays $H^0_1 \to H^0_3 Z^0$ and $H^0_3 \to H^0_1 Z^0$ are kinematically forbidden (see eq. (13)), whereas the production of $H^0_3 H^0_2$ by virtual Z^* exchange and the decay $H^0_3 \to H^0_2 Z^0$ are suppressed in amplitude by $\cos(\beta - \alpha)$. The latter decay rate, and branching ratios into $\gamma\gamma$, γZ^0, and $\gamma\Theta$ are too small to be viable signatures.[3] (Of course, if $m_{H_3^0}$ or $m_{H_1^0}$ happens to be smaller than $2m_t$, then detection in the $\gamma\gamma$ mode is likely to be possible.) The difficulty of charged Higgs detection is already apparent before the introduction of supersymmetry. As discussed in Section B, in the minimal supersymmetric model, certain rare decay modes which might have provided a useful signature are even further reduced due to the appearence of suppression factors like $\cos(\beta - \alpha)$ (*e.g.*, see eq. (8)).

However, before concluding that the picture is totally bleak, it is important to realize that a new feature is present. Because we are discussing a supersymmetric theory, there are new supersymmetric particles in the spectrum, which can couple to the Higgs bosons. In particular, it is possible that new Higgs decay modes into supersymmetric final states will be available which will radically alter Higgs phenomenology at the SSC. In a supersymmetric model, Higgs bosons can decay either into squark and slepton pairs, or into charginos and neutralinos (these are the mass eigenstates comprising the gauginos and higgsinos). If the relevant decays are kinematically allowed, then the corresponding branching ratios can be large. Indeed, over a fairly large region of the supersymmetric model parameter space, one finds total branching ratios into supersymmetric final states which are larger than 10%, and can easily approach 100%. As an example, we show the Higgs branching ratios into chargino and neutralino final states as a function of the supersymmetric parameters, taken from ref. 19. in fig. 4. For reasonable choices of certain supersymmetric model parameters, M and μ, described in detail in refs. 17, 18, and 19, we see in fig. 4 that a charged Higgs with mass of order 500 GeV can decay more than 80% of the time into chargino and neutralino modes, even when the tb decay channel is kinematically allowed.

The relatively large branching ratios into charginos and neutralinos can be explained by the fact that the relevant mass parameter which scales the Higgs couplings to these particles is m_W, m_Z, and the parameters of the neutralino and chargino mass matrices, which are presumably of the same order. However, unlike the coupling of H_1^0 to WW and ZZ, the Higgs couplings to the neutralinos and charginos are not suppressed, in general. As a result, it is not surprising that these final states can be dominant. The decay into squarks and sleptons can, in principle, also be an important fraction of the total Higgs widths. In evaluating various possible scenarios, we note that it seems more probable that some light charginos or neutralinos exist which would be accessible to Higgs decay. On the other hand, the general mass scale which controls the squark and slepton masses (and is *a priori* unrelated to the neutralino and chargino parameters) may be large enough so that Higgs decay into squark and sleptons would be forbidden. Clearly, no one can definitively predict, at present, which supersymmetric final states (if any) will dominate.

Supersymmetric decays of the Higgs present the possibility of completely novel signatures for Higgs searches. In particular, events with substantial missing transverse energy will now play an important role in the search for Higgs bosons. Previously, missing transverse energy was relevant in Higgs searches only in the search for W bosons in the final state which decayed leptonically or Z bosons

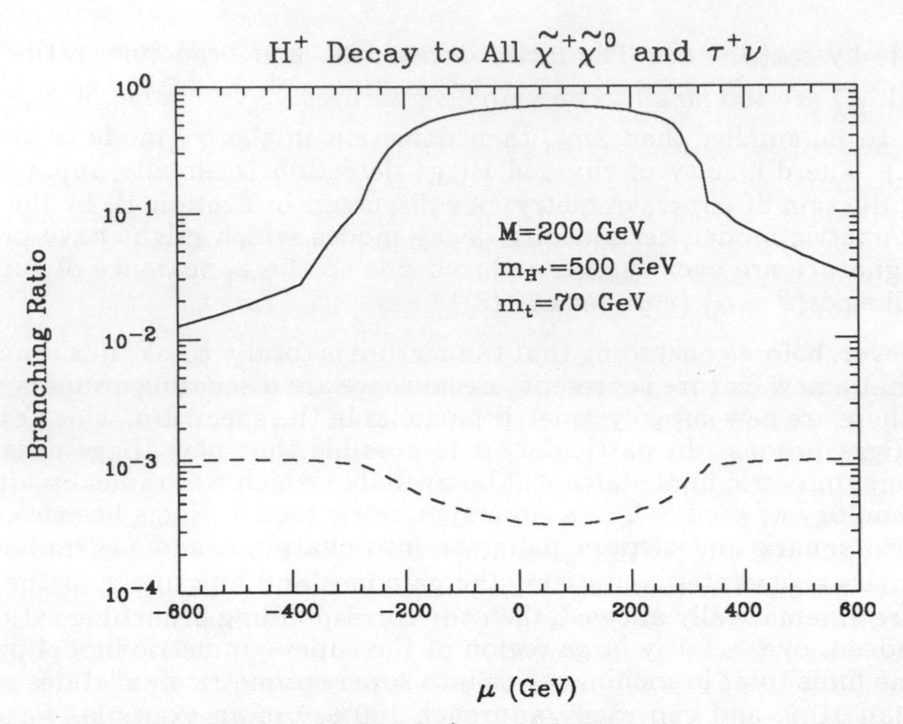

Figure 4: The branching ratio for $H^{\pm}$ to decay to any channel containing a neutralino-chargino pair, compared to the $\tau\nu$ branching ratio. We take $m_t = 70\ GeV$, $\tan\beta = 1.5$ and $m_{H^{\pm}} = 500\ GeV$, and have chosen a reasonable value of $M = 200\ GeV$ for the gaugino mass parameter of the model. We plot the branching ratio as a function of the higgsino mass parameter, μ. The curves are: solid, sum over all neutralino+chargino channels; and dashes, $\tau\nu$.

which decayed to $\nu\bar{\nu}$. In supersymmetric decays, it is very easy to generate large missing transverse energy without having a high-p_T lepton in the event. To fully assess the possibility of detecting various supersymmetric final states of the Higgs will require a substantial Monte Carlo effort. A sample study [20] was performed at Snowmass 1986, for a specific choice of parameters for which the neutralinos are very light, with encouraging conclusions. However, a systematic survey of the minimal supersymmetric model parameter space, along with appropriate Monte Carlo studies, is highly desirable. This is work which we hope will be undertaken in the near future. Clearly, if low-energy supersymmetry is correct, Higgs physics will become a branch of supersymmetry phenomenology.

F. Conclusions of the Non-Standard Higgs Working Group

In summary, we have examined the phenomenological consequences of the two Higgs doublet model relevant for the SSC. In the framework of the Standard Model, we have come to the conclusion that it may be very difficult to find evidence for a non-minimal Higgs sector. Among the neutral scalar Higgs, only those which couple strongly to WW and ZZ can be easily discovered. The techniques for discovery are identical to those used to detect the minimal Higgs. The neutral pseudoscalar Higgs does not couple to vector boson pairs at tree level. Thus its detection presents problems analogous to those encountered when looking for the "intermediate mass" minimal Higgs. In general, no "general purpose" strategy exists at present for such Higgses at a hadron collider. Only if the t quark is heavy, and the $t\bar{t}$ decay mode is forbidden, is there some hope for detection via rare decay modes. Primary among these is the the $\gamma\gamma$ mode discussed at length in ref. 3. For neutral Higgs bosons that have weak or vanishing couplings to vector boson pairs, the $\gamma\gamma$ mode should be usable for any Higgs boson with mass $\gtrsim 100\ GeV$ and $\leq 2m_t$. The general problem of detecting such neutral Higgs has much overlap with the work of the Intermediate Mass Higgs Working Group.[5] When a neutral Higgs has mass above $2m_t$, and its decay is dominated by the $t\bar{t}$ channel, we have been unable to develop a technique for discovering it at the SSC. The charged Higgs boson presents similar problems to those encountered for the neutral Higgs. Again, the t-quark mass is crucial. If $m_t > m_{H^\pm} + m_b$, then t-quark decays will provide a copious source of charged Higgs, and the $\tau\nu$ decays of the $H^\pm$ will have substantial branching ratio. We have seen that charged Higgs detection is possible in this case. But, if $m_{H^\pm} > m_t + m_b$, then the $H^\pm$ has a smaller production cross section and its decays will be dominated by the tb final state. The possibility of using rare decays to reduce the large backgrounds can still be considered; again, the most plausible decay of this type which might be observable at the SSC is the decay $H^+ \to \tau\nu$. Our Monte Carlo analysis of this scenario is not encouraging.

However, it is natural to go beyond the Standard Model framework when considering an extended Higgs sector. Indeed, it is probably true that the only sensible theoretical framework in which weakly coupled elementary Higgses can exist is "low-energy" supersymmetry. We have examined the consequences of the supersymmetric two-Higgs-doublet model. There are three general predictions. First, a neutral scalar Higgs boson with mass less than $\mathcal{O}(m_Z)$ almost certainly exists. Second, the heavy scalar neutral Higgs couples very weakly to vector boson pairs; in the absence of non-Standard Model decay modes this implies that its observability at the SSC is problematical. Finally, the widths of the charged and the heavy neutral Higgses may be dominated by the decay into supersymmetric final states. Large regions of the supersymmetric parameter space exist where the decays into neutralinos and charginos are dominant. This presents us with the possibility of new search strategies, involving missing transverse energy signatures, for the heavy Higgses at the SSC. A detailed appraisal and Monte Carlo study of such scenarios awaits future work.

Acknowledgements

This work was partially supported by the Department of Energy. We would like to thank the members of the Madison conference working group (see ref. 6) who participated in the earlier stages of this effort.

REFERENCES

1. S. Glashow and S. Weinberg, *Phys. Rev.* **D15**, 1958 (1977).
2. J.A. Grifols and A. Mendez, *Phys. Rev.* **D22**, 1725 (1980); G. Keller and D. Wyler, *Nucl. Phys.* **B274**, 410 (1986).
3. J. Gunion, G.L.Kane and J. Wudka, preprint UCD-87-28 (September, 1987).
4. R. Cahn, M. Chanowitz, M. Gilchriese, M. Golden, J. Gunion, M. Herrero, I. Hinchliffe, F. Paige, and E. Wang, to appear in *Proceedings of the 1987 Berkeley Workshop on "Experiments, Detectors and Experimental Areas for the Supercollider"*.
5. D.M. Atwood, J.E. Brau, J.F. Gunion, G.L. Kane, R. Madaras, D.H. Miller, L.E. Price, and A.L. Spadafora, "Intermediate Mass Higgs Boson(s)", to appear in *Proceedings of the 1987 Berkeley Workshop on "Experiments, Detectors and Experimental Areas for the Supercollider"*.
6. J.F. Gunion and H.E. Haber, "Probing the Higgs Sector at the SSC: the Standard Model and Beyond", *Proceedings of the 1987 Madison Workshop on "From Colliders to Supercolliders"*, *Int. J. Mod. Phys.* **A2**, 957 (1987).
7. H.E. Haber, D.E. Soper and R.M. Barnett, in *Physics Simulations at High Energies*, Proceedings of the 1986 Madison Workshop, edited by V. Barger, T. Gottschalk, and F. Halzen (World Scientific, Singapore, 1987) p. 425, and Oregon preprint, OITS-365 (1987).
8. J.F. Gunion, H.E. Haber, F.E. Paige, Wu-Ki Tung, and S.S.D. Willenbrock, *Nucl. Phys.* **B294**, 621 (1987).
9. J.F. Gunion and D. Millers, work in progress.
10. J.A. Grifols, J.F. Gunion, and A. Mendez, *Phys. Lett.* **197B**, 266 (1987).
11. See the reports by: J.F. Gunion *et al.*, p. 103, *Proceedings of the 1984 Snowmass Workshop on the Design and Utilization of the Superconducting Super Collider*, edited by R. Donaldson and J. G. Morfin; F.J. Gilman and L. Price, in *Proceedings of the 1986 Snowmass Workshop on the Design and Utilization of the Superconducting Super Collider*, edited by R. Donaldson and J. Marx, p. 185; and the Intermediate Mass Higgs group, ref. 5.
12. H-U Bengtsson, S. Komamiya, and H. Yamamoto, preprint SLAC-PUB-5369 (1987), to appear in *Proceedings of the 1987 Madison Workshop on "From Colliders to Supercolliders"*.
13. T. Barklow, in *Proceedings of the Second MARK-II Workshop on SLC Physics*, SLAC-Report-306 (1986), edited by G. Feldman, p. 189.
14. T. Sjostrand and M. Bengtsson, *Compt. Phys. Commun.* **43**, 367 (1987).
15. J.F. Gunion and H.E. Haber, 'τ Decay Spectra at the SSC', in *Proceedings of the 1984 Snowmass Workshop on the Design and Utilization of the Superconducting Super Collider*, edited by R. Donaldson and J. G. Morfin, p.150.

16. F. Halzen and C.S. Kim, *Proceedings of the 1987 Madison Workshop on "From Colliders to Supercolliders"*, have also computed the $gb \to W^{\pm}t$ cross section, including some QCD corrections.

17. J.F. Gunion and H.E. Haber, *Nucl. Phys.* **B272**, 1 (1986).

18. J.F. Gunion and H.E. Haber, *Nucl. Phys.* **B278**, 449 (1986).

19. J.F. Gunion, H.E. Haber, *et al.*, 'Decays of Higgs Bosons to Neutralinos and Charginos in the Minimal Supersymmetric Model: Calculation and Phenomenology', *Proceedings of the 1987 Madison Workshop on "From Colliders to Supercolliders"*, *Int. J. Mod. Phys.* **A2**, 1035 (1987); and J.F. Gunion and H.E. Haber, in preparation.

20. R.M. Barnett, J.A. Grifols, A. Mendez, J.F. Gunion, and J. Kalinowski, 'Detection of a Heavy Neutral Higgs Boson in a Higgsino-Neutralino Decay Mode', in *Proceedings of the 1986 Snowmass Workshop on the Design and Utilization of the Superconducting Super Collider,* edited by R. Donaldson and J. Marx, p. 188.

Detection of a Charged Higgs via the Decays $H^{\pm} \to \Upsilon W^{\pm}$ and $H^{\pm} \to \Theta W^{\pm}$

J.A. GRIFOLS

Dept. Fisica, UAB, 08193 Bellaterra (Barcelona), Spain

J.F. GUNION

Department of Physics, U.C. Davis, Davis CA 95616

A. MENDEZ*

Lawrence Berkeley Laboratory, Berkeley CA 94720

Abstract

We compute the branching ratios for a charged Higgs to decay to the final states $\Upsilon W^{\pm}$ and $\Theta W^{\pm}$. We assess signal and background rates for a high energy hadron collider.

* Permanent address: Dept. Fisica, UAB, 08193 Bellaterra (Barcelona), Spain

A great deal of effort has been focused on the detection of the single physical neutral Higgs boson arising in the one-doublet version of the standard $SU(2) \times U(1)$ model. However, even in the context of the SM, the minimal choice of one Higgs doublet is somewhat arbitrary. Additional doublets can be introduced without conflicting with the ρ parameter measurement of $\rho \approx 1$. The addition of a second $SU(2)$ Higgs doublet increases the physical Higgs boson degrees of freedom to 5: two neutral scalar Higgs; one neutral pseudoscalar Higgs; and a charged Higgs boson pair. It is the latter charged Higgs bosons that we wish to focus on here. Further motivation for the introduction of a second Higgs doublet is provided by the minimal supersymmetric extension of the SM, which requires two Higgs doublets—one to give masses to the up-type quarks and the second to give masses to the down-type quarks. In general, observation of a charged Higgs boson would be a definitive signature for a unified gauge theory that goes beyond the Standard Model.

However, detection of a charged Higgs boson at a hadron collider is non-trivial. First, we note that there are no tree level couplings for the decays $H^\pm \to W^\pm Z$ and $H^\pm \to W^\pm \gamma$.[1] Thus, even for large values of $m_{H^\pm}$, the charged Higgs will always decay predominantly to the heaviest fermionic channel available. (We neglect in this paper the possibility of supersymmetric decay modes that could compete with a heavy SM fermion channel.[2][3]) However, the QCD backgrounds to such a channel are expected to be very large. Thus, one must turn to rare decay modes of the charged Higgs. The first obvious possibility is the channel $H^\pm \to \tau^\pm \nu$. This was first considered in ref. 3. Here, we wish to consider the possibility of observing the charged Higgs through its decays to $\Upsilon W^\pm$ or $\Theta W^\pm$. These branching ratios are very sensitive to the top quark mass, as we shall see. There is also an important parameter $\tan\beta = v_2/v_1$ (where v_2 and v_1 are the vacuum expectation values of the Higgs doublets that give mass to the up and down quarks, respectively) that enters into the coupling of the charged Higgs to the tb channel. The $H^\pm tb$ coupling takes the form:

$$\frac{g}{2\sqrt{2}m_W}\left[m_t(1+\gamma_5)\cot\beta + m_b(1-\gamma_5)\tan\beta\right]. \tag{1}$$

For the purposes of our numerical calculations here we shall take $\tan\beta = 1$, but it should be kept in mind that many models yield values of $\tan\beta$ that are larger than 1, though rarely larger than 2. Using non-relativistic bound state techniques[4] we find the formulas for the two decay widths given below.

$$\Gamma(H^\pm \to \Upsilon W^\pm) = |R_\Upsilon(0)|^2 \frac{3\alpha^2\cot^2\beta}{x_W^2} \frac{m_\Upsilon m_t^4 \eta^{1/2}}{m_{H^\pm}^3 m_W^2\left[2m_{H^\pm}^2 + 2m_W^2 - m_\Upsilon^2 - 4m_t^2\right]^2} \times$$
$$(3 + \frac{\eta}{4m_\Upsilon^2 m_W^2}), \tag{2}$$

where $\eta = \lambda(m_{H^\pm}^2, m_\Upsilon^2, m_W^2)$, with $\lambda(a,b,c) = (a-b-c)^2 - 4bc$, $x_W = \sin^2\theta_W$, and $R_\Upsilon(0)$ is the value at the origin of the $b\bar{b}$ bound state wave function and is related to the experimental value of the width of $\Upsilon \to e^+e^-$ by

$$|R_\Upsilon(0)|^2 = \frac{9}{4\alpha^2} m_\Upsilon^2 \Gamma_{\Upsilon \to e^+e^-}. \tag{3}$$

$$\Gamma(H^\pm \to \Theta W^\pm) = |R_\Theta(0)|^2 \frac{3\alpha^2 \cot^2\beta}{64 x_W^2} \frac{m_\Theta \kappa^{1/2}}{m_{H^\pm}^3 m_W^2 \left[2m_{H^\pm}^2 + 2m_W^2 - m_\Theta^2\right]^2} \times$$
$$\left[16(m_{H^\pm}^2 - m_W^2)^2 - m_\Theta^2(10m_{H^\pm}^2 + 7m_W^2 - 2m_\Theta^2) + \frac{m_\Theta^2}{m_W^2}(m_{H^\pm}^2 - m_\Theta^2)^2\right], \tag{4}$$

where $\kappa = \lambda(m_{H^\pm}^2, m_\Theta^2, m_W^2)$, and $R_\Theta(0)$ is now the $t\bar{t}$ bound state wave function value at the origin. For numerical estimates, we take the Coulomb type expression

$$|R_\Theta(0)|^2 = 4\left(\frac{m_\Theta \alpha_s}{3}\right)^3. \tag{5}$$

The resulting decay widths are presented in Figs. 1 and 2. For the $\Upsilon W^\pm$ channel we have plotted the branching ratio for $H^\pm \to \Upsilon W^\pm$ relative to that for $H^\pm \to \tau^\pm \nu$, where the width for the latter decay is given by:

$$\Gamma(H^\pm \to \tau^\pm \nu) = \frac{G_F}{4\pi\sqrt{2}} m_\tau^2 m_{H^\pm} \left(1 - \frac{m_\tau^2}{m_{H^\pm}^2}\right)^2 \tan^2\beta. \tag{6}$$

We consider only $m_{H^\pm}$ and m_t values such that the $t\bar{b}$ channel is kinematically disallowed, so that the $\tau^\pm\nu$ branching ratio will be substantial. (The latter is, however, model dependent; when the $t\bar{b}$ channel is not allowed it varies from $\sim 30\%$ in the two-doublet SM extension at $\tan\beta = 1$ to as high as 50-90% in the minimal supersymmetric model.[3]) From the figure we see that in this situation the $\Upsilon W^\pm$ decay mode can be significant. In contrast, the $H^\pm \to \Theta W^\pm$ branching ratio is always extremely small, since the $H^+ \to t\bar{b}$ channel is always open if the former decay is allowed. Thus we do not feel that this latter decay can be used for detection of the charged Higgs.

Returning to the case in which the $H^+ \to t\bar{b}$ channel is not allowed, and the $H^\pm \to \Upsilon W^\pm$ channel has a reasonable branching ratio, it is of interest to determine if this decay can be used to detect the charged Higgs at a hadronic collider. There are two means by which the $H^\pm$ can be produced in hadronic collisions. These are:

1. $gg \to t\bar{t}$ followed by t decays to $H^\pm$. The $t \to H^+b$ decay mode is generally fully competitive with the $t \to W^+b$ channel,[5] with $BR(t \to H^+b)/BR(t \to W^+b) \to \cot^2\beta$ at large t mass. There is even a possibility that the t could be lighter than the W^+ but heavier than the H^+, in which case the t would decay almost exclusively to H^+. The event rates for $t\bar{t}$ production are, of course, very large for t masses below 250 GeV. For instance, at $m_t = 250\ GeV$ there will be $> 10^8$ $t\bar{t}$ pairs produced at the SSC for standard $L = 10^4 pb^{-1}$ yearly luminosity at $\sqrt{s} = 40\ TeV$. For purposes of illustration let us also adopt a definite value of $m_{H^\pm} = 150\ GeV$ and take $\tan\beta = 1$. Then the t decays 70% of the time to a W^+ and 30% of the time to a H^+.[6] From Fig. 1, assuming the SM-like $\tan\beta = 1$ result of $BR(H^\pm \to \tau^\pm\nu) \approx .3$, we find a $H^\pm \to \Upsilon W^\pm$ branching ratio of .0008. Let us also demand that the Υ decay to e^+e^- or $\mu^+\mu^-$ with net branching ratio of .06, and that the $W^\pm$ on both sides of the event decay to $e\nu$ or $\mu\nu$, with branching ratio of 2/9. The result is > 100 $t\bar{t}$ events with a final state containing two b jets, two leptonically decaying W's, and a leptonically decaying Υ. It is difficult to imagine significant backgrounds to such a final state.
2. $gb \to tH^\pm$, which is the two-to-two process for which the leading pole approximation can be thought of as tb fusion to the $H^\pm$. The cross section for this process was computed in ref. 3, and found to be quite significant at low to moderate $H^\pm$ masses. We have computed the event rates of $H^\pm$ production in this mode, followed by $H^\pm$ decay to $\Upsilon W^\pm$, with Υ decaying to e^+e^- or $\mu^+\mu^-$ and $W^\pm$ decaying to $e\nu$ or $\mu\nu$, for a few representative m_t and $m_{H^\pm}$ cases. Typically, when the top mass is near the minimum value for which the tb decay channel of the $H^\pm$ is closed, the event rate is of order 75 to 150 per SSC year. The event rate decreases as m_t increases at fixed $m_{H^\pm}$, becoming of order 30 to 40 events at $m_t = 100, 175, 250\ GeV$ for $m_{H^\pm} = 100, 150, 200\ GeV$, respectively. Since it is likely[5] [3] that the t quark that is a spectator to this $H^\pm$ production mode can be triggered on, backgrounds must derive from mixed QCD/Electroweak processes that produce a $t\Upsilon W^\pm$ final state. We have computed the irreducible background arising from the process $gb \to t\Upsilon W^\pm$ and found that it is completely negligible (typically $\lesssim .01$ event per SSC year).

Thus, we conclude that $H^\pm$ detection in the exotic $\Upsilon W^\pm$ mode will be quite free of backgrounds in all cases where it is useful. In contrast, the $\Theta W^\pm$ decay does not appear to provide a signficant event rate since the $H^\pm \to tb$ channel is always open if the $\Theta W^\pm$ decay is allowed.

Acknowledgements

We would like to thank the LBL Workshop on "Experiments, Detectors, and Experimental Areas for the SSC" for hospitality during the course of this work. This work was partially supported by the Department of Energy, by CAICYT of Spain, and the U.S.-Spain Joint Committee for Scientific and Technological

Cooperation. One of us (A.M.) acknowledges financial support from the theory group at LBL.

REFERENCES

1. J.A. Grifols and A. Mendez, *Phys. Rev.* **D22**, 1725 (1980).
2. J.F. Gunion and H.E. Haber, *Nucl. Phys.* **B272**, 1 (1986); *ibid.* **B278**, 449 (1986).
3. J.F. Gunion, H.E. Haber, F.E. Paige, Wu-Ki Tung, and S.S.D. Willenbrock, preprint UCD-86-15, submitted to *Nucl. Phys.*.
4. B. Guberina, J.H. Kuhn, R.D. Peccei and R. Ruckl, *Nucl. Phys.* **B174**, 317 (1980); R. Baier and R. Ruckl, *Z. Phys.* **C19**, 251 (1983); R. Gastmans, W. Troost and T.T. Wu, *Phys. Lett.* **B184**, 257 (1987).
5. See J.F. Gunion and H.E. Haber, "Probing the Higgs Sector at the SSC: the Standard Model and Beyond", preprint UCD-87-16, to appear in *Proceedings of the 1987 Madison Workshop on Colliders to Supercolliders.*
6. An expression for the relative decay rates appears in ref. 5.

FIGURE CAPTIONS

1) We plot the branching ratio for $H^\pm \to \Upsilon W^\pm$ relative to the branching ratio for $H^\pm \to \tau^\pm \nu$ for several $m_{H^\pm}$ masses, as indicated on the figure, as a function of the top quark mass, m_t. We have taken $\tan\beta = 1$. We consider only $m_{H^\pm}$ and m_t values such that the $t\bar{b}$ channel is not kinematically allowed. Otherwise the overall branching ratio for $H^\pm \to \Upsilon W^\pm$ will be very small and this channel not useful.
2) We plot the branching ratio for $H^\pm \to \Theta W^\pm$ for several $m_{H^\pm}$ masses, as indicated on the figure, as a function of the top quark mass, m_t. In this case, the $t\bar{b}$ channel is always open and dominates the $H^\pm$ decay width. We have taken $\tan\beta = 1$.

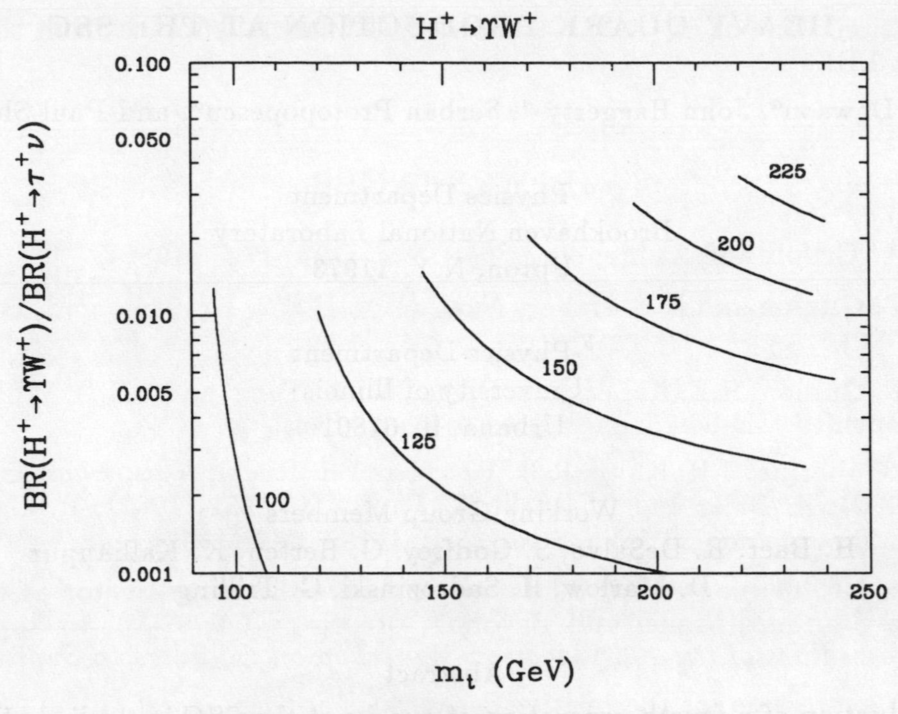

Figure 1

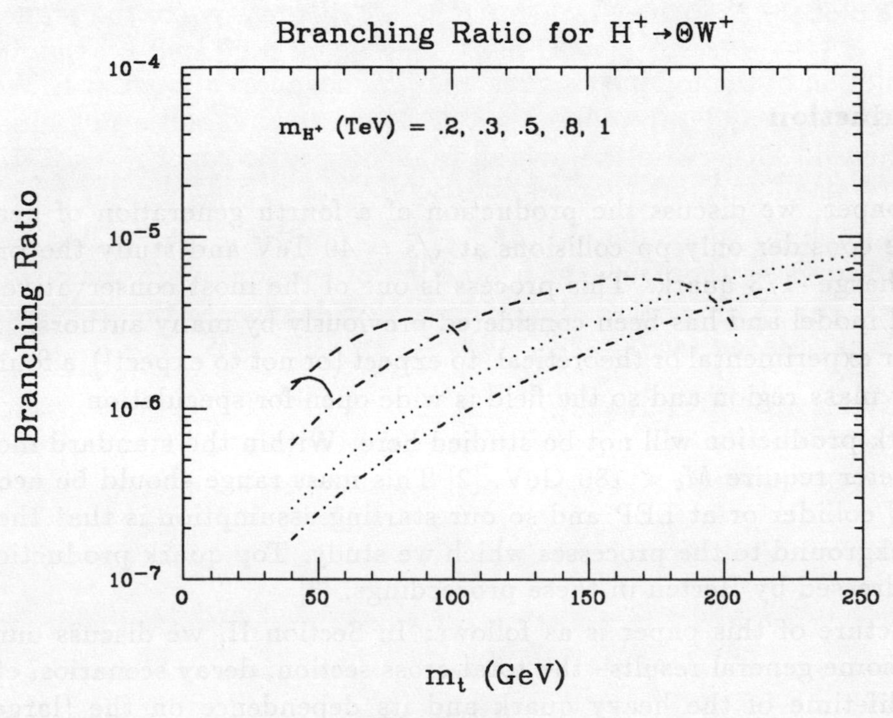

Figure 2

HEAVY QUARK PRODUCTION AT THE SSC

Sally Dawson[a], John Haggerty [a], Serban Protopopescu[a], and Paul Sheldon[b]

[a] Physics Department
Brookhaven National Laboratory
Upton, N.Y. 11973

[b] Physics Department
University of Illinois
Urbana, Il. 61801

Working Group Members
H. Baer, R. DeSalvo, S. Godfrey, G. Herten, K. Kallianpur
D. Marlow, H. Sadrozinski, G. Trilling

Abstract

The production of a fourth generation of quarks at the SSC is studied. We emphasize the properties of the events which are relevant for detector design. Quarks up to masses of 1 TeV should be observable via their two lepton decay modes without placing stringent demands on the detector.

I. Introduction

In this paper, we discuss the production of a fourth generation of heavy quarks at the SSC. We consider only pp collisions at $\sqrt{s} = 40$ TeV and study the production and decay of a charge -1/3 quark. This process is one of the most conservative extensions of the standard model and has been considered previously by many authors. [1] There is no reason, either experimental or theoretical, to expect (or not to expect!) a fourth generation quark in any mass region and so the field is wide open for speculation.

Top quark production will not be studied here. Within the standard model, limits on the ρ parameter require $M_t < 180$ GeV. [2] This mass range should be accessible at the TEVATRON collider or at LEP and so our starting assumption is that the top quark is simply a background to the processes which we study. Top quark production at the SSC has been addressed by Herten in these proceedings. [3]

The structure of this paper is as follows: In Section II, we discuss our assumptions and present some general results– the total cross section, decay scenarios, etc. We briefly discuss the lifetime of the heavy quark and its dependence on the (largely unknown) parameters of the 4 x 4 Kobayashi-Maskawa mixing matrix. Section III focuses on the experimental signals and the detector requirements for the case where the heavy quark decays yield two high transverse momentum leptons in the final state. In Section IV, we discuss the background processes and the reconstruction of the quark mass in this scenario.

We have chosen to focus our attention on final states with two leptons because it appears experimentally feasible to detect this process and it does not seem to place extraordinary requirements on the detector. Decay scenarios with a single high p_T lepton or with multi-muons in the final state are briefly discussed in Section V and in Section VI we present our conclusions.

II. Lifetimes and Cross Sections

We consider a pair of heavy quarks in an $SU(2)_L$ doublet,

$$\begin{pmatrix} U \\ D \end{pmatrix} \tag{2.1}$$

in which U (D) is a fourth generation charge 2/3 (-1/3) quark. The mass difference $| M_U - M_D |$, is restricted by measurements of the ρ parameter in the standard model to be less than 180 GeV at the 90 % confidence level. [2] The absolute values of the masses are limited only by unitarity of the tree-level scattering amplitude. [4] This limit is about 500 GeV (quarks can exist with masses above this value, but tree level scattering amplitudes cannot be calculated reliably in perturbation theory).

By analogy with the third generation of quarks we assume $M_U >> M_D$ and consider only the production of the D quark. If M_U is comparable to M_D, then there is an additional source of D quarks from the decay,

$$U \to W^+ D \tag{2.2}$$

Our results should thus be considered as conservative estimates of the production rates.

Since U and D quarks decay weakly via the real or virtual emission of a W boson, their decays are governed by the KM mixing matrix. Very little is known about the entries in a 4 x 4 KM matrix, with the restrictions coming primarily from unitarity requirements. At the 90 % confidence level[2,5],

$$\begin{aligned} | V_{uD} | &\leq .07 \\ | V_{cD} | &\leq .63 \\ | V_{Ud} | &\leq .14 \\ | V_{Us} | &\leq .62 \end{aligned} \tag{2.3}$$

There are two possible forms for the KM matrix which are experimentally acceptable. [5] In the first case, the KM matrix is approximately diagonal,

$$V_{ij} \simeq \begin{pmatrix} 1 & \theta & \theta^3 & \theta^5 \\ \theta & 1 & \theta^2 & \theta^4 \\ \theta^3 & \theta^2 & 1 & \theta^3 \\ \theta^5 & \theta^4 & \theta^3 & 1 \end{pmatrix} \tag{2.4}$$

and in the second case, the KM matrix has the approximate form,

$$V_{ij} \simeq \begin{pmatrix} 1 & \theta & \theta^3 & \theta^2 \\ \theta & 1 & \theta^2 & \theta \\ \theta^2 & \theta & \theta & 1 \\ \theta^3 & \theta^2 & 1 & \theta \end{pmatrix} \tag{2.5}$$

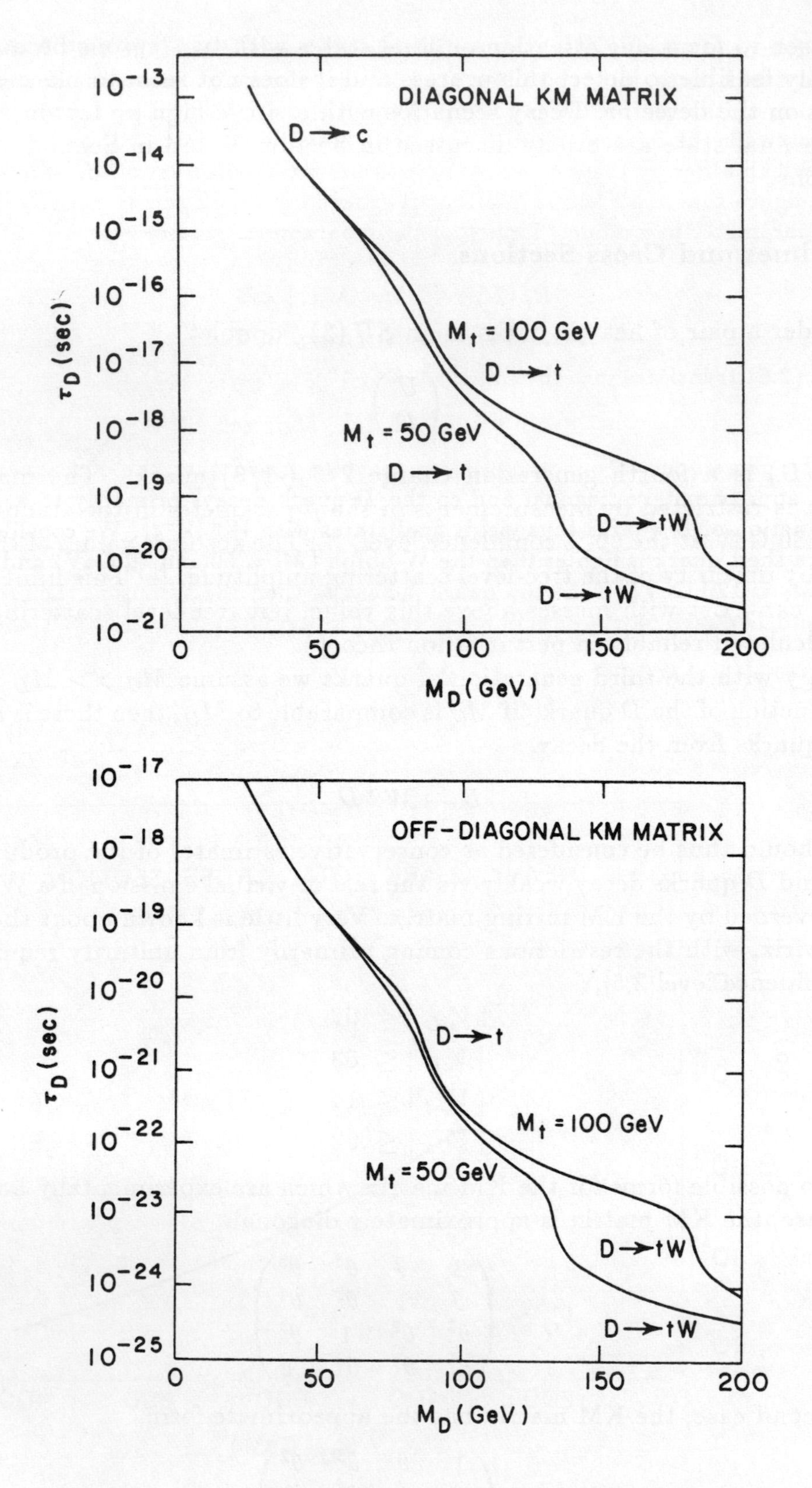

Figure 2.1. D quark lifetime.

The lifetime of a heavy quark depends on the mixing angles of the 4 x 4 KM matrix and on the top quark mass, but typically $\tau < 10^{-15}$ sec. Figure 2.1 shows the D quark lifetime as a function of its mass for the two forms of the KM matrix discussed above, (with central values chosen for the mixing angles). Note that the different forms of the KM matrix give lifetimes which differ by three orders of magnitude for fixed values of M_t and M_D. We see from this figure that the lifetime of a fourth generation quark can essentially be regarded as a free parameter in a model. Limits on the ρ parameter require, (for $M_U, M_D >> M_t$), [5]

$$\begin{array}{c} \Sigma_j \mid V_{jD} \mid^2 \mid M_D/M_W \mid^2 < 5 \\ \Sigma_j \mid V_{Uj} \mid^2 \mid M_U/M_W \mid^2 < 5 \end{array} \tag{2.6}$$

Equation (2.6) translates into the limit,

$$\tau > 5 \times 10^{-26} \, (1 \text{ TeV}/M_D) \text{ sec.} \tag{2.7}$$

For the remainder of our analysis, we have assumed that the Kobayashi-Maskawa matrix is approximately diagonal and so the D quark decays primarily to a W^- t pair. Pair production of D quarks thus yields final states with $W^+W^-t\bar{t}$. We consider both the case where the t quark is lighter than the W boson ($M_t = 50$ and 70 GeV) and where it is heavier ($M_t = 120$ GeV). If the top quark is heavier than the W boson, then it will decay to W^+b and D quark pair production will result in a final state with 4 W's and a $b\bar{b}$ pair.

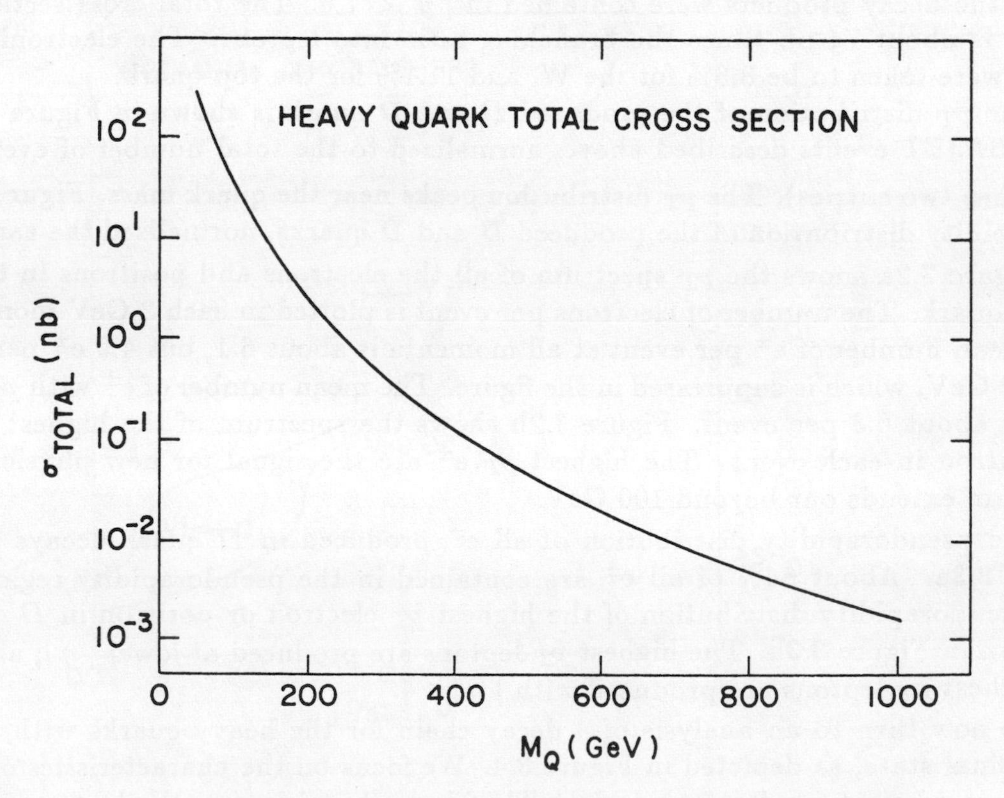

Fig. (2.2) Total Cross section for $pp \to D\bar{D}$ at $\sqrt{s} = 40$ TeV.

In hadronic interactions, heavy quarks can be produced by three mechanisms,
(i) Hard parton interactions, $q\overline{q} \rightarrow Q\overline{Q}$ and $gg \rightarrow Q\overline{Q}$,
(ii) Flavor excitation, $qQ \rightarrow qQ$ and $gQ \rightarrow gQ$, and
(iii) Jet fragmentation: $g \rightarrow Q\overline{Q}$ in a jet.

At the SSC, the dominant mechanism for heavy ($M_Q > 100$ GeV) quark production is gluon fusion. This is in contrast to charm production where jet fragmentation is a major source of $c\bar{c}$ pairs at high p_T.[1] The total cross section for heavy quark pair production is shown in Figure 2.2 and is quite large. For example, for $M_Q = 1$ TeV, there are approximately 10^4 events in an integrated luminosity of $10^{40} cm^{-2}$. (The cross section is the same for both D and U quark production.) The cross section falls rapidly with increasing quark mass. Unfortunately, the cross section becomes flatter when cuts on the momenta of the outgoing particles are imposed, so rate alone will be insufficient to determine the quark mass.

III. Event Characteristics

Heavy quark decays can be characterized by the number of leptons in the final state. The decays can yield a cascade of leptons, most of which are at low transverse momentum. ISAJET version 5.31 [6] was used to generate 2500 D decays to Wt with a D mass of 400 GeV and a top quark mass of 70 GeV at a total center of mass energy of 40 TeV. We generated events for which the D quark had transverse momenta greater than 20 Gev and which the decay products were contained in $| y |< 7.6$. The total cross section for these events is about 74 pb times the branching ratio into leptons. The electronic branching ratios were taken to be 8.3% for the W, and 11.1% for the top quark.

The p_T distribution of the produced D and $\overline{D}$ quark is shown in Figure 3.1a for the 2500 ISAJET events described above, normalized to the total number of events (so each event has two entries). The p_T distribution peaks near the quark mass. Figure 3.1b shows the rapidity distribution of the produced D and $\overline{D}$ quarks, normalized the same way.

Figure 3.2a shows the p_T spectrum of all the electrons and positrons in the decay of the D quark. The number of electrons per event is plotted in each 2 GeV momentum bin. The mean number of $e^{\pm}$ per event at all momenta is about 6.1, but 4.9 $e^{\pm}$ per event have $p_T < 2$ GeV, which is suppressed in the figure. The mean number of $e^{\pm}$ with $p_T > 40$ GeV is only about 0.3 per event. Figure 3.2b shows the spectrum of the highest p_T electron or positron in each event. The highest p_T $e^{\pm}$ are the signal for new physics, and their spectrum extends out beyond 100 GeV.

The pseudorapidity distribution of all $e^{\pm}$ produced in D quark decays is shown in Figure 3.3a. About 86% of all $e^{\pm}$ are contained in the pseudorapidity region $| \eta |< 4$. The pseudorapidity distribution of the highest p_T electron or positron in D quark decay is shown in Figure 3.3b. The highest p_T leptons are produced at lower $| \eta |$; about 96% of the highest p_T leptons are produced with $| \eta |< 4$.

We now turn to an analysis of a decay chain for the heavy quarks with two leptons in the final state, as depicted in Figure 3.4. We focus on the characteristics of the events which are relevant for detector design. The idea is to trigger on the leptons coming from one heavy quark decay and to reconstruct the quark mass using the hadronic decays of the other heavy quark. This scenario has been previously studied by Glover and Morris. [7]

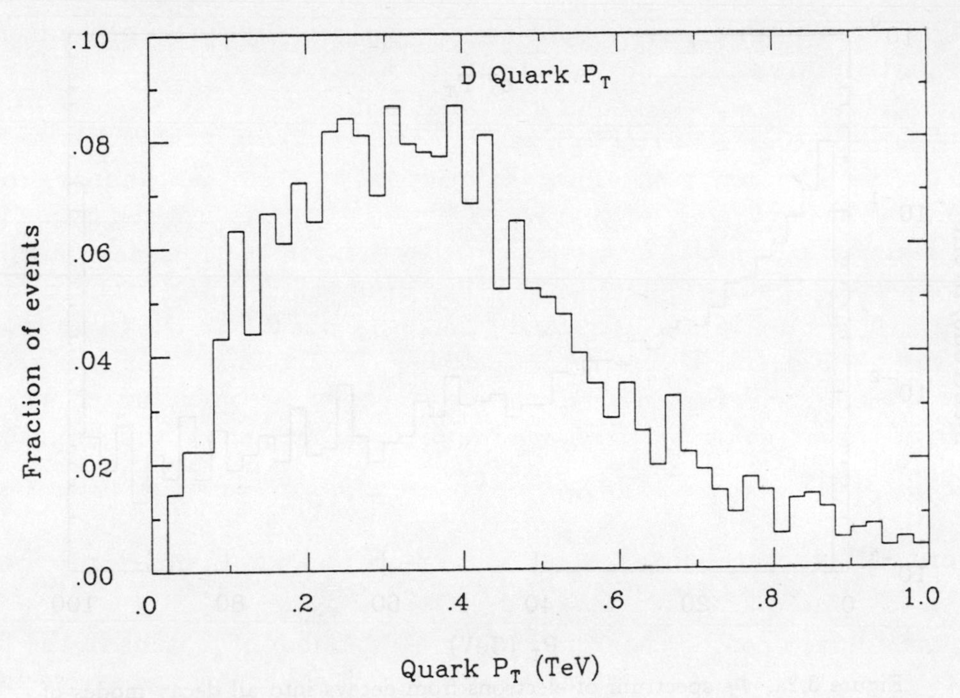

Figure 3.1a. P_T distribution for D and $\overline{D}$ quarks at $\sqrt{s}$ = 40 TeV for M_D = 500 GeV from ISAJET.

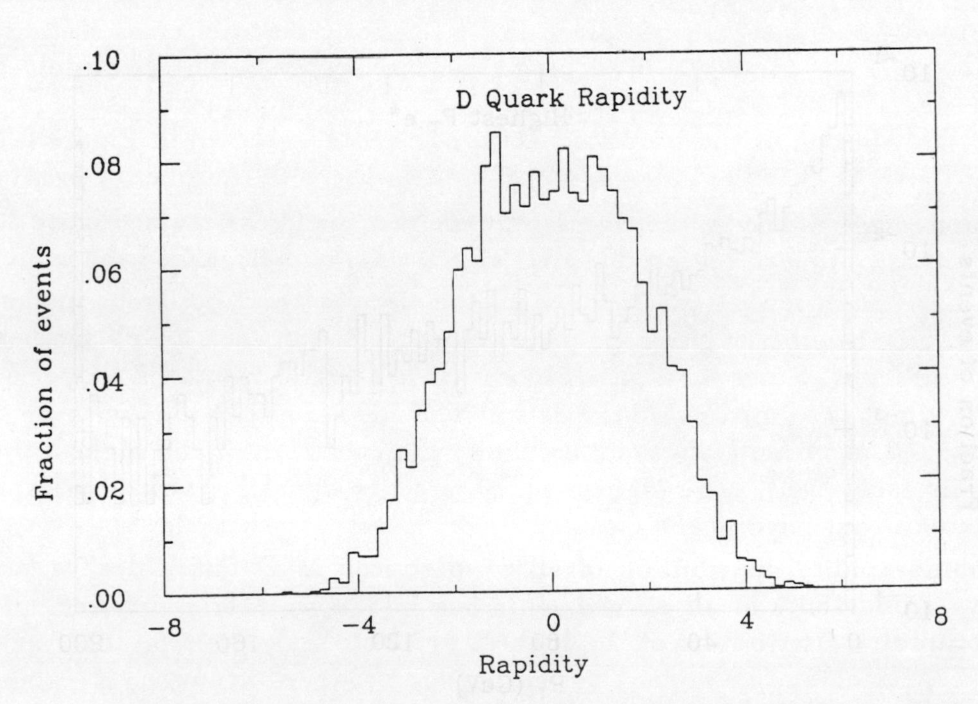

Figure 3.1b. Rapidity distribution for D and $\overline{D}$ quarks at $\sqrt{s}$ = 40 TeV for M_D = 500 GeV from ISAJET.

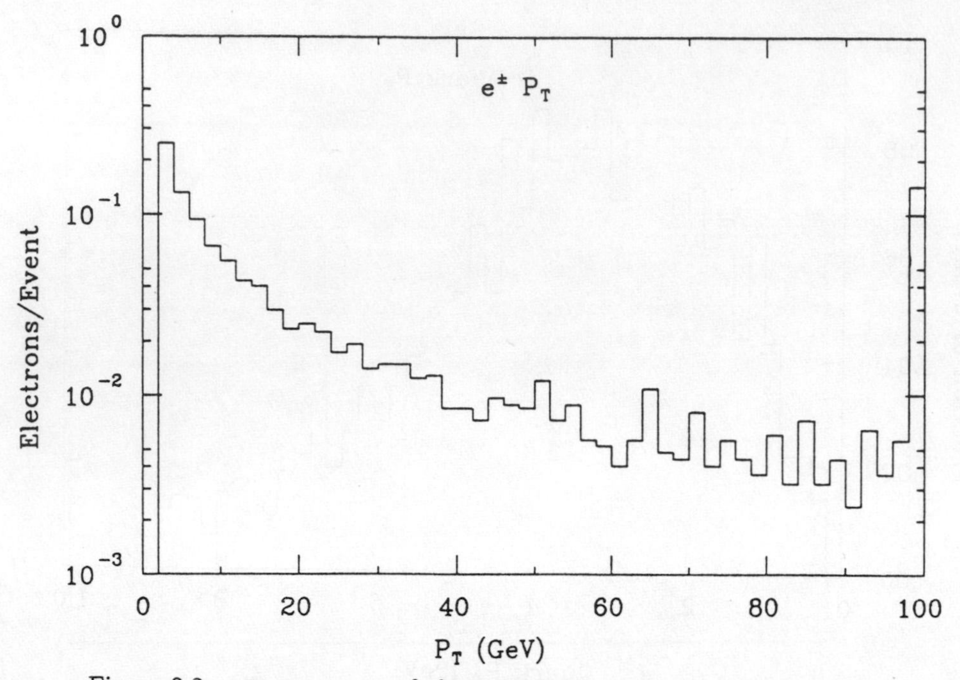

Figure 3.2a. P_T spectrum of electrons from decays into all decay modes of the D quark at $\sqrt{s}$ = 40 TeV for M_t = 70 GeV and M_D = 500 GeV from ISAJET. Both electrons and positrons with $p_T > 2$ GeV are plotted. Note the unsuppressed overflow bin of leptons at very high p_T.

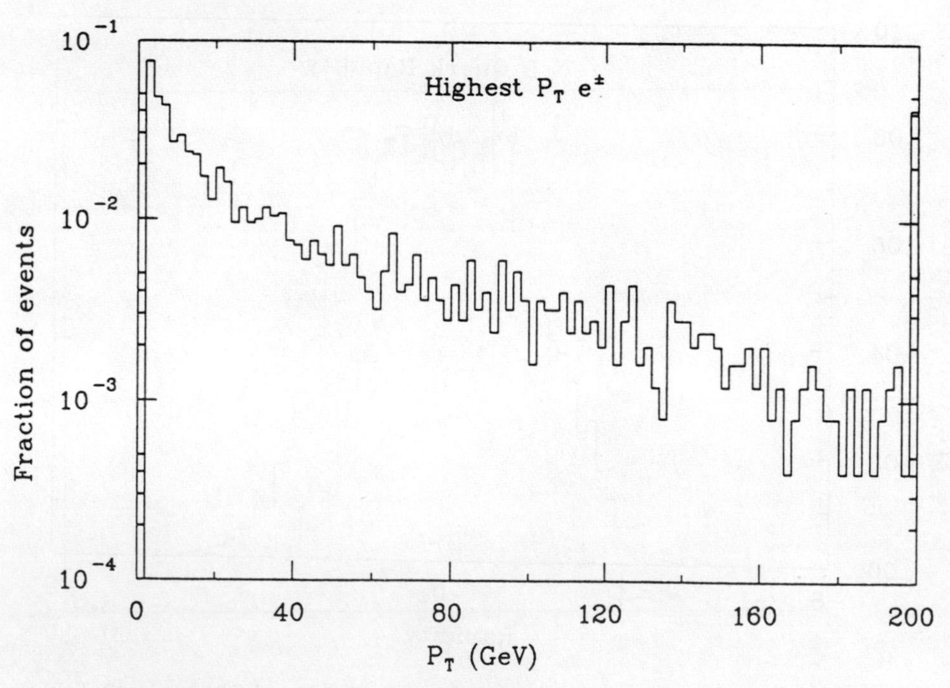

Figure 3.2b. P_T spectrum of the highest p_T electron or positron from decays of the D quark into all decay modes at $\sqrt{s}$ = 40 TeV from ISAJET.

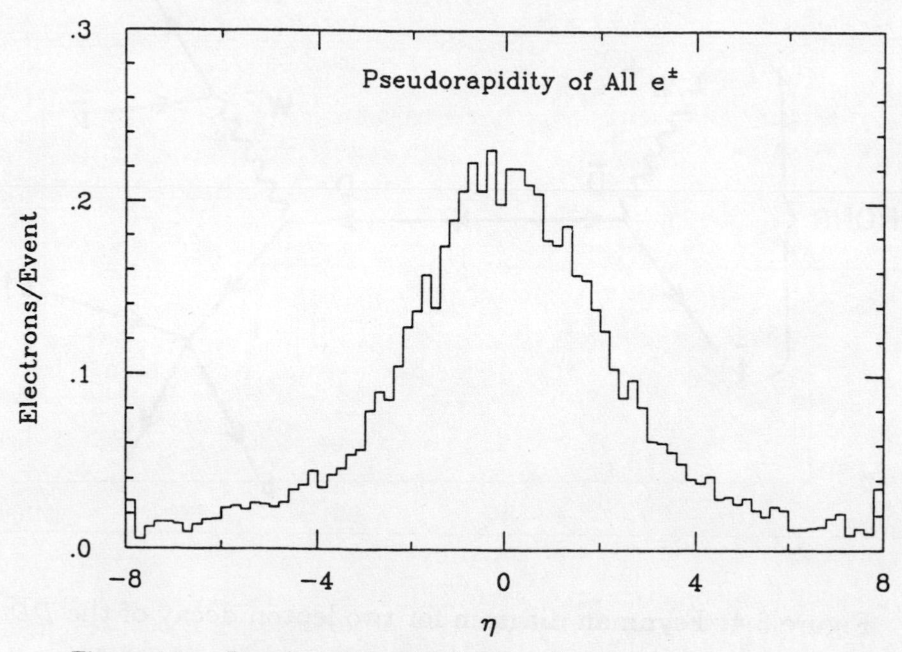

Figure 3.3a. Pseudorapidity distribution of all electrons or positrons from decays of the D quark into all decay modes at $\sqrt{s} = 40$ TeV for $M_t = 70$ GeV and $M_D = 500$ GeV from ISAJET.

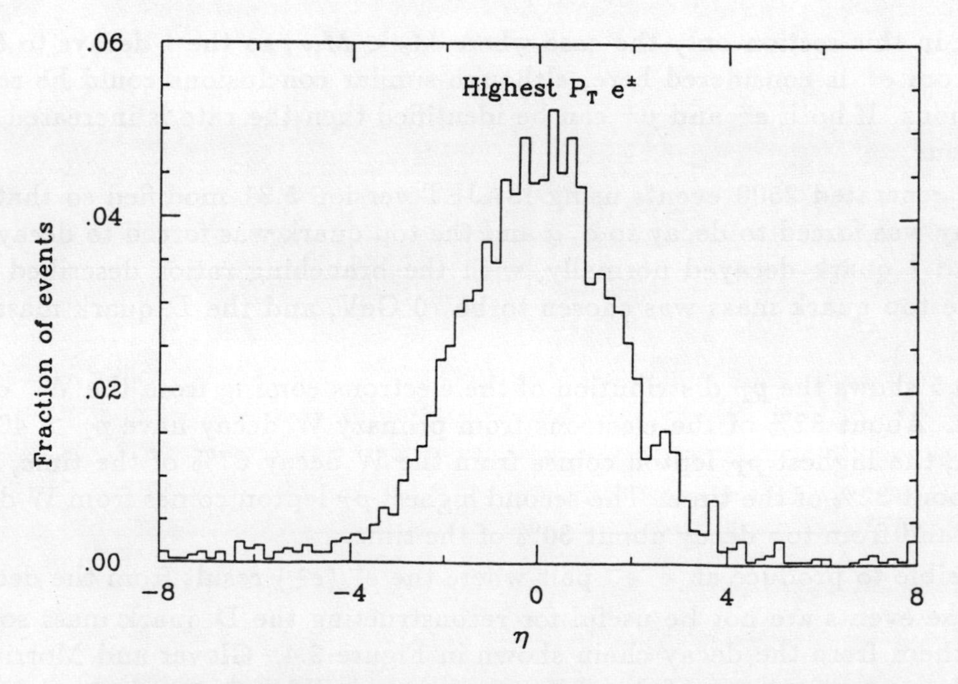

Figure 3.3b. Pseudorapidity distribution of the highest p_T electron or positron from decays of the D quark into all decay modes at $\sqrt{s} = 40$ TeV for $M_t = 70$ GeV and $M_D = 500$ GeV from ISAJET.

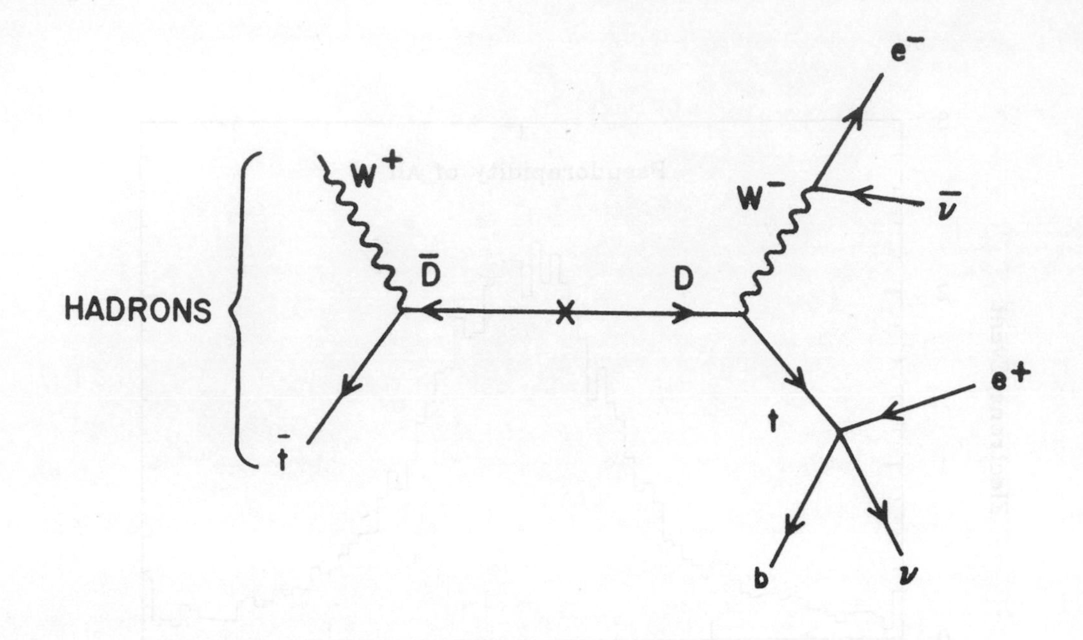

Figure 3.4. Feynman diagram for two lepton decay of the $D\overline{D}$.

We consider in this section only the case where $M_t < M_W$, so the t decays to $b\bar{l}\nu$. Only the signal from $e^{\pm}$ is considered here, although similar conclusions could be reached by studying muons. If both $e^{\pm}$ and $\mu^{\pm}$ can be identified then the rate is increased by about a factor of four.

We also generated 2500 events using ISAJET version 5.31 modified so that the W^- from D decay was forced to decay to $e^-\overline{\nu}$ and the top quark was forced to decay to $be^+\nu$. The W^+ and $\bar{t}$ quark decayed normally, with the branching ratios described above to leptons. The top quark mass was chosen to be 70 GeV, and the D quark mass was 500 GeV.

Figure 3.5 shows the p_T distribution of the electrons coming from the W^- decays for these events. About 82% of the electrons from primary W decay have $p_T > 40$ GeV. In these events, the highest p_T lepton comes from the W decay 67% of the time, and from top decay about 32% of the time. The second highest p_T lepton comes from W decay 62% of the time, and from top decay about 30% of the time.

It is possible to produce an e^+e^- pair where the e^- (e^+) result from the decay of the $D\left(\overline{D}\right)$. These events are not be useful for reconstructing the D quark mass so we must distinguish them from the decay chain shown in Figure 3.4. Glover and Morris [7] have calculated the transverse mass formed between the possible e^+e^- pairs and have found that by making the appropriate cuts on the transverse mass of the e^+e^- pair it is possible to obtain a sample of events where 70 to 80 % of the e^+e^- pairs results from the desired decay. (Note that this figure has a logarithmic scale). This method of selecting the signal requires determination of the electron sign up to momenta on the order of 300 GeV.

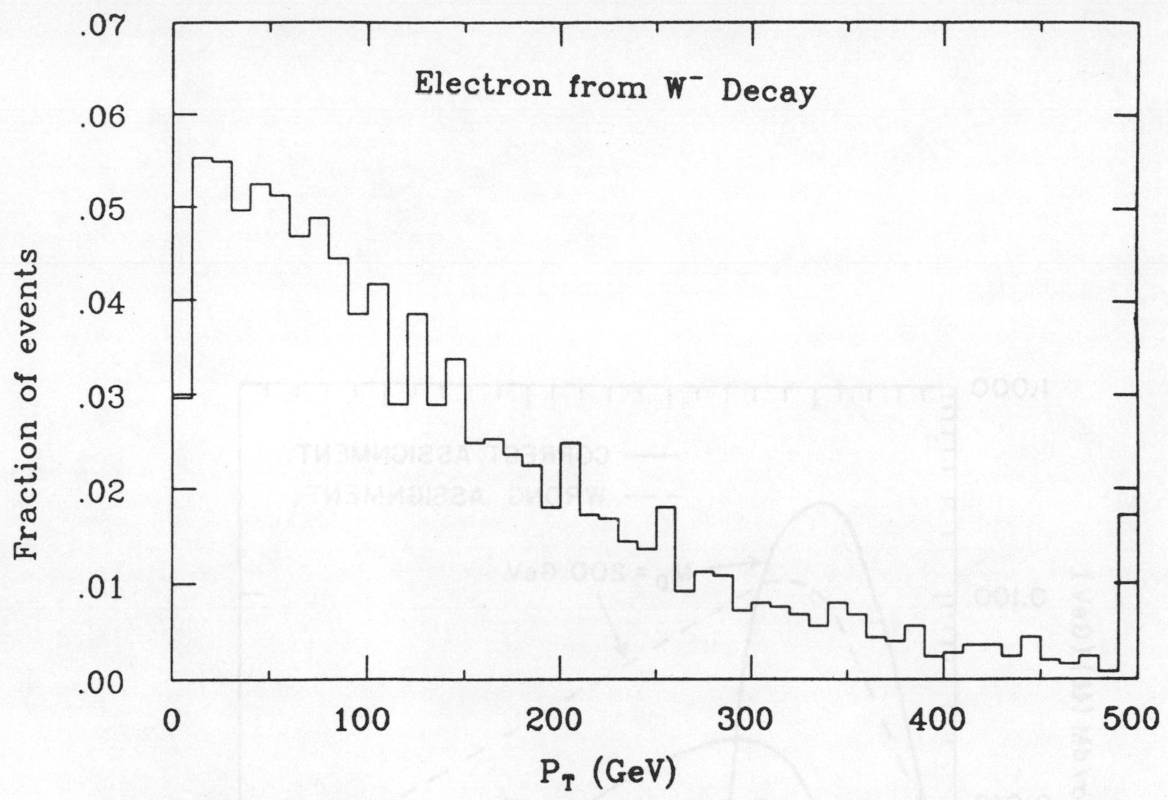

Figure 3.5. P_T distribution of electrons from $W^- \rightarrow e^-\overline{\nu}$ in decays of the D quark at $\sqrt{s} = 40$ TeV for $M_t = 70$ GeV and $M_D = 500$ GeV from ISAJET.

A simple calorimeter model was modified from ISAJET and used for studying hadronic jets and missing energy. The calorimeters have a cell size of $\Delta\eta = 0.1$ and $\Delta\phi = .087$ and extend over the pseudorapidity range $|\ \eta\ | < 6$. Electromagnetic energy was smeared with a resolution of 20% $/\sqrt{E}$; hadronic energy was smeared with a resolution of 80% $/\sqrt{E}$. No attempt was made to model real electromagnetic or hadronic showers, nor was any attempt made to model in detail physical processes in the calorimeters.

Jets were found in the hadron calorimeter using an algorithm modelled on the UA1 jet finding algorithm.[8] The process consists of finding the largest energy deposition in any calorimeter cell greater than 1 GeV and summing surrounding cells within ΔR of 1 with $E_T > 0.5$ GeV, where $\Delta R = \sqrt{\Delta\eta^2 + \Delta\phi^2}$. Jets with $E_T > 20$ GeV are kept, and the process is repeated with the remaining unused calorimeter hits until the maximum E_T in any calorimeter cell is less than 1 GeV.

The E_T distribution for jets in these events is shown in Figure 3.7. The mean number of jets per event was found to be about 6.6, with about 35% having $E_T > 100$ GeV. The pseudorapidity distribution of the jets is shown in Figure 3.8.

We used the jets found in the model calorimeter to study the distribution of hadronic energy around the high p_T leptons. Figure 3.9 shows the distance between either of the two highest p_T leptons and the nearest energy deposition (hadronic or electromagnetic) larger than 20 GeV. Included in this plot are all other electrons and any jets. Electrons are seen to be quite well isolated from other large energy deposits.

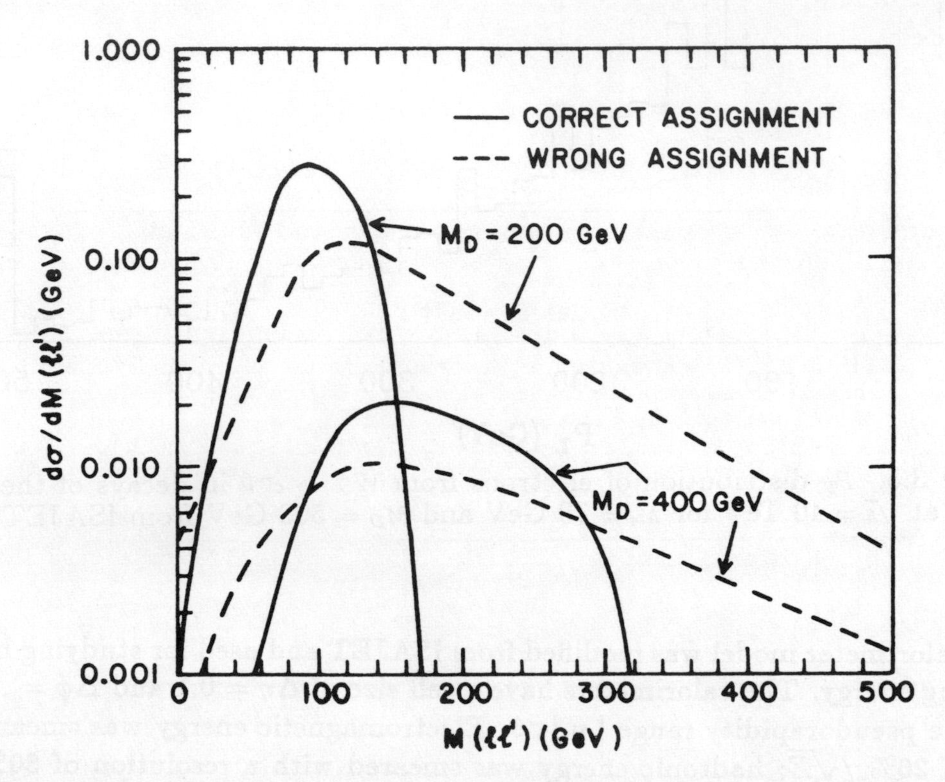

Figure 3.6. Transverse mass formed between lepton pairs from the decay of $D\bar{D}$. The curve marked "correct assignment" corresponds to the decay scenario of Figure 3.4, while the curve marked "wrong assignment" corresponds to the case where the leptons come from decays of different heavy quarks. This figure is from reference [7].

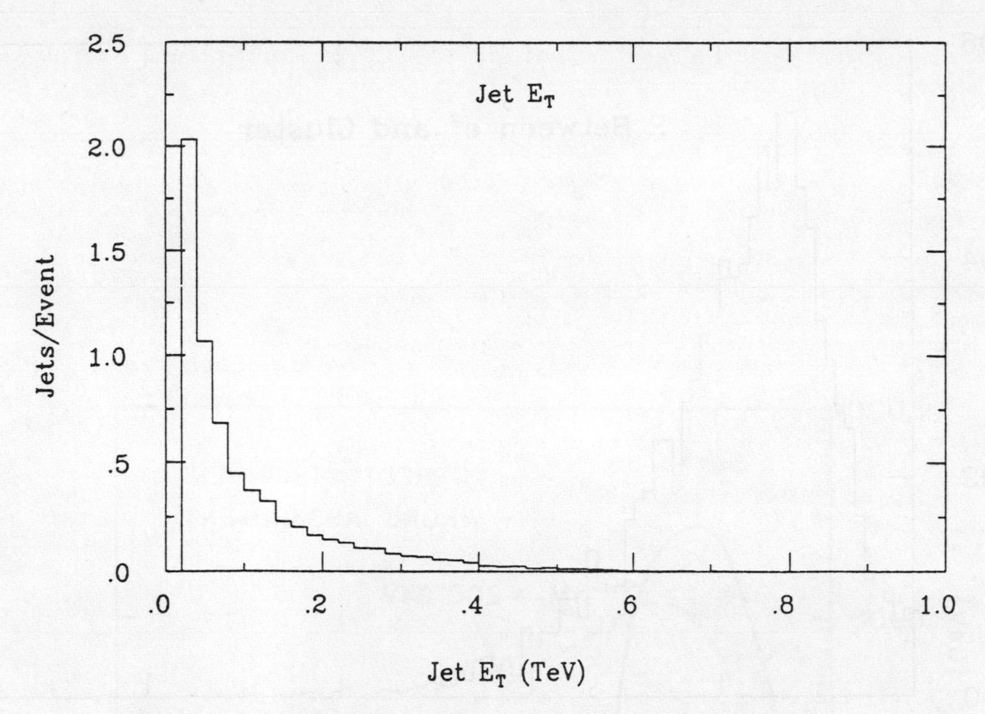

Figure 3.7. E_T distribution for jets found using the algorithm described in the text for decays of the D quark as shown in Figure 3.4 at $\sqrt{s} = 40$ TeV for $M_t = 70$ GeV and $M_D = 500$ GeV from ISAJET.

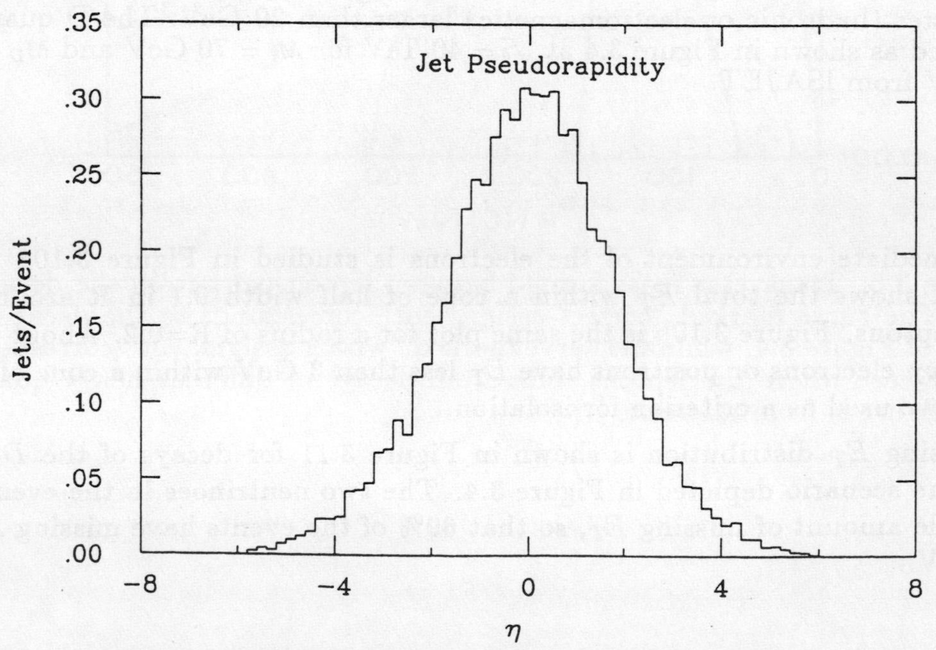

Figure 3.8. Pseudorapidity distribution of jets found using the algorithm described in the text for decays of the D quark as shown in Figure 3.4 at $\sqrt{s} = 40$ TeV for $M_t = 70$ GeV and $M_D = 500$ GeV from ISAJET.

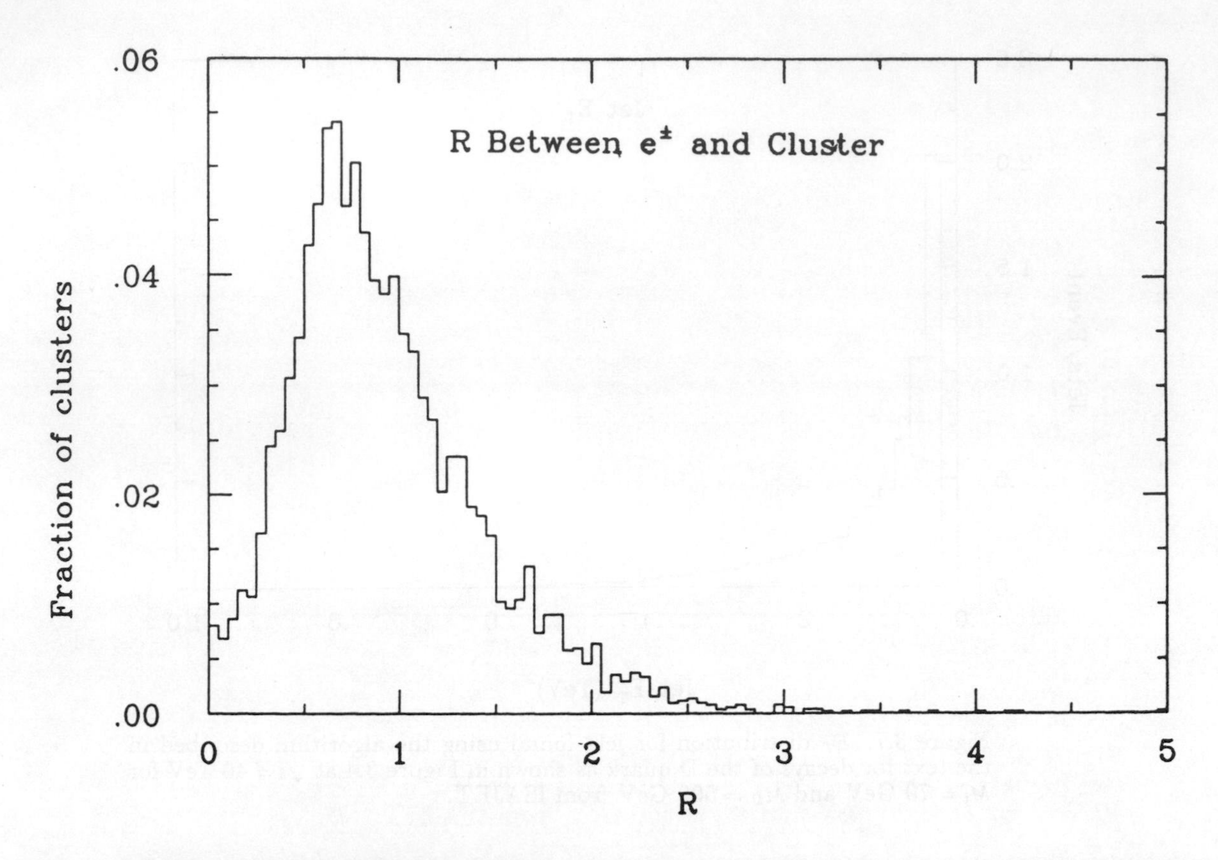

Figure 3.9. ΔR between either of the two highest p_T leptons and the nearest cluster (hadronic or electromagnetic) larger than 20 GeV. The D quark decayed as shown in Figure 3.4 at $\sqrt{s} = 40$ TeV for $M_t = 70$ GeV and $M_D = 500$ GeV from ISAJET.

The immediate environment of the electrons is studied in Figure 3.10a and 3.10b. Figure 3.10a shows the total E_T within a cone of half width 0.1 in R around the two highest p_T leptons. Figure 3.10b is the same plot for a radius of R=0.2. About 60% of the two highest p_T electrons or positrons have E_T less than 3 GeV within a cone of radius 0.2 in R, which we used as a criterion for isolation.

The missing E_T distribution is shown in Figure 3.11 for decays of the D quark according to the scenario depicted in Figure 3.4. The two neutrinoes in the event carry off a considerable amount of missing E_T, so that 69% of the events have missing E_T greater than 100 GeV.

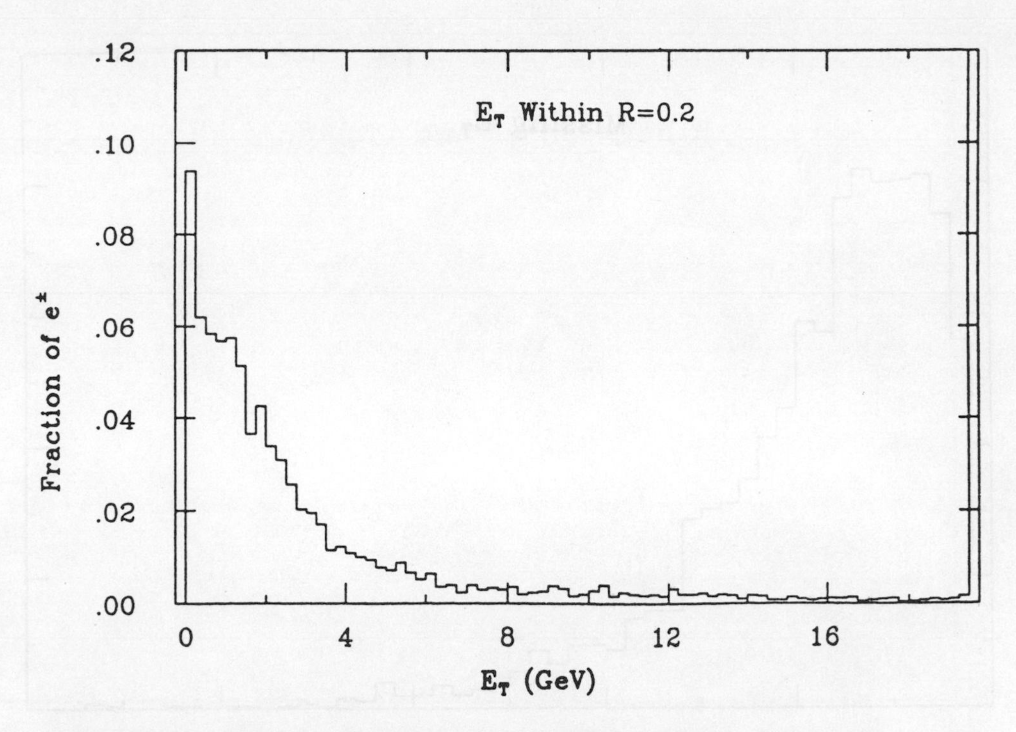

Figure 3.10a. Total E_T in a cone of half width ΔR=0.1 around either of the two highest p_T leptons. The D quark decayed as shown in Figure 3.4 at $\sqrt{s}$ = 40 TeV for M_t = 70 GeV and M_D = 500 GeV from ISAJET.

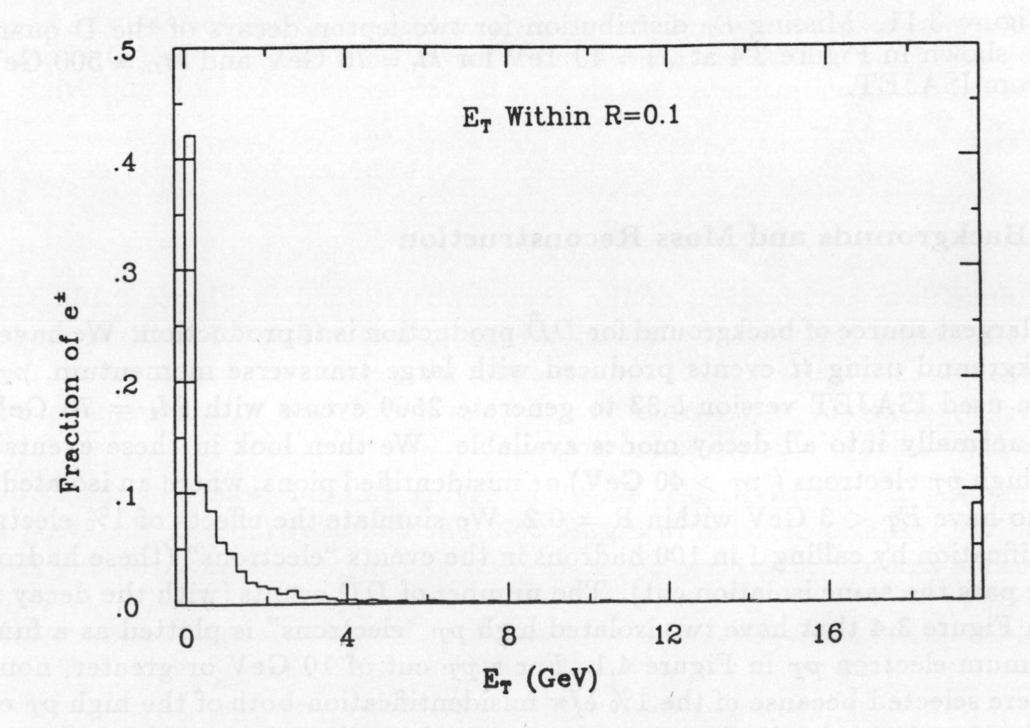

Figure 3.10b. Total E_T in a cone of half width ΔR=0.2 around either of the two highest p_T leptons. The D quark decayed as shown in Figure 3.4 at $\sqrt{s}$ = 40 TeV for M_t = 70 GeV and M_D = 500 GeV from ISAJET.

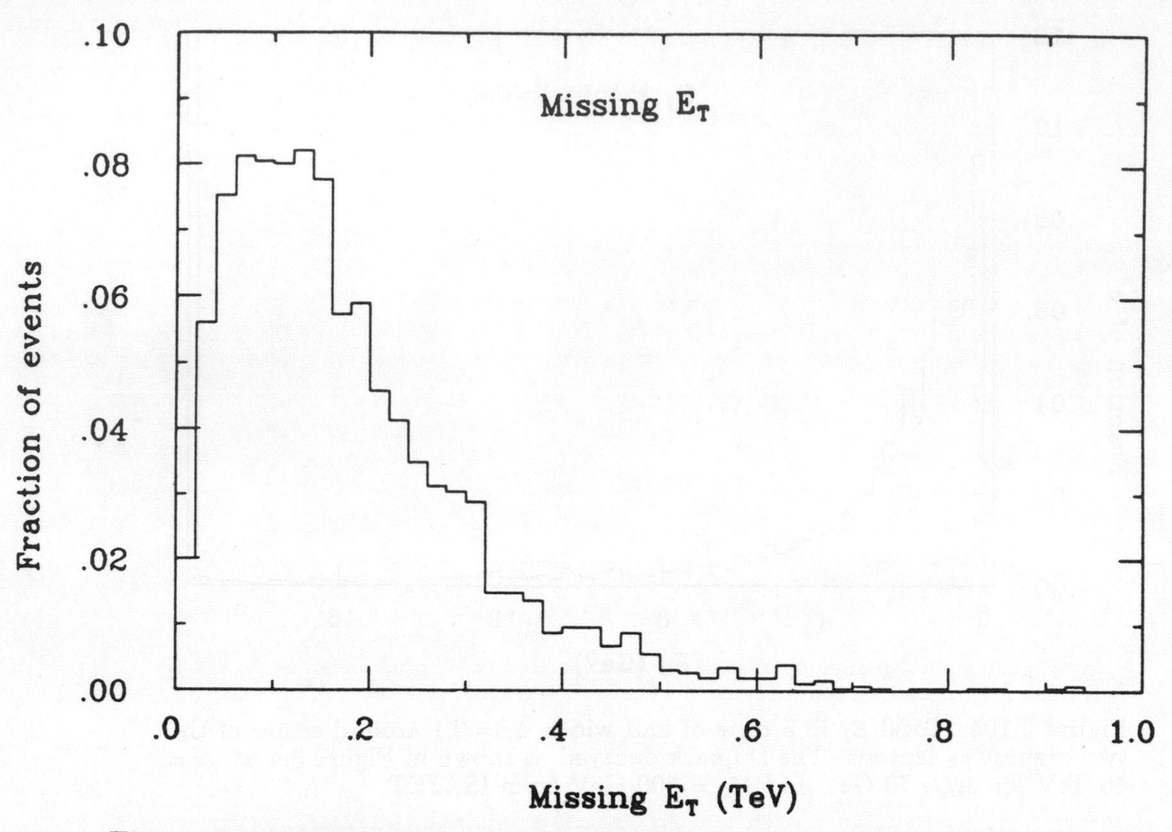

Figure 3.11. Missing E_T distribution for two lepton decays of the D quark as shown in Figure 3.4 at $\sqrt{s} = 40$ TeV for $M_t = 70$ GeV and $M_D = 500$ GeV from ISAJET.

IV. Backgrounds and Mass Reconstruction

The largest source of background for $D\overline{D}$ production is $t\bar{t}$ production. We have studied this background using $t\bar{t}$ events produced with large transverse momentum, $p_T > 400$ GeV. We used ISAJET version **5.33** to generate 2500 events with $M_t = 70$ GeV which decayed normally into all decay modes available. We then look in these events for two isolated high p_T electrons ($p_T > 40$ GeV) or misidentified pions, where an isolated track is defined to have $E_T < 3$ GeV within R = 0.2. We simulate the effects of 1% electron-pion misidentification by calling 1 in 100 hadrons in the events "electrons" (these hadrons must of course pass the same isolation cut). The number of $D\overline{D}$ events (with the decay scenario shown in Figure 3.4 that have two isolated high p_T "electrons" is plotted as a function of the minimum electron p_T in Figure 4.1. For a p_T cut of 10 GeV or greater, none of the events were selected because of the 1% e/π misidentification-both of the high p_T electrons were real. Also shown in the figure are the expected number of events from $t\bar{t}$ production. Because of low statistics, the error bars (statistical errors only) on the expected number of background events are large. For $p_T^{electron} > 10$ GeV, the number of events from D decay is about equal to the number of events from t quark decay even though we have used extremely simple event selection cuts. Just as for the signal events, all the background

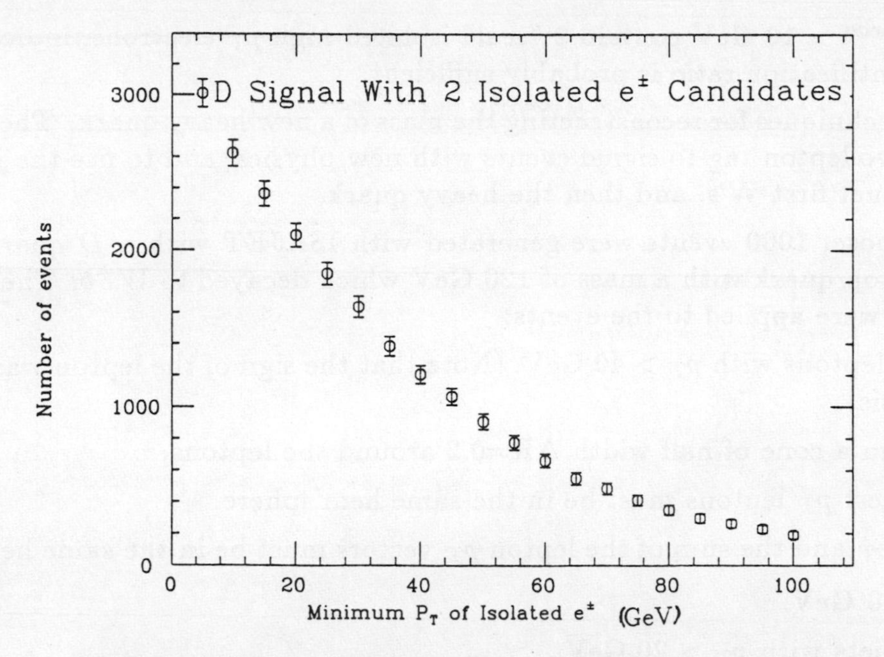

Figure 4.1a. Number of events with two isolated "electrons" from decays of the D quark as shown in Figure 3.4 for an integrated luminosity of 10^{40} cm^{-2} at $\sqrt{s}$ = 40 TeV for M_t = 70 GeV and M_D = 500 GeV from ISAJET. Pions were randomly misidentified as electrons 1% of the time.

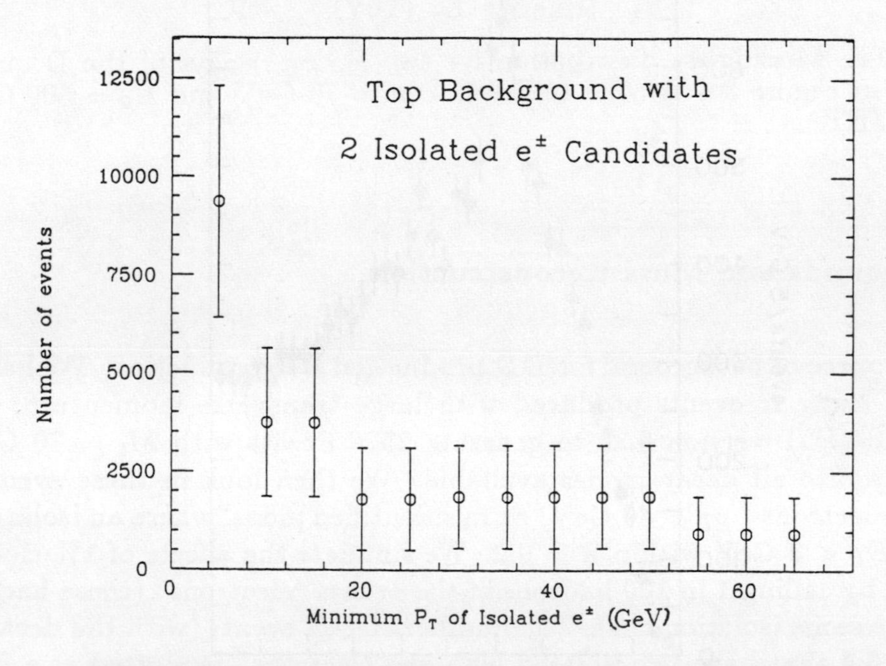

Figure 4.1b. Number of events with two isolated "electrons" from decays of the top quark for an integrated luminosity of 10^{40} cm^{-2} at $\sqrt{s}$ = 40 TeV for M_t = 70 GeV from ISAJET. Top quark p_T was required to be greater than 400 GeV. Pions were randomly misidentified as electrons 1% of the time.

events with $p_T^{electron} > 10$ GeV contain 2 "real" isolated high p_T electrons, indicating that a 1% e/π misidentification ratio is probably sufficient.

We studied techniques for reconstructing the mass of a new heavy quark. The approach was to use the two lepton tag to signal events with new physics, and to use the jets in the event to reconstruct first W's, and then the heavy quark.

For this purpose, 1000 events were generated with ISAJET with a D quark mass of 500 GeV, and a top quark with a mass of 120 GeV which decayed to W^+b. The following selection criteria were applied to the events:

1) Two high p_T leptons with $p_T > 40$ GeV. (Note that the sign of the lepton was not used in this analysis).
2) $p_T < 2$ GeV in a cone of half width ΔR=0.2 around the leptons.
3) The two highest p_T leptons must be in the same hemisphere.
4) The missing p_T and the sum of the lepton p_T vectors must be in the same hemisphere.
5) $p_T^{Missing} > 100$ GeV.
6) At least four jets with $p_T > 20$ GeV.

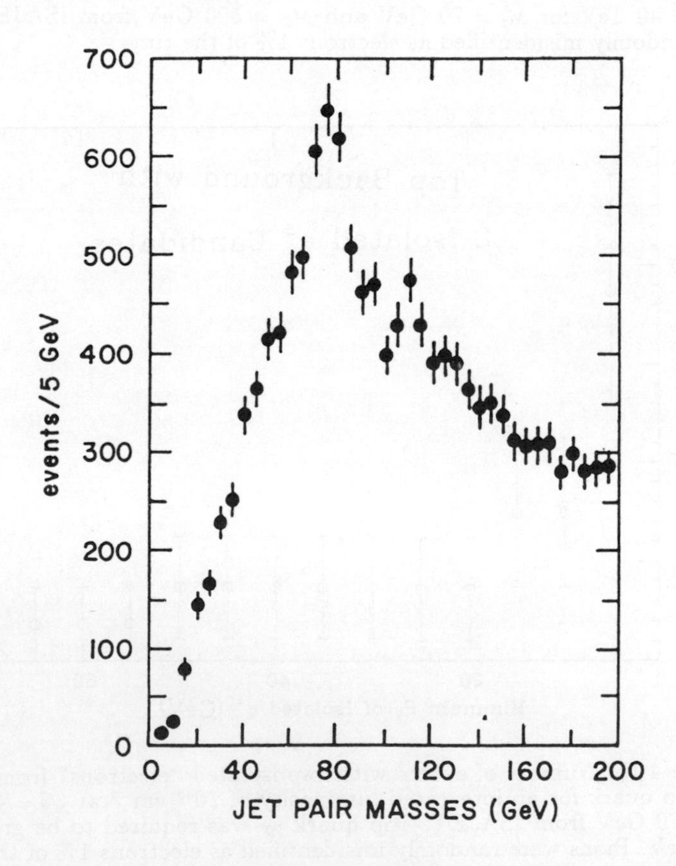

Figure 4.2. Jet pair mass from decays of the D quark as shown in Figure 3.4 at $\sqrt{s} = 40$ TeV for $M_t = 120$ GeV and $M_D = 500$ GeV from ISAJET W candidates were chosen to have $65 < M_{jet-jet} < 95$ GeV.

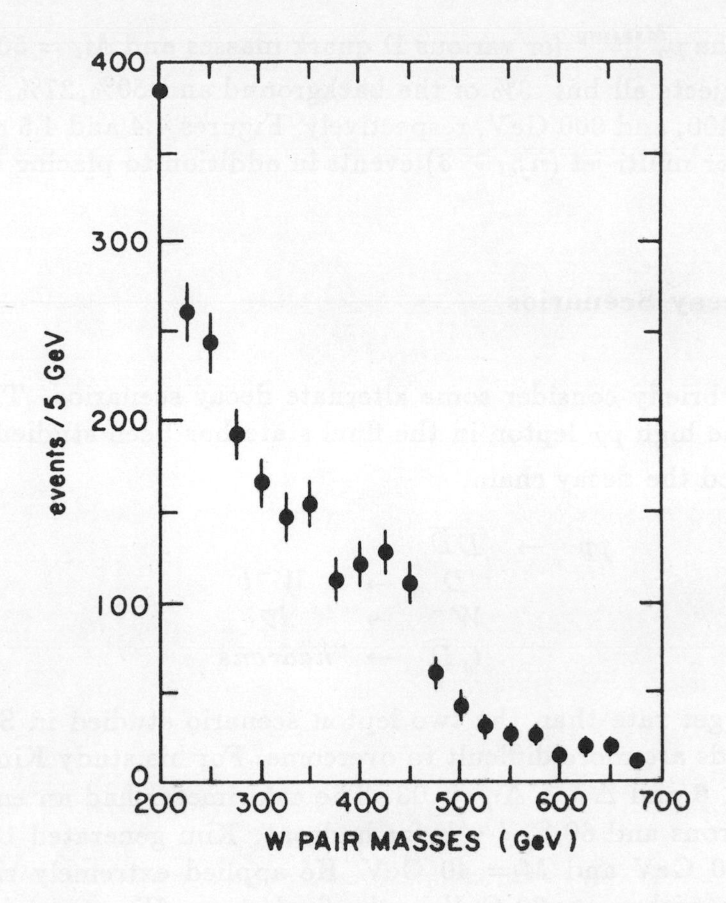

Figure 4.3. W-W mass from decays of the D quark at $\sqrt{s}$=40 TeV for M_t=120 GeV and M_D=500 GeV from ISAJET . The D quark decayed as shown in Figure 3.4.

These cuts retain about 0.2% of all $D\overline{D}$ events, which includes two semileptonic branching ratios of about 10% each and efficiently reject the top quark background. (The top quark background is dominated by 2 jet events, so cut 6) is the most restrictive). The jet-jet transverse mass distribution is shown in Figure 4.2. A clear W peak is visible above the combinatoric background. The pair mass of W candidates with mass between 65 and 95 GeV is shown in Figure 4.3, which shows an excess of events near 425 GeV. (The W pairs do not reconstruct to the D quark mass because the D follows the decay chain, $D \to W^- t \to W^+ W^- b$).

To study the effects of varying the D quark mass, we generated 60000 events using a simple Monte Carlo which included only the hard subprocesses ($gg \to D\overline{D}$, $q\overline{q} \to D\overline{D}$, etc.). For M_D = 400 GeV, the results were compared with the events described above from ISAJET and good agreement was found. Figure 4.4 shows the p_T distribution of the e^- from the W^- decay (plus the e^+ from theW^+) when the D quark decays as in Figure 3.4. This figure also includes the two lepton background from $t\bar{t}$ production for M_t = 50 GeV and from W^+W^- production. We see that a simple cut on p_{Te} will not be sufficient to reject the background.

Figure 4.5 shows the $p_T^{Missing}$ for various D quark masses and $M_t = 50$ GeV. A cut on $p_T^{missing} > 100$ GeV rejects all but .3% of the background and 50%, 27%, and 17% of the signal for $M_D = 200$, 400, and 600 GeV, respectively. Figures 4.4 and 4.5 demonstrate the necessity for looking for multi-jet ($n_{jet} \geq 3$) events in addition to placing cuts on p_{Te} and $p_T^{Missing}$.

V. Alternate Decay Scenarios

In this section we briefly consider some alternate decay scenarios. The heavy quark decay channel with one high p_T lepton in the final state has been studied extensively by Kim [9], who considered the decay chain,

$$\begin{array}{ccccc} pp & \rightarrow & D\overline{D} & & \\ & & D & \rightarrow & W^-t \\ & & W^- & \rightarrow & l\overline{\nu} \\ & & t,\overline{D} & \rightarrow & hadrons \end{array} \tag{5.1}$$

This process has a larger rate than the two lepton scenario studied in Sections III and IV, but the backgrounds are more difficult to overcome. For his study Kim used a simple calorimeter with $|\eta| < 6$ and $\Delta\phi = \Delta\eta = .03$. The calorimeter had an energy resolution of 15 % $/\sqrt{E}$ for electrons and 60 % / $\sqrt{E}$ for hadrons. Kim generated 100,000 ISAJET events with $M_D = 500$ GeV and M_t= 40 GeV. He applied extremely rigid cuts which required 6 or more jets with $p_T > 20$ GeV in the final state. His cuts rejected all but .4 % of the events (this corresponds to 380 events in an integrated luminosity of 10^{40} cm^{-2}) and the quark mass could be cleanly reconstructed. Kim considered the background from W^+W^- pair production and found that it could be efficiently rejected with his cuts.

Unfortunately, Kim did not investigate the dominant source of background which is $t\bar{t}$ production. The background to this process from $t\bar{t}$ production has been considered by Dawson and Godfrey [10]. They found that the top quark background could be cleanly rejected by looking for events with a single high p_T lepton and more than three jets. Their analysis, however, required the ability to clearly identify jets coming from b quarks. The other possible background to the one lepton decay channel of the D quark is from W plus jet production. However, looking for greater than 3 jets should eliminate this background.

The cascade decays of heavy quarks to final states with multi-muons has been considered by Baer [11] at this workshop. He looked at events with 5 or 6 muons of either sign or with 3 or 4 same sign muons in the final state. By requiring multi-muons in the final state, the top quark background can be almost completely rejected. For heavy quarks with a mass near 200 GeV, event rates are in the .1 pb. range. Because of the large number of neutrinos in the final state, the heavy quark mass cannot be reconstructed from the missing p_T of the leptons. Multi-muons may, however, be a clean signal for the onset of new physics.

Heavy quarks could also be searched for in their totally hadronic mode. An analysis of the 6 jet signal from heavy quarks has been compared with the 6 jet QCD background for pp collisions at $\sqrt{s}$= 17 GeV. The conclusion was that even with very high p_T cuts on the jets ($p_T > 50$ GeV) the signal was swamped by the QCD background. [12]

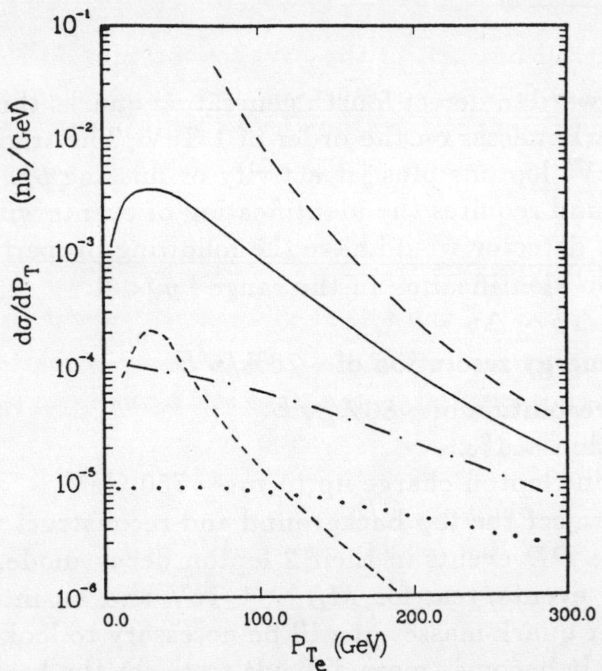

Figure 4.4. p_T distribution of $e^\pm$ from $W^\pm \to e^\pm \nu$ in the decays of the D ($\overline{D}$) quark as shown in figure 3.4 at $\sqrt{s} = 40$ TeV. This figure includes all branching ratios. The long-dashed line is the $t\bar{t}$ background for $M_t = 50$ GeV. The solid, dot-dashed, and dotted lines have $M_D = 200$, 400, and 600 GeV, respectively. The short-dashed curve is the background from W^+W^- production.

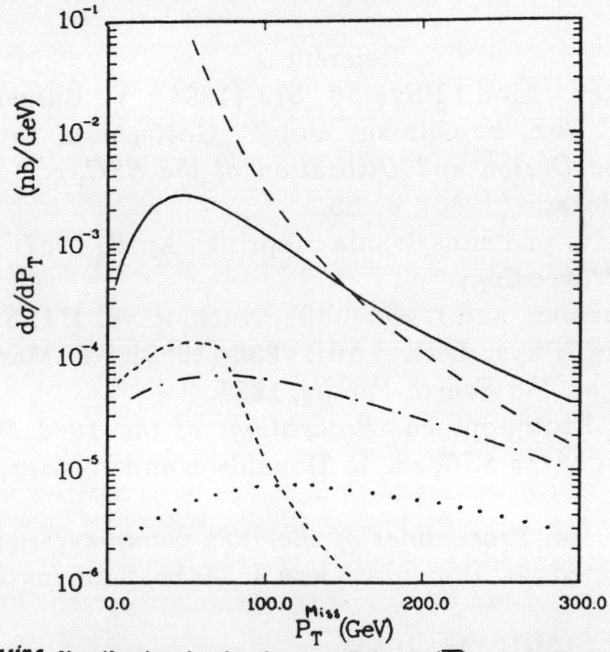

Figure 4.5. $p_T^{Missing}$ distribution in the decays of the D ($\overline{D}$) quark as shown in figure 3.4 at $\sqrt{s} = 40$ TeV. This figure includes all branching ratios. The long-dashed line is the $t\bar{t}$ background for $M_t = 50$ GeV. The solid, dot-dashed, and dotted lines have $M_D = 200$, 400, and 600 GeV, respectively. The short-dashed curve is the background from W^+W^- production.

VI. Conclusions

It appears straightforward to detect fourth generation quarks through their two lepton decay channels up to quark masses on the order of 1 TeV. The trigger for these events is two high p_T ($p_T > 40$ GeV) leptons plus jet activity or missing p_T. An efficient rejection of the top quark background requires the identification of events with greater than 3 or 4 jets. A good heavy quark detector would have the following properties:

1.) Electron and muon identification in the range $|y| < 4$.
2.) Calorimetry with $\Delta\phi \sim \Delta y \sim 0.1$.
3.) Electromagnetic energy resolution of $\sim 20\%/\sqrt{E}$.
4.) Hadronic energy resolution of $\sim 80\%/\sqrt{E}$.
5.) e/π misidentification $\sim 1\%$.
6.) Ability to determine lepton charge up to $p_T \sim 750$ GeV.

The cuts imposed to reject the top background and reconstruct the heavy quark mass retained about .2% of the $D\overline{D}$ events in their 2 lepton decay mode. With this efficiency, there would be about 20 events/year for $M_D = 1$ TeV and an integrated luminosity of $10^{40}/\text{cm}^2$. To reach higher quark masses, it will be necessary to look for them in their one lepton decay mode where it becomes more difficult to reject the background.

References

[1] E. Eichten *et. al.*, Rev. Mod. Phys **56** ,579 (1984); V. Barger *et. al.*, Phys. Rev. **D30**, 947 (1984); B. Cox, F. Gilman, and T. Gottschalk, *Proceedings of the 1986 Summer Study on the Design and Utilization of the SSC*, ed. R. Donaldson and J. Marx, Snowmass, Colorado (1986), p. 33.

[2] U. Amaldi *et. al.*, Univ. of Pennsylvania preprint , April, 1987.

[3] G. Herten, in these Proceedings.

[4] M. Chanowitz, M. Furman, and I. Hinchliffe, Nucl. Phys. **B153**, 402 (1979).

[5] S. Pakvasa and E. Ma, Phys. Lett. **156B**, 236 (1985), W. Marciano, *Proceedings of the UCLA Workshop on the Fourth Family*, 1987.

[6] F.E. Paige and S.D. Protopopescu, *Proceedings of the 1986 Summer Study on the Design and Utilization of the SSC*, ed. R. Donaldson and J. Marx, Snowmass, Colorado (1986), p.320.

[7] E. Glover and D. Morris, *Proceedings of the 1986 Summer Study on the Design and Utilization of the SSC*, ed. R. Donaldson and J. Marx, Snowmass, Colorado (1986), p. 238.

[8] G. Arnison,Phys.Lett., **139B**,115 (1984).

[9] S. Kim, *Proceedings of the 1986 Summer Study on the Design and Utilization of the SSC*, ed. R. Donaldson and J. Morfin, Snowmass, Colorado (1986), p. 241.

[10] S. Dawson and S. Godfrey, in these Proceedings.

[11] H. Baer, in these Proceedings.

[12] Z. Kunszt, *Proceedings of the La Thuile Workshop on Physics of Future Accelerators*, March 1987.

SINGLE LEPTONS FROM HEAVY QUARK PRODUCTION

Sally Dawson and Stephen Godfrey
Physics Department, Brookhaven National Laboratory, Upton, NY 11973

ABSTRACT

We consider the single lepton decay mode of heavy quarks as a signature of heavy quark pair production at the SSC. We find that events with multi-jets ($n_{\rm jet} \geq 3$) and a single high p_T lepton are a good signal for heavy quark production.

1. Introduction

In this note we consider fourth generation heavy quark pair production at the SSC. At $\sqrt{s} = 40$ TeV, heavy quarks are copiously produced from gluon fusion[1]. The heavy quarks then decay weakly to multi-jets and some number of leptons. The resulting events can be categorized by the number of leptons. Here we discuss the possibility of discovering heavy quark pair production from final states containing a single lepton. This scenario has been previously discussed by Kim[2]. Other scenarios are discussed elsewhere in these proceedings[3].

In analogy with the third generation, we assume that the charge $-\frac{1}{3}$ quark is the lightest member of the fourth generation and study only its production and decay. We consider the decay chain,

$$pp \rightarrow Q\overline{Q}$$
$Q \rightarrow W^+\bar{t}$
- $\bar{t} \rightarrow$ hadrons
- $W^+ \rightarrow \bar{e}\nu$

$\overline{Q} \rightarrow$ hadrons (1)

The rapidity distribution of the leptons is relatively flat and extends out to $|y| \lesssim 3$.

The primary background to this process is from $t\bar{t}$ production. We consider two top quark masses, $m_t = 50$ GeV and $m_t = 100$ GeV, and show that the problem of extracting the signal from the background is quite different in the two cases. The former case corresponds to the t-quark decaying via a virtual W-boson while the latter case corresponds to the t-quark decaying via a real W-boson.

In this note we do not consider the more difficult problem of reconstructing the heavy quark mass. A detailed discussion of our results can be found in ref. 4.

2. $m_t = 50$ GeV

We begin by discussing the case with $m_t = 50$ GeV. Here, the top quark will decay as follows;

$$\begin{aligned} t &\rightarrow bq\bar{q}' \\ t &\rightarrow b\bar{l}\nu \end{aligned} \tag{2}$$

The lepton coming from the W decay of Equation (1) can easily be distinguished from a lepton coming from a top quark decay by using the tranverse mass, $M_T\,(l\not{p}_T)$, formed from the lepton tranverse momentum and the missing p_T. $M_T\,(l\not{p}_T)$ is defined by $M_T\,(l\not{p}_T) = \sqrt{p_{lT}\not{p}_T\,(1-\cos\theta_T)}$ where θ_T is the angle separating the lepton and missing momentum in the plane tranverse to the beam axis. In Figure 1 we show the transverse mass distribution for leptons coming from W's and t-quark decays[5]. The $M_T\,(l\not{p}_T)$ distribution coming from $t \rightarrow b\bar{l}\nu$ is quite different from the $M_T\,(l\not{p}_T)$ distribution coming from W decay and offers a means of eliminating the t-quark background.

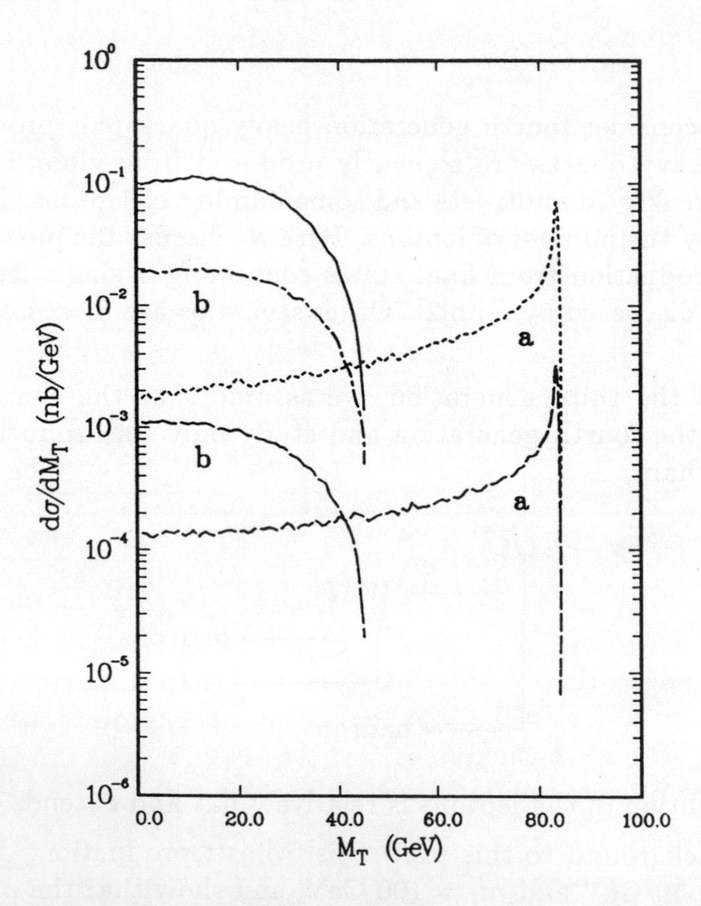

Fig. 1: $d\sigma/dM_T\,(l\not{p}_T)$ for $Q\overline{Q}$ production with $m_t = 50$ GeV at $\sqrt{s} = 40$ TeV. In this figure we have summed over single $e^{\pm}$ and $\mu^{\pm}$ production. The solid line is the lepton spectrum from $pp \rightarrow t\bar{t}$, $t \rightarrow l\nu b$, $\bar{t} \rightarrow hadrons$. The curves marked (a) are the signal with $pp \rightarrow Q\overline{Q}$, $Q \rightarrow Wt$, $W \rightarrow e\nu$, $t \rightarrow bq\bar{q}'$, $\overline{Q} \rightarrow hadrons$. The curves marked (b) are from the process $pp \rightarrow Q\overline{Q}$, $Q \rightarrow Wt$, $W \rightarrow q\bar{q}'$, $t \rightarrow be\nu$, $\overline{Q} \rightarrow hadrons$. (In all cases we have included the charge conjugate decay.) The dotted and dot-dashed curves are for $M_Q = 200$ GeV and the short-dashed and the long-dashed curves are for $M_Q = 400$ GeV. All of our figures include the appropriate branching ratios.

In Figure 2, we show the p_{l_T} distribution of the lepton resulting from the W decay along with the lepton background from t-quark decay. Fig 2a shows the lepton spectrum without cuts and fig 2b shows the p_{l_T} spectrum with a cut of $M_T\ (l\not{p}_T) > 50$ GeV demonstrating that the background from $t\bar{t}$ production can be completely eliminated by the cut on M_T. Note that the momentum of the lepton extends out to greater than 500 GeV.

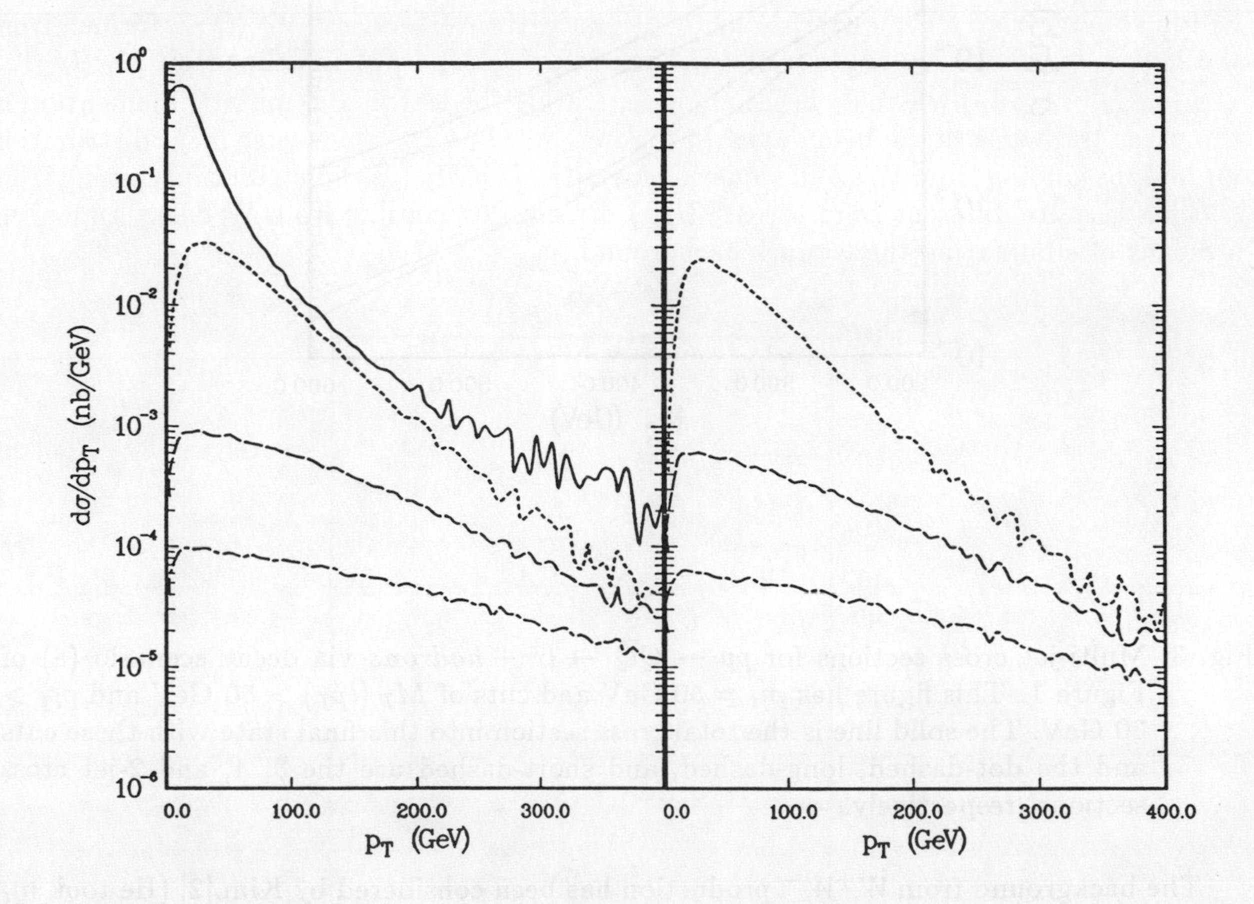

Fig. 2: $d\sigma/dp_{lT}$ for $pp \rightarrow Q\overline{Q} \rightarrow l^{\pm} + hadrons$ via decay scenario (a) of Figure 1. This figure has $m_t = 50$ GeV. The solid line is the $t\bar{t}$ background, while the short-dashed, long-dashed, and dot-dashed lines have $M_Q = 200$, 400, and 600 GeV respectively. Figure 2a has no cuts, while Figure 2b has $M_T\ (l\not{p}_T) > 50$ GeV.

The decays of heavy quarks tend to produce events with a large number of jets. Our jets are defined such that p_T (jet) > 30 GeV and Δ R $= \sqrt{\Delta\phi^2 + \Delta\eta^2} < 1$. Figure 3 shows the jet cross sections for $Q\overline{Q}$ production with a single lepton coming from the W decay. This figure has a cut on $M_T\ (l\not{p}_T) > 50$ GeV and $p_{lT} > 30$ GeV to eliminate the top quark background. For $M_Q = 600$ GeV, the 3-jet plus lepton cross section is 1.2×10^{-2} nb which corresponds to approximately $10^{\,5}$ events per year with an integrated luminosity of $10^{40}\mathrm{cm}^2$. The 3-jet plus high p_T lepton background from $t\bar{t}$ production has been totally eliminated by these cuts.

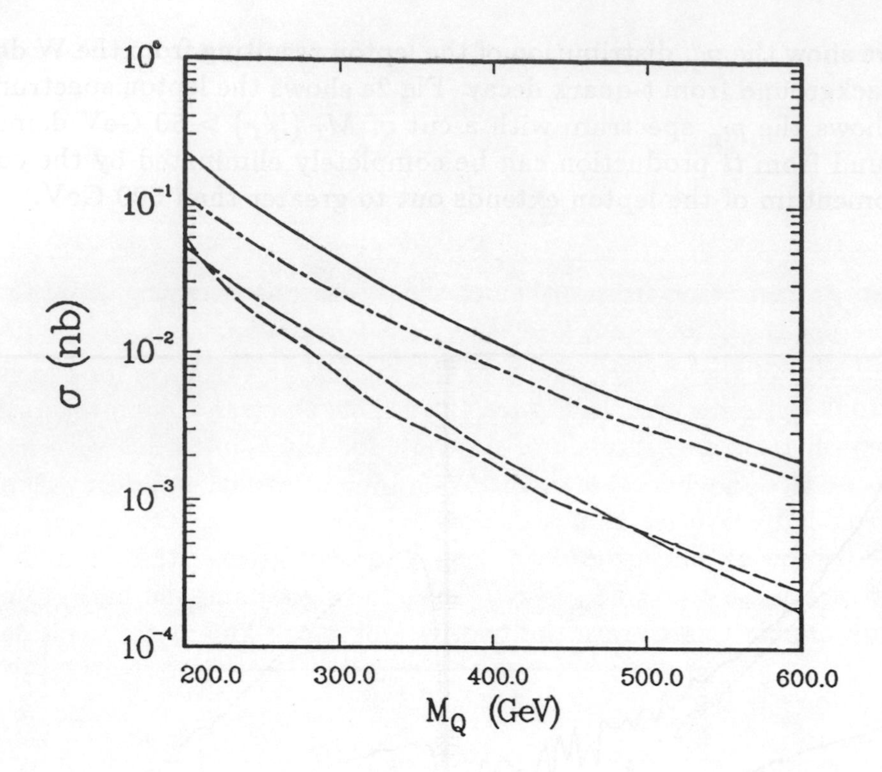

Fig. 3: Multi-jet cross sections for $pp \to Q\overline{Q} \to l^{\pm} + hadrons$ via decay scenario (a) of Figure 1. This figure has $m_t = 50$ GeV and cuts of $M_T\,(l\not{p}_T) > 50$ GeV and $p_{lT} > 30$ GeV. The solid line is the total cross section into this final state with these cuts and the dot-dashed, long-dashed, and short-dashed are the 3, 4, and 2-jet cross sections, respectively.

The background from W^+W^- production has been considered by Kim.[2] (He took $m_t = 40$ GeV). By looking for events with 6 jets in the final state and by applying extremely stringent cuts he was able to completely reject the W^+W^- background and to cleanly reconstuct the heavy quark mass. With his cuts, only 0.4% of the signal remained. With an efficiency of 0.4%, there would be about 400 3-jet plus lepton events per year for $M_Q = 600$ GeV.

There is also a large background from the partonic process $q + gluon \to W + q'$. The W decay here can yield a high p_T lepton which will mimic the signal. At a p_T of 200 GeV, $\frac{d\sigma}{dp_T dy}|_{y=0}$ for $pp \to W^{\pm}X \to \ell^{\pm}\nu X$ is approximately 3×10^{-4} nb. This process will yield primarily 1 and 2 jet events, so requiring events with multi-jets should cleanly reject this background.

3. $m_t = 100$ GeV

If the top quark is heavier than the W boson it becomes much more difficult to find a fourth generation quark. In this case the heavy quark decay chain is,

$$\begin{aligned} Q &\to tW^- \\ &\quad \to W^+ b \end{aligned} \tag{3}$$

Thus, the single lepton can come from either of the W-bosons decaying leptonically. We are interested in the decay chain where $t \to bW^+ \to b + hadrons$.

Due to the similarity of the decays, many of the kinematic variables for the signal and the $t\bar{t}$ background look quite similar. In Figure 4, we show the tranverse mass formed from the b-quark and lepton tranverse momenta, $M_T\,(lb)$, for the t-quark background and for the two possible decay chains where one of the W-bosons in equation 3 decays leptonically and the other hadronically. We see that a cut on $M_T\,(lb) > 80$ GeV can cleanly separate the signal from the top quark background for quark masses greater than 200 GeV. When the top quark mass is 100 GeV and M_Q is 200 GeV, the signal and the background are so similar that we were unable to separate the two by looking at kinematic variables.

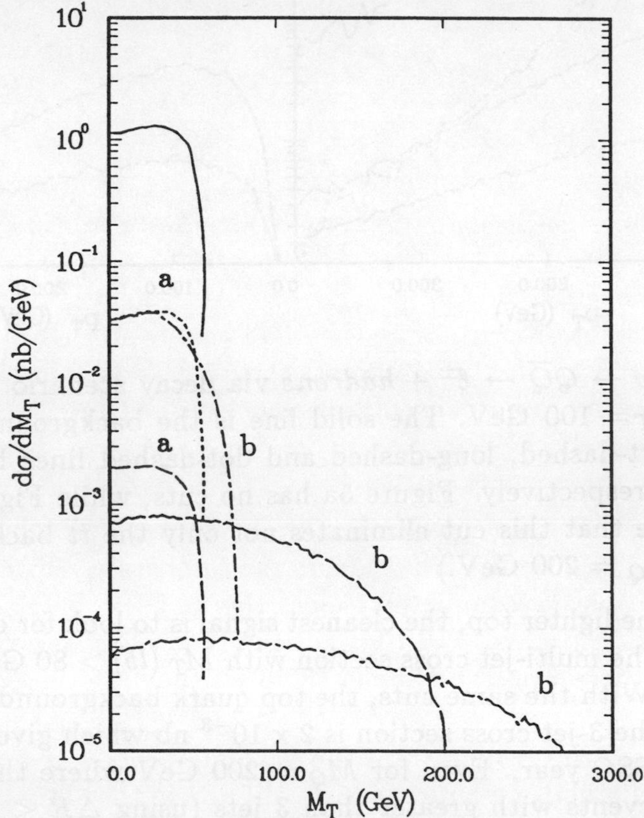

Fig. 4: $d\sigma/dM_T\,(lb)$ for $pp \to Q\overline{Q} \to \ell^\pm + hadrons$ with $m_t = 100$ GeV. The solid curve is the background from $t\bar{t}$ production with one t decaying semi-leptonically and the other hadronically. The curves marked (a) are for $Q \to W^- t,\, t \to W^+ b,\, W^+ \to l\nu$, $(\overline{Q}, W^-) \to hadrons$. The curves marked (b) are for $Q \to W^- t,\ W^- \to l\nu$, $(\overline{Q}, t) \to hadrons$. (In all cases we have included the charge conjugate process.) The dotted and short dashed-dot curves are for $M_Q = 200$ GeV, the short-dashed and long-dashed curves are for $M_Q = 400$ GeV, and the long dashed-dot curve is for $M_Q = 600$ GeV.

In Figure 5, we show the lepton p_T spectrum from heavy quark decay and from $t\bar{t}$ production with and without a cut on $M_T\ (lb)$. The p_T spectrum of the signal is considerably softer than for the lighter top quark and peaks at p_{l_T} near 100 GeV.

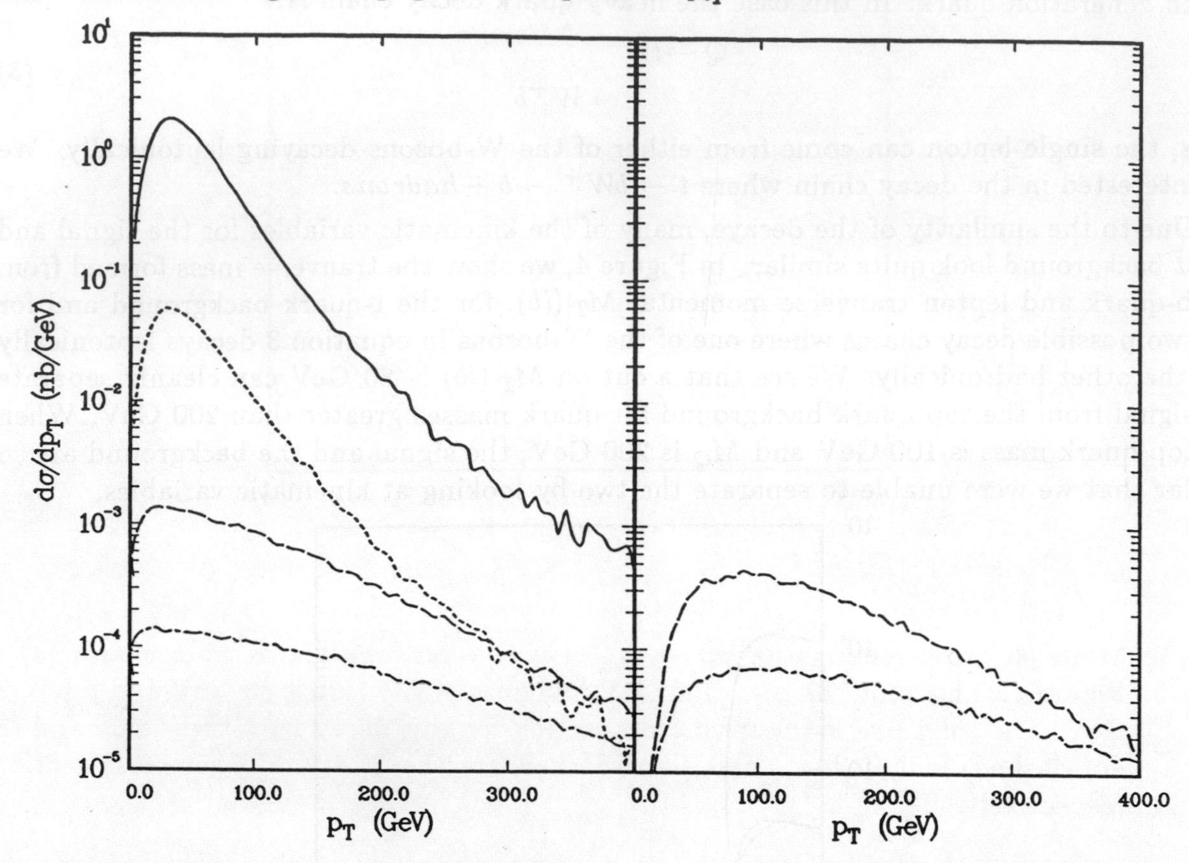

Fig. 5: $d\sigma/dp_{lT}$ for $pp \rightarrow Q\overline{Q} \rightarrow \ell^{\pm} + hadrons$ via decay scenario (a) of Figure 4. This Figure has m_T= 100 GeV. The solid line is the background from $t\bar{t}$ production while the short-dashed, long-dashed and dot-dashed lines have M_Q = 200, 400, and 600 GeV respectively. Figure 5a has no cuts, while Figure 5b has $M_T\ (lb) >$ 80 GeV. (Note that this cut eliminates not only the $t\bar{t}$ background, but also the signal from M_Q = 200 GeV.)

As in the case of the lighter top, the cleanest signal is to look for events with multi-jets. In Figure 6, we show the multi-jet cross section with $M_T\ (lb) > 80$ GeV and $p_{lT} > 30$ GeV as a function of M_Q. With the same cuts, the top quark background is totally eliminated. For M_Q = 600 GeV, the 3-jet cross section is 2 x 10^{-3} nb which gives approximately 2000 events in a standard SSC year. Even for M_Q = 200 GeV where the cut on $M_T\ (lb)$ was ineffective, requiring events with greater than 3 jets (using $\Delta R < 1$) cleanly rejects the top quark background.

4. Conclusions

Applying simple kinematic constraints to events with a single high p_T lepton and multi-jets allows the separation of events resulting from heavy quark decay from the background from top production. Clearly, it is *crucial* to be able the find multi-jet events although a segmentation of $\Delta R = 1$ appears to be adequate to reject the $t\bar{t}$ background.

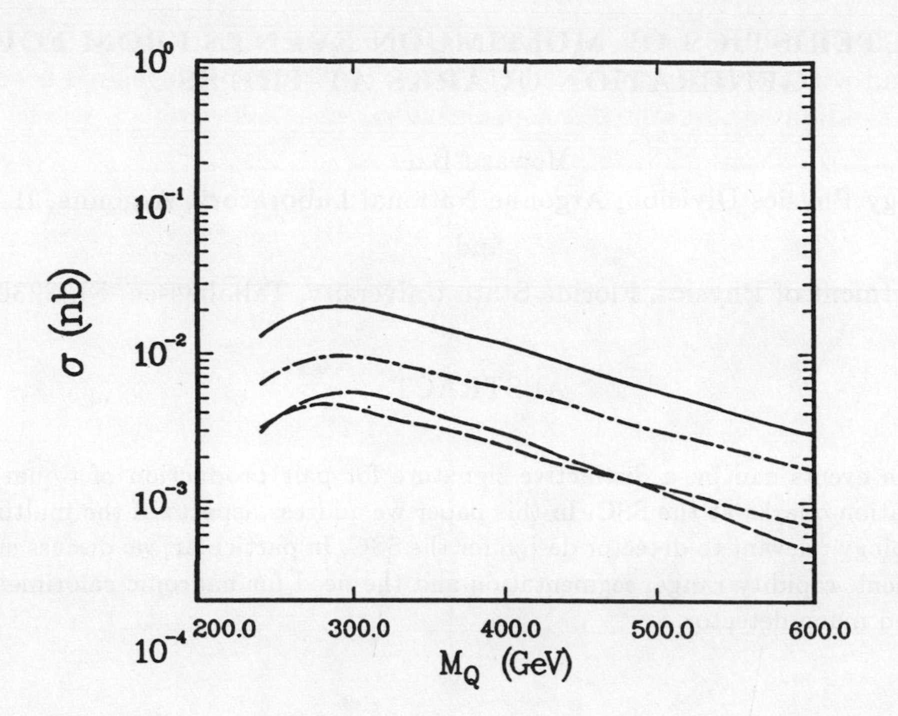

Fig. 6: Multi-jet cross sections for $pp \to Q\overline{Q} \to l^{\pm} + hadrons$ via decay scenario (a) of Figure 4. This plot has m_t= 100 GeV and cuts of $M_T\,(lb) > 80$ GeV and $p_{lT} > 30$ GeV. The solid line is the total cross section for this decay with these cuts and the dot-dashed, long-dashed, and short-dashed are the 3, 4, and 2-jet cross sections respectively.

Acknowledgments

This work has been supported in part by the United States Department of Energy under contract number DE-AC02-76CH00016.

References

1. See for instance. E. Eichten, I. Hinchliffe, K. Lane, and C. Quigg, Rev. Mod. Phys. **56**, 579 (1984).
2. S. Kim, in Proceedings of the Superconducting Supercollider Workshop, Snowmass, 1986, p. 241.
3. S. Dawson and S. Godfrey, Brookhaven National Lab. preprint (in preparation).
4. Summary of the Heavy Quark Physics Subgroup, these proceedings; H. Baer, these proceedings.
5. In calculating the cross sections we used the structure functions of ref. 1 (set 1).

CHARACTERISTICS OF MULTIMUON EVENTS FROM FOURTH GENERATION QUARKS AT THE SSC

Howard Baer
High Energy Physics Division, Argonne National Laboratory, Argonne, IL 60439
and
Department of Physics, Florida State University, Tallahassee, FL 32306

ABSTRACT

Multimuon events can be a distinctive signature for pair production of t-quarks or 4th generation quarks at the SSC. In this paper we address aspects of the multimuon event topology relevant to detector design for the SSC. In particular, we discuss energy measurement, rapidity range, segmentation and the need for hadronic calorimetry in a dedicated muon detector.

1. Introduction

An important test of one's capability to discover new physics at the SSC lies in the ability to distinguish a new heavy quark signal from background. Studies have been performed on the one or two lepton plus jets signal for new heavy quarks [1]. Although cross sections tend to be large there are substantial background problems in these channels. Convincing evidence for a new quark species might come through a complementary signal in another channel, which may be cleaner, but have lower rates. Recent studies indicate that the multimuon signal from 4th generation quarks is such a signal[2], and the purpose of this report is to expose characteristics of the multimuon event topology relevant to detector design for the SSC.

The underlying idea is that one can produce very many muons from a heavy quark as it decays through a cascade to lighter quark flavors. The fourth generation down-type quark decays mainly via

$$v \to t \to b \to c \to s$$

while the up-type quark decays via

$$a \to b \to c \to s$$

if $m_a < m_v + m_w$, or

$$a \to v \to t \to b \to c \to s$$

if $m_a > m_v + m_w$. As a result of these cascades, each $v\bar{v}$ event can produce a maximum of 16 muons; likewise, each $a\bar{a}$ event could produce up to 14 or 26 muons depending on its mass. Although the cross sections for so many muons in the final state are tiny, there nevertheless exists a substantial cross section for events with up to 5 or 6 muons in the final state, or 3 or 4 same sign muons, even after applying p_T and rapidity cuts.

In our calculations, we compute $Q\bar{Q}$ production through gg and $q\bar{q}$ fusion, and then implement the full cascade of weak decays down to final state u, d and s quarks and electrons, muons and neutrinos. At each stage we implement fragmentation according to the prescription of Peterson[3] *et al.*. We then apply p_T and rapidity cuts, and count the number of final state muons. The results given here and in Ref. 2 are for nearly degenerate a and v masses, with $m_a \approx m_v \approx 140$ or 240 GeV, values favored by GUT's[4] or SUSY GUT's[5] where the Yukawa couplings are evolved from the GUT scale to fixed points at the weak scale with no intermediate mass scales, and with the simplest Higgs sectors.

2. Results

Among the topics to be addressed at this workshop relevant to multimuon physics are luminosity, energy resolution, rapidity range, segmentation and hadronic calorimetry. I address each of these in turn.

Luminosity: A dedicated muon detector has an advantage over conventional detectors in that they are able to withstand a higher luminosity, due to lower trigger rates. Triggering on a muon with $p_T > 10$ GeV has little effect on the signal in 5 muon events, as shown in Fig. 1. Table I contains cross-sections for multimuon events down to the .1 pb range, which would correspond to 1000 events/year for the design luminosity of 10^4 pb^{-1}/year. This is clearly enough luminosity to detect fourth generation quarks in the mass range indicated. Raising the luminosity past its design value will only be beneficial for searching for new heavy quarks: one may then be sensitive to muon multiplicities even greater than six, corresponding to cross sections below .1 pb.

Table I. Cross-sections at 40 TeV for various multimuon configurations resulting from production and decay of heavy fourth generation $Q\bar{Q}$ pairs and of $t\bar{t}$ pairs, subject to cuts of $p_{T_{muon}} \geq 3$ GeV (10 GeV) and $\eta_{muon} \leq 1.75$ (3). Cross sections below .1 pb are indicated by a dash —.

Muons/Event	Cross Section (pb)		
	$v\bar{v} + a\bar{a}$		$t\bar{t}$
	$m_a \approx m_v$ (GeV)		m_t GeV
	140	240	60
Both Signs			
3	473 (171)	111 (49)	1325 (165)
4	57 (10)	15 (4)	60 (3)
5	4.3 (.24)	1.6 (.3)	1.3 (—)
6	0.8 (—)	0.2 (—)	— (—)
Same Sign			
2	1525 (778)	271 (174)	5295 (1350)
3	81 (15)	19 (7)	78 (5)
4	1.5 (—)	0.66 (—)	— (—)

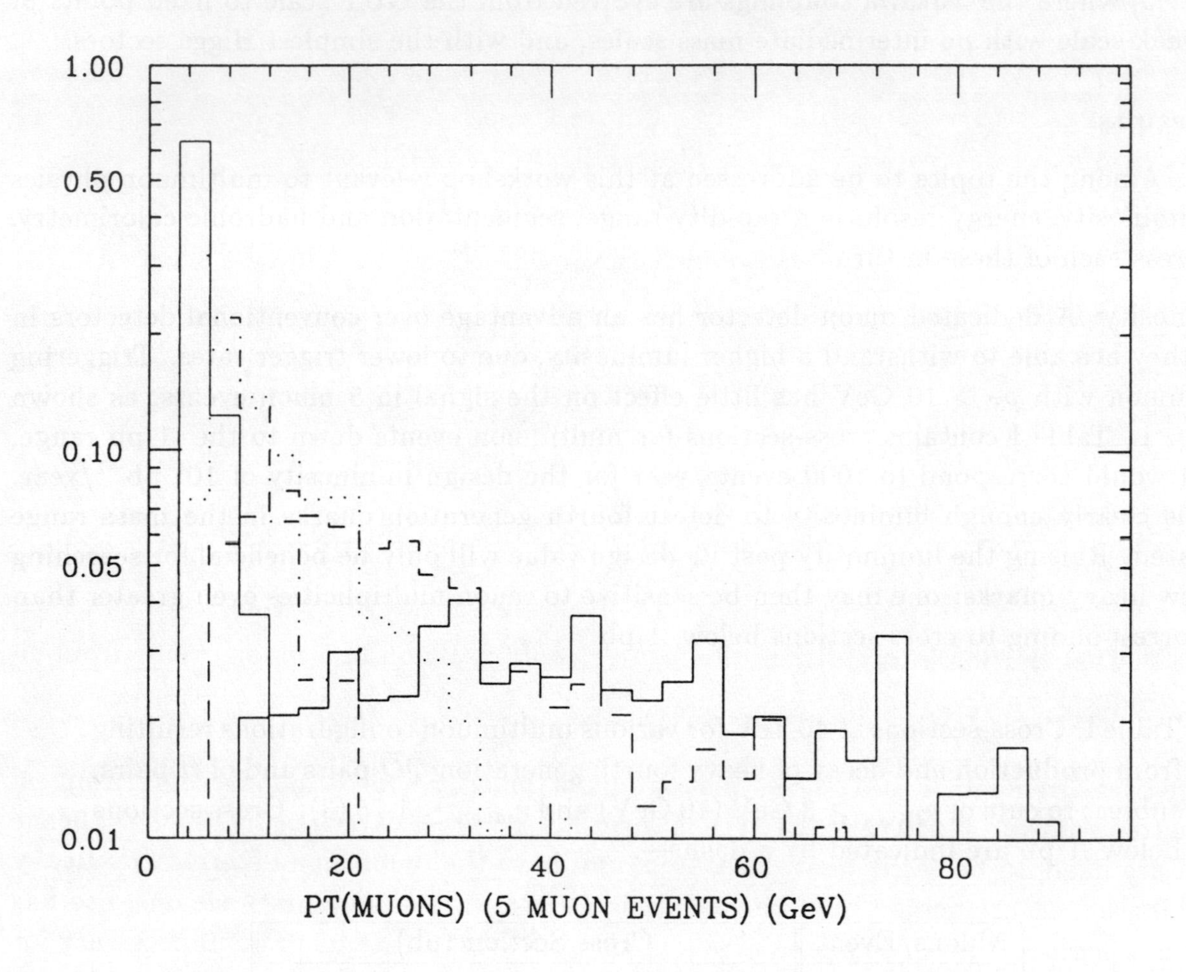

Figure 1: Histograms of p_T distributions of the 5 muons in 5 muon events from $v\bar{v}$ (240) production. In order of increasing p_T, the histograms are solid, dotdashed, dotted, dashed and solid. The only cut invoked in these figures is the requirement that $p_{T_\mu} > 3$ GeV.

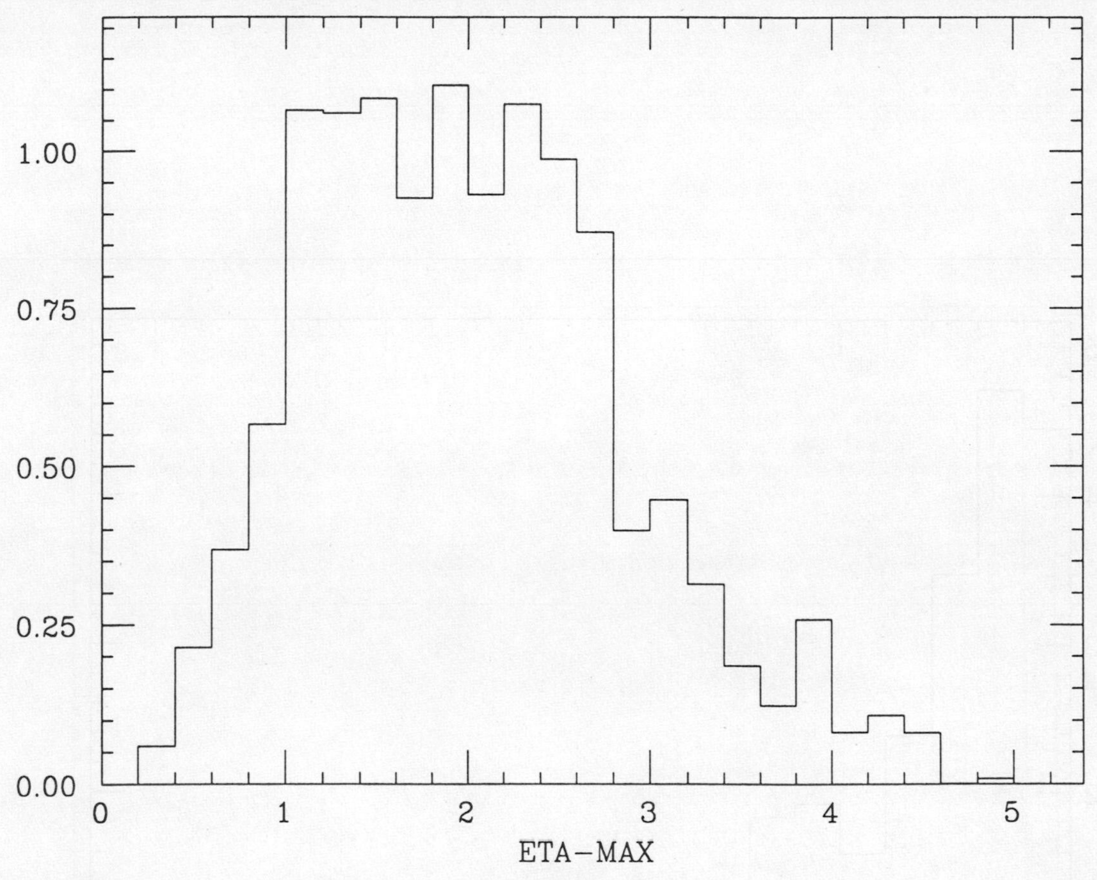

Figure 2: Distribution in rapidity of the highest rapidity muon in 5 muon events from $v\bar{v}$ (240) production.

Energy Measurement: Fig. 1 shows histograms of muon p_T for 5 muon events from $v\bar{v}$ production with $m_v = 240$ GeV. Even the fastest muons never exceed $p_T = 100$ GeV, hence tracks are likely to be easily bent, measured, and have the muon sign determined. Heavy quark mass measurements would be very difficult in multimuon events since one also has many missing neutrinos. Very precise muon energy resolution seems to be unnecessary for looking for heavy quarks via multimuon detection. In Table I we have plotted multimuon cross sections for p_T cuts of 3 and 10 GeV. Increasing the p_T cut from 3 to 10 GeV reduces the multimuon signal, but reduces the $t\bar{t}$ background even more. In this case one might concentrate on 4 and 5 muon events rather than 5 and 6 muon events.

Rapidity Range: In Fig. 2, we have plotted the rapidity of the highest rapidity muon in 5 muon events. The bulk of the signal occurs for $|\eta| < 3$; the multimuons tend to inhabit the central region.

Segmentation: Fig. 3 is a plot of $\Delta R = \sqrt{\Delta\eta^2 + \Delta\phi^2}$, where we have calculated all ΔR's between all pairs of muons in 5 muon events, and taken the minimum value. Only a small fraction of the multimuon events have a pair of muons with $\Delta R < .1$, so in general, the multimuons should be well separated and have distinct tracks.

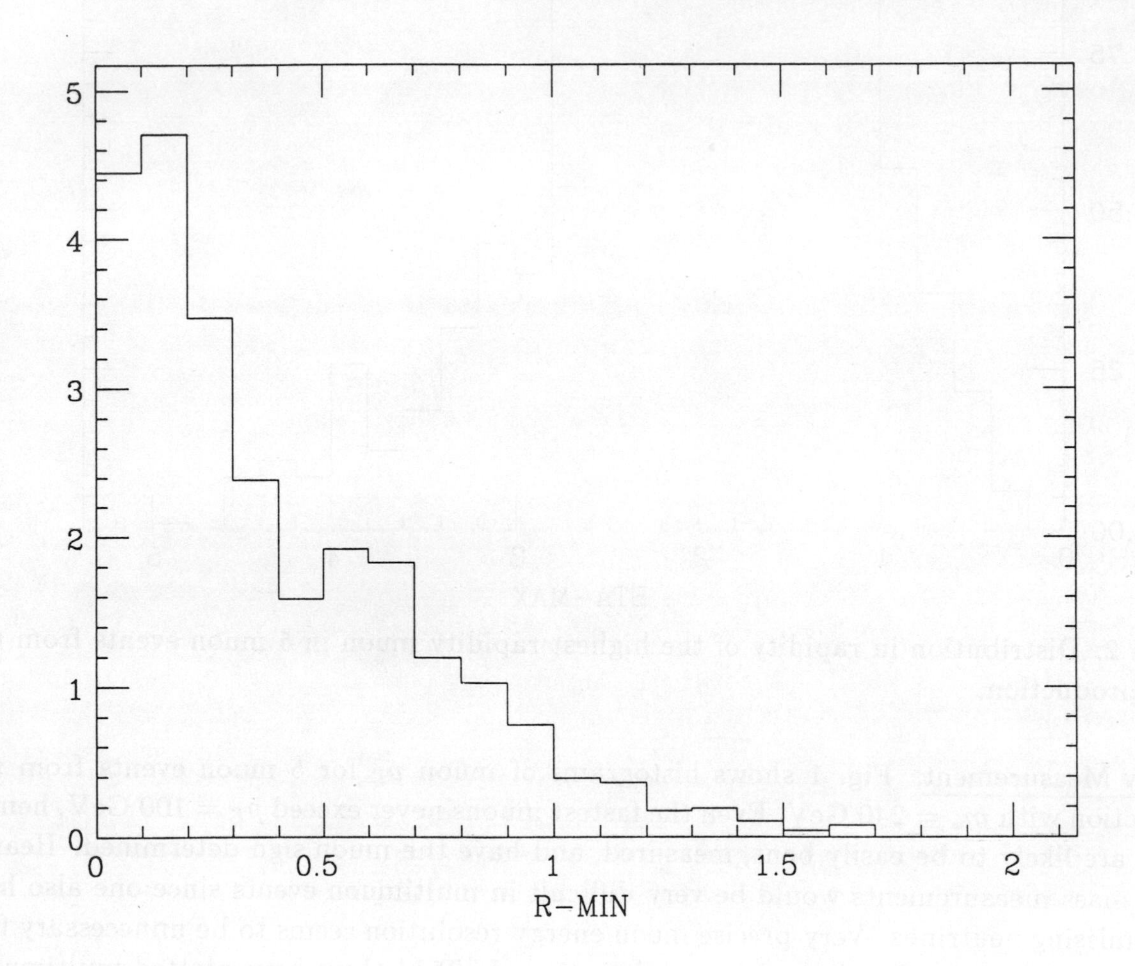

Figure 3: Minimum separation of any two muons in $\Delta R = \sqrt{\Delta\eta^2 + \Delta\phi^2}$ in 5 muon events from $v\bar{v}$ (240) production.

Hadronic Calorimetry: There will naturally be substantial jet activity in multimuon events from 4th generation quarks, and this information is very important for separating signals from backgrounds. The highest p_T muon from 4th generation quarks will often be isolated from substantial jet activity, allowing one to invoke isolation cuts to eliminate backgrounds from such sources as $b\bar{b}b\bar{b}$ events, or multiple scattering events.

3. Conclusion

In this report we have examined multimuon events from 4th generation quarks with an eye towards minimal detector design. An adequate dedicated muon detector would have muon detection with a rapidity range $|\eta| < 3$ and segmentation in $\Delta\eta$ and $\Delta\phi \leq .1$. Very precise energy measurement would not be needed. Muon sign determination is very important. Charged particle tracking and hadronic calorimetry at some level would be highly desirable for vetoing background events.

I thank V. Barger and H. Goldberg for the collaboration which led to this report,and also the workshop organizers for a stimulating environment to work in. This work supported by the U.S. Department of Energy, Division of High Energy Physics, Contract W-31-109-ENG-38.

References

1. V. Barger, H. Baer, K. Hagiwara and R. J. N. Phillips, Phys. Rev. **D30**, 947 (1984); S. Kim, Proceedings of SNOWMASS 86, pg. 241; E. W. N. Glover and D. A. Morris, Proceedings of SNOWMASS 86, pg. 238; J. Haggerty *et al.*, these proceedings; S. Dawson and S. Godfrey, to be published.

2. H. Baer, V. Barger and H. Goldberg, NUB-2725 T/E (to be published in Phys. Rev. Letters).

3. C. Peterson *et al.*, Phys. Rev. **D27**, 105 (1983).

4. C. T. Hill, Phys. Rev. **D24**, 691 (1981); N. Cabbibo, L. Maiani, G. Parisi, and R. Petronzio, Nucl. Phys. **158B**, 295 (1979); M. Pendelton and G. G. Ross, Phys. Lett. **98B**, 291 (1981); M. Machacek and M. T. Vaughn, Phys. Lett. **103B**, 247 (1981); E. A. Paschos, Z. Phys. **C26**, 235 (1984); J. W. Halley, E. A. Paschos, and H. usler, Phys. Lett. **155B**, 107 (1985); J. Bagger, S. Dimopoulos, and E. Masso, Nucl. Phys. **B253**, 397 (1985).

5. H. Goldberg, Phys. Lett. **165B**, 292 (1985); M. Cvetič and C. R. Preitschopf, Nucl. Phys. **B272**, 490 (1986).

Techniques for Finding Supersymmetry at the SSC*

R. MICHAEL BARNETT, CHRIS KLOPFENSTEIN, EDWARD WANG
Lawrence Berkeley Laboratory, University of California, Berkeley CA 94720

FELICITAS PAUSS
CERN ,Geneva, Switzerland

JOHN F. GUNION
Department of Physics, University of California, Davis CA 95616

HOWARD E. HABER
Department of Physics, University of California, Santa Cruz CA 95064

HOWARD BAER
Department of Physics, Florida State University, Tallahassee, FL 32306

MANUEL DREES, XERXES TATA
Department of Physics, University of Wisconsin, Madison WI 53706

KAORU HAGIWARA
KEK, Tsukuba, Ibaraki 305, JAPAN

Abstract

We examine the signatures at the SSC for supersymmetry for much of the (minimal) supersymmetric model parameter space. In particular, we survey the decay modes and signatures of gluinos and squarks. Gluinos (squarks) decay to two (one) jets and a chargino or neutralino ($\widetilde{\chi}$). This $\widetilde{\chi}$ may be the (stable) lightest supersymmetric particle, LSP (and lead to missing energy). Or $\widetilde{\chi}$ may have a two-body decay to another $\widetilde{\chi}$ plus a W, Z or Higgs boson. Finally, it may have a three-body decay to the LSP plus $q\overline{q}$, $e\nu$, $\mu\nu$, ee or $\mu\mu$. Only for very light gluinos and squarks is the decay mode containing the LSP dominant. In fact, for gluinos and squarks over 500 GeV, the decays to W and Z bosons dominate for much of parameter space. Finding such particles depends on our ability to detect W and Z bosons which decay to leptons. We discuss the efficiency for lepton detection in such events, and we estimate the backgrounds. The decays of gluinos and squarks which go directly to the LSP lead to very large missing energy. We report the initial results of a study of the backgrounds for this process.

* Work supported by Director, Office of High Energy and Nuclear Physics, Division of High Energy Physics of the US Department of Energy under contract nos. DE-AC03-76SF00098, DE-AA03-76-SF00010, and DOE-76ER-70191-MODA33, and by the National Science Foundation under agreement no. PHY83-18358.

1. Introduction

Attention has been focused on supersymmetry by subgroups at several Supercollider workshops during the last few years. New theoretical and experimental developments now permit a more sophisticated analysis of techniques for finding supersymmetry than was possible then. The motivation for searching for supersymmetry remains high. The development of superstring theories provided a deeper understanding of the expectations for supersymmetry. For some time theorists have given plausible arguments for why the masses of supersymmetric particles should be 1 TeV or less. Experimentally the lower bounds have now risen to 30-80 GeV depending on the particle.

The primary experimental lessons we can now take advantage of are those learned in analyzing the UA1 data[1,2] for monojets and dijets with missing energy and in searching for the t quark[3,4]. The UA1 Collaboration made considerable progress in finding techniques for separating signals from backgrounds. One can also make use of the reported mass limits on supersymmetric particles. Theoretically we now understand that it is essential to allow full mixing in the gaugino-higgsino sector, and to avoid making any assumptions concerning the mass of the lightest supersymmetric particle (we do not assume it is almost massless). In this paper our conclusions are not based on a single scenario, but rather on a complete survey of all of parameter space for the minimal supersymmetric model.

At a hadron collider the most copiously produced supersymmetric particles (kinematics allowing) will be the gluino and squarks, which are produced via the strong interactions. For $\widetilde{q}$ and $\widetilde{g}$ masses in the range that have been looked for at the CERN $Sp\bar{p}S$ collider ($M \lesssim 60\ GeV$) the gluino would decay primarily via

$$\widetilde{g} \rightarrow q\bar{q}\widetilde{\chi}_1^0,$$

and the squark (if $M_{\widetilde{q}} < M_{\widetilde{g}}$) via

$$\widetilde{q} \rightarrow q\widetilde{\chi}_1^0.$$

$\widetilde{\chi}_1^0$ is the lightest neutralino, which we take to be the lightest supersymmetric particle (LSP). Squark or gluino production, in this case, would be signalled[5] by events with jet(s) and missing transverse energy (E_T^{miss}), due to the LSP which escapes detection. The UA1 Collaboration,[2] after an analysis of the E_T^{miss} data sample has recently reported the limits $M_{\widetilde{g}} > 53\ GeV$ and $M_{\widetilde{q}} > 45\ GeV$. These limits are valid for $m_{\widetilde{\chi}_1^0} < 20-30\ GeV$. It has also been argued[6] that gluinos and squarks with masses up to $150 - 200\ GeV$ can be searched for at the Fermilab Tevatron. Above this mass range, searches can best be carried out at the next generation of hadron colliders such as the Superconducting Super Collider (SSC). Indeed, it is very possible that no evidence of physics beyond the Standard Model will be seen at the Tevatron, SLC or LEP. Such a result need not be in conflict with supersymmetry as the explanation of the origin of the electroweak scale. Certainly, sensible models of "low-energy supersymmetry" are easily constructed which have squarks, sleptons and gluinos with masses of the order of 500 GeV and beyond.[7]

In this paper, we are primarily interested in heavy gluinos and squarks which could be discovered at a future supercollider. For each particle we shall study its phenomenology only for the case that it is the lighter of the two particles. The reason for this is simple: if (say) gluinos are heavier than squarks, then they will decay into squarks (via a two-body decay). To study supersymmetry in this case, it would be far more efficient to study squarks which are produced directly. The detection of gluinos would then be a "second generation" type of experiment. Thus, for the rest of this paper, we assume when we study gluinos that $M_{\tilde{g}} < M_{\tilde{q}}$ and assume $M_{\tilde{q}} < M_{\tilde{g}}$ when we study squarks. With these two assumptions, the two-body decays $\tilde{g} \to \tilde{q}q$ and $\tilde{q} \to \tilde{g}q$ are forbidden. For gluinos one must consider all kinematically allowed decays of the type

$$\begin{aligned} &\tilde{g} \to q\bar{q}'\tilde{\chi}_i^{\pm} \\ &\tilde{g} \to q\bar{q}\tilde{\chi}_i^0 \ . \end{aligned} \tag{1}$$

and for squarks one must consider all kinematically allowed decays of the type

$$\begin{aligned} &\tilde{q} \to q'\tilde{\chi}_i^{\pm} \\ &\tilde{q} \to q\tilde{\chi}_i^0 \ . \end{aligned} \tag{2}$$

where the index i runs over all charginos or neutralinos.

We focus on the predictions for the above decays in the minimal supersymmetric extension of the Standard Model, specified in detail in refs. 8 and 9. In this model the spin-1/2 superpartners of the two Higgs doublets combine with the superpartners of the $W^{\pm}$ and of the γ,Z to yield two chargino mass eigenstates, $\tilde{\chi}_1^{\pm}$ and $\tilde{\chi}_2^{\pm}$, and four neutralino mass eigenstates, $\tilde{\chi}_1^0$, $\tilde{\chi}_2^0$, $\tilde{\chi}_3^0$, and $\tilde{\chi}_4^0$; the labelling is according to increasing mass. The importance of allowing neutralinos and charginos in the final state with arbitrary mixing angles must be stressed. One often finds analyses presented where special assumptions have been made (*e.g.* that the lightest neutralino is a pure photino). Apart from the fact that such assumptions are arbitrary, one can sometimes be led to wrong conclusions.

In general there are many allowed decay channels for gluinos and squarks. In particular, it is important to note that several of the charginos and neutralinos are usually substantially lighter than the gluino and squark. Thus the probability that a heavy gluino or squark will decay directly into jets and the LSP may be rather small; decays to a heavier chargino or neutralino, which eventually cascades down into the LSP could be (and, in fact, are) dominant.[10] Of special interest are decays to charginos and neutralinos that are heavy enough that they will, in turn, decay into a lighter chargino or neutralino plus a W or Z[11,12,13]. The resulting signature for gluino production is striking. In this paper, we will assess the relative importance of such decays as a function of the gluino (or squark) mass and other parameters of the supersymmetric theory.

Let us now discuss the parameters of the minimal supersymmetric model. The mass matrices for the $\widetilde{\chi}^{\pm}$ and $\widetilde{\chi}^{0}$ sectors depend on three unknown mass scales—μ, M_2, and M_1—in addition to the Higgs vacuum expectation values to be discussed shortly. Here μ is a supersymmetric Higgs mass parameter and M_2 and M_1 are gaugino mass parameters associated with the soft breaking of supersymmetry in the $SU(2)$ and $U(1)$ sectors. We will follow the common practice of reducing the parameter freedom by assuming that these latter two mass parameters are related to the gaugino mass of the $SU(3)$ subgroup, M_3 (which is equal to the gluino mass, $M_{\tilde{g}}$), by requiring that the three mass scales are equal at some grand unification scale. Using the notation of refs. 8 and 9, where $M_2 \equiv M$ and $M_1 \equiv (3/5)M'$, this implies

$$\begin{aligned} M &= \frac{g^2}{g_s^2} M_{\tilde{g}} \\ M' &= \frac{5g'^2}{3g_s^2} M_{\tilde{g}}. \end{aligned} \tag{3}$$

Turning to the Higgs sector of the minimal model, we emphasize that it contains exactly two doublets, H_1 and H_2. The vacuum expectation values of the neutral members of these two doublets, v_1 and v_2, give masses to the down and up-type quarks respectively. Of the eight degrees of freedom, three are absorbed in giving mass to the $W^{\pm}$ and Z, leaving two neutral scalar Higgs bosons, H_1^0 and H_2^0, a neutral pseudoscalar Higgs boson, H_3^0, and a pair of charged Higgs bosons, $H^{\pm}$. In the minimal model there are strong constraints upon the tree–level masses of these various Higgs bosons. Using the notation

$$\tan\beta = v_2/v_1, \tag{4}$$

one finds that by fixing $\tan\beta$ and one of the Higgs masses (say, $m_{H^{\pm}}$), all the other tree-level Higgs masses are determined:[14]

$$m_{H_3^0}^2 = m_{H^+}^2 - m_W^2 \tag{5}$$

$$m_{H_1^0,H_2^0}^2 = \frac{1}{2}\left(m_{H_3^0}^2 + m_Z^2 \pm \sqrt{(m_{H_3^0}^2 + m_Z^2)^2 - 4m_Z^2 m_{H_3^0}^2 \cos^2 2\beta}\right). \tag{6}$$

Note that H_2^0 is always lighter than the Z and is particularly light if $\tan\beta$ is near 1 or if $m_{H^{\pm}}$ is near m_W. Hence, one would expect H_2^0 to play a central role in the phenomenology of chargino and neutralino decays. Depending upon the choice of $m_{H^{\pm}}$ some, or all, of the remaining Higgs bosons may also be light enough to be important in neutralino and chargino decays (note though that $H^{\pm}$ is always heavier than the W boson). These considerations become important for the discussion of gluinos which do not decay directly into the LSP.

Our analysis always accounts for the existence of a region of μ for which the mass of the lightest chargino is less than the experimental lower bound which we take to be ~ 30 GeV

(the boundaries of this μ region depend on $M_{\tilde{g}}$). We use the above bound as a conservative limit based on the PETRA bound of 23 GeV[15] and the limit of ref. 16, inferred from UA1 data, of $\sim 40\ GeV$. In what follows we shall only present results for μ values that do not violate this bound.

2. Gluino Decays

As stated in the Introduction, when studying gluinos, we assume that $M_{\tilde{g}} < M_{\tilde{q}}$. For simplicity, we shall take six generations of $\tilde{q}_L$ and $\tilde{q}_R$ to be degenerate in mass. The formulae for gluino decay widths we use in obtaining branching ratios are given in Ref. 12. The only approximation that we make is to take the quarks which appear in the final state to be massless. The overall effect of finite quark mass on gluino branching ratios is small. Without loss of generality we may restrict β to lie between 0 and $\pi/2$. We make the assumption of CP invariance in the neutralino and chargino sectors, so M,M' and μ can be taken as real. The parameter M is taken as positive whereas μ can have either sign. We assume that the lightest supersymmetric particle (LSP) is the lightest neutralino, $\tilde{\chi}_1^0$; it is assumed to be stable and will escape collider detectors as missing energy. Finally, for our numerical work we have taken $M_{\tilde{q}} = 1.5 M_{\tilde{g}}$, but our results are insensitive to this choice.

In presenting our results for gluino decays to neutralinos and charginos, we generally consider two representative values for $\tan\beta$: 1.5 and 4. Results for $\tan\beta = 1$ are always very similar to those at $\tan\beta = 1.5$. Further, all results are unchanged if $\tan\beta \to \cot\beta$. Finally, recall that we shall only plot results corresponding to μ values that yield $m_{\tilde{\chi}_1^+} > 30\ GeV$.

We begin by considering the branching ratio for gluino decay to the LSP, $\tilde{\chi}_1^0$. The gluino searches at the CERN $Sp\bar{p}S$ have relied on this branching ratio being large for light gluinos. In figs. 1 and 2 the branching ratio for $\tilde{g} \to q\bar{q}\tilde{\chi}_1^0$ is plotted as a function of μ for a series of $M_{\tilde{g}}$ values ranging from $M_{\tilde{g}} = 50\ GeV$ to $M_{\tilde{g}} = 1\ TeV$, taking $\tan\beta = 1.5$ and $\tan\beta = 4$. We see that for $M_{\tilde{g}} = 50\ GeV$ there is a range of μ over which the branching ratio for this decay is unity (when $\tan\beta$ is not too large). However, even for this low $M_{\tilde{g}}$ value, the branching ratio for this decay decreases rapidly for $|\mu| \gtrsim 250\ GeV$. The branching ratio to the LSP also vanishes for $|\mu|$ very near 0. This is easily understood, since in this region the LSP is dominantly higgsino.

As we move to higher $M_{\tilde{g}}$ values, figs. 1 and 2 make it clear that the branching ratio to $\tilde{\chi}_1^0$ decreases very rapidly, especially in the case of $\tan\beta = 4$. Indeed, once $M_{\tilde{g}} \gtrsim 400\ GeV$ this branching ratio is essentially zero in the vicinity of $\mu = 0$ and rises to around 0.14 at large $|\mu|$.

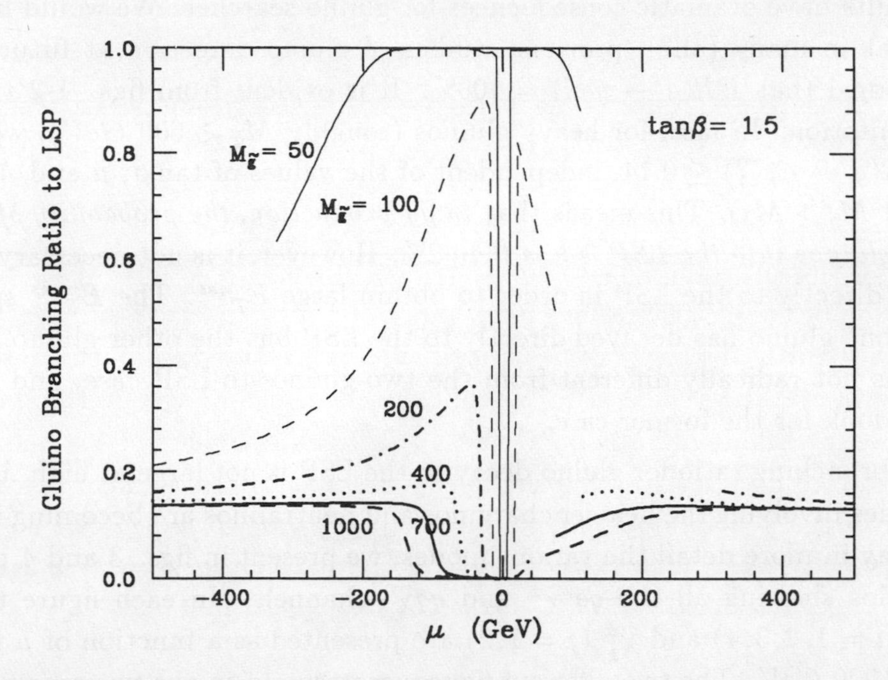

Figure 1: The branching ratio for $\tilde{g} \to q\bar{q}\tilde{\chi}_1^0$ as a function of μ for a series of $M_{\tilde{g}}$ values (in GeV units), where $\tilde{\chi}_1^0$ is the lightest supersymmetric particle (LSP). For this figure we take $\tan\beta = 1.5$. Sections of the curves that are not plotted, both here and in all succeeding graphs, correspond to parameter choices which yield $M_{\tilde{\chi}_1^+} < 30\ GeV$.

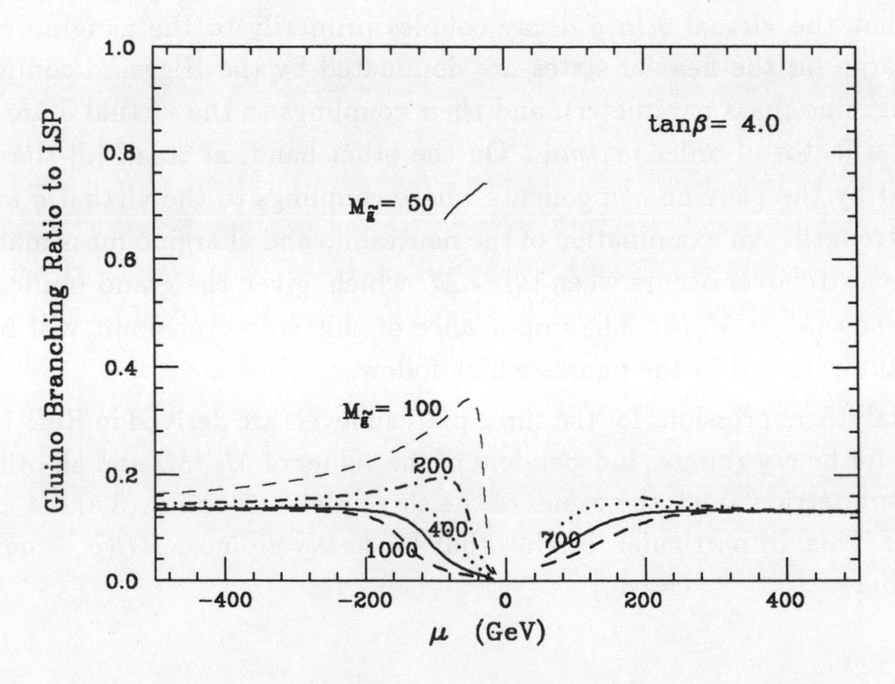

Figure 2: We present the same plots as in fig. 1, but for $\tan\beta = 4$.

These results have dramatic consequences for gluino searches. We would like to caution the reader that in many (all?) previous studies of gluino detection at future colliders, it has been assumed that $BR(\tilde{g} \to q\bar{q}\tilde{\gamma}) = 100\%$. It is evident from figs. 1-2 that this is an incorrect assumption. In fact, for heavy gluinos (roughly $M_{\tilde{g}} \gtrsim 600\ GeV$), we find a strict inequality $BR(\tilde{g} \to q\bar{q}\tilde{\chi}_1^0) \leq 0.14$, independent of the values of $\tan\beta$, μ and $M_{\tilde{q}}$ (assuming, of course, that $M_{\tilde{q}} > M_{\tilde{g}}$). This means that *in $\tilde{g}\tilde{g}$ production, the probability of having direct decay of both gluinos into the LSP is less than* 2%. However, it is not necessary to have both gluinos decay directly to the LSP in order to obtain large E_T^{miss}. The E_T^{miss} spectrum from events where one gluino has decayed directly to the LSP but the other gluino has any other decay mode, is not radically different from the two-gluinos-to-LSP case, and it is far more productive to look for the former case.

Since the branching ratio for gluino decay to the LSP is not large at high $M_{\tilde{g}}$ values, it is clear that modes involving the heavier charginos and neutralinos are becoming important. In order to display in more detail the various modes, we present in figs. 3 and 4 plots of gluino branching ratios showing all the $q\bar{q}'\tilde{\chi}^{\pm}$ and $q\bar{q}\tilde{\chi}^0$ channels. In each figure the branching ratios for $\tilde{\chi}_i^0$ $(i = 1, 2, 3, 4)$ and $\tilde{\chi}_j^{\pm}$ $(j = 1, 2)$ are presented as a function of μ for $M_{\tilde{g}} = 120$, 300, 700 and 1000 GeV. The two different figures correspond to our two representative $\tan\beta$ choices: $\tan\beta = 1.5$ and 4. Most apparent is the presence of three very distinct branching ratio levels. At large $|\mu|$ these correspond to $\tilde{g}$ decay to $\tilde{\chi}_1^{\pm}$, $\tilde{\chi}_2^0$ and $\tilde{\chi}_1^0$ in order of decreasing magnitude. At small $|\mu|$ these same plateau values emerge, but correspond to $\tilde{g}$ decay to $\tilde{\chi}_2^{\pm}$, $\tilde{\chi}_4^0$ and $\tilde{\chi}_3^0$, again in order of decreasing magnitude. That the dominant modes should switch from the heaviest states at small $|\mu|$ to the lightest states at large $|\mu|$ is easily explained by the fact that the virtual $\tilde{q}$ in $\tilde{g}$ decay couples primarily to the gaugino components of the $\tilde{\chi}$'s. At large $|\mu|$ the heavier states are dominated by the Higgsino components (recall that μ is a Higgsino mass parameter), and their couplings to the virtual $\tilde{q}$ are suppressed in amplitude by a factor of order m_q/m_W. On the other hand, at small $|\mu|$ the heavier states are dominated by the gaugino components whose couplings to the virtual $\tilde{q}$ are of standard electroweak strength. An examination of the neutralino and chargino mass matrices makes it clear that this switchover occurs when $|\mu| \sim M$, which, given the grand unification relations of eq. (3), means $|\mu| \sim M_{\tilde{g}}/4$. The importance of this crossover point will be apparent in much of the analysis and in the figures which follow.

Simple analytic expressions for the three plateau levels are derived in Ref. 12. From these one finds that for heavy gluinos, independent of the values of $M_{\tilde{g}}$, $M_{\tilde{q}}$ and all other parameters of the supersymmetric model, the values of the three plateaus of figs. 3 and 4 are about 0.58, 0.28 and 0.14. This, in particular, implies that for heavy gluinos, $BR(\tilde{g} \to q\bar{q}\tilde{\chi}_1^0) \leq 0.14$, as remarked earlier.

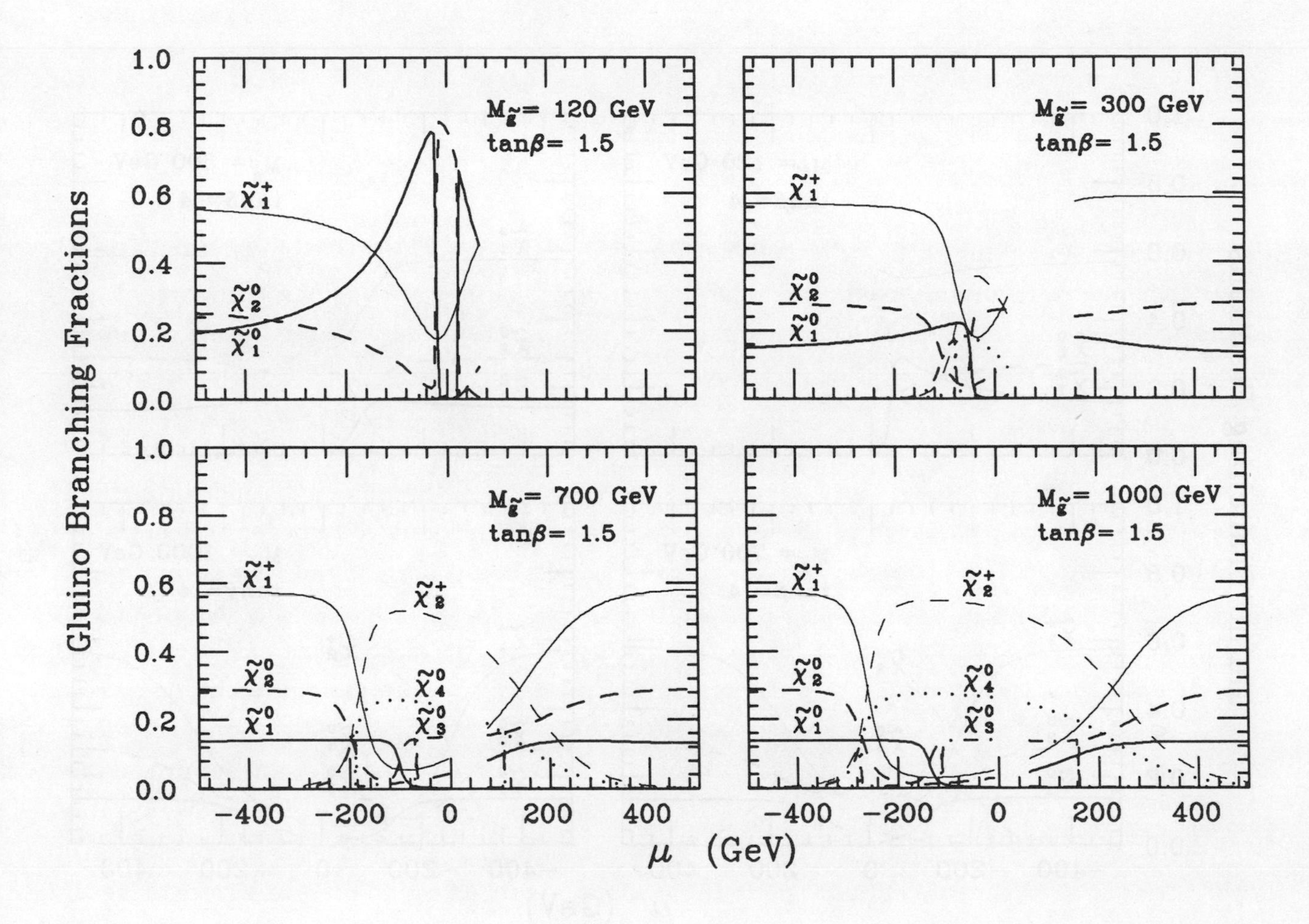

Figure 3: We give the branching ratios for $\widetilde{g} \to q\bar{q}\widetilde{\chi}_i^0$ and $\widetilde{g} \to q\bar{q}'\widetilde{\chi}_j^{\pm}$ $(i = 1,2,3,4$ and $j = 1,2)$ as a function of μ for four $M_{\widetilde{g}}$ values and $\tan\beta = 1.5$. The various curves correspond to: light solid line $= \widetilde{\chi}_1^{\pm}$; light dashed line $= \widetilde{\chi}_2^{\pm}$; heavy solid line $=\widetilde{\chi}_1^0$; heavy dashed line $= \widetilde{\chi}_2^0$; heavy dash-dot line $=\widetilde{\chi}_3^0$; and heavy dotted line $=\widetilde{\chi}_4^0$.

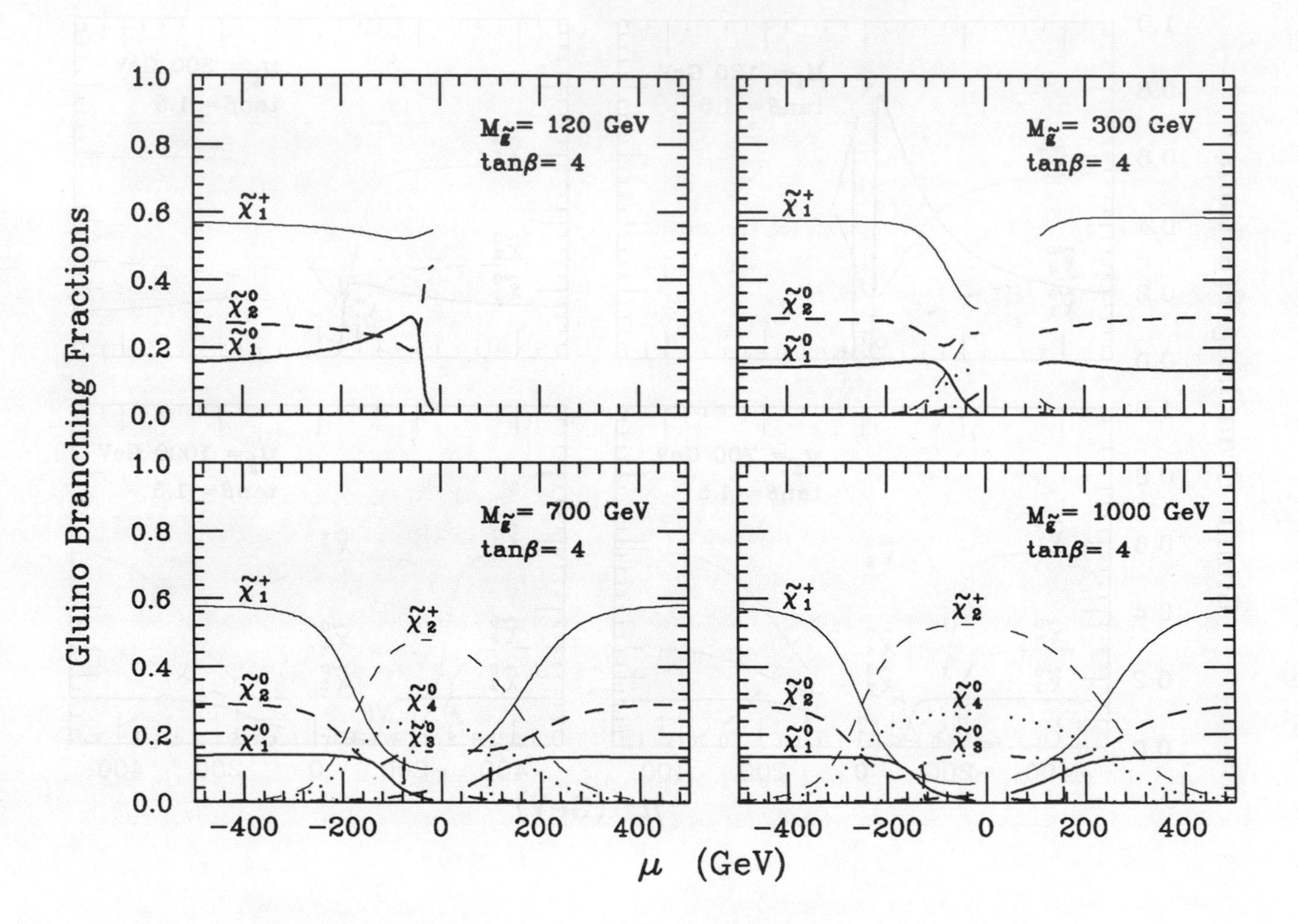

Figure 4: The same as for fig. 3, but with $\tan\beta = 4.0$.

Next we discuss the decay modes of the $\widetilde{\chi}^0$ and $\widetilde{\chi}^{\pm}$ which are produced in the decay of the gluino. Full results and formulae for these decays appear in a paper by Gunion and Haber.[17] The following two-body decays are allowed:

$$\begin{aligned} \widetilde{\chi}_i^0 &\rightarrow \widetilde{\chi}_j^0 + Z^0 \\ \widetilde{\chi}_i^0 &\rightarrow \widetilde{\chi}_j^{\pm} + W^{\mp} \\ \widetilde{\chi}_i^{\pm} &\rightarrow \widetilde{\chi}_j^0 + W^{\pm} \\ \widetilde{\chi}_i^{\pm} &\rightarrow \widetilde{\chi}_j^{\pm} + Z^0. \end{aligned} \tag{7}$$

$$\begin{aligned} \widetilde{\chi}_i^0 &\rightarrow \widetilde{\chi}_j^0 + H_k^0 \\ \widetilde{\chi}_i^0 &\rightarrow \widetilde{\chi}_j^{\pm} + H^{\mp} \\ \widetilde{\chi}_i^{\pm} &\rightarrow \widetilde{\chi}_j^0 + H^{\pm} \\ \widetilde{\chi}_i^{\pm} &\rightarrow \widetilde{\chi}_j^{\pm} + H_k^0 \end{aligned} \tag{8}$$

(where $k = 1$, 2, or 3).

If any of these two-body processes is allowed, they will certainly dominate any three-body decays mediated by virtual squark (or slepton) exchange. It is important to realize that two-body decays into a Higgs boson (especially the lightest Higgs) will, in general, be competitive with the production of vector bosons. By specifying $m_{H^\pm}$ in addition to those parameters already delineated for the neutralino/chargino sector, the widths for the decays to Higgs bosons may be computed. The constraints on the Higgs masses described above imply that H_2^0 is very light if either $\tan\beta$ is near 1 or $m_{H^\pm}$ is near m_W. In such cases, the decays $\widetilde{\chi}_i^0 \rightarrow \widetilde{\chi}_j^0 + H_2^0$ and $\widetilde{\chi}_i^+ \rightarrow \widetilde{\chi}_j^+ + H_2^0$ are certain to be important modes over nearly all of the supersymmetric parameter space. (If $m_{H^\pm}$ is near m_W, then H_3^0 is also light and can be similarly produced.)

In order to gain a more complete understanding of the phase space, both for $\widetilde{g}$ decays to the $\widetilde{\chi}^{\pm}$'s and $\widetilde{\chi}^0$'s and for $\widetilde{\chi}^{\pm}$ and $\widetilde{\chi}^0$ decays in the modes (7) and (8), we present in fig. 5 results for the masses of the various charginos and neutralinos for the four $M_{\widetilde{g}}$ values considered in figs. 3 and 4, taking $\tan\beta = 1.5$. (The mass spectra for $\tan\beta = 4$ are almost indistinguishable.) There are a number of features of these results that will be useful in the following discussions.

1. For all choices of $M_{\widetilde{g}}$ there are regions near $\mu = 0$ where the $\widetilde{\chi}_1^0$ and $\widetilde{\chi}_1^{\pm}$ are very light. Generally, the $\widetilde{\chi}_1^0$ is the LSP but there is always a narrow region of small positive μ for which $\widetilde{\chi}_1^{\pm}$ is the LSP. However, the bounds discussed earlier[15,16] imply that $m_{\widetilde{\chi}_1^+}$ must be $\gtrsim 30\ GeV$. From fig. 5 we see that this always rules out a set of small positive μ values, including those for which $m_{\widetilde{\chi}_1^+} < m_{\widetilde{\chi}_1^0}$. For these μ values we do not plot branching ratios in our various graphs.

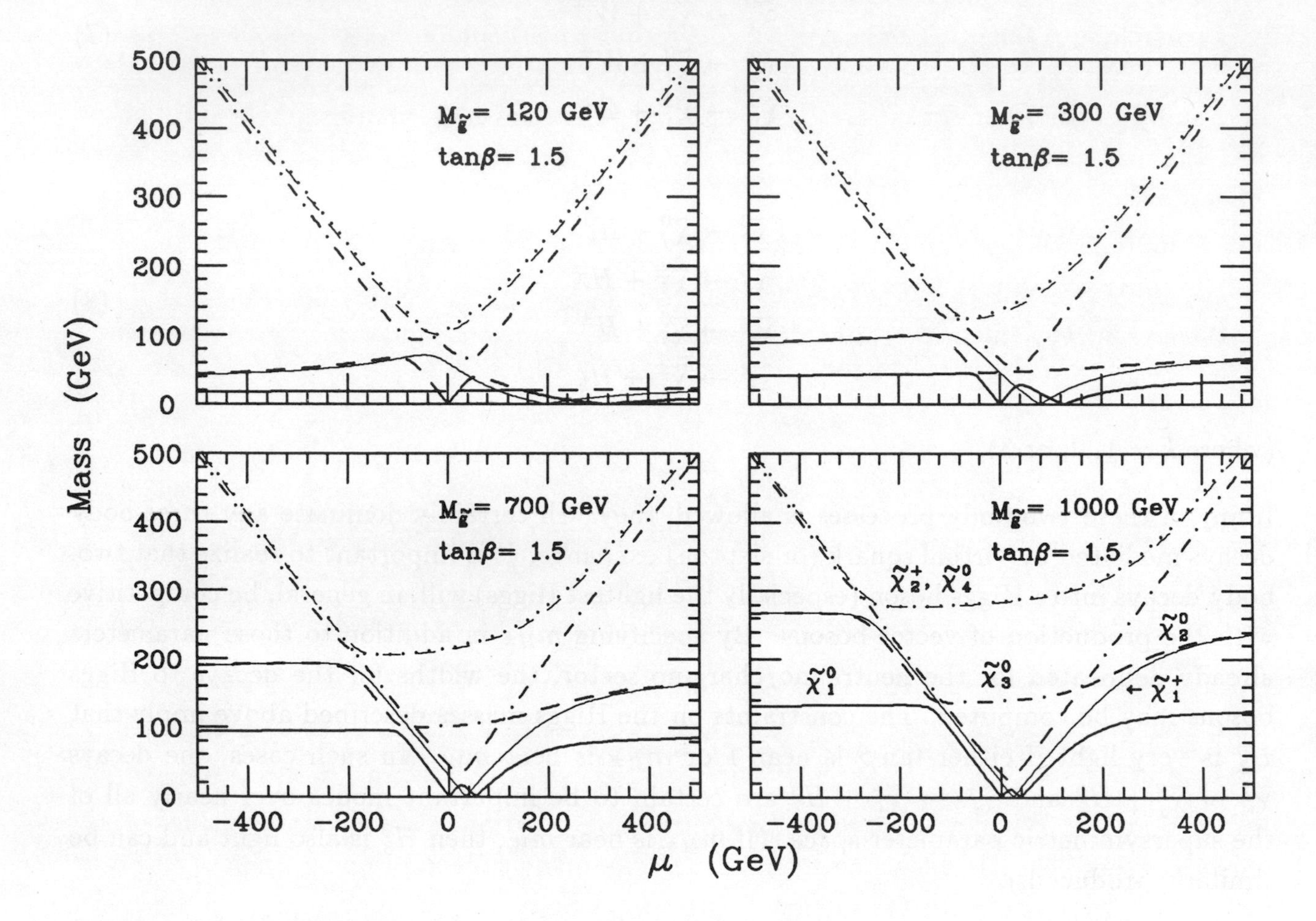

Figure 5: We present, using notation completely parallel to that of figs. 3 and 4, the masses of the $\widetilde{\chi}^0$'s and $\widetilde{\chi}^{\pm}$'s, for the same four gluino mass values at $\tan\beta = 1.5$.

2. When $M_{\tilde{g}} \gtrsim 700\ GeV$, fig. 5 shows that at large $|\mu|$ not only are $\widetilde{\chi}_1^\pm$ and $\widetilde{\chi}_2^0$ heavier than the Z, but so is the LSP ($\widetilde{\chi}_1^0$), contrary to old "theorems".[18]
3. Even for relatively low $M_{\tilde{g}}$ values, the $\widetilde{\chi}_2^\pm$, $\widetilde{\chi}_4^0$ and $\widetilde{\chi}_3^0$ can have masses that are larger than the $\widetilde{\chi}_1^\pm$, $\widetilde{\chi}_2^0$ and $\widetilde{\chi}_1^0$ masses by at least m_Z, so that the potential for the two-body decays listed in eq. (7) exists for decays with an initial $\widetilde{\chi}_4^0, \widetilde{\chi}_3^0, \widetilde{\chi}_2^\pm$ and a final $\widetilde{\chi}_2^0, \widetilde{\chi}_1^0, \widetilde{\chi}_1^\pm$.
4. The $\widetilde{\chi}_4^0$ and $\widetilde{\chi}_2^\pm$ are approximately degenerate, implying that chain decay of either particle into the other plus a W or Z is forbidden. A similar statement also applies to the $\widetilde{\chi}_3^0$ in the region $|\mu| \gtrsim M_{\tilde{g}}/4$, where its mass is only slightly less than those of $\widetilde{\chi}_4^0$ and $\widetilde{\chi}_2^\pm$. However, for $|\mu| \lesssim M_{\tilde{g}}/4$ and large $M_{\tilde{g}}$ the $\widetilde{\chi}_3^0$ mass can be significantly below the masses of $\widetilde{\chi}_4^0$ and $\widetilde{\chi}_2^\pm$, and the decays $\widetilde{\chi}_4^0 \to Z\widetilde{\chi}_3^0$ and $\widetilde{\chi}_2^\pm \to W^\pm\widetilde{\chi}_3^0$ are phase space allowed.
5. Finally, for large gluino mass, $M_{\tilde{g}} \gtrsim 800\ GeV$, the decays $\widetilde{\chi}_1^\pm \to \widetilde{\chi}_1^0 W^\pm$ and $\widetilde{\chi}_2^0 \to \widetilde{\chi}_1^0 Z$ become possible at large $|\mu| \gtrsim M_{\tilde{g}}/4$.

We now survey the basic gluino decay chains that lead to a signature of great interest:[19]

$$\tilde{g} \to jet(s) + W(\text{or } Z) + E_T^{miss}. \tag{9}$$

We present our results by plotting

$$BR[\tilde{g} \to \widetilde{\chi}_{1,2}^\pm(\to W^\pm)] + BR[\tilde{g} \to \widetilde{\chi}_{2,3,4}^0(\to W^\pm)] \tag{10}$$

and

$$BR[\tilde{g} \to \widetilde{\chi}_{1,2}^\pm(\to Z)] + BR[\tilde{g} \to \widetilde{\chi}_{2,3,4}^0(\to Z)] \tag{11}$$

in figs. 6 and 7. We plot these branching ratios as a function of μ for various values of $M_{\tilde{g}}$ and $m_{H^\pm}$ (there is little dependence on $\tan\beta$). The many sudden jumps in the curves occur due to two physical effects: (1) sudden changes in identity of a given neutralino or chargino, as mass eigenstates undergo level crossing and switch from being dominantly higgsino to dominantly gaugino; (2) the sudden onset or disappearance of the various 2-body decay modes, as determined by phase space.

The results are easily summarized. When $|\mu| \lesssim M_{\tilde{g}}/4$, and the heavier chargino and neutralino states dominate the $\tilde{g}$ decays, the branching ratios to W's and Z's become very significant. This is true so long as $M_{\tilde{g}} \gtrsim 300\ GeV$. A light $\widetilde{\chi}_1^\pm$, $\widetilde{\chi}_2^0$ or $\widetilde{\chi}_1^0$ is produced along with the W or Z—a $\widetilde{\chi}_3^0$ is essentially never produced by $\widetilde{\chi}_2^\pm$ and $\widetilde{\chi}_4^0$ decays. (Even though the decays $\widetilde{\chi}_2^\pm \to W^\pm\widetilde{\chi}_3^0$ and $\widetilde{\chi}_4^0 \to Z\widetilde{\chi}_3^0$ are phase space allowed at large $M_{\tilde{g}}$ and $|\mu| \lesssim M_{\tilde{g}}/4$, these decays are strongly suppressed by neutralino mixing angle factors.)

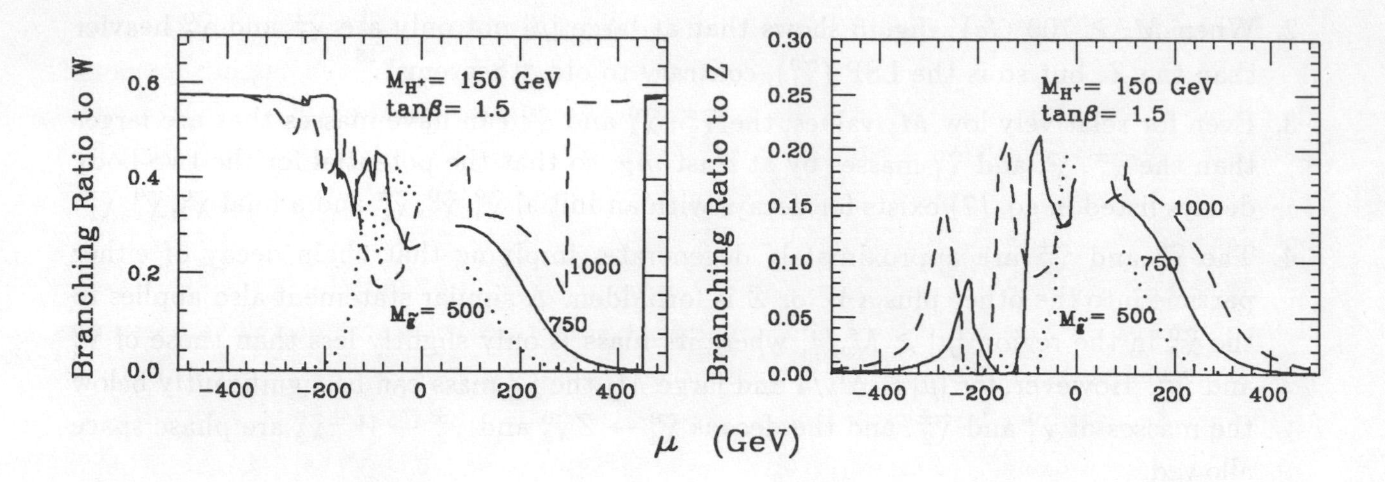

Figure 6: The gluino branching ratios to W and Z (see eqs. (10) and (11)) as a function of μ for $M_{\tilde{g}} = 500$ (dotdashes), 750 (solid), and 1000 GeV (dashes) at $\tan\beta = 1.5$. We have taken $m_{H^\pm} = 150$ GeV. As in previous figures, omitted portions of the curves correspond to parameter regions where $M_{\tilde{\chi}_1^+} < 30$ GeV.

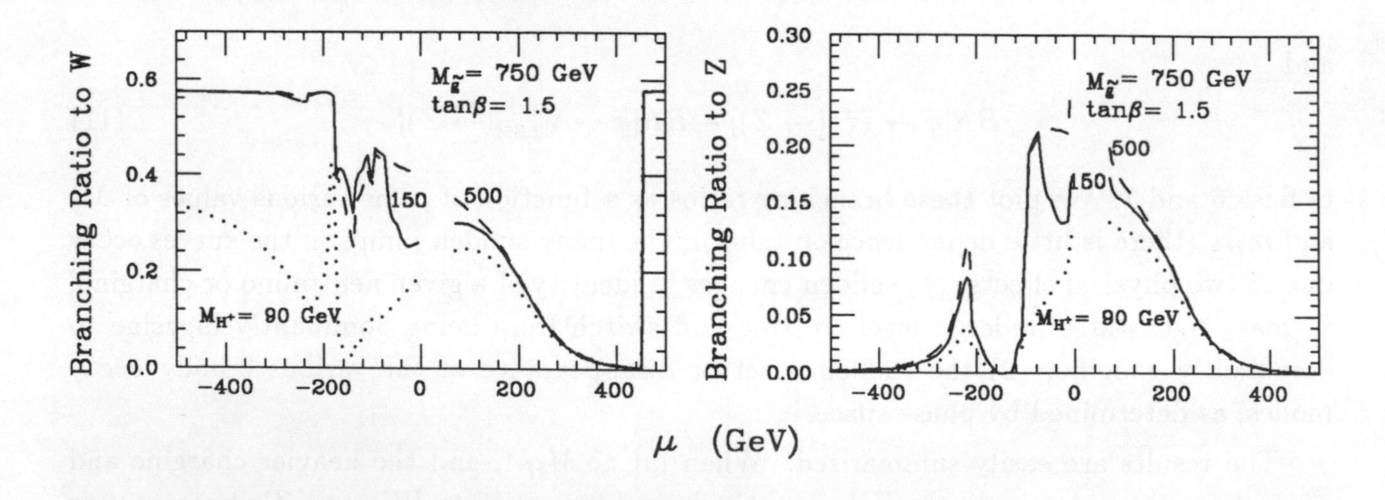

Figure 7: The gluino branching ratios to W and Z (see eqs. (10) and (11)) as a function of μ for $M_{\tilde{g}} = 750$ GeV and $\tan\beta = 1.5$ for 3 values of $m_{H^\pm}$: 90 (solid), 150 (dashes), and 500 GeV (dotdashes). As in previous figures, omitted portions of the curves correspond to parameter regions where $M_{\tilde{\chi}_1^+} < 30$ GeV.

For $|\mu| \gtrsim M_{\tilde{g}}/4$ we have seen that the lighter neutralinos and charginos dominate the $\tilde{g}$ decays, and it is only for very large $M_{\tilde{g}}$ ($\gtrsim 600\ GeV$) that $\tilde{\chi}_1^{\pm}$ and $\tilde{\chi}_2^0$ become heavy enough that they can decay to $W^{\pm}\tilde{\chi}_1^0$ and $Z\tilde{\chi}_1^0$, respectively. However, the $\tilde{\chi}_2^0 \to Z\tilde{\chi}_1^0$ decay is severely suppressed relative to the $\tilde{\chi}_2^0 \to H_2^0\tilde{\chi}_1^0$ mode. Thus, for large $M_{\tilde{g}}$ and $|\mu| \gtrsim M_{\tilde{g}}/4$ we obtain the result, apparent in the figures, that gluino decay *will* typically contain a W but *not* a Z.

Of course, gluinos will not always decay either to W's, Z's, and Higgs (using the two-body $\tilde{\chi}$ decays) or *directly* to the LSP. Some of the time they will decay to two SM fermions plus a $\tilde{\chi}$ which decays in a three-body mode, yielding five particles in the $\tilde{g}$ final state (prior to further decays). We term such decays 'five-body' modes.

We now summarize the possible signatures for gluino decay. In the discussion below, we sometimes refer to the quark and lepton modes. These result from the secondary (and tertiary) decays of the chargino or neutralino produced in the gluino decay chain. Their relative branching ratios are determined by the gaugino content of the particular $\tilde{\chi}$ state involved (since we neglect all final state quark and lepton masses). When we refer to leptonic modes, we are summing only over electrons and muons. We have not considered final states involving tau leptons since these lead to more complicated signatures.

Many of our results on gluinos signatures are summarized in fig. 8. The "five-body" gluino decays include:

$$q\bar{q}\ q\bar{q}\ \tilde{\chi}_1^0$$
$$q\bar{q}\ e^+e^-\tilde{\chi}_1^0$$
$$q\bar{q}\ \mu^+\mu^-\tilde{\chi}_1^0$$
$$q\bar{q}'e^+\nu\ \tilde{\chi}_1^0$$
$$q\bar{q}'\mu^-\bar{\nu}\ \tilde{\chi}_1^0.$$

The curves in fig. 8 labelled "LSP" refer to decays directly to the $\tilde{\chi}_1^0$ (the LSP):

$$q\bar{q}\ \tilde{\chi}_1^0.$$

Decays to W and Z bosons include:

$$q\bar{q}'\ W(\text{or } Z)\ \tilde{\chi}_1^0.$$
$$q\bar{q}\ q\bar{q}\ Z\ \tilde{\chi}_1^0$$
$$q\bar{q}\ e^+e^-Z\ \tilde{\chi}_1^0.$$
$$q\bar{q}'\ \mu^+\nu\ Z\ \tilde{\chi}_1^0.$$

The first mode occurs when the initial gluino decay product was $\tilde{\chi}_1^{\pm}$ or $\tilde{\chi}_2^0$. The others occur when the initial product is $\tilde{\chi}_2^{\pm}$ or $\tilde{\chi}_4^0$, and these cascade via another $\tilde{\chi}^0$ or $\tilde{\chi}^{\pm}$.

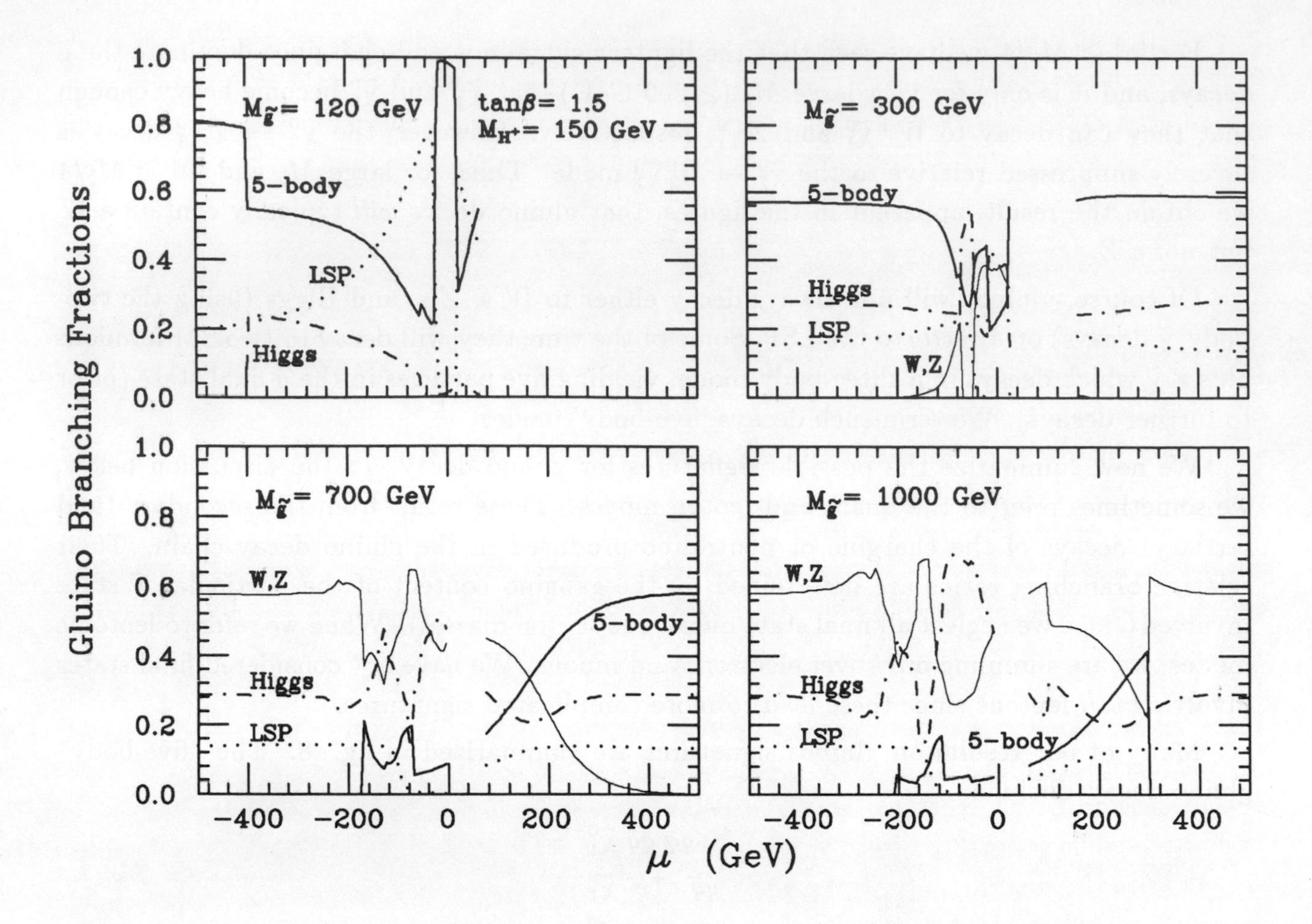

Figure 8: The branching ratios for gluino decay into the four different categories of tree-level accessible final states: 1) 5-body modes with no real W's, Z's, or Higgs; 2) the LSP ($\widetilde{\chi}_1^0$) directly produced in association with $q\bar{q}$; 3) any state with a real W or real Z; and 4) any state with a Higgs of any type. The branching ratios are presented for four different $M_{\tilde{g}}$ values as a function of μ, taking $\tan\beta = 1.5$ and $m_{H^\pm} = 150$ *GeV*. Modifications arising from varying these parameters are described in the text.

Among the gluino decays containing a Higgs boson are:

$$q\bar{q}\ H^0(\text{or } H^+)\ \widetilde{\chi}_1^0$$

$$q\bar{q}\ q\bar{q}\ H^0(\text{or } H^+)\ \widetilde{\chi}_1^0$$

$$q\bar{q}\ e^+e^-\ H^0(\text{or } H^+)\ \widetilde{\chi}_1^0.$$

Again the first mode occurs when the initial gluino decay product was $\widetilde{\chi}_1^{\pm}$ or $\widetilde{\chi}_2^0$. The others occur when the initial product is $\widetilde{\chi}_2^{\pm}$ or $\widetilde{\chi}_4^0$, and these cascade via another $\widetilde{\chi}^0$ or $\widetilde{\chi}^{\pm}$.

Next, we must remember that gluinos are actually pair produced at a hadron collider and that the above final state structure applies to the decay of each gluino. An incredible variety of search modes emerges. While many spectacular signatures with reasonable event rates are possible, the typical gluino-gluino event is extremely complex and probably cannot be distinguished from background. The exact scenario, of course, depends upon both $M_{\tilde{g}}$ and μ. We list below some promising modes as a function of $M_{\tilde{g}}$ and μ, along with their event rates at the SSC. Our purpose in this section is *not* to study these modes (or their backgrounds), but to identify some signatures deserving of further examination. The event rates shown do *not* take into account either efficiencies or cuts. Large numbers of events do not necessarily mean the signal is observable. It is fairly straightforward to transcribe the discussion below to any other supercollider, once the total $\widetilde{g}\widetilde{g}$ cross-sections have been computed. The following results are found in detail in Ref. 12

The total cross-sections at the SSC for gluinos of masses 120, 300, 700 and 1000 GeV are 217, 5.2, 0.10 and 0.015 nanobarns, respectively. We estimate event rates by assuming design luminosity which leads to 10^7 events per year per nanobarn. For $M_{\tilde{g}} = 120\ GeV$ there would be 2×10^9 events per year, but the separation of these events from background would be extremely difficult. For most values of μ there would be about 10^9 events in which one or both gluinos decayed directly to the LSP ($\widetilde{\chi}_1^0$) giving 2-4 jets plus E_T^{miss}. Unfortunately the magnitude of E_T^{miss} will not be that different from that of many backgrounds so that the Tevatron collider (where backgrounds will be much smaller) is a more logical place to search for such masses. Other signals for a 120 GeV gluino come from one gluino going directly to the LSP while the other has a 5-body decay, or from both gluinos undergoing a 5-body decay, and include:

$$q\bar{q}\ q\bar{q}\ l^+l^-\ E_T^{miss}$$

$$q\bar{q}\ q\bar{q}'\ \mu^+\ E_T^{miss}$$

$$q\bar{q}'\ q\bar{q}'\ \mu^+e^+\ E_T^{miss}$$

In the first two cases E_T^{miss} would be comparable to the double LSP case, whereas the latter has very little E_T^{miss}. The invariant mass of the l^+l^- pair in the first channel would be 45-70 GeV in most cases. Again backgrounds are probably prohibitive.

For $M_{\tilde{g}} = 300\ GeV$ the signals are very similar to those just described, but there may now be adequate E_T^{miss} to distinguish signal from background (although this requires study). For most μ values there would be ten million events/year of the single or double LSP signature (unless μ is very small in which case the event rate goes to zero). Our studies indicate that the E_T^{miss} spectrum from events in which one gluino goes directly to the LSP and the other gluino goes to anything is very similar to that in which both gluinos decay directly to LSP's (see Sec. 4 also). Placing a very high E_T^{miss} cut on the data leads to a suppression of 1.5 on the single-LSP case relative to the double-LSP case. But since there is a factor of 12 advantage in the branching ratio, it is the single-LSP case which is most important. The signals with leptons resulting from one or two five-body decays lead to several million events/year. Here we expect $M(l^+l^-) = 90-100\ GeV$ for $\mu < -50\ GeV$ and $M(l^+l^-) = 30-70\ GeV$ for $\mu > -50\ GeV$. It should be emphasized that all of these signatures are expected simultaneously independent of μ (although backgrounds may be more severe in some cases). At this gluino mass we see for the first time the possibility of the signal with W or Z in the decay products. For $-100 < \mu < 25\ GeV$ we expect 5-20 million events/year with a W or Z, *e.g.*

$$q\bar{q}\ q\bar{q}\ q\bar{q}\ q\bar{q}\ W\ E_T^{miss}.$$

A better signature is presumably found from the million or so events which would contain two vector bosons. The Z contribution begins only for $\mu > -30\ GeV$, but can be as much as 12% of the branching ratio. For these μ values we expect 500-2500 events/year in which two Z bosons result and both decay to lepton pairs.

For $M_{\tilde{g}} = 700\ GeV$ the nature of the signals is somewhat different depending on the value of μ. For $160 < \mu < 620/GeV$, the 5-body modes remain large at the expense of the W and Z modes, whereas for all other μ values the W and Z modes have replaced the 5-body modes. The single and double LSP modes (one or two gluino decays directly to the LSP) lead to 300,000 events/year with substantial missing energy and 2-4 hard jets (unless $|\mu| < 80\ GeV$). The number of jets depends on how many are coalesced by the jet-finding procedure. For high gluino masses (as here) we expect 3-4 jets most of the time. For $\mu < 160\ GeV$ there will be 100,000-300,000 events with two vector bosons. However, for $\mu < -100\ GeV$ these are always W bosons. For $-100 < \mu < 160\ GeV$, there will be 40-80 events/year with two Z bosons both of which decay to l^+l^-. In addition there are mixed modes where the two gluinos have different decays. One gluino may decay directly to the LSP ($\tilde{\chi}_1^0$) while the other goes to a Z giving:

$$q\bar{q}\ q\bar{q}\ q\bar{q}\ Z\ E_T^{miss}.$$

Thus one would find large E_T^{miss} and a Z boson plus 3-4 hard jets. Of these events, 1500 events/year with the Z decaying to leptons would occur if $60 < \mu < 200\ GeV$. (In this range, branching ratios for $\tilde{g} \rightarrow LSP$ and $\tilde{g} \rightarrow Z$ are both sufficiently large.) Far more common would be the equivalent process with W bosons (decaying to leptons). 10,000-30,000 such events would occur for most μ below 200 GeV. For large positive μ one finds events with one gluino going directly to the LSP and one having a 5-body decay. These 130,000 events would have large E_T^{miss} and very large total scalar E_T. Here, for the l^+l^- pair associated with 5-body decays, we expect $M(l^+l^-) = 100 - 160\ GeV$.

At $M_{\tilde{g}} = 1000\ GeV$ the 5-body decays are relevant for only limited μ values. The W boson modes are now substantial for all parameters, and the Z boson for $|\mu| < 300\ GeV$. However, with this large gluino mass the cross-section has dropped so that one can no longer look for events with small branching fractions. The number of events in which both gluinos decay to Z bosons (assuming $|\mu| < 300\ GeV$) and each boson decays to leptons is now only 10-15 events (although we expect backgrounds to be even smaller). The number of events/year with two W bosons each decaying leptonically is 800-3000; these events would have 3-4 jets with about 150-300 GeV each and the total scalar energy would be over 2000 GeV. There would be about 170 events with a W and a Z boson (if $|\mu| < 300\ GeV$) in which both decayed leptonically. Except for the region $-150 < \mu < 50\ GeV$, one or both gluinos could decay directly to the LSP giving 40,000 events with enormous E_T^{miss}. If the two gluinos have different decay modes, we can find 4000 events with a leptonically decaying W boson and a direct LSP (giving very high E_T^{miss}), or 300 events with $Z \rightarrow l^+l^-$ and the direct LSP.

In conclusion we wish to again emphasize that the discussion above is intended to indicate signatures deserving further consideration. No attempt was made here to account for backgrounds, efficiencies or cuts. Discussion of the role of the Higgs decay modes will appear in a forthcoming paper.[20] In the above we have assumed $\tan\beta = 1.5$ and $m_{H^+} = 150\ GeV$. For smaller $\tan\beta$ little change occurs. For larger $\tan\beta$ most changes are not significant, but decays to the 5-body channels are enhanced at the expense of the Higgs decay mode. If m_{H^+} is increased, there is little overall impact except for $M_{\tilde{g}} = 1000\ GeV$, where the W and Z modes are somewhat enhanced at relatively small $|\mu|$. However, if m_{H^+} is reduced to 90 GeV, we see a substantial change for $M_{\tilde{g}} > 500\ GeV$. The branching fractions for the W and Z bosons modes drop by a factor of 2-3 for negative μ, with the Higgs modes making up the difference. The impact is much smaller for positive μ.

Finally, we can compare the ability of the SSC and the LHC to find many of the signals discussed here. This comparison[12] appears in table I.

Table 1

Approximate Event Rates for SSC vs. LHC

Event rates for an integrated luminosity of 10^{40} cm^{-2} from $\widetilde{g}\widetilde{g}$ production, before cuts and efficiencies. Rates are given for those regions of μ where a given signal is most significant. These μ regions are indicated (in GeV units) by the parentheses. Note that there is always a gap in μ (near $\mu = 0$) due to our elimination of μ values for which $M_{\tilde{\chi}_1^+} < 30\ GeV$. We assume $\sqrt{s} = 40\ TeV$ and $\sqrt{s} = 17\ TeV$ for the SSC and LHC, respectively.

Signal	$M_{\tilde{g}} = 300\ GeV$	$M_{\tilde{g}} = 700\ GeV$	$M_{\tilde{g}} = 1000\ GeV$
1 direct LSP+any; $\Rightarrow$jets+E_T^{miss}	**10^7 vs. 10^6** $(-\infty,-40);(160,\infty)$	**3×10^5 vs. 2×10^4** $(-\infty,-100);(70,\infty)$	**4×10^4 vs. 1700** $(-\infty,-140);(60,\infty)$
Two 5-body decays with leptons	**10^7 vs. 10^6** $(-\infty,30);(160,\infty)$	**6000 vs. 400** $(70,620)$	**240 vs. 10** $(200,300)$
Two Z's; $Z\to l^+l^-$	**1000 vs. 100** $(-40,30)$	**50 vs. 3** $(-200,0);(70,200)$	**10 vs. 0.4** $(-300,-10);(60,250)$
Z+ direct LSP; $Z\to l^+l^-$	**100 vs. 14** $(-40,30)$	**2000 vs. 130** $(-200,-100);(70,200)$	**300 vs. 13** $(-300,-140);(60,300)$
5-body+direct LSP; 5-body $\to$ leptons	**10^6 vs. 10^5** $(-\infty,-40);(160,\infty)$	**20,000 vs. 1400** $(-180,-100);(70,620)$	**1500 vs. 60** $(100,300)$
W+ direct LSP; $W\to$ leptons	**5×10^5 vs. 8×10^4** $(-100,-70)$	**20,000 vs. 1400** $(-\infty,-100);(70,250)$	**4000 vs. 170** $(-\infty,-140);(60,\infty)$
$W+Z$; $W,Z\to$ leptons	**3×10^4 vs. 4×10^3** $(-40,30)$	**1000 vs. 70** $(-200,0);(70,200)$	**200 vs. 8** $(-300,-140);(60,300)$

3. Squark Decays

The assumptions made for the squark analysis are very similar to those made for the gluino analysis. The results discussed here are described in greater detail in Ref. 13 which also appears in these proceedings. Here of course we choose $M_{\tilde{q}} < M_{\tilde{g}}$. For definiteness in the results quoted, it was assumed that $M_{\tilde{g}} \approx M_{\tilde{q}}$. Some such assumption is essential since it fixes the relationship between $M_{\tilde{q}}$ and the parameter M (see eqn. (3)). It was checked that the conclusions remain qualitatively the same if a larger $M_{\tilde{g}}$ is chosen. In this section as in the gluino section, we will ignore $\widetilde{q}\widetilde{g}$ production and concentrate on $\widetilde{q}\widetilde{q}$ production.

Consideration of squarks is somewhat more complicated than that of gluinos since there are two flavors ("up" and "down") and since there are supersymmetric partners of the left-handed quarks ($\widetilde{q}_L$) and partners of the right-handed quarks ($\widetilde{q}_R$). (Other flavors are treated as being the same as u and d). Furthermore the cross-section for $\widetilde{q}_L\widetilde{q}_L$ production (or $\widetilde{q}_R\widetilde{q}_R$ production) is not the same as that for $\widetilde{q}_L\widetilde{q}_R$. When $M_{\tilde{q}} < M_{\tilde{g}}$, then the following two-body decays will dominate (if they are kinematically allowed):

$$\widetilde{q}_L \rightarrow q'_L\widetilde{\chi}_i^{\pm} \tag{12}$$

$$\begin{aligned} \widetilde{q}_L &\rightarrow q_L\widetilde{\chi}_i^0 \\ \widetilde{q}_R &\rightarrow q_R\widetilde{\chi}_i^0 \ . \end{aligned} \tag{13}$$

where i=1,2 for $\widetilde{\chi}_i^{\pm}$ and i=1,2,3,4 for $\widetilde{\chi}_i^0$. Note that there are no couplings of $\widetilde{q}_R$ to $\widetilde{\chi}_i^{\pm}$ for the same reason that q_R do not couple to W bosons. The branching ratios of all these modes can be found in Ref. 10 . These papers and fig. 9 show that left-handed squarks with mass $M_{\tilde{q}} > 150\ GeV$ rarely decay directly to the LSP ($\widetilde{\chi}_1^0$). However, the dominant decay of right-handed squarks *is* directly to the LSP *if* $|\mu| \gtrsim M_{\tilde{q}}/3$. When this condition is obeyed, the heavier neutralinos are dominated by their higgsino components (or by their neutral $SU(2)$ gaugino components) and therefore have a very small coupling to squarks.

The decays of the charginos and neutralinos (produced in squark decay) were discussed in the previous section. As for gluinos one of the products of cascading decays are W and Z bosons. In fig. 10 (from ref. 13) we show the branching fractions for the various types of squarks into real W and Z bosons. For the case $\mu \lesssim M_{\tilde{q}}/3$ $(= M_{\tilde{g}}/3)$ we see that left-handed squarks have branching ratios into W and Z bosons of 50% and 20% , respectively. Right-handed squarks of course have a small (3%) branching ratio to W and Z bosons, since they usually decay directly to the LSP. With design luminosity we expect about 8×10^5 $\widetilde{q}_L\widetilde{q}_L$ pairs annually at the SSC if $M_{\tilde{q}} = 0.5\ TeV$ or 4×10^4 pairs if $M_{\tilde{q}} = 1\ TeV$. We now find (for $|\mu| \lesssim M_{\tilde{q}}/3$) that the number of events in which both squarks have a cascade decay yielding a Z boson and both of the resulting Z bosons decay to l^+l^-, is 120 per year for $M_{\tilde{q}} = 0.5\ TeV$ and 6 events per year if $M_{\tilde{q}} = 1\ TeV$. These events would contain two hard jets from the primary decay in addition to the two leptonically decaying Z bosons.

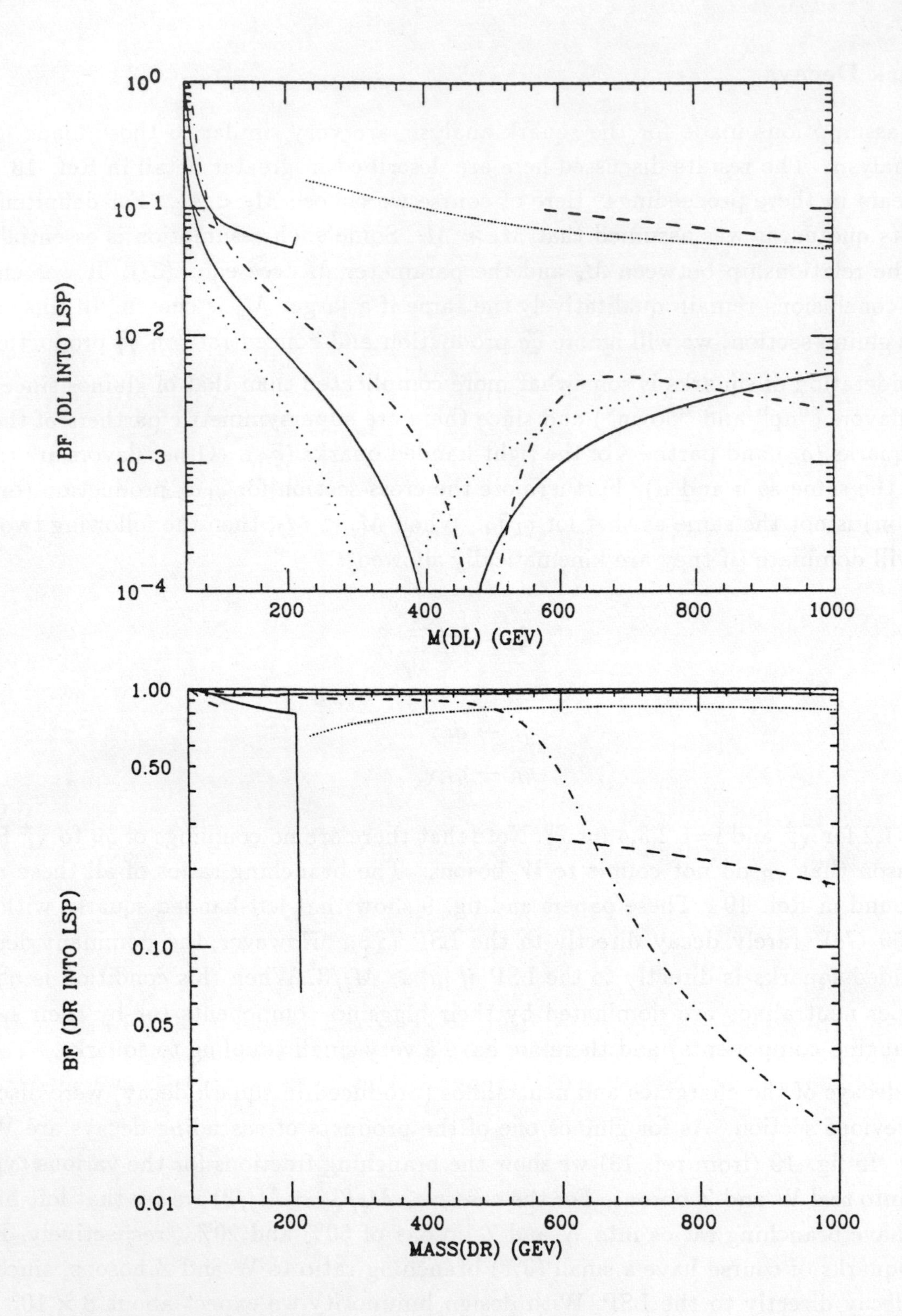

Figure 9: The branching ratios for direct decays of $\tilde{d}_L$ and $\tilde{d}_R$ to the LSP as a function of $M_{\tilde{q}}$ (with $M_{\tilde{q}} = M_{\tilde{g}}$) for $\mu = -3m_W$ (close dots), $-m_W$ (dashed), $-m_W/2$ (solid ending at 200 GeV), m_W (dot-dashed), $2m_W$ (solid continuing past 200 GeV), and $3m_W$ (dotted) at $\tan\beta = 1.5$. As in previous figures, omitted portions of the curves correspond to parameter regions where $M_{\tilde{\chi}_1^+} < 30\ GeV$.

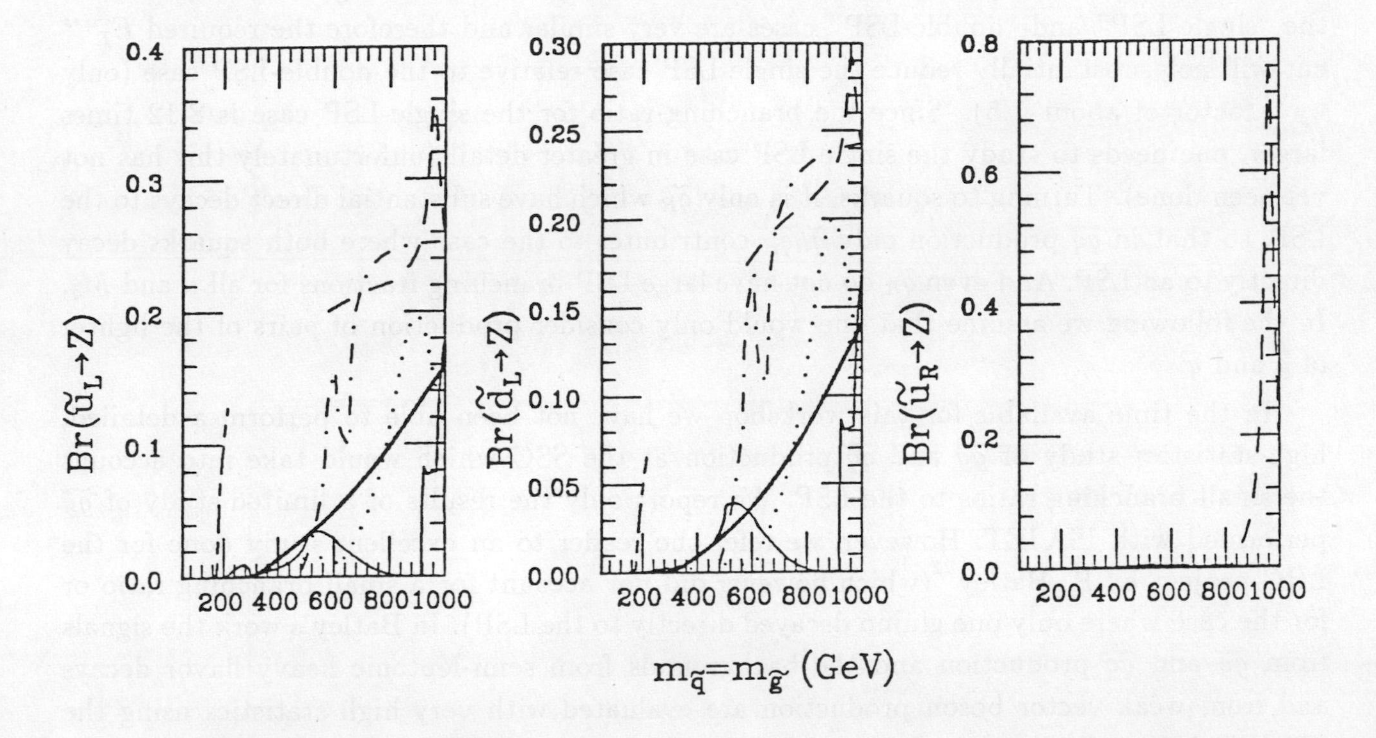

Figure 10: The squark branching ratios to W and Z as a function of $M_{\tilde{q}}$ (with $M_{\tilde{q}} = M_{\tilde{g}}$) for $\mu = -3m_W$ (thick solid), $-m_W$ (long dashed), $-m_W/2$ (long-short dashed), m_W (dot-dashed), $2m_W$ (solid), and $3m_W$ (dotted) at $\tan\beta = 1.5$. As in previous figures, omitted portions of the curves correspond to parameter regions where $M_{\tilde{\chi}_1^+} < 30\ GeV$.

4. Direct Decays of Gluinos and Squarks to the LSP

When most people think of supersymmetry, it is the jets plus large missing energy signal which comes to mind (where the missing energy originates in LSP's coming from decays of the gluinos or squarks). However, we have shown that the maximum branching ratio for direct decay of gluinos to the LSP ranges from 13 to 20% for masses of interest at the SSC. Therefore, in $\widetilde{g}\widetilde{g}$ production only 2 to 4% of the events have both gluinos decaying directly to the LSP. As discussed previously, this motivates us to look at the "single-LSP" case (where only one of the LSP's has decayed directly to the LSP). As seen in Fig. 11 the spectra for the "single-LSP" and "double-LSP" cases are very similar and therefore the required E_T^{miss} cut will not substantially reduce the single-LSP case relative to the double-LSP case (only by a factor of about 1.5). Since the branching ratio for the single-LSP case is 8-12 times larger, one needs to study the single-LSP case in greater detail (unfortunately this has not yet been done). Turning to squarks, it is only $\widetilde{q}_R$ which have substantial direct decays to the LSP, so that in $\widetilde{q}\widetilde{q}$ production only $\widetilde{q}_R\widetilde{q}_R$ contributes to the case where both squarks decay directly to an LSP. And even $\widetilde{q}_R$ do not have large LSP branching fractions for all μ and $M_{\widetilde{q}}$. In the following we assume that one would only consider production of pairs of the lighter of $\widetilde{g}$ and $\widetilde{q}$.

In the time available for this workshop we have not been able to perform a detailed, high-statistics study of $\widetilde{g}\widetilde{g}$ and $\widetilde{q}\widetilde{q}$ production at the SSC which would take into account the small branching ratios to the LSP. We report only the results of a limited study of $\widetilde{g}\widetilde{g}$ performed with ISAJET. However, we refer the reader to an excellent study done for the LHC project by R. Batley[21](which however did not account for a small branching ratio or for the case where only one gluino decayed directly to the LSP). In Batley's work the signals from $\widetilde{g}\widetilde{g}$ and $\widetilde{q}\widetilde{q}$ production and the backgrounds from semi-leptonic heavy flavor decays and from weak vector boson production are evaluated with very high statistics using the ISAJET Monte Carlo program (version 5.25). In his study the LHC energy was taken to be $\sqrt{s} = 17\ TeV$ and $M_{\widetilde{g}}$ and $M_{\widetilde{q}} = 600 - 1000\ GeV$ were considered. The backgrounds were dominated by $t\bar{t}$ events with t decaying semi-leptonically ($m_t = 40$ and $200\ GeV$ were considered). In calculating QCD jet production, ISAJET uses only the leading-order $2 \to 2$ matrix elements, but initial-state and final-state radiation is included.

The $t\bar{t}$ pair can be produced in the original hard scattering ("direct") or can occur in the evolution of one of the two gluon jets of the original hard scattering ("indirect"). Although direct production is 10-30 times larger than indirect, the indirect production is far more efficient at passing an E_T^{miss} cut so that the direct mode ends up being comparable to indirect. To pass a large E_T^{miss} cut presumably requires both t and $\bar{t}$ to undergo a semi-leptonic decay and requires both of the resulting neutrinos to go the in same direction (otherwise the missing energy cancels). In direct production the t and $\bar{t}$ tend to go in opposite directions whereas in indirect production both t and $\bar{t}$ tend to go opposite the gluon.

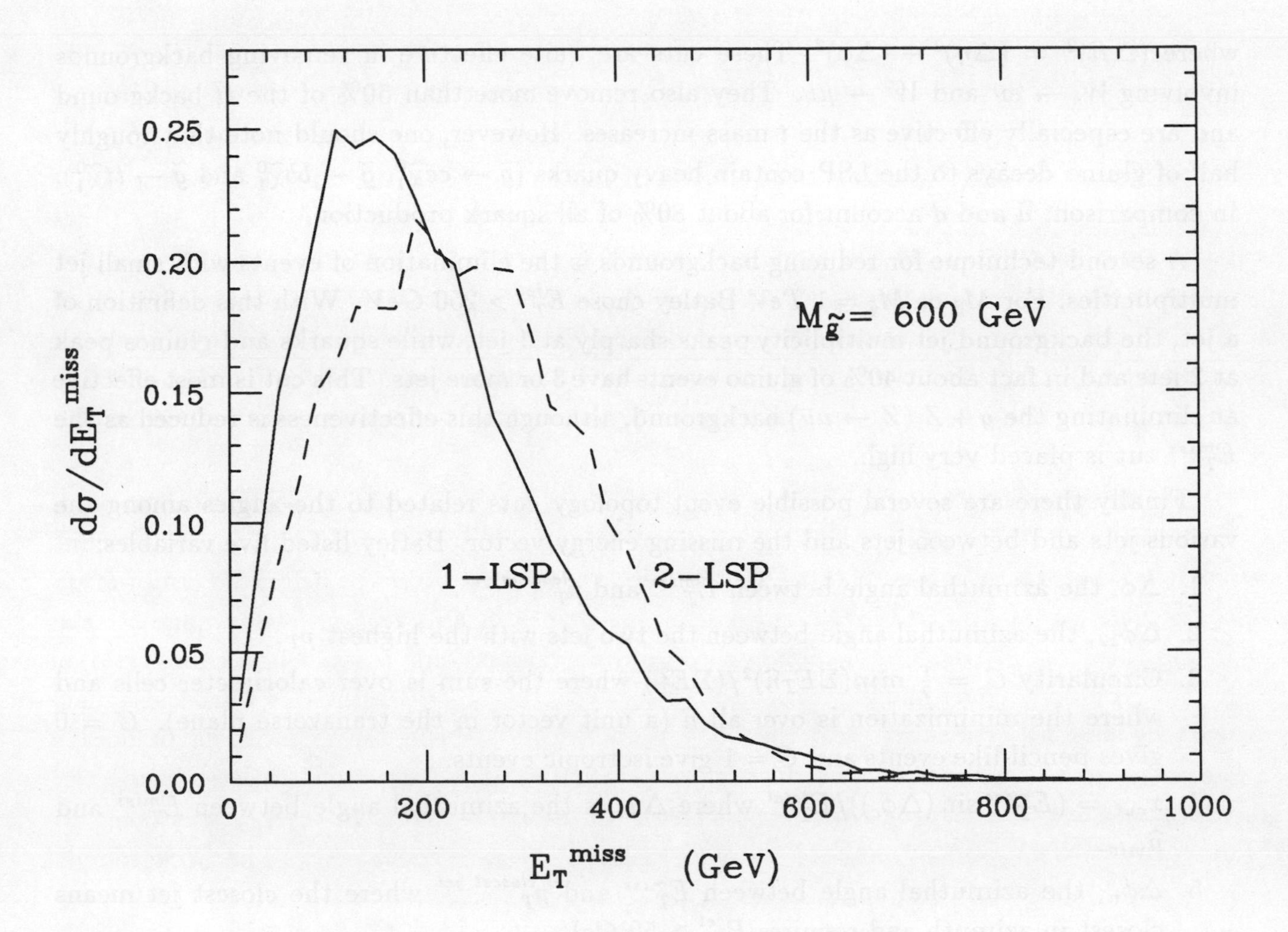

Figure 11: The E_T^{miss} spectra for $\widetilde{g}\widetilde{g}$ production ($M_{\tilde{g}} = 600\ GeV$) where one or two of the gluinos decay as $\widetilde{g} \to q\bar{q}\widetilde{\chi}_1^0$ (where $\widetilde{\chi}_1^0$ is the LSP). The double-LSP case has units of pb/GeV while the single-LSP case is normalized to have the same area.

Batley used several techniques to reduce backgrounds. First, events containing muons with $p_T^\mu > 15\ GeV/c$ were removed. Similarly, events containing an *isolated* electron passing this cut were eliminated. "Isolated" was defined as

$$\Sigma p_T^{hadron}(\Delta R < 0.4)\ /\ p_T^e\ <\ 0.1 \qquad (14)$$

where $(\Delta R)^2 = (\Delta\eta)^2 + (\Delta\phi)^2$. These cuts are quite effective in removing backgrounds involving $W \to e\nu$ and $W \to \mu\nu$. They also remove more than 50% of the $t\bar{t}$ background and are especially effective as the t mass increases. However, one should note that roughly half of gluino decays to the LSP contain heavy quarks ($\tilde{g} \to c\bar{c}\tilde{\chi}_1^0$, $\tilde{g} \to b\bar{b}\tilde{\chi}_1^0$ and $\tilde{g} \to t\bar{t}\tilde{\chi}_1^0$). In comparison, $\tilde{u}$ and $\tilde{d}$ account for about 80% of all squark production.

A second technique for reducing backgrounds is the elimination of events with small jet multiplicities. For $M_{\tilde{g}}$ or $M_{\tilde{q}} = 1\ TeV$ Batley chose $E_T^{jet} > 250\ GeV$. With this definition of a jet, the background jet multiplicity peaks sharply at 1 jet, while squarks and gluinos peak at 2 jets and in fact about 40% of gluino events have 3 or more jets. This cut is most effective at eliminating the $g + Z$ ($Z \to \nu\bar{\nu}$) background, although this effectiveness is reduced as the E_T^{miss} cut is placed very high.

Finally there are several possible event topology cuts related to the angles among the various jets and between jets and the missing energy vector. Batley listed five variables:

1. $\Delta\phi$, the azimuthal angle between E_T^{miss} and $p_T^{leading\ jet}$.
2. $\Delta\phi_{12}$, the azimuthal angle between the two jets with the highest p_T.
3. Circularity $C = \frac{1}{2}\ min(\Sigma E_T\hat{n})^2/(\Sigma E_T^2)$ where the sum is over calorimeter cells and where the minimization is over all $\hat{n}$ (a unit vector in the transverse plane). $C = 0$ gives pencil-like events and $C = 1$ give isotropic events.
4. $x_{out} = (E_T^{miss} \sin(\Delta\phi_c))/E_T^{total}$ where $\Delta\phi_c$ is the azimuthal angle between E_T^{miss} and $\hat{n}_{min}$.
5. $\Delta\phi_n$, the azimuthal angle between E_T^{miss} and $p_T^{closest\ jet}$ where the closest jet means closest in azimuth and requires $P_T^{jet} > 50\ GeV$.

Not all of these cut variables were used, in part, because even with the enormous statistics used in Batley's study, it would have become difficult to generate any events (especially for the $t\bar{t}$ backgrounds which unfortunately are the largest backgrounds). He therefore commented that it was likely that backgrounds could be further reduced. His results were presented in a table and included the assumption that the branching fractions for gluino and squark decays directly to the LSP were 100%. To give the reader a rough idea of the impact of branching ratios on Batley's analysis we have multiplied his gluino rates by 0.26 ($= 2 \times 0.14 - 0.14^2$ since either gluino can decay to an LSP) with the crude assumption that there would have been no difference in his analysis if he had studied the single-LSP case instead of the double-LSP case. We show in Table 2 a small portion of his table comparing the gluino signal with backgrounds.

Table 2

Event Rates after Selection Cuts
including Branching Ratio of $\tilde{g} \to LSP$

Event rates for an integrated luminosity of 10^{40} cm^{-2} from various backgrounds and $\widetilde{g}\widetilde{g}$ production, after the following cuts: $p_T^{jet} > 250$ GeV, $N_{jet} \geq 3$, $E_T^{miss} > 500$ GeV and circularity $C > 0.25$. Events with identified muons and isolated electrons were eliminated (see text). The LHC energy was taken as $\sqrt{s} = 17$ TeV. These results were taken from Ref. 21, but the gluino numbers have been multiplied by the branching ratio 0.26.

Process	Number of Events
QCD ($m_t = 40$ GeV)	**167 ± 48**
QCD ($m_t = 200$ GeV)	**64 ± 17**
$Z \to \nu\bar{\nu}$	**7 ± 2**
$W \to \tau\nu$	**7 ± 2**
other	**3 ± 1**
total bgd. ($m_t = 40$ GeV)	**184 ± 48**
total bgd. ($m_t = 200$ GeV)	**80 ± 17**
$\widetilde{g}\widetilde{g}$ ($M_{\tilde{g}} = 600$ GeV)	**494 ± 143**
$\widetilde{g}\widetilde{g}$ ($M_{\tilde{g}} = 800$ GeV)	**403 ± 52**
$\widetilde{g}\widetilde{g}$ ($M_{\tilde{g}} = 1000$ GeV)	**195 ± 26**
$\widetilde{g}\widetilde{g}$ ($M_{\tilde{g}} = 1500$ GeV)	**26 ± 1.3**

Clearly further analysis of cuts would be required for the higher $M_{\tilde{g}}$ in order to get an acceptable signal to background ratio. The event rates shown are for the LHC ($\sqrt{s} = 17$ TeV); at the SSC the signal would be increased by a factor of more than 10 (for the gluino masses given) while the background would increase at a somewhat slower rate (about a factor of 4 conservatively estimated, see Ref. 22.) We have not calculated the impact of branching ratios on Batley's results for squarks (we note, however, that he chose to study the worse case scenario in which the gluino is extremely heavy and in which squark-gluino scattering is ignored).

From the above discussion we have learned that severe cuts are needed to eliminate backgrounds. Unfortunately with these cuts, such a large fraction of the Monte Carlo generated background is eliminated that with plausible numbers of events generated, one frequently finds that no events have passed the cuts. Despite this, the resulting upper limits on the background rate can still be considerably larger than the expected signal. Of course the signal is not a problem since many signal events pass the cuts.

Work done in our subgroup (by Chris Klopfenstein) examined these same questions at SSC energies (compared with Batley's work which was at LHC energies) for the case of $\widetilde{g}\widetilde{g}$ production where $M_{\tilde{g}} = 300$ or $600\ GeV$. As mentioned above, in the time available for this workshop we have not been able to perform a detailed, high-statistics study. As was the case for Batley, this study examined the case in which both gluinos decayed directly to the LSP. We outline below the procedure which was followed:

1. Events were generated using ISAJET version 5.34.
2. A crude calorimeter simulation was used assuming perfect calorimetric coverage over $|\eta| < 5$ and all ϕ. Segmentation in η and ϕ was 0.05/cell. Smearing of energy with Gaussian resolution was taken to be:

$$\begin{aligned} (\sigma/E)^2 &= (0.15/\sqrt{E})^2 + (0.01)^2 \qquad for\ EM\ energy \\ (\sigma/E)^2 &= (0.35/\sqrt{E})^2 + (0.01)^2 \qquad for\ hadronic\ energy. \end{aligned} \tag{15}$$

3. Jets were found using the following algorithm (which is part of the ISAJET package): Find the cell with the highest E_T. If this exceeds E_{cut}^{cell} (= 5 GeV), then continue and include in this jet all cells within $\Delta R < R_{jet}$ (= 1.) with $E_T^{cell} > E_T^0$ (= 1 GeV). If the resulting E_T^{jet} has $E_T^{jet} > E_{cut}^{jet}$ (= 20 GeV), then keep the jet. This procedure is then repeated but ignoring all cells now in a jet.
4. Finally we applied a number of cuts to these events:
   - $N_{jets} \geq 3$.
   - $\Delta\phi < 150$ *degrees* (see Batley's definition (1) above).
   - $E_T^{miss} > M_{\tilde{g}}/2$.
   - $E_T^{jet} > 50\ GeV$.

The results of our study of $\widetilde{g}\widetilde{g}$ production in which both gluinos decay directly to the LSP are summarized in Table 3. In order to give the reader an idea of the impact of the single-LSP mode, the $\widetilde{g}\widetilde{g}$ cross sections have been multiplied by the branching ratios appropriate for the single-LSP mode (where only one of the gluinos has decayed directly to an LSP) instead by the branching ratios squared. The cross sections shown are obtained by summing separate ISAJET runs for different p_T ranges of the initial $2 \rightarrow 2$ processes. Each of these runs contained 10,000 to 40,000 events; however, for backgrounds there were often no events passing the cuts. When the initial cross sections were large, the resulting cross section limits after cuts were sometimes quite substantial despite having no events. The upper limits in Table 3 occur when p_T bins with no events are included. The lower limits occur when it is assumed that bins with no events make no contribution to the cross section.

Table 3

Signal and Background Cross Sections
including Branching Ratio of $\tilde{g} \to LSP$

Cross sections times branching ratios ($2 \times 0.2 - 0.2^2$ for $M_{\tilde{g}} = 300\ GeV$ and $2 \times 0.14 - 0.14^2$ for $M_{\tilde{g}} = 600\ GeV$) from $\widetilde{g}\widetilde{g}$ production and various backgrounds after the four cuts shown in text; the first box below uses $E_T^{miss} > 150\ GeV$ whereas the second has $E_T^{miss} > 300\ GeV$. Events with muons and electrons were *not* eliminated. The SSC energy was taken as $\sqrt{s} = 40\ TeV$.

Process	Cross Section (pb)
$gg\ (m_t = 70\ GeV)$	$\mathbf{419 > \sigma > 239 \pm 86}$
$t\bar{t}\ (m_t = 70\ GeV)$	$\mathbf{16 \pm 2}$
$Z \to \nu\bar{\nu}$	$\mathbf{0.5 \pm 0.1}$
$W \to \tau\nu$	$\mathbf{1 \pm 0.2}$
total bgd. $(m_t = 70\ GeV)$	$\mathbf{436 > \sigma > 256 \pm 86}$
$\widetilde{g}\widetilde{g}\ (M_{\tilde{g}} = 300\ GeV)$	$\mathbf{261 \pm 9}$

Process	Cross Section (pb)
$gg\ (m_t = 70\ GeV)$	$\mathbf{234 > \sigma > 1.5 \pm 0.3}$
$t\bar{t}\ (m_t = 70\ GeV)$	$\mathbf{5.9 > \sigma > 0.3 \pm 0.2}$
$Z \to \nu\bar{\nu}$	$\mathbf{0.15 > \sigma > 0.03 \pm 0.01}$
$W \to \tau\nu$	$\mathbf{0.39 > \sigma > 0.03 \pm 0.02}$
total bgd. $(m_t = 70\ GeV)$	$\mathbf{241 > \sigma > 1.8 \pm 0.5}$
$\widetilde{g}\widetilde{g}\ (M_{\tilde{g}} = 600\ GeV)$	$\mathbf{1.8 \pm 0.07}$

Clearly even when the lower limits are used, the signal to background ratios with this limited set of cuts are not yet adequate. Note, however, that if the branching ratio for $\tilde{g} \to LSP$ had been taken as 100% (as in previous studies), the signal would have been four times larger. The real problem with these analyses is that the cuts we used were inadequate (did not separate signal and background sufficiently) and yet we had difficulty generating enough background events so that a few events passed the cuts. Once we choose a stronger set of cuts the problem will be magnified. This situation has resulted from the fact that the signal cross sections have decreased by a factor of 3-4, necessitating more severe cuts. A complete study with very high statistics is needed to gain any real insight into this problem. Such a study should focus on the single-LSP case where only one of the gluinos decays directly to the LSP.

5. Decays of Gluinos and Squarks to W and Z Bosons

We have shown in Sec. 2 that heavy gluinos and squarks ($M_{\tilde{g}} > 300\ GeV$) have substantial probabilities for cascade decays resulting in W or Z bosons. However, the important question is whether or not a clear signal can be identified and whether there are large backgrounds. The problem is, of course, greater with W bosons even though there are roughly twice as many W bosons as Z bosons and even though the leptonic decay modes of W bosons are about three times those of Z bosons. We have therefore focussed on events in which both gluinos have decayed into Z bosons and both Z bosons have decayed to either e^+e^- or $\mu^+\mu^-$. Preliminary discussion of these questions first appeared in Ref. 19 where a number of figures were shown which displayed a variety of distributions for the signal. Similar distributions for squarks decaying to Z bosons were shown in Ref. 13 which appears in these proceedings. We do not have the space here to reproduce either set of figures, but we will summarize the conclusions. Other signatures are as useful as these but in the time available for this workshop were not considered.

In the work of Ref. 19 and in work reported here, we have examined the signal in which both gluinos have the decays:

$$\begin{aligned} \tilde{g} &\to q\bar{q}\tilde{\chi}_i \\ \tilde{\chi}_i &\to \tilde{\chi}_j Z \\ Z &\to l^+l^- \end{aligned} \tag{16}$$

The final-state from $\tilde{g}\tilde{g}$ production and decay is therefore $qq\bar{q}\bar{q}\ l^+l^-l^+l^-$ plus additional particles from the decays of the two remaining $\tilde{\chi}_j$. Thus, the signature is 4 or more hard jets plus two Z bosons. We have examined two cases: $M_{\tilde{g}} = 500\ GeV$ and $M_{\tilde{g}} = 750\ GeV$. We find that typically two of the jets are especially hard (more than 100-200 GeV). A third jet has at least 60-70 GeV. The total scalar energy coming from the three leading jets plus the two Z bosons is greater than 600-700 GeV.

Our studies indicate that the largest backgrounds for this signal is likely to be from the processes: a) $pp \to gq\bar{q}$ (or gqq) where each quark then radiates a Z boson, b) $pp \to ZZ$ where three additional jets occur as initial-state radiation and c) $pp \to qZ$ where the quark radiates a Z boson and two additional jets occur as initial-state radiation. (These three processes are not really unrelated, of course). We will show the first two backgrounds to be small and assume by analogy that the third is small. Starting with process b), we have calculated the signal and this background using ISAJET 5.34 and Pythia 4.9 (this work was done by Edward Wang). ZZ production was calculated with $p_T > 40\ GeV$. The cuts employed follow from the discussion in the previous paragraph and for $M_{\tilde{g}} = 750\ GeV$ ($500\ GeV$) were:

- $E_T^{leading\ jet} > 200\ GeV$ ($150\ GeV$).
- $E_T^{second\ jet} > 150\ GeV$ ($100\ GeV$).
- $E_T^{third\ jet} > 70\ GeV$ ($60\ GeV$).
- $E_T^{total\ scalar} > 700\ GeV$ ($600\ GeV$).
- $\phi^{leading\ jet} - \phi^{second\ jet} < 170\ deg.$

In doing this calculation it was assumed that $M_{\tilde{\chi}_i} = 200\ GeV$, $M_{\tilde{\chi}_j} = 80\ GeV$, and $M_{\tilde{\chi}_1^0} = 20\ GeV$. The Z bosons had perfect identification and reconstruction (for signal and background), the calorimetry was perfect out to $|\eta| < 5.5$, calorimeter granularity was $\Delta\phi = \Delta\eta = 0.05$, jet size was $\Delta R < 1$ and $E_T^{jet} > 25\ GeV$ was used.

Wang then found that the signal was 70-80% efficient at passing the cuts while only a tiny part of the background passed. Using the results of Sec. 2, we took the branching ratio of gluinos decaying to Z bosons to be 15% for $M_{\tilde{g}} = 750\ GeV$ and 10% for $M_{\tilde{g}} = 500\ GeV$. The number of events per year (integrated luminosity of $10^{40}\ cm^{-2}$) containing 3 jets + 2 Z bosons and passing the above cuts were:

	$M_{\tilde{g}} = 750\ GeV$	$M_{\tilde{g}} = 500\ GeV$
Signal	56 events	166 events
Background (b)	0.5 ± 0.3 events	2.2 ± 1.0 events

where background (b) is defined above. Clearly this background is not important. We should point out that there is an additional cut which we did not employ which we found would remove all the remaining backgound events generated with little impact on the signal. This variable is the total scalar transverse energy in the events (not just the three jets and two Z bosons). Since we had very few events surviving the above cuts (and no additional cuts were needed), we decided not to use this cut. This cut is effective because the signal involves the production of very heavy particles and because we ignored some of the decay products in the signal. One should be able to make use of the invariant mass of the three jets plus two Z bosons to make an estimate of gluino mass should such a signal be observed.

Turning to background (a), we are not able to perform an ISAJET calculation equivalent to the one done for background (b), since ISAJET does not radiate Z bosons from jets. We can use two techniques to estimate this background. The crudest method is to argue that background (a) is closely related to background (b) but the direct production of two Z bosons (background (b)) requires both incoming partons to be quarks whereas the direct production of a gluon and two quarks (or antiquarks) can originate in gluon scattering. This gives a factor of ten advantage to background (b), but (looking at the above table) we can afford that factor. If needed, the cuts employed could have been improved. The second method to estimate background (a) was to use a parton Monte Carlo program with a $2 \to 3$

squared matrix element for gluon-quark-quark production (provided by Ian Hinchliffe). In this program we employ the cuts described above. The results generated, of course, ignore the production of the two Z bosons. The additional effect of the radiation of transversely polarized Z bosons can then be estimated using the following factor for each radiation:

$$(g_L^2 + g_R^2)\frac{\alpha \ln \hat{s}/M_Z^2}{4\pi \sin\theta_W \cos\theta_W} f(x)$$

where x is p_Z/p_{quark} and $f(x)$ is a calculable function. This factor (squared) together with the output of the Monte Carlo program yields 1.3 $f(x)^2$ events per year using the cuts for $M_{\tilde{g}} = 750\ GeV$. While a full calculation has not been done, the function $f(x)$ is not likely to be large (especially when appropriate cuts are applied to the Z bosons). It therefore appears that there is no significant background to the signal in which both gluinos decay to Z bosons (plus other particles).

The work of Ref. 13 finds distributions similar to those described for gluinos. They conclude that for relatively small values of μ, one would expect 120 events/year for $M_{\tilde{q}} = 500\ GeV$ and 6 events/year for $M_{\tilde{q}} = 1000\ GeV$.

Acknowledgements

We wish to thank Ian Hinchliffe and Frank Paige for their valuable assistance in this work. We thank the organizers of the Workshop on Experiments, Detectors and Experimental Areas for the Supercollider for their hospitality.

REFERENCES

1. UA1 Collaboration, C. Albajar *et al.*, *Phys. Lett.* **185B**, 233 and 241 (1987); (Add.: **191B**, 462 and 463 (1987).
2. UA1 Collaboration, C. Albajar *et al.*, CERN-EP/87-148 (1987).
3. UA1 Collaboration, I. Wingerter *et al.*, to appear in the proceedings of the 1987 SLAC Summer Institute on Particle Physics: Looking Beyond the Z, August 1987.
4. UA2 Collaboration, B. De Lotto *et al.*, to appear in the proceedings of the 1987 SLAC Summer Institute on Particle Physics: Looking Beyond the Z, August 1987.
5. J. Ellis and H. Kowalski, *Nucl. Phys.* **B259**, 109 (1985); R.M. Barnett, H.E. Haber, and G.L. Kane, *Nucl. Phys.* **B267**, 625 (1986); E. Reya and D.P. Roy *Phys. Lett.* **166B**, 223 (1986).
6. H. Baer and E. Berger, *Phys. Rev.* **D34**, 1361 (1986); (E: **D35**, 406 (1987)); E. Reya and D.P. Roy, *Z. Phys.* **C32**, 615 (1986).
7. M. Claudson, L.J. Hall, and I. Hinchliffe, *Nucl. Phys.* **B228**, 501 (1983).
8. H.E. Haber and G.L. Kane, *Phys. Rep.* **117**, 75 (1985).

9. J.F. Gunion and H.E. Haber, *Nucl. Phys.* **B272**, 1 (1986); *Nucl. Phys.* **B278**, 449 (1986).
10. G. Gamberini, *Z. Phys.* **C30**, 605 (1986); H. Baer, V. Barger, D. Karatas, and X. Tata, *Phys. Rev.* **D36**, 96 (1987).
11. H. Baer *et al.*, preprint MAD/PH/357, to appear in *The Proceedings of the 1987 Madison "From Colliders to Supercolliders" Workshop.*
12. R.M. Barnett, J.F. Gunion, and H.E. Haber, reports nos. UCD-87-26 (also called LBL-24083 and SCIPP-87/104) and UCD-87-31 (also called LBL-24084 and SCIPP-87/105).
13. H. Baer, M. Drees, D. Karatas and X. Tata, report no. MAD/PH/362 (also called ANL-HEP-CP-87-88.
14. K. Inoue, A. Kakuto, H. Komatsu, and S. Takeshita, *Prog. Theor. Phys.* **68,** 927 (1982) (E: **70,** 330, (1983)); **71,** 413 (1984); R. Flores and M. Sher, *Annals of Physics (NY)* **148,** 95 (1983).
15. W. Bartel *et al.*, *Z. Phys.* **C29**, 505 (1985); H. J. Behrend *et al.*, *Z. Phys.* **C35,** 181 (1987).
16. H. Baer, K. Hagiwara, and X. Tata, *Phys. Rev.* **D35**, 1598 (1987).
17. J.F. Gunion and H.E. Haber, preprint UCD-87-24, (1987).
18. S. Weinberg, *Phys. Rev. Lett.* **50**, 387 (1983); A.H. Chamseddine, R. Arnowitt and P. Nath, *Phys. Rev. Lett.* **49**, 970 (1982).
19. H. Baer *et al.*, preprint MAD/PH/357, to appear in *The Proceedings of the 1987 Madison "From Colliders to Supercolliders" Workshop.*
20. R.M. Barnett, J.F. Gunion and H.E. Haber, in preparation.
21. R. Batley, to appear in the *Proc. of the 1987 La Thuile Workshop on Physics at Future Accelerators*, vol. 2.
22. E. Eichten *et al.*, Rev. Mod. Phys. **56**, 579 (1984).

Squark Signals at the SSC

Howard Baer[1,2] Manuel Drees[3] Debra Karatas[1,4] and Xerxes Tata[3,5] *

[1] High Energy Physics Division, Argonne National Laboratory
Argonne, IL 60439
[2] Physics Department, Florida State University, Tallahassee, FL 32306
[3] Physics Department, University of Wisconsin, Madison, WI 53706
[4] Illinois Institute of Technology, Chicago, IL 60616
[5] Theory Group, KEK, Tsukuba, Ibaraki 305, JAPAN

Abstract

We discuss the signals that emerge from the production of heavy squark pairs at the SSC. We show that left-handed squarks can be detected via their decay into real Z bosons unless the supersymmetric higgsino mass is too large, leading to typically 120 (6) $2j+4\ell+\not{p}_T$ events/year for $m_{\tilde{q}} = m_{\tilde{g}} = 0.5$ (1) TeV. Right-handed squarks can be detected via their direct decay into the lightest neutralino, which yields large missing p_T.

Squarks ($\tilde{q}$) and gluinos ($\tilde{g}$) will be copiously produced[1] at hadron colliders provided they are not too heavy. It has been shown[2] that these particles can be detected at the Tevatron if their masses do not exceed 150–200 GeV; if they are even heavier they can best be searched for at hadron supercolliders such as the SSC. In this report we focus on signals that emerge from the pair production of heavy squarks, $m_{\tilde{q}} \gtrsim 500$ GeV. A similar analysis for the pair production of heavy gluinos can be found in Ref. 3.

In our analysis we will assume that gluinos are not lighter than squarks. In this case squarks decay via

$$\tilde{q}_L \to \tilde{W}_{+/-}\, q'_L\,, \qquad\qquad \tilde{q}_{L,R} \to \tilde{Z}_i q_{L,R}\,, \tag{1}$$

where $\tilde{W}_+$ ($\tilde{W}_-$) is the heavy (light) chargino and $\tilde{Z}_i$, $i = 1,\ldots,4$ are the four neutralinos, $\tilde{Z}_1$ being the lightest one which we assume to be the lightest supersymmetric particle (LSP). Of course not all the decay channels of (1) need to be open. In minimal supergravity models[4] the masses and mixings of charginos and neutralinos are determined[5] by the values of three parameters: the gluino mass μ_3, the supersymmetric higgsino mass $2m_1$, and the ratio ω of the vacuum expectation values of the two neutral Higgs fields $\bar{H}$ and H, where $\langle \bar{H} \rangle$ gives masses to $I_{3L} = +\frac{1}{2}$ fermions. Therefore the branching ratios for the decay modes (1) depend on the gluino mass even if the gluino is heavier than the squark. For definitiveness we chose $m_{\tilde{g}} = m_{\tilde{q}}$. We have checked that our conclusions

* Spokesperson

remain qualitatively the same if a larger $m_{\tilde{g}}$ is chosen. We are also aware of the fact that for this choice of masses the associated gluino plus squark production is a sizeable or even dominant part of the total inclusive squark production cross section. In this first report we will, however neglect this additional source of squarks. We will see that squark pair production alone is sufficient to provide a clear signal over a wide range of parameters.

The breanching ratios of the various decay modes (1) for left- and right-handed up and down squarks can be found in Ref. 5. It is striking to note that left-handed squarks with $m_{\tilde{q}_L} \gtrsim 150$ GeV almost never directly decay into the LSP, whereas right-handed squarks dominantly decay into the LSP if $|2m_1| \gtrsim m_{\tilde{q}_R}/3$. The reason for this difference is that $\tilde{d}_R$, $\tilde{u}_R$ are SU(2) singlets and do thus not couple to charginos or the heavy neutralinos, which are dominated by either their higgsino or their neutral SU(2) gaugino components.

In Fig. 1 we show the total squark pair production cross section at $\sqrt{s} = 40$ TeV for $m_{\tilde{q}} = m_{\tilde{g}}$ (solid curve). We also show the cross sections for $\tilde{q}_L\tilde{q}_L$ (dashed) and $\tilde{q}_L\tilde{q}_R$ (dot-dashed).

Of course the signal for the production of a squark pair depends on how the charginos and heavy neutralinos produced via (1) decay. One very distinctive signal emerges from the decay of the heaviest gauginos $\tilde{Z}_4^0$ and $\tilde{W}_+$ into real W and Z bosons plus lighter gauginos[5,6]. In Figs. 2 and 3 we show the branching fractions for the various types of squarks to decay into real W and Z bosons. We see that this decay chain is very unlikely for right-handed squarks ($BR \lesssim 3\%$), whereas left-handed squarks have branching ratios into W's and Z's of 50% and 20%, respectively, if $|2m_1| \lesssim m_{\tilde{q}}/3$. Since we expect about $8 \cdot 10^5$ $(4 \cdot 10^4)$ $\tilde{q}_L$ pairs to be produced annually at the SSC if $m_{\tilde{q}} = 0.5$ (1) TeV, we can concentrate on the clean leptonic decay modes of the gauge bosons. Events with two hard jets from the primary squark decays and two leptonically decaying Z bosons offer an especially striking signature. For small $2m_1$ we expect 120 (6) of these events per year at the SSC for $m_{\tilde{q}} = 0.5$ (1) TeV.

However, the event rates are much smaller for larger $|2m_1|$, and in any case we cannot expect to see right-handed squarks in this decay mode. We therefore also investigated the signals that emerge if both squarks decay directly into the LSP, leading to events with 2 hard jets and missing p_T ($\not{p}_T$), or one squark decays directly into the LSP while the other cascades into a real Z, leading to a $2j + 2\ell + \not{p}_T$ signature.

In Fig. 4 we show the missing p_T spectra for these three cases. Here and in all subsequent figures we chose $m_{\tilde{q}} = 750$ GeV, a heavy gaugino mass of 200 GeV and a light gaugino mass of 25 GeV; this corresponds to $\mu_3 = 750$ GeV, $2m_1 = M_W$, and $\omega = 1.5$. The shown $\not{p}_T$ is only due to the two LSP's which escape detection; we neglected all contribution from particles that vanish in the beam pipe as well as from mismeasurements. The latter contribution would smear out our $\not{p}_T$ distribution, whereas the former would shift it towards higher $\not{p}_T$ values. We see, however, that the real $\not{p}_T$ is large enough to allow for a cut $\not{p}_T > 200$ GeV; even for the ZZ events (solid curve) this reduces the signal by only 40%, whereas the standard model 2 jet + $2Z$ background, which can produce $\not{p}_T$ only by mismeasurements and particles in the beam pipe, should be sharply reduced by this cut.

Even stronger $\not{p}_T$ cuts can be employed to isolate the events with only one Z (dashed) or with both squarks directly decaying into the LSP (dot-dashed curve).

In Fig. 5 we show the p_T distribution of the fastest and the second fastest jet from the primary squark decays, both for the ZZ events (solid) and for the events without Z. It might be somewhat surprising that the jets in the events that contain two heavy gauginos ($\tilde{Z}_4$ or $\tilde{W}_+$) are only slightly softer, although the available phase space in these squark decay modes is about 20% smaller. The shown p_T distribution of the hardest jet in $\tilde{q}\tilde{q}$ events is also very similar to the corresponding distribution in $\tilde{g}\tilde{g}$ events[3] for gluinos of the same mass. However, $\tilde{g}\tilde{g}$ events usually contain[3] at least three hard jets which should allow one to distinguish them from the $\tilde{q}\tilde{q}$ events under discussion.

In Fig. 6 we show the p_T distribution of the leptons that come from the decay of the Z bosons. This distribution should be very similar for events where one or both Z's are replaced by W's. We also show the corresponding distribution for the slowest of the four leptons. Even if the lepton identification requires a cut $p_{T\ell} > 10$ GeV we only lose about 10% of the signal.

In Figs. 7 and 8 we show the total scalar sum of transverse energy and invariant mass of the two jets and four leptons of the ZZ events (solid) and the two jets and two leptons of the $Z +$ LSP events (dashed). Obviously we can apply the cuts $E_T(2j + 4\ell) > 800$ GeV and/or $M(2j + 4\ell) > 900$ GeV without losing any signal. Figure 9 shows that the two hard jets still tend to be back-to-back; unfortunately this is also to be expected for QCD dijet events where 2 Z's are radiated off the final or initial state, which is one of the most severe background sources. Other potentially dangerous backgrounds are semileptonically decaying top quarks where the invariant mass of the dilepton systems happens to be close to M_Z. In principal this background can be eliminated by requiring the smallest invariant mass of any of the leptons together with any of the two hard jets to be larger than m_t. Figure 10 demonstrates that this cut would reduce the signal only slightly provided the top is not very heavy ($m_t < 100$ GeV).

Unfortunately a calculation of the most serious $2j + 2Z$ standard model background is not yet available. In Ref. 3 it has however been estimated that it can be reduced to the level of a few events per year by suitably chosen cuts on the p_T of the jets and the invariant mass of the $2j + 2Z$ system. For the $2j + 4\ell$ events this background can be reduced even further by cutting on $\not{p}_T$. The $2j + 2Z$ background, with one Z decaying invisibly, might however, severely constrain the possibility to extract the mixed signal where one squark cascades into a Z boson while the other directly decays into the LSP. The reason is that in most regions of parameter space this configuration requires a $\tilde{q}_R\tilde{q}_L$ pair in the initial state, which has a smaller cross section than $\tilde{q}_L\tilde{q}_L$ or $\tilde{q}_R\tilde{q}_R$ pairs if $m_{\tilde{g}} \geq m_{\tilde{q}}$. On the other hand preliminary studies[7] show that the signal from direct decay of both squarks into the LSP is detectable provided the branching ratio for this decay mode is not much smaller than 30%.

We thus conclude that heavy squarks with $m_{\tilde{q}} \gtrsim 500$ GeV should be detectable at the SSC either via the $2j+4\ell$ signal, if $|2m_1| \lesssim m_{\tilde{q}}/3$, or via the $2j+\not{p}_T$ signal if $|2m_1| > m_{\tilde{q}}/3$.

Acknowledgements

We thank the organizers of this workshop for their hospitality. This research was supported in part by the University of Wisconsin Research Committee with funds granted by the Wisconsin Alumni Research Foundation, and in part by the U. S. Department of Energy under contracts DE-AC02-76ER00881 and W-31-109-ENG-38.

References

1. P. H. Harrison and C. H. Llewellyn-Smith, Nucl. Phys. **B213**, 223 (1983); erratum, Nucl. Phys. **B223**, 542 (1983).
2. H. Baer and E. Berger, Phys. Rev. **D34**, 1361 (year); erratum, Phys. Rev. **D35**, 406 (1987); E. Reya and D. P. Roy, Z. Phys. **C32**, 615 (1986).
3. H. Baer *et al.*, to appear in the proceedings of the workshop *From Colliders to Super Colliders*, Madison, May 1987.
4. For reviews see H. P. Nilles, Phys. Rep. **110**, 1 (1984); H. E. Haber and G. L. Kane, Phys. Rep. **117**, 75 (1985).
5. See *e.g.* H. Baer, V. Barger, D. Karatas, and X. Tata, Phys. Rev. **D36**, 96 (1987), and references therein.
6. J. F. Gunion *et al.*, to appear in the proceedings of the workshop, *From Colliders to Super Colliders*, Madison, May 1987.
7. F. Pauss, these proceedings.

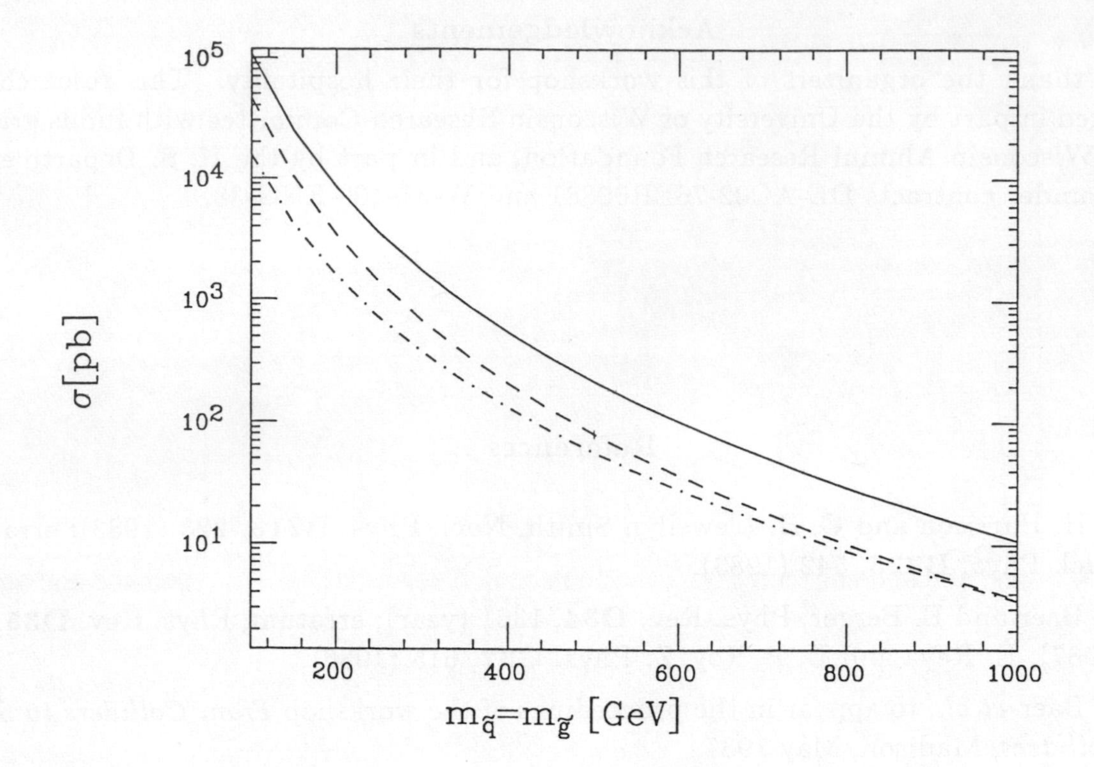

1. Total squark pair production cross section at $\sqrt{s} = 40$ TeV for $m_{\tilde{g}} = m_{\tilde{q}}$ (solid curve). We also show the $\tilde{q}_L\tilde{q}_L$ or $\tilde{q}_R\tilde{q}_R$ cross section (dashed) and the $\tilde{q}_L\tilde{q}_R$ cross section (dot-dashed).

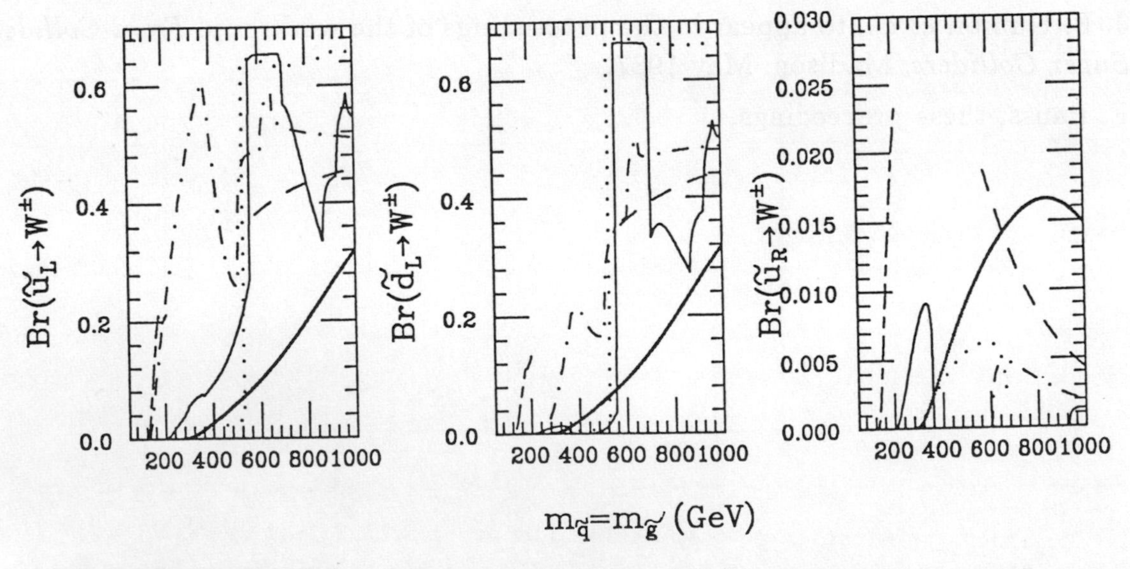

2. Branching ratios of the various squarks into real W bosons for $m_{\tilde{q}} = m_{\tilde{g}}$ and $\omega = 1.5$. The thick solid, long dashed, long-short dashed, dot-dashed, solid and dotted curves are for $2m_1 = -3M_W$, $-M_W$, $-\frac{1}{2}M_W$, M_W, $2M_W$, and $3M_W$, respectively. The gaps in the curves indicate that $m_{\tilde{W}_-} \leq 30$ GeV.

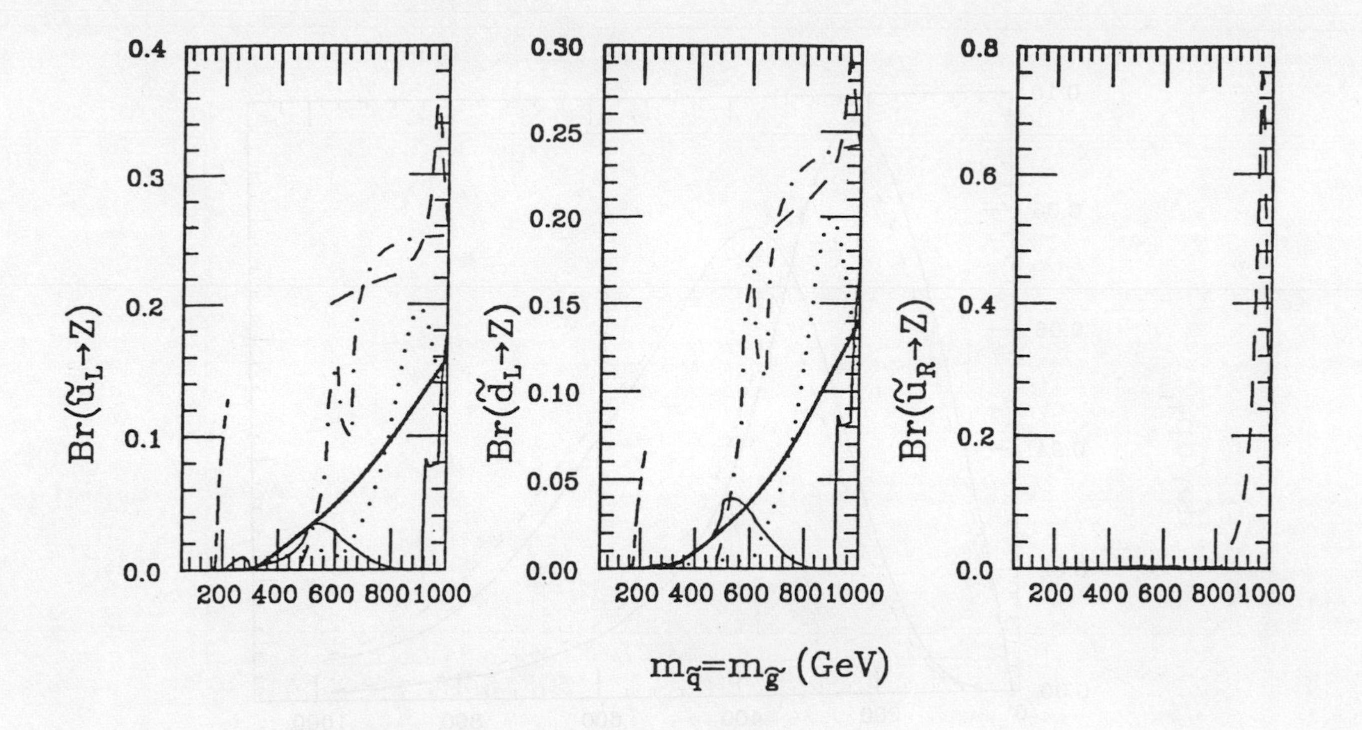

3. Branching ratios of the various squarks into real Z boson. Parameters and notations are as in Fig. 2.

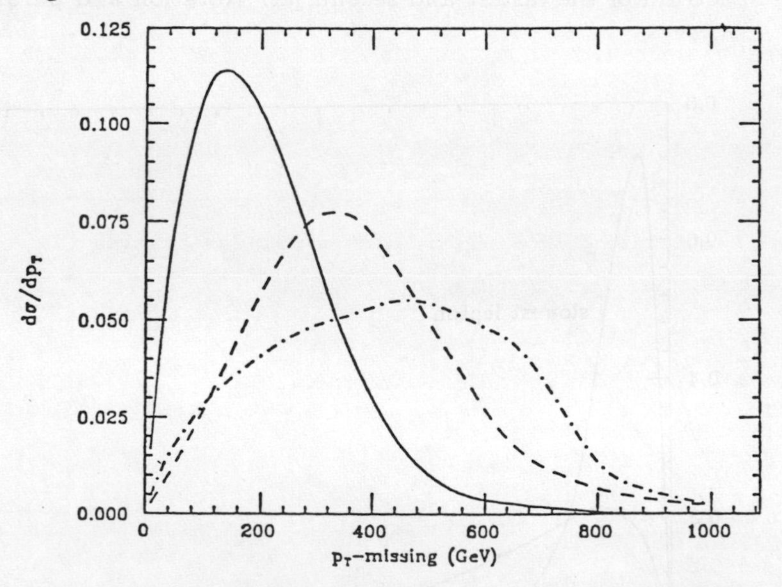

4. The $\not{p}_T$ spectrum from escaping LSP's for events with $2j+4\ell+\not{p}_T$ (solid), $2j+2\ell+\not{p}_T$ (dashed), and $2j+\not{p}_T$ (dot-dashed). Here and in the remaining figures the curves are normalized to yield the same total cross sections. The mass of the squarks, heavy gauginos and light gauginos is taken to be 750 GeV, 200 GeV and 25 GeV, respectively.

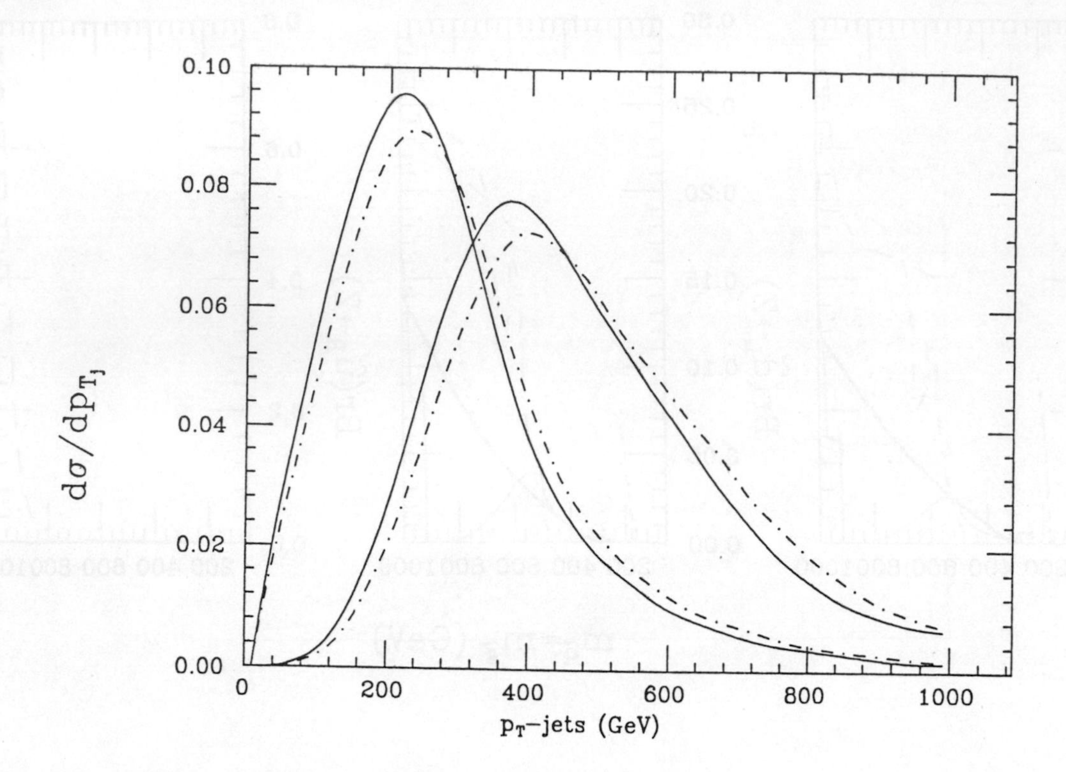

5. The p_T spectrum of the fastest and second jet. Notation and parameters are as in Fig. 4.

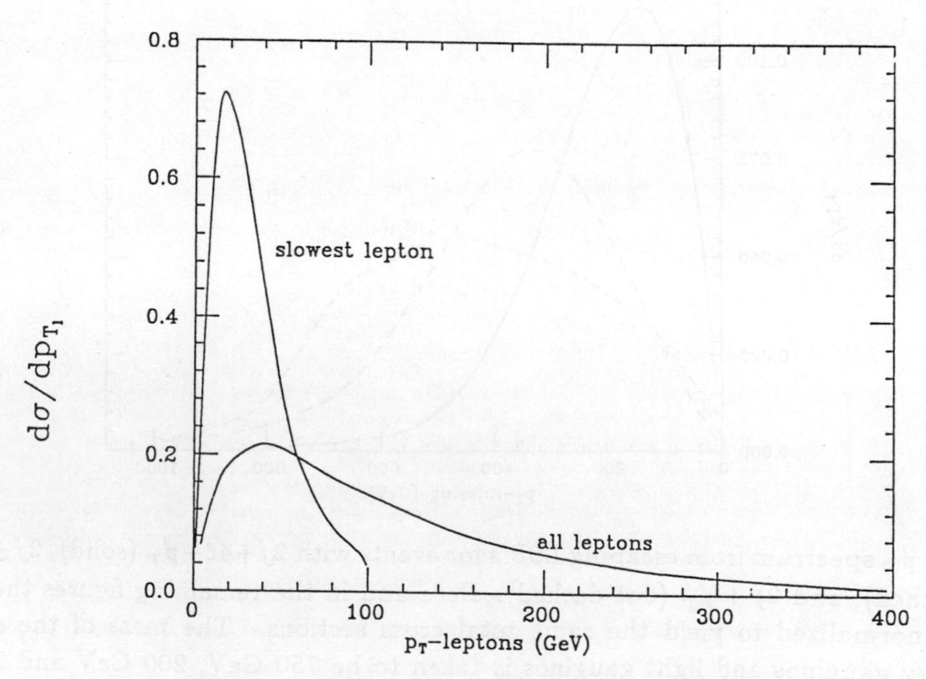

6. The p_T spectrum for all leptons as well as the slowest lepton in the $2j + 4\ell + \not{p}_T$ events.

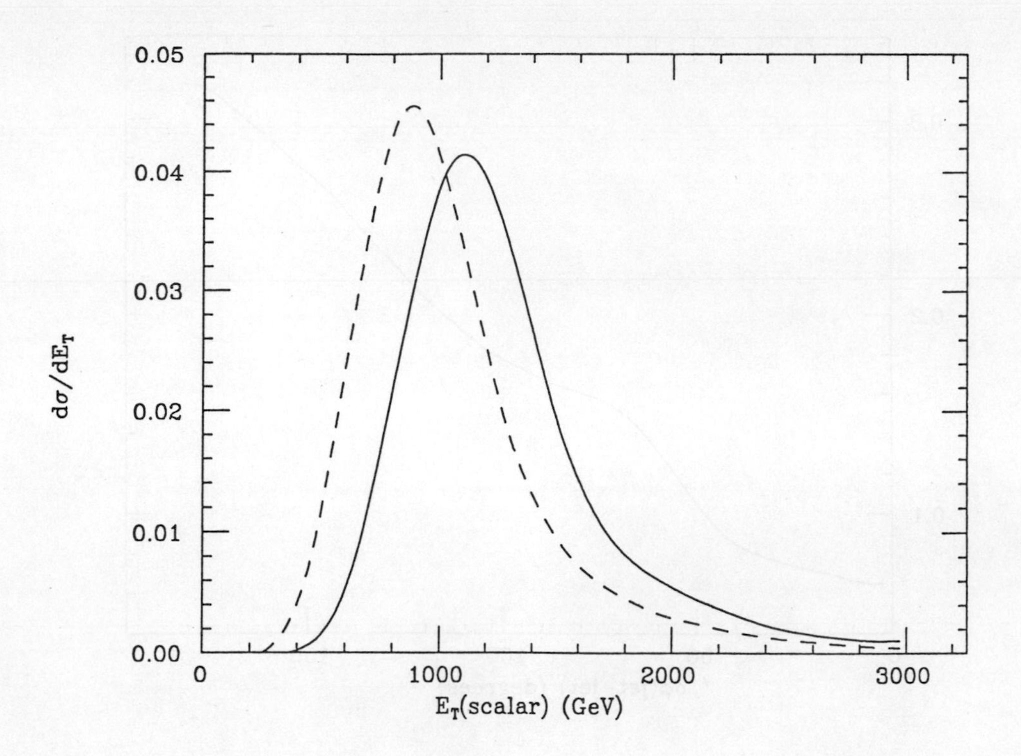

7. The scalar E_T spectrum from the two hard jets and the two Z bosons in the $2j + 4\ell + \not{p}_T$ events (solid), and from the two hard jets and one Z boson the events where one squark decays directly into the LSP (dashed).

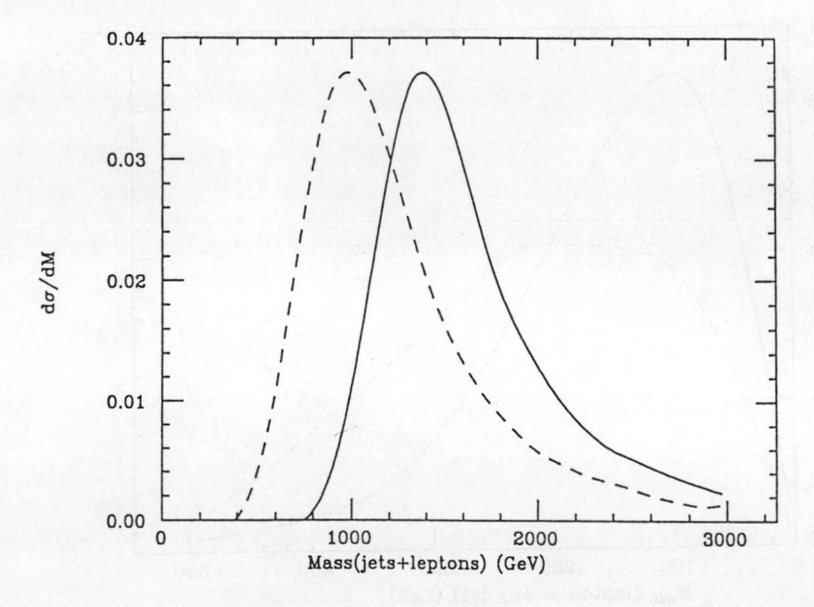

8. The invariant mass distribution of the two hard jets and two Z bosons (solid) in the $2j+4\ell+\not{p}_T$ events as well as of the two hard jets and one Z boson in the $2j+2\ell+\not{p}_T$ events (dashed).

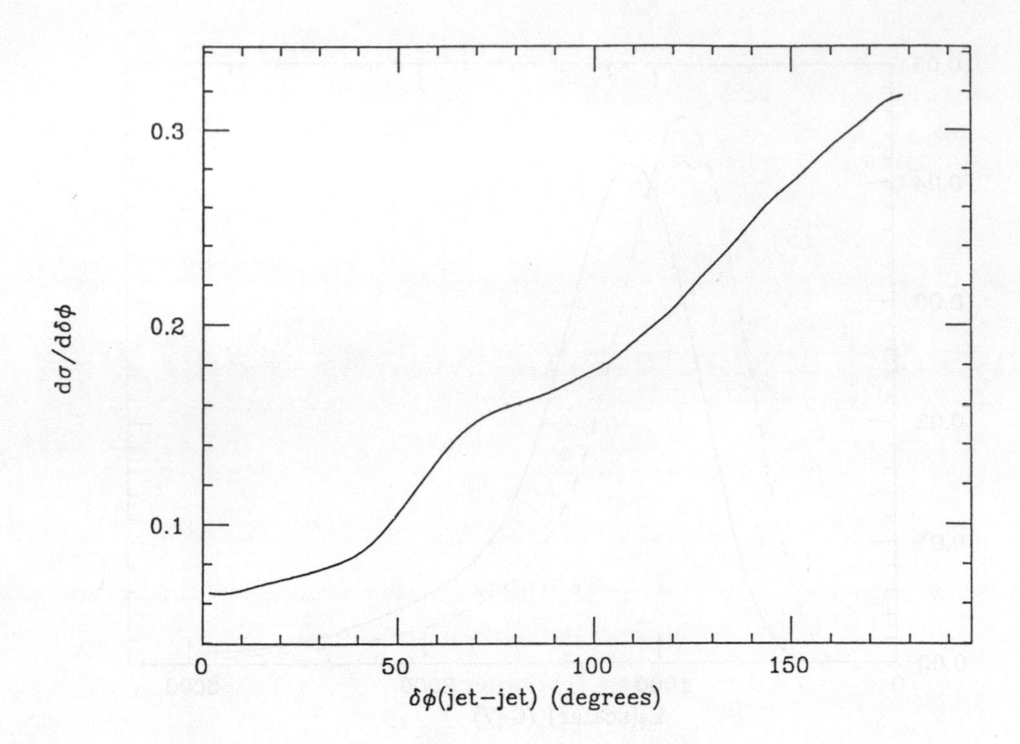

9. The distribution of opening angle in the transverse plane between the two hard jets from squark pairs where both squarks decay into a real Z boson.

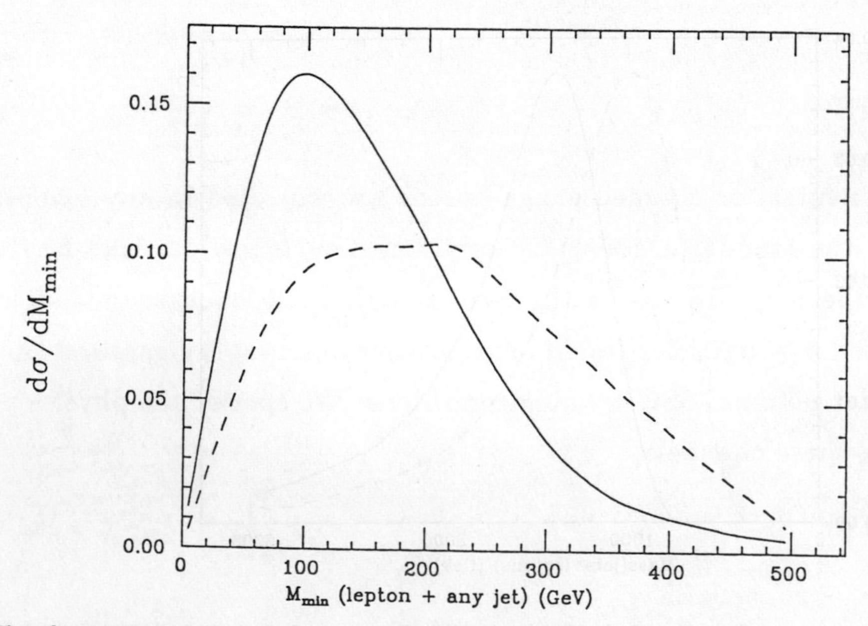

10. The distribution of the minimum invariant mass that results from combining a given lepton with any of the two hard jets. The solid and dashed curve is for the ZZ and $Z+\text{LSP}$ events discussed in the text.

Physics Parameters for New W's and Z's

J. S. Whitaker
Physics Department
Boston University
Boston, MA 02215

N. Deshpande
Institute of Theoretical Sciences
University of Oregon
Eugene, OR 97403

Abstract

New massive neutral or charged gauge bosons are expected in any extension of the gauge structure of the Standard Model. Observation of such new particles by their decays to quark jets will be very difficult in the face of high QCD backgrounds. The leptonic decay modes of such new particles would provide clear observable signatures up to masses around 5 TeV under nominal SSC running conditions. We specify the physics parameters which characterize these channels.

Introduction

The primary physics goal of the SSC is to explore the energy regime where high energy symmetries break down to form the gauge structure observed at low energies. Extension of the low energy group structure with local gauge symmetries predicts additional massive vector particles which are the gauge bosons of the extended symmetry. Properties of these new vector particles, designated generically as $W'^{\pm}$ and $Z^{0\prime}$, will depend on the specific model (see below). Observation of $W'^{\pm}$ and/or $Z^{0\prime}$ will be be the key in establishing higher symmetries, just as the observation of the $W^{\pm}$ and the Z^0 at the S$\bar{p}$pS was striking confirmation of the gauge structure of the Standard Model. The charge of this Study Group was to identify the physics parameters that characterize the new W's and $Z^{0\prime}$s in order to provide input for detailed detector designs. We have drawn heavily on previous results, especially the work of EHLQ [1] and the reports contained in the proceedings of the Snowmass 1984 and Snowmass 1986 workshops [2], [3].

Generic $W'^{\pm}$s and $Z^{0\prime}$s

The number and characteristics of possible $W'^{\pm}$s and $Z^{0\prime}$s depend on exactly how the gauge structure is to be expanded. Generally, a $Z^{0\prime}$ will appear when an additional $U(1)$ symmetry is added. The literature on this is very extensive, especially in the context of superstrings (see [4], [5], and references therein). If the added group is more complicated, for example a $SU(2)_R$ as in right-left symmetric theories [6] then both $W'^{\pm}$s and $Z^{0\prime}$s will arise. More exotic cases are possible. For example, "horizontal" gauge symmetry may be introduced [7] leading to particles which mediate interfamilial transitions and which may decay to *e.g.* an electron and a muon.

The general intent in these extensions of the Standard Model is that the groups unify at some large energy scale. This implies a relation between the coupling constants of the new group and those of the present low energy theory. Thus, despite a broad array of choices for the extended group, the gross properties of the $W'^{\pm}$ or $Z^{0\prime}$ are similar to the familiar $W^{\pm}$ and Z^0. Specifically, the new particles can be expected usually to couple to the familiar quarks and leptons with more-or-less standard couplings, and the widths of the new particles should scale as

$$\Gamma_{W'} \simeq \Gamma_W \times \frac{M_{W'}}{M_W}.$$

Although there are certainly exceptions to this generalization, we shall adopt this picture for this report. Details of specific models for production and decays of $Z^{0\prime}$ can be obtained from Reference [8]. The $W'^{\pm}$ and $Z^{0\prime}$ will then have widths which are a few percent of their mass. This is the crucial feature which makes these new particles potentially observable. The branching fractions for the $W'^{\pm}$ and $Z^{0\prime}$ in this picture will be roughly the standard values:

$BR(Z^{0\prime} \rightarrow l^+l^-) \simeq 3\%$ for $l = e,\ \mu,\ \tau$

$BR(Z^{0\prime} \rightarrow \nu\bar{\nu}) \simeq 6\%$ for each family

$BR(W'^{\pm} \rightarrow l^{\pm}\bar{\nu}_l) \simeq 8\%$ for $l = e,\ \mu,\ \tau$

with the remainders being decays to $q\bar{q}$.

The situation with $W'^{\pm}$ is complicated by the possibility that the the $W'^{\pm}$ may decay to a charged lepton plus a heavy Majorana neutrino. Such a particle arises for example in some left-right symmetric theories. The N^0 would presumably be unstable, and its decay products could confuse the signature for the $W'^{\pm}$. Alternatively, N^0 decays including a charged lepton could provide a distinctive signature for this process.[6]

Present lower bounds for the masses of $W'^{\pm}$ or $Z^{0\prime}$ are derived from the non-observation of flavor-changing neutral currents and from the success of Standard Model predictions for muon decay parameters, for asymmetries in e^+e^- annihilation, and for $W^{\pm}$ and Z^0 production at the S$\bar{\text{p}}$pS. Demonstrating the model dependence, the lower bounds for $M_{W',Z'}$ range from 100 GeV to over 1 TeV.

$Z^{0\prime}$ Detection

The search for the two-body or quasi-two-body decays of the $Z^{0\prime}$ will be the prototypical SSC experiment. We consider several possible decay modes:

<u>$Z^{0\prime} \rightarrow$ *jet jet*</u>

The $Z^{0\prime}$ is expected to decay roughly 70% to quark-antiquark, resulting in a two-jet topology. UA2 has recently reported [9]the observation of W' and Z^0 in the $q\bar{q}$ mode. At higher energies however, the gluon luminosity increases and the signal for $Z^{0\prime}$ in the the jet-jet channel will will be overwhelmed by the glue-glue jet background.

<u>$Z^{0\prime} \rightarrow W^+W^-$</u>

The decay $Z^{0\prime} \rightarrow W^+W^-$ has a branching ratio comparable to e^+e^- and is potentially observable. This process would provide information on the gauge group giving rise to the $Z^{0\prime}$. This decay mode is discussed in detail in the paper by N. Deshpande and F. Zwirner in these proceedings.

<u>$Z^{0\prime}\rightarrow l^+l^-$</u>

The decays of a narrow $Z^{0\prime}$ to e^+e^- or to $\mu^+\mu^-$ will be clear signals with essentially no background. In any reasonable detector, background from jets faking leptons will be effectively eliminated by the requirement of two leptons for the signal. Assuming a $Z^{0\prime}$ with standard couplings, the background continuum from the Drell-Yan process is at the level of 10^{-4} of the resonant signal (estimated using the EHLQ curves [1]) for masses in the vicinity of 1 TeV. Observation of a $Z^{0\prime}$ will hinge upon sufficient acceptance and sufficient luminosity to provide a few well-reconstructed events in this very clean channel.

The decay $Z^{0\prime}\rightarrow \tau^+\tau^-$ would be interesting as a test of lepton universality and for asymmetry studies, as discussed by Haber ([2], page 157.) The leptonic decays are likely to be very difficult to dig out of the backgrounds from charm production and from $W^\pm$ pair production at lower masses. The decays of the tau to single pions may provide a handle on this channel, but backgrounds due to QCD jets and due to the processes mentioned above will be very large.

$W'^\pm$ Detection

<u>$W'^\pm\rightarrow$ *jet jet*</u>

The situation here is the same as the case for $Z^{0\prime}\rightarrow$ *jet jet*. The decay of $W'^\pm$ to two jets will be a dominant decay mode but will be subject to a very background from standard QCD processes.

<u>$W'^\pm\rightarrow$ *leptons*</u>

The $W'^\pm$ is expected to decay to $l\nu_l$, where $l = e$ or μ, with a significant branching fraction (8% per generation in the Standard Model). This channel, the classic one by which the $W^\pm$ was originally observed, is characterized by an isolated high p_T lepton and large missing transverse energy $\not{E}_T$. The lepton transverse momentum distribution has the famous "Jacobian peak" at approximately half the mass of the $W'^\pm$. The question of backgrounds has been studied by Carr and Eichten ([2], page 703). They find that a factor of order 10^3 in pion rejection for lepton identification will suffice to isolate a $W'^\pm$ signal above backgrounds for masses in the 1 TeV range.

The decay $W'\rightarrow \tau\nu_\tau$ is potentially interesting for the opportunity it offers to analyze the W' polarization (see the paper of Haber previously cited). This channel is potentially observable by identifying the tau decay to three pions, but backgrounds are likely to be large enough from QCD jets and from other sources of τ's that establishing this decay mode would be very difficult.

In some models, the W' will decay $W' \rightarrow l^{\pm}N$, where N is a new heavy neutrino. The N may in turn decay, to a light neutrino plus jets for example. In such a case one loses the missing E_T signal for W' production, but the coincidence of an isolated lepton and a hadronic system with approximately balancing p_T may provide a sufficiently characteristic signal.

Luminosity Requirements and Mass Reach

The cleanliness of the W' or $Z^{0\prime}$ events leads us to conclude that the the design luminosity of $10^{33}\text{cm}^{-2}\text{sec}^{-1}$ should be useable in the search for heavy vector particles. The high p_T lepton should provide a clear trigger signal. It is difficult to imagine a detector operating at a higher luminosity and providing the necessary clean trigger information unless it was specialized to study only decays to muons, as in the detector concept described by R. Thun in these proceedings. We note that even at design luminosity, single beam crossings in which there are ten interactions will occur at 125 Hz! Detectors will have to be able to distinguish rare high-p_T leptons from the pileup of copious lower-p_T particles.

Discovery of the W' will depend on sufficiently copious production of the particle and on sufficiently large branching fractions into decay modes with sufficiently low backgrounds. Production of $W'^{\pm}$ or $Z^{0\prime}$ in proton-proton collisions would be dominantly by the Drell-Yan process of quark-antiquark annihilation. This implies that the transverse momentum of the W' will be small (negligible except for gluon radiation effects) and the longitudinal momentum of the W' will depend as $p_z = E_{beam} \times (x_a - x_b)$ where $x_{a,b}$ are the momentum fractions of the annihilating quark and antiquark. EHLQ [1] have calculated the production cross section for $W'^{\pm}$ and for $Z^{0\prime}$, convoluting the parton-parton cross sections with the structure functions of the incident protons. In terms of the W' rapidity y the production cross section scales as $\frac{d\sigma}{dy} \sim \frac{1}{M_{W'}^2}$ and is roughly flat in the central region, falling off as $|y|$ approaches $y_{max} \simeq ln(\frac{\sqrt{s}}{M_{W'}})$. One point to note is that due to the larger number of u quarks compared to d quarks at high x the cross section for producing W'^{+} is roughly twice that for W'^{-} at high mass. Observation of this charge asymmetry would be an important signature of $W'^{\pm}$ production.

Given the (mass-dependent) production cross sections and the branching fractions for decay into observable channels, and guessing at the statistics required to establish a positive observation, one can calculate the "mass reach" of an experiment as a function of the integrated luminosity. Figure 1, taken from EHLQ [1], shows the mass reach versus luminosity assuming that 10^3 particles produced with $|y| < 1.5$ would constitute a positive signal. An integrated luminosity of $\int Ldt = 10^{38}\text{cm}^{-2}$ implies a mass reach to about 2 TeV, while $\int Ldt = 10^{40}\text{cm}^{-2}$ will probe up to masses around 6 TeV.

Vertex Detection

Reconstruction of the primary vertex will be useful in the rejection of background from multiple interactions. There is no clear need for secondary vertex reconstruction in the search for the W' or $Z^{0\prime}$, aside from the possibility of studying $Z^{0\prime} \to \tau^+\tau^-$. Secondary vertex reconstruction would only partially solve the potential background problem from $W^\pm$ pair production.

Tracking

Tracking charged particles will be necessary for lepton identification, by pointing to calorimeter elements, and momentum determination by tracking in a magnetic field would assist in the energy measurement and in lepton identification. The leptons from W' decays are quite isolated. Figure 2 is a plot of the mean number of charged particles with momentum above 2 GeV as a function of the separation from one of the electrons in a $Z^{0\prime} \to e^+e^-$ event, calculated using ISAJET. The mass of the $Z^{0\prime}$ was 1 TeV, and the separation is expressed in terms of the parameter $R = \sqrt{(\Delta\eta)^2 + (\Delta\phi)^2}$, where η is the particle pseudorapidity. For each electron, The average number of charged particles with $R \leq .5$ is 4.5. If a lower cut of 2 GeV is applied, an average of 1.2 particles are within this cone.

Charged particle tracking should be extended to the smallest practicable angles from the beam lines, in order to achieve the best acceptance for the $W'^\pm$ and $Z^{0\prime}$ decays. Figure 3 shows, as a function of the rapidity of the $Z^{0\prime}$, the probability that the two electrons from $Z^{0\prime} \to e^+e^-$ will have angles from the beam lines θ greater than a minimum value θ_{min} that characterizes the low angle cutoff for lepton reconstruction. Folding these curves with the rapidity distribution for a given mass $Z^{0\prime}$ gives the total acceptance. Using ISAJET to simulate the rapidity distribution for a 1 TeV $Z^{0\prime}$, a detector with $\theta_{min} = 10°$ has an acceptance of 60% for reconstructing both electrons. The acceptance is higher at larger masses according to the scaling of y_{max} discussed above. On the basis of this study and of similar work[10] we conclude that coverage to $\theta_{min} \sim 5°$ to $10°$, corresponding to coverage to $|y| \leq 2.5$ to 3, is vital to the search for the W'.

Determination of the charge of the lepton is important to verify the opposite-sign character of the dilepton signal for the $Z^{0\prime}$ and to check the expectation that $\sigma(W'^{+}) \simeq 2\times\sigma(W'^{-})$. Charge determination is also important for the study of the forward-backward asymmetry, which can help determine the couplings of the W'. This has been studied in some detail[11]. Higher masses for the W' or $Z^{0\prime}$ imply lower statistics. For $M_{W'} \geq 2$ TeV or so ($p_T \geq 1$ TeV) the statistics for the W' will be too low in a reasonable run to allow asymmetry studies, and so charge determination is slightly less vital.

There is an interesting point to consider in a comparison of the electron channel and the muon channel. It may be that the most practical strategy will be to establish $W'^{\pm}$ or $Z^{0\prime}$ signals using the electron channel in the central region, where good energy and mass determination should be straightforward, and to study asymmetries and charge correlations using the muon channel at small angles, where asymmetry effects are larger and where the instrumentation for charge determination may be more feasible.

Electron Acceptance and Identification Requirements

In order to observe the decays of a W' or a $Z^{0\prime}$ the detector must be able to identify electrons and muons with high efficiency and low hadron contamination and must be able to reconstruct the four-vectors with good resolution.

There is a kinematic relationship between the energies of the two leptons for the case of a two-body decay of a particle which has only longitudinal momentum. Denoting the higher and lower energies as E_1 and E_2, then

$$\frac{E_2}{E_1} \geq sin(\theta_{min})$$

where θ_{min} is the smallest angle a detected particle will make with the beam lines. Since $E_1 + E_2 \geq M_{W',Z'}$, the reconstruction of $Z^{0\prime}$'s will require tracking of leptons in the rough momentum range $p \geq tan(\theta_{min}) \times M/2$. Since the mass range below several hundred GeV is already excluded or will be explored by experiments at the SLC and at LEP, electron identification will be important for momenta above 100 GeV or so. The upper limit to the range of interest is just set by the mass reach as defined by the integrated luminosity: electrons up to 4 TeV will be important for the W' search given nominal SSC operations. Electron identification will be vital over the range $|y| \leq 3$ or so with jet rejection at the 10^{-3} level as discussed above.

Calorimetry

Hadronic calorimetry will be important for establishing the missing E_T signal for the $W'^{\pm}$ search. A resolution of 100 GeV in E_T should be sufficient. Electromagnetic calorimetry will be crucial to the $W'^{\pm}$ and $Z^{0\prime}$ physics. The goal of measuring the width of the $Z^{0\prime}$ at the per-cent level will require energy resolution $\sigma(E)/E \simeq 1\%$ and angular resolution of $\sigma(\theta) \simeq 10$ mrad over the central region $|y| \leq 3$.

Muon Identification and Reconstruction Requirements

Muons will be very important for the $W'^{\pm}$ and $Z^{0\prime}$ physics. The rapidity range which should be covered is the same as for the electrons, $|y| \leq 3$ or so, with larger $|y|$ of interest for asymmetry studies. Energy resolution at the per-cent level would be required to study the $Z^{0\prime}$ width; this is likely to be very hard to accomplish without compromising other features of the detector. The energy resolution of 10–15% achieved in typical detectors is adequate for observing the Jacobian peak characteristic of $W'^{\pm}$ decays. Adeva *et. al.*([3], p. 257) have studied the energy resolution required to be able to distinguish between various $Z^{0\prime}$ models by studying the forward-backward asymmetry, and they conclude that energy resolution of $< 10\%$ at large rapidity would be required.

Triggering

The high-p_T, isolated leptons from $W'^{\pm}$ or $Z^{0\prime}$ decays will provide clear, straightforward trigger signals. It may be desirable to require a pair of high-p_T leptons in the trigger, and this should include the possibility of $e - \mu$ pairs to be sensitive to possible $Z^{0\prime}$'s from horizontal gauge symmetries. The ability to trigger on missing $E_T \geq 100$ GeV would be useful in the $W'^{\pm}$ search but might lose a $W'^{\pm} \rightarrow l^{\pm} N^0$ signal.

Conclusions

The decays of high-mass $W'^{\pm}$'s or $Z^{0\prime}$'s to electrons or muons will provide signatures that are clear, robust, and nearly background-free. Other decay channels, specifically to tau pairs or to W pairs, would be interesting but will be challenging to reconstruct. Decays to hadronic jets will be very difficult to separate from QCD multi-jet background. Coverage to $|y| \leq 3$ for lepton reconstruction is vital for good acceptance. Production cross sections should be large enough that a fiducial SSC run will provide sensitivity to masses up to over 5 TeV. The physics of new $W'^{\pm}$'s and $Z^{0\prime}$'s is of very fundamental interest and should be among the most accessible at the SSC.

References

[1] E. Eichten, I. Hinchliffe, K. Lane, C. Quigg, Rev. Mod. Phys. 56, October 1984.

[2] Snowmass 84: Proceedings of the 1984 Summer Study on the Design and Utilization of the SSC, Snowmass, CO, 1984; R. Donaldson and J. Morfin, editors.

[3] Snowmass 86: Proceedings of the Summer Study on the Physics of the Superconducting Supercollider, Snowmass, CO, 1986; R. Donaldson and J. marx, editors.

[4] L. S. Durkin and P. Langacker, Phys. Lett. 166B, 436 (1986).

[5] V. Barger, N. G. Deshpande, and K. Whisnant, Phys. Rev. Lett. 56, 30 (1986).

[6] See J. F. Gunion and B. Kayser, Snowmass 84, p. 153, and references therein.

[7] C. Albright *et. al.*, Snowmass 84, p. 144.

[8] V. Barger, N. G. Deshpande, J. Rosner, and K. Whisnant, Snowmass 84, page 224.

[9] R. Ansari *et. al.*, CERN preprint CERN-EP-87-04, 1987.

[10] D. Carlsmith, D. Hedin, and B. Milliken, Snowmass 86, p. 431.

[11] See e.g. H. Haber, Snowmass 84, p. 125, and references therein, and B. Adeva *et.al.*, Snowmass 86, p.257.

fig. 1

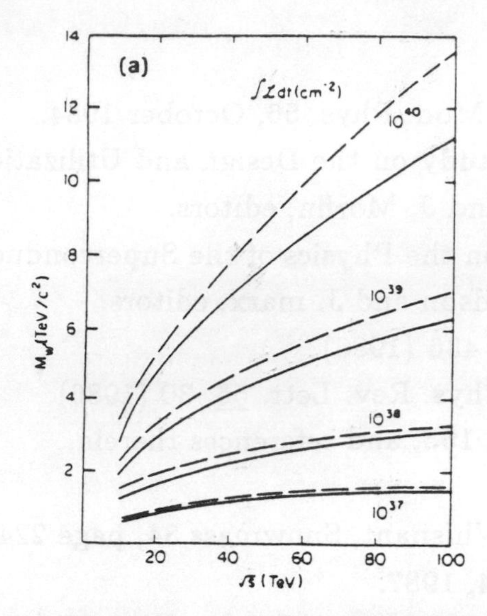

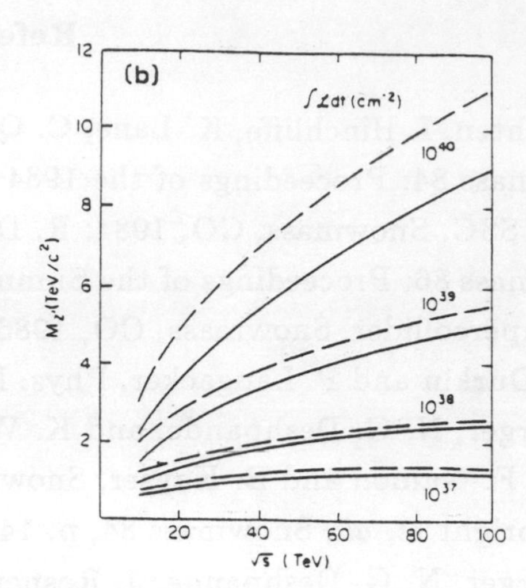

Maximum mass of a new charged intermediate boson for which 10^3 events are produced with $|y_W| < 1.5$ at the stated integrated luminosities in proton-proton collisions (solid lines) and in proton-antiproton collisions (dashed lines). Taken from EHLQ (1).

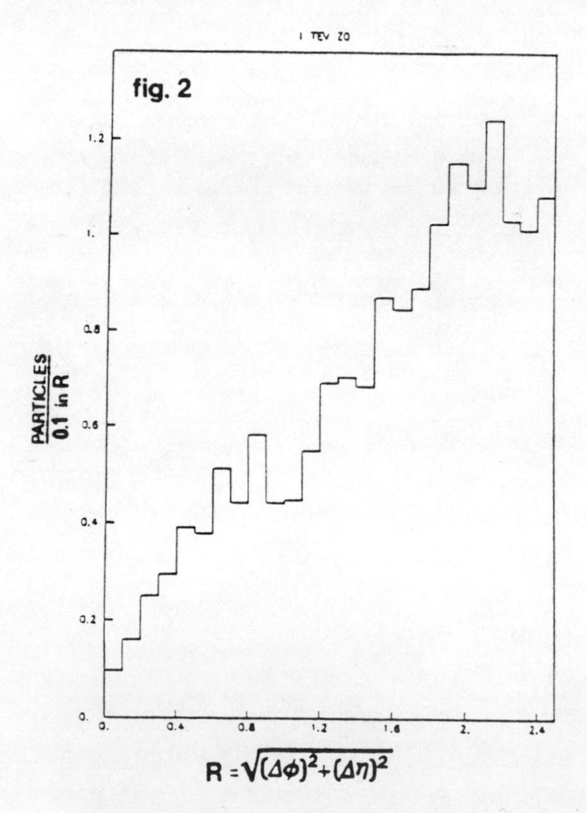

Number of charged particles with momentum > 2 GeV/c included within a cone of size R in azimuth-pseudorapidity space, about one of the leptons from the decay of a 1 TeV $Z^{o\prime}$ to e^+e^-.

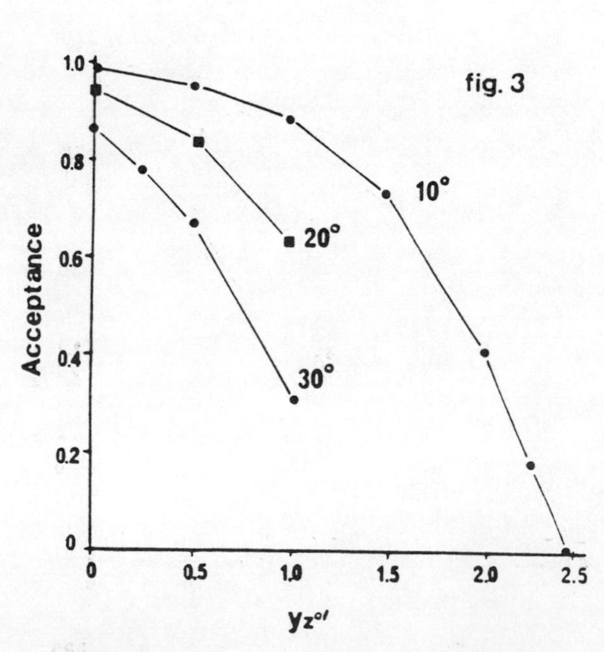

Acceptance for both leptons in $Z^{o\prime} \rightarrow l^+l^-$ at $\theta > \theta_{min}$ for several choices of θ_{min}

DETECTION OF A NEW Z' IN THE $Z' \to W^+W^-$ MODE AT THE SSC

N. G. Deshpande
Institute of Theoretical Science, University of Oregon, Eugene OR 97403

J. F. Gunion
Department of Physics, U. C. Davis, Davis CA 95616

and

F. Zwirner
Lawrence Berkeley Laboratory, Berkeley CA 94720

ABSTRACT

If a new Z' exists with mass in the TeV region, the decay rate for the mode $Z' \to W^+W^-$ is expected to be of the same order as the ones for $Z' \to e^+e^-$ or $Z' \to \mu^+\mu^-$. This mode can be detectable at the SSC via the secondary decays $WW \to (jet\ jet)(l\ \nu)$, $(l = e, \mu)$. We compare the expected signal with the backgrounds coming from continuum WW production and $W\ jet\ jet$ production in the standard model. Using for this decay selection criteria analogous to the ones proposed for the corresponding decay of a heavy Higgs, we conclude that the signal/background ratio should be considerably larger in the Z' case. This is primarily because, for masses in the TeV range, the Z' width is very much smaller than the Higgs width.

1 Introduction

If there is some 'beyond the standard model' physics in the TeV region which can be probed at the SSC, one attractive possibility is another massive neutral gauge boson (Z') in addition to the known Z [1]. The existence of additional neutral currents can be easily incorporated in non-minimal models of grand unification like $SO(10)$ or E_6, and this possibility arises naturally in the presently fashionable superstring-inspired models.

The most promising decay channel for the discovery of a new Z' is obviously the charged lepton-antilepton pair mode (e^+e^- or $\mu^+\mu^-$). Decays into ordinary quark-antiquark pairs appear to be undetectable, due to the huge QCD background. Most other possible channels must rely on final states involving particles that have not yet been discovered. At present the only other channel containing only standard model particles that looks promising for detection is $Z' \to W^+W^-$ [2]. Despite the fact that the $Z'W^+W^-$ coupling arises only through the mixing between Z and Z', which is forced to be small, in most models the Z' branching ratio into W pairs is naturally of the same order as the one into e^+e^- or $\mu^+\mu^-$. In this contribution we give a quantitative discussion of the possibility of detecting this last decay mode at the SSC.

2 The decay $Z' \to W^+W^-$

As pointed out in ref. [2], the decay $Z' \to W^+W^-$ can occur in any model with an extra Z', provided that $m_{Z'} > 2m_W$ and there is non-zero $Z - Z'$ mixing. The (unpolarized) decay rate for the process is given by

$$\Gamma(Z' \to W^+W^-) = f\,\frac{\alpha}{48}\tan^2\theta_W \sin^2\theta, \tag{1}$$

where α is the (running) electromagnetic fine structure constant, θ_W is the electroweak mixing angle, θ is the $Z - Z'$ mixing angle and

$$f = (1-4\eta)^{3/2}(1+20\eta+12\eta^2)\eta^{-2}, \qquad \eta \equiv \left(\frac{m_W}{m_{Z'}}\right)^2. \tag{2}$$

It is important to note that, in the limit $m_{Z'} \gg m_W$, $\sin^2\theta$ scales like $(m_W/m_{Z'})^4$ and f like $(m_{Z'}/m_W)^4$, so that one can have a significant branching ratio for $Z' \to W^+W^-$ even for very small values of the mixing angle θ.

In any model where $m_{Z'} > m_Z$ and only $SU(2)_L$ doublets and singlets have non-vanishing vacuum expectation values, one has

$$\sin^2\theta = \frac{m_{Z^0}^2 - m_Z^2}{m_{Z'}^2 - m_Z^2}, \tag{3}$$

where $m_{Z^0} = m_W/\cos\theta_W$ is the standard model prediction for the Z mass and m_Z, $m_{Z'}$ are the two eigenvalues of the $Z - Z'$ mass matrix. In particular, in the rank-5 superstring-inspired model [3] based on the group $SU(3)_C \times SU(2)_L \times U(1)_Y \times U(1)_{Y'}$, which will be taken here as an illustrative example,

$$\tan^2\theta = \frac{\sin^2\theta_W}{9}\left(\frac{m_{Z^0}^2}{m_{Z'}^2 - m_{Z^0}^2}\right)^2\left[\frac{(\overline{v}/v)^2 - 4}{(\overline{v}/v)^2 + 1}\right]^2, \tag{4}$$

where $\overline{v}$ and v are the vacuum expectation values of the Higgs doublets which give mass to the d-type and to the u-type quarks, respectively. On general grounds one expects $|\overline{v}/v| \leq 1$, while model calculations favour values of $|\overline{v}/v|$ between 0.2 and 0.6. As an example, for $|\overline{v}/v| = 0.5$ and three generations of ordinary quarks and leptons kinematically available for Z' decays, in the limit $m_{Z'} \gg m_W$ one gets

$$\Gamma(Z' \to W^+W^-) \simeq \Gamma(Z' \to e^+e^-) \simeq 0.035\,\Gamma_{Z'}^{TOT}. \tag{5}$$

For $m_{Z'} = 1\ TeV$, the above value of $|\overline{v}/v|$ is consistent with the limits on $Z - Z'$ mixing that will presumably be established from precise measurements of the Z mass at SLC and LEP [4].

Finally, before discussing the detectability of the mode $Z' \to W^+W^-$ at the SSC it is crucial to note that, for $m_{Z'} \gg m_W$, the outgoing W's are predominantly longitudinally polarized:

$$\begin{aligned} \Gamma(Z' \to W_L^+W_L^-) \quad &: \quad \Gamma(Z' \to W_L^+W_T^-, W_T^+W_L^-) \quad : \quad \Gamma(Z' \to W_T^+W_T^-) \\ = (1+2\eta)^2 \quad &: \quad 16\eta \quad : \quad 8\eta^2. \end{aligned} \tag{6}$$

For example, taking $m_{Z'} = 1\ TeV$ one gets

$$\frac{\Gamma(Z' \to W_L^+ W_L^-)}{\Gamma(Z' \to W^+ W^-)} \simeq 0.9. \tag{7}$$

3 Signal and background

The signal for $Z' \to W^+W^-$ is rather similar to the one for a heavy Higgs, $H^0 \to W^+W^-$, since in both cases the outgoing W's are predominantly longitudinally polarized. The main difference is that the Z' is expected to be much narrower than the Higgs, its typical width being 1−3% of its mass. Possible decay chains for the outgoing W's are ($l = e, \mu, \ldots; j = jet$)
a) $WW \to (l\nu)(l\nu)$;
b) $WW \to (j_1j_2)(j_3j_4)$;
c) $WW \to (j_1j_2)(l\nu)$.
The first one is problematic because of the small branching ratio and due to the difficulty of reconstructing the two missing-p_T signals associated with the two neutrinos. The second one appears to be overwhelmed by the 4-jet QCD background. The only promising one is the third, which will be studied in the following for $l = e, \mu$ and jets associated with light quarks (u,d,c,s).

In the rank-5 superstring-inspired model, for the values of the parameters given above and the standard cut $|y_W| < 2.5$, the cross section $\sigma(pp \to Z' \to W^+W^- \to l\nu j_1j_2)$ at $\sqrt{s} = 40\ TeV$ is shown in Fig. 1 as a function of $m_{Z'}$ (solid line). To take into account the branching ratio of W's into $e(\mu)\ \nu$ and light quark-antiquark pairs, the cross section for $Z' \to W^+W^-$ production has been multiplied by a factor of 1/6. For example, for $m_{Z'} = 1\ TeV$ one gets $\sigma(pp \to Z' \to W^+W^- \to l\nu j_1j_2) \simeq 0.1\ pb$, which corresponds to 10^3 events/year at the design SSC luminosity, $\mathcal{L} = 10^4\ pb^{-1}/year$.

The two main backgrounds to the signal are the continuum WW production and the single W production in association with two jets. The techniques to overcome these backgrounds are very similar to the ones proposed in the case of a heavy Higgs.

From EHLQ [5] we estimate that

$$\left(\frac{d\sigma}{dM}\right)_{WW}^{M=1\,TeV} \simeq 2 \times 10^{-3}\ pb/GeV \quad (|y_W| < 2.5). \tag{8}$$

Assuming a mass resolution $\Delta M = 0.05M$ in reconstructing the invariant mass of the WW system, then $(d\sigma/dM)\Delta M = 0.1\ pb$, which becomes a factor of 6 below the signal after including the branching ratios. Further signal/background discrimination is possible using the fact that the continuum WW's are predominantly transversely polarized, as will be discussed below.

The most serious background is however Wj_1j_2 arising from standard model interactions [6]. This has been widely studied in past years in connection with heavy Higgs searches at large hadron colliders. To compare this background to the signal of interest we use the results of ref. [7] for the electroweak/QCD processes yielding $pp \to Wj_1j_2 \to (l\nu)j_1j_2$, adapted to the present situation. We impose rapidity cuts on the charged lepton and outgoing jets of

$|y| < 4$ (these cuts are very similar in efficiency to those we impose on the signal, $|y_W| < 2.5$) and assume that the W can be reconstructed from the $j_1 j_2$ invariant mass with a $\sim$ 10% mass resolution. If we assume, in addition, that the Z' mass can be reconstructed with a $\sim$ 5% mass resolution in the $l\nu j_1 j_2$ mode, then we obtain the effective background cross section for $l\nu j_1 j_2$ given in Fig. 1 (dashed line). For $m_{Z'} = 1$ TeV we find a background

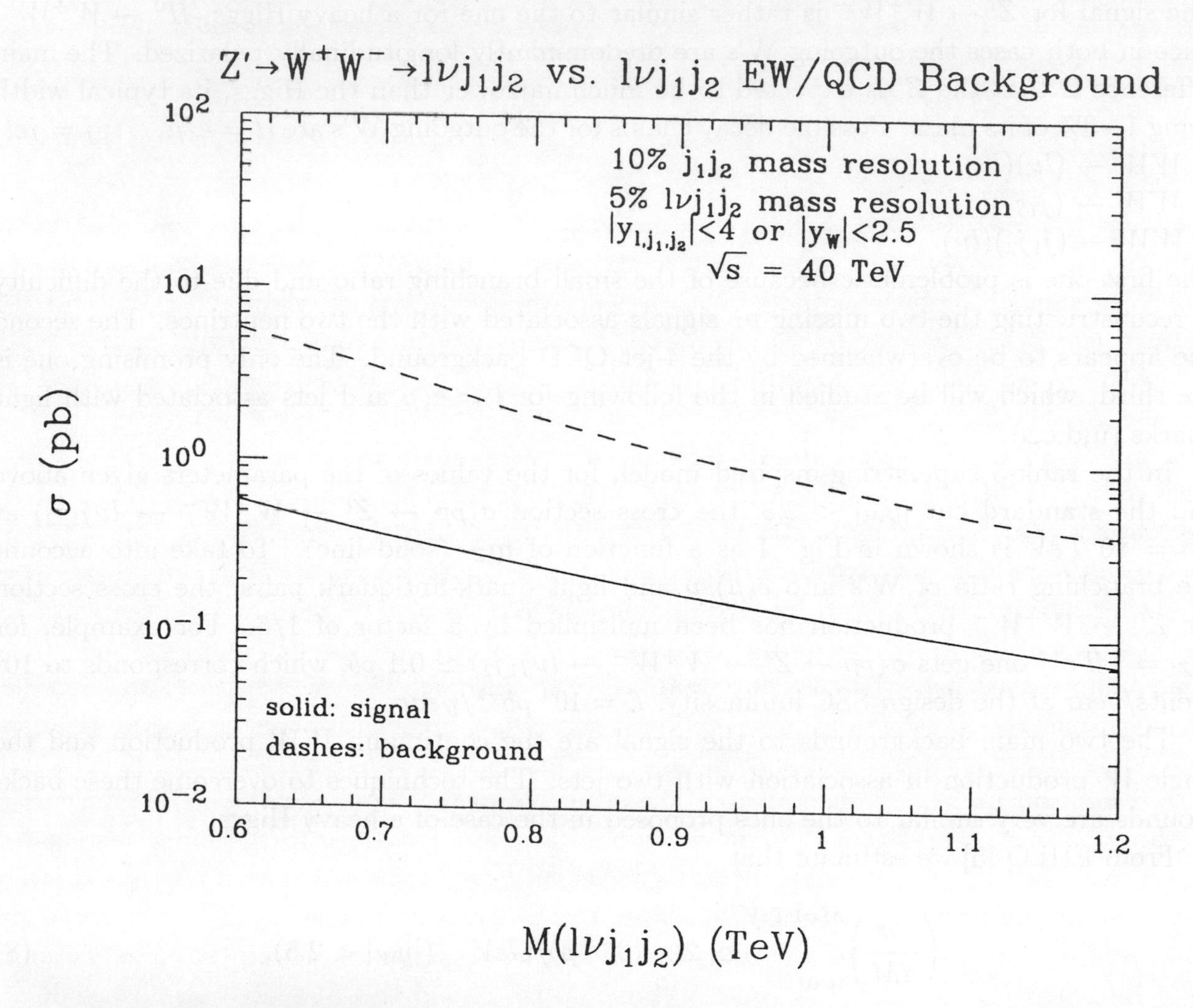

Figure 1: Signal and background for $pp \to Z' \to W^+W^- \to l\nu j_1 j_2$ at the SSC.

of 0.75 pb, to be compared with a 0.1 pb signal, while at $m_{Z'} = 0.7$ TeV we find 3.73 pb for the background and 0.37 pb for the signal. While the signal to background ratio is small, the signal is statistically significant and the $l\nu j_1 j_2$ background coming from the mixed electroweak/QCD processes can be normalized by moving the $j_1 j_2$ invariant mass off the m_W peak. Certainly the situation is much better than the one encountered in searching for a heavy Higgs boson in this same WW final state.(For instance, at $m_H = 1$ TeV, for the $j_1 j_2$ mass resolution stated above and with a WW mass resolution given by the large

Higgs width, Γ_H, at this $m_H = m_{WW}$ mass value, we obtain a Higgs signal in the $l\nu j_1 j_2$ mode of 4.5×10^{-2} *pb*, to be compared with a background of 8 pb. The much larger size of the background in the Higgs case is due to the large Higgs width at these high masses). Nonetheless additional cuts to enhance the signal/background ratio would be desirable. To this end, we remind the reader that the Z' and Higgs decay situations have in common the feature that both decay predominantly to longitudinally polarized W's. In contrast the $j_1 j_2$ system of the mixed electroweak/QCD background has a very 'transverse' nature for the $j_1 j_2$ angular distribution in the $j_1 j_2$ center of mass. A pair of variables which dramatically exhibits these differences was considered in ref. [7]. They are

$$r_{min} = \frac{p_T^{j_{min}}}{m_{WW}}, \quad r_{max} = \frac{p_T^{j_{max}}}{m_{WW}}, \tag{9}$$

where $p_T^{j_{min}} = min(p_T^{j_1}, p_T^{j_2})$ and $p_T^{j_{max}} = max(p_T^{j_1}, p_T^{j_2})$. They are most effective when computed in the WW center of mass. The mixed electroweak/QCD processes produce events with very small r_{min} over a range of r_{max}, whereas a longitudinally decaying W tends to produce jets perpendicular to the overall W three-momentum and thus with $r_{min} \sim r_{max}$ and both substantially away from 0. In the case of a 1 TeV Higgs, the cuts $r_{min} > 0.12$, $r_{min} + r_{max} > 0.35$ and $|\cos\theta^*_{e\nu}| < 0.5$ (where $\cos\theta^*_{e\nu}$ is the decay angle of the leptonically decaying W in the W rest frame), coupled with the previously mentioned $j_1 j_2$ and m_{WW} mass cuts, reduce the signal by a factor of ~ 2.5 but the background by a factor of ~ 80, yielding a signal/background ratio of $\sim 1/2$ with ≥ 300 signal events. While the partonic level results are not fully realized in detailed Monte Carlo studies which include detector simulation, fragmentation and QCD radiation, the results of ref. [8] show that the effect of these complications is not overwhelming and the Higgs may well be detectable in the $l\nu j_1 j_2$ mode. The situation in the case of a 1 TeV Z' can be much better. In the example under consideration, the $Z' \to l\nu j_1 j_2$ cross section is somewhat higher (a factor of 3) than the corresponding Higgs cross section, and, more important, the Z' is narrower and a smaller (5%) mass resolution in $M_{l\nu j_1 j_2}$ encompasses the entire Z' peak. Applying the same efficiencies for the $r_{min} - r_{max}$ and $|\cos\theta^*_{e\nu}|$ cuts as quoted in the Higgs case to the cross sections of Fig. 1, a 1 TeV Z' would yield ~ 500 signal events, while the background would contain ~ 100 events. Of course, as in the Higgs case, this partonic level result will be weakened by detector, fragmentation and QCD radiation effects, but it appears certain that the Z' signal could be observed. In fact, it should be possible to extend the Z' search in this mode to substantially higher masses. A 1.2 TeV Z' has a $l\nu j_1 j_2$ cross section of ~ 0.7 *pb*, yielding after all cuts ~ 280 signal events over a background of ~ 50 events.

4 Conclusions

If a new Z' exists in the TeV range, the detection of the $Z' \to W^+W^-$ mode is eminently possible and gives valuable information on $Z - Z'$ mixing. It might be the only mode other than $Z' \to e^+e^-, \mu^+\mu^-$ that is visible. Since the discovery of this mode would be a byproduct of heavy Higgs searches, we encourage further study of this Z' mode using realistic detector simulations.

This work is supported by the U. S. Department of Energy. One of us (F. Z.) is supported by a Fellowship of the Istituto Nazionale di Fisica Nucleare, Italy.

References

1. See, for example: V. Barger, N. G. Deshpande, J. L. Rosner and K. Whisnant, in Proceedings of the 1986 Summer Study on the Physics of the Superconducting Supercolliders (R. Donaldson and J. Marx eds.), p.224; B. Adeva, F. del Aguila, D. V. Nanopoulos, M. Quirós and F. Zwirner, ibidem, p.257; F. del Aguila, M. Quirós and F. Zwirner, preprint LBL-23694, to appear in Vol. II of the Proceedings of the Workshop on Physics at Future Accelerators, La Thuile (Val d' Aosta) and CERN; references therein.

2. F. del Aguila, M. Quirós and F. Zwirner, Nucl. Phys. B284, 530 (1987); see also: P. Kalyniak and M. K. Sundaresan, Phys. Rev. D35, 75 (1987); R. Najumia and S. Wakaizumi, Phys. Lett. 184B, 410 (1987).

3. E. Cohen, J. Ellis, K. Enqvist and D. V. Nanopoulos, Phys. Lett. 165B, 76 (1985); J. Ellis, K. Enqvist, D. V. Nanopoulos and F. Zwirner, Nucl. Phys. B276, 14 (1986) and Mod. Phys. Lett. A1, 57 (1986).

4. J. F. Gunion, L. Roszkowski and H. E. Haber, Phys. Lett. 189B, 409 (1987) and references therein.

5. E. Eichten, I. Hinchliffe, K. Lane and C. Quigg, Rev. Mod. Phys. 56, 579 (1984).

6. J. F. Gunion, Z. Kunszt and M. Soldate, Phys. Lett. 163B, 389 (1985) + (E) 168B, 427 (1986); S. D. Ellis, R. Kleiss and W. J. Stirling, Phys. Lett. 163B, 261 (1985).

7. J. F. Gunion and M. Soldate, Phys. Rev. D34, 826 (1986).

8. G. Alverson et al., in Proceedings of the 1986 Summer Study on the Physics of the Superconducting Supercolliders (R. Donaldson and J. Marx eds.), p.114.

Compositeness and QCD at the SSC

V. Barnes[a], B. Blumenfeld[b], R. Cahn[c], S. Chivukula[d],
S. Ellis[e], J. Freeman[f], C. Heusch[g], J. Huston[h], K. Kondo[i]
J. Morfín[f], L. Randall[c], D. Soper[j]

ABSTRACT

Compositeness may be signaled by an increase in the production of high transverse momentum hadronic jet pairs or lepton pairs. The hadronic jet signal competes with the QCD production of jets, a subject of interest in its own right. Tests of perturbative QCD at the SSC will be of special interest because the calculations are expected to be quite reliable. Studies show that compositeness up to a scale of 20 – 35 TeV would be detected in hadronic jets at the SSC. Leptonic evidence would be discovered for scales up to 10 – 20 TeV. The charge asymmetry for leptons would provide information on the nature of the compositeness interaction. Calorimetry will play a crucial role in the detection of compositeness in the hadronic jet signal. Deviations from an e/h response of 1 could mask the effect. The backgrounds for lepton pair production seem manageable.

[a]Purdue University, Lafayette, Indiana 47907
[b]Johns Hopkins University, Baltimore, Maryland 21218
[c]Lawrence Berkeley Laboratory, University of California, Berkeley, California 94720
[d]Boston University, Boston, Massachusetts, 02215
[e]University of Washington, Seattle, Washington 98195
[f]Fermi National Accelerator Laboratory, Batavia, Illinois 60510
[g]University of California at Santa Cruz, Santa Cruz, California 95064
[h]Michigan State University, East Lansing, Michigan 48824
[i]University of Tsukuba, Tsukuba, Japan
[j]University of Oregon, Eugene, Oregon 97403

1 Introduction

The standard model of the strong and electroweak interactions is in excellent agreement with all current experimental data. There is no reason, however, to believe that the standard model is correct at arbitrarily high energies. One possibility is that the quarks and leptons, which are assumed to be fundamental in the standard model, are in fact composite particles (for a general review of the subject see [1]). Current experimental limits show that the leptons are point-like down to distances of order (1 TeV)$^{-1}$ [2] and that quarks are point-like down to distances of order (few hundred GeV)$^{-1}$ [3]. The SSC is an ideal place to probe the structure of the quarks (and the leptons if they share common constituents with the quarks) at shorter distance scales.

Compositeness may provide the key to understanding flavor. In the standard fundamental doublet Higgs model, flavor is introduced by hand in the different values of the Yukawa couplings of the different quarks and leptons to the Higgs. It is to be hoped that there is a *dynamical* explanation for the different masses of the observed flavors of quarks and leptons (e.g. see [4]). In general, such dynamics give rise to dangerous flavor-changing neutral currents that contribute to the K_L - K_S mass difference. The smallness of this mass difference implies that the scale of the dynamics responsible for differences between flavors must be very high - of order 100 to 1000 TeV[5]. It is extremely difficult to construct a model in which one can account for the very large differences in the quark masses (in particular to account for the t quark mass) without inducing unacceptable flavor-changing neutral currents [6] [7].

In a model of composite quarks and leptons, however, it is possible that the different masses of the quarks and leptons are due to different masses of the constituent fermions and that flavor-changing neutral currents (which might arise from constituent fermion four-fermion interactions) are sufficiently suppressed [8] [9] [10]. In models of this type, the scale of compositeness is of order several TeV, and signs of compositeness should appear at the SSC.

As noted by Eichten, Lane, and Peskin [11], at energies below the scale of compositeness, the effects of compositeness may be summarized by a set of four-fermion operators arising from the exchange of constituent particles. We are mostly concerned with searching for the first signs of compositeness and therefore will concentrate on the effects of these four-fermion operators. An analysis of effects above the compositeness scale might proceed along the lines suggested by Hinchliffe and Bars [12].

Below the energy at which the constituents of quarks could be produced, compositeness would induce effective interactions among the quarks. These interactions would lead to increased quark-jet production at very high transverse momentum. As a result, tests of QCD and searches for compositeness are inevitably linked.

The problem of QCD jets is two-fold. On the theoretical side, jets are analyzed at the partonic level where quarks and gluons are the quanta, while real jets are composed of hadrons. On the experimental side, jets must be defined by an algorithm appropriate to the detector at hand, utilizing calorimetric information and possibly information from tracking. The relation of the algorithmically defined jet to the actual jet is a crucial issue. Important work remains to be done in bridging the gap between the theorists' partonic jet and the experimentalists' detector-dependent jet signature. These problems are addressed in Sections 2, 3, and 4 below.

Once the QCD jet production is understood, it will be possible to search for deviations from the expected cross sections. These deviations could be a sign for compositeness. The search for compositeness in hadronic jet production is discussed in Sections 4 and 5, following the treatment of EHLQ [13]. The most important results are the determination of the influence of calorimeter characteristics on the search for compositeness, explored in Section 4.

It may be that leptons as well as quarks are composite. If the leptons and quarks have common constituents, interactions of the form $q\overline{q} \rightarrow l^+l^-$ will be induced. At very high invariant mass, lepton pair production from these interactions could exceed that of the conventional Drell-Yan process. This manifestation of compositeness is discussed in Sections 6, 7, and 8. An important conclusion is that if compositeness is observed in lepton pair production, the study of the charge asymmetry will provide important information on the structure of the currents responsible for the interaction.

2 QCD Jet Cross Sections

At larger values of momentum transfer, the predictions of QCD become increasingly precise. The SSC is the natural place to test this theory. How precise can these predictions can be? At the momentum scale $Q \approx 1$ TeV appropriate to the SSC, hadronization effects, which produce corrections of order 1 GeV/Q in properly chosen inclusive processes, are negligible. Furthermore, $\alpha_s(Q) \approx 0.1$ is about half of its value at $Q = 10$ GeV, so that the effects of uncalculated higher order diagrams are smaller.

The main sources of error in the theoretical predictions will be:

1. Uncalculated higher order terms in the parton-parton hard-scattering cross section.
2. Errors in extrapolating the structure functions from the measured domain into the SSC energy region.
3. Measurement errors in determining the parton distribution functions.

The two cases in which three terms of QCD perturbation theory have been calculated provide guidance to the importance of higher order terms. The first is the beta function, which controls the evolution of $\alpha_s(Q)$. The result is [14]

$$\frac{d\ln\alpha_s(Q)}{d\ln Q^2} = -0.88\alpha_s\left\{1 + 0.74\alpha_s + 0.82\alpha_s^2 + \cdots\right\}. \tag{2.1}$$

where we have chosen the case of zero flavors in order to give definite numerical values for the coefficients.

Three terms have also been calculated for the total cross section for electron-positron annihilation to hadrons [15]. Here we quote the ratio R of this cross section to the cross section for producing a muon pair. We choose 4 quark flavors, $\Lambda = 150$ MeV in the $\overline{\mathrm{MS}}$ scheme, and c.m. energy $Q = 30$ GeV for purposes of illustration. Sensitivity to the choice of renormalization scale, μ^*, is minimized if $dR(\mu)/d\mu = 0$ at $\mu = \mu^*$. One finds μ^*=24.5 GeV and

$$R = 3\sum_i e_i^2\left\{1 + 0.32\alpha_s(\mu^*) + 0.069\alpha_s(\mu^*)^2 + \cdots\right\} \tag{2.2}$$

On the basis of Eqs. (2.1) and (2.2) we surmise a conservative rule: the coefficient of α^n is of order 1.

Let us now examine the cross section $d\sigma/dQ^2dy$ for producing muon pairs with dimuon mass Q and rapidity y in proton-proton collisions. We will take $\sqrt{s} = 40$ TeV, $Q = 1$ TeV, and $y = 0$. For the present illustrative purposes, we have included virtual photon decay but not virtual Z decay. We define the parton-level hard-scattering cross section using the $\overline{\mathrm{MS}}$ definition of parton distribution functions, which helps to improve the behavior of perturbation theory somewhat.

We show the cross section in Fig. 2.1, using the calculations of [16,17] for the one-loop corrections. The cross section depends on the choice of the factorization scale μ, so we display both the Born cross section and the cross section including the one-loop correction as functions of μ. In the graphs, we have normalized the cross sections to the Born cross section as conventionally defined with the scale μ set equal to Q. In order to fix the value of μ, we choose the value μ^* at which the derivative of the calculated cross section with respect to μ vanishes, $\mu^* \approx 700$ MeV. (The simple choice $\mu = Q$ would do about as well.) We see that the first order term provides about a 15% correction at the Born cross section at $\mu = \mu^*$. The value of $\alpha_s(\mu^*)$ is approximately 0.09. Thus the coefficient of $\alpha_s(\mu^*)$ is about 1.7. This is in agreement with the expectation that the coefficients in perturbation theory should be of order 1.

How big will the next term, which is uncalculated, be? The general estimate that the coefficients are of order 1 gives $1 \times \alpha_s(\mu^*)^2 \approx 0.008$. It seems safe then to conclude that the uncalculated higher order terms will give corrections of 5% or less, even allowing for a coefficient several times larger than the value 1 we have guessed.

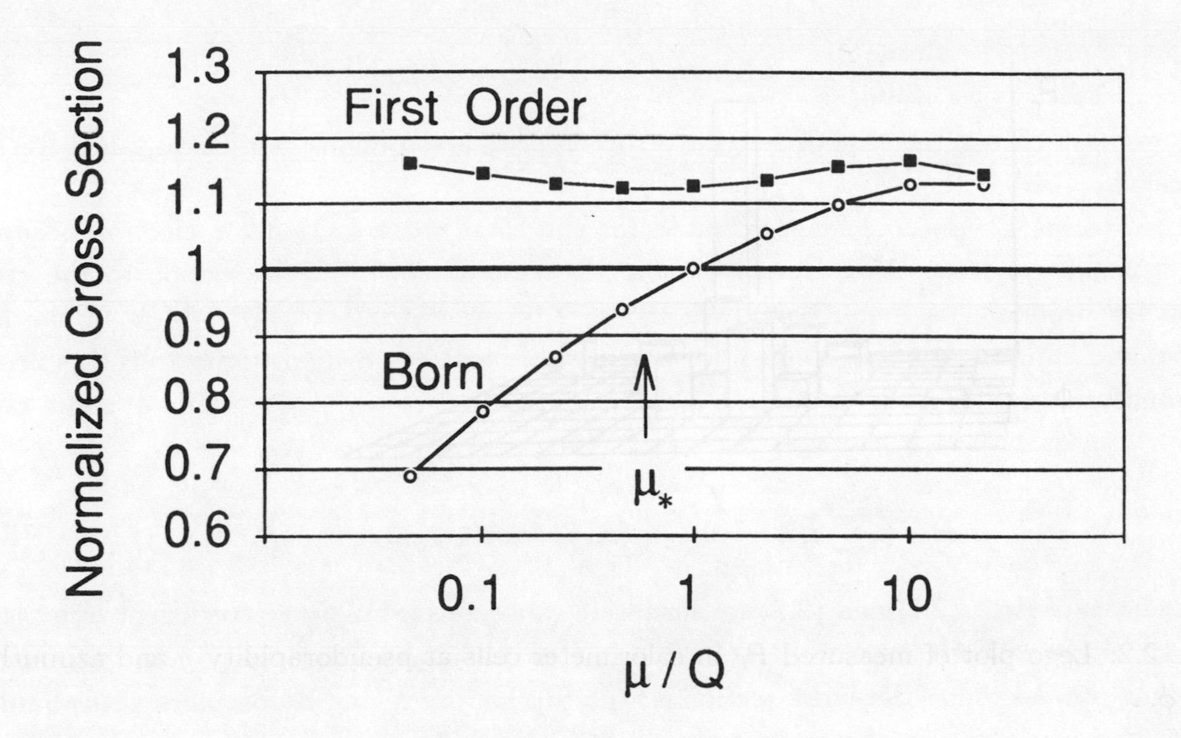

Figure 2.1: Drell-Yan cross section $d\sigma/dQ^2dy$ at $\sqrt{s} = 40$ TeV, $Q = 1$ TeV, and $y = 0$. The lower curve gives the Born cross section, while the upper curve gives the cross section including one-loop QCD corrections. The cross sections are plotted against the arbitrary factorization scale μ and normalized to the Born cross section at $\mu = Q$. We use $\Lambda = 50$ MeV in the 6 flavor $\overline{\text{MS}}$ convention, which corresponds to $\Lambda = 160$ MeV in the 4 flavor $\overline{\text{MS}}$ convention and gives $\alpha_s(\mu) = 0.091$ at $\mu = 1$ TeV.

A second source of error is that induced in the structure functions by using approximate kernels in the Altarelli-Parisi equations. We expect the residual uncertainty from this source to be quite small because evolution over the range 10 GeV $< \mu <$ 1 TeV for values of x near 1 TeV/20 TeV $= 0.05$ is itself a rather small effect. For instance, the gluon distribution at $x = 0.1$ changes only by about 30% between $\mu = 100$ GeV (Tevatron) and $\mu = 2$ TeV (SSC). Thus a 5% error in this evolution would produce a 1.5% error in the gluon distribution function.

Experimental errors in the measurement of the parton distributions also induce errors into theoretical predictions for the SSC. Between now and the time experiments at the SSC are underway, there will be additional measurements at HERA and at the Tevatron collider. In particular, measurements of the one jet inclusive cross section at the Tevatron collider can give a good measurement of the gluon distribution if the quark distributions are known from

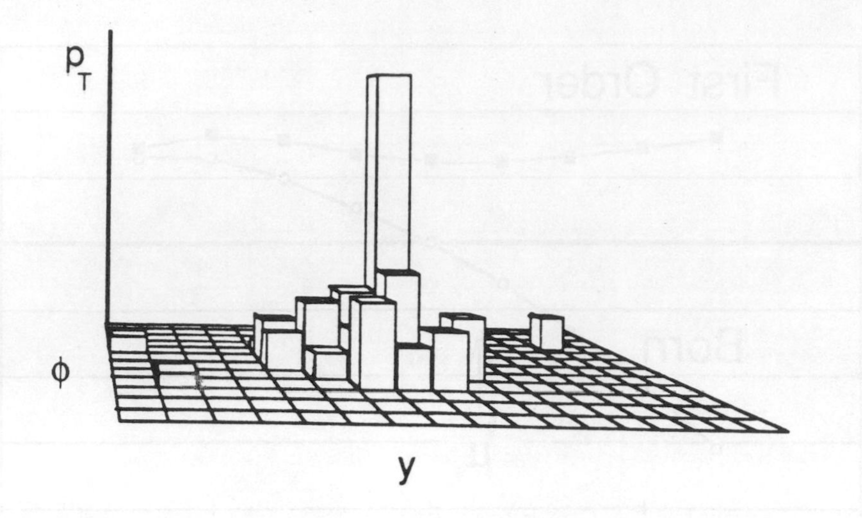

Figure 2.2: Lego plot of measured P_T in calorimeter cells at pseudorapidity y and azimuthal angle ϕ.

other sources. We have not been able to estimate the expected errors in any systematic way, but guess that a 10% uncertainty in the parton distribution functions is likely. This would result in a 20% error in the theoretical prediction for the cross sections, and would be the dominant source of error.

In order to test QCD beyond the Born level, one must adopt a definition of what a jet is, calculate the jet cross section based on that definition, and compare the prediction to a measurement using that definition. This has not been necessary in collider experiments up until now because there has been no jet cross section calculated beyond the Born level. In order to be suitable for this purpose, a jet definition must be simple enough to facilitate theoretical calculations at the parton level, and must have the property that hadronization effects and perturbative effects beyond the one-loop level yield small corrections to the prediction. We sketch one possible jet definition here. (The basic ideas follow those of Sterman and Weinberg [18].)

We imagine that the (y, ϕ) solid-angle space about the interaction point is covered by calorimeter cells that are capable of measuring the transverse momentum entering each cell. Here y denotes the pseudorapidity, $y = -\ln \tan \theta/2$. Then a typical jet event produces a Lego plot, as depicted in Fig. 2.2. Let the index j label the calorimeter cells. Let C denote the cone, depicted in Fig. 2.3, centered about the jet axis having angles (y, ϕ),

$$(y_j - y)^2 + (\phi_j - \phi)^2 < (\Delta R)^2 \tag{2.3}$$

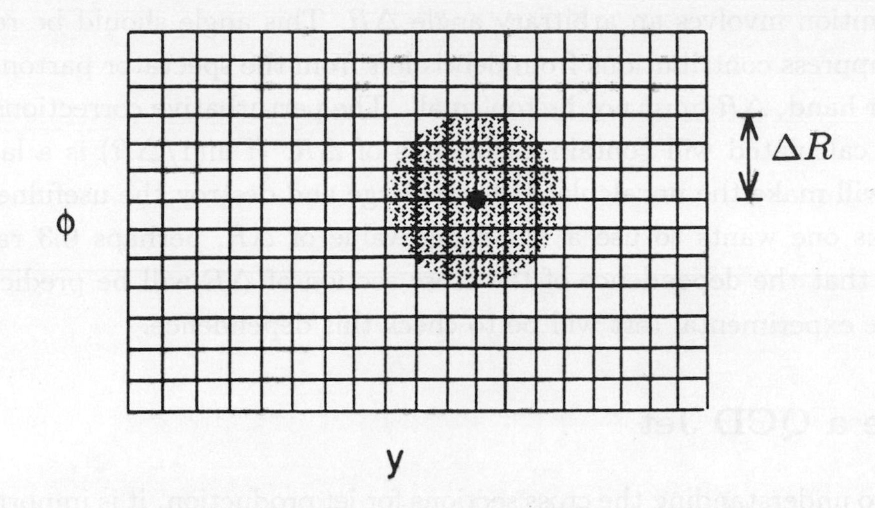

Figure 2.3: Jet cone in (y, ϕ)-space.

Here ΔR is a fixed quantity, effectively an angular acceptance. The jet P_T is defined as

$$P_T = \sum_{j \text{ in } C} P_{T,j}, \tag{2.4}$$

where $P_{T,j}$ is the absolute value of the transverse momentum in cell j. We define the jet axis (y, ϕ) as the P_T-weighted average of the angles in the cone:

$$y = \sum_{j \text{ in } C} P_{T,j} y_j / P_T, \tag{2.5}$$

$$\phi = \sum_{j \text{ in } C} P_{T,j} \phi_j / P_T. \tag{2.6}$$

Since the definition of the cone C depends on the angles (y_j, ϕ_j), this definition has to be applied self consistently: one must find a cone such that the average angle turns out to be located at the center of the cone. In certain cases the centers of two smaller jets can fall within the cone defining a single larger jet. In such cases, it seems best not to count the smaller sub-jets as legitimate jets in the jet definition.

One should note that with this inclusive jet definition, one does not try to determine how many jets there are in a given event. Nor is there any clustering algorithm beyond eliminating subjets if they fall within the defining cone of a larger jet. Evidently, this is not the definition to use if one wants to detect W-bosons decaying into two jets. Nor is it a good definition for examining delicate questions of jet substructure, which are washed out in the definition. However, it is insensitive to the details of shower development and hadronization.

The definition involves an arbitrary angle ΔR. This angle should be reasonably small so as to help suppress contributions from debris left from the spectator partons of the beam jets. On the other hand, ΔR must not be too small. The perturbative corrections beyond the order that can be calculated will contain logarithms of ΔR. If $\ln(1/\Delta R)$ is a large number, these logarithms will make the uncalculated terms large and destroy the usefulness of perturbation theory. Thus one wants to use a mid-range value of ΔR, perhaps 0.3 radians or so. One should note that the dependence of the cross section of ΔR will be predicted by QCD. One aspect of the experimental test will be to check this dependence.

3 Inside a QCD Jet

In addition to understanding the cross sections for jet production, it is important to understand the nature of the jets themselves for at least three reasons:

1. QCD "predicts" various features of the structure of jets. A careful study of these features at the SSC can constitute a test of QCD.

2. Jets will be a dominant feature of the interesting hadronic final states at the SSC and will play an essential role in the analysis of the physics. Thus it is important to have a sound knowledge of the structure of jets.

3. The role of jets will be central to understanding and carefully evaluating the standard model backgrounds to the experimental signals for new physics, e.g., compositeness, SUSY, etc. Without this type of analysis, it will be very difficult to identify reliably the appearance of new physics, a lesson already (painfully) learned at the CERN Collider.

These points, especially the third, seem to constitute a powerful argument for having a "jet spectrometer" at the SSC capable of analyzing a "typical" jet consisting of 50 charged particles all contained within an angular region of order $\Delta R = \sqrt{(\Delta y)^2 + (\Delta\phi)^2} \leq 0.1$! [See the earlier studies on this jet spectrometer question in earlier SSC workshops, [19,20]] From the work of EHLQ[13] we expect to be able to obtain an enormous data set of large energy jets at the SSC (of order 10^3 events per SSC year in a 10 GeV bin at a jet pair mass of 5 TeV/c^2 – see Fig. 3-22 of EHLQ). It in order to obtain a perspective on the problem let us review some of the expected features of jets based on QCD.

The perturbative techniques of QCD predict that jets should exhibit several features. However, it is important to keep in mind that these are *theorists' jets.* These features may or may not be exhibited by more "realistic" jets as generated by Monte Carlo simulators or as measured and identified by real detectors. It is, in fact, the conclusion of this review that a careful study should be undertaken to determine to what extent the very process of detecting and identifying jets in real detectors will obscure or change the theoretically anticipated structure.

First let us define a theorist's jet. It is the "spray" of hadrons generated by the interactions of QCD as a parton (quark or gluon) that is initially isolated in momentum space (as the result of some hard process) evolves to produce a color neutral final state (no isolated partons) on the 1 fermi scale. Thus the jet might be called the "smile of the Cheshire parton"; jets are how we actually *see* partons. We then characterize the approximately collinear (along the jet direction) distribution of hadrons within the jet in terms a fragmentation function $D_{h/j}(z)$, where z is essentially the fraction of the jet's energy carried by the hadron.

The first thing QCD says is that $D(z)$ is really $D(z,Q^2)$ where the scale Q characterizes the "size" of the parton, i.e., the momentum scale at which it is scattered out of the initial hadron, $Q \sim E_j$. Actually perturbative QCD really talks about $D_{p/j}(z,Q^2)$ not $D_{h/j}(z,Q^2)$ where p stands for a parton within the perturbative evolution of the jet. The mapping of perturbatively understood partons onto the final hadrons involves the magic (and mystery) of confinement and remains quantitatively uncertain at the moment. QCD tells us about the evolution of D with a variation in Q. We must input some initial conditions at some Q_o [where perturbative QCD is already valid] from experiment. Further, the leading order QCD predictions are truly reliable only in the limit of large Q where this means that $\ln(Q/\Lambda_{QCD}) >> 1$ with $\Lambda_{QCD} \sim 100$ MeV. With these warnings in mind we can proceed to briefly review the predictions of QCD.

1. Large Q jets will exhibit large gluon multiplicities (and therefore large hadronic multiplicities).

2. Jets "broaden" in P_T with increasing Q:

$$< P_T > \quad \propto \quad Q\alpha_s(Q^2) \tag{3.1}$$

due to "hard" emissions in the perturbative evolution.

3. In terms of angles the jet gets "narrower" with increasing Q:

$$\frac{< P_T >}{Q} \propto \alpha_s(Q^2) \tag{3.2}$$

The angle of a cone χ containing a fraction of the jet's energy ϵ is

$$\chi(\epsilon) \propto \exp\left[-C\ln(1/\epsilon)\ln\left(Q^2/\Lambda^2_{QCD}\right)\right] . \tag{3.3}$$

where C is a constant depending on the type of jet.

4. Quark initiated jets are harder than gluon jets:

$$\frac{D_G(z,Q^2)}{D_Q(z,Q^2)} \propto_{z\to1,Q^2\to\infty} \frac{(1-z)}{\ln(1/(1-z))}, \tag{3.4}$$

5. Gluon jets have more gluons than quarks jets:

$$\frac{<n>_{G/G}}{<n>_{G/Q}} \sim \frac{C_A}{C_F} = \frac{9}{4}, \tag{3.5}$$

6. Gluon jets are broader than quark jets:

$$\frac{<P_T>_{G/G}}{<P_T>_{G/Q}} \sim \frac{C_A}{C_F}, \tag{3.6}$$

$$\frac{\ln\left(\chi^G(\epsilon)\right)}{\ln\left(\chi^Q(\epsilon)\right)} \sim \frac{3C_F/2}{7C_A/5 + N_f/10} \approx \frac{C_F}{C_A} < 1, \tag{3.7}$$

7. Stimulated by the initially surprising results from UA1 on charm production in jets ("the great charm scare" which is now over) much more is now understood about the theory of the heavy flavor content of jets [21,22]. This recent work strongly suggests that perturbative predictions for heavy flavor production should be reliable.

Assuming that these predictions can be applied without essential modification to *hadrons* within jets we can proceed to look at data. In fact, the existing data on jets at both hadronic and e^+e^- colliders seem to exhibit, at least qualitatively, all of the above features [see, in particular, the review at the 1986 Berkeley Conference by G. Thompson][23].

However, there still remains an essential issue. Individual quarks and gluons cannot fragment directly into only hadrons in complete isolation, as implied by the naive jet picture. At the very least such a process would not conserve color, energy-momentum or electric charge. Thus jet formation must always involve some collaboration of the main jet initiating parton with at least one other parton in the event. Hence a real jet cannot, in principle, be precisely defined. On the other hand, like true love, even though a jet cannot be defined precisely, you know it when you see it. In practice jets are experimentally defined via a *jet algorithm* which in detail is experiment dependent (see the previous Section). As an example consider the UA1 jet algorithm, circa 1983. To identify and reconstruct the jet one uses the information on the energy deposited in the various segments of the calorimeter, the Lego plot. These energy elements are then clustered into jets according to a specific algorithm. In the UA1 case one adds (vectorially) the energies of all calorimeter cells with transverse energy $E_T > 2.5$ GeV within a cone defined by $\Delta R = \sqrt{(\Delta y)^2 + (\Delta\phi)^2} < 1$ and then further adds the energy of any other cells inside a cone angle of 45° with respect to the effective jet direction if the energy of the cell *transverse* to the jet direction is < 1 GeV. Thus the very process of finding and defining a jet will, to some extent constrain the structure of the jet: the dreaded trigger bias. In particular, such jet algorithms are likely to exclude from the jet both hadrons that are emitted at large angles and soft hadrons, even though these hadrons are, in some fundamental sense, associated with the jet. This can be expected to influence the average P_T and multiplicity (dominated by soft hadrons) experimentally obtained for jets defined this way. We are left with the concern that the very process of defining the jet will dictate it properties, independent of QCD or other expectations. [One can already see some of this effect in the very different $< P_T >$ values obtained by the AFS collaboration at the ISR depending on whether the jet direction is obtained from the calorimeter algorithm or directly from the charged tracks used in the $< P_T >$ measurement.]

Clearly hadronic jets will be a central feature of the physics at the SSC. However, since there is not a unique a priori definition of a jet either theoretically or experimentally, there will be an interplay between the theoretical and experimental definitions of jets. There is, therefore, the potential for distortion of results. Three topics seem the most important for further analysis in this regard:

1. Do the various Monte Carlo simulators of jets reproduce QCD? [Webber seems to be "ahead" here],

2. Jet defining algorithms to be used at the SSC [can QCD really be tested?],

3. SSC detector properties [can jets really be used as signatures for compositeness and other phenomena?].

4 Measurement of Jets at the SSC

The detection of compositeness through jet production requires the measurement of jets at the highest values of transverse momentum accessible at the SSC. The stable particle multiplicity distribution from 1000 hard-scattering events with a p_T of 5 TeV/c (as generated by ISAJET 5.34 [24]), is shown in Fig. 4.1. Although a mean of 1000 particles per event is probably an overestimate [25], the particle multiplicity in jet events will be considerably greater than in current experiments. Most of these particles have low energy, although a significant amount of the energy is carried by high energy particles. The particle $p_T(E)$ distribution for 1000 jets of approximately 5 TeV/c is shown In Fig. 4.2 (4.3). (There are approximately 200,000 entries in the first bin, not shown, from 0–100 GeV/c.) About 50% (70%) of the jets have a particle with $p_T(E) > 1$ TeV/c. Thus, essentially all of the events in this range will have particles with 1 TeV or greater. Clearly, the calorimetry must be capable of accurate measurements

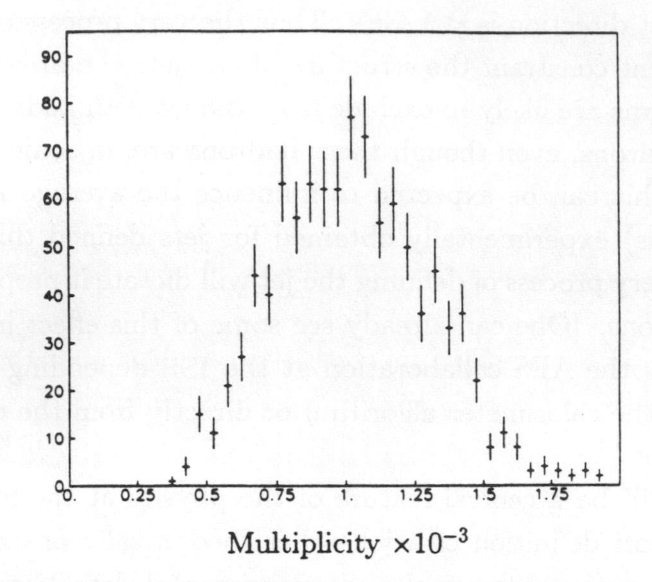

Figure 4.1: Multiplicity distribution for 1000 events generated by ISAJET 5.34. Each event contains a hard parton-parton scattering with p_T near 5 TeV/c. The horizontal axis gives the multiplicity in units of 10^3.

of particle energies from a few GeV to a few TeV. Another way of making this point is by showing the cross section $d\sigma/dP_T$ in Fig. 4.4 for particles generated by hard scattering in the range 0.5 TeV/c< p_T < 6.0 TeV/c and $|y| \leq 5.5$. Again it is evident that single particle production extends past $p_T \geq 2.5$ TeV/c.

Adequate containment of a few-TeV hadron requires a calorimeter thickness of the order of $10 - 12\lambda$. (For a more detailed review of the requisite detector properties at the SSC see, for example [26]) The resolution for both electromagnetic and hadronic calorimeters can be parameterized in the form $\sigma/E = A/\sqrt{E} + B$. The first term describes the contribution to the resolution from sampling fluctuations, while the second term represents contributions from systematic effects such as calibration, stability, and, as shall be discussed below, $e/h \neq 1$.

In existing large detectors, the constant term B is on the order of 2 to 3 % in electromagnetic calorimeters and 4 to 6 % in hadronic calorimeters. At the SSC we should aim for devices with constant terms of the order 1–2%. The energies being measured in this particular application are so large that the exact value of A is not so important, but the final resolution should be in the range of 10–15% $/\sqrt{E}$ for electromagnetic calorimeters and 35–55%$/\sqrt{E}$ for hadronic calorimeters.

One important source of the constant term B is the difference in the response of the calorimeter to electrons and hadrons, i.e. $e/h \neq 1$. In Fig. 4.5, the value of this constant term

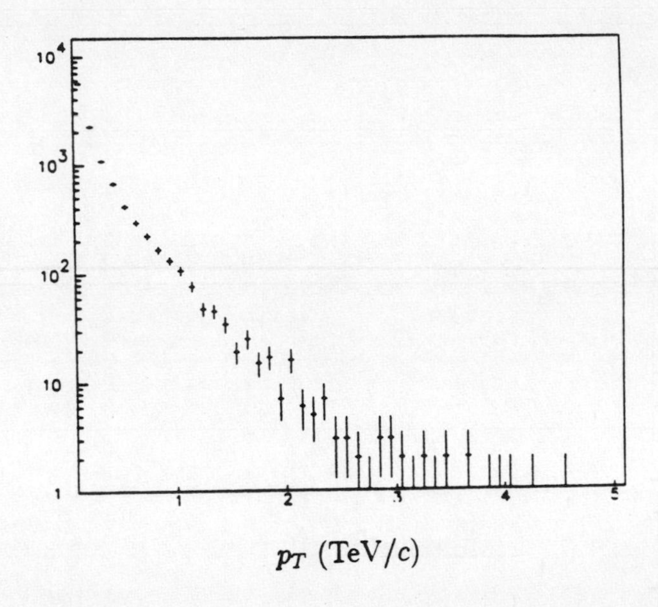

Figure 4.2: The distribution in p_T of particles produced in 1000 jets from hard collisions with p_T approximately 5 TeV.

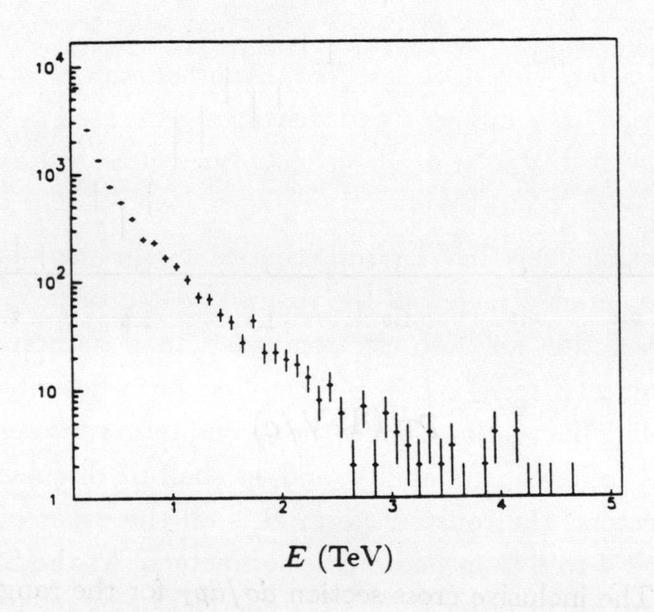

Figure 4.3: The distribution in E of particles produced in 1000 jets from hard collisions with p_T approximately 5 TeV.

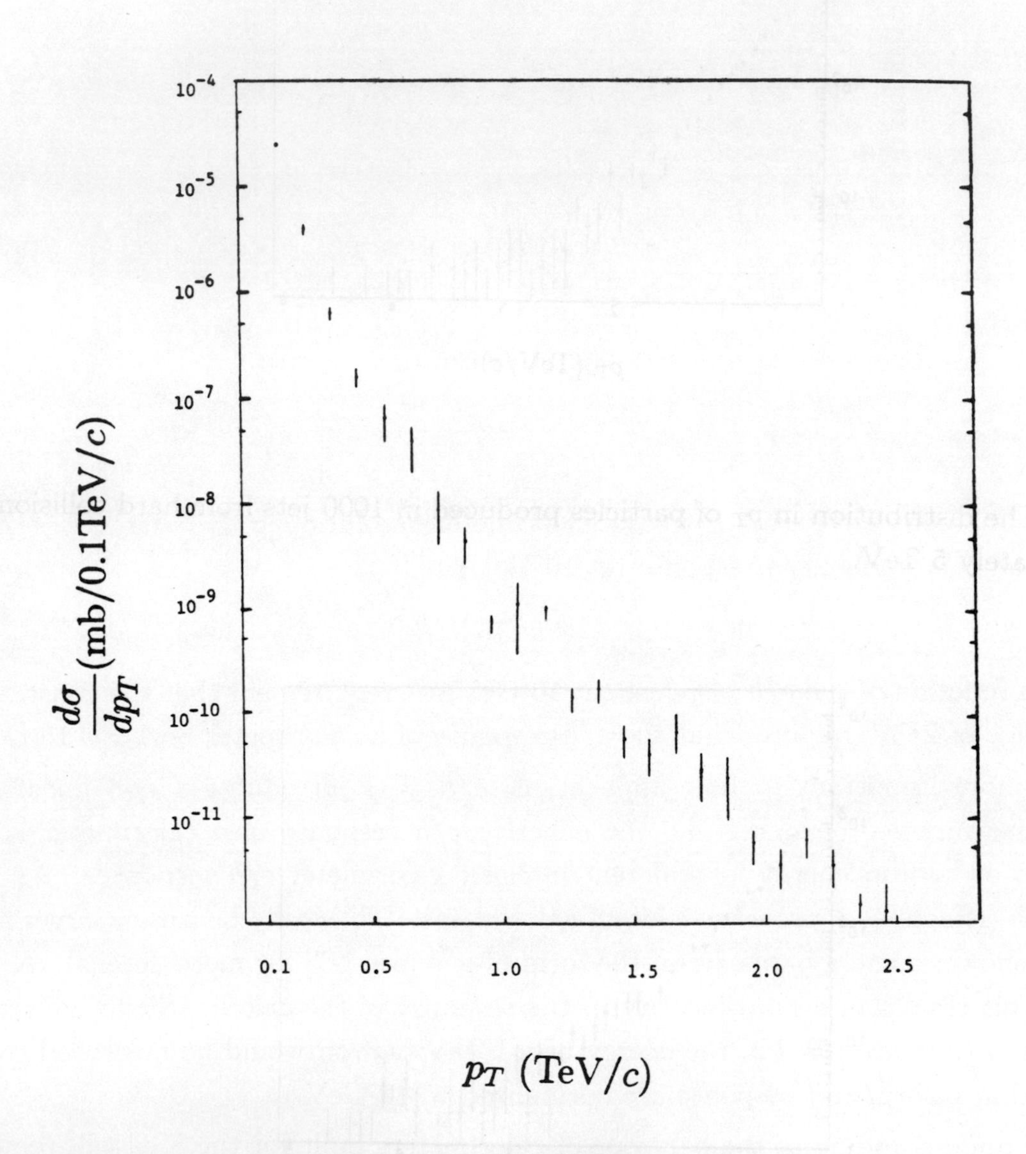

Figure 4.4: The inclusive cross section $d\sigma/dp_T$ for the range $|y| \leq 5.5$.

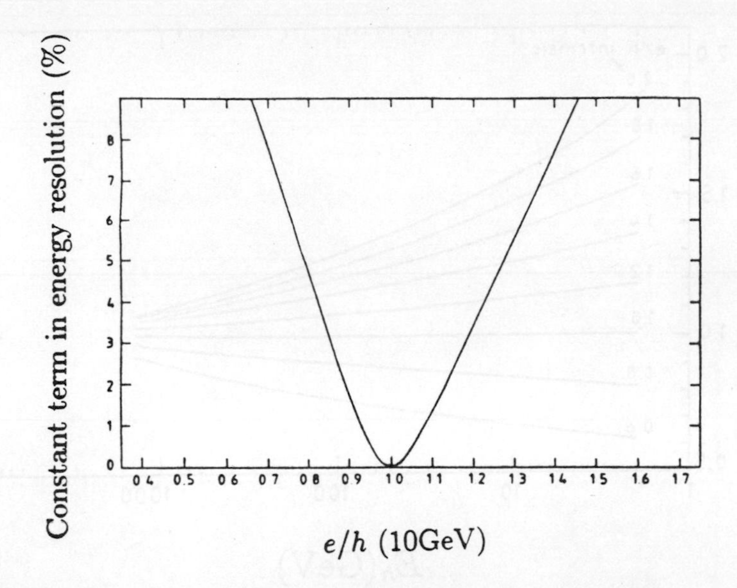

Figure 4.5: The constant term, B in the calorimeter resolution $\sigma/E = A/\sqrt{E} + B$, as a function of the calorimeter e/h response (at 10 GeV).

is shown as a function of the e/h response (at 10 GeV). A requirement that the constant term be in the range of 2–3% requires that the e/h response of a calorimeter be 1 ± 0.15 [27].

Perhaps more important for our application, $e/h \neq 1$ introduces a non–linearity into the energy measurement. In Fig. 4.6 the effective e/h response of a calorimeter is shown as a function of hadron energy for different intrinsic calorimeter e/h responses. As E_{hadron} increases, e/h (effective) approaches 1 for all e/h (intrinsic), basically because a larger fraction of a hadron shower's energy appears in the form of π^0s (see [27] for more details). As shown in Fig. 4.7, this results in a non–linearity in the response of the calorimeter to hadrons. For example, if e/h (intrinsic) = 1.4, the energy of a 1 TeV hadron would be measured to be 1.2 TeV. (Note that calorimeter response are normalized to 10 GeV.)

A proper understanding of the behavior of calorimeters built for the SSC will require the extensive use of test beams. As discussed earlier, there is a strong motivation for understanding the response of calorimeters to particles of energies on the scale of 1 TeV. A test beam of 1 TeV (e.g. Fermilab) probably would be adequate, but particles of higher energy will be encountered frequently and a complete understanding of a detector's response at these higher energies would certainly be useful.

We studied the detection of quark compositeness in production of high p_T jets by modifying ISAJET and including flavor diagonal contact terms as in EHLQ [13]. The p_T range from 0.5 - 6.0 TeV/c was considered. Jets were reconstructed using a simple calorimeter simulation with

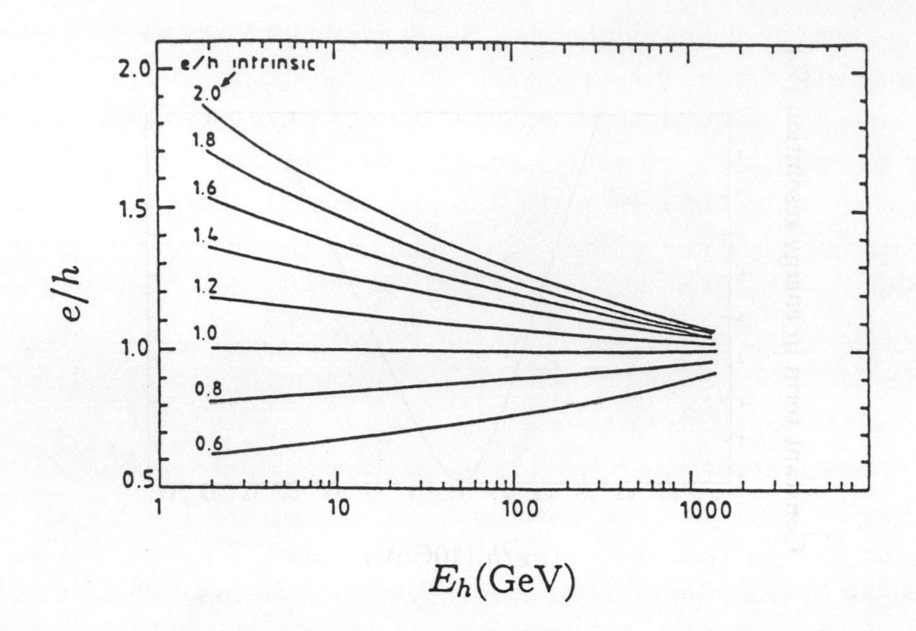

Figure 4.6: The effective e/h response of a calorimeter as a function of hadron energy for different intrinsic values of e/h for the calorimeter.

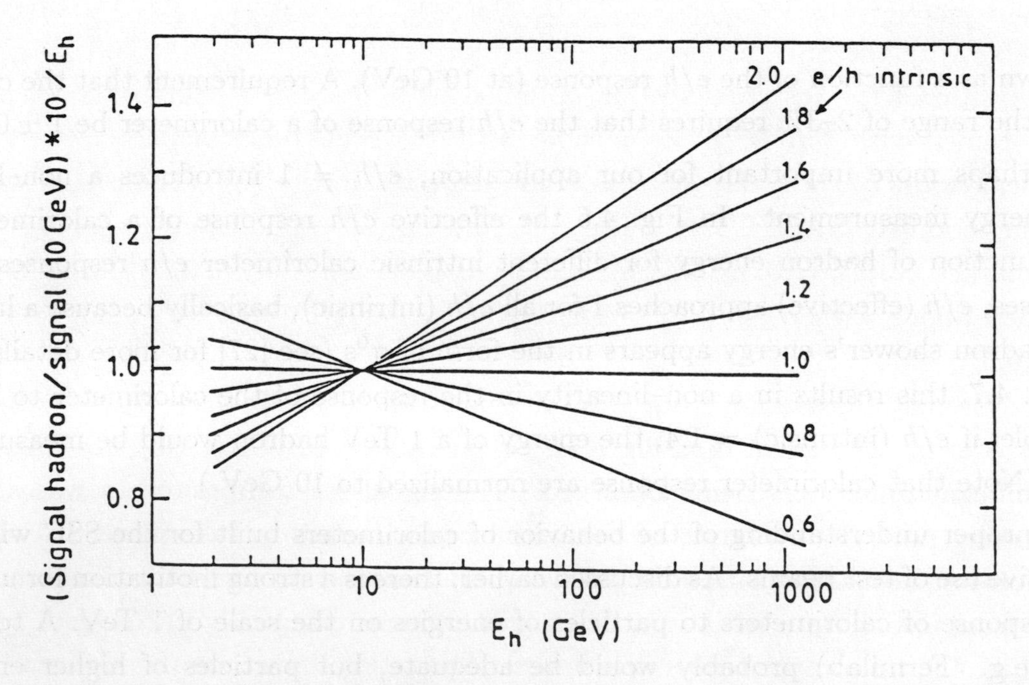

Figure 4.7: Response of a calorimeter as a function of hadron energy for various intrinsic values of e/h.

the calorimetry covering the pseudorapidity range, $|y| = \pm 5.5$ and with a lateral segmentation (both electromagnetic and hadronic) of 0.09×0.09 ($\Delta y \times \Delta\phi$). The energies of electrons and photons were smeared by a gaussian resolution terms with a σ typically of $0.15\ \sqrt{E}$. Hadron energies were smeared by a gaussian resolution term with a σ typically of $0.50\sqrt{E}$. For a calorimeter with different intrinsic responses to electrons and hadrons, i.e. $e/h \neq 1$, an appropriate constant term was added to the resolution (for example, 0.06 for $e/h = 1.4$; see [27]) In addition, each hadron's energy was corrected by the non–linearity appropriate to the e/h response of the calorimetry. (Note that this simulation lacks one aspect of realism in that hadron energies are deposited only in the hadron calorimeter. However, there are still useful lessons to be learned from this exercise.)

Jets were reconstructed using a UA-1 type cluster algorithm summing all calorimeter cells with energy greater than 1.5 GeV within a certain radius ($\Delta R = \sqrt{\Delta y^2 + \Delta\phi^2} = 0.5$ for most of the study) of a "seed cell" to form the jet energy. Typically 90% of the $\sqrt{\hat{s}}$ of an event is contained in the 2 jet mass formed using this algorithm. The jet cross section thus reconstructed is within 50% of the EHLQ jet cross section over the p_T range from $0.5 - 6.0$ TeV/c.

The major purpose of the study was to examine whether experimental or theoretical concerns could mask a signal for quark compositeness (see Section 5). The case considered below, a compositeness scale $\Lambda^* = 20$ TeV and destructive interference ($\eta = +1$), is near the limit of detectable with the SSC, as discussed in Section 5.

The ratio of the jet cross section obtained with the above compositeness parameters to that obtained with "QCD" alone is shown in Fig. 4.8. (The error bars are reflections of the statistical errors of the study, not the errors expected in the SSC. The events were generated with the modified ISAJET 5.34.) The "baseline" for all comparisons to come is the measurement marked RJET = 0.5. This measurement is obtained with the calorimeter stimulation described above with $\sigma_{EM} = 0.15\sqrt{E}$, $\sigma_{HAD} = 0.50\sqrt{E}$, $e/h = 1$ and with a jet radius $\Delta R = 0.5$.

No systematic deviation from 1 is seen until $p_T > 3.5$ TeV/c. At $p_T = 5.5$ TeV/c the ratio is over 2. The effect of using a larger jet radius (RJET = 1.0) was investigated with no significant differences being observed. The signal for compositeness was also examined (a) allowing no QCD evolution of the initial or final state to take place, (b) allowing QCD evolution but no hadronization, (c) allowing QCD evolution and hadronization but ignoring energy from initial state radiation. The signal remains essentially unchanged.

In Fig. 4.9, no compositeness is assumed. The ratios shown are those obtained by varying either the calorimeter response or the input structure functions. (The denominator in every ratio is the "standard QCD" jet cross section obtained with EHLQ structure functions RJET =0.5 and $e/h = 1$.) Using Duke–Owens structure functions instead of EHLQ causes a no-

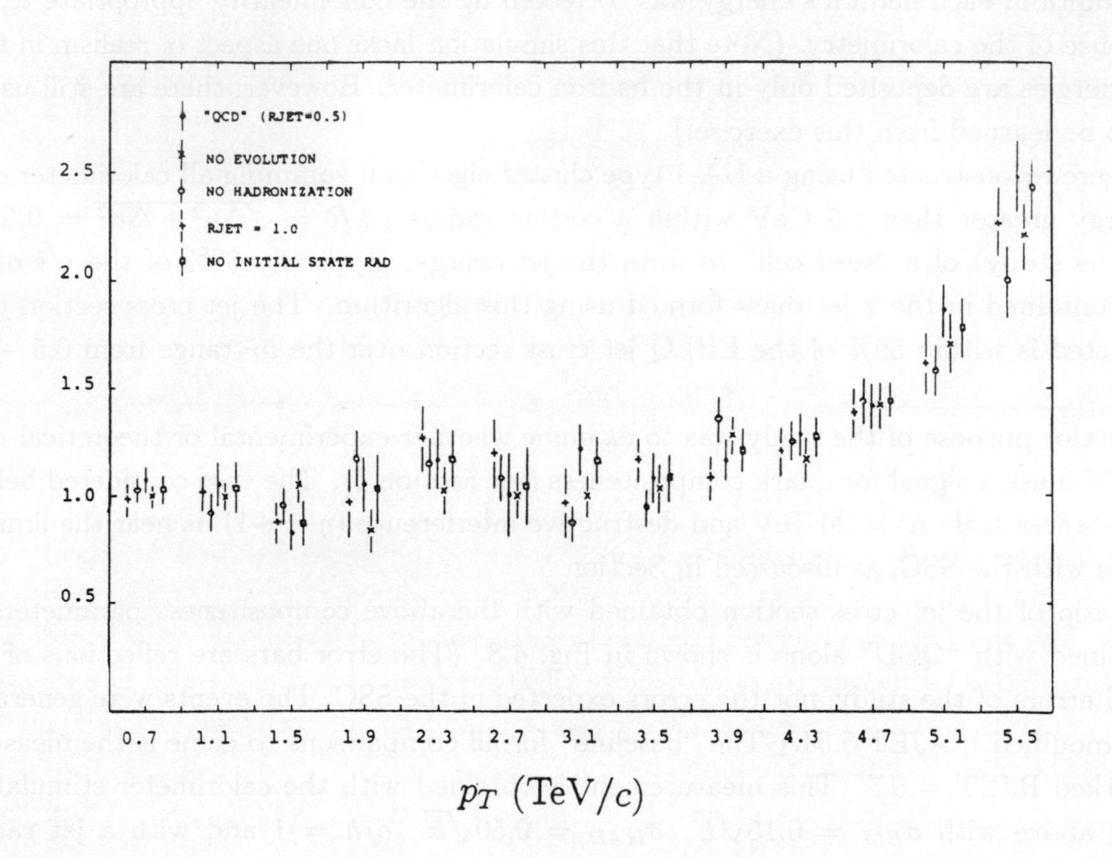

Figure 4.8: The ratio of the jet cross section with compositeness scale $\Lambda^* = 20$ TeV and destructive interference ($\eta=1$) to that obtained with just QCD interactions. The calorimeter is assumed to have $e/h = 1$, and hadronic calorimetry with $\sigma_E = 0.5\sqrt{E}$.

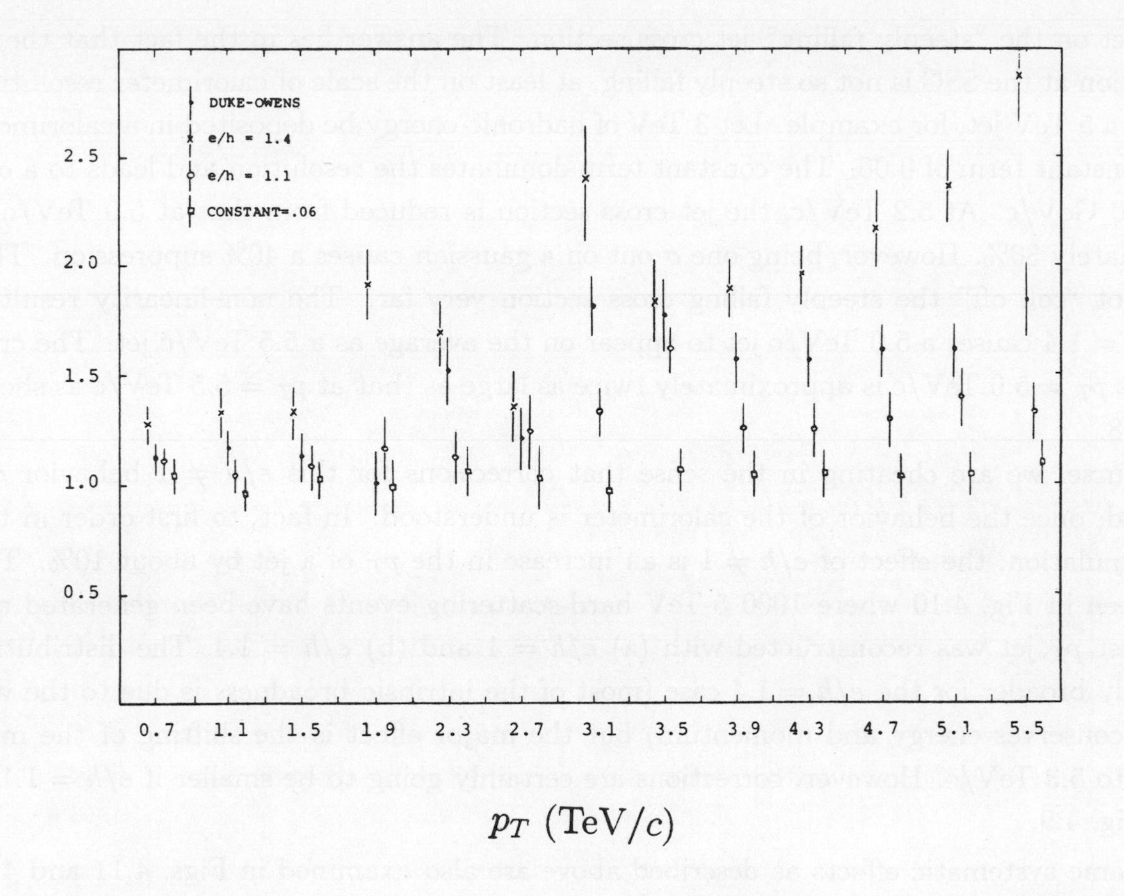

Figure 4.9: The ratio of the observed cross section to that for a "standard" calorimeter with $e/h = 1$, using EHLQ structure functions. Only QCD production is considered. The effects of $e/h \neq 1$, a constant term in the calorimeter resolution, and the use of Duke-Owens structure functions in the place of EHLQ are indicated.

ticeable increase in the jet cross section. The ratio is approximately 1.1 at $p_T = 1.0$ TeV/c and increases to approximately 1.8 at $p_T = 5.5$ TeV/c. The shape is not exactly the same as that expected for compositeness but a misunderstanding of the structure functions of this order (between EHLQ and Duke–Owens) could cause a great deal of confusion in extracting a compositeness signal. The ratio of the jet cross section obtained when the e/h response of the calorimetry is 1.4 to that for $e/h = 1$ is also shown. As stated above, the effect of $e/h \neq 1$ is the addition of a constant term to the resolution (of order 0.06) and a non–linearity in the energy measurement. As can be easily seen, the deviation of the ratio from 1 this causes is larger than that due to the compositeness signal itself. The bulk of this deviation is due to the non-linearity, not the constant term. This is seen in Fig. 4.9 where the effect of a constant term 0.06 in the energy resolution (but with e/h=1) is shown.

It may seem surprising at first that the addition of a constant term of 0.06 has such little effect on the "steeply falling" jet cross section. The answer lies in the fact that the jet cross section at the SSC is not so steeply falling, at least on the scale of calorimeter resolution. Consider a 5 TeV jet, for example. Let 3 TeV of hadronic energy be deposited in a calorimeter with a constant term of 0.06. The constant term dominates the resolution and leads to a σ of about 200 GeV/c. At 5.2 TeV/c, the jet cross section is reduced from that at 5.0 TeV/c by approximately 30%. However, being one σ out on a gaussian causes a 40% suppression. Thus jets do not "roll off" the steeply falling cross section very far. The non–linearity resulting from $e/h = 1.4$ causes a 5.0 TeV/c jet to appear on the average as a 5.5 TeV/c jet. The cross section at $p_T = 5.0$ TeV/c is approximately twice as large as that at $p_T = 5.5$ TeV/c as shown in Fig. 4.8.

Of course, we are cheating in the sense that corrections for this $e/h \neq 1$ behavior can be applied, once the behavior of the calorimeter is understood. In fact, to first order in this simple simulation, the effect of $e/h \neq 1$ is an increase in the p_T of a jet by about 10%. This can be seen in Fig. 4.10 where 1000 5-TeV hard-scattering events have been generated and the highest p_T jet was reconstructed with (a) $e/h = 1$ and (b) $e/h = 1.4$. The distribution is certainly broader for the $e/h = 1.4$ case (most of the intrinsic broadness is due to the way ISAJET conserves energy and momentum) but the major effect is the shifting of the mass from 4.8 to 5.3 TeV/c. However, corrections are certainly going to be smaller if $e/h = 1.1$ as seen in Fig. 4.9.

The same systematic effects as described above are also examined in Figs. 4.11 and 4.12 but as a function of the two-jet mass (for those jets produced between 60° and 120° in the jet–jet center of mass). The behavior is essentially the same. It is worth pointing out that the presence of composite terms in addition to increasing the jet cross section, also modifies the angular distribution, making it flatter. This modification of the angular distribution is a powerful tool for the detection of compositeness, especially since it is less sensitive to detector resolution effects.

It is also worth noting that measurements of this type (high p_T single or dijet production) are relatively insensitive to the effects of event pile-up. Even in the case of a relatively slow calorimeter such as liquid argon, the σ of the pile-up 'noise' from minimum bias events is only on the order of 300–400 MeV/c from $\mathcal{L} = 10^{33}\text{cm}^{-2}\text{s}^{-1}$ for a $\Delta y \times \Delta\phi = 0.09 \times 0.09$ cell [28],[29]. For $\mathcal{L} = 10^{34}\text{cm}^{-2}\text{s}^{-1}$ this increases to approximately 1 GeV/c. A jet defined with a radius of 0.5 incorporates about 100 cells, leading to a σ for the pile-up noise per jet of approximately 10 GeV/c.

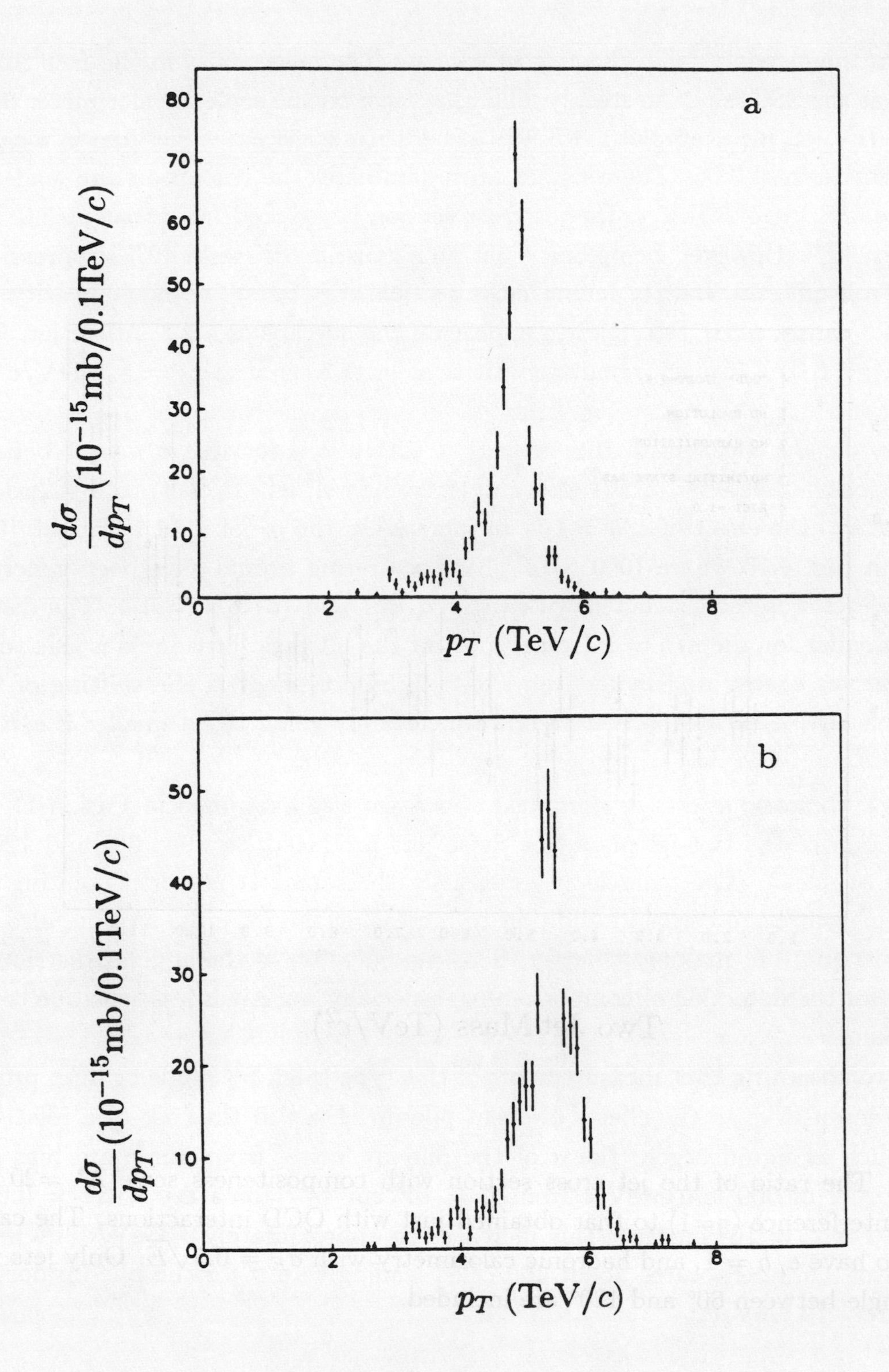

Figure 4.10: The p_T distribution of the highest p_T for hard scattering at $p_T = 5$ TeV/c. In (a) the calorimeter has $e/h = 1$. In (b) the calorimeter has $e/h = 1.4$.

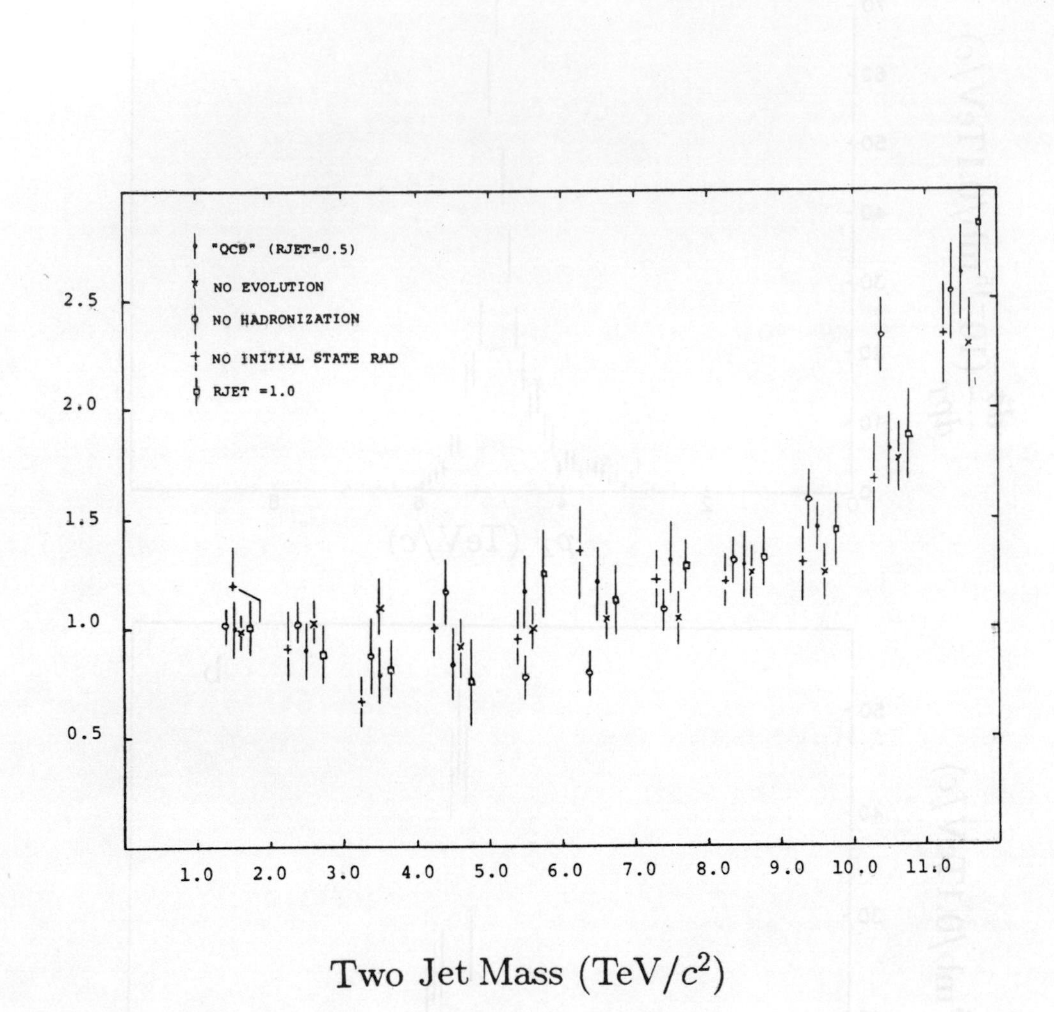

Figure 4.11: The ratio of the jet cross section with compositeness scale Λ^* =20 TeV and destructive interference (η=1) to that obtained just with QCD interactions. The calorimeter is assumed to have $e/h = 1$, and hadronic calorimetry with $\sigma_E = 0.5\sqrt{E}$. Only jets with c.m. scattering angle between 60° and 120° are included.

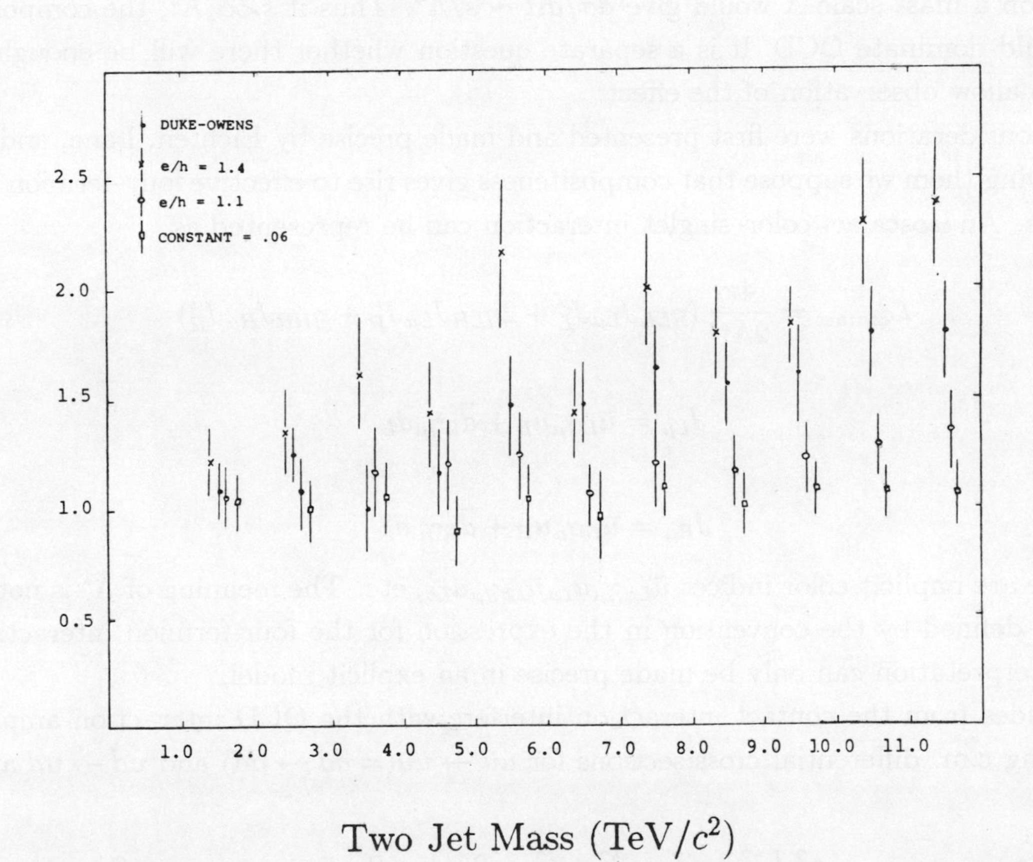

Figure 4.12: The ratio of the observed cross section to that for a "standard" calorimeter with $e/h = 1$, using EHLQ structure functions. Only QCD production is considered. The effects of $e/h \neq 1$, a constant term in the calorimeter resolution, and the use of Duke-Owens structure functions in the place of EHLQ are indicated. Only jets with c.m. scattering angle between 60° and 120° are included.

5 Compositeness Signal in Quark Jets

The effects of compositeness relative to QCD will be greatest for large values of $\sqrt{\hat{s}}$, the quark-quark c.m. energy, and for scattering near 90° in the quark-quark c.m. frame. Quark–quark scattering by QCD processes at 90^0 must have $d\sigma/d\Omega \sim \alpha_s^2/\hat{s}$ on dimensional grounds. Compositeness on a mass scale Λ would give $d\sigma/d\Omega \sim \hat{s}/\Lambda^4$. Thus if $\hat{s}\gtrsim\alpha_s\Lambda^2$, the compositeness effects should dominate QCD. It is a separate question whether there will be enough events at this $\hat{s}$ to allow observation of the effect.

These considerations were first presented and made precise by Eichten, Lane, and Peskin [11]. Following them we suppose that compositeness gives rise to effective four–fermion contact interactions. An isoscalar, color–singlet interaction can be represented as

$$\mathcal{L}_{contact} = \frac{4\pi}{2\Lambda^{*2}}\left(\eta_{LL}J_{L\mu}J_L^\mu + 2\eta_{LR}J_{L\mu}J_R^\mu + \eta_{RR}J_{R\mu}J_R^\mu\right) \tag{5.1}$$

where

$$J_{L\mu} = \overline{u}_L\gamma_\mu u_L + \overline{d}_L\gamma_\mu d_L \tag{5.2}$$

$$J_{R\mu} = \overline{u}_R\gamma_\mu u_R + \overline{d}_R\gamma_\mu d_R \tag{5.3}$$

where there are implicit color indices $\bar{u}_{La}\gamma_\mu u_{La}\bar{u}_{Lb}\gamma_\mu u_{Lb}$, etc. The meaning of Λ^* is not immediate. It is defined by the convention in the expression for the four-fermion interaction. Its physical interpretation can only be made precise in an explicit model.

Amplitudes from the contact interaction interfere with the QCD interaction amplitudes. The resulting c.m. differential cross sections for $uu \to uu (= dd \to dd)$ and $ud \to ud$ are

$$\begin{aligned}\left(\frac{d\sigma}{d\Omega}\right)_{uu\to uu} &= \frac{\alpha_s^2}{9\hat{s}}\left[\frac{\hat{s}^2+\hat{u}^2}{\hat{t}^2} + \frac{\hat{s}^2+\hat{t}^2}{\hat{u}^2} - \frac{2\hat{s}^2}{3\hat{t}\hat{u}}\right] + \frac{2\alpha_s\hat{s}}{9\Lambda^{*2}}(\eta_{LL}+\eta_{RR})\left(\frac{1}{\hat{u}}+\frac{1}{\hat{t}}\right) \\ &+ \frac{1}{\Lambda^{*4}\hat{s}}\left[\frac{2}{3}\hat{s}^2(\eta_{LL}^2+\eta_{RR}^2) + \frac{1}{2}\eta_{LR}^2(\hat{u}^2+\hat{t}^2)\right]\end{aligned} \tag{5.4}$$

$$\left(\frac{d\sigma}{d\Omega}\right)_{ud\to ud} = \frac{\alpha_s^2}{9\hat{s}}\frac{\hat{u}^2+\hat{s}^2}{\hat{t}^2} + \frac{1}{4\hat{s}\Lambda^{*4}}\left[(\eta_{LL}^2+\eta_{RR}^2)\hat{s}^2 + (\eta_{LR}^2+\eta_{RL}^2)\hat{u}^2\right] \tag{5.5}$$

where $\hat{s}$ is the center of mass energy squared and

$$\hat{t} = -\hat{s}(1-\cos\theta)/2 \tag{5.6}$$

$$\hat{u} = -\hat{s}(1+\cos\theta)/2. \tag{5.7}$$

Here θ is the c.m. scattering angle.

Table 5.1: The luminosity, $d\mathcal{L}/d\tau$, is parameterized as $d\mathcal{L}/d\tau = e^a\sqrt{\tau}^b(1-\sqrt{\tau})^c$. The luminosity is calculated with structure functions evaluated at $Q^2 = \tau s$, using structure functions from EHLQ.

	a	b	c
uu	-0.56	-2.5	8.7
dd	-2.7	-2.8	10.
ud	-0.48	-2.3	9.9
gg	-3.4	-4.2	15.
ug	-1.00	-3.4	12.
dg	-2.2	-3.5	13.

To see the effect of the contact interaction it is necessary to look at pp events with two balanced high p_T jets. By measuring the jets it is possible to reconstruct $\hat{s}$, the center of mass energy squared for the partonic collision, and θ, the scattering angle in the partonic center of mass. Only at large $\hat{s}$ can the contact interaction compete with QCD and in fact, it is not the process $qq \to qq$ that is largest, but $qg \to qg$ because of the large number of gluons in the proton. Generally, the luminosity for collisions with $\hat{s} = \tau s$ is, for $q_i q_j \to X$, $i \neq j$

$$\frac{d\mathcal{L}}{d\tau} = \int \frac{dx}{x}\left[q_i(x)q_j(\tau/x) + q_j(x)q_i(\tau/x)\right]) \tag{5.8}$$

and for $q_i q_i \to X$

$$\frac{d\mathcal{L}}{d\tau} = \int \frac{dx}{x} q_i(x)q_i(\tau/x) \tag{5.9}$$

The cross section as a function of τ and θ is

$$\frac{d\sigma}{d\tau d\Omega} = \frac{d\sigma}{d\Omega}(\hat{s} = s\tau)\cdot\frac{d\mathcal{L}}{d\tau} \tag{5.10}$$

For the QCD process $qg \to qg$ we have

$$\left(\frac{d\sigma}{d\Omega}\right)_{qg} = \frac{\alpha_s^2}{4s}\left(\frac{\hat{s}^2+\hat{u}^2}{\hat{s}^2}\right)\left(\frac{\hat{s}^2}{\hat{t}^2} - \frac{4\hat{s}}{9\hat{u}}\right) \tag{5.11}$$

and for $gg \to gg$

$$\left(\frac{d\sigma}{d\Omega}\right)_{gg} = \frac{9\alpha_s^2}{8\hat{s}}\left[3 - \frac{\hat{t}\hat{u}}{\hat{s}^2} - \frac{\hat{s}\hat{u}}{\hat{t}^2} - \frac{\hat{s}\hat{t}}{\hat{u}^2}\right]. \tag{5.12}$$

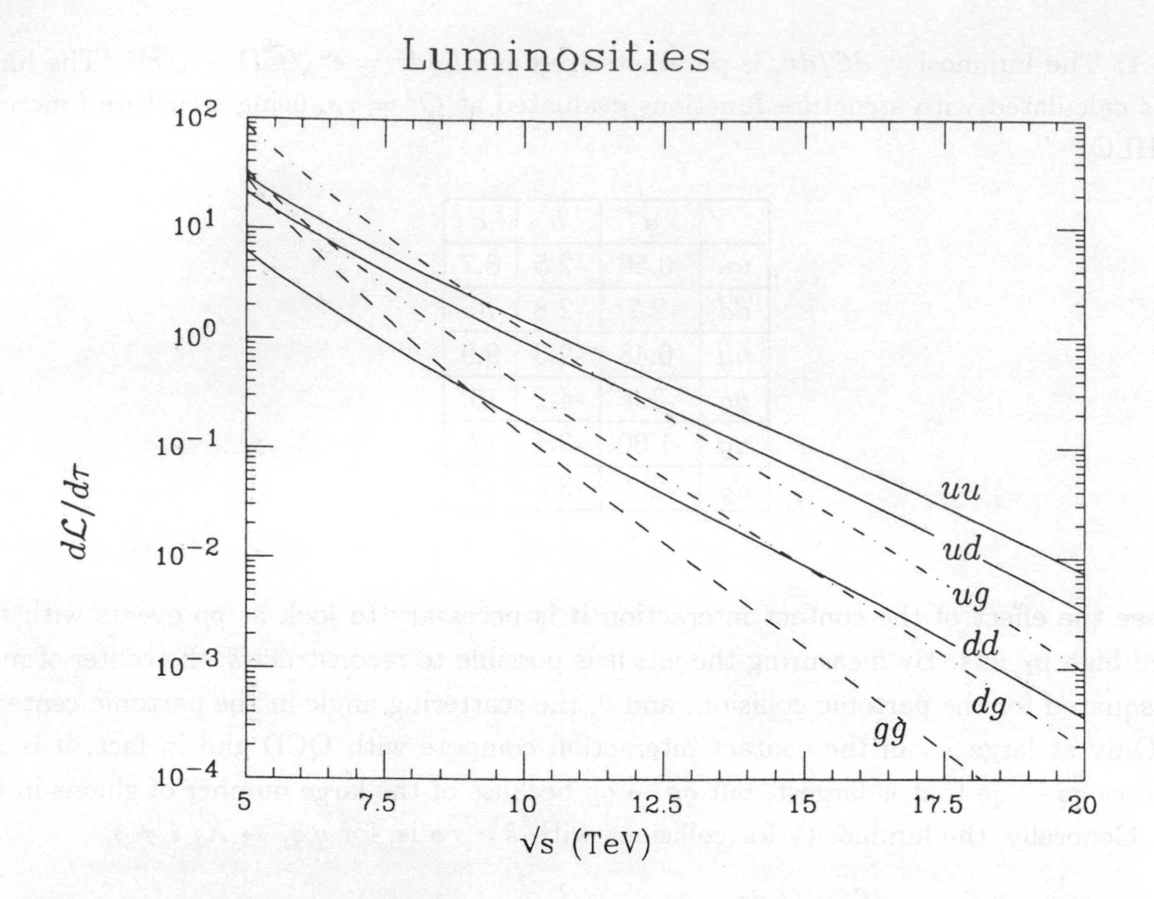

Figure 5.1: Luminosities, $d\mathcal{L}/d\tau$, at $\sqrt{s} = 40$ TeV as a function of $\sqrt{\hat{s}} = \sqrt{\tau s}$. The structure functions were taken from EHLQ.

In Table 5.1 we give a simple parameterization of $d\mathcal{L}/d\tau$. These luminosities are shown in Fig. 5.1. From the luminosities and the cross sections, Eqs.5.4, 5.5, 5.11, 5.12 we can compute the cross section to produce a jet pair with invariant mass greater than $\sqrt{\tau_0 s}$ and with c.m. scattering angle within $\pm\theta$ of 90°:

$$\sigma(\tau > \tau_0,\ |\cos\theta| < \cos\theta_0) = \int_{\tau_0}^{1} d\tau \frac{d\mathcal{L}}{d\tau} 2\pi \int_{-\cos\theta_0}^{\cos\theta_0} d\cos\theta \frac{d\sigma}{d\Omega}. \tag{5.13}$$

The results for $\theta_0 = 60^0$ and $\theta_0 = 30^0$ are shown in Figs. 5.2 and 5.3 and Tables 5.2, 5.3, 5.4, and 5.5. An integrated luminosity of $\int \mathcal{L} d\tau = 10^{40}\text{cm}^{-2}$ is assumed. The contact interaction has been taken to be a purely LL interaction, i.e. $\eta_{LL} = \pm 1, \eta_{LR} = \eta_{RL} = \eta_{RR} = 0$. For $\eta_{LL} = +1$, the interference with QCD is destructive, while for $\eta_{LL} = -1$ it is constructive.

We propose two criteria that should be satisfied for an "observable signal".

a. The contact plus QCD interactions must give twice as many events as anticipated from QCD alone.

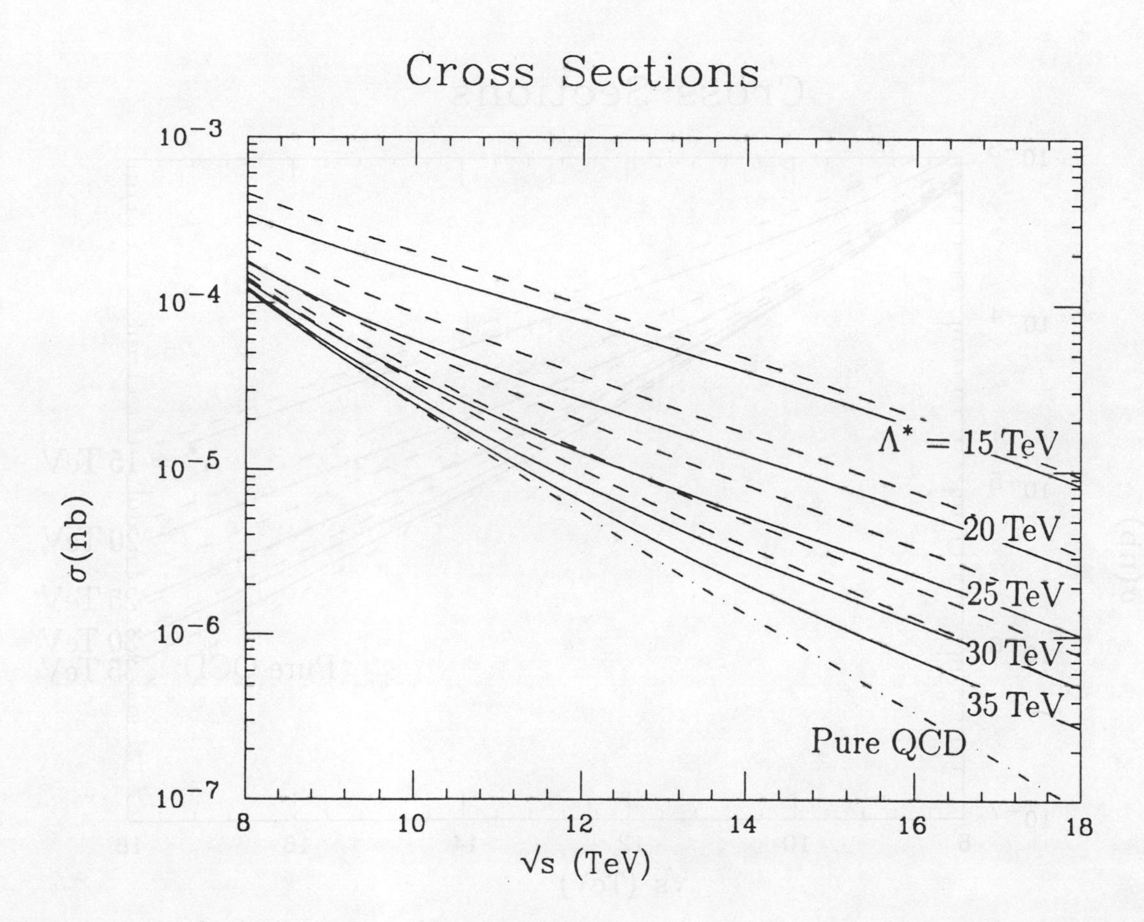

Figure 5.2: The cross section at the SSC to produce a jet-jet pair with invariant mass greater than $\sqrt{\hat{s}}$, with a scattering angle between 60° and 120° in the jet-jet c.m., as a function of $\sqrt{\hat{s}}$. The dashed line is for production by QCD only. The cross section increases as the value of Λ^* is decreased progressively, with the values 35, 30, 25, 20, 15 TeV. The dotted curves are for constructive interference between QCD and the compositeness interaction, the solid curves for destructive interference.

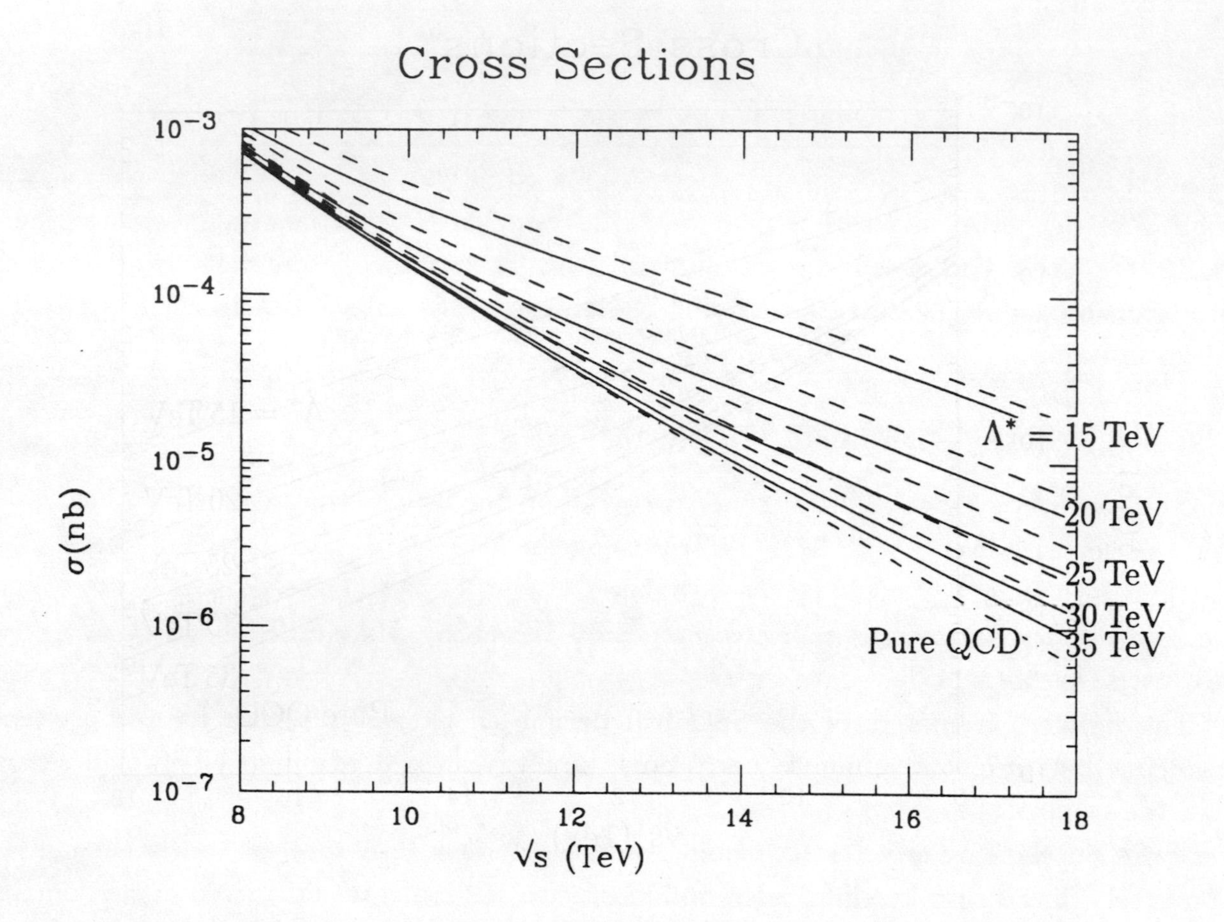

Figure 5.3: The cross section at the SSC to produce a jet-jet pair with invariant mass greater than $\sqrt{\hat{s}}$, with a scattering angle between 30° and 150° in the jet-jet c.m., as a function of $\sqrt{\hat{s}}$. The dashed line is for production by QCD only. The cross section increases as the value of Λ^* is decreased progressively, with the values 35, 30, 25, 20, 15 TeV. The dotted curves are for constructive interference between QCD and the compositeness interaction, the solid curves for destructive interference.

$\sqrt{s}$ (TeV)	8	10	12	14	16	18
$\Lambda^* = 15$	*3331*	*1536*	*780*	*390*	*185*	82
$\Lambda^* = 20$	1734	*591*	*260*	*123*	57	25
$\Lambda^* = 25$	1347	352	*126*	53	23	10
$\Lambda^* = 30$	1229	275	81	30	12	5
$\Lambda^* = 35$	1190	247	64	20	7	3
Pure QCD	1207	239	54	13	3	1

Table 5.2: Jet-jet events with $60^o < \theta < 120^o$ and jet-jet invariant mass greater than 8, 10, 12, etc. TeV, for various values of Λ^* in TeV. The contact interaction gives destructive interference ($\eta = +1$). Entries satisfying the observability criteria described in the text are shown in italics.

b. The number of events must be at least 100.

For the case of destructive interference we see from the Tables that a limit of $\Lambda^* > 25$ TeV could be set with events with $M > 12$ TeV using the $\pm 30^0$ data. Using the $\pm 60^0$ data, a limit a little less than 24 TeV could be set with data for $M > 14$ TeV. For constructive interference, the $\pm 30^0$ data would permit a limit of about 35 TeV using $M > 12$ TeV and about $\Lambda^* = 30$ TeV with the $M > 14$ TeV data.

The design parameters of the SSC will permit an extensive search for compositeness. Reducing instantaneous luminosity and thus the likely integrated luminosity would seriously limit the scope of the search. Tables 5.2 – 5.5 show that $\int \mathcal{L} d\tau = 10^{39}\text{cm}^{-2}$, which would reduce the event numbers by a factor 10, would lessen the domain of Λ^* probed to less than 20 TeV. Conversely, increasing the integrated luminosity to 10^{41}cm^{-2} would increase the domain to beyond 30 TeV for destructive interference and 35 TeV for constructive interference.

$\sqrt{s}$ (TeV)	8	10	12	14	16	18
$\Lambda^* = 15$	10657	*3596*	*1548*	*718*	*329*	*143*
$\Lambda^* = 20$	8030	2014	*670*	*264*	*111*	46
$\Lambda^* = 25$	7425	1628	448	148	54	21
$\Lambda^* = 30$	7258	1510	377	109	35	12
$\Lambda^* = 35$	7212	1470	351	94	28	9
Pure QCD	7302	1481	345	87	23	6

Table 5.3: Jet-jet events with $30^o < \theta < 150^o$ and jet-jet invariant mass greater than 8, 10, 12, etc. TeV, for various values of Λ^* in TeV. The contact interaction gives destructive interference ($\eta = +1$). Entries satisfying the observability criteria described in the text are shown in italics.

$\sqrt{s}$ (TeV)	8	10	12	14	16	18
$\Lambda^* = 15$	*4506*	*2015*	*975*	*468*	*215*	93
$\Lambda^* = 20$	*2395*	*860*	*370*	*167*	74	31
$\Lambda^* = 25$	1770	*525*	*196*	81	34	14
$\Lambda^* = 30$	1523	395	*130*	49	20	8
$\Lambda^* = 35$	1406	335	100	35	13	5
Pure QCD	1207	239	54	13	3	1

Table 5.4: Jet-jet events with $60^o < \theta < 120^o$ and jet-jet invariant mass greater than 8, 10, 12, etc. TeV, for various values of Λ^* in TeV. The contact interaction gives constructive interference ($\eta = -1$). Entries satisfying the observability criteria described in the text are shown in italics.

$\sqrt{s}$ (TeV)	8	10	12	14	16	18
$\Lambda^* = 15$	13367	*4701*	*1999*	*899*	*399*	*169*
$\Lambda^* = 20$	9555	2636	*924*	*366*	*150*	61
$\Lambda^* = 25$	8400	2026	611	*213*	80	30
$\Lambda^* = 30$	7935	1786	490	154	53	19
$\Lambda^* = 35$	7710	1673	434	128	41	13
Pure QCD	7302	1481	345	87	23	6

Table 5.5: Jet-jet events with $30^\circ < \theta < 150^\circ$ and jet-jet invariant mass greater than 8, 10, 12, etc. TeV, for various values of Λ^* in TeV. The contact interaction gives constructive interference ($\eta = -1$). Entries satisfying the observability criteria described in the text are shown in italics.

6 Lepton Pair Production as a Signal for Compositeness

If quarks and leptons are composite objects with common constituents, there will be two–quark two–lepton contact interactions below the scale of compositeness in addition to the four-quark operators discussed above. These interactions would affect quark-antiquark annihilation into lepton pairs. Following the analysis of Eichten, Hinchliffe, Lane, and Quigg [13], we consider the effect of these contact interactions on the lepton pair production process, pp→ $l^+ \, l^- X$.

If contact operators between quarks and leptons are detected, questions about their form will remain. The best test for the presence of compositeness is the total integrated cross section for the production of high invariant mass lepton pairs. However, if compositeness were found, further information would be obtained from the study of the differential cross section as a function of invariant mass of the lepton pair, rapidity, and the angle of the negatively charged lepton relative to the line-of-flight of the lepton pair.

In [30], similar questions were posed about the form of the coupling of an additional U(1) gauge boson, where the front-back asymmetry of the outgoing negatively charged lepton was studied. We perform a similar analysis below for contact interactions. Unlike the situation with only jets in the final state, the sign of the outgoing leptons can be identified, so there is a measurable asymmetry.

We use notation analogous to that used in Section 5 to describe the contact operators and cross sections. We restrict our attention to flavor-diagonal gauge singlet interactions which can be described by the Lagrangian:

$$L_{contact} = \frac{4\pi}{\Lambda_l^{*2}} \left(\eta_{LL} J_{L\mu} J_L^{l\mu} + \eta_{LR} J_{L\mu} J_R^{l\mu} + \eta_{RL} J_{R\mu} J_L^{l\mu} + \eta_{RR} J_{R\mu} J_R^{l\mu} \right) \tag{6.1}$$

where

$$J_{L\mu} = \bar{u}_L \gamma_\mu u_L + \bar{d}_L \gamma_\mu d_L \tag{6.2}$$

$$J_{R\mu} = \bar{u}_R \gamma_\mu u_R + \bar{d}_R \gamma_\mu d_R \tag{6.3}$$

$$J_{L\mu}^l = \bar{l}_L^- \gamma_\mu l_L^- \tag{6.4}$$

$$J_{R\mu}^l = \bar{l}_R^- \gamma_\mu l_R^- \tag{6.5}$$

where the implicit color indices (e.g. $\bar{u}_{La} \gamma_\mu u_{La}$) are summed. Here, Λ_l^* is the scale of the contact operator appropriate to the lepton channel being studied. For similar values of Λ^* and Λ_l^*, the cross section for jet pair production exceeds the cross section in pp collisions for lepton pair production (at the same pair invariant mass) because the lepton pair production requires an incident anti-quark while the jet production does not. The differential forward and backward cross sections are:

$$\left(\frac{d\sigma_F}{d\hat{s}dy}\right)_{q_i\bar{q}_i\to l\bar{l}} = \frac{\alpha_s^2}{24\hat{s}s} \left\{\left[\frac{7}{3}A_i(\hat{s}^2)+\frac{1}{3}B_i(\hat{s}^2)\right] q_i(x_1)\bar{q}_i(x_2) + \left[\frac{1}{3}A_i(\hat{s}^2)+\frac{7}{3}B_i(\hat{s}^2)\right] q_i(x_2)\bar{q}_i(x_1)\right\} \tag{6.6}$$

$$\left(\frac{d\sigma_B}{d\hat{s}dy}\right)_{q_i\bar{q}_i\to l\bar{l}} = \frac{\alpha_s^2}{24\hat{s}s} \left\{\left[\frac{7}{3}B_i(\hat{s}^2)+\frac{1}{3}A_i(\hat{s}^2)\right] q_i(x_1)\bar{q}_i(x_2) + \left[\frac{1}{3}B_i(\hat{s}^2)+\frac{7}{3}A_i(\hat{s}^2)\right] q_i(x_2)\bar{q}_i(x_1)\right\} \tag{6.7}$$

where $x_1 > x_2$ and

$$A_i(\hat{s}) = \left[Q_i - \frac{L_iL_l}{4x_W(1-x_W)}\frac{\hat{s}}{\hat{s}-M_Z^2+iM_Z\Gamma_Z} - \frac{\eta_{LL}\hat{s}}{\alpha\Lambda_l^{*2}}\right]^2 + \left[Q_i - \frac{R_iR_l}{4x_W(1-x_W)}\frac{\hat{s}}{\hat{s}-M_Z^2+iM_Z\Gamma_Z} - \frac{\eta_{RR}\hat{s}}{\alpha\Lambda_l^{*2}}\right]^2 \tag{6.8}$$

$$B_i(\hat{s}) = \left[Q_i - \frac{R_iL_l}{4x_W(1-x_W)}\frac{\hat{s}}{\hat{s}-M_Z^2+iM_Z\Gamma_Z} - \frac{\eta_{RL}\hat{s}}{\alpha\Lambda_l^{*2}}\right]^2 + \left[Q_i - \frac{L_iR_l}{4x_W(1-x_W)}\frac{\hat{s}}{\hat{s}-M_Z^2+iM_Z\Gamma_Z} - \frac{\eta_{LR}\hat{s}}{\alpha\Lambda_l^{*2}}\right]^2 \tag{6.9}$$

Here, $\hat{s}$ is the parton center of mass energy squared, s is the proton center of mass energy squared,

$$L_i = \tau_3 - 2Q_ix_W \tag{6.10}$$

$$L_l = -1 - 2Q_lx_W \tag{6.11}$$

$$R_i = -2Q_ix_W \tag{6.12}$$

$$R_l = -2Q_lx_W, \tag{6.13}$$

$x_W = \sin^2\theta_W$, and τ_3 is twice the weak isospin projection of fermion i.

These formulae are obtained by integrating the differential cross section over θ, the angle of the negatively charged lepton in the parton-parton pair frame, where the forward direction is defined to be the direction of motion of the center of mass system.

In Tables 6.1, 6.2, 6.3, and 6.4 we present our results for the total number of events that are expected for an integrated luminosity of 10^{40}cm^{-2} with center of mass energy of 40 TeV for several values of Λ_l^*. The data are binned in 500 GeV bins in $M = \sqrt{\hat{s}}$. The total cross section is integrated over the full rapidity range (up to a cutoff at $|y| = 3$), while the data for the forward and backward cross sections have only been integrated down to rapidity $|y| = 1$. We have not integrated over the small y data in this case because the asymmetry will be negligible near the symmetrical point $y = 0$.

Table 6.1 [6.2] presents the results for an "L-L" contact interaction, where $\eta_{LL} = 1[-1], \eta_{LR} = \eta_{RL} = \eta_{RR} = 0$. In Table 6.3 [6.4], we present the results for a "V-V" interaction, with $\eta_{LL} = \eta_{LR} = \eta_{RL} = \eta_{RR} = 1[-1]$.

We take the following criteria for "observability" of compositeness:

a. For at least one 500 GeV bin of M, there are twice as many events as predicted by the standard model .

b. There are at least 100 events in this bin.

With these criteria, we can test for an "L-L" composite interaction up to Λ_l^*=15 TeV when there is constructive interference with standard model Drell Yan processes (η=-1), and up to 12.5 TeV for destructive interference. Because of the increased cross section for "V-V" interactions (because both left and right-handed helicity states contribute to the cross section), compositeness can be tested up to scales of 20 [17.5] TeV for constructive [destructive] interference.

Assuming these criteria have been satisfied, we then test whether four-fermion operators with left-handed and vector currents are distinguishable. We do a χ^2 fit with 14 degrees of freedom to test how well the data for left-handed [vector] currents could mimic the prediction for vector [left-handed] currents. The 15 data points were the total cross section integrated over $|y| < 1$, the forward cross section integrated over $1 < |y| < 3$, and the backward cross section integrated over $1 < |y| < 3$ for the five values of M in Tables 6.1 – 6.4, and there was a single parameter, Λ_l^*. We required:

c. $\chi^2 > 50$ for discriminating against left-handed [vector] interactions fits to vector [left-handed] current data.

With this criterion, we find that for all compositeness scales within the discovery limits, the L-L and V-V contact operators could be distinguished. This indicates that there may be sufficient discriminating power in the data to test the predictions of a particular model. Although in practice such fits will be limited by the systematics, the situation for distinguishing among composite scenarios is encouraging.

M(TeV)	0.5	1.0	1.5	2.0	2.5
$\Lambda_l^* = 2.5$ TeV, Total	*170000*	*140000*	*120000*	*100000*	*86000*
F-B	47000	42000	34000	27000	21000
Total, y>1	110000	81000	61000	46000	34000
$\Lambda_l^* = 5.0$ TeV	36000	*10000*	*8000*	*6600*	*5500*
etc.	9400	3300	2400	1800	1300
	25000	6200	4200	3000	2200
$\Lambda_l^* = 7.5$ TeV	27000	*3100*	*1800*	*1400*	*1100*
	6500	980	550	390	280
	19000	1900	970	640	450
$\Lambda_l^* = 10.0$ TeV	25000	*1700*	730	*500*	*390*
	5700	530	230	140	97
	17000	1000	390	230	160
$\Lambda_l^* = 12.5$ TeV	25000	1200	*410*	*240*	*180*
	5500	380	130	68	45
	17000	770	220	110	71
$\Lambda_l^* = 15.0$ TeV	24000	1100	*280*	*140*	96
	5300	320	86	41	25
	16000	660	160	69	40
$\Lambda_l^* = 17.5$ TeV	24000	980	230	*100*	61
	5200	280	66	28	16
	16000	610	120	48	25
$\Lambda_l^* = 20.0$ TeV	24000	940	190	77	43
	5200	260	56	21	11
	16000	570	110	37	18
$\Lambda_l^* = \infty$	24000	820	130	34	12
	5000	210	32	8	2
	16000	490	68	16	5

Table 6.1: Number of lepton pair events with different lepton pair invariant masses (500 GeV bins) for different values of Λ_l^*. First row for each Λ_l^* is total number of events with $y<3$, second is number of forward events minus number of backward events (with $1< y <3$), and third row is total number of events with $1< y <3$. The data are for an LL contact interaction with constructive interference ($\eta = -1$). Entries satisfying the observability criteria in the text for total production are shown in italics.

M(TeV)	0.5	1.0	1.5	2.0	2.5
$\Lambda_l^* = 2.5$ TeV, Total	*130000*	*130000*	*110000*	*99000*	*84000*
F-B	29000	36000	32000	26000	20000
Total, y>1	82000	73000	58000	44000	33000
$\Lambda_l^* = 5.0$ TeV	27000	*7600*	*6800*	*6000*	*5100*
etc.	4800	2000	1900	1500	1200
	17000	4300	3400	2700	2000
$\Lambda_l^* = 7.5$ TeV	23000	*1800*	*1300*	*1100*	*980*
	4400	400	330	280	230
	15000	1000	640	500	380
$\Lambda_l^* = 10.0$ TeV	23000	980	*430*	*350*	*300*
	4600	200	99	82	68
	15000	550	210	150	110
$\Lambda_l^* = 12.5$ TeV	23000	800	220	*150*	*120*
	4700	170	45	32	26
	15000	460	110	61	45
$\Lambda_l^* = 15.0$ TeV	23000	760	150	77	58
	4800	170	30	15	12
	16000	440	73	32	21
$\Lambda_l^* = 17.5$ TeV	23000	760	130	50	33
	4900	180	25	9	6
	16000	450	63	21	12
$\Lambda_l^* = 20.0$ TeV	23000	760	120	39	21
	4900	180	25	7	4
	16000	450	60	16	8
$\Lambda_l^* = \infty$	24000	820	130	34	12
	5000	210	32	8	2
	16000	490	68	16	5

Table 6.2: Number of lepton pair events with different lepton pair invariant masses (500 GeV bins) for different values of Λ_l^*. First row for each Λ_l^* is total number of events with $y < 3$, second is number of forward events minus number of backward events (with $1 < y < 3$), and third row is total number of events with $1 < y < 3$. The data are for an LL contact interaction with destructive interference ($\eta = +1$). Entries satisfying the observability criteria in the text for total production are shown in italics.

M(TeV)	0.5	1.0	1.5	2.0	2.5
$\Lambda_l^* = 2.5$ TeV, Total	*580000*	*540000*	*470000*	*400000*	*340000*
F-B	15000	3000	1100	500	260
Total, y>1	380000	320000	240000	180000	130000
$\Lambda_l^* = 5.0$ TeV	70000	*38000*	*31000*	*26000*	*22000*
etc.	7500	910	300	130	66
	47000	22000	16000	12000	8600
$\Lambda_l^* = 7.5$ TeV	37000	*9100*	*6600*	*5300*	*4400*
	6100	520	150	63	30
	25000	5500	3400	2400	1700
$\Lambda_l^* = 10.0$ TeV	30000	*3900*	*2400*	*1800*	*1500*
	5700	390	100	39	18
	20000	2400	1200	820	580
$\Lambda_l^* = 12.5$ TeV	27000	*2300*	*1100*	*810*	*630*
	5400	320	76	28	13
	18000	1400	610	370	250
$\Lambda_l^* = 15.0$ TeV	26000	*1700*	*670*	*430*	*330*
	5300	290	62	22	9
	17000	1000	360	200	130
$\Lambda_l^* = 17.5$ TeV	25000	*1400*	*460*	*270*	*190*
	5200	270	54	18	8
	17000	840	250	130	78
$\Lambda_l^* = 20.0$ TeV	25000	1200	*350*	*180*	*120*
	5200	260	49	16	6
	17000	740	190	86	51
$\Lambda_l^* = \infty$	24000	820	130	34	12
	5000	210	32	8	2
	16000	490	68	16	5

Table 6.3: Number of lepton pair events with different lepton pair invariant masses (500 GeV bins) for different values of Λ_l^*. First row for each Λ_l^* is total number of events with $y < 3$, second is number of forward events minus number of backward events (with $1 < y < 3$), and third row is total number of events with $1 < y < 3$. The data are for a VV contact interaction with constructive interference ($\eta = -1$). Entries satisfying the observability criteria in the text for total production are shown in italics.

M(TeV)	0.5	1.0	1.5	2.0	2.5
$\Lambda_l^* = 2.5$ TeV, Total	*460000*	*510000*	*460000*	*400000*	*340000*
F-B	-4806	-2592	-1054	-487	-249
Total, y>1	290000	290000	230000	180000	130000
$\Lambda_l^* = 5.0$ TeV	39000	*29000*	*27000*	*24000*	*21000*
etc.	2600	-488	-239	-115	-60
	25000	17000	14000	11000	8100
$\Lambda_l^* = 7.5$ TeV	23000	*5400*	*5100*	*4600*	*4000*
	3900	-99	-88	-46	-25
	15000	3100	2600	2000	1600
$\Lambda_l^* = 10.0$ TeV	22000	*1800*	*1500*	*1400*	*1200*
	4400	37	-35	-22	-12
	14000	1000	750	610	470
$\Lambda_l^* = 12.5$ TeV	22000	1000	*600*	*540*	*480*
	4600	100	-10	-11	-7
	15000	550	290	230	180
$\Lambda_l^* = 15.0$ TeV	22000	760	*300*	*250*	*220*
	4800	130	2	-5	-4
	15000	430	140	110	84
$\Lambda_l^* = 17.5$ TeV	23000	700	190	*130*	*120*
	4800	160	10	-1	-2
	15000	400	88	55	43
$\Lambda_l^* = 20.0$ TeV	23000	690	140	79	65
	4900	170	15	0	-1
	15000	400	65	33	24
$\Lambda_l^* = \infty$	24000	820	130	34	12
	5000	210	32	8	2
	16000	490	68	16	5

Table 6.4: Number of lepton pair events with different lepton pair invariant masses (500 GeV bins) for different values of Λ_l^*. First row for each Λ_l^* is total number of events with y <3, second is number of forward events minus number of backward events (with 1< y <3), and third row is total number of events with 1< y <3. The data are for a VV contact interaction with destructive interference ($\eta = +1$). Entries satisfying the observability criteria in the text for total production are shown in italics.

7 Backgrounds for the Drell-Yan Process

The signature of the Drell-Yan process in the lowest order is simple and clean, but the higher order effects of initial and final bremsstrahlung diminish the clarity. In addition, the finite energy resolution and the limited particle identification of realistic detectors allow other processes to be misidentified as the Drell-Yan process. A Monte Carlo study was done to evaluate such physics backgrounds.

Among the backgrounds are (a) production of a pair of W's each of which decays leptonically $W \to \ell + \nu$, and (b) production of a pair of heavy quarks $Q\overline{Q}$, each of which decays semileptonically $Q \to q + \ell + \nu$. It is also possible that the abundant QCD jets combined with the detector limitations mimic the Drell-Yan events if each of a pair of back-to-back jets (c) contains a $\pi^0\pi^+(\pi^0\pi^-)$ overlap that simulates $e^+(e^-)$, or (d) contains a punch-through pion which imitates a muon, or (e) includes a leptonically decaying pion.

Events were generated for the processes of W-pair, t-quark pair, and two jet production using ISAJET 5.20. A t-quark mass of 50 GeV/c^2 was assumed. The background study was made for an invariant mass of the lepton pair of 2 TeV/c^2. Table 7.1 shows results of the simulation. The cross sections for the invariant mass W of 2 TeV for the intermediate states in various processes are shown in the first column. To get the cross section at $W = 2$ TeV, a rapidity interval of $2|\Delta y| = 4.0$ was multiplied by $(d\sigma/dMdy)_{y=0}$ from EHLQ[13]. The real background for the lepton pair mass of 2 TeV comes from a higher value for the invariant mass of the produced background system, and was estimated from Monte Carlo event generation as $(d\sigma/dM)$ in Table 7.1. The cross sections for some of the background processes are larger than those for the Drell-Yan processes, if no further cuts are made.

Two kinematical parameters of the dilepton system were studied for the signal and backgrounds. One is the normalized P_T sum of two leptons, defined by

$$x = |\mathbf{p}_{T1} + \mathbf{p}_{T2}| / \sqrt{|\mathbf{p}_{T1}| \cdot |\mathbf{p}_{T2}|},$$

where $\mathbf{p}_{T1}$ and $\mathbf{p}_{T2}$ are transverse momentum vectors of two leptons. The case $x = 0$ corresponds to an exact balance of transverse momentum. The other parameter is the sum of transverse energy flow around a lepton, defined by

$$E_b = (\Sigma E_T)_{\Delta R<1.0},$$

where ΔR represents the distance from a lepton in the $y - \phi$ space. A smaller value of E_b corresponds to a better isolation of the lepton. The distribution of parameters x and E_b for signal and background events are shown in Fig. 7.1 and Fig. 7.2.

The rejection efficiencies for cuts with $x < 0.2$, and $E_b < 100$ GeV (for $M_{ll} = 2$ TeV), are shown in Table 7.1. The overall cross sections with these cuts are shown in the last column

	EHLQ		ISAJET SIMULATION			
Process	$d\sigma/dM$ at $W = 2$ TeV (nb/GeV)	$(p_T)_W$ (TeV)	$d\sigma/dM$ at $M = 2$ TeV (nb/GeV)	p_T balance cut	Isolation cut	$d\sigma/dM$ after cuts (nb/GeV)
Drell-Yan	5.2×10^{-9}	0.5–1	5.5×10^{-9}	$\times 0.51$	$\times 0.72$	2.0×10^{-9}
W^+W^-	2.7×10^{-7}	1–5	1.0×10^{-11}	$\times 0.14$	$\times 0.053$	7.4×10^{-14}
tt	2.7×10^{-5}	1–5	1.8×10^{-9}	$\times 0.020$	$\times 0.0060$	4.7×10^{-13}
$j_1 j_2$	4.0×10^{-3}					
$\rightarrow (\pi^0\pi^+)(\pi^0\pi^-)$		2–6	1.6×10^{-9}	$\times 0.14$	$\times 0.038$	8.5×10^{-12}
$\rightarrow (\pi^+)(\pi^-)$		2–8	2.9×10^{-7}	$(\times 0.14)$	$(\times 0.038)$	1.5×10^{-9}

Table 7.1: Comparison of the Drell-Yan cross section and various backgrounds, showing the efficacy of cuts on transverse momentum balance and lepton isolation.

of the Table. The backgrounds are substantially reduced by these cuts. To estimate the backgrounds from (d) and (e), the cross section for a pair of QCD jets that contain a pair of pions with the appropriate invariant mass satisfying the isolation and P_T balance conditions is shown in the last column of the table. It is easy to show that $\pi\mu$ decay and punch-through are negligible if one takes into account the pion life-time and a reasonably deep muon filter, so the last entry in Table 7.1 is not a major concern.

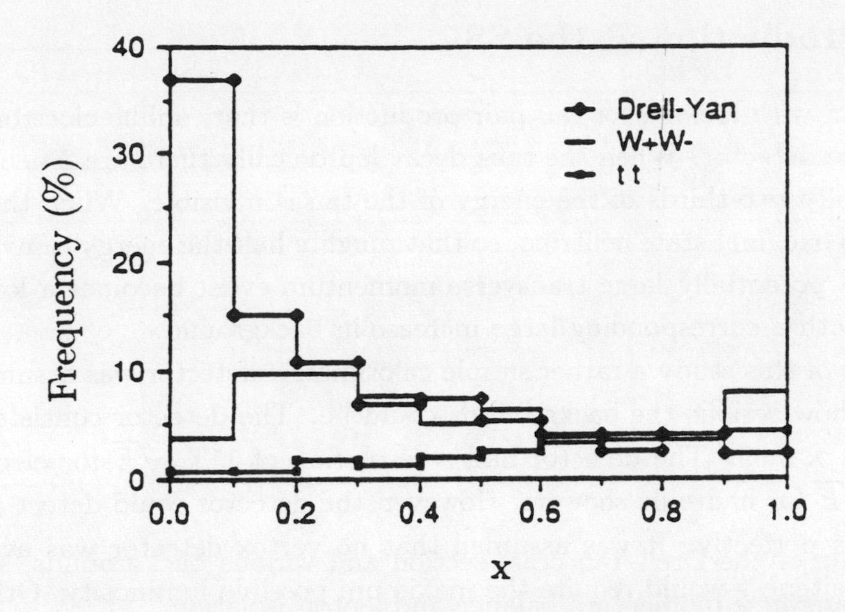

Figure 7.1: Distribution in the normalized transverse momentum imbalance x for Drell-Yan pairs and two sources of background.

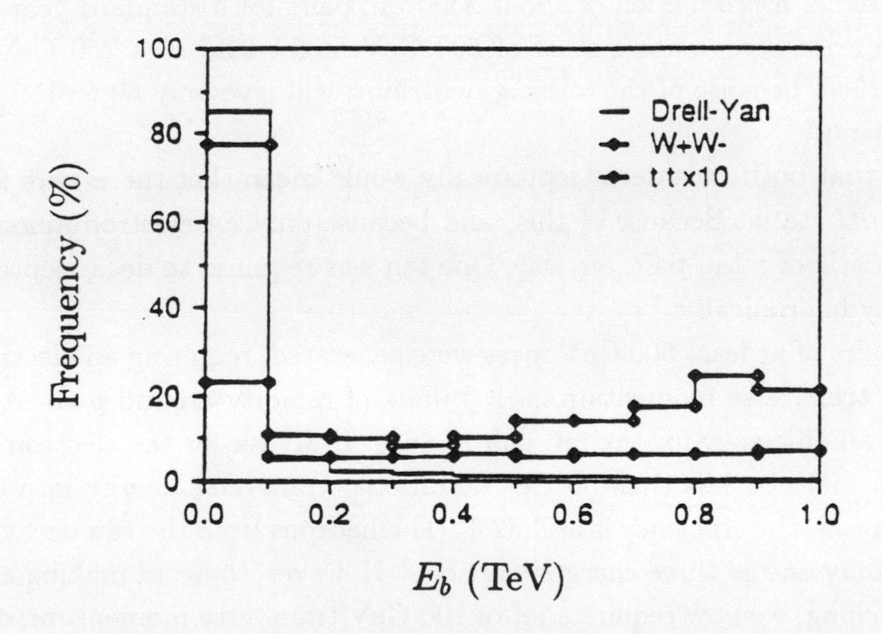

Figure 7.2: Distribution in E_b, showing the utility of an isolation cut on leptons.

8 Tau Pair Production at the SSC

The obvious problem with looking for tau pair production is that, unlike electrons or muons, taus decay inside the detector. When the taus decay leptonically there are 2 neutrinos in the final state, so typically two-thirds of the energy of the tau is invisible. When the tau decays hadronically there is one final state neutrino, so that roughly half the energy is invisible. What this means is that a potentially large transverse momentum event becomes a low transverse momentum event, with a corresponding large increase in background.

For the purposes of this study a rather simple calorimetric detector was assumed, in order to get a feeling for how serious the backgrounds could be. The detector consisted of cells of size $\Delta y \times \Delta\phi = 0.1 \times 0.09$. The detector had a resolution of $15\%/\sqrt{E}$ for electromagnetic showers and $50\%/\sqrt{E}$ for hadronic showers. However, the detector could detect and measure electrons and muons perfectly. It was assumed that no vertex detector was available since searching for compositeness would require the maximum possible luminosity. Other than the perfection for electrons and muons, no other assumptions were made about charged particle tracking or resolution. The event generator used was ISAJET 5.34, with the top quark mass set to 60 GeV.

Since compositeness, even for taus, should be evident only at higher mass scales, only production of tau pairs with an invariant mass above 500 GeV was considered. Standard Drell-Yan would predict a production of about 3300 tau pairs for a standard year at the SSC. Problems begin to arise here, since a mass of 500 GeV might lead to a 250 GeV transverse momentum tau, which, because of the missing neutrinos, will give only 50 to 150 GeV visible transverse momentum.

A requirement that both taus decay leptonically would mean that there were four missing neutrinos in the final state. Because of this, and because only an electron-muon final state would be characteristic of a tau pair, anyway, One tau was required to decay leptonically and the other one decay hadronically.

Standard tau pairs of at least 500 GeV mass were generated, requiring an electron or muon of at least 50 GeV transverse momentum in $\pm$ 3 units of rapidity around y=0. According to ISAJET this gives an efficiency for the tau pair of 36%. If we ask for the electron or muon to be isolated, that is there be less than 50 GeV additional transverse energy in a cone of ΔR =1.0 around the lepton, the efficiency is still 32%. The hadrons from the tau decay will form a very narrow jet of only one or three charged particles. However, to avoid making assumptions about hadronic tracking, we now require a jet of 100 GeV transverse momentum, defined only in the calorimeter in a cone of $\Delta R = 0.11$, to be in the hemisphere opposite the lepton. This narrow jet is also required to have less than 50 GeV additional transverse energy in a cone of $\Delta R = 1.0$ around it. This leads to an overall efficiency for detecting the tau pairs of 19%.

P_T(TeV)	Lepton	Clean Lepton	Clean Jet	Factor	Events
0.1-.25	0.24%	0.08%	–	$2.3 \cdot 10^7$	–
0.25-.50	2.5%	0.6%	0.16%	$6.9 \cdot 10^5$	$5.5 \cdot 10^6$
0.50-1.0	6.3%	1.1%	0.24%	$3.6 \cdot 10^4$	$4.3 \cdot 10^4$
1.0-2.0	10.6%	1.3%	0.10%	1300	6555
2.0-4.0	14.9%	1.8%	0.34%	29.5	501

Table 8.1: Background to tau pairs arising from QCD jets. For each transverse momentum bin, 5000 events were generated by Monte Carlo. The final column is an estimate of the number of fake Drell-Yan pairs per SSC year generated by QCD jets, subject to the cuts discussed in the text.

Given the standard Drell-Yan rate, this leads to a prediction of 620 events per year with the required signature.

The real problem in considering signals at the SSC is in trying to estimate the backgrounds. As a first step 5000 ordinary QCD jets in each of several transverse momentum bins were generated to see if any of them simulated the signature defined above. The results are summarized in Table 8.1.

The first column gives the transverse momentum range of the jets considered. The second column gives the fraction of the 5000 events that have an electron or muon of more than 50 GeV transverse momentum. The third column gives the fraction of the 5000 where the lepton is relatively isolated, as defined above. The fourth column gives the fraction of the 5000 where there is a narrow, isolated jet, of at least 100 GeV transverse momentum, opposite the lepton. The column labeled "Factor" gives the relative weight that the 5000 Monte Carlo events must be multiplied by to correspond to one standard year at the SSC. The last column gives the number of background events per year at the SSC with the required signature. Clearly a prediction of six million or so background events, with a signal of 600 events is rather disturbing. The cuts used to identify the tau pairs will have to be much tighter and there will need to be further cuts on the charged tracks if tau pairs are to be found.

However, this is only part of the story. Consider the generated events with P_T between 0.25 TeV and 0.50 TeV. Even if cuts were created to eliminate 7 of the 8 surviving events, the intrinsic rate at the SSC is so high that this one event would lead to a background prediction a thousand times the signal. Generating orders of magnitude more events, even if practical, would not really be satisfying, since even with the present statistics, we are looking at jet fragmentation effects below the 0.1% level.

The conclusions are that in order to try to see Drell-Yan production of taus, a detector would have to have very good charged particle tracking in addition to the calorimetry and probably a vertex detector, and that to understand the potential backgrounds we would need extremely reliable simulation techniques, probably far beyond what is available today.

9 Summary

The study of hadronic jets at the SSC will be pursued vigorously since they will be the manifestation of perturbative QCD. Since deviations from QCD predictions could signify the existence of compositeness below the level of quarks, reliable QCD predictions will be crucial. The precision of theoretical predictions for QCD cross sections will be limited primarily by uncertainties in the measured parton distributions to perhaps $\pm 20\%$. Care will be necessary in adopting a definition of jets that will be appropriate simultaneously to theoretical calculations and experimental measurement. The detailed study of the structure of QCD jets will be possible if suitable detectors are developed. Much of the dogma concerning the differences between quark and gluon jets might be subjected to experimental test with such a detector.

It is apparent that calorimetry will be of central importance in SSC detectors. For measuring very high energy jets, the constant term in the energy resolution, rather than the $1/\sqrt{E}$ term will be dominant. This size of this term is strongly dependent on e/h, the ratio of the response of the calorimeter to the electromagnetic part of the shower to the hadronic part. Test beams of 1 TeV may be adequate for calibration, but higher energy beams would be valuable. Monte Carlo studies show that compositeness signals for Λ^* could be detected at the SSC, but uncertainties in structure functions or deviations from e/h=1 could partially mask the effect. The deviation from $e/h = 1$ is especially important because the non-linearity results in a shift in the measurement of very high transverse momentum jets.

Within the context of the standard phenomenological formalism for compositeness, the SSC should be capable of detecting structure at the level of $\Lambda^* = 20 - 35$ TeV, depending on the detailed nature of the interaction, in the production of high transverse momentum hadronic jets. If leptons are composite as well, and share constituents with the quarks, compositeness could be visible in the production of lepton pairs. The reach in Λ_l^* is not quite so large for leptons since pair production requires the collision of a quark and an anti-quark and is thus concentrated at lower values of $\sqrt{\hat{s}}$. However, since the signs of the leptons will be measured, it will be possible to study asymmetries that cannot be observed for hadronic final states. In this way, if compositeness is found in lepton pair production, the structure of the currents responsible could be studied.

Backgrounds for lepton pair production have been studied using Monte Carlos. The sources include W pairs, heavy quark pairs, and hadrons simulating electrons or muons. The backgrounds seem manageable, so lepton pair production should continue to be a clean technique at the SSC. On the other hand, the study of tau pair production is fraught with difficulties.

Acknowledgments

The authors wish to thank G. Ballochi, F. Paige, S. Protopopescu and K. Tesima for useful discussions.

References

[1] M. Peskin, in Proceedings of the 1985 International Symposium on Lepton and Photon Interactions at High Energies, Edited by M. Konuma and K. Takahashi.

[2] JADE Collaboration, W. Bartel et. al., Z. Phys. **C30**, 371(1986).

[3] UA1 Collaboration, G. Arnison et. al., Phys. Lett. **B177**, 244(1986).

[4] H. Georgi, D. Kosower, and L. Randall, Phys. Lett. **B194**, 87(1987).

[5] R. Cahn and H. Harari, Nucl. Phys. **B176**, 135(1980).

[6] K. Lane and E. Eichten, Phys. Lett. **B90**, 125 (1980).

[7] I. Bars, Nucl. Phys. **B208**, 77(1982) .

[8] R. S. Chivukula and H. Georgi, Phys. Lett. **B188**, 99(1987).

[9] R. S. Chivukula, H. Georgi, and L. Randall, Nucl. Phys. **B292**, 93(1987).

[10] R. S. Chivukula and H. Georgi, Harvard University Preprint HUTP-87/A036, Phys. Rev. D to appear.

[11] E. Eichten, K. Lane, and M. Peskin, Phys. Rev. Lett. **50**, 811(1983).

[12] I Bars and I. Hinchliffe, Phys. Rev. **D33**, 704(1986).

[13] E. Eichten, I. Hinchliffe, K. Lane, and C. Quigg, Rev. Mod. Phys. **56**, 579(1984) and Errata, Fermilab Preprint Fermilab-Pub-86/75-T.

[14] O.V. Tarasov, A.A. Vladimirov and A.Yu. Zharkov, Phys. Lett. **93B**, 429 (1980).

[15] K.G. Chetyrkin, A.L. Kataev and F.V. Tkachev, Phys. Lett. **85B**, 277 (1979); M. Dine and J. Sapirstein, Phys. Rev. Lett. **43**, 668 (1979); W. Celmaster and R.J. Gonsalves, Phys. Rev. Lett. **44**, 560 (1980).

[16] G. Altarelli, R.K. Ellis, and G. Martinelli, Nucl. Phys. **B157**, 461 (1979).

[17] J. Kubar, M. Le Bellac, J.L. Meunier and G. Plaut, Nucl. Phys. **B175**, 251 (1980).

[18] G. Sterman and S. Weinberg, Phys. Rev. Lett. **39**, 1436 (1977).

[19] G. J. Van Dalen and J. Hauptmann, "A Specialized High p_T Jet Spectrometer for the SSC," in *Proc. of the 1984 Summer Study on the Design and Utilization of the Superconducting Super Collider*, R. Donaldson and J. G. Morfín, editors, p. 659.

[20] G. E. Theodosiou to the *Workshop on Triggering, Data Acquisition and Computing for High Energy/High Luminosity Hadron-Hadron Colliders* at Fermilab in 1985.

[21] A. H. Mueller and P. Nason, *Phys. Lett.* **157B**, 226 (1985).

[22] T. D. Gottschalk, "Remarks on the Discrimination of Prompt and Shower Bottom Production in Hadronic Jets", *Physics of the Superconducting Supercollider*, Snowmass, 1986. R. Donaldson and J. Marx, editors, p. 67.

[23] G. Thompson, *Proc. of the XXIII Int. Conf. on High Energy Physics, Berkeley, 1986*, p. 1148.

[24] F. E. Paige and S. P. Protopopescu, ISAJET 5.34. See F. E. Paige and S. P. Protopopescu, "ISAJET 5.30: A Monte Carlo Event Generator for pp and $p\bar{p}$ Interactions" in *Physics of the Superconducting Supercollider*, Snowmass, 1986. R. Donaldson and J. Marx, editors, p.320.

[25] See, for example, R. K. Ellis "New Results in Perturbative QCD," in *Supercollider Physics, Proc. of the Oregon Workshop on Super High Energy Physics*, D. E. Soper, editor, World Scientific, Singapore, 1986, p. 77.

[26] C. Baltay, J. Huston, and B. G. Pope, "Calorimetry for SSC Detectors", in *Physics of the Superconducting Supercollider*, Snowmass, 1986. R. Donaldson and J. Marx, editors, p.355.

[27] R. Wigmans, "On the Energy Resolution of Uranium and other Calorimeters", CERN/EF 86-18.

[28] A. Yamashita and K. Kondo, "Physics Noise to Calorimetry at SSC" in *Physics of the Superconducting Supercollider*, Snowmass, 1986. R. Donaldson and J. Marx, editors, p.365.

[29] G. O. Alverson and J. Huston, " Estimation of Background Noise in LAr Detectors Due to Pileup" *Physics of the Superconducting Supercollider*, Snowmass, 1986. R. Donaldson and J. Marx, editors, p.368.

[30] P. Langacker, R. Robinett, and J. Rosner, Phys. Rev. **D30**, 1470 (1984); D. London and J. Rosner, Phys. Rev. **D34**, 1530 (1986) ; V. Barger, N. Deshpande, J. Rosner and K. Whisnant, *Physics of the Superconducting Supercollider*, Snowmass, 1986. R. Donaldson and J. Marx, editors, p. 224. F. del Aguila, M. Quiros and F. Zwirner, Nucl. Phys. **B287**, 419 (1987).

HEAVY LEPTONS AT THE SSC *

G. Anderson and I. Hinchliffe
Lawrence Berkeley Laboratory
Berkeley, California 94720.

Abstract
We comment on heavy lepton searches at the SSC.

The cross-sections for the production of heavy leptons at the SSC are rather small. We shall discuss only a 4th generation lepton doublet (L, ν_L) and will assume that the neutrino ν_L is stable. There are three relevant production mechanisms; a L^+L^- pair can be produced from $q\overline{q}$ annihilation via an intermediate photon or Z boson; a $L\nu_L$ pair can be produced from $q\overline{q}$ annihilation via an intermediate W boson;[1] and a L^+L^- pair can be produced from the annihilation of a pair of gluons with an intermediate quark loop[2] (see figure 1). The rates from these processes are shown in figure 2. From this figure, it is clear the rates are rather small and that, if ν_L is very light compared to L, the rate from the process with a $L\nu_L$ final state is dominant over most of the relevant mass region. We shall assume that $\nu_L = 0$ in what follows.

The signals and backgrounds for the $L\nu_L$ final state were discussed at the La Thuile study[3] where it was concluded that the signal was fairly easy to extract. This conclusion has been challenged by Barger *et al.*[4] who carried out an exhaustive study (including one for the final state L^+L^-). They claim that all of the signals are obscured by backgrounds. In order to understand the controversy, we have studied the signals and background for the $L\nu_L$ final state.

The heavy lepton will decay to $W\nu_L$. The leptonic decay of the W will produce an isolated lepton (l) and missing transverse momentum. This signal is completely obscured by the process $q\overline{q} \to l\nu$.[4] We must therefore consider the hadronic decays of the W. The background now arises from processes which can make a W at large transverse momenta. There are two dominant possibilities; the final state $Z + W$ followed by the decay $Z \to \nu\overline{\nu}$; and $Z + jets$ where the jet system has an invariant mass near the W and the Z decays to $\nu\overline{\nu}$. The former is smaller than the signal if L is lighter than 200 GeV or so, hence we will concentrate on the latter.

The lowest order QCD processes which can produce $Z+$"fake" W occur at order α_s^2 and include $gluon + gluon \to Z + q + \overline{q}$. The relevant partonic matrix elements are known[5] and

*This work was supported by the Director, Office of Energy Research, Office of High Energy and Nuclear Physics, Division of High Energy Physics of the U.S. Department of Energy under contract DE-AC03-76SF00098.

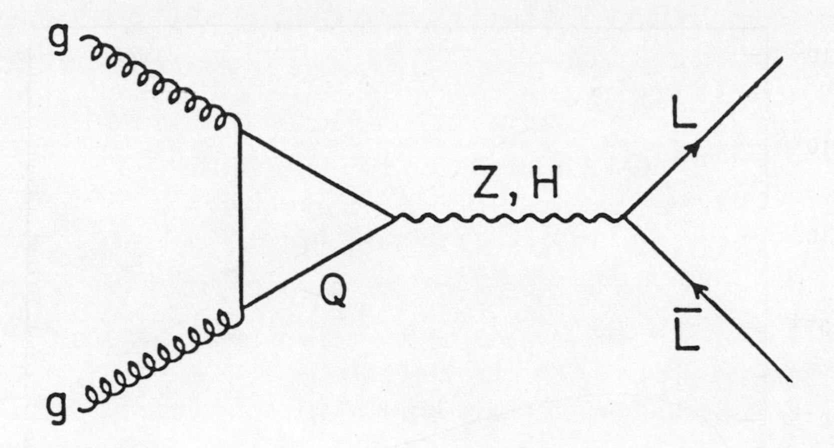

Figure 1: Diagram showing the parton model process $gg \to L^+L^-$.

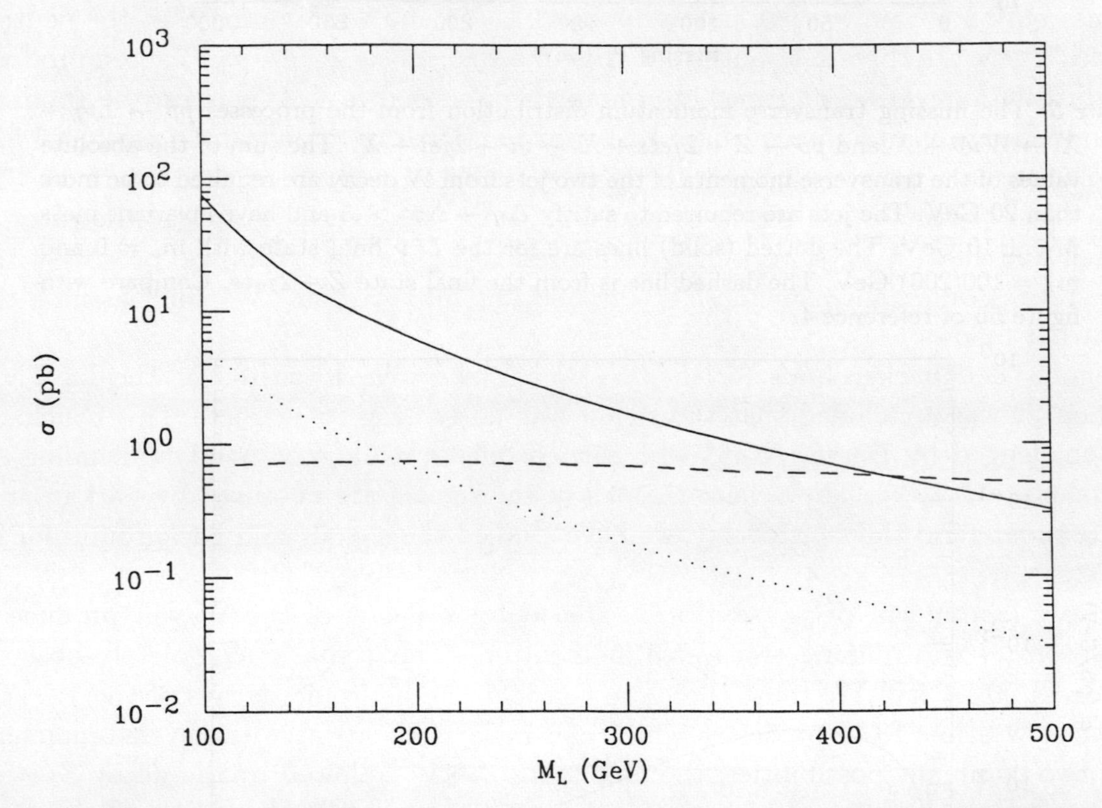

Figure 2: The cross section for heavy lepton production at the SSC as a function of the mass m_L of the charged lepton, for the three contributing partonic processes: $q\bar{q} \to L^+L^-$ (dotted line); $gg \to L^+L^-$ (dashed line); and $q\bar{q} \to L^{\pm}\nu$ (solid line);. The mass of the neutrino ν_L has been assumed to be zero. The gluon gluon rate[2] depends upon the Higgs mass, taken to be 100 GeV and upon the masses of the quarks in the fourth generation, taken to be m_L and $m_L + 250$ GeV for the charge 1/3 and 2/3 members of the doublet.[4]

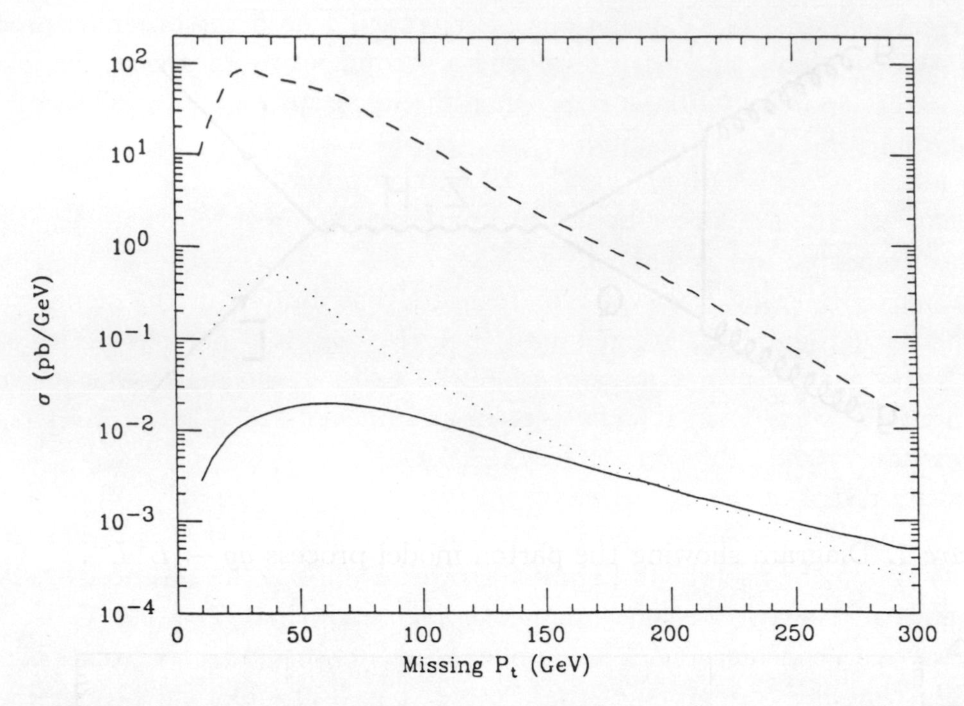

Figure 3: The missing transverse momentum distribution from the processes $pp \to L\nu_L + X \to W\nu\bar{\nu} + X$ and $pp \to Z + 2jets + X \to \nu\bar{\nu} + 2jet + X$. The sum of the absolute values of the transverse momenta of the two jets from W decay are required to be more than 20 GeV. The jets are required to satisfy $\Delta\eta^2 + \Delta\phi^2 > .5$ and have invariant mass $M_W \pm 10$ GeV. The dotted (solid) lines are for the $L^{\pm}\nu$ final state with $m_\nu = 0$ and $m_L = 100(200)$ GeV. The dashed line is from the final state $Z + 2jets$. Compare with figure 5b of reference 4.

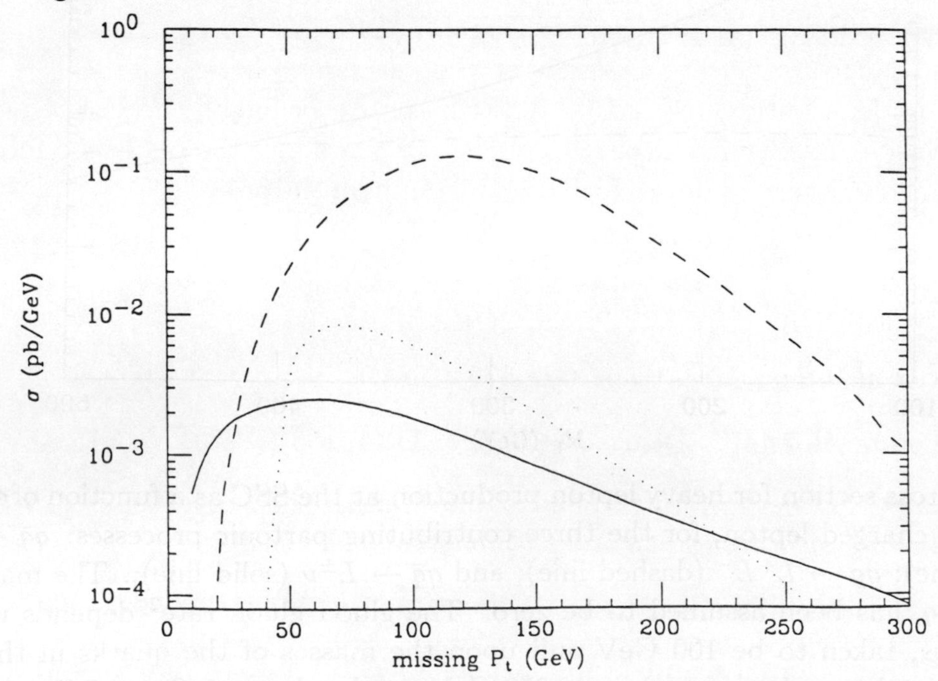

Figure 4: As figure 3 except each jet has transverse momentum of at least 50 GeV

are used in our calculation. The calculations of reference 4 used the order α_s processes of the type $q + \overline{q} \rightarrow Z + gluon$, and then produced a second jet by means of the leading log approximation†. This approximation is good when the invariant mass of the jet pair is much smaller than the partonic center of mass energy, so its accuracy in the region of interest where the two jets have an invariant mass of 80 GeV is not clear. The rates produced by our exact matrix elements are larger by approximately a factor of ten than those from the approximate form reported in reference 4.

Figure 3 shows the missing transverse momentum distribution in the signal and background. For the background we have required that the two final partons be separated by $\Delta\eta^2 + \Delta\phi^2 > .5$ and have invariant mass of $M_W \pm 10$ GeV. The signal to background ratio shown in this figure is worse than that of reference 4, since the $Z + jets$ rate is larger (see above) and reference 4 took a dijet mass of $M_W \pm 5$ GeV.

The momentum distributions of the jets from the real and "fake" W can be different so that techniques similar to those adopted in attempts to reject the $W + jets$ background to Higgs decay to WW may be useful.[8] In Figure 4 (5) we have made the additional requirement that the two jets each have transverse momentum greater than 50 (100) GeV. This cut reduces the background considerably. The number of events surviving these cuts is not great and it is unlikely that heavy leptons of mass more than 150 GeV will be observable. In addition the missing transverse momentum is not large so that detector hermeticity will be crucial.

In conclusion, detection of heavy leptons at the SSC seems to be very difficult, but perhaps not impossible. The feasibility depends critically upon the ability to identify events with W's decaying hadronically and missing transverse momentum.

Acknowledgment

This work was supported in part by the Director, Office of Energy Research, Office of High Energy and Nuclear Physics, Division of High Energy Physics of the U.S. Department of Energy under Contract DE-AC03-76SF00098. Accordingly, the U.S. Government retains a nonexclusive, royalty-free license to publish or reproduce the published form of this contribution, or allow others to do so, for U.S. Government purposes.

References

1. E. Eichten *et al.*, *Rev. Mod. Phys.* **56**, 179 (1984).

2. S.S.D. Willembrock and D. Dicus *Phys. Rev.* **D34**, 155 (1986).

3. D. Froidevaux, "Experimental Studies in the Standard Theory Group", in the *Proc. of the Workshop on Physics at Future Colliders*, CERN-87-07.

4. V. Barger, T. Han and J. Ohnemus, Madison preprint MAD/PH/331 (1987).

5. K. Ellis and R.J. Gonsalves, in "Supercollider Physics" Ed. D. Soper World Scientific Publishing, Singapore (1986).

---

†This is the method used by the Monte-Carlo event generators PYTHIA[6] and ISAJET[7]

6. H-U. Bengtsson and T. Sjostrand in *Proc. of the UCLA on Observable Standard Model Physics at the SSC: Monte Carlo Simulation and Detector Studies*,Ed., H.-U. Bengtsson *et al.* World Scientific Publishing, Singapore (1986).

7. F. Paige and S. Protopopescu in *Proc. of the UCLA on Observable Standard Model Physics at the SSC: Monte Carlo Simulation and Detector Studies*,Ed., H.-U. Bengtsson *et al.* World Scientific Publishing, Singapore (1986).

8. J.F. Gunion and M. Soldate, *Phys. Rev. D* 34:826 (1986).

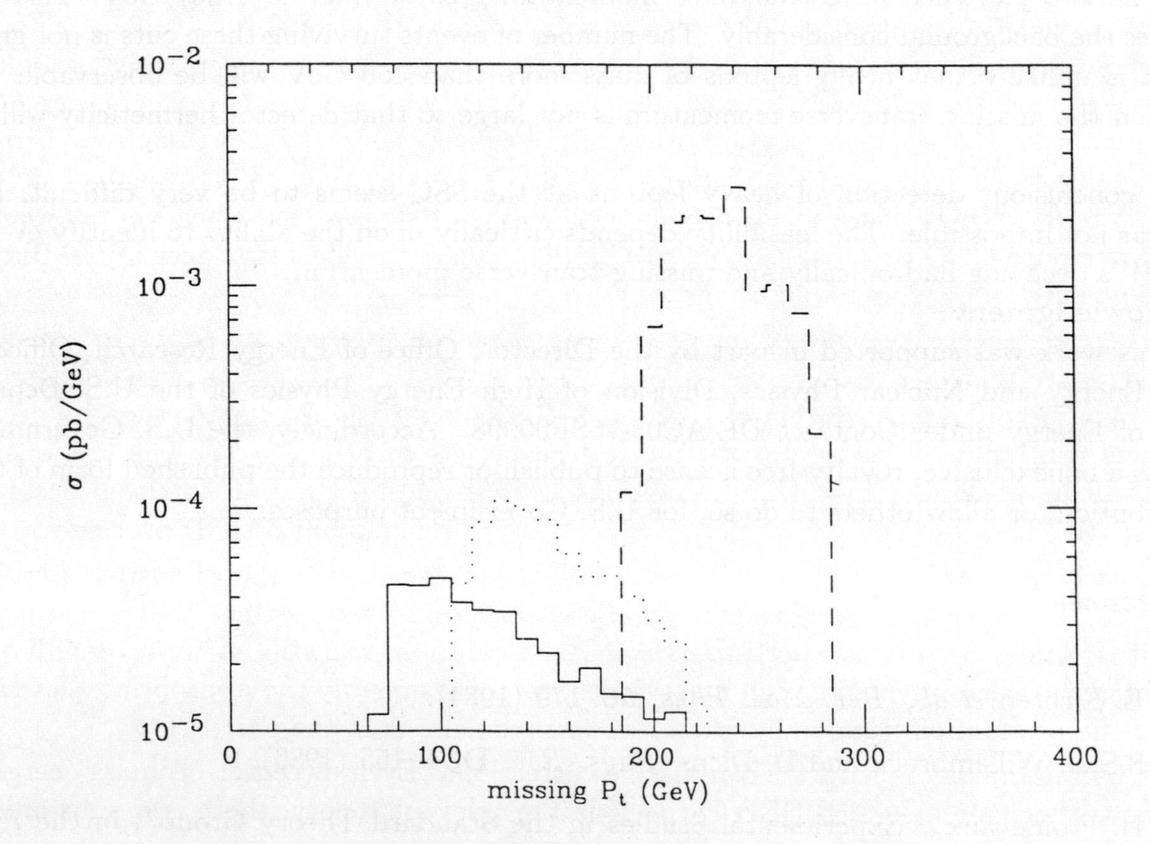

Figure 5: As figure 3 except each jet has transverse momentum of at least 100 GeV

RELIABILITY OF QCD MONTE-CARLO EVENT GENERATORS

John C. Collins

Department of Physics, Illinois Institute of Technology
Chicago, IL 60616, U.S.A.

ABSTRACT

I examine the extent to which Monte-Carlo simulations reproduce the predictions of perturbative QCD especially in the case of very high energy hadron-hadron scattering. Although the Monte-Carlos have great success in reproducing most of the qualitative features of the theory, they do not fully incorporate even the leading logarithmic approximation. Work is needed to give a systematic method for inclusion of both leading and non-leading effects.

1. INTRODUCTION

It is clear from the work at this and other SSC workshops, not to mention from current experimental technique, that calculations from QCD are necessary to analyze and understand data at all high energy hadron colliders. This is obviously true whenever one wants to analyze pure QCD processes like jet production, where the ideal is to test QCD. But it is also equally true for any kind of new physics.

There are several reasons for this: (a) The basic production process for a new particle is to generate it in a parton-parton collision; the cross section in a hadronic collision is given by a factorization theorem of QCD. (b) If a new particle is colored, then its decay products will include quarks and gluons, and hence jets of hadrons. (c) The rest of the event will be generated by QCD interactions; this is one background to the process of interest. (d) The competing background processes are generally also QCD processes.

Clearly one needs to do the QCD right, and this in fact means almost entirely perturbative QCD. Even if the new physics is completely different from anything that is discussed now, it is almost certain that perturbative QCD calculations will still be important: If one creates new objects involving high mass scales, then presumably the appropriate description of the production from and decay to known particles is at the parton level.

A useful analogy is with QED radiative corrections. Even at center-of- mass energies far above the scale at which QED by itself is the relevant theory, there is a substantial range of momentum transfers where it provides the dominant physics. In the past the discussion was a hot topic for theorists. Now the subject is so well-understood that the corrections are made by experimentalists as a matter of routine, and barely rate a mention in their published papers. In the case of QCD, the coupling is much bigger, the partons are confined (so that the radiative corrections have an infinite qualitative effect), and the theory is not so well understood.

If the signal to background ratio for some new physics is 1 : 1, then a factor of two error on the background calculation removes the signal. Since lowest order and leading logarithm calculations using perturbative QCD are often in error by this kind of ratio, there is a clear danger of losing signals. Of course, in practice one adjusts the background calculation to fit the data, so I have overstated the case. Even so the danger is clear.

These issues would not be worth emphasizing were it not that Monte-Carlo event generators are already so prevalent in the physics analysis. An important source of systematic error is often the systematic errors coming from our lack of knowledge of precise QCD predictions, particularly as expressed in the Monte Carlos. (Skeptics are reminded to look at the UA1 analysis of bottom production or of monojets etc.)

With this in mind, I will review the current status of the Monte Carlos and explain what needs to be done to improve them. We do not need all the improvements now, but we will need them when the experiments come online.

Of course, experimentalists should not take theorists at their word. We are often wrong. It will be important to test QCD at the SSC. Extrapolations over a factor of 20 in energy are fair game for demolition.

2. POWER LAW BEHAVIOR

The predictions that we are trying to make with the Monte Carlos are for a center-of-mass energy 20 times higher than at the Tevatron. This is a substantial extrapolation. Traditionally, perturbative QCD is supposed to give cross sections that vary logarithmically with energy, so that QCD physics at the SSC would not be very different from that at present energies. In fact many of the soft gluon effects that are responsible for detailed event characteristics are power law behaved. There is potential for substantial error in the predictions if they are not done well.

Recall the formula for the ratio of hadronic to muonic cross section in e^+e^- annihilation:

$$R_{e^+e^-} = 3\sum q_i^2[1 + \frac{\alpha_s}{\pi}(Q) + ...],$$

where $\alpha_s/\pi = \text{constant}/\ln(Q^2/\Lambda^2)$. At $Q = 5\text{GeV}$, α_s/π is .05, whereas at 5000GeV it is .02. Clearly the cross section changes little with energy, and extrapolations are appropriate. But many quantities involving soft gluons do not behave like this. Examples are multiplicities, cross sections at low x and the Drell-Yan transverse momentum distribution.

The small-x phenomena deserve mention. Traditionally, one assumes, on the grounds of Pomeron dominance, that parton distributions behave like $f(x) \propto 1/x$ at small x. This would imply for example that the cross section $d\sigma(W)/dy$ for making W's goes like s^0 as $s \to \infty$. However, in practical ranges of x, we find that $f(x) \propto 1/x^J$ is a better approximation, with J around 1.5. This gives $d\sigma(W)/dy \propto s^{J-1}$.

3. COMPARISON OF MONTE CARLO AND ANALYTIC CALCULATIONS

There are two schools of calculator in perturbative QCD: those who use the 'analytic' [1] methods and those who use the Monte-Carlo [2] methods. We may caricature these by saying that the analytic methods provide accurate calculations of (almost) nothing, while the Monte-Carlo calculations provide inaccurate calculations of everything. Particularly because they provide calculations of everything the Monte Carlos have become a favorite tool of experimentalists and many theorists for the comparison of QCD and the real world.

Let us remind ourselves of the main differences between the analytic and the Monte-Carlo methods.

1 The analytic calculations start from explicit factorization formulae for specific cross sections. Only certain kinds of cross section can be treated, mostly inclusive ones with large virtuality or transverse momentum. The Monte Carlos start from a probabilistic algorithm that generates complete events given a hard scattering as a trigger.

2 There is a more-or-less complete proof of the analytic factorization formula for many processes. But at best the Monte-Carlo algorithms are proved at the leading logarithm level. In certain aspects, the current versions of these algorithms do not even correctly include the leading logarithm approximation, as we will see later.

3 The perturbatively calculable parts of the analytic formulae, viz. the hard scattering cross sections and the evolution kernels, have systematic expansions in powers of the strong interaction coupling α_s. Thus analytic calculations can be improved in accuracy by using more terms in the expansions. However, there is no *systematic* method yet known for including higher order corrections in the Monte Carlos.

4 In many cases, large logarithms of ratios of kinematic variables mess up the convergence of perturbation expansions. Such a case is the Drell-Yan cross section at low transverse momentum. In the best cases, like Drell-Yan, the analytic formulae can be improved so that the large logarithms are effectively resummed; one really generates a new factorization formula in which the perturbative kernels no longer have large logarithms; the logarithms are generated by integrations over the kernels. By contrast, the Monte Carlos provide a leading logarithm resummation of such large logarithms. It is not always manifest that they do this correctly, and certainly one does *not* in general know how to make systematic improvements.

5 Non-perturbative phenomena appear in the analytic calculations in well-defined factors consisting of parton distribution and fragmentation functions. But the Monte Carlos rely on much modelling.

4. USES

It is clear, certainly, that the Monte Carlos are very useful, especially when one needs a simulation of the typically complicated events that are produced in high energy collisions. Qualitatively they reflect important features of QCD. They are needed in the design of detectors [3,4], to see whether proposed designs will properly perform its required function when presented with actual events. A second use is to provide input to a simulated detector so that efficiency corrections, etc, can be estimated. Provided that the corrections are insensitive to the precise details of the events (e.g., multi-particle correlations), and that the Monte Carlo used is tuned to fit data, this is a legitimate use. But if the data is a decade away in time and an order of magnitude away in energy, one must be careful if the proposed experiments depend significantly on getting the calculations correct.

But when a Monte Carlo is more intimately involved in the physics analysis, as so often happens now, the Monte Carlo must accurately reflect the real QCD theory.

5. WHY MONTE CARLOS WORK

First let us consider a model field theory (e.g., $(\phi^3)_6$, but not QCD) that has no gauge fields. Then it is relatively easy [5,6,7,8,9] to derive a factorization theorem for many interesting processes – Drell-Yan, jet production, etc. The structure of a typical factorization theorem is illustrated in Fig. 1. One parton out of each incoming hadron enters a hard scattering. Out of the hard scattering emerges some number of high p_T partons. These partons fragment into jets of hadrons. Let Q be a scale characteristic of the virtualities and transverse momenta of the hard scattering. Then the hard scattering can be calculated in a power series in the effective coupling $\alpha_s(Q)$. The distributions of partons in the incoming hadrons and of hadrons in the outgoing jets are non-perturbative. These distributions depend on the scale Q, and their Q-dependence is given by a renormalization group equation called the Altarelli-Parisi equation [10,11,12]. The kernel of this equation has a useful perturbative expansion in powers of $\alpha_s(Q)$. Corrections to the factorization formulae are higher twist, that is, suppressed by a power of Q.

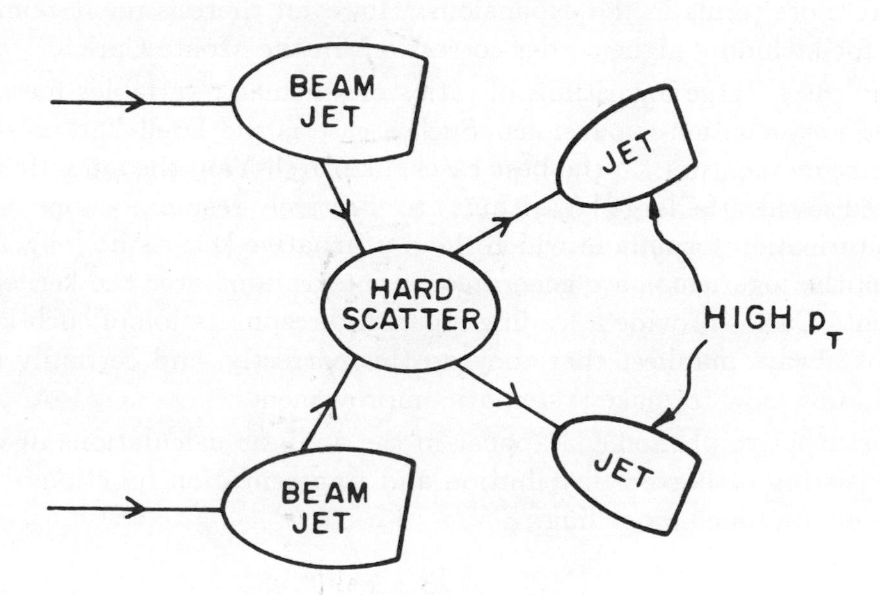

Fig. 1. Simple factorization theorem.

The standard factorization theorems are for inclusive processes: one sums over many unobserved particles. However, one can derive further factorization theorems that apply to the jets. The result is the jet calculus [13], which gives the picture summarized in Fig. 2. The partons coming into, or going out of, the hard scattering have lower virtualities than that of the hard scattering. One factors the associated jets into high and low virtuality pieces, and keeps repeating the process. Thus in Fig. 2, the lines have decreasing virtuality as one goes away from the hard scattering. The correct picture for the jets is of a cascade decay.

This picture leads immediately to a Monte-Carlo setup for the calculation [2], where one works with probabilities for decays. The jet calculus ensures that the kernels, the decay probabilities, etc can be expanded in powers of α_s at an appropriate scale.

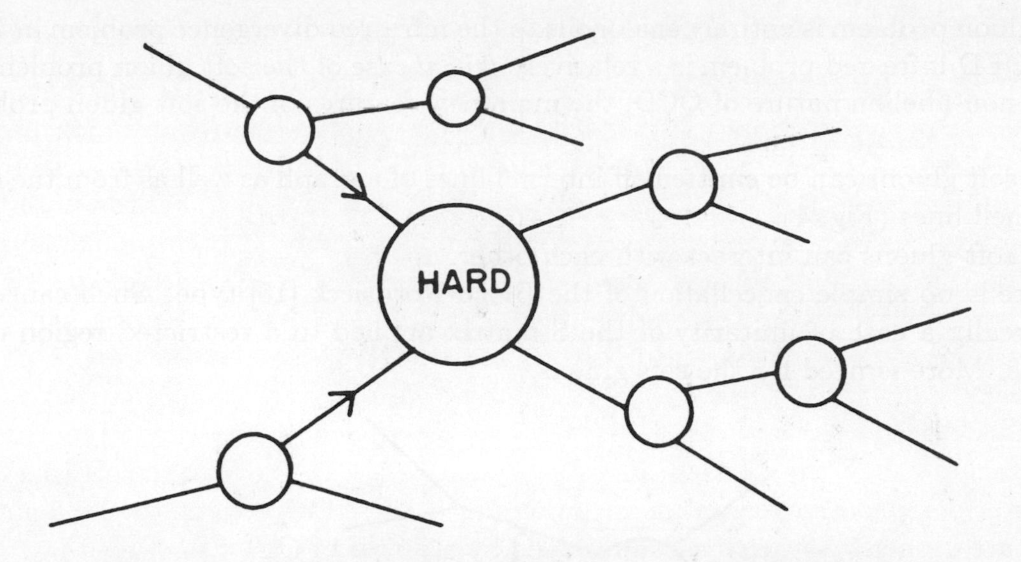

Fig. 2. Multiple factorization for jet calculus.

5.1 QCD and Soft Gluons

In QCD (or in any other field theory with elementary gauge fields), the basic hard scattering picture, Fig. 1 gets considerably modified [8,14]. At the leading twist level, soft gluons can couple any of the jet factors in a graph (Fig. 3). What is meant by a soft gluon is one all of whose momentum components are much less that the total center- of-mass energy. Thus a 10 GeV gluon is soft when one is considering production of $\frac{1}{2}$ TeV jets at the Tevatron ($\sqrt{s} = 2\,\text{TeV}$).

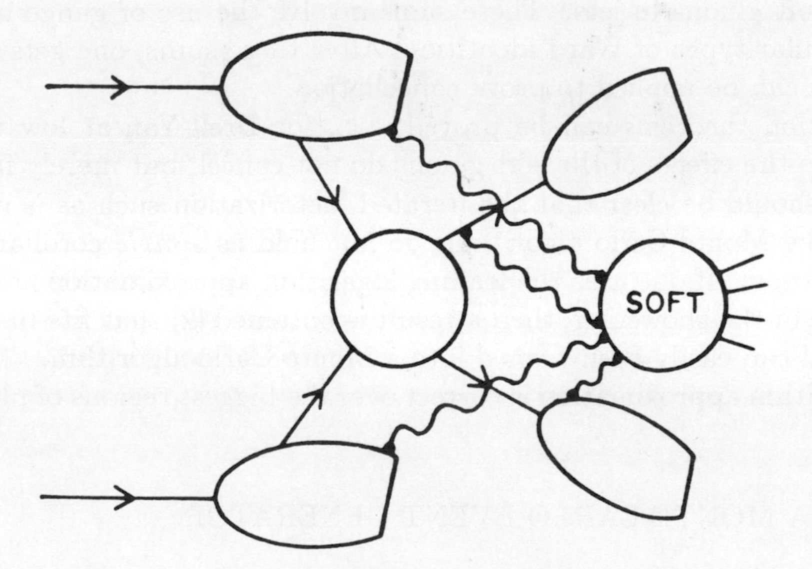

Fig. 3. Typical leading region for hard process in QCD.

A common misconception, inspired by some of the early papers on factorization, is that the soft gluon problem is entirely analogous to the infra-red divergence problem in QED. In fact the QED infra-red problem is a relatively *trivial* case of the soft gluon problem. Aside from the non-abelian nature of QCD, the main new features of the soft gluon problem are that

1 The soft gluons can be emitted off internal lines of a graph as well as from the external on-shell lines (Fig. 4).
2 The soft gluons can interact with each other.
3 There is no simple cancellation of the Bloch-Nordsieck [15] type. Such cancellations are really a case of unitarity of the S-matrix applied to a restricted region of phase space. More is need for the soft gluons.

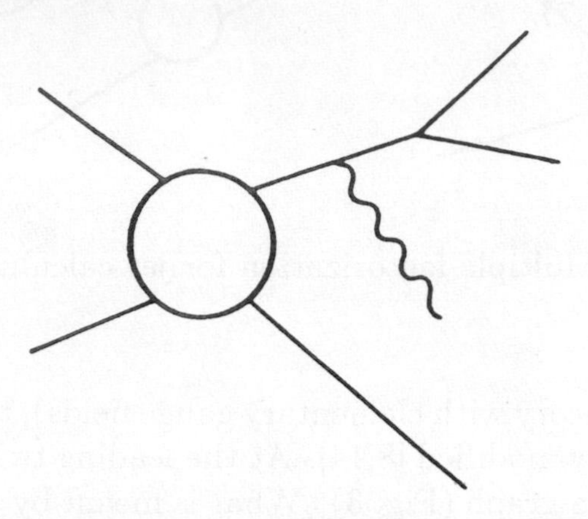

Fig. 4. Emission of soft gluons can be off internal lines.

The factorization theorems for hard processes can indeed be proved in QCD, but only after much labor. Most of the complete results are recent [16,17,18]. One needs to sum over attachments of soft gluons to jets. These sums involve the use of gauge invariance, as expressed in particular types of Ward identities. After these sums, one gets factors to which a unitarity sum can be applied to prove cancellation.

Fancier factorization theorems can be proved, e.g., for Drell-Yan at low transverse momentum [19], where the effects of the soft gluons do not cancel, but merely factorize.

At this stage, it should be clear that the iterated factorization such as is needed for the jet calculus and the Monte-Carlo algorithms do not hold as *simple* corollaries of the basic factorization theorem. If one uses the leading logarithm approximation and neglects azimuthal dependence in the showering, then a result is obtained [2] that fits in with a jet calculus structure, and can easily been turned into a Monte-Carlo algorithm.

The leading logarithm approximation is correct over the biggest regions of phase space for each graph.

6. STRUCTURE OF A MONTE-CARLO EVENT GENERATOR

The basic structure of a Monte-Carlo event generator can be summarized by Fig. 1, which represents a typical factorization formula for a *non-gauge* theory.

First one specifies a 'trigger' condition, that states what kind of hard scattering one is going to work with. This condition might be, for example, that one has a jet of greater than some specified transverse energy, E_{Tmin}. Or it might be that one has a Drell-Yan pair of invariant mass, Q, bigger than some value.

Then one generates the basic hard scattering with a probability distribution given by the hard scattering cross section convoluted with some parton distributions.

The partons associated with the beam jets and with the outgoing high p_T jets are generated by a probabilistic algorithm that implements the multiple-factorization that is the result of the jet calculus. This gives the showering associated with the initial- and final-state evolution.

The showering is stopped at some stage, and the partons in the final state are turned into hadrons. There are several methods used for the hadronization: independent fragmentation [20], string models [21], cluster algorithms [22].

Finally, some provision must be made for the fragmentation of the beam jets, ideally so that when no hard scattering occurs, a reasonable approximation to normal 'minimum bias' events is generated.

Monte Carlos do not always incorporate the state of the art in these areas. For example, independent fragmentation may be used, even though it is generally agreed that it is not a correct method in QCD [23]. Showering in the initial state has not been commonly included until recently [3,4]. Over a limited range of energy, it is often possible to tune up a Monte Carlo to agree with a set of data, even if the algorithm used is not correct.

7. WHERE THE MONTE CARLOS DON'T INCLUDE THE LLA

It can easily found out from, say, Webber's review [2] that there are two areas where none of the existing Monte Carlos do not even agree with the leading logarithm approximation. These both concern the azimuthal distributions of the partons produced in the showering. The first is concerned with the collinear fragmentation of partons, while the second concerns the physics of the soft gluons.

Even when the numerical effects of these correlations are small, they must be taken into account if one is ever to find a systematic method of including higher order corrections into the Monte-Carlo algorithms. If the azimuthal correlations are not treated, the higher order terms will have the infra red and mass divergences that come from incorrect approximations to the soft and collinear regions.

7.1 Collinear Decays

The problem of collinear decays can be explained by the example illustrated in Fig. 5 of a system of zero angular momentum that decays to two transversely polarized particles. Each of these then decays to two other partons. Let us work in the center-of-mass frame of the two gluons. Then the fact that the gluons have spin 1 restricts the azimuthal distribution of the final particles to be:

$$\frac{dN}{d\phi_a d\phi_b} \propto 1 + A\cos 2(\phi_a - \phi_b). \tag{1}$$

The coefficient A may have any value between $+1$ and -1. This happens even if the two decays are completely independent. This example embodies the physics of the 'paradox' of Einstein, Podolsky and Rosen. That A is in general non-zero is a purely quantum mechanical effect, and occurs even if the flight times are long so that the sequential decays can otherwise be thought of purely classically.

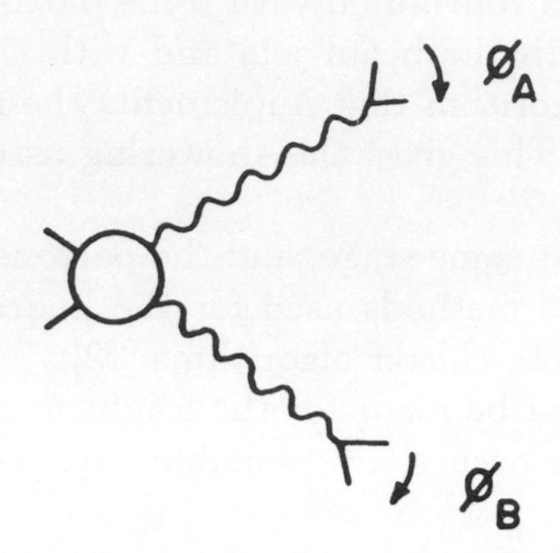

Fig. 5. Sequential decay that allows azimuthal correlations.

If one works in a field theory, then even in a kinematic region where the standard factorization theorem applies, A can be nonzero; indeed it can even take one of the limiting values ± 1. The effect is specific to spin-1 gluons: if the gluons were replaced by quarks, then the value of A would be of higher twist. Most current Monte Carlos replace A by zero. This approximation can be dangerously bad whenever one uses the correlations between several particles to get information on the spin of a parent particle, as in the process:

$$\text{Higgs} \rightarrow W^+ + W^- \rightarrow 4 \text{ jets}. \tag{2}$$

(To detect this decay of the Higgs, one obviously must subtract out the background due to purely QCD processes that create four jets, and one would typically wish to discriminate against the background on the basis of spin information.)

I have recently proposed a solution [24] to this problem, and Webber is currently working on incorporating it into the Webber- Marchesini program for e^+e^- annihilation.

7.2 Soft Gluon Emission

A second case where azimuthal correlations are neglected is in the emission of soft gluons. Suppose we have an initial parton P decaying into two collinear partons A and B, with an angle θ between them. Now consider a soft gluon emitted off the $A + B$ system (Fig. 6); we let the angles between the soft gluon and the partons A and B be θ_A and θ_B. Now the soft gluon cannot be uniquely associated with A or B, so that a priori one cannot use the simple showering algorithm that would be correct in a nongauge theory. However, [2] the probability P for the process may written as the sum of two terms, P_A and P_B, which have the following property after azimuthal averaging: The average of P_A (P_B) is the probability for collinear emission of the gluon off A (B) with θ_A (θ_B) restricted to be less than θ. This result [2] enables the soft gluon emission, with the neglect of azimuthal correlations, to be incorporated in a standard Monte Carlo very easily.

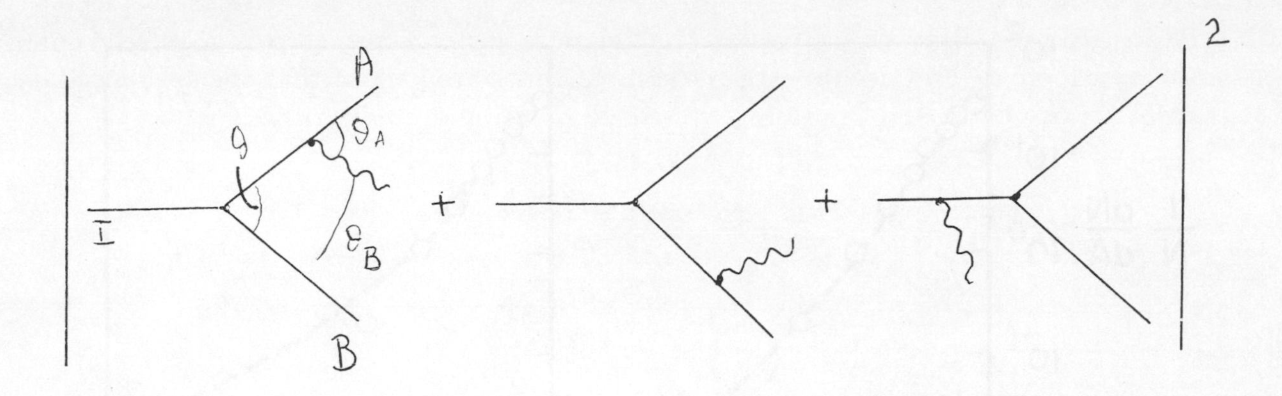

Fig. 6. Soft gluon emission

8. BEYOND THE LLA

8.1 Higher Order Corrections

Consider the integral:

$$\int_0^\infty dx \, \frac{Q}{(x+m)(x+Q)} = \int_m^Q \frac{dx}{x} + \text{finite (as } Q \to \infty). \tag{3}$$

This is a model of a momentum integral in a typical Feynman graph for a hard process, with x representing the virtuality of one of the lines. The first term on the right is the leading logarithm approximation. Since the finite remainder arises from the endpoints of the integration, one expects to be able to absorb the remainder into higher order corrections to the hard scattering and to the evolution kernels.

These corrections will evidently be particularly significant when one goes out on the tails of distributions and thereby samples the endpoints of the integrations. Consider for example Fig. 7 [25], which gives the aplanarity distributions in e^+e^- annihilation at 34 GeV. Both the data and two calculations are plotted. One calculation is the full α_s^2 result for this distribution, while the other is obtained from a Monte Carlo calculation, using only the LLA. The LLA is evidently substantially inaccurate.

Since the large aplanarity tail is where one might look for new physics (e.g., a top quark), one must include the higher order corrections. Similarly, if one wishes to use a Monte Carlo for measuring Λ from data, one must include non-leading terms.

The problem in incorporating higher order corrections is that one must avoid double counting. In the analytic methods, given a high order graph, there is a well-defined procedure for subtracting off the contribution that the graph makes in the LLA. Before obtaining the final result there are many cancellations of soft gluon effects.

There is no known *systematic* method of doing this for the Monte-Carlo algorithms. One problem is that after subtraction the remainder is not always positive. A negative probability cannot be directly incorporated into a Monte Carlo.

8.2 Soft Gluons

Since soft gluons cannot really be uniquely assigned to jets, except by convention, one

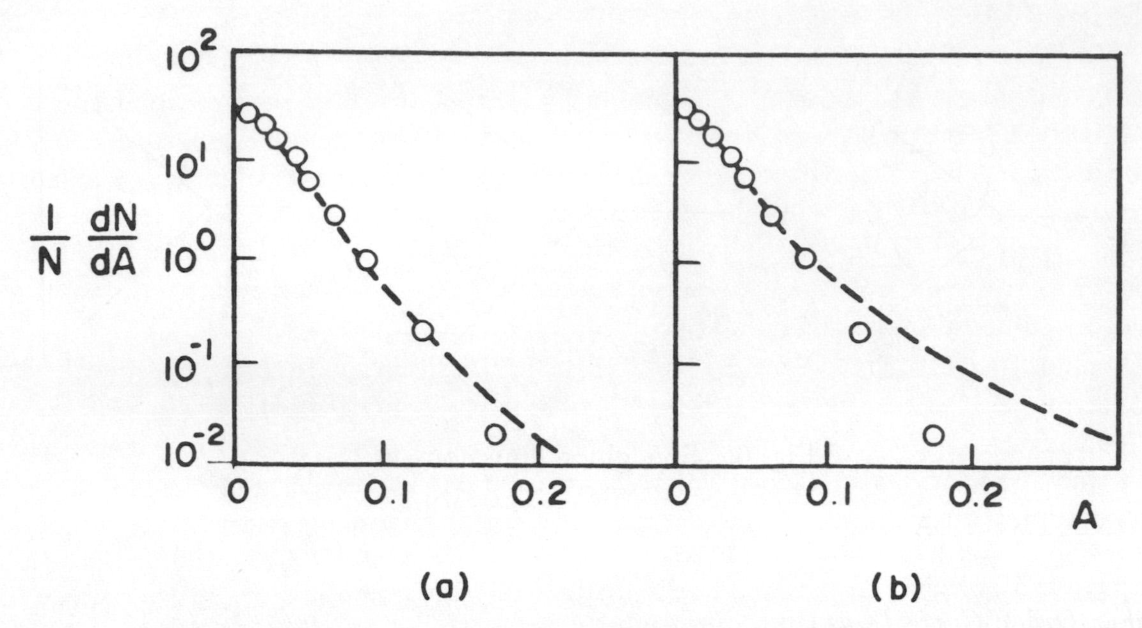

Fig. 7. Aplanarity distribution in e^+e^- annihilation: theory and experiment. The dots are the data at 34 GeV. The dashed line in graph (a) is the result of a full $O(\alpha_s^2)$ calculation; the dashed line in graph (b) is the result of a LLA shower calculation.

really needs a better treatment than the one usually given [2]. There are issues here that transcend the area of higher order corrections just discussed. In the analytic methods, even when they are applied to soft gluon emission, one works with an inclusive cross section that is not sensitive to the fate of the soft gluons, but at most to their effects on the hard partons (e.g., their transverse momentum distributions). One cannot afford this luxury in the Monte Carlos. One particular problem is that after emission, the soft gluons can interact among themselves.

This and related issues were raised by Brodsky [26].

Since there is a substantial amount of technology already available for treating soft gluons, as in the Drell-Yan process, one should be able to make progress here. It is sometimes said that, since there are many interference effects associated with soft gluons, it is impossible to treat them with a Monte Carlo. But cross sections are always positive, so it is not a priori impossible to develop a sophisticated Monte Carlo that correctly includes interference.

8.3 Hadronization of Jets

Inevitably one must choose a cutoff Q_0 below which one does not use perturbative methods, but has some parameterization of the nonperturbative physics. Ideally, the results of a calculation should be independent of this cutoff. It follows that models of the nonperturbative physics should be compatible with the perturbative physics. For example, in the analytic methods, the simplest factorization theorems force the nonperturbative physics to be in the form of parton distributions and fragmentation functions.

Certain models are thereby excluded, for example independent fragmentation [23]. The structure of perturbative QCD gives results that are much more like a string model.

8.4 Beam Jets

One must convert the remnants of the incoming hadrons into some part of the final state. This is mostly nonperturbative physics, but not entirely, as we will see. A substantial contribution to event structure is made by the beam jets. A simple way of making them is to generate a distribution of particles that is uniform in rapidity and is Gaussian in transverse momentum, with a fixed width. Such a parameterization (which is conveniently labelled as single Pomeron exchange) disagrees with experiment. To get the large fluctuations observed, one must allow a variable number of exchanged Pomerons.

One can get a perturbative model [27] for the multi-Pomeron exchange by permitting hard scattering into jets to be used down to low transverse momentum (around 1.5 GeV). There is a large enough probability of production of such minijets that one must allow for more than one hard scattering occurring simultaneously in the same hadron-hadron collision. The ladder graphs for hard scattering to minijets provide a perturbative model for the Pomeron that has many good properties [28]. As Sjöstrand [27] showed, the multiple hard scattering model is able to reproduce many features of the data – e.g., forward-backward correlations, KNO multiplicity distribution.

However, it appears that to fit the data, the probability of a double hard scattering must be rather higher than one's first guess; that is, the partons are confined in a rather smaller space in a hadron than the naïve radius. The current version of ISAJET [29] incorporates a phenomenologically similar fudge in its multiple Pomeron model to fit the activity associated with hard scattering.

There is in fact a reasonable rationalization of this effect. To get a hard scattering, the incoming hadrons must have a smaller impact parameter than normal. Thus the condition that there be a hard scattering biases the event towards more overlap of the hadrons and thus to a large number of exchanged Pomerons.

Clearly, further work should have the aims both of improving the perturbative treatments and of meshing these with the nonperturbative models for the Pomeron. These perturbative and nonperturbative treatments will be compatible in the sense defined earlier.

9. RECOMMENDATIONS

1 One should always use at least two different Monte Carlos in a calculations. This will provide a lower bound on systematic errors.

2 If the resulting uncertainties are excessive, then theoretical work to reduce them must be done. Also one must try to work with observables that are less sensitive to the uncertainties.

3 For precise QCD physics, one should aim at working with observables that are calculable by analytic methods.

4 Particular danger areas where the Monte Carlos are known to be inaccurate are: multi-particle correlations, scatterings involving wide angles between jets.

5 A more systematic derivation of the Monte Carlo algorithms is needed, as is the systematic treatment of non-leading logarithms.

6 It must be worked out how to deal with azimuthal correlations.

7 It would be useful to match the beam jet physics to perturbative Regge physics.

ACKNOWLEDGMENTS

This work was supported in part by the Department of Energy and by the John Simon Guggenheim Memorial Foundation. I would like to thank many colleagues for extensive discussions on the subject matter treated, especially T.D. Gottschalk, A.H. Mueller, F.E. Paige, D.E. Soper, and B.R. Webber.

LITERATURE CITED

1 E.g., Mueller, A.H., Phys. Rep. 73, 237 (1981).
2 Webber, B., in Annual Reviews of Nuclear and Particle Science, (1986), and references therein.
3 Proceedings of 1984 DPF Summer Study of the Design and Utilization of the Superconducting Super Collider, Donaldson, R., Morfin, J., (eds.),
4 Proceedings of 1986 DPF Snowmass Summer Study, Donaldson, R., (ed.).
5 Efremov, A.V. and Radyushkin, A.V., Teor. Mat. Fiz. 44, 17 (1980) [Eng. transl.: Theor. Math. Phys. 44, 573 (1981)].
6 Efremov, A.V. and Radyushkin, A.V., Teor. Mat. Fiz. 44, 157 (1980) [Eng. transl: Theor. Math. Phys. 44, 664 (1981)].
7 Amati, D., Petronzio, R., and Veneziano, G., Nucl. Phys. B140, 54 (1978) and B146, 29 (1978).
8 Libby, S. and Sterman, G., Phys. Rev. D18, 3252, 4737 (1978).
9 Ellis, R.K., Georgi, H., Machacek, M., Politzer, H.D. and Ross, G.G., Nucl. Phys. B152, 285 (1979).
10 Gribov, V.N. and Lipatov, L.N., Sov. J. Nucl. Phys. 46, 438 (1972).
11 Altarelli, G. and Parisi, G., Nucl. Phys. B126, 298 (1977).
12 Johnson, P.W. and Tung, W.-T., Phys. Rev. D16, 1769 (1977).
13 Konishi, K., Ukawa, A. and Veneziano, G., Nucl. Phys. B157, 45 (1979).
14 Efremov, A.V. and Radyushkin, A.V., Teor. Mat. Fiz. 44, 327 (1980). [Eng. transl.: Theor. Math. Phys. 44, 774 (1981)].
15 Bloch, F. and Nordsieck, A., Phys. Rev. 52, 54 (1937); Yennie, D., Frautschi and S.C., Suura, H., Ann. of Phys. 13, 379 (1961); Kinoshita, T., J. Math. Phys. 3, 650 (1962); Lee, T.D. and Nauenberg, M., Phys. Rev. 133, 1549 (1964).
16 Collins, J.C. and Sterman, G., Nucl. Phys. B185, 172 (1981).
17 Bodwin, G., Phys. Rev. D31, 2616 (1985).
18 Collins, J.C., Soper D.E. and Sterman, G., Nucl. Phys. B250, 199 (1985).
19 Collins, J.C., Soper, D.E. and Sterman, G., Nucl. Phys. B261, 104 (1985).
20 Field, R.D. and Feynman, R.P., Nucl. Phys. B136, 1 (1978).
21 Andersson, B., Gustafson, G., Ingelman G. and Sjöstrand, T., Phys. Rep. 97, 33 (1983).
22 Webber, B.R., Nucl. Phys. B238, 492 (1984).

23 Cox, B., Proceedings of the Oregon Workshop on Super High Energy Physics, 120 (1985); Sjöstrand, T., Z. Phys. C26, 93 (1984); Phys. Lett. 142B, 420 (1984); Gottschalk, T.D., Proceedings of the Oregon Workshop on Super High Energy Physics 94 (1985); Corcoran, M.D., Phys. Rev. D32, 592 (1985).
24 Collins, J.C., "Spin Correlations in Monte-Carlo Event Generators", IIT preprint 87/11.
25 Gottschalk, T.D., in Ref. 4.
26 Brodsky, S.J., in Ref. 3.
27 Sjöstrand, T., Fermilab preprint FERMILAB-PUB-85/119-T (1985).
28 Gribov, L.V.,Levin, E.M. and Ryskin, M.G., Phys. Rep. 100, 1 (1983).
29 Paige F.E. and Protopopescu, S.D., Proceedings of the UCLA Workshop on SSC Physics (1986).

High p_T Detector Configurations

Summary and Comparison of High p_T Detector Concepts

R. J. Cashmore
Department of Nuclear Physics
Oxford, UK

S. Ozaki
KEK, Tsukuba, Japan

G. Trilling
Lawrence Berkeley Laboratory
Berkeley, California 94720

I. Introduction

In this summary we have attempted to draw together the major requirements for high p_T detectors, an understanding of the present states of calorimetry and particle measurement and identification as they pertain to the SSC, some of the regions where R&D is required and finally an evaluation of some of the concepts presented at the workshop. Clearly, these concepts are only the beginnings and many will undergo substantial revision and detailed evaluation before becoming proposals for detectors at the SSC. The substantial R&D which will be performed over the next few years may have a dramatic effect on these proposals and ensure that sound detectors will be designed for experimentation in the area of high p_T phenomena at the SSC.

We first summarise the requirements for the detectors, particularly points that have emerged from the physics parameterization subgroups [1] [2] [3] [4] [5] [6] [7]. We then follow this with a discussion of the implications for detector design and compare the strengths of the various detectors which have been discussed. Finally we draw some general conclusions.

2. Detector Requirements for High p_T Physics

In considering what is required in a detector for the SSC we must remember that this machine provides a totally new energy regime, which may contain physics of which we are at present totally unaware. This implies that one should not be bound by projections from our present understanding and that designs of detectors should be based on sound general ideas. The high p_T detectors at the SSC should be capable of recognizing surprises and then exploring them.

It is, however, valuable to look at the requirements that are implied by our present understanding of physics and this has been pursued in the various physics parameterization subgroups. These studies suggest that luminosities in excess of 10^{33} may be desirable, the mass ranges and sensitivity to "new" phenomena being obviously extended. Such luminosities ($\leq 10^{34}$) present a distinct challenge to the experimentalists! The parametrization groups have studied the following areas:

(i) Higgs production and decay. This includes conventional Higgs particles of both intermediate ($M_H < 2M_W$) and heavy ($M_H > 2M_W$) mass as well as Higgs particles beyond the minimal standard model (e.g., charged and other neutral Higgs). The decay modes of heavy Higgs which have been considered and which lead to some of the detector requirements are

$$H \to ZZ, WW$$

with

$$Z \to l^+l^-, \nu\overline{\nu}, q\overline{q}$$

$$W \to l\nu, q\overline{q}$$

while for intermediate Higgs emphasis has been given to

$$H \to \gamma\gamma$$

(ii) New Quarks. In this case the features of heavy charge $-1/3$ quarks (D) were investigated in the decay mode

$$D \to tW$$

with subsequent decays of the t and W leading to leptons.

(iii) Supersymmetry - SUSY. Two alternative modes are considered which lead to different decay chains with different signatures.

$$\tilde{q} \to q\,\tilde{\gamma},\ \tilde{g} \to q\overline{q}\,\tilde{\gamma}$$

or

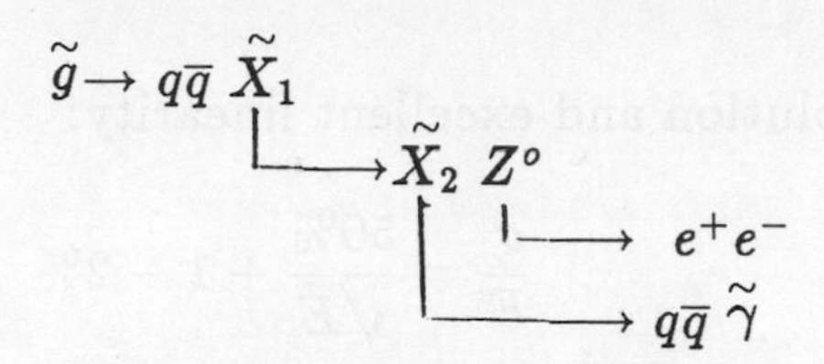

The first leads to large missing energy while in the second scheme missing energy is not as large but there exist multiple leptons in the final state.

(iv) New W', Z'. These new bosons are assumed to decay in a manner similar to the known W and Z.

(v) Drell-Yan, Jets and Compositeness. In this case the signature for compositeness lies in abnormally large cross-sections at high jet (or lepton) transverse momenta.

In summarizing the requirements for a high p_T detector in the areas of calorimetry, tracking, and lepton identification we give examples drawn from the various parametrization studies to illustrate the origin of the requirement. The individual group reports contain much more detailed discussions of the various processes and the corresponding implications.

2.1 Calorimetry

(i) Hadronic energy resolution and excellent linearity:

$$\frac{\sigma}{E} = \frac{50\%}{\sqrt{E}} + 1 - 2\%$$

This linearity (as we will see later) and the small constant term in the resolution imply $e/h = 1$. As an example in Fig. 2.1 we see the expected jet cross-section[7] divided by the pure QCD cross-section for the case of a contact interaction with $\Lambda^* = 20\ TeV$ and for the case of a calorimeter with $e/h = 1.4$ (and with no contact interaction). Unravelling the presence of a contact interaction will clearly be difficult if such a poor calorimeter is chosen. The energy spectrum of particles in the 4 TeV jets also indicate that calorimeters capable of containing $\sim 1\ TeV$ particles are essential. Figure 2.2 shows that these 4 TeV jets frequently ($\geq 10\%$ of the time) contain $\sim 0.5\ TeV$ particles. From previous SSC studies[8] this implies that calorimeters $10 - 12\ \lambda$ long are required.

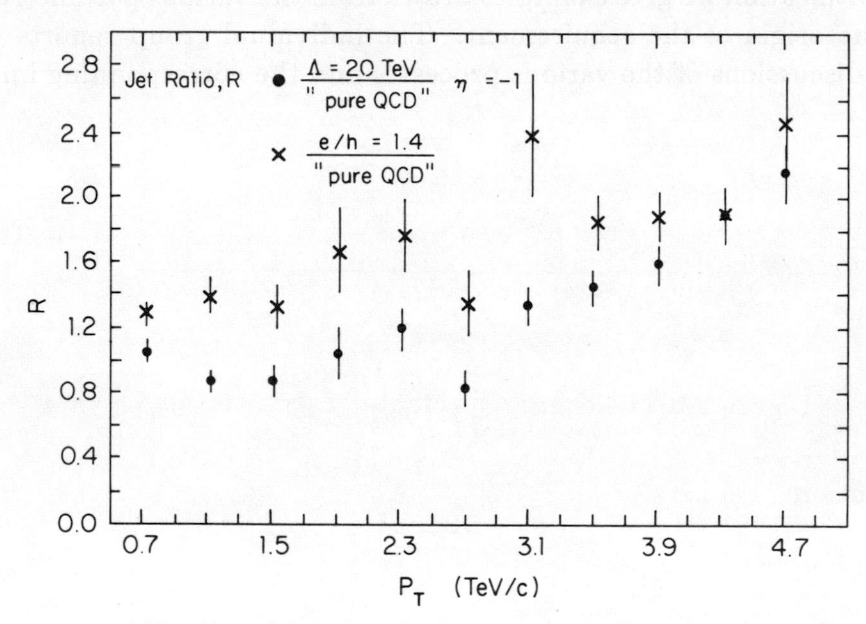

Figure 2.1 The ratio of jet cross-sections compared to QCD predictions for the presence of a contact interaction ($\Lambda^* = 20\ TeV$) or poor calorimeter performance ($e/h = 1.4$)

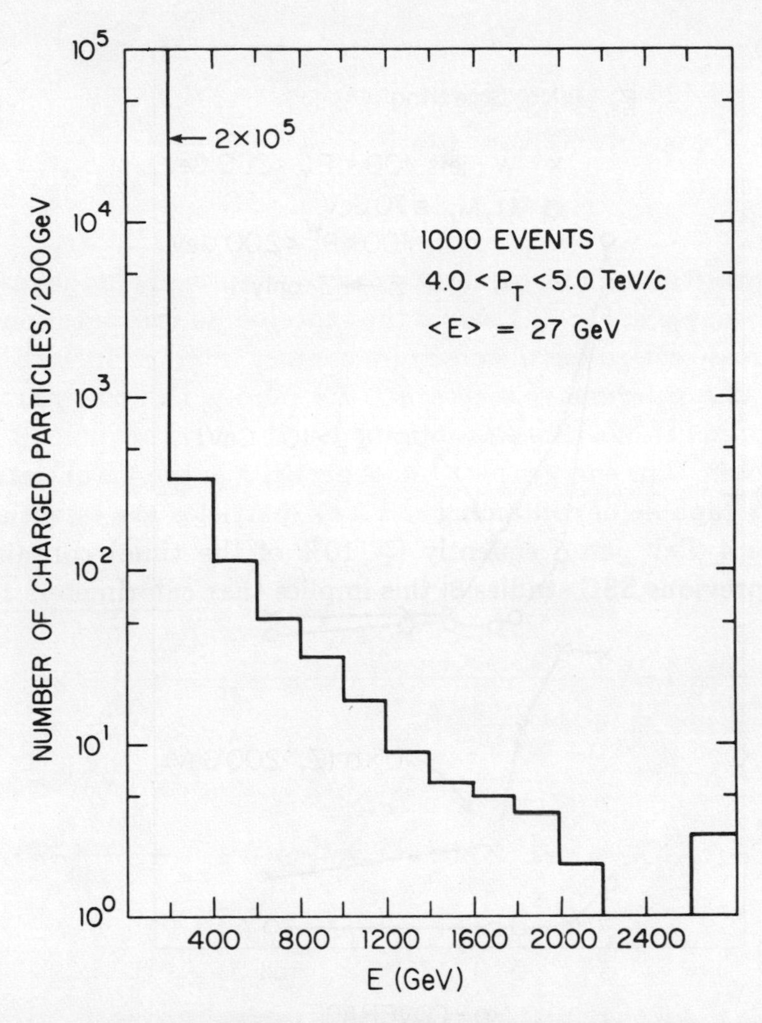

Figure 2.2 The energy spectrum of particles in 4 TeV jets

(ii) EM resolution:

$$\frac{\sigma}{E} \sim \frac{10-15\%}{\sqrt{E}} + 1\%$$

This type of requirement is derived from many of the subgroups.

(iii) Hermeticity:

Reliable measurement of missing $E_T \geq 100\,GeV$.

To do this requires a linear scale in measuring transverse energies which implies $e/h = 1$ and coverage to rapidities $|\eta| \leq 5 - 5.5$. In Fig. 2.3 we see the cross-sections from various background processes relevant to SUSY identification. Coverage beyond about $|\eta| \sim 5$ brings no further advantage in reducing the background cross-section. However coverage to $|\eta|$ values of only ~ 3 would bring substantial increases in background. These conclusions are, however, based on perfect resolution for the particles and may need to be reviewed when realistic energy and angular resolutions are included.

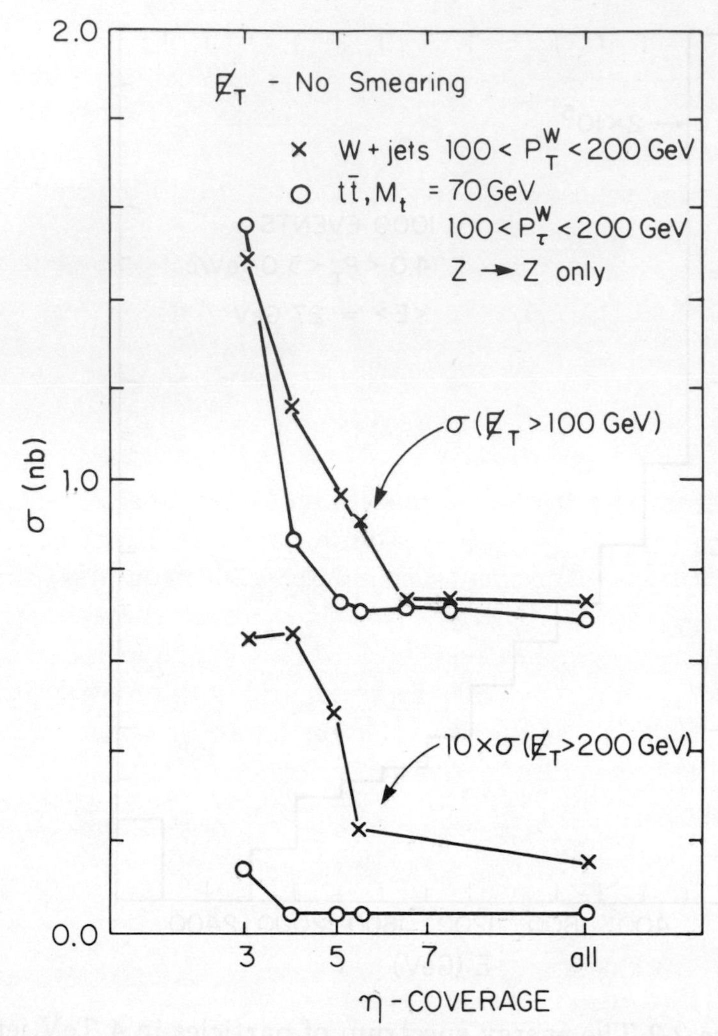

Figure 2.3 The missing E_T cross-sections from 2 background processes to SUSY production, as a function of calorimeter coverage. No resolutions are included.

(iv) Transverse Granularity:

$$EM \quad \Delta\phi \times \Delta y = (.03 - 0.05) \times (.03 - .05)$$

$$HAD \quad \Delta\phi \times \Delta y = (.06 - 0.10) \times (.06 - .10)$$

These are partly set by lepton isolation requirements [1] [2] [3] [4] [5] i.e., it is desirable to have empty towers (EM and Had) around any given isolated electron shower or muon signal in the calorimeter.

(v) Longitudinal Segmentation:

$$EM \sim 3 \text{ sections}$$

$$HAD \sim 3 \text{ sections}$$

These are set mainly by the need to achieve $\sim 10^{-3}$ hadron rejection in identifying electrons. [1] [2] [3] [4] [5] With this type of lateral and longitudinal segmentation such rejection can be obtained from the calorimeters.

2.2 Tracking

There were few specific requirements for tracking. In particular, there appears to be little specific need for high precision vertex detectors, with the possible exception of τ tagging for intermediate Higgs.[2] There was little study of the importance of being able to identify two or more separate z vertices in an event. We suspect, however, that it will be important to have this capability of multiple vertex identification, particularly when looking for the rare events at the SSC which are the signal of new physics. The major requirement was for

Momentum Resolution:

$$\frac{\sigma(p_\perp)}{p_\perp} \sim 0.5 p_\perp (TeV)$$

This comes mainly from requiring electron charge determination at the highest momenta (in Drell Yan processes [7], Z' decay asymmetries [6], Higgs and W^+W^+ sectors [1]). Of course a momentum measurement also enhances electron identification, since the equality of momentum (p) and energy deposition (E) can be checked. Finally, the presence of a magnetic field, even comparatively weak, is beneficial in reducing the confusion surrounding any track which might potentially be identified as an electron. In Fig. 2.4 we see the improvement in isolation of electrons [5] which can be achieved if charged tracks with $p_T < 5\ GeV/c$ can be removed from consideration.

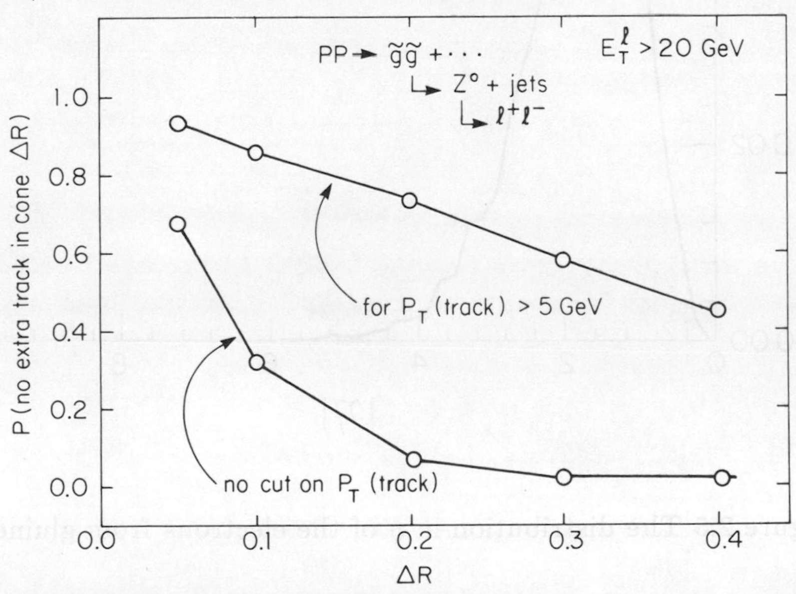

Figure 2.4 Isolation of tracks showing the effects of a p_T cut of $5 GeV/c$

2.3 Electron Identification

(i) Rapidity Coverage.

$$|\eta| \leq 2-3$$

To illustrate this requirement we show the destribution of the highest rapidity electron in $\tilde{g}$ decays [5] in Fig. 2.5, and the acceptance as a function of rapidity for different Higgs particles [1],in Fig. 2.6, where the Higgs has decayed to ZZ with subsequent decay to leptons. Similar coverage is required for heavy quarks [4] and new W', Z' [6].

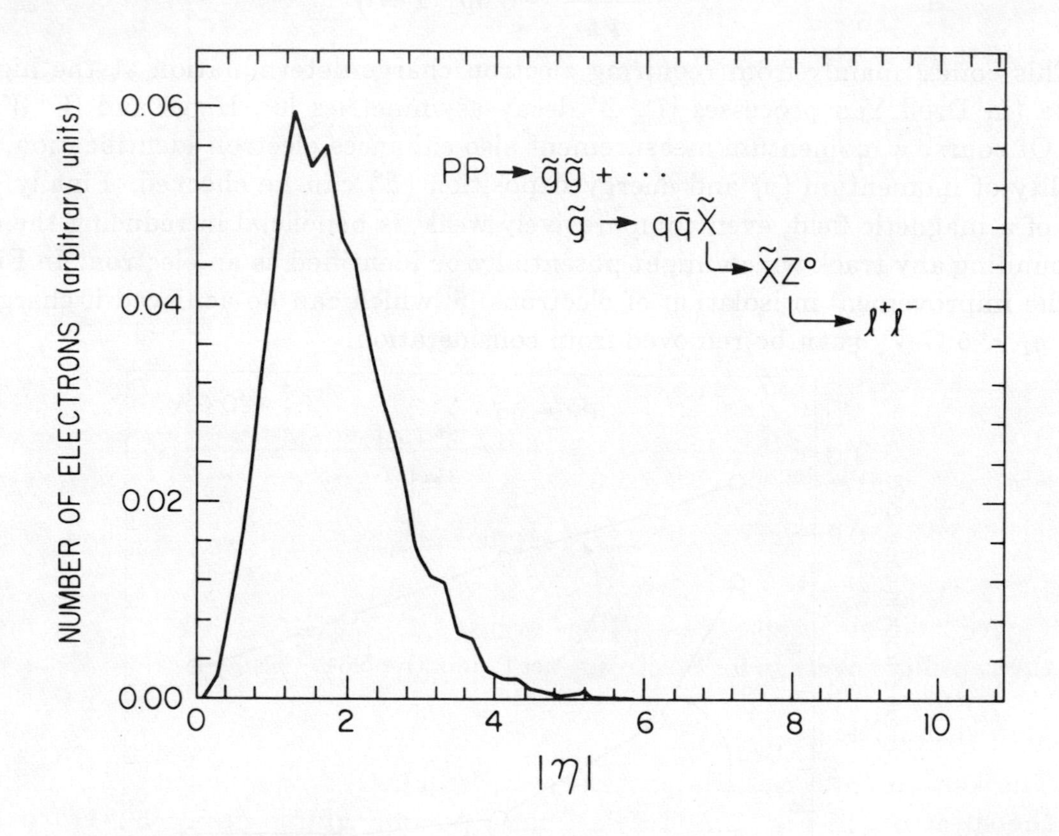

Figure 2.5 The distribution in η of the electrons from gluino decay

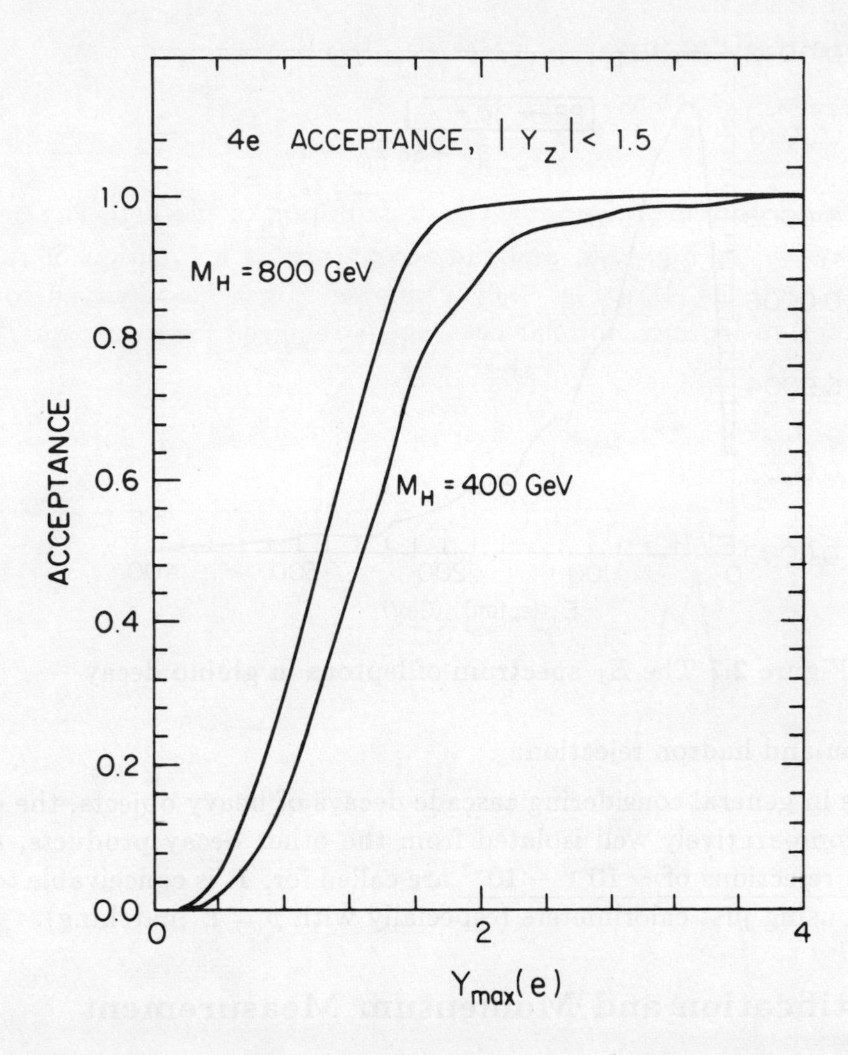

Figure 2.6 The acceptance of Higgs events ($H \to ZZ,\ \ Z \to e^+e^-$) as a function of the rapidity coverage for electrons

(ii) Momentum Range.

The leptons from cascade processes are frequently low momentum as indicated by the spectrum in Fig. 2.7, drawn from a possible gluino decay [5]. Of course, high E_T electrons are expected from W', Z' decay but these will be comparatively easy to identify.

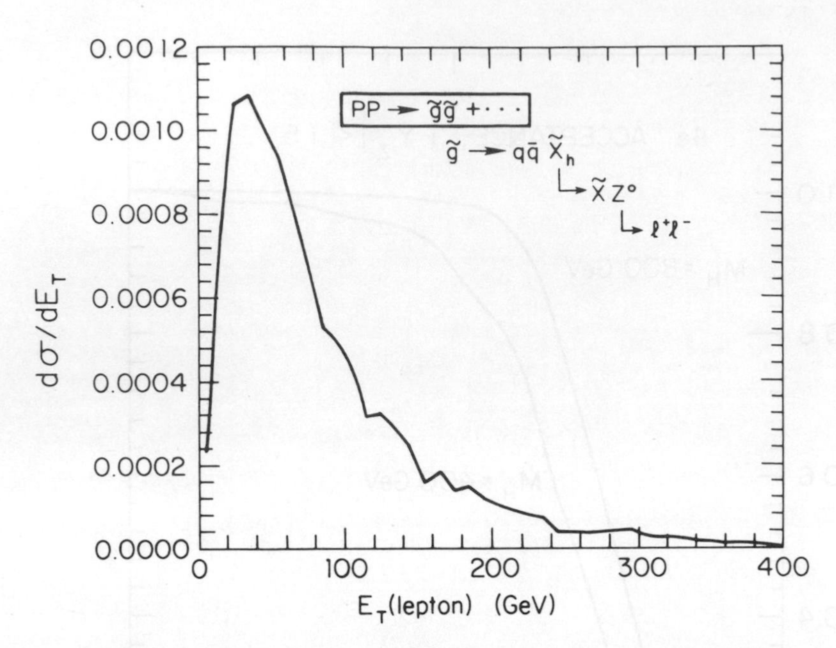

Figure 2.7 The E_T spectrum of leptons in gluino decay

(iii) Lepton isolation and hadron rejection:

Because we are in general considering cascade decays of heavy objects, the electrons (leptons) are comparatively well isolated from the other decay products, and only modest hadron rejections of $\sim 10^{-2} - 10^{-3}$ are called for. It is conceivable to achieve such rejections using just calorimeters (especially with $p = E$ matching).

2.4 Muon Identification and Momentum Measurement

Clearly the requirements for muon identification are identical to those for electrons. However, experience at colliders to date has indicated that it may be possible to go somewhat further with muons than electrons. For example, the muon η coverage can be extended to $3 <| \eta |< 5$ and muons even searched for within jets. These are desirable because it may then be possible to get a better understanding of the conventional physics backgrounds.

The muon momenta can be measured with a tracking + magnet system [9] [10] ($\frac{\sigma(p_\perp)}{p_\perp} < 0.5p_\perp$), with iron toroids ($\frac{\sigma(p_\perp)}{p_\perp} \sim .10$) [9] [10] [13] [14] or with an air core solenoid [15] ($\frac{\sigma(p_\perp)}{p_\perp} \sim .04p_\perp$). The benefits of a better momentum resolution are clearly seen in Figs. 2.8(a) and 2.8(b) where the reconstructed Z mass (from 1 TeV Higgs decay) is shown superimposed on the background estimated from $t\bar{t}$ pairs which have decayed to give leptons ($M_t = 200\ GeV$). This improved momentum measurement makes the muons far more comparable with the electrons, where the energy is derived from calorimetric techniques ($\frac{\sigma(E)}{E} \sim \frac{.15}{\sqrt{E}}$). Finally, the measurement of muon momentum in the range $| \eta |< 5$ may improve missing E_T measurements.

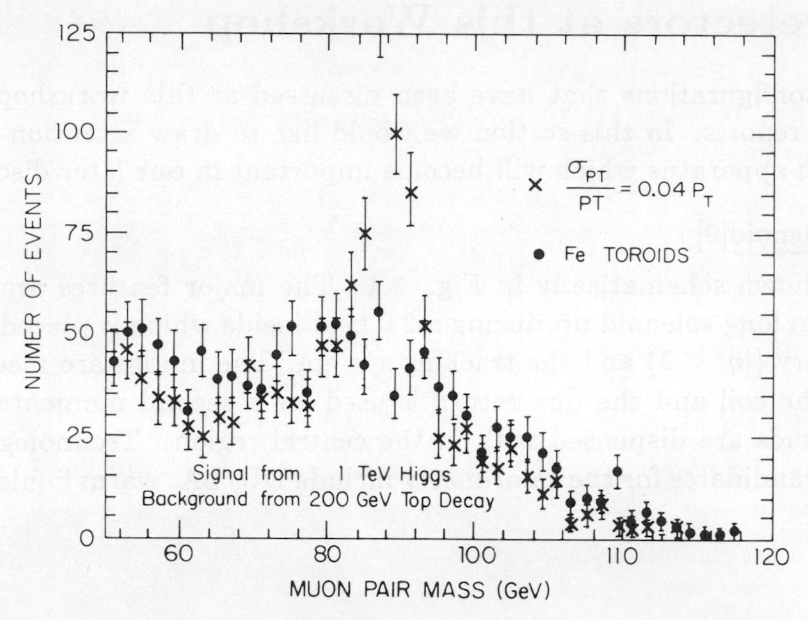

Figure 2.8 The $\mu^+\mu^-$ mass spectrum for different momentum resolutions. The Z originates from Higgs decay and the background is due to $t\bar{t}$ pair production $(M_t = 200GeV)$

2.5 General Remarks

Before leaving these requirements derived from the physics parametrization subgroups it is worth making a few general remarks and comments. Firstly, it appears to us that many of the requirements (e.g., hadron rejection for lepton identification) are not stringent enough to push experimental techniques to their state of the science possibilities. In other words, it is "fairly" easy to achieve the desired properties. This may, of course, be a dangerous way to begin doing physics in a totally new domain. For example, new phenomena revealed only by lepton identification in dense particle environments might be entirely missed. Thus, when finally designing experiments for the SSC, it may be sensible to go beyond these requirements. Secondly, much of the study has concentrated on the signals and not so much on understanding the backgrounds. In future studies these backgrounds should be looked at again, and particularly what experiments may be necessary to better define and measure these backgrounds. In doing this different criteria may emerge which are more stringent than the ones we have seen so far. In this respect it is worth reminding ourselves that many of the new phenomena to be hoped for at the SSC (Q, $\tilde{q}$, $\tilde{g}$ ...) will not appear as mass peaks in some distribution (due to missing ν, $\tilde{\gamma}$ ...) but rather as "excessive" rates for particular topologies and categories of event (e.g., multi-lepton, multi-jet final states). A good knowledge of the backgrounds will therefore be essential.

Finally, it is clear that at the SSC excellent calorimetry (for parton energy and direction measurement) and lepton identification will be vital and essential ingredients in any high p_T detector.

3. The Detectors at this Workshop

The detector configurations that have been discussed at this workshop are reviewed in detail in other reports. In this section we would like to draw attention to the particular features of each apparatus which will become important in our later discussions.

(i) Large Solenoid[9]

This is shown schematically in Fig. 3.1. The major features are the large radius (4m), 16m long solenoid producing a 2T field inside which is placed all of the central calorimetry ($|\eta| < 3$) and the tracking system. The muons are measured inside and outside the coil and the flux return is used for a second momentum measurement. Thus toroids are dispensed with in the central region. Technologies considered as possible candidates for the calorimetry include DU/LA, warm liquids, Pb/fibers, and DU/Si.

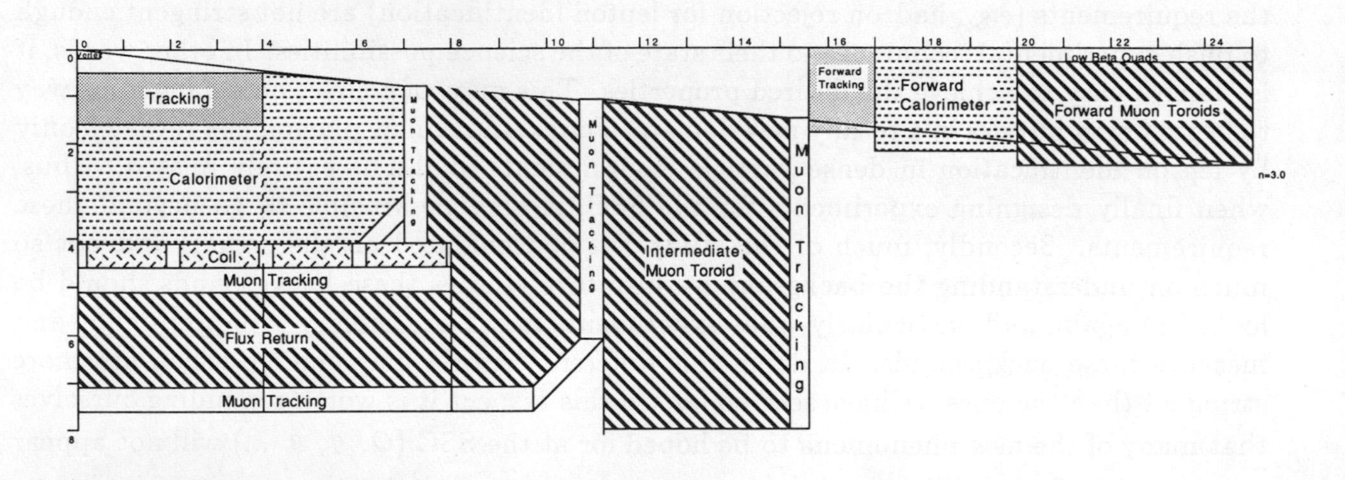

Figure 3.1 The Large Solenoid Detector

(ii) The Compact Solenoid[10]

This detector comes in two varieties. SSB: This consists of a 4T ~ 2m radius solenoid which contains a silicon strip tracking device, and a DU/Si calorimeter of 5-6 λ depth. The remainder of the calorimetry is outside the coil and muon measurement is primarily done with the toroid system (Fig. 3.2).

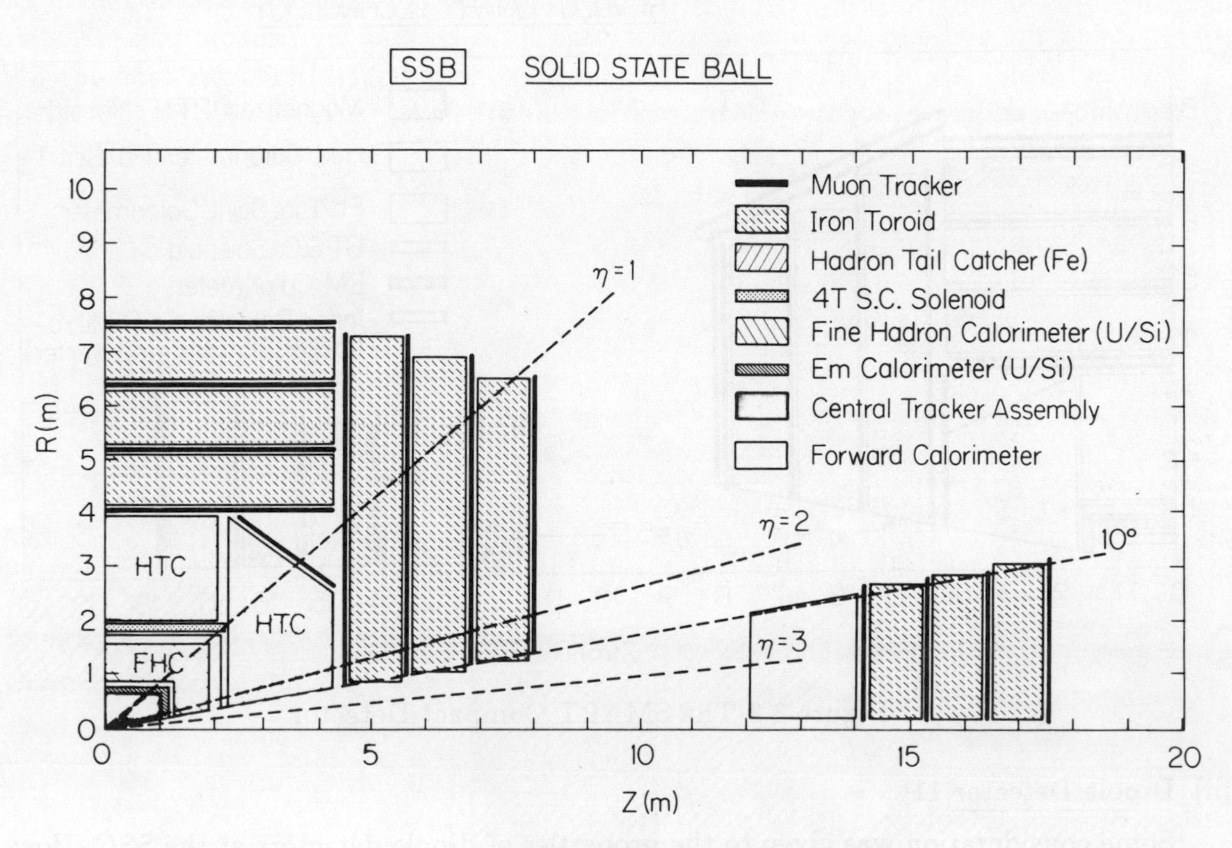

Figure 3.2 The SSB Compact Detector

SMART: This consists of a 6T, $\sim 1m$ radius solenoid containing a tracking device and electromagnetic calorimeter (+ preconverter + synchrotron radiation detector for electron identification) all based on scintillating fiber techniques. The hadronic calorimeter lying outside the coil is again based on lead and scintillating fibers.

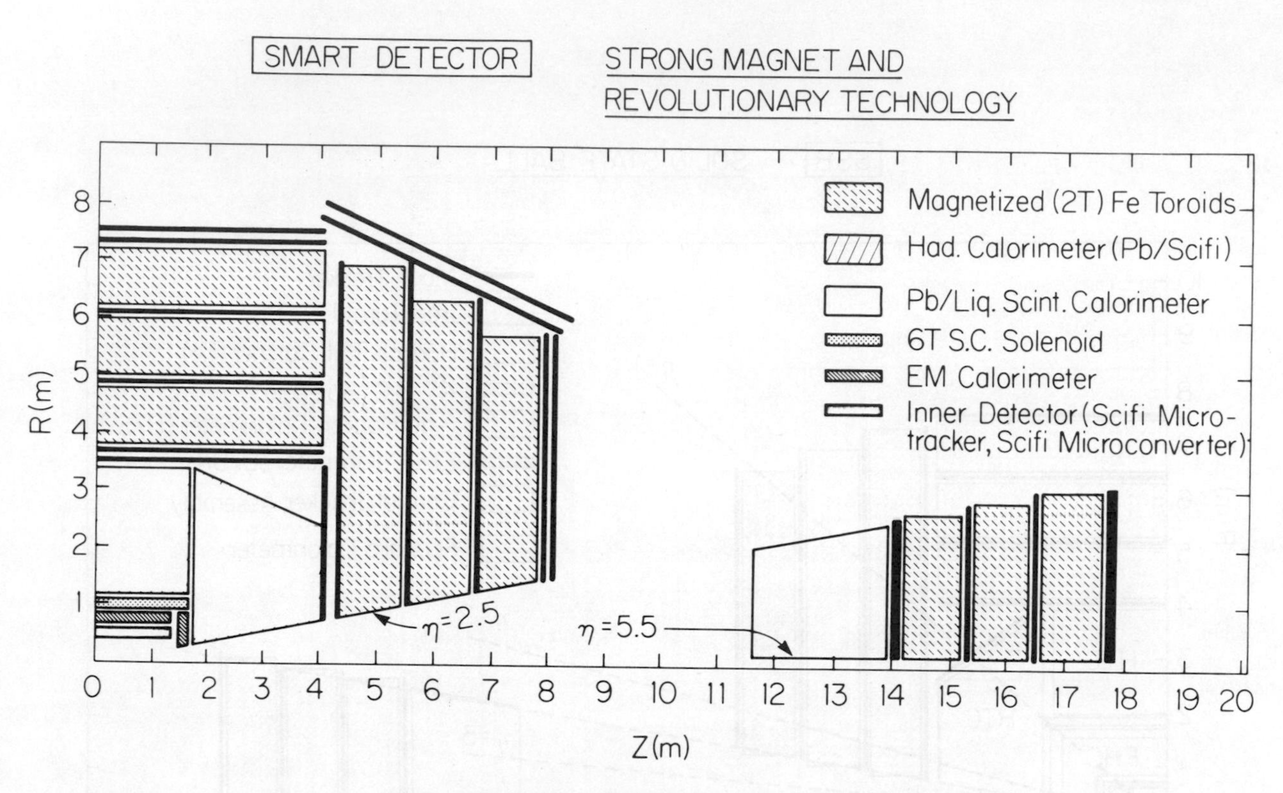

Figure 3.3 The SMART Compact Detector

(iii) Dipole Detector[11]

Some consideration was given to the properties of dipole detectors at the SSC. However, the study of high mass, high p_T processes, which are basically produced centrally, are more appropriately studied with a solenoid geometry. If a dipole design is pursued then it will be somewhat similar to UA1 [12] in conception.

(iv) Non Magnetic Detector[13]

This is shown schematically in Fig. 3.4. It is based around a high precision calorimeter, with tracking and TRD's inside to enhance electron identification. Muon measurement is totally performed in a toroid system. There exists the possibility of introducing a coil outside the precision calorimetry to produce a modest magnetic field. As in the large solenoid detector, several calorimeter technologies are considered possible candidates.

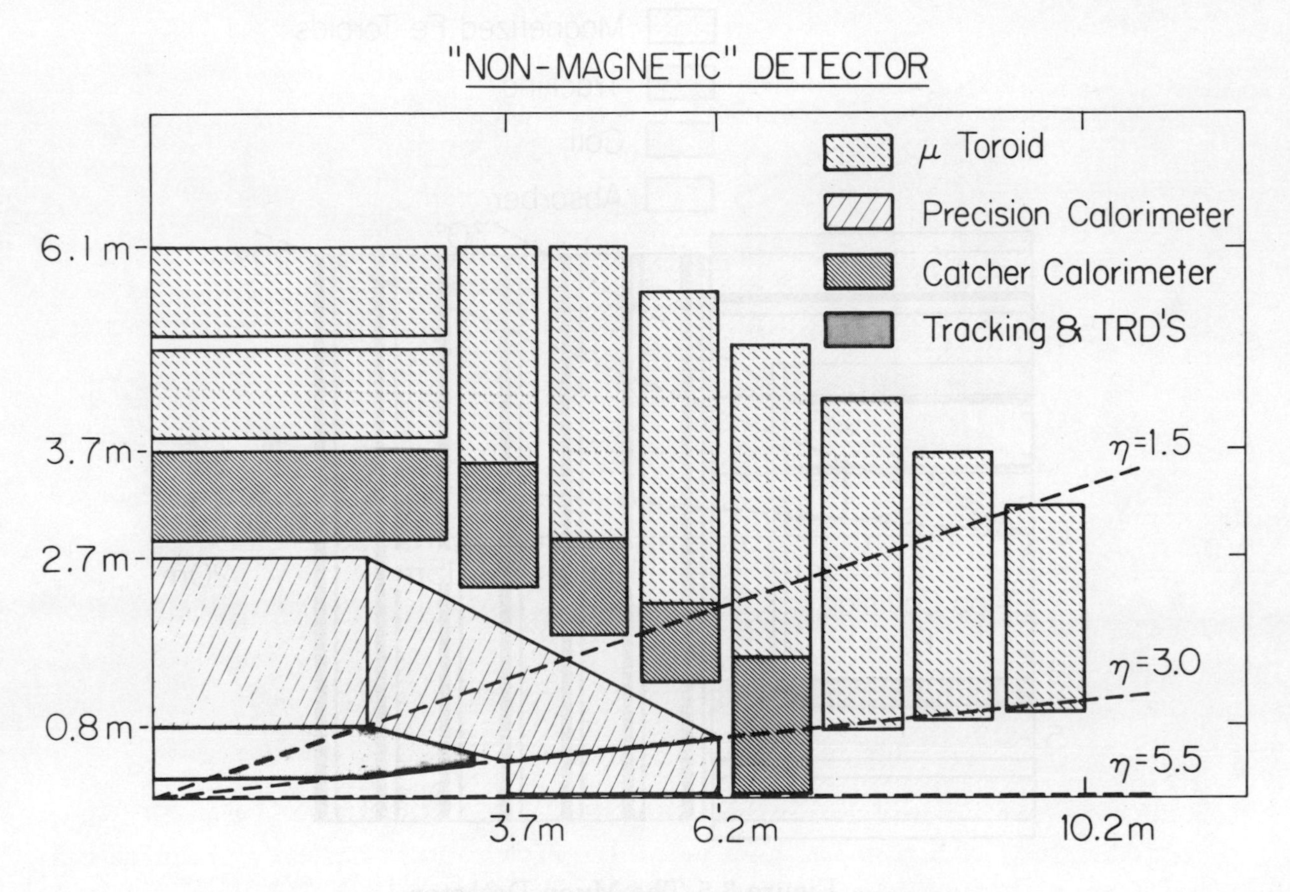

Figure 3.4 The Non-Magnetic Detector

(v) Muon Detector[14]

The design of solely a muon detector for the SSC was considered and is shown in Fig. 3.5. The principal features are a thick absorber close to the interaction point, followed by μ tracking in a solenoid and toroids. This detector is essentially blind to everything except muons, but may be able to handle very high luminosities.

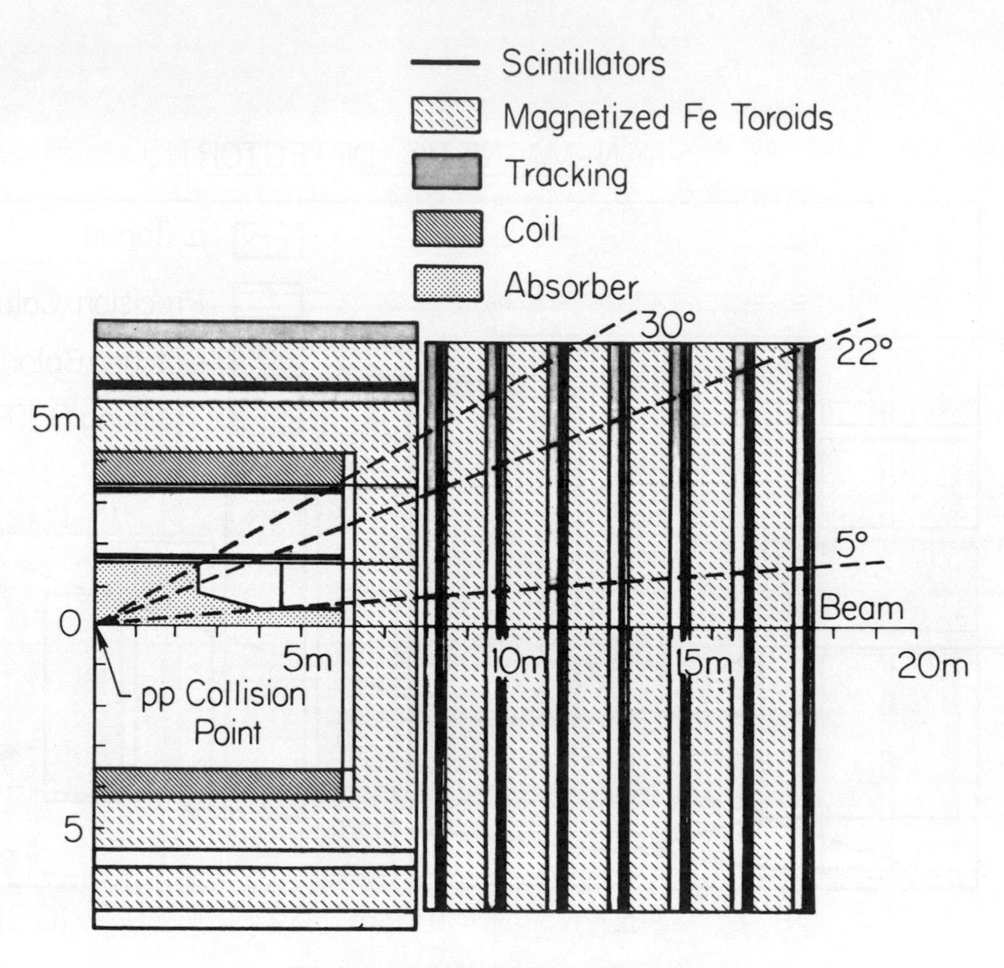

Figure 3.5 The Muon Detector

(vi) High Precision Muon Detector[15]

The layout of this detector is shown in Fig. 3.6a, with the calorimeter (a liquid Xe ball together with DU/TMS hadronic calorimetry) shown in Fig. 3.6b. This detector emphasizes high precision muon momentum measurement in the central region, $|\eta| < 1 - 1.5$, together with high precision electron energy measurement. The large radius solenoid produces a field of 7.5 Kg. Bearing a resemblance to the L3 detector, this has been referred to as L3 + 1.

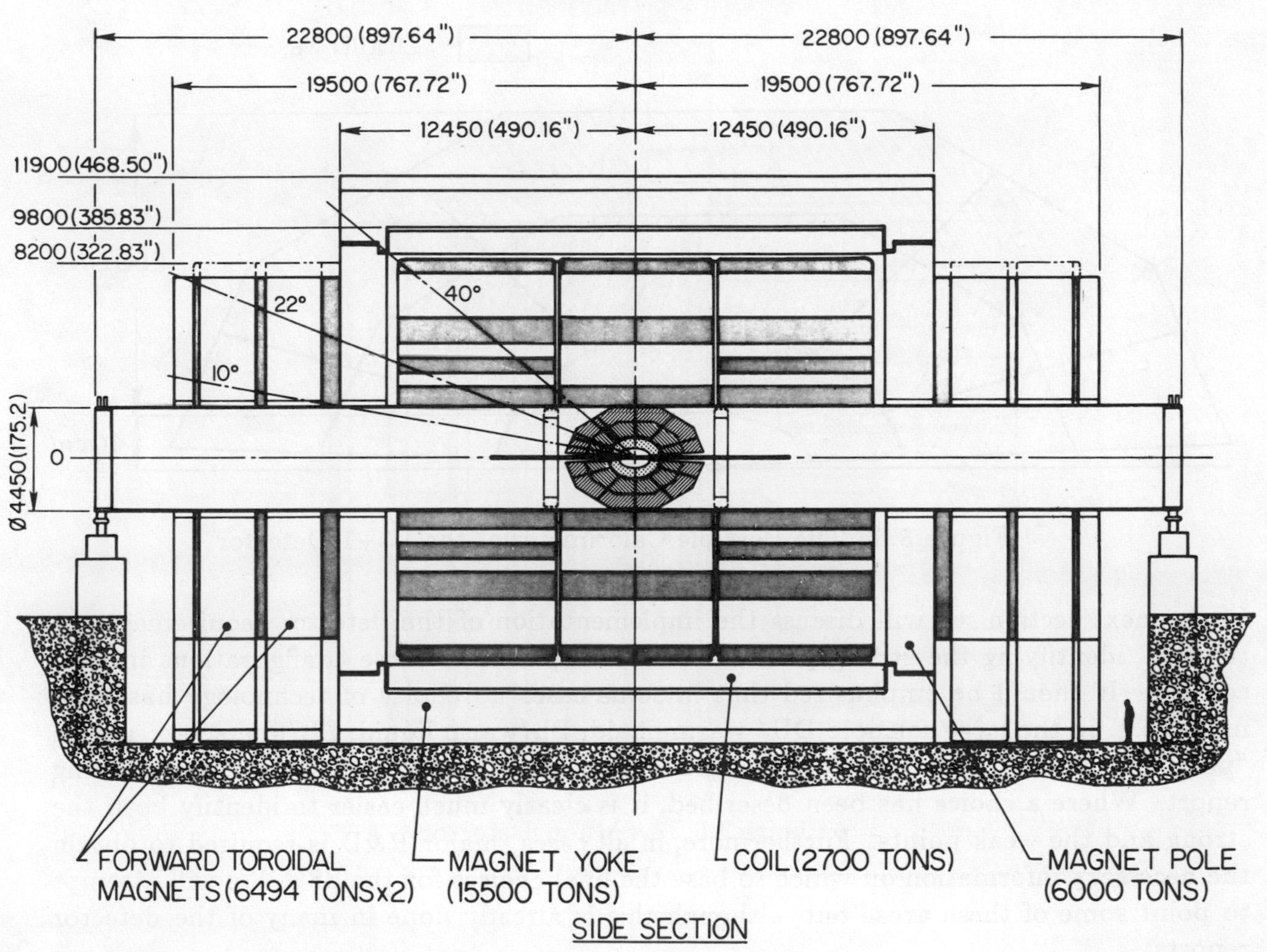

Figure 3.6a The L3+1 Detector

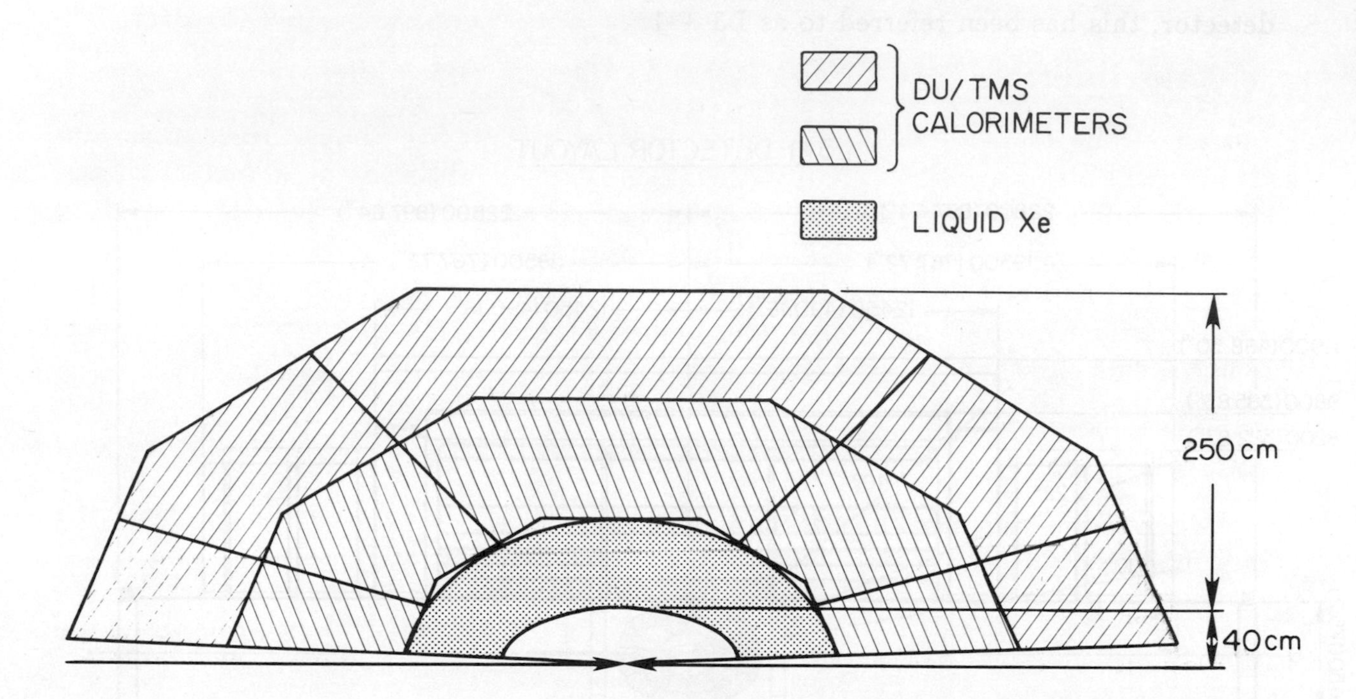

Figure 3.6b The Possible Calorimeter of the L3+1 Detector

In the next section we will discuss the implementation of the detector requirements at the SSC identifying the strong points and weaker points of these configurations in that context. It should be emphasized that in some cases no choice of technology has been made (e.g. in the large solenoid DU/ warm liquid, Pb/warm liquid, Pb/fibers, etc. are all "possible") and the factors affecting a final choice are often detailed in the corresponding report. Where a choice has been described, it is clearly much easier to identify both the strong and the weak points. Furthermore, in all cases, major R&D is required to obtain the necessary information on which to base the final choices for the SSC. We will attempt to point some of these areas out, although this is already done in many of the detector reports.

4. Implementation of Detector Requirements at the SSC

In this section we will make some general observations about detector performance at the SSC and comment, where appropriate, on the implementation in the preliminary designs at this workshop.

4.1 Calorimetry

From all of the studies made, it is clear that calorimetry, both electromagnetic and hadronic, will play an essential role at the SSC. There should be at least one experiment with outstanding calorimetry. Fortunately, because of tremendous advances in experiments and understanding[16][17] over the last two years or so, we now know the ingredients for building good hadron calorimeters. In fact there is now no reason to build an inadaquate hadron calorimeter at the SSC. If a poor hadron calorimeter is proposed, there must be strong benefits elsewhere to support approval.

The Ingredients for Good Calorimetry:

(i) $e/h = 1$

Much of the understanding of the benefits associated with this choice have been discussed by Wigmans at this conference[18] and elsewhere[16]. From the experimental data in Fig. 4.1 we see that the energy resolution retains the form

$$\frac{\sigma}{E} = \frac{const}{\sqrt{E}}$$

providing that $e/h = 1$. Values above or below result in additional systematic terms. In Fig. 4.2 we see another result, probably even more important, that the energy scale remains linear over the measured range to 200 GeV. This will, of course, be of major significance at the SSC where single particle momenta more than cover this range. These two points are further emphasized in Fig. 4.3 where the linearity and constant resolution term are shown as functions of e/h. Finally Fig. 4.4 compares the resolution functions for calorimeters with different e/h. Choosing $e/h = 1$ gives a good symmetric Gaussian curve[20], whereas $e/h = 1.3$ leads to asymmetric tails in the resolution[19].

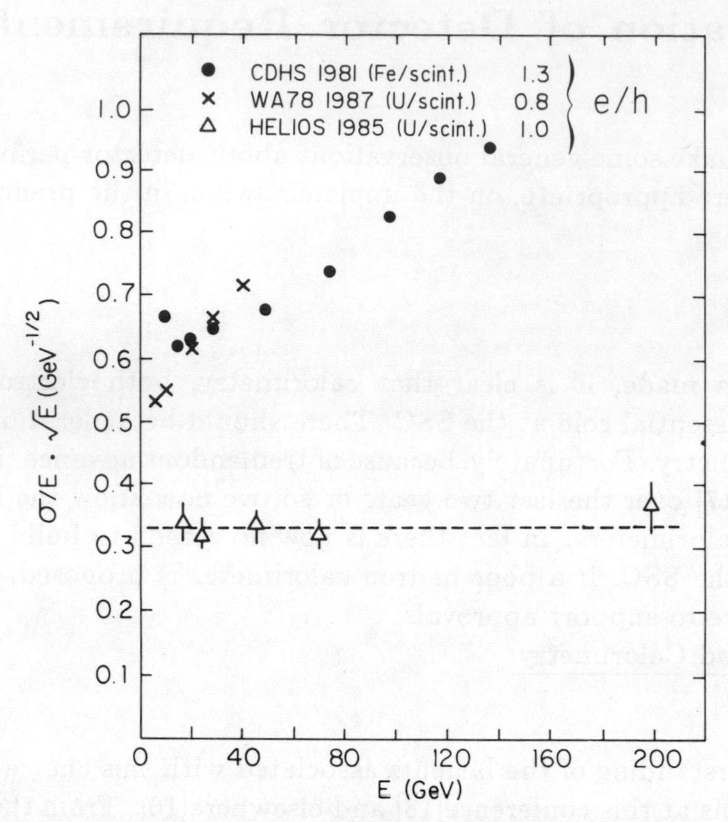

Figure 4.1 Resolution as a Function of Energy for Hadron Calorimeters with Different e/h.

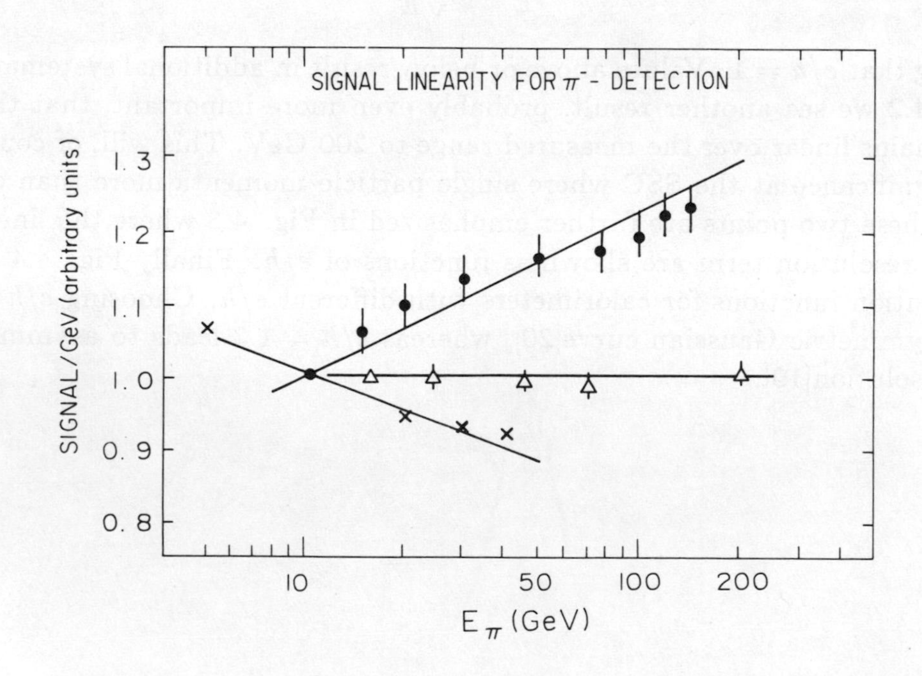

Figure 4.2 Linearity of Response for Different Hadron Calorimeters. (Data as in Fig. 4.1)

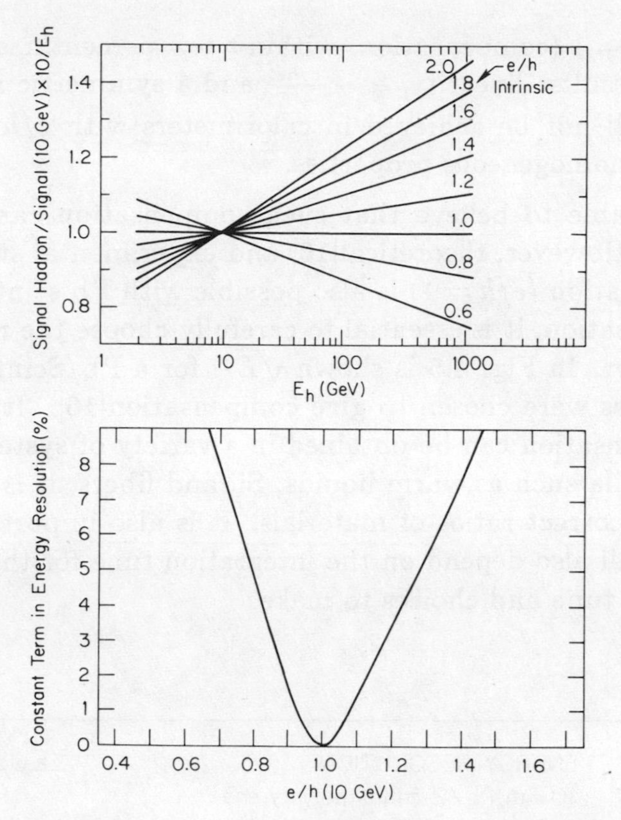

Figure 4.3 Calculated Resolution Constant Term and Linearity as a Function of e/h.

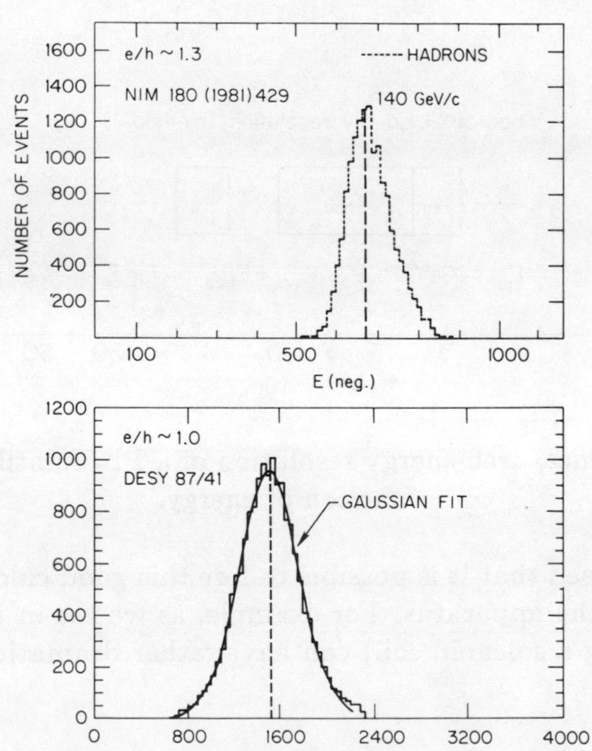

Figure 4.4 Shape of Experimental Energy Resolution Function for Different e/h.

Thus the choice $e/h = 1$ (compensation) within a few percent has these three fundamentally desirable results: linearity, $\frac{\sigma}{E} \sim \frac{const}{\sqrt{E}}$ and a symmetric resolution function. These properties will not be achieved in calorimeters with $e/h \neq 1$ or composite calorimeters of non-homogeneous properties.

It has been fashionable to believe that such compensation can only be obtained with DU absorber. However, theoretical[16] and experimental studies[20] have now shown that compensation ($e/h = 1$) is also possible with Pb scintillator systems. To obtain such compensation, it is essential to carefully choose the relative thickness of absorber and sampler. In Fig. 4.5 is shown $\sqrt{E}\frac{\sigma}{E}$ for a Pb/Scintillator calorimeter, where the thicknesses were chosen to give compensation[16]. It is now confidently believed that compensation can be obtained in a variety of systems. However, with "new" sampling media such as warm liquids, Si, and fibers, it is essential to do the R&D to ensure the correct ratios of materials. It is also important to remark that the compensation will also depend on the integration time for the pulses. There are many parameters to tune and choices to make.

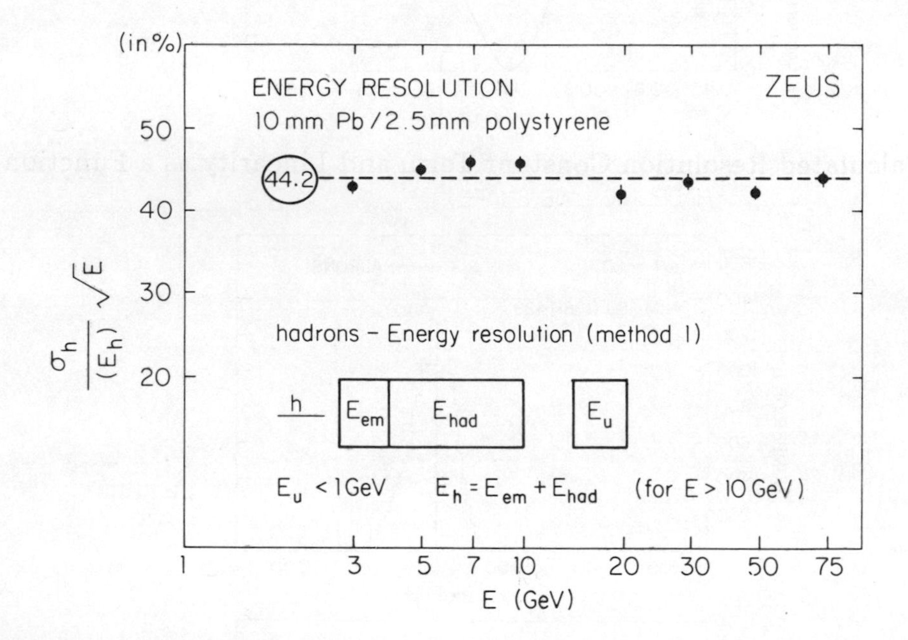

4.5 Experimentally measured energy resolution of a Pb/Scintillator calorimeter as a function of energy.

It must also be stressed that it is possible to lose this good calorimetric performance by other choices in the apparatus. For example, as we see in Fig. 4.6, the presence of inert material (e.g a solenoid coil) can have rather dramatic consequences on the value of e/h. This calculation is for normally incident $5GeV/c$ π and it is important to pursue these calculations further[21]. Moreover it should be remembered that many particles will be incident at shallow angles so that the effective thickness is much larger and the consequences even more undesirable.

Finally it is also important to point out that mixing calorimeters of different responses will also degrade e/h, with consequent effects on the resolution.

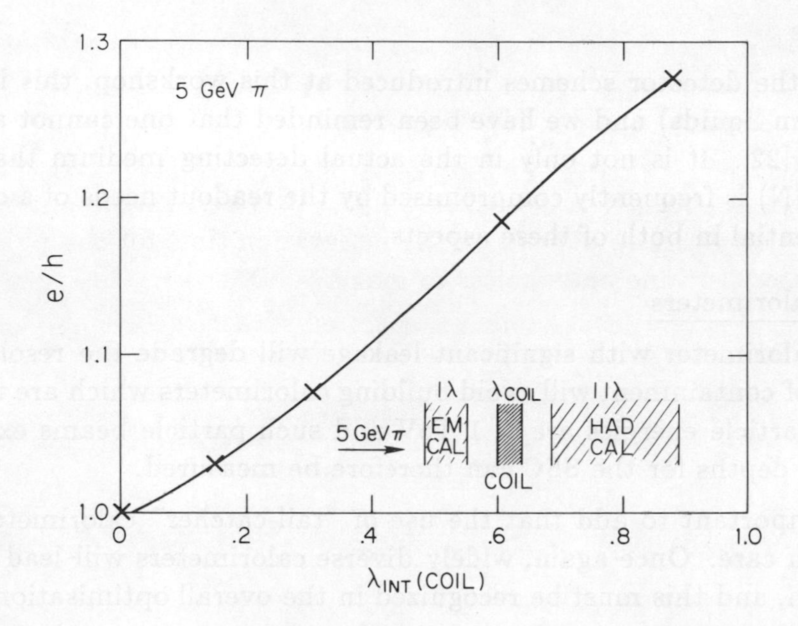

Figure 4.6 Effective value of e/h for 5 Gev/e π's in a calorimeter containing a S/C coil (of different thicknesses)placed after 1 absorption length of the calorimeter.

(ii) Calibration

As we have already remarked, this is a most important feature, and may be the limiting factor in calorimeter operation at the SSC. Liquids and silicon are generally regarded as being more desirable as the deposited charge is directly collected. However, this subject of calibration is always hard and it is essential for operation at the SSC that any scheme be clearly specified, sound, and well tested. R&D is again essential in this area.

(iii) Radiation Hardness

The calorimeters at the SSC will operate in regions of high radiation levels. R&D is required on this topic for almost all detecting media (Si, TMP, fibers) as well as for the associated electronics. The fluxes of neutrons inside one of the SSC detectors will probably be enormous[18] and should be treated with concern.

(iv) Segmentation

The transverse and longitudinal segmentations are mainly dictated by the needs of electron identification. A good transverse segmentation is clearly helped by a small Moliere radius (r_m). The requirement of $\Delta\phi \sim 0.03$ clearly implies that $r_m < .03\ r_{cal}$ where r_{cal} is the "radius" at the front face of the calorimeter. Furthermore, longitudinal segmentation of the EM calorimeter is essential to reach hadron rejections of $\sim 10^{-3}$.

(v) Signal/Noise

In many of the detector schemes introduced at this workshop, this is may be poor (e.g. in warm liquids) and we have been reminded that one cannot afford to throw charge away[22]. It is not only in the actual detecting medium that the problem arises, it (S/N) is frequently compromised by the readout needs of a calorimeter[22]. R&D is essential in both of these aspects.

(vi) Depth of Calorimeters

Having a calorimeter with significant leakage will degrade the resolution. A good knowledge of containment will avoid building calorimeters which are too thin. As we have seen particle energies are ≤ 1 TeV and such particle beams exist today. The appropriate depths for the SSC can therefore be measured.

It is also important to add that the use of "tail-catcher" calorimeters needs to be studied with care. Once again, widely diverse calorimeters will lead to degradation in resolution, and this must be recognized in the overall optimisation.

(vii) Mechanical and Geometrical Structure

All calorimeters require mechanical structures, but such structures are inert and absorb energy. This results in non linearities as a function of energy and angle, with consequent lack of hermiticity (to be hermetic a calorimeter has to be linear and have uniform response in η and ϕ). In Fig 4.7 we see a calculation of the response of a real calorimeter[23] as a function of η, the loss of energy being due to the presence of cryostat walls. In such circumstances, efforts are being made to recover this energy but it is never trivial to avoid these difficulties.

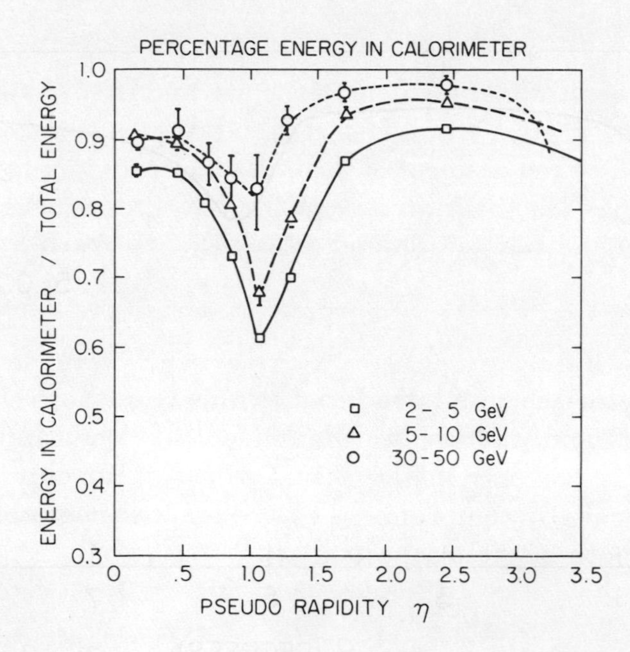

Figure 4.7 Response of a calorimeter as a function of η. The presence of inert material (cryostat walls) can lead to substantial losses.

It is also worth remarking that the way in which small angle calorimetry is achieved can have unpleasant and unexpected consequences. For example, in many of the designs the calorimeters have a cone surface at $\eta \sim 3$ in order to place forward calorimeters at greater distances from the intersection point and hence achieve more sensible granularity. However, leakage from this surface can deposit energy at substantially different azimuths leading to mismeasured E_T's. In Fig. 4.8 we see calculations of $\frac{p_T(measured)}{p_T(true)}$ for two configurations, one with a cone at 5° and the other with no cone.

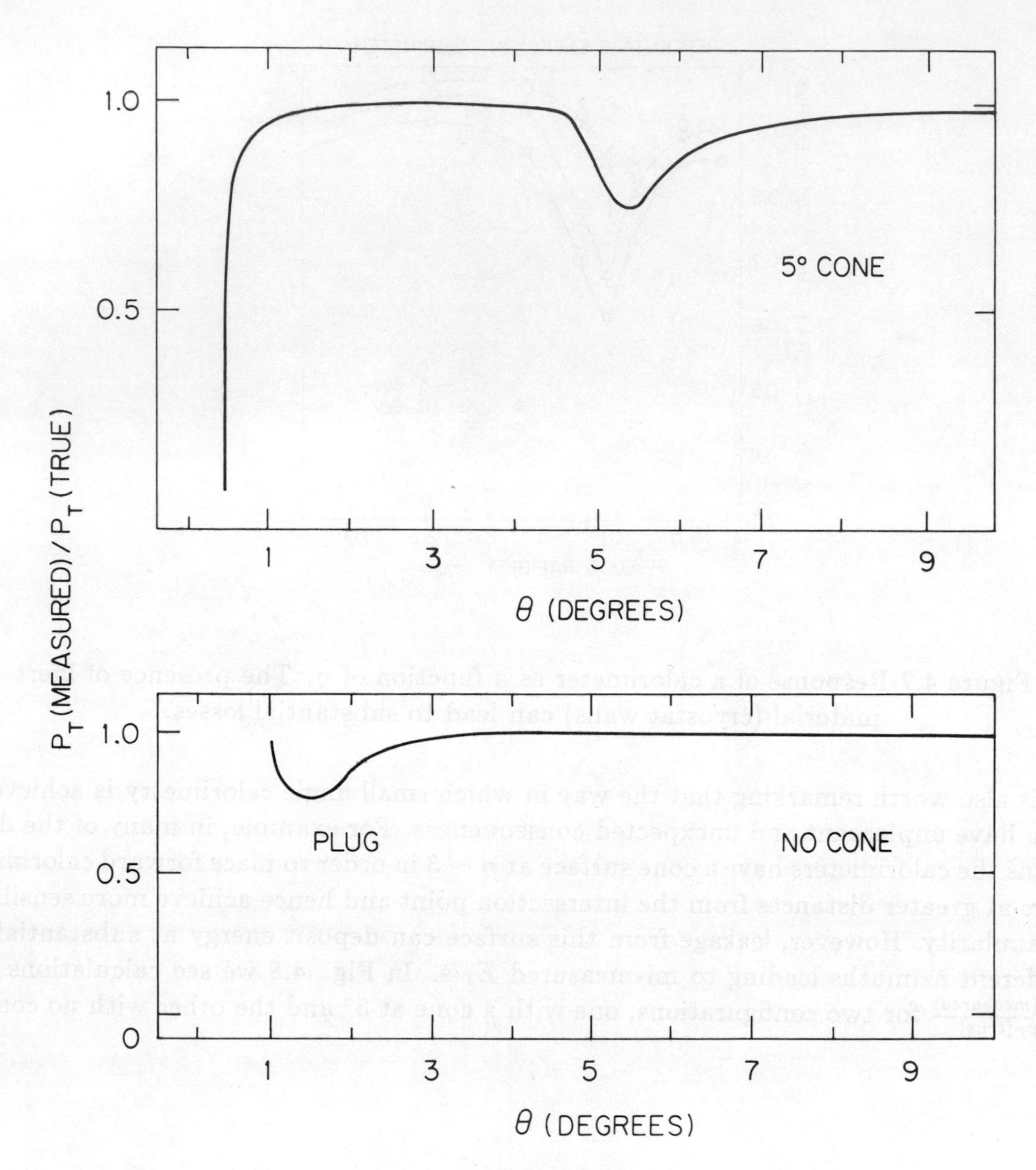

Figure 4.8 $\frac{p_T(measured)}{p_T(true)}$ **in calorimeters with and without a cone at 5°.**

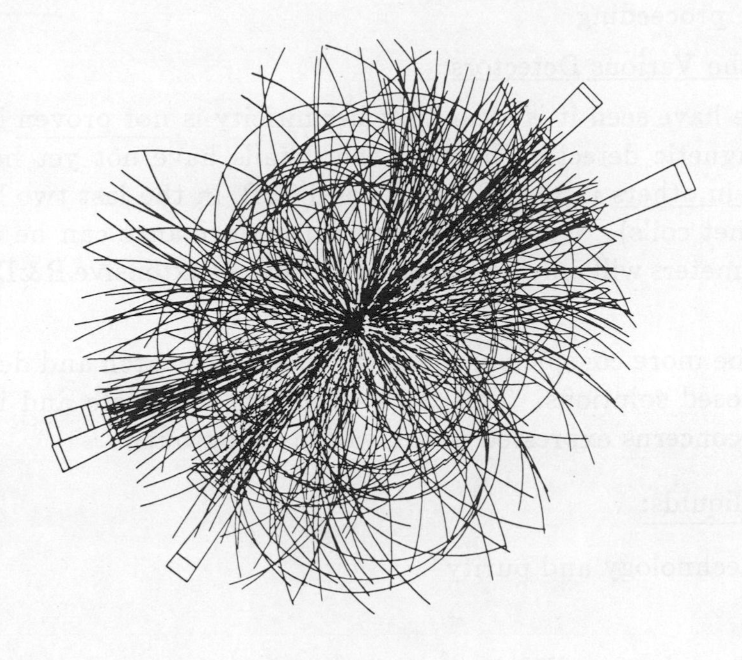

Figure 4.10 An SSC Event

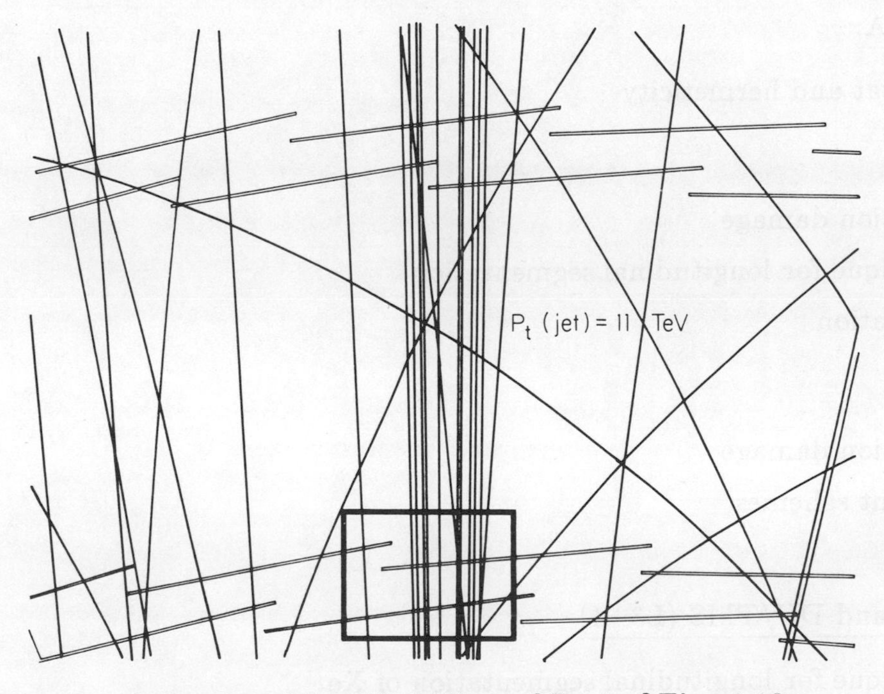

Figure 4.11 Magnified View of Part of Fig. 4.10.

The moral to be drawn is that achieving hermiticity is never easy and any calorimeter layouts must be mechanically and geometrically sound and subjected to detailed MC simulation before proceeding.

Calorimeters in the Various Detectors:

In the designs we have seen it is clear that hermiticity is not proven in some (large solenoid, non-magnetic detector) because the details have not yet been addressed and non existent in others (L3+1, SMART, and SSB, in the last two because of the intrusion of magnet coils). Whether the desired performance can be achieved from the various calorimeters will only be known as a result of extensive R&D, and detailed design.

We list some of the more conspicuous areas which need research and development for each of the proposed solutions. This is by no means complete and in many cases only repeats the concerns expressed in the various subgroups.

(i) Pb (DU)/Warm liquids:

- large scale technology and purity
- signal/noise
- need for high operating HV
- radiation damage

(ii) Pb(DU)/LA:

- cryostat and hermeticity

(iii) Pb/Fibers:

- radiation damage
- technique for longitudinal segmentation
- calibration

(iv) DU/Si

- radiation damage
- readout schemes
- cost

(v) Liquid Xe and DU/TMS (L3+1)

- technique for longitudinal segmentation of Xe

- calibration of Xe
- effect of large Xe Moliere radius and inhomogeneous hadron calorimeter on performance.

There are many further factors which are important in the various concepts. Size will not be an insignificant factor in their operation, and of course, with increased size, inevitably the costs grow. These will be major items in any choice for these detectors.

Choosing a Calorimeter for the SSC:

Building a calorimeter for the SSC is going to take ≥ 5 years and it will then be operated for ≥ 10 years. A substantial amount of detailed design and R&D will be necessary. However there are $\geq 3/4$ years in which to do this work and moreover there now exists a sound basis of calorimetry and much research already in progress.

Given the time scales, it appears to us that it will be possible to choose outstanding and reliable calorimeters for high p_T experiments at the SSC.

4.2 Magnets

The large solenoid and compact detectors call for designs more advanced than the current superconducting solenoids either in operation or under construction.

The large solenoid calls for a 2T magnet, 4m in radius, and $\sim$ 16m long which would probably be realized as a series of lumped coils. Although conceivable, such a magnet has ≥ 1 Gigajoule of stored energy, approximately a factor 10 larger than the current LEP magnets.

The designs for the "small" solenoids are even more adventurous calling for fields of 4T to 6T. These coils will again have stored energies in the region of 1 Gigajoule and, as Fig. 4.9 indicates, present designs would require them to be rather thick, $\sim 0.8\lambda$ at normal incidence. As we have seen this implies a rather dramatic effect on hadron calorimetry. Furthermore, the forces involved can become very large and particular care must be taken to cope with the inevitable quenches.

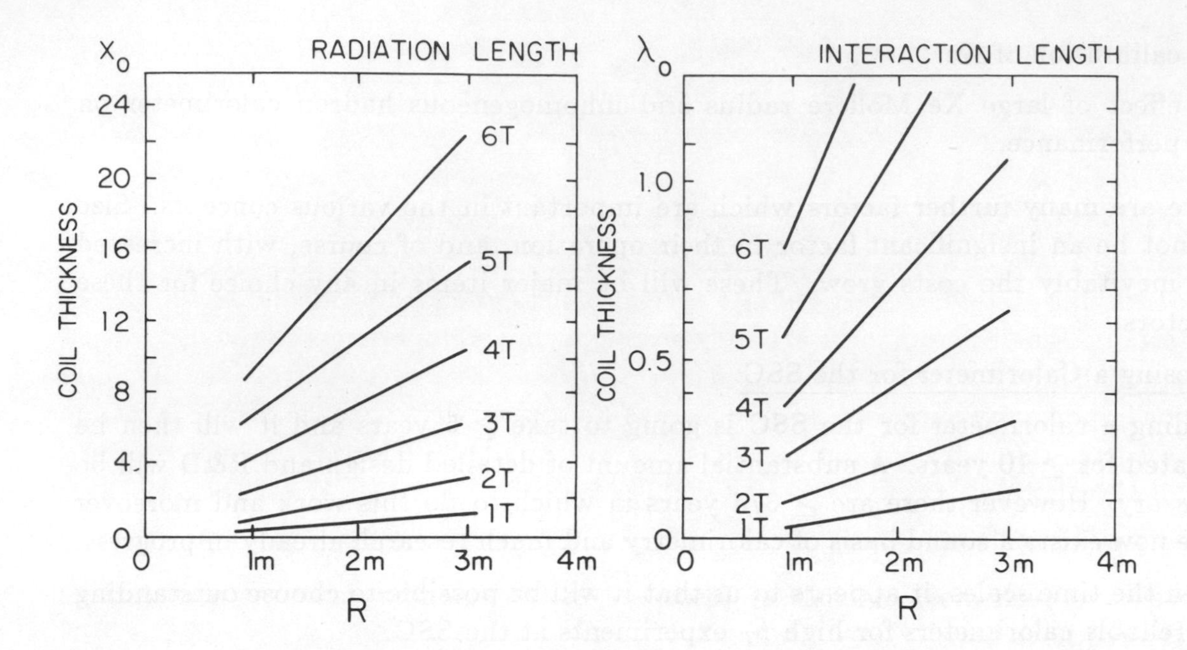

Figure 4.9 Thickness of high field superconducting magnets.

We believe that any of these designs, going substantially beyond present technology, require rather extensive R&D, the "smaller" solenoids requiring a particularly "aggressive" approach to the problem!

4.3 Charged Particle Tracking

Charged particle tracking has not been given a high priority in the deliberations and conclusions of the parameterization groups. Its principal function has been to facilitate or improve lepton identification. For electrons, a track is required to match the energy deposit in a calorimeter, and the identification (i.e. the hadron rejection) is clearly improved by being able to match the momentum (p) and the energy (E) with good resolution. The sign of electrons is clearly important in W' and Z' decays and is certainly essential if studies of W^+W^+ final states are to be compared with W^+W^- at the 1 TeV mass scale, (revealing perhaps a more complicated Higgs sector). In the case of muons, momentum is usually measured in an external system, e.g. toroids. The ability to measure the momentum close to the interaction point and subsequently match with the second measurement will bring a substantial improvement in muon identification.

The parameterization groups thus indicate that isolated tracks should be measured in

$$|\eta| \leq 2.5 - 3.0$$

and that a momentum resolution

$$\frac{\sigma(p_T)}{p_T} \sim 0.5 p_T \; (TeV)$$

would be satisfactory.

However we are entering a new energy domain. The power of tracking detectors in identifying secondary vertices and revealing new unexpected phenomena may be important. Furthermore there are some purely technical functions that have not been given much study at this workshop. Correct (or multiple) event vertex identification will be important when searching for rare events and while we expect the calorimeters to be better than any of their predecessors, the ability to check their behavior may well be invaluable.

Some Proposed Technologies

These have been most extensively discussed in the solenoid detectors and options considered have been:

(i) Drift or straw chambers operating in a field of 2T (large solenoid)

(ii) Silicon strip detectors (of 25 μ pitch) operating in a field of 4T (SSB) and

(iii) Scintillating plastic fibers operating in a field of 6T (SMART)

We have also seen presentations of silicon pixel devices[24] with sparse readout as a possible new vertex detector technique.

While options (i), (ii) and possibly (iii) (because the track finding is more limited in this case) can all give the required resolution and coverage required by the parameterization groups, only option (ii) allows the hope of reconstructing all the tracks in high p_T jets, should this be desirable. This is illustrated in Figs. 4.10, 4.11 and 4.12 where we show an event at 40 TeV together with successively closer views of a $\sim$ 1 TeV jet in the event. Although the density of tracks is high (!), a strip detector with 25μ pitch might well be able to resolve the various tracks. Of course further detailed studies of pattern recognition from layer to layer will be essential before these possibilities can be treated with any great confidence.

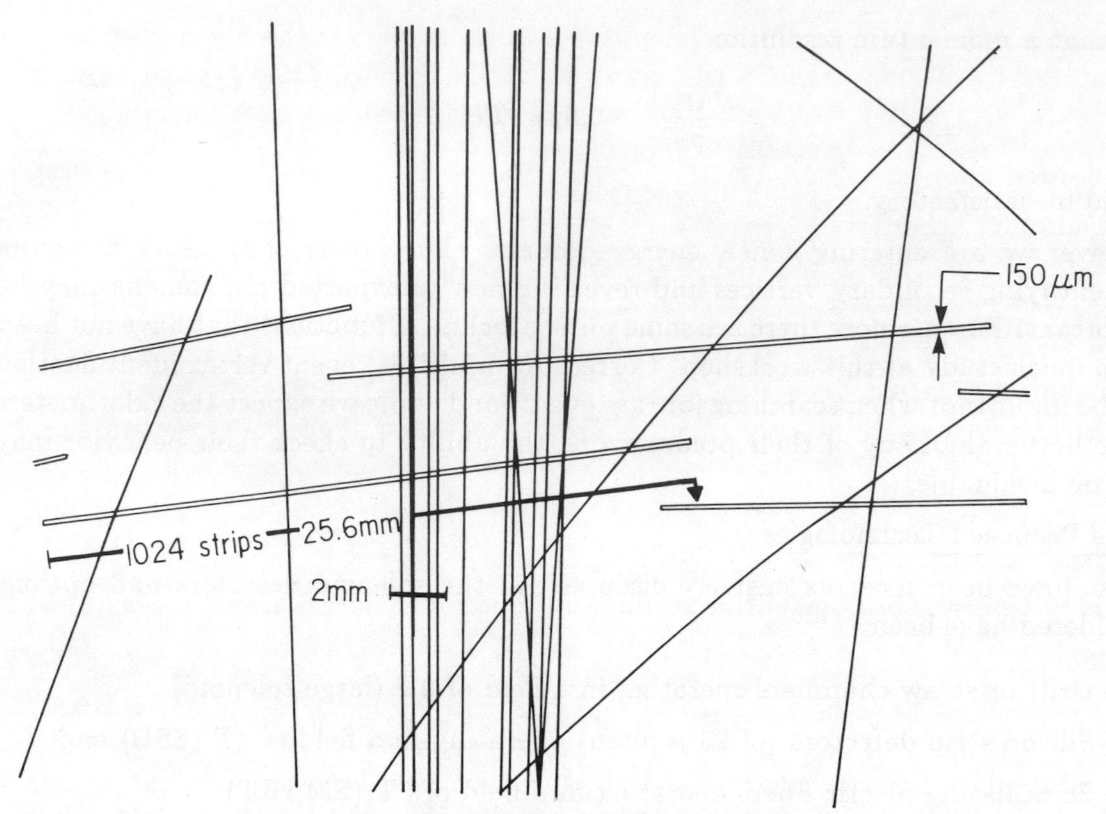

Figure 4.12 Magnified View of Part of Fig. 4.11.

Applicability of Various Technologies at the SSC

(i) Drift Chambers

This is clearly a known technology, but serious questions remain as to whether a luminosity of even 10^{33} can be survived. It has also yet to be proven that pattern recognition is possible in the environment of ≤ 10 superimposed events, although we believe that finding isolated electron or muon tracks should be possible given the regions and p_T's that will be defined by a calorimeter or external muon detector. What is clear is that drift chambers are unlikely to resolve particles in the cores of > 100 GeV jets.

Substantial R&D is required on at least the following topics:

- mechanical stability
- lower power local electronics
- radiation damage of chambers and electronics

before performance at the SSC can be predicted.

(ii) Silicon Strips or Fiber Tracking

These detectors offer a number of attractive features including rapid response (≤ 30 *nsecs*), a more compact detector (with consequent cost savings in detectors at larger radii) and even the possibility of reconstructing tracks in jet cores. On the other hand, the costs may be prohibitive and there are certainly doubts on the radiation hardness of detector and electronics. In this respect the fibers offer certain advantages in placing readout well away from the beams. However to have any confidence, very active R&D must be pursued in a number of areas including:

- desired fiber properties
- radiation hardness
- mechanical structures (so that the $\sim 5\mu$ accuracy can be explored)
- high field superconducting magnets

We include the latter since a high field is still essential to achieve the desired momentum resolution, despite the much improved accuracy of the measurements.

4.4 Lepton Identification

In this section we consider both electron and muon identification and measurement.

Electrons

It is essential to have a charged track pointing at an electromagnetic cluster for electron identification. Furthermore it is necessary to have both good longitudinal segmentation ($\sim$ 3 measurements) and good transverse segmentation, which matches the shower radial spread. Clearly, as we have already emphasised, a small Moliere radius helps in achieving this isolation.

If we take literally the requirement of $10^{-2} - 10^{-3}$ for hadron rejection then additional devices, over and above the calorimeter, are not essential. However we believe that redundancy is desirable and it would appear that the following rejections can be achieved:

(i) Calorimeter and momentum measurement	$\leq 10^{-3}$
(ii) Calorimeter and TRD	$\leq 10^{-4}$
(iii) Calorimeter and tracking and converter/synchrotron radiation	$< 10^{-4}$

We illustrate in Fig. 4.13 the rejection that can be achieved with a TRD detector[13] as a function of angle from the jet axis. Clearly a combined system should reach $\leq 10^{-4}$. The benefits associated with detecting synchroton radiation from detectors will be studied in the AMY detector at TRISTAN, and this experience will aid future designs of SSC detectors.

Once again it is essential to pursue an active R&D program and it is possible to identify, at least, the following areas which deserve particular interest:

- operation of TRD's in an SSC-like environment

— segmentation in liquid Xe and Pb/fiber calorimeters

Much of the R&D is of course covered in the section on calorimetry.

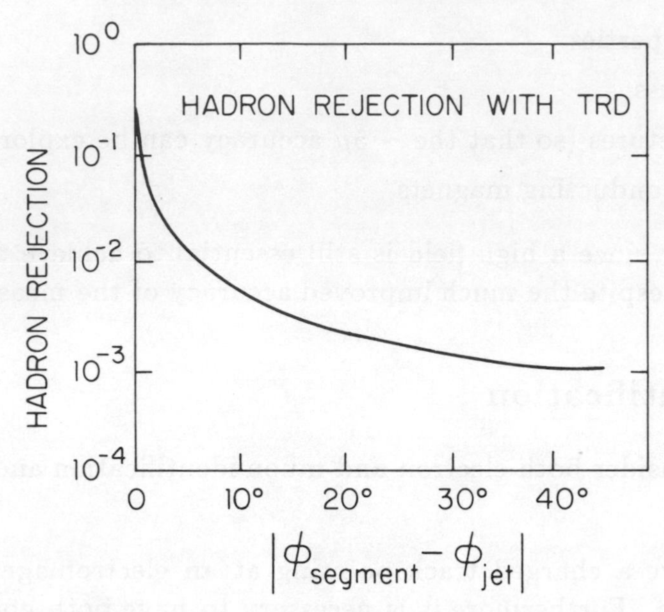

Figure 4.13 The hadron rejection in electron identification with a TRD that can be obtained as a function of the angle of a track from the jet axis.

Muons

In the various detectors the muon identifier is rather similar. In general it is based on the use of toroids[9][10][13][14] [15] in the forward direction and either toroids[10][13] or the solenoid return field[9][10] in the central region. These devices give a muon momentum resolution of

$$\frac{\sigma}{p} \sim 10 - 20\%$$

depending on the precise details.

In the one exception[15] to this, the use of an air cone solenoid leads to a resolution

$$\frac{\sigma}{p} \sim 0.04 \times p(TeV)$$

over a limited η range, $|\eta| \leq 1$ in the central region. Such a resolution is closer to that achievable with electrons $\left(\frac{\sigma}{E} \sim \frac{0.15}{\sqrt{E}}\right)$ and will clearly bring some advantages in background suppression.

The one attractive feature of the solenoid detectors (both large and small) is that they give two momentum measurements, at least in the central region, which is undoubtedly a good idea, leading to background suppression and most likely less confusion.

In general these muon detectors are expected to be comparitively long drift chambers not requiring too much R&D. However to achieve the very good momentum resolution[15] excellent knowledge of the mechanical stability is essential. The L3 experiment at LEP should provide valuable information on exactly this point.

4.5 Trigger

The problems of triggers at the SSC has not been discussed in much detail, the problems being common to all detectors. However, this will be a serious problem and deserves a great deal of further study. It is clearly very strongly coupled to the design and implementation of the read out electronics. The difficulties will be particularly significant for long drift space detectors (e.g. muon chambers) and schemes for identifying the crossing in which the interaction occurred will present an intriguing problem.

Clearly when the detector concepts are better refined it will be essential to return to this question of triggers and rates.

5. Brief Comparison of the Detectors

It is clearly difficult at this stage and inappropriate to make deep comparisons. We can only make some rather superficial comments.

(i) It appears to us that the large solenoid, compact solenoids and non magnetic detector are basically addressing the same physics issues. At present on paper and benefitting from much enthusiasm they appear somewhat comparable in performance, although there are some particular strengths. The compact solenoids will, however, require particularly large R&D progress to support detailed designs.

(ii) A dipole design appears to add nothing to calorimetry, its strength lying in a good charged track momentum resolution as a function of η although this is achieved at the expense of non-uniform track performance over the azimuth. If large mass scale phenomena (centrally produced) are important at the SSC this will be a disadvantage.

(iii) The muon momentum resolution in the $L3 + 1$ detector is unique, even though this performance is limited to the range $|\eta| \leq 1$. However in the present conception the calorimetry and electron identification are clearly compromised.

(iv) The proposed advantage of the 4π muon detector [14] is its ability to run at $L \sim 10^{34}$. However this luminosity is achieved by ignoring everything except for muons. It is important to point out that for many of the processes with which we are concerned at the SSC the ability to measure both electons and muons brings a factor of 4 in efficiency over muons alone. Thus the advantage of running this detector at 10^{34} over more general detectors at 10^{33} is not 10 but only ~ 2.5. In this context we must also remind ourselves that the various detectors must be made to work at 10^{33}, otherwise the physics potential of the SSC will certainly be reduced.

(v) In comparing the detectors we stress again that the SSC will open new domains, where we must be ready for surprises. Although the work of the physics parameterization groups is important it cannot and must not be the only input in any comparisons.

All of the detector designs require substantial R&D projects, which are both challenging and exciting and need to be pursued vigorously in the next few years. This R&D and new ideas, which will emerge, will lead to progress in the designs for detectors at the SSC.

6. General Conclusions

It does not appear on the basis of this very brief workshop that any very particular advantages of the large solenoid, compact solenoid, and non magnetic detector have emerged. This is probably due to the following reasons;

(i) There were no very stringent requirements from the physics parametrization groups that demanded compromises in the detectors to achieve excellent performances.

(ii) The Real Advantages of detectors will only appear with Detail Optimization.

(iii) Perspectives of what can be achieved will change with the R&D being pursued.

We can expect changes in all of these items in the future.

It appears to us that ≥ 1 (no will argue with at least 1!) high p_T detectors will be required to do the physics. The exact number and choices of detector will depend on the weight given to the following items (as well as others).

(i) Versatility during the exploratory phase of the SSC

(ii) The need for differing strengths

(iii) The need to confirm results at this unique machine

(iv) The costs

It will not be an enviable task to make this choice. However, with the time available, the experience being gained currently with large detectors, and an active R&D program (as indicated), we can be confident that whatever the choice(s) the detectors should be capable of addressing both the "expected" and "unexpected" physics at the SSC.

Bibliography

[1] Heavy Higgs Physics Parametrization Subgroup. These proceedings .

[2] Intermediate Higgs Physics Parametrization Subgroup. These proceedings.

[3] Non standard Higgs Physics Parametrization Subgroup. These proceedings.

[4] New Quarks or Leptons Physics Parametrization Subgroup. These proceedings.

[5] Supersymmetry Parametrization Subgroup. These proceedings.

[6] New $W'Z'$ Parametrization Subgroup. These proceedings.

[7] Compositeness, Drell-Yan, and Jets Parametrization Subgroup. These proceedings.

[8] C. Baltay, J. Huston, B.G. Pope, Snowmass (1986), p. 355.

[9] Large Solenoid Detector Group. These proceedings .

[10] Compact Detector Group. These proceedings .

[11] Dipole Detector Group. These proceedings.

[12] The UA1 Experiment, C. Tao, Int. J. Mod. Phys. A1(1986)749.

[13] Non magnetic Detector Group. These proceedings.

[14] Muon Detector Group. These proceedings.

[15] L3 + 1 Detector, Muon Detector Group. These proceedings.

[16] R. Wigmans, NIKHEF 87-08.

[17] B. Anders, et al., DESY 86-105, M.G. Catanesi, et al., DESY 87-027.

[18] R. Wigmans. These proceedings .

[19] H. Abramowicz, et al., NIM 80(1981), 429.

[20] E. Bernardi, et al., DESY 87/27.

[21] R. Brau, Compact Detector Group. These proceedings.

[22] C. Fabjan, These proceedings.

[23] R. Raja, Private Communication.

[24] D. Nygren. These proceedings.

REPORT OF THE LARGE SOLENOID DETECTOR GROUP*

G. G. Hanson
Stanford Linear Accelerator Center, Stanford University, Stanford, California 94305

S. Mori
Institute of Applied Physics, University of Tsukuba
Sakura-mura, Niihari-gun, Ibaraki 305, Japan

L. G. Pondrom
Physics Department, University of Wisconsin, Madison, Wisconsin 53706

H. H. Williams
Physics Department, University of Pennsylvania, Philadelphia, Pennsylvania 19104

B. Barnett
Johns Hopkins University

V. Barnes
Purdue University

R. Cashmore
Oxford University

M. Chiba
Tokyo Metropolitan University

R. DeSalvo
Cornell University

T. Devlin
Rutgers University

R. Diebold
U.S. Department of Energy

P. Estabrooks
Carleton University

S. Heppelmann
University of Minnesota

I. Hinchliffe
Lawrence Berkeley Laboratory

J. Huston
Michigan State University

T. Kirk
Fermi National Accelerator Laboratory

A. Lankford
Stanford Linear Accelerator Center

D. Marlow
Princeton University

D. Miller
Purdue University

P. Oddone
Lawrence Berkeley Laboratory

S. Parker
University of Hawaii

F. C. Porter
California Institute of Technology

N. Tamura
Kyoto University

D. Theriot
Fermi National Accelerator Laboratory

G. Trilling
Lawrence Berkeley Laboratory

R. Wigmans
NIKHEF, Amsterdam

N. Yamdagni
University of Stockholm

* Work supported in part by the Department of Energy, contract DE–AC03–76SF00515.

ABSTRACT

This report presents a conceptual design of a large solenoid detector for studying physics at the SSC. The parameters and nature of the detector have been chosen based on present estimates of what is required to allow the study of heavy quarks, supersymmetry, heavy Higgs particles, WW scattering at large invariant masses, new W and Z bosons, and very large momentum transfer parton-parton scattering. Simply stated, the goal is to obtain optimum detection and identification of electrons, muons, neutrinos, jets, W's and Z's over a large rapidity region. The primary region of interest extends over ± 3 units of rapidity, although the calorimetry must extend to ± 5.5 units if optimal missing energy resolution is to be obtained. A magnetic field was incorporated because of the importance of identifying the signs of the charges for both electrons and muons and because of the added possibility of identifying τ leptons and secondary vertices. In addition, the existence of a magnetic field may prove useful for studying new physics processes about which we currently have no knowledge. Since hermeticity of the calorimetry is extremely important, the entire central and endcap calorimeters were located inside the solenoid. This does not at the moment seem to produce significant problems (although many issues remain to be resolved) and in fact leads to a very effective muon detector in the central region.

1. Introduction

The main motivation for the SSC is the expectation that new physics in the form of new heavy particles, such as Higgs bosons, supersymmetric particles, heavy fermions, heavy W's or Z's, or composite particles, will be discovered in the TeV mass range. Such particles would be produced in the central rapidity region and would decay to high-p_T electrons, muons, or jets, often in events with large missing transverse energy ($\not{E}_T$) due to undetectable neutrinos.

The Large Solenoid Detector Group has studied a large 4π detector in a solenoidal magnetic field from two aspects:

1. Detector characteristics needed to look for the new physics
2. Improvements on the design of a large 4π solenoidal detector over previous designs [1–3].

Our large solenoid detector was conceived as being built with more-or-less "conventional" technology, although in practice such a detector would require a great deal of research and development to build, particularly for the calorimetry and electronics. Such a detector must be capable of operating at the SSC design luminosity of 10^{33} cm^{-2}s^{-1}. We also considered operation at lower and higher luminosities. The detector characteristics are dictated by the desire to detect and identify jets, electrons, muons, and neutrinos (via the missing momentum). Particular emphasis is placed on the identification of electrons and muons; backgrounds should be reduced to a level that is small compared to the rate for prompt real leptons. Since rates for interesting events will be small and the background processes complex, high priority was also given to the determination of the sign of the electric charge for both electrons and muons.

To accomplish electron identification and charge sign determination, tracking in the presence of a magnetic field is required in the inner volume surrounded by the calorimeter. In addition, there are many other motivations for tracking. Our summary of the most important reasons includes:

1. Identification of electrons.
2. Separation of multiple interactions within the same bunch crossing.
3. Matching electrons, muons, and jets to the correct vertex.
4. Electron charge sign determination.
5. Improving e/π separation.
6. Identification of secondary vertices.
7. Identification of τ leptons.
8. Invariant mass or momentum cuts.
9. Establishing the credibility of new physics and providing redundancy.

We note that items 4–9 require a magnetic field. While some of the items are of a higher priority than the others, we are convinced that a great deal of flexibility and power for addressing the physics issues is lost if a magnetic field is not incorporated.

There are additional arguments that may be given, e.g., that with a magnetic field one can verify calorimeter measurements and improve the hermeticity of the overall detector if it is not possible to build a "crackless" calorimeter. However, we do not give these arguments as large a weight since it is our goal to provide good charged particle tracking without compromising compensating and hermetic calorimetry.

2. Physics Requirements

We met with members of the Physics Parametrization Groups on Heavy Higgs, Intermediate Mass Higgs, Nonstandard Higgs, Supersymmetry, Heavy Quarks and Leptons, New W's and Z's, and Jets and Compositeness [4]. From these discussions and from the summary talks of these groups the physics requirements for detectors which will look for high-p_T physics at the SSC were determined. We should keep in mind that the new physics actually found at the SSC may be something other than what was expected. The Parametrization Groups provided models in terms of our present understanding of what physics the detectors should be able to deal with.

The basic requirements can be summarized as follows:

1. Electron and Muon Identification
   - $|\eta| < 2.5 - 3$ (5 for muons)
   - Sign of charge to $0.5 - 1$ TeV/c
   - $e, \mu/\pi$ rejection at least 10^{-3}
   - Mostly isolated tracks

2. Calorimetry
   - Missing $E_T > 100$ GeV
   - Hermeticity crucial!
   - $|\eta| < 5.5$ ($\sim 0.5°$)
   - Electromagnetic: $\sigma_E/E = (0.10 \text{ to } 0.15)/\sqrt{E} + 1\%$
   - Hadronic: $\sigma_E/E = 0.50/\sqrt{E} + 2\%$
   - Electromagnetic $\Delta\eta \times \Delta\phi$ segmentation: $0.02 \times 0.02 - 0.03 \times 0.03$
   - Hadronic $\Delta\eta \times \Delta\phi$ segmentation: 0.06×0.06
3. Tracking
   - $|\eta| < 2.5 - 3$
   - $\sigma_{p_T}/p_T \sim 0.3 - 0.5\ p_T$ (TeV/c)
   - Mostly isolated tracks
   - Useful to check missing energy
4. Microvertex Detector
   - Useful to tag b's and τ's

Probably the most difficult requirements to meet are those for an intermediate-mass Higgs decaying into $b\bar{b}$ or $\tau^+\tau^-$. For this physics one needs to tag b's or τ's using a microvertex detector at the design luminosity of 10^{33} cm^{-2}s^{-1}. In addition, this group wants to identify electrons in b jets.

3. Overview of Detector

In an effort to meet the physics requirements, we paid special attention to optimizing tracking and calorimetry together. The outer radius for tracking was reduced from the Snowmass 86 value in order to allow for at least 10 interaction lengths of hermetic compensating calorimetry entirely inside the magnet coil. In addition, we added intermediate tracking to cover $1.2 < |\eta| < 3$. The iron flux return was then available for use in the central muon detector. We discussed ways to optimize electron identification, possibly including transition radiation detectors, especially in the forward regions. We studied the measurement of muon momentum for muons in the central region by using only the bending angle measured just outside the coil and with central tracking included. We also discussed how such a large solenoid detector might operate at a high luminosity of 10^{34} cm^{-2}s^{-1}. The new design for a large solenoid detector is shown schematically in Figs. 1(a) and 2. Model A from the DCMAP report [2] is shown for comparison in Fig. 1(b). The new improved version has shrunk in outer radius from 10.5 m to 7.6 m. The detector components are described in detail in the following sections.

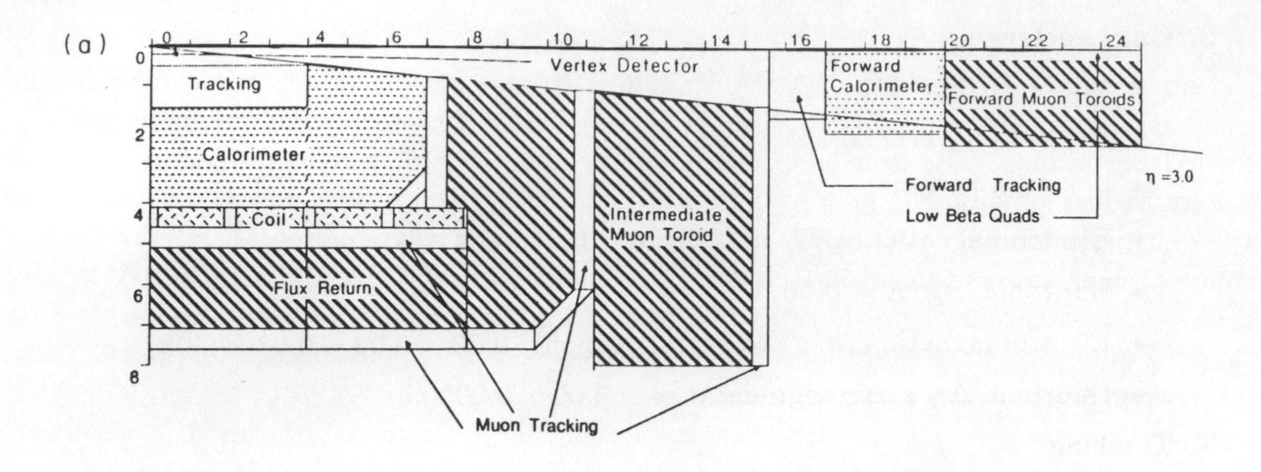

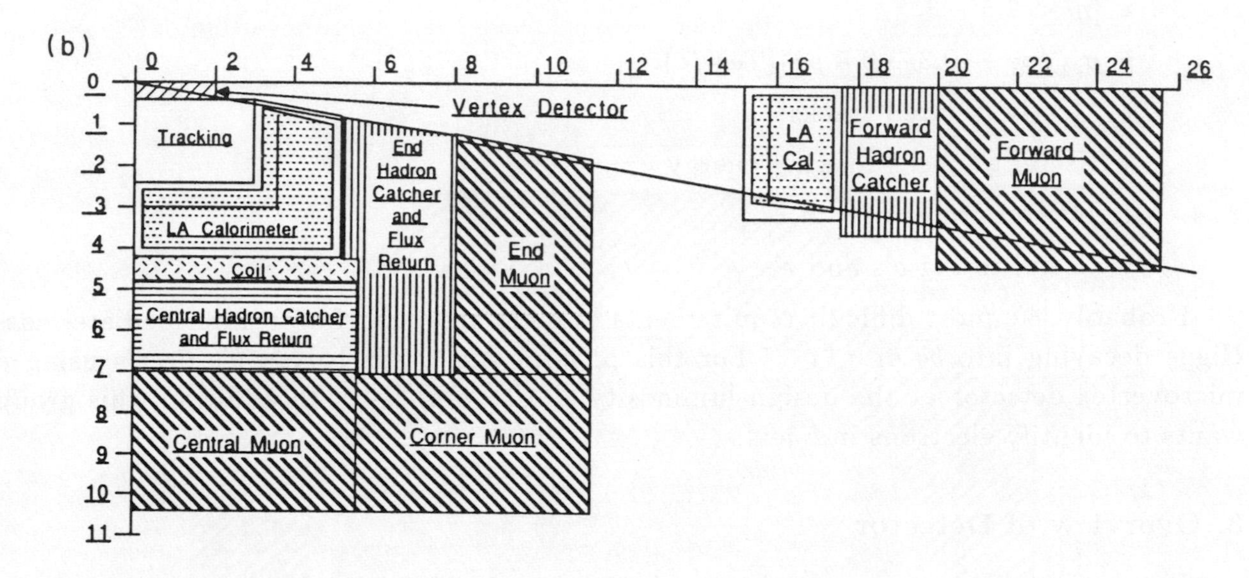

Fig. 1. (a) Schematic view of the Large Solenoid Detector. (b) Model A from Reference 2.

4. Magnet

The decision to locate all of the calorimetry inside the solenoid coil has profound consequences on the solenoid coil. The coil must be large (8.7 m $\times$ 16 m in this case) and must be able to carry a large weight inside the bore (5000 tons). However, restrictions on the thickness of the coil can be relaxed because the coil will no longer interfere with the energy measurement or affect the e/h signal ratio. Muon tracking chambers can be located just outside the coil because there is sufficient absorber to reduce the hadron shower. The flux return can now be utilized not only for flux return but for muon identification and a redundant momentum measurement of the muon as well.

4.1. Magnet Design

As is shown in Fig. 1(a), the coil extends from a radius of 4.1 m to 4.6 m. The magnetic field inside the coil is 2 Tesla providing over 8 Tesla-meters of magnetic analysis for exiting muons. The first layer of muon tracking chambers, extending from 4.6 m to 5.1 m, is just outside the coil. This tracking chamber is used to measure the angle of the exiting muon very accurately in order to be utilized as the p_T cutoff for the Level 1 muon trigger and as the primary momentum measurement for high energy (> 1 TeV/c) muons. The steel flux return, extending from 5.1 m to 7.1 m, is located outside the first layer of muon tracking chambers. The calculated magnetic field in this steel is 1.5 Tesla. The outer layer of muon tracking chambers, extending from 7.1 m to 7.6 m, is located outside the flux return. The angle measurement in this layer will be used in the Level 1 or Level 2 muon trigger to point back to the interaction vertex and as a redundant momentum measurement since the flux return has 3 Tesla-meters of magnetic analysis capability. An end view of the detector is shown in Fig. 2.

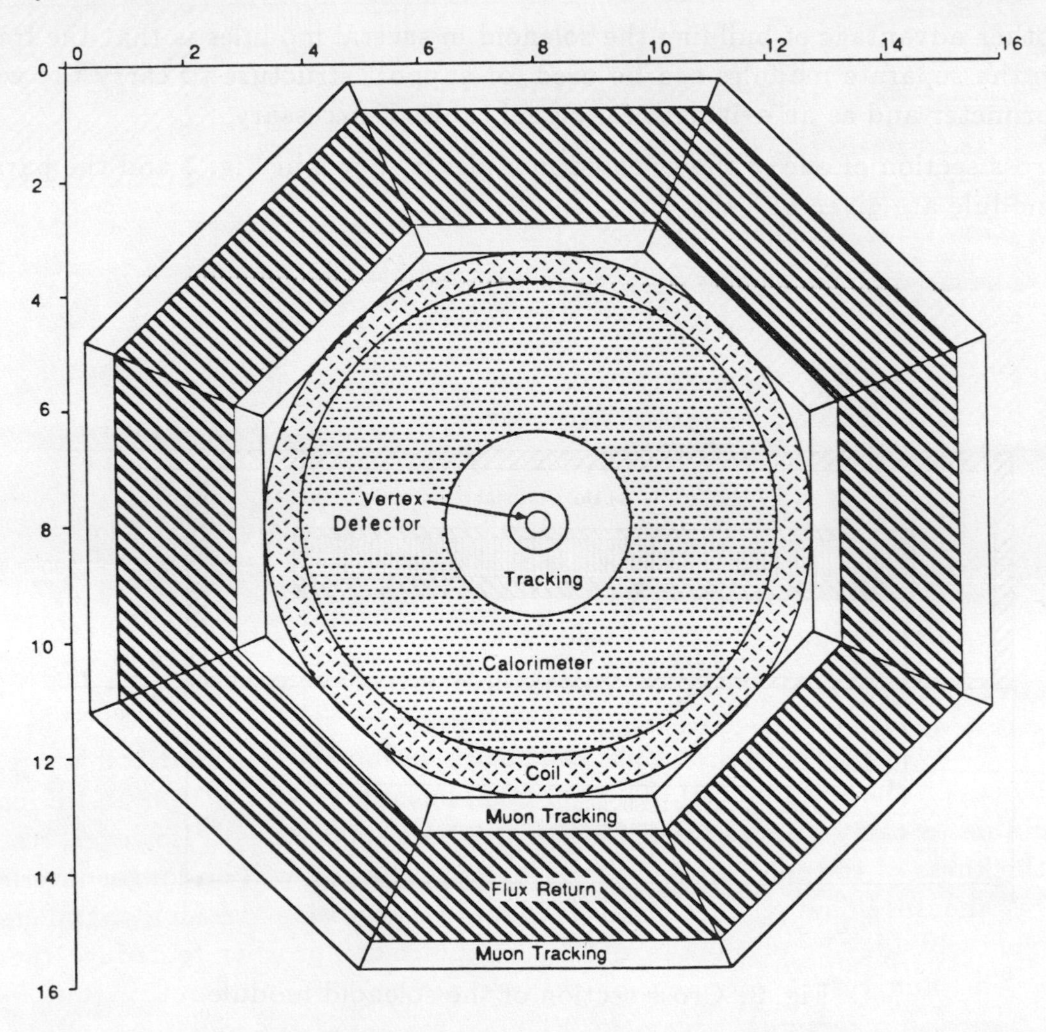

Fig. 2. End view of the Large Solenoid Detector.

The parameters of the solenoid are given in Table I. Because of the large size, it is proposed that the solenoid be constructed in several modules.

Table I. Main Parameters of the Super Solenoid

Bore Diameter	8.2 meters
Total Length	16 meters
Central Field	2 Tesla
Number of Modules	8
Free Space Between Adjacent Modules	0.34 meters
Total Stored Energy	1 GigaJoule
Overall Inductance	80 Henries
Total Weight	450×10^3 Kilograms

Another advantage of building the solenoid in several modules is that the free space between the separate modules can be used for support structure to carry the weight of the calorimeter and as an exit path for signal cables if necessary.

A cross section of one of the solenoid modules is shown in Fig. 3 and the parameters of the module are given in Table II.

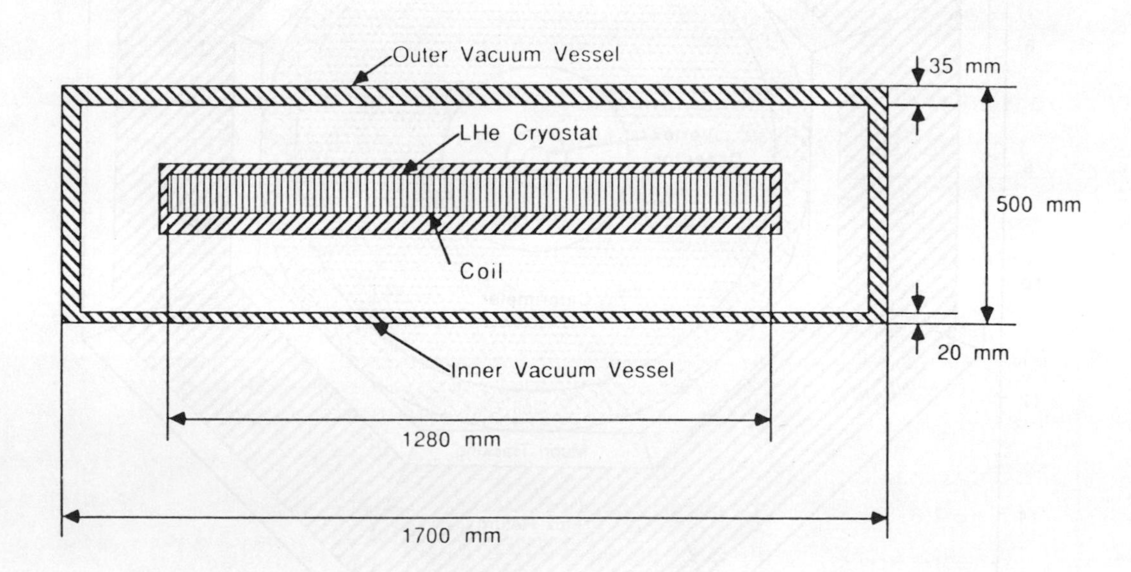

Fig. 3. Cross section of the solenoid module.

Table II. Parameters of the Solenoid Module

Length	1.7 m
Diameter (inner/outer)	8.2 m/9.2 m
Coil Length	1.28 m
Current	5000 A
Central Field	2.0 T
Winding	
Number of Layers	6
Total turns of winding	642
Stored Energy	125 MJ
Cryostat: He Vessel	
Material	SUS 304N
Wall Thickness (inner/outer)	3.5 cm/2.5 cm
Length	1.4 m
Conductor	AMY-type
Weight	56×10^3 Kg
Cold Mass	33×10^3 Kg
Temperature rise for 125 MJ	75°K from 4.2°K

Since the coil is now outside the calorimeter, one is no longer constrained to build a "thin" solenoid such as is currently used in CDF or the LEP detectors. A more conservative thick design is chosen. The conductor chosen here is the AMY-type conductor [5], cross section shown in Fig. 4 and parameters listed in Table III. The AMY conductor is made of hard copper and has an allowable stress of about 20 kg/mm^2. A stainless steel support structure can be used to reduce the conductor thickness. A detailed cross section of the liquid helium cryostat and the superconducting coil is shown in Fig. 5. The pool boiling method of cooling is used. Both the cryostat and vacuum chamber are made of stainless steel. The magnet is cryostable, but for safety reasons is designed to allow quenches to occur.

Table III. Parameters of the AMY Conductor

Superconductor	Nb-46.6 wt% Ti
Cross Section	9.8×10.2 mm^2
Strand Diameter	1.35 mm
Number of Strands	7
Number of Filaments (30 μmø) per strand	1025
Critical Current:	
At 4.4 T (4.2°K)	11760 A
At 6.0 T (4.2°K)	8610 A

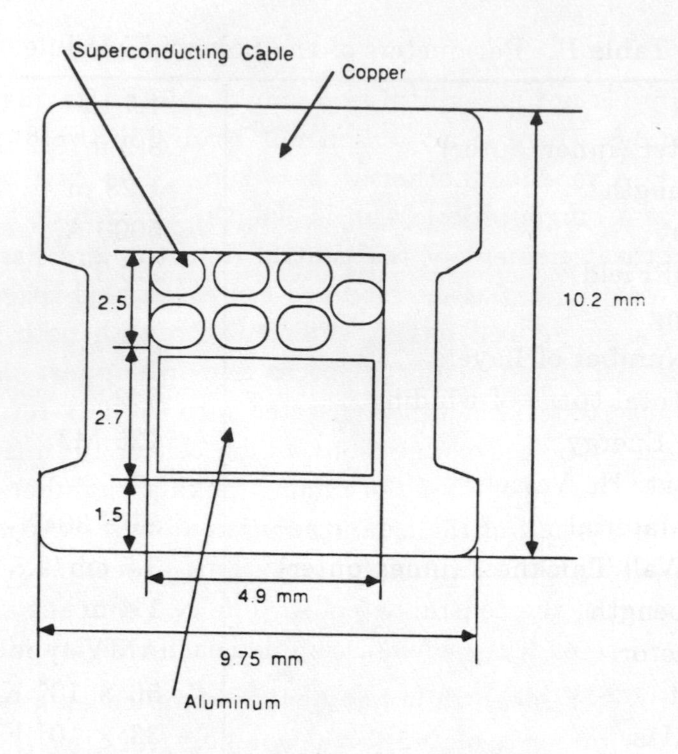

Fig. 4. Cross section of the AMY conductor.

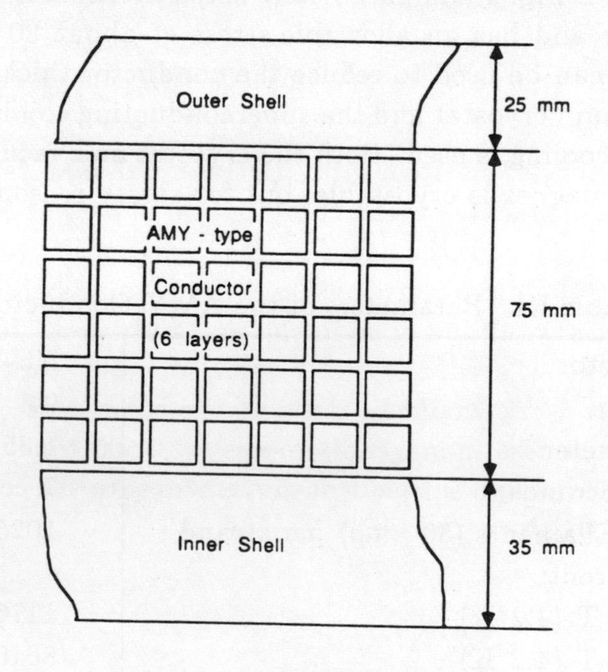

Fig. 5. Cross section of the liquid helium cryostat and superconducting coil.

4.2. Detector Assembly

A magnet this large is not assembled as a single object. Having already divided the coil into several modules, there is no reason not to divide the flux return as well. A possible cut is at $z = 4$ m and another at $z = 8$ m. The ring of the octagonal flux return is assembled as a single object, 4 m in length, 15 m wide and 15 m high. This forms the main structural element of the central detector and has a weight of almost 2500 metric tons. Two of the magnet modules are assembled together with additional structural support as a single unit having sufficient strength both to hold the modules themselves with their associated magnetic forces and to support the calorimeter which will subsequently be inserted. This unit is inserted into the flux return ring and literally hung from it by means of bolted connections which center the magnet modules, space them off the ring, and allow space for the muon tracking chambers. These connections will take up approximately half of the azimuthal area allowing the muon chambers to be inserted and aligned through the remaining gaps. This process is repeated four times to produce the entire barrel of the central detector. The two endcaps are solid steel and are assembled separately. The endcaps when assembled each weigh almost 3400 metric tons.

The central calorimeter meanwhile has been assembled as a separate object. The details of how this calorimeter is mated to the solenoid/flux return are dependent upon the calorimeter technology chosen. If the calorimeter is made of uranium-liquid argon, the number of cryostats would probably be kept to a minimum in order to minimize hermeticity problems. One possible choice is to cut it at 90°. This would give only two cryostats each weighing approximately 2500 tons. Each cryostat would be mated to two of the solenoid/flux return rings. If the calorimeter is made with either lead-scintillating fiber technology or lead-TMS technology, the calorimeter would probably be divided into four pieces, two barrels each weighing 1425 tons and two endcaps weighing 1075 tons each. Each of these pieces would then be mated with one of the solenoid/flux return rings.

After the calorimeter is inserted into the solenoid/flux return rings, the entire solenoid/flux return/calorimeter/endcap combination is assembled on the beam line to form the final central detector. If the experimental area consists of an assembly area and a separate collision hall, the flux return/solenoid work would be done in the collision hall and the calorimeter work would be done in the assembly area. If the experimental area consists of only the collision area, the flux return/solenoid work would be done at one end and the calorimeter work would be done at the other end. Since the detector has been cut up into more or less manageable pieces (< 7500 tons), assembly of complete sections in an assembly area and subsequent movement into the collision area is not ruled out entirely if required by the schedule.

5. Tracking and Vertex Detector

Tracking and microvertex detectors have been discussed rather extensively at previous workshops [6–10]. At the Snowmass 86 Workshop [7] a rather detailed design for a central tracking system was outlined, and we refer to that report for discussions of radiation damage, rates, and occupancy. Occupancy was found to be the limiting criterion, and cell widths were chosen so that the occupancy was $\leq 10\%$ (not including bending in the magnetic field or photon conversions), although even that level may pose difficulties for pattern recognition.

5.1. Microvertex Detector

We have not worked on a new design for a microvertex detector at this Workshop, but we assumed that a large solenoid detector might include one. A microvertex detector will be useful for tagging b's and τ's, studying heavy quarks, and measuring lifetimes of new particles. A large solenoid detector is a natural place to put a microvertex detector since momentum measurement is needed for interpretation of the microvertex detector data. Low momentum tracks can acquire large impact parameters due to multiple scattering and cannot be rejected without accurate momentum measurement. We refer to the design of Snowmass 86 [6]. Such a device could be made of silicon microstrips or pixels. At present, a microvertex detector is not considered possible for luminosities greater than 10^{32} cm^{-2}s^{-1}, but improvements in radiation-hardened electronics may make operation at higher luminosities possible by the time the SSC is running.

5.2. Central Tracking

The concepts for central tracking are essentially the same as in the Snowmass 86 reports. However, we have reconsidered the requirements for momentum resolution based on the physics. We would like to measure the sign of the charge of electrons for p_T up to 0.5–1.0 TeV/c. The most severe requirements come from the measurements of W^+W^+ and W^-W^- as signs of symmetry breaking at mass scales higher than 1 TeV. The momentum resolution is given by [11]

$$\frac{\sigma_{p_T}}{{p_T}^2} = \sqrt{\frac{720}{1+5/N}} \left(\frac{\sigma_x}{0.3\, q\, B\, L^2 \sqrt{N}} \right) , \tag{1}$$

where p_T is the transverse momentum of the particle in GeV/c, q is the charge in units of the electron charge, σ_x is the spatial resolution in m, B is the magnetic field in Tesla, L is the track length in m, and N is the number of measurements, assumed to be equally spaced. We have assumed at the present time a relatively uniform distribution of wires throughout the available volume; however, it is quite possible that the final design might employ a rather different distribution.

We have assumed that the outer radius for tracking is 1.6 m, as compared with the 2.35 m used at Snowmass 86, in order to allow for all of the calorimetry to be inside the magnet coil. In calculating momentum resolution we have used 150 μm spatial resolution instead of the 200 μm used previously; this is probably reasonable since a major contribution to spatial resolution is alignment errors and these may be less severe over a smaller radius. In addition, we have assumed a 2 T magnetic field instead of the 1.5 T used in the Snowmass 86 report. This gives a momentum resolution of $0.54p_T$ (TeV/c) for the 104 measurements assumed in our central tracker beyond a radius of 50 cm. If one uses the constraint that particles come from the interaction region the momentum resolution is improved to $0.26p_T$. The momentum resolution is improved by about the same factor for particles which come from decays of long-lived particles if a microvertex detector is used at small radius.

The tracking detector design is divided into central tracking ($|\eta| \lesssim 1.2$) and intermediate tracking ($1.2 \lesssim |\eta| < 2.5$). The central tracking chambers are assumed to have an inner radius of 40 cm. We do not expect the inner layers at radii less than 50 cm to survive at the design luminosity, but we also do not expect the SSC to begin operation at the full design luminosity. The inner layers can be removed or turned off when they are overwhelmed by the increased luminosity. Straw tube chambers are a natural candidate for a small cell design. The straws can be made small enough. They confine the sense wires to their own cell in case of breakage. They do not require a multitude of field wires resulting in large forces on the endplates. They provide mechanical support so that the chambers can be self-supporting. They provide a much better method of support than in conventional drift chambers for the long sense wires in order to achieve electrostatic stability. They can be pressurized to give better spatial resolution.

The central tracking system is assumed to be built of straw tubes of radii from 2 to 3.5 mm parallel or nearly parallel to the beam direction using a design similar to that given in Reference 12. The straws are made of aluminized polyester film (Mylar) or polycarbonate (Lexan) with wall thicknesses of about 30 μm. The straws are assumed to be at atmospheric pressure. Eight layers of straws are glued together to form superlayers, as shown in Fig. 6. Within each superlayer the layers are staggered by half the cell width in order to allow hits from out-of-time bunch crossings to be rejected and resolve left-right ambiguities as illustrated in Fig. 7. By dividing the chamber into eight-straw-thick superlayers we can obtain locally identifiable track segments with a high level of redundancy. Every other superlayer is small-angle stereo ($\sim 3°$) in order to measure the coordinate along the wire. Azimuthal cathode pad or strip readout is needed for bunch assignment since the propagation time along the wires is 16 ns for the outer layers. It is also useful to help in reducing stereo ambiguities. Cathode pad readout is included on the outer layers of the superlayers. There are 15 superlayers in all for a total of 120 measurements. The total number of cells is 122,368. The total number of radiation lengths is 8% for a particle traversing the central tracking chambers at 90°, as shown in Table IV. Thirteen superlayers are located at radii larger than 50 cm and are expected to be operable at the design luminosity. The central tracking system geometry is summarized in Table V.

The limiting factor in the momentum resolution will probably be knowledge of the relative positioning of the wires. It is difficult to see how one could use lasers to determine the alignment of straw tube chambers or chambers with many wires. One technique which could be used is mapping the wires with the help of real tracks as has been done in the

CLEO detector. Although spatial resolutions of about 50 μm have been achieved with straw tube chambers at 3 atmospheres absolute pressure, the problems of aligning the individual wires could well result in an effective spatial resolution much larger than 50 μm. If the alignment problems can be solved, it would be worthwhile to consider pressurizing the straws.

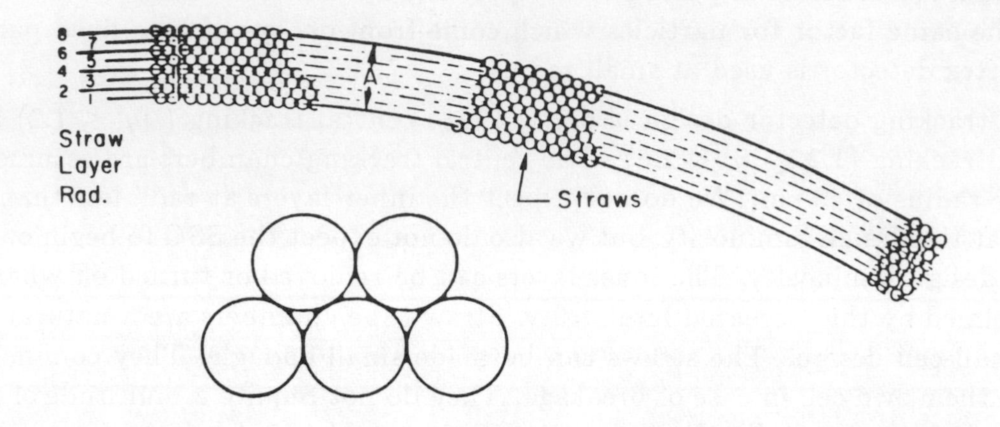

Fig. 6. Schematic drawing of a sector of a central tracking superlayer.

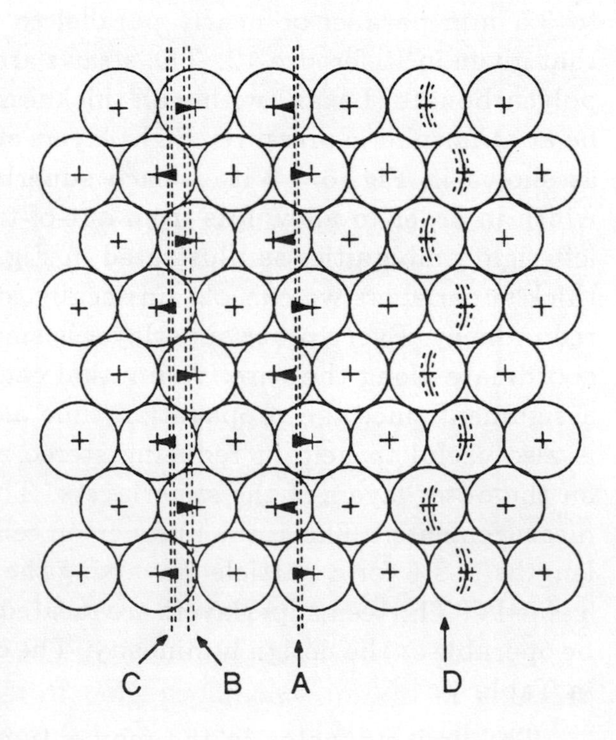

Fig. 7. Layers of straws in a superlayer with straws staggered by half the cell width. A single in-time track (A) will appear as a series of eight hits on the wires on alternate sides of the track. The left-right ambiguity is easily resolved locally. If two tracks are very close together (B and C), they will appear as a wide track to a single-hit readout; such a situation can be detected with a χ^2 fit. A track from an out-of-time bunch crossing is easily sorted out because the left and right drift times do not add up to the maximum drift time.

Table IV. Material in the Central Tracking System

Material	Thickness (cm)	Radiation Length for Material (cm)	Radiation Length (%)
Mylar	1.12	28.7	3.9
Glue	0.062	35.0	0.1
Stainless Steel Wires	0.044	1.76	2.5
Argon	50	17,800	0.3
Ethane	50	32,450	0.2
Pads on Mylar	0.15	28.7	0.5
Epoxy Foam	9.0	1,720	0.5
Total	73.5	–	8.0

Table V. Summary of Central Tracking Parameters

Superlayer Number	Inner Radius (cm)	Module Thickness (cm)	Half Length (cm)	Straw Diameter (mm)	Rapidity Range	Cell Occupancy (%)
1	40	2.7	85.2	3.92	1.50	9.7
2	48	2.7	85.2	3.92	1.34	7.3
3	56	2.7	119.0	3.92	1.50	7.0
4	64	2.7	119.0	3.92	1.38	5.6
5	72	4.1	119.0	5.89	1.28	10.4
6	80	4.2	170.0	6.04	1.50	11.6
7	88	4.2	170.0	6.17	1.41	10.3
8	96	4.3	170.0	6.28	1.34	9.3
9	104	4.4	170.0	6.38	1.27	8.4
10	112	4.5	238.5	6.47	1.50	9.5
11	120	4.5	238.5	6.55	1.44	8.7
12	128	4.6	238.5	6.61	1.38	8.0
13	136	4.6	238.5	6.68	1.33	7.4
14	144	4.6	238.5	6.73	1.28	6.8
15	152	4.7	238.5	6.78	1.23	6.3

5.3. Intermediate Tracking

In order to provide momentum measurement for $1.2 \lesssim |\eta| < 2.5$, we have added tracking in the intermediate region to take over where the central tracking ends. The intermediate tracking extends to ± 4 m along the beam line. Charged particles can be detected up to $|\eta| \leq 3.0$. The tracking chambers consist of several superchambers (at each end of the central tracking chamber), each with position measurements at several closely-spaced z values. We have considered two different designs for these chambers.

Option A. One alternative is to build planes of parallel wires between self-supporting plates consisting of 10 μm thick Mylar foils sandwiching 2 mm thick plastic foam sheets. The coordinate perpendicular to the wires is obtained in the usual fashion from the drift time. Alternate planes are offset by one half-cell to permit simple rejection of tracks from out-of-time bunch crossings. The anode wires would be every 4 mm, corresponding to a maximum drift distance of 2 mm and a sensitive time of 3 bunch crossings. The worst-hit drift cell would have an occupancy of only 2.1%. (This number should probably be doubled to account for the effects of photon conversions and for low momentum tracks which curl up in the magnetic field.)

Determination of the distance along the drift wires would be by means of 0.2×5 cm^2 cathode pads resistively chained together with every tenth pad being read out. The spatial resolution of such a system might be as good as 200 μm. The occupancy of a single pad cell would be less than 1.5% (3% after correction for photon conversions and curlers). The pads are arranged so that signals from two face-to-face pads can be locally correlated to depress noise and reduce the number of readout channels.

This design has 13 superchambers, each consisting of two half-moon modules with 8 anode planes each, on either side of the interaction point for a total of 104 measurements. All the wires in a superchamber are parallel to each other. The wires in each superchamber are at angles of 60° or 90° to the wires in neighboring superchambers. This option has quite low mass. The total thickness of the 13 modules is only about 7% of a radiation length for perpendicular incidence. It has the advantage of simple rejection of tracks from out-of-time bunch crossings. Local track segment finding for fairly stiff tracks should also be straightforward and might be useful for triggering purposes. On the other hand, it requires 64,000 anode wires and 250,000 cathode pad channels to be read out at each end.

Option B. A more natural geometry for a detector in a solenoidal field is one with a high degree of azimuthal symmetry. Such a geometry exists in the CDF forward radial tracking chambers [13] and in the radial chambers described by Saxon at La Thuile [9]. These chambers would have radial anode and field wires separated by stretched mylar foils with cathode pads. In order to keep the occupancy below 10%, there must be at least 800 azimuthal segments. Each superchamber consists of two sections, each with 6 anodes. The two halves are offset azimuthally by one half-cell so that, with information from 5 cm wide cathode pads, tracks from other bunch crossings can easily be rejected. There are 9 superchambers at each end of the interaction point, resulting in a total of 86,400 anode wires and 146,880 cathode pads to be read out. A track passing through the entire intermediate tracking system would have 108 measurements. Local track segment finding and momentum measurement should be straightforward. The cathode pad segmentation is also well matched to the calorimeter segmentation ($\Delta\eta = 0.03$ at the outer radius).

Local track segment finding and momentum measurement should be straightforward for either option. For example, the simple requirement that a track be radial to within 50 mrad can give one a trigger that the p_T of the track is greater than about 10 GeV/c.

The central and intermediate tracking systems are shown in Fig. 8. The momentum resolution as a function of polar angle and rapidity is shown in Fig. 9.

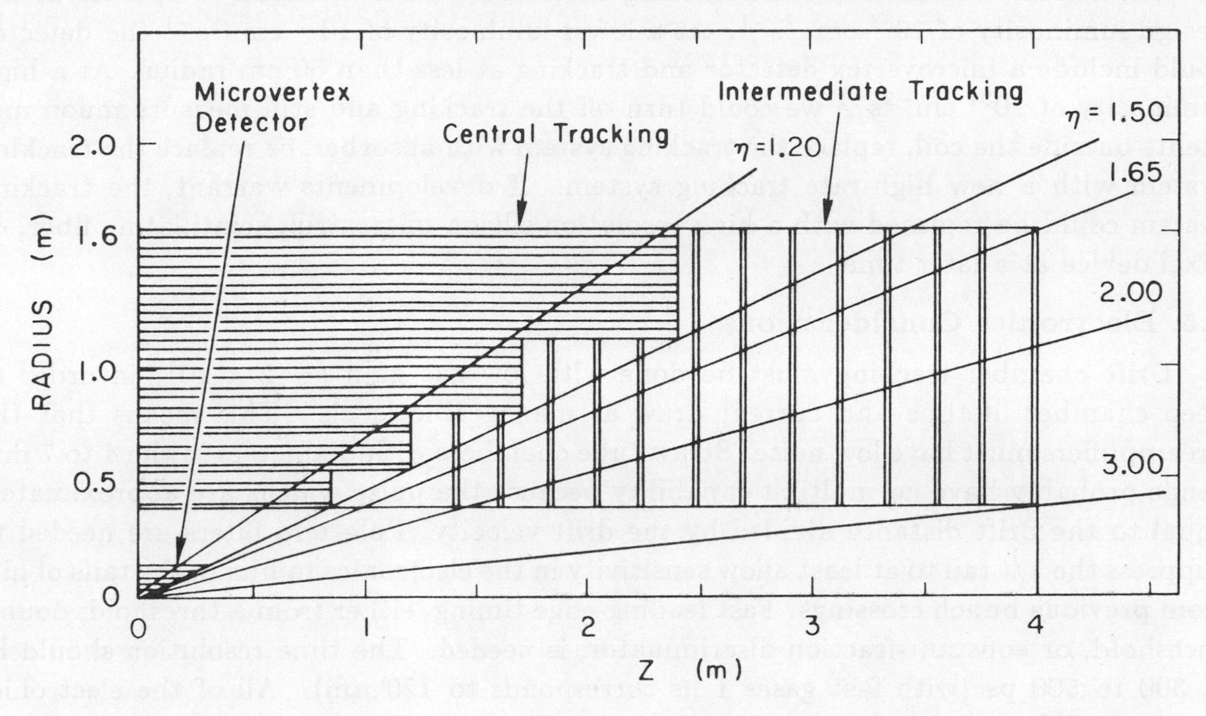

Fig. 8. Schematic view of central and intermediate tracking systems in the Large Solenoid Detector.

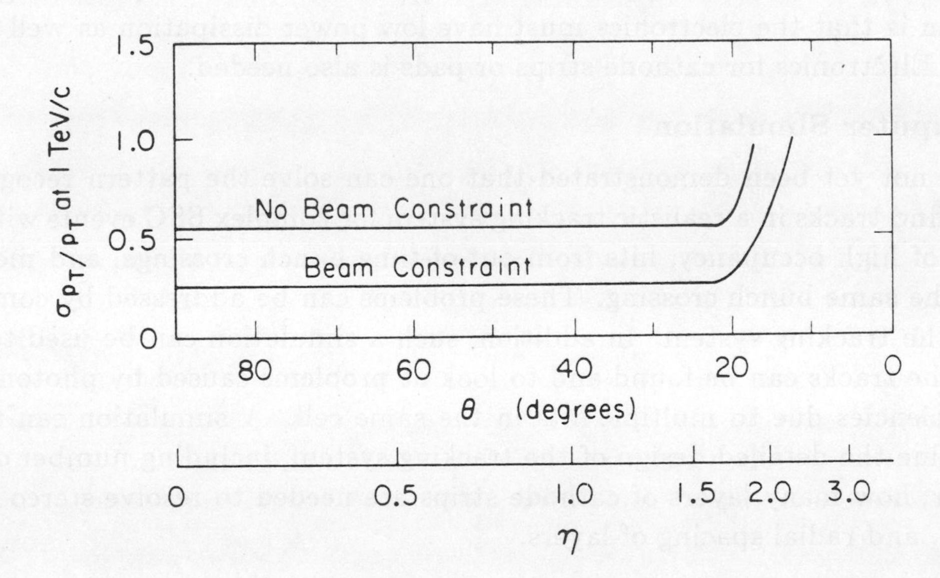

Fig. 9. Momentum resolution as a function of polar angle and rapidity in the Large Solenoid Detector for the 13 superlayers at radii > 50 cm in the central tracking system and intermediate tracking Option A.

5.4. Options at Lower and Higher Luminosities

The central and intermediate tracking described here is intended to operate at the design luminosity of 10^{33} $cm^{-2}s^{-1}$. At a lower luminosity of 10^{32} $cm^{-2}s^{-1}$ the detector could include a microvertex detector and tracking at less than 50 cm radius. At a high luminosity of 10^{34} $cm^{-2}s^{-1}$ we could turn off the tracking and still measure muon momenta outside the coil, replace the tracking system with absorber, or replace the tracking system with a new high-rate tracking system. If developments warrant, the tracking system could be replaced with a high-resolution silicon microstrip, scintillating fiber, or pixel device at a later time.

5.5. Electronics Considerations

Drift chamber tracking must be done with low gas gain ($\sim 2 \times 10^4$) in order to keep chamber lifetime and current draw at manageable levels. This means that the preamplifiers must have low noise. Straw tube chambers with diameters in the 4 to 7 mm range probably have no multihit capability because the pulse widths are approximately equal to the drift distance divided by the drift velocity. Pole-zero filters are needed to suppress the $1/t$ tail to at least allow sensitivity in the electronics to hits in the tails of hits from previous bunch crossings. Fast leading-edge timing, either from a threshold, double threshold, or constant-fraction discriminator, is needed. The time resolution should be $\leq$ 300 to 500 ps (with fast gases 1 ns corresponds to 120 μm). All of the electronics — preamplifiers, pulse shapers, discriminators, TDCs or TVCs, track processors to find track segments, and digital or analog pipelining — is expected to be located on the tracking detector in order to reduce the number of cables and processing time. The implication is that the electronics must have low power dissipation as well as radiation hardness. Electronics for cathode strips or pads is also needed.

5.6. Computer Simulation

It has not yet been demonstrated that one can solve the pattern recognition problems and find tracks in a realistic tracking system for complex SSC events with the added problems of high occupancy, hits from out-of-time bunch crossings, and more than one event in the same bunch crossing. These problems can be addressed by computer simulation of the tracking system. In addition, such a simulation can be used to study how many of the tracks can be found and to look at problems caused by photon conversions and inefficiencies due to multiple hits in the same cell. A simulation can also be used to determine the detailed design of the tracking system, including number of layers in a superlayer, how many layers of cathode strips are needed to resolve stereo ambiguities, cell width, and radial spacing of layers.

6. Calorimetry

The conceptual design of the calorimetry follows in a fairly straightforward way from the requirement that one be able to identify efficiently, and with as high a rejection of backgrounds as is possible, electrons, neutrinos (and other weakly interacting particles), quarks and gluons, and W's and Z's. These detection requirements specify the desired rapidity coverage and calorimeter thickness, transverse and longitudinal segmentation, and energy resolution. We first discuss these general properties of the calorimeter; the required values of the relevant parameters, though not optimized, are reasonably well understood. Following that we discuss which types of calorimeter construction, in particular which kind of absorber and sampling medium, are consistent with these goals.

6.1. Thickness

The overall thickness of the calorimeter is set both by the desire to obtain rather good energy resolution (of order a few per cent) for jets in the several TeV range and by the requirement of optimal missing E_T resolution. Extrapolation of existing measurements indicates that to contain 98% of the energy of a 1 TeV hadron requires a calorimeter thickness of 12 absorption lengths. Because the leading particle in a jet will have a longitudinal momentum fraction of order 0.2–0.25, energy resolutions of a few per cent may be obtained for jets of a few TeV with slightly lower thicknesses, e.g., 10–11 absorption lengths (λ). It should also be noted that measurements by different groups of the thickness required for containment of 98 or 99% of the energy of a hadron shower are not in very good agreement with one another [14]. Better measurements will be required before the thickness can be optimized. In the meantime, we assume a thickness of 10–12 λ at 90° and 13–14 λ in the forward direction.

As far as the thickness of the electromagnetic calorimeter is concerned, Monte Carlo studies [15] for the 1984 Snowmass Workshop indicated that even for an electron of 1 TeV, 26 radiation lengths will contain 98% of the energy; the thickness of the electromagnetic portion of the calorimeter was therefore taken to be 25 X_0.

6.2. Segmentation

The transverse segmentation of the electromagnetic calorimeter is determined both by the necessity for optimal identification of electrons and by the goal of detecting $W \to q\bar{q}$. Fine segmentation for electron identification is required to reject backgrounds from single hadrons and from single hadrons coincident with photons, and to enable reasonable efficiency for identifying an isolated shower in the high multiplicity environment. While it will be very difficult, if not impossible, to identify electrons arbitrarily close to a jet axis, a study [16] in the 1984 Workshop concluded that top quarks with a transverse momentum of 500 GeV/c could be identified via their semileptonic decays with an efficiency of 82% if the segmentation of the EM calorimeter was 0.02×0.02; the efficiency deteriorated significantly for coarser segmentations. Studies of $W \to q\bar{q}$ decays indicated that a similar segmentation was required to obtain optimal effective mass resolution if a small number of longitudinal samples were employed; good effective mass resolution for these decays is essential if one is to discriminate effectively against background from ordinary processes [17]. However, it has also been demonstrated that good effective mass resolution for quark-antiquark decays of W's may be obtained with coarser rapidity-phi segmentation if a very large number of longitudinal readouts are employed [18]. We did not consider this option; such an approach would appear to be most relevant if one desired to build a very compact calorimeter in which case the finite shower size prevents one from obtaining as fine segmentations. The segmentation of 0.02×0.02 corresponds to tower sizes of order 3 cm $\times$ 3 cm in the central region, a size compatible with a determination of the shower position to an accuracy of order 2–3 mm.

The question of what segmentation is required to obtain optimal electron/pion discrimination was addressed both in the 1986 Snowmass Workshop [19] (based primarily on test results for the CDF Endplug calorimeter) and at this meeting (see Section 7 below). These studies indicate that e/π rejection ratios of 10^{-3} may be obtained with transverse segmentations of a few centimeters and with 3–4 longitudinal segments. Given these facts, and bearing in mind the desire to keep the number of electronics channels from growing too large, the EM segmentation was assumed to be of order 0.02×0.02 in the central region and of order 0.03×0.03 in the more forward regions of rapidity. In the very forward region the finite size of the electromagnetic showers renders it pointless to utilize tower sizes smaller than 1 cm $\times$ 1 cm. In this region the size of the towers in $\eta - \phi$ then increases. Each EM tower was assumed to be subdivided longitudinally into three sections; a smaller number would compromise the electron/pion discrimination given the very large range of electron energies of interest (30 GeV to 2–3 TeV) and the concomitant change in the shower length. An optimization of the segmentation, both longitudinal and transverse, for electron/pion discrimination will probably require additional studies in a test beam.

The segmentation of the hadronic calorimeter in η and ϕ was assumed to be 0.06×0.06. This corresponds to tower sizes smaller than the typical hadronic shower size (although "hadronic" showers which are largely electromagnetic will have a narrower core) and very much smaller than the typical spread of the jet. No convincing argument was made to go to finer sizes, and the consensus was that this size would probably be adequate.

In rapidity the calorimeter is assumed to cover $|\eta| < 5.5$. Numerous studies, both at this meeting and at previous workshops, have emphasized the necessity of coverage over this interval to minimize the probability of initial state gluon radiation simulating events with large missing transverse energy.

6.3. Calorimeter Composition

If the thickness and segmentation of the calorimeter seem relatively straightforward, its composition is considerably more problematic. Many combinations of absorber (uranium, lead, iron) and sampling technique (scintillator, gas, liquid argon, warm liquid, silicon) were considered. The basic criteria were that the chosen technology must support the desired segmentation, must survive the high counting rates and radiation level (with sufficiently low noise), and must provide energy resolution of a few per cent at several TeV as well as excellent missing E_T resolution.

6.3.1. Compensation

Numerous studies have indicated that in order to obtain the goals for energy resolution and sensitivity to missing E_T, the ratio of electron (or photon) response to pion response, or alternatively the ratio of the response to the EM and non-EM components of hadron showers, must be of order 1.0 ± 0.1 [20–21]. Several Monte Carlo estimates based on the measured responses to electrons and charged pions indicate that the energy resolution of calorimeters constructed with iron or copper absorber will be of order 9–11% for 400 GeV jets due to the fact that the electron/pion response is 1.4–1.6 for low energy particles [22]. In contrast, uranium and lead calorimeters are estimated to have a resolution of order 2–4% under the same conditions. In addition, the very detailed study of calorimeter performance by Wigmans [21], which includes comparison with many experimental results, enables one to predict with considerable confidence the e/h signal ratio that will be obtained with almost any combination of absorber and active medium. One of the important conclusions of this study is that while it is possible to construct a compensating calorimeter (e/h signal ratio near 1.0) with iron or copper, the very thick plates required yield a very poor energy resolution, particularly for electrons. On the other hand, it is possible to obtain compensation with excellent energy resolution with either uranium or lead. Because of the lower cost and ease of handling, lead is strongly preferred although uranium is not completely excluded. Another important conclusion of this study is that no totally active calorimeter, such as might be constructed out of liquid scintillator or BaF_2, will be compensating. We therefore focus our attention on sampling calorimeters using lead or uranium as the absorber.

Calorimeters for which excellent compensation has been demonstrated include uranium-scintillator and lead-scintillator. In addition, it is anticipated, though not yet experimentally demonstrated, that one may obtain excellent compensation with silicon or warm liquid readout [21]. Experimental results on uranium-liquid argon and lead-liquid argon indicate $e/h = 1.1$ and 1.2, respectively, but are not conclusive at the present time [22]. Theoretically, it is estimated that these combinations will not yield excellent compensation. There is some evidence that uranium-gas calorimeters can achieve compensation, although gas sampling has significant other problems as is discussed below. It should be noted that the use of uranium does not, in general, allow a denser or more compact calorimeter than the use of lead; this follows because of the different fractions of light readout material required in order to obtain compensation. For example, uranium-scintillator and lead-scintillator calorimeters each have an effective nuclear absorption length of approximately 20 cm. An exception is that silicon readout does allow a denser calorimeter to be obtained with uranium than with lead.

6.3.2. Sampling Media

We now discuss each of the possible sampling media in the light of the requirements on segmentation, speed, calibration, and stability.

Scintillator. Scintillators that are relatively radiation hard should, in principle, survive the radiation from interactions in the central region and perhaps down to angles of 10–20° [22]. For very small angles the high radiation probably requires the use of another technique, or of a liquid scintillator which is either very radiation hard or which could be frequently exchanged. A big advantage of scintillator is the very fast response; signal collection times of tens of nanoseconds should be achievable. However, because much of the compensation relies on slow neutrons, it may be necessary to integrate for of order 100 ns in order to achieve the optimal e/π response. It has been suggested that one may be able to achieve optimal compensation at very short collection times by effectively designing the calorimeter to be overcompensated and then attenuating the compensation to the appropriate value with the short shaping time, but a systematic study of this approach has not yet been carried out.

Aside from radiation hardness, the biggest potential problems with scintillator calorimeters are the issues of calibration uniformity, stability and segmentation. Large scintillator calorimeters have been built which have been calibrated to 1% and experiments have successfully tracked the calibration over years to $< 2\%$. No one has yet demonstrated the capability to track a large system over several years to an accuracy of 1%, although systems have been implemented which should enable one to achieve this accuracy [23].

It is difficult, if not impossible, to obtain segmentations of order 2 cm $\times$ 2 cm with the "conventional" construction which uses scintillator plates and wavelength shifters. However, considerable success has been obtained constructing lead calorimeters with scintillating fibers as the active medium. This technique should clearly allow very fine transverse segmentation. It is more difficult to achieve simultaneously the desired longitudinal segmentation in the EM calorimeter. However, schemes have been suggested which may allow one sufficient longitudinal segmentation to attain excellent electron identification, which is the primary criterion. In addition, it is possible that with very fine transverse segmentation, it may not be necessary to have such fine longitudinal response. Figure 10 presents one design for a lead-scintillating fiber calorimeter which looks very promising and is described in the contribution of Wigmans to these Proceedings. The important issues of uniformity, stability, radiation hardness, and e/π rejection ratio should be evaluated in the next couple of years.

Gas Sampling. Calorimeters which employ sampling with wire chambers or proportional tubes, which we will often refer to simply as gas calorimetry, do enable fine segmentation in both the transverse and longitudinal directions in a simple and straightforward way (by means of pad readout). In addition, such calorimeters are relatively easy to construct and reasonably inexpensive. Cracks or dead spaces in the calorimetry occupy a rather small fraction of the volume, although they certainly exist.

Disadvantages of gas calorimetry include the fact that it is less dense and it yields poorer energy resolution than liquid ionization or scintillator calorimetry. The gaps between the absorber plates are typically of order 1 cm, though at least one electromagnetic calorimeter has been constructed with gaps as small as 5 mm [24]. The energy resolution is typically 1.5–2 times worse for electromagnetic showers and 1.5 times worse for hadronic showers (for compensating calorimeters). Of greatest concern, however, are the facts that (1) it is difficult to maintain the mechanical tolerances required for the

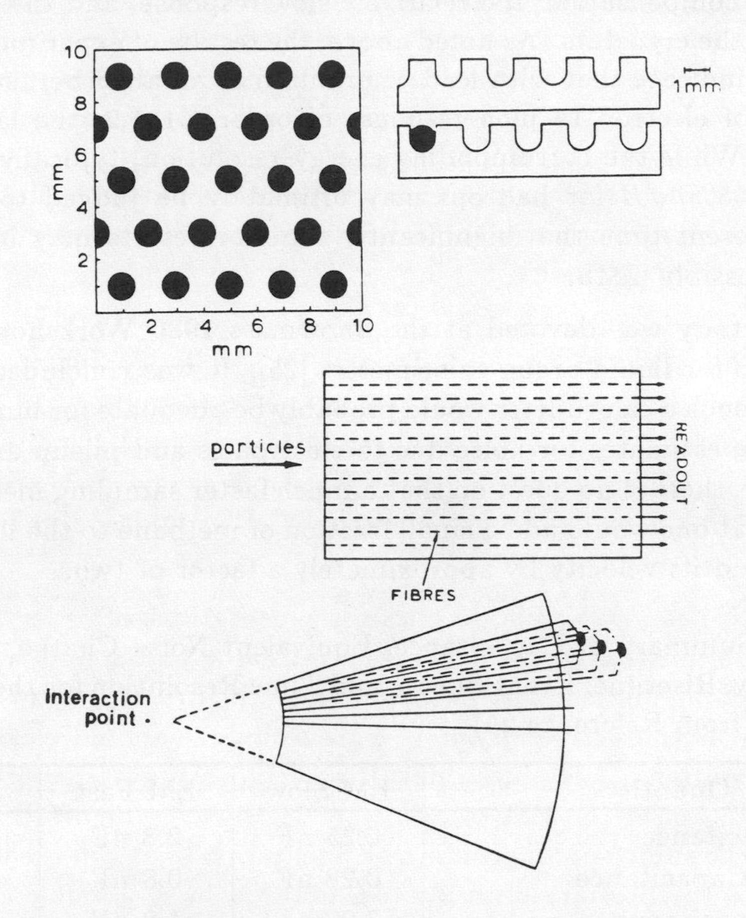

Fig. 10. Schematic design for a lead-scintillating fiber calorimeter.

response to be uniform to 1–2% over the entire calorimeter, (2) the gain will vary with changes in the temperature and pressure, (3) it is highly unlikely that gas calorimeters will sustain the instantaneous rate and radiation levels in the forward direction, and (4) the very low sampling fraction allows low energy neutrons scattering from protons to simulate very large energy depositions (as large as 50 GeV). This latter effect, which has proven to be a significant problem in some of the CDF gas calorimeters, could present a very serious problem for missing E_T measurements. It is possible that this latter effect can be strongly reduced by designing calorimeters with very low hydrogen content, and that each of the other concerns can be overcome and dealt with in a satisfactory manner. However, at the moment the problems seem sufficiently difficult and fundamental that gas calorimetry does not seem a likely candidate for the primary calorimeter system.

Liquid Argon. In comparison with scintillator and gas sampling, the use of liquid argon has a long list of advantages. The uniformity of response, both as a function of position and as a function of time, is excellent. The goal of a systematic error of 1% should be able to be obtained with a modest amount of effort. There is also no fundamental problem with achieving the degree of segmentation required.

Drawbacks of the use of liquid argon include the fact that one may not be able to attain excellent compensation, the relatively slow response, and the problems of non-hermeticity due to the cryostats. As noted above, the results of experimental and theoretical investigations indicate that with lead or uranium as an absorber, such a calorimeter will yield a ratio of electron to pion response of order 1.1–1.2, the latter value corresponding to lead. While the corresponding energy resolution, typically 12–15%/$\sqrt{E}$ for electrons and 45–55%/$\sqrt{E}$ for hadrons may ultimately be judged to be adequate, it appears at the present time that significantly superior results may be obtained using scintillator, and possibly TMS.

Considerable study was devoted at the Snowmass 1986 Workshop to the question of noise and pileup for liquid argon calorimeters [25]. It was concluded that the charge collection times of such a calorimeter would probably be adequate for luminosities of order 10^{33} cm^{-2}s^{-1}. The estimates for noise due to electronics and pileup are summarized in Table VI. However, there is no question that a much faster sampling medium is desirable. It is quite likely that one would add a small fraction of methane to the liquid argon which would increase the drift velocity by approximately a factor of two.

Table VI. Summary of Capacitance, Equivalent Noise Charge, Equivalent Noise Energy, Risetime, Event Pileup, and Time Resolution for the Calorimeter System (from Reference 25)

Quantity	EM Slow	EM Fast	Hadronic
Pad Capacitance	0.25 nF	2.8 nF	51 nF
Stripline Capacitance	0.28 nF	0.8 nF	
Cable Capacitance	0.90 nF	1.8 nF	9 nF
Total Capacitance	1.4 nF	5.4 nF	60 nF
Risetime	325 ns	65 ns	365 ns
Measurement or Peaking Time	400 ns	100 ns	500 ns
Equivalent Noise Charge	4800 e	25000 e	30000 e
Equivalent Noise Energy	15 MeV	65 MeV	120 MeV
Uranium Noise Energy	5 MeV	7 MeV	170 MeV
Event Pileup Noise	115 MeV	120 MeV	113 MeV
Time Resolution ($E = 5$ GeV)	4 ns	2 ns	8 ns

Perhaps the most difficult problem for liquid argon calorimeters is that of attaining minimal cracks and excellent hermeticity. Of the three large liquid argon calorimeters under construction, only that for the H1 experiment at HERA approaches the uniform coverage required. The H1 calorimeter covers only approximately 60–70% of the solid angle; whether a similar approach could be utilized to cover the entire solid angle remains to be seen. Access to the detectors inside such a calorimeter is a difficult problem. One may conclude that while it may be possible to construct a hermetic liquid argon calorimeter, no one has yet demonstrated how to do it.

Warm Liquids. A sampling medium which potentially offers great promise, although it is not without difficulties, is TMS (or a similar warm liquid). It should allow excellent transverse and longitudinal segmentation. As for compensation, calculations have shown that the e/h signal ratio sensitively depends on the recombination properties of the liquid (Birk's constant, see contributed paper by Wigmans), which determine the calorimeter response to the densely ionizing particles that dominate the non-EM signal. An evaluation of existing experimental data suggests that Birk's constant is considerably larger than for liquid argon or plastic scintillator (0.04–0.05 g/MeV-cm^2), at least for low fields. Therefore, it may turn out to be difficult to achieve sufficient compensation with lead absorber of acceptable thickness. Of course, this remains to be experimentally verified. Difficulties with TMS, as are well known, are the flammability, very high purity required, and relatively low ionization yield at moderate field strengths. While the safety issue remains to be answered, it has been suggested that when used together with lead, rather than uranium as was originally suggested, it should be possible to design a safe system. Techniques have been demonstrated which allow sufficient purity to be obtained in very small systems; it remains to be demonstrated that this purity can be maintained over long periods of time in very large systems. While each of these issues is a significant technical challenge, there is no *a priori* reason why each may not be solved. Perhaps the ultimate determining factor will be whether a system can be designed which simultaneously achieves the desired segmentation, the required purity, and the large drift fields at a tolerable cost.

There is considerable disagreement at the present time as to whether TMS is inferior or superior to liquid argon in terms of signal-to-noise ratio. For fields on the order of 15 kV/cm and for shaping times long enough that all the ionization is collected (several hundred nanoseconds) liquid argon is clearly superior since the total collected charge/cm exceeds that for TMS by more than a factor of five. For very short shaping times (e.g., tens of nanoseconds) the situation is substantially altered since the total "useful" induced charge is proportional to the peak induced current; in this case the much higher drift velocity of TMS, approximately 15×10^5 cm/s at 15 kV/cm *vs.* 5×10^5 cm/s for liquid argon, implies that the peak induced current for TMS is nearly 60% that for liquid argon, if one assumes the free ionization yields quoted by Gonidec *et al.*[26]. Furthermore, the signal for TMS increases rapidly with electric field since the ionization yield and the drift velocity both increase as the electric field increases. Strovink, based on a study for the D0 experiment, estimates that for fields in excess of 25 kV/cm, the induced current for TMS will exceed that for liquid argon [27]. On the other hand, a comparison by Radeka [28] concludes that TMS will yield a signal-to-noise ratio about 2/3 that of liquid argon even for very short shaping times and 25 kV/cm. A significant part of this discrepancy is probably due to the fact that the ionization yields reported by Gonidec are more than 50% larger than earlier results reported by Engler and Keim [29].

Even if one assumes the larger values reported by Gonidec, it is important to note the following:

1. It may be difficult to reach shaping times significantly less than 100 ns due to the capacitance and inductance of connections in a real calorimeter.

2. The signal-to-noise ratio achievable for shaping times of several tens of nanoseconds may not be acceptable for TMS or liquid argon (thereby forcing the use of longer shaping times).
3. Addition of small amounts of methane to liquid argon increases the drift velocity, and hence the signal relative to TMS, by a factor of two.

Our conclusion is that while the obvious warm-liquid advantages of TMS give it a very high priority for R&D, the potential signal-to-noise advantages of liquid argon are such that R&D is clearly warranted to solve the cryogenic and hermeticity problems.

Silicon Sampling. The use of silicon wafers to sample the ionization energy would allow very fine segmentation, fast response, and excellent calibration and stability. It has been estimated that by including sheets of polyethylene, or similar hydrogen rich material, next to each silicon layer, silicon sampling calorimeters may attain an e/h signal response near 1.0 [21]. Whether or not silicon is sufficiently radiation hard to be used in the calorimeter has not been conclusively demonstrated. It is known that sufficient bulk damage occurs at the radiation levels at the SSC that leakage currents will be significantly increased. However, it is argued that at the very short shaping times that would be utilized, the contribution to the noise due to this leakage current would be insignificant. On the other hand, it is currently estimated that very large fluxes of neutrons will exist within the calorimeters, and the damage due to this source has not yet been thoroughly investigated.

The greatest obstacle to the use of silicon as the primary sampling medium is cost. It has been estimated that if the cost per wafer can be reduced by a factor of ten, and there is some optimism that this can be achieved, it would be possible to build compact calorimeters for use at the SSC [18]. The use of expensive high-density absorber materials (tungsten, uranium) would then be justified by the amount of money saved on silicon and on the overall size of the detector. It should be mentioned that fission neutrons from uranium absorber would increase the radiation sensitivity problems for silicon. In any case, we have concluded that for the size calorimeter envisioned for the Large Solenoid Detector, silicon is too expensive even if the goal of a times ten reduction in cost/cm^2 is achieved. It seems more likely that silicon calorimeters may be used for specialized applications.

6.3.3. Summary of Calorimeter Composition

It is quite apparent that at this point in time, one cannot choose which calorimeter type will prove to be optimal. The leading candidates together with the primary technical problems that need to be answered are:

1. Lead-scintillating fibers

    Radiation hardness

    Uniformity of response

    Long-term stability

    e/π rejection

2. Lead-TMS

    Purity (including long term stability)

    Measurements of Birk's constant, e/h signal ratio, and hadronic energy resolution

    Safety

    High drift fields

    Are materials required for finely-segmented readout (e.g., printed circuit board techniques) consistent with required purity?

3. Lead-liquid argon

    Hermeticity

    Further calculations on adequacy of speed

    Further measurements on e/π ratio

6.4. Size and Layout

It has often been pointed out that a barrel calorimeter which is infinitely long, and has uniform segmentation along the beam direction, gives approximately the uniform sampling in rapidity that is desired. Needless to say, such a geometry is not practical both because of the overall size, and because of the acute angle between the calorimeter face and the incident particle direction for large values of rapidity. More realistic approximations to an ideal calorimeter have been proposed which begin with a barrel calorimeter, and gradually make the transition to the small angle calorimeter with plates perpendicular to the beam direction. While not explicitly indicated in the schematic diagram, we anticipate that if a plate geometry is ultimately utilized, such as for liquid argon or TMS sampling, a compromise between mechanical simplicity and the optimal performance would probably result in three plate orientations: parallel to the beam in the barrel region, perpendicular in the forward region, and at an intermediate angle in the middle region. If lead-scintillating fiber calorimeters prove feasible, it should be possible to make a very uniform and gradual change in the orientation as a function of rapidity.

The inner radius of the calorimeter in the present design is set by the space allowed for the tracking; it is well matched to the calorimeter requirements, however. The inner radius of 1.6 m at 90° allows for excellent angular information on jets, and the 4 m distance between the face of the calorimeter at 5° and the vertex results in a minimum tower size of 1.05 cm $\times$ 1.05 cm and in a counting rate per tower of 10^5 per second (the latter number neglects shower spreading).

In the 1986 Snowmass Workshop, the detector design located the coil in the middle of the calorimeter, with an internal "precision" calorimeter of 6 λ and an external "catcher" calorimeter. Because of concern that the calorimeter performance might be significantly compromised if the magnet coil were inserted in the middle, it was decided during this study to explore the possibility of placing the entire calorimeter inside the magnetic field. While the resulting magnet is quite large, such an approach appears to be quite feasible and may greatly improve the uniformity of the calorimeter. This approach results in a more compact calorimeter, and the savings in the size of the muon system and the overall calorimeter volume in the central region may more than compensate for the larger volume of "precision" calorimetry.

7. Electron Identification

The electron identification requirements for high p_T physics at the SSC, as determined by the physics parameterization groups at this Workshop, may be summarized as follows:

1. p_T(electron) > 10 GeV/c.
2. $|\eta| < 3$.
3. Hadron misidentification probability $\sim 10^{-3}$, principally for isolated particles, or jet misidentification probability $\lesssim 10^{-4}$.

Most of the high p_T-electron physics studied involves isolated electrons, with varying isolation criteria which are all coarse on the scale of the calorimeter segmentation. A notable exception is the intermediate mass Higgs particle, decaying to $b\bar{b}$. In this case, to select the b in its $b \rightarrow e\bar{\nu}_e X$ decay mode, one needs to be able to identify electrons in the b-jet, again at a $\sim 10^{-3}$ hadron rejection level. It should be noted that the requirements specified here are less severe than those stated in the Snowmass 1986 report [19,30]. In this report, we assume the goals as stated by the parametrization groups and provide evidence that an electron/pion discrimination of 10^{-3} may be obtained solely with the calorimeter and that rejections of order 3×10^{-4} may be achieved if E/p cuts are also included. An example of an "existence proof" that 10^{-3} performance can be obtained with a calorimeter is provided by the CDF beam tests of a lead-scintillator prototype shower counter [31]. We have therefore not considered in detail the use of auxiliary systems such as transition radiation devices (TRDs). However, as is discussed in greater detail at the end of this section, it is quite possible that better pion rejection will be required; in this case an auxiliary system such as TRDs should be included. We note that a nice review of the existing literature on π/e separation appears in the Report of the SSC Detector R&D Task Force [32]. We will not attempt to repeat that review here, concentrating instead on a few specific issues of central significance. We also do not discuss electron identification in the forward detector ($|\eta| > 3$).

7.1. Calorimeter Segmentation

We start by discussing the electron/pion discrimination which can be achieved with a segmented calorimeter alone, without momentum measurements, TRDs, etc. A useful test-beam study has been performed using an array of 60 lead glass bars [33]. In this study, the bars ($6.5 \times 6.5 \times 140\,\text{cm}^3$) were stacked in an array 5 across by 12 deep (total depth $\approx 24\ X_0$). Test beams of 15 to 47 GeV/c electrons and pions were measured with this array. By adding signals from various bars, the investigators were able to study the dependence of π-rejection on segmentation. With the finest segmentations, they obtained a pion rejection factor of 10^{-3}, independent of momentum in the range studied, with an electron efficiency of 90%. This result is obtained solely by the shower information in the bars – no knowledge of the true momentum is required. We may summarize their results on the performance of various transverse and longitudinal segmentations in Table VII.

Table VII. Performance of Various Transverse and Longitudinal Segmentations

Transverse Segmentation	Longitudinal Segmentation	Pion Rejection
$5 \times 2\ X_0$	$12 \times 2\ X_0$	1×10^{-3}
$5 \times 2\ X_0$	$4 \times 6\ X_0$	1×10^{-3}
$5 \times 2\ X_0$	$4\ X_0,\ 6\ X_0,\ 6\ X_0,\ 8\ X_0$	2×10^{-3}
$5 \times 2\ X_0$	$2\ X_0,\ 6\ X_0,\ 8\ X_0,\ 8\ X_0$	2×10^{-3}
$5 \times 2\ X_0$	$3 \times 8\ X_0$	2×10^{-3}
$5 \times 2\ X_0$	$6\ X_0,\ 8\ X_0,\ 10\ X_0$	2×10^{-3}
$5 \times 2\ X_0$	$2\ X_0,\ 6\ X_0,\ 16\ X_0$	4×10^{-3}
$4\ X_0,\ 6\ X_0$	$4 \times 6\ X_0$	7×10^{-3}
$10\ X_0$	$4 \times 6\ X_0$	60×10^{-3}

We see that no advantage was obtained using longitudinal segmentations finer than 4 total segments. In the transverse direction, anything coarser than the finest available resulted in significant performance loss. Thus, this study does not reveal the ultimate limit achievable with transverse segmentation – it is possible that better than 10^{-3} rejection can be obtained with finer division, though we know of no study which demonstrates this.

It must be kept in mind when applying these results to our present problem that there are several differences between the test apparatus and the sort of calorimeter we have in mind for the SSC: (1) The SSC device will presumably have a smaller energy sampling fraction, resulting primarily in a poorer energy resolution. (2) The SSC device will have transverse segmentation in two dimensions, rather than just one. Thus, better performance may be expected, but it is unknown by what factor. The factor is unlikely to be large, because of the high correlation (approximate circular symmetry) between the two dimensions. (3) The SSC electromagnetic calorimeter will be followed by hadron calorimetry, yielding information beyond the first $\sim 25\ X_0 \sim 1\ \lambda$. (4) The test beams used covered only the lower end of the energy scale we are interested in. Since no energy dependence was observed, and since shower shape properties tend to depend logarithmically on energy, a fairly substantial extrapolation may be valid. On the other hand, exclusive charge-exchange processes tend to decrease rapidly with increasing energy, suggesting that our extrapolation might err on the conservative side. (5) A lead-glass array is principally a Čerenkov radiation device, while the calorimeter for a large solenoid SSC detector will probably be ionization-sensitive. Thus, the response to hadronic showers may be somewhat different.

Keeping the above issues in mind, we expect our calorimeter design to have a pion rejection performance (for $\sim$ 90% electron efficiency) of order 10^{-3} for isolated particles, using the shower shape information. In particular, we choose a 3-part longitudinal segmentation in the electromagnetic calorimeter of 6 X_0, 8 X_0, and 11 X_0, with the hadronic calorimeter serving as a fourth longitudinal segment. This is under the assumption that a fourth segmentation within the electromagnetic portion will not yield significant improvement because of the information provided by the hadronic section. The transverse segmentation in the electromagnetic calorimeter is nominally $\Delta\eta \times \Delta\phi = 0.02 \times 0.02 - 0.03 \times 0.03$, providing a sufficiently fine grain to separate isolated particles from jets. We may calculate the approximate segmentation dimensions in units relevant to the shower at various angles, as shown in Table VIII.

Table VIII. Segmentation Dimensions*

η	θ (degrees)	s_ϕ (cm)	$s_\phi(X_0)$	s_η (cm)
0	90	3.4	3.0	3.4
1.64	22	3.4	3.0	9.1
2.0	15.4	2.2	2.0	2.3
2.0	15.4	3.3	2.9	3.4
3.0	5.7	1.2	1.1	1.2

* For $\eta \leq 1.64$, dimensions are calculated at a nominal 170 cm radius; for $\eta > 1.64$ at a nominal 400 cm $|z|$. For the first three rows, the segmentation is $\Delta\eta \times \Delta\phi = 0.02 \times 0.02$, and for the last two rows it is 0.03×0.03. It should be noted that the distance quoted for s_η in the barrel region is the distance parallel to the beam line. In fact, the most relevant distance is the size of the tower perpendicular to the direction of the particle. Hence, for example, for $\eta = 1.64$, the effective transverse size of the tower perpendicular to the direction of the particle and in the η direction is 3.4 cm.

The dimensions in X_0 are approximate, based on 50% of the volume occupied by lead, and the active material neglected. Once the corner is turned (from barrel geometry to endcap geometry – see Fig. 1(a)) at $|\eta| \sim 1.64$, the transverse segmentation becomes finer, compared to shower dimensions. To prevent the segmentation from becoming unmanageably small, and to reduce the channel count, we propose to change from the 0.02×0.02 segmentation to 0.03×0.03 at $|\eta| = 2$. For all of the barrel region, and some of the end region, the segmentation may be coarser than the above discussions suggest would be useful. If further study verified this, we could segment the middle longitudinal segment (containing shower maximum) more finely, by a factor of 2, in the ϕ coordinate for rapidities $|\eta| \lesssim 2.0$, at a cost of $\sim$ 60k additional channels. The total number of channels in the electromagnetic calorimeter with the longitudinal and transverse segmentation as described is $\sim$ 231k, for coverage to $|\eta| = 3$. An additional $\sim$ 81k channels are needed for forward electromagnetic calorimetry.

7.2. Measurement of E/p

The ability to measure the momentum of a particle with tracking in a magnetic field allows for a comparison with the energy deposited in the calorimeter. For an electron, we expect the momentum and energy measurements to be equal within error, while for a hadron, the energy deposited in the electromagnetic calorimeter will usually be substantially less than the measured momentum. This suggests both additional hadron rejection power over the calorimeter alone, as well as a redundant check of the calorimeter information. We have investigated these issues for devices with our design resolutions.

To get an idea of the additional power obtained by adding the measurement of E/p, it is again convenient to look at the work of Reference 33. They show a graph (see Fig. 11) of the probability that a 15–47 GeV/c π^- will survive shower shape cuts, as a function of E/p. Within statistics, this probability distribution is independent of momentum over the measured range. We may use this distribution to get an estimate of the additional rejection achievable by adding a momentum measurement. For this analysis we make the simple parameterization of the distribution:

$$f(E/p) = \begin{cases} a_1 e^{b_1 E/p} & 0.0 < E/p < 0.92 \\ c & 0.92 < E/p < 0.97 \\ a_2 e^{-b_2 E/p} & 0.97 < E/p \end{cases} \tag{2}$$

(normalized to $\int_0^\infty f(x)\,dx = 1$, with $a_1 = 0.0129$, $b_1 = 6.3$, $c = 4.23$, $a_2 = 2.03\times 10^{16}$, and $b_2 = 37.2$). The peak of the distribution is near $E/p = 1$, with a very steep fall-off above and a slower fall-off below. Thus, the shower information tends to pick hadronic showers in the electromagnetic calorimeter which contain most of the pion's initial energy, and we shouldn't expect a very large additional rejection from the added momentum information.

Use of the momentum information can be incorporated as a cut in E/p, which must be based on the resolution in E/p. For our estimates, we parameterize the momentum and electromagnetic calorimeter resolutions as:

$$\sigma_p/p = \sqrt{(a_{meas} p_T)^2 + a_{mcs}^2} \tag{3}$$

and

$$\sigma_E/E = \sqrt{(a_{samp}/\sqrt{E})^2 + a_{cal}^2 + (a_{noise}/E)^2} \quad . \tag{4}$$

Figure 12 shows the additional improvement attained by adding our E/p measurement to the shower shape rejection. The prediction above 50 GeV/c depends on the untested assumption that $f(E/p)$ remains invariant.

Reference 33 also contains a distribution for the energy deposited in the lead glass for 47 GeV/c pions before shower shape cuts. This distribution has a high peak below 1 GeV deposited energy, a broad peak near 20 GeV, and a tail extending up to the beam energy. We may apply a $\pm 2\sigma_{E/p}$ cut around $E/p = 1$ on this distribution to estimate the hadron rejection possible with a momentum-energy comparison alone (the energy

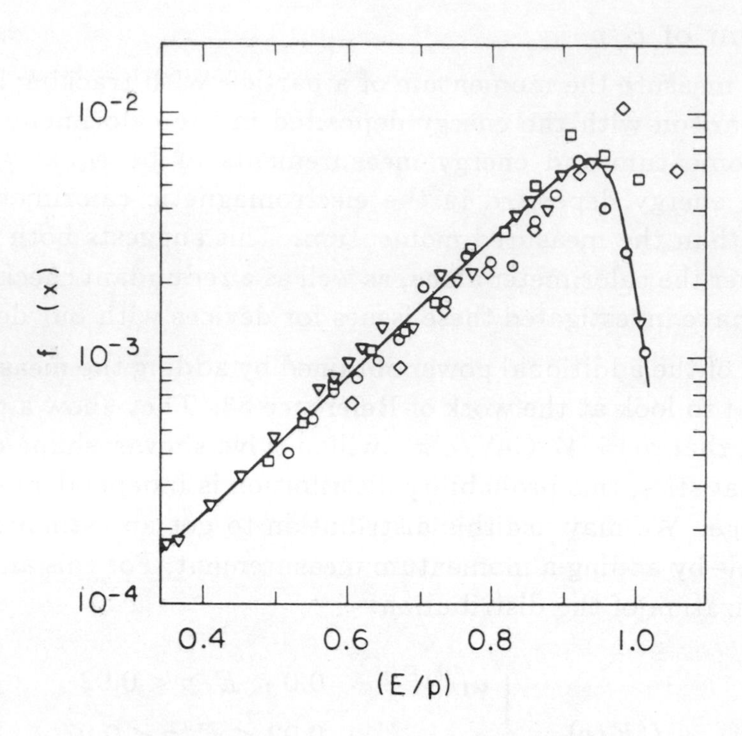

Fig. 11. Probability that a 15–47 GeV/c π^- will survive shower shape cuts, as a function of E/p (from Reference 33).

Fig. 12. Graph showing the estimated improvement factor for π rejection as a function of momentum obtained by adding a momentum measurement to the shower information from a well-segmented calorimeter which achieves 10^{-3} rejection before the momentum information. A $\pm 2\sigma$ cut in E/p is used. The energy and momentum resolutions (see text) are calculated with $a_{mcs} = 0.01$, $a_{cal} = 0.01$, and $a_{noise} = 0.5$ GeV. The solid curves are with $a_{meas} = 0.00026\,(\mathrm{GeV/c})^{-1}$, as appropriate for a beam-constrained momentum measurement, and the dashed curves are for $a_{meas} = 0.00054\,(\mathrm{GeV/c})^{-1}$, corresponding to no beam-constraint. For each pair of curves, the top one is with $a_{samp} = 0.10\,\mathrm{GeV}^{\frac{1}{2}}$ and the bottom one is with $a_{samp} = 0.15\,\mathrm{GeV}^{\frac{1}{2}}$.

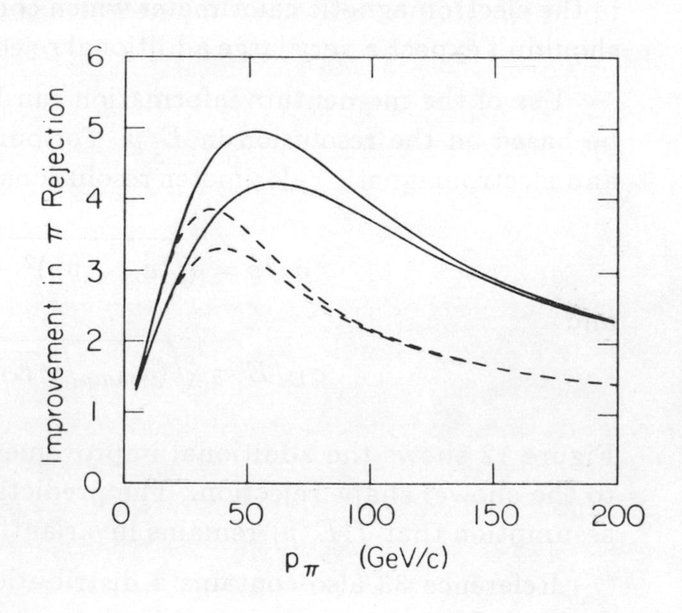

resolution of the lead-glass array is good enough to be neglected in this estimate). The result is a rejection of $\sim 2\text{–}3\times10^{-3}$, with an additional uncertainty of perhaps 50% from residual electron contamination in the pion beam.

We conclude this discussion with the following observations: Using the calorimeter information alone, a pion rejection of $\sim 10^{-3}$ is possible, and using E/p alone a rejection of a few $\times\ 10^{-3}$, at least for 50 GeV/c pions. Adding the E/p information to the shower rejection improves the rejection power by factors of a few. There is substantial correlation between the two methods, so they are primarily redundant rather than independent methods. Having both approaches means that they can be used as cross checks on each other. Furthermore, one method, E/p, is most powerful for $|\eta| \lesssim 1.6$, where the momentum resolution is best, and the other method is most powerful for $1.6 \lesssim |\eta| \lesssim 3.0$, where the transverse segmentation is best. Thus, we expect to be able to meet the desired specification for all $|\eta| \lesssim 3.0$ by using the two methods. While we have concentrated our discussion on truly isolated particles, it is worth noting that the E/p information is extremely valuable in rejecting situations with photons overlapping $\pi^{\pm}$ tracks [19]. It also should be noted that tracking in a magnetic field enhances our ability to reject backgrounds from Dalitz pairs and converted photons. The separation, s, of the e^+ and e^- from the symmetric conversion of a photon of energy E_γ, after traversing 1 meter in radial distance from the conversion point is given by $s \approx [1.2/E_\gamma(\text{TeV})]$ mm, assuming a 2 Tesla field. Thus, the separation is readily detected up to quite high energies. It should finally be stressed that we have based our estimates heavily on the detailed information available at $\sim$ 50 GeV/c – studies with realistic and flexible prototypes at higher test beam energies will be a crucial step in any rational design process for an SSC calorimeter.

7.3. Transition and Synchrotron Radiation Devices

Given the preceding discussion, and the specifications put forward by the physics parameterization subgroups, it can be concluded that an appropriately segmented calorimeter, plus some momentum measurement capability, will be sufficient to do the physics. However, as noted above, we have varying degrees of discomfort on this matter for the following reasons: First, there is some concern that the conclusions of the parameterization groups may understate the requirements. Perhaps the background studies have so far been insufficiently comprehensive or realistic. This concern is fueled partly by the more stringent requirement (10^{-4} or even 10^{-5} hadron misidentification) stated at the Snowmass 86 Workshop [19,30], although those numbers do not appear to have been the result of an exhaustive study. Second, even if the parameterization studies are accurate, there may be other physics not considered. A more conservative approach to the question of how well we might ultimately wish to do at hadron rejection is to ask what rejection is required to reduce the hadron background below the level of real "prompt" electrons. Thus, for isolated particles, we first ask how often an isolated hadron occurs in the background processes (as a function of p_T), and then what further rejection is needed to get the level below the rate of electrons from b, t and W decays. A careful investigation along these lines has not been completed, but is essential in order to understand what level of hadron rejection is optimal. Third, the calorimetric methods of electron identification are difficult when in the environment of a jet. Fortunately, this is not the typical situation of interest, but one example does exist – the decay of an intermediate mass Higgs to $b\bar{b}$. In this case, the b-jet has a fairly small p_T, so it is not hopeless, but it is marginal whether the required performance can be achieved without additional devices. Fourth, even if the 10^{-3} rejection level is adequate at $10^{33}\,\mathrm{cm}^{-2}\mathrm{s}^{-1}$, a higher-luminosity upgrade would probably be looking for rarer signals, requiring proportionately better rejection.

Because of these concerns, we feel it is imperative to keep the option available of installing additional devices for electron identification. This need for flexibility is a major argument for not shrinking the size of the central cavity any further. For example, a TRD design from Snowmass 86 requires about 60 cm in radius for four TRDs. Currently, the candidate devices with the most promise are TRDs and SRDs. Both types of device are being used in existing experiments. These possibilities, along with the difficulties requiring further study for application at the SSC, are described in References 19, 30, and 32. There is also a recent general review of TRDs by Dolgoshein [34].

An interesting idea for using TRDs in a high-luminosity environment is to turn the hardware threshold up above $\sim 8\,\mathrm{keV}$ (we imagine the Snowmass 86 design, consisting of TRDs with 100 layers of 40 micron polypropylene foil followed by a 50-50% Xe-C_2H_6 X-ray wire chamber detector). Then the many uninteresting pions below $\sim$ 300 GeV/c are effectively invisible in this device, electron identification below this momentum is available, and the TRDs may be used for tracking both pions and electrons above $\sim$ 300 GeV/c.

8. Muon Identification and Momentum Measurement

The goal of the muon detector system in the Large Solenoid Detector is to identify and measure the vector momenta of muons between about 10 GeV/c and 2 TeV/c and over a rapidity range of $\pm$ 5 units. The system chosen in this detector takes strong advantage of the fully integrated functional nature of the Large Solenoid Detector, integrating the muon momentum measurements with the central tracking detector, the large solenoid magnetic flux return yoke and the hadron calorimetry (all of which is located within the solenoid coil). Muons at small angles (less than about 30°) are measured by means of conventional magnetized iron toroids placed around the beam pipe and shaped to permit muon momentum measurements at all angles down to 0.8°. The relationship of the detector elements is shown in Fig. 1(a). The particular subsystem aspects of the muon detector are now discussed in some detail.

8.1. Momentum Measurement of Large-Angle Muons

Large-angle ($\theta > 30°$) muons leaving the interaction region are picked up by the central and intermediate tracking systems, allowing a partial orbit determination to be made. They then pass through the calorimeter where they undergo multiple scattering in the absorber material. As they exit the calorimeter and after they pass through the magnet coil, their positions and slopes are measured by modules of drift tubes located between the solenoid coil in its cryostat and the iron flux return yoke of the magnet. It is assumed that the spatial location precision of points along the orbit in the bending direction can be held within a combined statistical and systematic error of 50 μm at this position together with an angular measurement lever arm of 50 cm. It is further assumed that the longitudinal position parallel to the beam can be determined within an error of a few centimeters using current division or induced cathode charge measurements. These achievable error assumptions relate directly to the momentum resolution capability of the muon system (for high momenta especially).

For large-angle muon momenta *below* about 500 GeV/c, a classic orbit sagitta measurement can be accomplished *entirely within* the central tracking devices; the muon momentum, therefore, is well measured using data only from these detectors. The muon *particle identification function* is, of course, established in all cases by its subsequent penetration through the calorimeter and magnetized iron material outside the tracker. For muon momenta *above* 500 GeV/c, the added magnet track length gained by measuring the muon's magnetic trajectory *through the calorimeter* (with the drift tube modules) allows a very significant gain to be made in $B\,L^2$, hence in muon momentum resolution. This gain in resolution is made because the spatial resolution in sagitta measurement takes over from multiple scattering as the dominant source of error as muon momenta increase above this value.

The large-angle muons then pass through the magnetized iron return yoke where a third, largely independent, momentum measurement can be obtained by deflection in the yoke iron. The muon positions are measured once more on exiting the yoke and the deflection angle determined to an accuracy of 0.14 mrad. This angular precision assumes, as before, that the outside muon drift tube modules can be maintained in space with a statistical and systematic location error not exceeding 50 μm and that a 0.5 meter lever arm for measuring the exit angle is available. These assumptions present severe technical challenges and appropriate technology will need to be identified and developed to insure that these tolerances can be met.

Under the conditions noted in the paragraphs above, the muon momentum is measured at large angles with the fractional precision shown in Fig. 13. Even if the final exit angle from the magnetized iron yoke is not measured, the momentum precision is not degraded for most conditions (it is largely dependent on the measured angle of deflection in the central solenoid field); the valuable particle identification constraint of having a third, backup momentum determination is lost, however.

The momentum resolution graph for large-angle muons (displayed in Fig. 13) is dominated by two sources of measurement uncertainty: multiple Coulomb scattering and spatial resolution. Spatial resolution in the central tracking detector dominates the overall momentum resolution for large-angle muon momenta below about 400 GeV/c and a combination of both error sources contributes significantly in the full coil measurement path for momenta above this value. Shown in Fig. 13 is a significant gain in resolution realized by combining the two measurement methods available for large-angle muons. Of special interest is the ability of the adopted muon system to generate an electric charge sign determination for muon momenta up to about 2 TeV/c. Since the momentum measurement errors are Gaussian for the *reciprocal* of the momentum, a second graph is presented which shows the reciprocal momentum measurement error (Fig. 14). This graph allows an easier assessment of the probability of making large fractional momentum errors, or of errors in determining the electric charge sign for very high momentum muons. In the example shown, it is seen that a 2 TeV/c muon has a 2 standard deviation probability for misdetermination of the electric charge sign. To assess the magnitude of measurement errors from multiple Coulomb scattering, the calorimeter was assumed to have lead absorber plates distributed uniformly over the radial zone from 1.6 to 3.6 m in the central region, and to have a total thickness of 1900 gm/cm^2 of lead (300 X_0, 10 λ). As the angle of incidence decreases from 90°, the effective absorber thickness will increase, as will the amount of multiple scattering. There will also be a decrease in momentum measurement precision for muons below an angle of about 30° as the full turning angle in the solenoid is decreased.

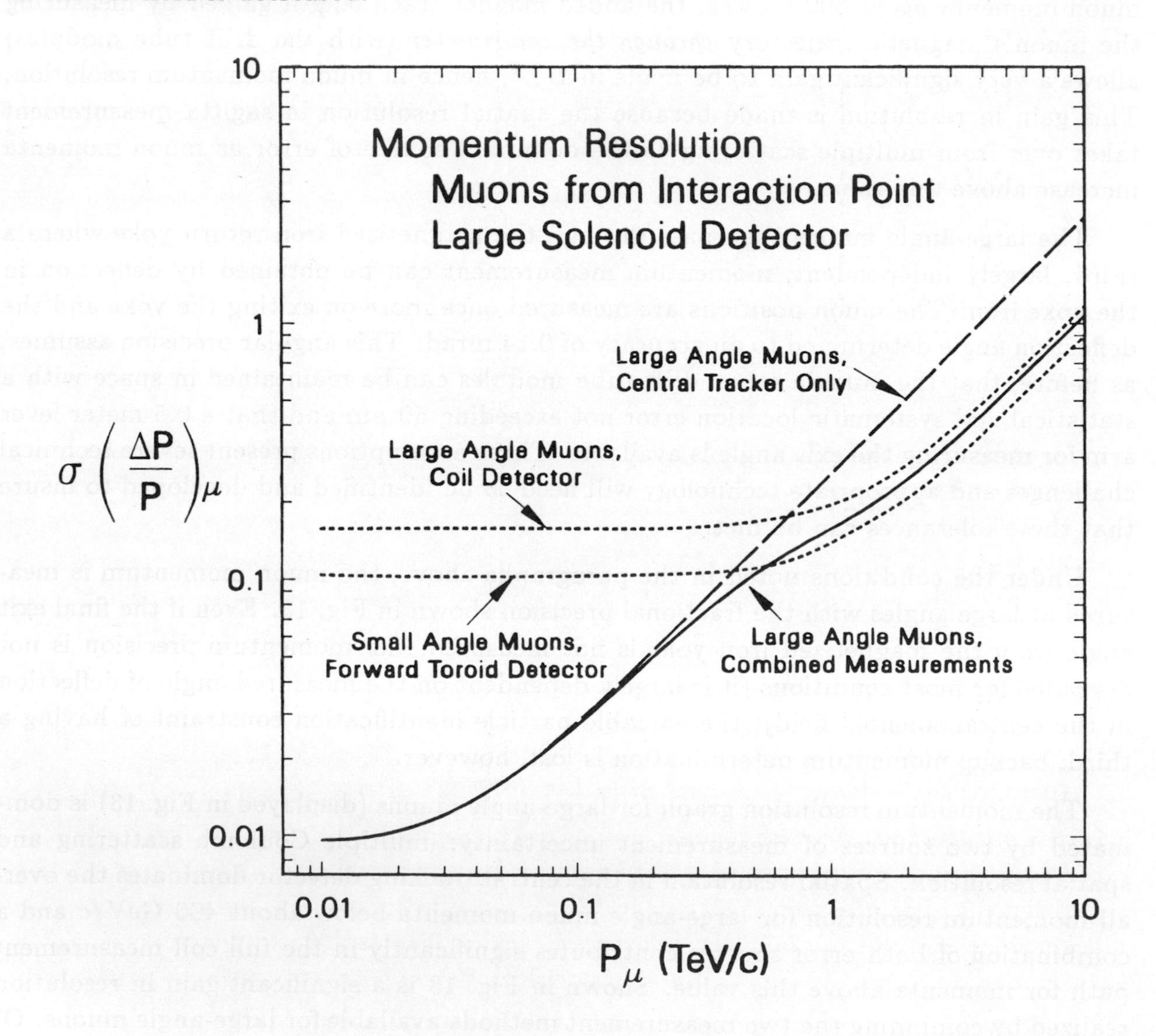

Fig. 13. Momentum resolution for measurement of large-angle and forward-going muons in the Large Solenoid Detector.

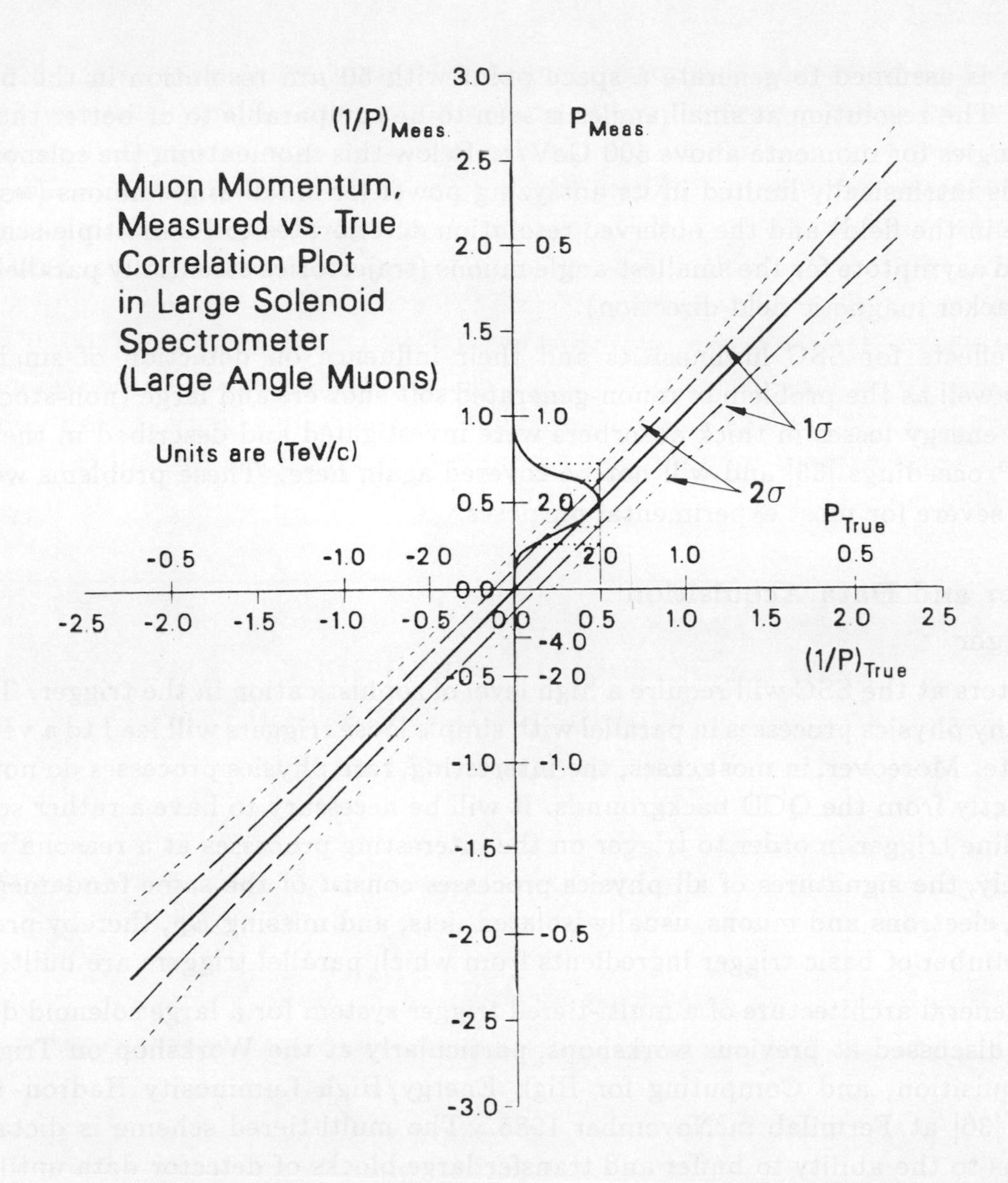

Fig. 14. Gaussian error contours for reciprocal muon momentum measurements in the Large Solenoid Detector.

8.2. Momentum Measurement of Small-Angle Muons

In the small-angle forward regime, muons are measured by deflection through magnetized iron toroids (see Fig. 1(a)). To suppress loss of measuring ability through the track-obscuring effects of soft electromagnetic showers in the toroids, redundant samples are taken of the muon position at two points within the toroids and a final angle is measured behind them. The entering muon angle is determined by the forward tracking system.

The muon momentum resolution for small-angle muons is shown in Fig. 13 for comparison with the resolution for large-angle muons. The small-angle detector elements are multi-layer muon proportional tube modules interspersed with magnetized iron toroids.

A module is assumed to generate a space point with 50 μm resolution in the bending direction. The resolution at small angles is seen to be comparable to or better than that at large angles for momenta above 500 GeV/c. Below this momentum, the solenoid field topology is intrinsically limited in its analyzing power for small-angle muons (less rotation angle in the field) and the observed resolution deteriorates to its multiple-scattering dominated asymptote for the smallest-angle muons (trajectories essentially parallel to the central tracker magnetic field direction).

Rate effects for SSC luminosities and their influence on detection of small-angle muons, as well as the problem of muon-generated soft showers and large (non-stochastic) fractional energy losses in thick absorbers were investigated and described in the Snowmass 86 Proceedings [35] and will not be covered again here. These problems were not felt to be severe for most experimental purposes.

9. Trigger and Data Acquisition

9.1. Trigger

Detectors at the SSC will require a high level of sophistication in the trigger. Triggering on many physics processes in parallel with simple loose triggers will lead to a very high trigger rate. Moreover, in most cases, the interesting, rare physics processes do not stand out distinctly from the QCD backgrounds. It will be necessary to have a rather sophisticated on-line trigger in order to trigger on the interesting processes at a reasonable rate. Fortunately, the signatures of all physics processes consist of the same fundamental ingredients, electrons and muons, usually isolated, jets, and missing E_T, thereby providing a small number of basic trigger ingredients from which parallel triggers are built.

The general architecture of a multi-tiered trigger system for a large solenoid detector has been discussed at previous workshops, particularly at the Workshop on Triggering, Data Acquisition, and Computing for High Energy/High Luminosity Hadron–Hadron Colliders [36] at Fermilab in November 1985. The multi-tiered scheme is dictated by limitations to the ability to buffer and transfer large blocks of detector data until a final trigger decision is complete. Each tier rejects a large fraction of the event candidates received from the previous tier, thereby affording the next tier more time to make a more sophisticated decision to reject further events. The first tier, or Level 1, is usually considered to make the most rapid decision possible, using analog sums of calorimeter data and at least some clustering to obtain a rejection of 10^3–10^4 in about 1 μs. Level 2 is then allowed about 10 μs to reject an additional factor of 10^2 using more detailed considerations, such as the distribution and clustering of energy and the association of charged tracks with electromagnetic energy. Finally, Level 3, which is itself multi-tiered, has available for the first time all the data from all parts of the detector. It then uses a large farm of microprocessors to reject an additional factor of about 10^2, reducing the final event rate to 1–10 Hz. Table IX, borrowed from the report of the Physics Signatures Working Group [37] at the Fermilab Triggering Workshop, illustrates how one path in a parallel trigger might select Higgs events of the sort:

$$H \rightarrow W + W \rightarrow e + \nu + jet + jet \quad .$$

Table IX. Summary of Trigger Strategy and ISAJET Results for $H \rightarrow W^+ + W^-$. Note that the rejection factors apply to individual cuts while the efficiencies are cumulative.

Trigger Selections	Rejection Factor	Remaining Cross Section (nb)	$H \rightarrow WW$ Efficiency
First Level:			
(a) Select electron candidate as calorimeter cell with $E_T > 25$ GeV and with at least 80% of the energy electromagnetic	700	3×10^4	0.86
Require $p_T^{miss} > 40$ GeV/c for the event	4	7500	0.43
Second Level:			
Require the electron candidate to be isolated, with a surrounding region of ± 5 calorimeter cells in both η and ϕ containing less than 20% of the E_T of the electron candidate cell	1.3	5700	0.37
(b) Make a jet requirement of either 1 jet having $E_T > 80$ GeV or 2 jets each having $E_T > 40$ GeV	20	290	0.32
Third Level:			
Require that tracking show a charged particle with $p_T > 10$ GeV/c pointing to the candidate electron calorimeter cell	255	1.1	0.32

Note that the "Third Level" trigger selection of the previous example required a charged track with $p_T > 10$ GeV/c and pointing to the candidate electron calorimeter cell. It is therefore very important that one have some minimum cut on charged track momentum in the trigger; this could be accomplished either by reconstructing the momenta of stiff tracks or by use of TRD information. One method for determining the p_T of a charged track associated with a cluster of electromagnetic energy in the calorimeter is to measure the angle of the track in the outer superlayer (group of 8 layers) of the central tracking system. For moderate momenta, sufficient angular resolution exists, even without realizing at the trigger level the full spatial resolution of the drift chamber. Thus, the magnetic field provides an additional tool at the trigger level to reducing backgrounds to electron candidates.

The above example of Higgs decay into W-pairs is but one example of many interesting physics processes upon which to trigger. It appears to be a manageable example as described by the Fermilab Triggering Workshop; however, it is a relatively easy example. For each other physics process of interest a set of analysis cuts must be defined in order that a set of realistic on-line trigger cuts can be chosen. Then a conceptual design for trigger algorithms, including appropriate hardware, can be developed for representative detector topologies. At that stage, two crucial issues can be addressed. First, since the time required for trigger decisions will depend on detector topology, are there important implications for the detector topology? For instance, does matching charged tracks to electromagnetic showers prefer a particular geometry of tracking chambers, or is there an appropriate segmentation for a transition radiation detector to match it to the calorimeter? Second, how sharp must cuts be at the trigger level; for instance, what p_T resolution is needed at the trigger level to efficiently trigger on electrons from $W \rightarrow e\nu$ without excessive trigger rates from background? This consideration will impact the uniformity of response necessary from detectors and electronics and will determine what calibration corrections will be needed at the trigger level. Since details of the trigger affect the details of detector design in such ways, the trigger must be realistically included in simulations of detector designs. The difficult triggering environment at the SSC demands that the trigger be considered as part of the interplay of physics goals and detector design.

9.2. Data Acquisition

The general aspects of data acquisition for a large solenoid detector were described by the report of the Data Filtering/Acquisition Group at the November 1985 Workshop on Triggering, Data Acquisition, and Computing for High Energy/High Luminosity Hadron-Hadron Colliders at Fermilab [38]. The data acquisition electronics includes detector-mounted custom VLSI circuits to amplify, shape, and sample the detector signals and to buffer the samples during Level 1 and Level 2 trigger decisions. During the Level 1 decision, samples from all beam crossings must be buffered, and during the Level 2 decision, samples from all Level 1 triggers must be buffered. An example of this "front-end" electronics was also presented at the Fermilab Workshop [39]. More detailed examples of electronics for drift-time measurement and calorimetric measurement were described in the report of the Triggering and Electronics Group at the Snowmass 1986 Summer Study [40]. These workshops highlighted the concerns of power dissipation and radiation hardness and the need for VLSI R&D for both the amplifying/shaping and the sampling/buffering functions. The front-end electronics must also preprocess and sparse scan the data for each Level 2 trigger in order to limit the required bandwidth at the output of the front-end electronics.

Trigger requirements will impact the design of the front-end electronics; however, the implications have not yet been explored. Note that the quantity of data in the front-end buffers is tremendous and the portion of this data which can be used in the trigger may be limited by the bandwidth of the connections and busses linking the front-end electronics to the trigger processors. Locating a large part of the trigger electronics at the front-end electronics and segmenting the trigger electronics in a geographical way will ease this bandwidth issue and at the same time provide parallelism in the trigger processing which is needed for prompt trigger decisions. Examples of local and geographic trigger processing include local shower clustering and local track segment finding.

Conceptually, all sparse-scanned data from Level 2 triggers is transferred via event builders to the Level 3 trigger processors, which are a farm of general-purpose microprocessors. This portion of the data acquisition system perhaps looks more conventional than the front-end electronics; however, the necessary bandwidths and processing power far exceed systems currently being implemented. We should be encouraged by the experience now being gained, and soon to be gained, with microprocessor farms in several existing detectors. Nonetheless, new hardware and, perhaps more importantly, new software tools will be needed to manage this data and processing.

The task of producing particle four-vectors, which has traditionally been performed off-line, may be efficiently performed on-line at the SSC. The basic tasks of reconstructing energy clusters and charged tracks and of associating particle identification attributes will be performed to a large extent in the trigger and the Level 3 processors. Retaining this information, even if further refinement in on-line algorithms is necessary, will lead to economies in computing. Furthermore, reducing the raw data to reconstructed tracks and parameters on-line will lead to economies in data handling and bandwidth, which if used judiciously can increase the acceptance of interesting physics events and processes.

In summary, the detailed architecture of the data acquisition system is yet to be developed. It will to a large extent be determined by considerations of power and of bandwidth. Economies in how much data is buffered, transferred, and processed will alleviate these concerns. Thus, the details of the distribution of processing within the architecture sketched above will be significant.

10. Research and Development Requirements

A considerable amount of research and development will be required before a large solenoid detector for the SSC can be designed and built. The time needed for R&D, design, and construction for this type of detector is probably commensurate with the time scale for building the SSC itself since the detector concepts are rather well understood. The research and development requirements for the various detector components are described in the following sections.

10.1. Solenoid Magnet

The main work needed here is not so much R&D since we have chosen an existing superconductor but an honest engineering conceptual design to determine the feasibility of producing the proposed solenoid and flux return and the cost of such an undertaking.

10.2. Tracking and Vertex Detection

There are many areas in which research and development are required for tracking in a large solenoid detector. The high-rate and high-radiation environment provides considerable challenges.

Vertex Detectors and High-Resolution Tracking Devices. While vertex detectors were not discussed in detail by this group, we assumed that a large solenoid detector would include one. For completeness, we will mention some of the R&D required, although more detail can be found in other reports in these Proceedings. The main problem for silicon microstrip and pixel detectors is radiation sensitivity of the detector itself and the electronics, particularly since they need to be located close to the beam line. In addition, the electronics needs to have low power dissipation. The main areas for R&D

for scintillating fibers are short attenuation length in small-diameter glass fibers and long fluorescence decay times and readout times.

Central and Intermediate Tracking. Some of the major requirements for research and development for central and intermediate tracking for a large solenoid detector are listed here.

1. There are many mechanical and electrical problems involved in building a central tracking system out of long straw-tube chambers. What is the minimum wall thickness required? How can long straws be handled, held straight, and positioned? How well can the wires be positioned? How can the wires be supported for electrostatic stability? Can a chamber be built with stereo straws? How can cathode pads be implemented? Are pressurized straws a realistic possibility? Feasibility studies are needed.
2. Much more work is needed even at the conceptual design stage for intermediate tracking. The two options described here, planes of parallel wires and radial tracking chambers, are possibilities, but a more attractive solution may be found after more work. The two options described require a large number of cathode pads for reading out the coordinate along the wire. Their pattern recognition capabilities need study. Either option also involves mechanical design problems.
3. "Fast" drift chamber gases, such as mixtures of CF_4, could be very useful in reducing the occupancy for a fixed cell size or reducing the number of cells by allowing a larger cell for a fixed occupancy. More research is needed on these gases to determine their radiation resistance, spatial resolution, double-hit resolution, and operation characteristics in moderate magnetic fields.
4. Efforts are needed in understanding how to align or measure the position of the wires. Systematic errors in wire position will probably be the limiting factor in the momentum resolution obtained with any wire chamber system.
5. At this time it is not clear whether we will need to record multiple hits for each wire or digitize the pulses. Straw-tube chambers probably do not have multi-hit capability. R&D is needed in this area.
6. Computer simulation is needed to study pattern recognition in the high-rate SSC environment. For central and intermediate tracking the dominant constraint is the combination of cell occupancy and double-hit resolution. It is crucial to determine what tracks can actually be found for SSC events given the high multiplicity and density of tracks and the added hits from out-of-time bunch crossings. Pattern recognition studies are also needed for high-resolution tracking devices such as silicon microstrip and pixel devices. Computer simulation can then be used to determine suitable mixes of pixel devices, silicon strip devices, high-precision drift chambers, and large straw tube or drift chamber systems for a large solenoid detector. Finally, computer simulation can be used to help determine the detailed cell and tracking system designs.

10.3. Calorimetry

Fundamental questions must be answered by research and development and by experience with running or proposed detectors before a choice can be made from among the three most attractive options for calorimetry in a large solenoid detector. The major areas for R&D are listed below.

Lead/Scintillating Fibers. The possible problems for this type of calorimeter are radiation hardness, calibration uniformity and stability of response, long-term stability, and sufficient longitudinal segmentation to obtain e/π rejection of at least 10^{-3}.

Lead/TMS. Operational experience with warm liquid calorimeters is needed to determine whether the conditions for safety and purity can be met. High drift fields are needed to increase the induced signal current to at least the level of liquid argon. Birk's constant and the e/h signal ratio need to be determined experimentally; it may turn out to be difficult to achieve sufficient compensation. Hadronic energy resolution should be measured. Although there are important technical problems to be solved, warm liquid calorimeters may ultimately prove to be better able to meet the hermeticity requirement for calorimeters at the SSC.

Lead/Liquid Argon. Liquid argon calorimeters have been used quite successfully in the past; however, the requirements for calorimetry for a large solenoid detector at the SSC may be difficult to meet with liquid argon. Hermeticity is probably the hardest problem to solve. The cryostat must be designed so that cracks are minimized. Compensation is also a problem which needs experimental investigation. Further calculations are needed to determine whether the charge collection time is adequate for the SSC.

Beam Tests. For any calorimeter design beam tests will be needed to measure the e/π ratio and hadronic and electromagnetic energy resolution.

10.4. Electron Identification

In addition to R&D on calorimetry, outlined in the previous section, more development of TRD technology relevant to the high energies at the SSC is needed to prove the feasibility of some of the ideas that have been proposed, in particular to push the energy range up to 300 GeV or more. Likewise, studies of realistic synchrotron radiation devices would be useful to determine feasibility in the SSC environment.

10.5. Muon Identification and Momentum Measurement

Specific areas for R&D for muon detection include spatial alignment of the tracking detectors and stability as a function of time and temperature. In addition, integration of the muon detector wire geometry into the processing required for low-level triggering should be studied. R&D is also needed concerning the effects of interactions of high-energy muons with material on muon identification and triggering.

10.6. Electronics

The high interaction rates at the SSC lead to an overall data acquisition scheme (see Section 9.2) based upon highly integrated and sophisticated systems of front-end electronics mounted on the detector. Development and construction times for electronics for present detectors already frequently exceed times for mechanical systems. Moreover, the front-end electronics designs will be integrated into the mechanical designs and the data acquisition system. These front-end circuits will be the most challenging R&D problem in electronics. Timely R&D of front-end electronics is essential to SSC physics.

Some areas for R&D for front-end electronics were listed in the report of the Snowmass 86 Triggering and Electronics Group [40]. In some respects, custom VLSIs developed for current experiments, such as analog memory devices for SLD, microplex readout of silicon strips for Mark II and DELPHI, and pipelined readout of calorimeters for ZEUS, serve as models for SSC electronics. However, these circuits are in general not adequate for the SSC with respect to scale of integration, readout times, power consumption, radiation hardness, and system design.

R&D for front-end electronics for a large solenoid detector should take two directions. The first direction should be generic studies. For instance, different integrated circuit technologies should be examined for their appropriateness to the SSC. In addition, relative advantages of various pipeline structures should be studied, such as CCDs *vs.* switched capacitors or digital *vs.* analog, with respect to speed, charge resolution, power and calibration aspects. Techniques for improving radiation hardness should also be investigated. The second direction should be the development of front-end electronics circuits for specific detector components. Prototypical circuits for drift-time and charge measurement should be built in order to demonstrate the principle of the SSC data acquisition scheme and to gain experience with the techniques. The circuits could be tested in an actual experimental environment. The detailed design of these circuits, however, depends on the detailed design of the detector, including signal risetime, detector capacitance, detector and cabling impedance, packaging, etc., and on the details of the overall data acquisition system. Circuits for third coordinate readout in tracking chambers – delay lines, charge division, and cathode pads – must be developed. In addition, circuits for simultaneous measurement of time and pulse height will be developed. Front-end circuits for finely-segmented silicon devices, much different from existing multiplexed designs, will be needed. In light of the long lead times involved in developing and producing front-end circuits and their dependence on system concerns, R&D on some of the details of the overall data acquisition system would now be timely. Of particular concern are (1) control and management of the front end, such as clocking and labeling, (2) determination and application of calibration constants, (3) outputs needed for the trigger, (4) data processing needed and (5) test features.

In the area of off-detector data acquisition electronics, that is, high-speed busses, event builders, and microprocessor farms, much of the necessary development will occur naturally through implementation of on-line farms for existing experiments, such as CDF and D0. Some further R&D into high-speed busses and event builders is warranted.

The trigger processors will most likely depend on custom VLSI circuits for reasons of speed and the advantages of detector mounting. Consequently, R&D in this area is needed. The final designs will depend on, and help determine, details of system integration. Trigger circuits of certain usefulness include analog sum and discrimination circuits (including E_T sums), shower cluster finders, and track segment finders. Also of interest is the study of on-line track finders using specialized, fully-custom, or general-purpose processors.

10.7. Detector Simulation

Computer simulation of the components of the Large Solenoid Detector and of the detector as a whole will be a very important part of the design of the detector. The ratio of events from interesting physics to events from background processes is very low, so detailed understanding of detector response to the backgrounds is needed. Any large detector for the SSC will be very complex and expensive and every effort must be made to design a detector which will have excellent performance and will not have to undergo major rebuilding. Some areas in which computer simulation is particularly needed are pattern recognition in tracking detectors, discussed in previous sections, and development of shower simulation code which can be trusted as an aid in the design and optimization of the calorimeter, including its electron identification performance. Existing codes require too much computer time to be practical and need to be compared with test beam data at energies of several hundred GeV. Computer simulation of the processing of the data, including electronics response to the signals from the detector components, processing of the data by microprocessors on the detector, trigger, and data acquisition will be required.

11. Conclusions

The physics at the SSC will require high resolution hermetic calorimetry and excellent electron and muon identification. These needs are met by the large solenoid detector discussed here. Hermetic calorimetry is obtained in the presence of a magnetic field by placing the entire unit inside the solenoid coil. Good momentum resolution in the tracking has been preserved by choosing a 2 Tesla magnetic field (although moderately lower fields would not substantially change the performance of the detector). Some thought has been given to intermediate angle tracking although the designs outlined are very preliminary. Electron identification with a pion misidentification probability of less than 10^{-3} can be obtained with only calorimetry and conventional tracking. This meets the goals outlined by the physics study groups. However, there was a strong feeling that better rejection than this should be provided if possible; some kind of TRD system, integrated with the tracking, is a likely candidate. The solenoid coil, 8 meters in diameter and 16 meters long, has been designed in cryogenically separate modules for ease of construction and to allow access to the calorimetry. The large integral $B\,d\ell$ of the solenoid (8 Tesla-meters) allows adequate momentum resolution for muons to 1–2 TeV/c and charge sign measurement to momenta exceeding 5 TeV/c.

The design parameters of the proposed detector are summarized in Table X.

Table X. Summary of the Detector Design Parameters

SOLENOID COIL	
Inner diameter	8.2 meters
Length	16 meters
Central field	2 Tesla
Weight (including flux return)	16,450 metric tons
CENTRAL TRACKING	
Inner radius	0.40 meters
Outer radius	1.6 meters
Number of superlayers	15
Number of cells	122,368
$\|\eta\|$ coverage	< 1.2
INTERMEDIATE TRACKING (OPTIONS A & B)	
$\|\eta\|$ coverage	$1.2 < \|\eta\| < 3.0$
z position	$\|z\| < 4.0$ meters
Total number of chambers	26 (A) or 18 (B)
Total anode wires	128,000 (A) or 172,800 (B)
Total cathode pad channels	500,000 (A) or 293,760 (B)
ELECTROMAGNETIC CALORIMETER	
Depth	25 X_0
Transverse segmentation	$(\Delta\eta \times \Delta\phi)$
$\|\eta\| < 2.0$	$.02 \times .02$
$2.0 < \|\eta\| < 4.5$	$.03 \times .03$
$4.5 < \|\eta\| < 5.5$	$.03 \times .03$ to $.08 \times .08$
Longitudinal segmentation	6 X_0, 8 X_0, 11 X_0
Total number of towers	104,000
Total number of electronics channels	312,000
Weight	
Central	200 metric tons
Forward	35 metric tons
HADRONIC CALORIMETER	
Depth	10–12 λ
Transverse segmentation	$(\Delta\eta \times \Delta\phi)$
$\|\eta\| < 2.0$	$.06 \times .06$
$2.0 < \|\eta\| < 4.5$	$.06 \times .06$
$4.5 < \|\eta\| < 5.5$	$.06 \times .06$ to $.08 \times .08$
Longitudinal segmentation	2 segments
Total number of towers	19,100
Total number of electronics channels	37,200
Weight	
Central	4800 metric tons
Forward	965 metric tons
MUON SYSTEM	
Total number of electronic channels	$\sim$ 100,000
Weight of toroids	13,000 metric tons

A very substantial amount of R&D must be performed if the detector is to be optimally and efficiently constructed, and specific R&D areas were identified and discussed in the preceding section. In addition to basic research and development, prototype work must be done on all components of the detector. The choice of sampling medium for the calorimeter is especially critical, but testing is needed for all aspects of the design, partly because of the scale of the construction task. However, it is believed that the amount of R&D required is consistent with the length of time available so that the detector can be built and ready at the turn-on of the SSC.

Relatively little time was spent at this Workshop discussing the problems of electronics, triggering, rate and pile-up effects, and data handling and analysis. For some of these subjects reasonably careful studies were performed during Snowmass 1986 (or at previous workshops), but all require further study and must be brought under control in the early stages of detector design.

In spite of the rather large amount of R&D which remains to be done, there was considerable optimism that there are no fundamental obstacles to the design, construction, and operation of a detector with the excellent detection of electrons, muons, jets, and missing energy that is required for analyzing the exciting physics that awaits at 40 TeV.

References

1. Proceedings of the 1984 Summer Study on the Design and Utilization of the Superconducting Super Collider, edited by R. Donaldson and J. G. Morfin, Snowmass, CO (1984).
2. Cost Estimate of Initial SSC Experimental Equipment, SSC–SR–1023, SSC Central Design Group, June 1986.
3. Proceedings of the 1986 Summer Study on the Physics of the Superconducting Supercollider, edited by R. Donaldson and J. Marx, Snowmass, CO (1986).
4. See the Reports of the Parametrization Subgroups in these Proceedings.
5. K. Tsuchiya *et al.*, KEK Preprint 86-63, October 1986.
6. T. Kondo, "Report of the Microvertex Detector Group," Proceedings of the 1986 Summer Study on the Physics of the Superconducting Supercollider, edited by R. Donaldson and J. Marx, Snowmass, CO (1986), p. 743.
7. D. G. Cassel and G. G. Hanson, "Report of the Central Tracking Group," *ibid.*, p. 377.
8. R. DeSalvo, "A Proposal for an SSC Central Tracking Detector," *ibid.*, p. 391.
9. D. H. Saxon, "Vertex Detection and Tracking at Future Accelerators," Proceedings of the Workshop on Future Accelerators at High Energies, La Thuile/CERN (1987), RAL 87–011.
10. E. Elsen and A. Wagner, "Tracking Detectors for Large Hadron Colliders and for e^+e^- Linear Colliders," *ibid.*
11. R. L. Gluckstern, Nucl. Instr. and Meth. **24**, 381 (1963).
12. R. DeSalvo, "A Proposal for an SSC Central Tracking Detector," CLNS 87/52.
13. M. Atac, T. Hessing and F. Feyzi, IEEE Trans. Nucl. Sci. **33**, 189 (1986).

14. H. A. Gordon and P. D. Grannis, "Calorimetry for the SSC," Proceedings of the 1984 Summer Study on the Design and Utilization of the Superconducting Super Collider, edited by R. Donaldson and J. G. Morfin, Snowmass, CO (1984), p. 541.
15. T. Kondo, H. Iwasaki, Y. Watanabe, and T. Yamanaka, "A Simulation of Electromagnetic Showers in Iron-, Lead-, and Uranium–Liquid Argon Calorimeters Using EGS and Its Implications to e/h Ratios in Hadron Calorimetry," *ibid.*, p.556.
16. R. Partridge, "Calorimeter Requirements for Tagging the Semi–Leptonic Decays of Top Quarks," *ibid.*, p. 567.
17. E. Fernandez *et al.*, "Identification of W Pairs at the SSC," *ibid.*, p. 107.
18. T. Akesson *et al.*, "Detection of Jets with Calorimeters at Future Accelerators," Proceedings of the Workshop on Future Accelerators at High Energies, La Thuile/CERN (1987), CERN/EP/87-88.
19. G. Brandenburg *et al.*, "Identification of Electrons at the SSC," Proceedings of the 1986 Summer Study on the Physics of the Superconducting Supercollider, edited by R. Donaldson and J. Marx, Snowmass, CO (1986), p. 420.
20. P. Jenni *et al.*, "Report of the Jet Group," Proceedings of the Lausanne Workshop on the LHC.
21. R. Wigmans, CERN-EP/86–141 (1986), submitted to Nucl. Instr. and Meth. See also the references contained therein.
22. C. Baltay, J. Huston, and B. G. Pope, "Calorimetry for SSC Detectors," Proceedings of the 1986 Summer Study on the Physics of the Superconducting Supercollider, edited by R. Donaldson and J. Marx, Snowmass, CO (1986), p. 355.
23. S. R. Hahn *et al.*, "Calibration Systems for the CDF Central Electromagnetic Calorimeter," submitted to Nucl. Instr. and Meth., August 1987.
24. Technical Proposal for Aleph, CERN/LEPC/83-2, LEPC/P1 (1983).
25. T. J. Devlin, A. Lankford, and H. H. Williams, "Electronics, Triggering, and Data Acquisition for the SSC," Proceedings of the 1986 Summer Study on the Physics of the Superconducting Supercollider, edited by R. Donaldson and J. Marx, Snowmass, CO (1986), p. 439.
26. A. Gonidec *et al.*, Proceedings of the International Conference on Advances in Experimental Methods for Colliding Beams, March 9–13, 1987, Stanford, California, to be published in Nucl. Instr. and Meth.
27. W. Wenzel, D0 Internal Note #524, Fermilab.
28. V. Radeka, "Fundamental Limits on Ionization Calorimetry," Proceedings of the International Conference on Advances in Experimental Methods for Colliding Beams, March 9–13, 1987, Stanford, California, to be published in Nucl. Instr. and Meth.
29. J. Engler and H. Keim, Nucl. Instr. and Meth. **223**, 47 (1984).
30. H. H. Williams, "Detector Summary Report", Proceedings of the 1986 Summer Study on the Physics of the Superconducting Supercollider, edited by R. Donaldson and J. Marx, Snowmass, CO (1986), p. 327.
31. L. Nodulman *et al.*, Nucl. Instr. and Meth. **204** 351 (1983).

32. Report of the Task Force on Detector R&D for the Superconducting Super Collider, SSC–SR–1021, SSC Central Design Group, June 1986.

33. R. Engelmann *et al.*, Nucl. Instr. and Meth. **216**, 45 (1983).

34. B. Dolgoshein, Nucl. Instr. and Meth. **A252**, 137 (1986).

35. D. Carlsmith *et al.*, "SSC Muon Detector Group Report," Proceedings of the 1986 Summer Study on the Physics of the Superconducting Supercollider, edited by R. Donaldson and J. Marx, Snowmass, CO (1986), p. 405.

36. Proceedings of the Workshop on Triggering, Data Acquisition and Offline Computing for High Energy High/High Luminosity Hadron–Hadron Colliders, edited by B. Cox, R. Fenner and P. Hale, Fermilab, Batavia, IL (1985).

37. G. Kane, F. Paige, L. Price *et al.*, "SSC Physics Signatures and Trigger Requirements," *ibid.*, p. 1.

38. A. J. Lankford and G. P. Dubois, "Overview of Data Filtering/Acquisition for a 4π Detector at the SSC," *ibid.*, p. 185.

39. P. Cooper *et al.*, "A Feasibility Design for the Readout of a 4π SSC Detector," *ibid.*, p. 200.

40. T. J. Devlin, A. Lankford, and H. H. Williams, "Electronics, Triggering, and Data Acquisition for the SSC," Proceedings of the 1986 Summer Study on the Physics of the Superconducting Supercollider, edited by R. Donaldson and J. Marx, Snowmass, CO (1986), p. 439.

Report of the Compact Detector Subgroup

J. Kirkby
CERN, Geneva 23, Switzerland
T. Kondo
KEK, Tsukuba-gun, Ibaraki-ken 305 Japan
S. L. Olsen
University of Tsukuba, Ibaraki-ken 305 Japan
and University of Rochester, Rochester, NY 14627 U.S.A.

M. Asai	Hiroshima Institute of Technology
R. E. Blair	Argonne National Laboratory
J. E. Brau	University of Tennessee
R. Cashmore	Oxford University
R. Darling	Rutgers University
C. Newman-Holmes	Fermi National Laboratory
V. P. Kenny	U.S. Department of Energy
D. Koltick	Purdue University
T. Ohsugi	Hiroshima University
S. Ozaki	KEK
R. A. Partridge	Brown University
P. Peterson	Rockefeller University
L. E. Price	Argonne National Laboratory
P. G. Rancoita	INFN Milano
J. L. Ritchie	SLAC
R. W. Rusack	Rockefeller University
H. Sadrozinski	UC Santa Cruz
F. Sannes	University of Queensland
A. Seiden	UC Santa Cruz
T. Tauchi	KEK
G. S. Tzanakos	Columbia University
G. Valenti	CERN
W. Vernon	UC San Diego
A. Weinstein	UC Santa Cruz
R. J. Wilson	Boston University

1. Introduction

1.1 General Approach

At the outset we decided to focus the design of our detector on the detection of the Higgs (H^o) particle of the Standard Model. This choice follows since we consider the discovery of the H^o to be the highest priority among the experiments conjectured for the SSC. Since we do not know the H^o mass, we require sensitivity over a broad range: from a mass $\sim$ 50 GeV/c^2, corresponding to the upper mass sensitivity of current machines such as the SPS Collider and TeV I, up to the physics and SSC mass limits, which coincide at $1 \sim 2$ TeV/c^2. We see immediately (section 7) that an optimized H^o detector must have:

1. High rate capability. Typical detected signal rates are ~ 1000 year^{-1} for low Higgs' masses down to a few$\times$10 year^{-1} for the highest accessible masses for $\int \mathrm{Ldt} = 10^{40}\mathrm{cm}^{-2}$.
2. Excellent sensitivity to a broad range of final states. These include $H^o \rightarrow \gamma\gamma$ (good γ/π^o separation) and $\tau^+\tau^-$ at low masses, $H^o \rightarrow b\bar{b}$ (good e/μ identification in jets, good secondary vertex detection) at intermediate masses, and $H^o \rightarrow Z^oZ^o/W^+W^-$ (good e/mu identification and resolution, good jet measurement for ν and W/τ identification) at the highest mass range.

Taken together, these requirements imply a **general purpose detector with precise charged particle tracking in a magnetic field and high quality calorimetry.** The detection of the H^o imposes a stringent test of detector performance after which other physics is 'easy'.

1.2 General Tracking Considerations

We consider precise magnetic tracking to be vital in these studies. [By "precise magnetic tracking" we imply the ability to track **all** charged particles inside a jet with high efficiency and with a momentum resolution $\sigma_p/p \leq 25\%$ p(TeV/c).] The reasons are as follows:

1. To provide internal detector redundancy. This provides internal cross-checks on the validity and nature of a signal, increases rates for rare signals and helps substantially in the early development of detector algorithms and in the checking of calibrations during data-taking. The redundancy includes charged tracking vs calorimeter energies (to identify poor calibrations or external backgrounds), μ energy measurements in the inner vs outer systems, e energy measurements in the inner tracker vs electromagnetic calorimeter and, e vs μ rates.
2. To determine electron and μ signs.
3. To provide σ_p/p $\sim$ few% for μ's over most of the momentum range. The most accurate measurement of σ_p/p is typically provided by the inner

tracker. In addition, the inner tracker will guard against an anomalous high-momentum tail created by scattering or interactions in the massive outer muon filter.

4. To measure the internal structure of jets, something that cannot be resolved by calorimetry. Here we wish to observe the low charged multiplicities associated with τ orW/Z jets which will be an important tool, at high p_T, for distinguishing them from QCD jets. Another example where the internal stucture of jets may play a vital experimental role is in the search for signs of compositeness in very high $p_T(\geq 5$ TeV/c) jets. Finally, we wish to be sensitive to any new physics that may lead to jets with an anomalous internal structure.
5. Searches for heavy $H^o \rightarrow Z^oZ^o$ with the using subsequent decays $Z^o \rightarrow l^+l^-$, have a natural figure of merit for the l^+l^- invariant mass resolution, i.e., the $\Gamma_{Z^o}/2.35$, where Γ_{Z^o} is the natural width of the Z^o. For 200 GeV/c μ's, this implies $\sigma_p/p \sim 5\% \times p$, unattainable in a conventional magnetized iron spectrometer.

Our requirements of precise magnetic tracking fundamentally influence the design of the detector. We list three available technologies for tracking (along with simplified performance figures):

1. Wire drift chambers: σ(2 hit) $\sim$ 1mm; σ(sagitta) $\sim$ 100μm.
2. Silicon microstrips: σ(2 hit) $\sim$ 100μm; σ(sagitta) $\sim$ 10μm.
3. Scintillating fibers (SCIFI): σ(2 hit) $\sim$ 100μm; σ(sagitta) $\sim$ 10μm.

Simulations of TeV gluon and quark jets using QCD evolution models such as ISAJET[1] indicate a mean angle between particles inside the cores of jets of about 0.5mrad. Therefore, in order to track efficiently inside jets, a two-hit resolution of 100 μm is required at a radius of 25 cm (see Fig.1.1). The **minimum** radius where this can be achieved for drift chambers is at r $\sim$ 250 cm compared to 25 cm for Silicon microstrips or SCIFI. For a given magnetic field (**B**) the requirement of momentum resolution of $\sigma_p/p \leq 25\%$ p(TeV/c) specifies the lever arm (**L**) of the tracking system. Given the large size of the drift chamber tracking system, practical considerations limit the field strength B to about 1.5 T, implying a lever arm L$\sim$ 250 cm. We argue that the small size of solid state trackers and their insensitivity to magnetic fields allow for much stronger magnets; for a field of B = 6 T, a level arm of only 42 cm will give the design momentum resolution. We immediately see from these two considerations that **only solid state tracking devices** (Si microstrips or SCIFI) **can potentially meet the necessary tracking requirements** in general purpose detector.

1.3 Calorimetry Requirements

Previous SSC workshops have stressed that the calorimeter is the key system for most SSC experiments.[2,3] Many of the physics signatures will require excellent calorimetry. The detection and measurement of jets should be somewhat analogous to detection of hadrons at present e^+e^- collider detectors. This is best done with good energy and direction measurements from calorimetry since the high multiplicities involved reduce substantially the effectiveness of other techniques.

The design of the electromagnetic calorimeter is driven by a few potentially very important physics channels where large backgrounds must be overcome by good resolution. One such process is the decay $H^o \to \gamma\gamma$, which the intermediate Higgs working group has stressed as an important process to explore.[4] For an electromagnetic resolution of $15\%/\sqrt{E}$, which should provide the necessary power, a calorimeter depth of $25X_o$ is necessary.[5] The transverse segmentation of $\Delta\eta \times \Delta\phi = 0.03 \times 0.03$ is required for the high multiplicities that are expected.[3]

In order to achieve the best possible jet resolution it is important for the combined electromagnetic/hadronic calorimeter system to have nearly equal response to hadrons and photons. This is important for two reasons. First one wants to equalize the response of the detector to fluctuations of the jet composition between electromagnetic and hadronic components.[6] However, even when these fluctuations are not contributing to the jet resolution one still needs good basic hadronic resolution.[7] An inequality in this response will mean that fluctuations in the ratio of EM and hadronic components of a shower will distort the calorimeter's response. This can be avoided by making the sampling medium sensitive to the neutrons that are produced within the hadronic shower cascade. Since, on average, the energy content of these neutrons complements the amount of electromagnetic energy in a given shower, they can be exploited to equalize the electromagnetic and hadronic response. The success of this scheme is intimately connected to the choice of sampling material as hydrogenous materials are more easily compensating.[8] In addition to the role of neutron detection to achieve compensation, one can suppress the electromagnetic response through the use of high Z materials as radiators such that the electromagnetic energy is preferentially deposited in the radiator rather than the sampling medium.[9] These issues have lead one naturally to a calorimeter with a high Z radiator and a readout that is sensitive to neutrons. Suitable sampling media include scintillator for its hydrogenous nature, and silicon covered with a thin hydrogenous layer for its potential source of knock out protons into the non-saturating silicon.[10] A thickness of 12 lambda is required to achieve negligible leakage fluctuations in the TeV regime. A transverse segmentation of $\Delta\eta \times \Delta\phi = 0.06 \times 0.06$ has been required by most of the physics working groups.[3]

While the resolution of the electromagnetic and hadronic calorimeters are key to their success, other important properties must be optimized. It is very important to be able to measure not only visible energy, but also missing E_T. The importance of this has been proven at the $Sp\bar{p}S$ and is now universally recognized. Since any losses of energy from the detector are interpreted as real missing energy, one must minimize cracks and dead spaces in the detector and extend the sensitive coverage to as small an angle as possible. Angles down to $0.5 \sim 1^\circ$ are required.[3] Another effect that tends to wash out the missing transverse energy measurement is the pileup of many events within the sensitive time of the calorimeter. The multiple interactions within one beam crossing make some pileup unavoidable, but one must try to minimize this by keeping the detector sensitive time as close to the beam crossing interval of about 30 nanoseconds as possible.[11]

Many other issues must be considered in the final choice of calorimeter design. The ease and stability of the calibration is crucial to obtain the very few percent resolution demanded by the Higgs search at the SSC.[12] Radiation resistance is very important as the backgrounds from the machine will be large and, in addition, event generated fluences of neutrons and other particles are significant from the enormous number of interactions that the detector must survive to find the small signals that the new physics will produce. Finally, the cost issue will play a role as the ideal solutions are more expensive than compromised ones. A realized system is likely to require some compromises.

1.3 The Advantages of Compactness

Consideration of the above requirements led us to the concept of the **compact solenoid detector**, which is based on:

1. A precise compact inner tracker.
2. An ultra-high-field magnet.

This detector concept has several important advantages in addition to high resolution tracking. In particular, the overall detector volume and cost is substantially reduced. The inner radius of the calorimeter is $\sim$ 1m compared with $\sim$ 2.5m in the conventional ("Model A") general purpose SSC detector; the volume of the calorimeter is approximately 1/4 the Model A design. This reduction in size will make the provision of deep and uniform calorimetry affordable, leading to an improved performance. Moreover, the high field introduces a significant new technique for electron identification: the detection of hard synchrotron γ rays.

In order to explore this detector concept as fully as possible, we developed detector designs based on both technologies:

1. **SSB** (**S**olid **S**tate **B**ox). Characterized by an inner tracker of Si microstrips and emphasizing extremely precise σ_p/p and secondary vertex measurements.
2. **SMART** (**S**trong **M**agnet **a**nd **R**evolutionary **T**echnology). Characterized by an inner SCIFI tracker and emphasizing the best e identification and operation at the highest possible luminosities ($\geq 10^{33}cm^{-2}s^{-1}$).

These detectors employ essentially identical outer muon systems and forward detectors but differ in their magnetic tracking and calorimetry. In the following sections we first describe the distinct components of each of our two detectors, and then their common detector components. The physics performance of the detectors is then evaluated. This is followed by a summary of studies of certain key aspects of the detectors: the design of high field magnets, the effect of the magnet coil material on the performance of the calorimeter, the possible role of BaF_2 calorimeter and radiation hardness. Finally we present cost estimates and an evaluation of the R&D required to establish these detectors from concepts into reality.

REFERENCES

1) Proceedings of the ECFA-CERN Workshop on a Large Hadron Collider in the LEP Tunnel, pgs. 200 and 503 (1983).

2) M.G.D. Gilchriese, Snowmass '84.

3) C. Baltay, J.Huston, and B.G. Pope, Snowmass '86.

4) Intermediate Mass Higgs Working Group Report, these proceedings.

5) T. Kondo and K. Niwa, Snowmass '84.

6) T. Akesson *et al.*, Nucl. Inst. & Meth., **A241**, 17 (1985).

7) C.W. Fabjan *et al.*, Nucl. Inst. & Meth., **A141**, 61 (1977).

8) J. Brau and T.A. Gabriel, Nucl. Inst. & Meth., **A238**, 489 (1985).

9) J. Brau, Proceedings of the Caltech Workshop on Compensated Calorimetry (1985); H. Breukmann, ibid.

10) T. Akesson *et al.*, CERN-EP/87-88; T.A. Gabriel *et al.*, Workshop on Detector Simulation for the SSC, Argonne, 1987.

11) J.O. Alverson and J. Huston, Snowmass '86; A. Yamashita and K. Kondo, ibid.

12) High Mass Higgs Working Group Report, these proceedings.

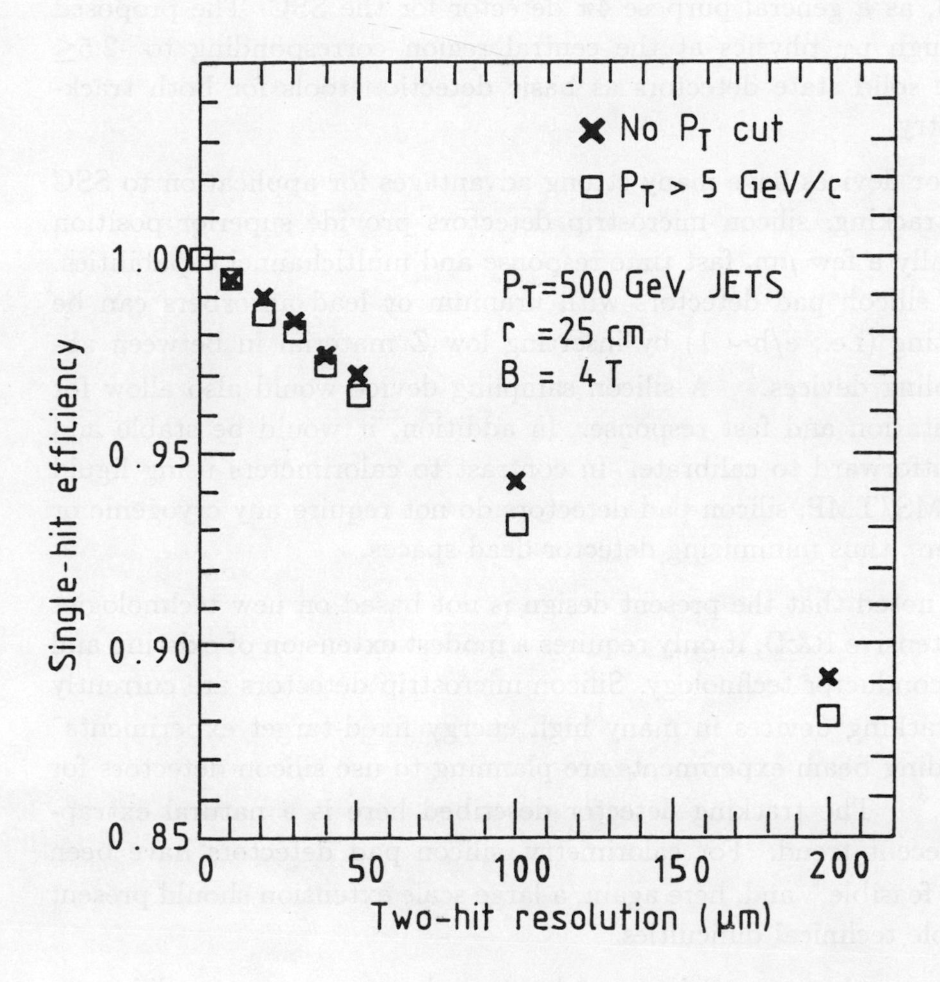

Fig.1.1 The single hit efficiency as a function of two track resolution for an SSC tracking detector.

2. The Solid State Box (SSB) Detector

2.1 Design Concept of SSB

We propose here a conceptual design of a compact solenoid detector, Solid State Box, **SSB**, as a general purpose 4π detector for the SSC. The proposed detector is for high p_T physics at the central region corresponding to $-2.5 \leq y \leq 2.5$. We use solid state detectors as basic detection tools for both tracking and calorimetry.

Semiconductor devices have many strong advantages for application to SSC detectors. For tracking, silicon microstrip detectors provide superior position resolution, typically a few μm, fast time response and multichannel capabilities. For calorimetry, silicon pad detectors with uranium or lead absorbers can be made compensating (i.e., $e/h \sim 1$) by inserting low Z material in between absorber and sampling devices.[1] A silicon sampling device would also allow for very fine segmentation and fast response. In addition, it would be stable and relatively straightforward to calibrate. In contrast to calorimeters using liguid Argon and/or TMS/TMP, silicon pad detectors do not require any cryogenic or vacuum containers, thus minimizing detector dead spaces.

It should be noted that the present design is not based on new technologies which require extensive R&D; it only requires a modest extension of existing and established semiconductor technology. Silicon microstrip detectors are currently being used as tracking devices in many high energy fixed-target experiments[2] and several colliding beam experiments are planning to use silicon detectors for vertex detection.[3] The tracking detector described here is a natural extrapolation of this recent trend. For calorimetry, silicon pad detectors have been demonstrated as feasible[4] and, here again, a large scale extension should present no insurmountable technical difficulties.

It is quite reasonable to anticipate at least modest progress on solid state detectors over the next few years, this would be sufficient to make the suggestions presented here realistic. This optimism derives from the following facts:

1. There are no fundamental problems remaining with the notable exception that a better understanding of the effects of radiation damage is desperately needed (see below).
2. Recent progress in microelectronics makes it possible to adopt VLSI technology for readout electronics extensively.
3. Semiconductor detectors have been developed for use in a wide variety of fields, including space science, nuclear fusion, astronomy, medical science,

etc. As a result, there is an existing industrial technological base for mass production, quality control and thus, the potential for cost reduction.

The one major problem that remains is the current lack of understanding of the susceptability to radiation damage. At the SSC, radiation levels due to charged particles has been calculated[5] but, as was pointed out in the present Workshop, the number of ~1MeV neutrons, generated by hadron showers inside the calorimeters, might be significantly higher.[6,7] Neutrons cause dislocations in the silicon crystal lattice more efficiently than charged particles, making it vitally important to evaluate, both theoretically and experimentally, the neutron radiation levels and the radiation damage effects caused by neutrons.

2.2 Design of a Tracking System for 1 TeV

2.2a Basic Considerations

We propose a tracking system with very fine segmentation in order to resolve tracks, even those inside of tightly collimated jets. To avoid ending up with an enormous number of channels, we are forced to use infrequent sampling. In addition, we have to go out to a moderately large radius to provide for precision momentum measurements.

Here we look in some detail at the organization of a tracker made entirely of silicon strips. They offer extremely good resolution in space and time. In addition, it is an area where much progress is being made and the detectors, for the scale discussed below, are still affordable in price.

We employ vector tracking ideas described in the report presented by A. Seiden at this meeting.[8] These are similar to those used by Mark III, AMY, Mark II, CDF, SLD and ZEUS. The use of a vector tracking chamber allows a crude local calculation of the transverse momentum using track segments, allowing the efficient selection, during reconstruction, of high p_T tracks only.[9] This vastly reduces the computer time required for track reconstruction for those events in which only high p_T tracks are desired.

2.2b. Tracker Parameters

The parameters of the tracker are as follows:

1. We propose 16 layers of silicon strips organized into 8 pairs (superlayers) of detectors. The detectors are assumed to be double sided (one side axial and one side stereo) so the total number of measurements is 32. Thus, the detector can be thought of as 8 axial and 8 stereo superlayers, although the axial and stereo measurements are grouped together locally.
2. The tracker ranges from r=8cm to r=50cm. The inner radius is determined

by the amount of radiation damage that can be tolerated. For the SSC operating at a luminosity of $10^{33}cm^{-2}sec^{-1}$, this value for the inner radius corresponds to about 0.4 Mrad per year.[5] Clearly, the design of sufficiently radiation-resistant electronics is critical.

3. We assume each layer is 150μm thick and has double sided readout. This corresponds to 2.5% X_o for the silicon and probably about 4-5% X_o in total if we include a foam support structure. It is of great importance to keep the amount of material in this range to avoid creating many soft electron-positron pairs from γ conversions. In addition, B meson identification by means of vertex reconstruction, as required, for example, for a light Higgs search, involves lower momentum tracks and is sensitive to multiple scattering errors.

2.2c Pattern Recognition Requirements

For pattern recognition we require:

1. Different tracks give distinct hits.
2. Hits in the paired layers, separated by some radial distance δ, can be locally associated into correct vector segments.
3. Vector segments from different pairs, with radial separation Δ, can be correctly linked into tracks.

Using 0.5mrad as the minimum separation between particles in a jet core in conjunction with a detector resolution of 5μm, results in optimized performance when δ=5mm and Δ=50mm.[8].

We have looked in some detail at the ability of such a system to do tracking in TeV jets. Fig.2.1 shows a high p_T event in a 3 Tesla magnetic field. A region near the core of the tight 1.5 TeV jet in the upper left quadrant of the event, at a radius of 26cm, has been blown up in Fig.2.2. The scale is indicated by the size of a single strip detector containing 1024 elements of transverse dimension 25μm.

To quantify this further, we have performed the first two steps of the pattern recognition for events with 1 TeV jets. Using a simple algorithm for linking hits into segments, and considering only hit pairs which can come from tracks with $p_T \geq .4$GeV, we evaluate the efficiency vs p_T for finding tracks if we require at least 4,6 or 8 linked segments for a good track. The results are shown in Fig.2.3, where the last bin includes overflows. The efficiency approaches 100% for high momentum tracks. In contrast to conventional drift chamber trackers, which are only efficient for tracks that are outside of jets, the silicon tracker is quite efficient for high p_T tracks in the core of jets.

2.2d z-coordinate Measurements

The optimum way of z coordinate determination is a complicated question. Here we look at the use of stereo measurements. We assume each pair of strips is double-sided and organized as shown in Fig.2.4. The goal is to measure z coordinates locally within each pair. Each of the 8 pairs making up the full detector would then have space coordinates and tangent vectors assigned locally to each track. These would then be matched from layer to layer.

The main problem for matching locally in z comes from accidental matches caused by the high multiplicity of tracks. We minimize this by using a very small stereo angle, trading off resolution in z for reducing the track pairing problem. This does not significantly degrade the momentum or mass resolution in a solenoidal spectrometer. For example, for a stereo angle of 5mrad, the excellent transverse (5μm) resolution of the strips will yield a z resolution of 1mm. With strip lengths of 10cm, tracks tranversely separated by $\geq 500\mu$m will not interfere. This is probably small enough to guarantee that most stereo hits can be associated with only a small number of altenative assignments.

2.2e Momentum Resolution

We assume a vertex constrained fit as would be appropriate for the high momentum leptons from the heavy Higgs decay. We also assume that the four measurements in a superlayer give an effective measurement with an error σ, and that the eight superlayers are approximately equally spaced out to a maximum radius r_M, we obtain

$$\sigma_{p_T}/p_T^2 = 5.5\sigma/(.3\ B\ r_M^2).$$

Taking $\sigma = 3\mu$m for the average over the four measurements in a superlayer, and $r_M = .5$m gives:

$$\sigma_{p_T}/p_T^2 = 6.5\%/(.3B) \quad (\mathrm{TeV}^{-1}).$$

For B=4T, we get:

$$\sigma_{p_T}/p_T = 5.5\% p_T(\mathrm{TeV}).$$

For the intermediate mass Higgs search, as discussed earlier, we wish to reconstruct Z^o masses, using leptons, with a resolution better than the natural width limit, implying:

$$\sigma_{p_T}/p_T^2 \leq 2\% \quad (\mathrm{TeV}^{-1}),$$

at about 200 GeV. Thus a field $\geq$2.5 T would be adequate.

With such good momentum resolution multiple scattering will dominate up to rather high momentum. Assuming 5% of a radiation length in total, distributed over a 50cm path length, the contribution is $\sigma_{p_T}/p_T = 0.8\%$ for B=2.5 T.

Fig.2.5 shows a possible arrangement for the tracking detector. This arrangement provides:

a. Tracking with $\geq$5 superlayers for $|y| \leq$2.3.

b. Tracking with 8 superlayers for $|y| \leq$1.75.

c. Tracking over a fixed radius of 50cm for $|y| \leq$1.44.

d. Tracking with axial strips for $|y| \leq$1.1.

Fig.2.6 shows σ_{p_T}/p_T^2 vs rapidity for p_T in TeV, assuming $\sigma_{p_T}/p_T = 8\% \times p_T(\text{TeV})$ at 90° and ignoring multiple scattering.

In order to exploit the excellent resolution provides by the use of silicon and the high magnetic field, it will be essential to align the silicon to an accuracy of a few microns.

2.2f Curling Tracks

The high magnetic field tends to trap particles, resulting in large occupancy rates and increased radiation damage. In this section we look at this problem as a function of the B field, tracker radius and rapidity coverage.

Fig.2.7 shows the ISAJET momentum spectrum for minimum bias events. A track will be trapped in the field for $2\rho = r_M$, where ρ is the track's radius of curvature. Thus, for example, trapping will occur for $p_T \leq$ 250MeV for r_M=50cm and B=3.3 T in a compact solenoid; or $p_T \leq$ 450MeV for r_M=165cm and B=2 T for a conventional solenoidal detector. A large fraction of the tracks in minimum bias events lie in this momentum range, making trapping an important consideration.

To estimate the effect we have calculated the effective track length seen in the detector per minimum bias event, comparing a compact (r_{min} = 8cm, r_{max} = 50cm) and large solenoid (r_{min} = 50cm, r_{max} = 160cm). We use as a unit the distance a straight track requires to traverse the tracking volume, counting as fractions those tracks that only pass through a part of the detector (i.e, with $2\rho \leq r_{max}$, or for tracks that exit in z). Tracks with $\rho \leq r_{min}$ make no contribution. The maximum number of spirals allowed is 100 based on estimates of energy loss and multiple scattering. We have repeated the calculation with this number reduced to 30 to study the effects of this cutoff. The change in total track length is 20%, which provides an estimate of the uncertainty in the calculation. The results are presented as a function of the B field and rapidity in Tables 2-I and 2-II.

Table2-I

Average number of track lengths for the compact solenoid per minimum bias event.

B Field (Tesla)

Y range	0	1.5	2.0	2.5	3.0	3.5	4.0
$\|y\| \leq 1.0$	8.6	9.7	10.0	12.0	16.5	17.2	19.3
$\|y\| \leq 1.5$	13.2	15.8	16.4	20.0	26.2	28.3	32.0
$\|y\| \leq 2.0$	17.8	22.6	24.0	30.0	38.0	42.6	49.3
$\|y\| \leq 2.5$	22.5	30.7	33.7	42.0	54.0	62.9	72.9

Table 2-II

Average number of track lengths for the large solenoid per minimum bias event.

B Field (Tesla)

Y range	0	1.5	2.0	2.5	3.0	3.5	4.0
$\|y\| \leq 1.0$	8.6	17.1	18.5	16.6	14.1	12.1	9.6
$\|y\| \leq 1.5$	13.2	28.4	30.9	28.3	24.5	20.8	16.8
$\|y\| \leq 2.0$	17.8	43.8	47.4	44.6	39.6	33.5	27.2
$\|y\| \leq 2.5$	22.5	65.1	71.9	68.8	61.7	52.3	42.9

These results indicate that the occupancy is increased by more than a factor of two, as compared to B=0, for the compact solenoid with 3.5 T≤B≤4 T or the large solenoid with 1.5 T≤B≤2.5 T. For the compact solenoid the occupancy is not a problem because of the large segmentation and excellent time resolution. Even at the innermost radius of 8cm, with a luminosity of 10^{33}, the occupancy is 2×10^{-3} in a two-hit resolution interval of 100μm(4 strips). This contrasts to an occupancy of ~25%, in a 4mm cell at a 50cm radius, in a large solenoidal detector.

2.3 The Silicon Sampling Calorimeter

2.3a Basic Structure

The calorimeter system is divided into three major parts, an electromagnetic calorimeter, a fine hadron calorimeter and a hadron tail catcher, as shown in

Fig.2.8. The first two are located inside the solenoid, while the tail catcher is placed radially outside of the magnet coil. The inner calorimeter segments use silicon detectors for sampling and uranium as the absorber. The tail catcher consists of scintillation counters (or gas sampling counters) and iron absorber plates which also serve as a magnetic flux return. We expect that the presence of superconducting solenoid and the transition of absorber material (U→Fe) and sampling media (Si→scintillator or gas) at a depth of 4.8 λ_o in the calorimeter will not degrade the energy resolution or the e/h ratio appreciably, since most of hadronic shower stops before a depth of $5\lambda_o$. This tendency is more notable for jets since high energy jets tend to be composed of many relatively low energy particles. This is illustrated in Fig.2.9, in which the hadron shower development of an 1TeV jet made out of ten 100GeV pions, simulated by GHEISHA, is displayed.

2.3b Calorimeter Dimensions

The EM and fine hadron calorimeters are designed to be as compact as possible, in order to minimize the total area of silicon detectors, which are expensive. On the other hand, the calorimeters must be at some distance from the interaction point in order to maintain the separation between jets. Considering the transverse shower size in Uranium/Si calorimeters and the minimum tracking length for a good momentum resolution, we tentatively decided to set the inner radius of the EM calorimeter to 60 cm. However, more studies are needed for optimization.

2.3c Sampling Frequency

The sampling frequency is a compromise between energy resolution and cost (detector area). The major parameters of EM and hadron calorimeters are summarized in Table 2-III. The e/h ratio is controlled by adjusting the thickness of low-Z insertions between the absorber plates.[1] This technique needs more study, both theoretically as well as experimentally. The SICAPO group has observed a large suppression of electromagnetic components in prototype W/silicon calorimeter, due to the addition of G-10 plates of ~1mm before and after the silicon sampling layers, as shown in Fig.2.10.

2.3d Energy Resolution

Three prototype electromagnetic calorimeters have been assembled and tested in electron beams.[4] These tests demonstrated good linearity and an energy resolution given by $\sigma_E/E = 17\% \times \sqrt{(\tau/E)}$, where τ is the sampling frequency in units of radiation length. There are no experimental results on hadron calorimeters. However, calculations indicate that the energy resolution is expected to be $45 \sim 50\%/\sqrt{E}$ with the sampling frequency suggested here.[1]

2.3e Feasability of Construction

Prototype EM calorimeters have been assembled with little difficulty. Hadron calorimeters can be made using the same basic technique. There are no fundamental problems foreseen as far as detector construction concerns. Silicon detectors do not require vacuum containers nor special light guides.

2.3f Segmentation

Silicon pads ranging from 4 to 10 cm^2 in area are taken as the basic elements. Therefore, each calorimeter cell can be as small as 2×2cm by attaching separate readout electronics to each elements. More detailed shower profiles can be obtained by inserting 1mm strip detectors at around EM shower maximum, at a depth of about $5X_o$. The longitudinal segmentation is rather arbitrary in the present design; we think 4 divisions in the EM part and 2 in the hadron part would be adequate for reasonable e/π discrimination.

2.3g Electron Identification

The following techniques can be exploited for electron identification:

a. E/p selection; in Fig.2.11 we show the momentum resolution of charged particles together with the EM energy resolution. One sees that both are excellent in the interesting energy range of 20-200 GeV indicating that the E/p method will function effectively.

b. Longitudinal Shower Profile; in the present design, there are 6 (4 in the EM section and 2 in fine hadron section) pulse height measurements along the depth of a shower. This configuration will not only provide the usual HAD/EM ratio for e/π separation, but also longitudinal correlations among EM sections. These have been demonstrated to be quite effective in enhancing e/π discrimination.[10]

c. Transverse Profile; we propose to place silicon strip detectors with strip widths of about 1mm at depths of, for example, $2X_o$, $5X_o$ and $10X_o$. Pulse heights from strips are read out individually to obtain detailed profiles of the electromagnetic showers. The proposed fine measurement of transverse energy profiles of EM showers (narrow and dense) would be very powerful for discriminating EM from hadronic showers.[11] The advantages of the present scheme are (a) the position resolution of sub-mm for EM showers, (b) the ability to identify multiple EM showers even when they are close to each other, and (c) the immunity against existence of nearby particles since the central cores of energy deposition are directly measured.

d. Track-Shower Matching; as described above, we can measure the center position of an EM shower with $\leq$1mm resolution. A requirement of track-shower connections in x and y within such a resolution will reduce backgroud from coalescing π^o-charged particles.

These identifications are not completely independent each other, correlations will be important. Nevertheless, although detailed calculations have not yet been done, we anticipate 10^{-3} to 10^{-4}, or possibly better, rejections against hadrons by the combination of the four methods. It should be noted that the proposed electron identification will work not only for isolated electrons, but also for electrons moderately close to a jet core because of the fine segmentation of the tracking and EM shower profile measurements.

Table 2-III Characteristics of a Silicon Sampling Calorimeter

	EM calorimeter	Fine hadron calorimeter
absorber	2 mm Uranium	6.6 mm Uranium
no. of samplings	40	60
sampling frequency	$0.63X_o$	$2X_o$
total thickness	$25X_o$, 0.8 λ_o	$120X_o$, $4.0\lambda_o$
sampling device	4 kΩcm silicon	4 kΩcm silicon
pad size	2×2 cm^2	2×2 cm^2
tower size	2×2 cm^2	6×6 cm^2
$\Delta\eta \times \Delta\phi$ at η=0	0.03×0.03	0.06×0.06
$\Delta\eta \times \Delta\phi$ at η=2	0.065×0.066	0.14×0.15
longitudinal segm.	4	2
total no. of pads	1.025×10^6	4.125×10^6
total detector area	410 m^2	1650 m^2
no.of readout chan.	102 K	15 K
energy resolution	13% $/\sqrt{E}$	45~50% $/\sqrt{E}$

2.3h Readout Electronics

The readout electronics should be mounted directly on the calorimeters, in order to minimize the cable length. For each readout channel, there must be an analog time-sliced memory followed by analog multiplexiers and FADC/memories for handling multiple channels. Prompt analog signals are summed and sent separately for triggering. Front-end electronics must be developed to minimize the electronic noise caused by relatively high capacitance of the silicon detectors. Transformer couplings cannot be employed because of the high magnetic field. Pad detectors can be serially connected to reduce the capacitance as long as the effect of stray capacitances are kept small.

2.3i Costs

Silicon detectors use high resistivity crystals, which are not widely available and therefore expensive. Current prices are \$15~\$20 for a 2cm×2cm pad. This must be reduced an order of magnitude in order to provide 2000m^2 of detector area within a reasonable budget. More detailed cost estimates are provided in section 11 of this report.

REFERENCES

1) R.Wigmans, Signal Equalization and Energy Resolution for Uranium/ Silicon Hadron Calorimeters, Contribution paper to this workshop.

2) For example, NA32,WA75 at CERN, E653,E691 at Fermilab.

3) For example, Mark II, CDF, SLD, Delphi, Aleph.

4) G. Barbiellini *et al.*, Nucl. Instr. and Meth., **A235**, 55 (1985) and **A236**, 316 (1985); A. Nakamoto *et al.*, Nucl. Instr. and Meth., **A251**, 275 (1986); M. Bormann *et al.*, Nucl. Instr. and Meth., **A240**, 63 (1985).

5) Report of the Task Force on Detector R&D for the Superconducting Supercollider, SSC-SR-1021(1986).

6) D.Nygren, Talk at the present Workshop.

7) R.Wigmans, Talk at the present Workshop.

8) A.Seiden, Note on tracking at the SSC.

9) W.Atwood, SLAC SLD experiment, Internal Note no.135.

10) Y.Fukui et al., Proceedings of the Physics of the Superconducting Supercollider, Snowmass, 1986, p.417.

11) T.Kondo and K.Niwa, Proceedings of the Design and Utilization of the SSC, Snowmass, 1984, p.559.

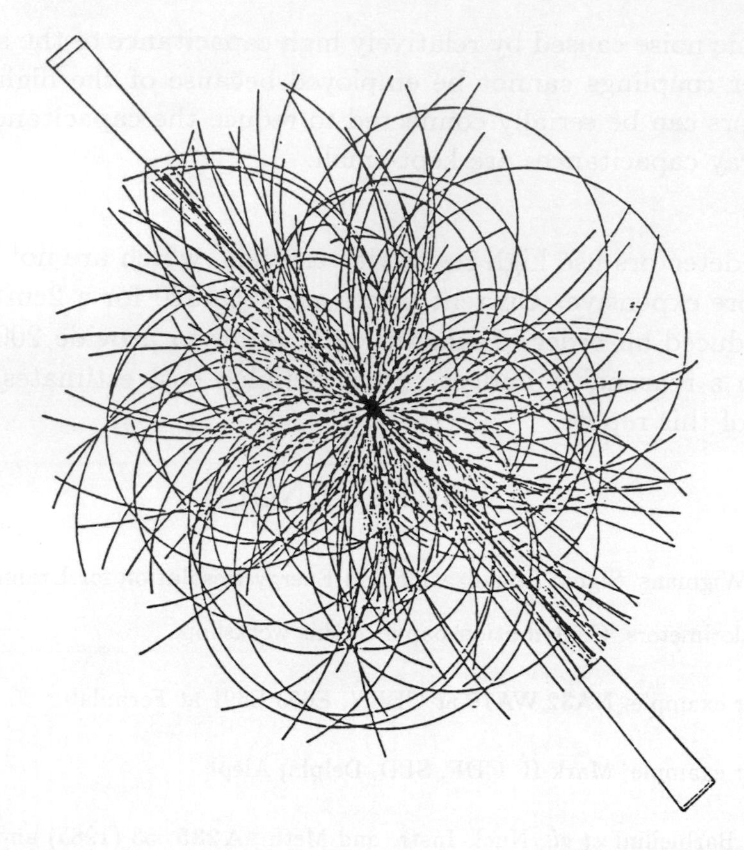

Fig.2.1 Charged tracks from a high p_T event in a 3 Tesla magnetic field.

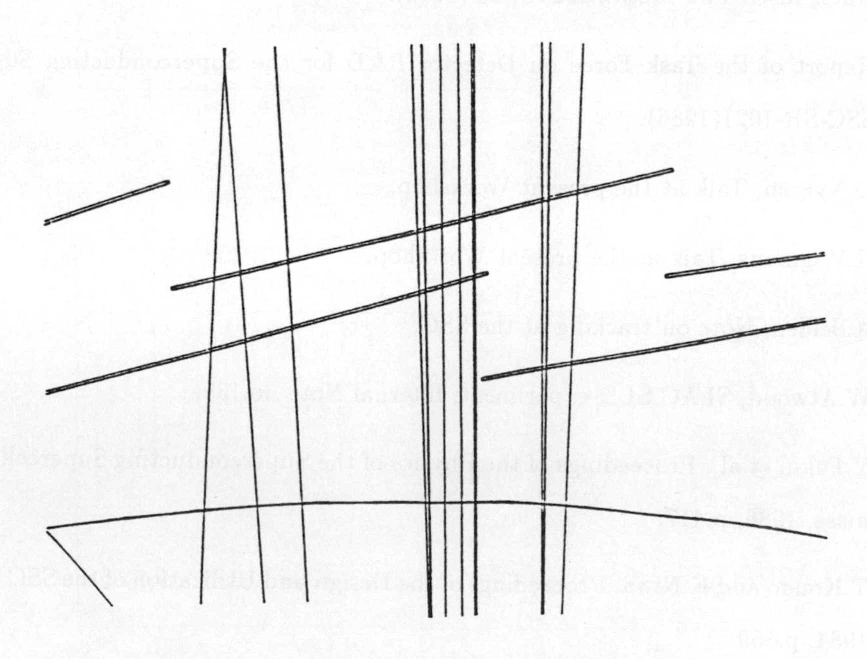

Fig.2.2 A blow up of the event, showing the jet core.

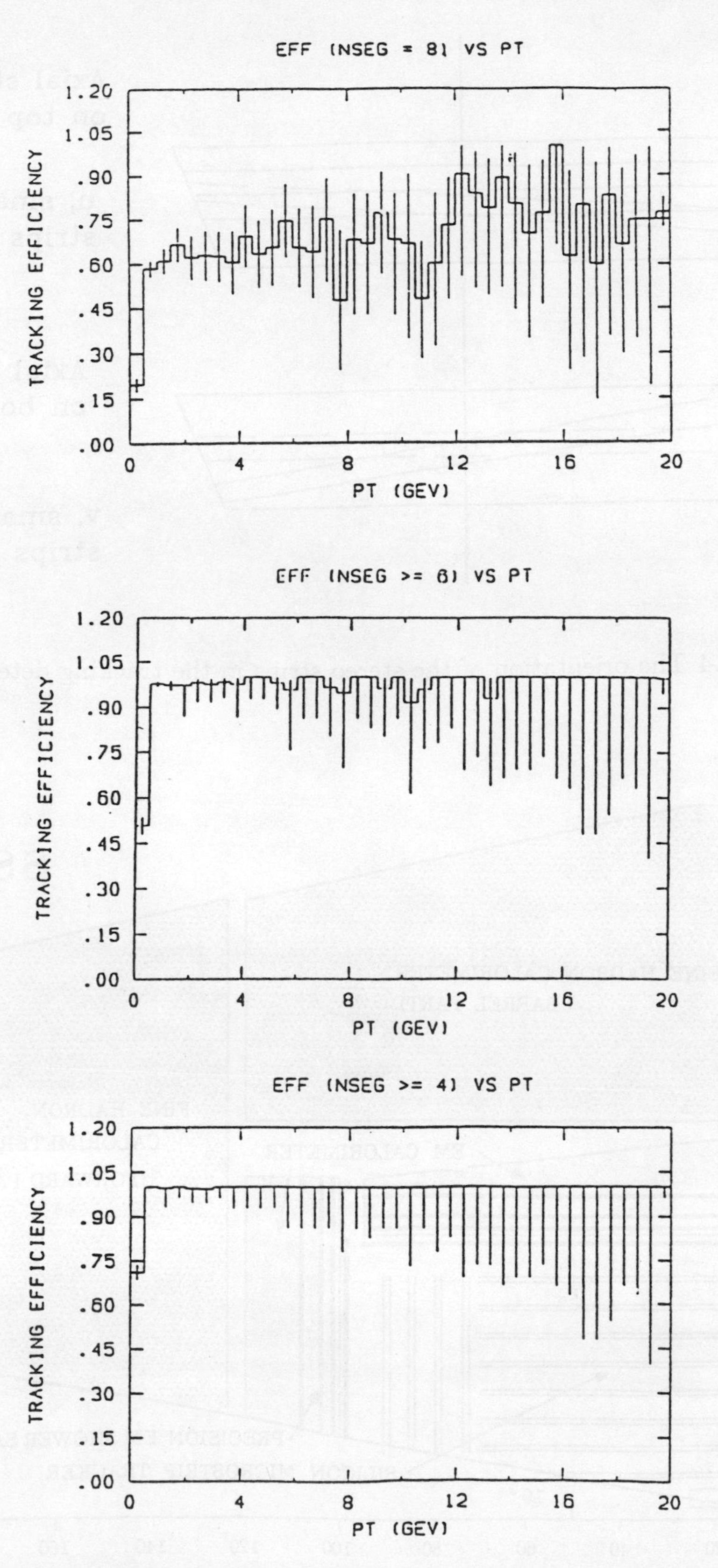

Fig.2.3 The tracking efficiency vs p_T for the cases when 8, 6 and 4 track segments are required.

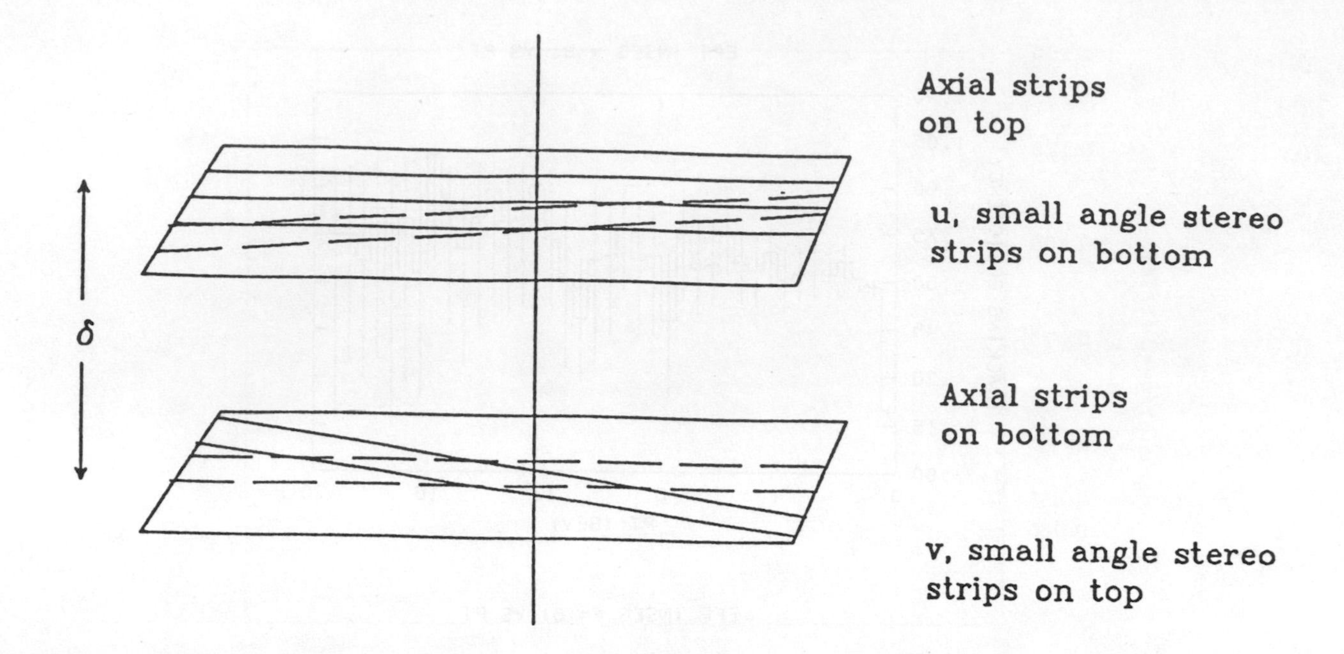

Fig.2.4 The orientation of the stereo strips in the tracking detector.

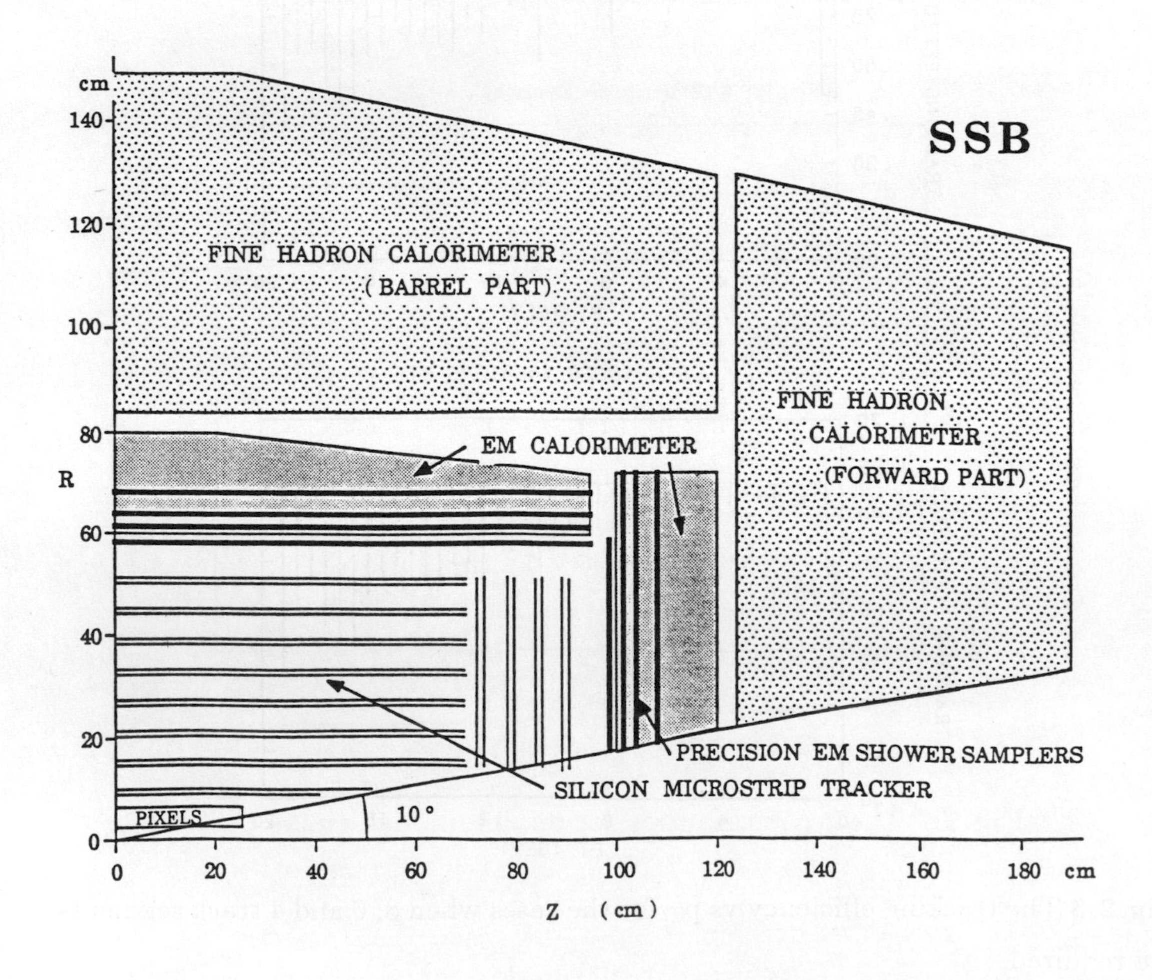

Fig.2.5 The inner components of the SSB detector.

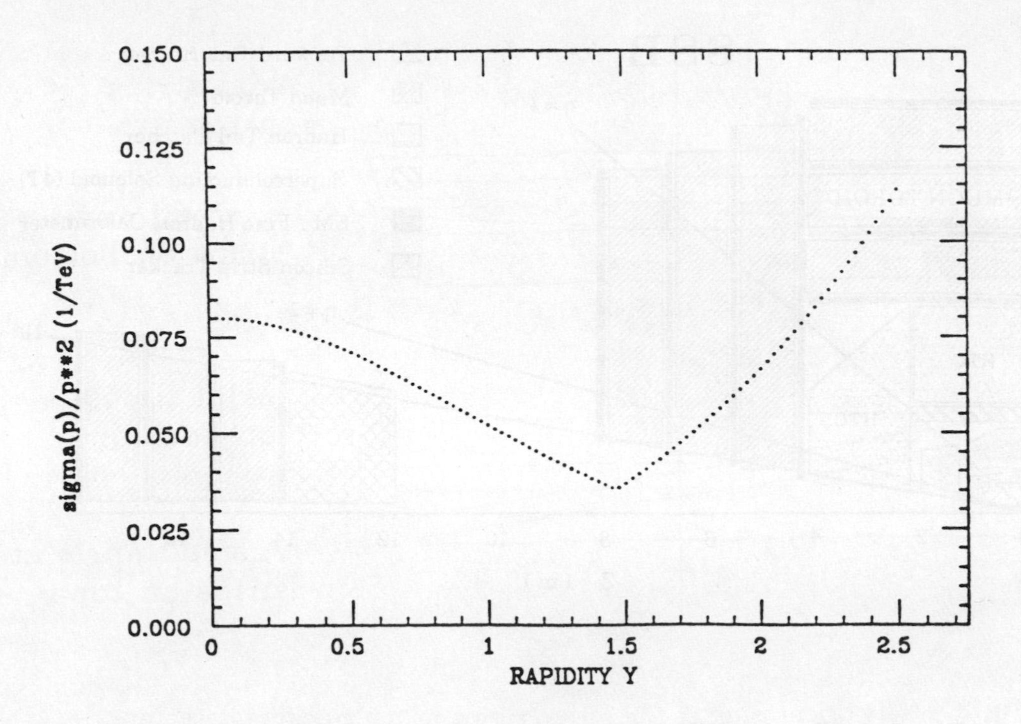

Fig.2.6 The momentum resolution for the **SSB** tracking system.

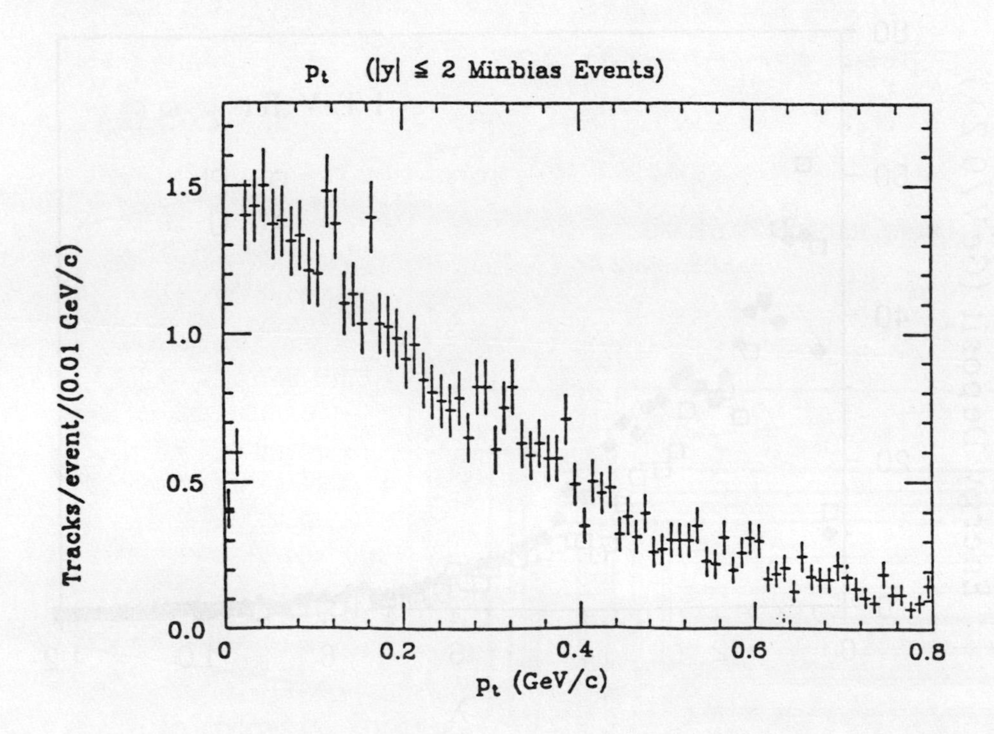

Fig.2.7 The p_T distribution for minimum bias events.

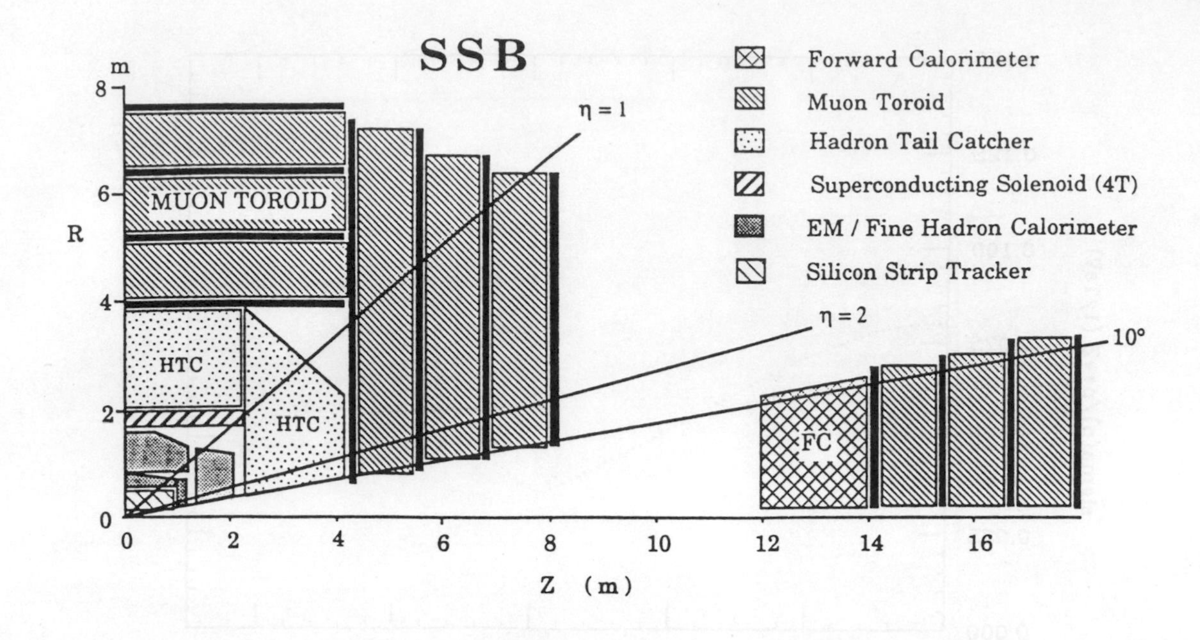

Fig.2.8 The overall configuration of the **SSB** detector.

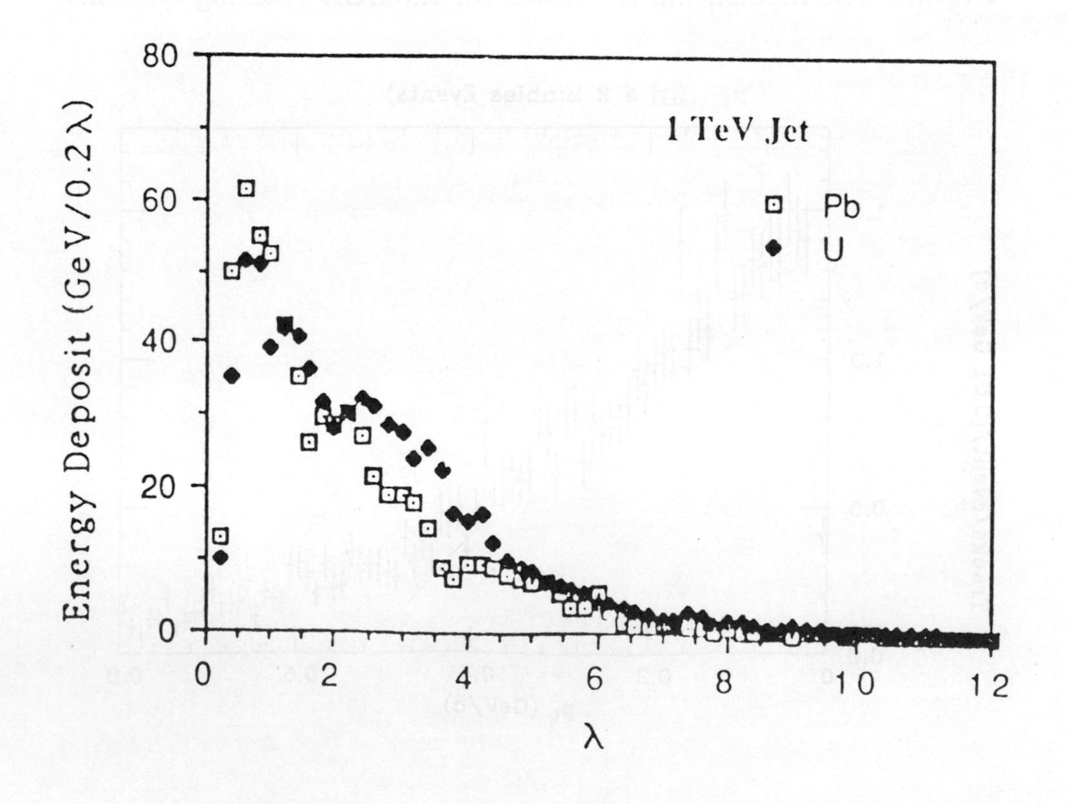

Fig.2.9 The longitudinal shower development of a a 1 TeV jet composed of 10 100 GeV pions.

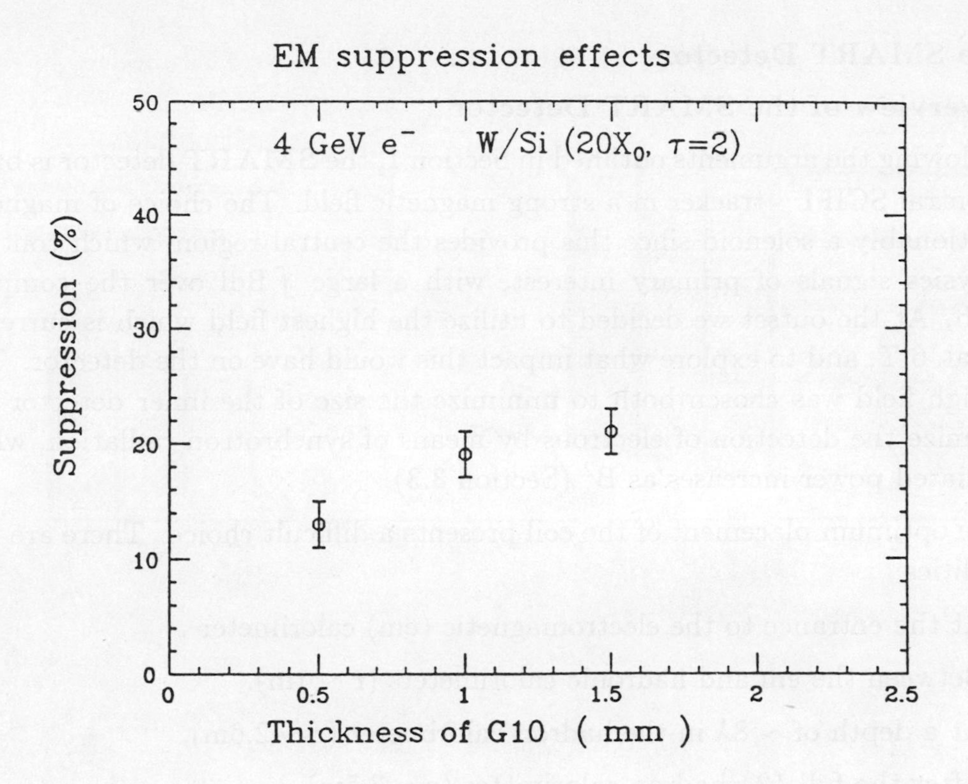

Fig.2.10 The suppression of the EM response of a Uranium-Silicon calorimeter by the insertion of G10.

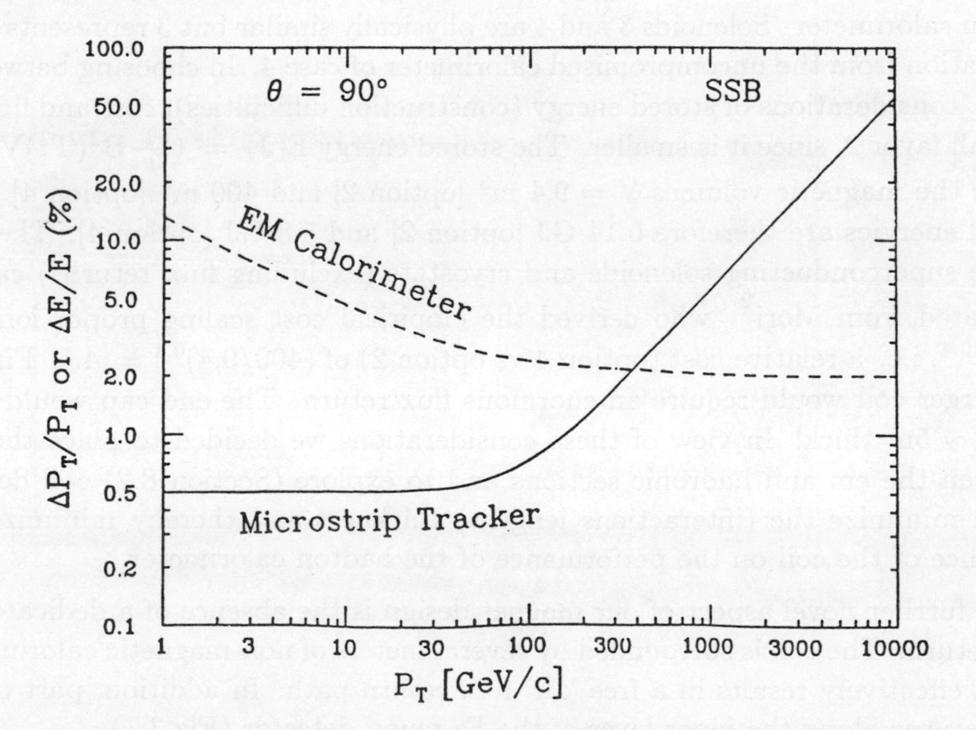

Fig.2.11 A comparison of the momentum resolution in the tracking chamber to the energy resolution in the EM calorimeter of the SSB detector.

3. The SMART Detector

3.1 Overview of the SMART Detector

Following the arguments outlined in Section 1, the **SMART** detector is based on a central SCIFI[1] tracker in a strong magnetic field. The choice of magnet is unquestionably a solenoid since this provides the central region, which contains the physics signals of primary interest, with a large $\int$ Bdl over the complete azimuth. At the outset we decided to utilize the highest field which is currently practical, 6 T, and to explore what impact this would have on the detector. This ultra-high field was chosen both to minimize the size of the inner detector and to optimize the detection of electrons by means of synchrotron radiation, where the radiated power increases as B^2 (Section 3.3).

The optimum placement of the coil presents a difficult choice. There are four possibilities:

1. At the entrance to the electromagnetic (em) calorimeter .
2. Between the em and hadronic calorimeters (r $\sim$ 1m).
3. At a depth of $\sim 8\lambda$ in the hadron calorimeter (r $\sim$ 2.6m).
4. After the full 12λ hadron calorimeter (r $\sim$ 3.5m).

It is possible immediately to narrow the choice to 2 and 4. Physically, solenoids 1) and 2) are similar but 1) substantially deteriorates the performance of the em calorimeter. Solenoids 3 and 4 are physically similar but 3 represents a deterioration from the uncompromised calorimeter of case 4. In choosing between 2 and 4, considerations of stored energy (construction difficulties), cost and flux return all favor 2, since it is smaller. The stored energy $E(J) = (\frac{1}{2\mu_o}B^2(T^2)V(m^3)$ where the magnetic volumes $V = 9.4\ m^3$ [option 2] and 400 m^3 [option 4]. The stored energies are therefore 0.14 GJ [option 2] and 5.8 GJ [option 4]. The cost of the superconducting solenoids and cryostats (excluding flux returns) can be estimated from Mori[2] who derived the empirical cost scaling proportional to $B \times V^{0.4}$, i.e. a relative cost (option 4 vs option 2) of $(400/9.4)^{0.4} = 4.5$. Finally, the larger coil would require an enormous flux return. The end cap would need to be $\sim$ 5m thick! In view of these considerations we decided to place the coil between the em and hadronic sections and to explore (Section 8.2) coil designs which minimize the (interactions lengths) thickness and thereby minimize the influence of the coil on the performance of the hadron calorimeter.

A further novel aspect of our magnet design is the absence of a dedicated Fe flux return. The coil is surrounded by several meters of non-magnetic calorimetry, which effectively results in a free 'air' flux return path. In addition, part of the flux returns along the inner layer of the Fe muon detector (Fig.3.1).

After defining the magnet and inner tracker, the rest of the detector follows in a fairly straightforward way. The complete detector (Fig.3.2) involves:

1. A compact inner detector which includes a SCIFI micro-tracker and SCIFI micro-converter.
2. A compact 6 T superconducting solenoid.
3. A Pb/SCIFI electromagnetic and hadronic calorimeter of total depth 12 λ.
4. An outer magnetized muon detector of Fe thickness 3m.
5. A separate forward detector for particles emitted at polar angles less than 10°. This separation is made to simplify the construction and readout of the detectors in the harsh forward region and to ensure good detection down to very small polar angles ($\sim 1^o$).

We will now describe in the following sections each of these systems in more detail.

3.2 Tracking: SCIFI micro-tracker

Following the discussion in Section 1, the inner tracker is arranged into a 'superlayer' geometry, i.e. condensed regions where many track measurements are made over a small radial distance. There are two distinct arrangements for the superlayers:

1. Three superlayers which are spaced at equal intervals to measure the sagitta of curved tracks.
2. Two superlayers which are placed at the outer edge of the tracking volume. These effectively measure the total magnetic bending angle to determine the track momentum. This 'beam constrained fit' makes use of the very small ($\sigma \sim 10\mu$m) transverse dimension of the interaction point.

In order to reach the design momentum measurement accuracy, $\sigma_p/p \leq$ 25%p(TeV/c), arrangement 1) would involve three superlayers at radii of 20, 40, and 60 cm, in the field B = 6 T. For arrangement 2) we consider two superlayers set at radii of 50 and 60 cm, with a relative systematic alignment error of 10 μm. The magnetic impact parameter at the origin, $b = L^2/2\rho$ where the track curvature ρ(cm) = 100 p(GeV/c)/[.3 B(T)] = 55.6 p(GeV/c). Therefore b (μm) = 225/p (TeV/c). The measurement error on the magnetic impact parameter $\sigma_b \sim 5 \times 10\mu m = 50\mu m$. Hence the momentum resolution $\sigma_p/p = \sigma_b/b$ = 22% p(TeV/c). Note that the beam size has a negligible effect on the precision of this measurement.

The second 'minimal tracking' approach is very attractive from the viewpoint of high luminosity operation: the innermost layer at a radius of 50 cm receives 'only' $\sim 2 \times 10^4$ rads/year and the occupancy and double-hit problems are substantially reduced relative to operation at smaller radii. The price that must be paid is a loss of the ability to measure secondary vertices such as from b decay. We point out that the momenta of the particles from b decay are still well measured over essentially the full kinematic range; the *magnetic* impact parameter for tracks below 100 GeV/c is $\geq$2 mm, compared with a *decay* impact parameter of typically $\leq 100\mu$m. With these considerations we chose to incorporate a high-luminosity 'minimal tracking' geometry for the SMART inner detector (Fig.3.3). Our approach would be to install an inner SCIFI vertex detector for a dedicated experiment (at lower luminosity) involving detection of secondary vertices.

The SCIFI micro-tracker (Fig.3.4) is built of two concentric cylindrical 'superlayers' constructed from coherent arrays of narrow (25 μm) diameter plastic fibers. Each superlayer consists of four layers, each 2 mm thick, made from fibers arranged in the directions z-u-v-z. The u and v fibers form a narrow angle stereo of $\pm$34 mrad with respect to the z fibers, which lie parallel to the beam axis. The perpendicular material in each superlayer is 2% X_o. With a yield of 5 hits per mm there are a total of 40 hits per track in each superlayer.

This geometry results in excellent track reconstruction since each superlayer measures local track vectors in space (three-dimensional position and angle) and in momentum (by impact parameter relative to the beam axis). For example, inside a superlayer the azimuthal angle precision, $\sigma_\phi = 1.1$ mr, and the momentum precision, $\sigma_p/p \leq 0.1\%$ p(GeV/c). Unambiguous linkage of the track vectors between superlayers can therefore be made. In addition the vector stubs of (uninteresting) low energy tracks can be rejected at an early stage of event reconstruction. In association with the ultra high field (6 T) solenoid, this compact SCIFI tracking system achieves remarkable performance (Table 3-I).

The SCIFI micro-tracker is read out by an image intensifier/CCD system (Fig.3.5). The high-speed operation of this system is paced by the CCD. The maximum rate at which a CCD may be exposed is several$\times 10^5$Hz. This is set by the fast clear time of 1 μs which would, for example, lead to a 50% dead time at an exposure rate of 10^6Hz. Readout times are typically 10 ms, reducing to 1 ms for the highest speed contemporary CCDs. With these slow readout times (which are due to the high degree of multiplexing) it is clear that detailed CCD information cannot participate in the trigger until level 3 ($\leq 10^3$Hz) where the events are analysed in parallel microprocessors. It is possible, however, to obtain fast OR'd analogue SCIFI pulses which can participate in level 2 triggers ($\leq 10^5$Hz). These pulses are generated at the anodes of the final image intensifier stages and correspond to the sums of the total SCIFI pulse height in regions of area ~ 1 cm^2.

Table 3-I SCIFI micro-tracker characteristics

Number of hits per track	80
σ_{xy}hit^{-1}	15 μm
Two-hit resolution	50 μm
σ_{ϕ}per superlayer	1 mr
σ_p/p per superlayer	0.1% × p (GeV/c)
σ_{xy}per superlayer	3 μm
$\sigma_z$per superlayer	140 μm
σ_z(vertex)track^{-1}	1 mm
σ_p/p^2	$[22\% \times p\ (TeV/c)]^2 + [0.7\%]^2$

At the high rates of the SSC it is necessary to provide an optical delay line upstream of the gated image intensifier stage (Fig.3.5). The first two stages are proximity focussed photodiodes mounted directly at the edge of the SCIFI detector. Since these devices do not have a micro-channel-plate they have a high quantum efficiency and operate in longitudinal magnetic fields of any strength. The photodiodes serve to amplify the optical signal by a factor of 100 and to shift the light to a longer wavelength for improved optical transmission. The fastest currently available phosphors (P47, τ = 35 ns) are too slow for our purposes. Following discussions with industry we expect a new fast ($\tau \leq$10 ns) and efficient phosphor can be developed. This phosphor would be used in each of the first two stages of image intensifier. The output is connected to a flexible coherent fiber bundle which acts as the optical delay line. The length of this delay line must be sufficient to allow the level 1 trigger to open the gate of the final stage before the arrival of the optical data, e.g. a level 1 decision time of 350 ns would require a 60 m optical delay line. This situation is shown in Fig.3.6 for a period of 10 ns between crossings, which corresponds to the minimum system integration time. In order to preserve the lifetime of the final stage image intensifier, the device and CCD may be cooled and the gain reduced. (Suitably cooled, CCD'S can record and shift signals of $\leq$10 electrons!)

The ability of a SCIFI micro-tracker to handle the high rates of the super-collider is due to its extremely high granularity: the full system comprises $\sim 10^8$ fibers or $\sim 10^7$ 'cells' (defined by the 50 μm image intensifier spot size). During the integration time of 10 ns approximately one event will occur, with a total multiplicity of 100 (high p_T event) charged particles. The resulting occupancies are very low. For example, in a single fiber layer at r = 50 cm there are 6×10^4

cells, and so the occupancy per trigger is 0.2% per fiber layer.

The calibration and monitoring of the SCIFI micro-tracker is accomplished as follows. Charged particles (field off) check the mechanical assembly, the fiber coherency and radiation damage to the SCIFI. Illuminated slit fiducials placed at the opposite end of the superlayers check for drifts during data taking and for any field on/field off effects. We note that SCIFI is well-suited to stable operation in high B fields since the data is almost purely optical (no E×B effect in the detector) and the device is inherently stable to short term drifts (no gas, no time-to-distance relationship, etc.).

The SCIFI micro-tracker involves 137 kgm of SCIFI (total fiber length 2×10^5 km). The readout requires 550 image intensifier/CCD units, assuming the use of CCD's with area 1 cm^2, and unit magnification. We discuss the radiation hardness and the cost of this detector in Sections 10.2 and 11.4, respectively.

3.3 Imaging preconverter: SCIFI micro-converter

Immediately following the micro-tracker there is an imaging preconverter for electrons and photons (Fig.3.4). The 'micro-converter' (Fig.3.7) is made from plastic SCIFI of diameter 100 μm embedded in a medium of thin high-Z converters, e.g. Pb foils, of a total thickness $\sim 3X_o$. The SCIFI/Pb volume ratio is 1/4 to preserve the same ('compensated') composition as the calorimeter. We estimate the minimum detectable photon energy to be ~5 MeV (equivalent to 10 pe).

The first application, illustrated in Fig.3.7a, for this device is to identify the presence of a nearby π^o and γ which could lead to false electron candidates in the relatively coarse-grained cells of the electromagnetic calorimeter. In addition to rejection of this background, the SCIFI micro-converter can make a positive electron identification in two ways (Fig.3.7b):

1. Initiation of the electron shower in the thin pre-converter.
2. Identification of hard synchrotron photons radiated in the ultra-high magnetic field.

The characteristics of the synchrotron light radiated by an electron in a 6 T magnetic field over a path length of 60 cm are as follows: number of photons (independent of E_e) = 22, total synchrotron energy E (keV) = $28E_e^2$, critical energy E_c(keV) = $8E_e^2$ and mean photon energy $E_\gamma = 1.3E_e^2$, where E_e is expressed in GeV. This shows the potential for a very powerful technique (in addition to calorimetry) for the identification of high-energy electrons, namely the detection of hard (MeV) synchrotron photons which are located in a well-defined 'fan' along the tangent to the electron track. For a γ threshold of 5 MeV, we can see (Fig.3.8) the threshold for electron identification is $\sim$ 30GeV;

lowering of the γ threshold would lead to a corresponding reduction in the electron identification threshold. The SCIFI tracking layers will be essentially transparent to photons of these energies ($\leq$1 MeV) and so most will reach the micro-converter. Photons which convert in the tracking layers will of course be detected with good efficiency and will aid the identification of electrons. This technique for electron identification is of particular importance since most of the alternative techniques, such as transition radiation or matching between the calorimetry energy and incident momentum, are of little benefit above $E_e \sim 100$ GeV. At these key energies, the only positive experimental signatures of an electron are longitudinal shower profile and production of hard synchrotron photons. Finally we comment that the radiation of photons in the material of the SCIFI tracker will increase the apparent synchrotron flux and aid the identification of the electron.

The final application of the SCIFI micro-converter (Fig.3.7c) concerns the identification of single photons. Here the major background is from π^o or η production which leads to two or more nearly-coincident photons. A two-shower resolution of 150 μm will allow π^o identification up to $p_T \sim 1$ TeV/c.

We have done a Monte Carlo simulation of a simplified model of the micro-converter. Here, we have assumed 200 μm diameter fibers interspersed between 0.1 X_o lead radiators, for a total of 3 X_o. In Fig.3.9a,b,c and d are examples of a 100 GeV electron, a 100 GeV single γ, a 100 GeV charged pion and a 100 GeV charged pion with a 10 GeV γ 9 mm away, respectively. If we declare a γ to be 'detectable' if it converts in the micro-converter and lies more than 750 μm away from the primary electron track, we get an average number of 'detectable' γ's as a function of E_e shown in Fig.3.10. From 30GeV $\leq E_e \leq$ 350GeV there are $\geq$5 'detectable' γ's/e.

The micro-converter is read out analogously with the micro-tracker. The only difference involves a linear demagnification, by a factor of 2, using fiber optic reducers, in order to match better the fiber and CCD pixel sizes. The full system involves about 200 image intensifier/CCD units.

3.4 Calorimeter: Pb/SCIFI

We will briefly review our arguments in favor of a Pb/SCIFI calorimeter for SMART. The primary requirement is to achieve a relative electron/hadron response e/h $\sim$ 1. (In fact, due to unavoidable smearings caused by the jet physics itself, the practical requirement on the calorimeter is e/h = 1 $\pm$0.15). This condition was demonstrated originally in uranium calorimeters but now also it has been predicted[3] and measured[4] in Pb calorimeters which utilize a read-out which is rich in hydrogen (plastic scintillator) and therefore sensitive to the neutron component of the hadronic shower. Up to the present there have been two main arguments supporting uranium-based calorimeters: one is compensation and the other, high density. These arguments are now considerably weaker since we know 'compensation' is not unique to uranium and since the overall thickness, including readout, of practical compensating hadronic calorimeters is not substantially different between the various dense absorbers. Moreover uranium poses strong disadvantages with respect to cost, toxicity and in particular to the irradiation of the entire detector with thermal neutrons, both from the natural fission and from hadronic showers. The latter source, which appears to be a severe contributor of damaging radiation, is at least a factor of 4 higher in the case of a uranium calorimeter compared with a Pb device.

We have therefore based our calorimeter on Pb absorber. The sampling is chosen to be plastic SCIFI of 1 mm diameter, with a Pb/SCIFI volume ratio of 4/1 throughout, in order to achieve e/h = 1 (in the absence of coil material).

The advantages of this calorimeter are as follows:

1. High density. For a Pb/SCIFI volume ratio of 4/1 the radiation length X_o = 7.0 mm and the interaction length, $\lambda = 20.3$ cm.
2. Fast. The signal integration time is $\leq$20 ns and the individual cell times can be determined to $\leq$1 ns.
3. Hermetic and uniform. Internal dead regions or 'cracks' are absent.
4. Inexpensive materials and relatively simple construction compared with other calorimeters such as liquid argon or TMP.
5. Good electromagnetic and hadron energy resolution.
6. Readout system which is compact, simple, flexible and stable in operation.
7. 'Compensated', i.e. relative electron/hadron response = 1 (in the absence of coil material).
8. Radiation hard to a level sufficient for operation in a supercollider detector.

The calorimeter is divided longitudinally (Fig.3.10) by cutting fiber blocks of the appropriate thickness, followed by machining and polishing of the entrance and exit faces. In the electromagnetic section (28 X_o, including 3 X_o micro-converter) these blocks have thicknesses of 5 X_o and 20 X_o in order to follow the shower profile. The hadronic section (an additional 11 λ, for a total of 12 λ) is segmented into three longitudinal sections [3 λ, 5 λ, 3 λ]. The transverse granularities are $\Delta\eta \times \Delta\phi = 0.03 \times 0.03$ (em) and 0.06×0.06 (hadronic).

The fiber directions lie approximately radially in the barrel region and along z in the end cap region. In this geometry the fibers are approximately, but not exactly, pointing to the interaction regïon and so particles do not travel along a single fiber and avoid the absorber material. The readout (Figs.3.3, 3.10 and

3.11) is accomplished with small waveshifter plates, each with a clear polystyrene fiber ribbon attached to one edge face, which are arranged in a mosaic at the outer surface of each layer. With this approach the transverse granularity and cell shapes can be simply chosen or later modified, since the calorimeter medium itself has no fixed internal boundaries. The ribbons, of length 10 m, guide the light outside the detector to multi-anode phototubes. The pulse-height of each calorimeter cell is recorded on an individual anode pixel. There are a total of 100 K cells which are read out using 1.5K multi-anode phototubes, each having 64 anodes.

We calculate a light yield of 250 pe GeV^{-1}, i.e. there is essentially zero electronics noise per tower ($\leq$20 MeV). The calculated performance of the calorimeter, in the absence of coil material, is $\sigma_E/E = 17\% \times E^{-1/2}(GeV^{-1/2})$ [em] and $31\% \times E^{-1/2}(GeV^{-1/2})$ [hadrons], with e/h = 1. The effect of the coil on the performance of the hadronic calorimeter is discussed in Section 8.1 of this report.

The calibration and monitoring of the calorimeter is effected by means of an optical flasher system, which illuminates the fibers at the entrance of each tower, and also by means of radioactive sources. The systematic calibration accuracy which has been achieved in practice, by the UA2 scintillator sampling calorimeter is a cell-to-cell spread of $\pm$1.5% (em) and $\pm$3% (hadronic) and an absolute energy uncertainty of $\pm$1% (em) and $\pm$2% (hadronic). We are therefore encouraged that the overall systematic uncertainties could be reduced to $\pm$1%, after further development.

The radiation dose at the entrance to this calorimeter is 10^4 rads/year (3×10^{11} minimum ionizing particles $cm^{-2}year^{-1}$). If we assume the mean photon energy is ~1 GeV then at shower maximum there will be ~10 particles per incident photon and the dose will be ~ 10^5 rads $year^{-1}$. On this basis we see that plastic SCIFI must perform reliably up to doses of 10^6 rads. This is a reasonable expectation based on present performances and new materials (Section 10.2) and considering the short fiber lengths involved. The radiation dose in the hadron calorimeter will be approximately an order of magnitude less (i.e. $\leq 10^4$rads $year^{-1}$) because of the larger radial distance and the reduction in mean number of particles at shower maximum.

REFERENCES

1) For a review of SCIFI detectors and other references, se J. Kirkby, Today and Tomorrow for Scintillating Fiber (SCIFI) Detectors, INFN Eloisation Project, Workshop on Vertex Detectors, Erice, CERN-EP/87-60 (1987).

2) S. Mori, Report of the Task Force on R& D for the Superconducting Super Collider, SSC-SR-1021, 251 (1986).

3) R. Wigmans, CERN-EF/86-18 (1986), submitted to Nucl. Inst. & Meth.

4) E.Bernardi *et al.*, Performance of a Compensating Lead-Scintillator Hadronic Calorimeter, DESY 87-041 (1987).

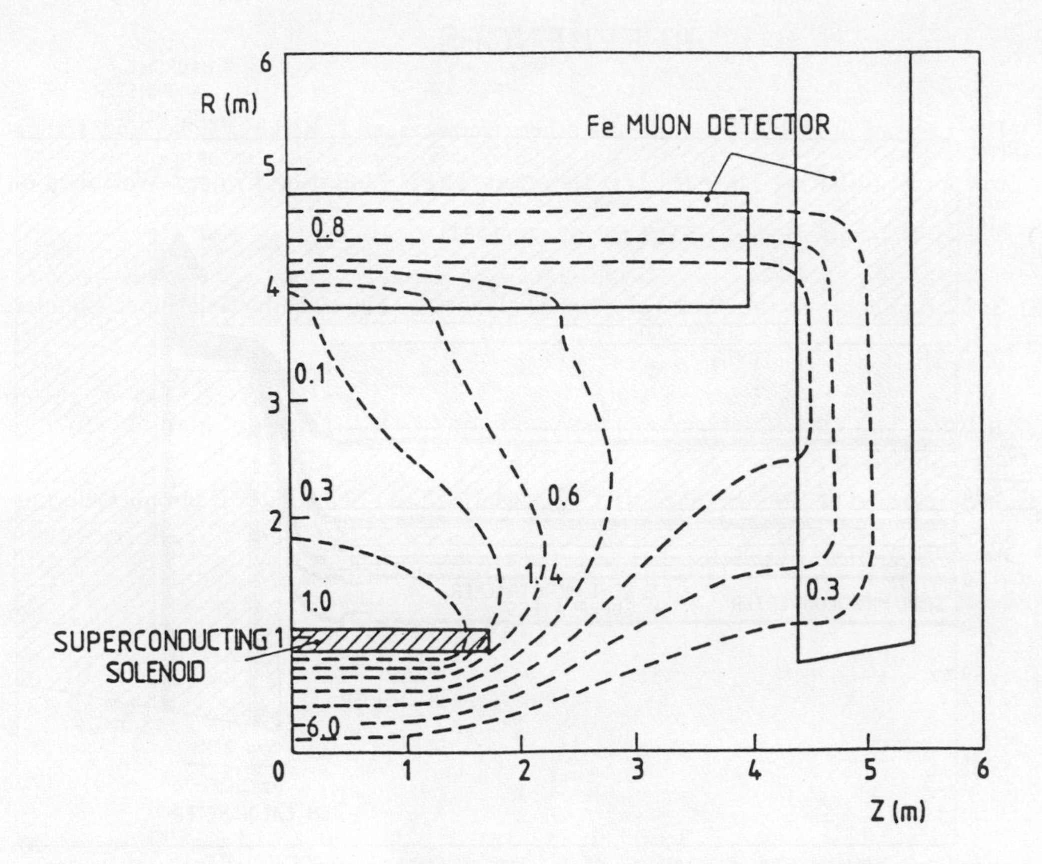

Fig. 3.1 A field map of the **SMART** solenoid.

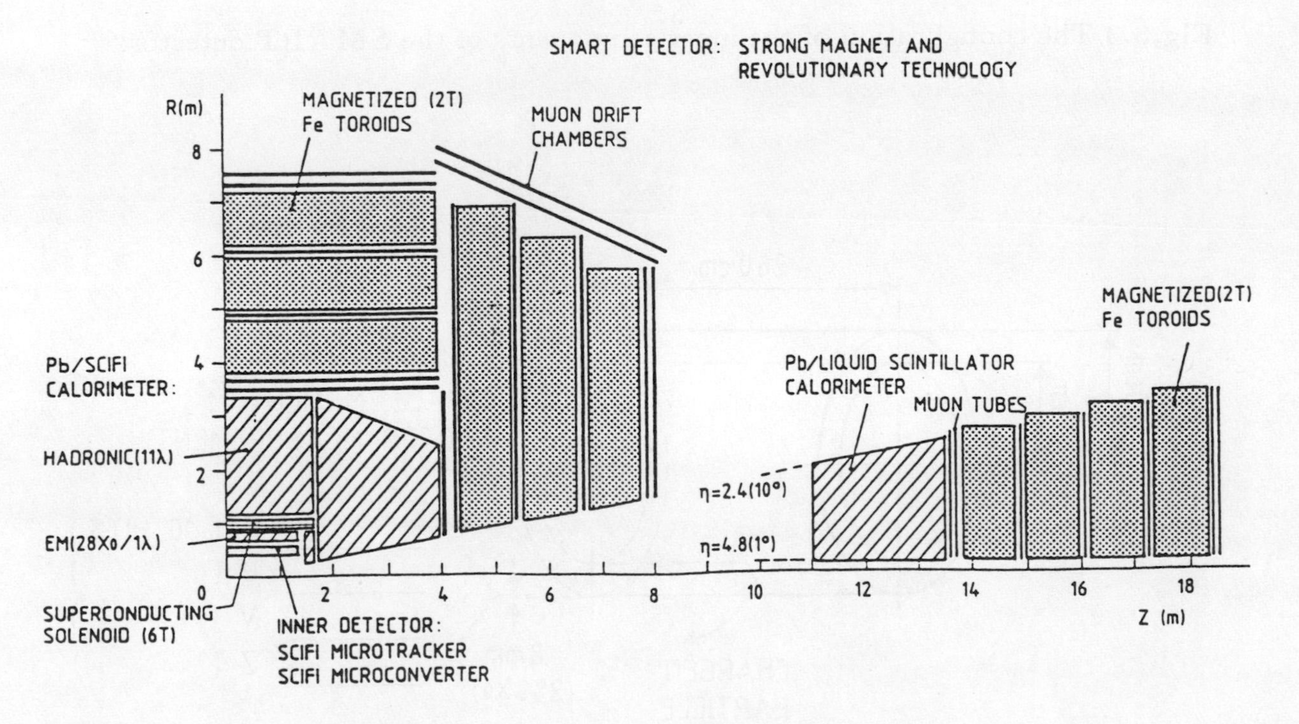

Fig. 3.2 An overview of the **SMART** detector.

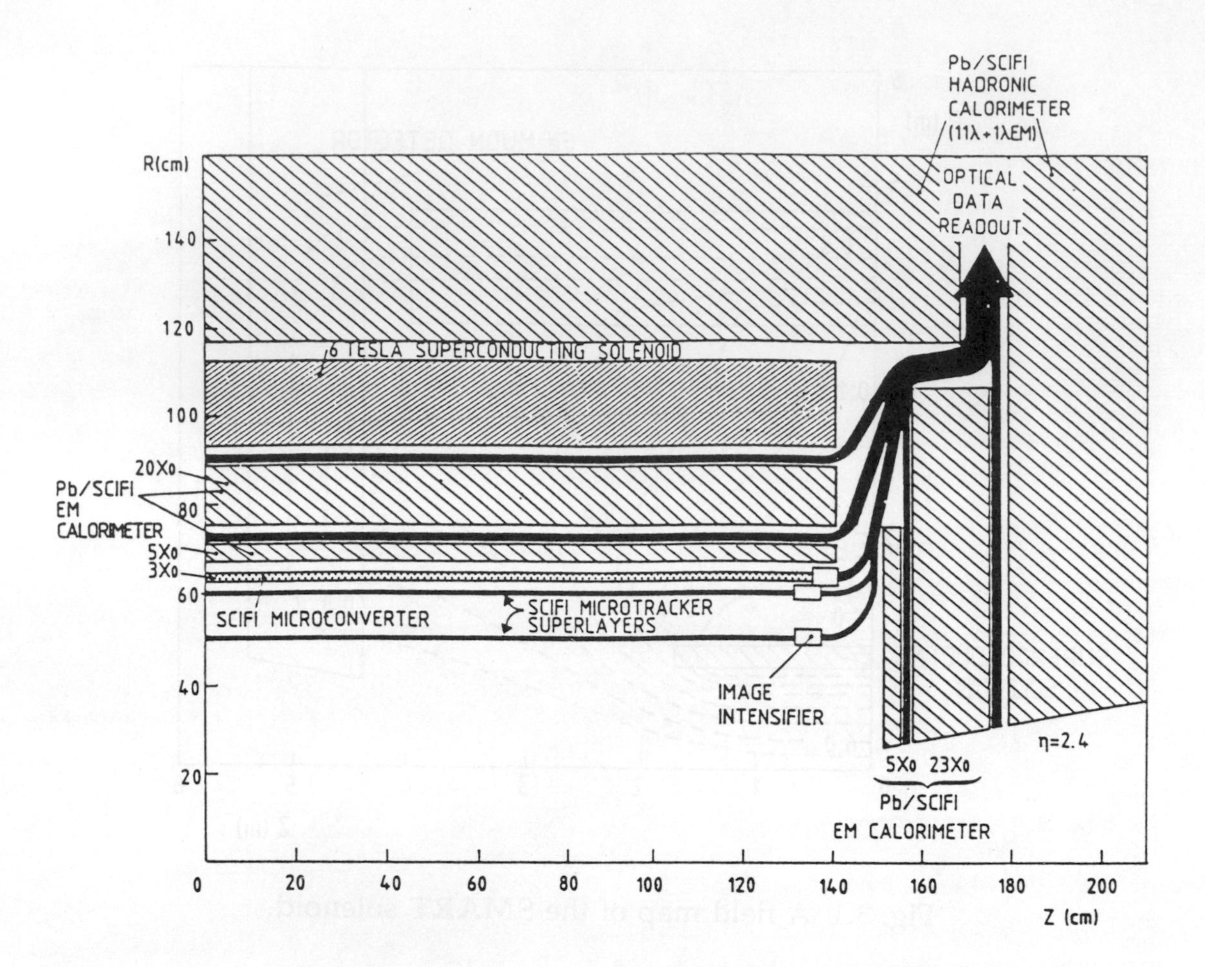

Fig.3.3 The configuration of the inner components of the **SMART** detector.

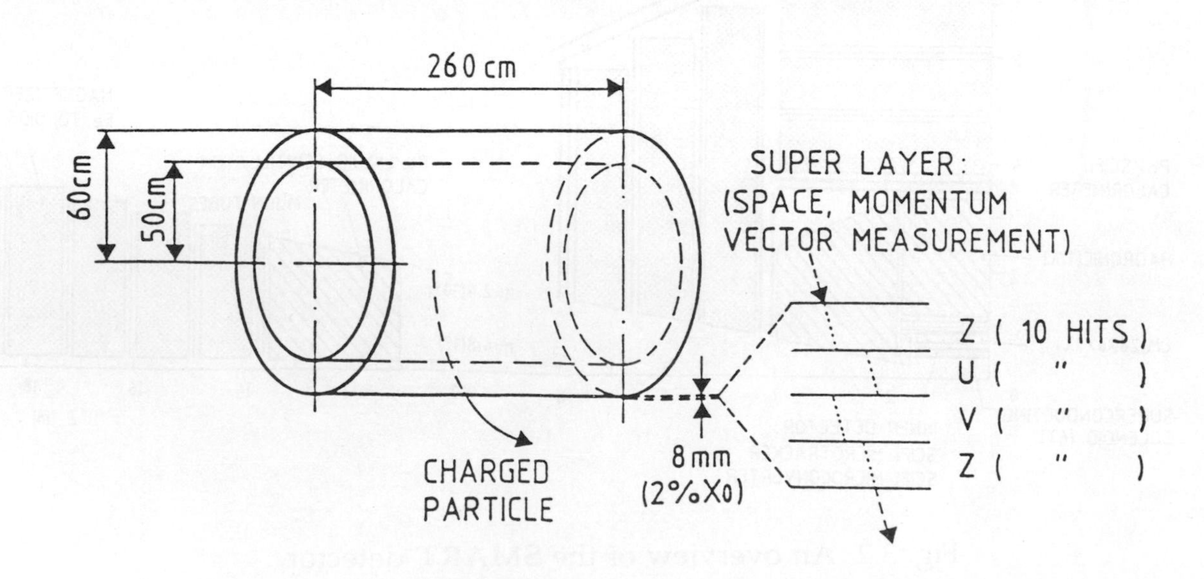

Fig.3.4 A SCIFI microtracker superlayer.

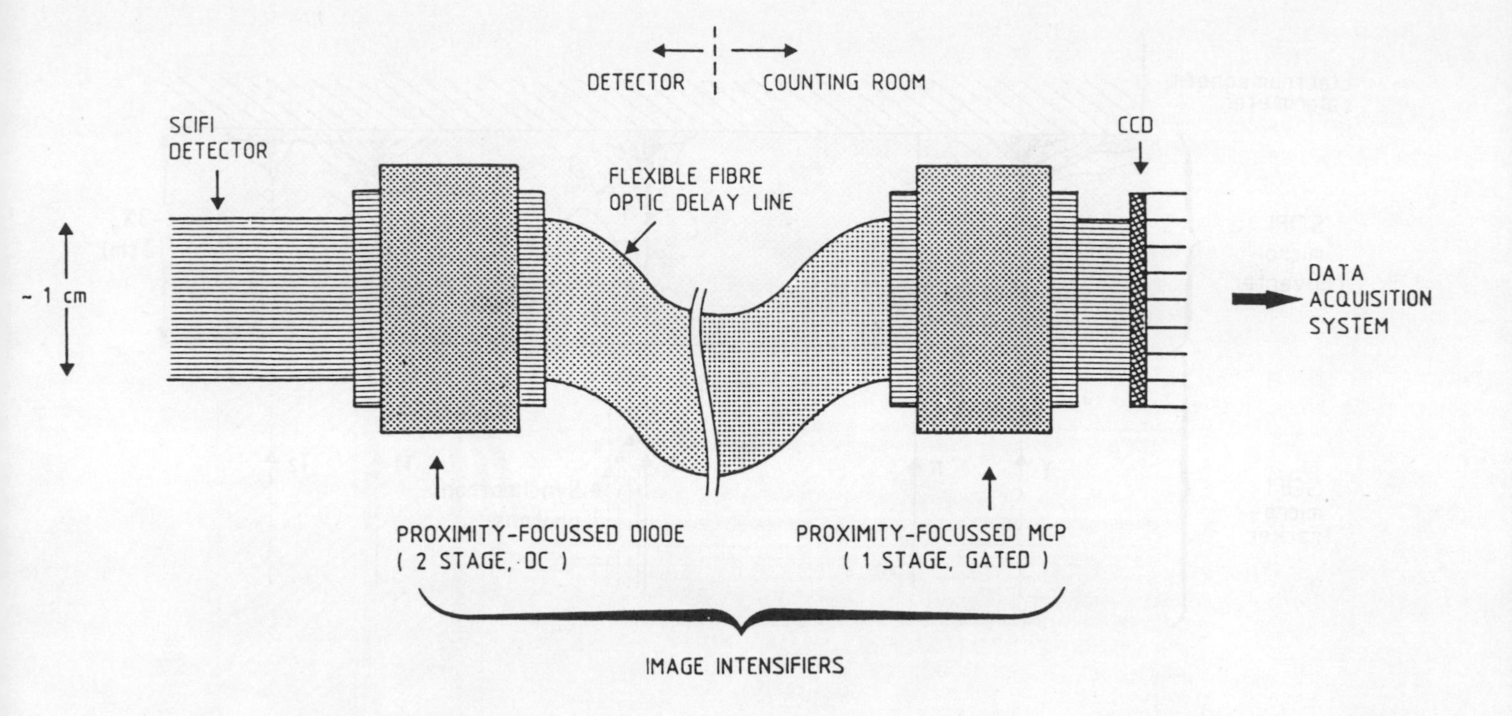

Fig.3.5 The image intensifier, optical delay line readout system.

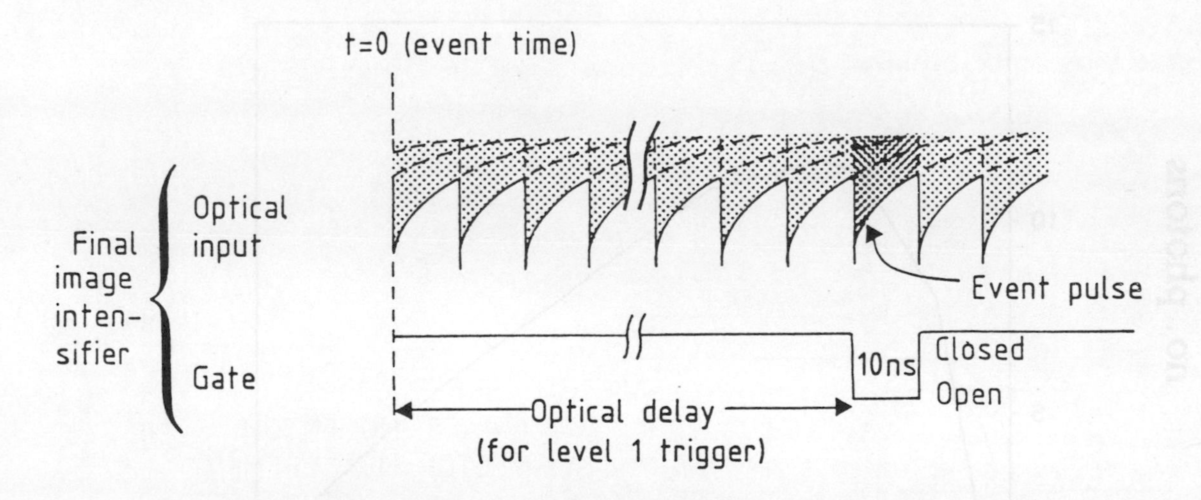

Fig.3.6 The signal structure at the gated image intensifier.

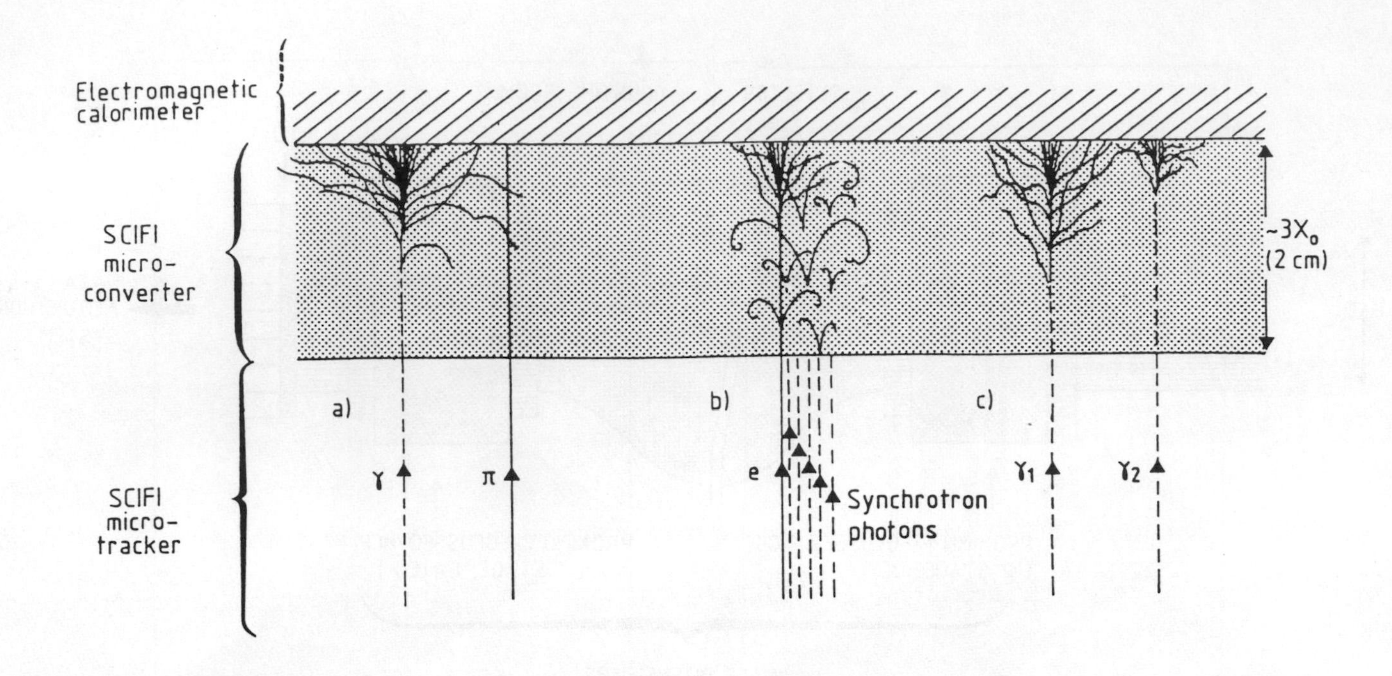

Fig.3.7 The operation of the micro-converter for: a) $\pi^{\pm}/\gamma$ overlap suppression; b) electron identification; c) π^{o}/γ discrimination.

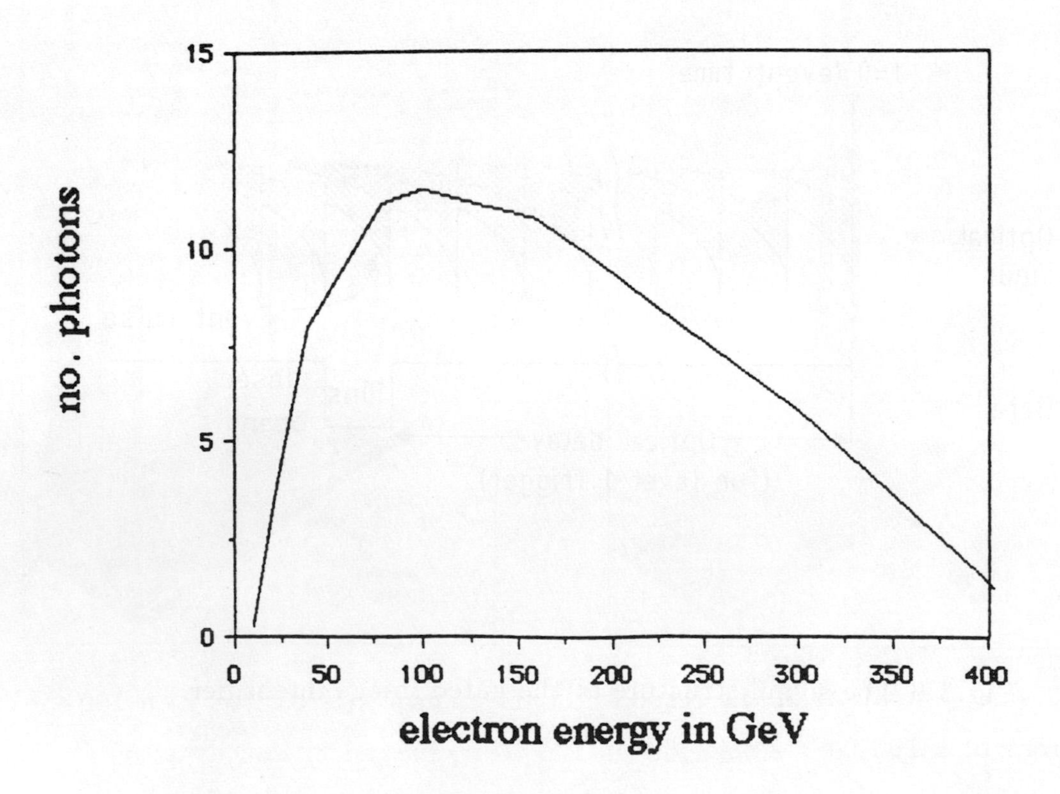

Fig.3.8 The number of identifiable synchrotron γ conversions vs E_e.

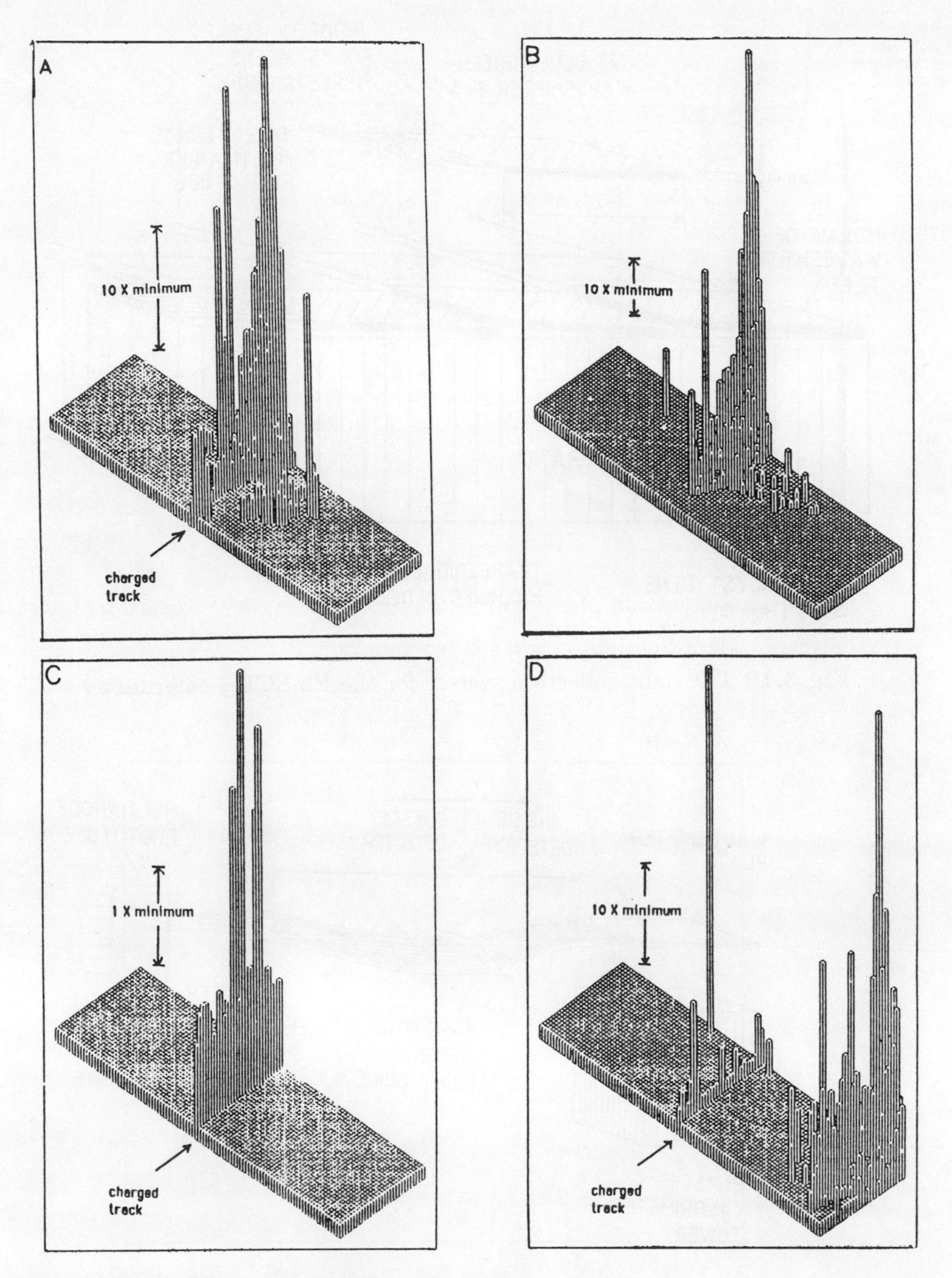

Fig.3.9 A simulation of the response of the microconverter to: a) a 100 GeV electron; b) a 100 GeV single γ; c) a 100 GeV charged π; and d) a 100 GeV π with a 10 GeV γ that is 9 mm away.

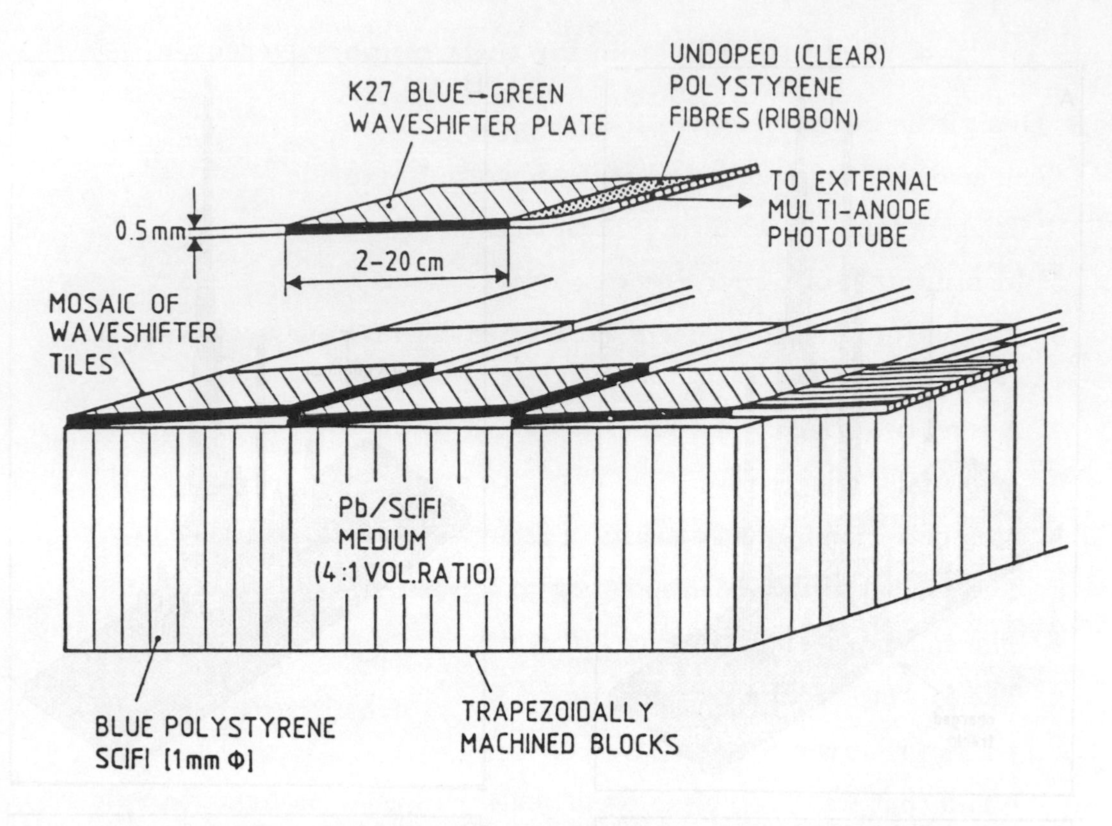

Fig.3.10 The light collection system for the Pb-SCIFI calorimeter.

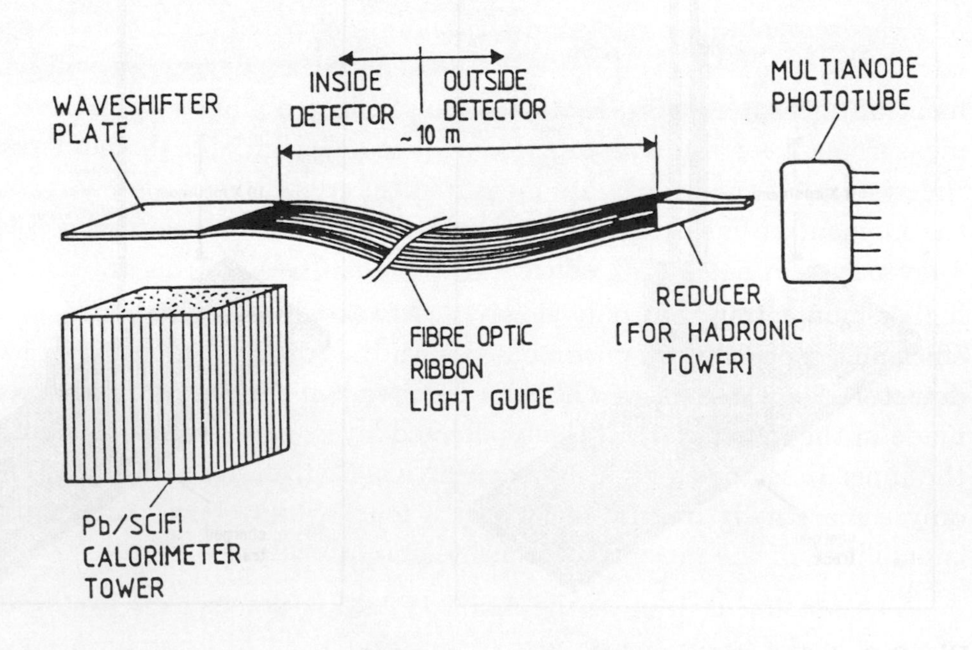

Fig.3.11 The readout scheme for the Pb-SCIFI Calorimeter.

4. The Muon System for the Compact Detector

4.1 Design Criteria

The goals of the muon detection system are to provide:

1. Triggers from μ's with $p_T \geq 20$ GeV/c.
2. Identification of μ's for $\theta \geq 1°$.
3. A momentum measurement dominated by multiple scattering ($\Delta p/p \sim 10\%$) for μ momenta up to ~1 TeV/c.
4. A level two trigger that associates the μ to the correct RF bucket.
5. A low false trigger rate.

The muon detection system consists of 3 main sections (see Fig.4.1).

a. The central section corresponding to $\theta \geq 60°$.

b. The end cap section covering $10° \leq \theta \leq 60°$.

c. The forward section covering $1° \leq \theta \leq 10°$.

4.2 The Four Layer System

We find that we can achieve these goals with a muon detection system consisting of 4 detection planes separated by 1 m thick magnetized iron absorbers, as shown in Fig.4.2. The number of layers in the system was chosen both to achieve good momentum resolution, high trigger efficiency as well as to minimize the loss of measurements due to Brehmsstrahlung and pair production. As will be seen later, these processes will not adversely effect the trigger, but will harm the momentum measurements. Table 4.1 shows the probability to make a momentum measurement for a μ as a function of μ momentum and the number of layers in the system. If the coordinate measurement from 2 layers are corrupted, we lose the momentum measurement in a 4 layer system. We see from Table 4.1 that a 4-layer system has a 95% efficiency up to 500 GeV/c; this efficiency worsens at higher momentum, but only slowly.[1] It was felt that this represented excellent efficiency, especially when combined with the tracking information in the central detector. For those cases where an independent momentum measurement is not made in the muon system, the unaffected layers can be used to identify tracks in the inner detector as high momentum μ's. A five layer system would represent only a marginally improvement over a four-layer system, particularly when the capabilities of the inner tracker are taken into consideration.

The amount of iron in the muon system is determined by the desired level of precision of the momentum measurement used to form the trigger. The EM and hadron calorimeters, corresponding to 2 meters of iron equivalent, provide the initial punch-through rejection. Each detection plane is assumed to provide coordinates with $\sigma = 100\mu m$ in the bend plane and $\sigma \sim 1cm$ in the non-bend plane.

Table 4.1

Probability to Make an Independent Momentum Measurement

μ Momentum (TeV/c)	Probability of a single lost layer	3 layers	4 layers	5layers
.10	.05	95%	99%	100%
.20	.10	90%	99%	100%
.50	.23	77%	95%	99%
1.0	.37	63%	86%	95%
2.0	.55	45%	70%	83%
5.0	.66	34%	56%	71%

The system has a total absorber of 5 meters of equivalent iron. Estimates indicate that the trigger rate and misidentification due to punch-through are negligible.[2]

4.3 Triggering Strategy

The measuring elements of the muon system consist of 1 cm thick drift cells that are 2 cm wide and 4 m long. These cells would have a maximum drift time of 200 ns, a resolution of 150 μm in the drift coordinate. The coordinate along the wire could be measured by cathodes or by current division to yield a resolution of $\sim$1.5 cm. Two layers of such cells are combined in a staggered fashion to form a single measuring layer of the system, yielding the required precisions. A scintillator system, placed outside of the entire system, consists of elements 20 cm wide and 2 m in length. The counter size is chosen to provide the capability of tagging the correct RF bucket for a given μ, at the 63 MHz crossing rate.

The trigger functions as follows. Roads are formed that vary from approximately 7 cells wide at $\theta \sim 90^o$ to 1.5 cells wide at $\theta \sim 10^o$. These roads yield a transverse momentum cut-off of 20 GeV/c. Tracks of lower momentum fall outside of these roads and are rejected. Each road points to the interaction region as indicated in Fig.4.3. A fast compare of the signals from each of the chambers is done in a fast memory look-up table.[3] Because of the integration time of the chambers, a road will persist for ~100 ns (longer persistence is easily acvhieved). Once a road is formed, a coincidence with the scintillator system is made, yielding a μ trigger identified with a specific crossing.

This system has the distinct advantage that the transverse momentum trigger can easily be altered by changing the contents of the memory look-up tables. To lower the cut-off, simply increase the road size. Because the cells are formed by two layers, the road widths can be altered in steps of size 1 cm, corresponding to $\Delta p_T \sim 2$ GeV/c at $\theta \sim 90°$.

4.4 Momentum Resolution

The momentum resolution[1] for tracks with at least three useful layers is shown in Fig.4.4. Here it can be seen that the resolution is dominated by multiple scattering up to 500 GeV/c. The sign of the μ is well measured up to 5 TeV/c.

4.5 Singles Rates and False Triggers

In order to find an upper limit on the false trigger rate from the muon system, the scenario illustrated in Fig.4.5 was considered. Two low momemtum tracks cross such that a roadway has hits on all layers. In the angular region $10° \leq \theta \leq 45°$ the rate for particles exiting from the iron[1] is 80 KHz. This results in a counting rate in each scintillator of 130 Hz (including the cosmic ray rate of $\sim$ 100 Hz/m^2). If the multiple tracks are uncorrelated, this translates to an accidental trigger rate of .04 Hz/road or 150 Hz total for the ~4000 roads in the entire detector. Detailed calculations need to be done to determine the effects of two-particle correlations on the accidental trigger rate.

The singles rates due to π and K decays-in-flight are shown in Fig.4.6 and Fig.4.7, for two configurations of the inner detector.

REFERENCES

1) SSC Muon Detector Group Report, Snowmass, Physics of the Superconducting Supercollider, Editors R. Donaldson and J. Marx, pg. 405 (1986).

2) D. Green *et al.*, Nucl. Instr. and Meth., **A244**, 356 (1986).

3) Y. Arai and T. Ohsugi, Snowmass, Physics of the Superconducting Supercollider, Editors R. Donaldson and J. Marx, pg. 455 (1986).

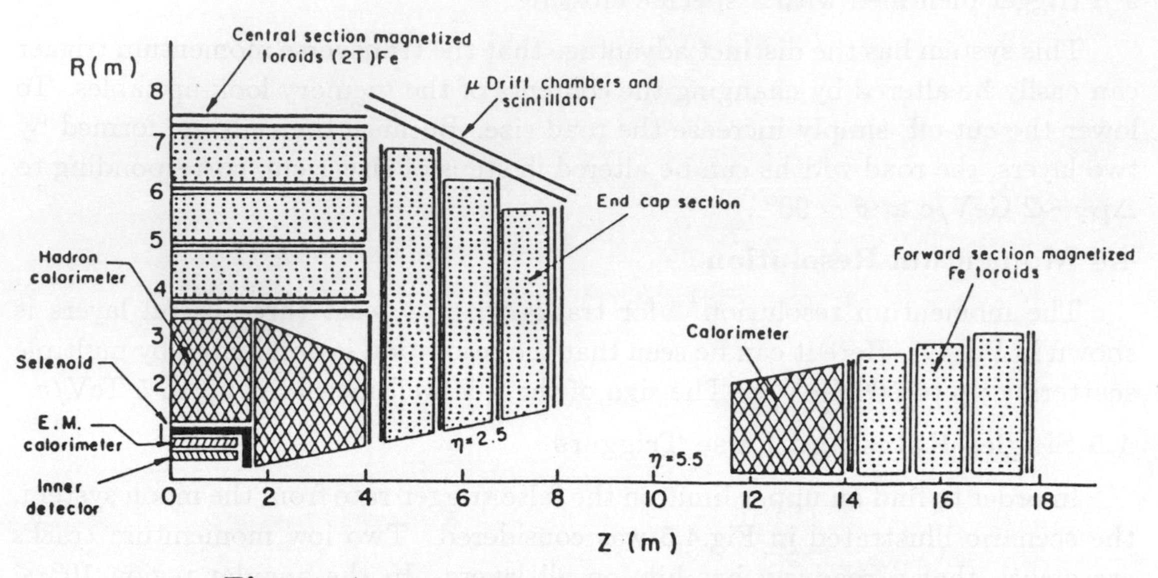

Fig.4.1 An overview of the muon detection system.

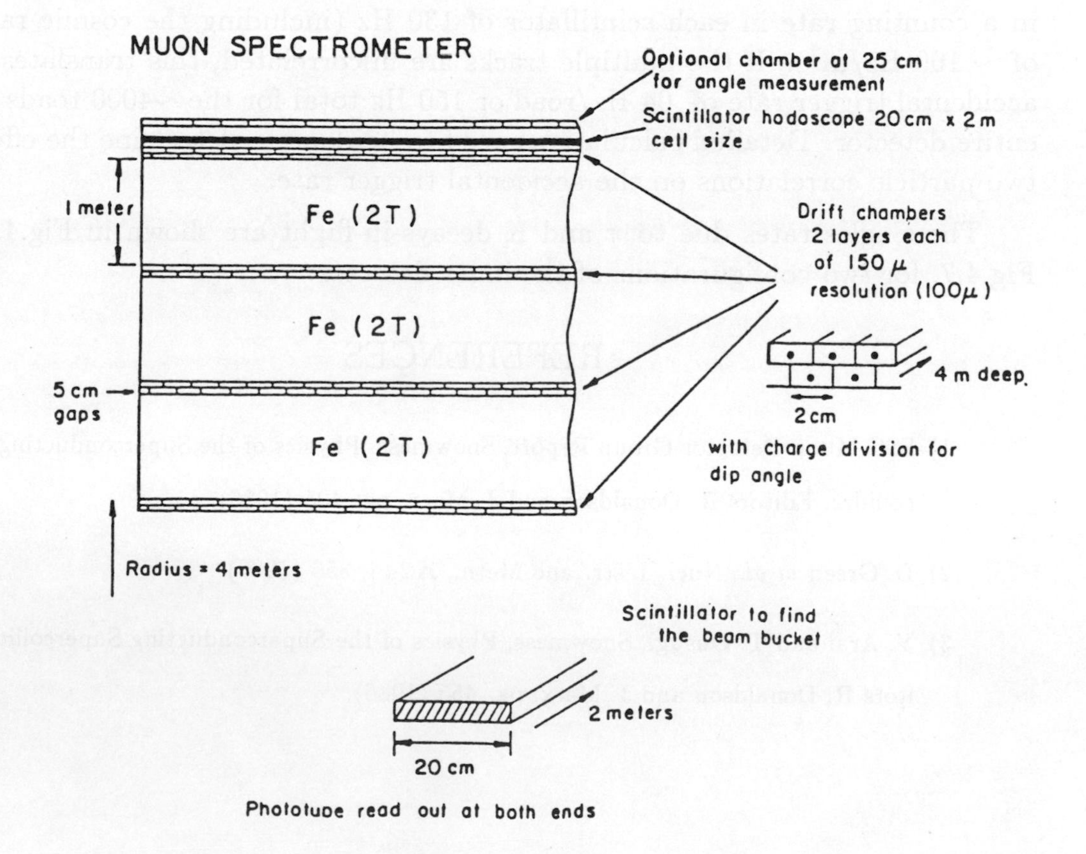

Fig.4.2 The 4-layer detector arrangement in the muon spectrometer.

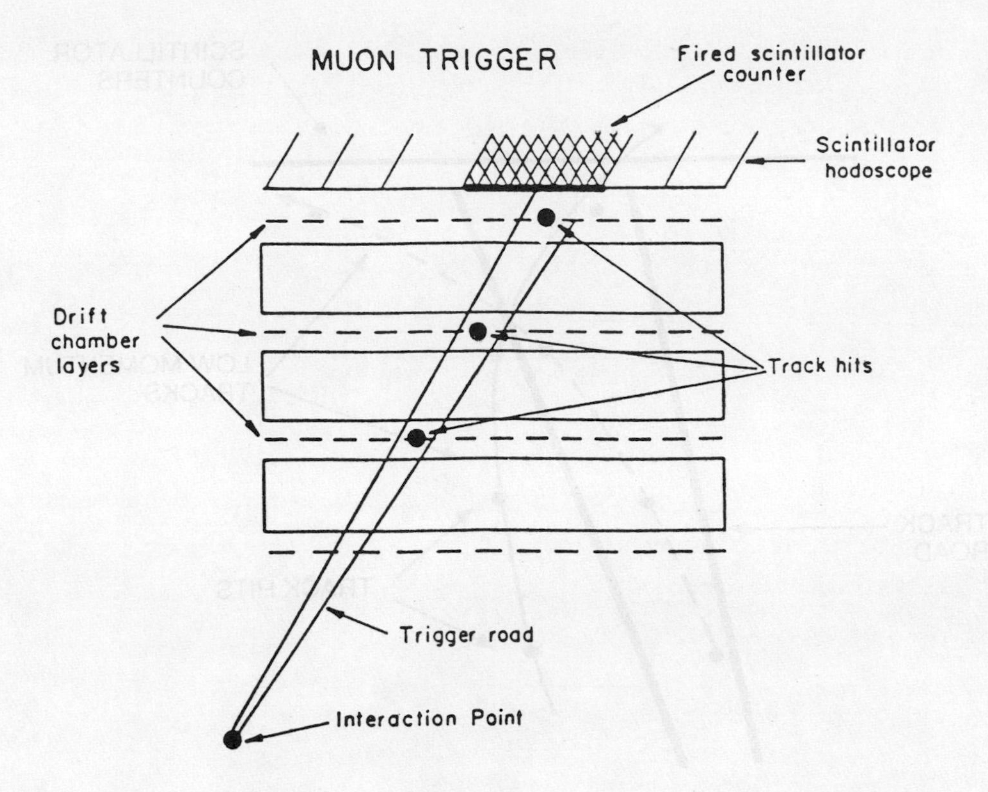

Fig.4.3 The road system used for triggering on high p_T muons.

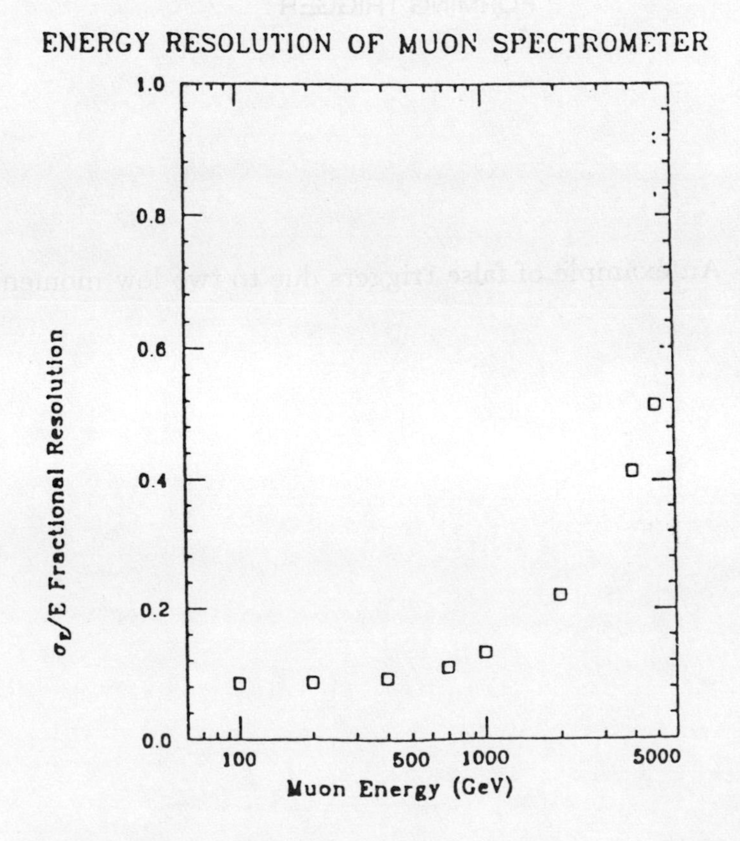

Fig.4.4 The energy resolution expected for the muon system.

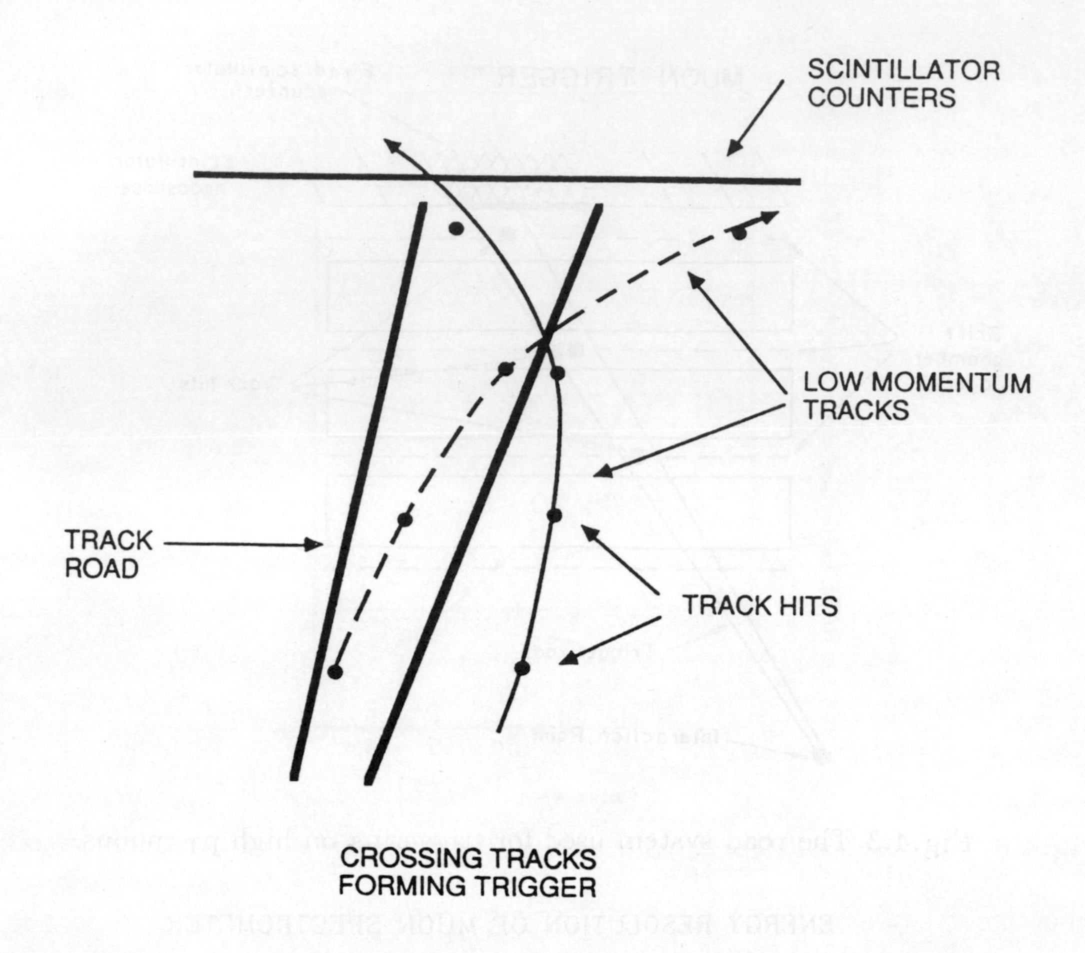

Fig.4.5 An example of false triggers due to two low momentum tracks.

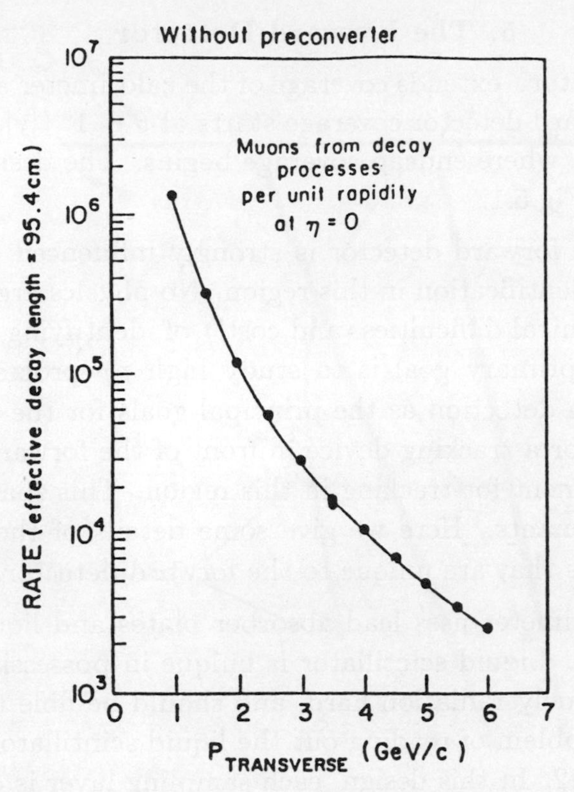

Fig.4.6 The singles rate due to π and K decays-in-flight without preconverter (effective decay length=95.4cm).

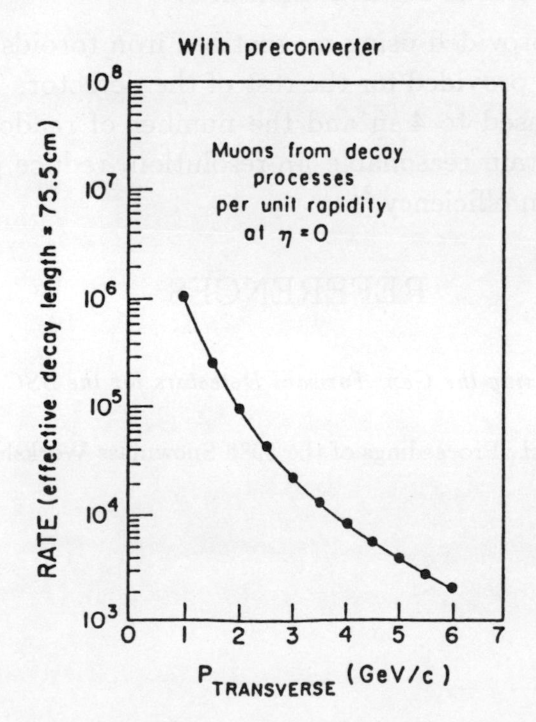

Fig.4.7 The singles rate due to π and K decays-in-flight with preconverter (effective decay length=75.5cm).

5. The Forward Detector

The forward detector[1] extends coverage of the calorimeter and muon systems to small angles. Forward detector coverage starts at $\theta \sim 1^o$ ($|y| \sim 5$) and extends to $\theta \sim 10^o$ ($|y| \sim 2.4$), where endcap coverage begins. The design of the forward detector is shown in Fig.5.1.

The design of the forward detector is strongly influenced by a decision not to attempt electron identification in this region. No physics argument was found that justified the technical difficulties (and cost!) of identifying forward electrons in a detector whose primary goal is to study high p_T processes. This leaves hermeticity and muon detection as the principal goals for the forward detector. While there is room for a tracking device in front of the forward calorimeter, we found no strong argument for tracking in this region. This space is available for rent to responsible tenants. Here we give some details of the calorimeter and muon detector designs that are unique to the forward detector.

The forward calorimeter uses lead absorber plates and liquid scintillator as the sampling medium. Liquid scintillator is unique in possessing very fast time response, being extremely radiation hard, and should be able to acheive an e/π response of 1. The problem of reading out the liquid scintillator is solved by the design shown in Fig.5.2. In this design, each sampling layer is divided into ~ 32 wedges, with each wedge readout using by a photomultiplier tube. The liquid scintillator vessel is etched to provide a $\sin\theta$ response. Thus, the calorimeter measures E_T directly with no segmentation in θ.

Muon coverage is provided using magnetized iron toroids and drift chamber readout similar to that provided for the rest of the detector. The total thickness of iron has been increased to 4 m and the number of readout layers has been increased to 5 to maintain reasonable p_T resolution, reduce punchthrough, and increase muon detection efficiency.[2]

REFERENCES

1) R. Partridge, *Closing the Gap: Forward Detectors for the SSC*, in these proceedings.

2) D. Carlsmith *et al.*, Proceedings of the 1986 Snowmass Workshop, p. 405.

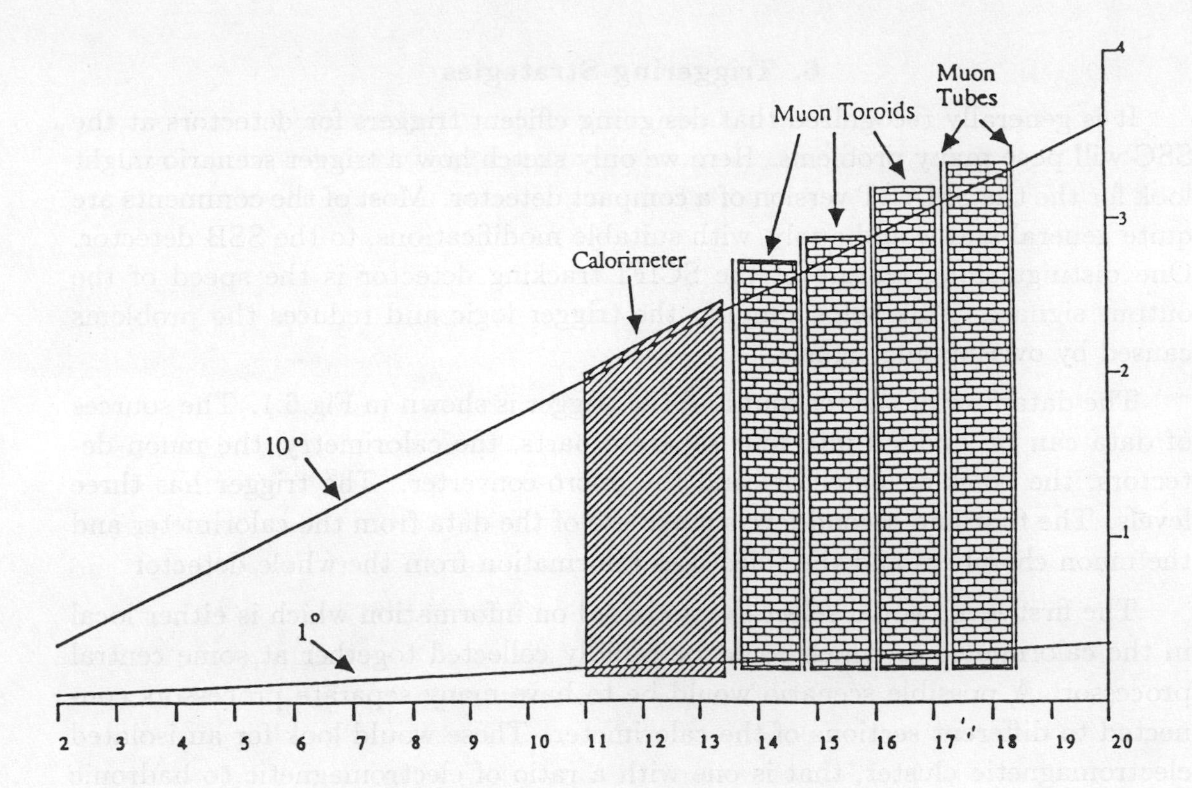

Fig.5.1 A schematic view of the forward detetector.

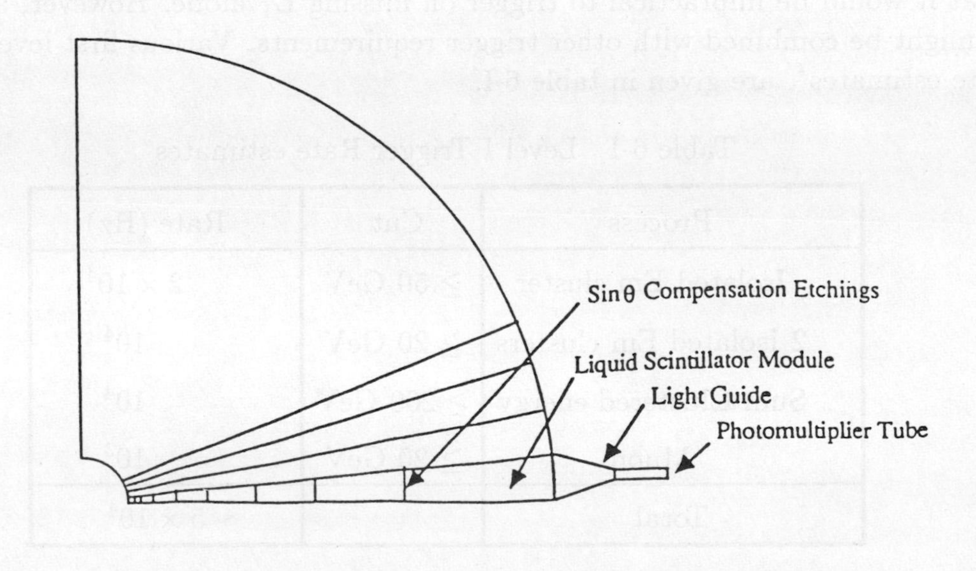

Fig.5.2 The liquid scintillator sampling module used for E_T sampling in the forward detector.

6. Triggering Strategies

It is generally recognized that designing efficent triggers for detectors at the SSC will pose many problems. Here we only sketch how a trigger scenario might look for the the SMART version of a compact detector. Most of the comments are quite general and would apply, with suitable modifications, to the SSB detector. One distinguishing feature of the SCIFI tracking detector is the speed of the output signals. This helps simplify the trigger logic and reduces the problems caused by overlapping signals.

The data flow of the readout and the trigger is shown in Fig.6.1. The sources of data can be divided into four separate parts, the calorimetry, the muon detectors, the tracking detectors and the micro-converter. The trigger has three levels. The first and second levels make use of the data from the calorimeter and the muon chambers and the third uses information from the whole detector.

The first level would select events based on information which is either local in the calorimeter or which could be rapidly collected together at some central processor. A possible scenario would be to have many separate processors connected to different sections of the calorimeter. These would look for an isolated electromagnetic cluster, that is one with a ratio of electromagnetic to hadronic energy greater than 80% , or for a high energy hadronic cluster. They would also generate for each bunch the vector and scalar energy deposited in each section of the calorimeter. This would be passed to a central processor so that sum E_T, sum vector E_T and sum missing E_T could be tested. At the first level it appears that it would be impractical to trigger on missing E_T alone. However, a test on it might be combined with other trigger requirements. Various first level trigger rate estimates[1] are given in table 6-I.

Table 6-I Level I Trigger Rate estimates

Process	Cut	Rate (Hz)
Isolated Em cluster	≥ 50 GeV	2×10^4
2 Isolated Em clusters	≥ 20 GeV	10^4
Sum Clustered energy	≥ 200 GeV	10^4
Muon	≥ 20 GeV	10^3
Total		$\sim 5 \times 10^4$

Due to the speed requirement, the implementation of the first level trigger would most likely be in analog electonics and it would have to be pipelined. The output of the this trigger would be used to gate the micro-channnel plate image intensifiers from the pre-shower and the tracking detectors and to initiate the second level of the trigger.

The second level trigger would have an approximate input rate of 5×10^4Hz at a luminosity of 10^{33}cm^{-2}sec^{-1} and select events at a rate of 5×10^2Hz and it would probably be implemented in digital electonics. The required speeds would imply that a specialized processor making extensive use of look-up tables and fast special purpose circuits rather than micro-processors would be used. At this level triggers requiring combinations or correlations between different parts of the apparatus could be formed. For example, the trigger of two isolated showers with a large invariant mass together with the additional requirement that there be no high p_T jet present in the rest of the apparatus, a trigger suggested for selecting the decay of the Higgs to two gammas, could be implemented at this level. Rate estimates for such a trigger are given in Table 6-II.

Table 6-II Level II Trigger Rate estimates

Requirements	Rate (Hz)
2 Em clusters $E_\gamma \geq 20\ GeV$ + Max. jet $p_T \leq$ 100Gev	100
2 Em clusters $E_\gamma \geq 20\ GeV$ + missing $E_T \leq$ 100Gev	100

The output of the second level trigger would clear the CCD's if the event is rejected or, if accepted, initiate the data transfer to the farm of processors where the third level of the trigger is made.

The third level trigger would make use of a farm of processors. Each element would run code which would select events based on the full detector information. Besides the muon detectors and the calorimetry the tracking and the pre-shower detectors would enter into the decision at this stage. The input bandwidth and the processing power limit the input rate to the farm and complexity of the algorithms which can be applied to the data.

REFERENCES

1) F.Paige, Proceedings of the Workshop on Triggering, Data Acquisition and Computing for High Energy/High luminosity Hadron-Hadron Colliders. FNAL Nov 1985

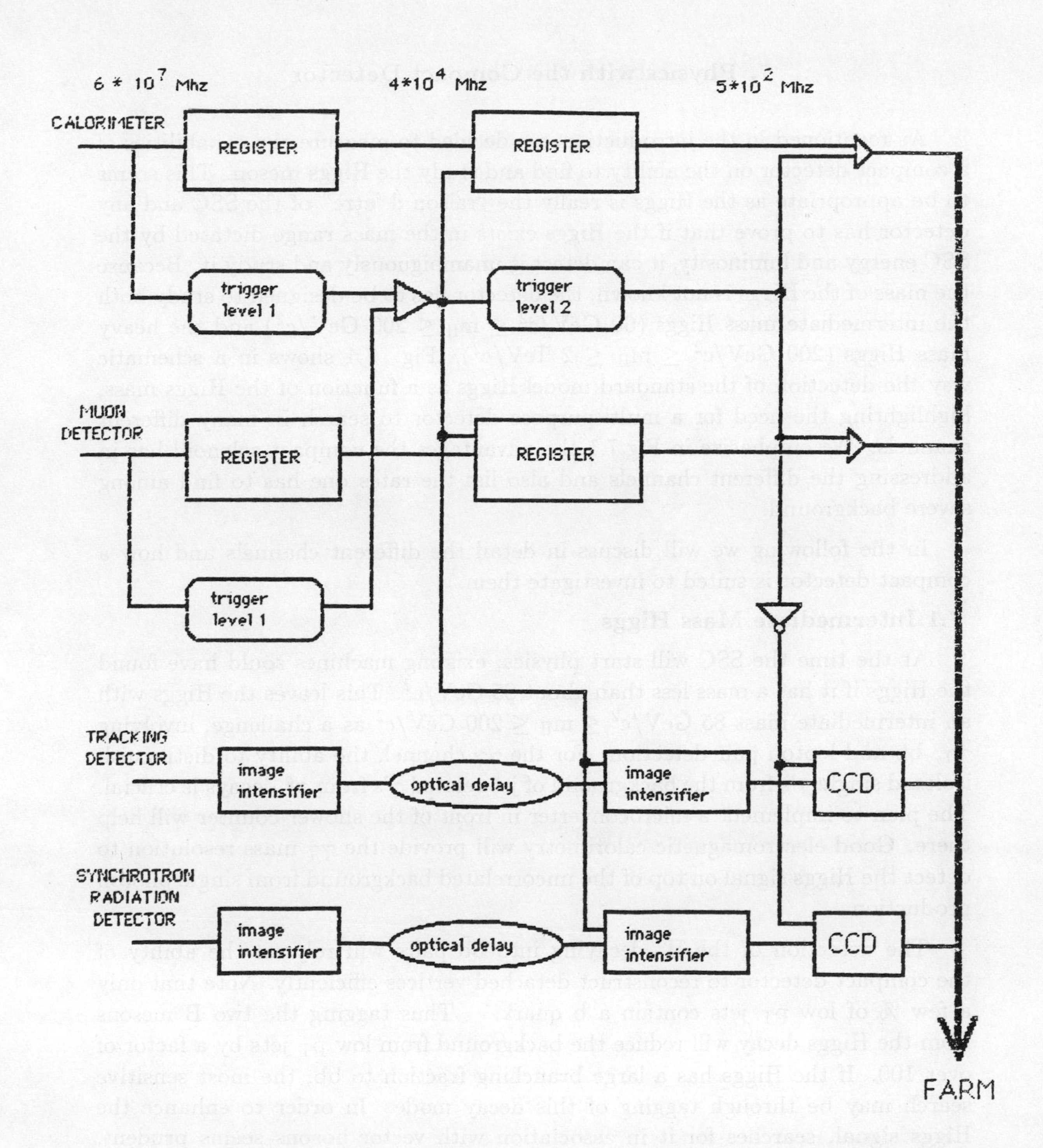

Fig.6.1 The three trigger levels in the **SMART** detector.

7. Physics with the Compact Detector

As mentioned in the introduction, we decided to measure the capabilities of a compact detector on the ability to find and study the Higgs meson. This seems to be appropriate as the Higgs is really the "raison d 'etre" of the SSC and any detector has to prove that if the Higgs exists in the mass range dictated by the SSC energy and luminosity, it can detect it unambiguously and study it. Because the mass of the Higgs is not known, the detector has to be designed to study both the intermediate mass Higgs (60 GeV/c^2 $\leq m_H \leq$ 200 GeV/c^2) and the heavy mass Higgs (200 GeV/c^2 $\leq m_H \leq$ 2 TeV/c^2). Fig. 7.1 shows in a schematic way the detection of the standard model Higgs as a function of the Higgs mass, highlighting the need for a multi-purpose detector to search in many different channels. We emphasize in Fig.7.1 the advantages the compact solenoid has in addressing the different channels and also list the rates one has to find among severe background.

In the following we will discuss in detail the different channels and how a compact detector is suited to investigate them:[1]

7.1 Intermediate Mass Higgs

At the time the SSC will start physics, existing machines sould have found the Higgs if it has a mass less than about 85 GeV/c^2. This leaves the Higgs with an intermediate mass 85 GeV/c^2 $\leq m_H \leq$ 200 GeV/c^2 as a challenge, involving $\gamma\gamma$, $b\bar{b}$ and lepton pair detection. For the $\gamma\gamma$ channel, the ability to distinguish isolated single γ's from the background of jet related γ's from π^o decays is crucial. The plan to implement a microconverter in front of the shower counter will help there. Good electromagnetic calorimetry will provide the $\gamma\gamma$ mass resolution to detect the Higgs signal on top of the uncorrelated background from single photon production.

The detection of the H^o decaying into $b\bar{b}$ pairs will rely on the ability of the compact detector to reconstruct detached vertices efficiently. Note that only a few % of low p_T jets contain a b quark.[2] Thus tagging the two B mesons from the Higgs decay will reduce the background from low p_T jets by a factor of over 100. If the Higgs has a large branching fraction to $b\bar{b}$, the most sensitive search may be through tagging of this decay mode. In order to enhance the Higgs signal, searches for it in association with vector bosons seems prudent. Since the associated production with a Z^o suffers from large backgrounds derived from radiative processes, W associated production appears as the most promising avenue. The process $pp \to W$ + Higgs, where the Higgs decays to $b\bar{b}$, has been studied by the Intermediate Higgs group.[3] They required that the transverse momentum in the fundamental parton scattering exceed 50 GeV/c. Furthermore they tagged on a lepton with a high jet transverse momentum, meaning one of the b jets decayed semi-leptonically. Both jets had to have decay vertices with impact distances in the transverse plane of 50 μm or more. It is significant that the surviving background events are largely pairs of b quarks. In one year of SSC running one expects, for a Higgs mass of 100 GeV/c^2, about 400 signal events over a background of 2100 from gluon/quark and quark/anti-quark production, which is an 8 standard deviation effect. The ability to see vertices at the level that permits this 50 μm cut is crucial; without the vertex cuts the signal represents a much less significant excess over the background (about 700 events out of 54,000).

7.2 Heavy Higgs

In the clean heavy Higgs channels like $H \to Z^o Z^o$, with the subsequent decays $Z^o \to l^+ l^-$, we will be rate-limited,and parameter optimization enters fully. For example, in a detector where only electrons can be used (as would be the case for a detector that relied only on iron toroids for μ momentum measurements), a Higgs with a mass of 300 GeV/c^2 will yield 140 events/SSC year; this number drops to 29 at a Higgs mass of 500 GeV/c^2. Using both electrons and μ's with comparable momentum resolution, as proposed for the compact detector, gives 560 and 116 events, respectively. Additionally, combinatorial backgrounds (such as those from heavy top) can be measured in μe channels and like-sign channels—another reason to track both μ and e precisely. As an example, we have generated with Isajet[4] the process $pp \to H^o \to Z^o Z^o$ where both Z^o's subsequently decay as $Z^o \to e^+e^-$, for the compact detector, for Higgs' masses varying from 200 GeV/c^2 to 2 TeV/c^2 (we used 60 GeV/c^2 for the top quark mass). A figure of merit is the mass resolution of the Z^o which should be smaller than the natural width, $\Gamma_{Z^o}/2.35$. We show in Fig.7.2 the invariant mass of all leptons with $p_T \geq$ 10 GeV/c and $|y| \leq 2.5$ for a Higgs mass of 500 GeV/c^2 and compare it with the natural width of the Z^o (dotted line). Note that the resolution in the SSB both in the electrons and μ's, contributes very little to the apparent width of the Z^o but that for a toroidal μ tracker the resolution is considerably worse than Γ_{Z^o}.

In Fig.7.3 we show the FWHM resolution of the Z^o mass as a function of Higgs mass for μ (Fig.7.3.a) and electron pairs (Fig.7.3.b), respectively, for the compact detector and for an electromagnetic calorimeter alone and toroid's alone. We also indicate the natural width Γ_{Z^o} of the Z^o. We can conclude that the compact detector allows accurate reconstruction of Z^o's from H^o decay with a resolution comparable or better than the Z^o width up to Higgs masses of 2 TeV/c^2. Fig.7.4 shows the resolution in the Higgs mass from the $Z^o \to e^+e^-$ channels, compared with the Higgs width Γ_H. It should be pointed out that the resolution of the compact detector is below the Higgs width even for low Higgs masses.

For Higgs masses close to 1 TeV/c^2, one of the W's or Z^o's in the Higgs decay has to be detected in its hadronic mode (Fig.7.1). As pointed out by Protopopescu in Snowmass '86,[5] the background from W + jet, where the mass of the jet is close to the W mass, is so large that signal to background is 1:8 when only the calorimeter is used. A compact detector with precision tracking inside of the jet will provide additional discrimination between W jets and quark/gluon jets by determination of the number of charged particles inside the jet. Using Pythia 4.8,[6] we generated either Higgs $\rightarrow$ WW or a 2-jet system, both with invariant mass of 1 TeV/c^2, and counted the number of charged tracks pointing toward the jets with invariant mass of the W. In Fig.7.5 we show the distribution of charged multiplicities m in (a) quark/gluon jets (mean = 30) and (b) W jets (mean = 18). A cut at m $\leq$ 20 GeV/c^2 leaves 66% of the W jets while cutting out 93% of the quark/gluon jets. The ability to resolve the charged tracks in the core of jets is increasing the signal to background in the above example from 1:8 to better than 1:1, thus making a difficult measurement almost trivial.

REFERENCES

1) A more detailed discussion can be found in: H.F.W. Sadrozinski, A. Seiden and A. Weinstein, SCIPP Preprint 87/100.

2) B. Cox *et al.*, Snowmass, Proceedings of the Physics of the Superconducting Supercollider, Editors R. Donaldson and J. Marx (1986).

3) J. Brau, L. Price, G. Kane *et al.*, These Proceedings.

4) F.E. Paige and S.D. Protopopescu, Isajet 5.34.

5) S.D. Protopopescu, Snowmass, Proceedings of the Physics of the Superconducting Supercollider, Editors R. Donaldson and J. Marx (1986).

6) H.-U. Bengtsson and T. Sjoestrand, Pythia 4.8, UCLA-87-001.

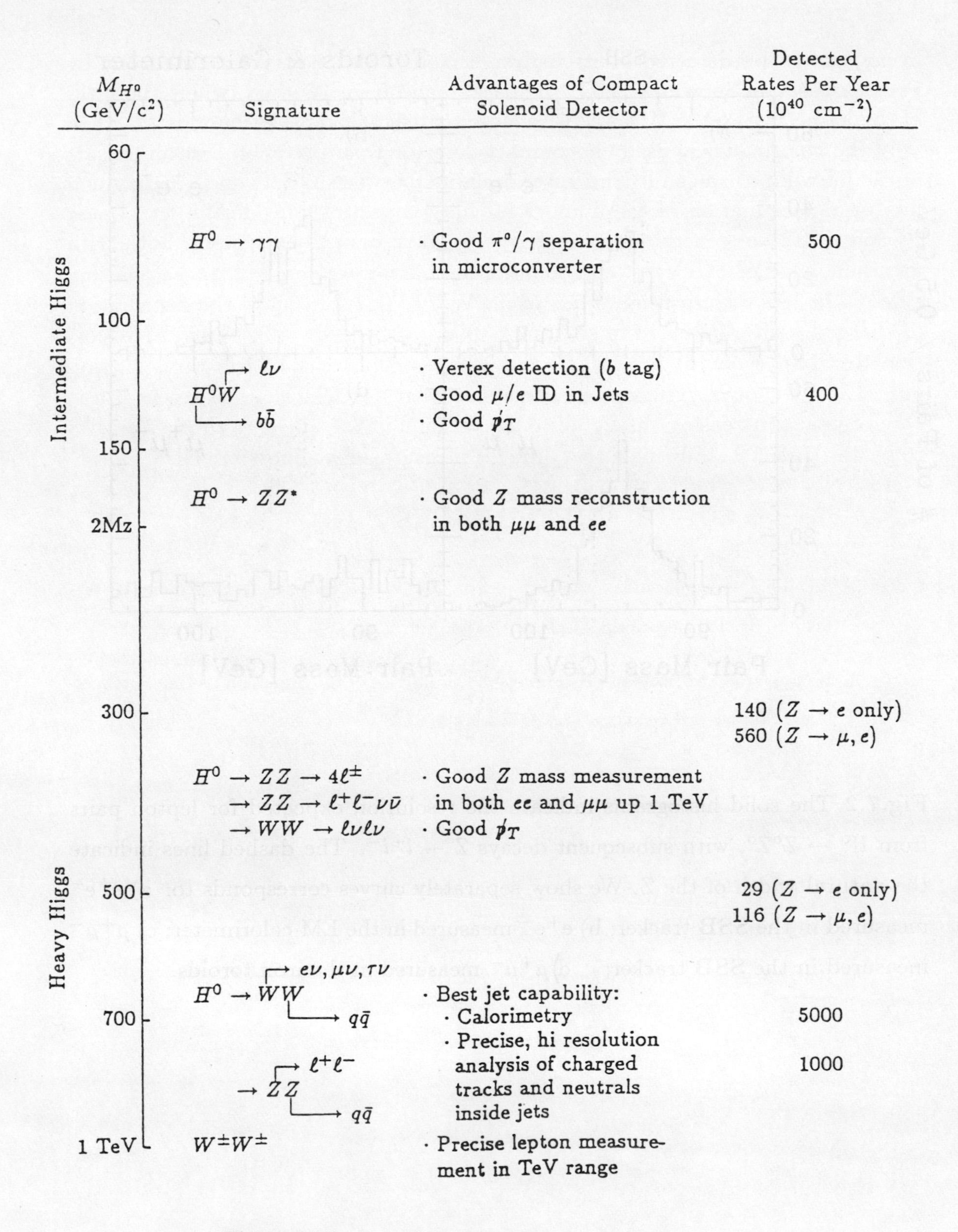

M_{H^o} (GeV/c^2)	Signature	Advantages of Compact Solenoid Detector	Detected Rates Per Year (10^{40} cm^{-2})
60			
Intermediate Higgs	$H^0 \to \gamma\gamma$	· Good π^o/γ separation in microconverter	500
100			
	H^0W ($W \to \ell\nu$, $H^0 \to b\bar{b}$)	· Vertex detection (b tag) · Good μ/e ID in Jets · Good $\not{p}_T$	400
150			
	$H^0 \to ZZ^*$	· Good Z mass reconstruction in both $\mu\mu$ and ee	
2Mz			
300			140 ($Z \to e$ only) 560 ($Z \to \mu, e$)
Heavy Higgs	$H^0 \to ZZ \to 4\ell^\pm$ $\to ZZ \to \ell^+\ell^-\nu\bar{\nu}$ $\to WW \to \ell\nu\ell\nu$	· Good Z mass measurement in both ee and $\mu\mu$ up 1 TeV · Good $\not{p}_T$	
500			29 ($Z \to e$ only) 116 ($Z \to \mu, e$)
	$H^0 \to WW$ ($\to e\nu, \mu\nu, \tau\nu$; $\to q\bar{q}$)	· Best jet capability: · Calorimetry	5000
700	$\to ZZ$ ($\to \ell^+\ell^-$; $\to q\bar{q}$)	· Precise, hi resolution analysis of charged tracks and neutrals inside jets	1000
1 TeV	$W^\pm W^\pm$	· Precise lepton measurement in TeV range	

Fig.7.1 H° detection in the compact detector.

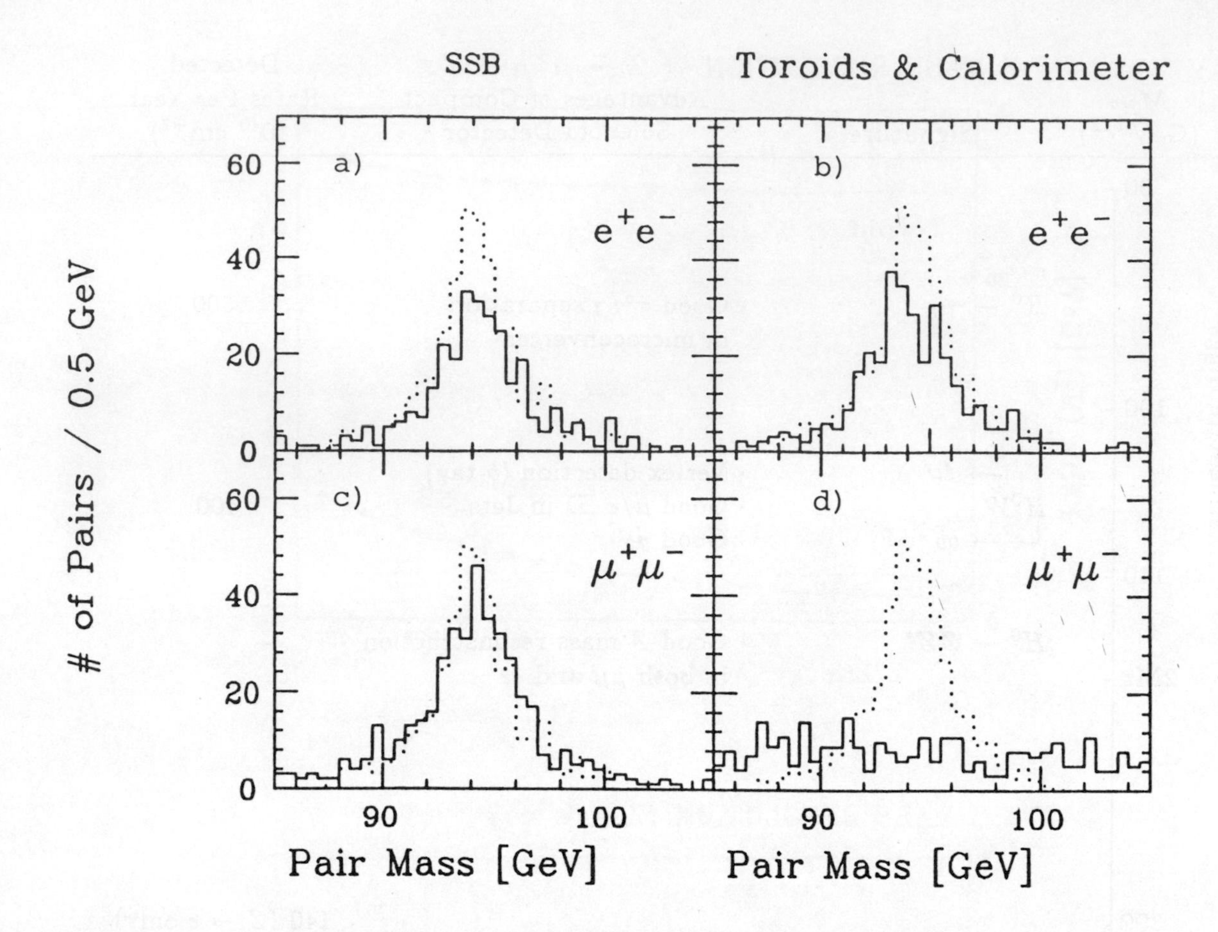

Fig.7.2 The solid histograms indicate the resolution expected for lepton pairs from H° → Z°Z°, with subsequent decays Z → l^+l^-. The dashed lines indicate the natural width of the Z. We show separately curves corresponds to: a) e^+e^- measured in the **SSB** tracker; b) e^+e^- measured in the EM calorimeter; c) $\mu^+\mu^-$ measured in the **SSB** tracker; d) $\mu^+\mu^-$ measured in the iron toroids.

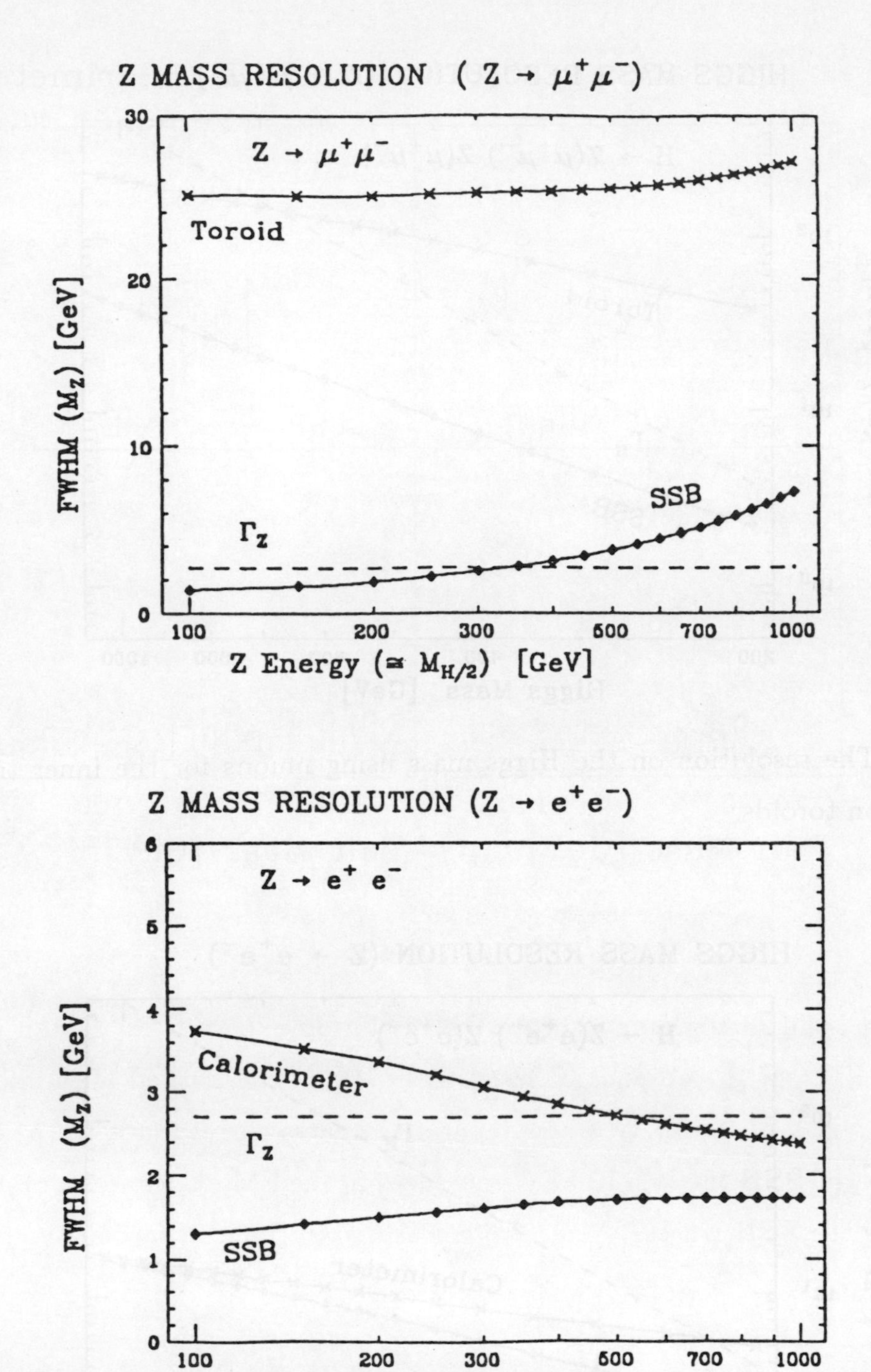

Fig.7.3 The FWHM resolution of the Z° mass as a function of Higgs mass for a)μ pairs measured in the inner tracker and in the toroids; b) e pairs measured in the inner tracker and calorimeter.

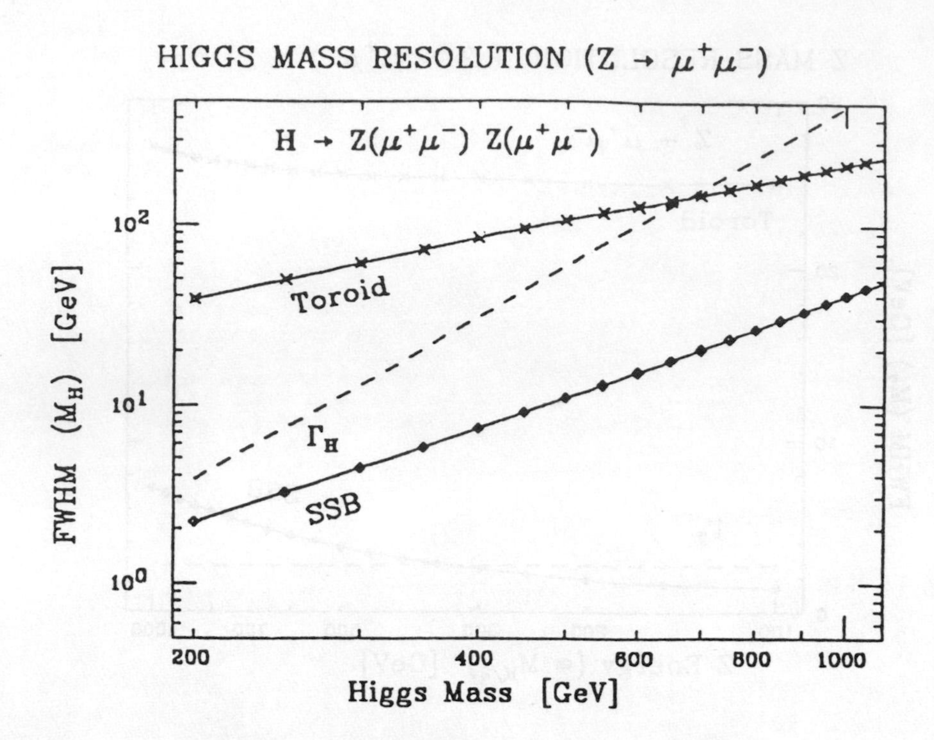

Fig.7.4a The resolution on the Higgs mass using muons for the inner tracker and the iron toroids.

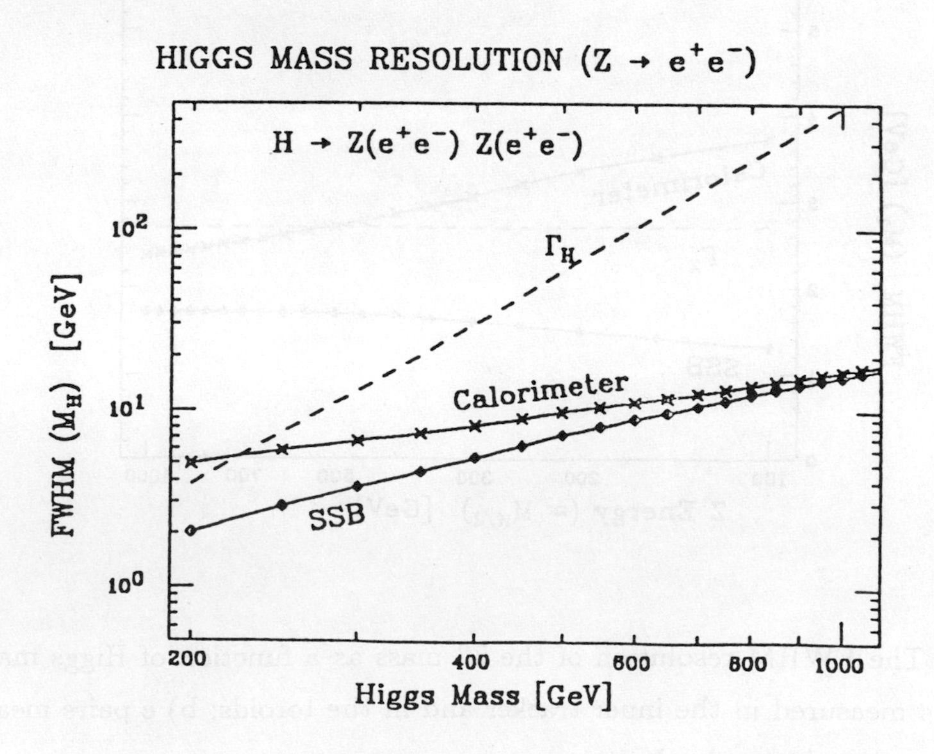

Fig.7.4b The resolution on the Higgs mass using electrons in the inner tracker and the calorimeter.

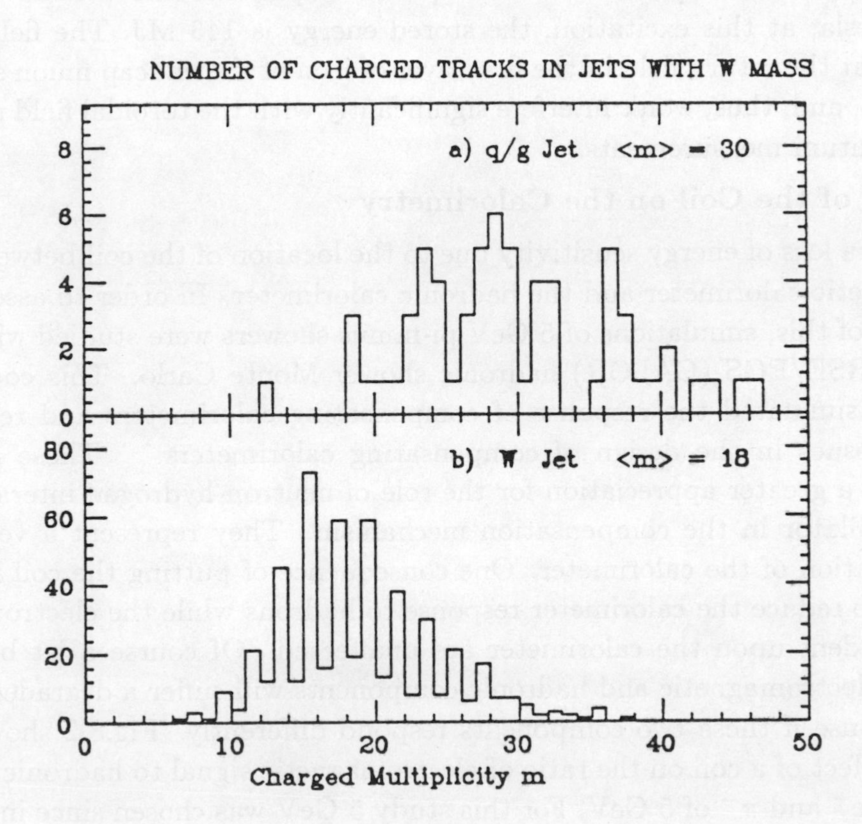

Fig.7.5 A comparison of the expected charged multiplicity distribution for quark-gluon jets with that from W decay.

8. Considerations on High Field Solenoids

The requirement for a 6 T field, essential for many aspects of the SMART detector, pushes the state-of-the-art for magnet technology. As described in section 3.1 of this report, it is necessary to position this coil between the EM calorimeter and the hadron calorimeter, at r$\sim$1 m. In this case, the nearest iron for returning the flux is at a distance of 4.5 m in z and 3.8 m in r. As a result the magnet is very nearly an air-core solenoid. A field map is shown in Fig.8.1. Approximately 15M ampere-turns are required to produce the desired central field of 6 Tesla; at this excitation, the stored energy is 143 MJ. The field map indicates that the return field in the first layer of iron of the endcap muon system is only 0.3 T, and, thus, won't interfere significantly with the toroidal field needed for μ momentum measurements.

8.1 Effects of the Coil on the Calorimetry

There is a loss of energy sensitivity due to the location of the coil between the electromagnetic calorimeter and the hadronic calorimeter. In order to assess the significance of this, simulations of 5 GeV pi-minus showers were studied with the HETC/MORSE/EGS (CALOR) hadronic shower Monte Carlo. This code has successfully simulated the response of compensating calorimeters and revealed important issues in the design of compensating calorimeters.[1] These results have lead to a greater appreciation for the role of neutron-hydrogen interactions within scintillator in the compensation mechanism. They represent a very detailed simulation of the calorimeter. One consequence of putting the coil in this location is to reduce the calorimeter response to hadrons while the electrons and photons incident upon the calorimeter are unaffected. Of course a jet being a mixture of electromagnetic and hadronic components will suffer a degraded integrated response if these two components respond differently. Fig.8.2 shows the calculated effect of a coil on the ratio of electromagnetic signal to hadronic signal for incident e^- and π^- of 5 GeV. For this study 5 GeV was chosen since in many applications the majority of the particles in a high energy jet are in the range of 0-10 GeV. From this Figure we see that a coil thickness of less than 0.4 interaction lengths is required to keep the electron to hadron ratio at 5 GeV below 1.1. Clearly such a variation in different response to photons and pions will broaden the resolution on jet energies as the jet composition fluctuates between different fractions of electromagnetic and hadronic components. As will most calorimeters, response of the device is not constant with energy, even without this effect. Fig.8.3 shows the expected relative response of incident hadrons compared to the signal for incident electrons of the same energy (the conventional e/h ratio) as a function of energy for the coil thickness of 0.3 interaction lengths.

8.2 Coil Designs

Using recent experience with the three superconducting solenoids for TRISTAN experiments, i.e. VENUS, AMY and TOPAZ, we can estimate how thin a superconducting magnet coil can be made using existing technology. There is a good summary of conventional field strength superconducting solenoids ($B \leq 2T$) given in ref.2. Here we extrapolate to magnet designs for higher field ($B \geq 2T$).

For a "thin" magnet, the coil and cryostat thickness, t_{mag} can be split into two parts, namely an aluminum-equivalent part and a copper-equivalent part:

$$t_{mag} = t_{Al} + t_{Cu}.$$

The quantities t_{Al} and t_{Cu} are further divided as:

$$t_{Al} = t_{IV}(r) + t_{RS}(r) + t_{ST}(r, B^2) + t_{OV}(r),$$

and

$$t_{Cu} = t_{SC}(B^2),$$

in which IV, RS, ST, OV and SC represent the inner vessel, radiation shield, aluminum stabilizer, outer vessel and superconducting wire, respectively, while r is the magnet radius and B is the magnetic field. Evaluating each term, based on the past experiences on large superconducting solenoids, gives[3]

$$t_{Al} = 2 \times 10^{-3} r + 2 \times 10^{-3} r + 7.4 \times 10^{-3} B^2 r + 10^{-2} r,$$

and

$$t_{SC} = 5.3 \times 10^{-4} B^2 / \cos\theta,$$

where r is in meters, B is in Tesla, and $\cos\theta$ takes into account the effect of finite magnet length; θ is the polar angle to the edge of the coil. It should be noted that, in the present estimate, we assumed the ratio of stored energy to cold mass, E/M, to be 10 Joules/gram. Recent magnets have $E/M \sim 5$ J/g; we are assuming a factor 2 improvement in future designs of quench protection systems. Fig.8.4 shows the total radiation length as a function of coil radius. It is obvious from the figure that the B^2 term for stabilizer dominates at higher magnetic field.

There remain, however, two major technical problems for high field solenoids:

1. The electromagnetic pressure to the solenoid coil, given by $P = B^2/2\mu_o$, is$\sim 10^7 N/m^2$ at $B = 5T$. This corresponds to a tensile strength beyond the yield strength of regular aluminum. There may be three ways to solve this:

i) use of a distributed support structure using high strength aluminum; ii)use of a pre-stressed aluminum cylinder; or iii) using NiTi/Cu wires embedded in a stainless steel conduit. In the last case, the total radiation length may be doubled, though the interaction length remains in the same order as shown in Fig.8.5.

2. The best cooling method is most likely indirect Helium cooling. So far, there has been no experience with this in the case of high field magnets.

In summary, we think a high field compact solenoid such as that required for **SSB** can be built with a modest extrapolation of current technology. On the other hand, making a coil for the **SMART** detector with $t_{mag} \sim 0.4\lambda$ will require substantial progress; R& D has to be pursued, especially in the areas of providing low mass mechanical support systems to contain the electromagnetic hoop stresses, and the development of quench protection systems that will allow for the safe operation of coils with $E/M \geq 10$ J/g.

REFERENCES

1) J. E. Brau and T. A. Gabriel, Nucl. Inst. & Meth., **A238**, 489 (1985).

2) S. Mori, contribution to the Report of the Task Force on Detector Research and Development for the Superconducting Supercollider, SSC-SR-10, pg. 251 (1986).

3) A. Yamamoto and K. Maehata, private communication

Fig.8.1 The flux map of the **SMART** solenoid. It is very nearly an air core magnet.

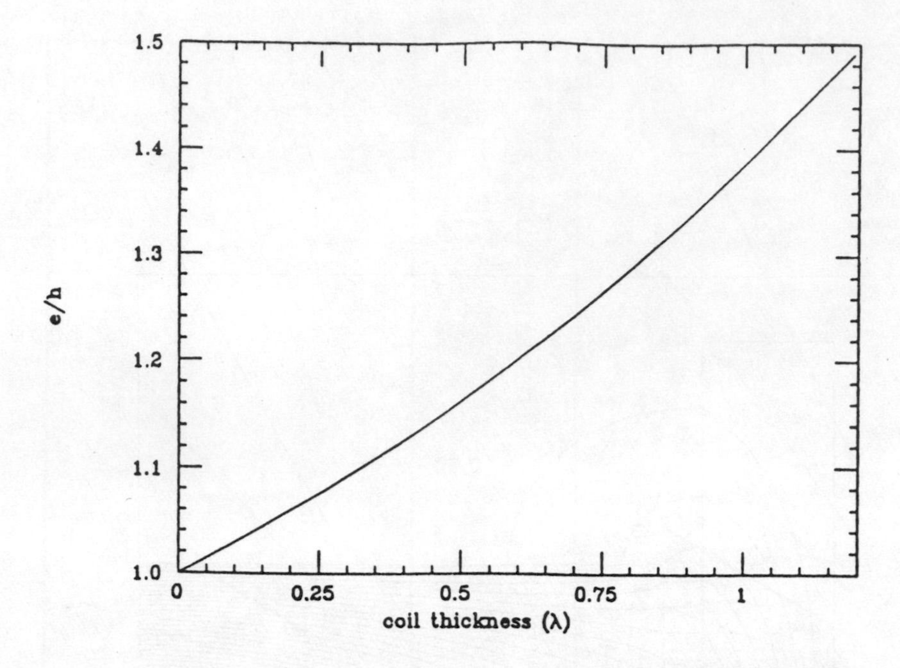

Fig.8.2 e/h for a compensated calorimeter with a coil placed between the EM and hadron sections vs thickness of the coil in interactionlengths.

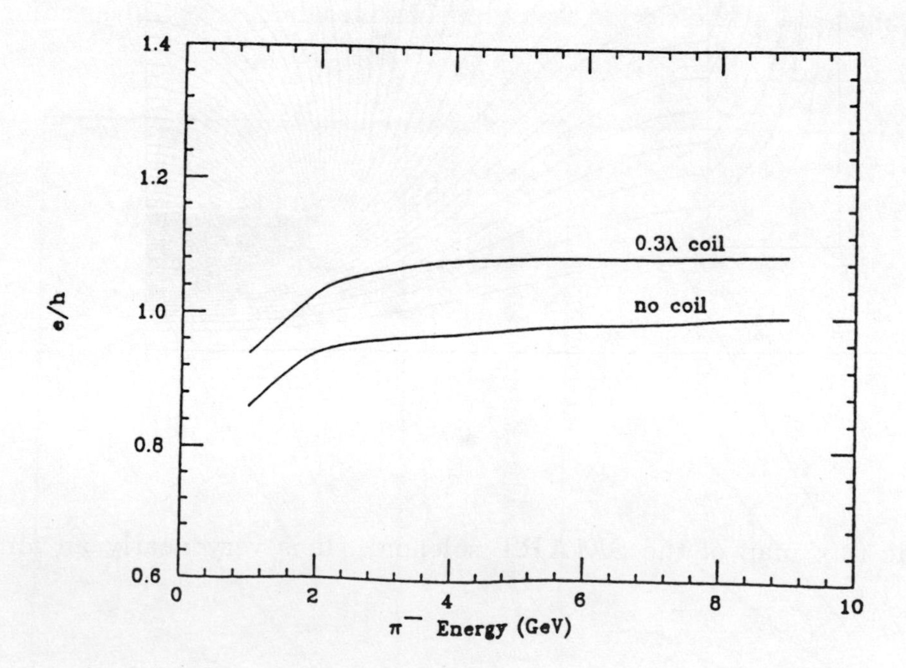

Fig.8.3 e/h for π's incident on a calorimeter vs π energy for no coil and a 0.3λ coil between the EM and hadron sections.

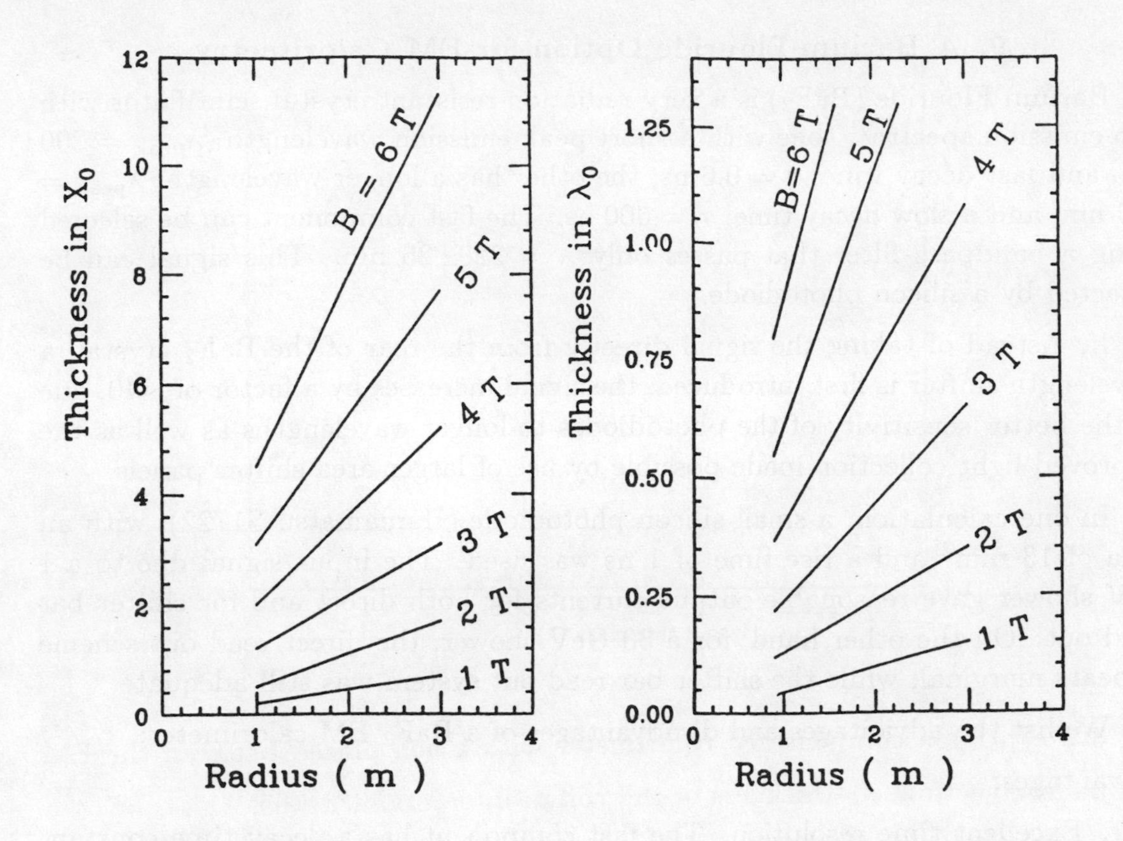

Fig.8.4 Estimated thicknesses for aluminum stabilized coils, vs inner radius, for various central fields.

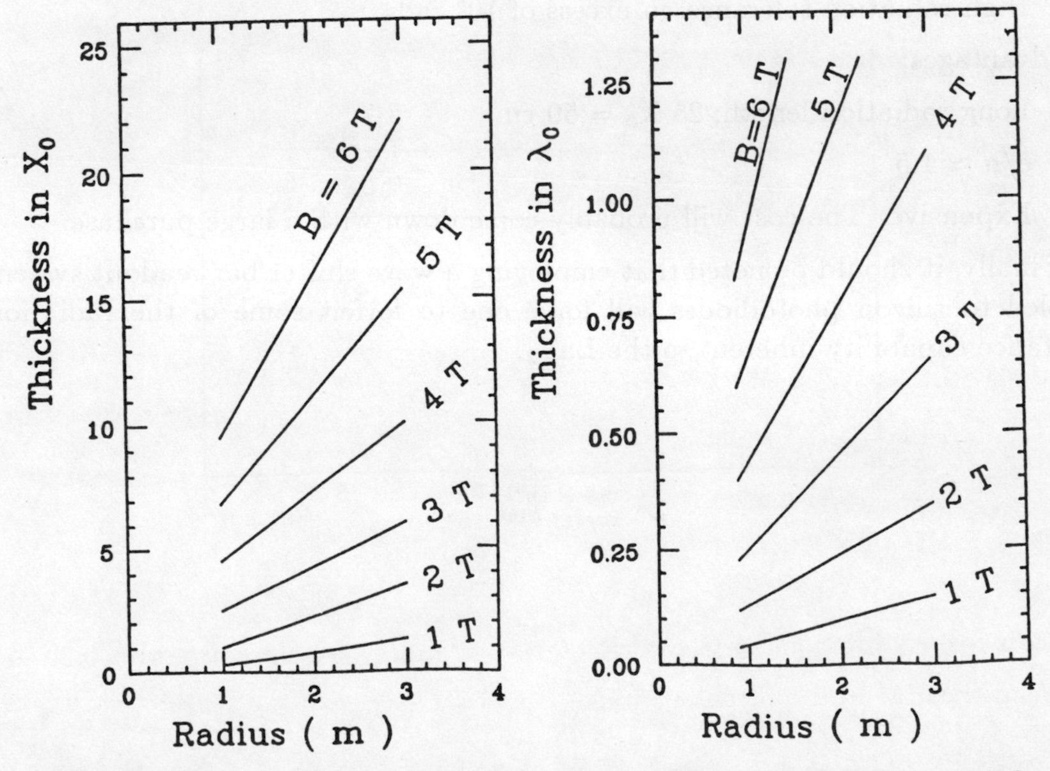

Fig.8.5 Thickness estimates for coils made using superconductor in stainless stell conduit.

9. A Barium Flouride Option for EM Calorimetry

Barium Flouride (BaF_2) is a very radiation resistant crystal scintillator with two emission spectra:[1] one with a short peak emission wavelength, $\lambda_{\text{peak}} = 200$ nm, and fast decay time, $\tau = 0.6$ ns; the other has a longer wavelength, $\lambda_{\text{peak}} = 310$ nm, and a slow decay time, $\tau = 600$ ns. The fast component can be selected using a bandpass filter that passes only $\lambda = 220 \pm 25$ nm. This signal can be detected by a silicon photodiode.

If, instead of taking the signal directly from the rear of the BaF_2 crystal, a wavelength shifter is first introduced, the signal increases by a factor of $\sim$40, due to the better sensitivity of the photodiodes to longer wavelengths as well as the improved light collection made possible by use of larger area shifter panels.

In one calculation, a small silicon photodiode (Hamamatsu S1722), with an area of 13 mm^2 and a rise time of 1 ns was used. The input signal due to a 1 TeV shower gave reasonable output currents for both direct and for shifter bar read out. On the other hand, for a 30 GeV shower, the direct read out scheme appears marginal, while the shifter bar read out system was still adequate.

We list the advantages and disadvantages of a BaF_2 EM calorimeter:

Advantages:

1. Excellent time resolution. The fast component has a decay time constant of $\tau = 0.6$ ns.
2. Good energy resolution;[2] $\Delta E/E \approx 2.5\%/\sqrt{E}$.
3. High radiation tolerance; in excess of 10^8 rads.

Disadvantages:

1. Long radiation length; 25 $X_o = 50$ cm.
2. $e/h \approx 1.5$
3. Expensive. The cost will probably come down with a large purchase.

Finally, it should be noted that employing a wave shifter bar readout system coupled to silicon photodiodes will force one to forfeit some of the radiation resistance capability inherent to the BaF_2.

REFERENCES

1) M. Laval *et al.*, Nucl. Instr. and Meth., **A206**, 169 (1983).

2) D. Anderson *et al.*, Nucl. Instr. and Meth., **A217**, 217 (1983).

10. Considerations on Radiation Damage

10.1 Effects of Neutron Damage on the Silicon Calorimeter

10.1a The Neutron Damage Coefficient for Silicon

Neutron damage is parametrized by the damage coefficient K, defined as follows;[1]

$$1/\tau = 1/\tau_o + \Phi/K$$

where τ_o and τ are the carrier lifetimes before and after irradiation, respectively; Φ is the integrated neutron flux.

The carrier lifetime is closely related to current generation in the depletion region of p-n junction diode, which results in leakage current of a silicon detector. The generation current, I_g, is given by:[2]

$$I_g = \frac{qn_iW}{\tau}$$

where q is the electron charge, n_i is the intrinsic carrier concentration and W is the depth of the depletion region.

If we consider damage for the case where the term $1/\tau_o$ is negligible compared to $1/\tau$, the generation current becomes

$$I_g = \frac{qn_iW}{K} \times \Phi = \alpha_n \Phi.$$

This formula shows that the smaller the coefficient K, the more sensitive the material to damage.

The damage coefficient for 14 MeV neutrons and for reactor neutrons, as measured by J.R. Srour,[1] is shown in Fig.10.1. The dominant process for neutron production is the evaporation of neutrons from excited nuclei. The evaporation neutron energy spectrum has the approximate form[3]

$$N(E_n) = \sqrt{E_n}\exp(-E_n/T)$$

where T=1.3 MeV for fissionable elements, and is more or less independent of the exciting mechanism. The primary neutrons in a reactor exhibit this spectrum. The energy dependence of displacement damage due to neutrons is shown in Fig.10.2, which is taken from ref.1. It shows a gradual increase between 0.2 to 10 MeV with some fine structure superimposed. Although the value of the damage coefficient for 14 MeV neutrons is slightly smaller than that of fission neutrons, it is still a good approximation to estimate the radiation damage of silicon detector in the hadron calorimeter considering the effects of all neutrons to be similar.

Typical silicon crystals in use as radiation detectors have resistivities of a few $K\Omega - cm$. In the Fig.10.1, we can see that the damage coefficient for $3K\Omega - cm$ p-type silicon is $\sim 1.3 \times 10^6$ sec/cm. From this we evaluate the damage coefficient for leakage current to be

$$\alpha_n = 1.8 \times 10^{-15} \text{amp/cm},$$

where we have used $n_i = 1.4 \times 10^{-10} cm^{-3}$, and a unit depletion depth.

In contrast, measurements of proton-induced damage expressed using the same parameterization yield[3]

$$\alpha_p = 3.0 \times 10^{-17} \text{amp/cm}.$$

Thus, one can conclude that the effects of neutron damage are more than an order of magnitude more severe than that for minimum ionizing protons. A similar conclusion can be derived from Fig.10.3, which was taken from reference 1.

10.1b The Neutron Production Rate in Hadron Showers

Fig.10.4 shows a energy distribution of neutrons produced by protons incident on aluminum.[4] The energy spectrum has a peak below 10 MeV, corresponding to evaporation neutrons. For heavy elements, this evaporation peak is expected to shift slightly to lower energy due to lower excitation temperature of excited nucleus. Neutrons with energies more than 10 MeV are produced by cascade (or spallation) processes and are expected to have an energy spectrum that is independent of the atomic number of the absorber element. The fission neutron spectrum from fissionable elements, whether induced by charged particle interaction or by neutron capture, can be presented by the evaporation spectrum given above.

The proton-induced neutron yield has been discussed in detail in texts on neutron scattering.[5] The measured neutron yield, shown in Fig.10.5, can be parametrized as

$$Y(E_{inc}, A) = 0.1(E_{inc} - 0.12) \times (A + 20)$$

for non-fissionable materials and

$$Y(E_{inc}, A) = 50(E_{inc} - 0.12)$$

for ^{238}U. Here $Y(E_{inc},A)$ is the number of neutrons per incident proton of kinetic energy E_{inc}, in units of GeV, and A is the target mass number. This formula gives a neutron yield of 44 neutrons/GeV/proton for a depleted uranium absorber, in good agreement with the data point at 1 GeV and also with the measured value of ~45 n/GeV/proton.[5] In the case of lead, the formula for non-fissionable element gives 20 n/GeV/proton, which reproduces well the 1 GeV data point in Fig.10.5. At high energies, a significant fraction of the incident energy goes into pion production and subsequently into electromagnetic showers originating from π^o's. Photon and electron-induced neutron yields have been measured to be an order of magnitude below than that for incident protons.[6] This is the reason that the neutron yield indicates a gradual saturation for incident energies above 1 GeV. To our knowledge, there are no similar measurements for incident $\pi^{\pm}$'s. One would expect the $\pi^{\pm}$-induced neutron yield to be equal or less than that of protons at the same kinetic energy. Because the pion has more inelastic channels for producing π^o's, it tends to dissipate a larger fraction of its energy electromagnetically. On the hand, the rest-mass energy of the $\pi^{\pm}$ contributes to the excitation of nuclei, thus liberating neutrons. In the following, we use the results for incident protons and assume that the yield from $\pi^{\pm}$'s is the same.

10.1c The Neutron Yield in the Hadron Calorimeter

At the SSC, the dominent radiation source will be minimum-bias events. Assuming an event rate of 10^8 Hz, 6 charged hadron per unit of rapidity with an average particle momentum of $1/\sin\theta$ (GeV), we find the energy flow per second per cm^2 at a surface of 1 m from the interaction point to be

$$1.5 \times 10^6 \text{ GeV/sec/cm}^2 \quad \text{at } \eta = 2.50, \ (\theta = 10^o);$$

$$1.0 \times 10^4 \text{ GeV/sec/cm}^2 \quad \text{at } \eta = 0, \quad (\theta = 90^o).$$

The total neutron yield from this hadron flux inside of solid uranium can then be calculated. Using 40 n/GeV/proton and the energy flow estimated above, we find

$$6 \times 10^7 \text{ n/sec/cm}^2 \ = \ 6 \times 10^{14}\text{n/yr/cm}^2 \quad (\text{at } \theta = 10^o);$$

$$4 \times 10^5 \text{ n/sec/cm}^2 \ = \ 4 \times 10^{12}\text{n/yr/cm}^2 \quad (\text{at } \theta = 90^o);$$

here one year is assumed to correspond to 10^7 seconds of operation.

Those neutrons with energies greater than about 1 MeV are likely to dissipate their energy through inelastic (n,γ) reactions on heavy nuclei; for energies of 100 keV or less they are eventually captured by nuclei. Comparing the longitudinal distribution of the fission and spallation products with the neutron capture products as shown in Fig.10.6, taken from ref.5, one can conclude that the neutrons produced inside the uranium absorber travel a distance of 0.5~1 interaction length before they are captured. Taking into account of the path lengths of the neutrons and the longitudinal distribution of the production of fission products, we estimate that at most one fifth of produced neutrons, i.e., at $\theta = 10^{o}$, $\sim 10^{14}$n/yr/cm^2 pass through the silicon planes near shower maximum.

10.1d Radiation Damage to the Silicon in the Calorimeter

Radiation damage effects on the solid state detectore can be stated as the level of increase of the leakage current. The leakage current can kill signals and cause problems associated with the total power dissipation. At the position exposed to the most severe radiation in the silicon calorimeter, the leakage current/cm^2, due to neutron damage, can be estimated as

$$I_g = \alpha_n \times \Phi \times W = 3.5 \times 10^{-3} \text{amp/cm}^2.$$

A value is about two orders of magnitude larger than that required for stable operation. Clearly, considerable effort must be put into understanding this problem.

10.2 Radiation Properties of Scintillating Fibers

The experimental data (see ref.7) on radiation damage to plastic scintillators and SCIFI are incomplete and in some cases contradictory. This is probably due to the fact that there are many parameters which influence the amount of radiation damage, amongst which are:

1. Total dose.
2. Dose rate.
3. Type of radiation: photons, electrons, hadrons, neutrons or heavily ionizing particles.
4. Surrounding chemical environment, especially the presence of oxygen.
5. Composition of the material, in particular the presence or not of a UV generating component. In the case of fibers, the composition of both the core and the cladding is relevant.
6. Presence in the material of small concentrations of impurities.
7. Time between irradiation and measurement (recovery), and the surrounding chemical environment during this interval.
8. Heat cycling (annealing).

We can find the following qualitative consensus among the experimental measurements.

1. All data agree that acrylic scintillators are more readily damaged by radiation than polystyrene or PVT scintillators. Equivalent damage (in the region of the scintillation spectra) occurs in the latter scintillators at about an order of magnitude higher doses.
2. Under irradiation, there is a preferential deterioration in transmission at shorter wavelengths ('yellowing' of the material). This argues for the use, where practical, of green-emitting rather than blue-emitting fibers.
3. The amount of damage does not *seem* to depend strongly on the type of ionizing particle which deposits a given total energy.
4. After periods of a few days in air following exposure, substantial recoveries in transmission are observed.

The data are in less clear agreement over the following issues:

1. The influence of an oxygen atmosphere. Some measurements show an improvement in an inert (N_2) atmosphere; others show the opposite occurs.
2. The relative contribution to the observed loss of light from a deterioration of the scintillation efficiency (by destruction of the scintillation molecules), from preferential absorption of the primary scintillation light, and from a decrease in the attenuation length of the final waveshifted light in the bulk material.

In summary, it seems that PVT or polystyrene scintillators and their fibers will operate with tolerable losses up to several 10^5 rads, and perhaps up to 10^6 rads. This is just at the level where they are viable for supercollider detectors. Furthermore it is likely that new waveshifters can be introduced which will substantially improve the radiation hardness.

10.3 Conclusions on Radiation Damage

What is now clearly needed is a new comprehensive and systematic study to be made of radiation damage to high resitivity silicon, plastic scintillators and plastic scintillating fibers. These tests should be made under conditions (exposure rate, particle composition, especially n content, etc.) which simulate as closely as possible those expected in a supercollider detector. Moreover, a clear chemical and physical model of radiation damage effects in silicon and plastics must be developed if we wish to make systematic progress beyond the somewhat empirical approach at present.

REFERENCES

1) J.R. Srour, D.M. Long, D.G. Millward, R.L. Fitzwilson and W.L. Chadsey, **Radiation Effects on and Dose Enhancement of Electronic Materials**, NOYES PUBLICATIONS, Park Ridge, New Jersey (1984); J.R. Srour, IEEE Nucl. Sci. **20**, 190 (1973).

2) S.M. Sze, **Physics of Semiconductor Devices**, 2nd edition, JOHN WILEY and SONS, New York, p.90 (1981).

3) T. Ohsugi *et al.*, KEK preprint 87-22, May 1987.

4) H.W.Patterson and R.H.Thomas, **Accelerator Health Physics**, ACADEMIC PRESS, New York, p.128 (1973).

5) Claude Leroy, Yves Sirois and Richard Wigmans, Nucl. Instr. and Meth. **A252**, 4 (1986).

6) John M. Carpenter and William B. Yelon, Methods of Experimental Physics **23A**, Neutron Scattering, ACADEMIC PRESS, p.99 (1986).

7) J. Kirkby, *Today and Tomorrow for Scintillating Fiber (SCIFI) Detectors*, INFN Eloisation Project, Workshop on Vertex Detectors, Erice, CERN-EP/87-60 (1987).

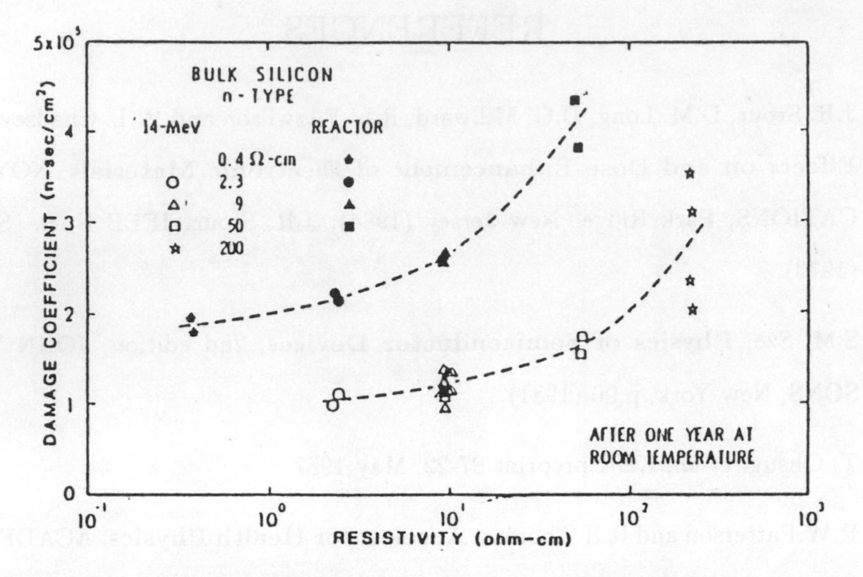

Illustration of the resistivity dependence of the low-injection-level value of K_r for neutron-irradiated n-type bulk silicon.[21]

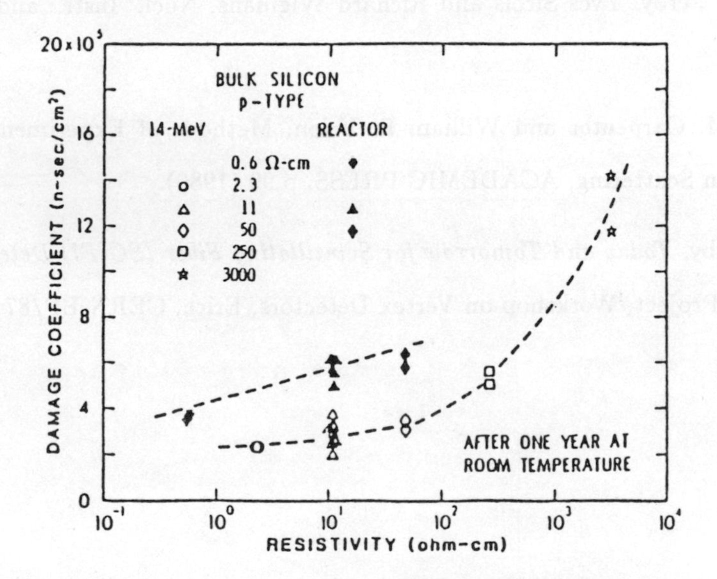

Illustration of the resistivity dependence of the low-injection-level value of K_r for neutron-irradiated p-type bulk silicon.[21]

Fig.10.1 Radiation damage coefficients for n-type and p-type bulk silicon irradiated with neutrons.

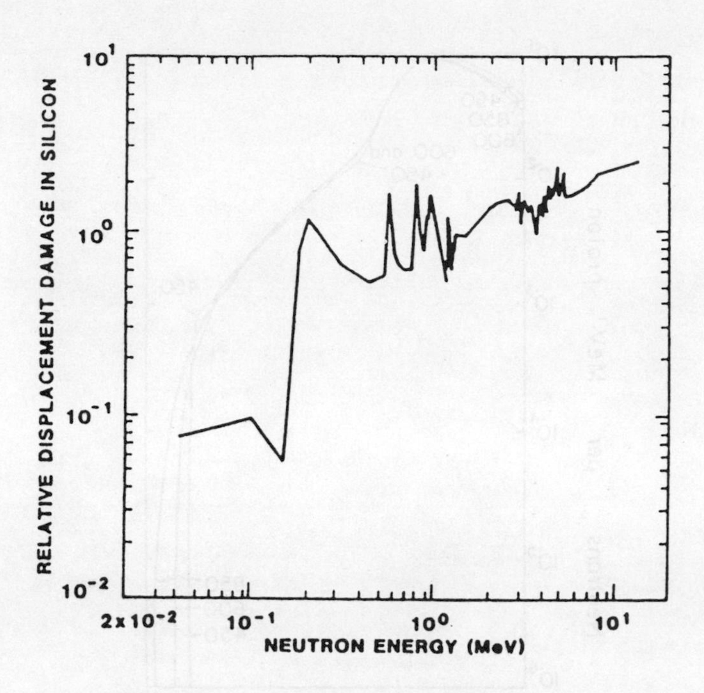

Fig.10.2 The displacement damage in Silicon for different neutron energies.

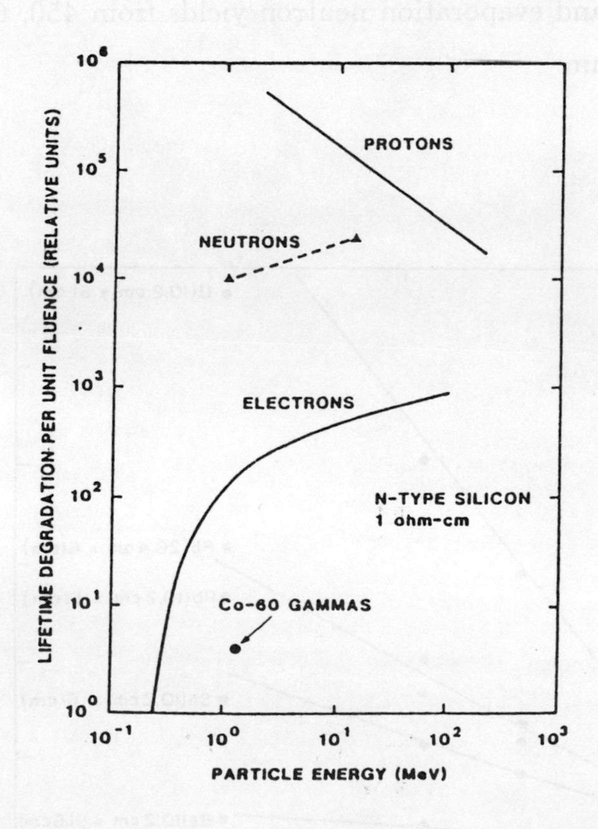

Fig.10.3 A comparison of the effectiveness of various types of radiation for degrading the carrier lifetime in Silicn.

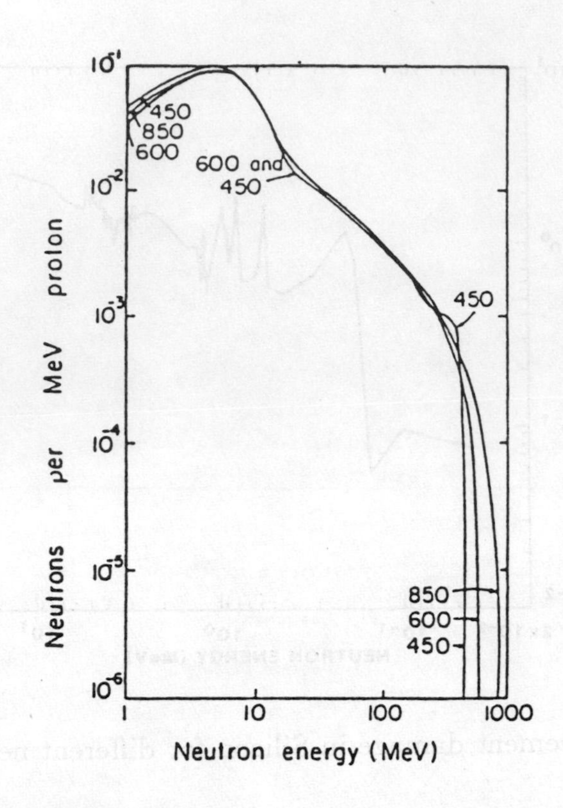

Fig.10.4 Cascade and evaporation neutron yields from 450, 600, and 850 MeV protons on Aluminum.

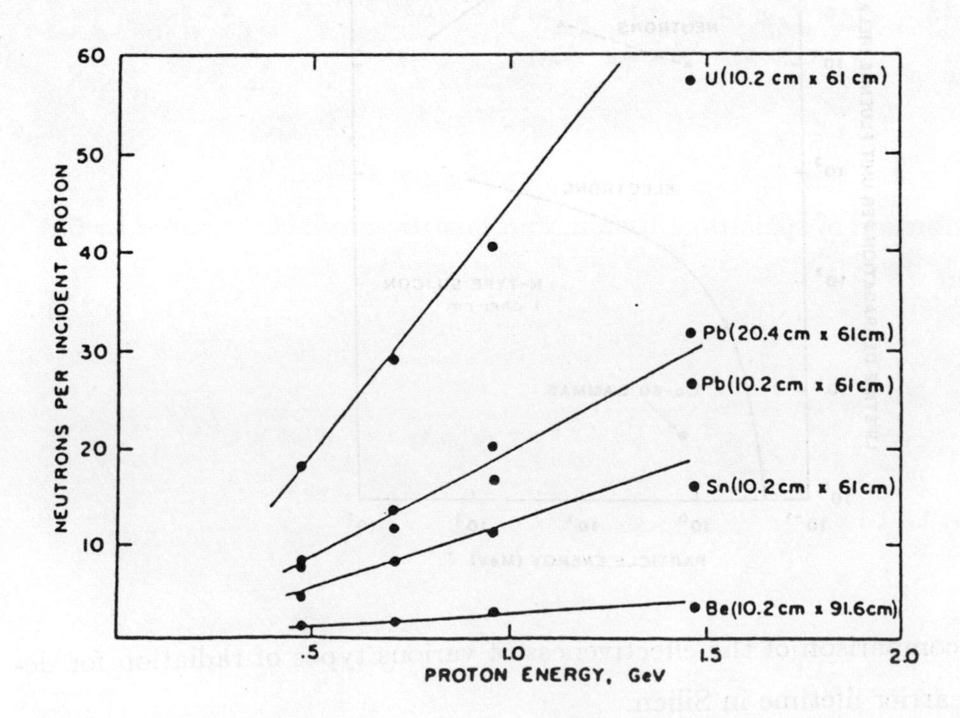

Fig.10.5 The neutron yield vs proton energy for various targets.

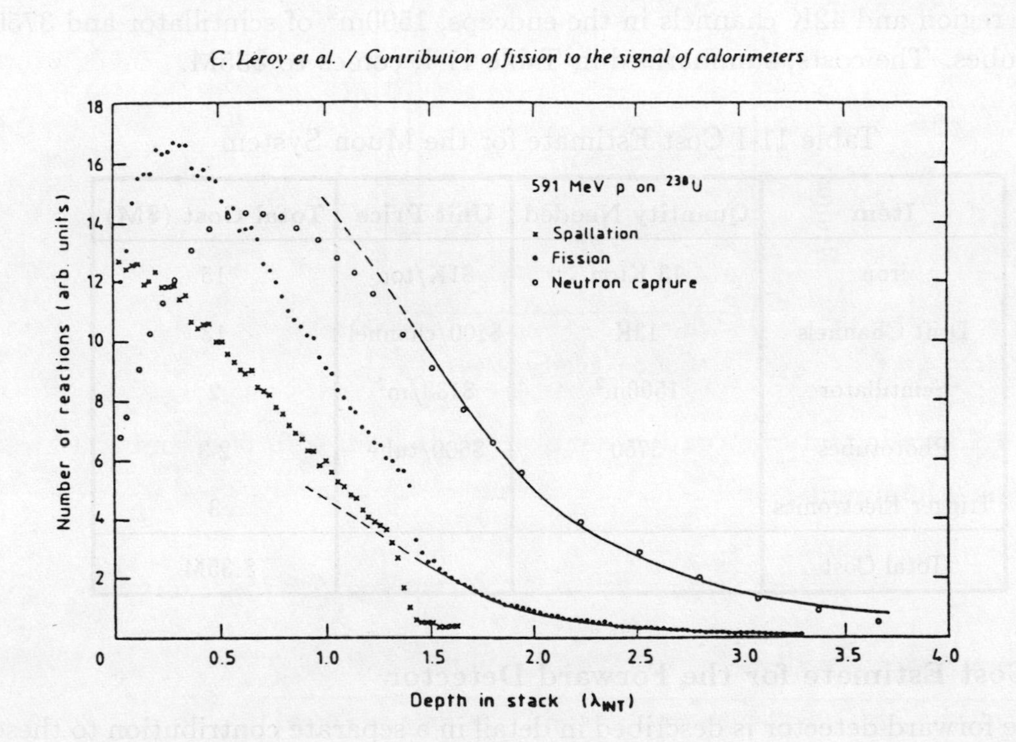

Fig.10.6 Distribution of spallation, fission, and capture products, created by 591 MeV protons in ^{238}U.

11. Cost Estimates

We have made preliminary cost estimates for the various items described in this report using the guidelines suggested in ref.1. We first give estimates for the muon system and the forward detector, items common to both the **SSB** and **SMART** designs, and subsequently estimate the total cost of each option.

11.1 Cost Estimate for the Muon System

The major elements of the muon system consist of iron, chambers, scintillator and phototubes. There are 6000 tons of iron, 90K drift chamber channels in the central region and 42K channels in the endcaps, 1500m^2 of scintillator and 3750 phototubes. The costs, summarized in Table 11-I, comes to $35M.

Table 11-I Cost Estimate for the Muon System

Item	Quantity Needed	Unit Price	Total Cost ($M)
iron	13 Kton	$1K/ton	15
Drift Channels	13K	$100/channel	13
Scintillator	1500m^2	$130/m^2	2
Phototubes	3750	$600/tube	2.3
Trigger Electronics			3
Total Cost			$ 35M

11.2 Cost Estimate for the Forward Detector

The forward detector is described in detail in a separate contribution to these proceedings.[2] There, a total cost estimate of $12M is given.

11.3 Cost Estimate for the Complete SSB Detector

Here, for purposes of cost estimating, we assume a modest factor of two reduction in the cost of high-resistivity silicon from the current small quantity price. The costs are summarized in Table 11-II.

Table 11-II Cost Estimate for SSB

Item	Quantity	Unit Price	System Cost ($M)
microstrip detectors	12000pcs	$1K/pc	12
strip detectors for EM	$100m^2$	$\$10/cm^2$	10
pads for calorimeters	$2100m^2$	$\$2/cm^2$	42
uranium for EM	20 tons	$16K/ton	0.3
uranium for hadron	200 tons	$16K/ton	3
hadron tail catcher	2000tons	$3.5K/ton	7
magnet(B=4T;r=1.8m)	$170Tm^3$		10
Helium system			2
microstrip readout	14K chips	$2K/chip	28
calorimeter readout	115K channels	$150/channel	17
hadron catcher readout	20K channels	$150/channel	17
muon detector			35
forward detector			12
online computer			5
Total			$200M
Total + 15%			$230M

11.4 Cost Estimate for the Complete Smart Detector

Here, we use raw material costs and an estimate of the cost of a special purpose fabrication facility for estimating the cost of the SCIFI. Total costs are given in Table 11-III.

Table 11-III **SMART** Cost Estimate

Detector System	Sub-system	Sub-cost ($ M)	System Cost ($ M)
SCIFI micro-tracker	mechanical	2	
			18
	readout	16	
SCIFI micro-converter	mechanical	3	
			10
	readout	7	
6 T magnet	coil + cryostat	10	
			12
	Helium system	2	
pb/SCIFI calorimeter	mechanical	26	
			44
	readout	18	
Muon System			35
Forward Detectors			12
Miscellaneous			10
Total			$141M
Total + 15%			$162M

REFERENCES

1) R. Schwitters *et al.*, Cost Estimate of Initial SSC Experimental Equipment, SSC-SR-1023, 114 (1986).

2) R. Partridge, *Closing the Gap: Forward Detectors for the SSC,* in these proceedings.

12. R&D Requirements

It is evident that compact solenoid detectors require a substantial advance beyond the current state-of-art in detector technology. We were led to this conclusion after finding current technologies, especially wire chamber inner trackers, were inadequate. We argue that this is an **asset** in favor of our detectors; if we do not aim for significant advances in experimental techniques then the full physics potential of the SSC will not be realized.

An important area for R&D that is central to both compact detector concepts is the development of ultra-high-field solenoids that are relatively thin. Here the aim should be for coils that are as thin as $\sim .5\lambda_o$ and capable of operating as high as 6 Tesla. The possibilities for use of warm superconducting ceramics should be carefully explored.

We list, in addition, a number of items that are specific to the two designs.

12.1a R&D for the SSB Tracking System:

a. VLSI readout for silicon microstrips with time slice capability. It is very important to have VLSI that maintain high speed with low power dissipation.

b. Radiation damage studies of silicon microstrip detectors and VLSI readout chips.

c. Double sided silicon microstrips.

d. Mechanical techniques for stable and accurate mounting.

e. Development of efficient pattern recognition techniques.

f. Studies of the effects of very high magnetic fields on the operation of the microstrip detectors, their readout electronics and the support and allignment system.

12.1b R&D for the SSB Calorimeter

a. Studies of radiation damage due to neutrons, charged particles and γ-rays.

b. Construction of prototype hadron calorimeters.

c. Development of compensation techniques.

d. Design of readout electronics for detectors with high capacitance.

e. Low-cost production techniques for large area silicon pad detectors.

f. Optimization studies of the segmentation needed for jet reconstruction.

g. Hermeticity studies.

h. Experimental study of electron ID using strip sampling.

Here radiation damage questions require actual beam tests. It is also important to understand the dose rate effects as well as self- and forced-annealing characteristics. Studies b through h require real prototype electromagnetic and hadron calorimeters with silicon sampling detectors. It is therefore strongly recommended to initiate construction of prototype calorimeters as soon as possible.

12.2a R&D for the SCIFI tracker and micro-converter:

a. Efficient narrow diameter (25 μm) plastic SCIFI.

b. Development of new radiation-hard plastic scintillators.

c. Radiation damage tests.

d. Development of high speed image intensifiers and CCD'S.

e. Use of coherent optical delay lines or devices.

f. Prototyping of precision mechanical assemblies.

g. Development of calibration techniques.

h. Prototype tests of the synchrotron photon/shower micro-converter.

12.2b R&D for the SCIFI calorimeter:

a. Construction of prototypes and the development of assembly procedures.

b. Studies of radiation damage sensitivity.

c. Development of segmentation techniques.

d. Large scale applications of low-noise readouts for multi-anode phototubes.

e. Development of calibration techniques.

13. Summary

We advocate the concept of a general-purpose detector based on a compact, precise inner tracker in association with an ultra-high solenoid. This detector requires a significant R&D effort, especially with regard to solid state tracking detectors: Si microstrips and SCIFI. The natural scenario for such a detector would be to approve it among the 1st round proposals after the **feasibility** of the compact tracker had been demonstrated. The full R&D program of the novel inner tracker need not be completed at the time of approval and, if necessary, the detector could commence physics and the tracker could be phased in at a later stage. We argue this concept will result in the **best-performing general purpose detector at the SSC,** with unmatched capabilities for tracking, muons, electrons, photons and the capability of operations at extreme luminosities. Finally, at ~1/2-2/3 the cost of the conventional dinosaur, it is a bargain!

REPORT OF THE NON-MAGNETIC DETECTOR GROUP*

T. Åkesson† and C. W. Fabjan
CERN, CH-1211 Genève 23, Switzerland
R. N. Cahn, C. Klopfenstein, R. J. Madaras, M. T. Ronan,
M. L. Stevenson, M. Strovink,† and W. A. Wenzel
Lawrence Berkeley Laboratory, Berkeley, California 94720
J. M. Dorfan‡ and G. J. Feldman†
Stanford Linear Accelerator Center, Stanford University, Stanford, California 94305
P. Franzini and P. M. Tuts‡
Columbia University, New York, New York 10027
D. Hedin‡
Northern Illinois University, Dekalb, Illinois 60115
C. A. Heusch
University of California, Santa Cruz, California 95064
M. D. Marx
State University of New York, Stony Brook, New York 11794
H. P. Paar‡
University of California, La Jolla, California 92093
F. E. Paige and S. D. Protopopescu
Brookhaven National Laboratory, Upton, New York 11973
R. Raja
Fermi National Accelerator Laboratory, Batavia, Illinois 60510
P. G. Rancoita
INFN, I-20133 Milano, Italy
D. D. Reeder‡
University of Wisconsin, Madison, Wisconsin 53706
J. S. Russ‡
Carnegie-Mellon University, Pittsburgh, Pennsylvania 15213
J. P. Rutherfoord
University of Washington, Seattle, Washington 98195
L. R. Sulak and J. S. Whitaker‡
Boston University, Boston, Massachusetts 02215
A. P. White‡
University of Florida, Gainsville, Florida 32611

---

* Work supported in part by the Department of Energy, contracts DE–AC03–76SF00515 and DE–AC03–76SF00098.

† Group coordinator
‡ Subgroup coordinator

1. Introduction: Why a Non-Magnetic Detector?

Why should anyone consider a non-magnetic detector for the SSC? Other things being equal, a detector does not improve by removing a feature.

The question in this case is whether the requirement of measuring the momenta of 1 TeV/c electrons hinders higher priority goals, such as optimum calorimetry. The orientation of the non-magnetic group is that calorimetery and lepton detection are the most important characteristics of an optimized SSC detector. Thus, a "non-magnetic" detector does not necessarily mean the absence of a magnetic field in any particular region; it merely means that the magnetic field does not play a major role in the optimization of the detector.

What are the direct physics advantages of magnetic analysis? At the parton level, the only advantage is a determination of the electron sign. (In any detector, the sign of the muon will be determined by magnetic analysis and the sign of a quark jet is not measurable.) There were only a few physics processes reported to us by the parameterization groups which benefited from a central magnetic field. These were like-sign W pair production and asymmetry measurements from new W and Z bosons. In both cases the physics can be done with muons alone at a cost of two or four in rate.

Are there other advantages to having a central magnetic field? We considered several possibilities which are listed below with comments:

1. Redundancy: The argument is that cracks or inefficiencies in calorimeters can be found by seeing stiff tracks pointing to them. This argument is based on experience with present day "hermetic" detectors. Considering the precision required by SSC calorimeters and the fact that approximately half of all energy is carried by neutral particles, we do not find this a compelling argument in the SSC environment. Calorimeters with cracks or inefficiencies large enough to make this redundancy check useful at the SSC will simply be incapable of measurements of adequate precision.
2. Secondary vertex detection: Secondary vertex detectors will not work without some level of momentum measurement because there will be too much confusion from multiple scattering of soft particles. A modest magnetic field which might be inserted into a "non-magnetic" field would probably be sufficient for this purpose. However, the usefulness of secondary vertex detection for high-p_t physics seems minor:
   (a) b Jets: Monte Carlo simulations have shown that on the average, every 500 GeV/c p_t jet has a fairly stiff B meson in it. Thus, at high p_t the identification of a b jet does not seem to be a very powerful signature.
   (b) τ's: In almost all cases, τ's will be lost in the multihadronic jet background. In cases in which the τ is well isolated, most physics process will also yield electrons and muons, which can be identified with higher efficiency.
3. p/E for electron identification: This is probably useful, but we are not sure to what extent. We will return to this point when we discuss electron identification.

2. The Detector

Table 1 shows a list of requirements for a non-magnetic detector that match well the needs specified by the physics subgroups.

Table 1: Detector Requirements

Electrons:	$\lvert y \rvert \leq 3$
	$\Delta E/E \leq 15\%/\sqrt{E}$
	$\Delta E/E \leq 1\%$ (systematic)
	hadron misidentification $\ll 10^{-3}$
Muons:	$\lvert y \rvert \leq 3$
	$\Delta p/p \approx 13\%$ at 1 TeV
Calorimetry:	$\lvert y \rvert \leq 5.5$ for hermeticity
	$\lvert y \rvert \leq 3$ for jet reconstruction
	$\Delta y = \Delta\phi \approx 0.05$ (hadronic towers)
	$\Delta E/E \leq 50\%/\sqrt{E}$
	$\Delta E/E \leq 2\%$ (systematic)
	hermetic design

Figure 1 shows a schematic drawing of the detector which we are proposing to meet the requirements listed above. In the next three sections, we will discuss the methods we propose to achieve the necessary calorimetry and lepton identification.

3. Tracking and Election Identification

As outlined in this section, we will accomplish charged-particle tracking and electron identification using transition radiation detectors (TRD's) in one compact device. It will be placed inside the calorimeter and will cover ± 3 units of rapidity. To cover this large rapidity range requires dividing the tracking volume into several geometrical regions — the central (or barrel) region and the forward regions — each requiring appropriate resolution. The exact placement of the transition between these regions and the best choice of coordinates for each region requires a careful study of structural integrity, readout, and minimization of dead space. In the absence of a detailed design we have chosen the representative set of parameters given below.

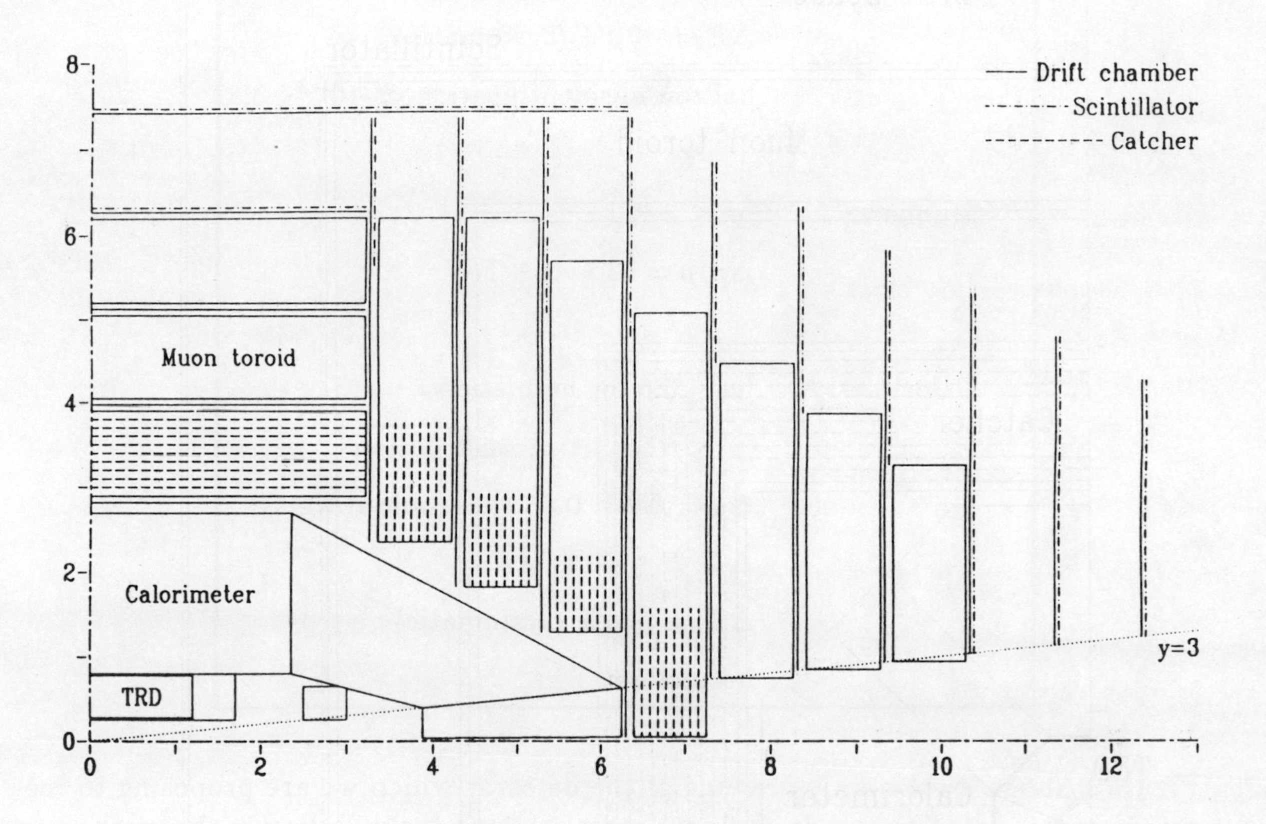

Fig. 1a. Elevation View of the "Non-Magnetic Detector."

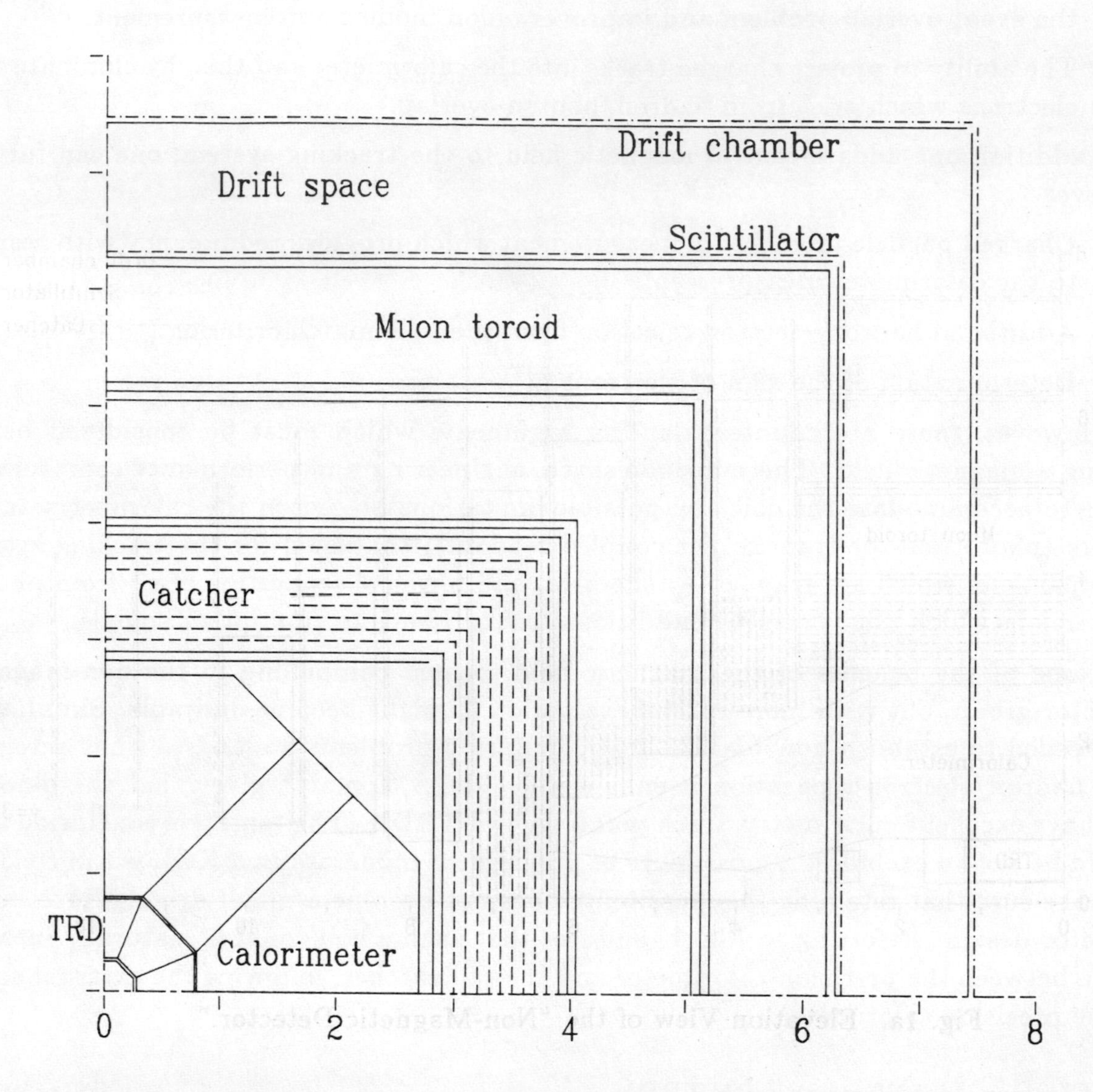

Fig. 1b. Cross Sectional View of the "Non-Magnetic Detector."

3.1. Tracking Considerations

The main functions of charged particle tracking in a non-magnetic detector are:

1. The measurement of the longitudinal position of the event vertex, which helps resolve the event overlap problem and improves muon momentum measurement.
2. The ability to project charged tracks into the calorimeter and thereby eliminate fake electrons which arise from hadron/photon overlap.

If in addition one adds a central magnetic field to the tracking system, one can further achieve:

3. Charged particle momentum measurement which provides redundancy with respect to the calorimeter measurements.
4. Additional hadron/electron rejection from the p/E match criterion.
5. Determination of the sign of electrons.

However, there are counterbalancing arguments which must be considered before adding a magnetic field. These include space, engineering and performance compromises made to accommodate the coil, the possible limitations placed on the calorimetry transducers (phototubes or transformer coupled readout), confusion for the tracking system from particles which are trapped by the field, and physical separation of $e^{\pm}$ from photon conversions which might result in additional backgrounds to prompt $e^{\pm}$ signals.

None of the benefits of the magnetic field seemed compelling to the non-magnetic detector group, but with more careful evaluation it might become desirable. Simulations are needed to establish how bad the track confusion problems are and to what extent, if any, hadron/electron separation is enhanced by the p/E match given that the detector will have excellent calorimetry, track pointing, and TRD's. The tentative conclusion then would be not to exclude the possibility of providing a moderate ($\approx$5 KG), warm coil, but to make sure that retaining this possibility does not become a major driving force in the detector design. Referring to Fig. 1, one sees that such a coil could fit naturally into the space between the precision calorimeter and the tail catcher, following the external shape of the precision calorimeter.

3.2. Electron Identification with TRD's

Efficient identification of electrons will be the key to much of the most interesting physics at the SSC. Background to electron identification in the SSC environment will arise both from single hadrons and from jets. The pattern of calorimetric energy deposition can provide single pion rejection in the range 10^{-2} to 10^{-3} depending on sampling granularity.[1] Tracking charged particles and comparing the extrapolated track with the centroid of electromagnetic energy deposition in the calorimeter will augment pion rejection. Additional independent electron identification systems will likely be required to beat down the backgrounds that arise in the study of isolated electrons, and, in the study of electrons in jets, from the limitations of calorimeter granularity and large multiplicity. A transition radiation detector (TRD) will provide additional pion rejection on a per track basis. The TRD in an SSC detector primarily will reject fake electron candidates that are selected by calorimetric criteria.

3.3. A Straw Design

In the present design the TRD and tracking functions are integrated in a straw wire chamber system.[2] This system will provide charged particle tracking to rapidity of 3. It will identify electrons with high efficiency and by itself will have pion rejection of better than 10^{-2}.

The design is shown in Fig. 2. The solid angle is divided into three regions: the central region, $|z| < 120$ cm; the forward region, 120 cm $< |z| <$ 170 cm; and the very forward region, 250 cm $< |z| <$ 300 cm. In each region there are 32 repetitions of a basic TRD unit. The TRD unit consists of 1 cm of radiator and two layers of straw chambers in close packing. The total thickness of a TRD assembly is 50 cm.

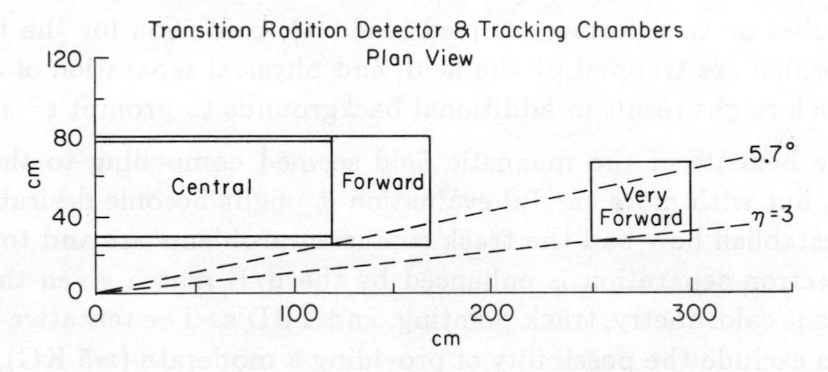

Fig. 2. Plan view of the transition radiation detector and tracking system.

In the central region, the TRD radiators form hexagonal cylinders with sides parallel to the z axis. The straw chambers form hexagonal "cobwebs" in the x-y plane. This geometry reduces the occupancy of each wire to an acceptable level and provides good dip angle measurement. The TRD occupies 28 cm $< r <$ 78 cm. Inboard from the TRD there are eight layers of straws running axially, at radius 25 to 28 cm. Outboard from the TRD are four more layers of axial straws.

In the forward and very forward regions, the TRD radiators form planes that are perpendicular to the z axis. Each TRD consists of a 50 cm package of 32 TRD units as in the central region. The straw chambers also lie in the x-y plane. They are arranged in parallel arrays; successive planes are rotated by 60° to give stereo information for track reconstruction. The forward region occupies 25 cm $< r <$ 80 cm and the very forward region occupies 25 cm$< r <$ 65 cm. The total system provides TRD coverage to $|\eta| = 3$.

This design incorporates a very large number of straw chambers. The straw diameter is set at 4 mm as part of the optimization of the TRD performance. In the TRD there are approximately 230,000 straws in the central region, 35,000 in each forward region, and 32,000 in each very forward region. The total number of TRD straws is 364,000. The axial straws number roughly 17,000, assuming that 240 cm long straws are feasible.

3.4. Tracking System

In the central region, charged particles are tracked in the r-ϕ plane by the axial straws and in the r-z plane by the TRD straws. The axial straws are equipped with charge division readout and multiple hit electronics to allow a match with tracks in the TRD. Particles at $\theta > 30°$ will be projected into the calorimeter with resolution of order 1 mm in both azimuthal and longitudinal directions. In the forward angles ($5° < \theta < 30°$) the 32 y-u-v measurements from the TRD straws (see below) will provide comparable resolution.

3.5. TRD Implementation

Transition radiation detectors can be operated in two different modes: total ionization measurement and cluster counting.[3] In the first case the TRD is optimized for high energy X-rays and one measures in the X-ray absorption gap the combined signal of ionization and X-rays. The Landau-distributed fluctuations in dE/dx diminish the electron identification power of the TRD. The second method counts the number of energy deposits above a given threshold. In this case one measures the transition radiation quanta in a large number of proportional chambers sensitive to low energy X-rays. The thickness of each proportional chamber is sufficiently small that the typical ionization energy deposition is significantly smaller than the energy deposition from an X-ray interaction. The identification power of the TRD in this case depends on the Poisson-distributed number of δ-rays. The smaller fluctuations in the Poisson distribution relative to the Landau distribution can give the cluster-counting method better controlled tails and better pion rejection.

The present design is based on the cluster counting method. The radiator is optimized for production of X-rays with energies around 5 keV by controlling the fiber diameter and packing.[4] The radiator is divided into 1 cm slabs, each followed by a double layer of straws. This radiator thickness is roughly equal to the X-ray attenuation length. This arrangement of the straws provides numerous uncorrelated samples for the measurement of the X-ray production probability and is well suited to their simultaneous use in charged particle tracking.

The straw chambers are filled with a xenon-rich gas mixture to enhance X-ray detection efficiency. The walls of the straws are thin enough that X-ray absorption is negligible compared to interactions in the gas. Straws of polycarbonate and mylar with 30 μm walls have been achieved;[5] these walls are just a few per cent of the X-ray absorption length. In a gas that is 60% xenon, the X-ray interaction probability in the double layer of straws is approximately 80%. For a high energy electron, there will be approximately 1/4 detected X-ray per double layer of straws.[3] The energy deposited by ionization by the primary particle is 2 keV per layer, a fraction of the energy deposited in the gas by a typical X-ray interaction.

Given the large number of straw chambers, the electronics per straw must be simple and inexpensive. One possibility is to use "two-bit" electronics: each straw would be instrumented with one channel consisting of an amplifier and two comparators along with associated readout. A low threshold would be used to flag the low energy deposit from the primary particle ionization; a higher threshold being exceeded would flag an X-ray interaction.

3.6. TRD Test Results

A detector similar to the present design has been described in Ref. 3. The detector, shown in Fig. 3, consists of a large number of radiator-proportional chamber sets. Each radiator has a length of 1 cm and consists of 40 polypropylene foils with a thickness of 18 μm each. The proportional chamber is 3 mm thick and has a wire spacing of 2 mm. The chamber gas is a mixture of 60% xenon, 35% helium, and has 5% methane as a quencher.

The single-particle response of such a detector is shown in Fig. 4. The relevant figure of merit is the rejection factor R, defined as the ratio of pion efficiency to electron efficiency. Figure 4(a) shows R versus detector length for a 50 GeV particle when 90% electron efficiency is required. Figure 4(b) shows the variation of rejection with particle energy for a 40 cm long detector with 30 proportional gaps. The number of clusters required for an electron signal corresponds to 90% electron efficiency at 50 GeV. The rejection has a minimum of 2×10^{-3} at 3 GeV and then rises slowly until about 200 GeV where the pions start radiating. The slow rise until 200 GeV is due to increased δ-ray production by the pions. Poor rejection below 3 GeV is due to the drop in electron efficiency. The excellent rejection of low energy pions will provide the improvement in electron identification necessary to supplement the calorimetric response.

3.7. TRD Performance in Jets

The performance of the detector described in the previous section has been simulated using ISAJET.[6] The geometry assumed 2-meter axial wires starting at 20 cm from the beam axis, so the results will be a "worst-case" bound on the performance of the present design. Two-jet events at 500 GeV per jet were generated assuming one interaction per beam crossing. Figure 5(a) shows the number of charged particles per azimuthal segment per event versus the azimuthal angle from the jet axis. An azimuthal segment is the angle subtended by one proportional cell at 20 cm radius, or 0.6° for the test detector. The level varies from 0.7 at the jet axis to 0.2 at 45° from the jet. The properties of the particles vary strongly over this region. This is clearly demonstrated in Fig. 5(b), which plots the probability per jet and per azimuthal segment that the hadrons fake an electron, requiring 90% electron efficiency. The distribution has a level of about 10^{-3} for angles

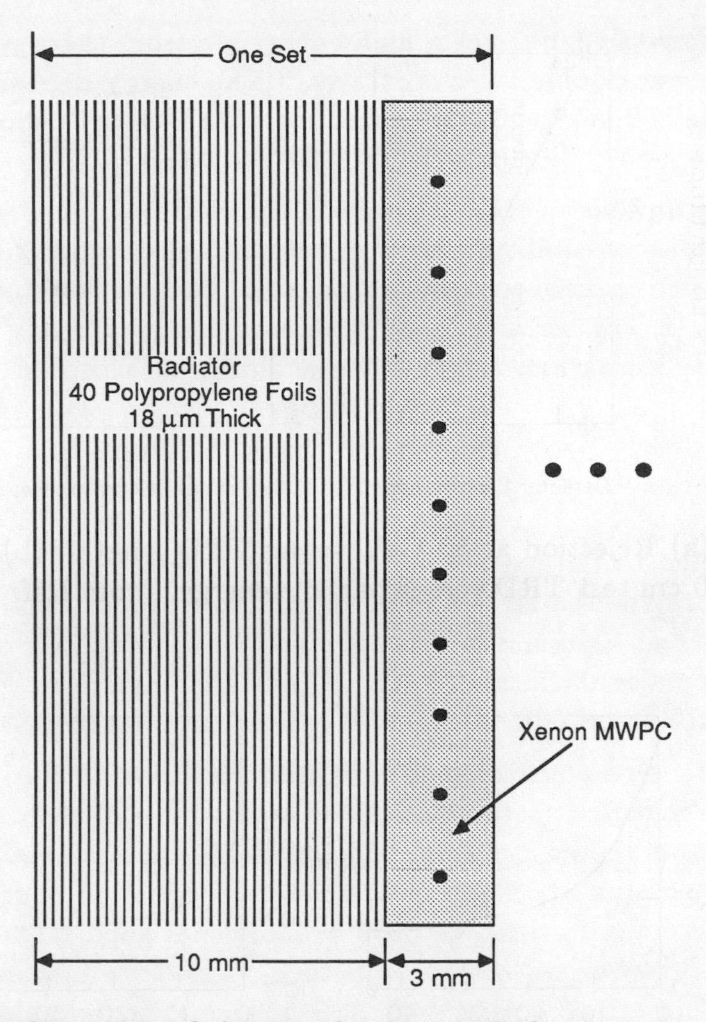

Fig. 3. Configuration of the test detector in Ref. 3.

larger than 5° from the jet axis and rises to about 15% at the jet core. The reason for this increase is twofold: the pions are energetic and start to radiate, and they have, of course, a strong angular correlation. Considering the role of the TRD response as a test of electron candidates selected calorimetrically, Fig. 5(b) demonstrates a TRD rejection factor of 1000 as close as 5° from the jet axis.

This same ISAJET study was used to estimate the effect on wire chamber performance of radiation from beam-beam interactions. Using data on gain shift versus integrated charge given by Walenta,[7] the innermost axial wire in the tracking system should experience a gain change of only about 5% in a year of running at the nominal luminosity of 10^{33} cm^{-2}s^{-1}.

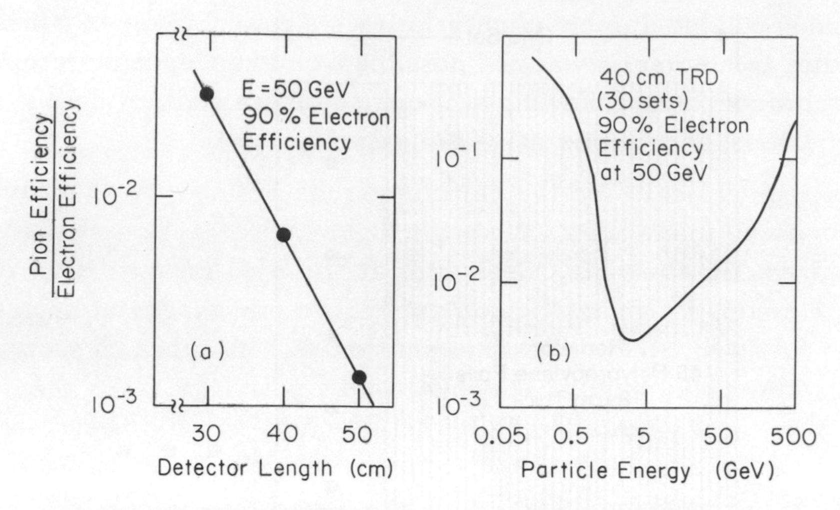

Fig. 4. (a) Rejection at 50 GeV versus TRD length. (b) Rejection for the 40 cm test TRD versus particle energy. From Ref. 3.

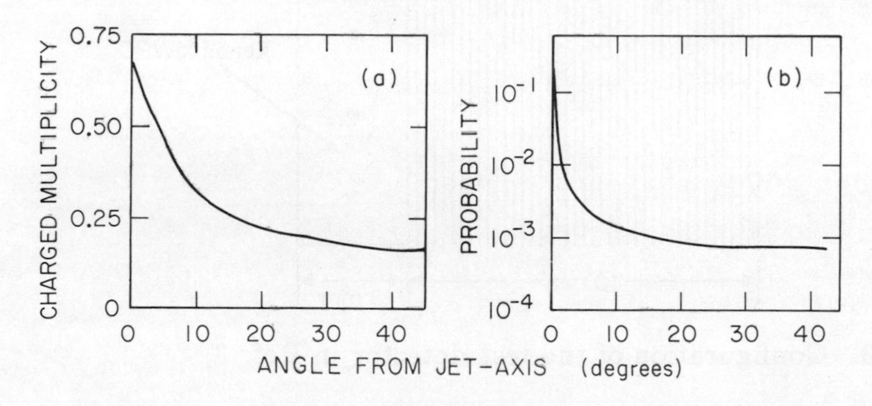

Fig. 5. (a) Number of charged particles per azimuthal segment and per jet versus azimuthal angle from jet axis, for 500 GeV ISAJET jets. (b) Probability per azimuthal segment and per jet that charged hadrons fake an electron, versus angle from the jet axis.

4. Calorimetry

4.1. Introduction

The design discussion for the non-magnetic detector calorimetry benefitted greatly from the experience gained on current detectors and prototypes, and particularly from recent work on the detailed understanding of the role of compensation in determining calorimeter performance.[8] This experience and understanding were applied to the evaluation of a variety of possible calorimeter designs for the SSC. Some designs have already received much attention at other workshops, but newer, innovative approaches were also considered. Because a large number of questions were raised, many of which need R&D or engineering studies to answer, it was not possible to make a specific recommendation for a calorimeter technique. The following sections therefore contain detailed consideration of a number of options, and comparative discussion.

The physics requirements for the calorimetry as defined by the physics subgroups have been summarized in Table 1. We begin below with a discussion of the additional operational requirements that any calorimeter at the SSC must satisfy. We then discuss the general calorimeter design in the context of the non-magnetic detector as a whole system. This is followed by detailed examination of a number of proposed techniques which represents the main part of the subgroup's activities. We then give a critical comparison of these techniques and end by listing questions and topics for future R&D.

4.2. Operational Requirements

There are a number of requirements that must be satisfied by any calorimeter system and some that arise specifically from the environment of an SSC experiment. Most of these requirements are not unique to the non-magnetic detector but were discussed in detail and form much of the basis for our critical comparisons later. Further, this list of requirements is not exhaustive but reflects the main areas of concern of the members of the subgroup.

4.2.1. Radiation hardness

It has been calculated that the radiation dose at the face of an SSC calorimeter will vary from 10^3 rads/year at $\eta = 0$ to 10^{10} rads/year in the $\eta > 5.5$ region at the 10^{33} design luminosity.[9] Depending on the design, the active calorimeter material, readout path, and electronics are all potentially at risk from this radiation. The resulting problems can be complex. In a piece of scintillator or scintillating fiber, the matrix, fluor, and wavelength shifter can all be damaged, resulting in changing efficiencies for light production, transmission, and output. The detailed study and separation of such effects is only now being undertaken in conjunction with the search for more radiation resistant media.

Apart from the direct radiation from the interaction point, there are also large numbers of neutrons produced in the passive material, particularly when this material has a large atomic weight (*e.g.*, uranium). The magnitude of this effect has only recently been estimated.[10] In view of this incomplete knowledge of radiation effects, it is not possible to give a definitive statement on the survival time of certain calorimeter materials beyond optimistic extrapolation from present knowledge. However, for some of the more radiation sensitive techniques it is already apparent that the high radiation regions at low polar angle will require a change in technology that could potentially introduce strange effects across boundaries.

4.2.2. Calibration and Stability

It is a fundamental requirement to be able to calibrate the calorimeter and know over what period it will be stable, so as to give a contribution to the energy resolution which is small compared to other unavoidable contributions. Sources of instability can lie in variations in active material density and purity (liquids), radiation damage effects, and drifts in electronics. To achieve a good calibration it must be possible to monitor each part of the calorimeter and readout path since, for instance, radiation effects will vary with polar angle and depth into the calorimeter. While it is relatively easy to arrange for testing of the readout path even with amplifiers mounted on the calorimeter, it is much more difficult, particularly in certain designs, to arrange to be able to create calibration "energy" deposits throughout the active volume of the calorimeter.

4.2.3. Speed of Response

The problems of calorimeter pile-up effects due to the high event rate at the SSC were examined quantitatively at Snowmass 86.[11] For cases in which the integration time is long compared to the beam crossing interval, significant residual energy ("physics noise") and missing E_t contributions can result. Generally the magnitude of such effects was found to depend on the square root of the signal development time. Thus an increase in the drift velocity of a liquid by a factor of two may be insufficient. Clearly, the best solution is to use an intrinsically fast response device.

4.2.4. Hermeticity

Much of the new physics predicted for the SSC requires the ability to be confident that a significant amount of energy has not been lost in a dead region of the detector. While any initially hermetic calorimeter design will become modified by the realities of structural engineering, some options are intrinsically less problematic in this area. However, there is still room for imaginative new approaches to the location of dead regions in established techniques. Careful and detailed simulation is needed, for instance, to decide the relative merits of a scheme with one major but thoughtfully situated dead region and a scheme with more uniformly distributed but individually less significant dead spots.

4.2.5. Calorimeter Depth and Shower Containment

The parametrization of Gordon and Grannis indicates that a calorimeter should have a depth of 10 ± 1 interaction lengths in order to contain a 1 TeV hadron shower at the 98% level.[12] The remaining few percent should be measured in several more interaction lengths of coarse "tail catcher" calorimetry to ensure a low level of energy leakage.

4.2.6. Electron Identification

Good transverse and longitudinal segmentation is required to provide hadron rejection and allow for electron isolation for decay tagging.

4.3. General Design

Given the absence of a magnetic field and the consequent removal of the need for long track lengths for momentum reconstruction, one can consider a compact calorimeter design to surround the limited volume of the TRD-tracker system. The inner radius of the calorimeter was thus set to be 0.8 meter.

The general approach to the calorimeter design was to use an adequately compensated "fine" calorimeter 8 to 9 interactions lengths deep, backed by a 4 to 5 interaction length magnetized iron tail catcher which also forms the first part of the muon system. The catcher would typically measure about 2% of the energy in a 1 TeV shower and can thus afford to be a device of inferior resolution. A significant cost saving is also achieved by the resulting limitation on the outer radius of the "fine" calorimetry.

The fine calorimeter has an electromagnetic section with three longitudinal segments, the second segment having a finer transverse segmentation than the others for improved electromagnetic shower position measurement. There is also the option in some designs to install a very fine position measurement detector behind the second segment to distinguish electrons from nearby photon showers. The fine hadronic calorimetry also has three longitudinal segments.

The general shape of the calorimeter has three main features, a barrel section, a sloping end section and a low polar angle plug section (Fig.1). The sloping end section gives an approximately uniform angle of incidence over its length for particles from the interaction region, and allows larger physical tower sizes (of fixed rapidity interval) at the lower angles than would an orthogonal end cap situated at the end of the barrel section.

In the low polar angle region we have avoided a scheme with a disconnected forward calorimeter since studies for D0 have shown that such an arrangement gives a worse missing-p_t resolution.[13]

4.4. Specific Techniques

Four possible techniques were considered in detail:

(*a*) Lead (or uranium)/liquid argon

(*b*) Lead (or uranium)/warm liquid

(*c*) Uranium/silicon

(*d*) Lead/scintillating fibers

The liquid argon option is by now a reasonably well understood technique although with potentially serious drawbacks in terms of dead areas that would require a very careful engineering study for the SSC. The alternative of using a warm liquid, which may go a long way towards solving the hermeticity problems, has been the subject of much recent development but a great deal still remains to be learned about actual operational use.

If a compact calorimeter design is possible then the third option of using silicon as the active medium becomes financially viable, although there are many unknowns since such a device has only been tested as an electromagnetic calorimeter; a hadronic prototype is still under construction.

The last option, lead/scintillating fibers, is the most speculative, but potentially offers the possibility of constructing a high performance calorimeter if questions of segmentation, calibration and radiation sensitivity can be satisfactorily resolved.

We shall now consider each of these options in detail. The degree of detail for the various designs varies according to factors such as the amount of prior work performed, the level of attention received at the Workshop, and the desire to avoid repeating well known facts about the more established approaches.

4.4.1. Lead (Uranium)/Liquid Argon

Of all the choices for sampling calorimetry at the SSC, liquid argon provides the best developed and understood choice. For this option we did not evolve a specific design as was done for the other options discussed below. We describe here only the main point of concern about liquid argon which was discussed at the Workshop.

While the understanding of liquid argon calorimetry extends to its inherent advantages (stability, radiation hardness, and signal size (relative to warm liquids), its disadvantage, as displayed in current experiments like D0, is also clearly indicated. This disadvantage is the lack of hermeticity introduced by segmenting the calorimetry into 3 angular intervals, each contained in its own double-walled insulation vessel.

In D0 this is the region between central and endcap calorimeters as shown in Fig 6. Both the endcap and central calorimeter vessels are made of stainless steel. This choice was dictated primarily by the desire to have code vessels which are welded closed after insertion of the calorimeter modules. The choice of welded vessels was dictated by both monetary and space constraints, as well as a prejudice against the complexity of cryo-liquid seals. It should be noted, that of the inert material in this transition region, almost one half comes from the support mechanisms for the calorimeter modules. This amount of material is likely to occur in any calorimeter — warm liquid, scintillator/lead, etc. — for support columns and for readout and access channels.

Another design for this transition region is shown in Fig. 7, the vessel designed for the H1 liquid-argon calorimeter at HERA.[14] This design makes use of the e-p kinematics, requiring hermeticity only in the proton direction. The single vessel encompasses both the central and endcap regions. In addition the vessel walls between tracking and calorimetry are aluminum, while the remainder of the vessel is stainless steal. Finally, the vessel is sealed using flanges and sliding seals between dissimilar metals. An additional complication in this design are the support feet, which must accommodate the 0.3% relative motion of the warm and cold vessels. Approximately four times more engineering resources were applied to the design of the H1 vessel than the D0 vessel.

Designs of liquid argon calorimetry for the SSC will have to further optimize the arrangement of modules and vessel designs. One can clearly make the design of a liquid argon system competitive with other choices (warm liquid, lead/scintillator) by placing the entire system within one large vessel. Since, as in the H1 design, the inner vessels can be aluminum, this design would be indistinguishable from other systems from the point of hermeticity. However, complications from this approach arise because the tracking system is sealed within the interior volume. One must solve the problems of cables and services, which can be accomplished with re-entrant holes through the calorimeter volume, and access to the tracking system. A possible solution is to build a "NASA style" tracking

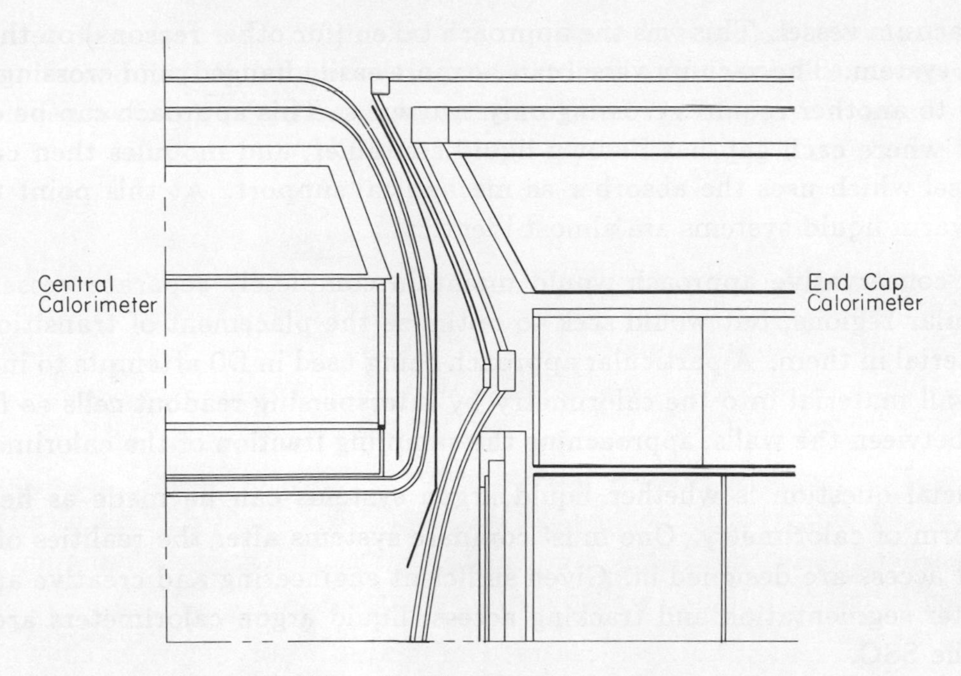

Fig. 6. D0 detector (Fermilab) central/endcap transition region.

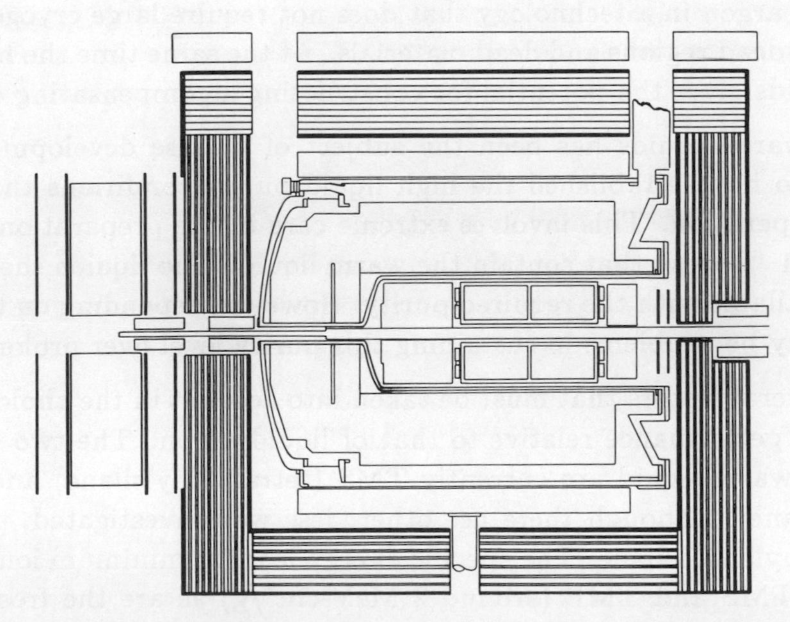

Fig. 7. H1 detector (HERA).

system. It would be built early enough to run in and debug and then buried with limited access for repairs. Given the size of the SSC calorimeters, one can even contemplate person sized access ports, judiciously placed to permit limited access to the tracker.

An alternate approach is to subdivide the argon vessels, and place them in a common insulating vacuum vessel. This was the approach taken (for other reasons) on the Mark II liquid argon system. The vacuum vessel can be more easily flanged, and crossing from one argon vessel to another requires crossing only two walls. This approach can be continued to the point where each gap has its own liquid container, and modules then can have a vacuum vessel which uses the absorber as mechanical support. At this point the liquid argon and warm liquid systems are almost identical.

A more conservative approach would maintain completely separate vessels for the various angular regions, but would seek to optimize the placement of transition regions and the material in them. A particular approach being used in D0 attempts to incorporate the vessel wall material into the calorimetry by interspersing readout cells as frequently as possible between the walls, approaching the sampling fraction of the calorimeter itself.

The crucial question is whether liquid argon systems can be made as hermetic as any other form of calorimetry. One must compare systems after the realities of support, services and access are designed in. Given sufficient engineering and creative approaches to calorimeter segmentation and tracking access, liquid argon calorimeters are a viable option for the SSC.

4.4.2. Lead (Uranium)/Warm Liquid

The use of warm liquids in calorimetry is partly an attempt to realize the desirable features of liquid argon in a technology that does not require large cryogenic vessels with their problems of dead regions and dead materials. At the same time the hydrogen content of the warm liquids gives the potential for constructing a compensating device.

The use of warm liquids has been the subject of intense development by the UA1 collaboration who have established the high liquid purity conditions that are necessary for satisfactory operation. This involves extreme care in the preparation and cleaning of the so-called UA1 "boxes" that contain the warm liquid. The liquids themselves are now commercially available with the required purity. However, depending on the details of the system, there may be problems in sustaining this purity level over prolonged periods.

There are several factors that must be taken into account in the choice of warm liquid and its expected performance relative to that of liquid argon. The two main contenders in the choice of warm liquid are currently TMS (tetramethylsilane) and TMP (2,2,4,4-tetramethylpentane), although there are other, less well investigated, possibilities such as hexamethylethylenedisilane. The specific energy loss for minimum ionizing particles is very similar for TMS and TMP (around 2 MeV-cm^2/g) as are the free electron yields. The yields are, however, significantly smaller (by about a factor of three) than that of liquid argon. This situation can be improved somewhat by operating at high electric fields. There are also benefits in terms of increased drift velocity from high field operation. These two factors taken together can lead to signal/noise ratios for TMS and TMP equal to or better than that for liquid argon depending on the value of the field used. This occurs when fast shaping electronics is used and the signal/noise ratio is determined only by the peak current. This current is proportional to the product of the amount of liberated charge and the drift velocity. As the electric field is increased a point is reached at which the lower charge yield of the warm liquid is compensated by the increase in drift velocity, which is faster for TMS and TMP than for liquid argon. The result is that the signal/noise ratio for TMS is equal to that for liquid argon at 25 kV/cm and the corresponding point is reached for TMP at 80 kV/cm. The price paid for this lies in the increased problem of high voltage breakdown at these high fields and is an area that needs to be studied.

The physical properties of the warm liquids lead to constraints on their use. Both TMS and TMP are highly flammable and thus have major safety problems. The use of TMS in combination with uranium is excluded because of the uranium fire hazard. However, after the recent work on the understanding of compensation, it appears to be feasible to design a compensating lead/warm liquid calorimeter thus widening the allowed choice of warm liquid. This is also important since the response time of a TMS calorimeter would be 2 to 3 times faster than a liquid argon device which would, for instance, lead to a reduction in the fake missing E_t signal from pile-up.

The use of warm liquids would also allow the possibility of putting the charge preamplifiers inside the liquid since there would be no strong restriction on their power dissipation. This gives very short cable delays and hence short signal rise times. Alternatively, the use of small modules containing warm liquid (as opposed to the large liquid argon systems) would also allow fairly short cables even if the preamps were mounted outside the modules.

One of the main original motivations for the use of warm liquids was the potential for a hermetic design. Clearly only single walls are needed to contain the warm liquid and not the double walls, with their attendant dead spaces and dead material, needed for liquid argon. Also the need to have a small number of large vessels is removed and this allows the smaller individual modules to have thinner walls. Although a careful engineering study is needed, it should be possible to support the warm modules using external braces, or, if necessary, supporting structural members could be bathed in the warm liquid since the thermal connection to the exterior is no longer a problem.

A wide variety of module shapes is possible. One possibility discussed at the Workshop would be to build "logs" of stacked thin boxes and absorber plates and then build up the calorimeter from these logs. The cracks between the logs can and should be non-projective and cables from the calorimeter and TRD/tracking system extracted at a number of distributed azimuthal locations. Even though there are still potentially troublesome cracks between the central and end sections of the calorimeter these should be much less severe than in an equivalent liquid argon design.

There should be no problem in satisfy the physics requirements of transverse segmentation since the situation is close to the well studied liquid argon case. Equally, longitudinal segmentation is achieved by the appropriate connections of the thin box layers.

The question of radiation hardness of the warm liquids is unclear. This question has to be studied in the context of whether one has a fill-and-seal system or a recirculating system for the warm liquid. A recirculating system would, apart from the obvious increase in complexity, add extra dead material in the form of pipes and manifolds. Such effects need careful modeling to be understood. While warm liquids are almost certainly not as radiation resistant as liquid argon, it may only be necessary to provide for liquid replacement in the region of the end plug, close to the beam axis. Calculations indicate that it should be possible to build even a non-uranium warm liquid calorimeter with a hadronic energy resolution of 30 to $35\%/\sqrt{E}$ and a small constant term if the device is correctly compensating and has the potential stability of calibration and operation foreseen for such devices.[8]

4.4.3. Uranium/Silicon

The uranium/silicon design described here is an evolution of the design originally outlined at the Workshop on Physics of Future Accelerators, La Thuile.[15] Because of this previous work this design is presented in rather greater detail than the other options considered. The advantage of silicon readout for calorimetry is its potential for absolute gain calibration and stability. It also offers the chance for very fine segmentation in both transverse and longitudinal directions with correspondingly good two jet separation and lepton isolation capabilities.

The disadvantages of silicon are its high unit cost and its limited tolerance to radiation damage, compared to liquid argon or warm liquid calorimeters.

This section discusses an extension of the La Thuile work, presenting a mechanical structure, a readout scheme, and a preliminary radiation damage estimate for a very compact device aimed at excellent lepton isolation and electron/hadron separation.

4.4.3.1 Electromagnetic Section

The electromagnetic section of the calorimeter is one of the most important parts of the detector. It must identify and measure the vector momentum of all electrons in the midst of high-density hadronic flux through the same region. Therefore, good lateral and longitudinal segmentation is needed to reject hadronic interactions in the electromagnetic section, as well as to provide a strong isolation capability to tag, for example, b-jet decays.

Fig. 8 shows the electromagnetic section in detail. The cell structure uses 2-mm (0.6 radiation length) uranium plates followed by a 2-mm gap containing the silicon readout layer (0.4 mm) and its backing layer (1.6 mm G-10). The readout electronics for each layer is embedded in the backing. As at La Thuile, the cell size is 2 cm × 2 cm in the silicon. After each 5 radiation length module, there would be a silicon strip detector (x-y readout) of 0.5 mm pitch to allow determination of the shower centroid to a precision of about 0.15 mm, as well as to supply a measure of the electron angle from the shower shape asymmetry. This additional readout data adds little to the overall cost or complexity of the detector but adds greatly to the ability to resolve electrons from nearby photon showers. It makes the effective two-shower separation in the detector roughly a factor of 10 better than the 2 cm square cells by themselves. The effective segmentation for electrons of this device is approximately 0.5 × 0.5 cm at a mean radius of 64 cm. This segmentation is much finer than is needed for trigger purposes, but it gives a very powerful handle on lepton isolation in the offline analysis.

The characteristics of the electromagnetic section are given in Table 2. The barrel region is made of "logs" assembled with non-pointing cracks. The importance of ensur-

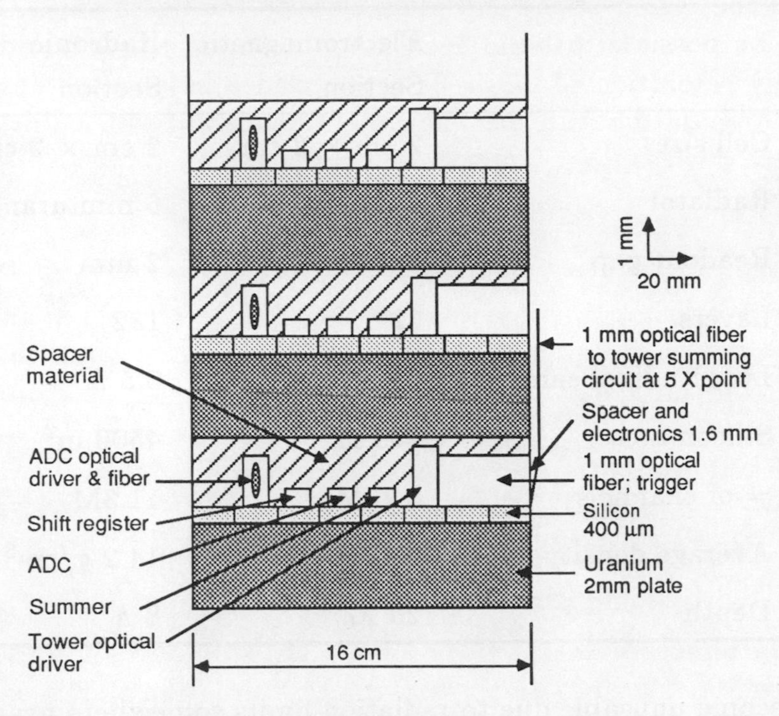

Fig. 8. Uranium/silicon calorimeter electromagnetic section detail.

ing full shower containment despite assembly breaks in the calorimeter was emphasized repeatedly at the Workshop by many people. This particular structure represents a mechanical assembly that is both practical and minimizes shower degradation. A real design will take careful engineering and physics analysis to optimize. We have imagined that the plates would remain parallel to the beam axis out to 45 degrees, then turn normal to the axis. This minimizes the effects of polar angle variations in the resolution and also makes a mechanical assembly using external support frames conceivable. Details are indicated in Fig. 9. Readout fibers from the longitudinal section would emerge in the matching region between longitudinal and vertical plates. For the endcaps, one must confront the difficult problem of a greatly-increased radiation damage rate as the distance off the beam axis is reduced. The $1/\sin\theta$ increase of $\Delta\eta$ with fixed $\Delta\theta$ in the forward direction requires a cell size in the electromagnetic towers that is not much larger than the 2 cm $\times$ 2 cm cells in the barrel region in order to maintain good electron isolation sensitivity for $\eta > 1.7$.

The question of how best to cover this forward region with good electron resolution is a very demanding one for any calorimeter technique and one for which we have no unique answer. Based on the La Thuile radiation damage calculations and the estimates of radiation tolerance given in that report, we conclude that silicon becomes unusable in the first few layers of the calorimeter for $\eta > 1.7$. This limit is soft; present studies of the question of radiation damage to silicon in a calorimeter environment will help to establish a more definite picture of the problem within the coming year . However, silicon

Table 2: Characteristics of the Uranium/Silicon Calorimeter

	Electromagnetic Section	Hadronic Section
Cell size	2 cm $\times$ 2 cm	2 cm $\times$ 2 cm
Radiator	2 mm uranium	6 mm uranium
Readout gap	2 mm	2 mm
Layers	32	132
Longitudinal sampling	5 X_0	0.5 Λ
Silicon area	300 m^2	4500 m^2
# of channels	0.75M	11.3M
Average density	9.45 g/cm^3	14.2 g/cm^3
Depth	20 X_0	8 Λ

is certain to become unusable due to radiation limits somewhere near $\eta = 2$. Can any of the radiation resistant media, in particular TMS, be made with sufficiently fine cell sizes to give good electron tags for rapidities up to 3? In order to achieve a tower size of .05 $\times$ .05 in $\Delta\eta \times \Delta\phi$, the detector element for $\eta = 2$ must be 2 cm $\times$ 2 cm in this design (z = 1.5m). At $\eta = 3$, this is reduced to 0.85 cm $\times$ 0.9 cm for the same resolution. How to achieve such small cell sizes without introducing breaks in the calorimeter structure or without extending the barrel region to such a length that there become serious weight/cost issue seems to be an important question for future SSC detector research.

We point out that a change of readout media from silicon to TMS in this sector of the calorimeter, in which plates are normal to the beam, would imply no noticeable change in detector resolution and no mechanical losses. A warm liquid device with thin container walls can be placed in the same 2 mm gap as the silicon readout. If the warm liquid has multiple plates in one liquid container, then as long as the plates are well aligned, the mechanical forces can be supported on the plates in the silicon region. From the shower resolution standpoint, if the effective radiation length of the two regions is the same, then the shower development is the same and only the cell segmentation matters for shower recognition. In U/Si or U/TMS detectors with the same gap, the radiation length is essentially identical. Therefore, one can think of mating these two different readout schemes in the transition region without sacrificing electron identification power. It may be the best way to handle this transition region.

The electronics to handle the readout for triggering and for offline analysis is discussed in the following section on readout and triggering.

4.4.3.2. Hadronic Calorimeter

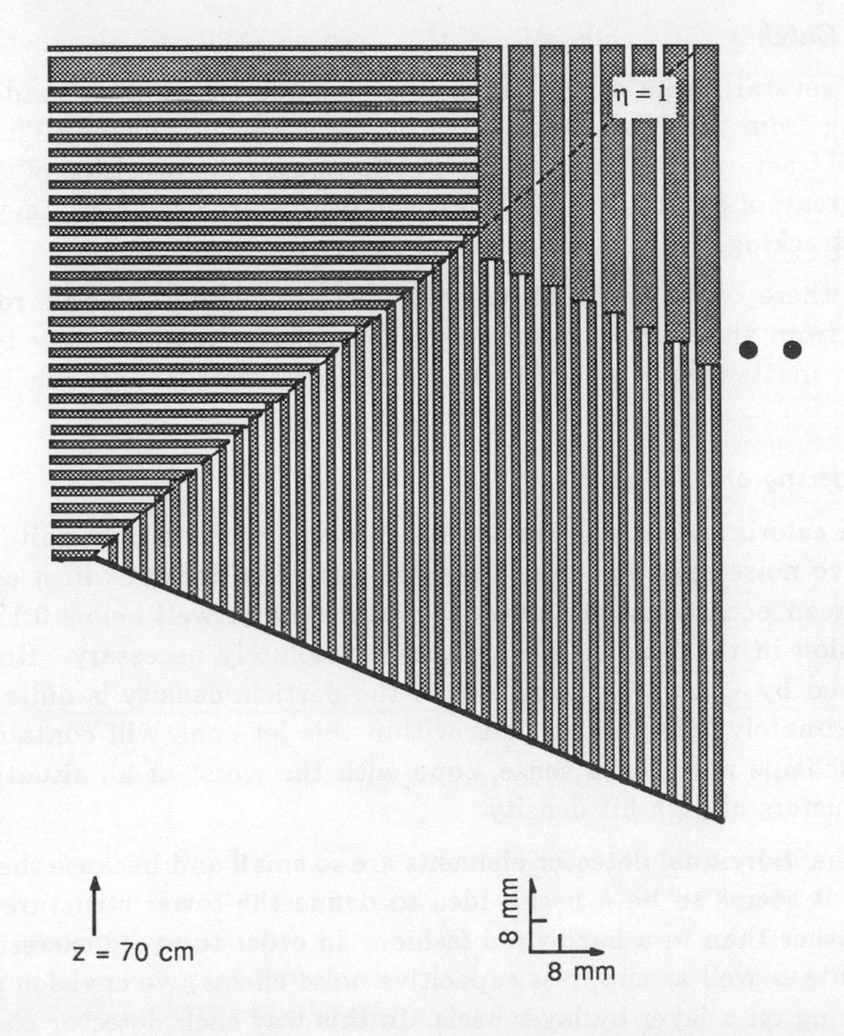

Fig. 9. Uranium/silicon electromagnetic calorimeter mechanical structure near $\theta = 45°$ where plates change orientation.

The hadronic section is built up with the same general structure as the electromagnetic section. The uranium plates are 6 mm thick, rather than 2 mm. The silicon structure is identical. The hadronic section is 8 interaction lengths deep, divided into 16 longitudinal samplings. As demonstrated in the shower resolution studies in the La Thuile proceedings, this sampling gives extremely good two-jet separation properties and will facilitate reconstructing events with W decays in the presence of other QCD background jets. The silicon required for the barrel region totals 3700 m^2.

As in the La Thuile study, this detector aims to have a balanced electromagnetic and hadronic response (e/h=1) by including polyethylene coatings as appropriate over the silicon detectors. The expected electromagnetic and hadronic resolutions should have a very small constant term. The cross-talk problem will be essentially zero, because signal processing is local and signals are converted to light directly. The anticipated resolutions, then, are $15\%/\sqrt{E}$ for the electromagnetic section and $55\%/\sqrt{E}$ for the hadronic section.

4.4.3.3. Tail Catcher

The first several layers of the muon filter include calorimeter readouts to measure energy leaking from the 8 interaction length precision calorimeter. The energy leakage will be a small fraction of the highest energy showers, so the precision of the "tail catcher" need not be great, of order $100\%/\sqrt{E}$. If gas detectors are used here, they contribute also to the muon tracking.

However, there could be problems with gaseous readouts in this region due to the neutron flux from the hadron calorimeter. Scintillator readout may be better in this regard. Such matters will be better understood as more experience is gathered from CDF.

4.4.3.4. Triggering and Readout

In a U/Si calorimeter the intrinsic segmentation is necessarily quite fine in order to keep capacitive noise suitably small. For the very fine divisions in η and ϕ mentioned earlier, the mean occupancy per silicon detector will be well below 0.1%. Hence, good zero suppression in the readout electronics is absolutely necessary. However, within a jet cone defined by $\sqrt{(\Delta\eta)^2 + (\Delta\phi)^2} = 0.4$ the particle density is quite high and many of the approximately 2000 silicon pads within this jet cone will contain energy. Thus, the readout scheme must, in a sense, cope with the worst of all situations — sparsely distributed clusters of high hit density.

Because the individual detector elements are so small and because the detector radius is also small, it seems to be a useful idea to define the tower structure of the detector in software rather than in a hardwired fashion. In order to avoid excessive cracks in the detector layout, as well as suppress capacitive noise effects, we envision doing the initial signal processing on a layer by layer basis. In this way each detector element in a given layer is available for combination with one or several elements in deeper layers to make a software-defined tower. Towers are readily modified, and tower structures can be changed in different η regions. This may be useful for special purpose studies, such as energy flow triggers or t-meson triggers.

The implementation of this idea is illustrated in the logic diagram in Fig. 10. Each detector has an on-chip preamplifier for charge/current conversion. Preamplified output signals from the 16 detectors would be loaded into a 16 channel parallel-load CCD shift register after each bunch crossing and then shifted into a 1.1 GHz flash ADC for processing. This ADC thus digitizes all the 16 detector signals in the interval between bunch crossings.

The 12-bit ADC output and 4-bit address, along with an 8-bit timing label to tag which bunch crossing out of 256 produced these data, would then go to a 2.4 GHz optical serial encoder. This encoder will not handle the high hit density that would arise from a jet between bunch crossings. One must insert a 24-bit wide by 256 bit deep 1 GHz shift register to buffer the parallel data stream coming into the serial encoder. Zero suppression is managed by imposing a digital threshold on ADC values and addresses fed into the digital delay line. In this way, the serial encoder for a given group of 16 pads sees only the average data rate over a 3.8 ms interval, rather than that from successive bursts.

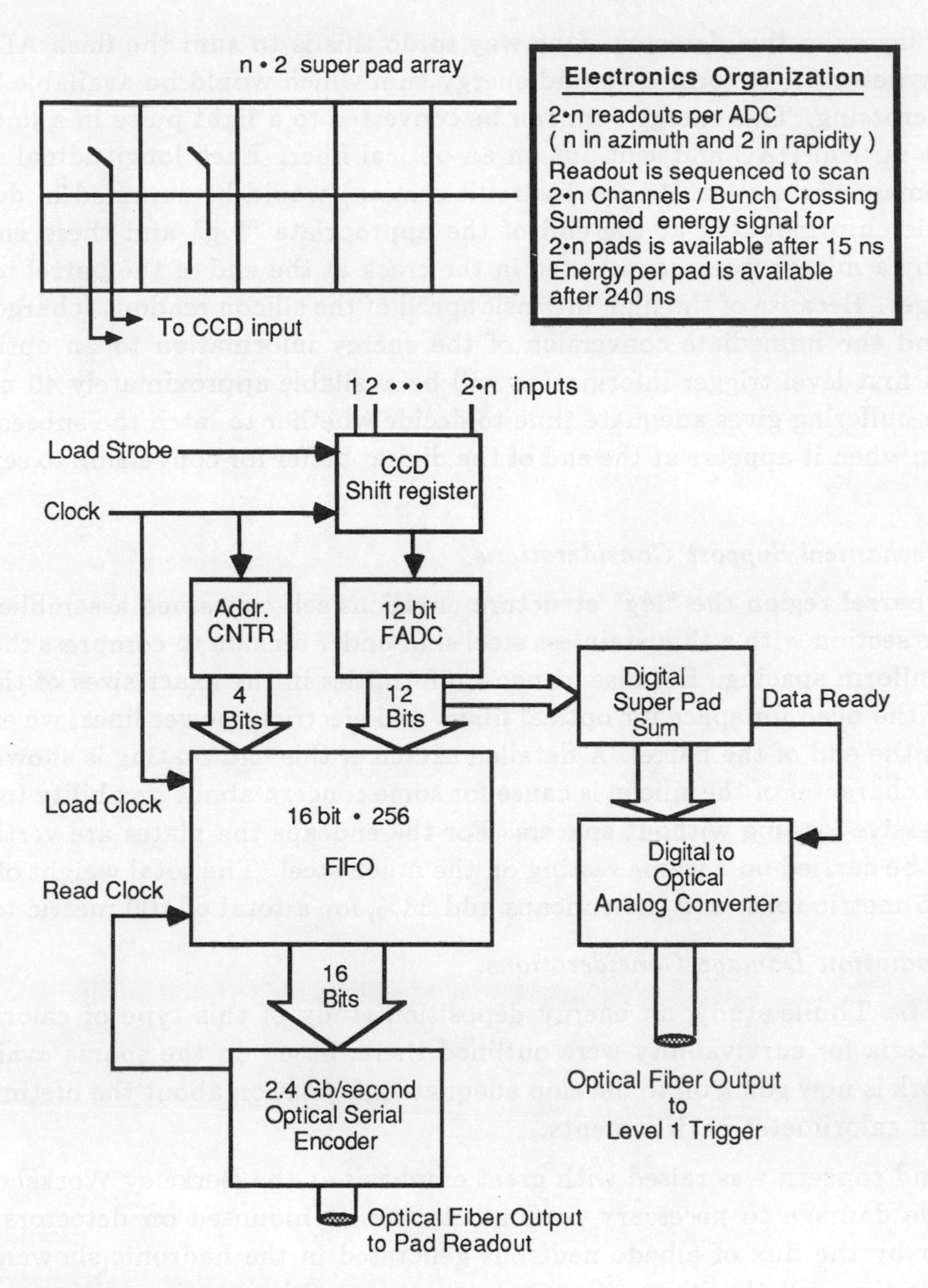

Fig. 10. Uranium/silicon calorimeter tower scheme read-out.

This shift register would have a 1GHz write clock to load data in from the ADC, but only a 40 MHz read clock to bring 24-bit words from the buffer into the serial converter input register. This kind of buffering would allow for two "full-occupancy" events, i.e., events in which all 16 detectors in one readout cluster were hit, for each 16 bunch crossings. Given the mean occupancy rate of 1.5% per crossing for any detector in the group, this seems to be sufficiently conservative to handle the design luminosity.

For trigger purposes, one must define local energy clusters. This requires summing over many layers in this detector. One way to do this is to sum the flash ADC output, detector by detector, to form a 16-fold energy sum which would be available 30 ns after the bunch crossing. This energy sum can be converted to a light pulse in a linear fashion in a 1 GHz optical DAC and sent out on an optical fiber. Each longitudinal segment (8 per electromagnetic section, 13 per hadronic section) would be summed in depth by an electro-optic chip mounted at the end of the appropriate "log" and these energy sums combined by a microprocessor mounted in the crack at the end of the barrel to go to the level 1 trigger. Because of the high intrinsic speed of the silicon readout (charge collection <10 ns) and the immediate conversion of the energy information to an optical analog signal, this first level trigger information will be available approximately 40 ns after the event. The buffering gives adequate time to decide whether to latch the subsequent ADC information when it appears at the end of the digital buffer for conversion to serial optical data.

4.4.3.5. Mechanical Support Considerations

In the barrel region the "log" structure envisions self-contained assemblies of trapezoidal cross section with a thin stainless steel skin under tension to compress the structure and give uniform spacing. Because of non-uniformities in the exact sizes of the uranium blocks and the need for space for optical fibers and electrical power lines, we envision a 3 mm gap at the end of the barrel. A detailed sketch of this end mating is shown in Fig. 9. The brittle character of the silicon is cause for some concern about its ability to withstand the compressive loading without spacers. For the endcaps the plates are vertical, so the weight can be carried on a frame resting on the muon steel. The total weight of the barrel region is 75 metric tons, and the endcaps add 33%, for a total of 100 metric tons.

4.4.3.6. Radiation Damage Considerations

In the La Thuile study, an energy deposition study of this type of calorimeter was made. Criteria for survivability were outlined there, based on the sparse available data. Further work is now going on to develop adequate information about the lifetime of silicon detectors in calorimeter environments.

A second concern was raised with great emphasis at the Berkeley Workshop, namely, the possible damage to necessary readout electronics mounted on detectors inside the calorimeter by the flux of albedo neutrons generated in the hadronic showers. It is not clear at this time whether the readout scheme proposed here, using microprocessors and digital logic inside the calorimeter volume can be realized with sufficiently radiation-hardened devices. However, it is clear that the lifetime of readout electronics within the large detectors is a major SSC issue. The CDG has formed a task force to pursue these questions both in model studies and experiments. Answers should be available within another one or two years and will greatly affect the ultimate design of SSC detectors.

4.4.4. Lead/Scintillating Fiber

This proposed design evolved directly out of work on the understanding of the benefits of compensating calorimetry, and how to achieve compensation by varying the relative fractions of active and passive material. The basic structure is shown in Fig. 11. Plastic fibers doped with fluor and wavelength shifter are laid in channels cut in lead sheets. Many sheets can then be stacked to build up a block. The thickness ratio of lead to fiber (about 4 to 1) is set by the requirement to obtain e/h = 1.0. This approach yields a fairly homogeneous medium and the possibility of making arbitrary shape calorimeter elements relatively easily. There may even be the possibility of constructing partially self-supporting structures from the blocks. This design overcomes the problems found in lead/scintillator sandwich designs in which one wants both thin lead plates to reduce the effect of sampling fluctuations and the correct lead/scintillator ratio for effective compensation. This leads to very thin scintillator plates which unfortunately have poor optical properties (short attenuation length). The fibers, on the other hand, have much better optical properties with attenuation lengths much longer than the effective nuclear interaction length of the lead/fiber combination. This is important for the hadronic calorimeter section since the depth profile of light production will vary from shower to shower. The problems of lateral variation in light yield across a slab of scintillator are also clearly avoided by the use of fibers.

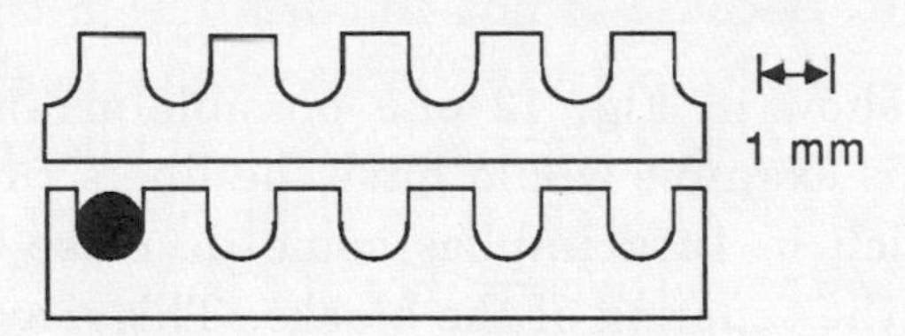

Fig. 11. Cross section of lead sheet/scintillating fiber module.

The fibers would have their axes pointing in the general direction of the interaction point, although offset by a few degrees to avoid the problem of "channeling" of particles down the fibers. There remain, however, problems associated with the finite spread of the interaction region and channeling of secondary particles produced in the calorimeter itself. More complex designs with "wiggled" fibers may offer a solution at the price of complicating the fabrication process.

The light yield of the wavelength shifter doped fibers is expected to be about an order of magnitude higher then in a system with separate scintillator and wavelength shifter. This may allow the use of solid state readout devices which would have merits in terms of calibration and stability, as well as being more compact than photomultipliers. With such a readout, longitudinal fibers and no external wavelength shifters it should be possible to achieve a design with very little dead space other that required by essential mechanical supports and signal paths to the exterior.

The attenuation induced in the fibers by radiation damage is an area of concern. It is known that, for instance, by shifting the light to longer wavelength (blue to green) it is possible to be less sensitive to the radiation damage effects. There still remain, however, unresolved issues of the effects of the large neutron flux generated in the calorimeter, and the detailed ways in which radiation of all types affects the various components of the doped scintillating fibers. Further, until there are answers to these questions it is not possible to define with any certainty the minimum polar angle at which this type of calorimeter could operate without rapid deterioration.

One of the main practical issues in this design is how to achieve satisfactory transverse and longitudinal segmentation. The longitudinal segmentation may be achieved either by having a number of layers with fibers in each layer running the full depth of the layer, or by having groups of fibers starting at a number of different depths through a single block. The first possibility introduces more material in the form of readout devices in between the longitudinal segments which may be important for the electromagnetic section. The transverse segmentation is, however, straightforward in this option.

The second possibility leads to a serious problem of calibration since the ends of many fibers would be buried inside blocks and would be inaccessible to calibration sources. Also the transverse segmentation scheme is more complicated since it would require sums over the separately extracted longitudinal segments.

However, in both schemes there exists the possibility of choosing an almost arbitrarily fine granularity limited ultimately by the cost of the electronics for the number of channels created.

As an example we show in Fig. 12 one possible arrangement of longitudinal and transverse divisions. This example would have the fibers running the full length of each longitudinal section which in turn implies some decrease in the fiber/lead ratio with increasing radius due to the tapering of the blocks. This may not be too serious, however, since contribution to the constant term in the energy resolution is expected to be small for values of the lead/fiber ratio between 3 and 5.

Resolutions of $30\%/\sqrt{E}$ for hadrons and $15\%/\sqrt{E}$ for electrons plus a very small constant term can be obtained due to the compensated design, assuming that the contributions to the constant term from instabilities and variations across the calorimeter can be held down to low levels. These estimates will be tested in prototypes currently planned for testing in about 1 to 2 years. It should also be noted that this technique has already been successfully used on a small scale in the CERN Omega inner (electromagnetic) calorimeter.

4.5. Comparison of Techniques

While a specific recommendation for the choice of calorimeter technique for the non-magnetic detector was not made during the Workshop, there was much discussion of the relative merits of the various options. We now give a summary of the main areas of those discussions.

There was considerable concern about the survivability of detector components in the SSC high radiation environment. Only liquid argon is without potential problems in this area and can be used for all parts of a calorimeter system. The effect of radiation on warm liquids is unclear although with a fluid there always exists the option to design a recirculating, or at least a rechargeable system. The radiation hardness of silicon and

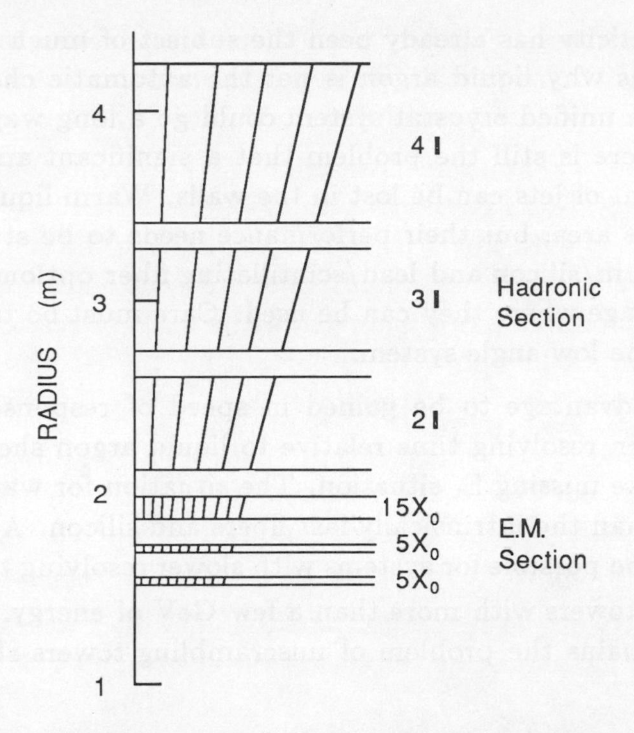

Fig. 12. Lead/scintillating fiber calorimeter segmentation.

plastic scintillating fibers is of greater concern. For silicon even "radiation hardened" versions are unlikely to be usable close to the beam axis. For the fibers there is some hope that a significant improvement in radiation hardness can be achieved with new doped scintillating materials now being developed. However, it still remains likely that there would be a change of technology in our detector in the region of 2 to 3 in η.

If such a break in the detector must occur then every effort must be made to have a smooth transition, particularly as our design aims to avoid the problems associated with a disconnected forward calorimeter section as discussed in section 4.3. In view of this, the advantages of a silicon or fiber calorimeter must be offset by the added complexity of building and operating two different types of calorimeter system. In the case of lead/scintillating fibers it was suggested that, at lower angles, it may be possible to use glass tubes containing liquid scintillator in place of the fibers. This still represents a somewhat different system, although not as fundamental as the suggested transition from uranium/silicon to warm liquid.

One problem common to all the techniques discussed arises if the amplifiers are situated on the detector. Besides the direct radiation there is expected to be a large flux of neutrons coming from the uranium or lead absorber plates. Lead may pose less of a problem but probably by less than an order of magnitude in neutron flux. The tolerance level of standard electronics to these neutrons is unclear. While it may be possible to consider radiation hardened electronics this would almost certainly increase the cost of the calorimeter.

The issue of hermeticity has already been the subject of much discussion and is one of the principal reasons why liquid argon is not the automatic choice for the SSC. An imaginative design of a unified cryostat system could go a long way towards solving the problem. However, there is still the problem that a significant amount of energy from the very soft component of jets can be lost in the walls. Warm liquids systems represent an improvement in this area, but their performance needs to be studied with a detailed simulation. The uranium/silicon and lead/scintillating fiber options are potentially very hermetic over the η range where they can be used. Care must be taken not to spoil this at the interface with the low angle system.

There is a clear advantage to be gained in speed of response by using silicon or fibers. The reduction in resolving time relative to liquid argon should give a significant improvement in the fake missing E_t situation. The situation for warm liquids is closer to that for liquid argon than the intrinsically fast fibers and silicon. At Snowmass 86 it was suggested that it may be possible for systems with slower resolving time to arrange to flag the beam crossing for towers with more than a few GeV of energy.[16] Even if this were possible there still remains the problem of unscrambling towers shared by events from different crossings.

In terms of energy resolution, the four techniques considered should all be capable of giving the required $0.5/\sqrt{E}$. Comparisons therefore focus on the constant term. With the possible exception of liquid argon it is predicted that it should be possible to design for $e/h = 1.0$, thus essentially eliminating one component of the constant term. The remainder of this term is driven by the stability of the calorimeter, its uniformity, and how well it can be calibrated.

Since the $\sqrt{E}$ term becomes less important at high energies, these factors must receive close attention. Good calibration and stability is indicated for all options except lead/scintillating fibers where the case has yet to be proved. There are also concerns for both the silicon and fiber systems if their performance is subject to change due to a measurable amount of radiation damage over the lifetime of the calorimeter.

The required segmentation, both transverse and longitudinal, seems no problem except in the lead/scintillating fiber case where a viable scheme has yet to be worked out in detail.

Finally, the cost of implementing the calorimeter in any of the four schemes should not be very different except perhaps in the case of the silicon where the unit cost for very large quantities needs clarification.

4.6. Questions and R&D Topics

We conclude by listing for each option the areas of concern for the focus of future R&D:

(*a*) Lead (uranium)/liquid argon

- cryostat design — needs an engineering study to look at the feasibility of a unified design.

- resolving time — increasing drift velocity
- electronics for event separation
- amplifier location, cable lengths
- how to make e/h closer to 1.0 (additives?)

(*b*) Lead (uranium)/warm liquids

- resolving time
- box design and dead material
- high field operation — breakdown problems
- maintaining purity in a large system
- effects of radiation on liquids and/or scheme for liquid replacement

(*c*) Uranium/silicon

- radiation effects
- silicon costs for large scale production

(*d*) Lead/scintillating fibers

- radiation effects
- liquid scintillator in glass tubes?
- practical longitudinal segmentation scheme(s)
- calibration scheme
- fiber orientation — channeling effects

(*e*) General

- support structures
- engineering breaks in technology
- use of radiation hard electronics

5. Muon Detection

5.1. Rationale for Muon Coverage and Resolution

Among the items studied by the muon subgroup were the importance of lepton coverage at small angles and the effects of muon resolution. Our conclusions are that muon coverage below 5° provides only marginal improvement in the physics capabilities and that, except for a few cases, muon momentum resolution of 10% is acceptable provided that it remains "reasonable" (less than 30%) to p_t's of ~ 2 TeV/c.

The necessary lepton coverage for high-p_t events has been discussed previously[17] and coverage down to 5° ($\eta \approx 3$) appeared sufficient. In the case of heavy Higgs production, an even smaller pseudorapidity cut would be made to increase the signal-to-background rate compared to the Z-pair continuum.[18] For new heavy Z production, the desire to measure forward-backward asymmetries requires good acceptance at high η of the Z.[19] In Fig. 13, the acceptance is shown as a function of the rapidity of the Z for events with muons having angles $\geq 5°$ and a Z mass of 1 TeV. The total geometric acceptance is about 90% and is about 50% at the highest η(Z).

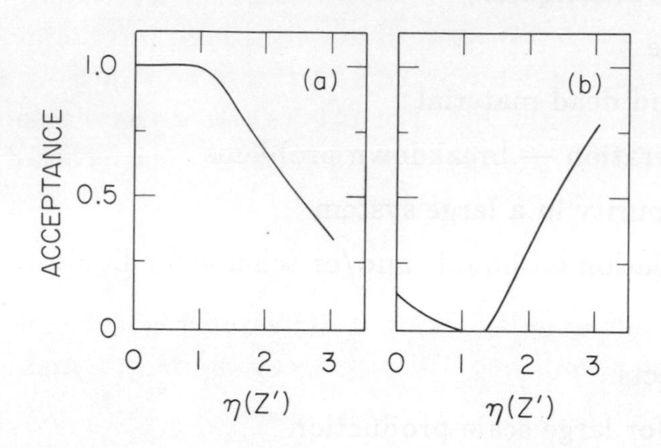

Fig. 13. The geometric acceptance versus η(Z) for a new Z with mass 1 TeV for a) a minimum angle of 5° on both leptons from $Z \to ll$ and b) requiring both leptons to be in the interval from $2.5° \leq \theta_l \leq 25°$.

Another argument which has been advanced in support of extended lepton coverage is that these muons signal the presence of neutrinos in events having missing p_t. Two cases were examined at this workshop. In the first the production and leptonic decay of the W ($W \to \mu\nu$) with $p_{tw} > 100$ GeV/c was examined. For a minimum angle of 5°, in 16% of the events the muon would not be accepted and of these, 40% would have a missing $p_t >$ 100 GeV/c (assuming perfect calorimeter coverage for $\eta \leq 5$). The contribution of these events to a missing-p_t sample is small compared to the ineradicable contribution from $W \to \tau\nu$ and $Z \to \nu\nu$ and therefore the gain obtained from extending the muon coverage below 5° would be of doubtful value.

Events containing Z + jet having $p_{tz} > 100$ GeV/c formed the second class of events with missing energy. These events are background to $H^0 \to ZZ \to ll\nu\nu$. A sample of 135,000 Z + jet events was generated using ISAJET with top masses in the range 30-85 GeV (the results were independent of the value of the top mass). As above, an almost perfect calorimeter was assumed and therefore events with missing p_t greater than 50 GeV/c were almost exclusively due to semileptonic decays of heavy quarks. The identification of extra leptons in an event would signal these quark decays. If only muons with $p_t > 10$ GeV/c and $\theta > 5°$ could be identified, then 25% of the events would be so flagged. This efficiency improved only slightly to 28% when the muon coverage was extended to 2°. A much improved efficiency could be obtained by reducing the muon p_t selection to 5 GeV/c (32%) or by the identification of both extra electrons and muons (53%). If both these selections were made 69% of the Z + jet events would be identified.

So again, additional muon coverage at small angles results in only a marginal improvement in background rejection.

Muon momentum analysis will effect the physics capability in a number of ways. The measurement of the width of a new Z depends on the resolution, but for this the electron channel will probably always be superior to the muon. Also, as Rosner has pointed out in Ref. 19, asymmetries in new Z production are not only a function of the rapidity of the Z but can also depend on the lepton-lepton mass through interference with the standard Z. Since the muon charge can be determined up to large p_t, it is the natural channel in which these asymmetries can be examined. For a 1 TeV Z, a mass resolution of 100 GeV may not be adequate to measure the asymmetries. A exhaustive study of this has not been done, but it would appear that a 25-50 GeV mass resolution at large $\eta(Z)$ would suffice. A possible upgrade to the muon detector should a new Z be discovered would be to increase the iron thickness in the 2.5° to 25° region from 4 m to 12 m. The $\eta(Z)$ acceptance for both leptons to be in this angular region is shown in Fig. 13. The momentum resolution for these muons will be about 5 to 8% with a corresponding mass resolution at 1 TeV of about 40 GeV.

Many high-p_t objects, such as the Higgs, will not have narrow widths. Even so, muon momentum resolution is an important consideration in the measurement of these events. Among the possible effects of poor momentum resolution are an increase in background due to the dimuons produced in $Z \rightarrow \mu\mu$ and a dilution in the ability to determine the polarization of W's or Z's arising from the Higgs decay. Only the first topic was studied during this workshop.

The predominant source of dimuon backgrounds is due to decay of heavy quarks. A heavy top decaying to W's and b's would be a great source of high-p_t muons. However, when we look at the Higgs decaying into Z's, with the Z's in turn decaying to muons, all four muons are well isolated. Imposing isolation cuts on the top events reduces that background to negligible levels simply because only a small fraction of the b decays have an isolated muon. The ability to impose these isolation cuts is clearly crucial and finely segmented calorimetry is indispensable. For the subsequent discussions we will assume that the calorimeter is segmented at least into cells of 0.1 in $\Delta\phi$ by 0.1 in $\Delta\eta$. We define a muon to be isolated if the sum of the transverse energy in the cells surrounding the one intersected by the muon plus that of the intersected cell is less than 2 GeV (a total of 9 cells). The only process we could find that could generate four isolated muons in significant numbers to be a potential background to a Higgs signal is the scenario involving a heavy fourth generation quark ($\approx$200 GeV) decaying to a heavy top ($\approx$100 GeV). If the heavy quark then decays via W + t and the top via W + b, pair production of these heavy quarks will produce events with four W's. The cross section for producing such heavy quark pairs is 4000 pb compared to the standard cross section of 4 pb for production of a 400 GeV Higgs. However, for this process the cross section for a 400 GeV Higgs is boosted to 40 pb (because of the coupling to the heavy quarks), so the background is less fierce than it might appear at first glance.

To study the importance of these heavy quark backgrounds for a Higgs signal we generated events for a 400 GeV Higgs and 200 GeV fourth generation b quark using ISAJET. Events were retained if they had four isolated muons (according to the above criteria) with $p_t > 30$ GeV. For an integrated luminosity of 10,000 pb^{-1} we are left with 3000 background and 80 signal events. Looking at the Z's from the Higgs events we find that, for 10% rms muon momentum resolution, the full-width half-maximum Z mass resolution is 14 MeV. If we then select events with 2 distinct opposite-sign μ pairs in the mass range 82 to 106 GeV, we are left with 60 signal and 100 background events. The invariant mass of the ZZ pair, however, peaks below 300 GeV for the background events. If we further select those events with the ZZ effective mass in the range 350 to 450 GeV, we have 50 signal/15 background events. Note that the four muon final state is the worst possible case. The inclusion of events with electron pairs will reduce the background because of the better electron energy resolution and the reduction of the combinatorial background.

In conclusion, although a heavy fourth generation quark could generate non-negligible backgrounds to a purely leptonic Higgs signal, the signal-to-background ratio is expected to be good enough so that it is not a serious problem. Muon momentum resolutions of better than 10% are not required, but, if the resolution were to get worse by a factor of 2, the signal-to-background ratio in the 4-muon final state would deteriorate from 10/3 to 5/6.

5.2. Muon Detector

The apparatus needed to identify and measure the muon is shown schematically in Fig. 1. Three meters of iron are arranged in roughly cylindrical shapes just outside the calorimeter. At each end are seven annular disks about the beam. The entire spectrometer is magnetized such that there is a toroidal magnetic field of 1.8 T in each iron section (of course, the field approaches 2.0 T near the inner radii of the disks).

The environment outside of the calorimeter is assumed to be quiet enough to allow the use of drift chambers. The choice of drift distance is clearly dependent on the noise rate and the readout cost. The chambers are presumed to have a 70 μm point resolution and have some vector capability as well. The inner chambers are presumed to measure the direction of the muon incident on the iron to an accuracy of 0.5 mrad. This is a significant requirement but can be achieved by measuring several points to a precision of 70 μm over 30 cm. Similarly, the muon direction is also measured after exiting the iron to the same accuracy. (The gain in resolution in measuring the exit angle much better is marginal unless the muon track can be identified in the central tracking device.) In addition there are scintillation counters deployed as shown which can be used in the trigger and to tag the particular bunch crossing. They also provide the time reference for the drift chambers.

The geometry is chosen so that a muon will traverse ≥ 5.4 Tm for $\theta \geq 28°$ and 7.2 Tm for $5° < \theta < 28°$. The fractional momentum resolution is calculated assuming point measurements of 70 μm accuracy between the iron segments and adjacent to each outer side, and independent angle measurements both before and after the iron to an accuracy of 0.5 mrad. The resolution is obtained from a calculation[20] including measurement errors and multiple coulomb scattering and is shown in Fig. 14 for the different paths (no allowance for the increase in path due to oblique incidence is made).

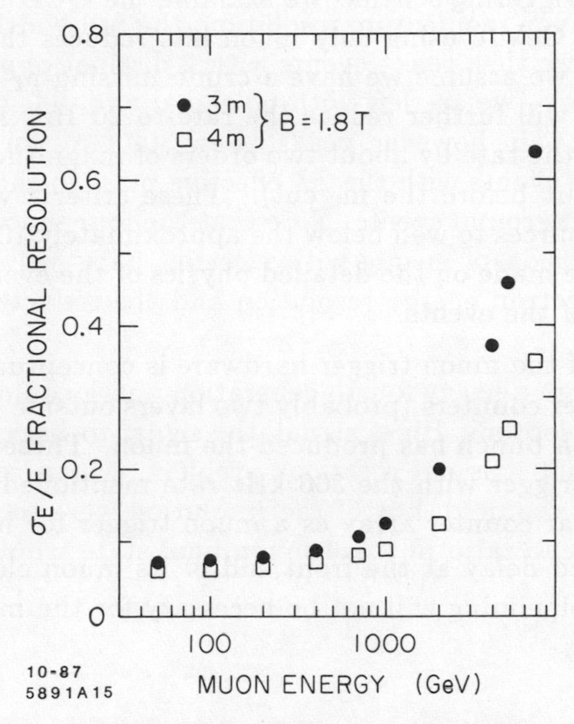

Fig. 14. The resolution of the muon spectrometer as a function of momentum. The muon traverses 3 m of magnetized iron for angles greater than 28° and for angles less than 28° and greater than 5° it traverses 4 m. Chamber resolutions are as noted in the text.

5.3. Muon Trigger Rates

The muon trigger rates for a generic SSC detector were calculated during Snowmass 86 and are given in the Muon Detector Group Report.[21] The geometry assumed was similar to that envisioned for the non-magnetic detector with precisely the same number of calorimeter absorption lengths and lengths of magnetized iron. The central detector was assumed to be a cylinder 5 m long with a 2.5 m radius (i.e. shorter but thicker then for the non-magnetic design). The size of the central detector affects the π/K decay rate. The trigger rate is dominated by decays at small angles and low p_t (below 15 GeV/c), but these decays make an almost negligible contribution to higher p_t single muon rates or to dimuon rates. For now, we will ignore the small differences in geometry and just summarize the Snowmass 86 numbers.

At the standard luminosity, the raw muon rates for angles greater than 5° will be about 300 kHz. In order to lower the rate to 1 Hz, one will have to impose a p_t cut of about 100 GeV/c. Most of the muons are in the forward region with only about a 10 kHz rate for angles greater than 20°. The dimuon rate is much lower with only a 1 kHz rate for no p_t cut and a 10 Hz rate if both muons are required to be have p_t greater than 25 GeV/c.

As an example of a triggering scheme, we examine the QCD rates for a W trigger. A first level cut of $p_t \geq 25$ GeV/c using only muon hits reduces the rate from 300 kHz to 600 Hz. If at this point we assume we have a crude missing-p_t vector, then a loose m_t cut between 35 and 120 will further reduce the rate to 20 Hz. Finally, an isolation cut on the muon will reduce the rate by about two orders of magnitude (it may be preferable to apply this isolation cut before the m_t cut). These criteria will reduce the $W \rightarrow \mu$ trigger rate from QCD sources to well below the approximately 10 Hz rate from $W \rightarrow \mu\nu$. Further selection must be made on the detailed physics of the event, *i.e.*, the p_t of the W, or other characteristics of the events.

To date the design of the muon trigger hardware is conceptual at best. It is believed necessary to install trigger counters (probably two layers outside the iron) to quickly and precisely determine which bunch has produced the muon. These counters would also be used as a level 0 muon trigger with the 300 kHz rate mentioned above. D0 is currently envisioning using a similar counter array as a muon trigger for high luminosity and will also add a 300 ns lumped delay at the front end of its muon electronics.[22] With this, they believe that other pipelining will not be necessary for the muon data.

6. Trigger

A major problem for any 4π SSC detector is how to reduce the 60 Mhz interaction rate to an approximate trigger rate of 1 Hz. We have discussed some aspects of triggering in previous sections (4.4.3.4 and 5.3); here we give a brief overview of a generalized trigger which would be tailored to the specific adopted technologies.

The trigger would contain four levels, each reducing the rate by approximately two orders of magnitude. The zeroth level would be a deadtimeless pre-trigger which works on pipelined signals from the calorimeter towers. Sums of electromagnetic and hadronic energy in $\Delta\eta \times \Delta\phi$ bins of 0.1×0.1 would be formed, and transverse energy cuts would be made at around 20 to 30 GeV. This trigger would probably take between 500 ns and 1 μs, requiring a pipeline of this length.

The level 1 and 2 triggers together would reduce the rate from about 1 MHz to about 100 Hz. They would do this by applying increasingly sophisticated analyses to the calorimeter data and by integrating information from the muon trigger (see Section 5.3). Isolation cuts and topology cuts could be applied at these levels.

The final level of trigger would be done by microprocessor farms doing a preliminary analysis of the entire event.

7. Detector Cost

We have estimated the cost of the non-magnetic detector using the cost schedule established by the SSC Detector Cost Evaluation Panel.[23] The result is shown in Table 3. We note that the total, 193 M$, is about two-thirds of the cost estimated for a large solenoidal detector.[23]

Table 3: Detector Cost Estimates

Item	Number	Units	Cost (M$)
Tracking & TRD's			
Mechanical	330	k wires	33.0
Electronics	340	k channels	40.8
Calorimeter			
Precision mechanical	1600	tons	25.6
Catcher mechanical	1500	tons	5.2
Electronics	200	k channels	24.8
Muon system			
Toroids	10500	tons	12.6
Chambers	3700	m^2	7.4
Electronics	120	k channels	12.2
Computing	1	system	5.0
EDIA	20%		32.2
Total			**193.0**

8. Conclusions

We have presented a model of a detector that meets all of the major requirements given to us by the physics parameters groups.

By employing a combination of track-matching, calorimetric longitudinal and transverse shower profile measurements, and TRD information, the detector is capable of superior electron identification. It has an excellent muon identification and measurement system. And it allows the use of an optimized calorimeter that is not compromised by the requirements of a large magnetic field.

Thus, we conclude that a "non-magnetic" detector is a strong candidate for a 4π SSC detector for the study of high-mass phenomena.

Acknowledgement

We are indebted to V. Chernjatin, B. Dolgoshein, and J. Schukraft for information on TRD design considerations and performance that has been included in this report.

References

1. C. Fabjan, in *Experimental Techniques in High Energy Physics,* T. Ferbel, ed., Addison-Wesley, 1987; Y. Fukui *el al.*, *Proceedings of the Summer Study on the Physics of the Superconducting Supercollider,* Snowmass 1986, p. 417.
2. R. DeSalvo, *Proceedings of the 1986 Summer Study on the Physics of the Superconducting Supercollider,* June 23–July 11, 1986, Snowmass, Colorado, p. 391; A. Odian, *ibid.*, p. 398.
3. B. Dolgoshein, *Nucl. Instrum. Methods* **A252**, 137 (1986).
4. S. P. Swordy *el al.*, *Nucl. Instrum. Methods* **193**, 591 (1982); H.-J. Butt *el al.*, *Nucl. Instrum. Methods* **A252**, 483 (1986).
5. J. Beatty, private communication.
6. F. E. Paige and S. D. Protopopescu, BNL-38034 (1986).
7. A. H. Walenta, *Nucl. Instrum. Methods* **217**, 65 (1983).
8. R. Wigmans, *Nucl. Instrum. Methods* **A259**, 389 (1987).
9. B. Pope, *Physics at the Superconducting Super Collider Summary Report,* Fermilab, June 1984.
10. R. Wigmans, these proceedings.
11. T. Kondo *el al.*and J. Huston *el al.*, *Proceedings of the 1984 Summer Study on the Design and Utilization of the Superconducting Super Collider,* June 23–July 13, 1984, Snowmass, Colorado.
12. H. Gordon and P. Grannis, *Proceedings of the 1986 Summer Study on the Physics of the Superconducting Supercollider,* June 23–July 11, 1986, Snowmass, Colorado.
13. A. Jonckheere, D0 internal note.
14. H1 Detector Design Report.
15. *Proceedings of the Workshop on Physics at Future Accelerators,* La Thuile, Val d'Aosta, Italy, CERN 87-07.
16. C. Baltay, J. Huston, B. Pope, *Proceedings of the 1984 Summer Study on the Design and Utilization of the Superconducting Super Collider,* June 23–July 13, 1984, Snowmass, Colorado.
17. D. Carlsmith, D. Hedin and B. Milliken, *Proceedings of the 1986 Summer Study on the Physics of the Superconducting Supercollider,* June 23–July 11, 1986, Snowmass, Colorado, p. 431.
18. Report of the Heavy Higgs Group, these proceedings.

19. J. L. Rosner, *Proceedings of the 1986 Summer Study on the Physics of the Superconducting Supercollider,* June 23–July 11, 1986, Snowmass, Colorado, p. 213.

20. C. Zupancic, CERN/EP/NA4 Note 81-8 (unpublished) and C. Zupancic (to be published)

21. D. Carlsmith *el al.*, *Proceedings of the 1986 Summer Study on the Physics of the Superconducting Supercollider,* June 23–July 11, 1986, Snowmass, Colorado, p. 405.

22. D. Green, D0 Note 557, Fermilab, 1987.

23. SSC Central Design Group, SSC-SR-1023 (1986).

4π Dipole Detector for the SSC

H. H. BINGHAM

Department of Physics, University of California, Berkeley, CA 94720

and

J. A. J. MATTHEWS*

Department of Physics, Johns Hopkins University, Baltimore, MD 21218

and

R. VAN KOOTEN

Stanford Linear Accelerator Center, Stanford, CA 94305

ABSTRACT

A 4π dipole detector is considered for the study of intermediate and high P_t physics at the SSC. Dipole detectors emphasize physics over a large rapidity interval, and are typically superior to solenoid detectors for the measurement of forward charged particles. The strengths of solenoid and dipole detectors are largely orthogonal, and suggest that these detectors form a complimentary pair for the study of high P_t phenomena at the SSC.

1. Introduction

Although the high luminosity options for the SSC emphasize "discovery" physics at large mass scales or transverse momenta, the actual scale of this new physics is not known. Previous detector studies have emphasized solenoidal central detectors.[1,2] Such detectors are most effective for the measurement of charged particles with pseudo-rapidities $|\eta| \leq 1.5$, and are well suited to the study of physics at the highest mass scales accessible at the SSC. In contrast other "new physics" topics, for example the study of intermediate mass scales, may require a detector that is optimized for a large coverage in rapidity. The possibility of using a large dipole central detector is discussed in this note.

* Co-ordinator

The plan of this note is as follows. In Section 2, we review the detector requirements for high-P_t physics,[3] and in particular the motivation for the measurement of charged particles over a large rapidity interval. The physics then sets some "natural" requirements on the momentum resolution of the detector as a function of $|\eta|$ and P_t. The dipole option for a central magnetic detector is addressed in Section 3. The emphasis here is on the performance of the charged particle tracking system, and on the comparison of the dipole versus the solenoidal choice of magnetic field. For concreteness two "generic" types of dipole tracking systems are considered: one based on straw chambers, and one based on scintillating optical fibers. A very schematic version of a complete 4π dipole detector is presented in Section 4 in order to emphasize specific features of the dipole system. In particular, we study the effects of the detector transverse magnetic field on the circulating beams in the SSC. Finally Section 5 presents our summary.

2. Physics Motivation

The detector requirements needed for the study of typical "high P_t" physics topics were reviewed for the 1987 SSC Detector Workshop.[3,4] Many other topics and associated detector requirements have been discussed in the literature. In general, it is found that detectors should have highly-segmented linear calorimetry over $|\eta| \leq 5.5$, and good electron and muon coverage over $|\eta| \leq 3.0$. In detail, different physics emphasizes one feature of the detector over another. Furthermore, different detector features are often important in the "discovery" phase and in the "understanding" phase of any measurement. For this discussion we will focus on physics at intermediate mass scales that benefit from charged particle "magnetic" analysis. Topics that benefit significantly from the knowledge of the charge of high P_t (electron) tracks include: search for Heavy Higgs, search and analysis of new W'or Z', and the study of WW, WZ, and ZZ scattering.

Heavy Higgs

The Higgs decay: Higgs $\rightarrow$ ZZ, followed by the decays $Z \rightarrow e^+e^-$ (or $\mu^+\mu^-$), provides a realistic means to search for the Higgs at the SSC.[5] If we require acceptance efficiencies of at least 50% the detector must have good lepton coverage for $|\eta| \leq 2.0$. The search for the Higgs in the decay channel: Higgs $\rightarrow$ WW is much more difficult, but can potentially result in better statistics than in the ZZ channel. The analysis of either Higgs decay channel would profit from a knowledge of the continuum WW, WZ, ZZ and the W+jets and Z+jets distributions. The distributions of W/Z from these "backgrounds" are typically much flatter in rapidity than from Higgs decays.[6,7] This suggests that detectors should have good lepton coverage to pseudo-rapidities of $|\eta| \approx 3.0$ in order to have a background "monitor" region with uniform detection capabilities. To be sensitive to Higgs as massive as 1 TeV, the leptons from the Higgs decay must be measured up to transverse momenta $P_t \leq 0.5$ TeV. Since the measurement of the sign of the charge of each lepton is critical for the study of the backgrounds from continuum pairs of vector bosons, and also minimizes combinatoric backgrounds in the Higgs decay channels, [8] the detector should also determine the lepton's charge for $|\eta| \leq 3.0$ and $P_t \leq 0.5$ TeV.

New W′or Z′

High energy pp colliders are excellent laboratories for the discovery of new W′ and Z′ bosons. The signals for these new bosons have been reviewed often.[9,10] For this discussion we assume that the bosons are studied only through their leptonic decays. The range in pseudo-rapidity of the leptons depends on the mass of the parent boson, but typically varies from about $|\eta| \leq 1.5$ for 5 TeV bosons to $|\eta| \geq 3.0$ for bosons below 1 TeV. Following discovery of a "new gauge boson", it can be a more challenging matter to determine their properties. At the highest accessible masses only "discovery" may be possible due to the very limited number of events. However detailed studies should be possible for masses in the 1-2 TeV range assuming 10^{33} $cm^{-2}sec^{-1}$operation of the SSC. For the W′ and the Z′, asymmetry measurements can discriminate among many of the possibilities.[11] This requires a measurement of the lepton charge, and in particular a measurement of the charge out to large rapidities and to transverse momenta of approximately $P_t = 0.5 - 1.0$ TeV. [12]

Single Boson and Multi-Boson Cross Sections

Although the production of "continuum" W and/or Z pairs is often discussed as a background to the discovery of new physics at the SSC, it should not be forgotten that these "backgrounds" are of interest in their own right. The precision measurement of the cross sections for the production of WW, WZ or ZZ pairs is an important test of the standard model [6,7,13] and should be one of the goals of any central detector. The cross sections at large VV-pair masses are small and limit this study to intermediate mass scales.[6,15] As a result the cross sections are significant out to several units of rapidity. This emphasizes the importance for good lepton detection capabilities to $|\eta| \approx 3$.

Another "background" for new physics is the production of W+jets and/or Z+jets.[5,7,14] This has stimulated interest in the calculation of QCD higher order corrections to W and/or Z production. As a result, the ability of QCD calculations to describe these data [15,16] may ultimately allow precision measurements of α_s in channels such as high mass Z+1-jet cross sections.[17] As in the measurement of pairs of vector bosons, the study of W+jets and/or Z+jets should be done out to several units of rapidity.

Physics Summary

In summary, many of the important physics topics at the SSC are not at the highest mass scales, but fall in an intermediate mass range of approximately 0.5 - 2.0 TeV. To study this physics detectors should have good lepton detection to pseudo-rapidities of $|\eta| \leq 3.0$ and to transverse momenta $P_t \approx 0.5 - 1.0$ TeV. In almost all cases, the sign of the charge of the lepton must also be determined. Assuming that the charge and momentum of muons will be measured in a magnetized external muon system, the measurement of the sign of very high-P_t electrons sets the required momentum resolution of the central detector as a function of $|\eta|$ and P_t.[18]

3. Central Tracking in a Dipole Detector

The emphasis of our study is on central tracking and on momentum resolution for a large 4π detector with a dipole magnetic field. To do the study we use two "generic" tracking systems: one based on straw chambers, and one based on scintillating optical fibers. These choices are motivated by the central tracking chamber in the UA1 detector [19] at the CERN $\bar{p}$ p collider. Although these tracking systems represent possible extensions of the UA1 design adapted to the high luminosity environment of the SSC, they should not be considered to be optimized or final designs. Rather they are chosen to illustrate certain strengths and weakness of the dipole geometry, and to provide a design from which acceptance and momentum resolution can be estimated as a function of a track's pseudo-rapidity and P_t.

Tracking Systems

We choose a coordinate system centered on the interaction region (and on the detector), with the x-axis in the horizontal plane, the y-axis up, and the z-axis along the beam direction. In this geometry the dipole magnetic field is along the x-axis, and charged tracks bend in the yz-plane. For the optimal measurement of track momentum, the wires (or optical fibers) should be parallel to the magnetic field. This would suffice if it were possible to obtain a "third" coordinate along the wire, for example by means of charge division as is done in the UA1 case. In practice there are problems in obtaining a third coordinate, [20] and we assume that a simple stereo system is used instead. Thus we consider a central tracking chamber composed of many "planes" of wires/fibers oriented perpendicular to the z-axis. These "planes" are stacked-up along the z-axis to fill the tracking volume. Alternative planes provide measurements along the "y" direction, or at $\pm 30°$ to the y-axis for example.

Although planes of wires/fibers perpendicular to the z-axis are a "natural" measurement system for tracks that are at moderate or small angles to the z-axis, planes of wires/fibers that are separated by finite distances along the z-axis will have "cracks" for tracks at large angles. Since there are solutions to the problem of "cracks", for the present study we will only consider planes of chambers oriented orthogonal to the z-axis throughout. In fact there is some merit in this choice: there is only one tracking system which should result in major advantages in pattern recognition, and possibly in momentum resolution. To solve the "crack" problem, the wires/fibers must be densely packed in the region of approximately $|z| \leq 50$ cm.[21] In this case there are no "cracks" for large angle tracks. There is one remaining problem: the number of measurements on tracks moving almost parallel to the x-axis is small. This occurs over a small solid angle and we ignore it. A solution based on using 5 mm diameter straw proportional tubes is sketched in Table I. Two features of the dipole geometry are: all wires are approximately of the same length, and the number of wires, but not their length, scales with the total chamber length along the beam axis. An analogous, but far more speculative, solution based on using scintillating optical fibers is also given in this table. Here we assume that bundles of fibers are drawn into thin strips of dimensions 1 cm $\times$ 400 μm for example, where the individual fibers are 100 μm $\times$ 100 μm "square".[22]

Table I. Tracking Chamber Parameters

	Straw Chamber	Scint. Fiber Chamber
Minimum tracking radius (r_{min})	50 cm	10 cm
Maximum tracking radius (r_{max})	150 cm	50 cm
Extent along the beam axis (z_{max})	±400 cm	±200 cm
Straw diameter/Fiber transverse dimensions	0.5 cm	$(100\ \mu m)^2$
Charge/light collection time (t_{max})	50 nsec	< 100 nsec
Maximum straw/fiber length	≈ 310 cm	≈ 100 cm
Orientation of successive "planes"	Y U V ...	Y U V...
Spatial resolution / measurement	150 μm	30 μm
Typical number of measurements / track	200	160
Magnetic field (B)	0.7 T	1.0 T

For the straw chamber we have used a model where the wires are supported only at the outer radius of the chamber. This results in a very low mass tracking system which is an important criterion for a dipole detector where the magnetic field tends to "extract" forward going particles from the beam region out into the detector. A "massive" UA1 wire support structure [19] could possibly be used for the very central "z" region of the chamber but would generate large backgrounds if used throughout. Wires are strung to some minimum distance to the beam, $r_0 \approx 5$ cm. This results in somewhat lower number of wires available to track particles that are near the xz-plane, where tracking is done primarily in the "U" and "V" stereo planes. The total channel count is given in Table II. In this "design" the density of measurements along the z-axis decreases as $|z|$ increases to minimize the total number of channels. Even with this reduction the total number of channels is very large. Finally, the wire occupancy,[23] shown in Fig. 1, is kept below 10% at SSC luminosities of 10^{33} cm^{-2}sec^{-1}, by starting the active part of the wire at a radius of 50 cm from the beam. This should be possible if the walls of the mylar straws are aluminized only at chamber radii ≥ 50 cm as is sketched in Fig. 2.

No detailed information will be provided for the scintillating fiber tracking chamber; the design is largely conceptual. For modelling the chamber performance we use the parameters of Table I. In this case we assume that the fibers physically occupy only the space between r_{min} and r_{max}. Fibers that would come closer than r_{min} to the beam, if extended to the full width of the chamber, terminate at r_{min}. Thus these fibers are in "two pieces", and two readouts are needed per coordinate measurement. Given the conceptual nature of this design we will not speculate on a channel count. One benefit of the small element size is that occupancies for individual fibers are $\leq 2\%$ at SSC luminosities of 10^{33}

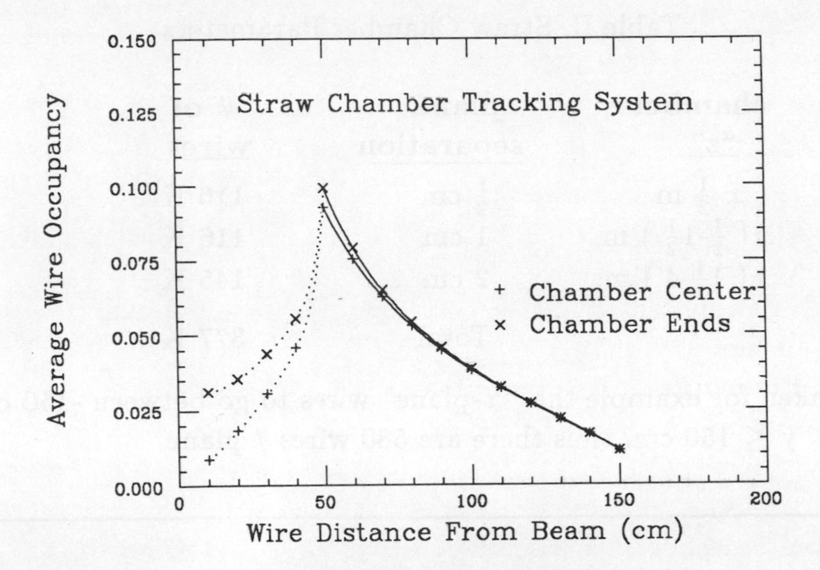

Figure 1: Wire occupancy versus wire closest distance to the beam. For the wires >50 cm from the beam the wires are live along their entire length, solid curves; for wires that approach to closer than 50 cm from the beam the wires are live only along part of their length, dotted curves.

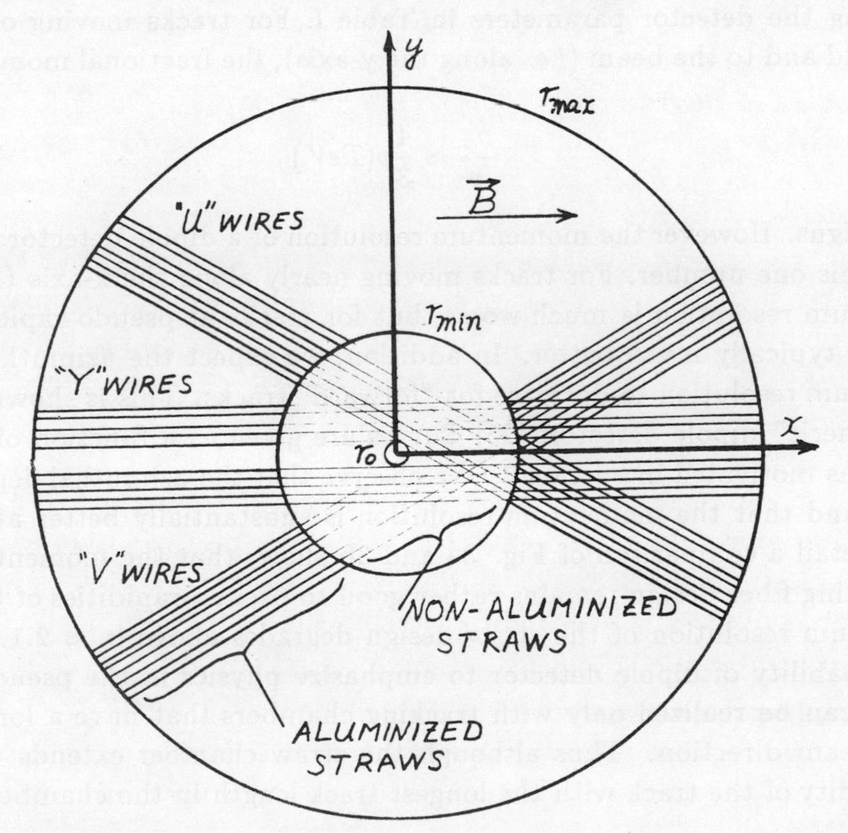

Figure 2: End view of straw chamber dipole tracking system showing orientation of Y U V stereos wires. The closest wire to the beam is at radius $r_0 \approx 5$ cm, and the the live region of the wires is from $r_{min} \leq r \leq r_{max}$, see Table I.

Table II. Straw Chamber Parameters

chamber "z"	plane separation	# of wires
$\pm\frac{1}{2}$ m	$\frac{1}{2}$ cm	116 K
$\pm(\frac{1}{2}$-$1\frac{1}{2})$ m	1 cm	116 K
$\pm(1\frac{1}{2}$-4$)$ m	2 cm	145 K
	Total	377 K

Note: we have taken for example the "Y-plane" wires to go between -150 cm $\leq$ y $\leq$ -5cm and from 5 cm $\leq$ y $\leq$ 150 cm, thus there are 580 wires / plane.

$cm^{-2}sec^{-1}$.

Momentum Resolution

The momentum resolution for the two "generic" dipole tracking systems is estimated [20,24] using the detector parameters in Table I. For tracks moving orthogonal to the magnetic field and to the beam (*i.e.* along the y-axis), the fractional momentum resolution is:

$$\frac{\sigma_p}{p} \approx \frac{4}{3}p(TeV)$$

for both designs. However the momentum resolution of a dipole detector is poorly characterized by this one number. For tracks moving nearly along the x-axis (B-field direction) the momentum resolution is much worse, but for tracks at pseudo-rapidities $|\eta| > 1$ the resolution is typically much better. In addition, we expect the azimuthal dependence of the momentum resolution to decrease for "forward" tracks. This is shown in Fig. 3a,b for the two "generic" dipole detectors.[25] Curves are given as a function of pseudo-rapidity at fixed P_t, as motivated in Section 2. We observe that the azimuthal dependence is small for $|\eta| > 1$, and that the momentum resolution is substantially better at $|\eta| > 1$ than for $\eta \approx 0$. In detail a comparison of Fig. 3a and 3b shows that the momentum resolution of the scintillating fiber design remains rather good to pseudo-rapidities of $|\eta| \approx 2.7$, whereas the momentum resolution of the straw design degrades above $|\eta| \approx 2.1$. This illustrates that the capability of dipole detector to emphasize physics in the pseudo-rapidity range $1 \leq |\eta| \leq 3$ can be realized only with tracking chambers that have a long "aspect ratio" along the beam direction. Thus although the straw chamber extends to larger $|z|$, the pseudo-rapidity of the track with the longest track length in the chamber is:

$$|\eta|^{longest}_{straw\ chamber} \approx 1.7$$

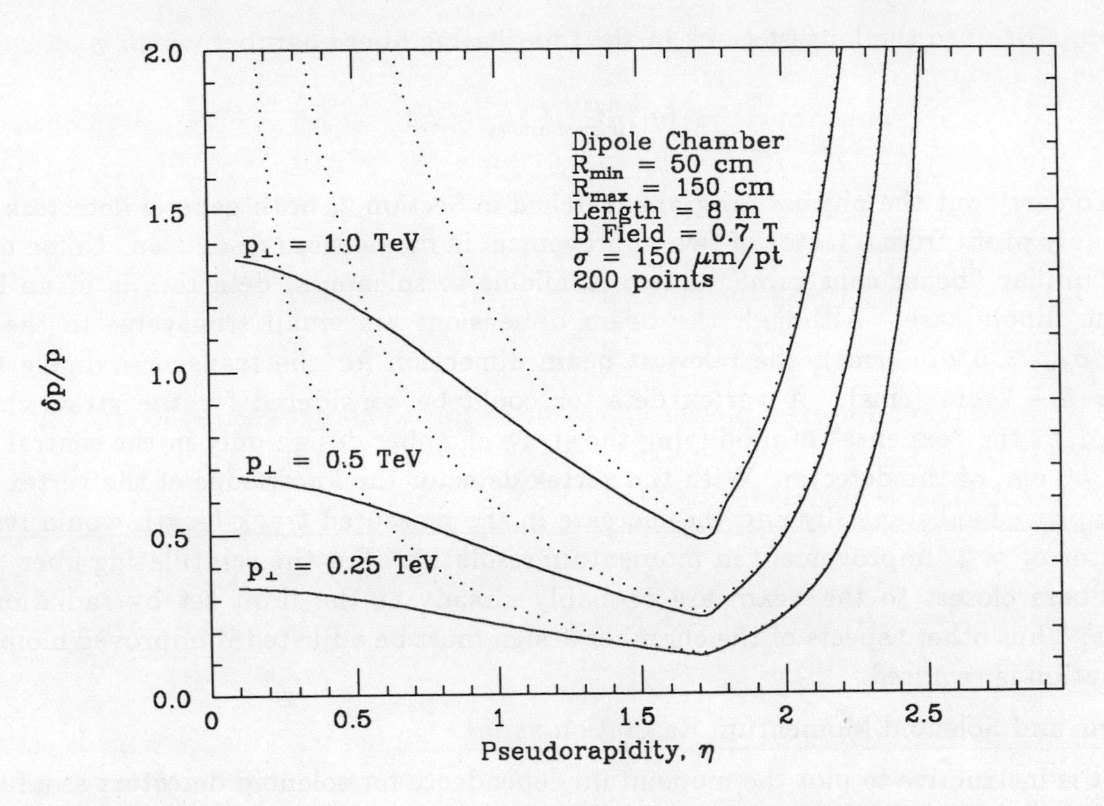

Figure 3a: Momentum resolution of the straw chamber dipole tracking system. Solid lines: track perpendicular to B field; Dotted lines: track parallel to B field.

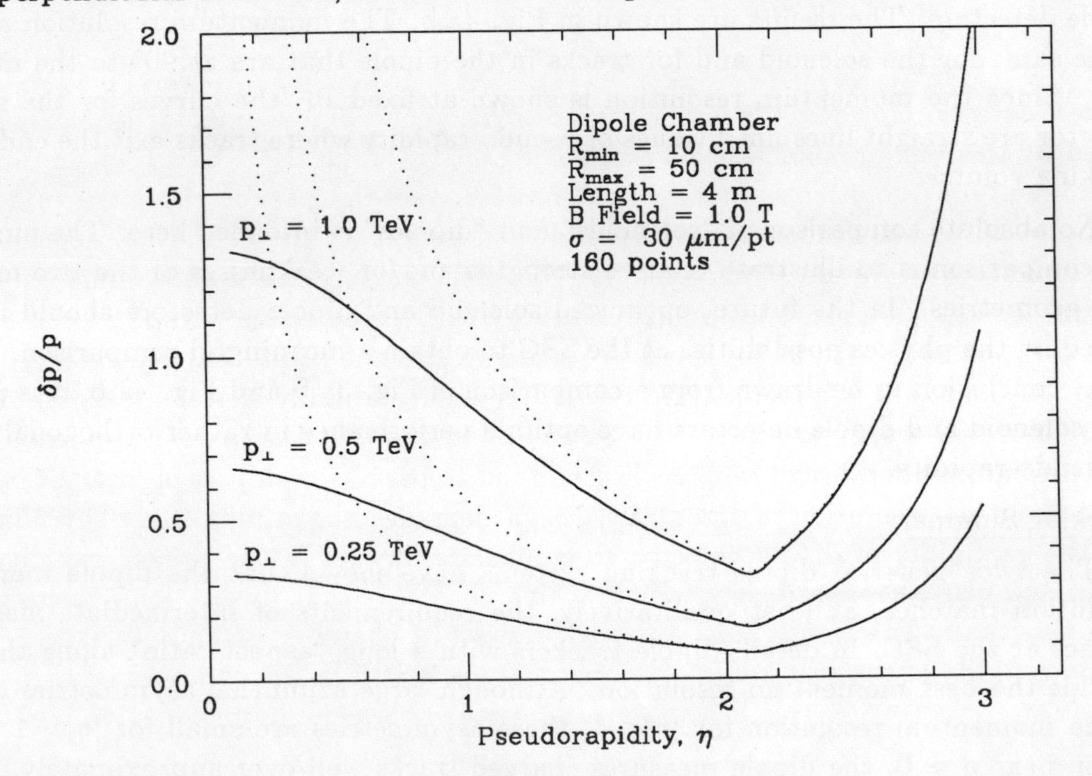

Figure 3b: Momentum resolution of the scintillating fiber tracking system. Solid lines: track perpendicular to B field; Dotted lines: track parallel to B field.

in comparison to the longest track in the scintillating fiber chamber which is at:

$$|\eta|^{longest}_{fiber\ chamber} \approx 2.1$$

To carry out the physics program sketched in Section 2, both generic detectors would probably profit from a factor of two improvement in momentum resolution. Unfortunately the familiar "beam constraint" that is available to solenoidal detectors is of no benefit in the dipole case. Although the beam dimensions are small transverse to the beam axis, $\sigma_{xy} \approx 5\ \mu$m (rms), the relevant beam dimension for the transverse dipole field is $\sigma_z \approx 5 - 7$ cm (rms). A vertex detector could be considered for the straw chamber design, at the "expense" of modifying the straw chamber design only in the central region $|z| < 50$ cm, of the detector. With the vertex detector the knowledge of the vertex would be improved substantially and the increase in the measured track length would result in a factor of ≈ 2 improvement in momentum resolution. For the scintillating fiber design, the fibers closest to the beam are probably already at the limit set by radiation dose levels. Thus other aspects of the chamber design must be adjusted if improved momentum resolution is required.

Dipole and Solenoid Momentum Resolution vs $|\eta|$

It is instructive to plot the momentum dependence for solenoid detectors as a function of $|\eta|$ at fixed P_t as is done above. To simplify the comparison with the dipole study, we use the same detector parameters for the solenoid detectors as given in Table I for the dipole detectors. The results are shown in Fig. 4a,b. The momentum resolution at $\eta \approx 0$ is the same for the solenoid and for tracks in the dipole that are at 90° to the magnetic field. Since the momentum resolution is shown at fixed P_t, the curves for the solenoid detector are straight lines until values of pseudo-rapidity where tracks exit the ends of the tracking volume.

No absolute comparison of "solenoids" and "dipoles" is intended here. The purpose of our comparison is to illustrate relative strengths and/or weaknesses of the two magnetic field geometries. In the future, optimized solenoid and dipole detectors should be challenged by the physics possibilities at the SSC to obtain a meaningful comparison. If there is any conclusion to be drawn from a comparison of Fig. 3a,b and Fig. 4a,b, it is possibly that solenoid and dipole detectors have optimal performance in rather orthogonal regions of pseudo-rapidity.

Tracking Summary

The two "generic" dipole tracking systems have shown that the dipole momentum resolution matches, at least qualitatively, the requirements of intermediate mass scale physics at the SSC. In detail, dipole trackers with a long "aspect ratio" along the beam provide the best momentum resolution. Although large azimuthal asymmetries do exist in the momentum resolution for $|\eta| < 1$, these asymmetries are small for $|\eta| > 1$. In the region near $\eta \approx 0$, the dipole measures charged tracks well over approximately one half the azimuth, and the part of the azimuth that is poorly measured is "backed up" in the full detector by good calorimetry.

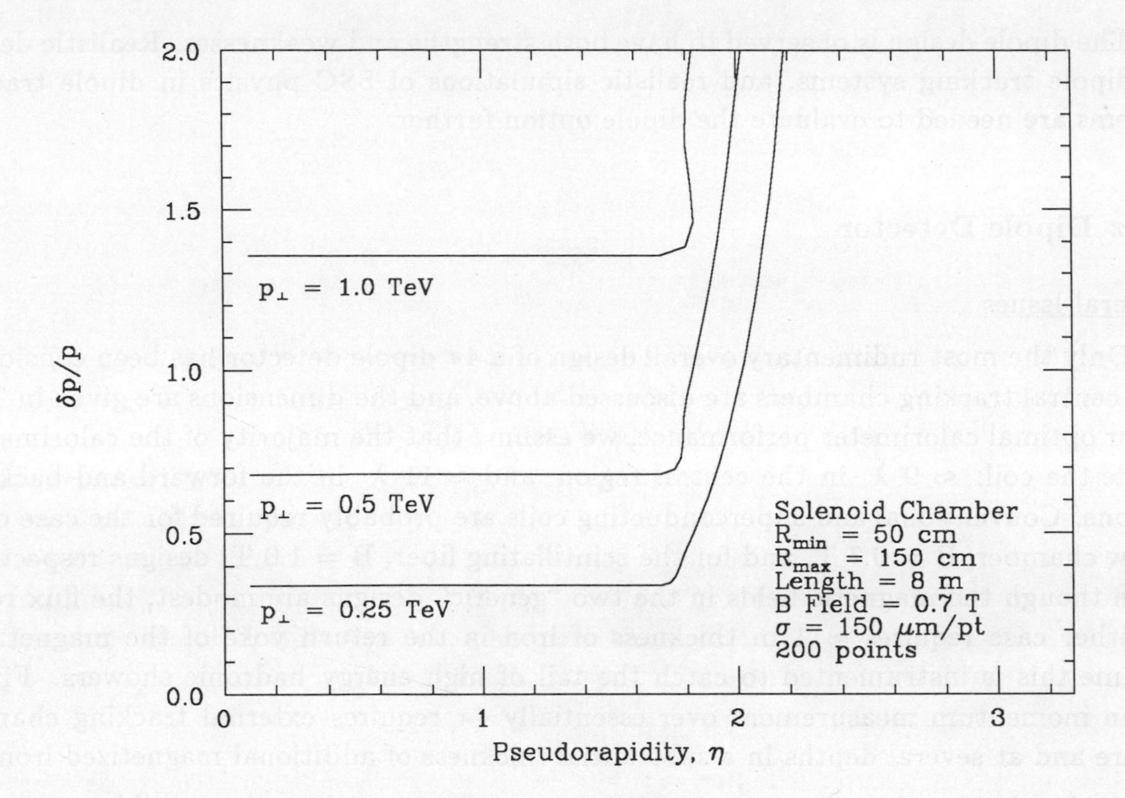

Figure 4a: Momentum resolution of a solenoidal tracking system with the same parameters as the straw chamber dipole system.

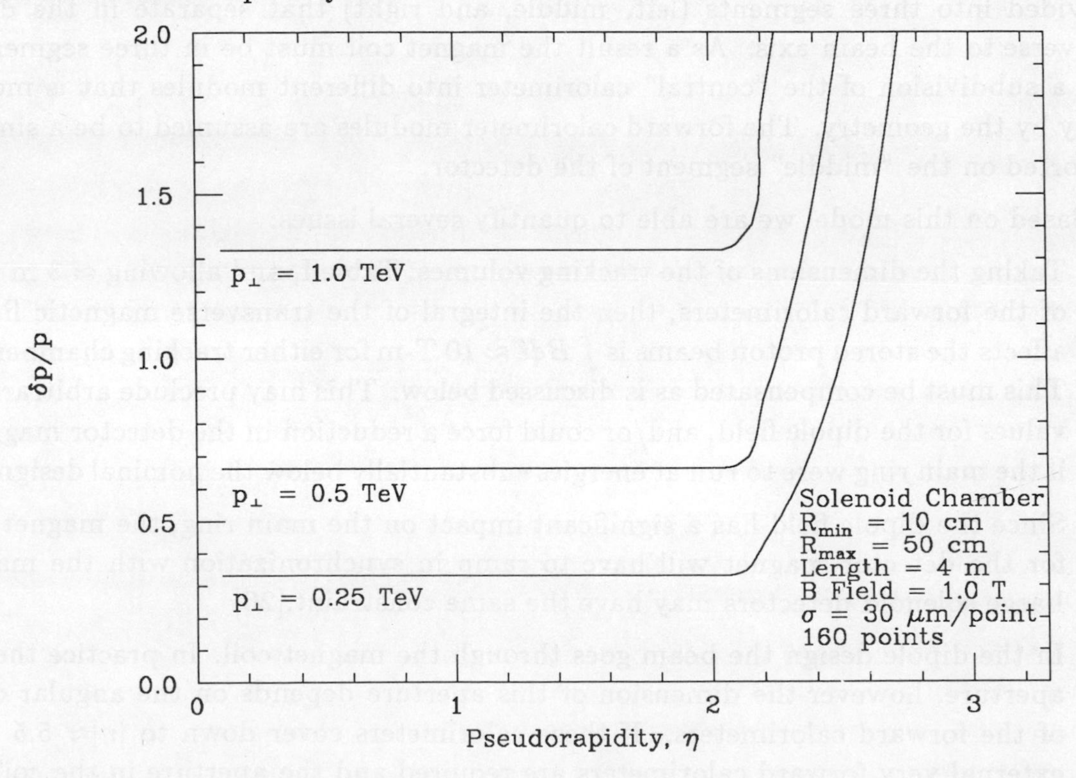

Figure 4b: Momentum resolution of a solenoidal tracking system with the same parameters as the scintillating fiber dipole system.

The dipole design is observed to have both strengths and weaknesses. Realistic designs for dipole tracking systems, and realistic simulations of SSC physics in dipole tracking systems are needed to evaluate the dipole option further.

4. 4π Dipole Detector

General Issues

Only the most rudimentary overall design of a 4π dipole detector has been considered. The central tracking chambers are discussed above, and the dimensions are given in Table I. For optimal calorimeter performance, we assume that the majority of the calorimeter is inside the coil: $\approx 9\ \lambda$ in the central region, and $\approx 11\ \lambda$ in the forward and backward regions. Conventional and superconducting coils are probably required for the case of the straw chamber, B = 0.7 T, and for the scintillating fiber, B = 1.0 T, designs respectively. Even though the magnetic fields in the two "generic" designs are modest, the flux return in either case requires $\approx$ 1 m thickness of iron in the return yoke of the magnet. We assume this is instrumented to catch the tail of high energy hadronic showers. Finally, muon momentum measurement over essentially 4π requires external tracking chambers before and at several depths in a substantial thickness of additional magnetized iron.

A sketch of such a detector, based on using the straw chamber tracking system, is shown in Fig. 5; the external muon system is not shown in this figure. The main detector is divided into three segments (left, middle, and right) that separate in the direction transverse to the beam axis. As a result the magnet coil must be in three segments. We show a subdivision of the "central" calorimeter into different modules that is motivated purely by the geometry. The forward calorimeter modules are assumed to be a single unit supported on the "middle" segment of the detector.

Based on this model we are able to quantify several issues:

1. Taking the dimensions of the tracking volumes, Table I, and allowing $\approx$ 3 m for each of the forward calorimeters, then the integral of the transverse magnetic field that affects the stored proton beams is $\int Bd\ell \approx 10$ T-m for either tracking chamber design. This must be compensated as is discussed below. This may preclude arbitrarily large values for the dipole field, and/or could force a reduction in the detector magnet field if the main ring were to run at energies substantially below the nominal design energy.
2. Since the dipole field has a significant impact on the main ring, the magnet current for the detector magnet will have to ramp in synchronization with the main ring. Large solenoid detectors may have the same constraint.[26]
3. In the dipole design the beam goes through the magnet coil. In practice there is an aperture, however the dimension of this aperture depends on the angular coverage of the forward calorimeters. If these calorimeters cover down to $|\eta| \approx 5.5$ then no external very forward calorimeters are required and the aperture in the coil for the beam can be very small. However, if the forward calorimeters cover only to values of $|\eta| \equiv \eta_{dividing}$ for example, then the aperture in the magnet coil should be adequate

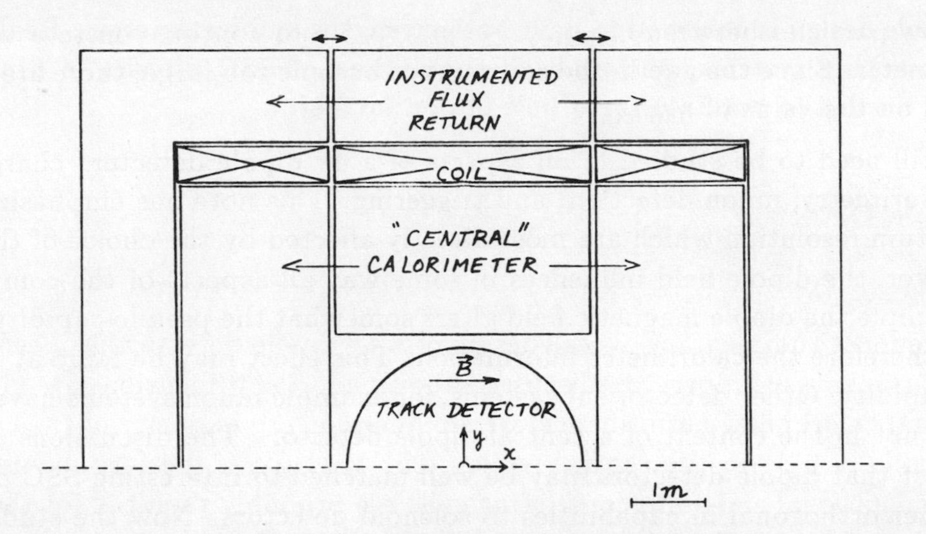

Figure 5a: "Half section" through the center of a schematic 4π dipole detector showing an end view of the tracking detector, calorimetry, magnetic coil, and instrumented flux return.

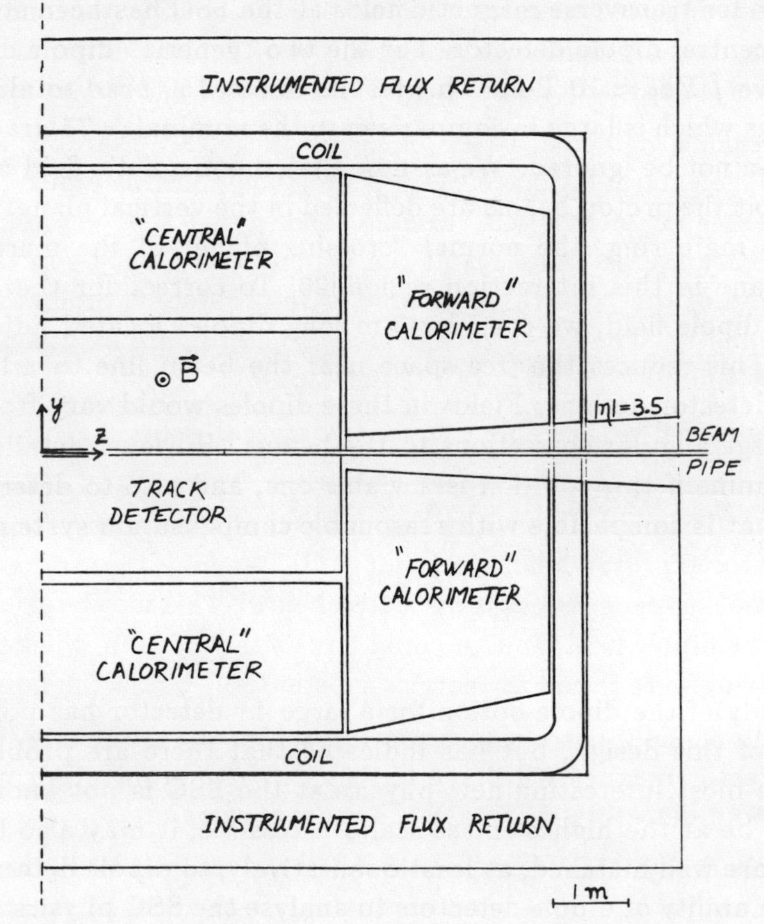

Figure 5b: "Half section" through the center of a schematic detector showing a side view of the detector elements.

to allow particles between $\eta_{dividing} \leq |\eta| \leq 5.5$ to continue on to a very forward calorimeter. Since the gap in the coil can not become too large, there are constraints placed on the value of $\eta_{dividing}$.

Issues still need to be studied in all aspects of a 4π dipole detector: charged particle tracking, calorimetry, muon detection, and triggering. This note has emphasized tracking and momentum resolution which are most directly affected by the choice of the magnetic field. However, the dipole field influences in some way all aspects of the complete detector. For example the dipole magnetic field alters somewhat the pseudo-rapidity of charged tracks and therefore the calorimeter information. This effect may be large at large values of pseudo-rapidity. Other detector subsystems, for example muon systems have been studied,[27] but not in the context of a central dipole detector. The discussions of Section 2 and 3 suggest that dipole detectors may be well matched to interesting SSC physics, and may be rather orthogonal in capabilities to solenoid detectors. Now the studies must be done to test this in detail.

Dipole Field Compensation

Compensation for transverse magnetic fields at the SSC has been studied,[26] but not for the case of a central dipole detector. For the two "generic" dipole detectors discussed above, the effective $\int Bd\ell \approx 10$ T-m. This results in a ≈ 150 μrad total angular deflection for 20 TeV protons which is large in comparison to the nominal ≈ 75 μrad crossing angle of the beams and can not be ignored. We assume that the magnetic field is in the horizontal plane and thus that the proton beams are deflected in the vertical plane. Then to minimize the effect on the main ring, the normal "crossing plane" of the machine should be in the horizontal plane in this interaction region.[26] To correct for the angular deflection imparted by the dipole field, we consider 5-m long dipoles located adjacent to the high-β quadrupoles. This reduces the free space near the beam line to ± 15 m, which could impact on some detector designs. Fields in these dipoles would vary from ≈ 0.5 to 2 T to provide rather large angular corrections to the beams. However detailed simulations are required to determine if this solution is a viable one, and also to determine the range of detector $\int Bd\ell$ that is compatible with reasonable compensation systems.

5. Summary

An initial study of the dipole option for a large 4π detector has pointed to a number of the strengths of this design, but has indicated that there are problems to be solved. What will be the most interesting new physics at the SSC is not known. Although the new physics may be at the highest mass scales accessible, it may also be at intermediate mass scales that are well matched, at least qualitatively, to dipole detectors. Quantitative evaluations of the ability of dipole detectors to analyse the SSC physics must still be done. Initial indications are that the dipole detector and the solenoid detector have optimal performance in rather complimentary regions of rapidity. However more studies must be done to determine if one dipole and one solenoid detector form the optimal pair of detectors for high-P_t physics at the SSC.

References

1. H. H. Williams, Proceedings of the 1986 Summer Study on the Physics of the Superconducting Supercollider, Eds. R. Donaldson and J. Marx, 327 (1986)

2. G. J. Feldman, M. G. D. Gilchreise, J. Kirkby *et al.*, Proceedings of the 1984 Summer Study on the Design and Utilization of the Superconducting Super Collider, Eds. R. Donaldson and J. Morfin, 623 (1984)

3. I. Hinchcliffe and P. Oddone, these proceedings

4. The "high P_t" physics topics included the discovery and study of: Heavy Higgs, Light Higgs, Non-Standard Higgs, Heavy Quarks and Leptons, SUSY, Jets and Compositeness, and New W′ or Z′ Gauge Bosons.

5. G. Alverson *et al.*, Proceedings of the 1986 Summer Study on the Physics of the Superconducting Supercollider, Eds. R. Donaldson and J. Marx, 114 (1986)

6. E. Eichten, I. Hinchliffe, K. Lane, and C. Quigg, *Rev. Mod. Phys.* **56**, 579 (1984)

7. W. J. Stirling, R. Kleiss, and S. D. Ellis, *Phys. Lett.* **163B**, 261 (1985)

8. M. Chanowitz, M. Gilchreise, J. Gunion, and E. Wang, these proceedings

9. H. E. Haber, Proceedings of the 1984 Summer Study on the Design and Utilization of the Superconducting Super Collider, Eds. R. Donaldson and J. Morfin, 125 (1984)

10. J. F. Gunion, Proceedings of the 1984 Summer Study on the Design and Utilization of the Superconducting Super Collider, Eds. R. Donaldson and J. Morfin, 147 (1984)

11. N. G. Deshpande, J. F. Gunion, and H. E. Haber, Proceedings of the 1984 Summer Study on the Design and Utilization of the Superconducting Super Collider, Eds. R. Donaldson and J. Morfin, 119 (1984);
J. L. Rosner, P. Langacker, and R. W. Robinett, $\bar{p}$ p Options for the Supercollider, Eds. J. E. Pilcher and A. R. White, 202 (1984);
J. L. Rosner, Proceedings of the 1986 Summer Study on the Physics of the Superconducting Supercollider, Eds. R. Donaldson and J. Marx, 213 (1986);
B. Adeva *et al.*, ibid, 257;
N. Deshpande, and S. Whitaker, these proceedings

12. We note that for large mass W′and Z′ there is a correlation between $|\eta|$ and the maximum transverse momentum of the decay leptons; see for example ref. 10. However this correlation is fairly weak for the intermediate mass scales accessible in asymmetry measurements.

13. M. Kuroda *et al.*, Supercollider Physics, Ed. D. E. Soper, 294 (1985)

14. J. F. Gunion *et al.*, Proceedings of the 1986 Summer Study on the Physics of the Superconducting Supercollider, Eds. R. Donaldson and J. Marx, 142 (1986)

15. Z. Kunszt *et al.*, Proceedings of the Workshop on Physics of Future Accelerators, 123, CERN report: CERN 87-07 Vol I, (1987)

16. S. D. Ellis, R. Kleiss, and W. J. Stirling, *Phys. Lett.* **154B**, 435 (1985)

17. C. Quigg, private communication

18. For good charge identification we require that the fractional momentum uncertainty $\sigma_p/p \leq 1/3$.

20. D. G. Cassel, G. G. Hanson *et al.*, Proceedings of the 1986 Summer Study on the Physics of the Superconducting Supercollider, Eds. R. Donaldson and J. Marx, 377 (1986)
21. Note that at the SSC the beam "diamond" has an extent along the beam axis of approximately ±15 cm.
22. S. Reucroft *et al.*, Plastic Scintillating Fibers, Talk at the Workshop on Experiments, Detectors and Experimental Areas for the Superconducting Super Collider, Berkeley, July 7-17, 1987
23. To estimate the wire occupancy we use the following: 1.6 interactions / crossing at luminosities of 10^{33} cm^{-2}sec^{-1}, a maximum drift chamber drift time of 50 nsec which means that the chambers are sensitive to ≈ 3.2 crossings, [20] and an average charged multiplicity of 6 / unit of rapidity. No factor(s) for curving tracks, or for charged secondaries have been included.
24. R. L. Gluckstern, *Nucl. Instrum. Methods* **24**, 381 (1963)
25. For this study the effective track length in the magnetic field varies with track direction to reflect the geometrical constraints of the tracking volumes, however the number of measurements on the track is taken to be a constant.
26. S.Peggs, Proceedings of the 1986 Summer Study on the Physics of the Superconducting Supercollider, Eds. R. Donaldson and J. Marx, 540 (1986)
27. D. Carlsmith *et al.*, Proceedings of the 1986 Summer Study on the Physics of the Superconducting Supercollider, Eds. R. Donaldson and J. Marx, 405 (1986)

MUON SPECTROMETERS
REPORT OF THE DETECTOR SUBGROUP

U. Becker, D. Bintinger, B. Blumenfeld, C.Y. Chang,
M. Chen, C.Y. Chien, P. Duinker, J.J. Eastman, K. Freudenreich,
M. Fukushima, G. Herten, H. Hofer, V. Innocente, P. LeComte,
R. Leiste, S.C. Loken, K. Luebelsmeyer, T. Matsuda, E. Nagy, H. Newman,
P. Piroue, J. A. Rubio, P. Seiler, P. Spillantini, R. Thun,
S.C.C. Ting, W. Wallraff, R. Yamamoto, C.P. Yuan and R.Y. Zhu

I. Introduction

This report summarizes the work of the Muon Spectrometer Group. The detection of leptons has in the past been a powerful tool for identifying new particles in hadronic interactions. For example, the J particle was observed in the di-electron spectra, the Υ in the dimuon spectra and the Z° was discovered simultaneously in electron and muon decay channels. The detection of leptons has also been a very useful tag of the weak decays of heavy fermions. The recent measurement by the UA1 Group of like-sign muon pairs gave the first indication of large mixing for B-mesons and is another example of using muons for the study of heavy fermions. Looking forward to the SSC physics program, we can already identify a number of areas where muon detection will again play an important role in understanding various physical processes. For example, the measurement of multi-gauge boson production is of fundamental interest and Z° bosons will be readily identified in the dimuon decay channel. A more speculative possibility is the existence of new, very heavy bosons which may be discovered by their decay into muon pairs. Even if some possible new signals are also observable in electron channels, it will be very important to check lepton universality by measuring the corresponding muon channels.

The detection of muons in SSC experiments provides a number of unique capabilities. Perhaps the most important of these is the potential for measurements at the highest conceivable luminosities. Muons can penetrate thick absorbers which shield the active detector elements from the tremendous flux of secondary particles. For those experiments with tracking and calorimetric detectors inside such an absorber, muon detection may provide the only reliable means of tagging leptons within dense particle jets. Finally, precise measurements of muon trajectories and momenta are useful in calibrating calorimeters *in situ* as has been demonstrated in a number of experiments including L3 at CERN.

In this report we examine in detail two possible detectors which emphasize muon measurements:

1. A "muons-only" spectrometer designed specifically for operation at the highest SSC luminosities.

2. A precision lepton-photon detector which utilizes a high-resolution muon spectrometer for $\theta > 25^o$ with a 4π solid angle high resolution calorimeter consisting of liquid xenon and TMS-uranium inside for precision measurement of electrons, photons and hadronic jets as well.

We note that other large general-purpose detectors described elsewhere in these proceedings also include muon detection capabilities. In those detectors the main emphasis is, however, on achieving excellent hadron calorimetry and the designs are not directed toward optimal muon detection.

This report continues with some general remarks on possible muon detector geometries and discuss the challenges and problems of muon measurements. It then describes the two detectors mentioned above. The conclusion gives some brief remarks about these detectors in the context of an overall SSC experimental program and enumerates topics for future studies and detector development.

II. Geometries for muon spectrometers

Previous workshops have investigated a broad range of possible detectors for high energy muons. These include non-magnetic detectors, magnetized iron spectrometers, and air-core spectrometers.

Non-magnetic techniques have been investigated at the LHC workshop [II-1]. They concluded that it is possible to determine muon momenta by measuring transition radiation or muon energy loss in dense material. In either case, the detectors are larger and more costly than conventional spectrometers and the precision is not competitive with magnetic measurement.

Iron spectrometers have been discussed extensively either as a dedicated muon spectrometer or as part of a general purpose detector. Both toroidal and solenoidal geometries have been considered for the central region. For the forward direction, the iron toroid seems to be the only possibility. In the next section, we consider some of the technical problems associated with the iron spectrometers. These include muon energy loss in the iron and difficulties of detector alignment.

Air core spectrometers can achieve excellent momentum resolution. The traditional choice has been to use a solenoid with a very long path length. A superconducting toroid is also a possible solution and has been considered by L.W. Jones elsewhere in these proceedings.

III. Challenges and problems

Solid iron spectrometers offer an attractive possibility for providing reasonably high magnetic field in a large volume for low cost. Such devices have been used

[II-1] C. Gößling and Č. Zupančič, *Unconventional Methods for Muon Momentum Measurements*, In M. Jacob, editor, *Large Hadron Collider in the LEP Tunnel: Proceedings of the ECFA-CERN Workshop*, page 223, 1984.

in fixed target experiments [III-1] and colliding beam experiments [III-2]. At SSC energies, however, it becomes increasingly more difficult to use iron spectrometers.

At very high energy, the energy loss of muons is dominated by radiative processes. These processes have two effects in a magnetized iron muon spectrometer. They introduce a significant low-energy tail on the muon momentum measurement and the electromagnetic shower resulting from the radiated photon or electron pair causes many additional hits in the chamber planes of the spectrometer.

The effect of the electromagnetic showers in iron has been estimated by Eastman and Loken. The Monte Carlo simulation uses an accurate model for all energy loss mechanisms and uses the EGS program to model the development of the resulting shower. The details of the calculation are described elsewhere in these proceedings.

The additional hits in each detector layer will make it extremely difficult to track muons through an iron spectrometer. The authors have assumed that a chamber plane will not contribute to a track measurement if there is more than a single hit in the chamber. Figure III-1 shows the probability of detecting only the muon in a single chamber plane downstream of a meter of iron. The probability of finding a clean chamber hit decreases significantly at high momentum. It is clear that multiple detector planes separated by magnetic field or thick absorbers will be necessary to track high energy muons with acceptable efficiency. A similar analysis has been done for muons in uranium where the loss of chamber information is even more severe (Fig. III-1).

At high energy, the radiative processes also contribute to the angular deviations of muons in the spectrometer. These deviations add a significant, non-gaussian tail to the scattering distribution and will cause some difficulty in fitting tracks. These effects have not yet been estimated quantitatively.

At very high energy, the problems of chamber alignment become more severe. In a large solid angle iron spectrometer, the alignment of hundreds of chambers separated by heavy iron yokes is very difficult and considerable effort will be necessary to achieve the necessary precision. These alignment problems can be alleviated using air magnets. The proponents of L3+1 have developed extensive expertise in aligning large chambers to very high precision in an air magnet. Their techniques use lasers to provide relative alignment between chamber layers.

All of these considerations have led the muon subgroup to design spectrometers which measure muon momenta in an air gap in the central region. We turn now to the description of two specific detectors which were discussed at the workshop.

[III-1] H.Y. Frisch *et al.*, Phys. Rev. D25(1982)2000; R. Kopp *et al.* (the BCDMS Collaboration). Z. Phys. C28, 171 (1985); P.D. Meyers *et al.* Phys. Rev. D34, 1265 (1986)

[III-2] ISR-R209, D. Antreasyan *et al.*, Phys. Rev. Lett. 45 (1980)863

IV. Dedicated Muon Spectrometer

A large detector which specializes in the measurement of muons is described by R. Thun *et al.* elsewhere in these proceedings. The design is shown in Fig. IV-1 and consists of an iron-toroid system at forward angles and a superconducting solenoid in the central region. The basic philosophy behind this design is to obtain a detector that can pursue a limited but very important physics program at the highest conceivable SSC luminosities with special emphasis on Higgs detection. For this reason priority in the detector design is given to the suppression of hadronic punch-thru, decay muons and low-pt prompt muons. This is achieved with a very thick, dense, close-in absorber. An equally high priority has been given to achieving a good muon trigger through the implementation of a highly segmented scintillator system and by demanding reasonably small cells in the wire tracking chambers. The detector design also emphasizes a broad rapidity coverage ($-3 \leq Y \leq 3$) and redundancy in the number of measuring planes to avoid problems from electromagnetic showers from bremsstrahlung and pair production. The momentum resolution that is obtained with such a detector is of order 10%. The only concession to particles other than muons is a rudimentary calorimeter in the absorber consisting of a small number of layers of proportional wires run at very low gain in order to survive the rates and radiation levels at high luminosity. The sole purpose of such a calorimeter is to tag the presence of energetic hadronic jets and thereby determine the degree of isolation of the muons. The main goals of such a "muons-only" detector are measurements of single and multiple Z^0 production and searches for heavy higgs and other new gauge bosons. The total cost of the detector is at the $100 million level. The answer to the question of whether a "muons-only" detector will produce physics not easily accessible to the other detectors depends on a number of factors. The most important of these are the luminosity to be actually provided by SSC and the ability of the "open" calorimetric detectors to handle particle rates and general radiation levels. A careful analysis of these factors has not been made. Experience at the ISR indicated that experiments could operate there at luminosities of 10^{32} cm^{-2} sec^{-1} while experience at the higher-energy SPS and TeV I proton-antiproton colliders has already shown some rate and radiation problems at luminosities below 10^{30} cm^{-2} sec^{-1}. The argument has been made that the calorimetric detectors will measure electrons as well as muons and therefore require a factor of two to four less luminosity for multi-lepton signals. However, high-quality electron detection demands good tracking for interaction point and momentum measurements. Charge identification requires, in addition, a fairly long track length in a magnetic field. It has not yet been demonstrated that the general-purpose detectors with these capabilities can really be operated at luminosities much above 10^{32} cm^{-2} sec^{-1}.

While the "muons-only" detector can probably withstand particle rates for luminosities up to 10^{34} cm^{-2} sec^{-1}, it has not yet been shown convincingly that reasonably low muon trigger rates can be achieved at very high luminosities. In this regard the calorimetric detectors have an advantage since a large energy deposit

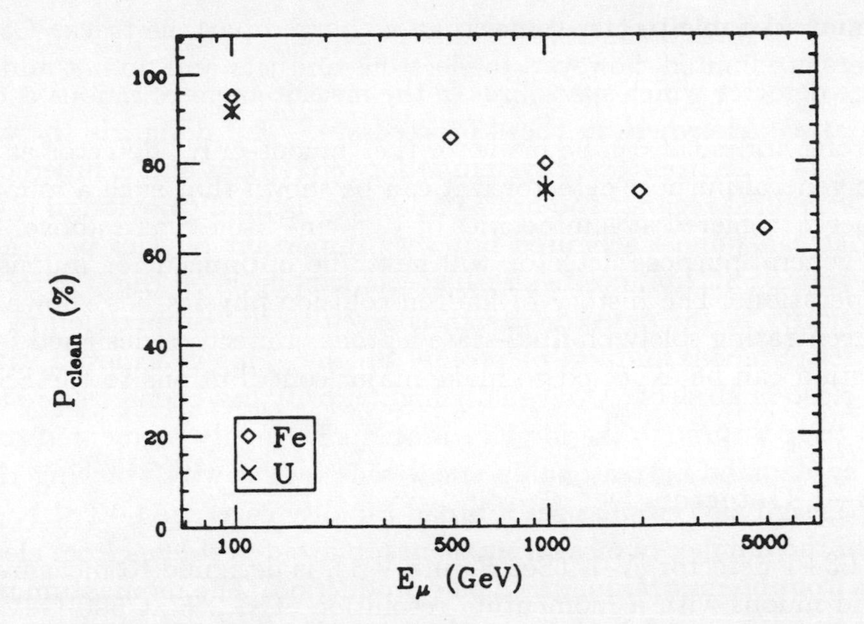

Fig. III-1 The probability of a clean hit (i.e. a muon track unaccompanied by any other charged particle) as a function of muon energy E_μ in a chamber downstream of 1 meter of iron. Also included are two points for a chamber downstream of 90 cm of uranium.

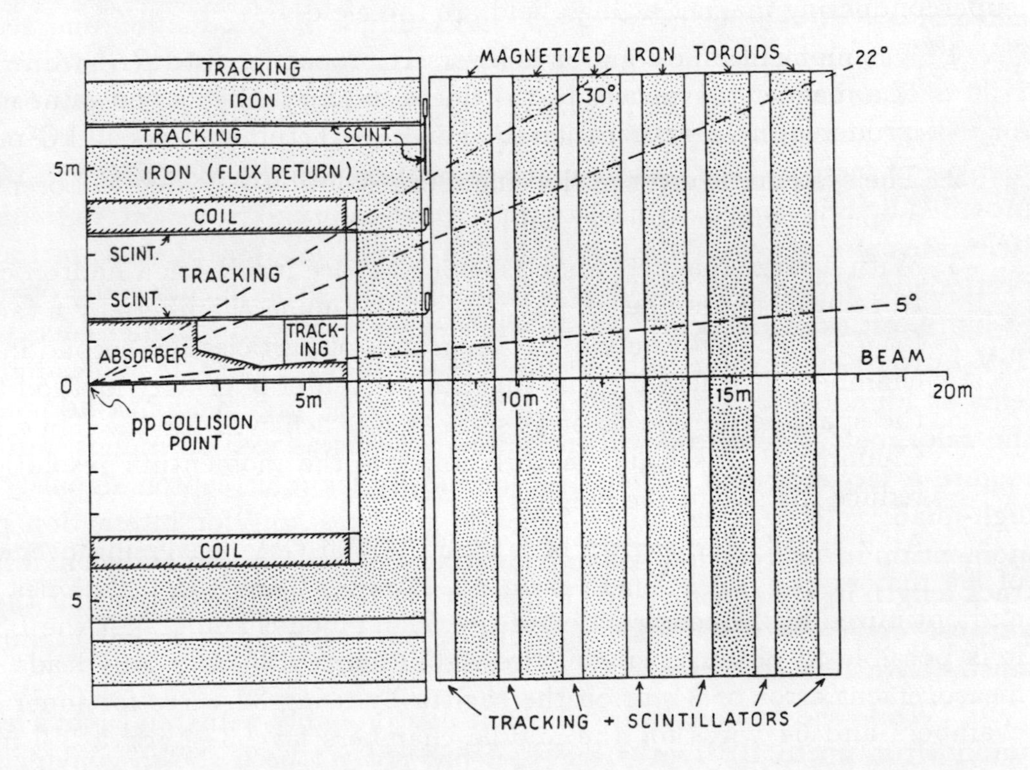

Fig. IV-1 A large detector, which specializes in the measurement of muons only, consists of an iron-toroid system at forward angles and a superconducting solenoid in the central region.

provides a more reliable trigger signal than a charged-particle track. Calorimeter-based triggers are limited, however, to electrons and jets and do not address muon physics directly.

A strong argument can be made for the "muons-only" detector as a complement to the general-purpose detector if it can be shown that such a muon detector can be properly triggered at luminosities of 10^{33} cm^{-2} sec^{-1} and above. The point is that the general-purpose detector will never be optimized for muons either in design or operation. The history of hadron-collision physics has shown the great value of concentrating solely on final-state leptons. Detectors designed for optimal lepton detection can be expected to make major contributions to the SSC physics program.

V. The L3+1 Detector

The L3+1 detector [V-1] (See Figure V-1), is designed to measure photons, electrons and muons with a momentum resolution of $< 2\%$ at 500 GeV. It is the result of many years of study on how to construct the most accurate lepton detectors based on the experience of constructing the L3 detector. The L3+1 muon detector has an air gap spectrometer in the central region and an iron toroid in the forward and backward region.

The reasons for using a large air magnet at low field rather than a smaller superconducting magnet at high field are threefold:

1. The momentum resolution is inversely proportional to BL^2 while the energy of a magnet is typically proportional to $(LB)^2$. Thus one gains much faster by increasing L rather than B. For L3 + 1, the BL^2 is 540 kG-m^2.
2. The systematic error in the chamber resolution, e.g. due to Lorentz bending angle, is smaller at lower field.
3. With a large magnet, there is enough space inside for a hadron calorimeter to reduce the number of particles at the muon chambers by a factor of 10^5 for $\theta > 30^o$. With only a few thousands of particles per second in the muon chambers, it is feasible that one can use center of gravity method to improve the space resolution to 40 μ such as was achieved for the Time Expansion Chamber. This could in turn improve the momentum resolution and/or reduce the size of the magnet.

At $\theta = 90^o$, the sagitta of a muon of 500 GeV traversing over a distance of 5.5 meters in a 7.5 kG field is only 1.72 mm. Using the techniques developed and optimized in the construction of precision muon chambers for the L3 detector, it is possible to obtain a single wire resolution of 136μm. This leads to a total measurement error of 34μm on the sagitta by using 32 wires for inner and outer chambers and 64 wires for the middle chambers for L3+1, or a 2% momentum

---

[V-1] U. Becker *et al.*, Nucl. Inst. and Meth. A253(86)15.

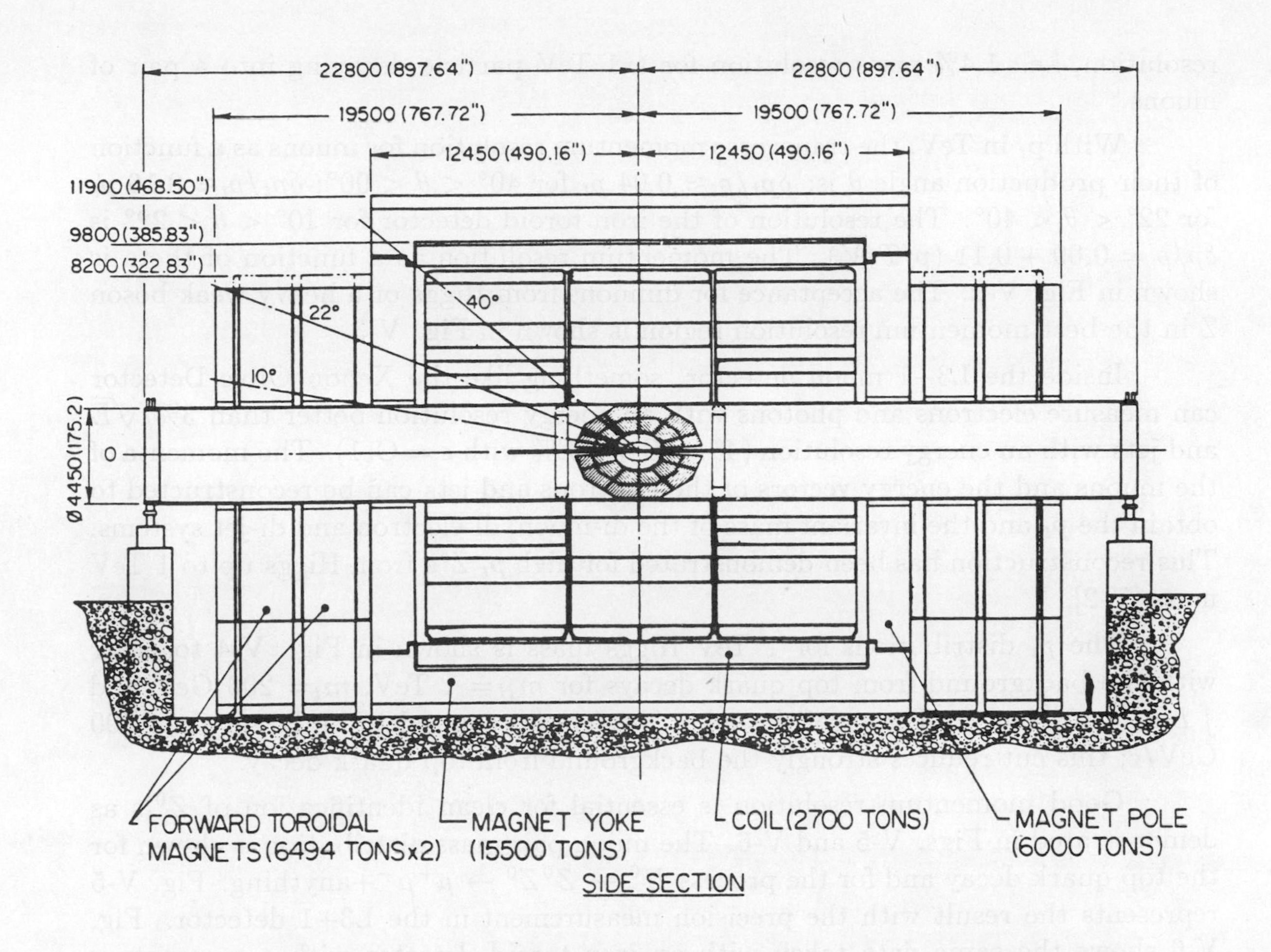

Figs. V-1 The side view of the proposed L3+1 detector showing the precision muon detector at $\theta > 22^o$ and the endcap toroidal magnets. It is designed to measure photons, electrons and muons with a momentum resolution of $< 2\%$ at 500 GeV. The calorimeter illustrated here is the Xenon Olive and TMS-U detector.

resolution, i.e. 1.4% mass resolution for a 1 TeV particle decaying into a pair of muons.

With p_t in TeV, the transverse momentum resolution for muons as a function of their production angle θ is: $\delta p_t/p_t \approx 0.04\ p_t$ for $40° < \theta < 90°$; $\delta p_t/p_t \approx 0.16\ p_t$ for $22° < \theta < 40°$. The resolution of the iron toroid detector for $10° < \theta < 22°$ is $\delta p/p = 0.09 + 0.11\ (p/TeV)$. The momentum resolution as a function of theta is shown in Fig. V-2. The acceptance for dimuons from Higgs or a heavy weak boson Z in the best momentum resolution region is shown in Fig. V-3.

Inside the L3+1 muon detector, something like the Xenon Olive Detector can measure electrons and photons with an energy resolution better than $3\%/\sqrt{E}$ and jets with an energy resolution $(45/\sqrt{E} + \epsilon)$E% with $\epsilon = O(1)$. The momenta of the muons and the energy vectors of the electrons and jets can be reconstructed to obtain the p_t and the invariant mass of the di-muon, di-electron and di-jet systems. This reconstruction has been demonstrated for high p_t Z^os from Higgs up to 1 TeV mass [V-2].

The p_t distributions for 1 TeV Higgs mass is shown in Fig. V-4 together with the background from top quark decays for m_H= 1 TeV, m_t= 200 GeV and $\int Ldt = 10^{40} cm^{-2}$. The p_t of the $\mu^+\mu^-$ system is required to be larger than 300 GeV/c; this cut reduces strongly the background from top quark decay.

Good momentum resolution is essential for clean identification of Z^0's as demonstrated in Figs. V-5 and V-6. The muon pair mass distribution is shown for the top quark decay and for the process $H^0 \to Z^0 Z^0 \to \mu^+\mu^-$+anything. Fig. V-5 represents the result with the precision measurement in the L3+1 detector. Fig. V-6 shows the same data taken with an iron toroid detector with a momentum resolution of $\delta p/p = 0.09 + 0.11$ (p/TeV). With high resolution as in the L3+1 detector a clear Z^0 signal from the Higgs boson decay is detected, whereas only a shoulder is vaguely visible for an iron toroid detector.

A detailed simulation of high energy muons going through an air-gap magnetic spectrometer has been carried out by Innocente (contribution to this workshop), who showed another advantage of an air gap magnet over magnetized iron, *i.e.* the contamination problem due to electromagnetic showers associated with high energy muons is negligible for air magnets owing to the presence of the magnetic field in the region of the chambers. This is in sharp contrast with the case of magnetized iron detectors where about 20% of the hits are ambiguous because of accompanying showers along with the muons. These accompanying showers not only spoil the muon momentum resolution, but can also artificially produce a significant tail of fake high momentum muons.

VI. Xenon Olive Detector for Electrons and Jets

[V-2] M. Chen *et al.*, MITLNS 162 and contribution to this workshop.

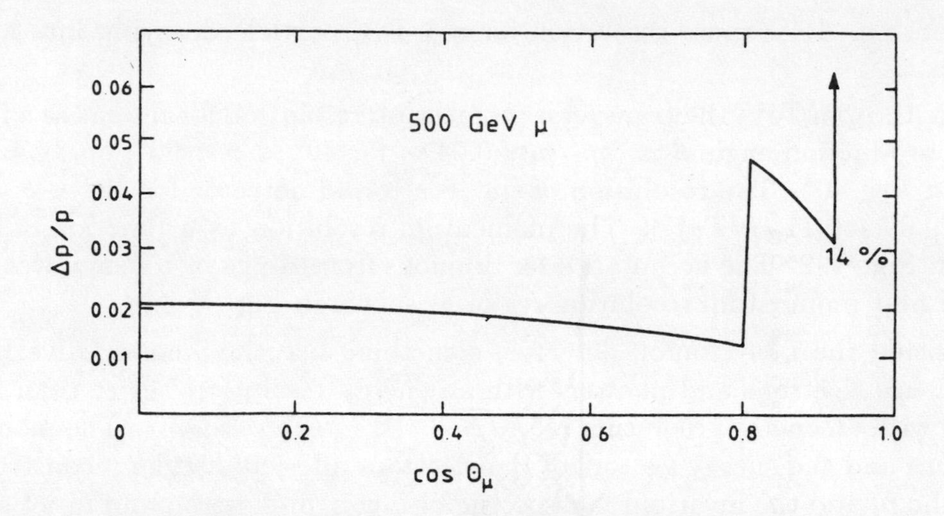

Fig. V-2 The muon momentum resolution of the L3+1 detector as a function of the production angle.

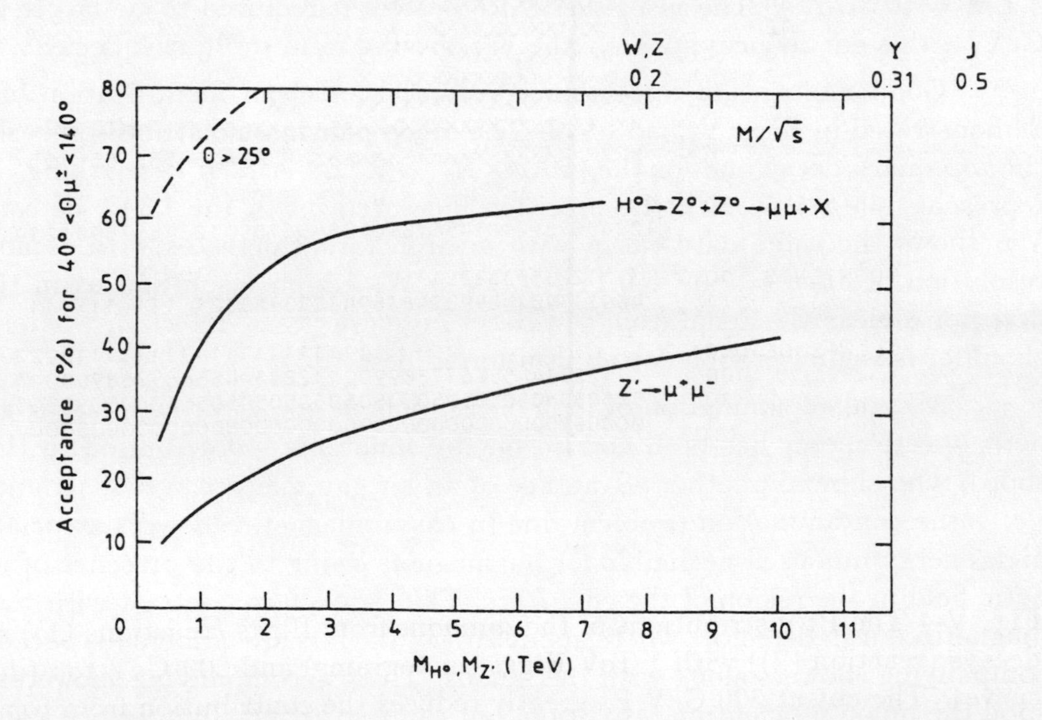

Fig. V-3 The di-muon acceptance in the good momentum resolution regions, i.e. $\theta > 40^o$ (solid) and $\theta > 25^o$ (dashed) of the L3+1 detector for the Higgs and for a new heavy Z'. The top scale shows the scaling variable x = the ratio of the mass of H^o or Z' to the cm energy of the machine. The J particle, the Υ and the W and Z were originally discovered at x= 0.5, 0.31 and 0.2 respectively, which illustrates the mass region SSC should in principle be sensitive to (with good ideas, detectors and unlimited luminosity etc.).

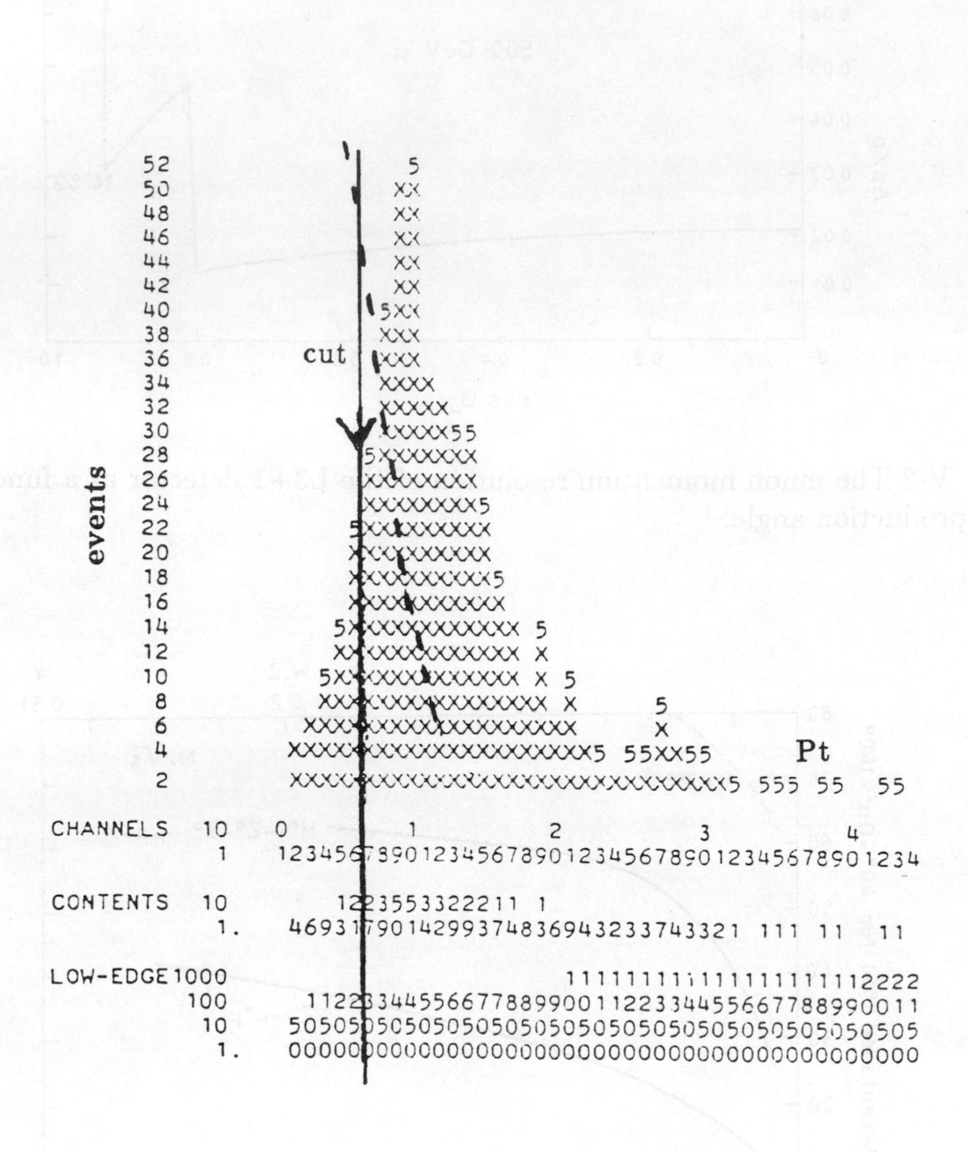

Fig. V-4 The Pt distributions of the dimuons from Higgs (reactions (1)) and top decays (reaction (2)) with 1 TeV Higgs (histograms) and 200 GeV top (dashed curve). The cut at 300 GeV Pt greatly reduces the contribution from top relative to the Higgs.

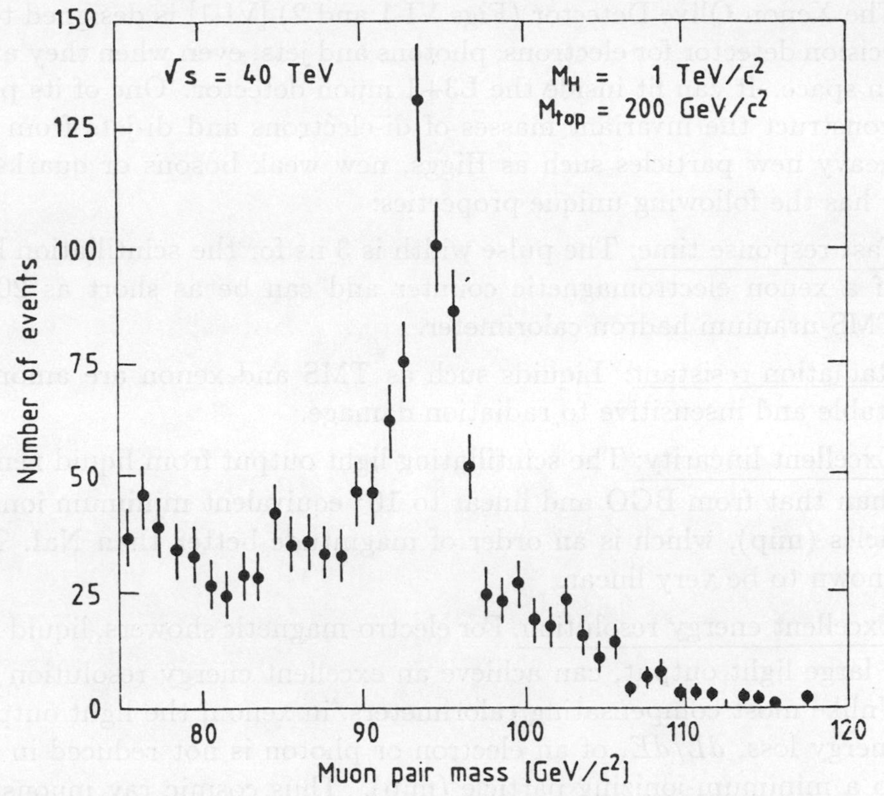

Fig. V-5 The reconstructed invariant mass distributions of the dimuons from reactions (1) and (2) with 1 TeV Higgs and 200 GeV top for the L3+1 detector. The Higgs signal is clearly visible above the heavy quark background.

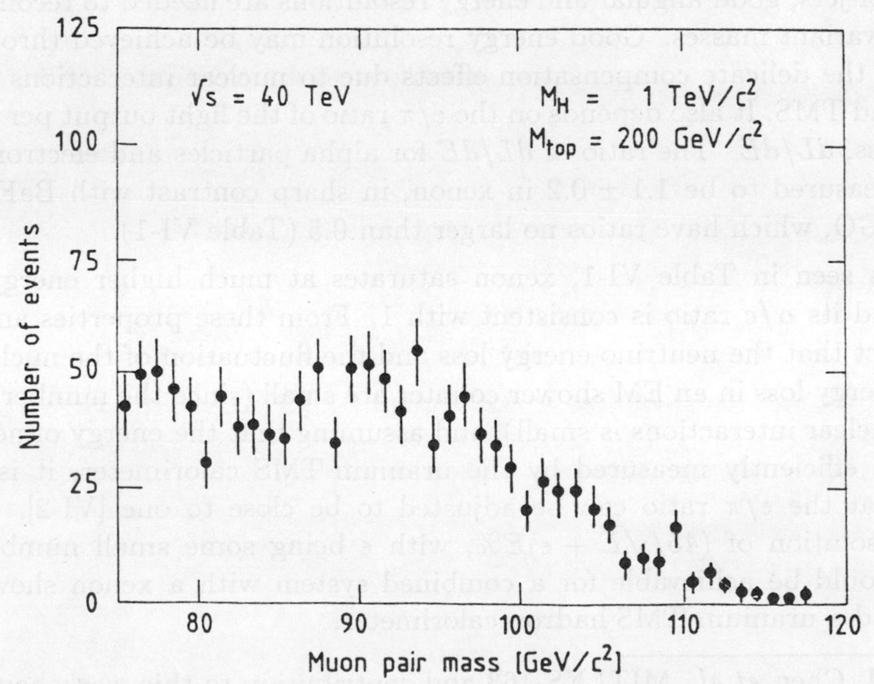

Fig. V-6 The reconstructed invariant masses distributions of the dimuons from reactions (1) and (2), same as in Fig. V-6, but with the momentum resolution of an iron toroid.

The Xenon Olive Detector (Figs VI-1 and 2) [VI-1] is designed to be a compact precision detector for electrons, photons and jets, even when they are produced closely in space. It can fit inside the L3+1 muon detector. One of its physics goals is to reconstruct the invariant masses of di-electrons and di-jets from high p_t Z's to tag heavy new particles such as Higgs, new weak bosons or quarks, etc. This detector has the following unique properties:

1. Fast response time: The pulse width is 3 ns for the scintillation light output of a xenon electromagnetic counter and can be as short as 20 ns for the TMS-uranium hadron calorimeter.
2. Radiation resistant: Liquids such as TMS and xenon are among the most stable and insensitive to radiation damage.
3. Excellent linearity: The scintillating light output from liquid xenon is larger than that from BGO and linear to 10^4 equivalent minimum ionization particles (mip), which is an order of magnitude better than NaI. TMS is also known to be very linear.
4. Excellent energy resolution. For electro-magnetic showers, liquid xenon, with a large light output, can achieve an excellent energy resolution of $3/\sqrt{E}\%$. Unlike most compensating calorimeters, in xenon, the light output per unit energy loss, dL/dE, of an electron or photon is not reduced in comparison to a minimum-ionizing particle (mip). Thus cosmic ray muons, with their trajectories and momenta measured by the precision muon detectors, can be used to calibrate the xenon counters *in situ*.

   For jets, good angular and energy resolutions are needed to reconstruct their invariant masses. Good energy resolution may be achieved through studies of the delicate compensation effects due to nuclear interactions in uranium and TMS. It also depends on the e/π ratio of the light output per unit energy loss, dL/dE. The ratio of dL/dE for alpha particles and electrons has been measured to be 1.1 ± 0.2 in xenon, in sharp contrast with BaF_2, NaI and BGO, which have ratios no larger than 0.5 (Table VI-1).

   As seen in Table VI-1, xenon saturates at much higher energy densities, and its α/e ratio is consistent with 1. From these properties and from the fact that the neutrino energy loss and the fluctuation of the nuclear binding energy loss in an EM shower counter are small (since the number of inelastic nuclear interactions is small), and assuming that the energy of neutrons will be efficiently measured by the uranium-TMS calorimeter, it is concluded that the e/π ratio can be adjusted to be close to one [VI-2]. An energy resolution of $(45/\sqrt{E} + \epsilon)E\%$, with ϵ being some small number = O(1), should be achievable for a combined system with a xenon shower counter and a uranium-TMS hadron calorimeter.

[VI-1] M. Chen *et al.*, MITLNS-163 and contribution to this workshop.
[VI-2] H.S. Chen *et al.*, MITLNS-164 and contribution to this workshop.

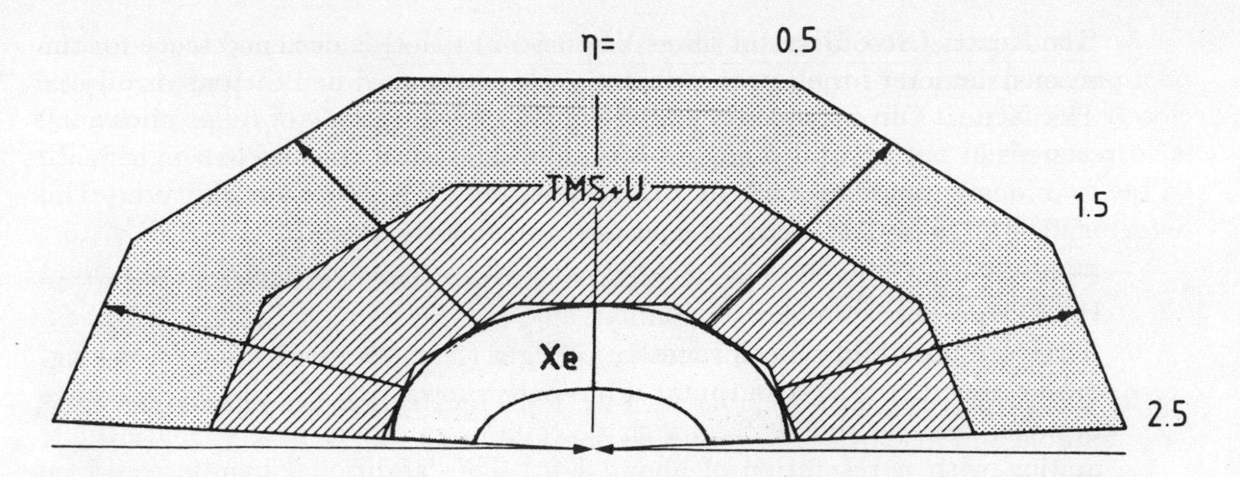

Fig. VI-1 The schematic drawing of the proposed Xenon Olive Detector. The xenon electro-magnetic shower counter is divided into 2 x 10^4 cells, about 2.2 degrees in θ at θ around 90° and 1 degree at θ around 10°, corresponding to a pseudo-rapidity interval of 0.03 at $\theta = 90^o$, 0.05 at $\theta = 45^o$ and 0.06 at $\theta = 15^o$

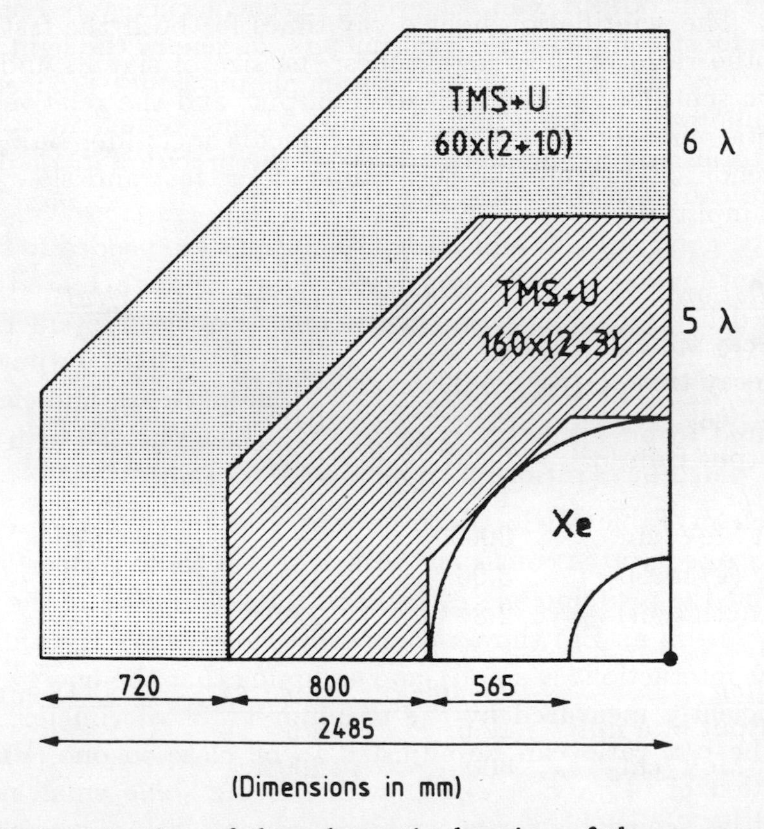

Fig. VI-2 The R-ϕ projection of the schematic drawing of the proposed Xenon Olive Detector, showing the sampling thickness of the hadron calorimeter counter, the dimensions and numbers of absorption lengths of the electro-magnetic and hadron calorimeter counters.

5. Good spatial resolution in three dimensional space. This is necessary for the reconstruction of the invariant mass of closely spaced di-electrons or di-jets. The Xenon Olive Detector is divided into about 2 x 10^4 cells, about 2.2 degrees in both θ and ϕ angles at θ around 90° and the size in θ is reduced to 1 degree at θ around 10°, corresponding to a pseudo-rapidity interval of 0.03 at $\theta = 90°$, 0.05 at $\theta = 45°$ and 0.06 at $\theta = 15°$. For $\theta < 15°$, liquid xenon could be sandwiched with lead plates to reduce the equivalent radiation length to about 5mm.
6. A remote possibility. Both scintillation light (fast) and ionization (slow) signals of xenon give large outputs. It has been shown that the timing difference between the two signals from a single counter can yield precise space information with a resolution of about 30μ. This additional handle could be valuable for the identification of high energy electrons (photons), where one needs as much independent information as possible to reject hadrons.

Including electronics, the cost for the xenon shower counter is estimated to be 25 million dollars and that for the calorimeter 40 million dollars.

Table VI-1. The scintillation light decay times for both the fast and slow components and the ratio of their amplitudes; the size of signals and the dynamical range for both scintillation and ionization output; and the relative light output of alpha and beta particles of several common media including BaF_2 in comparison with liquid xenon. T_1, T_2 are the decay time of the fast and slow component and A_1, A_2 their amplitudes.

Scintillation	*Liquid Xe*	*BaF_2*	*NaI*	*BGO*
fast decay time T_1 (ns)	3	0.6	250	60
slow decay time T_2(ns)	25-30	620	—	300
ratio A_1/A_2	25	0.09	—	0.05
rel. output from α	0.137	0.006	1.	0.08
ratio α/e	1.1 ± 0.2	0.34	0.5	0.23
K_B^{-1} in MeV/cm	30600	--	3670	500
density (g/cm^3)	3.06	4.89	3.67	7.1
rad. length (cm)	2.8	2.05	2.59	1.12

Ionization	*Liquid Xe*	*Liq. Ar*	*TMS*	*TMP*
rel. output of a mip	12.6	4.6	1	1.1
K_B^{-1} in MeV/cm	300	200	-	70

VII. R&D required

A. On high resolution muon detector:

To measure the momenta of high energy muons accurately one needs precise measurements of the trajectory in strong (>> 5 kG) and sometimes non-uniform magnetic fields (e.g. air toroids). This calls for drift chamber gases with the following properties:

1. high ionization;
2. low diffusion coefficients;
3. small multiple scattering;
4. and most important, small Lorentz bending angle.

To carry out this systematic search for a suitable gas satisfying the above criteria, the ionization processes and drift of electrons along the field direction within various candidate media can be simulated using computers. The response time of each candidate media should be computed and then measured.

B. On the xenon shower counter:

1. test fast and radiation resistant wave length shifters;
2. test fast devices to readout the light signals from xenon;
3. build and test an array of xenon shower counters, say consisting of 7 x 7 cells with a size of 2 x 2 x 60 cm^3. Find the optimum method to use both light and ionization signals for precise space and timing information.

C. On hadron calorimeter using TMS-like media:

1. carry out a systematic search for suitable condensed media,which is even faster, safer and with larger signals than TMS, by studying the ionization processes and drift of electrons along the field direction within various candidate media with the aid of computer simulation and test chambers;
2. develop fast, compact electronics for data acquisition and signal processing. Particularly for fast, noise free electronics. This is necessary since the requirement of fast read out times in practice often implies signals being cut off early, resulting in small electronics signals.

VIII. Conclusion

The two detector designs described above, as well as the general detectors discussed elsewhere, approached the task of muon detection from a number of different directions. A detailed comparison in the context of an overall SSC physics program is somewhat premature since very little work has been done in evaluating detector designs for their ability to get at the interesting physics in a realistic simulation of the SSC environment. A study of trigger rates, backgrounds and radiation problems is therefore an essential next step for the proper evaluation of the various designs.

The L3+1 detector is a precision muon, electron (photon) and jet detector. It differs from all other general purpose detectors by placing the main tracking devices outside the hadron absorbers thus making precision muon measurements possible even at very high luminosities. This arrangement also makes the size of the calorimeter small enough such that the rare noble gas, xenon, could be used at the

SSC as BGO is being used in L3 at LEP. The "muons-only" detector really belongs to a quite different class since it is designed for a more restricted but important physics program and has a much lower estimated cost. This detector should be able to operate at the highest SSC luminosities since all interacting particles other than high-momentum muons are absorbed. However, detailed trigger-rate studies must be done before one can be confident of handling various muon signals at such high luminosities.

It is, of course, clear that the detectors discussed by the muon sub-group as well as those described elsewhere in these proceedings are largely paper studies. To make further progress, work has to continue in three major areas:

a. Rate studies for physical processes and their backgrounds: we need a better understanding of the ability of large detectors to operate at high SSC luminosities.
b. Detector research and development: compact, fast, radiation-hard calorimeters are obviously desirable in any detector including those that emphasize muons. The liquid-xenon detector represents an idea in this direction.
c. Muon detection at small angles ($\theta < 25°$) presents special problems. Are magnetized toroids the only practical solution? If so, what is the optimal arrangement of magnets and detector components?

As these detector studies progress from workshop exercises to realistic proposals, we can expect many of these questions to be answered more fully.

To conclude, we have learned the following at this workshop:

1. It is advisable to measure either the initial or exit angles of muons in order to obtain reasonable resolution for magnetized toroids in the small angle regions.
2. Air gap magnets provide much better momentum resolution than magnetized iron and they are relatively free from contamination of electromagnetic showers associated with high energy muons, thus making the identification of high energy muons much more reliable.
3. Good lepton pair mass resolution (1%) is important to distinguish heavy Higgs from heavy quark decays.
4. Placement of hadron filters inside a muon detector significantly increases the maximum luminosity the detector can cope with. After most particles are absorbed, much better position resolution and thus also better momentum resolution for the muons may be possible by using the center of gravity method.
5. Precision energy measurements of electrons, photons and jets inside a large muon detector are possible using compact and fast calorimeters (e.g. xenon and TMS-U counters).

6. A high resolution muon spectrometer is a built-in precision calibration device for calorimeters, which is essential for precision electron and photon detection.

MUON ENERGY LOSS AT HIGH ENERGY AND IMPLICATIONS FOR DETECTOR DESIGN

J. J. Eastman and S. C. Loken
Lawrence Berkeley Laboratory
University of California, Berkeley, CA 94720

Abstract

We study the effects of energy loss and associated electromagnetic showers on muon tracking and momentum measurement in muon detectors operating in the energy range 100 GeV–5 TeV. A detailed Monte Carlo simulation tracks muons and shower particles through a detector structure and evaluates the charged-particle environment in chambers. We find that catastrophic energy loss events accompanied by energetic showers can pose serious problems to designers of muon spectrometers.

1 Introduction

In the energy regime to be explored by the SSC, traditional methods of muon identification and momentum measurement are complicated by the introduction of new energy loss mechanisms. At muon energies above a few hundred GeV, radiative processes dominate over ionization as sources of muon energy loss. These processes, bremsstrahlung and direct production of electron pairs, give rise to photons or electrons which can carry a significant fraction of the muon energy. In a muon spectrometer or calorimeter, the hard photons or electrons will initiate electromagnetic cascades which can obscure the muon track in active detector planes, making a precision position measurement in those planes impossible.

A number of groups have reviewed the mechanisms for energy loss by muons. This work has been done in connection with precision muon spectrometers for fixed target experiments [1,2,3] and in SSC studies [4]. The purpose of the study reported here is to investigate in detail the effects of showers on muon tracking. We have developed a Monte Carlo simulation that transports muons through material and models their energy loss and angular deflection arising from bremsstrahlung, direct e^+e^- pair production, ionization, multiple Coulomb scattering and photonuclear interactions. Bremsstrahlung photons and electron pairs are generated in discrete events along the muon path. The resulting electromagnetic showers are modeled

Figure 1: Contributions of several processes to the average energy loss of a muon in 5 m of iron as a function of muon energy. The processes are (p) direct e^+e^- pair production, (b) bremsstrahlung, (n) photonuclear interactions, and (i) ionization; t marks the total loss.

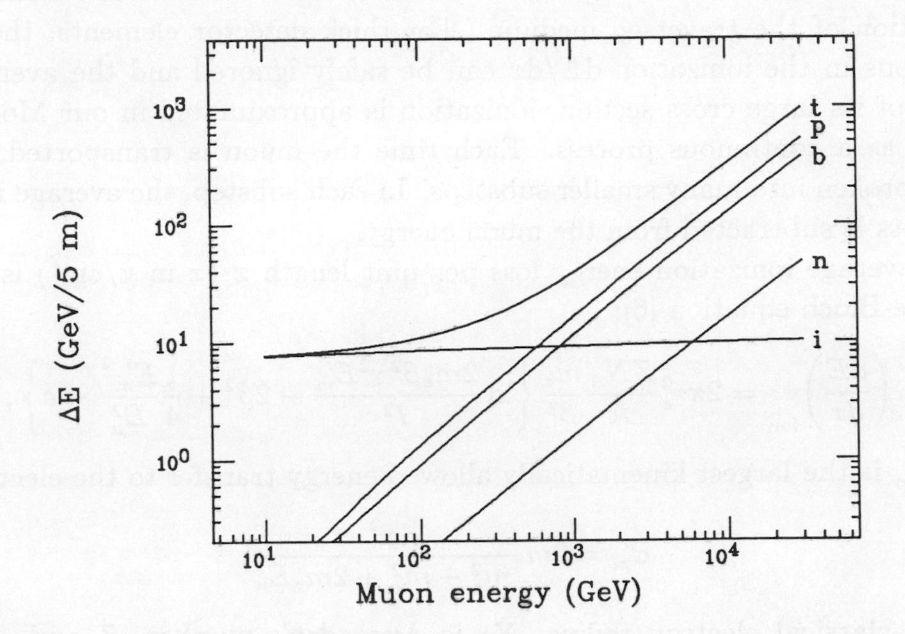

using the EGS4 Monte Carlo code [5] to determine occupancy rates and hit distributions in the active layers in two model detector configurations.

In the next section, we review the energy-loss processes and describe their parameterizations. We then discuss the behavior of hard electromagnetic showers in typical muon spectrometer configurations. Results are presented in Section 3 for muon energies from 100 GeV to 5 TeV.

2 Muon transport in matter

High energy muons passing through matter lose energy by ionization, bremsstrahlung, direct pair production, and photonuclear interactions. Our muon transport code simulates each of these processes. Figure 1 shows the average contribution of each process to the total average energy loss of a muon in 5 m of iron as a function of the muon energy.

In our treatment, ionization energy loss is considered to be continuous and uniform over the transport, and the hard processes, which occur relatively rarely in a meter of iron, are considered stochastically. These hard processes, characterized by sudden large energy losses and high-multiplicity showers, make muon tracking and momentum reconstruction difficult.

2.1 Ionization

The dominant source of energy loss at muon energies below a few hundred GeV is ionization of the traversed medium. For thick detector elements, the Landau fluctuations in the ionization $\mathrm{d}E/\mathrm{d}x$ can be safely ignored and the average used. Because of its large cross section, ionization is approximated in our Monte Carlo program as a continuous process. Each time the muon is transported, the step taken is broken into many smaller substeps. In each substep, the average ionization energy loss is subtracted from the muon energy.

The average ionization energy loss per unit length x (x in g/cm^2) is given by the Bethe-Bloch equation [6]:

$$\left(\frac{dE}{dx}\right)_{\mathrm{Ion}} = 2\pi r_e^2 \frac{ZN_A}{A}\frac{m_e}{\beta^2}\left\{\ln\frac{2m_e\beta^2\gamma^2E'_m}{I^2} - 2\beta^2 + \frac{1}{4}\frac{E'^{\,2}_m}{E_\mu^2} - \delta\right\}, \tag{1}$$

where E'_m is the largest kinematically allowed energy transfer to the electron:

$$E'_m = 2m_e\frac{p^2}{m_e^2 + m_\mu^2 + 2m_eE_\mu}, \tag{2}$$

r_e is the classical electron radius, N_A is Avogadro's number, Z and A are the atomic number and atomic weight of the medium, I is the medium's mean ionization potential, m_e and m_μ are the electron and muon masses, β and γ are the usual relativistic variables, and δ is the density effect correction. Sternheimer *et al.* [7] parameterize δ as a function of $X = \log_{10}(\beta\gamma)$ as

$$\delta = \begin{cases} 2X\ln 10 + a(X_1 - X)^m + C, & X_0 < X < X_1, \\ 2X\ln 10 + C, & X > X_1, \end{cases} \tag{3}$$

where X_0, X_1, a, m, and C are empirical material-dependant coefficients [8].

2.2 Stochastic processes

As it transports a muon, our Monte Carlo program determines when to generate a stochastic process by computing the total cross section for each of the processes simulated. The mean free path λ (in g/cm^2) is the inverse of the total cross section per gram. The muon is then transported a distance $-\lambda\ln(r)$, where r is a random number from 0 to 1, and the particular interaction to be simulated at the end of the transport is selected according to its contribution to the total cross section.

The energy fraction v lost by the muon in the interaction is drawn randomly from the distribution $\sigma^{-1}d\sigma/dv$. For each process, we use an unnormalized parameterization of the differential cross section derived by Van Ginneken [9] that is simple enough in form to be integrated analytically to obtain the total cross section and inverted explicitly for drawing v. The parameterization is normalized by carrying out a more detailed calculation of the differential cross section at one value of v, and scaling the parameterized function to match the detailed calculation at that point. The average energy loss to each process, computed either by explicit integration of $E_\mu v d\sigma/dv$ or by recording discrete energy loss events in a large number of Monte Carlo runs, matches the tabulated values of ref. [6] at the 5% level for muon energies in the range 50 GeV–10 TeV.

Photons or e^+e^- pairs are generated with the selected energy fraction v and are passed to EGS for propagation through the detector structure. Shower particles are followed until their energies fall below 1.5 MeV.

In the expressions that follow, energies and masses are expressed in GeV.

2.2.1 Bremsstrahlung

The total cross section for muon bremsstrahlung is not large – the probability for a 1 TeV muon to bremsstrahlung in 1 m of iron is 5% – but the probability $v d\sigma/dv$ for producing a photon of energy vE_μ is nearly flat across a wide range of values of v. This means that bremsstrahlung contributes a tail of catastrophic loss to the muon energy loss distribution, and that the treatment of this process as a continuous contribution to the average energy loss is a very poor approximation.

The detailed calculation in this case is that of Petrukhin and Shestakov [10], who provide the following expression for the differential cross section for bremsstrahlung, taking into account the effects of nuclear and atomic form factors:

$$\left(\frac{d\sigma}{dv}\right)_{\text{Brem}} = \alpha\left(2Zr_e\frac{m_e}{m_\mu}\right)^2\left(\frac{4}{3}-\frac{4}{3}v+v^2\right)\frac{1}{v}\ln\frac{\frac{2}{3}A_R\frac{m_\mu}{m_e}Z^{-2/3}}{1+\frac{A_R\sqrt{e}}{2}\frac{m_\mu{}^2}{m_eE_\mu}\frac{v}{1-v}Z^{-1/3}}, \tag{4}$$

for $Z > 10$. Here $e = 2.178\ldots$, and $A_R = 189$ is the radiation logarithm.

The parameterization of Van Ginneken of the differential cross section for muon bremsstrahlung from a nuclear target [1] is

$$\sigma^{-1}\frac{d\sigma}{dv} = \begin{cases} k_1 v^{-1}, & v_{\min} < v < 0.03, \\ k_2 v^{-m}, & 0.03 < v < v_2, \\ k_3(1-v)^n, & v_2 < v < 1. \end{cases} \tag{5}$$

Here k_1, k_2, and k_3 are fixed by continuity and by the condition that the differential cross section must match Equation 4 at a specific point v_{norm} in the k_2 domain; $m = 1.39 - 0.024\ln E_\mu$; $n = 1.32 - 0.12\ln E_\mu$; $v_{\min} = 0.001/E_\mu$; and $v_2 = \sqrt{E_\mu}/(\sqrt{E_\mu} + 4.5)$. The k_i depend in general on E_μ, Z and the normalization point v_{norm}, which itself varies with E_μ.

[1] We use Van Ginneken's parameterization for bremsstrahlung from nuclear targets, but normalize it as though it represented the differential cross section for all contributions to muon bremsstrahlung.

2.2.2 Direct pair production

At energies above $\sim$ 500 GeV in iron, and $\sim$ 150 GeV in uranium, direct e^+e^- pair production becomes the single most important source of muon energy loss. Unlike the bremsstrahlung case, the pair energy probability peaks at low energies, making high-energy pairs relatively unlikely.

For muon energies above $\sim$ 100 GeV both e^+e^- and $\mu^+\mu^-$ pair production are possible. $\mu^+\mu^-$ production by muons is a potentially important process that can lead to misassignment of the sign of the incident muon, but this mechanism contributes less than 0.01% to the total energy loss [6] and is not modeled in the present study.

Kel'ner and Kotov [11] give the following expression for the differential cross section for e^+e^- production by muons, which includes form factors:

$$\left(\frac{d\sigma}{dv}\right)_{\text{Pair}} = \frac{16}{v\pi}(Z\alpha r_e)^2[F_a(E_\mu, v) + F_b(E_\mu, v)], \tag{6}$$

with

$$\begin{aligned} F_a(E_\mu, v) = & \int_{1/E_\mu v}^{1/2} dx_+ \int_{t_{\min}}^{\infty} \frac{dt}{t^2} \left\{ \left[t\left(1 - v + \frac{v^2}{2}\right) - m_\mu^2 v^2 \right] \right. \\ & \times \left[\frac{1}{\gamma}\left(\frac{1}{2} - x_+ x_-\right) + \frac{x_+ x_-}{6\gamma^2}(2 - t) \right] \\ & \left. + \frac{t x_+^2 x_-^2}{3\gamma^2} \left[t\left(3(1-v) + \frac{v^2}{2}\right) - m_\mu^2 v^2 \right] \right\} \ln \frac{A_R Z^{-1/3}\sqrt{\gamma}}{\zeta b + 1}, \\ F_b(E_\mu, v) = & \frac{1}{12 m_\mu^2} \left[\left(\frac{4}{3} - \frac{4}{3}v + v^2\right)\left(\ln \frac{m_\mu^2 v^2}{1-v} - \frac{5}{3}\right) + 1 - \frac{1}{3}(1+v)^2 \right] \\ & \times \ln \left[\frac{A_R m_\mu Z^{-1/3}}{\frac{A_R\sqrt{e}}{2} Z^{-1/3} \frac{m_\mu v^2}{E_\mu(1-v)} + 1} \right], \end{aligned} \tag{7}$$

where $x_\pm$ are the fractions of the pair energy $\omega = vE_\mu$ carried by the positron and electron, $-t$ is the square of the four-momentum transferred to the muon, $\gamma = 1 + t x_+ x_-$, $\zeta = 100\gamma Z^{-1/3}/\omega x_+ x_-$, $b = A_R\sqrt{e}/200$, and $t_{\min} = m_\mu^2 v^2/(1-v)$. In practice the computationally expensive double integral was evaluated numerically at a number of points to form a lookup table; the value of the differential cross section at a given E_μ, v, and Z was found by interpolation as needed.

Van Ginneken provides the following approximation to the differential cross section:

$$\sigma^{-1}\frac{d\sigma}{dv} = \begin{cases} k_0, & 5m_e/E_\mu < v < 25m_e/E_\mu, \\ k_1 v^{-1}, & 25m_e/E_\mu < v < .002, \\ k_2 v^{-2}, & .002 < v < .02, \\ k_3 v^{-3}, & .02 < v < 1. \end{cases} \tag{8}$$

k_0, k_1, k_2 and k_3 are fixed by continuity and by requiring that the differential cross section match Eq. 6 at a point in the k_2 region, which is the region responsible for the bulk of the energy loss.

2.2.3 Photonuclear interactions

Photonuclear interactions account for about 5% of the total energy loss of high-energy muons in iron, and about 2% in uranium. The losses are concentrated in rare, relatively hard events. In this paper we consider these interactions as a contribution to the energy loss but we do not simulate hadronic showers.

Bezrukov and Bugaev [12] present the following expression for the differential cross section for photonuclear interactions of muons, including a correction for nucleon shadowing effects:

$$\begin{aligned}\left(\frac{d\sigma}{dv}\right)_{\mu A} &= \frac{\alpha}{2\pi} A\sigma_{\gamma N}(vE_\mu)v\left\{0.75G(x)\left[\kappa\ln\left(1+\frac{m_1^2}{t}\right)-\frac{\kappa m_1^2}{m_1^2+t}-\frac{2m_\mu^2}{t}\right]\right. \quad (9)\\ &\quad +0.25\left[\kappa\ln\left(1+\frac{m_2^2}{t}\right)-\frac{2m_\mu^2}{t}\right]\\ &\quad \left.+\frac{m_\mu^2}{2t}\left[0.75G(x)\frac{m_1^2}{m_1^2+t}+0.25\frac{m_2^2}{t}\ln\left(1+\frac{t}{m_2^2}\right)\right]\right\},\end{aligned}$$

where $t = m_\mu^2 v^2/(1-v)$, $\kappa = 1 - 2/v + 2/v^2$, $m_1^2 = 0.54\ \mathrm{GeV}^2$, $m_2^2 = 1.8\ \mathrm{GeV}^2$, $\sigma_{\gamma N}(E) \simeq 114.3 + 1.647\ln^2(0.0213E)\ \mu\mathrm{b}$ is the total photoabsorption cross section on nucleons, $x = 0.00282A^{1/3}\sigma_{\gamma N}(E_\mu v)$, and $G(x) = (3/x^3)(x^2/2 - 1 + e^{-x}(1+x))$.

Van Ginneken's parameterization is based not on this expression but on a simpler treatment that uses leptoproduction scaling and parameterizations of measured structure functions. The formula is

$$\sigma_{-1}\frac{d\sigma}{dv} = \begin{cases} k_1(E_\mu v)^2, & 0.144 < E_\mu v < 0.35, \\ k_2(E_\mu v)^{-1.11}(1-v)^2, & 0.35 < E_\mu v < E_\mu - m_\mu. \end{cases} \quad (10)$$

As before k_1 and k_2 are fixed by continuity and by matching the parameterized differential cross section to Equation 9 at a value of v in the k_2 domain.

3 Implications for detectors

Figure 2 illustrates a model spectrometer used in this study. The spectrometer is a stack of five unmagnetized iron blocks 1 meter thick. Chambers between the blocks are used to measure the muon position. Electromagnetic interactions in the iron produce showers that can penetrate the chambers. The 1-meter blocks are sufficiently thick to eliminate correlations in shower hits among the chambers.

Figure 2: A 1 TeV muon traverses a model detector used in the study. Five thin chambers alternate with five 1-meter slabs of iron. The muon (entering from the left) has undergone four substantial pair-production events and a number of smaller interactions in the iron. All charged particle paths are shown.

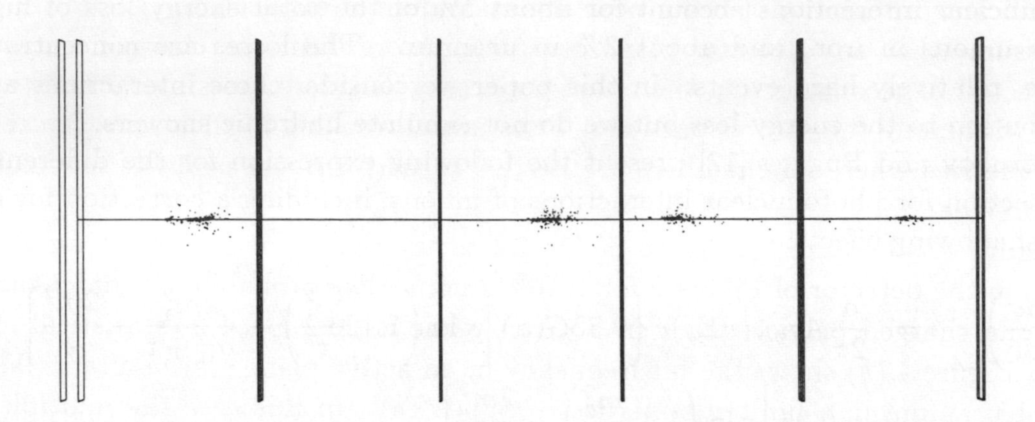

A second detector configuration used in limited-statistics runs was a 90 cm uranium slab followed by a single chamber. This configuration was used to study the charged particle environment at the exit of a hadron calorimeter.

3.1 Energy loss

Figure 3 shows various contributions to the energy loss spectrum per muon for 10000 1 TeV muons traversing 5 m of iron. The total energy lost in the 5 m by each muon to each hard process is histogrammed. The striking features of these distributions are the long tails associated with catastrophic energy losses.

Because of energy loss in the material of a muon spectrometer, the muon momentum measured in the spectrometer will be significantly less than the incident momentum. The measured momentum can be corrected on average and the fluctuations in energy loss are comparable to the typical resolution of a solid iron muon spectrometer. The fluctuations in energy loss are, however, significantly greater than the precision of an air-gap spectrometer. If the muon momentum is to be determined outside a thick calorimeter, as in the L3+1 proposal [13], it is necessary to reconstruct the lost energy for each event to achieve the full detector resolution.

3.2 Chamber illumination

The most significant result of the energy loss processes is the electromagnetic showers that are initiated by the photons and electron pairs. For a 1 TeV muon, the typical shower energy is of order 100 GeV and the shower will develop over approximately 50 cm of the muon track. Any detector plane in a muon spectrometer has a significant probability of being near the maximum of an electromagnetic shower. We assume that any detector plane containing more than one charged track will be rejected in reconstructing the muon trajectory as it is impossible to isolate the muon hit among the shower particles. Substantial tracking inefficiencies can result in geometries where all chambers are needed to perform a reasonable momentum measurement.

Figure 4 (a) is the histogram of hit frequency per plane per event for active planes in the detector of Figure 2 for 1 TeV muons. The probability of having more than one charged particle track in a given plane is $20.2 \pm 0.4\%$ (statistical error only). Figure 4 (b) shows the hit frequency in an active plane placed after a 90-cm slab of uranium in a limited statistics run at 1 TeV. In this case the multiple-hit probability is $26 \pm 3\%$. This shows that a muon plane placed at the exit of a uranium calorimeter can suffer serious inefficiencies.

The single-hit probability is shown in Figure 5 as it falls with increasing muon energy. At 1 TeV in iron, if three out of three chambers are required for a momentum measurement, the probability of finding a measurable track is only 51%. Clearly this sort of geometry is to be avoided. If three out of four chambers are required, the probability of a measurable track rises to 82%; a substantial inefficiency remains.

3.3 Multi-hit requirements for chambers

Some SSC muon detector designs [13,14] have featured muon position resolutions in the barrel region of on the order of 100 μ. If, however, the detector's two-track resolution is too large, planes containing multiple hits may not be tagged as such. In this case the muon position measurement is smeared by the size of the shower, which is typically much larger than the nominal resolution. It is therefore important that a detector be able to tag such multiply-hit planes and remove them from the momentum fit. We have used our simulation to examine the requirements for reasonable tagging.

In our model, an active detector plane is characterized by a two-track resolution σ_{2T}. We assume that a detector plane can be tagged as multiply hit by some means (eg. pulse width or height analysis) if the plane contains more than 20 hits. Otherwise, we tag a multiply hit plane only if it contains at least one pair of hits separated by more than σ_{2T}. Figure 6 shows the efficiency for tagging a multiply-hit detector plane by this method as a function of σ_{2T}. The residual $\sim 10\%$ efficiency at very large σ_{2T} comes from the 20-hit tagging capability. The tagging efficiency drops

Figure 3: Energy loss per muon from each of several processes for 10000 1 TeV muons in the detector of Figure 2. Shown are the contributions from *a*) bremsstrahlung; *b*) direct e^+e^- pair production; and *c*) deep inelastic scattering on nucleons. Note the substantial tails.

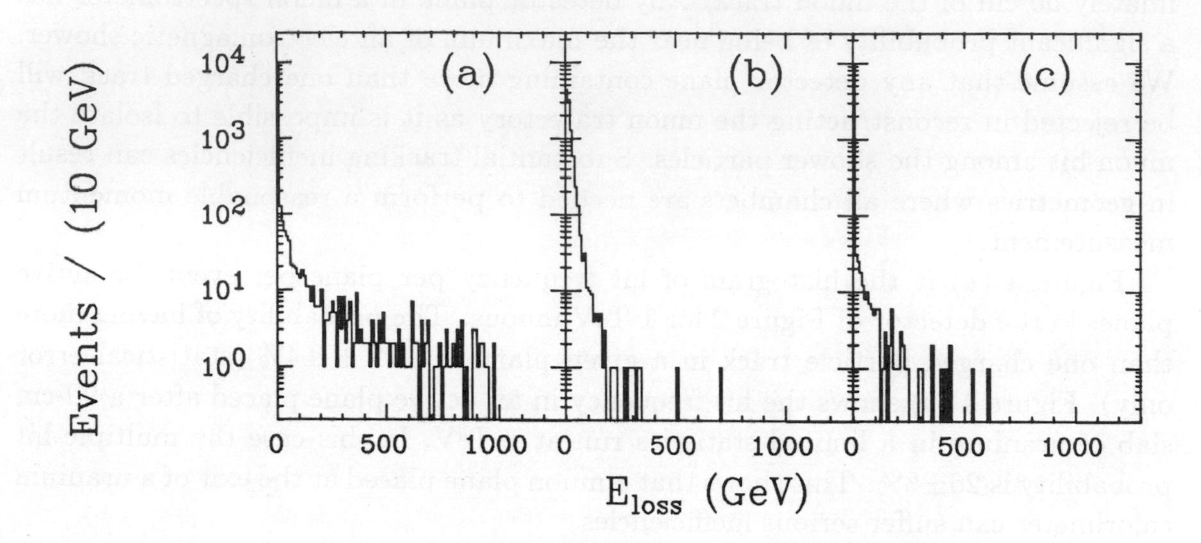

Figure 4: *a*) The hit multiplicity for 1 TeV muons in the active planes of the detector of Figure 2 and *b*) in an active plane after 90 cm of uranium.

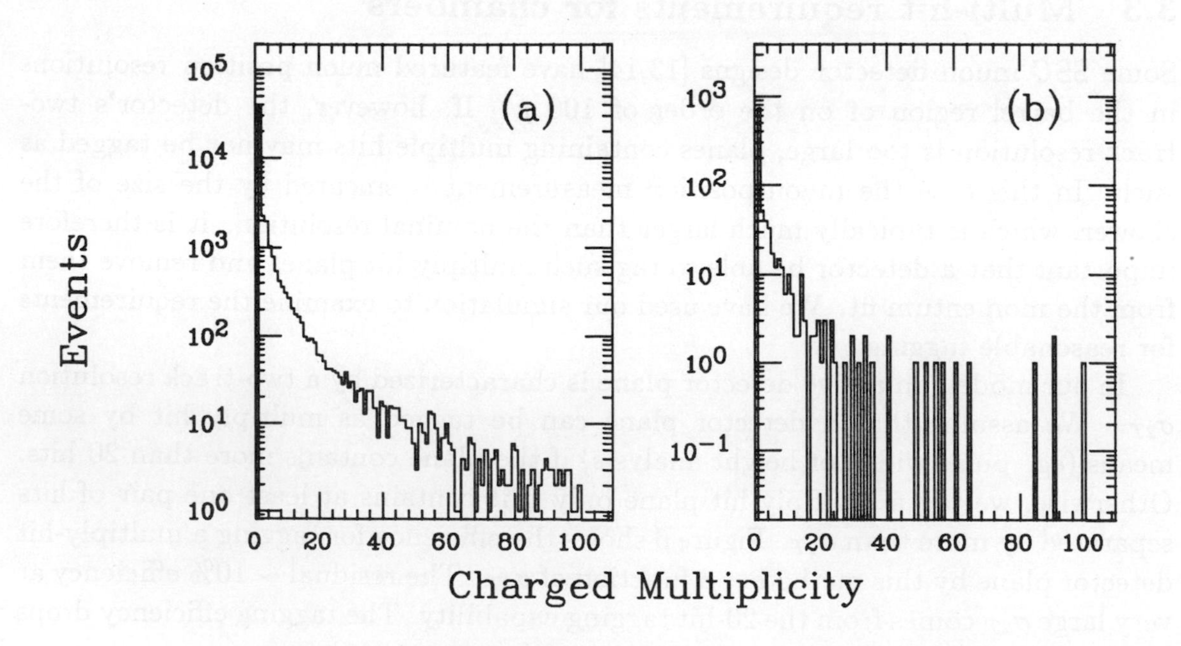

Figure 5: The probability of a clean muon hit (i.e., a muon track unaccompanied by any other charged particle) in an active layer in the detector of Figure 2 as a function of E_μ (diamonds). Also included are two points for an active layer placed behind 90 cm of uranium (crosses).

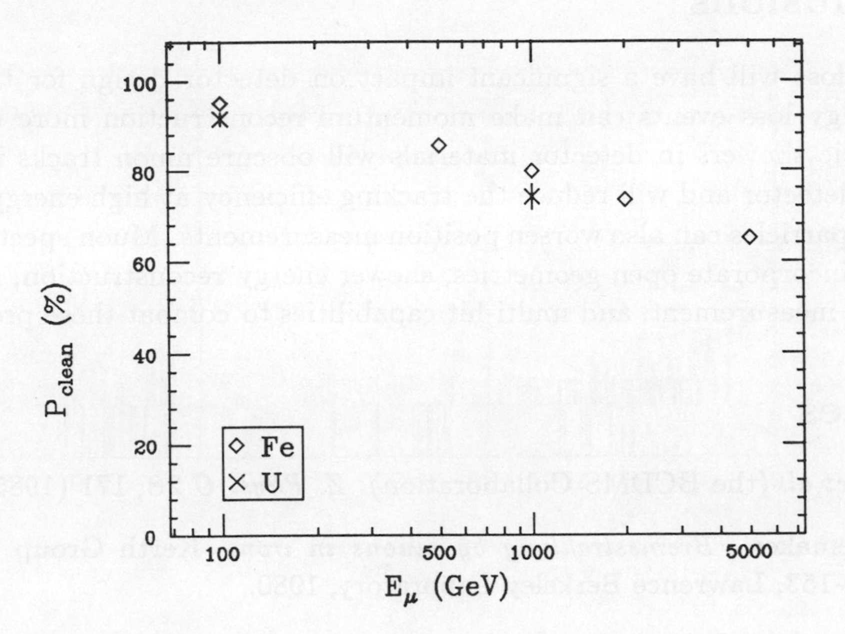

Figure 6: The probability of identifying a multiply hit chamber as a function of the chamber's two-track resolution for active planes in the detector of Figure 2 and for an active layer placed behind 90 cm of uranium. Assumptions are given in the text.

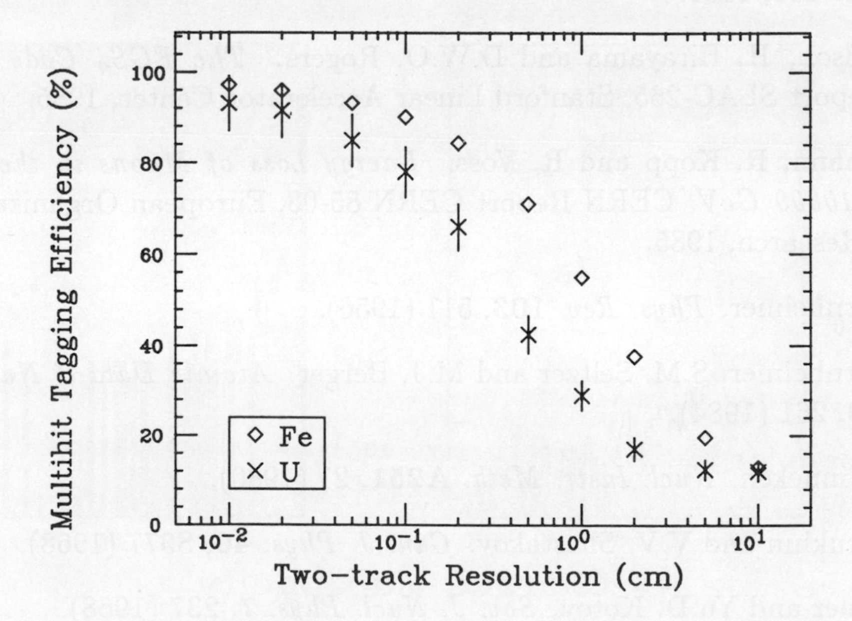

dramatically for σ_{2T} greater than about 1 mm. Obtaining two-track resolutions of this size in large chambers is a challenge for detector designers.

4 Conclusions

Muon energy loss will have a significant impact on detector design for the SSC. Localized energy loss events can make momentum reconstruction more difficult. Electromagnetic showers in detector materials will obscure muon tracks in many planes of the detector and will reduce the tracking efficiency at high energy. Unresolved shower particles can also worsen position measurements. Muon spectrometer designs could incorporate open geometries, shower energy reconstruction, multiple redundancy of measurement, and multi-hit capabilities to combat these problems.

References

[1] R. Kopp *et al.* (the BCDMS Collaboration). *Z. Phys. C* **28**, 171 (1985).

[2] F.C. Shoemaker. *Bremsstrahlung by muons in iron.* Kerth Group Internal Note GIN-153, Lawrence Berkeley Laboratory, 1980.

[3] F.C. Shoemaker. *Direct pair production by muons in iron.* Kerth Group Internal Note GIN-155, Lawrence Berkeley Laboratory, 1980.

[4] Muon Detector Subgroup. SSC muon detector group report. In R. Donaldson and J. Marx, editors, *Physics of the Superconducting Supercollider: Snowmass 1986*, page 405, 1986.

[5] W.R. Nelson, H. Hirayama and D.W.O. Rogers. *The EGS4 Code System.* SLAC Report SLAC-265, Stanford Linear Accelerator Center, 1985.

[6] W. Lohmann, R. Kopp and R. Voss. *Energy Loss of Muons in the Energy Range 1-10000 GeV.* CERN Report CERN 85-03, European Organization for Nuclear Research, 1985.

[7] R.M. Sternheimer. *Phys. Rev.* **103**, 511 (1956).

[8] R.M. Sternheimer, S.M. Seltzer and M.J. Berger. *Atomic Data & Nucl. Data Tables* **30**, 261 (1984).

[9] A. Van Ginneken. *Nucl. Instr. Meth.* **A251**, 21 (1986).

[10] A.A. Petrukhin and V.V. Shestakov. *Can. J. Phys.* **46**, S377 (1968).

[11] S.R. Kel'ner and Yu.D. Kotov. *Sov. J. Nucl. Phys.* **7**, 237 (1968).

[12] L.B. Bezrukov and É.V. Bugaev. *Sov. J. Nucl. Phys.* **33**, 635 (1981).

[13] U. Becker *et al.* *Nucl. Inst. Meth.* **A253**, 15 (1986). Also see submission to these proceedings.

[14] R. Thun, C.-P. Yuan *et al.* *Searching for Higgs* $\rightarrow Z^0Z^0 \rightarrow \mu^+\mu^-\mu^+\mu^-$ *at SSC.* Physics Department Report UM HE 86-32, University of Michigan, 1986. Also see submission to these proceedings.

Electromagnetic Showers Background Associated with High Energy μ's

Vincenzo Innocente
INFN - sezione di Napoli - Naples - Italy

ABSTRACT

We investigated the background, induced in an air gap magnetic spectrometer such as the L3 detector, due to electromagnetic showers produced by high energy muons interacting with the materials placed in front of it and conclude that, for 500 GeV muons, less than 1% of the muon tracks will be lost due to the overlap with spurious electron (or positron) tracks. This is in contrast with iron spectrometers where about 20% of the hits can be spoiled.

1. Introduction

Muons that interact with materials produce secondary particles as δ-rays, bremsstrahlung photons and direct produced pairs. If these particles will be produced in front of a muon spectrometer they can reach the muon chambers and, in principle, create so many spurious tracks to degrade the performances of the muon spectrometer itself. It has been reported [1] that about 20% of the hits in an iron spectrometer could be ruined in this manner.

We have studied these problems in an air gap spectrometer with the L3 experimental set-up [2] (fig.1) for 50 and 500 GeV muons, this last case being particularly interesting for muon detectors at the SCC.

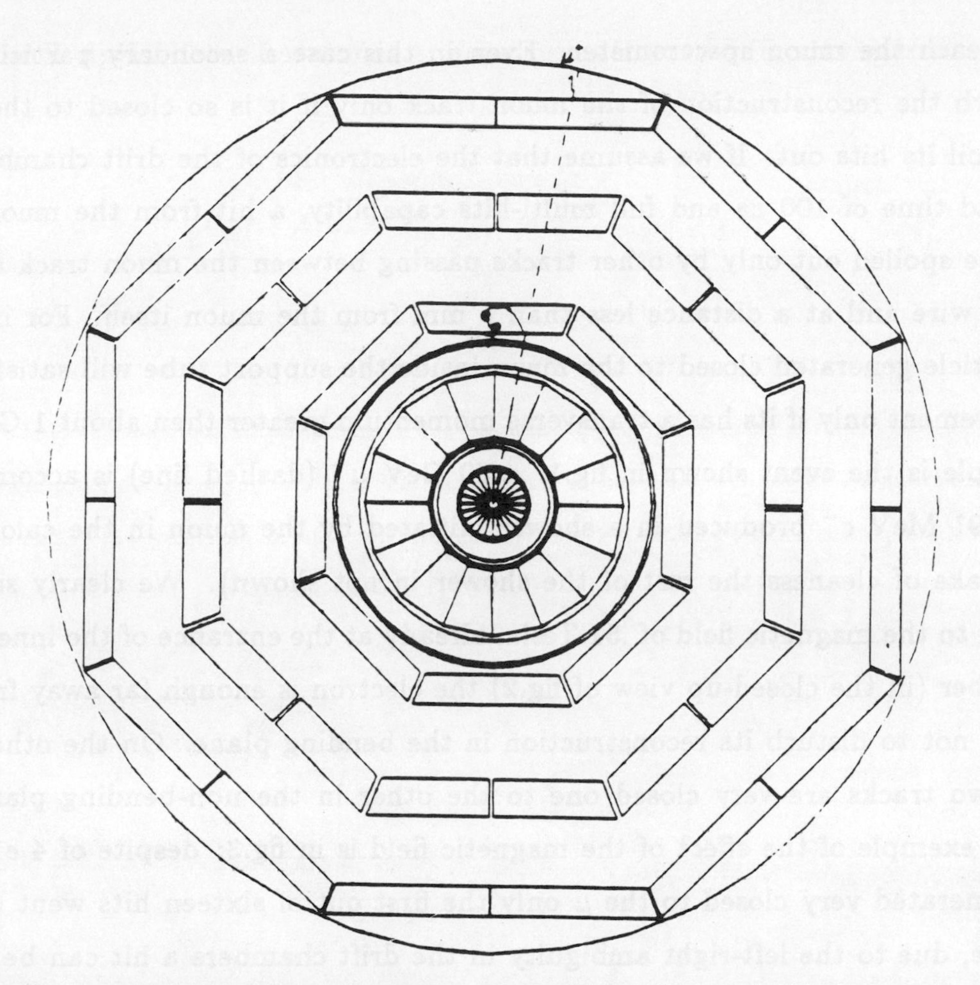

Fig.1 Frontal view of the L3 Detector. In the upper part a 50 GeV muon accompanied by a 91 MeV electron spiralizing in the one of the inner muon chambers is shown.

2. The L3 Muon Spectrometer

The L3 muon spectrometer [2] consist of three drift chamber layers arranged in 16 modules (octant) inside a .51 Tesla solenoidal magnetic field. The innermost and the outermost layers contain 16 wire planes (p-wires) measuring the radial coordinate (and then the transverse momentum) in the bending plane, and 4 wire planes (z-wires) measuring the longitudinal coordinate in the non-bending plane. The middle chambers are equipped with 24 p-wire planes.

The innermost p-chambers are 15.5 cm apart from the steel support-tube, that is the last dense material in front of them. This means that, due to the presence of a magnetic field, only charged particles with momentum greater than 12 MeV can reach the muon spectrometer. Even in this case a secondary particle could disturb the reconstruction of the muon track only if it is so closed to the muon to spoil its hits out. If we assume that the electronics of the drift chambers has a dead time of 100 ns and full multi-hits capability, a hit from the muon track can be spoiled out only by other tracks passing between the muon track and the sense wire and at a distance less than 5 mm from the muon itself. For instance a particle generated closed to the muon inside the support tube will satisfies this requirement only if its has a transverse momentum greater then about 1 GeV. An example is the event shown in fig.1: a 50 GeV μ^+ (dashed line) is accompanied by a 91 MeV e^- produced in a shower initiated by the muon in the calorimeter (for sake of cleaness the rest of the shower in not shown). We clearly see how, thank to the magnetic field of .51 Tesla, already at the entrance of the inner muon chamber (in the closed-up view of fig.2) the electron is enough far away from the muon not to disturb its reconstruction in the bending plane. On the other hand the two tracks are very closed one to the other in the non-bending plane. An other exemple of the effect of the magnetic field is in fig.3: despite of 4 e^- and 2 e^+ generated very closed to the μ only the first out of sixteen hits went lost. Of course, due to the left-right ambiguity in the drift chambers a hit can be spoiled out also by particles just on the other side of the sense wire as can be seen in the events in fig.4. Finally in fig.5 it is shown a photon stopped in the meddle chamber that creates an electron partially overlapping the muon track.

The conclusion is that an e^+ (or e^-) can overlap the muon track in the bending plane only if it has a high energy or if it is produced directly inside the drift chambers (or its walls) by a photon of the shower emerging from the calorimeter.

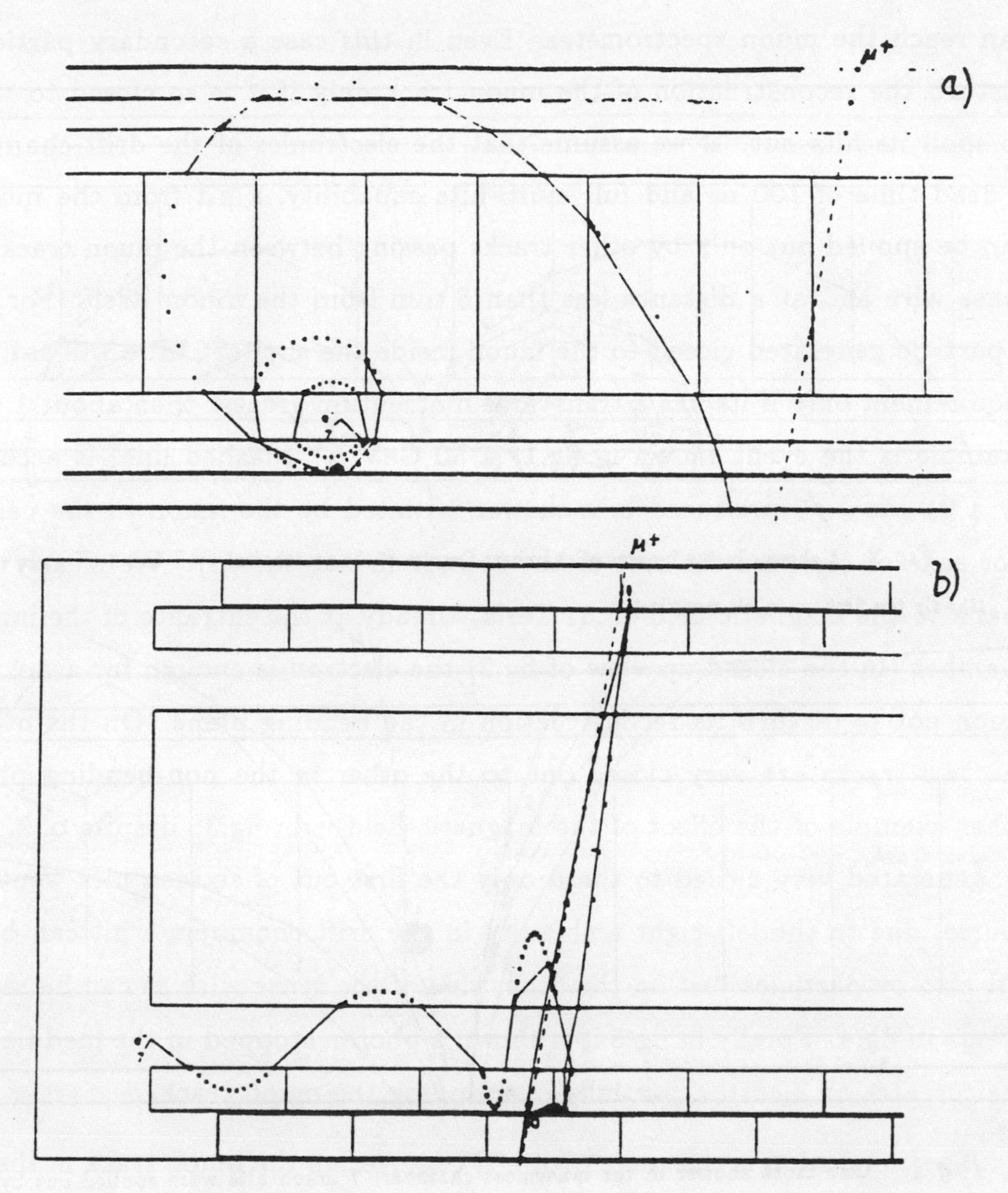

Fig.2 **Close-up view inside a inner muon chamber: *a)* front view (bending plane) *b)* side view (non-bending plane). Clearly seen are the muon (dashed line) and the 91 MeV electron. while the e^- is produced closed to the μ in the calorimeter, it is already well far away from it not to disturb its reconstruction in the radial coordinate. In the longitudinal coordinate, instead, the two track as very closed together and in at least one of the z-planes (the second from the bottom) the muon hit will be spoiled out by the electron hit. In both view the boxes represent the 10 cm wide drift cells**

3. Simulation of High Energy Muons in the L3 Detector

To simulate the passage of high energy muons through the L3 Detector we have used the L3 Monte-Carlo program based upon the GEANT program package

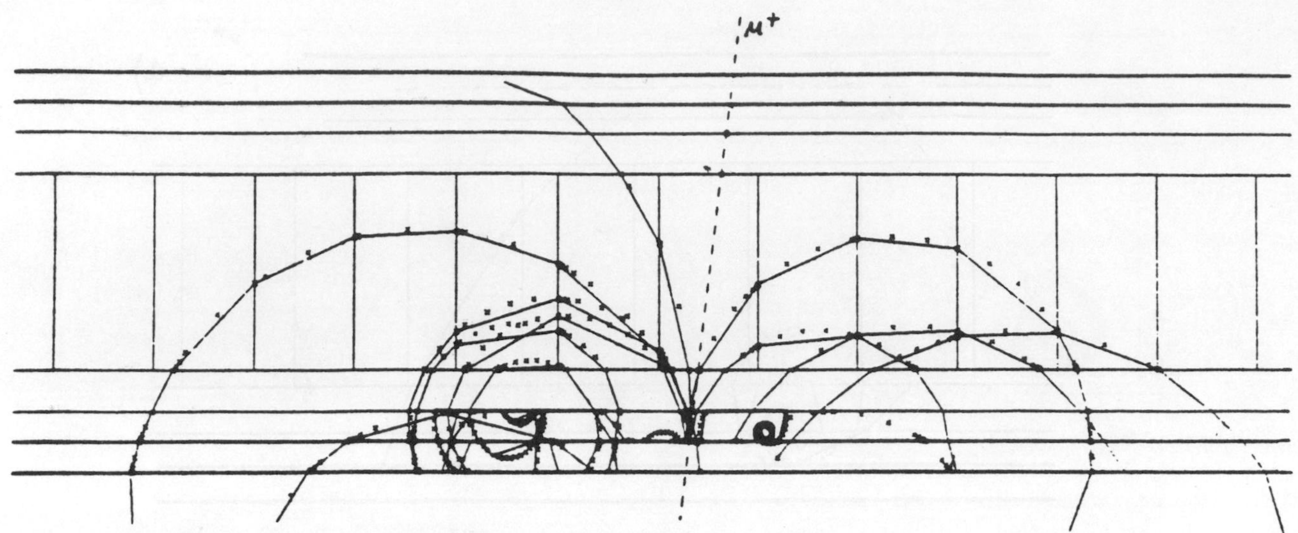

Fig.3 A shower in the innermost chamber: Despite of the presence of 4 e^- and 2 e^+ close to the μ, only the first muon hit went lost.

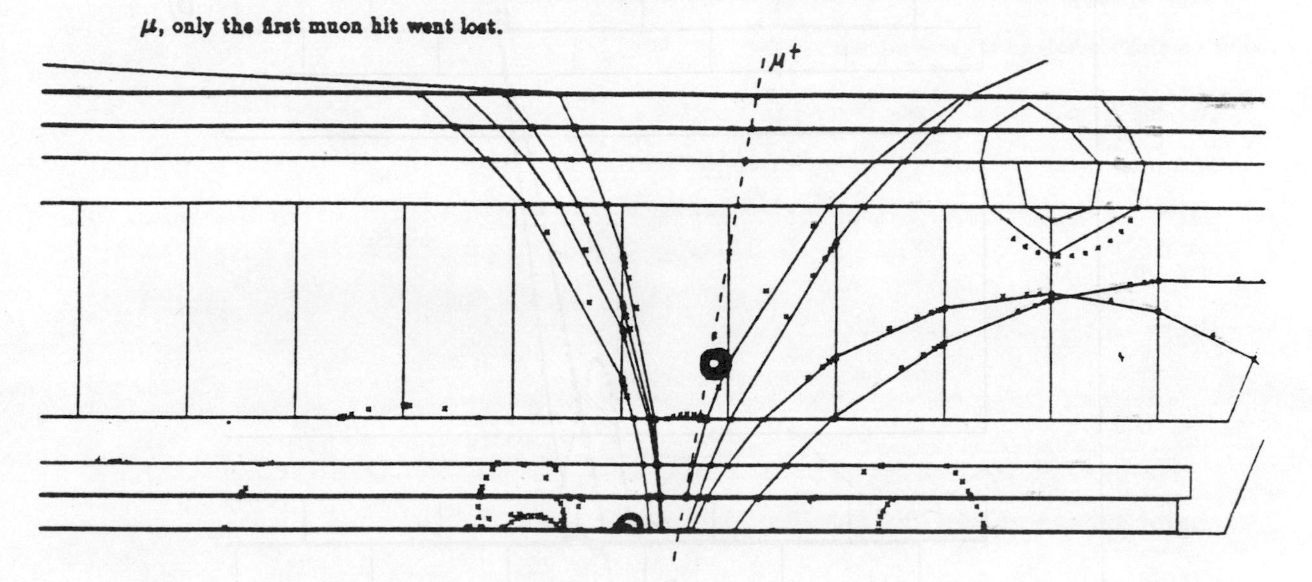

Fig.4 One more shower in the innermost chamber: 7 muon hits were spoiled out by one of the electrons passing on the other side of the drift cell

of CERN. [3] Muons are followed in their path through the detector and their interactions with materials recorded. Secondary particles are explicitely generated and tracked if their momentum is larger than a given cut-off. Finally the detector response to the passage of all these particles is simulated.

We have generated 1000 single muons starting from the interaction vertex. The cut-off for tracking electrons and photons was set at 1 MeV. Tests with lower

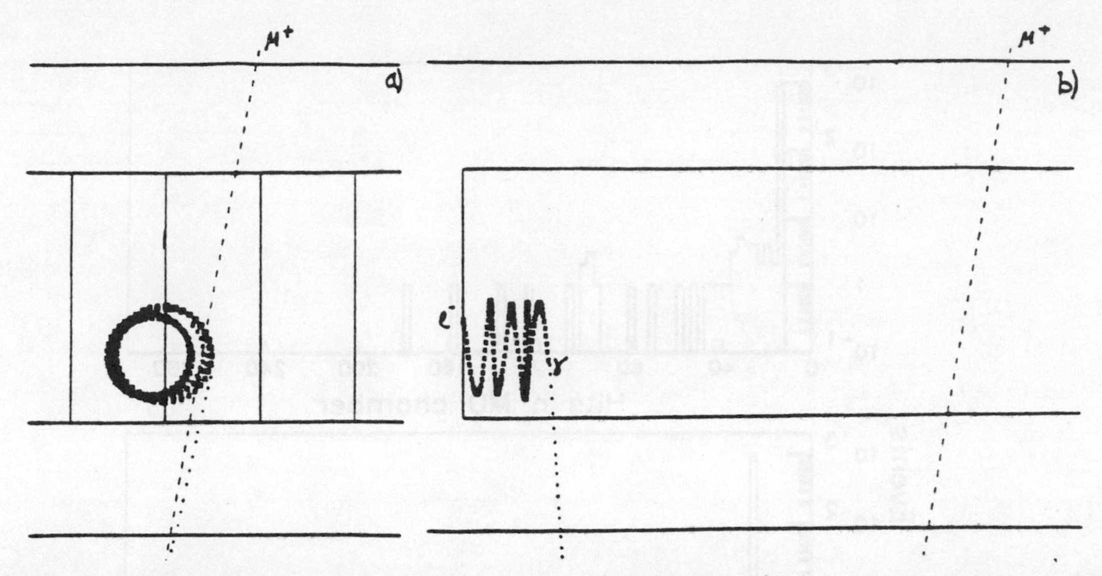

Fig. 5 **Photon stopped in the meddle chamber: *a)* front view *b)* side view. it created an electron that partially overlaps the muon track.**

cut-off at 50 GeV show that the results are not affected by this choice. Almost one half of the 50 GeV muons emerges from the calorimeter accompanied by other particles, mainly low energy photons. Not accompanied 500 GeV muons are only few per cent.

4. Results

In figure 6 the histograms of the number of hits per event recorded in the muon chambers are shown for 500 GeV muons. We see that in more then 90% of the events only the muon gave tracks in the chambers. It should be also pointed out that the number of additional hits in the outer chambers is not negligible at all. This is due to the fact that, in presence of large showers, some photons will interact in the drift chambers itself or in their frames giving electrons and positrons that, even having a small momentum, can produce several hits in the p-planes. Anyhow the number of events where muon hits are spoiled out is very limited. In fig.7 the histograms of the number of the muon hits lost in each chamber becouse of a time overlap with other tracks is shown. For each muon the maximum among them is histogrammed in fig.8. Assuming that a track is lost if more than one third of the hits in any of the chambers has been spoiled out, than we realize from fig.8 that only 7 out of 1000 generated 500 GeV muons will be lost. For

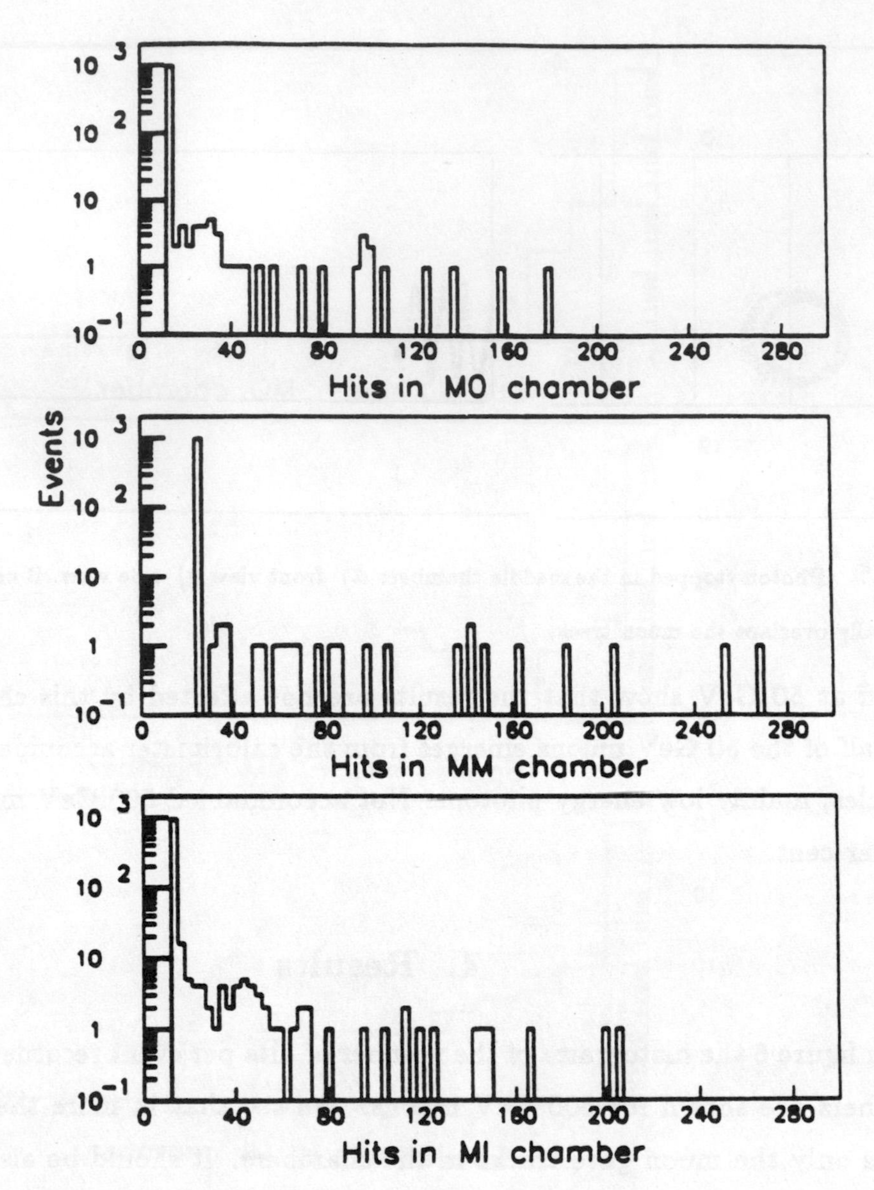

Fig.6 **Histograms of the number of hits per event recorded in each muon chamber for 500 GeV muons.**

50 GeV muons this number reduces to 5 per mille. The evident conclusion is that the background induced by electromagnetic showers in an air gap magnetic spectrometer is negligible, at least in the bending plane, even for 500 GeV muons.

In the non bending plane the situation is of course worse because of the absence of bending due to the magnetic field and also due to the limited (8) number of wire planes in the L3 detector. On the other hand, as can be seen in fig.2, the z-wires are staggered so that, if a hit is spoiled out in one plane, is very unlikely

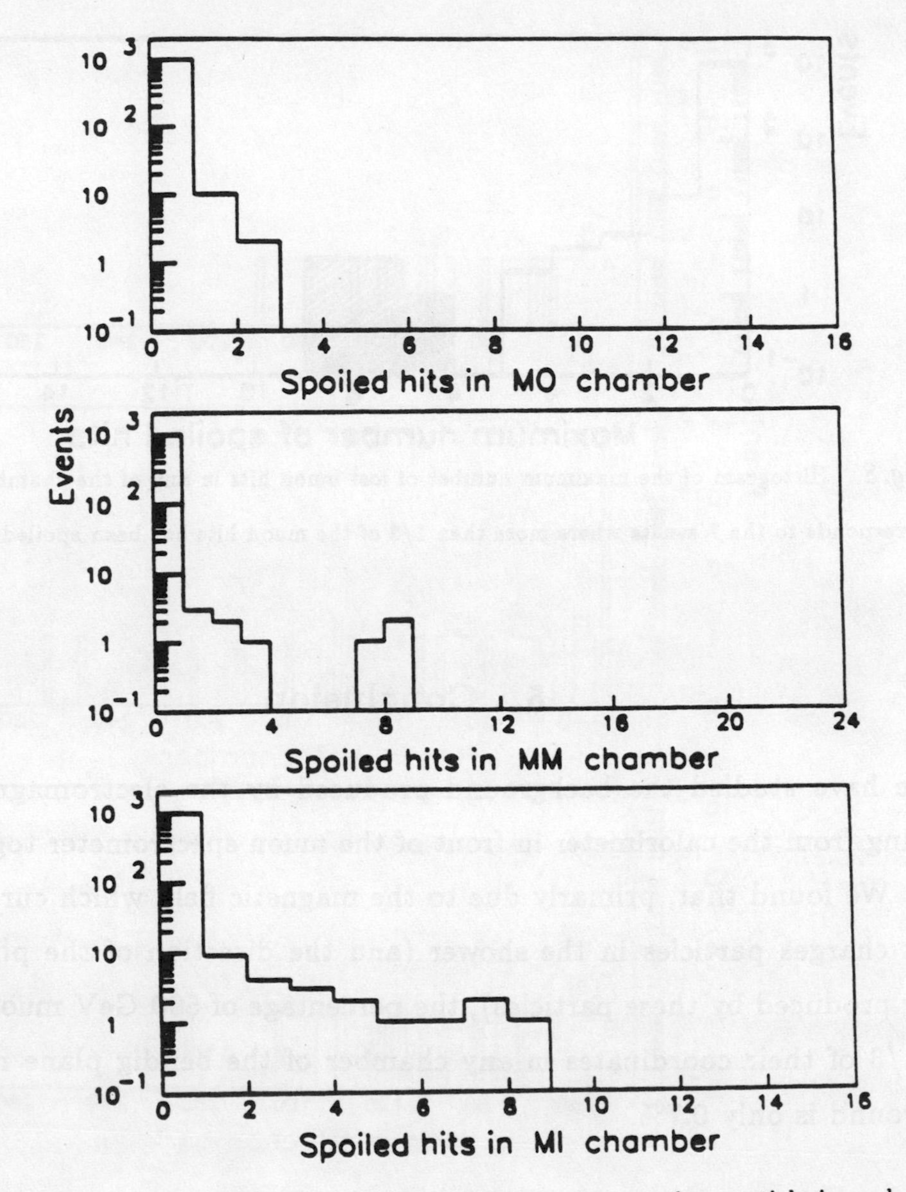

Fig. 7 **Histograms of the number of muon hits spoiled out by an other particle in each muon chamber for 500 GeV muons.**

that also the hit in the nearby plane will be spoiled out by the same track. For both 50 and 500 GeV muons we found that about 7 per cent of the muons have at least one hit spoiled out in the first two planes. This number reduces to 2 per cent in the other 6 planes.

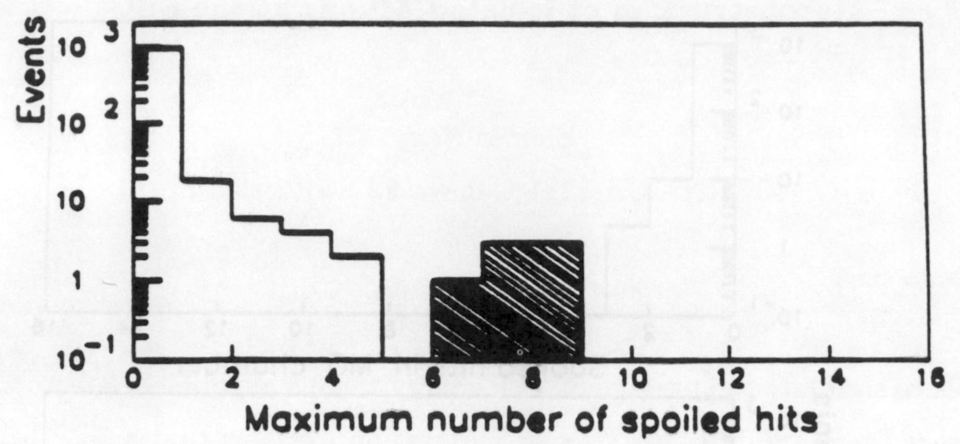

Fig.8 Histogram of the maximum number of lost muon hits in any of the chambers. The dashed area corresponds to the 7 events where more then 1/3 of the muon hits has been spoiled out.

5. Conclusion

We have studied the background produced by the electromagnetic showers emerging from the calorimeter in front of the muon spectrometer together with a muon. We found that, primarly due to the magnetic field which curls up the low energy charges particles in the shower (and the direction of the photons subsequenly produced by these particles), the percentage of 500 GeV muons with more than 1/3 of their coordinates in any chamber of the bendig plane ruined by the background is only 0.7%.

REFERENCES

1. S.Loken et al., contribution to this workshop
2. L3 Technical Proposal, 1982
3. R.Brun et al. GEANT Version 311, CERN Program Library.

Reconstruction of Isolated Muons at the SSC

Elemér Nagy
CERN - CH-1211 Geneva 23 Switzerland

ABSTRACT

Tools to simulate the passage of *TeV* energy muons in collider detectors and to reconstruct them are described. As an illustration we show that a 500 GeV muon detected in the barrel part of the "L3+1" Detector can be perfectly reconstructed.

1. Introduction

Detection and precize measurement of high energy muons probably remain among the most efficient tools to discover new physics at the SSC,[1] it is therefore of great importance to investigate the precision by which muons can be reconstructed at these energies. In this paper we study the reconstruction of *isolated* muons for two reasons:

i) **in many interesting physics processes (like e.g. $pp \longrightarrow H^0 + X$; $H^0 \longrightarrow Z^0 Z^0$; $Z^0 \longrightarrow \mu^+ \mu^-$ (Ref1)) the muons are indeed isolated;**

ii) **muons in the TeV energy range interact with the detector material in a much more complex way then one has to used to it in the GeV range – therefore the first step towards the understanding of its behaviour consists of the study by isolating the muons from the background of the other particles produced in the same interaction.**

2. Simulation of the Passage of High Energy Muons

To simulate the passage of high energy muons through the detector material we have used the GEANT program package of CERN.[2] The simulation includes the following physics processes contributing to the energy loss : *a*) Energy loss due to ionization, *b*) δ-ray production, *c*) Bremsstrahlung, *d*) direct e^+e^--production and *e*) nuclear interaction. For each discrete physics process the user can define a cut-off value of the energy loss above which the program explicitely generates the process and below which a mean energy loss is substracted from the muon energy in a given tracking step. These cut-off values are selected depending on how detailed simulation of the secondary tracks (δ-rays, pairs, bremsstrahlung, etc) is aimed at. E.g. a cut-off value for the pair-production, P_{cut}, near to its lower bound corresponds to generate explicitely e^+e^- pairs when the energy loss of the muon is caused by the pair-production process. On the other hand, if P_{cut} is set to the kinematically allowed maximum value, the energy loss due to the pair production is taken into account by substracting the avarage value of the energy loss without generating any pair. We call therefore this energy loss in the following as continuous energy loss. Obviousely, a high cut-off value can save computing time.

The avarage total energy loss, of course, should not depend on the chosen cut-off values. Based on 100 generated muons in the "L3+1" set-up (see Section 4) Table 1 shows that this is indeed the case. Moreover, the avarage total energy losses obtained for different muon energy and materials are in excellent agreement with the published values.[3] [4]

Table 1

Energy loss of a 500 GeV Muon in "L3+1"

	B_{cut}, P_{cut}					
Eloss [GeV]	*Minimum*	1 *GeV*	5 *GeV*	10 *GeV*	50 *GeV*	500 *GeV*
ΔE_{tot}	32 ± 4	32 ± 5	19.5 ± 3	21 ± 4	24 ± 5	21.3
ΔE_{brems}	4.5 ± 1.5	4.5 ± 1.5	1.0 ± 0.4	7.0 ± 2.5	9 ± 4	--
ΔE_{pair}	24.3 ± 3	21.5 ± 3	8.3 ± 2	2 ± 2	0 ± 1	--
ΔE_{cont}	3.5	6.2	10.2	11.9	14.7	21.3

3. Reconstruction of High Energy Muons

The momentum-vector of the muon is not always measured close to its production point where its value is of physical interest. This is the case e.g. for the L3 Detector[5] or for the "L3+1" detector proposed for the SSC,[6], where the muon traverses a dense calorimeter before its momentum is measured. Therefore in order to deduce the original momentum-vector of the muon from its measured value one has to

i) measure all the emitted energy by the muon in the form of bremsstrahlung, pair-production, etc. beyond a certain minimum energy, E_{min}, and to

ii) transport the track with proper *error propagation* back to the origin taking into account the physics processes the muons undergo. Since in the colliders the interaction point is usually well-known (especially its transverse spread is small), a correct error propagation can be used to exploit this information to improve the track parameters at the origin.

In this paper we deal only with requirement *ii*) assuming that requirement *i*) can be satisfied for $E_{min} \geq 1\ GeV$ for isolated muons. Our method is the following:

1) Choosing a value E_{min} we track the muon backward (inversing its direction and charge) from the detector plane to a so called *reference plane* which contains the interaction point and is approximately perpendicular to the muon direction. For the back-tracking we use the GEANT tracking facility (Ref2) calculating step-by-step the avarage energy loss setting $B_{cut} = P_{cut} = E_{min}$ where B_{cut} is the cut-off value of the bremsstrahlung and P_{cut} is the same for pair-production and nuclear interaction processes.

2) The error propagation is carried out at the tracking time by a program written by the author* which is based on a program package used by the European Muon Collaboration (EMC).[7] [8] It takes into account step-by-step the multiple Coulomb scattering of the muon and the avarage angular deflection due to the bremsstrahlung and pair-production below B_{cut} calculated by GEANT.

4. An Example: 500 GeV Muon in the L3+1 Detector

The method outlined above was tested on a 500 GeV muon sample generated in the proposed "L3+1" Detector (Ref6) (see Fig.1). The detector consists of a dense calorimeter at the central part capable to measure the total electromagnetic and hadronic energy produced in a pp collision and is followed in the barrel part by high precision drift chambers contained in a solenoid magnetic field of 0.75 T such that they are capable to measure the muon momentum with a resolution of

$$\delta p/p = 4 \cdot 10^{-5}\ p, \tag{4.1}$$

where p is the muon momentum in GeV. This "bench-mark" muon used in this analysis is also shown in the upper part of the detector displayed in Fig. 1. From the production point upto the entry point in the first muon chamber the muons traverse $\approx 2110\ g/cm^2$ material equivalent to uranium.

---

* Not yet published

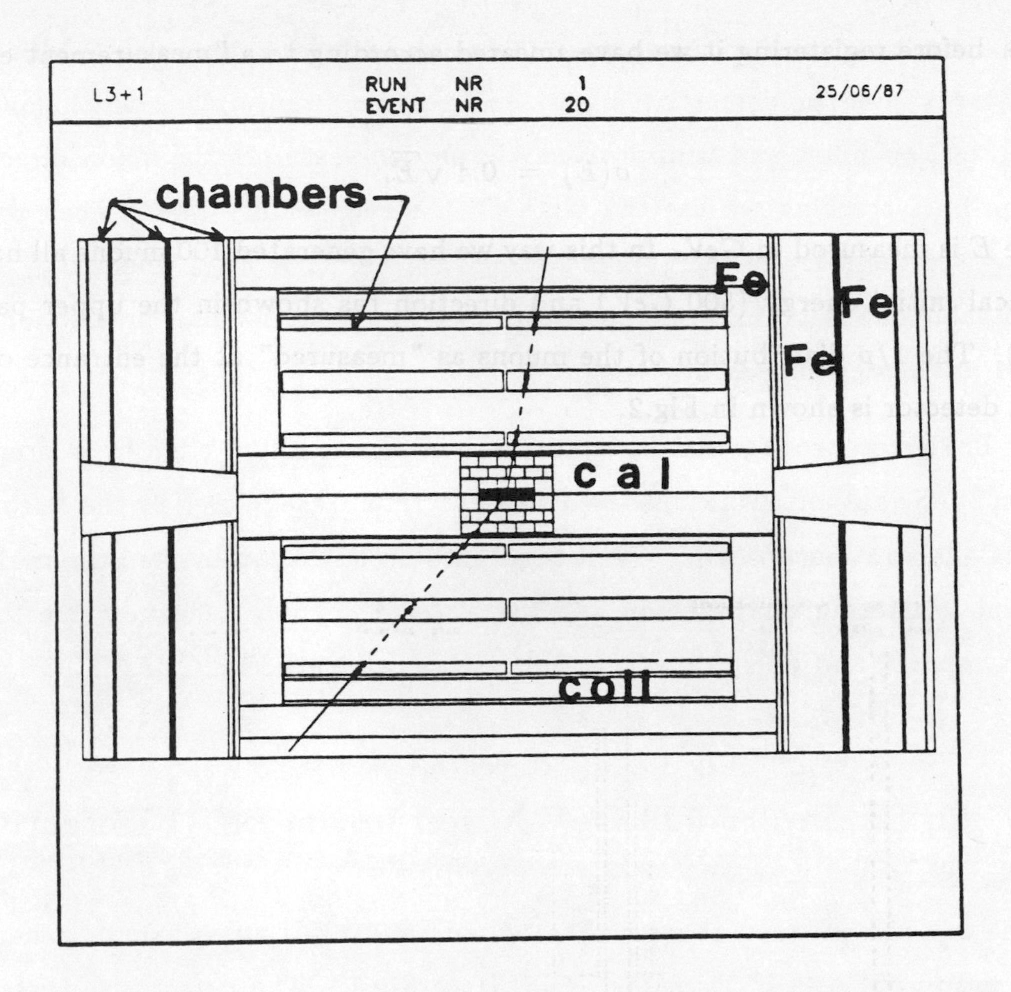

Fig.1 Side view of the "L3+1" Detector. In the upper part the "banch-mark muon" is shown.

Generation of the Bench-Mark Muons

In the generation process we have set

$$B_{cut} = P_{cut} = 1\ GeV, \tag{4.2}$$

i.e. we have generated explicitely bremsstrahlung photons and e^+e^- pairs whose total energy was always $\geq 1\ GeV$.

At the entrance of the muon chamber we have smeared the muon momentum according to (4.1) and its direction by a Gaussian distribution with r.m.s of 1 $mrad$. We have also simulated the measurement of the emitted electromagnetic energy,

E, i.e. before registering it we have smeared according to a "measurement error" of

$$\sigma(E) \;=\; 0.4\,\sqrt{E}, \tag{4.3}$$

where E is measured in GeV. In this way we have generated 100 muons all having identical initial energy (500 GeV) and direction (as shown in the upper part of Fig.1). The $1/p$ distribution of the muons as "measured" at the entrance of the muon detector is shown in Fig.2.

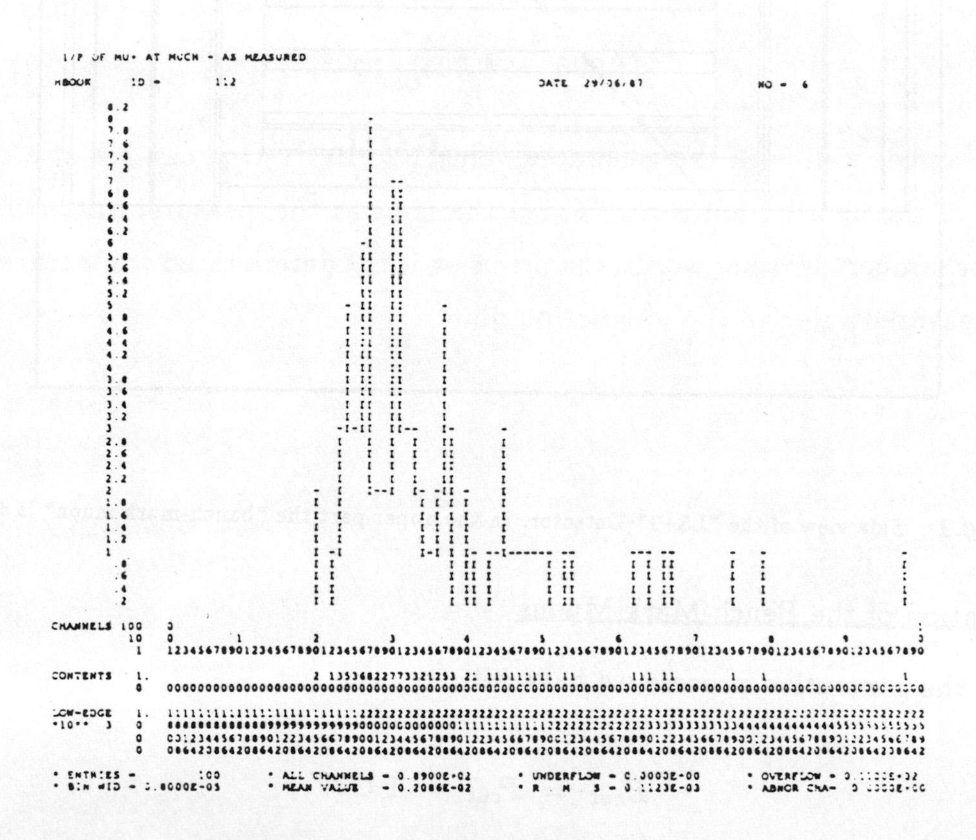

Fig.2 **$1/p$ distribution of the generated muons "as measured" at the entrance of the muon chamber.**

Reconstruction of the Bench-Mark Muons

First by fixing $P_{cut} \;=\; B_{cut} \;\geq\; E_{min}$, each generated muon was back-tracked as outlined in Section 3 to the reference plane, where the complete error-matrix of the trajectory was calculated. Next, we corrected the muon energy on the reference plane by the sum of all "measured" electromagnetic energy E_i, emitted by the muon *above* P_{cut}:

$$E_{corr} = \sum_{E_i \geq P_{cut}} E_i. \tag{4.4}$$

Finally, we have fitted the trajectory to the interaction point. For $P_{cut} = 1\ GeV$ the result is shown in Figs. 3-6. Fig. 3 shows the distribution of $1/p$, the reconstructed inverse momentum at the origin. The avarage value of this distribution is

$$(1.996 \pm 0.004) \cdot 10^{-3},$$

which agrees well with the generated value: $2 \cdot 10^{-3}$. One can see also that the width of the distribution is $\approx$ 2%, i.e. the same as the measurement error on the muon chamber. In other words, the precision is not deteriorated while transporting the measured value to the interaction point.

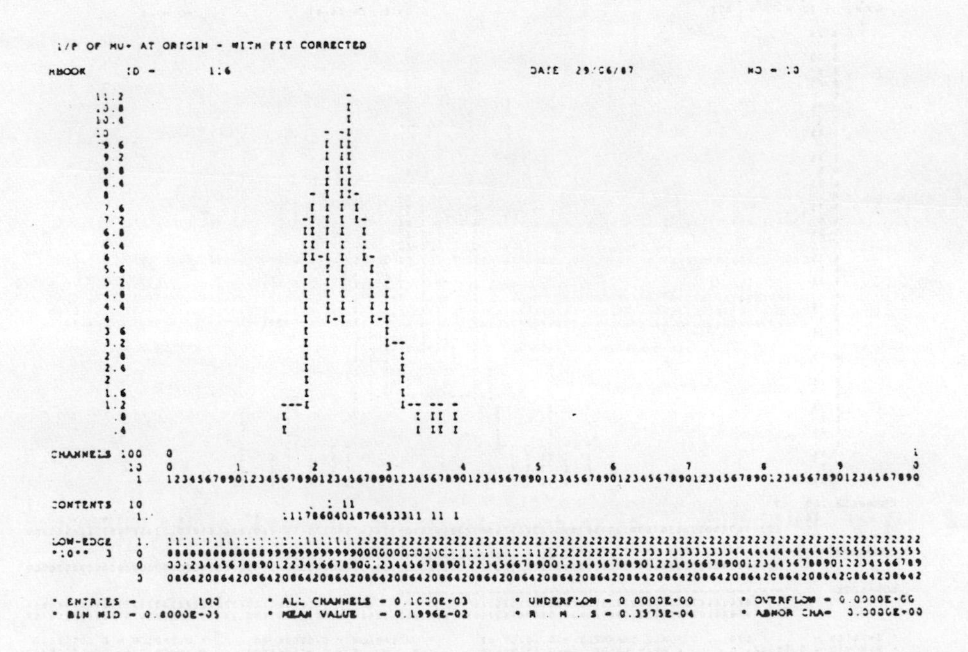

Fig.3 $1/p$ distribution of the reconstructed muons at their origin $P_{cut} = 1\ GeV$.

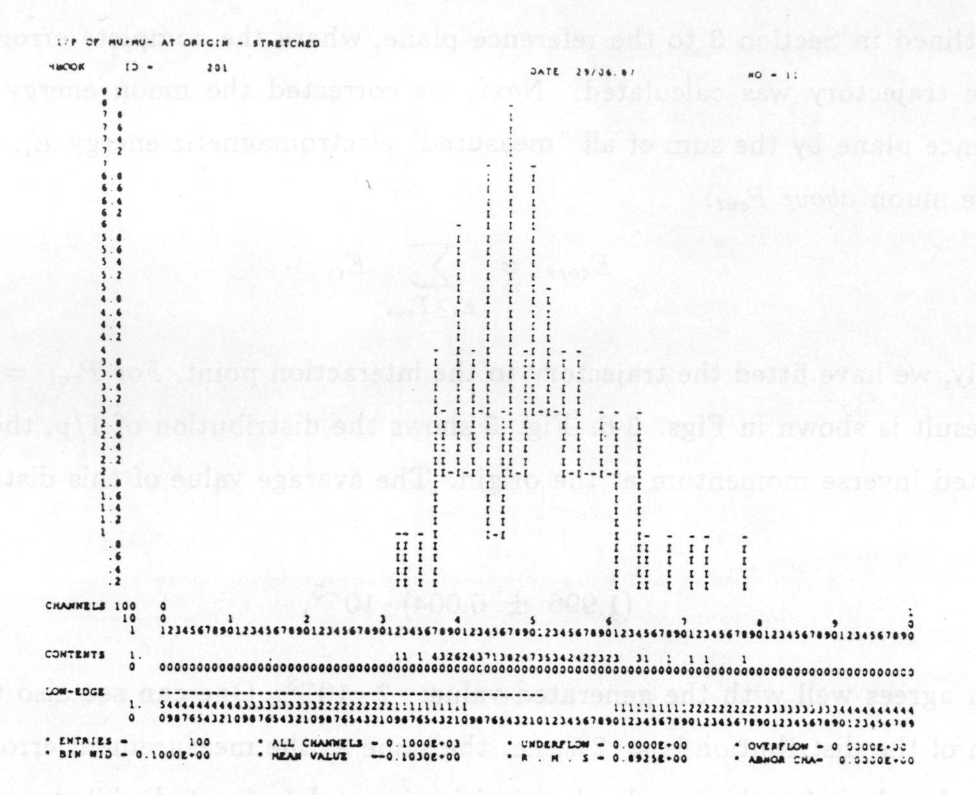

Fig.4 **"Strectched" distribution of $1/p$**

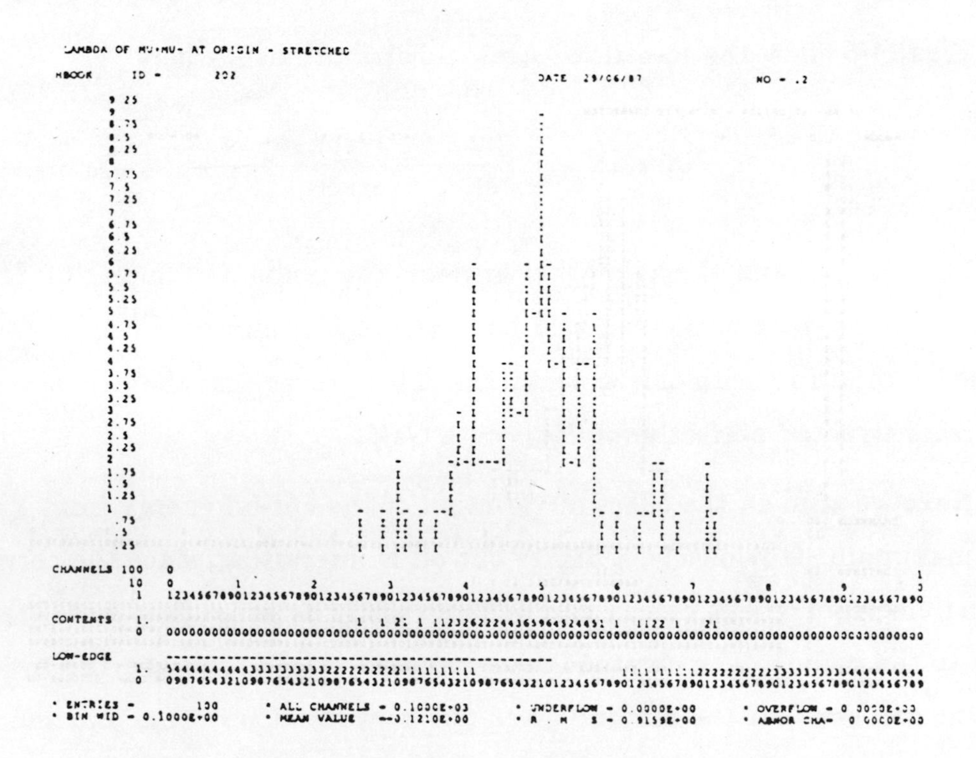

Fig.5 **"Strectched" distribution of the dip angle of the muon**

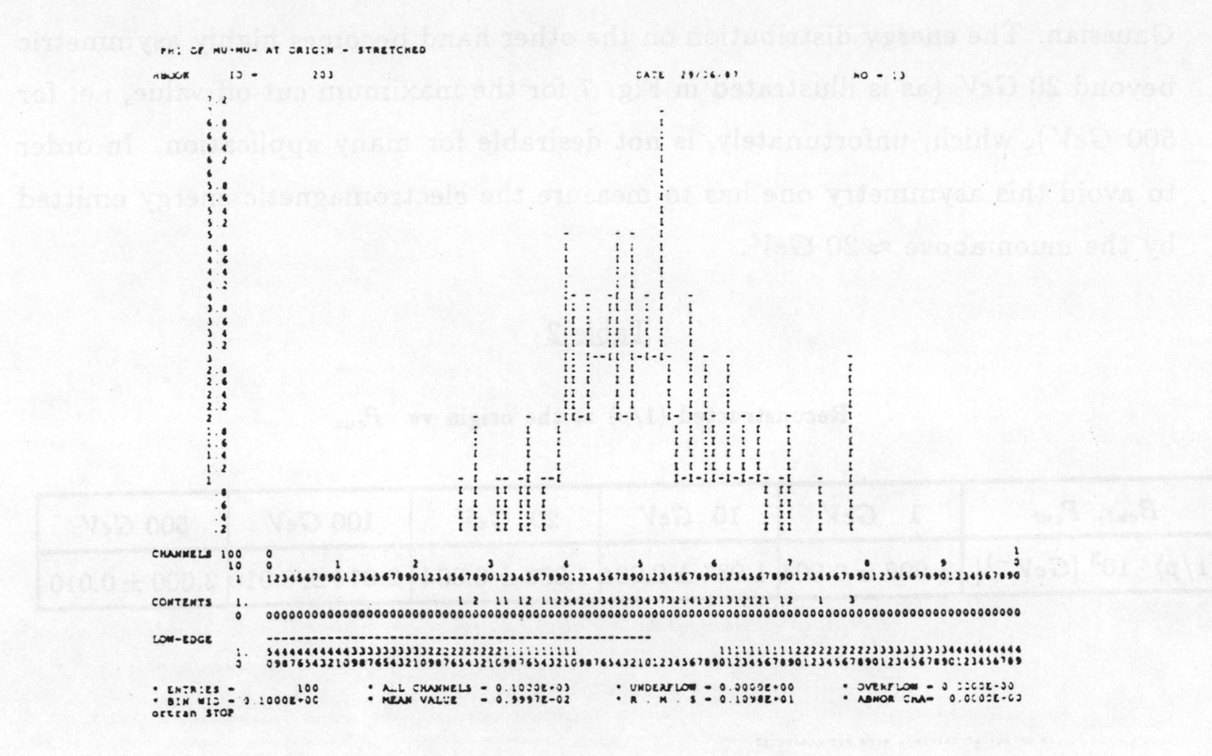

Fig.6 **"Strectched" distribution of the azimuthal angle of the muon**

Figs. 4-6 show the so-called "stretch-distributions", where

$$"stretch" \;=\; \frac{true\; -\; reconstructed}{error}, \tag{4.5}$$

for $1/p$, the dip and the azimuthal angle of the muon at the origin. In the ideal case all these distributions should have zero mean and unit r.m.s.; the result is indeed compatible with this expectation. This proves that the bench-mark muons are reconstructed perfectly for $P_{cut} \;=\; 1\; GeV$.

Next we studied the effect of variation of the cut-off energy from 1 GeV upto the maximum. Obviousely, a higher cut-off is interesting, since the corresponding higher electromagnetic energy can be more easily detected (i.e. more easily associated to the muon). The obtained reconstructed $\langle 1/p \rangle$ values are displayed in Table 2. One can see that the method gives *unbiased* result upto the maximum value of P_{cut}, i.e. even in the case when *no* electromagnetic energy deposit is measured *at all.* This result is encouraging for the application when the muon is *not isolated.* The angle reconstruction remains unbiased as well, the distribution is very close to

Gaussian. The energy distribution on the other hand becomes highly asymmetric beyond 20 GeV (as is illustrated in Fig. 7 for the maximum cut-off value, i.e. for 500 GeV), which, unfortunately, is not desirable for many application. In order to avoid this asymmetry one has to measure the electromagnetic energy emitted by the muon above $\approx 20\ GeV$.

Table 2

Reconstructed $\langle 1/p \rangle$ at the origin vs. P_{cut}

B_{cut}, P_{cut}	1 GeV	10 GeV	20 GeV	100 GeV	500 GeV
$\langle 1/p \rangle \cdot 10^3\ [GeV^{-1}]$	1.996 ± 0.004	1.984 ± 0.004	1.983 ± 0.004	2.014 ± 0.010	2.000 ± 0.010

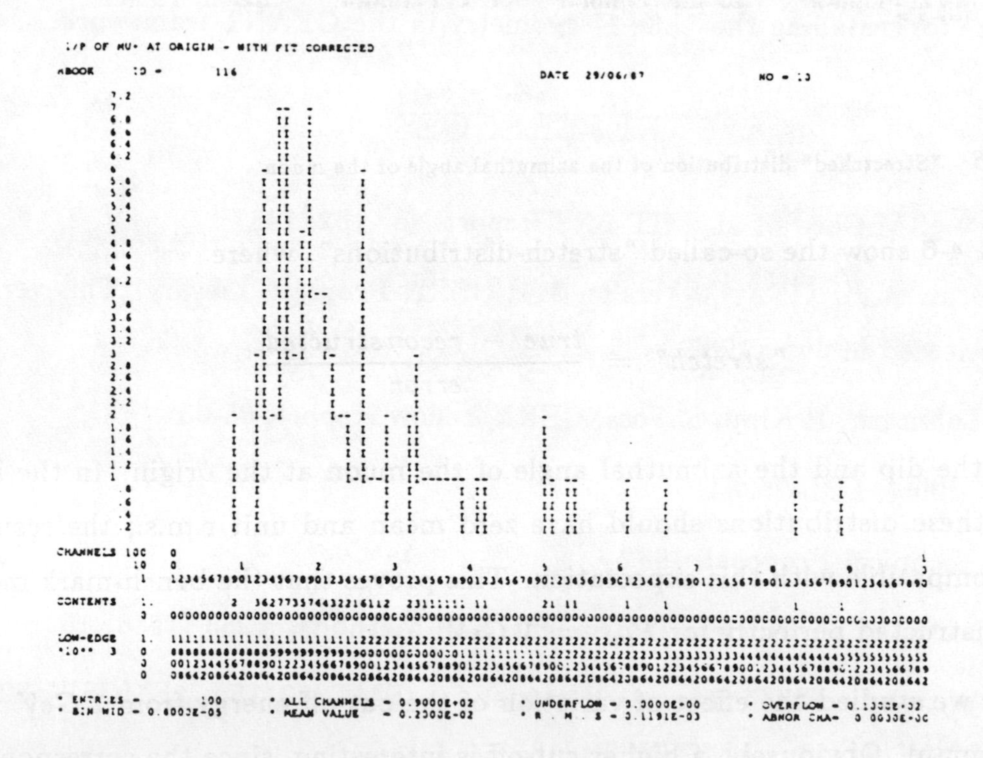

Fig.7 $1/p$ **distribution of the reconstructed muons at their origin** $P_{cut} = 500\ GeV$.

5. Conclusion

We have described tools, which can cope with the unusual simulation of the passage of *TeV* energy muons through the collider detectors and with their reconstruction, when they are assumed isolated. In the particular example of the "L3+1" Detector we have shown that a 500 *GeV* muon can be perfectly reconstructed and its parameters at the origin can be as precisely determined as at the point of the measurement even if these two points are separated by materials of hundreds of radiation length.

ACKNOWLEDGEMENTS

I am indebted to Prof. S.C.C.Ting for supporting this work, as well as to Prof. M.Chen for his constant interest and many fruitful discussions. I thank also Dr S.Banerjee for installing the "L3+1" geometry in the GEANT framework.

REFERENCES

1. **See e.g. M.Chen et al. MIT-LNS Report No. 162, 1987**
2. **R.Brun et al. GEANT Version 311, CERN Program Library. This version is released in June 1987**
3. **W.Lohmann, R.Kipp, R.Voss, CERN Yellow Report, 85-03**
4. **G.Trilling, LBL-22384**
5. **L3 Technical Proposal, 1982**
6. **U.Becker et al., Nuclear Instruments and Methods, A253 (1986) 15**
7. **W.Wittek, EMC Internal Reports: EMC/80/15, EMCSW/80/39, EMCSW/81/13, EMCSW/81/18**
8. **A.Haas, Contribution to the EMC UTIL42 package**

The Xenon Olive Detector:
A Fast and Precise Calorimeter with High Granularity for SSC

M. Chen*, C. Dionisi$, Yu. Galaktinov#, G. Herten*, P. LeCoultre%, Yu. Kamyshkov#, K. Luebelsmeyer+, W. Wallraff+, R. K. Yamamoto*

Abstract

We propose a fast detector with fine resolution and granularity, consisting of a shower counter and a hadron calorimeter. Both are linear, insensitive to radiation damage and with good spatial resolution in 3-dimensional space in order to measure di-electron and jet masses at the next generation high luminosity colliders.

New heavy particles such as the Higgs, top or fourth generation quarks or leptons with masses up to a few TeV could be produced abundantly at the next generation hadron colliders with luminosity $> 10^{33} cm^{-2}/s$. The main challenge has been to find a suitable signal which can stand out among huge QCD backgrounds. These heavy particles, when their masses are above 200 GeV, decay predominantly into at least one real weak boson [1]. The weak boson mass can be reconstructed from its final state leptons or jets to tag these high mass particles.

As the masses of these particles become larger ($>$ 0.5 TeV), the weak bosons from the decay of these particles become more and more energetic and their decay products also more and more collimated (with an opening angle of about 10 degress for 1 TeV Higgs). Experimentally, the situation is similar to the case of detecting neutral pions in the GeV region, i.e. to reconstruct their masses by using the two final state photons. As is well known, very fine electromagnetic shower counters are needed for neutral pions above a few GeV of energy. Here the two photons are replaced by a pair of charged leptons, or more frequently by two collimated jets from the weak boson decays. In order to reconstruct the invariant mass of the electrons or two jets which are closely spaced, one needs a calorimeter which can measure the energy distributions of the di-electrons or jets in 3-dimensional space precisely.

---

* MIT, Cambridge, Mass 02139,USA

$ Universita 'La Sapienza' and INFN , Rome, Italy

ITEP, Moscow 117259, USSR

% ETH, Zurich, Switzerland

+ I. Physikalishes Institut, RWTH Aachen, D-51 Aachen, Bundesrepublik Deutschland

[1] M. Chen, C. Dionisi, G. Herten, and E. Nagy MIT-LNS REPORT No.162, 1987

In the case of di-jets from high p_t weak boson decays, hundreds of particles are concentrated in a small cone of a few degrees of opening angle. It is difficult to use timing information or shower profiles to locate the position of individual tracks. Thus the calorimeter must be divided into many physical cells. For Higgs like particles with mass > 0.5 TeV, the spatial resolution of the calorimeter becomes more important than the energy resolution of the jets in reconstructing the invariant masses. Thus the mass resolution of the detector is to a large extent limited by the physical size of the cell as well as by the lateral spread of the shower.

From the above studies, we conclude that an ideal calorimeter at next generation colliders with high luminosity should meet the following criteria:

1. Fast response time, comparable to the beam crossing intervals of $10 - 20ns$, in order to cope with the high interaction rate ($> 10^8$/s). The drift velocity as a function of field strength of some known liquids are shown in Fig. 1, respectively. As seen, the fastest known ionization media at high field strength is the TMS liquids[2] ,[3] One could achieve a 20 ns response time at high field (50 kV/cm) and small drift distances (1mm). On the other hand, as shown in Table I and Fig. 2, the scintillation speed of liquid xenon (with a peak wave length of 168 nm) is even faster with a width of 3 ns for 96% of the light and 20 to 30 ns for the rest. The large pulse height together with such a sharp rise time makes it possible to obtain a resolution of much less than 1 ns. This is in sharp contrast with Ba F_2 which has only 9% of light for the fast component while 91% of the light is as slow as 620ns (Table I).

2. Particle sensitive media which are insensitive to radiation damages. The liquids such as TMS[3] and xenon[4] are among the most stable and insensitive to radiation damages. For liquid xenon we expect a tolerance of more than 10^7 rad.

3. Linear response over large dynamical range, as there are many particles in a small cone of no more than a few degrees. The sizes of signals and the dynamical range of some relevant media such as xenon, argon, NaI, BGO, TMS and TMP are shown in Table I. The scintillating light output from liquid xenon [5] is larger than that of BGO and linear to 10^4 equivalent minimun ionization particles (mip), which is an order of magnitude better than NaI and two orders of magnitude better than BGO. TMS is also known[3] to be very linear.

---

[2] Andries Hummel and Werner F. Schmidt, Radiation Research Review, 5 (1974) 199,R.A. Holroyd and D.F. Anderson, Nuc. Instr. and Meth. A236(1985)294-299; S. Ochsenbein SIN preprint (1987)

[3] D.F. Anderson, Fermilab, Batavia, IL; C. Rubbia and Werner F. Schmidt, UA1; and S. Ochsenbein, SIN, Private Communications.

[4] C.S. Wu and A. Sayres, Rev. Sci. Instr. 28(1957)758; R.A. Muller et al.,Phys. Rev. Lett. 27(1971)532.

[5] M. Mutterer Nucl. Inst. and Meth. 196(1982)73; A. Hitachi et. al., Nucl. Inst. and Meth. 196(1982)97.

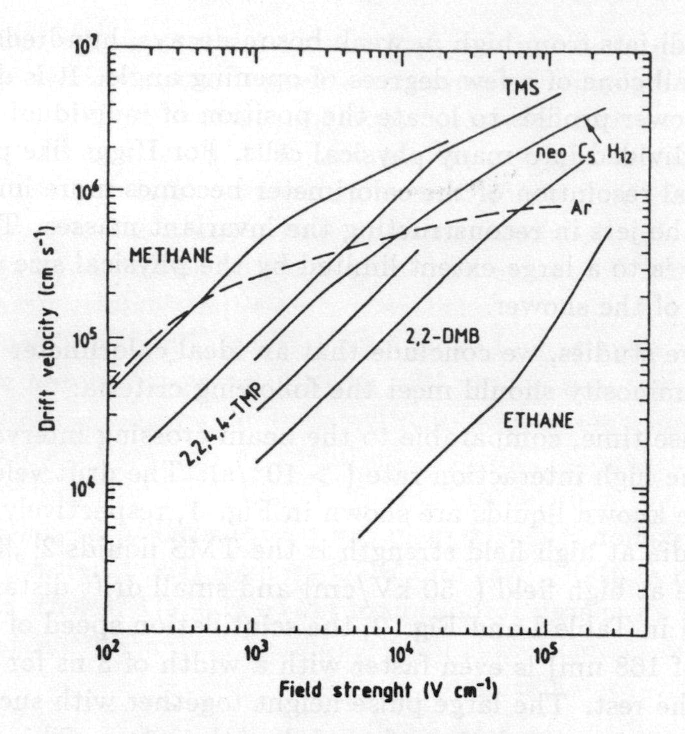

Fig. 1 The electron drift velocity as a function of field strength for some known liquids [2,3]. Liquid xenon is similar to liquid argon.

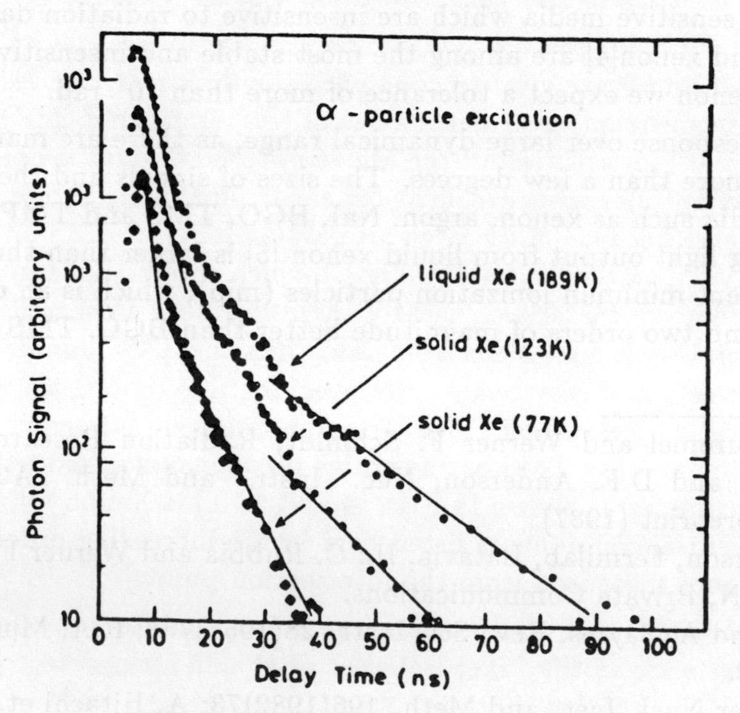

Fig. 2 The distribution of scintillation light of xenon to show speed of the faster and slower components[10].

4. Uniform acceptance and excellent energy resolution. For jets, it can be achieved through studies of the delicate compensation effects through nuclear interactions in Uranium and TMS. Some relevant physical properties, such as density and radiation length, of xenon are listed in Table I in comparison with NaI, BGO and BaF_2. For electromagnetic showers, liquid xenon, due to its high density, small radiation length (2.8 cm) and high light yield (750000 photo-electrons/GeV), could achieve an excellent energy resolution of $3/\sqrt{E}\%$ [6]. The large light output of the liquid xenon also eases the requirement on amplifiers and makes it insensitive to noises, particularly uranium noise. For jets, we need both excellent angular and energy resolutions. The latter depends on the e/pi ratio of the light output per unit energy loss, dL/dE, for electrons and pions. The ratio of the light output per unit energy loss, dL/dE, for alpha and beta particles, has been measured to be 1.1 $\pm$0.2 in xenon. This is again in sharp contrast with BaF_2, NaI and BGO etc. which have a ratio of about 0.5 (Table I). Thus for a linear and scintillating shower counter such as liquid xenon, the e/π ratio can be adjusted to be much closer to one than other scintillators such as BaF_2, NaI and BGO etc.(see H.S. Chen Et. al. , contribution to this workshop) and an energy resolution of $(45/\sqrt{E} + \epsilon)E\%$ with ϵ being some small number = O(1), should be achievable for a combined system with a xenon shower counter and a uranium-TMS hadron calorimeter.

5. Good spatial resolution in three dimensional space in order to reconstruct the invariant mass of closely spaced di-electrons or di-jets. This can be achieved only by dividing the calorimeter into very fine cells with a pseudo-rapidity interval of around 0.05. Also material with short radiation and absorption length is needed in order to limit the lateral spreading of shower sizes.

While items 1 and 2 are vital to cope with the high rate due to the machine, the rest is essential to reconstruct invariant masses of jets, and particularly for high P_t weak bosons. No single material satisfies all the above criteria. Liquid xenon, with its fast rise time, and high linearity, clearly is excellent for electro-magnetic shower counter but not for hadron calorimeter because of its large interaction length. TMS on the other hand has a low electron yield as can be seen in Fig. 3 which shows the large variation of electron yields of some known materials. Clearly there is a need to search for new ionization media even faster and with higher electron yield than the TMS liquid, and with properties consistent with safe and easy usage. However, for this report, we will discuss a hypothetic detector, called the Xenon Olive Detector, using liquid xenon as the electromagnetic shower detector and the TMS as the active media sandwiched with Uranium plates for the hadron calorimeter as shown in Figs. 4 and 5. This Xenon Olive Detector consists of the following essential components listed here in the increasing order of the radius R, defined as the distance measured from the intersection point:

1. For R between 5 and 40 cm, the space is reserved for some fast and high resolution tracking devices such as the Pixel Detector or silicon strip detectors.

[6] K. Masuda et al., Nucl. Inst. and Meth. 160(1979)247

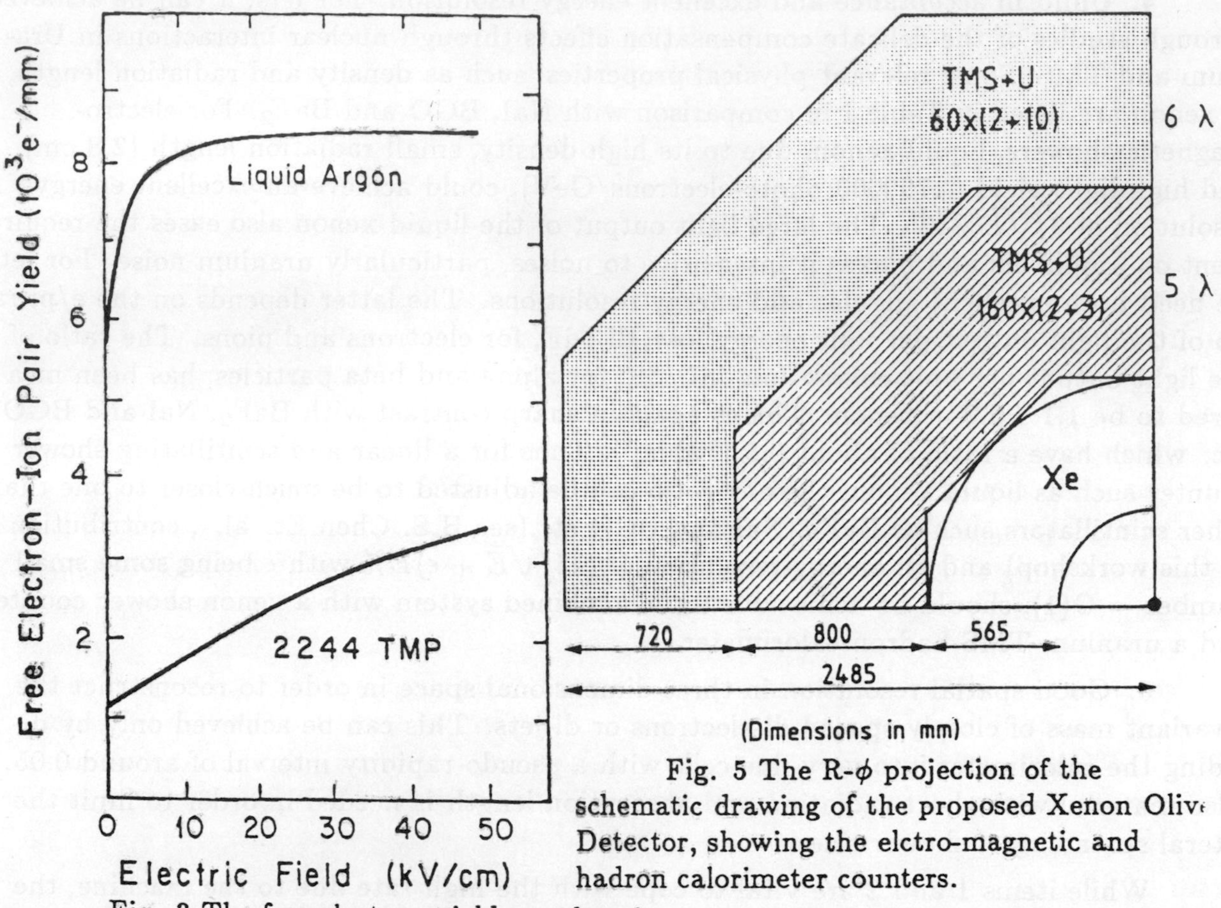

Fig. 3 The free electron yields produced by a minimun ionization particle in liquid argon and TMP as a function of the applied electric field[8]. TMS is about same as TMP while liquid xenon is about 2.5 times larger than liquid argon.

Fig. 5 The R-ϕ projection of the schematic drawing of the proposed Xenon Olive Detector, showing the eltro-magnetic and hadron calorimeter counters.

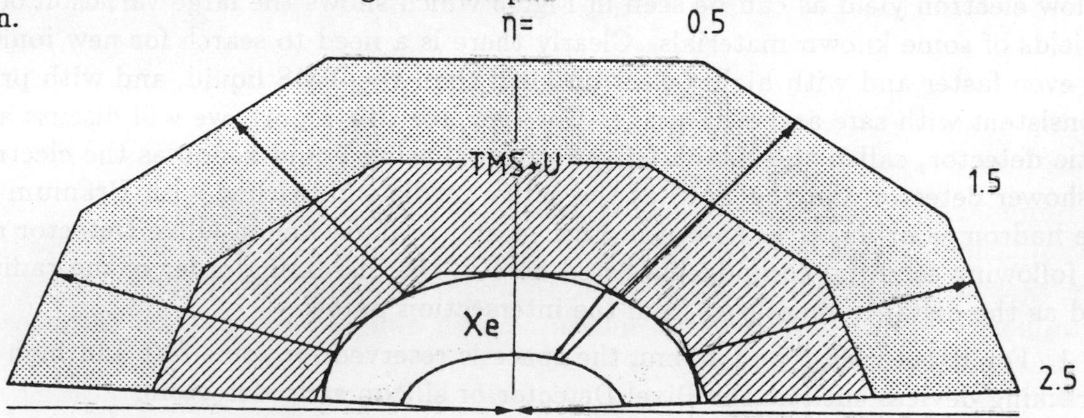

Fig. 4 The schematic drawing of the proposed Xenon Ball Detector, showing the cell arrangement of the xenon electro-magnetic shower counter.

2. For R between 40 and 100 cm, 20 radiation lengths of liquid xenon is used as electromagnetic shower counters. In order to provide good spatial resolution, it is divided into about 2 x 10^4 cells, about 2.2 degrees in both θ and ϕ angles at θ around 90° and the size is reduced in θ to 1 degrees at θ around 10°, corresponding to a pseudo-rapidity interval of 0.03 at $\theta = 90^o$, 0.05 at $\theta = 45^o$ and 0.06 at $\theta = 15^o$. For $\theta < 15^o$, Liquid xenon has to be sandwiched with lead plates to reduce the equivalent radiation length to about 5mm. The total volume of liquid xenon is about 5 cubic meters and its cost is between that of NaI and BGO or BaF_2 crystals (table IIIb). Fast wave length shifters (with rise time about 1ns) such as 2 mg/cm^2 of Sodium-Salicylate [6] or 30 $\mu g/cm^2$ of p-p' diphenylstilbene (D.P.S) [7] have been used to convert the ultraviolet light into visible light which in turn could be read out by fast photo multipliers or diodes at the far end of each cell. The effect of eliminating internal reflections against the wall should conpensate the focusing effect of the trapeziform geometry of the cell and yields uniform light collection efficiency as we have learned with the L3 BGO crystals.

3. For R between 100 and 250 cm, there are two sets of hadron calorimeters with different sampling thickness. First there are 160 layers of 2mm liquid TMS sandwiched with 3 mm uranium plates. This is followed by 60 layers of 2mm liquid TMS sandwiched with 10 mm uranium plates to yield a total of 11 absorption lengths. It is also divided into small cells to match the liquid xenon counters. The signals from many layers of the same cell can be linked together to form a tower. A similar though a bit slower media, TMP, has been demonstrated by the UA1 collaboration to have good spatial and energy resolution[8] when used in an electro-magnetic shower counter sandwiched with uranium.

4. for R > 250 cm, large drift chambers, such as these of the proposed L3+1 detector [9] (Fig 6), measure muon momenta precisely to obtain a momentum resolution of $4\% P_t^2$ with p_t in TeV. It has been demontrated [1] that it is important to measure both muon and electron momenta precisely in order to establish the Higgs signal against background.

In Fig. 7, we show the energy distribution in the xenon detector for one hundred GeV electrons as simulated by the Geant3 Monte Carlo program. Typically more than 62% of the energy concentrates in one or two cells, which allows a precise determination of the impact point of the electron position to 1 to 2 mm. The results of a detailed simulation of such a detector using the Geant3 Monte Carlo program to reconstruct di-electrons from heavy Higgs decays will be described in a subsequent paper.

Summary: We described a new type of fast calorimeter detector, which is insensitive to radiation damage and has very good energy and spatial resolution. These properties are vital for the measurement of di-electron and di-jet masses at the next generation high luminosity colliders. The xenon shower counter alone, due to its fast response time,

[7] John A Northrop, Rev. Sci. Inst. 29(1968)437; Peter Kozma and Cestmir Hradecny, Nucl. Inst. and Meth. A254 (87)639

[8] M. Albrow et. al., CERN-EP/87-55(1987).

[9] U. Becker et. al., Nucl. Inst. and Meth. A253(86)15.

excellent linearity and resolution, and high radiation damage tolerance, may also serve as ideal small angle luminosity counters for experiments at LEP, SLC, HERA and future colliders.

Acknowledgement. We would like to thank Profs. S. C. C. Ting, J. Friedman, A. Kerman and G. Gilchriese for their support and encouragement and Profs. D. F. Anderson, U. Becker, B. Borgia, G. Charpak, C. Fabjan, H. Fesefeldt, C. Kukowka, E. Lorenz, D. Luckey, J. P. Martin, U. Micke, H.U. Martyn, M. Mutterer, E. Nagy, H. Newman, S. Ochsenbein, P. Piroue, C. Rubia, K. Strauch, Werner F. Schmidt, L. Sulak, J. P. Theobald, and W. Van Donninck for many useful discussions.

Table I. The scintillation light decay times for both the fast and slow components and the ratio of their amplitudes; The size of signals and the dynamical range for both scintillation and ionization output; and the relative light output of alpha and beta particles of several common media including BaF_2 in comparison with liquid xenon. t_1, t_2 are the decay time of the fast and slow component and A_1, A_2 their amplitudes.

scintillation	*Liquid Xe*	*BaF_2*	*Na I*	*BGO*
fast decay time T_1 (ns)	2 to 3 [10]	1	250	300
slow decay time T_2	20-30	620	—	—
ratioA_1/A_2	25	0.09		
rel. output from α	0.137	0.006	1.	0.08
ratio of α/e [10,11]	1.1± 0.2	0.34	0.5	0.23[14]
K_B^{-1} in MeV/cm[12]	30600	− − −[13]	3670	1000
density (g/cm^3)	3.06	4.89	3.67	7.1
rad. length (cm)	2.8	2.05	2.59	1.12

ionization	*Liquid Xe*	*Liq. Ar.*	*TMS*	*TMP*
rel. output of a mip	12.6	46	1	1.1
K_B^{-1} in MeV/cm	300	200	–	70

references [10] [11] [12] [13] [14]

[10] S. Kubota et. al., Nucl. Inst. and Meth. 196 (82)101, and A242(1986) 291
[11] M.R. Farukhi et. al.,IEEE Trans. Nucl. Sci. NS-18(1971)200.
[12] The output/energy loss, dL/dE, reduces by 50% in Birk's law. See Ref.5.
[13] It is poorly defined since the size of the fast component critically depends the particle types, such as α,e or heavy ions etc., see refs. 10 and 11.
[14] J. P. Martin, private communication.

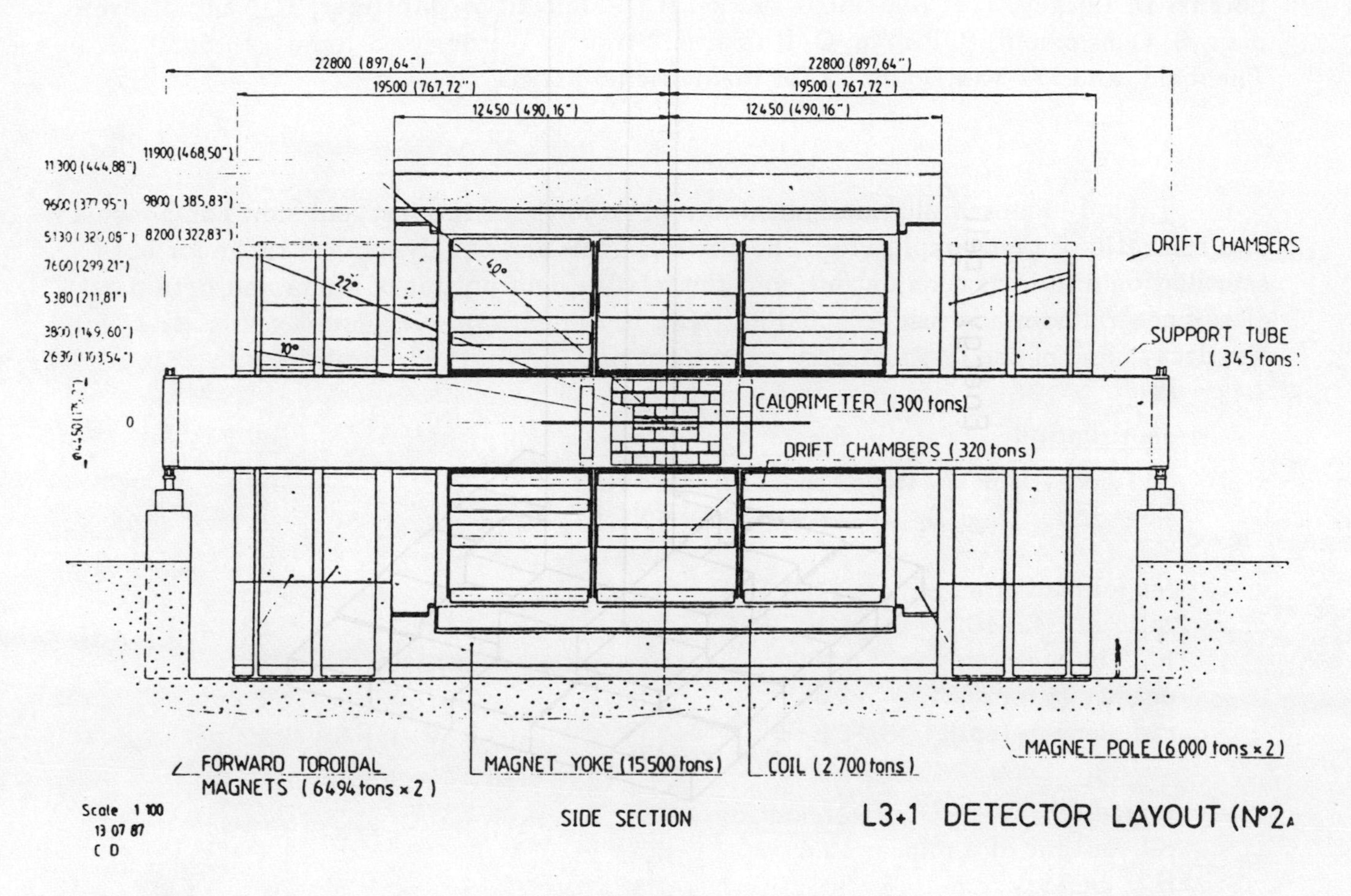

Fig. 6 The schematic drawing of the proposed L3+1 Detector showing the large muon detector and the endcap iron toroids. Cal can be the Xenon Olive and TMS-U calorimeter.

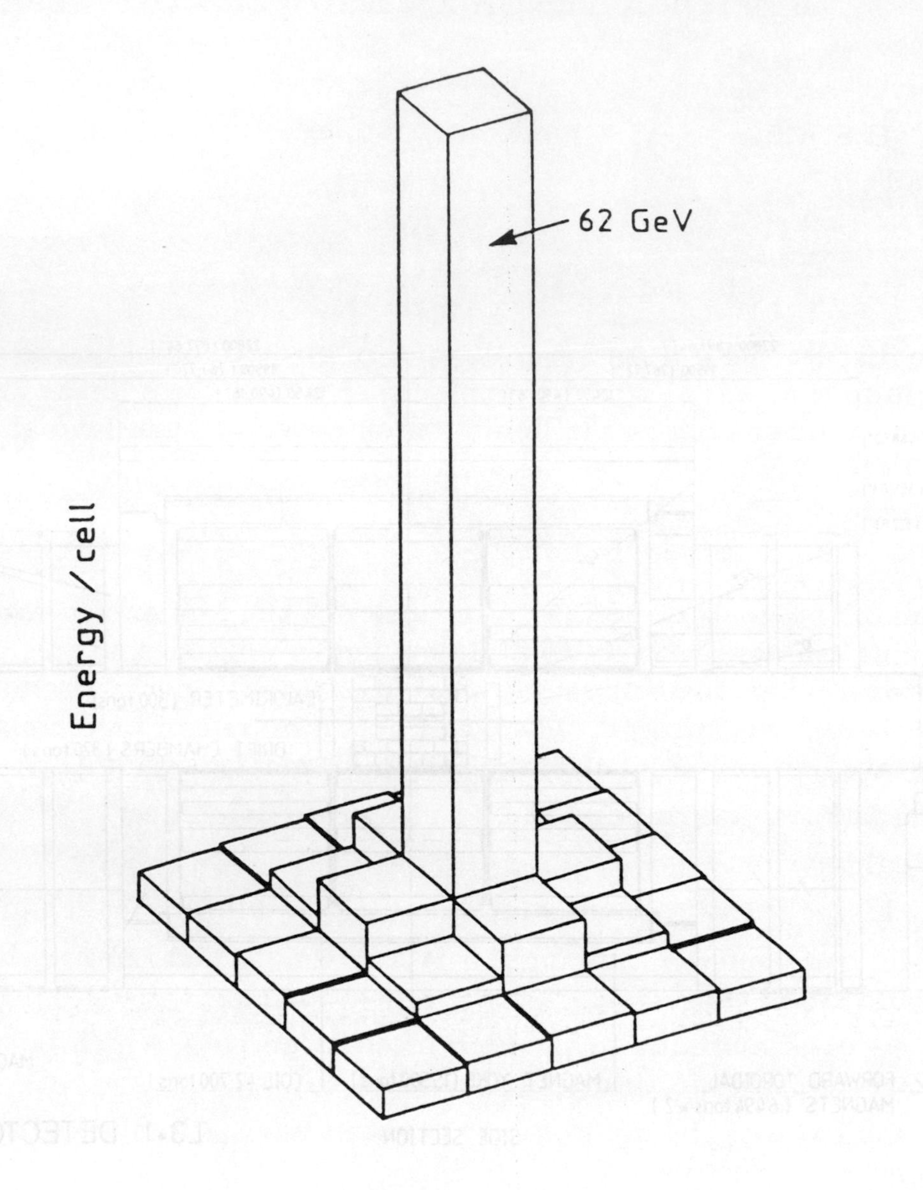

Fig. 7 The energy distribution in the xenon detector for a hundred GeV electrons as simulated by the Geant3 Monte Carlo program. Typically more than 62% of the energy concentrates in one or two cells.

e/pi FOR A XENON EM SHOWER COUNTER

H.S. Chen[+], M. Chen[*], C. Dionisi[$], G. Herten[*], C. Luci[$]

Abstract

We discuss the e/π for several types of electromagnetic shower counters and conclude that e/π should be much closer to one for a linear and scintillating shower counter such as the liquid xenon detector than lead glass, NaI, BaF_2 and BGO etc.

One of the goals of SSC is to reconstruct the jets and measure the jet-jet invariant masses precisely to find new heavy particles. Higgs, top and high mass leptons, when their masses are above 200 GeV, decay predominantly into at least one real weak boson. The weak boson mass can be reconstructed from its final state leptons or jets to tag these high mass particles[1].

The factors determining the jet mass resolution are:

1. calibration,
2. cluster identification, which requires small physical calorimeter cells,
3. intrinsic energy and angular resolution, which requires large signal outputs and small cells,
4. event pile up due to the high interaction rate, which favors detectors with fast response time, and
5. the ratio of the light output per unit energy loss, dL/dE, for electrons and pions, i.e. e/π ratio.

Since there is large fluctuation in the percentage of the neutral component of a jet, in order to get good jet energy resolution, it is often desirable to design a calorimeter with e/π close to be 1. It has been shown [2] that if the e/π value deviates from 1 by more than 0.1, it will produce a systematic error

---

[+] IHEP, Beijing, China

[*] Physics Department, MIT, Cambridge, Mass 02139,USA

[$] Universita 'La Sapienza' and INFN , Rome, Italy

[@] Invited paper to SSC-WORKSHOP, Berkeley, Calif. July7 to 17, 1987

[1] M. Chen, C. Dionisi, G. Herten, and E. Nagy MIT-LNS REPORT No.162, 1987; also submitted to this workshop

[2] R. Wigmans, contribution submitted to this workshop

in energy larger than 2%, which becomes significant for jets above 200 GeV. On the other hand, some non-compensating EM shower counter (such as NaI, BGO or rare gases like xenon) situated in front of a fully compensating hadron calorimeter such as uranium-TMS sandwich counter has the following advantages:

1. The electromagnetic component of the jets (about 40%) is measured with much higher precision than most compensating calorimeters;

2. In addition, this calorimeter has very good energy resolution for photons and electrons even at low energies. This can be essential for detecting new onium via the dielectron mode or a Higgs boson via the two photon decay mode in the intermediate mass region[3].

The liquid xenon detector was thus proposed[4] to be a precise and compact shower counter for high luminosity colliders because it is

1. fast with response time of 3ns,

2. radiation damage insensitive,

3. linear to 10^4 mip,

4. with large pulse height (more than 75000 photo-electrons/ GeV) and thus excellent signal to noise ratio,

5. excellent granularity to measure space point in real 3-D space.

Here we like to point out one more important property of xenon, i.e. there are a lot of evidence showing that its e/pi is close to 1 in contrast to lead glass, BaF_2, NaI and BGO etc.

For an EM shower counter, e/pi refers to the relative light output for the same amount of energy loss, E_{loss}, of a high energy electron or pion. E_{loss} is the difference of the incoming particle energy and the total energy of the outgoing charged particles, photons, and neutrons which presumably can be measured by a fully compensated hadron calorimeter situated behind the shower counter.

There are three mechanisms which make e/pi ratio of many common shower counters, such as lead glass, NaI, and BGO etc., to deviate from one:

1. saturation: hadrons produce heavy ions via nuclear interactions. dL/dE for Heavy ions can be significantly lower than that of electrons due to saturation and also due to the fact that sometime electrons and hadrons excite the same media to different energy levels;

2. nuclear binding energy loss;

3. energy loss due to neutrinos.

---

[3] Contribution submitted to this workshop by the intermediate Higgs group

[4] M. Chen et. al. , MIT-LNS REPORT No.163,June 30,1987, also submitted to this workshop

Qualitatively, one can understand why e/pi = 1 for xenon shower counter using the following simple arguments:

The light output for an electron in a xenon counter is:

$$L(e) = E_{loss}\ Eff(e)$$

and for a hadron,

$$L(h) = (E_{loss} - < n >< U >)\ Eff(h)$$

from which we have,

$$L(h) = (L(e)/\ Eff(e) - < n >< U >)\ Eff(h)$$

where $< U >$ = mean undetectable energy loss per nuclear interaction, including nuclear binding energy and energy loss due to neutrinos. It is feasible to have $< U > = 0$ for a compensated hadron calorimeter such as TMS-Uranium sandwich counter; $< n >$ = mean no. of nuclear collision of all the particles in the shower in the counter excluding elastic or quasi-elastic scatterings. $< n >$ = hundreds for a hadron calorimeter, but $< n > = O(10)$ for a hadron interacting in an EM shower counter; Eff(h,e) is the efficiency to convert ionization energy loss to scintillating light for h and e respectively.

We note that even though $< U >$ is small for high energy particles, $< n >< U >$ is not negligible for hadron calorimeters unless $< U > = 0$, i.e. for compensated calorimeters. However, as already described above, since $< n >$ is small for a scintillating EM shower counter like xenon, $< n >< U >$ is negligible even though $< U >$ does not exactly vanish. Furthermore, Eff(h) = Eff(e) for xenon since liquid xenon is known [5] to be linear to 10^4 MIP[5], [6] .

Therefore we have L(h) = L(e), or e/pi =1 for a xenon EM shower counter.

---

[5] M. Mutterer Nucl. Inst. and Meth. 196(1982)73; A. Hitachi et. al., Nucl. Inst. and Meth. 196(1982)97.

[6] Saturation: The output/energy loss, dL/dE, reduces by 50% like in Birk's law.

One may then ask why e/pi is greater than one for many commonly used shower counters such as lead glass, BaF2 [7] or BGO[8]? The e/pi of a lead glass counter is about 3 because EFF(e) is much greater than EFF(h) for the emmision of Cerenkov light. The e/pi of BGO in front of one hadron calorimeter module has been measured to be 1.6± 0.2 for 20 GeV π [13] partly because of side leakage of hadronic showers and partly because BGO saturates at 1 GeV/cm [8], which is two orders of magnitude lower than that of xenon[5]. After correcting the saturation effect (0.3), we expect the e/pi ratio in BGO should reduce to 1.3 ± 0.2, which includes side leakage and binding energy loss. The e/pi for BaF_2 must also deviate significantly from being one because its α/e ratio is only 0.34[11], [9] .

Experimentally, the ratio of the light output per unit energy loss, dL/dE, for alpha and beta particles, has been measured to be 1.1 ±0.2 [10] in xenon. This is in sharp contrast with that of BaF_2 (0.34[11]), BGO (0.23[12] ,[13]) or NaI(0.5). Table I summarizes some of the important properties of xenon together with several other media of shower counters:

Table I. Some important properties of xenon, including the saturation constant for both scintillation output; the relative light output of alpha and beta particles; and the scintillation light decay times for both the fast and slow components and their ratio of amplitudes; and the size of the fast signals etc. in comparison with several common media including BaF_2. T_1, T_2 are the decay time of the fast and slow component and A_1, A_2 their amplitudes.

scintillation	*Liquid Xe*	*BaF_2*	*Na I*	*BGO*
density (g/cm^3)	3.06	4.89	3.67	7.1
rad. length (cm)	2.8	2.05	2.59	1.12
absorpt. length (cm)	55	30	41	22
K_B^{-1} in MeV/cm[6]	30600[5]	− − −[7]	3670[5]	1000[8,12]
K_B^{-1} in mip	10^4	− − −[7]	10^3	120
fast decay time T_1 (ns)	3[10]	1	250	300

[7] It is poorly defined since the magnitude of its fast component critically depends the particle types, such as electrons, α particles,or heavy ions etc.

[8] E. Lorenz, private communication.

[9] M. Laval et. al., Nucl. Inst. and Meth. 206 (1983)169

[10] S. Kubota et. al., Nucl. Inst. and Meth. 196 (1982)101, and A242(1986) 291

[11] M.R. Farukhi et. al.,IEEE Trans. Nucl. Sci. NS-18(1971)200.

[12] J. P. Martin, private communication.

[13] C. Chen et. al., to be submitted to Nucl. Inst. and Meth.

slow decay time T_2	20-30	620	—	—
ratio A_1/A_2	25[10]	0.09[9]	—	—
rel. output from α	0.137	0.006	1.	0.08
ratio of α/e [10,11]	1.1± 0.2	0.34	0.5	0.23[12]

Conclusions. As seen in Table I, xenon stands out for the following important features,

1. it saturates at much higher energy densities,

2. its α/e ratio is consistent with 1.

From these properties and from the above argument

3 that the neutrino energy loss and the fluctuation of the nuclear binding energy loss in an EM shower counter are small,

and the assumption

4. that neutrons will be efficiently detected by the uranium-TMS calorimeter,

we believe that for a xenon shower counter together with a fully compensated hadron calorimeter, the e/π ratio can be adjusted close to one and the system should yield excellent energy resolution for jets as well as for electrons/photons.

Acknowledgement. We would like to thank Drs. R. Cashmore, S. Cooper, C. Fabjan, H. Fesefeldt, P. Franzini, G. Gilchriese, E. Longo, K. Strauch, V. Telegdi, G. Trilling, and W. Wallraff for many useful discussions.

A FANTASY ON TRACKING INSIDE THE XENON BALL

R. Leiste, R.K. Yamamoto

I. GOALS

Ideally, a tracking device for the Xenon-ball should be able to determine the direction, impact point, and momentum of isolated charged tracks with good resolution.

Due to the limited volume offered by the Xenon-ball and because of the high event rates at the SSC, typical tracking devices such as drift chambers are not practical. Toward this end, the pixel device discussed by D. Nygren at this workshop, and/or microstrip devices seem to be the best candidates for this application.

For this study we assume the use of microstrip devices with 25 µm minimal pitch for our tracking system. However, with the advent of 25 µm x 100 µm pixel devices, the microstrips could be replaced without any loss in momentum resolution and with the added advantage of two dimensional coordinate measurements resulting in fewer ambiguities and superior pattern recognition capabilities.

II. REMARKS ON MOMENTUM RESOLUTION

In order to study the momentum resolution as a function of the point measuring accuracy σ_i, the number of detector elements in the RΦ-projection N and the lever arm L, a three ring arrangement was considered. The rings are positioned at radii R_1 (= 5 cm), $R_2 = R_1 + L/2$ and $R_3 = R_1 + L$, respectively.

For a microstrip device, the measurement accuracy σ is in first approximation proportional to the strip pitch D ($\sigma^2 = D^2/12$) and hence inversely proportional to the number of strips in a ring. As we are interested only in the RΦ projection, no detector segmentation in Z is assumed.

The sagitta of the particle trajectory in the RΦ projection is

$$S = \frac{.3 \cdot B \cdot L^2}{8 \cdot P_\perp}$$

In the three point approximation used here the sagitta is measured as

$$S_e = (X_1 + X_3)/2 - X_3$$

with an RMS error of

$$\delta S = [(\sigma_1^2 + \sigma_3^2)/4 + \sigma_2^2]^{1/2} .$$

The maximum momentum $P_{\perp max}$ to define the sign of the particle charge is given by the condition

$$P_{\perp}/\delta P_{\perp} = S/\delta S_e \geq 3.$$

Three cases representing different R dependences of σ were considered:

$$\text{Case A : } \sigma_i = \sigma_1 = 5\ \mu m$$

$$\text{Case B : } \sigma_i = \sigma_1 \cdot (R_i/R_1)^{1/2}$$

$$\text{Case C : } \sigma_i = \sigma_1 \cdot R_i/R_1$$

The results are summarized in Tables 1 and 2 and plotted in figs.1-4.

Table 1: $P_{\perp max}$ ("Physics") and N ("Cost") as function of L

L [m]	$P_{\perp max}$ [GeV] for case A	B	C	N for case A	B	C
.2	57	47	18	113k	62k	38k
.4	230	145	42	188k	78k	38k
.6	515	275	66	264k	91k	38k
.8	915	431	90	340k	102k	38k
1.0	1430	609	115	415k	112k	38k
.2	104	85	33	753k	413k	253k
.4	417	262	76	1253k	520k	253k

Table 2: $P_{\perp max}$/N ("Physics/Cost") and $P_{\perp max}$/N•L ("Physics/Cost•Space") comparison

L [m]	$P_{\perp max}$/N [GeV/10^3] for case A	B	C	$P_{\perp max}$/N•L [GeV/10^3•m] for case A	B	C
.2	.5	.8	.5	2.5	3.8	2.4
.4	1.2	1.9	1.1	3.1	4.7	2.8
.6	2.0	3.0	1.7	3.3	5.0	2.9
.8	2.7	4.2	2.4	3.4	5.3	3.0
1.0	3.5	5.4	3.0	3.5	5.4	3.0

For L = .4 m (in the present design the inner radius of the Xenon-ball) a value of $P_{\perp max}$ ~ 200 GeV can be attained. In case B one reaches about 70 % of this momentum at less than half of the cost. In both cases, the L^2 dependence favors a bigger lever arm, so an optimization with the Xenon-ball concerning cost and performance would be useful.

Obviously, case C yields very little physical output and should be considered only if the cost of the tracking system is the limiting parameter.

In general, a three point measurement is certainly not sufficient for pattern recognition purposes. Usually, one assumes in the order of 20 or more points per track. In a microstrip arrangement with 20 rings the multiple scattering would cause some limitations at small momenta. In Table 3 the values of $P_{\perp min}$ are summarized, the momentum below which the coordinate uncertainty in the outermost detector from multiple scattering is bigger than the assumed measurement accuracy. In the calculations a detector thickness of 300 μm was assumed.

Table 3: $P_{\perp min}$ as function of L

L [m]	$P_{\perp min}$ [GeV] for case A	B	C
.2	36	16	7
.4	73	24	8
.6	109	30	8
.8	146	35	9
1.0	182	40	9

Of course, $P_{\perp min}$ is not a pricipal limitation, simply below this momentum the internal resolution of the tracking system is not completely used.

On the other side, the overall accuracy of the system and hence $P_{\perp max}$ raises roughly with the square root of the number of points, but N increases linearly. The recalculated numbers for L = .2 m and .4 m are added in Table 1 as separate rows. Although the numbers for $P_{\perp max}$ are impressive, we notice the huge increase of N and therefore of the cost of the system.

III. POSSIBLE SCHEME FOR THE TRACKING SYSTEM

As stated from the very beginning, the tracking system for L3+1 is not a general purpose tracking device. It is not foreseen to resolve complicated high multiplicity topologies, but rather as part of the EMC to identify isolated electrons, positrons and gammas against hadrons. Therefore pattern recognition capabilities should not play the dominant role in the design of the system.

From that and the calculations in Chapter II one can draw a few general conclusions:

- A tracking system composed of microstrip detectors in any of the cases A, B or C fulfills the requirements in the determination of the direction and impact point of a particle at the EMC entrance.

- Obviously, the lever arm should be chosen as big as tolerable from EMC design considerations.

- For a given lever arm, the number of detector rings should be chosen as small as possible. Although not the main task, some pattern recognition problems have to be solved, probably by treating the EMC and the tracking system as one unit. Besides that, the maximum measurable momentum as well as the cost of the sytem will strongly influence the choice.

As an example, we consider a five ring system drawn in fig.5. With this number of possible points per track one gets 99 % efficiency to measure at least three points even if 10 % of the strips are dead.

Assuming a constant pitch of 25 µm in all rings (case A in Chapter II) and a maximum size of an individual microstrip device of 10 cm x 10 cm, one can calculate the following numbers:

Lever arm at $\theta = 90°$		:	.3 m
$P_{\perp max}$	for $90° \geq \theta \geq 38°$	:	150 GeV
	for $38° \geq \theta \geq 24°$	:	150 - 66 GeV
	for $24° \geq \theta \geq 9.5°$	:	66 - 8 GeV
Rapidity range covered		:	≈ 2.5
Occupancy at the innermost ring		:	.03
Number of individual microstrip devices		:	540
Total number of strips (=N)		:	2.2×10^{6}

Before starting the real design and construction of such type of a tracking sytem, a number of problems have to be studied. Some of them are:

- Radiation hardness of microstrip detectors at very high doses, especially against neutrons.

- The charge collection time in microstrip detectors is of the order of 10 nsec, matching very well the SSC requirements. However, at present there is no fast high integrated readout electronics available. As this possible future electronics has to be mounted directly on the detector, it should be as hard against radiation as the detectors themselves.

- The maximum strip length (presently about 10 cm) limited by the production technology (wafer size) can be, in principle, enlarged using hybrid technologies. Which maximum length is tolerable with sufficient S/N?

- Positioning of hundreds of individual microstrip detectors with an accuracy of a few microns within a volume of one cubic meter and monitoring the positions in situ certainly is an exciting challenge.

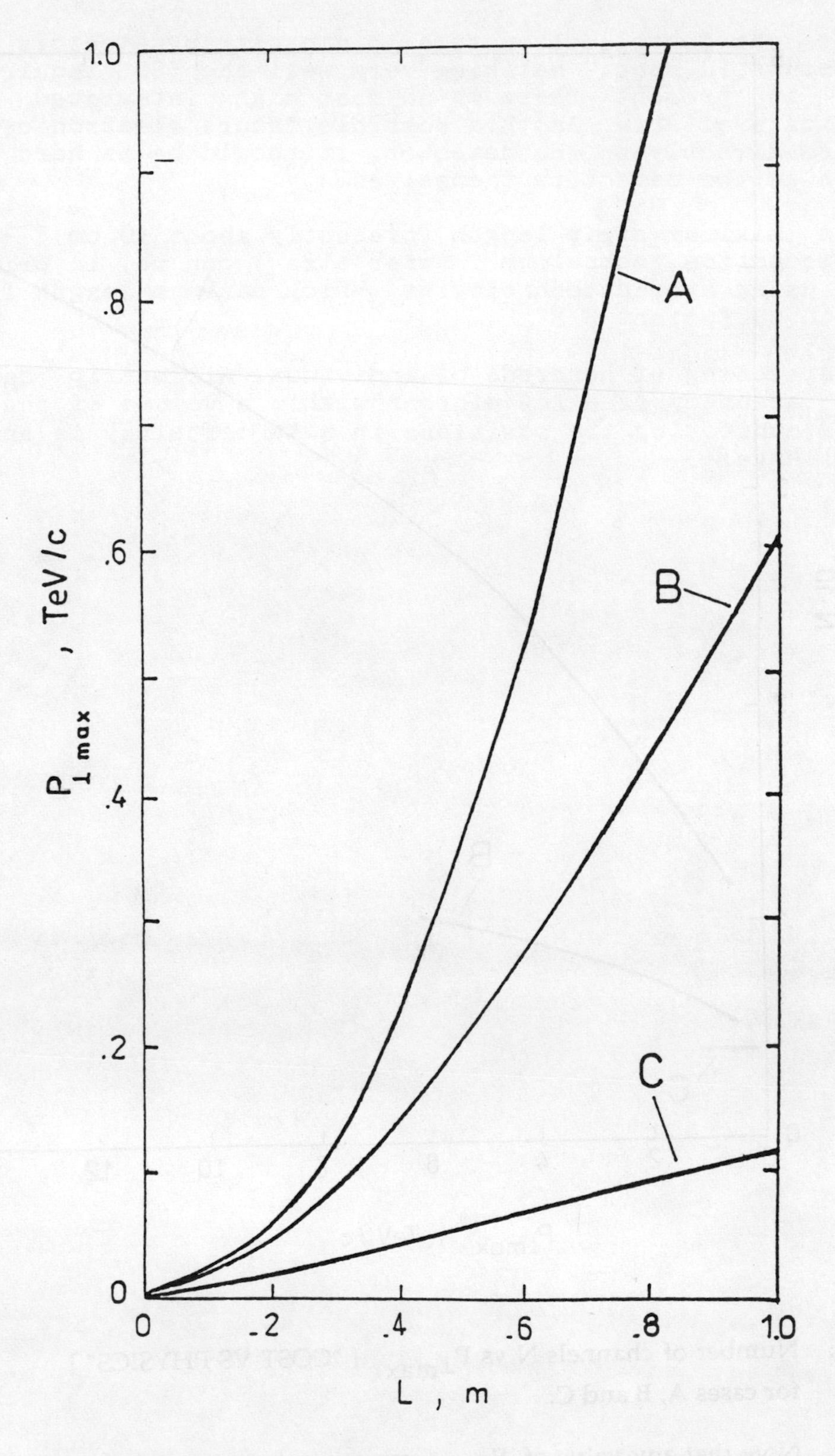

Fig. 1: Maximum transverse momentum $P_{\perp max}$ to define the sign the particle charge vs lever arm L for the three assumptions on the $\sigma = f$ (R) dependence (see text).

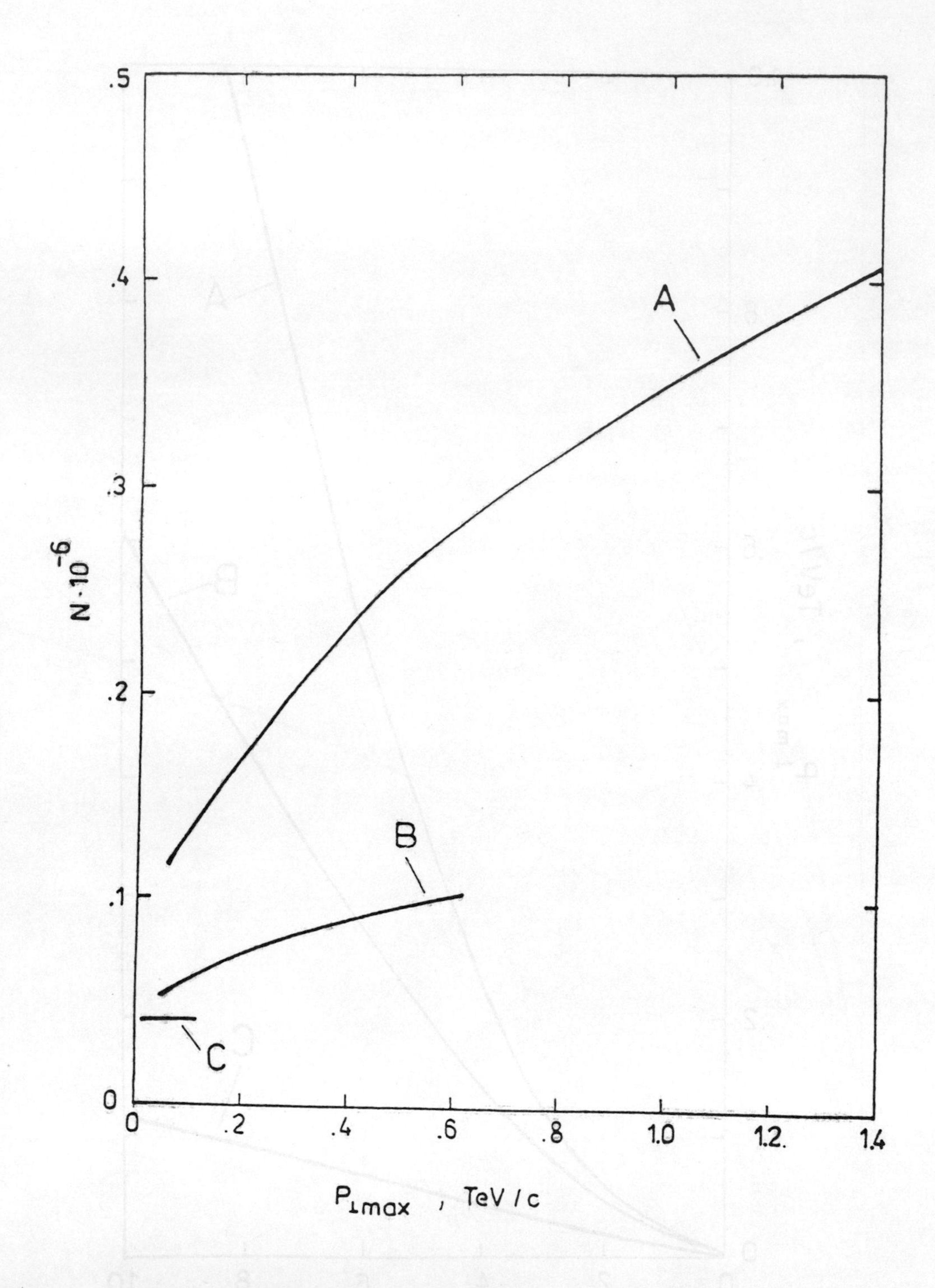

Fig. 2: Number of channels N vs $P_{\perp max}$ ["COST VS PHYSICS"] for cases A, B and C.

Note that any value of $P_{\perp max}$ corresponds to a certain lever arm L depending on the given $\sigma = f(R)$ assumption. We found it more informative to plot $P_{\perp max}$ instead of L.

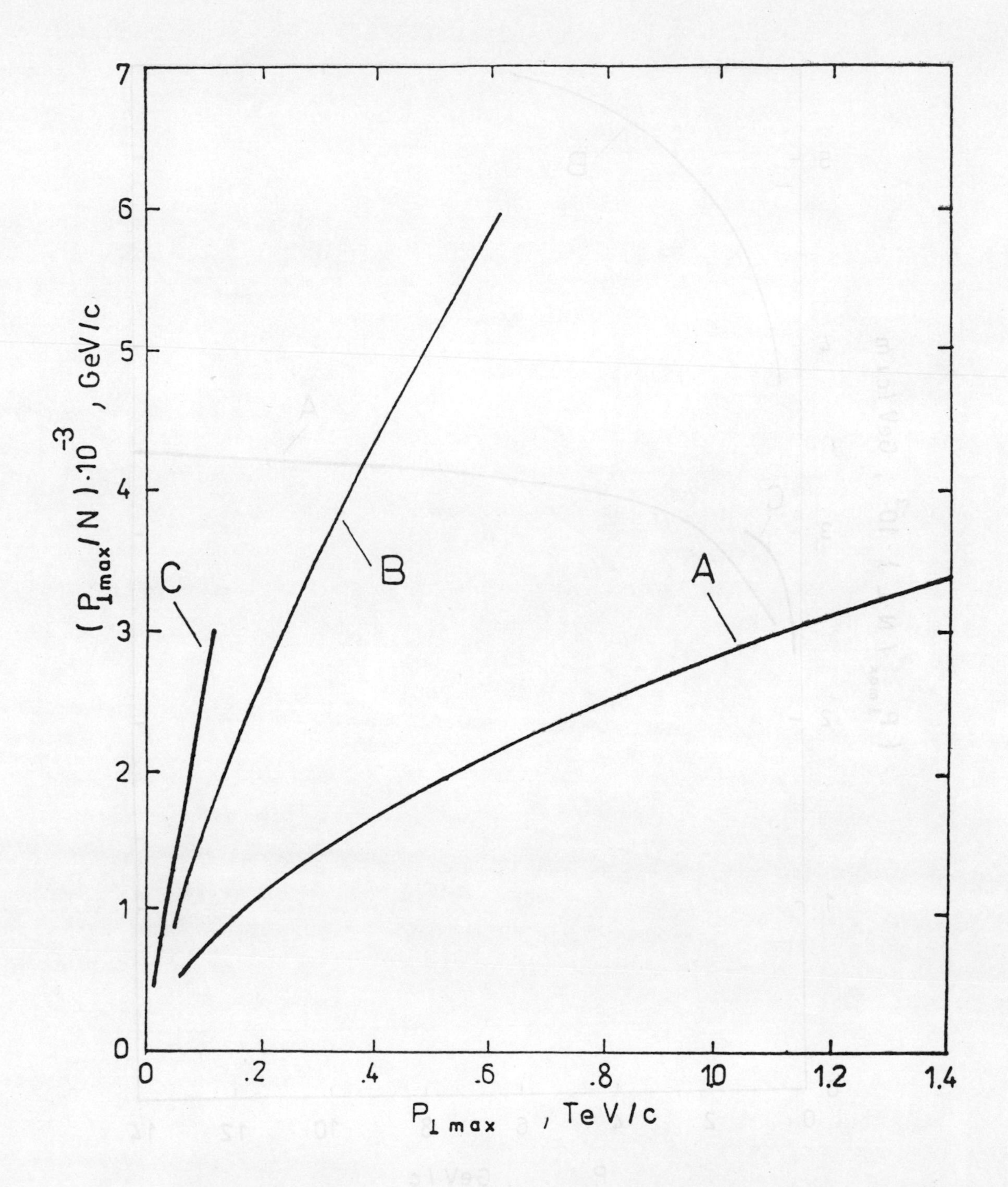

Fig. 3. $P_{\perp max}/N$ vs $P_{\perp max}$ ["PHYSICS PER COST VS PHYSICS"] for cases A, B and C.

Note that any value of $P_{\perp max}$ corresponds to a certain lever arm L depending on the given $\sigma = f$ (R) assumption. We found it more informative to plot $P_{\perp max}$ instead of L.

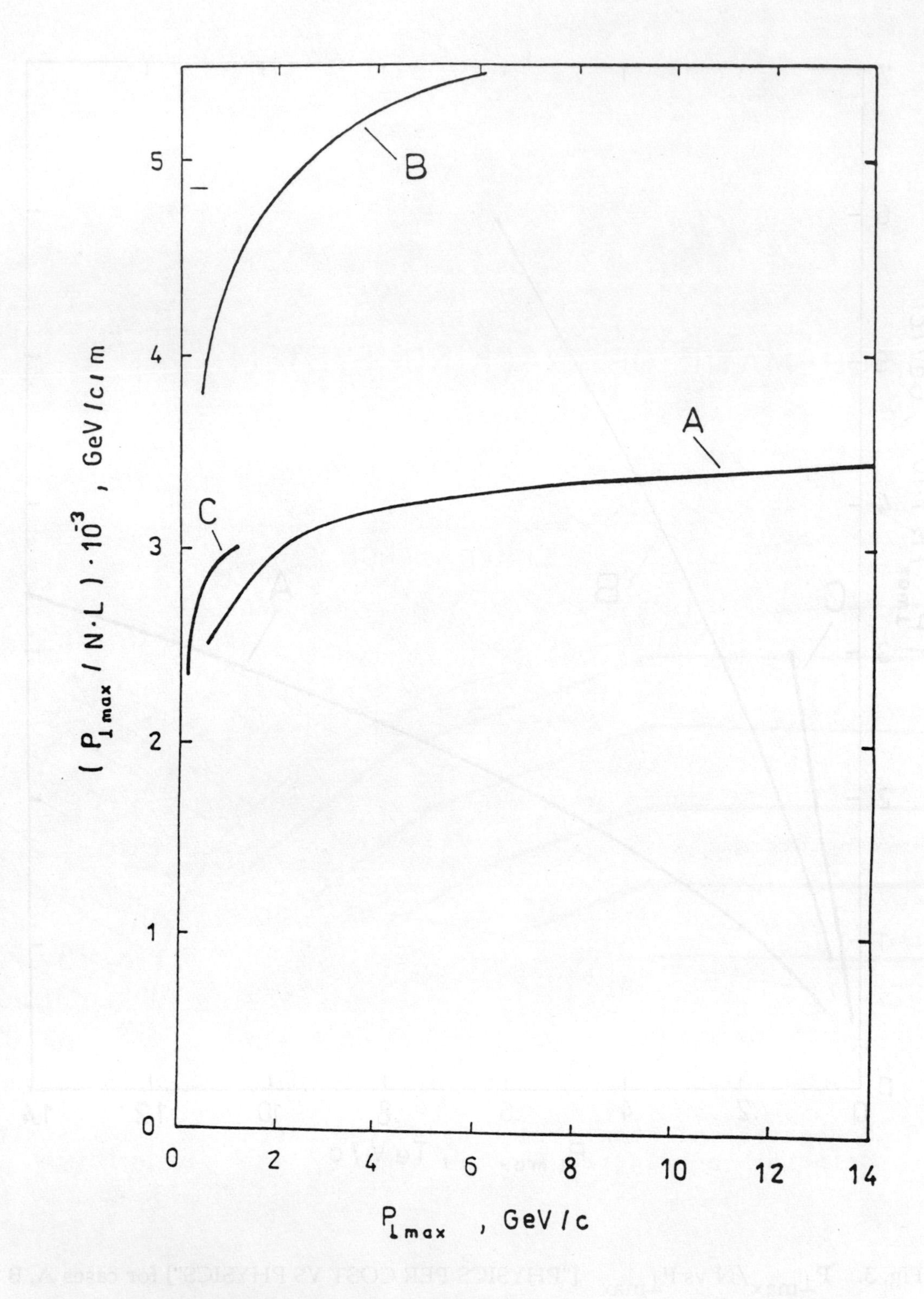

Fig. 4. $P_{\perp max}$ /N•L vs $P_{\perp max}$ ["PHYSICS PER COST AND SPACE VS PHYSICS"] for cases A, B and C.

Note that any value of $P_{\perp max}$ corresponds to a certain lever arm L depending on the given $\sigma = f$ (R) assumption. We found it more informative to plot $P_{\perp max}$ instead of L.

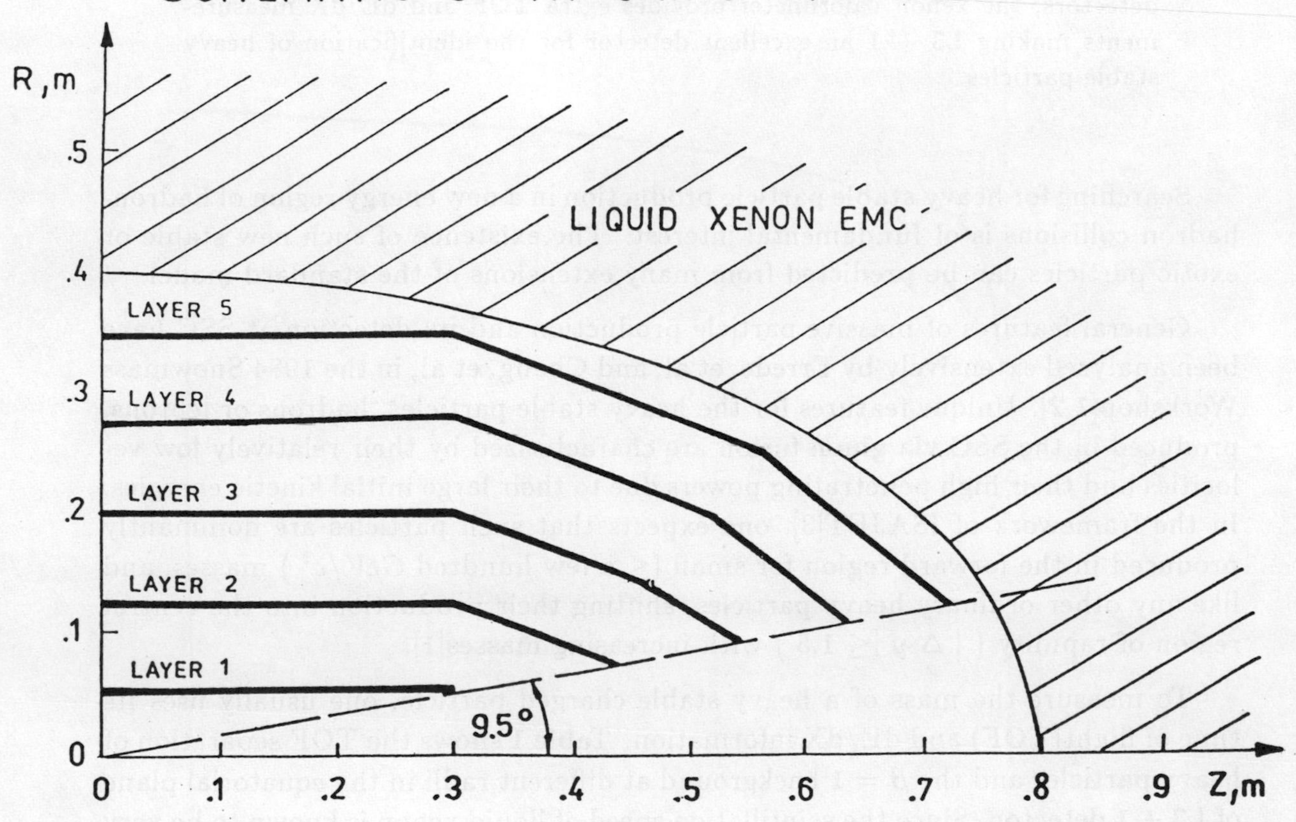

Fig.5: Schematic of the tracking system based on microstrip detectors considered in this study.

IDENTIFICATION OF HEAVY STABLE PARTICLES WITH L3 + 1

C. Y. Chang and D. Fong, University of Maryland

Abstract

In adition to the high resolution on momentum measurements of the muon detectors, the xenon calorimeter provides extra TOF and dE/dX measurements making L3 + 1 an excellent detector for the identification of heavy stable particles.

Searching for heavy stable particle production in a new energy region of hadron-hadron collisions is of fundamental interest. The existence of such new stable or exotic particles can be predicted from many extensions of the standard model.

General features of massive particle production and its detection at SSC have been analyzed extensively by Errede, et al, and Chang, et al, in the 1984 Snowmass Workshop[1,2]. Unique features for the heavy stable particles, hadrons or leptons, produced in the SSC via gluon fusion are characterized by their relatively low velocities and their high penetrating powers due to their large initial kinetic energies. In the framework of ISAJET[3], one expects that such particles are dominantly produced in the forward region for small ($\leq$ a few hundred GeV/c^2) masses, and like any other ordinary heavy particles, shifting their production into the central region of rapidity ($|\ \Delta\ y\ | \leq 1.5$) with increasing masses[1].

To measure the mass of a heavy stable charged particle, one usually uses its time of flight(TOF) and dE/dX information. Table 1 shows the TOF separation of heavy particles and the $\beta = 1$ background at different radii in the equatorial plane of L3 + 1 detector. Since the scintillation speed of liquid xenon is known to be very fast with a width of 3 nsec for 96% of the light and 20 to 30 nsec for the rest[4], the time information extracted from the liquid xenon chambers at the innermost and outermost sampling layers can provide two independent meassurements of the TOF for the charged stable heavy particles(straight tracks) with excellent accuracy. As shown in in Table 1, when the xenon TOF information is combined with two more such measurements obtained with plastic scintillation counters located before and after the L3 + 1 muon detectors, the time separations provide an identification of the heavy particles.

Figure 1. shows the dE/dX for charged heavy stable particles (Q=1) in liquid xenon. The density effect for the relativistic rising has been taken into account by using the parameters of solid Tin [5]. The dE/dX of charged pions and protons are also included in Fig. 1 for comparison. In Fig. 2, we show the dE/dX for Q=1/3 heavy particles vs their momenta. It is clear that with the high resolution muon detectors of L3 + 1, and the dE/dX measurements by the xenon chambers should provide an excellent identification of these supermomentum particles produced at the SSC. For completeness, we show in Fig. 3 the β of the heavy particles involed in these discussions. We see that we are not really dealing with very slow particles at 1000 GeV/c^2 of masses at SSC.

Foot-Notes and References:

1. S. Errede, et al, " Stable/Exotic Particle Productio at the SSC ", Proceedings of the 1984 Summer Study on the Design and Utilization of Superconducting Super Collider, Snowmass Colorado, page 175, (1984)

2. C Y Chang, et al, " TOF for Heavy Stable Particle Identi fication ", Proceedings of the 1984 Summer Study on the Design and Utilization of Superconducting Super Collider, Snowmass Colorado, page 202, (1984)

3. F E Paige, et al, BNL Report BNL-29777.

4. M Chen, et al, MIT-LNS Report No.163, June 30, 1987. See also: M Chen, et al, these Proceedings.

5. R M Sternheimer, Phys. Rev. Vol-103, 511 (1956) S Hayakawa, " Cosmic Ray Physics ", John Wiley & Sons, 1969.

Table 1: TOF separations between Heavy Stable Particles of integal charge and the $\beta = 1.0$ background.

Mass = 500. GeV/c^2

R= Counter	40 Xe	100 Xe	260 Sci	815 Sci (cm)
Mom.(GeV/c)	δ (TOF) in nsec.			
50.	12.066	30.166	78.432	245.853
100.	5.465	13.663	35.525	111.357
150.	3.307	8.267	21.494	67.376
200.	2.257	5.642	14.669	45.982
300.	1.258	3.146	8.178	25.636
500.	0.552	1.381	3.590	11.253
1000.	0.157	0.393	1.023	3.207

Mass = 1000. GeV/C^2

R= Counter	40 Xe	100 Xe	260 Scin	815 Scin (cm)
Mom.(GeV/c)	δ (TOF) in nsec.			
500.	1.648	4.120	10.713	33.580
600.	1.258	3.146	8.178	25.636
900.	0.660	1.649	4.289	13.443
1000.	0.552	1.381	3.590	11.253
2000.	0.157	0.393	1.023	3.207

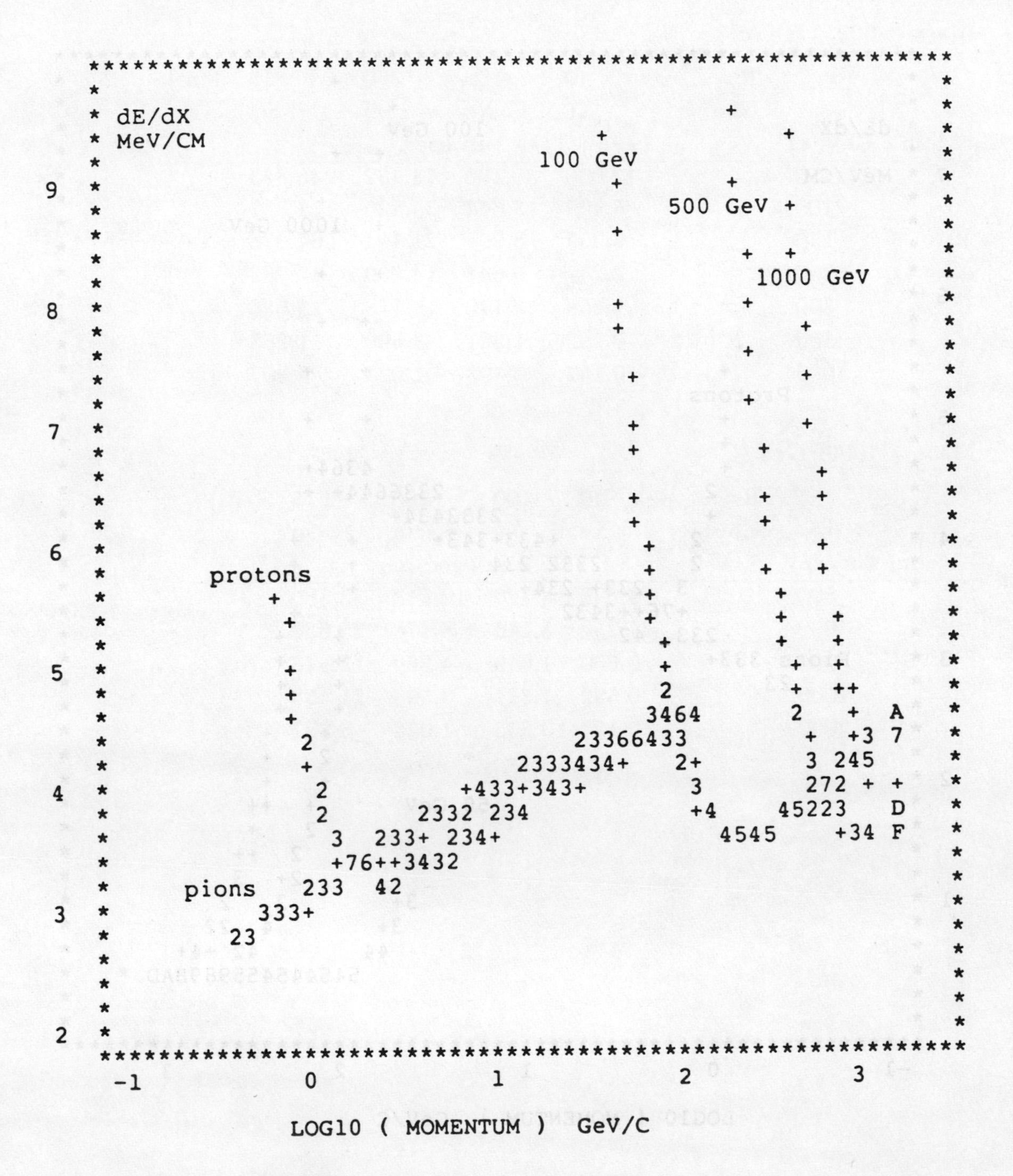

Figure 1. dE/dX for charge = 1 heavy particles against their Momenta.

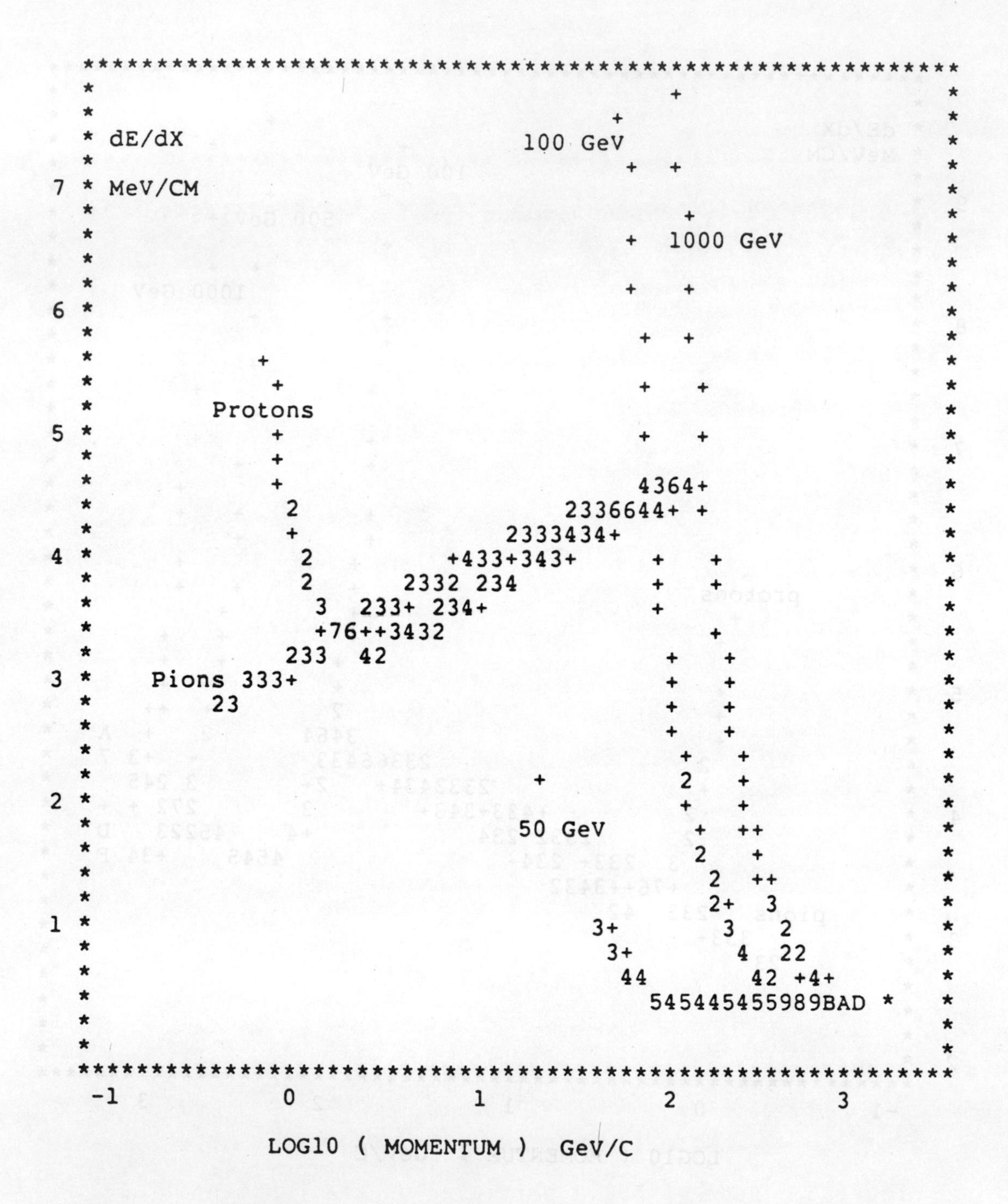

Figure 2. dE/dX for charge = 1/3 heavy particles against their Momenta.

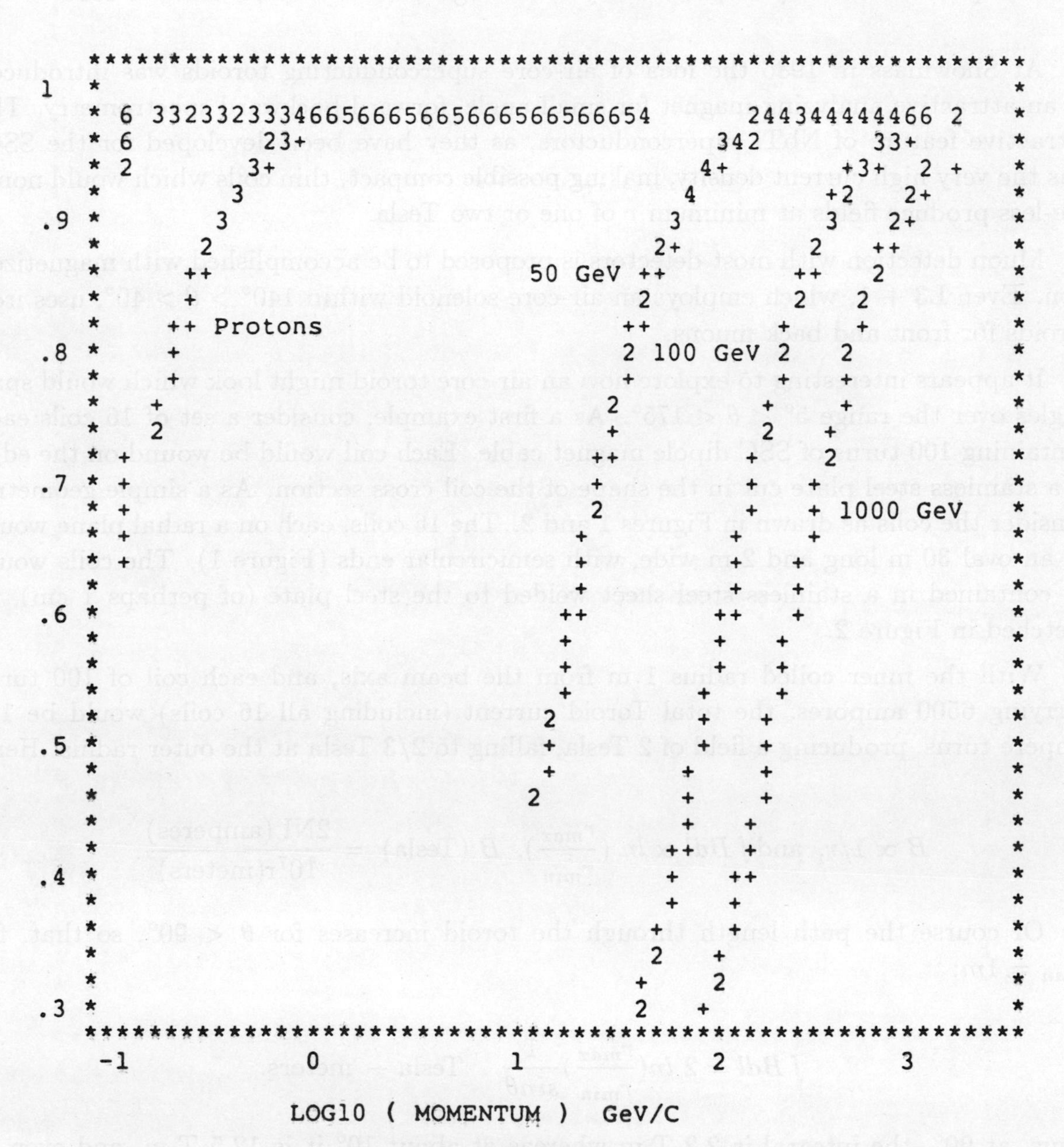

Figure 3. Beta of heavy particles vs their Momenta.

AIR CORE SUPERCONDUCTING TOROIDS AS MAGNETS AROUND INTERIOR POINTS

Lawrence W. Jones
Department of Physics, University of Michigan, Ann Arbor, Michigan 48109

At Snowmass in 1986 the idea of air-core superconducting toroids was introduced as an attractive analyzing magnet for small-angle, forward-backward spectrometry. The attractive feature of NbTi superconductors, as they have been developed for the SSC, was the very high current density, making possible compact, thin coils which would none-the-less produce fields at minimum r of one or two Tesla.

Muon detection with most detectors is proposed to be accomplished with magnetized iron. Even L3 + 1, which employs an air-core solenoid within $140° > \theta > 40°$, uses iron toroids for front and back muons.

It appears interesting to explore how an air-core toroid might look which would span angles over the range $5° < \theta < 175°$. As a first example, consider a set of 16 coils each containing 100 turns of SSC dipole magnet cable. Each coil would be wound on the edge of a stainless steel plate cut in the shape of the coil cross section. As a simple geometry, consider the coils as drawn in Figures 1 and 2. The 16 coils, each on a radial plane would be an oval 30 m long and 2 m wide, with semicircular ends (Figure 1). The coils would be contained in a stainless steel sheet welded to the steel plate (of perhaps 1 cm), as sketched in Figure 2.

With the inner coiled radius 1 m from the beam axis, and each coil of 100 turns carrying 6500 ampores, the total Toroid current (including all 16 coils) would be 10^7 ampere turns, producing a field of 2 Tesla, falling to 2/3 Tesla at the outer radius. Here,

$$B \propto 1/r, \text{ and} \int Bdl \propto ln\ (\frac{r_{max}}{r_{min}}),\ B\ (\text{Tesla})\ = \frac{2\text{NI (amperes)}}{10^7\text{r(meters)}}.$$

Of course the path length through the toroid increases for $\theta < 90°$, so that, for $r_{min} = 1m$:

$$\int Bdl = 2\ ln(\frac{r_{max}}{r_{min}})\frac{1}{sin\theta} \quad \text{Tesla} - \text{ meters.}$$

Thus, at 90°, the integral is 2.2 T-m whereas at about 10° it is 12.5 T-m, and even at 5°, where the trajectory leaves the toroid at r=2m, the integral is about 16 T-m.

The adjacent coils are separated by about 35 cm at the inner radius, corresponding to 90-95% open space azimuth.

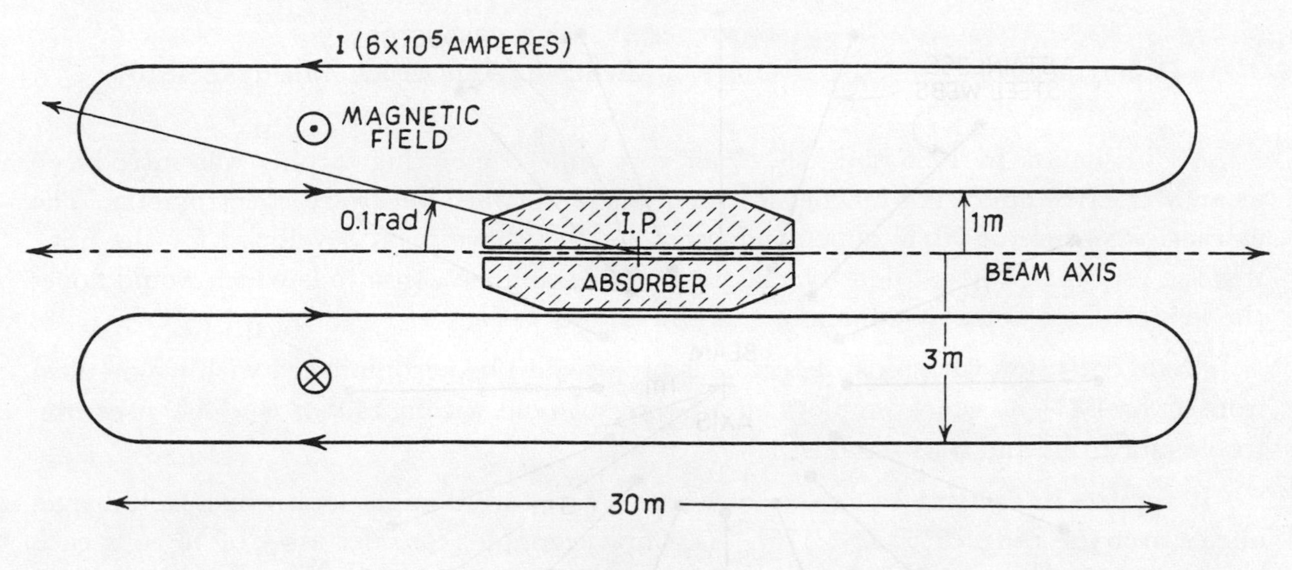

Figure 1

The greatest engineering problems in this toroid geometry are the mechanical forces on the coils. There are three forces; the forces within each conductor bundle, the forces internal to each coil, and the net force on each coil. The forces within the 100-conductor bundle are similar to those in the SSC dipoles and would be compressive, forcing the conductors together. No exceptional structure is required for this. The forces within each coil would be in a direction to drive the conductors apart. The sheet would then be under tension. This tension would be about a half ton per cm.

The net displacement force will try to force all coils into the beam axis, each with a radial force of about half a ton per cm. Substantial mechanical structures are thus necessary to support these forces. The cryogenics has not been considered here.

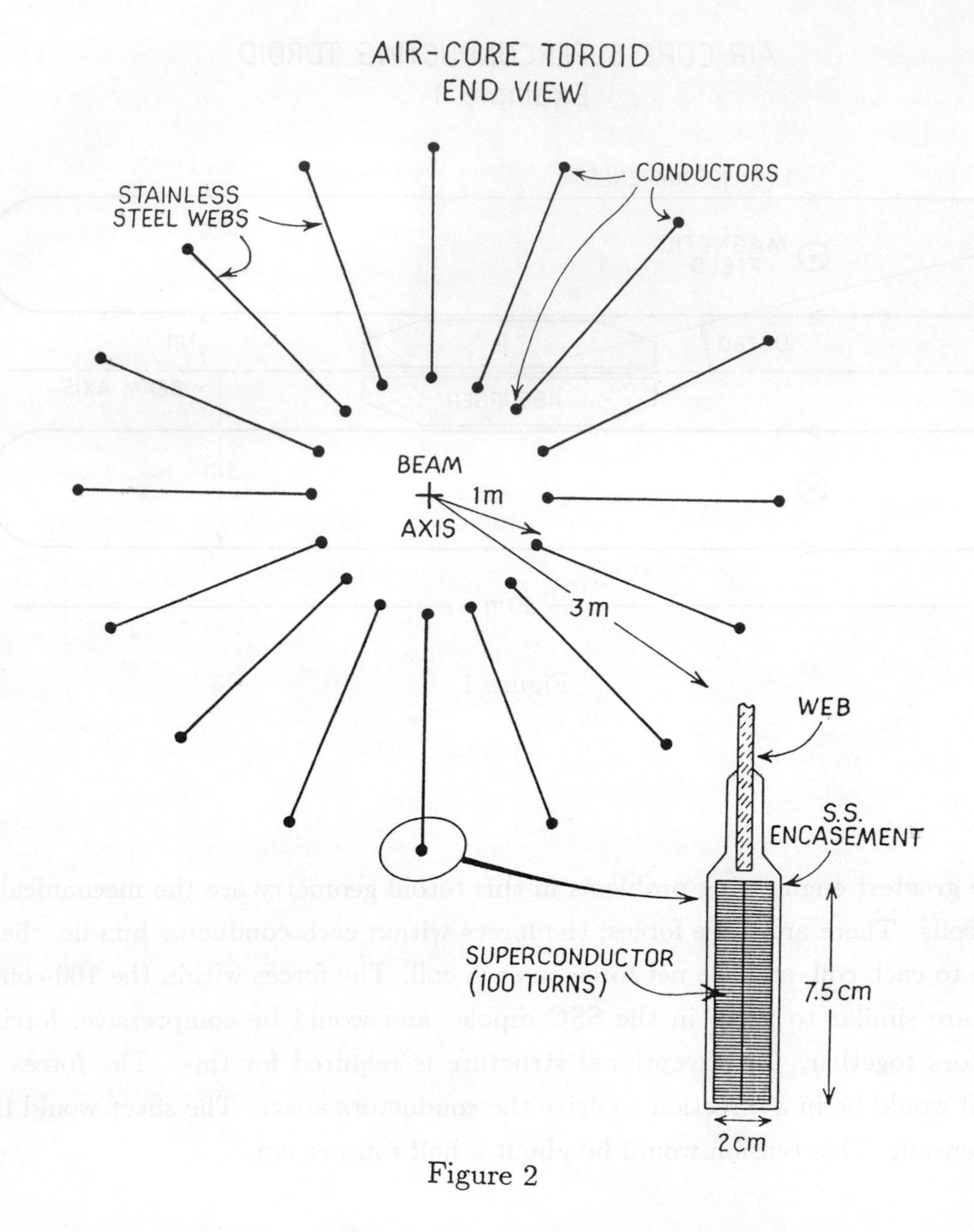

Figure 2

Obvious applications of this toroid concept would be to the dedicated muon detector as discussed by R. Thun. This would permit excellent momentum resolution of muons independent of angle, down to 5°. Such a toroid would also be applicable to detectors such as the "non-magnetic" detector, and the L3+1 concept. In R. Thun's detector, the central system has a radius of over two meters. Of course the shape and dimensions of Figure 1 are only the simplest example. In Figure 3 another example is sketched designed to surround a fatter central, non-magnetic detector and to also close to within 50 cm of the beam pipe forward and backward. In this case there may be 32 coils of 100 turns and/or more turns in each coil to provide at least 1.5 T at 3m.

Several observations may be made. Although the cryogenics and mechanical support are sophisticated, they are very light compared to the massive iron muon detectors they replace. Thus considerable simplification in rigging engineering, and access to the central detector system would be realized. Such coils may be ideal first applications for new warm superconductors. The geometry is simple, and the cryogenics is tedious. Given the improvement in muon momentum resolution of perhaps an order of magnitude which this approach may offer as compared with magnetized iron, this approach appears to merit further study.

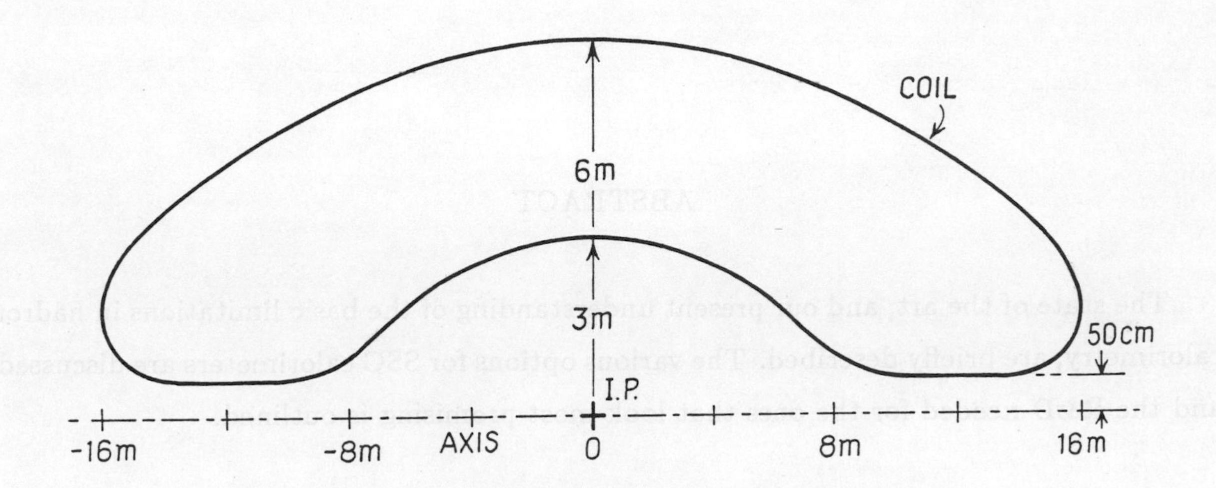

AIR-CORE SUPERCONDUCTING TOROID
Example 2.
(ONE OF 16-32 COILS ONLY SHOWN)

Figure 3

CALORIMETRY AT THE SSC

Richard Wigmans
NIKHEF-H, Amsterdam, The Netherlands

ABSTRACT

The state of the art, and our present understanding of the basic limitations in hadron calorimetry, are briefly described. The various options for SSC calorimeters are discussed, and the R&D needed for the ones that look most promising is outlined.

Invited talk given at the Workshop on Experiments, Detectors,
and Experimental Areas for the Supercollider
Berkeley, July 7 - 17, 1987

1. INTRODUCTION

It has become increasingly clear that the success of experiments at the SSC will depend to a large extent on the quality of the available calorimetry. The calorimeters will have to fulfill the following tasks:

1) Provide triggers that select the potentially interesting events. Since these represent only a tiny fraction of the total number of interactions ($< 10^{-8}$), extraordinary requirements on trigger selectivity, efficiency and reliability have to be met.

2) Provide the experimental information needed for the analysis of hadron production in the *pp* collisions (jet physics). In this respect, calorimeter data will be the only thing available.

3) Provide crucial information on lepton production. Both the identification of leptons (e, ν) and the measurement of their 4-vectors (e, ν, μ) will rely heavily on calorimeter data. This is also true for leptonically decaying particles, like intermediate vector bosons.

In order to be able to properly fulfill these tasks, SSC calorimeters should respond to showering particles with a signal that is fast, independent on the type of particle or its impact point, and narrowly distributed as a Gaussian around a mean value that is proportional to the particle energy. It should be possible to distinguish electromagnetically interacting particles (e,γ) from hadrons with a high degree of reliability by means of the shower profile, and the noise should be sufficiently low to measure muons.

The performance of SSC calorimeters will, therefore, crucially depend on (in random order of importance):

a) The hermeticity

b) The granularity

c) The energy resolution

d) The e/h signal ratio

e) The uniformity of the response

f) The signal formation time

g) The noise level

h) The signal stability

Because of the harsh environment in which the calorimeters will have to operate, their components and in particular the active material and the electronics mounted inside the sensitive volume, should be able to sustain high radiation levels. Although the calorimeters should be sufficiently large to make the effects of shower leakage negligible, they should be as compact as possible in order to limit the size of costly downstream equipment (magnet, muon detector).

The latter requirement calls for the use of high-density shower absorption materials, although also the cost and other properties of such materials have to be taken into account. The desire to do excellent electron identification naturally favours high-Z absorbers. The ratio between the nuclear interaction length and the radiation length, the relevant parameter in this respect, is for materials like uranium and lead a factor of three larger than for copper and iron.

In this talk I will briefly describe the state of the art, and our present understanding of the basic limitations in hadron calorimetry. I will review the various options for SSC calorimeters, and suggest R&D needed for the ones that I consider most promising.

2. RECENT PROGRESS IN HADRON CALORIMETRY

Thanks to the efforts by many people, considerable progress has recently been made in hadron calorimetry, and in our understanding of the mechanisms that determine the performance of these devices. I will illustrate that with three examples.

The HELIOS Collaboration at CERN, who operate a 3 mm U/2.5 mm plastic-scintillator calorimeter, amongst others for studying heavy-ion collisions with a fixed target, have measured hadrons with an energy resolution σ/E better than 2%[1] . This result, shown in fig. 1, was obtained at running conditions. Oxygen-16 ions at 3.2 TeV were sent on a tungsten target at a rate of 10^6 per burst (2.4 s). The trigger asked for highly inelastic collisions ($E_T > 100$ GeV). Figure 1 shows the total calorimeter signal distribution recorded for the events fulfilling this requirement. Apart from the beautiful energy resolution, also the signal linearity is remarkably good. Based on calibration with 8 – 70 GeV electrons the nominal ion beam energy, two orders of magnitude higher, is reproduced to within 2%.

The second example comes from ZEUS, who have tested a 10 mm Pb/2.5 mm plastic-scintillator calorimeter, as part of their R&D program[2] . This rather onorthodox detector (thick lead, thin scintillator) measured hadrons in the energy range 3 – 75 GeV with a resolution $\sigma/E = 44\%/\sqrt{E}$ (fig. 2) , much better than the results obtained with a "standard", more frequently sampling lead/scintillator calorimeter (5 mm Pb/5 mm plastic), or any other non-uranium device.

The third example illustrating the progress in this field is shown in fig. 3. This figure, taken from ref. 3, shows that the odd lead/scintillator result mentioned above, confirms the prediction that a resolution of $42\%/\sqrt{E}$ could be achieved for this configuration. Although the authors of ref. 2 do not mention this prediction, it was of course the one and only reason why this test calorimeter was built.

What can we conclude from from this?

The first result shows that it is not unrealistic to aim for hadron calorimetry at the 1% precision level in the SSC era. What this may mean for the quality of experimental results is illustrated by fig. 4, taken from the same HELIOS ^{16}O exposure. The figure shows the total energy recorded in the calorimeter for minimum bias events. The beam contains a very small contamination of other even-mass ions due to dissociated ^{16}O nuclei. Owing to the good energy resolution, the hadron calorimeter acts as a perfect mass spectrometer. And a perfect mass spectrometer is precisely what is needed to discover new particles and determine their properties. At SSC energies hadron calorimetry with the precision of nuclear γ-ray spectroscopy seems feasible, and it seems feasible with plastic-scintillator readout.

The second result suggests that one doesn't need uranium for such good performance, and the third result shows that we now understand *how* to achieve this goal. In the next section, the improved understanding of hadron calorimetry that lead to this prediction is briefly described[3,4] .

3. THE PHYSICS OF HIGH RESOLUTION HADRON CALORIMETRY

3.1 The role of the e/h signal ratio

In a given calorimeter, the energy resolution with which hadronic showers are detected is worse than for electromagnetic ones. This is, among others, due to the fact that in hadronic showers fluctuations occur in the fraction of the initial energy that is converted into kinetic energy carried by ionizing particles. Shower leakage and, much more importantly, the energy needed to release nucleons from the nuclei in which they are bound, may consume on average up to 40% of the incident energy, with large fluctuations about this average.

As a consequence, the signal distribution for pions is wider than for electrons at the same energy, and has in general a smaller mean value ($e/\pi > 1$). The calorimeter response to the electromagnetic (em) and non-em components of hadron showers is similarly different ($e/h > 1$). Since the event-to-event fluctuations in the fraction of the energy spent on π^0 production (f_{em}) are large and non-Gaussian, and since $< f_{\text{em}} >$ increases (logarithmically) with energy, the following effects have to be expected if $e/h \neq 1$:

i) The signal distribution for mono-energetic hadrons is non-Gaussian.

ii) The fluctuations in f_{em} give an additional contribution to the energy resolution.

iii) The energy resolution σ/E does not improve as $E^{-1/2}$ with increasing energy.

iv) The calorimeter signal is not proportional to the hadron energy (alinearity).

v) The measured e/π signal ratio is energy dependent.

Because of the latter effect I prefer the energy-independent quantity e/h. The difference between e/h and e/π is in practice small, and vanishes at low energies and for e/h close to 1. All these effects have been experimentally observed (fig. 5), and can be simulated with a simple Monte Carlo. At increasing energies, deviations from $e/h = 1$ (the *compensation* condition) become rapidly a dominating factor for the (lack of) calorimeter performance, *e.g.* through the constant term that they add to the energy resolution σ/E (fig. 5a). Signal alinearities of $\sim$ 20% over one order of magnitude in energy have been observed, both in overcompensating ($e/h < 1$) and undercompensating ($e/h > 1$) calorimeters (fig. 5b). But perhaps the most disturbing consequence of a non-compensating calorimeter for operating in an SSC experiment is the non-Gaussian line shape (fig. 5c), which will cause severe problems if one wants for example to trigger on transverse energy. It will be extremely hard to unfold the steeply falling E_T distribution and the response function. Moreover, severe trigger biases are likely to occur: If $e/h < 1$ (> 1) one will predominantly select events that contain little (a lot of) em energy from π^0's.

Several experiments have developed weighting schemes in an attempt to get rid of the mentioned consequences of $e/h \neq 1$, using the details of the measured shower profile. Although some improvement can be obtained in this way, it turns out not to be possible to optimize simultaneously on energy resolution, signal linearity and line shape. In addition, even the best energy resolutions obtained with such schemes are at least 50% worse than results obtained with compensating calorimeters.

In view of the calorimeter tasks and because of the large energy range covered by particles produced in collisions at the SSC, compensating calorimeters will turn out to be an essential advantage. It should be emphasized that other sources of experimental uncertainty, like calibration errors, will produce effects similar to the ones caused by deviations from $e/h = 1$. Therefore, it is not necessary that e/h be 1.0000. It has been estimated[3] that for resolutions at the 1% level, $e/h = 1 \pm 0.05$ is good enough.

3.2 The factors determining e/h

The response of a sampling calorimeter to a showering particle is a complicated issue, that depends on many details. This is particularly true for hadronic showers. It has become clear that showers can by no means be considered as a collection of minimum ionizing particles (mip's) that distribute their energy to absorber and active planes according to dE/dx. The calorimeter signal is to a very large extent determined by very soft particles from the last stages of the shower development, simply because these particles are so numerous. There are many observations that support this statement. Simulations of high energy em showers in lead- or uranium sampling calorimeters show that $\sim 40\%$ of the energy is deposited through ionization by electrons softer than 1 MeV. Measurements of pion signals in fine-sampling lead/plastic-scintillator calorimeters revealed that there is almost no correlation between the particles contributing to the signal of consecutive active layers. This proves that the particles that dominate the signal travel on average only a very small fraction of a nuclear interaction length indeed.

In order to evaluate the e/h signal ratio of a given calorimeter one must, therefore, understand in detail what goes on in the last stages of the shower development, *i.e.* analyse the processes at the nuclear and even the atomic level. It turns out that the particles that decisively determine the calorimeter response are soft photons in the case of em showers, and soft protons and neutrons from nuclear reactions in non-em showers. Since most of the protons contributing to the signal are highly non-relativistic, the saturation properties of the active material for densely ionizing particles are of crucial importance.

Apart from this, there are many other factors that affect the signals from these shower components, and thus e/h. Among these, there are *material* properties like the Z values of the active and passive components, the hydrogen content of the active media, the nuclear level structure and the cross section for thermal neutron capture of the absorber; and *detector* properties like the size, the signal integration time, the thickness of the active and passive layers and the ratio of these thicknesses.

For a given combination of active and passive material, the calorimeter response (signal per GeV) to the important shower components may be quite different. In calorimeters with high-Z absorber and low-Z active material, the response to em showers is suppressed relative to mip's by about $30-40\%$, since the low-energy γ's from the shower get almost completely absorbed by the absorber because of the photo-electric effect. On the other hand, the response to the spallation protons from hadron showers may be somewhat *larger* than for mip's because of the dE/dx characteristics for non-relativistic particles, although this effect is counterbalanced by self-absorption in the passive layers and signal saturation in the active ones.

Since neutrons are themselves not ionizing, the calorimeter response to these particles depends completely on nuclear physics details. Relative to mip's, the response to soft neutrons (*i.e.* those which do not produce charged particles through inelastic nuclear reactions) may be anything from very small to very large. A crucial ingredient in this respect is the presence of a significant fraction of hydrogen in the active material. Hydrogen is very effective in absorbing the kinetic energy of these neutrons, and the resulting recoil protons fully contribute to the calorimeter signal, although saturation may considerably reduce the size of this contribution. In the absence of hydrogen, neutrons will only contribute through the nuclear γ's that they generate by inelastic scattering or capture. As mentioned before, high-Z sampling calorimeters are inefficient in detecting such γ's.

For a given combination of absorber and readout materials, the relative calorimeter response to the various shower components and, therefore, the e/h signal ratio can be varied within certain limits, making use of the characteristics described above. For example, the π^0 response may be selectively reduced by inserting thin low-Z foils in between the high-Z absorber material and the readout layers. It is predicted that e/h can be reduced by at maximum $\sim 10\%$ in this way. Very interesting possibilities for tuning e/h would obviously be opened up if the saturation properties of the active material could be affected, *e.g.* by varying the strength of the electric field in ionization chambers.

The most important handle on e/h is provided by the neutrons, in particular for calorimeters with hydrogenous readout media. In this case, the fraction of the kinetic neutron energy transferred to recoil protons in the active layers varies much more slowly with the ratio of the amounts of passive and active material than does the fraction of the energy that mip's deposit in the active layers. Therefore, the relative contribution of neutrons to the calorimeter signal, and hence e/h, can be varied through the sampling fraction. A smaller sampling fraction enhances the relative contribution of neutrons. It is estimated that in compensating lead- or uranium-scintillator calorimeters neutrons make up for $\sim 40\%$ of the non-em signal, on average. The lever arm on e/h provided by this mechanism may be considerable. It depends on the energy fraction carried by soft neutrons (favouring high-Z absorbers), on the hydrogen fraction in the readout media, and on the signal saturation for densely ionizing particles (favouring materials with a low value of Birk's constant k_B).

Figure 6 shows an example of the results of calculations on e/h for uranium calorimeters. For hydrogenous readout materials (plastic scintillator, warm liquids) the e/h value sensitively depends on the relative amount of active material, and in any case a configuration can be found with $e/h = 1$. Experimental results clearly confirm the tendency predicted for scintillator readout. For non-hydrogenous readout (LAr, Si) the mechanism described above does not apply. Here the neutron response and hence the e/h ratio can be changed through the signal integration time, taking more or less advantage of the considerable energy released as γ's when thermal neutrons are captured by nuclei, a process that occurs at a time scale of $\sim 1\ \mu$s. The calculations predict a 15% *increase* of e/h if the signal integration time is reduced from 1 μs to 100 ns, for detectors sufficiently large to thermalise the neutrons. At the SSC one would certainly want to work with as fast a signal as possible.

Experimental results obtained so far seem to confirm the prediction that it will be hard to achieve compensation with liquid-argon readout. Ideas to bring e/h down closer to 1 include the insertion of low-Z foils in between the absorber and the argon, mixing methane into the argon, adding cadmium to enhance the γ-yield from neutron capture, and choosing the field polarity such as to maximise the signal from soft spallation protons stopping in the liquid (central anode).

The calculations indicate that compensation should not be too hard to achieve for uranium calorimeters with silicon readout (fig. 7). When each Si plane is sandwiched in between thin hydrogenous foils, e/h can be tuned over a comfortable range through the thickness of these foils, owing to two effects: Suppression of the em signal, as discussed before, and amplification of the non-em signal through detection of recoil protons from elastic neutron scattering in the foils. The absence of signal saturation in Si is essential in this respect.

Detectors with gaseous readout media (*e.g.* proportional wire chambers) offer a convenient way to tune e/h to the desired value, *i.e.* through the hydrogen content of the gas mixture. This has been experimentally demonstrated by the L3 Collaboration[7]

The curves for TMP calorimeters given in fig. 6 are based on the assumption that the signal saturation in this liquid is equal to either liquid argon or PMMA plastic scintillator. Experimental data, although not completely conclusive, indicate that the signal suppression for soft protons in warm liquids is considerably larger, and perhaps dependent on the electric field strength and the particle's angle with the field vector. Figure 8 shows how the e/h signal ratio depends on k_B for TMP calorimeters with either uranium or lead absorber. Clearly, a detailed knowledge of the recombination phenomena in warm liquids is absolutely crucial for predicting the parameters of compensating configurations.

In contrast to what has been thought for a long time, compensation is not a phenomenon restricted to uranium calorimeters. It has become clear that it is the readout medium rather than the absorber material that determines the possibilities in this respect. Compensation is easier to achieve with high-Z absorbers because of the large neutron production and the correspondingly large leverage on e/h. But even materials as light as iron allow compensation, if used in combination with e.g. plastic scintillator, be it with impractically thick absorber plates.

The neutron production in lead is considerably smaller than in uranium. In order to bring e/h to 1 for lead/scintillator detectors, the neutron signal has therefore to be more amplified relative to mip's than for uranium/scintillator; as a consequence, the optimal sampling fraction is smaller for lead. The calculations predicted e/h to become 1 for lead plates about 4 times as thick as the scintillator, while for uranium a thickness ratio of about 1:1 is optimal. This prediction was confirmed by the ZEUS tests mentioned in sect. 2. They found $e/h = 1.05 \pm 0.04$, a hadronic energy resolution scaling with $E^{-1/2}$ over the energy range $3 - 75$ GeV, and no deviations from a Gaussian line shape.

A final word should be said about fully sensitive hadron calorimeters. The mechanisms that are described in this section and that make compensating calorimeters possible, are based on the fact that we are dealing with *sampling* calorimeters: Only a small fraction of the shower energy is deposited in the active layers, and by carefully choosing parameters one may equalise the response to the em and non-em shower components. This does *not* work for homogeneous devices like the liquid xenon detector proposed at this workshop[8]. In the non-em shower part inevitably losses will occur that cannot be compensated for, *e.g.* the binding energy required to release protons and nucleon aggregates from Xe nuclei. Also the binding energy needed to liberate neutrons, which eventually might be gained back when these neutrons are captured again, will have to be considered as lost when one wants fast signals. Therefore, the e/h signal ratio will always be considerably larger than 1 for such detectors.

Measurements performed so far with homogeneous hadron detectors support this conclusion, both for what concerns the e/h ratio[9], and the resulting alinearity, non-Gaussian response and poor energy resolution ($\sigma/E > 10\%$ at 150 GeV)[10]. In the case of liquid Xe, the estimated nuclear binding energy loss amounts to $\sim 35\%$ of the non-em shower component. Therefore, the smallest possible value for e/h is $1/0.65 \sim 1.54$, which corresponds to $e/\pi \sim 1.37$ at 10 GeV. Any inefficiency in detecting the energy of the hadronic shower particles, *e.g.* due to saturation, albedo neutrons or decaying pions, will increase this value. Also a combination of a front section of this type with a compensating backup part will probably not perform anywhere close to what can be achieved for hadron detection with a compensating calorimeter of uniform structure.

3.3 The energy resolution

The deviation from a purely statistical improvement of the energy resolution at increasing energies induced by $e/h \neq 1$, was already discussed in sect. 3.1. It should be emphasized once more that similar effects, which limit the high energy performance, may result from uncertainties in the calibration, non-uniformities in the detector structure or in the signal formation, *etc.* It is believed that for well-designed detectors these effects can be limited to the 1% level, and the result shown in fig. 1 supports this belief. Nevertheless, this will be one of the main challenges for SSC calorimeters.

For perfect, compensating calorimeters the precision of the measured energy is determined by two factors: The sampling frequency and the intrinsic resolution. Since the fluctuations resulting from these sources add in quadrature, it does not make sense to reduce the contribution of sampling fluctuations far below the intrinsic limit. The intrinsic resolution of compensating calorimeters is dominated by the event-to-event fluctuations in the nuclear binding energy losses, a consequence of the large variety of possible nuclear reactions. Efficient neutron detection may considerably reduce the effect of these fluctuations, since the kinetic neutron energy is strongly correlated with the binding energy loss, particularly in high-Z materials. The calculations yield intrinsic limits close to $20\%/\sqrt{E}$ for compensating calorimeters with hydrogenous readout. This was recently experimentally confirmed for the compensating lead/plastic-scintillator calorimeter. Therefore, the already impressive ZEUS result (fig. 2) is completely dominated by sampling fluctuations, and a fine-sampling device should be able to reach $\sigma/E = 30\%/\sqrt{E}$ or better. Because of the observation that the calorimeter signal is largely dominated by particles that travel only a fraction of a centimeter, one should expect the sampling fluctuations to be determined to a large extent by the total surface of the boundary between the absorber and readout media. This favours structures with fibres surrounded by absorber rather than the standard sandwich geometries.

In compensating calorimeters where the neutrons deposit only a small fraction of their kinetic energy directly in the active layers, like U/Si, the damping of the nuclear binding energy fluctuations does *not* work; as a consequence, the intrinsic limit on the energy resolution is much larger ($\sim 40\%/\sqrt{E}$) for these devices, and one should not expect total resolutions much better than $\sigma/E = 50\%/\sqrt{E}$.

4. CALORIMETERS FOR THE SSC

In the previous section I have given arguments as to why detectors with $e/h = 1 \pm 0.05$ should be strongly preferred when it comes to choosing a calorimeter for an SSC experiment. As was outlined in sect. 1, this is by far not the only criterion that counts, and maybe not even the most important one. I will now briefly discuss the various options, limiting myself to compensating calorimeters, or calorimeters that expectedly can be made compensating, in view of the other criteria imposed by the physics and the circumstances at the SSC.

So far, compensation has been experimentally demonstrated for three types of sampling calorimeter:

a) Uranium/plastic scintillator

b) Lead/plastic scintillator

c) Uranium/gas

There is a reasonable expectation that it can be achieved with

d) Uranium/warm liquid

e) Lead/warm liquid

f) Uranium/silicon + polyethylene

g) Lead/gas

And if one is lucky with

h) Uranium/liquid argon + methane

There are of course more possibilities, involving other high-Z absorbers like Hg, Au, W, Pt, *etc.*, but for various reasons these are much less practical than the ones listed. Anyone who has ever built a uranium calorimeter will agree that it would be a major advantage if one could avoid this nasty material. If only because of this reason, the lead alternatives deserve a very serious study. It is interesting to notice that the effective nuclear interaction length is the same for compensating uranium/scintillator and lead/scintillator calorimeters (20 cm), due to the different optimal sampling fractions. So in spite of its much larger density, uranium does not allow constructing more compact calorimeters in this case.

Gas calorimeters

Gas sampling calorimeters employing wire chambers or proportional tubes for readout, enable a sufficiently fine three-dimensional segmentation. They are easy to construct, cheap, and allow a rather hermetic structure.

Disadvantages of this readout technique include the fact that it is very hard to make the response sufficiently uniform over the entire calorimeter volume, and to maintain the required stability over an extended period of time. In addition, present experience indicates that gas calorimeters will considerably suffer from radiation damage at the SSC.

Of greatest concern, however, are the effects introduced by the extremely small sampling fraction, typically $10^{-4} - 10^{-5}$ for mip's. Because of the virtual absence of saturation effects, densely ionizing particles may simulate energy deposits orders of magnitude beyond reality. For example, in a calorimeter with a sampling fraction of 10^{-4} a recoil proton from elastic neutron scattering in the gas, will appear as a 10 GeV local energy deposit if the proton stops in the chamber. Such effects ("Texas towers") have been observed at CDF and cause very serious problems. If only because of this reason, gas calorimetry does not seem a likely candidate for the SSC.

Silicon

Similar effects, due to the combination of a very small sampling fraction and the absence of signal saturation for densely ionizing particles, may also affect the performance of hadron calorimeters with silicon readout, be it at a lower level (sampling fraction $\sim 0.5\%$). Silicon offers some major advantages: Very fine segmentation, fast response, excellent uniformity, calibration and stability. In combination with uranium or tungsten it would make possible extremely compact calorimeters (seven interaction lengths per meter).

On the other hand, crystalline silicon is known to be particularly vulnerable to slow neutrons, which are definitely not going to be lacking in the SSC environment. Measurements have shown[11] that on average 40-50 neutrons per GeV are produced when non-em showers develop in uranium, depending on the detector configuration. It is feared that the small fraction of the neutrons that leak backwards through the front face of the calorimeter (albedo) might yield a major problem for any electronics installed in front of the calorimeter[12] ; yet another reason to prefer calorimeters with lead absorber, where the soft neutron production was measured to be a factor of 3 lower than in uranium[11] , owing to the absence of nuclear fission. Before embarking on a program for Si-calorimetry at the SSC, the radiation hardness should be seriously studied under realistic circumstances.

Although somewhat outside the scope of this review, the greatest obstacle to the massive use of silicon in SSC calorimeters may well turn out to be cost. Even if one is very optimistic in this respect, a calorimeter with Si readout for the Large Solenoid Detector seems at least an order of magnitude beyond affordability.

I will now discuss the more serious candidates, although it should be emphasized that none of these is without problems and that a vast R&D effort will be needed before one can justify any particular choice.

Plastic scintillator readout

As for silicon, a big advantage of plastic scintillator is the fast response and charge collection time. Since compensation relies on detecting the *kinetic* neutron energy rather than on the γ's released in the capture process, it can be achieved within 100 ns. Inspired by the beautiful ZEUS result mentioned before, Jenni *et al.*[13] have proposed a new type of hadron calorimeter, based on scintillating plastic fibres embedded in lead at a ratio appropriate for achieving compensation. The fibres are oriented (roughly) in the direction of the entering particles. The same technique has already successfully been applied in detectors for em showers.

Apart from the excellent energy resolution envisaged for such a detector ($30\%/\sqrt{E}$ for jets, $15\%/\sqrt{E}$ for em showers), this scheme offers some crucial advantages relevant to the SSC. Among these, the arbitrarily fine lateral segmentation possibilities and the excellent hermeticity achievable with this technique should be mentioned. On the other hand , longitudinal segmentation, important for electron identification, is not trivial. Several ideas exist on paper, but should be tested in practice. Because of the strong correlation between longitudinal and lateral shower information, very fine longitudinal segmentation is unlikely to be essential for achieving the desired degree of e/π separation, once a fine-grained lateral structure is available.

Other issues needing a careful evaluation are signal uniformity and stability. In this respect the result shown in fig. 1, obtained with plastic scintillator, is encouraging since the spaghetti calorimeter is expected to be more uniform and stable than the HELIOS detector.

Like for all proposed detectors, radiation damage should be a major worry. Radiation stability is certainly not the strongest point of this option. However, built-in redundancy like a long light attenuation length, semi-continuous calibration monitoring and transmitting at long wavelengths might basically eliminate the problem up to $\sim$ 1 MRad, a level which will never be exceeded in a large fraction of any SSC calorimeter. Close to the beam pipe one would either have to replace some modules regularly, or choose a different technique.

Liquid argon

Liquid argon is in many ways an ideal active medium for reading out sampling calorimeters, and several experiments have successfully used this beautiful technique for detecting (mainly em) showers. Among the advantages relevant to SSC experimentation can be mentioned the excellent uniformity of the response, both in space and in time, achievable with this technique. Also the segmentation needed to achieve the desired degree of electron identification from the shower profile should in principle be no problem.

Unfortunately, there are some serious disadvantages as well. First of all, there is the slow response; the charge collection time is typically $\sim 1\ \mu s$ in most calorimeters. At a luminosity of 10^{33}, with an event frequency of 100 MHz, this is at best marginal and, generally, faster sampling media are considered highly desirable. An interesting development comes from HELIOS, who considerably reduced the capacitance , and thus the pulse shaping time by placing the preamplifiers very close to the gaps, in the cryostat. In this way they obtain 100 ns signals.

This has two drawbacks. Firstly, it makes the detectors vulnerable to radiation damage (neutrons, see silicon), thus giving up the nice feature that the argon itself is highly radiation resistant. Secondly, according to the calculations it will further deteriorate the e/h ratio, which is already marginal for long pulse shaping times (see fig. 6).

Perhaps the most serious difficulty is that of attaining an acceptable hermeticity, arising from the fact that the detector has to be operated cryogenically. No one has yet been able to show what a 4π calorimeter that meets the SSC requirements on hermeticity and uniform solid angle coverage should look like.

Warm liquids

Warm liquids offer potentially similar advantages as liquid argon, in terms of segmentation, uniformity, stability and radiation hardness of the liquid, *without* the need of operating in a cryostat. Therefore, this option is in principle very attractive. However, the experience with these liquids is still limited, and serious problems have to be solved before one might design a large reliable detector based on this technique.

Compensation has yet to be demonstrated. As was pointed out in sect. 3.2, the signal saturation due to recombination is absolutely crucial in this respect. In my opinion, detailed measurements of Birk's constant and its dependence on the electric field vector should be the first priority of any R&D program on warm-liquid calorimeters.

As for all other ionization chambers, the signals are slow. At moderate drift fields (10 kV/cm) the electron mobility and also the ionization yield (signal/noise!) are considerably lower than in liquid argon. At high fields the situation may drastically improve, especially for TMS. Schemes to push the pulse shaping down to the 100 ns regime a la HELIOS, will have to face the jeopardy of radiation damage of the electronics by neutrons.

Because of the safety aspects, it would be a major advantage if lead absorber could be used. This question is closely linked to the compensation issue (fig. 8). Maybe there are other, safer liquids that offer the advantages of TMS with respect to electron mobility and ionization yield (TMSn?).

The signals from warm liquid chambers are extremely sensitive to electron-trapping impurities. The statement that this technique provides excellent signal uniformity and stability is only correct if such impurities can be controlled at the ppb level, which is far from trivial for a large system.

Perhaps none of the listed problems will turn out to be insurmountable, but clearly a lot of work is needed to find out.

5. SUMMARY

At this point in time it is clear that one cannot decide what is the best calorimeter composition for experiments at the SSC. Given the requirement that the detector should be sufficiently compensating, there remain only 2 serious candidates, and 2 that I'm willing to give the benefit of the doubt.

The most promising candidates are

a) Lead/scintillating fibre. The questions needing answers concern radiation hardness, e/π separation, signal uniformity and stability.

b) Lead (or uranium)/TMS (or other warm liquids). The R&D program should concentrate on compensation questions, purity control, high drift fields, safety and signal speed.

The other candidates are

c) Uranium/liquid argon, where primarily the hermeticity and compensation problems need satisfactory solutions, and

d) Uranium (or tungsten)/silicon, where the radiation damage and cost issues need clarification.

In any case a vast R&D program will be needed to find out to what extent the challenging requirements of SSC experimentation can be met.

REFERENCES

1) T. Akesson *et al.*, Performance of the Uranium/Plastic Scintilator Calorimeter for the HELIOS Experiment at CERN, CERN-EP/87-111.

2) E. Bernardi *et al.*, Performance of a Compensating Lead-Scintillator Calorimeter, preprint DESY 87-041 (1987).

3) R. Wigmans, Nucl. Instr. and Meth. **A259** (1987) 389.

4) R. Wigmans, High Resolution Hadron Calorimetry, preprint NIKHEF-H/87-8 (1987).

5) H. Abramowicz *et al.*, Nucl. Instr. and Meth. **180** (1981) 429.

6) M.G. Catanesi *et al.*, Hadron-, Electron-, and Muon Response of a Uranium-Scintillator Calorimeter, preprint DESY 87-027 (1987).

7) Y. Galaktionov *et. al.*, Nucl. Instr. and Meth. **A251** (1986) 258.

8) W. Walraff, Contribution to this Workshop.

9) E.B. Hughes *et al.*, Nucl. Instr. and Meth. **75** (1969) 130.

10) A. Benvenuti *et al.*, Nucl. Instr. and Meth. **125** (1975) 447.

11) C. Leroy, Y. Sirois and R. Wigmans, Nucl. Instr. and Meth. **A252** (1986) 4.

12) D. Nygren, Contribution to this Workshop.

13) P. Jenni *et al.*, The High Resolution Spaghetti Hadron Calorimeter, NIKHEF-H/87-07 (1987).

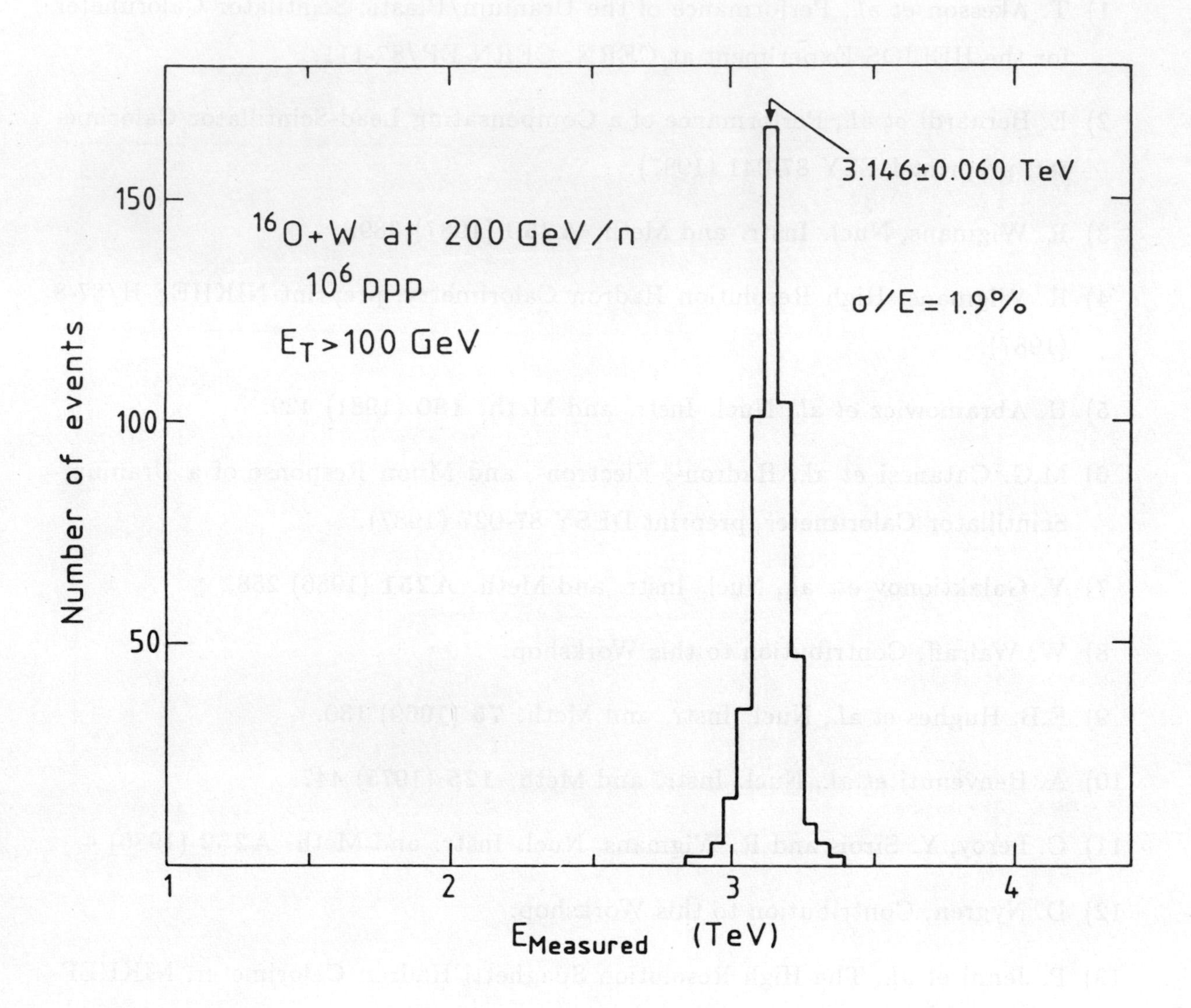

Figure 1

1. The total energy distribution for the reaction products from ^{16}O + W central collisions at 3.2 TeV, measured with the HELIOS uranium/plastic-scintillator calorimeter.

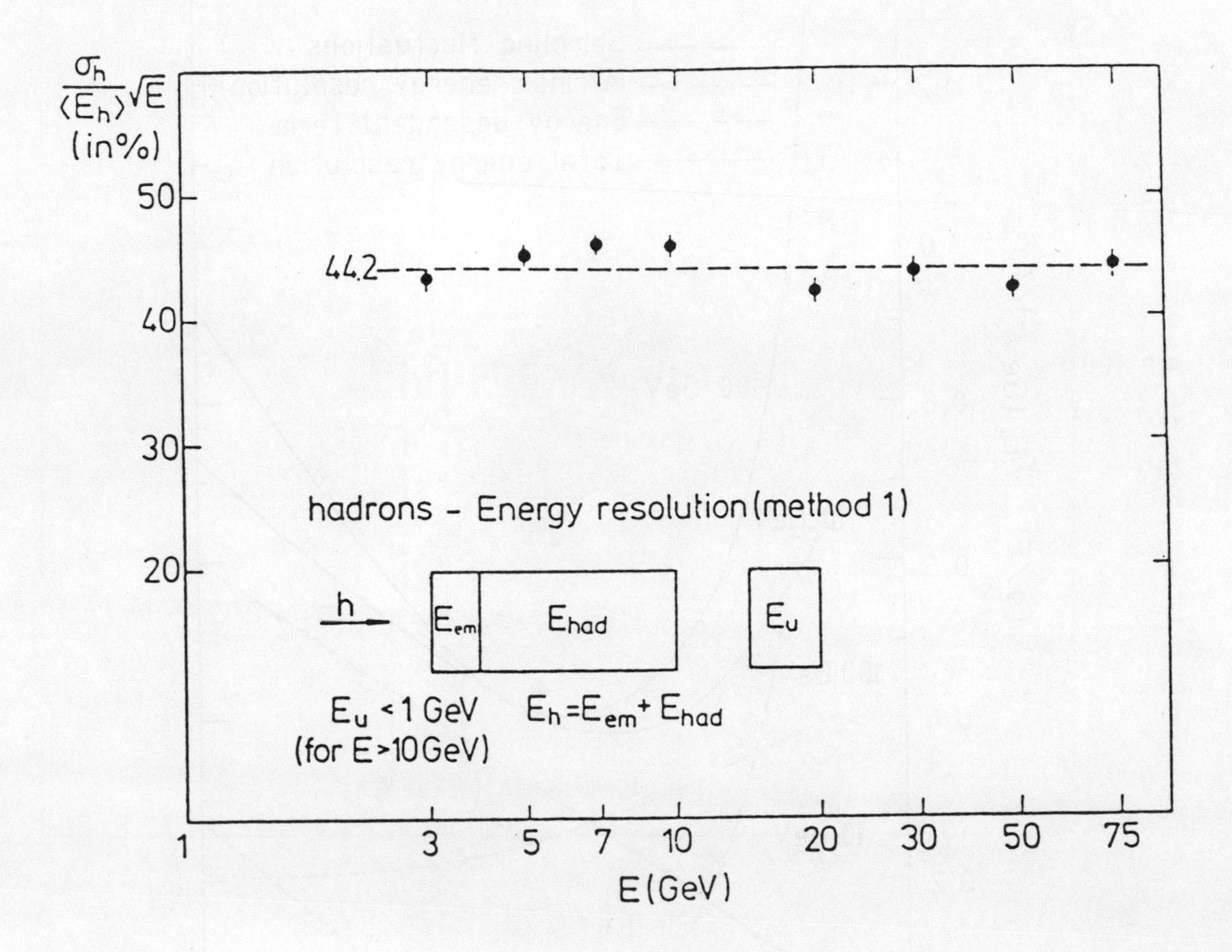

Figure 2

2. The hadronic energy resolution $\sigma/\sqrt{E}$ as a function of E, measured with a 10 mm Pb/2.5 mm plastic-scintillator calorimeter. Data from ref. 2.

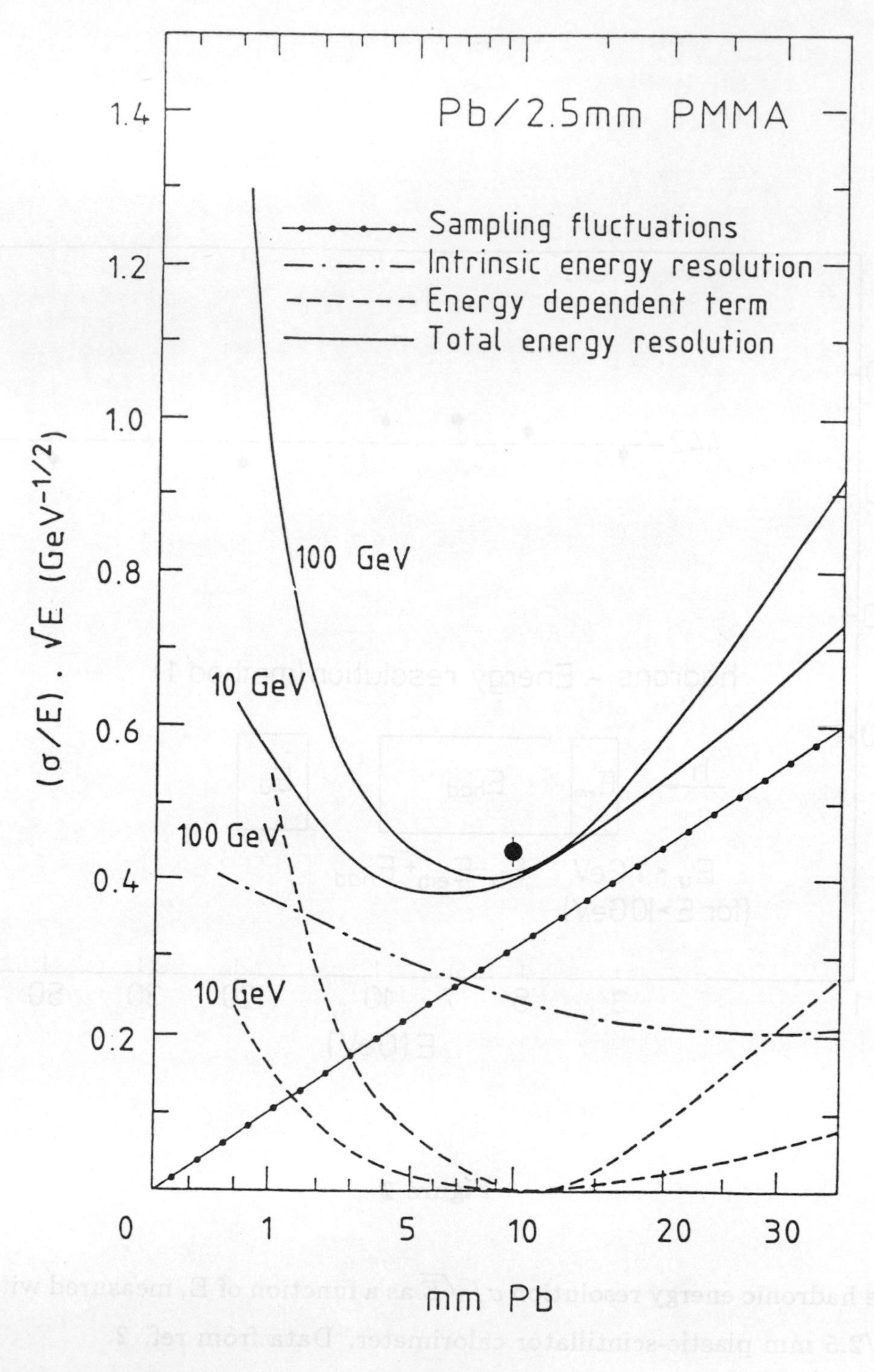

Figure 3

3. The total energy resolution and the various factors contributing to it, for detection of 10 and 100 GeV hadrons in Pb/PMMA calorimeters, as a function of the thickness of the lead plates. The thickness of the scintillator plates is 2.5 mm.

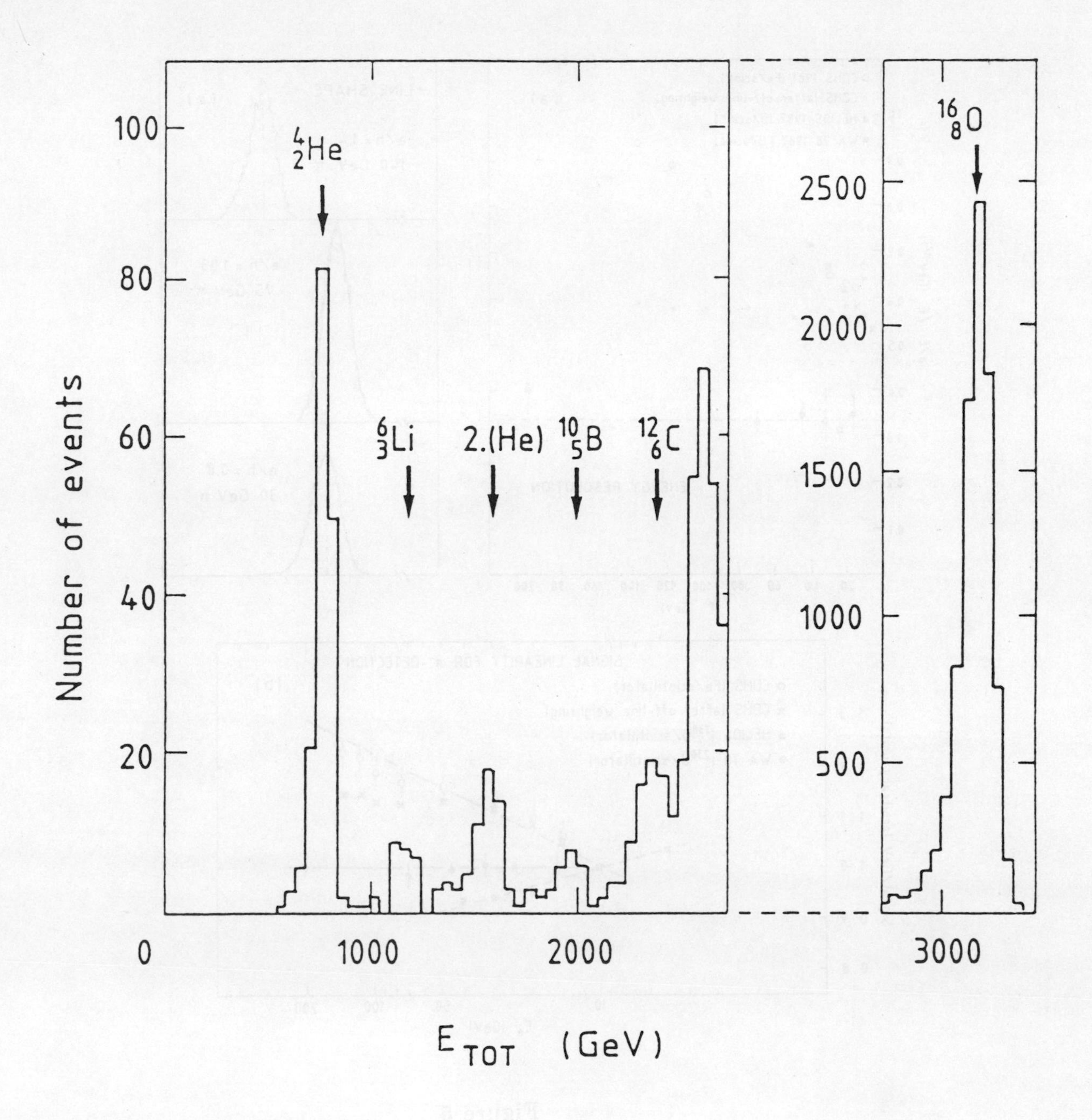

Figure 4

4. The HELIOS calorimeter as a high-resolution mass spectrometer. Total energy measured in the calorimeter for minimum bias events showing the composition of the SPS heavy-ion beam.

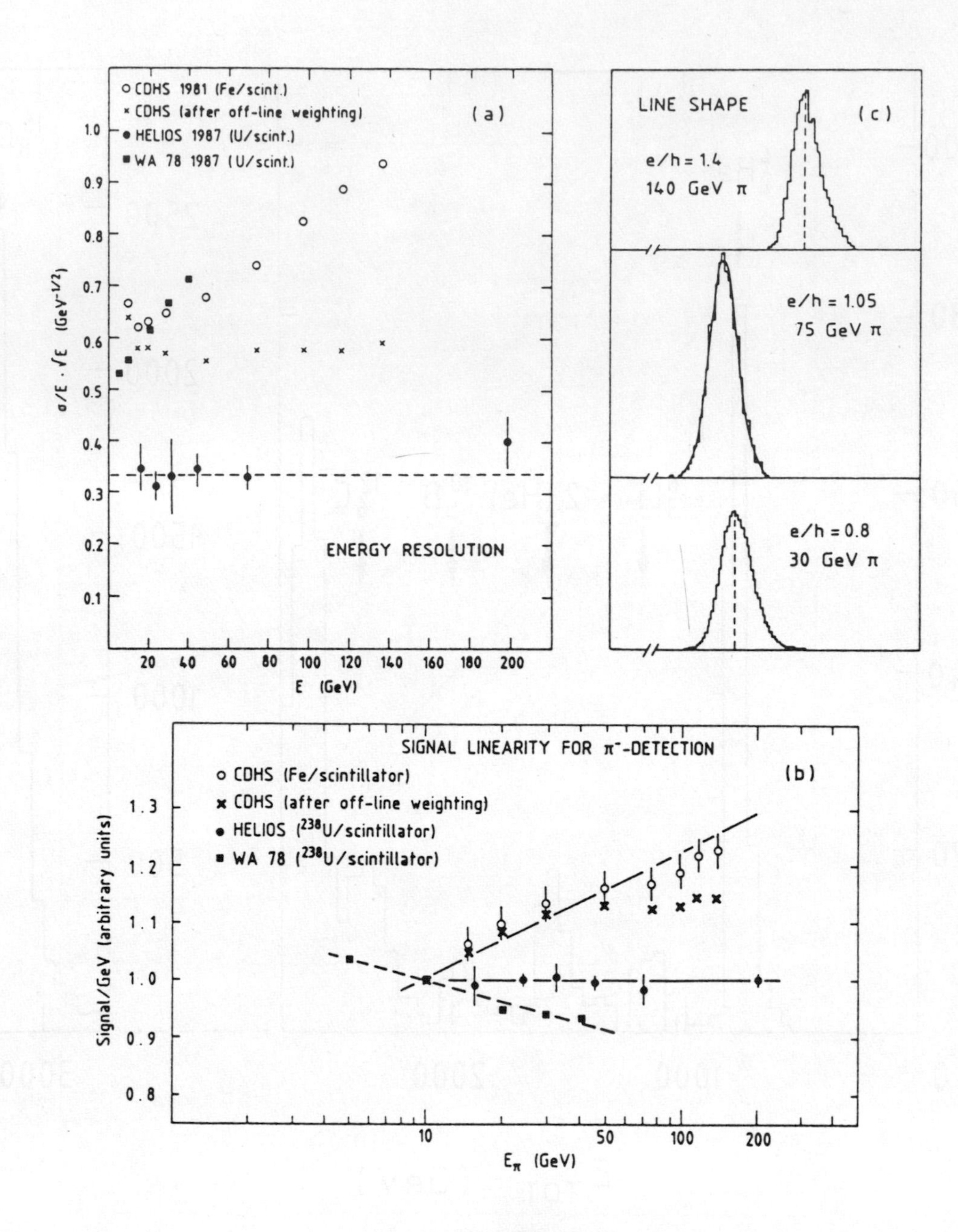

Figure 5

5. Experimental observation of the consequences of $e/h \neq 1$. Energy resolution (a), signal per GeV (b) and signal distribution (c) for pion detection in under-compensating (ref. 5), compensating (ref. 1) and overcompensating (ref. 6) calorimeters.

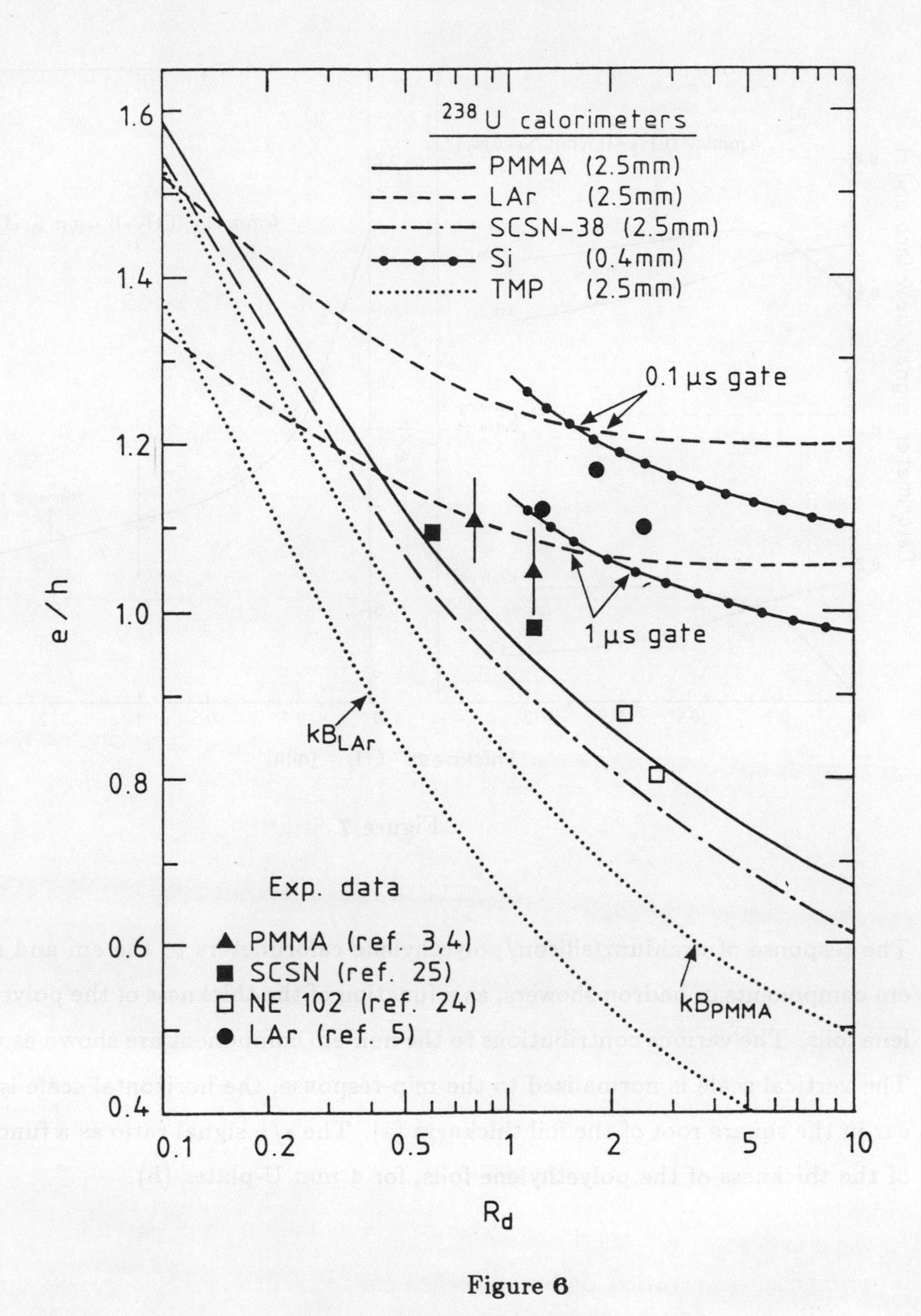

Figure 6

6. The signal ratio e/h for uranium calorimeters employing different readout materials, as a function of the ratio of the thicknesses of absorber and readout layers. Results of experimental measurements are included.

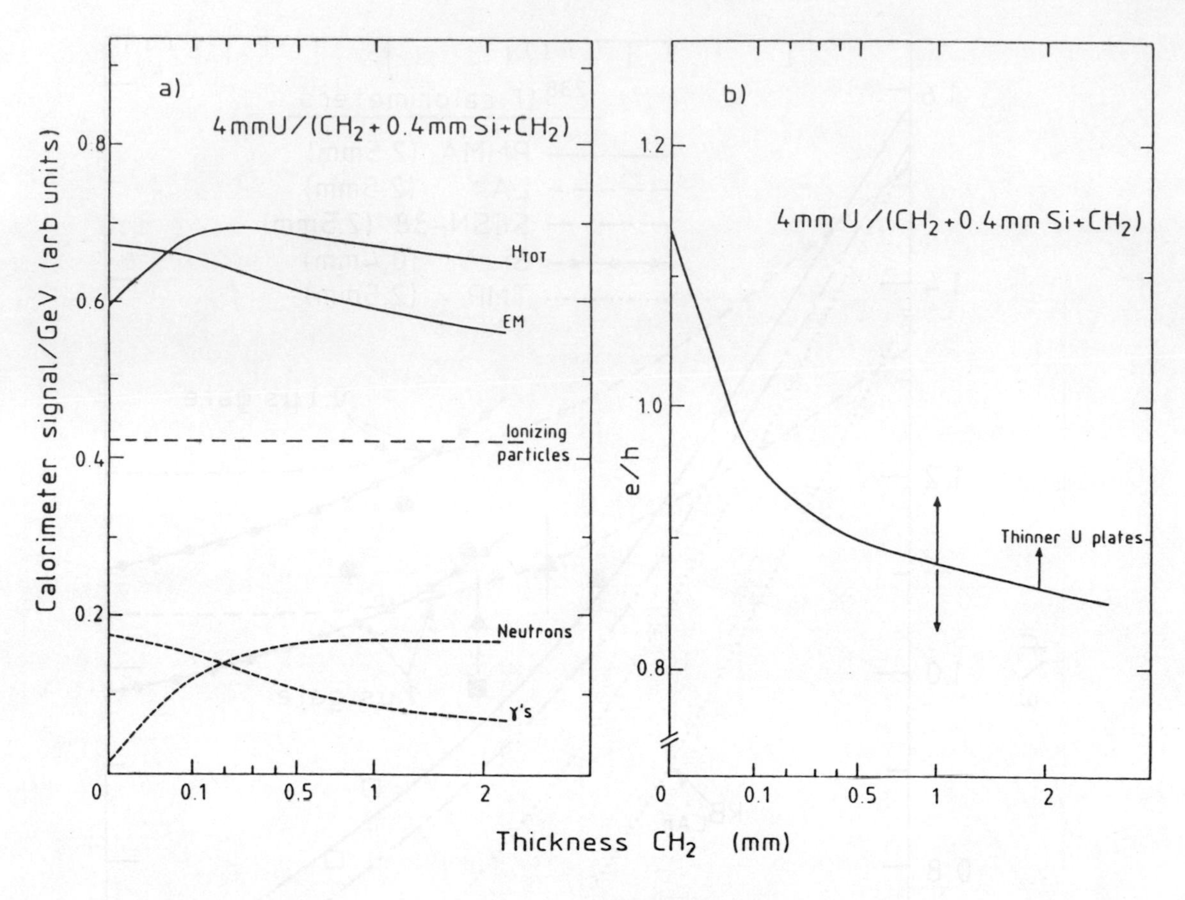

Figure 7

7. The response of uranium/silicon/polyethylene calorimeters to the em and non-em components of hadron showers, as a function of the thickness of the polyethylene foils. The various contributions to the non-em component are shown as well. The vertical scale is normalized to the mip-response, the horizontal scale is linear in the square root of the foil thickness (a). The e/h signal ratio as a function of the thickness of the polyethylene foils, for 4 mm U-plates (b).

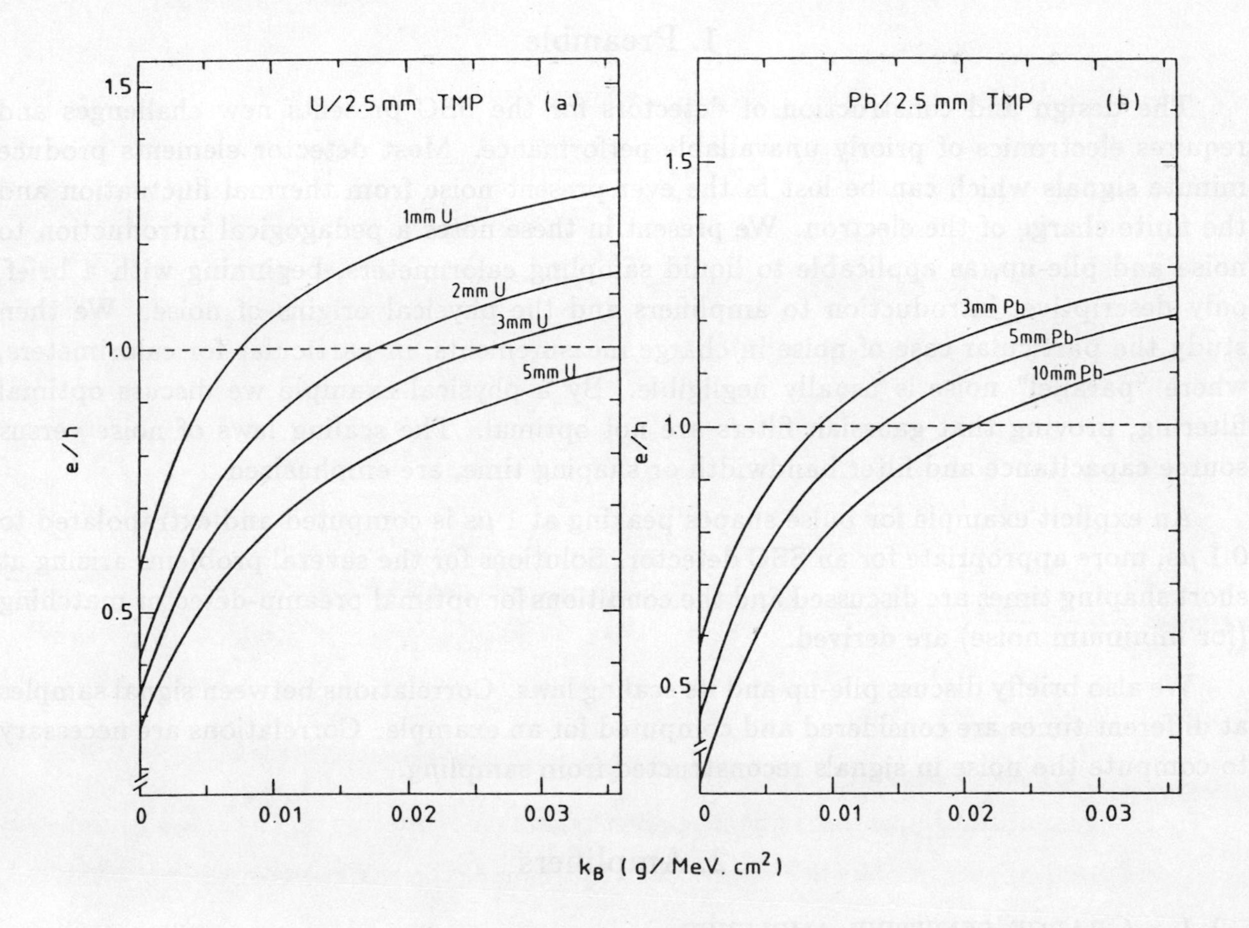

Figure 8

8. The e/h signal ratio as a function of Birk's constant k_B, for TMP calorimeters with uranium (a) or lead absorber (b). The liquid gaps are 2.5 mm wide.

Noise and Pile-up in Liquid Sampling Calorimeters

PAOLO FRANZINI
Columbia University
New York, NY

1. Preamble

The design and construction of detectors for the SSC presents new challenges and requires electronics of priorly unavailable performance. Most detector elements produce minute signals which can be lost in the ever present noise from thermal fluctuation and the finite charge of the electron. We present in these notes a pedagogical introduction to noise and pile-up, as applicable to liquid sampling calorimeters, beginning with a brief, only descriptive, introduction to amplifiers and the physical origins of noise. We then study the particular case of noise in charge measurements, in particular for calorimeters, where "parallel" noise is usually negligible. By a physical example we discuss optimal filtering, proving that gaussian filters are not optimal. The scaling laws of noise versus source capacitance and filter bandwidth or shaping time, are emphasized.

An explicit example for pulse shapes peaking at 1 μs is computed and extrapolated to 0.1 μs, more appropriate for an SSC detector. Solutions for the several problems arising at short shaping times are discussed and the conditions for optimal preamp-detector matching (for minimum noise) are derived.

We also briefly discuss pile-up and its scaling laws. Correlations between signal samples at different times are considered and computed for an example. Correlations are necessary to compute the noise in signals reconstructed from sampling.

2. Amplifiers

2.1 CHARGE SENSITIVE AMPLIFIER

We define in the following a few fundamentals of noise and amplifiers[1] especially for the so called "charge sensitive amplifier" (often called preamp), as used for charge measurements. We begin with an ideal amplifier, figure 1 A), with infinite, negative, gain and infinite bandwidth. Adding a negative feedback loop as shown in figure 1 B), the gain becomes well defined and is given by:

$$G \equiv \frac{V_{out}}{V_{in}} = 1 + \frac{Z_2}{Z_1} \simeq \frac{Z_2}{Z_1}. \tag{2.1}$$

while for the configuration in figure 1 C) the gain is

$$G = -\frac{Z_2}{Z_1}, \tag{2.2}$$

In eqs. (2.1) and (2.2), V_{in} and V_{out} are AC signals and Z_1 and Z_2 are complex impedances, ideally functions of the angular frequency ω, but not of the voltage across them or the current through them, i.e. Z_1 and Z_2 are linear circuit elements. Examples are resistance with $Z = R$, capacitance with $Z = 1/i\omega C$ and inductance with $Z = i\omega L$, where $i = \sqrt{-1}$. G is in general complex. Since we are interested in $|G|^2$, we will ignore the sign of G. From (2.2) the amplifier of figure 2 A) has a gain $G = 1/i\omega RC$, for AC signals. For any time dependent input signal $V_{in}(t)$, the output signal is given by $V_{out} = \frac{1}{RC} \int V_{in}(t)\, dt + const.$

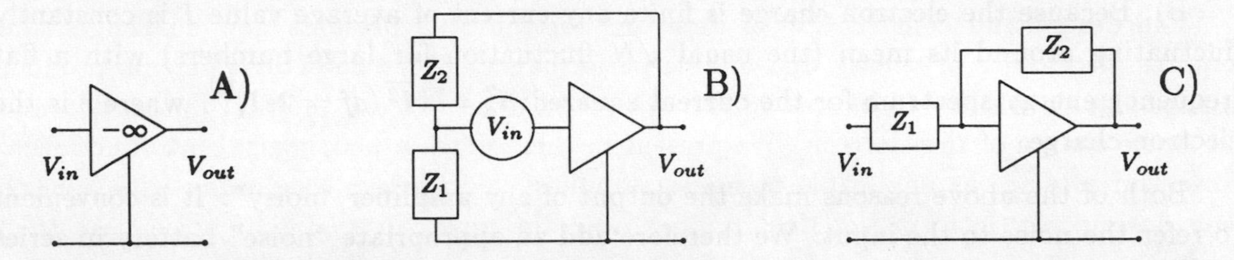

Figure 1. A): Ideal amplifier, gain=$-\infty$. B) and C): Negative feedback configurations.

We are mostly interested in the feedback configuration of figure 2 B). From the standard arithmetics of feedback it follows that (for the ideal amplifier defined at the beginning) the input impedance with feedback is $Z_{in} = 0$. We cannot anymore define a voltage gain but we can apply a current signal to the input and define a "conversion gain" as the ratio of V_{out} to I_{in}. The AC conversion gain is just Z_f, from which the name "transimpedance amplifiers".

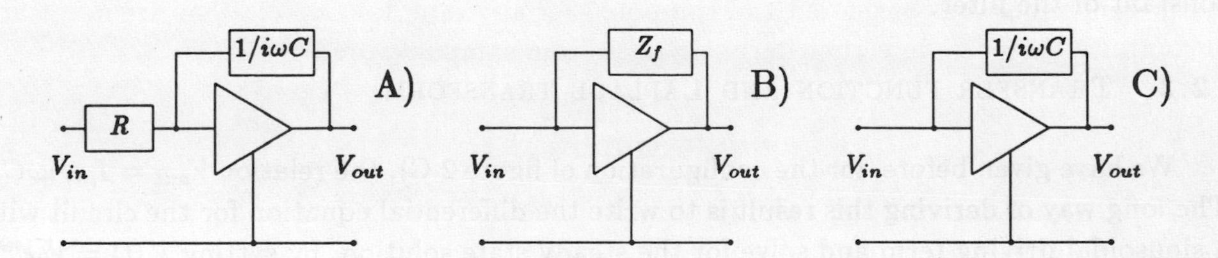

Figure 2. A) Voltage integrator. B) Transimpedance amplifier. C) Current integrator.

For the case of a feedback capacitor, figure 2 C), the output voltage is given in terms of the input current by $V_{out} = I_{in}/i\omega C_f$, for AC signals and in general by $V_{out} = \frac{1}{C_f} \int I_{in}(t)\, dt + const$, from which the name charge sensitive amplifier. We will come back to the limitations of practical charge sensitive amplifiers, for the following we will assume the ideal properties $Z_{in} = 0$ and $V_{out} = Q_{in}/C_f$.

2.2 SOURCES OF NOISE

There are two reasons why an amplifier always produces noise:

A). From thermodynamics we know that if a dissipative mechanism is present in a system, energy is dissipated with a flat frequency energy spectrum. From this follows that a resistor R at temperature T is equivalent to a noiseless resistor in series with a battery with a "voltage squared noise density" $e_n^2 = d\overline{V^2}/df = 4kTR$[2] .

B). Because the electron charge is finite any current of average value I is constantly fluctuating around its mean (the usual $\sqrt{N}$ fluctuation for large numbers) with a flat frequency energy spectrum for the current squared: $i_n^2 = d\overline{I^2}/df = 2eI,$[3] where e is the electron charge.

Both of the above reasons make the output of any amplifier "noisy". It is convenient to refer the noise to the input. We therefore add an appropriate "noise" battery in series with the input of our ideal amplifier. The magnitude of the noise source is determined by e_n^2, measured in V^2/Hz. This noise is called series noise. If the amplifier has feedback we can compute how much noise appears at the output. From e_n^2 we can estimate the rms fluctuation of V. V and I fluctuate without correlation in any frequency interval Δf, therefore $\mathrm{rms}(V) = \sqrt{\int_0^\infty e_n^2\, df} = e_n\sqrt{\int_0^\infty df},\ for\ e_n(f) = const.$, which is clearly divergent. A single RC integration removes this divergence: $\mathrm{rms}(V) = e_n\sqrt{\int_0^\infty a^2/(\omega^2 + a^2)\, df} = e_n\sqrt{a\pi/2}$, where $a = 1/RC$. We see a general property of series noise: for $e_n^2(f) = const$, the noise scales as the square root of the cut-off frequency or of the inverse of the RC time constant of the filter.

2.3 TRANSFER FUNCTION AND LAPLACE TRANSFORM

We have given before, for the configuration of figure 2 C), the relation $V_{out} = I_{in}/i\omega C_f$. The long way of deriving this result is to write the differential equation for the circuit with a sinusoidal driving term and solve for the steady state solution, by setting $V(t) = V_0 e^{i\omega t}$. This leads to the well known algebraic methods for solving AC circuits which we have used. Things are more complicated for non sinusoidal driving terms. Since most simple signals used for the analysis of circuit response are not in general square summable, use of the Laplace rather than the fourier transform is appropriate. There is moreover a very trivial way to write the Laplace transform of the solution of the differential equation for the circuit. By the substitution $i\omega \to s$ in the AC solution for V_{out} one obtains the Laplace transform $(\mathcal{L}\{V(t)\} or \tilde{V}(s))$ of the output voltage. For the AC solution above, the Laplace transform of the output signal is obtained in terms of the Laplace transform of the input current: $\tilde{V}(s) = \tilde{I}(s)/sC_f$. The inverse Laplace transform $(\mathcal{L}^{-1})$ of this equation is $V_{out} = \frac{1}{C_f}\int I_{in}(t)\, dt$. We will usually ignore constants since they can always be obtained trivially because of the linearity of the circuit equations.

We shall in the following refer to $g(\omega) \equiv (AC - Output)/(AC - Input)$ as the AC response of the circuit and to $g(s) \equiv Output(s)/Input(s)$ as the transfer function of the circuit; they are the same function with $i\omega \rightleftharpoons s$. For most circuits of interest the transfer function is the ratio of polynomials in s. Likewise the driving term is very simple for signals of interest. $\tilde{V}(s)$ is easily reducible to sum of terms whose inverse Laplace transform is well known. It is therefore straightforward to obtain $V(t)$. Essentially the same arguments apply to $g(\omega)$. It is usually easy to analytically compute $\int_0^\infty |g(\omega)|^2\, df$, although numerical integration on your home PC is faster.

2.4 NOISE CHARGE REFERRED TO THE INPUT

As discussed above, amplifiers always produce noise, in the form of a randomly fluctuating, normal distributed output voltage, with rms value V_{noise}. Output voltages are of practical relevance but do not convey in a simple way the relative importance of the amplifier noise as compared to the magnitude of the charge signal in which we are interested. Inverting the relation $V_{out} = Q_{in}/C_f$ for $V_{out} = V_{noise}$ we obtain the "input equivalent noise charge" or $Q_{noise} = C_f/V_{noise}$, which allows immediate comparison of amplifier noise and input signal. If we know the "responsivity" R_E of the detector defined by $Q_{out}^{Det} = R_E E_{in}^{Det}$, where E_{in}^{Det} is the energy deposited in the calorimeter, from Q_{noise} we also obtain the error in the energy measurement due to the amplifier noise: $E_{noise} = Q_{noise}/R_E$

3. Noise in charge sensitive amplifiers

3.1 SERIES NOISE

Since the response of a charge sensitive preamp to a charge signal is the same as that of the feedback capacitor, were we able to put the signal charge on it, one might ask why use them.

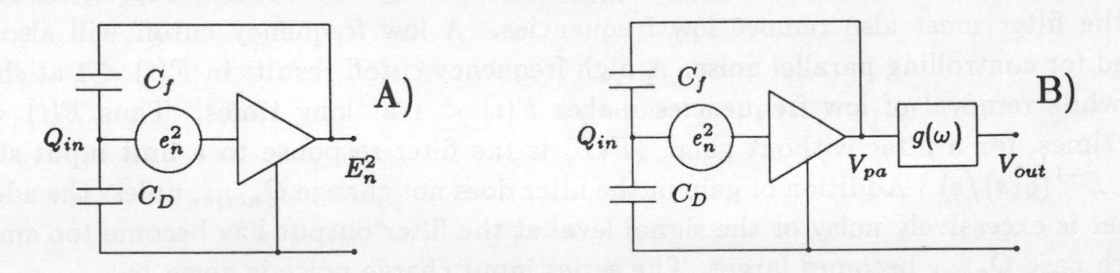

Figure 3. Detector and charge sensitive amplifier.

Figure 3 A) shows an approximation to the real world, with a signal charge Q_{in} coming from a detector of very large capacitance C_D, connected to a charge sensitive amplifier. Because of its zero input impedance (which is not quite true of a real preamp) all charge is transferred out from the detector and the output signal is $V = Q_{in}/C_f$, where C_f could be for instance 1 pF. The charge sensitive preamps gives the desired signal no matter how large is C_D. The latter however completes a feedback loop which amplifies the amplifier input noise. From eq. (2.1) the amplifier output voltage squared density is $E_n^2 = e_n^2(C_D/C_f)^2$. Without frequency cutoffs (filtering) the integral of E_n^2 diverges. However the input noise charge squared density is $q_n^2 = e_n^2 C_D^2$, therefore $Q_{noise,series} \propto e_n C_D$. If $g(\omega)$ is the AC frequency response of the filter in figure 3 B), then the rms voltage noise at the output of the filter is given by:

$$\mathrm{rms}(V) = e_n(C_D/C_f)\sqrt{\int_0^\infty |g(\omega)|^2\, df}. \tag{3.1}$$

The filter however changes the relation between Q_{in} and V_{out}. In fact we must first define input and output signals more precisely. We will for now assume that the input charge Q_{in} appears at the preamp input in ≈ 0 time. Then the preamp output is a step in time: $V_{pa}(t) = (Q_{in}/C_f)\theta(t)$, figure 4, where $\theta(t) = 0$ for $t < 0$, $= 1$ for $t > 0$.

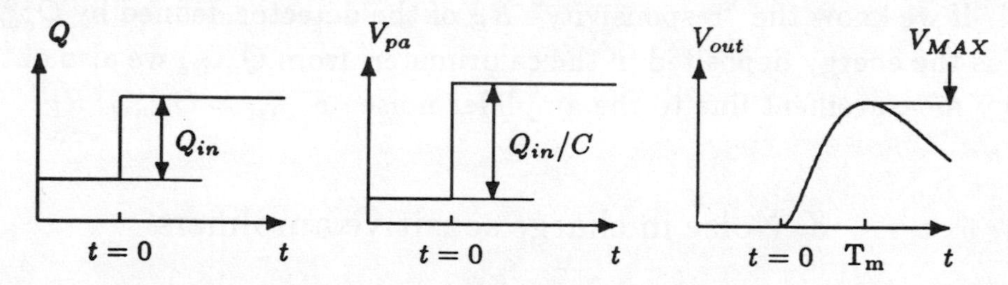

Figure 4. Definition of input and output signals.

If the filter removes only high frequencies (and has no gain), the filter output will after some time reach the value Q_{in}/C_f. In order to control pile-up, we require that the filter output be a function of time $V(t) = F(t)(Q_{in}/C_f)$, where, for a unit input voltage step, $F(t)$ rises to some value F_{MAX} for $t = T_m$ and returns rapidly to zero. Thus the filter must also remove low frequencies. A low frequency cutoff will also be required for controlling parallel noise. A high frequency cutoff results in $F(t) < 1$ at short times while removal of low frequencies makes $F(t) < 1$ at long times. Thus $F(t) < 1$ for all times, for a filter without gain. ($F(T)$ is the filter response to a unit input step, $F(t) = \mathcal{L}^{-1}(g(s)/s)$.) Addition of gain in the filter does not change Q_{noise}, unless the added amplifier is excessively noisy or the signal level at the filter output has become too small, in which case Q_{noise} becomes larger. The series input charge noise is given by:

$$Q_{noise,series} = (e_n C_D/F_{MAX})\sqrt{\int_0^\infty |g(\omega)|^2\, df}. \tag{3.2}$$

From either eq. (3.1) or (3.2) we obtain that the input noise charge is proportional to the detector capacitance and to the square root of an inverse time, for a given value of F_{MAX}, i.e. for a given type of filter. This time can be identified with T_m, as we will explain later. Later we shall also see, that while the basic proportionality of noise to capacitance is correct for each primitive element of the detector, the total detector noise scales as the square root of the total capacitance, when we take into account the subdivision of the detector into many channels.

3.2 PARALLEL NOISE

Dissipative elements across the preamp input result in a noise current into the charge sensitive preamp. Typical of these are bias/feedback resistors in the preamp, resistors to the H.V. power supplies, dielectric losses in capacitors or transformer core losses. In addition the preamp bias current and leakage currents also contribute to noise. Contrary to the case of series noise, which cannot be removed, parallel noise can in general be controlled. We should mention however that signal pile-up at high rate is in all respects a parallel noise, and the present discussion applies to its minimization. All sources of parallel noise can be combined into a single current source, in parallel with the input, with current squared density i_n^2 measured in A^2/Hz. Since $V_{out} = I_{in}/i\omega C_f$, the output voltage squared noise density is $E_n^2 = i_n^2/\omega^2 C_f^2$. Notice that $\int_{f_{lo}}^{\infty} i_n^2/\omega^2 \, df$ diverges for $f_{lo} \to 0$, requiring a low frequency cut-off such as an RC differentiation. After an appropriate filter, of response $g(\omega)$, the parallel noise is given by:

$$Q_{noise,parallel} = (i_n/F_{MAX})\sqrt{\int_0^{\infty} \frac{|g(\omega)|^2}{\omega^2} df}. \tag{3.3}$$

for $i_n^2(f) = const.$, from which $Q_{noise,parallel} \propto \sqrt{time} \propto \sqrt{T_m}$.

3.3 OTHER CONFIGURATIONS

As a short aside we also consider the case of an inductor across the charge sensitive preamplifier input, as is the case for the leakage inductance of a coupling transformer. The output voltage squared noise density is given by $E_n^2 = e_n^2/(\omega^2 LC)^2$, from which $\mathrm{rms}(V_{out}) = (e_n/LC)\sqrt{\int (1/\omega^4)\, df}$. At least 2 RC differentiations are required to remove this divergence, resulting in a so called bipolar pulse shape, *i.e.* pulses which after reaching a maximum amplitude of a given sign, return to zero and cross it, reach some maximum amplitude of the opposite sign and finally return to zero. Pulses of this kind will be briefly considered in section 4.3. For the case of an inductor across the preamp input, the noise increases as ${T_m}^{3/2}$.

4. Filters

4.1 INTRODUCTION

We have seen how $|g(\omega)|$ must vanish at both low and high frequencies. The natural question is: *IS THERE A BEST FILTER*? The answer is not unique, it depends on the relative amounts of series and parallel noise, where pile-up and other things such as uranium radioactivity, might or not be included. It also depends on practical considerations such as how much signal loss one can tolerate before noise contributions from the following gain stages become important. Also "EMI", [4] noise pickup from external sources, must be considered for true optimization. We derive here some properties of a simple class of filters, and prove wrong a common belief that filters producing exponential cusp or gaussian shaped pulses are best. They are physically realizable only in the limit in which $F_{MAX} = 0$, because they give an output from $t = -\infty$ and give *infinite noise*.

4.2 CR-RC FILTERS

These filters consist of a single RC differentiation section and n RC integrations, as shown in figure 5 with all sections having the same value of the time constant $RC = 1/a$.

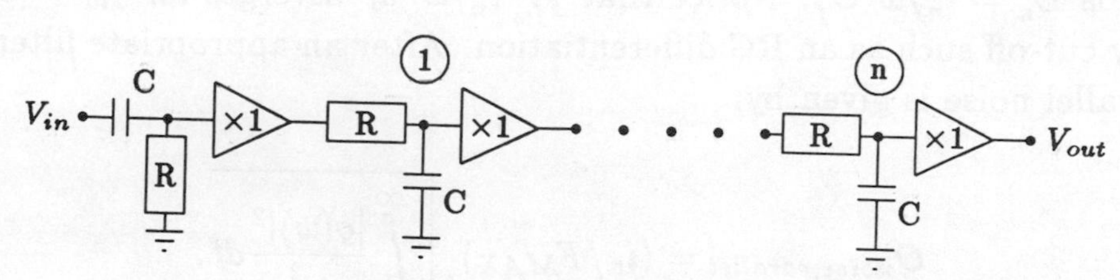

Figure 5. One differentiation plus n integrations filter.

The filter response is independent of the order of the sections. The amplifiers between RC stages are ideal unity gain amplifiers, inserted to make the response of each stage independent from the other ones. This is *not* the way one *usually* realizes such filters. The AC response of a differentiation is $i\omega/(i\omega + a)$ and of an integration is $a/(i\omega + a)$. Therefore we have:

$$|g_n(\omega)|^2 = \omega^2 a^{2n}/((\omega^2 + a^2)^{n+1}), \qquad a = 1/RC \tag{4.1}$$

$$g_n(s) = \tilde{V}_{out}(s)/\tilde{V}_{in}(s) = sa^2/(s+a)^{n+1}, \tag{4.2}$$

where $\tilde{V}_{in}(s) = 1/s$ for a unit input step at $t = 0$. From eq. (4.2) $\tilde{V}_{out,n}(s) = a^2/(s+a)^{n+1}$, and

$$F_n(t) \equiv V_n(t) = (at)^n e^{-at}/n!, \tag{4.3}$$

from which $F_{MAX,n} = n^n e^{-n}/n! \simeq (1/\sqrt{2\pi n})(1 + 1/12n)$, for $t = n/a \equiv \mathrm{T_m}$.

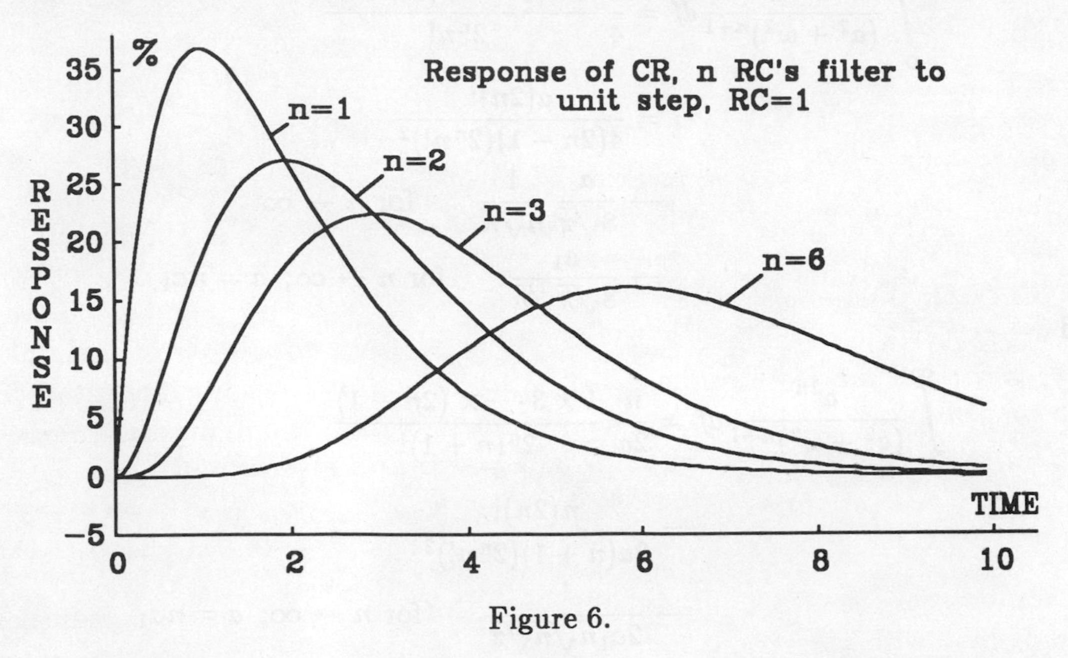

Figure 6.

The signals $V_{out,n}$ are shown in figure 6 for $a = 1$ (time^{-1}). Since signals which last longer and longer in time are not acceptable (because of pile-up, if nothing else) we choose a in such a way that all filters give signals peaking at '1', that is $a_n = na_1$, where $a_1 = 1$. Figure 7 shows these signals.

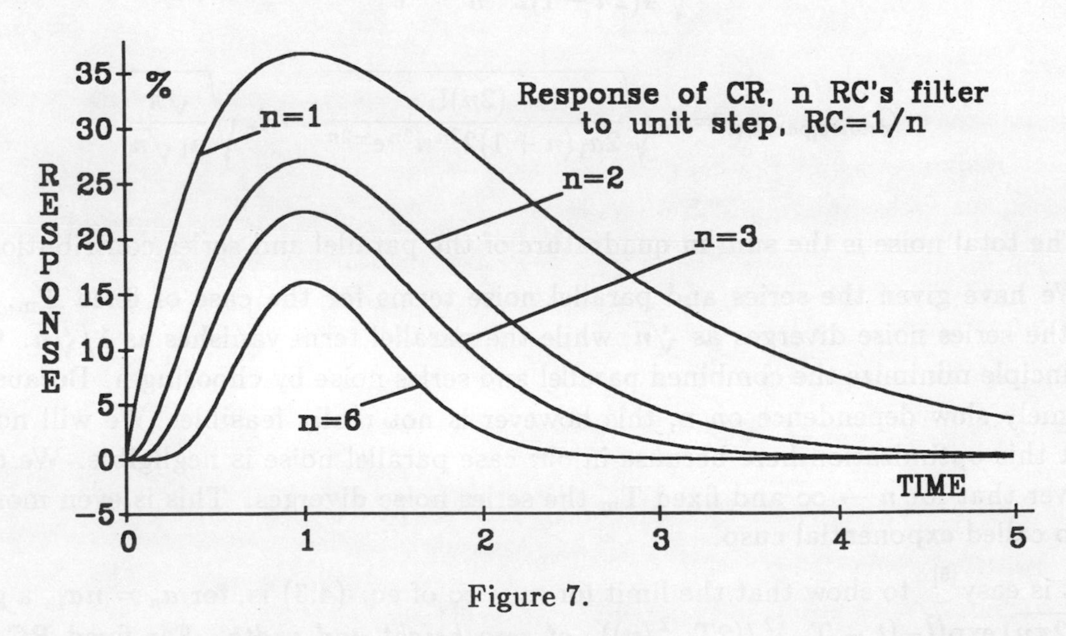

Figure 7.

Increasing n makes the signal resemble a gaussian. To compute series and parallel noise, we need the integrals $S_n = \int_0^\infty |g(\omega)|^2 \, df$ and $P_n = \int_0^\infty (|g(\omega)|^2/\omega^2) \, df$, (eqs. (3.2) and (3.3)), which are given by:

$$\int_0^\infty \frac{a^{2n}\omega^2}{(a^2+\omega^2)^{n+1}} df = \frac{a}{4}\,\frac{1\times 3\cdots\times(2n-3)}{2^n n!}$$

$$= \frac{a(2n)!}{4(2n-1)(2^n n!)^2}$$

$$\to \frac{a}{8\sqrt{\pi}}\,\frac{1}{n\sqrt{n}} \qquad \text{for } n\to\infty$$

$$\to \frac{a_1}{8\sqrt{\pi}\sqrt{n}} \qquad \text{for } n\to\infty;\ a = na_1$$

and

$$\int_0^\infty \frac{a^{2n}}{(a^2+\omega^2)^{n+1}} df = \frac{n}{2a}\,\frac{1\times 3\cdots\times(2n-1)}{2^n(n+1)!}$$

$$= \frac{n(2n)!}{2a(n+1)(2^n n!)^2}$$

$$\to \frac{1}{2a_1 n\sqrt{n}\sqrt{\pi}} \qquad \text{for } n\to\infty;\ a = na_1$$

Combining these results with $F_{MAX,n}$ from eq. (4.3) we finally have:

$$Q_{noise,series} = e_n C_D \sqrt{\frac{a_1(2n)!}{4(2n-1)2^{2n}n^{2n-1}e^{-2n}}} \to e_n C_D \sqrt{a_1\sqrt{\pi}\sqrt{n}} \tag{4.4}$$

$$Q_{noise,parallel} = i_n \sqrt{\frac{(2n)!}{2a_1(n+1)2^{2n}n^{2n}e^{-2n}}} \to i_n \sqrt{\frac{\sqrt{\pi}}{a_1\sqrt{n}}} \tag{4.5}$$

The total noise is the sum in quadrature of the parallel and series contributions.

We have given the series and parallel noise terms for the case of fixed T_m. In this case the series noise diverges as $\sqrt[4]{n}$, while the parallel term vanishes as $1/\sqrt[4]{n}$. One can in principle minimize the combined parallel and series noise by choosing n. Because of the extremely slow dependence on n, this however is not really feasible. We will not worry about this optimization here because in our case parallel noise is negligible. We do insist however that for $n \to \infty$ and fixed T_m the series noise diverges. This is even more so for the so called exponential cusp.

It is easy[5] to show that the limit for $n \to \infty$ of eq. (4.3) is, for $a_n = na_1$, a gaussian $(1/\sqrt{2\pi n})\exp((-(t-T_m)^2/(2{T_m}^2/n))$, *of zero height and width.* For fixed RC in each filter section, when $n \to \infty$, we find instead that the series noise vanishes and the parallel noise diverges. Also $T_m \to \infty$ as n and the limit of eq. (4.3) is again a gaussian of zero height and infinite width. These results are valid for *any causal filter* which produces a gaussian shaped output pulse for an input step, because of the uniqueness of time domain to frequency domain transformations (fourier transform!).

In table 1 we give values for the parallel and series noise for $T_m = 1\ \mu s$ and unit e_n/C_D and i_n. For $1 \leq n \leq 3$, the noise varies by less than 5%. An estimator of the noise to be expected after a simple, unipolar filter is therefore the time T_m, at which the output signal peaks, for an input step at t=0. For bipolar shaped output pulses, which are equivalent to sampling the filter output twice, once before the signal and once at the signal peak, and then taking the difference of the two samples, see section 7.2, the noise will in general be $\simeq \sqrt{2}$ times larger for the same value of T_m. The lowest series noise is obtained for n=2.

4.3 OTHER FILTERS

I have of course in no way proved that the 'n=2' filter above is the *best* filter. Other filters for which I have computed the noise integrals are those giving output pulses as shown in figure 8, for a step input.

n	F_{MAX}	$\sqrt{S_n}/F_{MAX}$	$\sqrt{P_n}/F_{MAX}$
1	0.37	961	961
2	0.27	924	924
3	0.22	966	882
4	0.20	1012	846
10	0.13	1217	715
100	0.04	2110	422
-	-	$\sqrt{Hz}$	$(10^{12} Hz)^{-1/2}$

Table 1. Series and parallel noise factor of CR- nRC filters for $T_m = 1\ \mu s$.

Peculiar as the idealized shapes might appear, *they are meant to be made of straight line segments*, such pulse shapes can be obtained by integration, followed by differences of appropriately scaled and delayed signals or by sampling techniques.

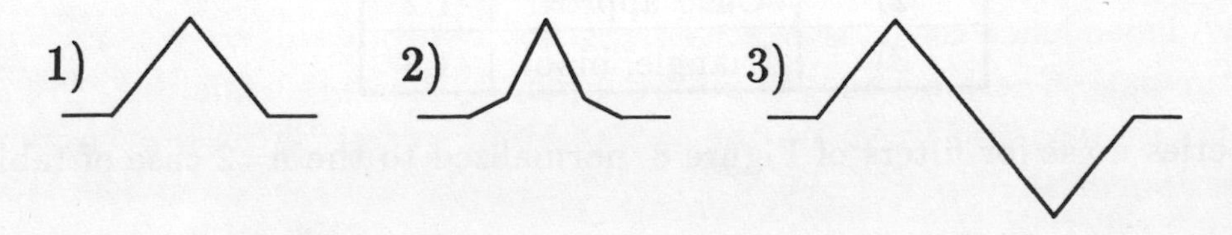

Figure 8. Output pulse shapes.

The noise analysis for signal sampling followed by manipulation of the samples, either in analog or after digitization, is equivalent to that of filters which use delays. Generating delays (*e.g.* with delay lines) is very expensive, while multiple sampling is a very economic and efficient technique. This is why I include these filters and give an example in this section of how to produce a triangle pulse and how to compute the filter response. The relative series noise, for the same value of T_m for all filters, is given in table 2, together with results for other shapes. The filter circuit given in block diagram form in figure 9 produces a triangle output. For a step input $V_{in} = V_a\theta(t)$ the output signal is $V_{out}(t) = V_b((t/\tau)\theta(t) - 2((t-\tau)/\tau)\theta(t-\tau) + ((t-2\tau)/\tau)\theta(t-2\tau))$. The transfer function is $g(s) = 1/s\tau - (2/s\tau)e^{-s\tau} + (1/s\tau)e^{-2s\tau}$ and the AC response $g(\omega) = \frac{1}{i\omega}(1 - 2e^{-i\omega\tau} + e^{-2i\omega\tau})$. The modulus squared of the transfer function is:

$$\begin{aligned} |g(\omega)|^2 &= \frac{1}{\omega^2}(6 - 8\cos(\omega\tau) + 2\cos(2\omega\tau)) \\ &= \frac{1}{\omega^2}(16\sin^2(\omega\tau/2) - 4\sin^2(\omega\tau)) \\ &\propto \omega^4 \text{ for } \omega \to 0 \end{aligned}$$

equivalent to two differentiations for low frequencies, as obvious from figure 9, while at high frequency $|g(\omega)|^2 \propto 1/\omega^2$, equivalent to a single integration, as expected.

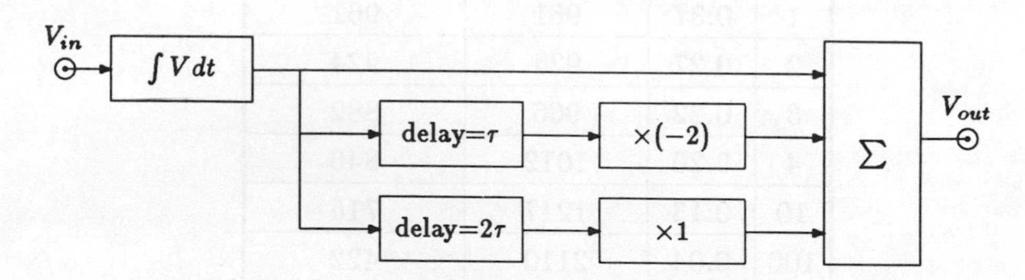

Figure 9. Filter for producing a triangle pulse

Filter	'Shape'	Noise
CR-2RC	-	1.0
1)	Triangle	1.08
2)	Cusp approx.	1.2
3)	Triangle, bipol.	1.5

Table 2. Series noise for filters of Figure 8, normalized to the n=2 case of table 1.

5. An Example

5.1 NOISE MAGNITUDE

We can now estimate the magnitude of the noise for a detector element. We will use in the following the n=2 filter but we still need values for e_n^2 and i_n^2. For charge measurements at moderately long times ($T_m \geq 20 - 50$ ns), the lowest noise is obtained with preamps using a junction field effect transistor (JFET or FET) as their input stage. For a good design, the preamp noise is essentially entirely due to the input voltage noise of the FET. A popular device is the type 2SK147 FET and similar ones, for which $g_m \equiv \partial i_{out}/\partial v_{in}]_{v_{out}=const} \simeq 0.05\ \Omega^{-1}$ at a drain current of $\simeq 6$ mA. The voltage and current noises of a FET are approximately (ref. 1) given by:

$$e_{n,FET} = \sqrt{4kT\frac{2}{3}\frac{1}{g_m}} = 0.46 \quad \text{nV}/\sqrt{\text{Hz}}$$

$$i_{n,FET} = \sqrt{2e \times 10^{-12}} = 5.6 \times 10^{-16} \quad \text{A}/\sqrt{\text{Hz}}$$

Choosing $C_D = 1$ nF and $T_m = 1\ \mu$s we obtain

$$Q_{noise}^2 = 923^2(\text{Hz}) \times e_n^2(\text{V}^2/\text{Hz}) \times C_D^2(\text{F}) \times (10^{-6}(\text{s})/T(\text{s})) + (744 \times 10^{-6})^2(\text{Hz}^{-1}) \times i_n^2(\text{A}^2/\text{Hz}) \times (T(\text{s})/10^{-6}(\text{s}))$$

which reduces to

$$Q_{noise}^2 = 1.84 \times 10^{-31} + 1.8 \times 10^{-37} \qquad \text{C}^2$$

The first term in both expressions above is the contribution from the series noise of the FET while the second term is the parallel noise. As anticipated, the parallel noise, at least that part of it due to the preamp, is negligible. A 10 MΩ resistor across the preamp input gives a noise $Q_{noise}^2 \sim 10^{-35}$. The noise for the above choice of T_m and C_D is finally: $Q_{noise} = 0.43$ fC or 2700 electrons. In the following we will always give the input noise charge in electrons: e. From the value of Q_{noise} we can estimate the error in energy measurements, due to amplifier noise. In argon one electron ion pair[6] is produced per $\simeq$26 eV of energy loss. Since however the positive ion drift velocity is extremely slow, the current in an argon gap where N electron-ions pairs are produced along the drift field, is given by $(Ne/T_{dr})(1 - t/T_{dr})$ for $0 < t < T_{dr}$, where T_{dr} is the time for electrons to drift across the full gap. The integral of this current is $Ne/2$, therefore the responsivity of a sampling liquid argon calorimeter is $\varrho/52\ e/\text{eV}$, where ϱ is the so called "sampling fraction", *i.e.* the fraction of the energy deposited in argon. A typical value for ϱ is 0.07 for which the above noise charge corresponds to an energy 'noise': $E_{noise} = 2$ MeV/nF. For the DØ detector we have $E_{noise} = 1.5$ MeV/nF, for $T_m = 2\ \mu$s.

6. Detector noise

6.1 NOISE SCALING

We now use the results above to estimate the noise in a realistic detector. In liquid argon, the capacitance of each gap is given, see for instance Jackson,[7] by $C(\mathrm{pF}) = .88 \times \epsilon_r \times area(\mathrm{cm}^2)/gap(\mathrm{mm})$, $\epsilon_r = 1.6$. In the DØ detector, between any two uranium plates we have a double gap of 2.3 mm of liquid argon. This configuration gives the smallest drift time for a given amount of argon. We assume that the electromagnetic part of the calorimeter consists of towers each having 2×20 gaps of 10×10 cm^2 in area. Therefore $C_D = 2.5$ nF and E_{noise}/tower $\simeq 5$ MeV. A realistic value of $\mathrm{T_m}$ for an SSC detector is $\mathrm{T_m} \simeq 100$ ns (or smaller). For $\mathrm{T_m} = 100$ ns the noise should be larger than the result above by $\sqrt{10}$. However the results derived above are valid only if $\mathrm{T_m} < T_{dr}$ and the preamp bandwidth is sufficiently large. Both conditions are not valid for $\mathrm{T_m} = 100$ ns. We can calculate the filter output for the input current signal, by decomposing the input as in figure 10. The Laplace transform of the input is therefore given by: $\tilde{I}(s) = I_0(1/s - 1/s^2 T_{dr} + (1/s^2 T_{dr})e^{-sT_{dr}})$. After the charge sensitive preamp and the filter (sec. 2.4 and 4.2, eq 4.2) the signal is:

$$V_{out}(t) = f(t) - g(t) + g(t - T_{dr})\theta(t - T_{dr}), \quad \text{with :}$$

$$f(t) = 1 - \frac{t^2 e^{-t}}{2} - \frac{te^{-t}}{2} - e^{-t}$$

$$g(t) = t - 3 + \frac{t^2 e^{-t}}{2} + 2te^{-t} + 3e^{-t}$$

where an overall normalization is ignored.

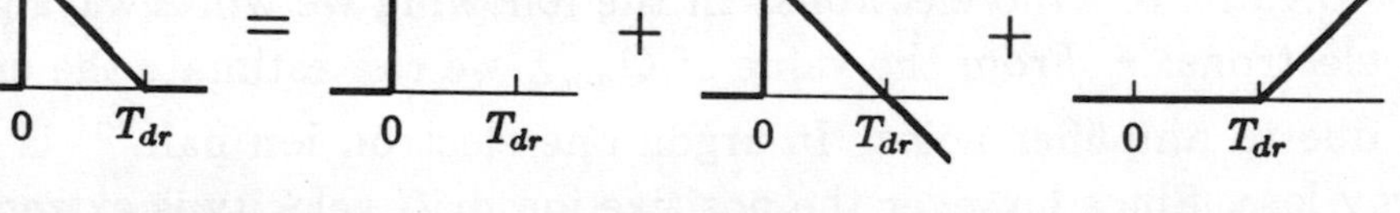

Figure 10. Decomposition of a triangular pulse.

$V_{out}(t)$ is shown in figure 11, for various values of T_{dr}/RC, where, for our example, $RC = \mathrm{T_m}/2 = 50$ ns.

From figure 11 we might conclude that a gap of 0.5 mm, for which $T_{dr} \simeq 100$ ns (T/RC=2) is acceptable. If the gap is reduced by a factor four, the absorber has to be made thinner and the total number of gaps increased by a factor four, to maintain the same sampling fraction.

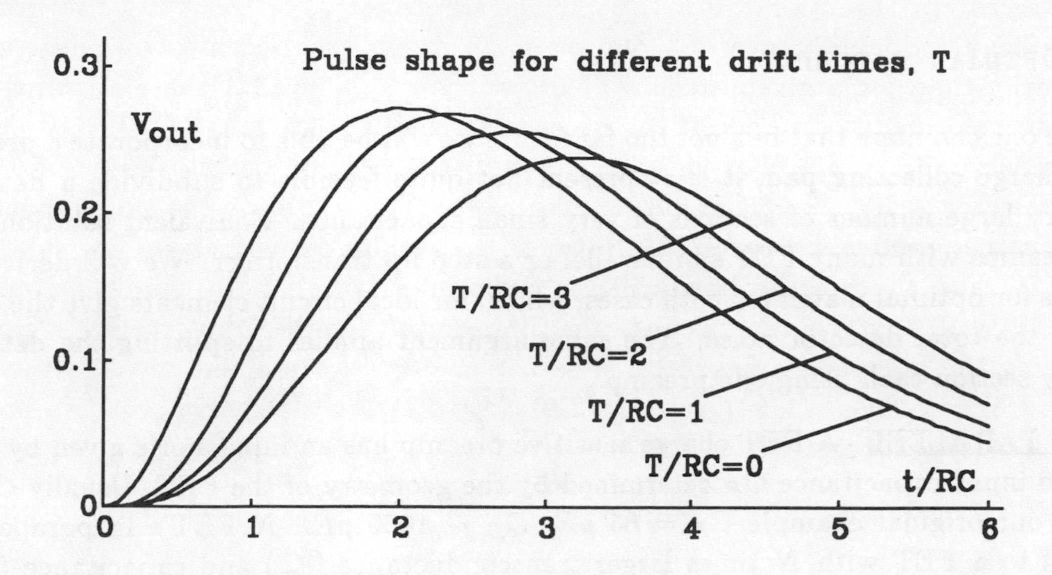

Figure 11. Dependence of filter output on drift time.

The tower capacitance is therefore increased by a factor 16 and

$$E_{noise} = 16 \times \sqrt{10} \times 5 = 253 \text{ MeV}.$$

The assumed infinite bandwidth of the preamp is the other severe limitation in achieving low noise at short times. A real preamp has a gain which at high frequencies [8] is $\propto \omega_0/i\omega$ where $\omega_0 = 1/(g_m C_c)$, where C_c controls the response of the amplifier. With capacitive feedback the preamp input impedance becomes purely ohmic, $R_{in} = (C_c/C_f)/g_m$, where for reason of stability $C_c/C_f > 1$. Thus in practice $R_{in} > 1/g_m \simeq 20\ \Omega$. Charge cannot be transferred from detector to preamp in times shorter than $R_{in}C_D$. The calculation of the output signal becomes long and tedious but straightforward. Instead of giving the answer, consisting of a couple dozen terms, we can analyze the effect of finite $R_{in}C_D$ in the simplified model that the detector is discharged through R_{in}. Then $Q(t) = Q(1 - \exp(-t/R_{in}C_D)) \simeq Qt/R_{in}C_D$. In our example $C_D = 40$ nf and $R_{in}C_D \simeq$ 1000 ns. Therefore only about 10% of the signal charge is collected at $t = \mathrm{T_m}$. This results in yet another factor ten increase in noise! The important lesson to learn here is that $Q_{noise} \propto C_D$ and $\propto \sqrt{\mathrm{T_m^{-1}}}$ only for $T_{dr}, R_{in}C_D < \mathrm{T_m}$. Otherwise $Q_{noise} \propto C_D^2$ and $Q_{noise} \propto \mathrm{T_m}\sqrt{\mathrm{T_m}}$. In scaling from $1\mu s$ to $0.1\mu s$ we have reached the situation where

$Q_{noise} \propto C_D^2$. This can be remedied by subdividing the tower into N sections, each connected to its own preamp. Not only do we go back to $Q_{noise} \propto C_D$, we additionally reduce the noise by a factor $\sqrt{N}$. The noise for each section is down by N and when the N signals are added together the noise is increased by $\sqrt{N}$, since each preamp noise is uncorrelated to the others. This in fact suggests that we reexamine the question of matching of preamp to detector.

6.2 OPTIMAL MATCHING

While one can hope that in a not too far future we will be able to incorporate a preamp in each charge collecting pad, it is at present not quite feasible to subdivide a detector into a very large number of sections of very small capacitance. Equivalent solutions are to use preamps with many FET's in parallel or a step up transformer. We will derive the conditions for optimal match for both cases, which, for ideal circuit elements give the same result for the total detector noise. The same argument applies to splitting the detector into many section each using one preamp.

FET's IN PARALLEL A FET charge sensitive preamp has an input noise given by $e_n^2 \propto g_m$ and an input capacitance C_F determined by the geometry of the FET. Usually $C_F \ll C_D$. For our original example $C_F = 65\ pF$, $C_D \sim 1000\ pF$. N FET's in parallel are equivalent to a FET with N times larger transconductance (g_m) and capacitance (C_F). The total noise charge for a preamp with N FET's is therefore proportional to $(C_D + NC_F)/\sqrt{N} = C_D/\sqrt{N} + C_F\sqrt{N}$ which is minimum for $C_D = NC_F$ or $N = C_D/C_F$. For given C_D the noise is reduced by a factor $2/\sqrt{N}$ when $N = C_D/C_F$ FET's in parallel are used. We have computed a noise of 253 MeV for 40000 pF, if $R_{in}C_D < 100\ ns$. A preamp with 40000/65=615 FET's in parallel would improve the noise by $\sqrt{615}/2 = 12.4$. Having increased g_m by ~ 600, Z_{in} is decreased by a similar value and we do indeed satisfy the condition $R_{in}C_D < 100\ ns$. Therefore, the noise per tower can be as low as 20 MeV.

In practice, physics considerations suggest subdividing the detector into sections with typical capacitance below 5 nF. Optimal matching would still require preamps with 5000/65 $\sim$ 100 FET's. Since the noise increases only as $\sqrt{N}$, we can obtain quite satisfactory results with significantly fewer FET's.

TRANSFORMER If a detector element of capacitance C_D is connected to a preamp input via a step up transformer[1] with turn ratio 1:N, the preamp sees at its input a capacitance C_D/N^2. The signal charge Q is however, reduced to Q/N at the preamp input. Referring all quantities to the transformer input, where we can compare directly to the signal, the noise is $\propto (C_D + N^2C_F)/N = C_D/N + NC_F$, which is minimum for $N = \sqrt{C_D/C_F}$. Using a transformer with turn ratio 1:$\sqrt{C_D/C_F}$, reduces the noise by a factor $2/\sqrt{C_D/C_F}$, just as with the optimal number of FET's in parallel. The effective input impedance is likewise reduced by C_D/C_F as before, satisfying the condition for fast charge transfer from detector to preamp.

WHICH IS BEST? The use of transformers is appealing, instead of 600 FET's in parallel all we need is a step up transformer with turn ratio of 1:25. Transformers, unfortunately, have many disadvantages. They produce a highly divergent noise component at low frequency, requiring two differentiations in the filter, resulting ultimately in about $\sqrt{2}$ more series noise. They are very sensitive to magnetic fields. Transformers can only be used in non magnetic detectors.

The real problem is however, to construct a good transformer with low core losses and tight coupling over a very large frequency range. A study of transformers by the UA1 collaboration[9] for $C_D = 6800\ pF$ and $C_F = 65\ pF$, for which one should achieve a noise reduction of $\sim \frac{1}{2}\sqrt{100} = 5$, only achieved a factor of 1.55. The authors of ref. 6 have reported significantly improved noise performance, when using noisier preamps. Radeka[10] has reported results obtained with a 1:11 transformer, which, for the case discussed here, would give $E_{noise} = 66$ MeV/tower. Transformers are very expensive and bulky while fabrication of dies with tens or hundreds of transistors is very easy with modern technology. A very serious problem with transformers is that the preamp sensitivity must be increased by N because of the reduction of the charge signal by N. This means in practice that all sources of external, non random noise must be kept at a much lower level, since, when adding signals from many channels coherent noise can very rapidly overcome preamp noise. FET's too, have a problem: increased power dissipation in the preamp. Incidentally, while I refer to FET's in parallel, this is of course equivalent to a single FET of appropriately scaled geometry.

6.3 Test Results

I have built preamps with four FET's in parallel, followed by a low noise amplifier and an "n=2" filter with RC= 50ns to achieve a T_m of 100 ns. An additional differentiation with $RC \simeq 250$ ns was used to reduce noise when using a transformer. The theoretical noise slope, from section 4.2 is 4270 e's/nF at $T_m = 100$ ns. The measured values were

$$Q/C_D = 4000 \text{ e's/nF}, \qquad R_{in} = 5\ \Omega$$

I then inserted a very high quality "delay line" 1:2 transformer (Varil, model LF-402, -1 dB bandwidth 10 KHz to 100 MHz, 0.5 dB insertion loss - and that is 12% power loss) which should have given a factor $\simeq 1.9$ improvement. The measured noise slope was

$$Q/C_D = 2500$$

or a factor of 1.6 improvement. There was in addition, a large noise at zero capacitance, notwithstanding two differentiations. Using only the four FET's preamp, results in a noise slope of 3 MeV/nF, which for a 40 nF tower divided in 5 nF sections gives a total noise of $3 \times 5 \times \sqrt{8} = 42$ MeV. This value, for $T_m = 100\ ns$, is probably acceptable for an SSC detector. Including a 1:2 transformer, would bring this value down to 22-26 MeV per tower. 22 MeV could also be obtained with 16 FET's per preamp.

6.4 Detector Noise

There is clearly room for improvement in the technology for charge measurements. The example we went through was for $T_m = 100$ ns, which is somewhat on the high side for an SSC detector. We also ignored the hadronic part of the calorimeter. We have however over estimated the area of the e.m. tower. In order to estimate the total calorimeter noise, we can imagine covering the canonical $\Delta\eta \times \Delta\phi = 10 \times 2\pi$ region with 5000 em towers with an optimistic noise of 22 MeV and 5000 hadronic towers with 44 MeV noise (4 times as much capacitance). Then the total noise in the calorimeter is

$$E_{noise} = \sqrt{5000 \times 22^2 + 5000 \times 44^2} = 3.5\ GeV.$$

This value can be taken as being terrible or very good, depending very much on one's attitude. Consider the case of a pure e.m. shower, typically confined to one tower, for which $E_{noise} = 20$ MeV. For the size of tower considered, a hadronic shower might span 50 towers, then $E_{noise} = 44\sqrt{50} = 310$ MeV! Clearly blindly adding all channels in the calorimeter is a silly thing to do. If, by pattern recognition and clustering, we avoid adding empty channels, we might need adding only a few hundred towers, in which case $E_{noise} \sim \sqrt{n}\sqrt{100}\times 50 \leq 1$ GeV, which means an E_T resolution of 1 GeV. If the resolution function is a Gaussian of 1 GeV width, that would be quite good.

Finally, I wish to remark that if one can remove all signals with $|E_i| < \mathcal{K} E_{noise}$, then the noise in the total energy sum is reduced by a factor $\mathcal{R}$, resulting in a reduced energy noise, given in table 3, which remains normal distributed.

$\mathcal{K}$	$\mathcal{R}$	E_{noise} (GeV)
2	1.96	1.8
2.5	3.19	1.1
3	5.85	0.6

Table 3. Noise reduction $\mathcal{R}$ after removal of signals with $|E| < \mathcal{K} E_{noise}$

7. Pile-up and more on noise

7.1 A simple estimate

Pile-up can produce an average shift in the zero of the energy scale in addition to random fluctuations. In general, because of the various AC couplings in the electronics chain and the signal filter, $\langle E \rangle = 0$. We are therefore left with the base line fluctuations due to accumulated energy deposits in the detector. In terms of the signal V(t) produced by a fixed energy deposit with frequency f, the mean square fluctuation of the signal is given by

$$\langle (V(t))^2 \rangle = \lim_{T\to\infty} \frac{fT\int_0^T (V(t))^2\,dt}{T} = f\int_0^\infty (V(t))^2\,dt.$$

We have derived the form of V(t) in section 4.2 for the CR- nRC filters and defined other shapes of V(t) in figure 8. The integrals are trivially performed, but in order to derive the rms energy fluctuation per tower we need values for f and the average energy deposit.

We assume for this calculation a luminosity $\mathcal{L} = 10^{33}\ s^{-1}\ cm^{-2}$, $\sigma_{inel} = 10^{-25}\ cm^2$, a multiplicity $\langle n \rangle$ =100 particles/interaction and a detector of 50000 towers each covering a constant $\Delta\eta\Delta\phi$ interval. Then $f = 2 \times 10^5$. Table 4 gives the rms fluctuation of the signal, normalized to $V_{MAX} = 1$ and the rms energy noise for $\langle E_T \rangle =$ 500 MeV/particle and $T_m = 0.1\ \mu s$. The rms error from pile-up scales as $\sqrt{f}$ and $\sqrt{T_m}$ and varies with pulse shape as parallel noise. The discussion of section 6.4 applies to pile-up 'noise' as well, in particular the results of table 3.

Shaping	$\sqrt{\langle V^2 \rangle}$	$\sigma(E_T)$ (MeV)
CR-2RC	0.13	68
CR-10RC	0.08	37
Triangle	0.08	40
Triangle, bipol.	0.11	58

Table 4. Energy error per tower from pile-up.

7.2 MULTIPLE SAMPLES, CORRELATIONS AND NOISE

Double and multiple sampling can be used to reduce errors from pile-up especially if T_m is shorter than the time for the signal to return to zero. If we define the first difference of the signal as:

$$V^I(t) = V(t) - V(t-\tau),$$

obtainable by 'delay line clipping' or sampling V(t) at t and t+τ, we have ($\langle V \rangle = 0$):

$$\langle (V^I(t))^2 \rangle = \langle (V(t))^2 \rangle + \langle (V(t-\tau))^2 \rangle - 2\langle V(t)V(t-\tau) \rangle$$

or

$$\sigma^2(V^I) = 2\sigma^2(V) - 2\langle V(\tau)V(0) \rangle.$$

If $\tau \ll T_m$, V(t) and $V(t-\tau)$ are fully correlated, and $\sigma(V^I) = 0$. If $\tau \simeq T_m$ or larger, the last term vanishes, giving $\sigma(V^I) \simeq \sqrt{2}\sigma(V)$ as anticipated in section 4.2. Similarly, for:

$$V^{II}(t) = V^I(t) - V^I(t-\tau) = V(t) - 2V(t-\tau) + V(t-2\tau),$$

$\sigma(V^{II}) \simeq \sqrt{6}\sigma(V)$. Thus, triple sampling, often considered the best way to ensure the fastest return of the signal to zero, increases the noise by $\sqrt{6}$. For different pulse shapes there are optimal weighted differences. For the 'n=2' filter the difference $V(t) - 0.6V(t - T_m)$ reduces the error due to pile-up by a factor 1.23, while the noise is increased by 1.17, which clearly is not a great improvement. This is an example of "pole-zero" cancellation by sampling and differences. Pole-zero cancellation is a fancy name for something which at long times is just a voltage divider and is also a very elegant way to get into trouble. It is not recommended for calorimeters and proportional chambers at hadron colliders.

In general, unless the electronics noise is very small, there is very little to be gained by using multiple sampling and shape fitting to reduce pile-up mistakes in measuring a signal of interest. If the samples are enough apart in time to aid in distinguishing signals from pile-up, they also have uncorrelated noise. Figure 12 shows the error on the first finite difference $V(t) - V(0)$ versus t, in units of T_m from a 'n=2' filter and the correlation factor $\langle V(t)V(0)\rangle$, both normalized to $\sigma(V)$.

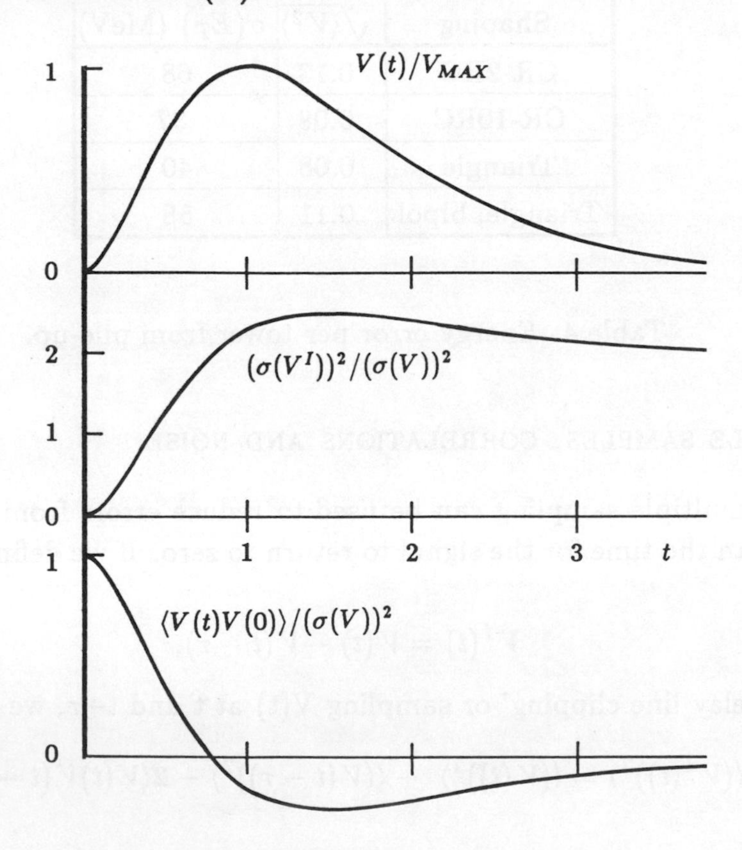

Figure 12. Signal, noise and correlation.

8. Postscript

8.1 SINS OF OMISSION

I have kept this presentation very narrowly confined to the topics in the title and therefore ignored several questions which are in fact intimately related to optimizing the performance of a calorimeter in the SSC environment. I feel I must at least mention that cross talk and channel to channel calibration, which are *almost trivial* at $T_m \sim 2\ \mu s$, become very major problems at $T_m \sim 0.1\ \mu s$. NA34, ref. 10, begins with 14 nF/m^2 in the argon gap, which is then increased to 180 nF/m^2 by ground planes between pads and traces to control cross talk. In the example developed here a similar approach would be 4 times less disastrous, since I have chosen a gap in argon of 0.5 mm, rather than $\simeq$2 mm as in ref. 10. The mechanical designers of the calorimeter stacks might not find that an easy thing to do. Connections from pads to preamp must be made with transmission lines of very low characteristic impedance, Z_0, satisfying the usual condition: $Z_0 C_D < T_m$. Typically the only way to obtain low impedance transmission lines is to increase the capacitance per unit length C of the line. Since $Z_0 \propto 1/\sqrt{C}$, adequate transmission lines have capacitance in the several hundred pF per meter. In ref. 10 the solution was to immerse the preamps in liquid argon thus avoiding this last problem.

A major R&D effort aimed to a high level of detector-electronics integration is a possible approach which gives optimal detector preamp match and allows both shielding and transmission not to become the overwhelming factors in the noise fight.

One of the major advantages of a calorimeter with no gain in its sampling medium, is that channel to channel response can be obtained to high accuracy, by charge injection at the preamp input. If $T_m \gg T_{dr}$, $C_D R_{in}$, no particular attention to pulse shape is required nor to where the charge is injected. Relative accuracy of 0.1% are easily achieved in DØ for T_m=2 μs whether charge is injected at the preamp, or at the pad which is connected to the preamp by $\sim$ 30 ft of cables, with C_D in the range 0.5 to 5 nf. If the appropriate conditions are not satisfied, response to charge injection at the preamp input and to charge freed in the detector gap become different and corrections involving geometry, capacitance, cable lengths and so on must be applied. A high level of integration of electronics and detector would again help this problem.

8.2 HOW IT ALL CAME TO BE

Most of the elementary discussion about noise above, comes from many years of frustration in trying to understand incomplete, obscure (or wrong) books and papers on the subject. My work with solid state detectors, NaI and BGO crystals and, in the past five years, on the design of the calorimeter electronics for the DØ detector has profited from a better understanding of noise and required many calculations as the ones discussed here. That they finally came to be in writing is the fault of Mark Strovink, who convinced me to participate in this workshop, and more so of Gil Gilchrist who convinced me to write them up. Unfortunately I cannot blame anybody but myself for errors and mistakes, despite Juliet Lee-Franzini's several perusals and correcting the ones she caught.

ACKNOWLEDGEMENTS

This work was supported in part by the National Science Foundation.

REFERENCES

1. P. Horowitz and W. Hill. The art of electronics. Cambridge University Press, Cambridge 1980. (The most entertaining introduction to electronics)
2. H. Nyquist, Phys. Rev. **32**, 110 (1928).
3. W. Schottky, Ann. Physik, **57**, 541 (1918).
4. For a practical definition of EMI see for example the Tektronix catalog, any year.
5. E. Fairstein and J. Hahn, Nucleonics, **23**, 50 (1965).
6. W.J. Willis and V. Radeka, Nucl. Inst. and Meth., **120**, 221 (1974).
7. J. D. Jackson. Classical Electrodynamics. Wiley, New York, 1962.
8. V. Radeka, IEEE Trans. Nucl. Sci. NS-**21**, 51 (1974).
9. Design report of a U-TMP calorimeter for the UA1 experiment, UA1 collaboration, UA1 technical note TN/86-112, 1986 (unpublished).
10. V. Radeka, 1987 SLAC Instrumentation Conference.

Closing the Gap: Forward Detectors for the SSC*

Richard Partridge
Department of Physics, Brown University, Providence, RI 02912

ABSTRACT

Forward detectors are likely to be part of any general purpose detector designed to study high-p_T processes at the SSC. The physics requirements and detector environment faced by forward detectors and their impact on detector design are examined.

General purpose detectors designed to study high-p_T processes invariably provide some form of forward coverage to close the gap between the central/endcap region and the beam pipe. Typically, forward coverage begins at $0.5^\circ - 1^\circ$ ($|\eta| \approx 5 - 5.5$) and extends to $5^\circ - 10^\circ$ ($|\eta| \approx 2.5 - 3$). While the rapidity intervals covered by the forward and central detectors are approximately equal, it is not necessary to provide equivalent detection capabilities in the forward and central regions since high-p_T processes typically have $|\eta| < 3$. The forward region also differs from the central region in having particle densities and radiation levels at small angles that will be 1-2 orders of magnitude greater than seen by similar components at 90°. This paper examines the physics requirements and detector environment faced by forward detectors and their impact on detector design.

The principal aim of the forward detector is to provide hermeticity for the calorimeter. Previous studies[1] indicate coverage out to a pseudorapidity of $5 - 5.5$ is required to maximize missing p_T resolution. Before examining the design of the forward calorimeter, it is useful to consider what other capabilities are desirable for the forward detector. In particular, we shall examine the benefits of electron and muon identification in the forward region.

Providing muon coverage in the forward region has several advantages. It has been shown that muon coverage in the forward region significantly improves the detector acceptance for multiple muon processes, such as $H^0 \rightarrow ZZ$, $Z^0 \rightarrow \mu^+\mu^-$.[2] If a heavy neutral gauge bosons is found, forward muon coverage will aid asymmetry measurements needed to determine the boson's couplings.[3] Finally, a high p_T muon in the forward region will have only a small fraction of its energy measured, degrading the hermeticity of the calorimeter. While none of these arguments mandates forward muon coverage, the benefits appear to be sufficient to justify including an iron toroid muon spectrometer in the forward region.

---

* This work supported in part by U.S. Department of Energy Contract DE-AC02-76ER03130

The design of such a detector has been studied[4] and will not be discussed further. While several of the above arguments apply equally to electron and muon identification, there are several reasons for not attempting forward electron identification.

- Electron identification requires a highly segmented calorimeter to isolate the electron's energy deposit,[5] substantially increasing the number of calorimeter channels in the experiment.
- Supplementing the calorimeter π/e separation with TRD's, momentum analysis with a dipole spectrometer, or preshower microconverters would be costly and pose considerable technical difficulties.
- Construction of a forward tracking device to track the electron will be non-trivial.
- There is no compelling physics argument for forward electron identification in high-p_T processes. While electron identification may be important for certain intermediate-p_T processes, such as CP violation in B decay, these processes are best studied by specialized detectors in dedicated interaction regions.
- The existence of muon detectors in the forward region allows the study of forward leptons and provides "insurance" against missing some exotic new physics process involving forward leptons.

While it may well be possible to overcome the technical difficulties listed above, the solution is likely to be costly and provide limited benefit. The remainder of this paper assumes that no attempt will be made to identify electrons.

The design of the forward detector is strongly influenced by the decision not to attempt forward electron identification.

- There is no longer any clear requirement for forward tracking. Previous studies[6] indicate that forward tracking, while feasible, will be difficult.
- The principal function of the forward calorimeter is to measure E_T and provide good missing p_T resolution. Coarse transverse segmentation should be adequate for these purposes. The required degree of segmentation is discussed below.
- There is no obvious advantage to separate electromagnetic and hadronic calorimeter sections since the overall calorimeter energy resolution is not significantly improved by fine longitudinal sampling in the electromagnetic section.

The required transverse segmentation in the forward calorimeter can be determined by examining the E_T resolution that results from various segmentations.

Fig. 1(a) shows the E_T resolution as a function of rapidity for particles with transverse energy of 1 GeV, 10 GeV, and 100 GeV in a calorimeter that has $\Delta\eta = 0.1$. The calorimeter is assumed to have an energy resolution of $0.5/\sqrt{E} + 0.02$. For this segmentation, the E_T resolution is dominated by the energy resolution for all rapidities. Fig. 1(b) shows that increasing the segmentation to $\Delta\eta = 0.2$ produces only small changes in the E_T resolution. However, increasing the segmentation to $\Delta\eta = 0.5$ substantially degrades the E_T resolution in the forward region. These results indicate that a segmentation of $\Delta\eta = \Delta\phi = 0.2$ is quite adequate for achieving good E_T and missing p_T resolution.

A segmentation of $\Delta\eta = 0.2$ requires the innermost segments to subtend a polar angle of a few milliradians, resulting in very small calorimeter cells. Since the hadronic shower size ultimately limits how fine a calorimeter can be segmented, there will be a minimum segmentation depending on the calorimeter design, absorber, and distance from the interaction point. Fig. 2 shows the E_T resolution for a segmentation of $\Delta\eta = 0.2$ with a minimum segmentation of 2, 5, and 10 mr. It can be seen that the resolution begins to degrade very quickly at large rapidities when the minimum segmentation exceeds 5-10 mr. This may be a problem for designs where the forward detector is brought in close to the interaction region to be integrated with the endcap calorimeter.

These studies indicate several features of forward calorimetry:

- At forward angles, energy resolution becomes less important than in the central region for a given E_T.
- For particles with E_T above a few GeV, segmentation dominates the E_T resolution in the forward region.
- Increasing rapidity coverage beyond 5.5 will not necessarily improve total E_T or missing p_T resolution and may, in certain cases, degrade the resolution.
- Forward calorimeters placed close to the interaction region must have a high density to minimize the transverse spread of hadronic showers.

The design of the forward calorimeter must take into account the extremely high radiation exposure this region is subject to. A detector coming within 10 cm of the beam will have a radiation dose of 300,000 Rad due to incident particles; this will be multiplied substantially as the particles shower in the detector. For this reason, it is natural to consider a liquid sampling medium such as liquid argon, warm liquids, or liquid scintillator. Other considerations for the calorimeter include obtaining an e/π response of 1, fast readout times, and maximizing the calorimeter density, especially if the forward detector is integrated into the central/endcap calorimeter.

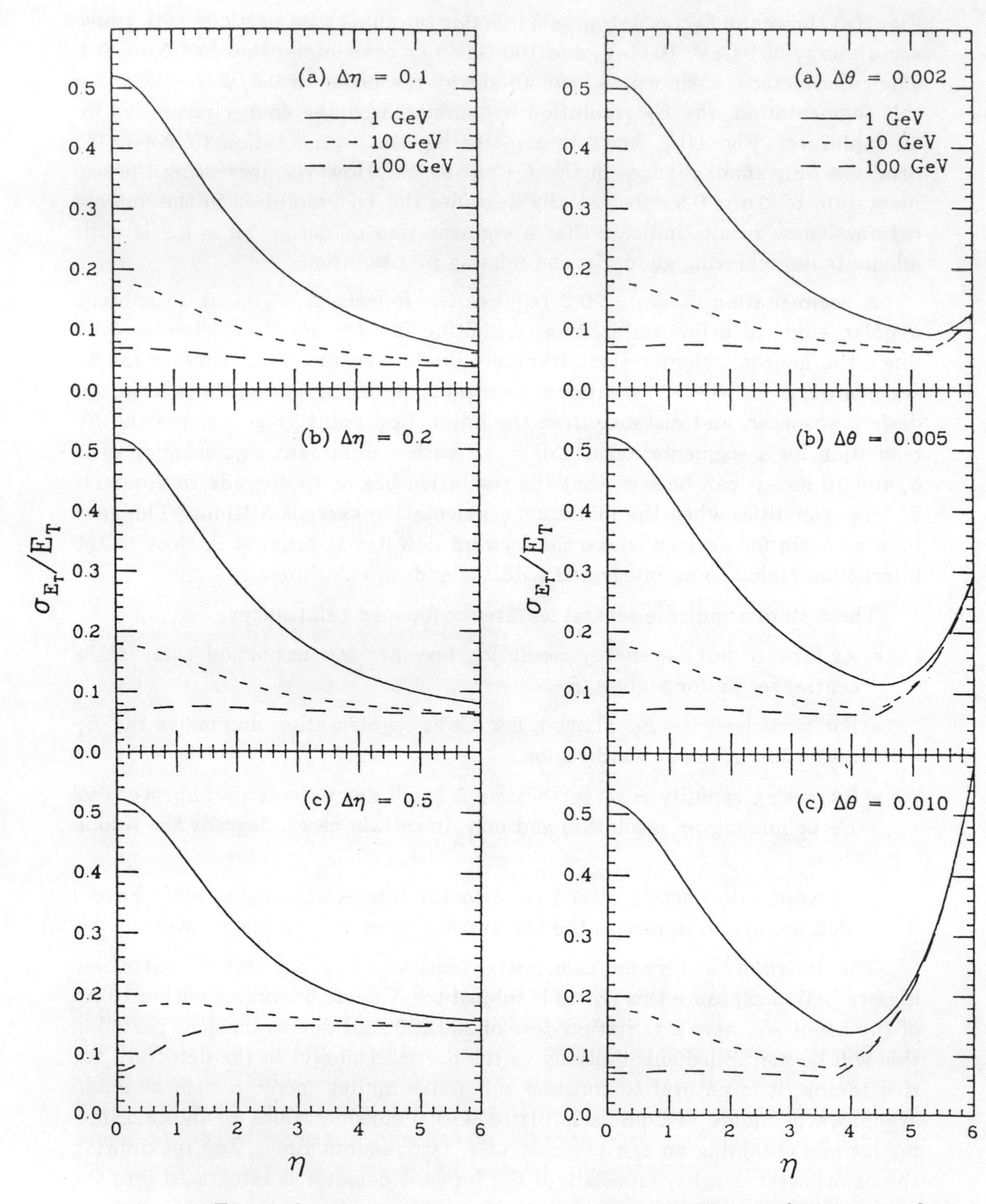

Fig. 1 E_T resolution as a function of pseudorapidity for various values of E_T. (a) $\Delta\eta = 0.1$; (b) $\Delta\eta = 0.2$; (c) $\Delta\eta = 0.5$.

Fig. 2 E_T resolution as a function of pseudorapidity for various values of E_T. (a) $\Delta\theta > 2$ mr; (b) $\Delta\theta > 5$ mr; (c) $\Delta\theta > 10$ mr.

A possible design for a forward calorimeter sampling module employing liquid scintillator is shown in fig. 3. Liquid scintillator is unique in possessing very fast time response, being extremely radiation hard, and should be able to achieve an e/π response of 1. The problem of reading out the liquid scintillator is solved by etching the scintillator vessel to yield a $\sin\theta$ response. Thus, the calorimeter measures E_T directly with no need to segment in θ. Segmentation in ϕ is easily achieved by dividing the detector into ≈ 32 wedges, with photomultiplier readout of each wedge.

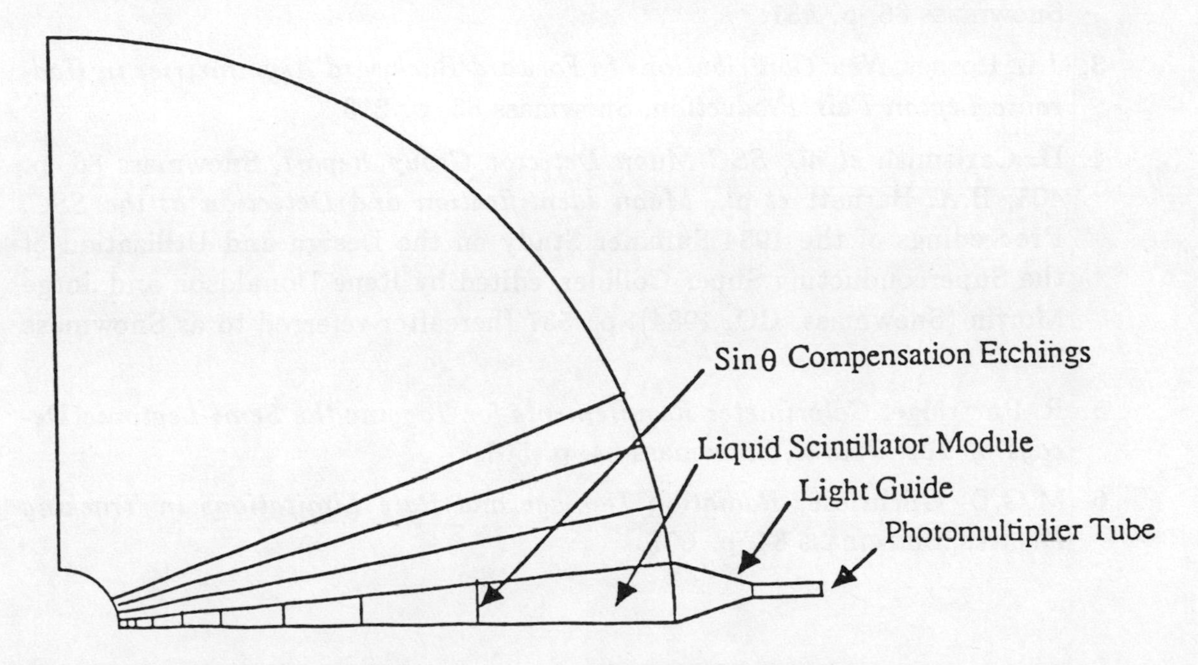

Fig. 3 Diagram showing construction of liquid scintillator sampling module for a forward detector calorimeter.

In conclusion, a number of aspects in the design of forward detectors have been examined. Such a detector should probably include the best possible calorimetry and full coverage for muons. Forward electron identification is difficult to achieve and has limited physics potential. The effect of calorimeter segmentation on E_T resolution was studied; a segmentation of $\Delta\eta = \Delta\phi = 0.2$ provides good resolution. Intense radiation exposure suggests the use of a liquid sampling medium. One promising detector technique is to use liquid scintillator compensated to directly measure E_T.

REFERENCES

1. H. Iwasaki, B. Milliken, S. Protopopescu, and R. Raja, *Hermeticity Studies in SSC Type Calorimeters*, Proceedings of the Summer Study on the Physics of the Superconducting Supercollider, edited by Rene Donaldson and Jay Marx (Snowmass, CO, 1986), p. 371 (hereafter referred to as Snowmass 86).
2. D. Carlsmith, D. Hedin, and B. Milliken, *Leptonic Angular Acceptances*, Snowmass 86, p. 431.
3. J.L. Rosner, *New Contributions to Forward-Backward Asymmetries in Hadronic Lepton Pair Production*, Snowmass 86, p. 213.
4. D. Carlsmith *et al.*, *SSC Muon Detector Group Report*, Snowmass 86, p. 405; B.A. Barnett *et al.*, *Muon Identification and Detection at the SSC*, Proceedings of the 1984 Summer Study on the Design and Utilization of the Superconducting Super Collider, edited by Rene Donaldson and Jorge Morfín (Snowmass, CO, 1984), p. 537 (hereafter referred to as Snowmass 84).
5. R. Partridge, *Calorimeter Requirements for Tagging the Semi-Leptonic Decays of Top Quarks*, Snowmass 84, p. 567.
6. M.G.D. Gilchriese, *Radiation Damage and Rate Limitations in Tracking Devices*, Snowmass 84, p. 607.

Kinematic Isolation and Reconstruction of Hadronic W and Z Decays at a Super Collider

John Hauptman
Physics Department and Ames Laboratory
Iowa State University, Ames, IA 50011

and

Aurore Savoy-Navarro
CEN Saclay, Gif-sur-Yvette, France

Abstract

A specific and detailed method is described for the kinematic reconstruction of hadronic decays of the W and Z gauge bosons, and statistical criteria for their efficient isolation from the much larger QCD background of single-jet and multi-jets are presented and evaluated. The FORTRAN 77 code is available upon request.

Introduction

We will consider one problem and its solution, originally formulated by Frank Paige[1]: the production of a high-mass (800 GeV/c^2) Higgs boson, its subsequent decay to two W bosons, followed by the decay of one W to $q\bar{q}$ and the other W to $\ell\nu$. As suggested by Frank Paige, an easy line of attack is to generate events with a QCD jet on one side and a W decaying hadronically on the other side (DRELLYAN in ISAJET, and processes 15 and 18 in Pythia). These processes,

$$pp \rightarrow W + QCDjet(s) \rightarrow q\bar{q} + QCDjet(s) \rightarrow jet + jet + QCDjet(s), \qquad (.1)$$

constitute the training sample from which we construct useful quantities for the statistical separation of the W two-jet system from the single or multiple jet QCD system. One useful quantity is the mass. Distributions of the invariant mass of the QCD jet and the $W \rightarrow q\bar{q}$ systems are shown in Figure 1. This mass and other discrimination quantities, described in section I, are then applied to two independent samples of events: the Higgs signal and two QCD jet backgrounds. The signal is the production of a high-mass Higgs, its subsequent decay to two Ws, and their subsequent decays to $\ell\nu$ and to $q\bar{q}$ (Pythia subprocesses 25 and 26),

$$pp \rightarrow H^0 \rightarrow W^+W^- \rightarrow \ell\nu + q\bar{q} \rightarrow \ell\nu + jet + jet. \qquad (.2)$$

The two background samples are the production of a single QCD jet accompanied by a W, with the subsequent decay of the W to $\ell\nu$ (Pythia subprocesses 15 and 18),

$$pp \rightarrow W + QCDjet \rightarrow \ell\nu + QCDjet, \qquad (.3)$$

and the production of multiple QCD jets in association with a leptonic W decay. (a selected subset [2] of Pythia subprocesses 15 and 18),

$$pp \rightarrow W + QCDjets \rightarrow \ell\nu + QCDjets. \qquad (.4)$$

---

[1] F.E. Paige, Report of the Theoretical Group on Experimental Signatures, in Physics at the Superconducting Super Collider. Summary Report, June 1984, p. 107.

[2] H.-U. Bengtsson multi-jet selection algorithm: the higher (lower) energy jet is required to have $P_T > 0.50\ (0.25)\ P_T^W$, and the invariant mass of the two highest energy jets between 40 and 200 GeV/c^2.

The object of this study is to find a set of selection criteria which accept a large fraction of reaction (2) while at the same time rejecting nearly all of reactions (3) and (4). As noted in ref. 1, the cross sections for the background processes are larger than the signal by about a factor of 200, so we require the relative acceptances to differ by this factor. These samples are clearly statisically independent, but systematically not independent, since we use Pythia for all three. However, by the time the SSC runs, we will understand jets in W decays from LEP II data.

Throughout this note we consider a non-magnetic Super Collider tower calorimeter detector with angular segmentation in rapidity, y, and azimuth, ϕ, of .02 radians for hadronic energy and .01 radians for electromagnetic energy. This calorimetric detector has no imposed magnetic field, has exact compensation of electromagnetic and hadronic energies, and has energy resolutions of $0.15/\sqrt{E}$ and $0.50/\sqrt{E}$ for electromagnetic and hadronic energies, respectively. The energy resolutions chosen in the simulation are achievable with many technologies available today. Rms noise is introduced at 1.5% per channel.

Although no extensive studies have been done (they can be) we have generally found that segmentation can be degraded by a factor of 2 or 3 with little loss of information content in the calorimeter. We have not studied the effects of a non-compensating calorimeter, although that could be done. We also have not introduced mis-calibrations into the event sample, nor have we considered event pile-up. Finally, the imposition of a magnetic field would complicate the analysis of events, requiring a solution of the track finding problem in dense track environments, and the subsequent matching of these tracks with energies in the calorimeter system, so that calorimetric energies of charged tracks could be put "back into the right place" in the calorimeter after having been bent by the field.

Events are generated either by the Pythia 4.7 or the ISAJET monte carlo, all stable particles are showered by a simple two-dimensional algorithm, and the energy in each (y, ϕ) tower is collected. The final output "raw data record" for an event is a list of the triplets (E_i, y_i, ϕ_i) for i = 1,...,Nactivated towers. A threshold of 0.1 GeV per tower is applied to reduce the size of each event with negligible loss of information.

I. Parameterization of the W $\rightarrow q\bar{q}$ energy pattern.

After a collection of calorimeter towers has been identified by a pattern recognition algorithm as belonging to a $W \rightarrow q\bar{q}$ hypothesis, [3] the most important characteristic of this collection of towers is its invariant mass. Taking each tower as an element of zero mass, and using the angles (θ_i, ϕ_i) in spherical coordinates pointing to the center of each tower from the interation vertex, we sum the four-vector of this collection of towers as

$$E = \sum_{i=1}^{n} E_i \quad (.5)$$

$$P_x = \sum_{i=1}^{n} E_i \ \sin\theta_i \ \cos\phi_i \quad (.6)$$

[3] This pattern recognition is, of course, an important step. We take care at this stage to isolate a concentration of activated energetic towers by the following procedure: (i) make a first guess at the energy of the W (for example, from the opposing leptonic W), then (ii) search for a calorimeter cluster inside a cone of half-angle $4 \cdot M_W / P_W$, (iii) re-estimate the W energy and direction using only towers inside this cone. Iterate (ii) and (iii) until convergence. Finally, the semi-major axis of the charge distribution is found, and an ellipse of eccentricity 0.5 is draw around the energy concentration. This tight ellipse reduces the probability of minimum bias hits from simultaneous events contaminating the W towers.

$$P_y = \sum_{i=1}^{n} E_i \ sin\theta_i \ sin\phi_i \qquad (.7)$$

$$P_z = \sum_{i=1}^{n} E_i \ cos\theta_i, \qquad (.8)$$

$$(.9)$$

where i runs over all activated towers. The invariant mass is calculated as $M = \sqrt{E^2 - P^2}$.

A W or Z decay to quarks is a two-body decay characterized in a calorimeter by two concentrations of energy, separated in angle by roughly $\alpha = M_W/P_W$. In detail this angular separation depends on the quark decay angle θ^* in the W frame, and reduces to M_W/P_W for a 90^0-decay to massless quarks. We form a pattern for these energy concentrations which is energy-independent and has minimum variance about the mean. In the (y,ϕ) coordinates of the calorimeter towers, we translate to the center-of-energy (y_0,ϕ_0) of the energy pattern, and rotate in (y,ϕ) to define an axis $\hat{s} = (s_y, s_\phi)$ which is aligned with the semi-major axis of the energy distribution in the calorimeter. We use a thrust algorithm in two dimensions, and align the positive $\hat{s}$ axis towards the higher energy concentration. For each tower energy E_i, the variable s_i is defined to be the distance along the $\hat{s}$ axis from the center-of-energy (y_0,ϕ_0) scaled by P_W/M_W,

$$s_i = [(y_i - y_0)s_y + (\phi_i - \phi_0)s_\phi]\mathrm{P}_W/\mathrm{M}_W. \qquad (.10)$$

The density of tower energies in s is accumulated and normalized to form the quantity

$$< \frac{1}{E}\frac{dE}{ds}(s) >= \frac{1}{\sum E_i}\sum \frac{dE_i}{ds} \qquad (.11)$$

in s-bins of $\Delta s = 0.20$.

For a 90^0-decay to massless quarks, this quantity will have two equal density peaks at s=+1 and s=−1, and for an asymmetric W decay the forward decaying quark will have a higher energy density at smaller s, while the backward decaying quark will have lower density at larger, but more negative, s. This is kinematically just like $\pi^0 \rightarrow \gamma\gamma$ decay. A contour plot of this normalized energy density as a function of the cosine of the quark decay in the W center-of-mass is shown in Fig. 2, with a scale of 1% per contour. Slices in cosine θ^* are shown in the left-hand column of Fig. 3.

We want to minimize the variance about this mean independent of the energy of the W, and that is why we normalize by 1/E, translate to (y_0,ϕ_0), rotate to align each W with $\hat{s}$, and form $< 1/EdE/ds >$ as a function of cosine θ^*. We also exclude b (and t) quarks since they will contribute too much variance to $< 1/EdE/ds >$. The rms variation about this mean is shown in the middle column of Fig. 3. There are bin-to-bin correlations in this plot, and the correlation coefficient $\rho_{i\,i-1}$ between the ith s bin and the $(i-1)^{st}$ s bin is shown in the right-hand column of Fig. 3. These plots will be useful in the statistical pattern recognition of hadronic W decays.

Both an estimate of cosine θ^* and the χ^2 for the consistency of a particular $W - q\bar{q}$ hypothesis with the expected average are readily obtained using the $< 1/EdE/ds >$ pattern and its variance. The χ^2 between the hypothesis and the average, including the correlation with the nearest bin, is calculated at several values of cosine θ^*, and assuming that this χ^2 is parabolic near its minimum, both the value of cosine θ^* and the value of χ^2 at the minimum are calculated.

Other parameters characteristic of W and Z decay are the elongation of the calorimeter energy pattern along the $\hat{s}$ axis, and the relative widths of the energy concentrations at positive and negative s. These are both apparent from Figs. 2 and 3. Relative to the centroid of the energy distribution (y_0, ϕ_0), let $\vec{e}$ be the vector to the ith tower located at (y_i, ϕ_i) of magnitude E_i. This elongation may be measured by $E_\perp/E_\parallel$, where

$$E_\perp = \frac{1}{E}\sum_{i=1}^{n} \vec{e}_i \times \hat{s} \tag{.12}$$

and

$$E_\parallel = \frac{1}{E}\sum_{i=1}^{n} \vec{e}_i \cdot \hat{s} \tag{.13}$$

Further characteristics of the W $\rightarrow q\bar{q}$ energy pattern are the relative widths of the +s and -s energy concentrations, σ^+/σ^-, where for positive $\vec{e_i} \cdot \hat{s}$,

$$\sigma^+ = \sqrt{1/E\sum(\vec{e}_i \cdot \hat{s})^2 - (1/E\sum \vec{e}_i \cdot \hat{s})^2}, \tag{.14}$$

(and similarly for σ^- for negative values of the dot product), and the sum of the root-mean-square of the deviations of the $\vec{e}$ from s=0, $s^+ + s^-$, where

$$s^+ = \sqrt{1/E\sum(\vec{e}_i \cdot \hat{s})^2} \tag{.15}$$

for positive $\vec{e_i} \cdot \hat{s}$, and similarly for s^-. Clearly, these quantities depend on cosine θ^*, as seen in Figs. 2 and 3.

Many of these variables are highly correlated. For example, if events with mass near M_W are selected out by a mass cut, then necessarily this sample is enriched in events with substantial transverse momentum with respect to the W laboratory direction, and this usually means two, or more, jets each with about a P_T of $M_W/2$. Hence $E_\perp/E_\parallel$ is small, and it is not too difficult to match the shape of the $< 1/EdE/ds >$ distribution for some value of cosine θ^*. As a result, subsequent cuts on cosine θ^* and χ^2 for the shape fit are less effective. In a previous note,[4] we took full advantage of these correlations, and achieved results superior to the present results. However, as noted at that time, these correlations may to some extent be artifacts of the monte carlo physics simulation, and contain no true discrimination. The simple calculations and cuts presented in this note have the singular advantage of being more easily understood.

II. Evaluation

The events of reactions (2), (3) and (4) are subjected to the following event analysis. The P_T of the missing ν is taken to be the negative of the overall momentum vector of the calorimeter system. We take the lepton (either e or μ) from the Pythia record with efficiency of 1., and in the case of background events from reactions (3) and (4) with no direct lepton, we take the highest energy hadron as the lepton. We solve the quadratic equation for the neutrino z-momentum, and choose the solution with the smaller absolute value of the W $\rightarrow \ell\nu$ z-momentum. The pattern recognition search for the hadronic W decay is initiated at (y, ϕ) opposite to this leptonic W. Misreconstructions and pattern

[4] E. Fernandez, et al., Proc. Snowmass Workshop 1984, p. 107. Also, p. 121.

recognition inefficiencies are included in the overall W $\rightarrow q\bar{q}$ efficiency. Thus, a more sophisticated algorithm at this point would improve our present results.

Apart from the invariant mass of the system (Fig. 1), distributions of $E_{\perp}/E_{\parallel}$ for W decays to light (u,d,s,c) quarks (solid line) and QCD jets (dashed line) are shown in Fig. 4(a), followed by χ^2 in Fig. 4(b), σ^+/σ^- in Fig. 4(c), and the eccentricity of the energy pattern in Fig. 4(d). The distributions of these features after a mass selection has been made about the W mass are shown in Figs. 5(a-d).

We define the W efficiency, ϵ_W , to be the probability that a W will survive the selection cuts. The probability for a QCD jet to pass the W selection cuts is called $\beta_{j\rightarrow W}$, and we define the QCD jet rejection factor to be the ratio ϵ_W / $\beta_{j\rightarrow W}$. One set of selections which results in a W acceptance (or efficiency) of about 10 % and a rejection against single QCD jets (relative to W's) of 100 is:

cosine θ^*	< 0.80
$E_{\perp}/E_{\parallel} + .048 \times \sigma^+/\sigma^-$	< 0.484
$\mid$ M(system)-M(W) $\mid + 32.6 \cdot E_{\perp}/E_{\parallel}$	< 10.9
$\chi^2 + 11.0 \cdot \sigma^+/\sigma^-$	< 44.4
$s^+ + s^-$	> 4.20

These cuts yield the solid points in Figure 6 showing both the W selection efficiency and the QCD jet rejection factor for both single jets (as generated by Pythia) and multi-jets intended to simulate the multiple jet processes suggested by Gunion, et al. [5]. Variation of each cut about its central value results in the efficiency *versus* QCD jet rejection curves shown. The selection criteria were optimized for W against single jets alone; presumably we could improved the multi-jet rejection at the expense of the single jet rejection.

III. Figure-of-Merit.

Any set of selection criteria will necessarily result in an efficiency *versus* rejection curve similar to those in Figure 6. Further and more stringent cuts lower the efficiency of the wanted ensemble and increase the rejection against the unwanted ensemble. As defined above, the rejection factor of jets relative to hadronic W decays is defined as

$$R_{jet\rightarrow W} = \frac{\epsilon_W}{\beta_{j\rightarrow W}}, \tag{.16}$$

which, for equal numbers of W's and jets initially, reduces to the ratio of the number of W's to the number of jets surviving the selection cuts. The W efficiency and jet rejection are not exactly inversely proportional, but they are nearly so. We suggest a figure-of-merit for this problem, by which different algorithms and methods may be compared, to be the product of the W efficiency and the jet rejection factor,

$$Figure-of-Merit = \epsilon_W \times R_{jet\rightarrow W} \,. \tag{.17}$$

This quantity is just $\epsilon_W{}^2/\beta_{j\rightarrow W}$, which is the square of a common quantity often used to estimate the significance of a signal S over background B,

$$\frac{S}{\sqrt{B}} \propto \frac{\epsilon_W}{\sqrt{\beta_{j\rightarrow W}}}. \tag{.18}$$

[5] J.F. Gunion, Z.Kunszt, and M. Soldate, SLAC-PUB-3709, and H-U. Bengtsson, private communication

For the set of characteristics and cuts presented here, this figure-of-merit is about 10 for single QCD jets from Pythia, and about 1.5 for multiple QCD jets (Refs. 2 and 5). [6]

IV. "You, too, can find W's and Z's! Just step right up!"

The code used to generate calorimeter "raw data tapes" of reactions 1, 2, 3 and 4 is available upon request, as are the tapes themselves. The code used to analyze these events, including the routine to calculate the features and to perform the selections cuts, is also available. In any particular problem, after the pattern recognition has identified the calorimeter towers believed to be primarily activated by the W decay products, the simple call

- Call Calculate _Vector(Mass, n, E _y _ ϕ, Vector)

will calculate the quantities used in the discrimination, where you specify the appropriate central mass (Mass) of the W or Z (which may depend upon the e/h of your calorimeter), the number of activated towers (n), and the array of energy-rapidity-azimuth (E _y _ ϕ) for each tower. A subsequent function call

- Passed = Passed _ W _ Cuts(Set, Degree, Vector)

returns a true value if this vector of features passed all selection cuts, and false if not. "Set" and "Degree" are integer variables which allow the user some flexibility: Set=1,2 are one-dimensional cuts, and Set=3,4 are two-dimensional cuts. Degree varies the severity of the cuts, from less (-5) to more (+5) severe. For example, Fig. 6 shows results for Set=4 and Degree=0, the nominal severity setting.

V. Summary

We have described a procedure for isolating hadronic W and Z decays from the much larger backgrounds of single and multiple QCD jets expected at the SSC. This procedure has been checked by direct simulation of events in a nominal calorimeter, and subsequent event analysis. We have found that discrimination against single QCD jets is adequate for subsequent physics analysis, but against multiple QCD jets it is not. It is important that physics generators be augmented to include at least some approximate generation schemes, together with cross section estimates, for W + multi-jet events

---

[6] The selection criteria of S. Protopopescu, Proc. on the Physics of the SSC, Snowmass, July 1986, pp. 180-184, yield a figure-of-merit of about 3.3 for ISAJET single jets.

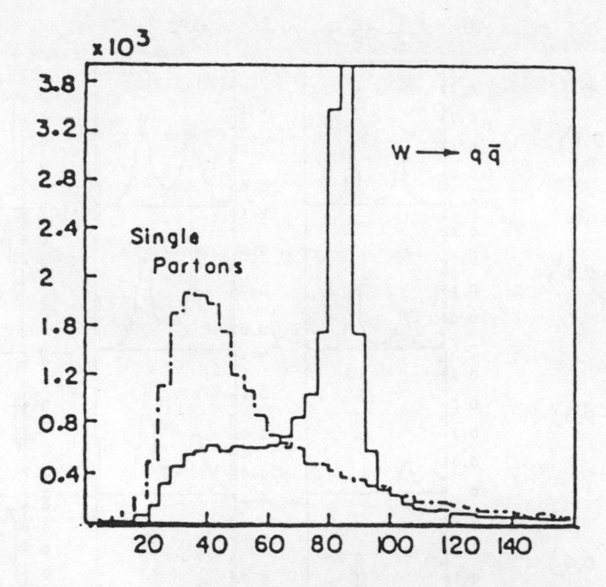

Figure 1. Mass of W $\rightarrow q\bar{q}$, all quarks flavors.

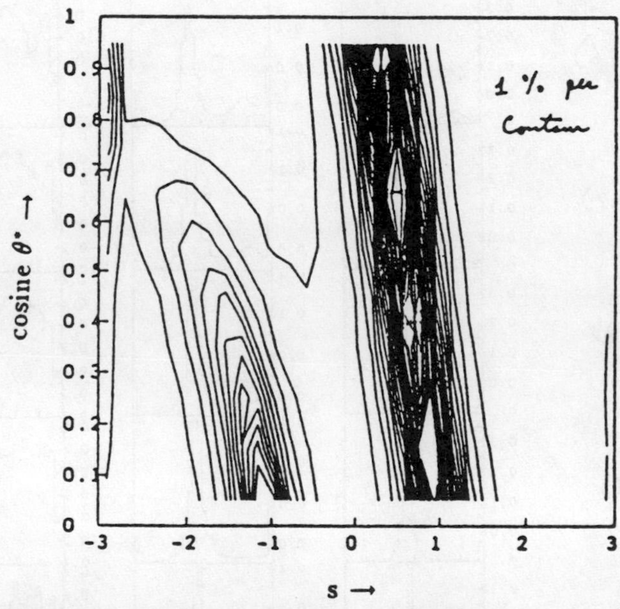

Figure 2. Contour plot of $< 1/EdE/ds >$ *vs.* s and cosine θ^*. Scale is 1% per contour line.

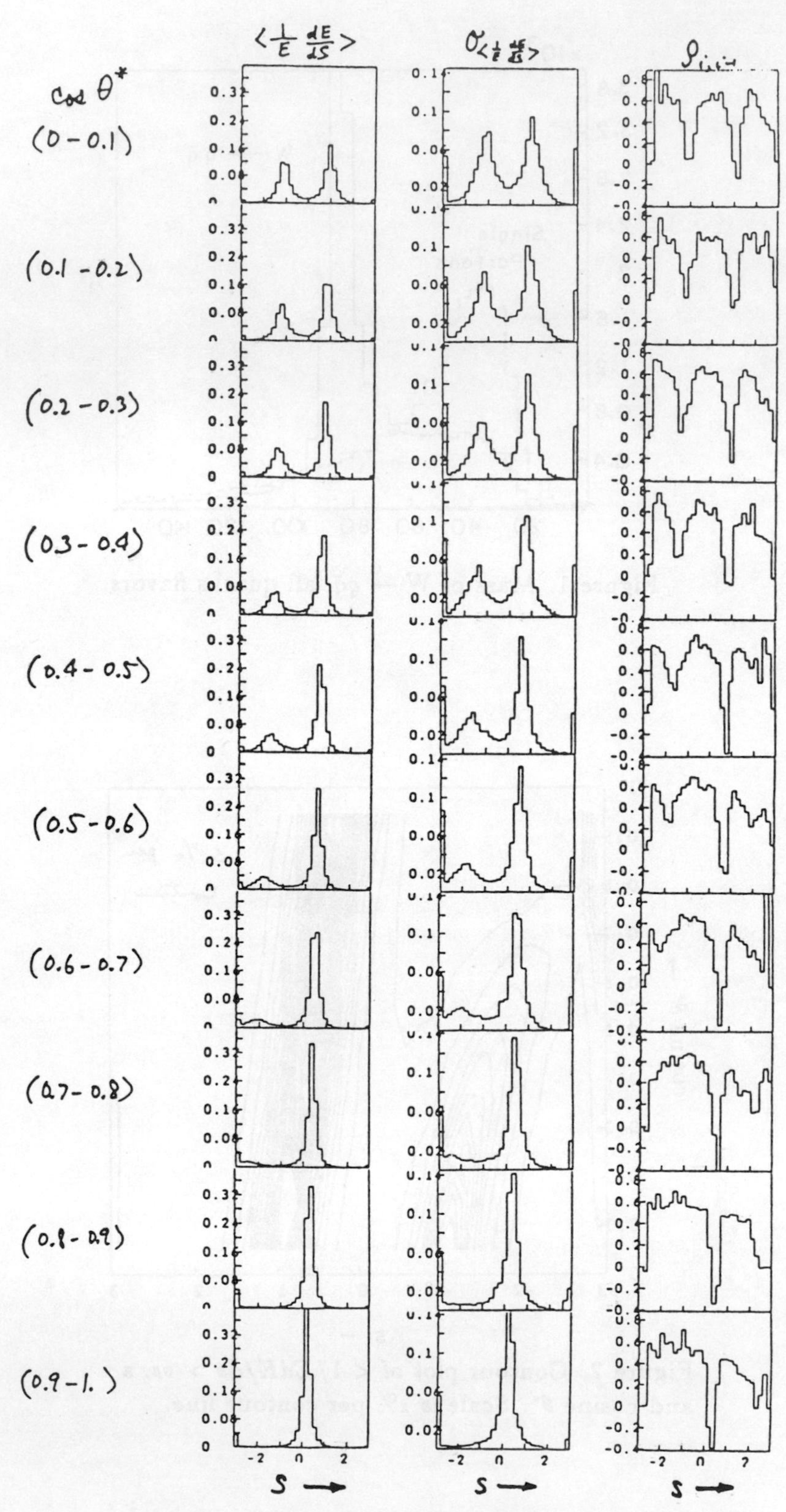

Figure 3. Slices in cosine θ^* for $< 1/EdE/ds >$ (left), rms variation of $< 1/EdE/ds >$ (middle), and nearest bin correlation coefficient (right).

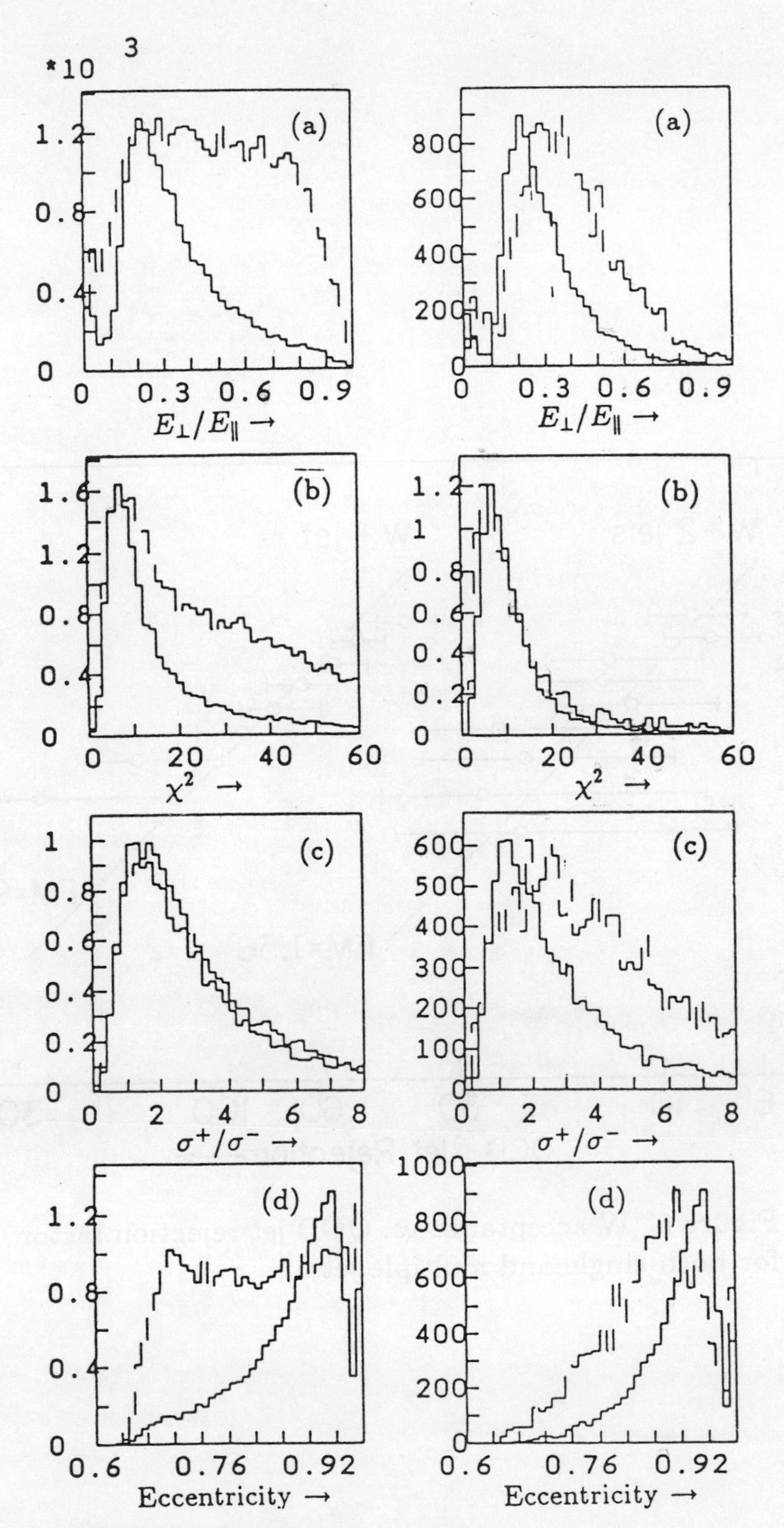

Figure 4. Distributions of (a) $E_\perp/E_\parallel$, (b) χ^2, (c) σ^+/σ^-, and (d) eccentricity of the calorimeter energy pattern for $W \rightarrow q\bar{q}$ (solid) and QCD jets (dashed).

Figure 5. Same as Fig. 4, except after a mass selection about the W mass.

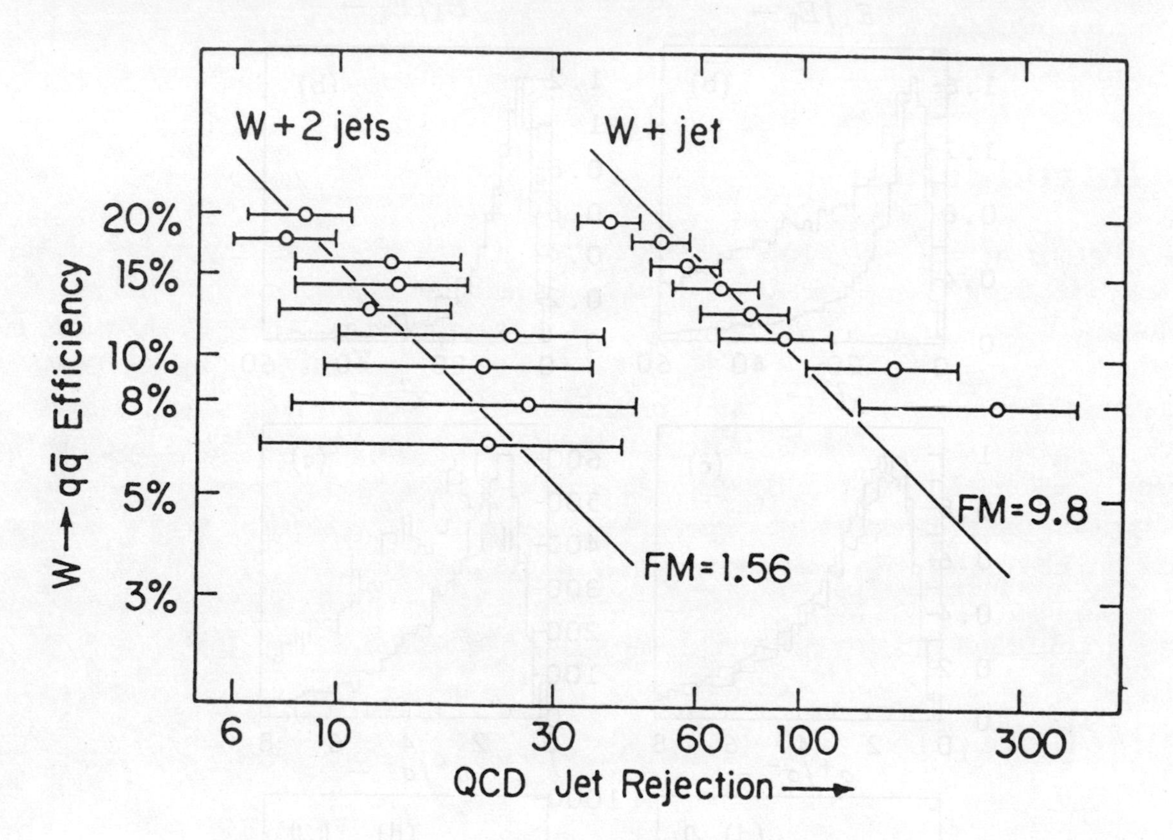

Figure 6. W acceptance *vs*. QCD jet rejection factor for both single and multiple jets.

DETECTOR DEPENDENT CONTRIBUTIONS TO JET RESOLUTION

J. Freeman and C. Newman-Holmes

Fermi National Accelerator Laboratory[1], Batavia, IL

Abstract

We present results of calculations undertaken to study detector dependent effects that contribute to the energy resolution of jets. To accomplish this study, we have developed a fast, easily varied simulation of a generic 4π calorimeter. Physics processes used for benchmarks of performance were the transverse energy resolution of 1 TeV jets and the mass resolution of W $\rightarrow$ 2 jets, for W's with 500 GeV transverse momentum.

1 Introduction

It is generally agreed that good measurement of jet energies is an important capability for any SSC detector. A variety of detector dependent effects contribute to jet energy resolution. These include calorimeter thickness, segmentation, energy resolution, electron/hadron response ratio and cracks or dead areas. The problems of energy carried off by neutrinos, overlapping events and limitations of jet reconstruction by clustering also affect jet energy measurements. Attempts to study a collection of effects such as these can easily become aimless rambles through a multi-dimensioned parameter space. Instead we approach the problem with a belief that many of the general features of these diverse effects can be studied with a relatively simple Monte Carlo program. We have written such a program incorporating parametrized showers and a very simple geometry. We describe the program below and present some results obtained with it.

2 Calorimetry Simulation

The calculations described herein were performed with a computer program interfaced to the widely used ISAJET event generation program [1]. A simple detector is simulated using several settable parameters. The detector consists of three regions: a spherical decay volume, an electromagnetic calorimeter and a hadronic calorimeter. Both calorimeters are spherical shells. The decay volume contains a cylindrical region where a solenoidal magnet field is present. The radius of the spherical decay volume is chosen to just enclose the cylinder, with the cylinder's size specified by the user. The radiation length, absorption length and thickness of each calorimeter are also settable.

The program goes through a list of particles made by ISAJET. For each particle, a distance to decay point, distance to electromagnetic conversion and distance to hadronic interaction are calculated using probablity distributions appropriate for the particle type. Particles are then tracked through the detector one by one. In the decay volume, a particle

[1]Operated by Universities Research Association under Contract with the U.S. Department of Energy

will decay if the distance to its decay point is less than the distance it travels through this volume. If a nonzero magnetic field has been specified, the particle's trajectory is appropriately changed.

When a particle reaches its predetermined shower or conversion point, parameters for its shower are generated. The shower parametrization has been described elsewhere [2]; longitudinal and transverse shower profiles as well as fluctuations are modelled. This parametrized shower is then integrated over the distance between the shower point and the calorimeter edge. If a shower starts in the electromagnetic calorimeter, the same shower is continued into the hadronic calorimeter. The electromagnetic/hadronic energy response is settable for each calorimeter. Note that in this model, a particle may not decay once its shower has begun. The total (electromagnetic + hadronic) energy deposited in the calorimeters is available for each particle both before and after smearing with a resolution function. The resolution is assumed to be of the form $\sigma_E/E = const/\sqrt{E} + 1\%$ where the constant is supplied by the user for each calorimeter (electromagnetic and hadronic) and the additional 1% is a systematic error associated with calibration. In addition, the energy is deposited in an η - ϕ array with specifiable segmentation. Energy is shared between the central tower (i.e., the one to which the particle track pointed) and its four nearest neighbors in η - ϕ space. The fraction of energy deposited in the central tower depends on the ratio of the shower size to tower size. The remaining energy is shared equally among the four nearest neighbor towers. Electromagnetic and hadronic calorimeter energies are saved separately in the η - ϕ array; only their sum is saved in the particle-oriented arrays. A simple clustering algorithm is used to find energy clusters in the η - ϕ array.

At a luminosity of 10^{33} cm^{-2}/sec, one expects an average of about 10 interactions in a 100 ns integration time. Our simulation includes an option to overlap minimum bias events with the generated events of interest. The average number of events to overlap may be varied. Then the actual number of overlapped interactions is determined for each event by sampling from a Poisson distribution with the specified mean.

3 Analysis

In this paper, we consider two benchmark measurements: The E_t distribution for 1 TeV jets, and the invariant mass distribution for 500 GeV P_t W's decaying into quarks which then hadronize into jets. The events were produced with version 5.2 of ISAJET. The jets were from TWOJET events with P_t constrained to be within the range 1000 - 1010 GeV, and θ 80° to 90°. The W's were produced by the WPAIR option, with θ between 80° and 90°, and the P_t range 450 - 550 GeV. After the events were created, the previously described simulation was used to generate calorimetry energy depositions. Clusters were then found in the calorimetry using a standard algorithm from CDF:

1. The set of towers with $E_t > 5.0$ Gev was determined. These were the seed towers for potential clusters.
2. For each seed tower not in a cluster, all nearest neighbor towers were searched, and any tower with $E_t > 1.0$ GeV was added to the cluster.

3. Neighbor towers of all towers in the cluster were searched, and any towers not in the cluster with $E_t > 1.0$ GeV were added. This step was repeated until no new tower was added to the cluster.

4. All clusters with centroids within $R < 0.7$ in η-ϕ space were merged into one large cluster, and the new cluster centroid was found. R is defined as:

$$R = \sqrt{((\phi_i - \phi_j)^2 + (\eta_i - \eta_j)^2)}$$

where i and j are cluster indices. For the jet E_t calculation, all energy within $R < 0.7$ of the cluster direction was summed.

The momenta of all final state partons (hadronizing) that fell within the 0.7 R cone were summed and the resultant E_t was calculated. This quantity was used as a normalization on an event-by-event basis. We felt that this choice of normalization separated the effects of gluon bremsstrahlung, and other "physics" processes from the effects of the actual detector performance.

It is interesting to see the difference between clustering, and summing all energy within the cone in η-ϕ space. Figure 1 shows the error in E_t measurement divided by the E_t of the hadronizing partons in the cone for two-jet events. Three curves are shown. The curve labelled "perfect" is obtained if one sums the energies deposited in the calorimeter for all daughter particles of the parton. This is clearly better than one can do in the real world where one doesn't know from which parton an observed particle is descended. The curve labelled "cluster" is obtained if one simply assumes the energies of the two highest P_t clusters are the parton energies. The curve labelled "cone" is obtained if one uses a reconstructed cluster to define a direction but then sums all energy within some distance in η - ϕ space to approximate the parton energy. The fact that "cone" is better than "cluster" indicates that there is energy from the parton which will not be included by a naive clustering algorithm.

To reconstruct the W invariant mass each tower within $R < 0.7$ was treated as a massless particle with all energy assumed to be deposited in the tower center, and the resultant invariant mass of this set of "particles" was calculated. We note that these calculations are very insensitive to the clustering algorithm used, since only the cluster direction is required. For the W analysis some additional cuts were applied to suppress problems in pattern recognition. The jets produced by the decay of 500 Gev W's are coalesced in our choice of calorimeter geometry. To ease pattern recognition, we chose events where no more than 25 GeV E_t of cluster energy was outside of the 2 leading clusters, and where the ratio of E_t's of the leading to the next to leading cluster was less than 1.25 . These cuts were typically 40 % efficient. The remaining events can in principle be used, but the pattern recognition is more difficult.

The curves shown below are from a sample of approximately 175 W's per case for the W analysis and 350 jets for the jet analysis.

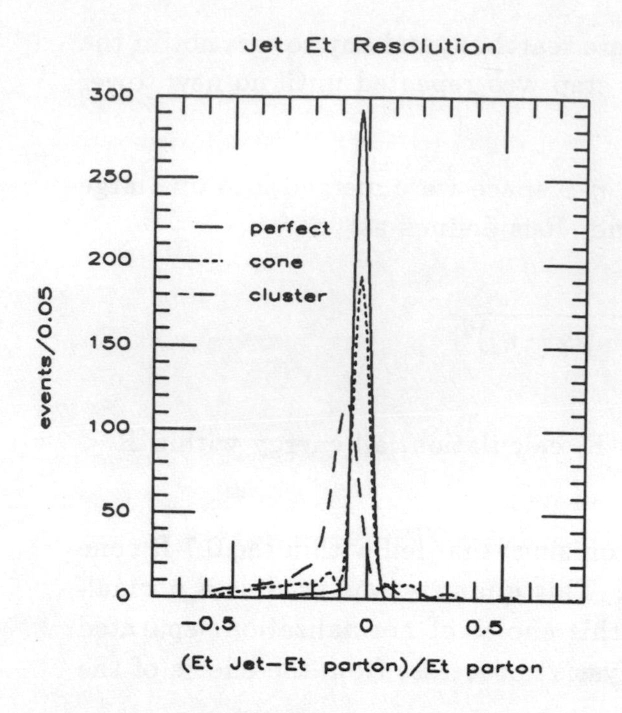

Fig. 1. Fractional Jet E_t resolution for various choices of pattern recognition.

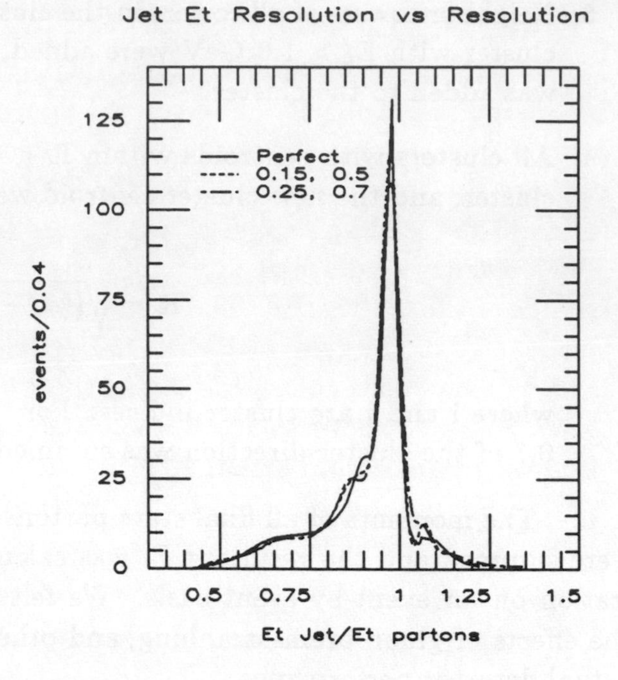

Fig. 2. Jet E_t resolution vs. calorimeter energy resolution.

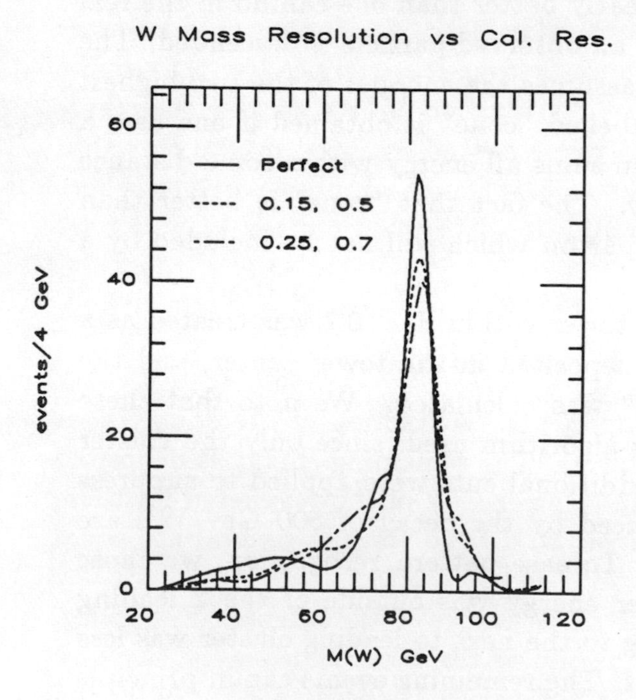

Fig. 3. W Invariant Mass resolution vs. calorimeter energy resolution.

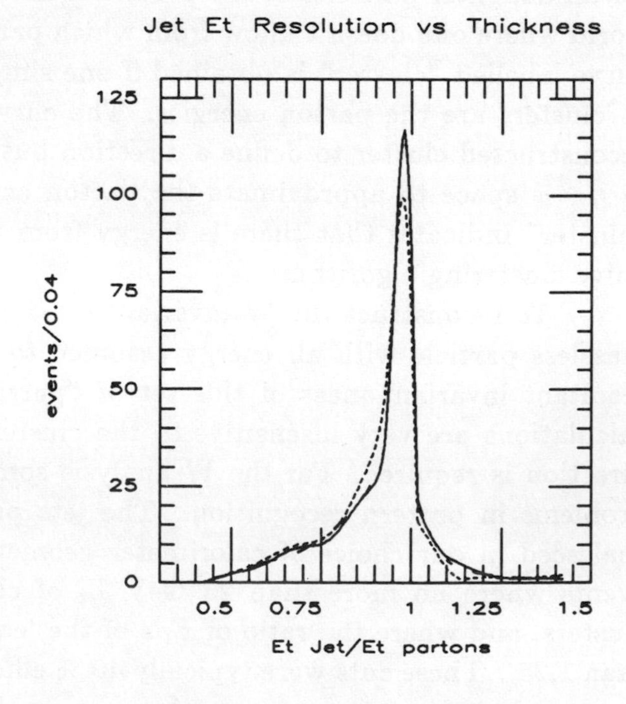

Fig. 4. Jet E_t resolution as a function of hadron calorimeter thickness in interaction lengths.

3.1 Dependence on Calorimetry energy resolution

Figure 2 shows jet E_t distributions for three choices of electromagnetic/hadronic calorimetry energy resolution. Numbers in the figure are to be divided by the $\sqrt{E}$ to determine the σ of the gaussian resolution for different energy depositions. There is no striking difference between the three cases, because the E_t resolution of a jet is determined by the E_t resolution of the leading particles, which, due to their high energy, are always well measured. The low energy tails of the distributions are due primarily to particles from the jet falling outside of the 0.7 R cone. The events in the tail may have a high energy parton lying close to the edge of the cone of $R < 0.7$. Particles from the hadronization of this parton often land outside of the cone. In a sense, this tail is an artifact of the jet definition.

Figure 3 shows W mass resolution for the three cases. Here we see that calorimeter resolution is more important than in the jet case. Low energy particles in the lab frame can make large contributions to the invariant mass calculation. It is important to measure these particles accurately.

3.2 Dependence on Hadron calorimeter thickness

Figure 4 shows the E_t distribution for 1 TeV jets as a function of the thickness in interaction lengths of the hadron calorimetry. In all cases the electromagnetic calorimeter in front of the hadron calorimeter is assumed to be 17 radiation lengths, and 0.8 interaction lengths thick. The curves for 10 and 15 interaction lengths are indistinguishable. The distribution for five interaction length thick hadron calorimetry shows a slightly worse resolution, and a slight low energy tail from leakage. The effect of calorimetry thickness would be more pronounced in the forward/backward regions, where the particle energies (for a fixed E_t) are larger. Figure 5 shows the results for the mass distribution of W's. In this case there is no difference between any of the three cases. This is partially caused by the cuts that define the event sample. The requirement of cluster E_t balance suppresses events where there is substantial leakage or punchthrough. In addition, the jets from the W decay are softer, again reducing the effect of leakage.

3.3 Dependence on Intrinsic E/H Response ratio

Hadronic shower energy deposition has two components. A fraction of the energy goes into π_0s and is deposited like electromagnetic showers. The remaining energy is deposited by ionizing hadrons. The ratio of calorimetry response per GeV of shower energy for these two types of energy deposition is called the intrinsic E/H ratio. It is in general not equal to 1.0 [3].

Figure 6 shows the jet E_t distributions for three values of the intrinsic E/H ratio. We note that there is a broadening of the distribution as the ratio varies away from 1.0 The curves for 0.8 and 1.2 are not very different. In general, E/H not equal to 1.0 causes a nonlinearity of response for different energy jets. Figure 7 shows the observed E_t fraction versus jet E_t for 2 choices of E/H, normalized to $E/H = 1$. Coincidentally, the curves intersect at about 1 TeV, explaining the lack of shift of the peaks of the distributions

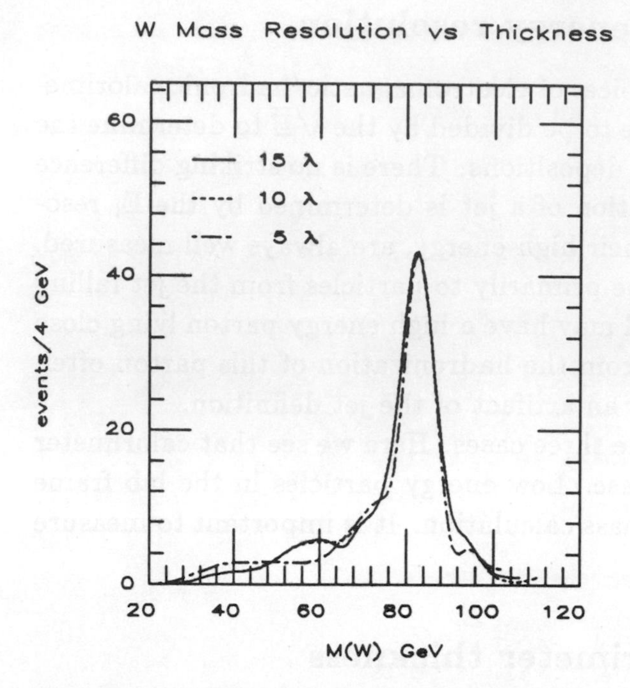

Fig. 5. W mass resolution as a function of hadron calorimeter thickness.

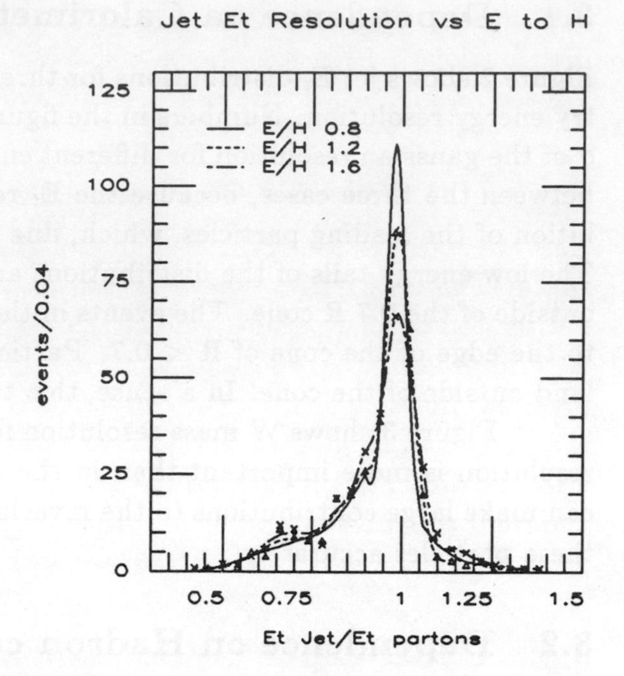

Fig. 6. Jet E_t resolution vs. intrinsic E/H ratio.

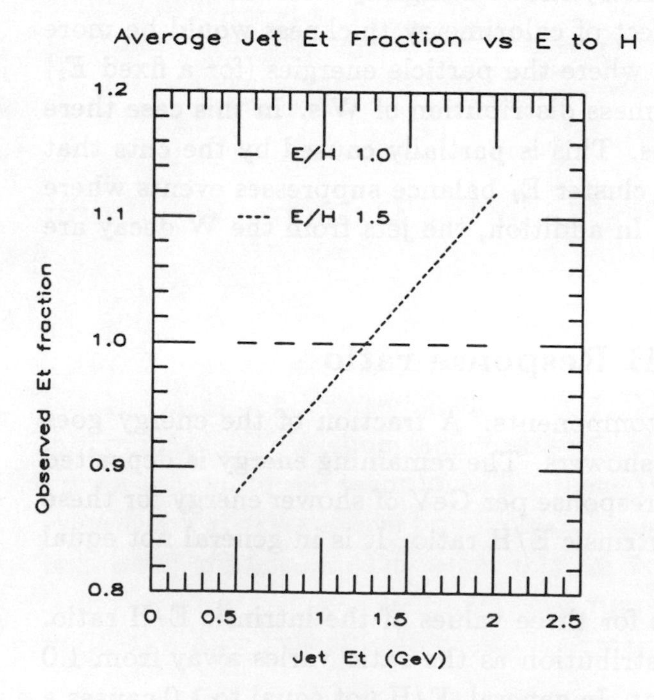

Fig. 7. Average jet E_t fraction as a function of jet E_t for different values of E/H.

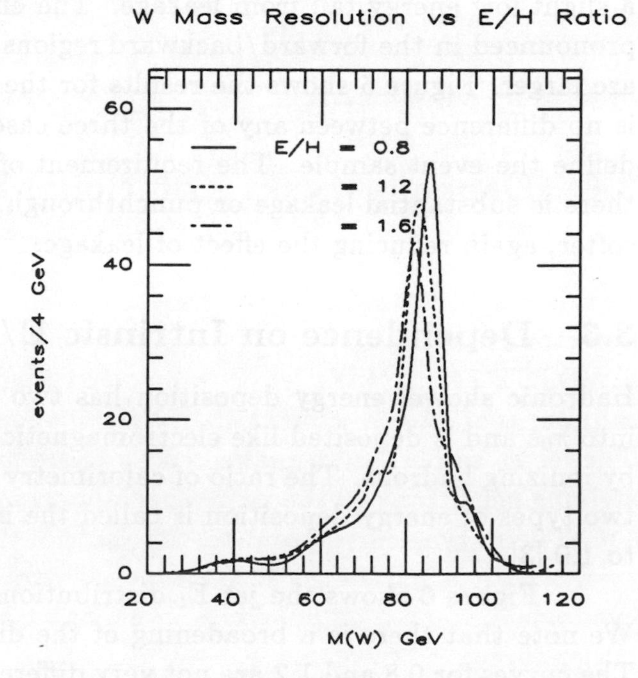

Fig. 8. W mass resolution vs. calorimetry E/H.

in Figure 6. This nonlinearity can be a serious problem for measurements of energy distributions. Figure 8 shows the effect of varying E/H on the W mass distribution. We see an increase of the width and a systematic shift of the peak energy.

3.4 Dependence on Calorimeter Tower Size

Here we consider the effect of different tower sizes. The towers are rectangular cells in η-ϕ space. The only effect tower size can have on jet E_t resolution is from mismeasure of the θ direction of energy flow, causing an error in the $\sin(\theta)$ weighting for the E_t calculation. At θ approximately 90°, this is a very small effect for any reasonable tower size. In our invariant mass algorithm, the tower size is very important. All energy deposited in a tower is assumed to come from the tower center, so the larger the tower, the larger the possible error in direction determination of the energy flow into the tower. Figure 9 shows W invariant mass distributions for three cases of tower size: $\delta R = 0.01$, 0.03, and 0.1 . The mass resolution is significantly worse with the largest tower size.

3.5 Dependence on Solenoidal Magnetic Field

Since jet E_t resolution is determined primarily by the stiff leading particles, magnetic field in the central tracking volume has very little effect. The W invariant mass distribution is somewhat affected by the applied magnetic field, with a broadening at larger fields. The central tracking volume was 1.4 meters in radius for these studies. Figure 10 shows the change in the shape of the W invariant mass distribution for 3 choices of magnetic field.

We note that many of the SSC detector designs under consideration have at least some of the calorimetry inside the magnetic field. Problems this may cause are not addressed by our simulation.

3.6 Dependence on Additional Events in the Detector Resolving Time

In this study, we consider the effect of additional background events which fall within the detector resolving time of the signal events. For the background events, we used ISAJET TWOJET events with jet P_t's between 3 and 15 GeV. This corresponds to about 150 millibarns of cross section at 40 TeV, so it should be a reasonable model for the background. The jet E_t resolution is unaffected by these soft events. The W mass resolution has a very striking sensitivity to the number of superimposed background events as shown in Figure 11. The curves correspond to the distribution for the signal events, and for the signal events with a number of superimposed background events. The number of background events is extracted on an event-by-event basis from a Poisson distribution with a mean of either three or 10 events.

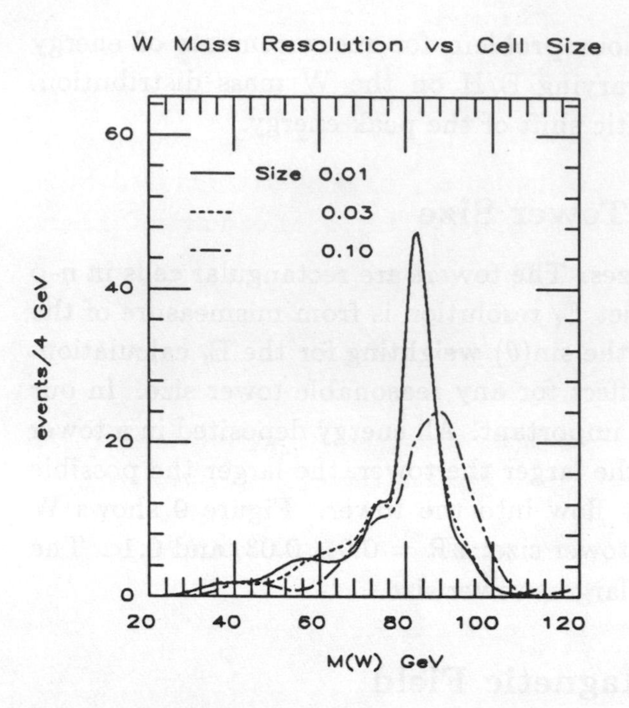

Fig. 9. W mass resolution vs. calorimeter tower size in $\eta - \phi$.

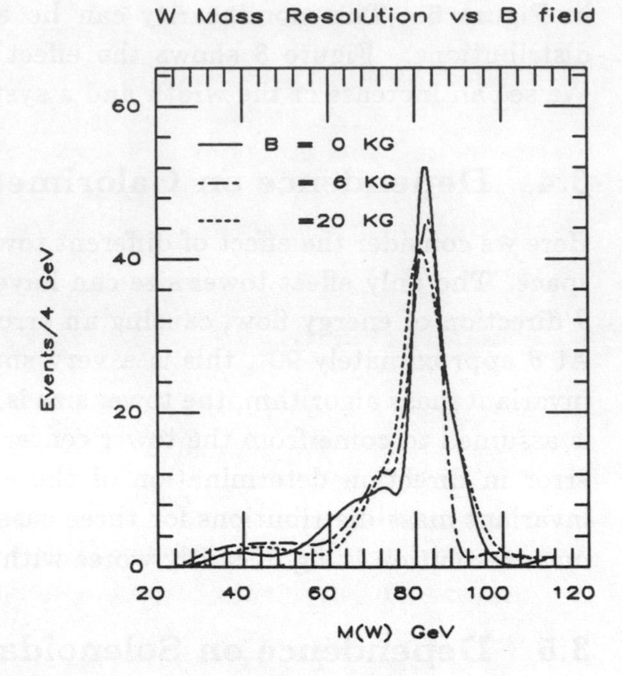

Fig. 10. W mass resolution as a function of magnetic field.

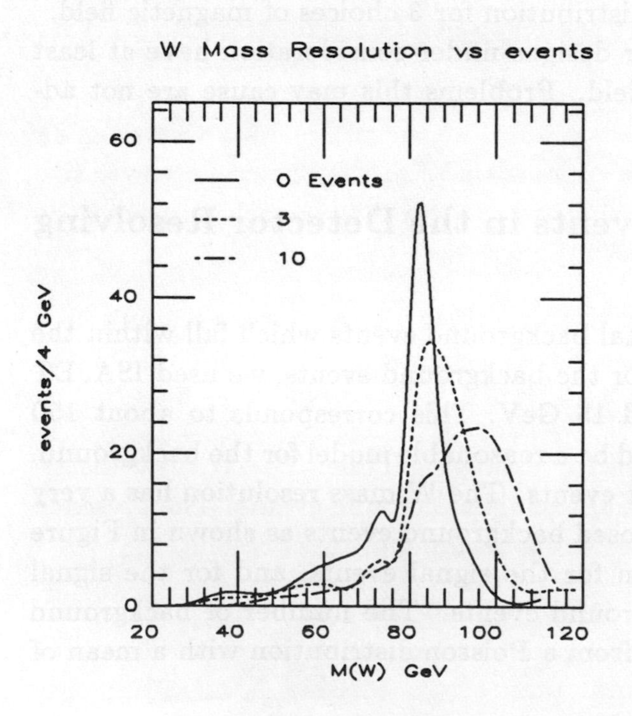

Fig. 11. W mass resolution vs number of superimposed background events.

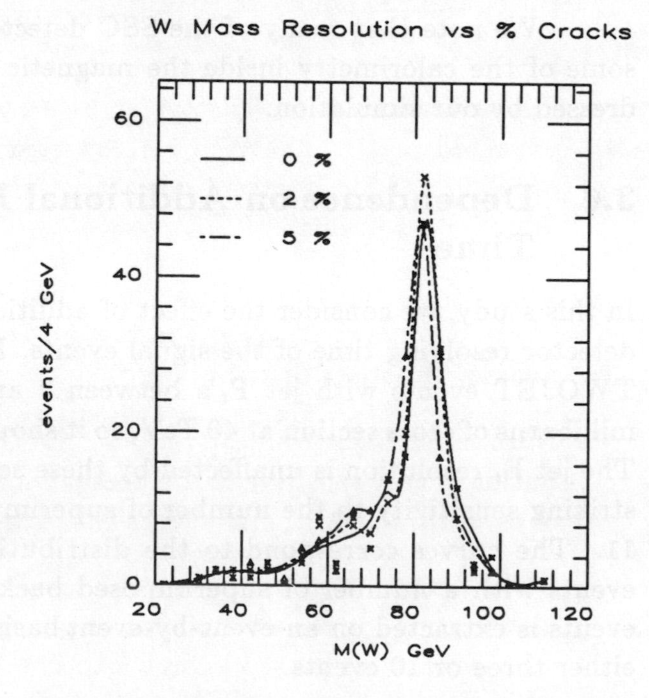

Fig. 12. W mass resolution for different percentages of cracks.

3.7 Dependence on Dead/Crack Regions

We adopted an ad hoc procedure to investigate the effect of dead material and cracks. Each tower was given a fixed percentage of dead area. Particles landing in the dead portion of the tower had the energy that they were to deposit into the calorimetry reduced by a factor of 2. Figure 12 shows W mass distributions for 0%, 2%, and 5% dead areas. There is surprisingly little effect.

4 Conclusions

We have used a simple Monte Carlo program to investigate several contributions to jet energy resolution. We summarize our conclusions about each of the effects considered:

1. Calorimetry Resolution: The jet E_t signal is independent of the calorimetry energy resolution. The W invariant mass signal however does have some dependence. Accurate measurement of low energy particles is important.

2. Calorimetry Thickness: Neither signal is particularly sensitive to the thickness of the hadron calorimeter in interaction lengths. Higher energy jets would have more degradation due to leakage. Missing E_t signatures are the most sensitive benchmark to determining the desired thickness.

3. Both signals are sensitive to the E/H ratio. Since there is an understanding of this ratio, and the ability to tune it, there is no reason to build calorimetry with E/H far from 1.0 . The range 1.1 to 0.9 is acceptable.

4. Tower Size in η-ϕ: The jet signal is almost totally insensitive to tower size, an expected result. The W invariant mass is very sensitive to the choice. A tower size of $0.03\eta \times 0.03\phi$ is acceptable. Increasing the size to 0.1 x 0.1 seriously degrades the W mass resolution.

5. B field: The jet E_t distribution is insensitive to the field. The W mass resolution is affected somewhat. 30 kG-m of field causes about a factor of 1.5 broadening of the peak of the mass distribution.

6. Number of Background Events in the Detector Resolving Time: The W mass distribution is very sensitive to the presence of additional background events. The presence of even three additional events causes a serious worsening of mass resolution.

7. Dead Areas. Dead areas at the level of a few percent have little effect on either distribution. It is very difficult to generalize about cracks, since the exact details of their properties must be known to determine their effect. Missing E_t measurements are probably most sensitive to the effects of cracks.

References

1. F.E. Paige and S.D. Protopopescu, 1982 DPF Summer Study, Snowmass, CO, June 1982, p. 471.

2. J. Freeman and A. Beretvas, 1986 DPF Summer Study, Snowmass, CO, June 1984, p. 482

3. R. Wigmans, CERN/EF 86-18, Sept.,1986

LUMINOSITY LIMITATIONS

D. Bintinger
SSC Central Design Group, c/o LBL 90-4040, Berkeley, California 94720

ABSTRACT

This paper investigates the limitations to luminosity that are contained in the present SSC design. Increase in luminosity is considered by varying beam current, emittance, β*, and the total number of beam bunches. Possible luminosity at different beam energies is also considered. The difference between initial luminosity and integrated luminosity is emphasized.

1. Introduction

Initial luminosity (L_o) is given by the formula

$$L_o = I_B^2 E_B C/4\pi m_p c^3 e^2 \epsilon_N \beta^* M \quad (1)$$

where

I_B = beam current in each ring
E_B = beam energy
C = ring circumference
m_p = proton mass
c = speed of light
e = electron charge
ϵ_N = normalized beam emittance
β^* = focusing parameter at the interaction point
M = number of proton bunches in each ring.

Increasing luminosity by varying each of the parameters I_B, ϵ_N, β^*, M and E_B is considered in this paper. The design limitations encountered are noted. The design [1] values for these parameters, giving an initial luminosity of $10^{33} cm^{-2}s^{-1}$, are

I_B = 73 mA
ϵ_N = 1×10^{-6} rad•m
β^* = 0.5 m
M = 17100
and E_B = 20 TeV.

The time evolution of luminosity involves a complex interplay between the number of protons in a bunch (N_B) and the emittance. A computer program is used to calculate luminosity as a function of time [2]. For the design values of the SSC, the calculated luminosity rises from the initial value, before eventually falling. Figure 1 shows this. The rise is due to emittance shrinkage caused by synchrotron radiation damping. This calculation is probably naive given the realities of an actual collider; however, it is based on our present best knowledge and provides a method for comparing integrated luminosity under various assumptions.

By knowing luminosity as a function of time and assuming a time to fill both rings (here, 2 hours), the time to maintain stored beams for physics that will maximize integrated luminosity can be calculated. Using this storage time, the maximized integrated luminosity for a given period is obtained. For one year (3.15×10^7 seconds) of continuous operation of the SSC at design parameters, the maximized integrated luminosity is $5.3 \times 10^{40} cm^{-2}$. This quantity will be recomputed as the various luminosity equation parameters are varied. It will be seen that increasing initial luminosity by a factor will in general increase integrated luminosity by a lesser factor.

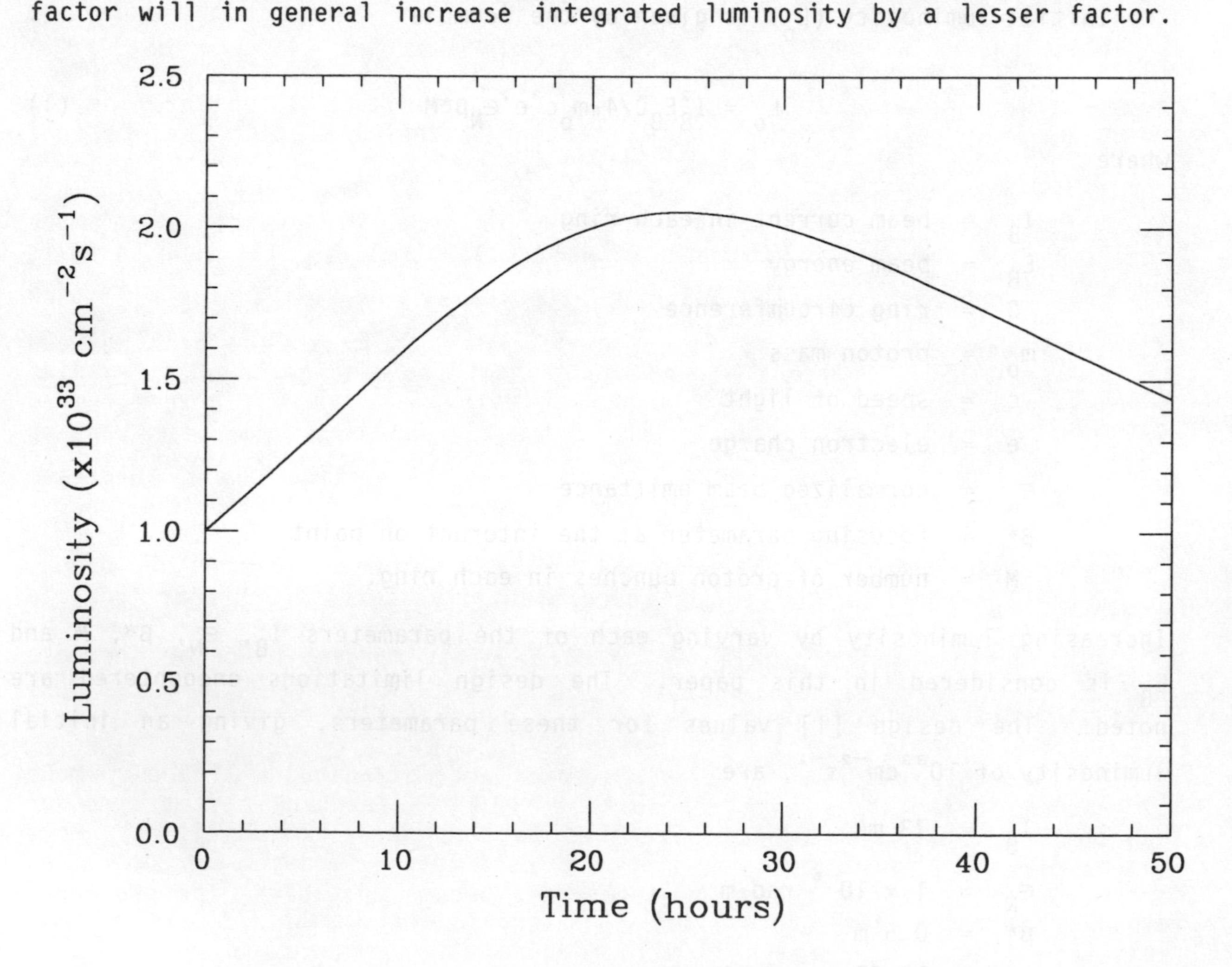

FIG. 1. Luminosity vs time for the design parameters.

The primary reason for this is that the number of protons is more quickly depleted at higher luminosities.

2. Beam Current Variation

In this section the effect of increasing the luminosity by increasing the beam current will be considered. All other parameters in Eq. (1) will be held constant. To be specific, an increase in initial luminosity of a factor of 10 will be considered, implying a beam current increase of $\sqrt{10}$. The number of protons in a bunch will be increased by this same $\sqrt{10}$ factor. The number of proton-proton interactions per bunch crossing increases proportionally to luminosity giving 14 interactions per crossing initially, versus 1.4 for design beam current and σ_{Inel} = 90 mb. Following the prescription to obtain integrated luminosity of the previous section, an increase in beam current by $\sqrt{10}$ results in an increase in integrated luminosity by a factor of 5.5, less than the initial luminosity increase of a factor of 10.

The remainder of this section will investigate the effects of beam current increased by $\sqrt{10}$ on the various subsystems of the SSC and on beam stability. The majority of the comments below are taken from a paper by R. Diebold [3].

a) Injector System. A limitation to the current injected into the Low Energy Booster (LEB) by the LINAC is given by the incoherent Laslett tune shift. This tune shift increases with beam current. Its current value is 0.09 [4] and it is desirable to keep it at or below this value. To do this for a beam current larger by $\sqrt{10}$ will require the energy of the LINAC to be increased from the design value of 600 MeV to 1600 MeV, as this tune shift is inversely proportional to $\beta\gamma^2$ of the LINAC. This increased energy will then cause premature stripping of the H^- ions in the transfer line bend between the LINAC and LEB. Either the bend will have to be eliminated or its radius increased. A larger beam current in the LEB will also require more turns to fill the LEB resulting in an increased emittance due to the beam passing through the H^- stripping foil. This increased emittance will decrease somewhat the luminosity gain from increased current.

b) Collider Stability. As noted above, the number of protons per bunch, N_B, is also increased. The single bunch stability limits are proportional to N_B^{-1}. The design limits have a factor of 6 safety margin; hence an increase in N_B by $\sqrt{10}$ is feasible.

The growth rates of the transverse and longitudinal dipole mode multi-bunch instabilities will increase by $\sqrt{10}$. A feedback system stronger than the present system will be required to damp these instabilities. More importantly, the quadrupole multi-bunch oscillation modes, for which no control is necessary at design beam current, become unstable. The preferred method at present for controlling this quadrupole instability is a fast feedback system. Such a system will require R&D to fabricate and apply to the SSC. The problem of multi-bunch instability is a potentially serious one for increased beam current.

The beam-beam tune shift is a beam stability parameter that is directly proportional to beam current, and whose value must be kept below a limit. The beam-beam tune shift has two components, the head-on and the long-range. Both are related to the crossing of the beam at interaction points. An optimistic upper limit for the beam-beam tune shift is that the head-on portion should be less than 0.005 per interaction point; a pessimistic upper limit is that the sum of head-on and long-range should be less than 0.005. With design value parameters, the tune shifts are 0.0008 for head-on and 0.002 for long-range. A beam current increase by $\sqrt{10}$ will exceed the pessimistic limit but not the optimistic. Some of the long-range tune shift can be cancelled out by having interaction point crossing planes alternate between horizontal and vertical. It is felt that the beam-beam tune shift can be brought safely below limit for a beam current increased by $\sqrt{10}$.

c) Heating in IR Quadrupoles. The final focus quadrupoles in the Interaction Regions will absorb radiation from the interaction point at rates up to 2 W/m, or 0.2 mW/gm, at design luminosity. The amount of radiation scales with the luminosity, hence a $\sqrt{10}$ increase in beam current will cause radiation depositions of up to 2.0 mW/gm. The level at which quenches will occur, extrapolated from Tevatron data, is estimated as 10 mW/gm. This implies a safety margin of 5 at the increased luminosity; however, there are large uncertainties in the data extrapolation. Better measurements and calculations are necessary.

d) Tunnel and IR Radiation. Radiation in both the tunnel and interaction halls will increase as the square of the beam current. In the interaction hall this is because radiation is proportional to luminosity. In the tunnel this is due to the fact that increased beam current also gives increased synchrotron light which in turn causes increased gas desorption from beam tube walls. Electronics in both areas will thus have to survive or be shielded from ten times the radiation if beams current is increased by $\sqrt{10}$. As outlined in the Invitation for Site Proposals [5], a factor of $\sqrt{10}$ increase in beam current will not result in environmental radiation beyond proscribed limits.

e) Vacuum. As explained in the above paragraph, beam gas scattering will increase with the square of the beam current. Beam lifetime will then decrease. It is assumed that operation with increased current will occur far into the life of the SSC when present information suggests that the beam lifetime at design current will be about 1000 hours. Increasing beam current by $\sqrt{10}$ will then degrade the beam lifetime to approximately 300 hours. This is well within tolerable limits.

f) Total Beam Energy. At the design current of 73 mA the total energy in one 20 TeV beam is 405 megajoule. This energy, of course, will scale with beam current. Thus all systems sensitive to being damaged by significant beam loss will have their jeopardy increased proportionally.

g) Beam Dump System. The current beam dump system can safely handle up to $\sqrt{10}$ times the design beam current. The beam dump consists of graphite pellets whose boiling point is 4000°C. A beam current $\sqrt{10}$ larger than design will cause a temperature rise in the beam dump of 770°C.

h) Cryogenics. A study by Mike McAshan [6] indicates that an increase in beam current, and therefore synchrotron radiation heat, by $\sqrt{10}$ will increase refrigeration load by a factor of 2.1. (The refrigeration load will not directly scale as there is a significant static load.) For extended high current operation, then, refrigerator capacity will have to be increased by this factor. The diameter of the helium gas return line in the magnet cryostats will also have to be enlarged to carry increased flow. Other cryostat and magnet cold mass modification may also be necessary due to increased heat flow outward from the beam tube. Doubling the number of recoolers in the tunnel may also be necessary. The cryostat and cold mass design and the number of recoolers must be decided before initial construction. Refrigerator capacity, however, can be increased in stages with some components, namely heat exchangers, best installed at initial construction. An alternative solution to handling increased synchrotron heat is to install a beam tube liner that would intercept the synchrotron heat at higher temperature (~20 K). This would make refrigerator size largely independent of synchrotron radiation load. However, a beam tube liner would increase the magnet coil size, significantly increasing cost. Additionally, the effects of a liner on aperture and impedance would have to be examined closely.

(i) RF System. The amount of RF power required increases from 1.74 (1.50) MW for design current acceleration (storage) to 3.74 (2.98) MW for current times $\sqrt{10}$ acceleration (storage). Doubling the number of klystrons should provide for this increase.

Summarizing this section, there are three areas where limitations to increasing luminosity by increasing beam current by $\sqrt{10}$ appear in the present design. They are the injector system, multi-bunch stability, and cryogenics. In none of these areas are the problems insurmountable. Increasing luminosity by increasing beam current, then, is feasible with additional design work.

3. Emittance Variation

The design emittance of the SSC is considered to be achievable under optimum conditions. Decreasing it further is problematic; however, for purposes of this article we will investigate a factor of 2 decrease in emittance. This decrease will give a factor of two increase in initial luminosity and a factor of 1.6 increase in integrated luminosity. The number of interactions per bunch crossing will increase initially from 1.4 to 2.8. As previously, the remainder of this section will investigate the effects of emittance decreased by a factor of 2 on subsystems of the SSC.

a) Injector System. Almost the same comments that apply to the case of increased beam current apply here. The incoherent Laslett tune shift doubles with halved emittance. It is desirable to keep this tune shift at or below its design value. An increase in LINAC energy can be used to offset the increase caused by decreased emittance. The problem of premature stripping of H^- in the transfer line bend between LINAC and LEB occurs again with the same cure of decreasing the curvature of or eliminating the bend.

b) Collider Stability. Neither single bunch nor multi-bunch stability is affected by decreased emittance. The head-on beam-beam tune shift increases with decreasing emittance, while the long-range beam-beam tune shift remains constant. A factor of 2 increase in head-on beam-beam tune shift is not serious.

c) Heating in IR Quadrupoles. The same comments for an increase in luminosity due to increased beam current apply here for an increase in luminosity due to decreased emittance. The IR quadrupoles will have to absorb more radiation. A factor of 2 more should produce no quenches.

d) IR Radiation. This also will increase proportionally with any increase in luminosity.

Summarizing this section, the main limitation to increasing luminosity by decreasing emittance is the ability to decrease emittance. As mentioned at the start of this section, a factor of 2 decrease is thinkable. If such a decrease is achieved, then the restrictions of the LINAC-LEB interface will be encountered.

4. β^* Variation

In this section the effects of increasing luminosity by decreasing β^* are investigated. The parameter β^* is directly related to the cross section of the beams at the interaction point and, hence, inversely proportional to luminosity. The design β^* is 0.5 m and the design free distance on either side of the interaction point is 20 m. (There also exist in the present design 2 medium luminosity interaction points with β^*s of 10 m and free distances of ±120 m. These will not be discussed.) A standard method of decreasing β^* in e^+e^- storage rings is to move the final focusing quadrupoles closer to the interaction point. This moves the points of maximum vertical and horizontal beam spread, referred to as β^{MAX}, closer to the interaction point to allow for tighter focusing. For the SSC, the final focusing quadrupole triplet is 55 m long, and β^{MAX} vertical and horizontal are 50 m and 80 m from the interaction point respectively. Even if the quadrupole triplet is moved the entire 20 m of free distance, not much of a fractional gain would be made in moving vertical or horizontal β^{MAX} closer to the interaction point. Additionally, to take advantage of moving β^{MAX} closer to the interaction point, either β^{MAX} must be increased or the gradient of the quadrupoles must be increased. At present β^{MAX} is approximately 8 Km which implies a beam cross section at β^{MAX} that is as large as is presently considered prudent in the aperture of the quadrupoles.

Increasing the gradient of the quadrupoles will decrease β^*. Achieving a gradient higher than the current design of 230 T/m is considered difficult. However, being optimistic concerning gradient, and simultaneously reducing free distance, a reduction in β^* by a factor of 2 is conceivable [7]. This reduction gives a factor of 2 in initial luminosity and, following the prescription outlined in the introduction, a factor of 1.5 in integrated luminosity. Again, as in previous sections, the number of interactions per bunch crossing will increase, here initially from 1.4 to 2.8.

There is a natural limit to increasing luminosity by decreasing β^* for a given proton bunch length. This occurs when the distance along the beam over which the beam is focussed becomes less than the proton bunch length. A measure of this is when the bunch length becomes order of the same length as β^*. The bunch length for the SSC is 7 cm. Hence a β^* of 0.5 m allows for a significant luminosity increase, although factors previously mentioned restrict this increase.

Other considerations for increasing luminosity by decreasing β^* are increased beam-beam long-range tune shift (head-on portion not affected), increased IR quadrupole heating, and increased IR radiation. The IR quadrupole heating will not necessarily be any worse due to quadrupoles in a reduced β^* configuration being closer to the interaction point. This is due to the fact that the point of highest radiation deposition, even with the 0.5 m β^*, is significantly away from the beginning of the closest quadrupole. The increase in radiation absorbed will come from the increased luminosity.

In summarizing increasing luminosity by decreasing β^*, the main point is that at present a factor of 2 decrease in β^* is the maximum practical. An auxiliary consideration is that free distance may be reduced to as little as 3 m. A concern related to β^* considerations is that IRs at the SSC come in pairs of matching β^*. With present β^*s this pairing is not a restriction. For example, one IR in a pair could have $\beta^* = 0.5$ m and its mate IR have $\beta^* = 10$ m. Whether this pairing becomes more restrictive if one β^* is less than 0.5 m is presently being investigated.

5. Number of Proton Bunches Variation

Keeping the current constant, the luminosity can be increased by decreasing the number of proton bunches. The number of protons in each bunch then increases. The luminosity increase is directly proportional to the decrease in number of bunches, M, as shown in Eq. (1). The number of interactions per bunch crossing, however, increases as the square of the decrease. For specificity a factor of 3 reduction in number of bunches will be considered. This gives a factor of 3 increase in initial luminosity, but only a factor of 1.7 increase in integrated luminosity. This method of increasing luminosity gives the poorest integrated luminosity return on increased initial luminosity of the methods studied in this paper.

Other systems or effects to be considered when reducing the number of bunches are the injector system, collider stability, and IR radiation. The injector system is again faced with the limitations imposed by the incoherent Laslett tune shift. However, the solution here is a new, lower frequency RF system for the LEB. For this case the more extensive injector modifications of sections 2 and 3 can be avoided. The single bunch stability limits will be lowered by a factor of 3 which still leaves a factor of 2 safety margin. Multi-bunch stability may be affected by resonances excited by the new bunch spacing; however, it should be possible to avoid these resonances by slightly altering machine tune. The head-on beam-beam tune shift will increase by a factor of 3. The long-range beam-beam tune shift will remain constant. The total beam-beam tune shift will stay below the limit of 0.005 given earlier. As for all the previous cases IR radiation will increase proportionally to luminosity.

In summary for this section, increasing luminosity by holding the current constant and decreasing the number of bunches has the major drawback that it gives the poorest return of integrated luminosity. It also increases the number of interactions per bunch crossing as the square of the initial luminosity increases.

6. Variation of Beam Current and Number of Bunches

The luminosity can be increased by adding more proton bunches to the beam. If the added bunches have the same number of protons as the original bunches, the current will increase proportionally to the increase in the number of bunches, and, according to Eq. (1), so will the initial luminosity. For this case the integrated luminosity will also increase proportionally. In this section we will consider a factor of three increase in the number of bunches and therefore in the beam current and initial and integrated luminosities. The number of interactions per bunch crossing will remain constant; however, the time between bunches and crossings will decrease from 16 ns to 5.3 ns.

The remainder of this section will investigate the effects of increasing beam current a factor of 3 by increasing the number of bunches. Most of these effects have been considered in the section on beam current variation, section 2, and those portions of section 2 will be referred to here.

a) Injector System. Again, as in section 2a, the incoherent Laslett tune shift limits increasing current in the LEB. Here, though, the limit is due to the tune shift being inversely proportional to bunch spacing. The solution to offsetting this limit is the same as in section 2a, namely to increase the LINAC energy and to decrease or eliminate the curvature in the LINAC-LEB transfer line. A consequence of decreasing the bunch spacing not encountered in section 2a is that the RF system frequency must be increased by a factor of 3 in all three booster rings. The emittance will increase in the LEB due to the increased current thus degrading the luminosity increase by a factor that is unknown at this time.

b) Collider Stability. Since the number of protons per bunch remains constant, there is no change to single bunch stability. However multi-bunch stability encounters the same problems as outlined in section 2b. The solution is, as in section 2b, a fast feedback system to control this instability. The head-on beam-beam tune shift will not increase. The long-range beam-beam tune shift increases since it is inversely proportional to the bunch spacing. Again, some of the long-range tune shift can be cancelled out by having interaction point crossing planes alternate between horizontal and vertical. The beam-beam tune shift should not be a serious problem.

The effects on the following topics or systems are the same as those considered in section 2 and will not be repeated here. These effects result from beam current or luminosity increase, regardless of how the increase is achieved. The topics or systems along with the corresponding portion of section 2 are: Heating in IR Quadrupoles (2c), Tunnel and IR Radiation (2d), Vacuum (2e), Total Beam Energy (2f), Beam Dump System (2g), Cryogenics (2h), and RF System (2i). Concerning the RF system, it should be noted that in the present design every sixth bucket is filled, hence there is room for a factor of three increase in number of buckets.

Summarizing this section, the same three areas where limitations appear in section 2 for a factor of $\sqrt{10}$ beam current increase also appear here. These areas are the injector system, multi-bunch instabilities, and cryogenics. Again, in none of these areas are the problems insurmountable, and, as in section 2, the conclusion is that, with additional design work, increasing luminosity by increasing both beam current and number of bunches is feasible.

7. Variation of Beam Energy

Since synchrotron radiation power varies as E_B^4, decreasing beam energy will quickly lessen the cryogenic load due to synchrotron radiation. At a lower beam energy, then, the excess cryogenic capacity can be used to absorb the power from increased beam current. If this is done, beam current will scale as E_B^{-4}, and initial luminosity, as shown by Eq. (1), will scale as E_B^{-7}. For a beam energy decrease from 20 TeV to 15 TeV, the initial luminosity will increase from 1×10^{33} $cm^{-2}s^{-1}$ to 7.5×10^{33} $cm^{-2}s^{-1}$. Table I lists beam energy, initial luminosity, and integrated luminosity for beam current scaled to keep total synchrotron radiation power constant. The last entry is for the beam current scaled in the same manner if beam energy is raised. All the considerations of section 2 apply for increased beam current, except of course the need for increasing cryogenic capacity. Avoiding the expense of increasing cryogenic capacity is in fact the major reason for increasing luminosity by decreasing beam energy.

8. Summary

Table II lists the methods by parameter for increasing luminosity that have been considered in this paper. The table lists a variation for the parameter that is considered possible and the resulting initial luminosity

Table I. Luminosity vs. Beam Energy.
Scaled by Keeping Synchrotron Power Constant.

Beam Energy (TeV)	Initial Luminosity (cm^{-2} s^{-1})	Integrated Luminosity for 1 Year (cm^{-2})
15.0	7.5×10^{33}	2.6×10^{41}
17.5	2.5×10^{33}	9.7×10^{40}
20.0	1.0×10^{33}	5.3×10^{40}
22.5	4.4×10^{32}	3.2×10^{40}

Table II. Parameter Variation vs. Luminosity

Parameter	Variation	Initial Luminosity Increase	Integrated Luminosity Increase	Major Limitations
Beam Current	x$\sqrt{10}$	x10	x5.5	Injector System Multi-bunch Stability Cryogenics
Emittance	x1/2	x2	x1.6	Emittance itself Injector System
β*	x1/2	x2	x1.5	β* itself IR Free Space
Number of Bunches	x1/3	x3	x1.7	Limited Integrated Luminosity
Beam Current and Number of Bunches	x3	x3	x3	Decreased Bunch Spacing Injector System Multi-bunch Stability Cryogenics
Beam Energy	x3/4	x7.5	x4.9	Cross Section Decrease Injector System Multi-bunch Stability

and integrated luminosity increases. Also listed are what are considered the major limitations to varying this particular parameter. It is seen from the table that significant luminosity increase is only achieved by increasing the beam current. However, as beam current is increased emittance may also increase lessening the gains.

9. Acknowledgements

This paper is an outgrowth of a group that studied luminosity limitations. The group consisted of Alex Chao, R. Diebold, Gil Gilchriese, Al Garren, Dave Johnson, Steve Peggs, Jack Peterson, W. Scandale, and the author. This paper would not have been possible without the contributions and criticism of the entire group.

References

1. Superconducting Super Collider Conceptual Design, SSC-SR-2020, March 1986.
2. The program that computes luminosity as a function of time includes beam-beam interaction proton loss, beam-gas interaction proton loss, and emittance change due to synchrotron radiation damping, intrabeam scattering, beam-beam elastic scattering, and beam-gas coulomb scattering.
3. R. Diebold, Extending the Reach of the SSC, ANL-HEP-CP-86-122.
4. M. A. Furman, Effect of the Space-Charge Force on Tracking at Low Energy, SSC-115, March 1987.
5. Invitation for Site Proposals for the Superconducting Super Collider, DOE/ER-0315, April 1987.
6. M. McAshan, Cryogenic Implications of Operation of the SSC at Substantially Increased Current, Sept. 11, 1986, uncirculated.
7. A. A. Garren, D. E. Johnson, and W. Scandale, Layout and Performance of a High-Luminosity Insertion for the SSC, these proceedings.

LAYOUT AND PERFORMANCE OF A HIGH-LUMINOSITY INSERTION FOR THE SSC

A. A. GARREN, D. E. JOHNSON AND W. SCANDALE†
SSC CENTRAL DESIGN GROUP*

ABSTRACT

An experimental insertion with the beta functions at the crossing point as low as a quarter of a meter has been designed for the SSC by using quadrupoles with 330 T/m gradients. A free space of ±3 m around the interaction point is available to install an experimental device dedicated to the observation of large transverse momentum reactions. A phase advance of 90° across the overall insertion ensures to some extent the mutual compensation of the chromatic aberrations of the quadrupoles triplets with those of the next experimental region.

Introduction

A possible increase of luminosity of the SSC above the present design is of obvious interest to the HEP community. We consider here one route to such an increase. In a hadron collider with round beams at the collision points, the luminosity is expressed by the following formula:

$$\mathcal{L} = \frac{kN_b^2 f\gamma}{4\pi\epsilon_N \beta^*} \tag{1}$$

with
- k = number of bunches per beam
- N_b = number of protons per bunch
- f = revolution frequency
- γ = relativistic factor
- $\epsilon_N = \gamma\sigma^2/\beta$ = normalized emittance
- σ = rms beam radius
- β^* = beta function at the crossing point

An increase in luminosity is possible by increasing the number of protons per bunch, by increasing the number of bunches, or by decreasing the beta function at the crossing point. The first issues have been extensively considered.[1] Here we focus our attention on the third issue.

In a given insertion, the beta at the crossing point can be reduced by using a triplet closer to the interaction point, with stronger gradient quadrupoles. The lower limit of β^* that produces a luminosity consistent with Eq. (1) is determined by the longitudinal rms bunch length, and is about 7 cm for the SSC. Such a low value is beyond present

† Visiting guest from CERN

⋆ Operated by Universities Research Association, Inc. under contract with the U.S. Department of Energy.

technical feasibility in the SSC, due to the excessively high gradient required in the triplet to maintain the associated peak beta values within a reasonable range. A more realistic goal is a beta of 25 cm, *i.e.* a factor of two below the nominal value. We assume that quadrupoles with a gradient of the order of 330 T/m and a bore winding diameter of 4 cm will be available. This gradient is generally consistent with that planned for the LHC, which uses 250 T/m quadrupoles with a 5 cm winding diameter.[2]

Geometry of the high-luminosity insertion

The geometrical layout of the high-luminosity insertion proposed here is derived from the nominal SSC insertion.[3] The layout and collision lattice functions for one half of this insertion are shown in Figure 1.

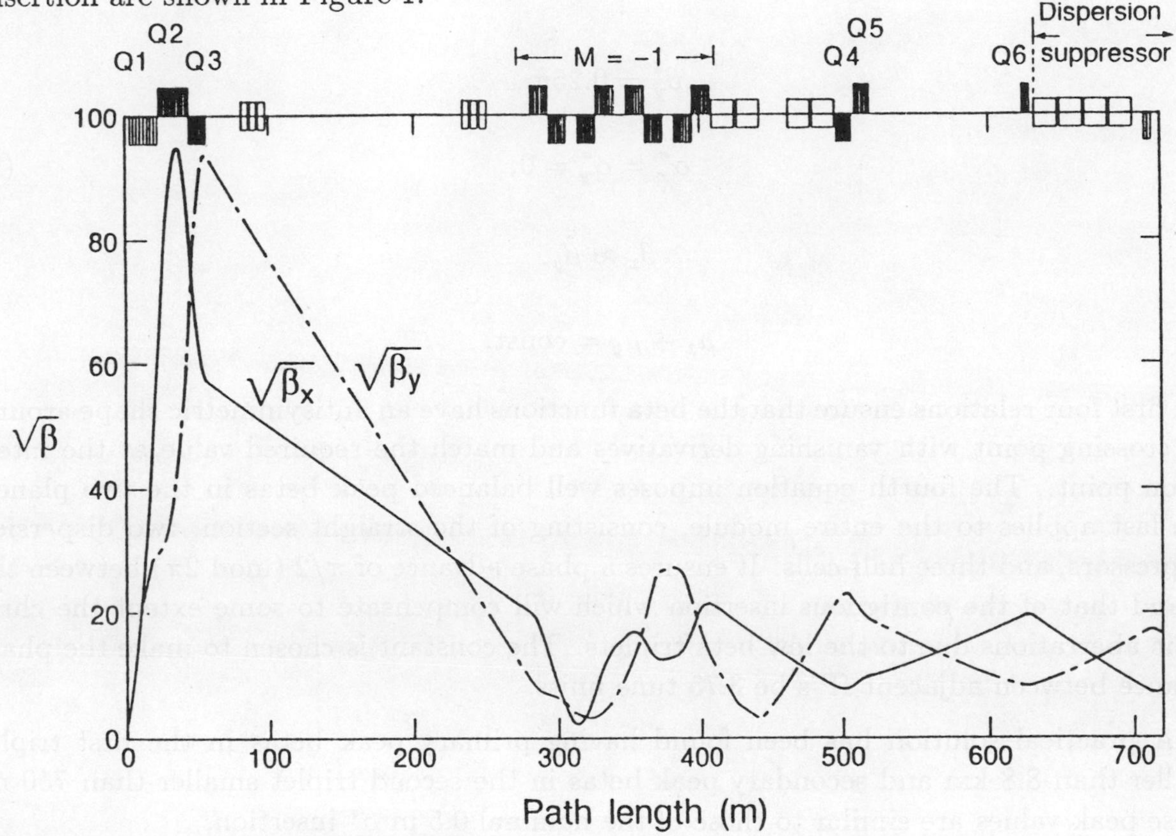

Figure 1 – Right half of the high-luminosity insertion.

Two triplets of quadrupoles separated by an M = -1 matrix vertical dispersion matching section and by the vertical crossing magnets are used on either side of the crossing point. The dispersion suppressor is unchanged. The first triplet starts 3 m from the crossing and is 55.5 m long. Its midpoint is 15.5 m closer to the interaction point, compared with the same triplet of the nominal insertion, which is 20 m from the crossing and 52.5 m long. The second triplet, next to the dispersion suppressor, has the same purpose of tuning the beta function between the injection and collision values and is about the same in the high-luminosity insertion as in the nominal one, 133.3 m compared to 135 m. The free space of ± 3 m available around the interaction point has to to accommodate the experimental apparatus as well as a cryogenic end connections and a set of collimators which will protect the first triplet of superconducting quadrupoles from radiation caused by the beam-beam interactions.

Optical properties

The high-luminosity insertion is required to obey six optical constraints, namely:

$$\begin{aligned} \beta_x^* &= 0.25\text{m}, \\ \beta_y^* &= 0.25\text{m}, \\ \alpha_x^* &= \alpha_y^* = 0, \\ \hat{\beta}_x &\approx \hat{\beta}_y, \\ \mu_x + \mu_y &= \text{const.} \end{aligned} \tag{2}$$

The first four relations ensure that the beta functions have an antisymmetric shape around the crossing point with vanishing derivatives and match the required value at the interaction point. The fourth equation imposes well balanced peak betas in the two planes. The last applies to the entire module, consisting of the straight section, two dispersion suppressors, and three half-cells. It ensures a phase advance of $\pi/2 \pmod{2\pi}$ between the IP and that of the contiguous insertion which will compensate to some extent the chromatic aberrations due to the low beta triplets. The constant is chosen to make the phase advance between adjacent IP's be 3.75 tune units.

A practical solution has been found having primary peak betas in the first triplet smaller than 8.8 km and secondary peak betas in the second triplet smaller than 750 m. These peak values are similar to those of the nominal 0.5 m β^* insertion.

By appropriate tuning of the six low-beta quadrupoles, mainly those in the outer triplet, a smooth path has been found to an injection configuration in which the maximum betas have been greatly reduced. This path resembles that of the nominal low-beta insertion. Along this path the primary peak betas in the first triplet decrease whilst the betas at the crossing point and the secondary peak betas in the second triplet increase. The cross-over point of the primary and the secondary peak beta happens when β^* is 2.826 m and the peak betas are 790 m. Plots of the quadrupole tuning gradients and peak betas as a function of β^* are shown in Figures 2 and 3.

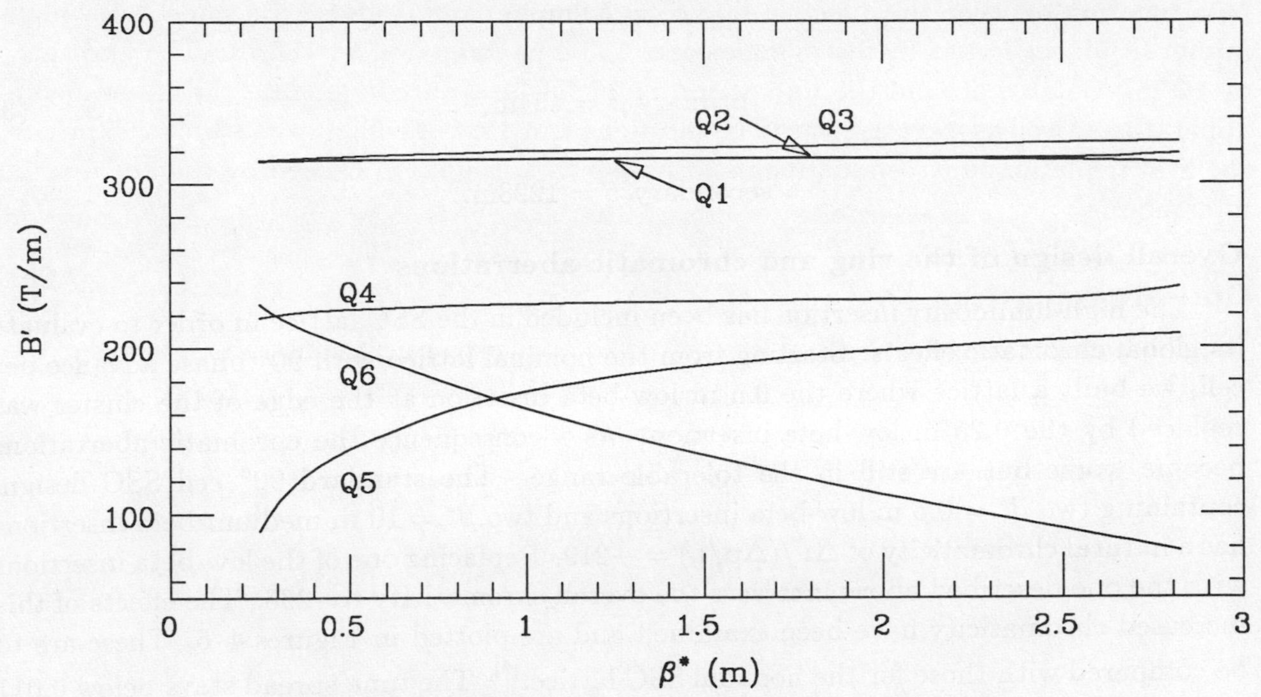

Figure 2 – Tuning curves for the high-luminosity insertion.

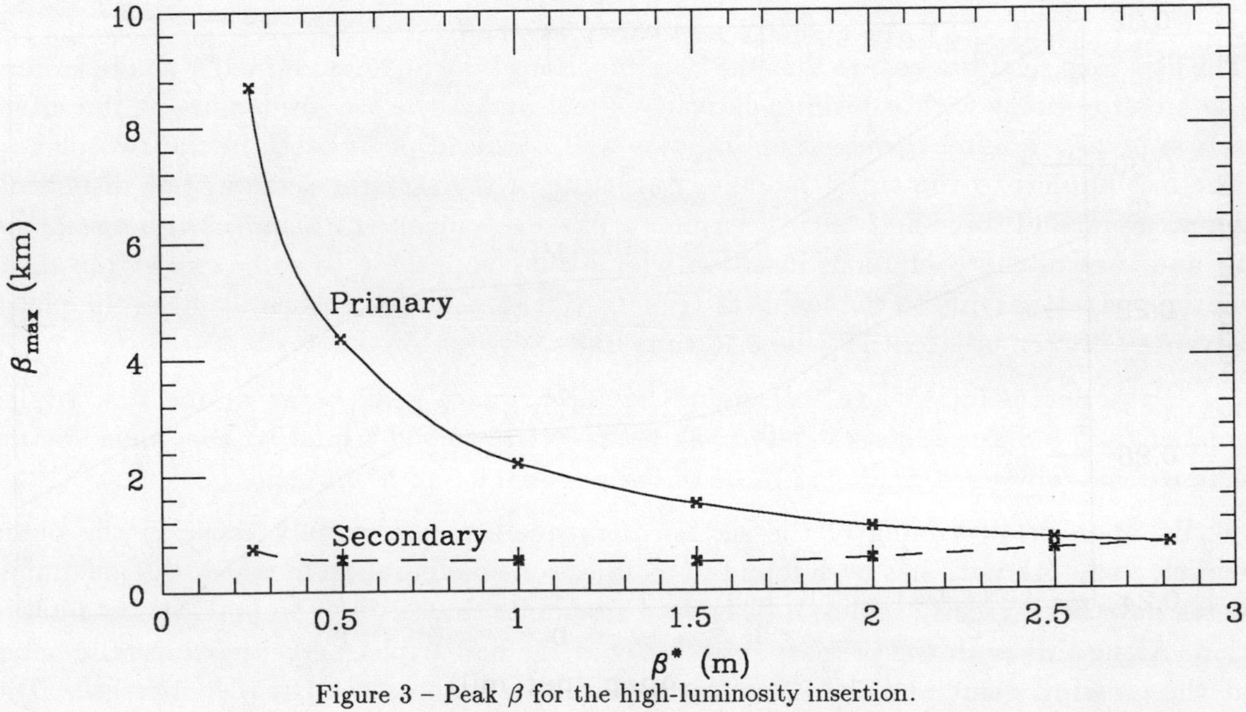

Figure 3 – Peak β for the high-luminosity insertion.

The injection path can be continued to provide configurations with a primary peak beta smaller than the secondary peak beta up to the following values:

$$\beta^* = 5.0\text{m},$$

$$\text{primary } \hat{\beta} = 454\text{m}, \tag{3}$$

$$\text{secondary } \hat{\beta} = 1298\text{m}.$$

Overall design of the ring and chromatic aberrations

The high-luminosity insertion has been included in the SSC lattice in order to evaluate its global chromatic effects. Starting from the nominal lattice with 90° phase advance per cell, we built a lattice where the 0.5 m low-beta insertion at the edge of the cluster was replaced by the 0.25 m low-beta insertion. As a consequence the chromatic aberrations become worse but are still in the tolerable range. The standard 90° cell SSC design, containing two $\beta^* = 0.5$ m low-beta insertions and two $\beta^* = 10$ m medium-beta insertions has a natural chromaticity of $\Delta\nu/(\Delta p/p) = -219$. Replacing one of the low-beta insertions with the one described above increases the overall chromaticity to -235. The effects of this increased chromaticity have been examined and are plotted in Figures 4–6. These are to be compared with those for the nominal SSC lattice.[4] The tune spread stays below 0.011 over a momentum spread range of $\Delta p/p = \pm 0.001$, which may be compared to 0.007 for the standard lattice.

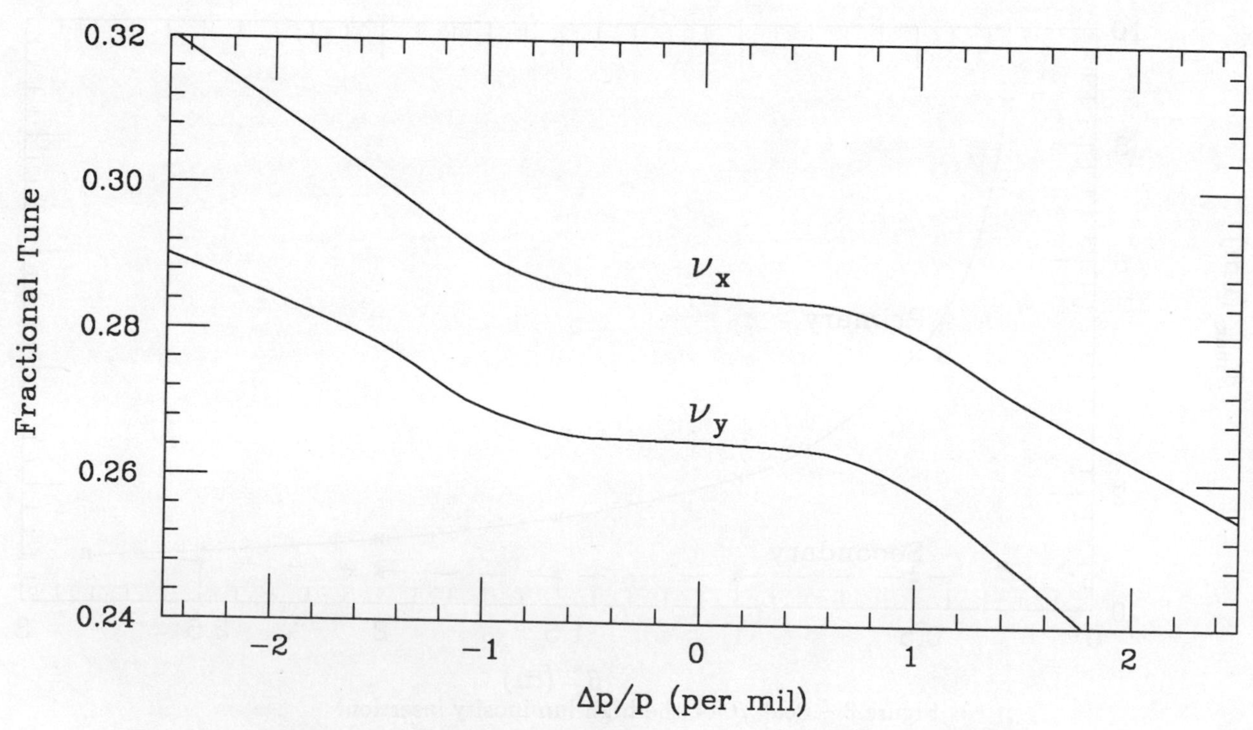

Figure 4 – Tune plot for the SSC with one $\beta^* = 0.25$m insertion.

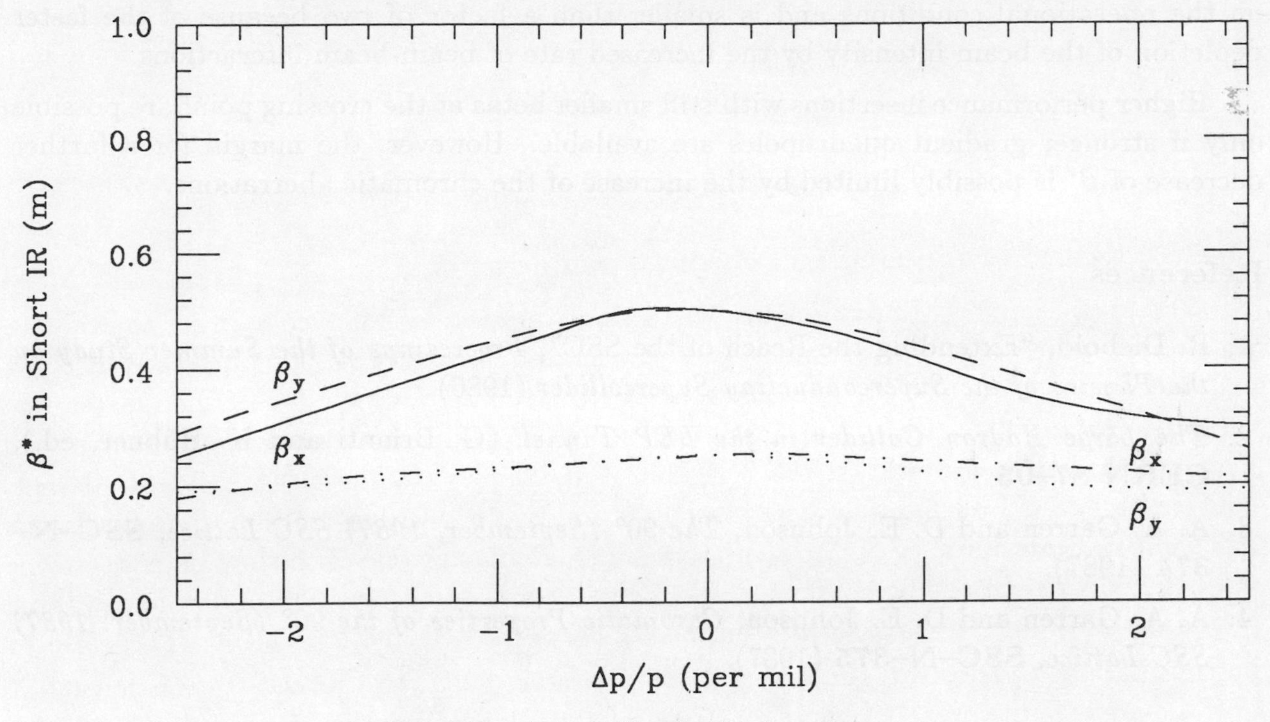

Figure 5 – β^* for 0.25m & 0.5m insertions.

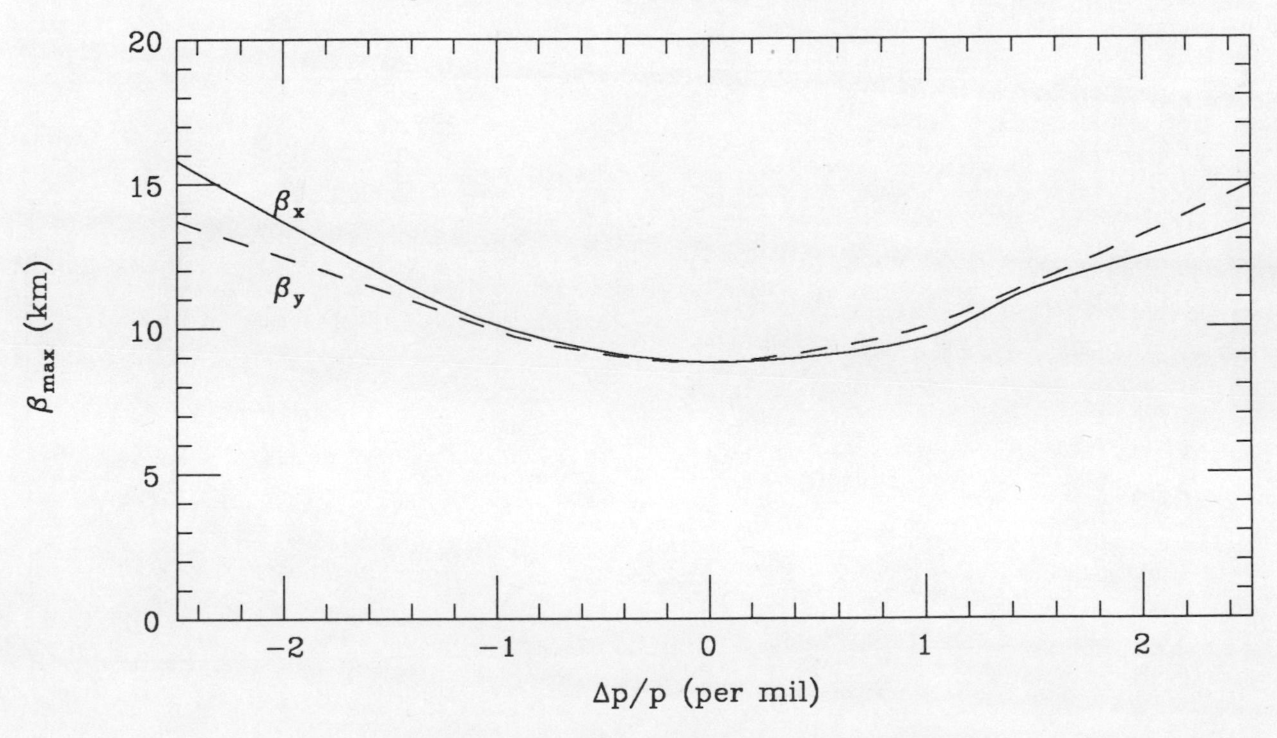

Figure 6 – Peak β for the SSC with one $\beta^* = 0.25$m insertion.

Conclusions

A high-luminosity insertion has been found for the SSC with 90° cells, by which the initial luminosity can be doubled. In terms of average luminosity the effective gain depends on the operational conditions and is smaller than a factor of two because of the faster depletion of the beam intensity by the increased rate of beam-beam interactions.

Higher performance insertions with still smaller betas at the crossing point are possible only if stronger gradient quadrupoles are available. However, the margin for a further decrease of β^* is possibly limited by the increase of the chromatic aberrations.

References

1. R. Diebold, "Extending the Reach of the SSC", *Proceedings of the Summer Study on the Physics of the Superconducting Supercollider* (1986).
2. *The Large Hadron Collider in the LEP Tunnel,* (G. Brianti and K. Hübner, ed.), **CERN 87–05**.
3. A. A. Garren and D. E. Johnson, *The* 90° *(September, 1987) SSC Lattice,* **SSC–N–374** (1987).
4. A. A. Garren and D. E. Johnson, *Chromatic Properties of the* 90° *(September, 1987) SSC Lattice,* **SSC–N–375** (1987).

Intermediate p_T Physics

BOTTOM AND TOP PHYSICS

K. J. FOLEY*
Brookhaven National Laboratory, Upton, Long Island, NY 11973

A. FRIDMAN
ELOISATRON Project and *Institut National de Physique Nucleaire et de Physique des Particules, France*

F. J. GILMAN†
Stanford Linear Accelerator Center, Stanford University, Stanford, CA 94305

G. HERTEN*
Department of Physics and Laboratory for Nuclear Science, Massachusetts Institute of Technology, Cambridge, MA 02139

I. HINCHLIFFE‡
Lawrence Berkeley Laboratory, University of California, Berkeley, CA 94720

A. JAWAHERY◇
Physics Department, Syracuse University, Syracuse, NY 13210

A. SANDA°
Physics Department, Rockefeller University, New York, NY 10021

M. P. SCHMIDT△
Physics Department, Yale University, New Haven, CT 06511

K. R. SCHUBERT
Institut fur Hochenergiephysik, Universitat Heidelberg, Heidelberg, Germany

---

*Work supported by the Department of Energy, contract DE-AC02-76CH00016.

†Work supported by the Department of Energy, contract DE-AC03-76SF00515.

*Work supported in part by the Department of Energy, contract DE-AC02-76ER3069.

‡Work supported by the Department of Energy, contract DE-AC03-76SF00098.

◇Work supported by the National Science Foundation.

°Work supported by the Department of Energy, contract DE-AC02-87ER-40325-Task B.

△Supported by an Alfred P. Sloan Research Fellowship and by the Department of Energy, contract DE-AC02-76ER03075.

ABSTRACT

The production of bottom quarks at the SSC and the formalism and phenomenology of observing CP violation in B meson decays is discussed. The production of a heavy t quark which decays into a real W boson, and what we might learn from its decays is examined.

1. Introduction

The physics of heavy flavors promises to be an important component of SSC physics and has been studied intensively in the Snowmass Workshops[1,2] in 1984 and 1986. The production of heavy quark flavors occurs primarily by the strong interactions and offers another arena in which to test QCD and to probe gluon distributions at very small values of x. Such quarks can also be produced as decay products of possible new, yet undiscovered particles, *e.g.*, Higgs bosons, and therefore are a necessary key to reconstructing such particles. The decay products of heavy quarks, especially from their semileptonic decays, can themselves form a background to other new physics processes. Perhaps most important of all, in their rare decays and in CP violating asymmetries formed by studying their weak decays, particles containing heavy quarks can give us further insight into the boundary between what is to be understood inside the standard model and what physics lies beyond.

This is particularly the case with respect to the B meson system. It has been apparent for some time that rare decays and CP violation are especially important to pursue[3] at the SSC where the number of produced B mesons will exceed by many orders of magnitude those produced at any existing or planned colliding beam (but not fixed target) facility.At Snowmass 86, a major development which increased the optimism about doing B physics at the SSC, and that drove much of the discussion in this area, was the success of silicon strip vertex detectors in cleanly extracting charm physics in fixed target experiments. During this past year there was another major development which makes the difficult measurement of a CP asymmetry in B decays much easier (although still very difficult)—the observation of large $B_d^\circ - \overline{B}_d^\circ$ mixing by the ARGUS collaboration.[4]

In the next Section we review the production of b quarks at the SSC, which has an interest of its own from the perspective of QCD. It is also a prerequisite to the study of CP violating asymmetries in B decays, a subject taken up in Section 3. There we lay out the various classes of CP violating asymmetries and re-evaluate the expectations for their magnitude (and hence the number of B mesons needed to establish a statistically significant effect) in the light of the ARGUS measurement.[4] Section 4 takes up the production of t quarks in the now not unlikely case that M_t is comparable to, or larger than M_W. Finally, in Section 5 some of the physics to be explored in t quark decays is considered.

2. The Production of b Quarks at the SSC

The total cross-section for the production of bottom quarks is controlled by the parton processes $gg \to b\bar{b}$ and $q\bar{q} \to b\bar{b}$. The former process dominates at SSC energies. In order to calculate the rate we must know; the quark and gluon structure functions at the appropriate values of x and Q^2; the bottom quark mass (m_b); and the partonic cross sections which are presently known only in lowest order,[5] *i.e.*, α_s^2. The values of the momentum fractions x_1 and x_2 of the incoming partons are given by,

$$x_1 x_2 > \frac{4m_b^2}{s} \quad , \tag{1}$$

where s is the center-of-mass energy squared of the proton-proton system. Since the structure functions are rapidly falling functions of x, the dominant regions of x_1 and x_2 occur where they are equal and of order 2.5×10^{-4}. This raises an immediate problem since the quark distributions are measured only for $x > 0.015$, and the gluon distribution, which must be inferred from the Q^2 dependence of the antiquarks, is known even less well.[6]

A conventional assumption is that $xf(x, Q_0^2)$, for either gluons or sea quarks, tends to a constant as $x \to 0$. Here Q_0 is some fixed scale at which measurements are made. Such an assumption is not consistent with perturbative QCD. As Q^2 is increased, $xf(x, Q^2)$ takes the following asymptotic form:[7]

$$x\ f(x, Q^2) \sim \exp\left\{\sqrt{\frac{144}{33-2n_f}\ log\left(\frac{1}{x}\right)\ log[log(Q^2)]}\right\} \tag{2}$$

at small x. Here n_f is the number of light quark flavors. Collins[8] has argued that a better form at Q_0^2 is obtained by assuming that $x\ f(x, Q_0^2) \sim x^{-1/2}$, which is more stable with respect to the Q^2 evolution. Figure 1 shows the difference in these two assumptions at $Q_0^2 = 5\ \text{GeV}^2$. The relevant Q^2 for the total bottom quark rate is of order $4m_b^2$, the minimum value of the center-of-mass energy squared in the parton system. Figure 1 also shows how the two different starting distributions have evolved at $Q^2 = 100\ \text{GeV}^2$. Notice that the large differences at $Q^2 = 5\ \text{GeV}^2$ have washed out to some extent. The differences at small x can be probed either at HERA or by measuring Drell–Yan dilepton production at low invariant mass at the Tevatron collider,[9] and should be resolved before the SSC starts running.

The total cross section for bottom pair production is shown in Figure 2, where a bottom quark mass of 4.9 GeV has been assumed. Two curves which are shown reflect the choices in Figure 1 for the distribution functions. It can be seen that the differences are larger at SSC than at the Tevatron since the values of x are smaller. Until we have better data on structure functions we cannot know where the true answer lies. However the upper curve is likely to be nearer the truth than the lower one.

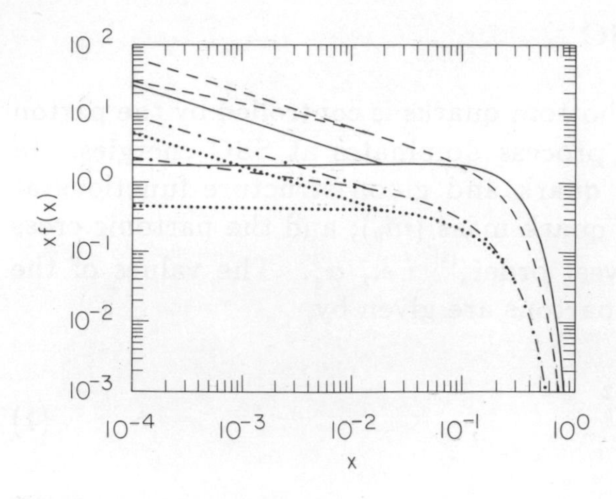

Fig. 1. The x dependence of $xf(x, Q^2)$ for gluons (dashed and solid lines) and antiquarks (dotted and dot-dashed lines) at $Q^2 = 5$ GeV^2 (solid and dotted lines) and $Q^2 = 100$ GeV^2 (dashed and dot-dashed lines). In each case two lines are shown, with the upper one corresponding to the choice $xf(x, 5) \sim x^{-1/2}$ and the lower one to $xf(x, 5) \sim const$ for $x < 0.02$.

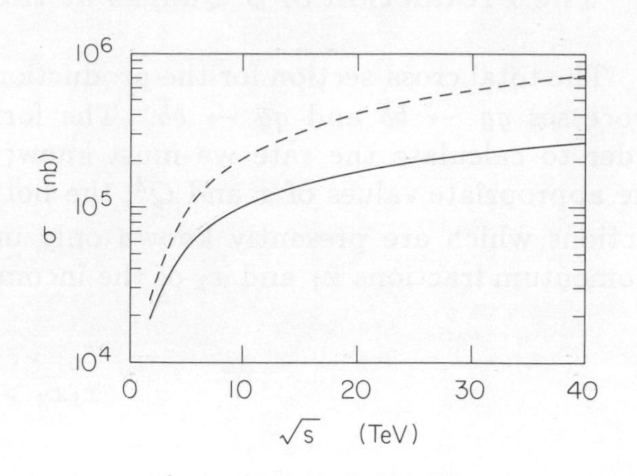

Fig. 2. The total cross section for $pp \to b\bar{b} + X$ as a function of $\sqrt{s}$.

Since the cross section varies roughly as m_b^{-3} over the relevant range, there is an uncertainty of order 50% associated with the choice of a bottom quark mass. The scale Q which appears in α_s (the partonic cross section is proportional to α_s^2) and in the structure functions, is unknown. It should be of order $4m_b^2$, but a better determination is possible only after the order α_s^3 contributions to the partonic cross section have been computed.[10] A bad choice of scale will result in large corrections.

The rapidity distribution of the produced b quarks is flat out to rapidities of ~ 4, as shown in Figure 3. It should be noted that the produced b and $\bar{b}$ are close in rapidity. This is because the rapidity separation is related to the invariant mass, $\sqrt{x_1 x_2 s}$, of the $b\bar{b}$ system, which likes to be small since the parton distributions fall rapidly with x. The transverse momentum distribution is shown in Figure 4. Notice that it is dominated by values of p_T of order m_b. The effect of the uncertainty in the structure functions is much less at large p_T since the relevant values of x are larger.

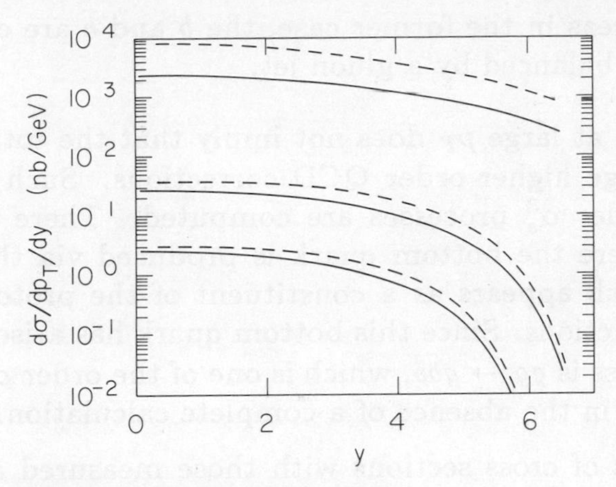

Fig. 3. The cross section $d\sigma/dp_T\ dy$ for the production of a b quark in pp collisions at $\sqrt{s} = 40$ TeV as a function of rapidity y at $p_T = 5$, 30, and 55 GeV. The solid (dashed) line corresponds to the choice $xf(x,5) \sim$ const. $(xf(x,5) \sim x^{-1/2})$. The scale Q^2 appearing in α_s and $f(x,Q^2)$ has been set to $4m_b^2 + p_T^2$.

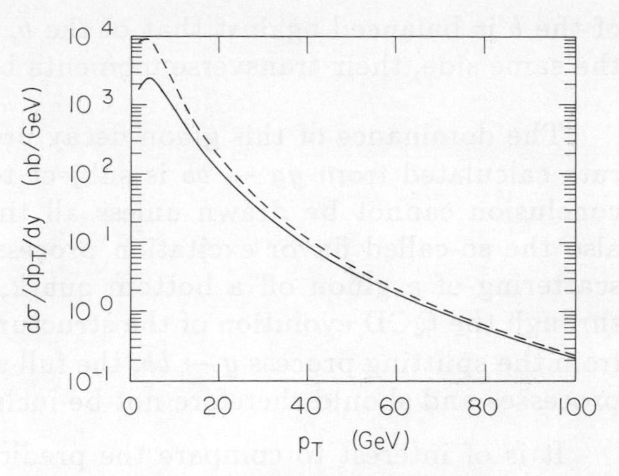

Fig. 4. The cross section $d\sigma/dp_T\ dy$ for the production of a b quark in pp collisions at $\sqrt{x}$=40 TeV as a function of p_T at $y =$ 0. The solid (dashed) line corresponds to the choice $xf(x,5) \sim$ const. $(xf(x,5) \sim x^{-1/2})$. The scale Q^2 appearing in α_s and $f(x,Q^2)$ has been set to $4m_b^2 + p_T^2$.

If one is interested in the production of b's with a transverse momentum much larger than their mass, then other partonic processes can become important. A gluon produced at large transverse momentum can "decay" into a quark anti-quark pair.[11] This decay probability can be computed in the leading logarithm approximation, with the result shown in Figure 5. Since the production of a gluon jet at large p_T occurs with a rate which is of order 200 times[12] that for the direct production of a b via $gg \to b\bar{b}$, this mechanism will dominate at large p_T. The rate is much larger than that shown in Figure 4 at very large p_T. Notice that the kinematics of the gluon decay process is quite different from that of $gg \to b\bar{b}$; in the latter case the transverse momentum

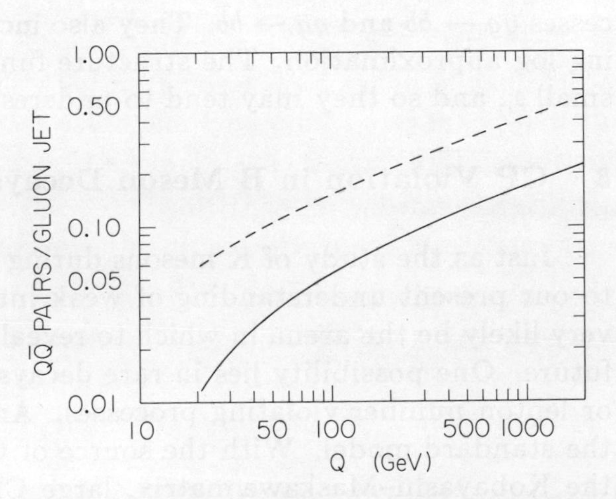

Fig. 5. The average number of heavy quark pairs resulting from the fragmentation of a gluon jet as a function of the off-shellness, Q, of the gluon jet. For jets produced in pp collisions, Q is of order p_T. (Solid: $b\bar{b}$, dashed: $c\bar{c}$.)

of the b is balanced against that of the $\bar{b}$, whereas in the former case, the b and $\bar{b}$ are on the same side, their transverse momenta being balanced by a gluon jet.

The dominance of this gluon decay process at large p_T does not imply that the total rate calculated from $gg \to b\bar{b}$ is subject to large higher order QCD corrections. Such a conclusion cannot be drawn unless all the order α_s^3 processes are computed. There is also the so-called flavor excitation process where the bottom quark is produced via the scattering of a gluon off a bottom quark, which appears as a constituent of the proton through the QCD evolution of the structure functions. Since this bottom quark has arisen from the splitting process $g \to b\bar{b}$, the full process is $gg \to gb\bar{b}$, which is one of the order α_s^3 processes and should therefore not be included in the absence of a complete calculation.

It is of interest to compare the predictions of cross sections with those measured at current energies. The rates predicted for charm production tend to be too low, unless a very small value (of order 1.1 GeV) is used for the charm quark mass. However, charm is not very heavy and the validity of QCD for such small scales is in doubt. There is a measurement of the $b\bar{b}$ rate in $p\bar{p}$ collisions at 630 GeV by the UA1 collaboration. The value[13] of $1.1 \pm .1 \pm .4$ μb for $p_T > 5$ GeV and $|y| < 2$ is in very good agreement with the expected cross section. There is also a measurement[14] in a pion beam at CERN. Here the theoretical uncertainties are larger since the pion structure function is not well known. Nevertheless, the observed value is in reasonable agreement with expectations.[15]

The Monte-Carlo event generators ISAJET[16] and PYTHIA[17] have the parton processes $gg \to b\bar{b}$ and $q\bar{q} \to b\bar{b}$. They also include the gluon "decay" contribution in the leading log approximation. The structure functions used are of the type $xf(x,5) \sim$ const. at small x, and so they may tend to underestimate the total rate.

3. CP Violation in B Meson Decays

Just as the study of K mesons during more than 30 years has contributed enormously to our present understanding of weak interactions, we believe that B meson decays will very likely be the arena in which to reveal various new weak interaction phenomena in the future. One possibility lies in rare decays—for example, flavor changing neutral currents or lepton number violating processes. Another, less revolutionary possibility, lies within the standard model. With the source of CP violation in this case being a single phase in the Kobayashi–Maskawa matrix, large CP violating asymmetries of ~ 10 to $\sim 30\%$ are predicted in certain decays of B mesons.

Thus, if we can attain the requisite sensitivity, we are faced with a situation in which, at the very least, we verify the dramatic predictions of the standard model. At the most, we could discover more interesting phenomena which point beyond the standard model. With an integrated luminosity of 10^{40} cm^{-2}, several times 10^{12} $b\bar{b}$ pairs are produced at the SSC, so that even with reduced luminosity or a moderate acceptance the 'raw' data rate to attempt this kind of experiment is there. The study of B physics at SSC is guaranteed to be very interesting.

Mixing and CP Violation

Mixing describes a situation where the mass eigenstates are a coherent superposition of a particle and antiparticle.[18,19] The time evolution of a meson that was produced as a $B^\circ(\bar{b}d)$ or $\bar{B}^\circ(b\bar{d})$ meson at time t = 0 is given by

$$\begin{aligned} |B^\circ(t)> &= g_+(t)|B^\circ> + \frac{q}{p}g_-(t)|\bar{B}^\circ> \\ |\bar{B}^\circ(t)> &= \frac{p}{q}g_-(t)|B^\circ> + g_+(t)|\bar{B}^\circ> \quad , \end{aligned} \tag{3}$$

where

$$g_\pm(t) = \exp\{-\frac{1}{2}\Gamma_1 t\}\ \exp\{im_1 t\}\Big(1 \pm\ \exp\{-\frac{1}{2}\Delta\Gamma t\}\ \exp\{i\Delta m t\}\Big)/2$$

and we have made the definitions:

$$\Delta\Gamma = \Gamma_2 - \Gamma_1; \quad \Delta m = m_2 - m_1; \quad \frac{q}{p} = \frac{1-\epsilon_B}{1+\epsilon_B} \quad . \tag{4}$$

Γ_i and m_i, $i = 1,2$ are the width and mass, respectively, of the two mass eigenstates B_i. In the following we set $\Delta\Gamma = 0$ for convenience; for B° mesons, one computes $\Delta\Gamma << \Gamma$ with considerable confidence, as we see below.

Using B mesons one can study rare decays, mixing, and CP violation, all of which will critically test the standard model, aside from the possibility of surprises which will point to new physics. Below we shall concentrate on CP violation and give short explanations for six classes of CP asymmetries in B decays which are relevant for SSC physics.

- Class I—The Charge Asymmetry in Same Sign Dileptons

The charge asymmetry in $B^\circ\bar{B}^\circ \rightarrow \ell^\pm\ell^\pm + X$ is given by[20]

$$\frac{\sigma(B^\circ\bar{B}^\circ \rightarrow \ell^+\ell^+ + X) - \sigma(B^\circ\bar{B}^\circ \rightarrow \ell^-\ell^- + X)}{\sigma(B^\circ\bar{B}^\circ \rightarrow \ell^+\ell^+ + X) + \sigma(B^\circ\bar{B}^\circ \rightarrow \ell^-\ell^- + X)} = \frac{|\frac{p}{q}|^2 - |\frac{q}{p}|^2}{|\frac{p}{q}|^2 + |\frac{q}{p}|^2}$$

$$= \frac{\mathrm{Im}(\Gamma_{12}/\mathrm{M}_{12})}{1 + \frac{1}{4}|\Gamma_{12}/M_{12}|^2} \tag{5}$$

where we define $< B^\circ|H|\bar{B}^\circ >= M_{12} + \frac{i}{2}\Gamma_{12}$. In order to estimate the size of this asymmetry consider

$$\left|\frac{\Gamma_{12}}{M_{12}}\right| = \left|\frac{\Gamma_{12}}{\Gamma}\right| \left|\frac{\Gamma}{M_{12}}\right| = \frac{2}{x}\left|\frac{\Gamma_{12}}{\Gamma}\right| , \tag{6}$$

where we used the definition

$$x = \frac{\Delta m}{\Gamma} = \frac{2|M_{12}|}{\Gamma} .$$

Now, to estimate Γ_{12}, note that it gets contributions from B° decay channels which are common to both B° and $\bar{B}^\circ$. For example, one may consider a Cabibbo-suppressed decay channel $B \to D\bar{D} +$ pions $\to \bar{B}$. If this is the only relevant process one might guess

$$\left|\frac{\Gamma_{12}}{\Gamma}\right| \sim 10^{-3} . \tag{7}$$

Putting these numbers into Eq. (6) with $x = 0.78$, the central value of ARGUS,[4] we obtain

$$\left|\frac{\Gamma_{12}}{M_{12}}\right| \sim 2 \times 10^{-3} . \tag{8}$$

Inevitably there are other decay channels which contribute, and there generally will be cancellations among those channels. Nevertheless, in the standard model the CP asymmetry in Eq. (5) should not exceed 10^{-2}, but at the other extreme, we can not be sure it is less than 10^{-3}.

- Class II—Mixing With Decay to a CP Eigenstate

  Since there is substantial $B^\circ - \bar{B}^\circ$ mixing, one can consider two decay chains:

$$\begin{matrix} B^\circ \to B^\circ \searrow \\ \qquad\qquad f \\ B^\circ \to \bar{B}^\circ \nearrow \end{matrix} ,$$

where f is a CP eigenstate. The amplitudes for these decay chains can interfere and generate nonzero asymmetries between $\Gamma(B^\circ(t) \to f)$ and $\Gamma(\bar{B}^\circ(t) \to f)$.

An analysis using Eq. (3), which parallels that for the K meson system, gives

$$\Gamma(B^0(t)\to f)\sim e^{-\Gamma t}\Big\{\Big(1+\cos[\Delta m\, t]\Big)|\rho|^2 + \Big(1-\cos[\Delta m\, t]\Big)|\frac{q}{p}|^2 + 2\sin[\Delta m\, t]\ \mathrm{Im}(\frac{q}{p}\rho^*)\Big\} \tag{9a}$$

$$\Gamma(\bar{B}^\circ(t)\to f)\sim e^{-\Gamma t}\Big\{\Big(1+\cos[\Delta m\, t]\Big) + \Big(1-\cos[\Delta m\, t]\Big)|\frac{p}{q}\rho|^2 + 2\sin[\Delta m\, t]\ \mathrm{Im}(\frac{p}{q}\rho)\Big\} \tag{9b}$$

Here $\rho = A(B\to f)/A(\bar{B}\to f)$. Noting that $|\frac{p}{q}|\sim 1+\frac{1}{2}\mathrm{Im}\frac{\Gamma_{12}}{M_{12}}$ and $|\mathrm{Im}\frac{\Gamma_{12}}{M_{12}}|\ll 1$, we can set $|\frac{p}{q}|=1$ to a very good approximation. Using CPT to set $|\rho|=1$, Eq. (9) simplifies to:

$$\Gamma(\bar{B}^\circ(t)\text{ or }B^\circ(t)\to f)\sim e^{-\Gamma t}\Big\{1\pm\sin[\Delta m\, t]\mathrm{Im}(\frac{p}{q}\rho)\Big\}\quad , \tag{10}$$

The study of CP violation in these modes requires information on the identity of the B, i.e., whether it is a B° or $\bar{B}^\circ$ at $t=0$. Since b and $\bar{b}$ quarks are pair produced at SSC, such information can be obtained by tagging the other particle as to its b or $\bar{b}$ content. The observable asymmetry in the case where a B^+ or B^- is used as the "tag" is

$$\frac{\Gamma(\bar{B}^\circ(t)\to f)-\Gamma(B^\circ(t)\to f)}{\Gamma(B^\circ(t)\to f)+\Gamma(\bar{B}^\circ(t)\to f)} = \sin[\Delta m\, t]\ \mathrm{Im}(\frac{p}{q}\rho). \tag{11}$$

If the "tag" is also a neutral B which can oscillate, the situation is slightly more complicated and oscillation of both B° and $\bar{B}^\circ$ must be taken into account. With a common final state f and a semileptonic tag of the associated neutral B, the decays of a $B\bar{B}$ pair in a coherent state of given charge conjugation are,

$$BR\Big(B(t)\bar{B}(\bar{t})\,|_{C=\mp1}\to f+(D\ell\bar{\nu}X)_{\mathrm{tag}}\Big) \propto e^{-\Gamma(t+\bar{t})}\{1-\sin[\Delta m\,(t\mp\bar{t})]\ \mathrm{Im}(\frac{p}{q}\rho)\} \tag{12a}$$

$$BR\Big(\bar{B}(t)B(\bar{t})\,|_{C=\mp1}\to f+(\bar{D}\bar{\ell}\nu\bar{X})_{\mathrm{tag}}\Big) \propto e^{-\Gamma(t+\bar{t})}\{1+\sin[\Delta m\,(t\mp\bar{t})]\ \mathrm{Im}(\frac{p}{q}\rho)\}\quad . \tag{12b}$$

Note that for $C=-1$, *i.e.*, $B\bar{B}$ in an odd relative angular momentum state, the potential asymmetry vanishes if the times t and $\bar{t}$ are treated symmetrically. For example, when

$C=-1$ Eqs. (12a) and (12b) become identical when $t=\bar{t}$. Furthermore, they become equal when the rates are integrated over time. This tends to reduce the value of the observable asymmetry slightly, but for the present purposes we shall ignore this effect (from production of $B^\circ\bar{B}^\circ$ pairs in a $C=-1$ state). Then the process where one B decays to a CP eigenstate and the other B is tagged gives an observable asymmetry which can be written as

$$\frac{\Gamma(B\bar{B}\to f+\bar{\ell}_{\rm tag})-\Gamma(B\bar{B}\to f+\ell_{\rm tag})}{\Gamma(B\bar{B}\to f+\bar{\ell}_{\rm tag})+\Gamma(B\bar{B}\to f+\ell_{\rm tag})}=\sin[\Delta m\ t]\ {\rm Im}(\frac{\rm p}{\rm q}\rho)\quad . \tag{13}$$

For example, possibilities for the final state f are ψK_s, $\psi K_s X$, $\psi\pi^+\pi^-$, $D\bar{D}K_s$, $\pi^+\pi^-$, D^+D^-, and $D^{0*}\bar{D}^\circ$ + c. c. . Obviously, the "lepton tag" shown above can be replaced by any decay mode which identifies the particle or antiparticle nature of the associated B.

We stress that *in this class of asymmetries the quantities* ${\rm Im}(\frac{\rm p}{\rm q}\rho)$ *can be predicted from the Kobayashi–Maskawa matrix.* The hadron dynamics cancel in the ratio of amplitudes ρ. Since $|\frac{p}{q}|\approx 1$, as described above, and for a CP eigenstate $|\rho|=1$, we can write

$$\frac{p}{q}\rho_f=e^{i\phi_q}\quad . \tag{14}$$

For a B_q° meson, it can be shown that the phase on the right-hand-side of Eq. (14) is

$$\phi_q=2\arg\left([U_{tb}\,U^*_{tq}]\begin{bmatrix}U^*_{cb}\,U_{cs}\\ U^*_{cb}\,U_{cd}\\ U^*_{ub}\,U_{ud}\end{bmatrix}\right)\quad\text{for}\quad\begin{matrix}b\bar{q}\to c\bar{c}s\bar{q}\\ b\bar{q}\to c\bar{c}d\bar{q}\\ b\bar{q}\to u\bar{u}d\bar{q}\end{matrix}\quad , \tag{15}$$

giving an explicit formulation in terms of KM matrix elements. In the Wolfenstein parameterization[21] of the KM matrix

$$U_{KM}=\begin{pmatrix}1-\lambda^2/2 & \lambda & \lambda^3A(\rho-i\eta)\\ -\lambda & 1-\lambda^2/2 & \lambda^2A\\ \lambda^3A(1-\rho-i\eta) & -\lambda^2A & 1\end{pmatrix}\quad . \tag{16}$$

An explicit example is provided by the process $b\bar{d}\to c\bar{c}s\bar{d}$, where substituting the Wolfenstein parametrization into Eq. (15) gives[21]

$${\rm Im}(\frac{\rm p}{\rm q}\ \rho)={\rm Im}(e^{i\phi_d})=\frac{2(1-\rho)\eta}{(1-\rho)^2+\eta^2}\quad . \tag{17}$$

Numerically, $\lambda=\sin\theta_C=.22$, $|U_{cb}|=.045\pm.008$ implies $A=.93\pm.17$, and $|U_{ub}/U_{cb}|<.19$ implies $\rho^2+\eta^2\lesssim.75$. The value of η is constrained by the measurement of ϵ, expressing

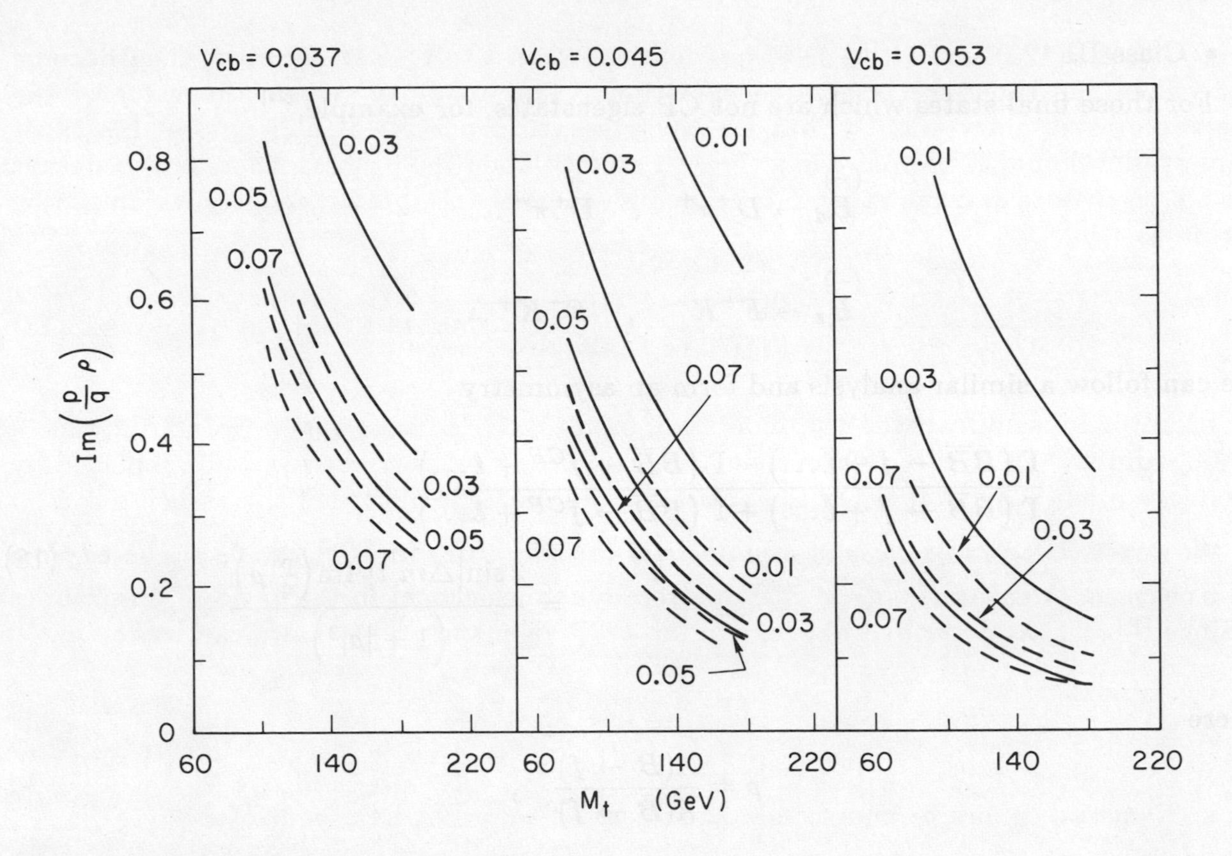

Fig. 6. Values of $\mathrm{Im}\left(\frac{\mathrm{p}}{\mathrm{q}}\rho\right)$ as a function of the mass of the top quark, M_t, for various values of the Kobayashi–Maskawa matrix element, U_{cb}, with dashed curves corresponding to the process $b\bar{q} \to c\bar{c}s\bar{q}$ and solid curves corresponding to the process $b\bar{q} \to u\bar{u}d\bar{q}$. The parameter B_K is taken to be unity, and the curves are labelled by the ratio of the rates, $\Gamma(b \to u)/\Gamma(b \to c)$.

the strength of CP violation in the K meson system. For given values of U_{cb}, of $\rho^2+\eta^2$, *i.e.*, fixed $\Gamma(b \to u)/\Gamma(b \to c)$, of M_t, and of the parameter B_K, which is equal to unity when the matrix element of the $\Delta S = 2$ operator giving rise to the $K_L - K_S$ mass difference has its vacuum insertion value, the constraint coming from ϵ gives two possible values for ρ and η. For those values of ρ and η, we compute Δm_B for the $B_d - \bar{B}_d$ system and choose the solution which gives a larger value of Δm_B. Finally, for the values of ρ and η which satisfy the above criteria, we compute $\mathrm{Im}(\frac{\mathrm{p}}{\mathrm{q}}\ \rho)$, which is shown in Figure 6. The alternative choice of ρ and η, which gives a smaller value of Δm_B, leads to larger values of $\mathrm{Im}(\mathrm{p}/\mathrm{q})\rho$. In what follows we shall take

$$0.1 < \mathrm{Im}\left(\frac{\mathrm{p}}{\mathrm{q}}\rho_{\mathrm{f}}\right) < 0.6$$

- Class III

For those final states which are not CP eigenstates, for example,[22]

$$\overset{(-)}{B}_d \to D^-\pi^+ \quad , \quad D^+\pi^- \ldots$$

$$\overset{(-)}{B}_s \to F^+K^- \quad , \quad F^-K^+ \ldots \quad ,$$

one can follow a similar analysis and form an asymmetry

$$\frac{\Gamma\left(B\overline{B} \to f + \bar{\ell} \ldots\right) - \Gamma\left(B\overline{B} \to f^{CP} + \ell \ldots\right)}{\Gamma\left(B\overline{B} \to f + \bar{\ell} \ldots\right) + \Gamma\left(B\overline{B} \to f^{CP} + \ell \ldots\right)} = \frac{2\sin[\triangle m\, t]\ \mathrm{Im}\left(\frac{p}{q}\,\rho\right)}{\left(1 + |\rho|^2\right)} \tag{18}$$

where

$$\rho = \frac{A(B \to f)}{A(\bar{B} \to f)} \quad ,$$

as before, but now ρ depends on hadron dynamics. For example, using a factorization ansatz to compute the hadronic matrix element,

$$\rho_f = \frac{A\left(B_d \to D^+\pi^-\right)}{A\left(\bar{B}_d \to D^+\pi^-\right)} \approx \frac{U^*_{ub}U_{cd}f_D m_B^4}{U_{cb}U^*_{ud}f_\pi\left(m_B^2 - m_D^2\right)^2} \sim (1-2)\lambda^2(\rho + i\eta) \quad , \tag{19}$$

where our ignorance concerning the correct value for f_D/f_π is the source of the uncertainty in the coefficient of $\lambda^2(\rho + i\eta)$. In addition, there are further uncertainties concerning the validity of the factorization ansatz, etc. In any case, the ratio ρ is not a unit vector in the complex plane, nor then is $\frac{p}{q}\,\rho$ when f is not a CP eigenstate, and in general considerable uncertainties surface. An analogous procedure for B_s decays yields with comparable uncertainties,

$$\rho_f = \frac{A\left(B_s \to F^+K^-\right)}{A\left(\bar{B}_s \to F^+K^-\right)} = \frac{U^*_{ub}U_{cs}}{U_{cb}U^*_{us}}\frac{f_F}{f_K}\frac{m_{B_s}^4}{\left(m_{B_s}^2 - m_F^2\right)^2} \sim 1.6\,(\rho + i\eta) \quad . \tag{20}$$

• Class IV

The two cascade decays shown in Figure 7 lead to the same final states

$$B^- \rightarrow D^\circ s\bar{u}$$
$$\quad \hookrightarrow K_s^\circ u\bar{u}$$

$$B^- \rightarrow \bar{D}^\circ s\bar{u}$$
$$\quad \hookrightarrow K_s^\circ u\bar{u}$$

and therefore their amplitudes can interfere. From the quark diagrams in Figure 7, one reads off the Kobayashi–Maskawa matrix element factors which are the coefficients of the two amplitudes,

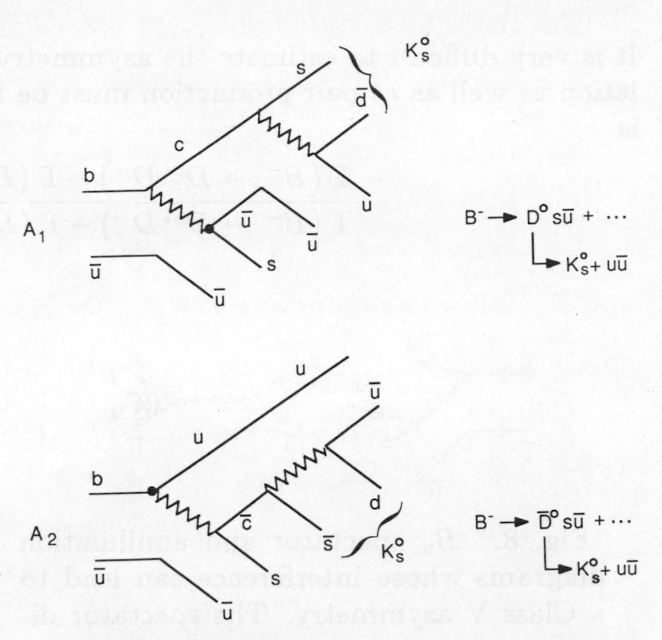

Fig. 7. B_u decay diagrams whose interference can lead to a Class IV asymmetry.

$$A\left(B^- \rightarrow K_s^\circ u\bar{u}s\bar{u}\right) \propto U_{cs}^* U_{cb} U_{us}^* U_{ud}\; A_1 + U_{ub} U_{cs}^* U_{cs} U_{ud}^*\; A_2 \quad . \tag{21}$$

Then the asymmetry can be obtained from

$$\begin{aligned} \Gamma\left(B^\pm \rightarrow (K_s + X)_D + Y\right) \propto \Bigg\{ 1 + \frac{|U_{ub}|^2 |U_{cs}|^2}{|U_{cb}|^2 |U_{us}|^2} |\bar{\rho}|^2 \\ +2Re\left(\frac{U_{ub} U_{cs} U_{ud}^*}{U_{cb} U_{us} U_{ud}}\right) Re\bar{\rho} \pm Im\left(\frac{U_{ub} U_{cs} U_{ud}^*}{U_{cb} U_{us} U_{ud}}\right) Im\bar{\rho} \Bigg\} \end{aligned} \tag{22}$$

where $\bar{\rho} = A_2/A_1$. Im$\bar{\rho}$ where $\bar{\rho} = A_2/A_1$. Im$\bar{\rho}$ comes from the phase shift difference between scattering of $D + (s\bar{u})$ states and $\bar{D} + (s\bar{u})$ states. Since the isospin structures of these states are different, we expect nonvanishing values of Im$\bar{\rho}$. Again, at least compared with a Class II type of asymmetry, the theoretical prediction is quite uncertain.

• Class V

B decays can also receive contributions from quark decay and from weak annihilation diagrams,[23] as shown in Figure 8. They can contribute coherently to modes such as

$$B_u \to D^{\circ *} D^- \quad . \tag{23}$$

It is very difficult to estimate the asymmetry for this process. The effect of weak annihilation as well as $c\bar{c}$ pair production must be included in an estimate. An optimistic guess is

$$\frac{\Gamma\left(B^- \to D^{\circ *} D^-\right) - \Gamma\left(B^+ \to \bar{D}^{\circ *} D^+\right)}{\Gamma\left(B^- \to D^{\circ *} D^-\right) + \Gamma\left(B^+ \to \bar{D}^{\circ *} D^+\right)} \sim 10\% \quad . \tag{24}$$

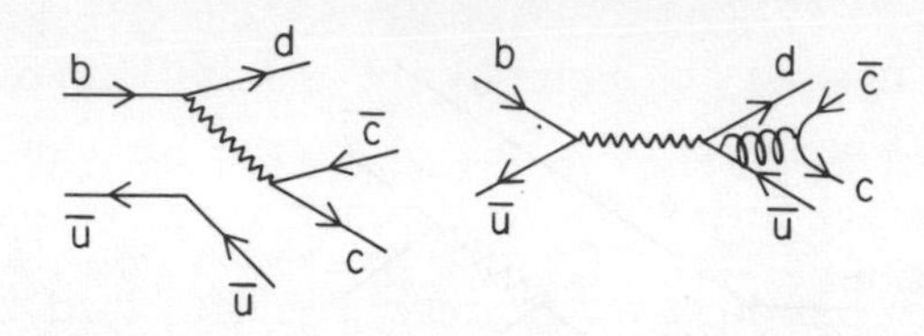

Fig. 8. B_u spectator and annihilation diagrams whose interference can lead to a Class V asymmetry. The spectator diagram is shown in a Fierz transformed manner to indicate more directly the identity of the final state quarks with those from the annihilation diagram.

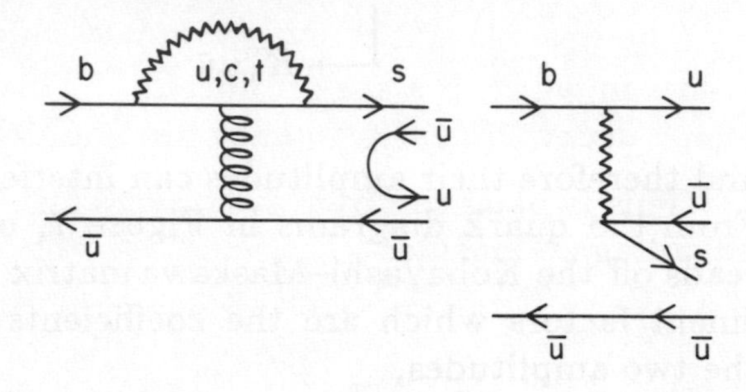

Fig. 9. B_u penguin and spectator decay diagrams whose interference can lead to a Class VI asymmetry.

- Class VI

Allowing for penguin operators, as shown in Figure 9, opens up another possibility for generating a CP violating asymmetry.[24] A particular channel of interest is

$$B_u^- \to K^- \rho^\circ \quad ,$$

since the signature is rather clear.

A rough estimate gives

$$\frac{\Gamma\left(B^- \to K^- \rho^\circ\right) - \Gamma\left(B^+ \to K^+ \rho^\circ\right)}{\Gamma\left(B^- \to K^- \rho^\circ\right) + \Gamma\left(B^+ \to K^+ \rho^\circ\right)} \sim 10\% \quad . \tag{25}$$

Again, the estimate contains uncertainties, including those due to long distance effects which supply strong interaction phases. Note that we can not only have differences in overall rates, but CP violating asymmetries in the differential rates, *i.e.*, in the Dalitz plot, for certain modes.[25]

We summarize this subsection by emphasizing again the fact that Class II asymmetries are predicted unambiguously in the KM model and are cleaner theoretically. But it is still very important to search for the other classes of asymmetries. They may be large, and some of them do not require tagging, possibly making them easier to observe experimentally. For example, although we may have eliminated the superweak model by the time SSC experiments begin, a single unambiguous observation of an asymmetry in charged B decays will immediately rule out the superweak model of CP violation, since in that case no asymmetry is expected in decays which do not involve mixing.

Time Dependence

The observation of the secondary vertex for B decay is very likely crucial in isolating events in which B's are produced.[26] This inevitably leads to a loss of events for those B's which decay too early to allow distinguishing the secondary from the primary vertex.

This is balanced by ways in which the time-dependent asymmetry can be used to our advantage. Consider

$$\Gamma\left(\overset{(-)}{B} \to f\right) \propto e^{-T}\left(1 \pm \sin[xT]\,\mathrm{Im}(\frac{\mathrm{p}}{\mathrm{q}}\,\rho)\right) \quad , \tag{26}$$

where $T = \Gamma t$, is the time in the units of lifetime. Comparing this with the time-integrated expression

$$\Gamma\left(\overset{(-)}{B} \to f\right) \propto \left(1 \pm \frac{x}{1+x^2}\,\mathrm{Im}(\frac{\mathrm{p}}{\mathrm{q}}\rho)\right) \quad , \tag{27}$$

it is quite clear that for B_s system in which

$$x_s > 6\ x_d \sim 5 \quad , \tag{28}$$

the oscillations tend to wash out the time-integrated asymmetry. Thus an asymmetry measurement for B_s decays must be accompanied by measurement of the secondary vertex with a spatial resolution in the transverse plane of approximately,

$$\ell = \frac{\pi}{10} \cdot \frac{1}{x_s} \cdot \left(100\mu m\right) \quad .$$

The time dependence of Eq. (26) for $\mathrm{Im}\left((\mathrm{p}/\mathrm{q})\rho\right) = .1$ and .6, the conservative and optimistic values, is shown in Figure 10 for the B_d system with $x = 0.78$. Note that the asymmetry vanishes at $T = 0$. Thus by cutting out the events with small values of T, the asymmetry will increase, even though the number of events has decreased.

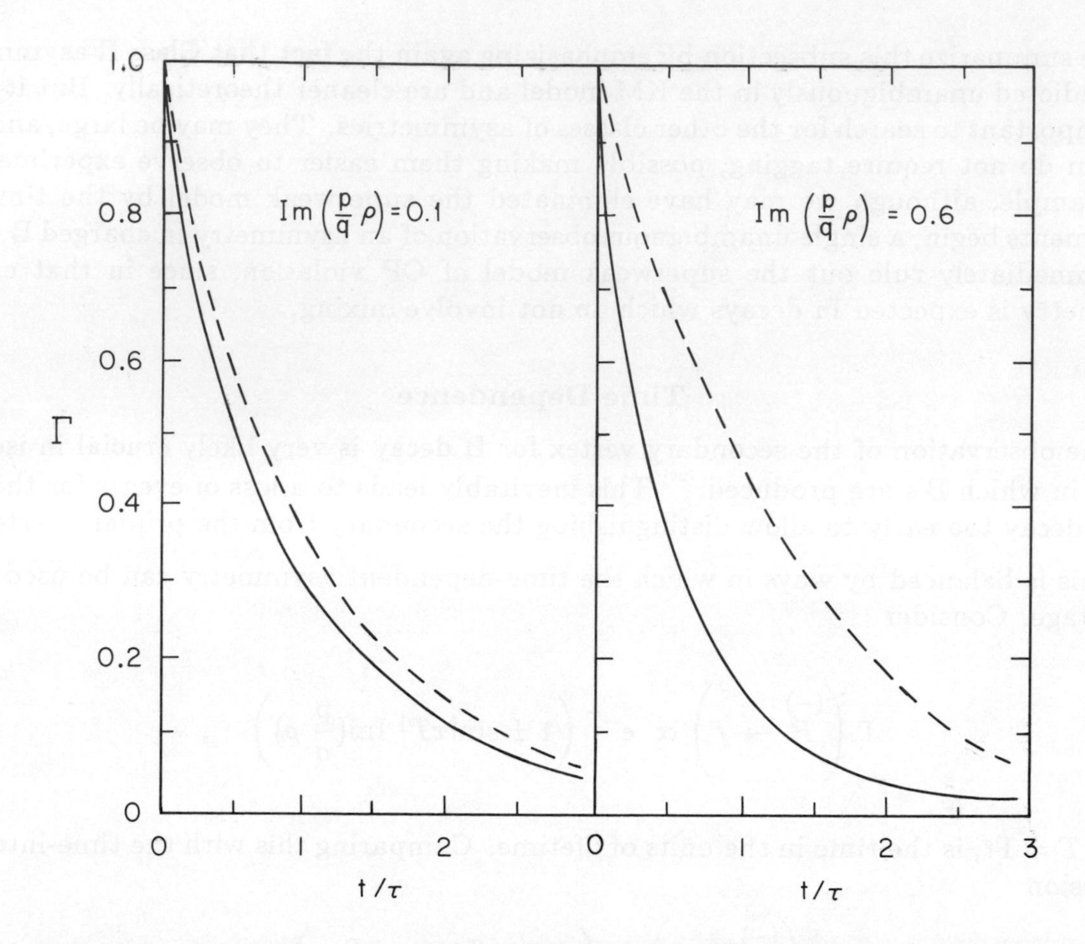

Fig. 10. The time dependence for the decay of a B_d or $\overline{B}_d$ meson to a CP eigenstate, as in Eq. (26), with $x_d = 0.78$, and $\mathrm{Im}\left(\frac{p}{q}\rho\right) = 0.1$ and 0.6.

A crude use of the full information contained in the time dependence is obtained by considering the asymmetry for decays [see Eqs. (9a) and (9b)] occurring between times T_1 and T_2 is given by

$$\begin{aligned} \text{Asymmetry} = \Big\{ & \left(e^{-T_1} \cos xT_1 - e^{-T_2} \cos xT_2\right) x \\ & + \left(e^{-T_1} \sin xT_1 - e^{-T_2} \sin xT_2\right) \Big\} \frac{\mathrm{Im}(\frac{p}{q}\rho)}{(1+x^2)\,(e^{-T_1} - e^{-T_2})} \ . \end{aligned} \tag{29}$$

For example, Figure 11 shows the net asymmetry as a function of T_1 when we set $xT_2 = \pi$. Note that the asymmetry indeed peaks at $T_1 = 1$. A best choice for T_1 can be extracted from the following:

T_1	0	.25	.5	.75	1
$\frac{\mathrm{Asym}(T_1)}{\mathrm{Asym}(O)} e^{-T_1}$	1	1.21	1.29	1.24	1.02

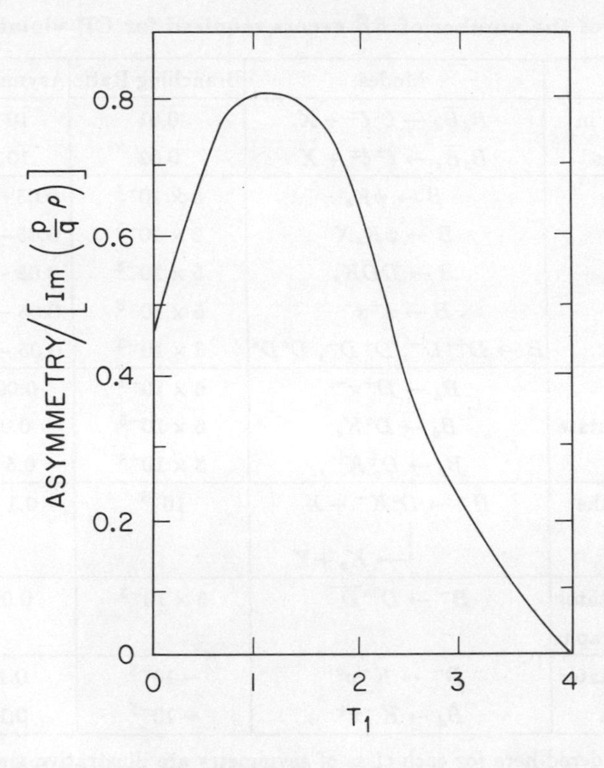

Fig. 11. The net asymmetry as a function of the lower cutoff in proper time (in units of the lifetime), T_1, with the upper cutoff set at $T_2 = \pi/x$.

Compared to a case in which there is no t cut ($T_1 = 0$), selecting $T_1 = 1$ reduces the number of available events, but almost exactly compensates for this by a larger asymmetry; such a cut will lead to a measurement with the same statistical significance. Similarly, a cut at $T_1 = 0.5$ will actually give a 30% reduction in the number of events required to obtain an asymmetry of a given significance when compared to an experiment without the T_1 cut.

Estimation of the Number of $B\bar{B}$ Events Required for CP Violation Studies at the SSC

Now we proceed to use the theoretical predictions for the CP violation asymmetries in B meson decays developed above to estimate the minimum number of B mesons which would be needed for observing CP violation in an experiment at the SSC. These predictions are divided into the six classes of asymmetries which we have discussed and are summarized in Table I. The estimated number of produced $B\overline{B}$ pairs necessary to yield a three standard deviation effect of a certain class, given a branching ratio, asymmetry, and tagging efficiency (explained below) are given in the last column.

The Table was constructed using input and assumptions as follows:

- The branching ratios given in the third column are in many cases below the level of sensitivity of experiments at the existing e^+e^- machines. For these unmeasured modes, the branching ratios given are purely theoretical estimates.

Table 1. Estimation of the number of $B\bar{B}$ events required for CP violation studies at the SSC*

Class	Modes	Branching Ratio	Asymmetry	# $B\bar{B}$ Events Required
I. Charge Asymmetry in Same Sign Dileptons	$B_d\overline{B}_d \rightarrow \ell^{\pm}\ell^{\pm} + X$	0.01	10^{-3}	6×10^9
	$B_s\overline{B}_s \rightarrow \ell^{\pm}\ell^{\pm} + X$	0.02	10^{-4}	2×10^{12}
II. Mixing with Decay to a CP Eigenstate	$B \rightarrow \psi K_s$	5×10^{-4}	$0.05 - 0.3$	$(1 - 34) \times 10^8$
	$B \rightarrow \psi K_s X$	2×10^{-3}	$0.05 - 0.3$	$(2 - 85) \times 10^7$
	$B \rightarrow D\overline{D}K_s$	5×10^{-3}	$0.05 - 0.3$	$(3 - 100) \times 10^7$
	$B \rightarrow \pi^+\pi^-$	5×10^{-5}	$0.05 - 0.5$	$(0.3 - 32) \times 10^8$
	$B \rightarrow D^{+*}D^-,\ D^+D^-,\ D^*D^*$	3×10^{-3}	$0.05 - 0.3$	$(0.7 - 26) \times 10^8$
III. Mixing with Decay to a CP Non–Eigenstate	$B_d \rightarrow D^+\pi^-$	6×10^{-3}	0.001	3×10^{11}
	$B_d \rightarrow D^oK_s$	6×10^{-5}	0.01	7×10^{11}
	$B_s \rightarrow D_s^+K^-$	3×10^{-4}	0.5 ?	5×10^7
IV. Cascade Decays to the Same Final State	$B^- \rightarrow D^oK^- + X$, $\quad\llcorner\!\rightarrow K_s + Y$	10^{-5}	0.1 ?	9×10^8
V. Interference of Spectator and Annihilation Graphs	$B^- \rightarrow D^{0*}\overline{D}$	3×10^{-3}	0.01	2×10^9
VI. Interference of Spectator and Penguin Graphs	$B^- \rightarrow K^-\rho^o$	$\sim 10^{-5}$	0.1	1×10^8
	$\bar{B}_d \rightarrow K^-\pi^+$	$\sim 10^{-5}$	0.1	1×10^8

***The specific channels considered here for each class of asymmetry are illustrative and not exhaustive. At this time we need to keep an open mind as to which channels will be best suited for CP violation studies. In this spirit we have included among the modes illustrating Class II asymmetries $\psi K_s X$ and $D\bar{D}K_s$, which are not necessarily CP eigenstates. There is some danger of a cancellation between the asymmetries produced by sub-channels with opposite CP quantum numbers, but a total cancellation is unlikely.**

- The asymmetries were estimated as described above in the discussion of each of the six classes of CP violating asymmetry.

- To observe an asymmetry, a = (N(B) - N($\bar{B}$))/(N(B)+N($\bar{B}$)) with S standard deviation confidence requires a minimum of $(S/a)^2$ B mesons. The number of $B\bar{B}$ events required is calculated from:

$$N(B\bar{B}) = (\frac{1}{2})(\frac{S^2}{a^2})(\frac{1}{\epsilon_{tag} \cdot BR(B \to f) \cdot BR(f \to X_{ch}) \cdot f(B)}) \quad , \tag{30}$$

where ϵ_{tag} is the tagging efficiency (if it is necessary to tag the initial identity of the B or $\bar{B}$), $BR(B \to f)$ is the branching ratio for the CP violating decay mode $B \to f$, $BR(f \to X_{ch})$ is the branching fraction for f decays into stable charged particles, and $f(B)$ is the fraction of B mesons in a jet of the type needed for a given asymmetry measurement. In the Table we use $S = 3$.

- Class I modes require B° - $\bar{B}^\circ$ mixing as well as the detection of both B mesons in their semileptonic decays. The branching ratios in Table 1 account for the semileptonic decay branching ratio of both B mesons (0.04) and the probability that one of the B° mesons mixes and decays to a "wrong" sign lepton. The mesons are then already "tagged" by the charges of the leptons and no other charged particles need be detected.

- When tagging is necessary (in Classes II and III), we propose using the semileptonic decay $B \to D + \ell + X$, where ℓ is an electron or a muon, as the tagging signature for the "other B". Provided that a common vertex is reconstructed for the D and the lepton, the sign of the lepton, aside from mixing, gives the identity of the initial b quark. The presence of a D meson in a common vertex with the lepton provides a powerful discrimination against leptons from semileptonic decays of charm hadrons not originating from B meson decays. The tagging efficiency is the product of the branching ratios for $B \to D + \ell + X$ (~ 0.2), for D decay into all charged final states (~ 0.2), and for a B_d or B_u in the jet (0.76, see below), which combine to make $\epsilon_{tag} \sim 0.03$.

- B meson decays that can produce asymmetries in Classes IV, V, or VI are "self-tagging", *i.e.*, do not require tagging the accompanying B meson. We take their tagging efficiency to be unity.

- Considering the complicated nature of SSC events in the forward region, we assumed that only charged stable particles will be used for reconstructing B meson decays. For the purpose of these calculations, we assumed J/ψ detection in the e^+e^- and $\mu^+\mu^-$ modes, D and D_s detection in all charged particle modes, and K_s detection in $\pi^+\pi^-$ with branching ratios of 0.14, 0.2, 0.1 and 0.67, respectively.

- For the relative population of the various species of B hadrons in a b quark jet we use the fractions f(B_d°) = 0.38, f(B^+) = 0.38, f(B_s) = 0.15 and f(bqq) = 0.09.

- The estimated numbers of $B\bar{B}$ events required for observing both the lower and the upper range of the predicted asymmetry in each mode are given in the last column of Table 1. A preliminary study shows that using the full proper life-time dependence of the asymmetry reduces the number of required B_d's by a factor of 1.3 compared to the numbers given in the table for the time-integrated asymmetries. For the B_s meson, where due to maximal mixing many oscillations can occur in one lifetime, we use the full time-dependence information to estimate the required number of events.

These estimates do not include the effect of detection efficiencies. Furthermore, since we have not accounted for background effects, they should only be considered as the minimum number of $B\bar{B}$ events required for CP violation studies at the SSC. Clearly, a realistic assessment of such an experiment requires a detailed study of the background effects as well as of detection efficiencies, including effects due to geometrical acceptance, tracking, momentum and energy resolution, vertex reconstruction, particle identification, and triggering efficiency.

4. The Production of t Quarks at the SSC

Recent limits[27] from UA1, indicating that the mass of the top quark is larger than 44 GeV, together with theoretical interpretation[28] of the ARGUS result[4] on $B-\bar{B}$ mixing have increased considerably the possibility that M_t is comparable to, or even greater than, M_W. The production and detection of the t quark when $M_t < M_W$ has been considered previously.[29,30] Now the question of the production and signatures of top quarks with $M_t > M_W$ at the SSC needs to be examined, together with the resulting lepton spectra from cascade decays of the top quark.[31] Heavy top quark production is interesting in another way in that it turns out to be a severe background in the search for a heavy Higgs boson in the W^+W^- decay mode.[31]

The results have been obtained with the ISAJET Monte Carlo[16] and concentrate on the measurement of muons from t quark decays, assuming a high resolution muon detector, like L3+1. The momentum resolution is taken to be $\delta p/p = 4 \times 10^{-5}$ p, with p in GeV/c. An angular coverage of $5° < \theta < 175°$ is assumed, and only muons with $p_T > 10$ GeV have been used for the analysis.

Top quarks with $M_t > M_W$ are mainly produced through gluon-gluon fusion. Figure 12 shows the total cross section for $t\bar{t}$ production as a function of the t mass for $p_T > 10$ GeV and $p_T > 100$ GeV. The cross sections have an uncertainty of about a factor 2, because of poor knowledge of the gluon density function at very small values of x. For $\int Ldt = 10^{40}$ cm^{-2} at $\sqrt{s} = 40$ TeV one expects 1.8×10^8 events (2.6×10^7) for $M_t = 100$ GeV (200 GeV) and $p_T^{top} > 100$ GeV. At $p_T > 100$ GeV the cross section for b quark production is 3 (21) times larger than for a top quark with mass $M_t = 100$ GeV (200 GeV).

Since the t quark decays into a W boson and a b quark, the final state in $gg \to t\bar{t} \to W^+ W^- b\bar{b}$ consists of two W bosons and two b jets. For $M_t = 100$ GeV, the mean

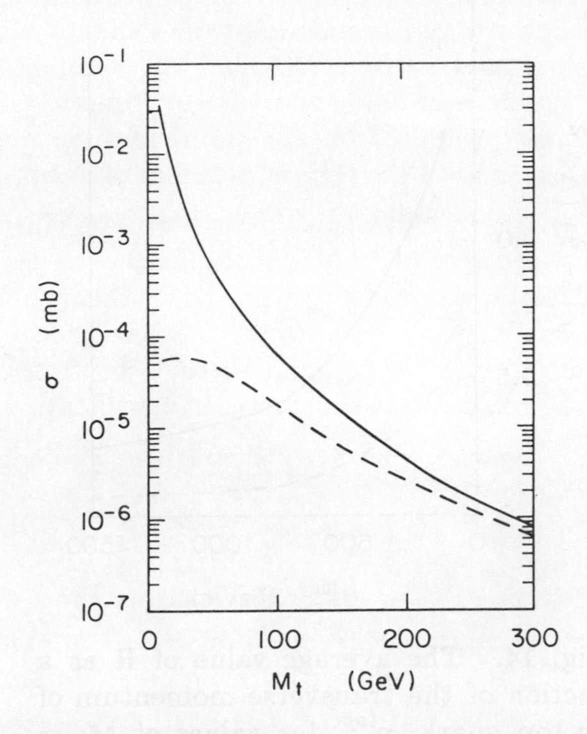

Fig. 12. The cross section for $t\bar{t}$ production in pp collisions at $\sqrt{s} = 40$ TeV as a function of the mass of the top quark, M_t, and $p_T > 10$ GeV (solid curve) and $p_T > 100$ GeV (dashed curve).

transverse momentum of the b jet is about 20 GeV. The p_T of the b jets is especially small if the mass of the top quark is only slightly larger than M_W. The decay of a heavy top quark represents a large source of W pairs at the SSC, being about two orders of magnitude larger than the continuum production.

A relatively clean signal for heavy t quark production is obtained if one W boson decays leptonically (into a muon plus a neutrino) and the other hadronically (into quarks). One then observes an isolated muon and a low p_T b jet, balanced on the other side by 2 jets from the W decay and a low p_T b jet. To further reduce QCD background from high p_T W production, one can study events with two muons produced in the cascade decay of the top quark:

$$t \to W^+ + b \to \mu^+\nu_\mu + \mu^-\bar{\nu}_\mu + c \quad .$$

The background from high p_T $b\bar{b}$ and $c\bar{c}$ production can be reduced by requiring that the high p_T muon be isolated. If one requires the energy in the calorimeter to be less than 10 GeV inside a cone of $\Delta R < 0.3$ around the high p_T muon, one can reduce the light quark background to a sufficiently low level. We are taking advantage of the signature of heavy top decay being an isolated high p_T muon plus a second low p_T muon inside the bottom quark jet. The mean p_T of both muons has a strong dependence on the top mass, with the ratio of the mean transverse momenta of the muons, $< p_T(\mu_W) > / < p_T(\mu_b) >$ being about 10 for $M_t = 100$ GeV. This ratio can be used to measure M_t.

Another method to determine the t quark mass is through the distribution of $R = \sqrt{(\Delta\eta)^2 + (\Delta\phi)^2}$, which measures the separation in space between the muons from the W and b decay ($\Delta\phi$ is the difference in azimuthal angle of the muons with respect to the proton beam, and $\Delta\eta$ the difference in their pseudo-rapidities.) This is shown in Figure 13 for $p_T = 300$ GeV. The distributions are clearly distinct for $M_t = 100$ GeV and 200 GeV. Thus a measurement of R at fixed p_T^{top} is a measure of the mass of the t quark, where p_T^{top} is measured from the total p_T of the jets opposite to the muon pair.

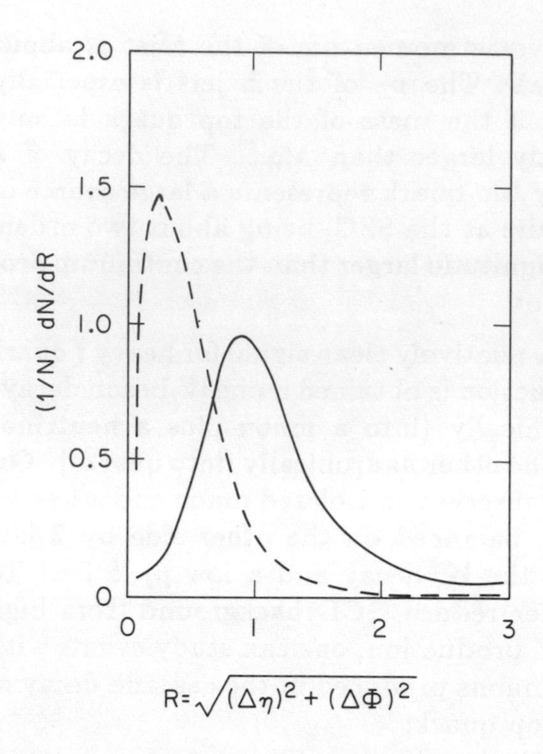

Fig. 13. The distribution in R between the muons coming from W and b decay for $M_t = 100$ (dashed curve) and $M_t = 200$ GeV (solid curve) and $p_T^{top} = 300$ GeV.

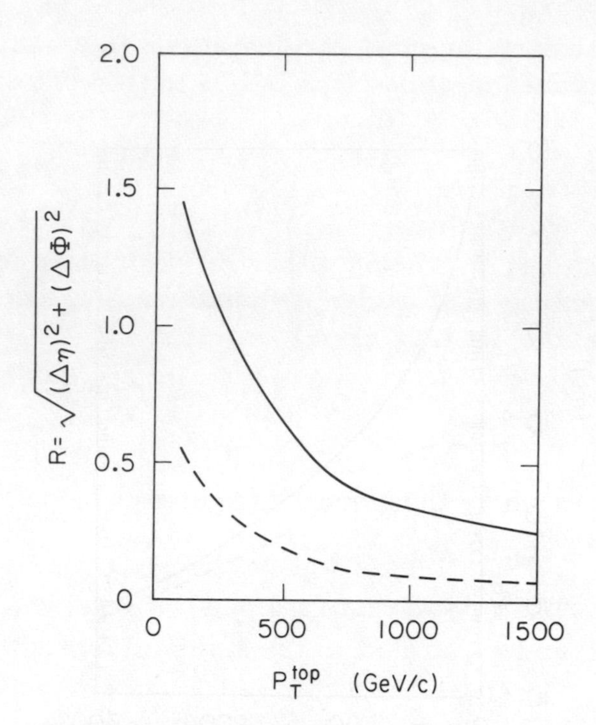

Fig. 14. The average value of R as a function of the transverse momentum of the top quark, p_T^{top}, for values of M_t = 100 (dashed curve) and M_t = 200 (solid curve) GeV.

Figure 14 shows the average R as function of the p_T of the primary t quark. This measurement can provide a fairly precise determination of the mass of the t quark. The event rate at the SSC is sufficiently high to allow for this measurement. Even for $p_T^{top} > 1$ TeV, from the gluon-gluon fusion production mechanism alone one still expects 1800 events with lepton pairs from the cascade decay of the t quark when the integrated luminosity, $\int Ldt = 10^{40} cm^{-2}$.

5. The Physics of t Quark Decays

If we limit our scope to three generations of quarks and leptons, then a great deal of the physics of t decays is fixed. The t mass is constrained to be above 25 GeV from TRISTAN,[32] above 44 GeV from UA1,[27] and above about 50 GeV from considerations of $B - \bar{B}$ mixing.[28] That means that M_b/M_t is certainly < 0.2, and is quite likely < 0.1. These are small numbers, smaller than the ratio of final to initial quark masses in the dominant charm or bottom decays, and small enough to be negligible to the accuracy necessary for most considerations (but see below for a special situation). For example, standard formulas then tell us that 99.7% of t decays will be of the form:

$$t \to b + W^+ \quad ,$$

with the W^+ being either real or virtual, depending on the t mass. Since $|U_{ts}| \approx |U_{cb}|$ when the KM angles are small (as they are known to be), and $|U_{cb}|$ is known from the B lifetime, we may already say that

$$t \to s + W^+$$

will only be a $\approx 0.2\%$ decay mode. It is convenient to break the discussion up into the cases where the t quark is lighter or heavier than the W.

- $M_t < M_W$.

The t decays into a b plus a virtual W, which materializes as $e^+\nu_e, \mu^+\nu_\mu, \tau^+\nu_\tau$, $u\bar{d}$, and $c\bar{s}$, so that in the standard formula

$$\Gamma(t \to \text{all}) = N\,\Gamma(t \to q_{-1/3}\; e^+\;\nu_e) = N\frac{G_F^2 M_t^5}{192\pi^3} \quad , \tag{31}$$

N, the total number of lepton flavors and quark flavors and colors, is equal to 9. This means a total width of ≈ 70 keV for $M_t = 50$ GeV, a number that needs to be corrected slightly upward for QCD, slightly downward for the final quark masses (in particular, the b quark mass, which has been neglected), and upward for the effects of the finite W mass. This latter is the biggest effect, and amounts to $\approx 25\%$ for the case ($M_t = 50$ GeV) cited above.[33] This corresponds to far too short a lifetime to allow separation of a production from a decay vertex, but also far too small a width to be within conceivable experimental resolution. The rate for weak decay of the constituent t quarks within possible hadrons is now comparable with that for electromagnetic and strong decays. Weak decays of toponium become a major fraction of, say, the $J^P = 1^-$ ground state, and even for the $T^*(t\bar{q})$ meson, weak decays can dominate the radiative magnetic dipole transition from this 1^- state to the 0^- ground state.[34]

A t quark in this mass range will very likely be discovered and its properties examined in detail well before the SSC is operating. Depending on its precise mass, TRISTAN, SLC, TEV I, LEP I, and LEP II all have a shot at the initial discovery. A detailed exploration of the properties of the t in its weak decays is the property of the e^+e^- machines. It should be possible, by finding the ground state of toponium to determine M_t to ± 0.2 GeV, or better.[35] In principle, the $V - A$ nature of the $t \to b$ transition can be checked by using longitudinally polarized electrons to form a polarized 1^- toponium state, and then examining the correlation of the momentum of the final lepton with the spin direction of the t quark when it decays semileptonically. Moreover, the fact that t quark weak decays are competitive with strong and electromagnetic decays of toponium can be used to our advantage to measure $\Gamma(t \to \text{all})$ through the chain:[36]

$$\Gamma(t \to \text{all}) = \left[\frac{\Gamma(t \to \text{all})}{\Gamma(t\bar{t} \to \text{all})}\right] \cdot \left[\frac{\Gamma(t\bar{t} \to \text{all})}{\Gamma(t\bar{t} \to \ell^+\ell^-)}\right] \cdot \Gamma(t\bar{t} \to \ell^+\ell^-) \quad . \tag{32}$$

The first factor is the fraction of weak t decays out of all toponium decays, the second is the inverse of the toponium leptonic branching ratio, and the third, the absolute width for toponium decay to $\ell^+\ell^-$, which can be obtained in the standard way by finding the area under the peak in the total cross section for toponium production in e^+e^- collisions. Using vertex detectors, one should also be able to verify that almost all t decays involve the transition $t \to b$. At the SSC, such a t quark is detectable in its semileptonic decays by using an "isolated lepton" cut.[29,30]

- $M_t > M_W$

In this case the t quark can decay into a real W with a width

$$\Gamma(t \to q_{-1/3} + W) = \frac{G_F}{8\pi\sqrt{2}} \frac{M_W^2(M_t^2 - M_W^2)^2}{M_t^3} \left(2 + \frac{M_t^2}{M_W^2}\right) \quad . \tag{33}$$

The width, being only first order in G_F, is much larger than one would obtain from (wrongly) extrapolating Eq. (31); for $M_t = 100$ GeV, the width is ≈ 80 MeV. It also grows asymptotically like M_t^3 rather than M_t^5.

In decays of the ground state T or $T^*(t\bar{q})$ mesons, individual exclusive channels should have very small branching ratios. In decays of heavy flavor mesons their branching ratios scale like $(f/M_Q)^2$, where f is a meson decay constant (like f_π or f_K), of order 100 MeV, and M_Q is the mass of the heavy quark. For D mesons individual channels have branching ratios of a few percent; for B mesons they are ten times smaller; and for T mesons they should be a hundred or more times smaller yet. It should be possible to treat T decays in terms of those of the constituent t quark, $t \to b + W^+$, with the b quark appearing in a b jet not so different than those already observed at PEP and PETRA.

There is one possible exception to these last statements, and that is when $M_t \approx M_b + M_W$. Then T and T^* will decay into a few exclusive channels with a real W plus a B, B^*, or slightly higher mass meson. If the W were a narrow resonance and M_t was slightly smaller than $M_b + M_W$, the $t \to b$ transition would be severely suppressed, allowing the $t \to s$ transition to be dominant in spite of its suppression by the KM factor, $|U_{ts}|^2$. However, the approximate 3 GeV width of the W smears out the threshold, allowing decays to B mesons through the lower tail of the W Breit–Wigner line shape even when $M_t \approx M_W$. The $t \to b$ transition is never suppressed by phase space compared to $t \to s$ by more than about[33] 30%.

Thus for $M_t > M_W$, the t quark is to be seen generally decaying into jets and discovered at TEV I or, if heavy enough, at the SSC. The production cross sections are discussed in the previous Section. The characteristic signature is obtained by looking for $t\bar{t}$ production from gluon fusion, with, say the W from the $\bar{t}$ decaying hadronically, $\bar{t} \to \bar{b} + W^- \to \bar{b}\bar{q}'q$, and the W from the t decaying leptonically, $t \to b + W^+ \to b\ell^+\nu_\ell$. The lepton should be isolated there being a missing momentum due to the neutrino, while both the $\bar{t}$ and t masses reconstruct within errors to the same value. The mass of the t quark can be determined from the momentum spectrum of the lepton relative to that of the b jet, or equivalently, as discussed in the last Section, the distribution in ΔR between the leptons from the W and the semileptonic decay of the b quark. It seems likely that in this way the gross properties of the t quark can be determined, but not much more.

ACKNOWLEDGEMENTS

We thank I. I. Y. Bigi and A. Soni for their participation in the discussions; J. D. Bjorken, both for his contributions during the Workshop and his advice on this manuscript; and I. Dunietz for his comments on the manuscript.

REFERENCES

1. *Proceedings of the 1984 Snowmass Summer Study on the Design and Utilization of the Superconducting Super Collider*, edited by R. Donaldson and J. G. Morfin (American Physical Society, New York, 1984), referred to as SNOWMASS 84.

2. *Proceedings of the 1986 Summer Study on the Physics of the Superconducting Super Collider*, edited by R. Donaldson and J. Marx (American Physical Society, New York, 1987), referred to as SNOWMASS 86.

3. See D. Loveless *et al.* in *$\bar{P}P$ Options for the Supercollider*, edited by J. E. Pilcher and A. R. White (University of Chicago, Chicago, 1984), p. 294; J. W. Cronin *et al.*, SNOWMASS 84, p.161; B. Cox, F. J. Gilman, and T. D. Gottschalk, SNOWMASS 86, p. 33.

4. H. Albrecht *et al*, Phys. Lett. 192B, 245 (1987).

5. See, for example, B. L. Combridge, Nucl. Phys. B151, 429 (1979).

6. See, for example, the discussion in E. J. Eichten, I. Hinchliffe, K. Lane, and C. Quigg, Rev. Mod. Phys. 56, 579 (1984).

7. V. Gribov, E. M. Levin, and M. G. Ryskin, Phys. Rep. 101C, 1 (1982).

8. J. C. Collins, in *Supercollider Physics*, edited by D. Soper (World Scientific, Singapore, 1986), p. 62.

9. F. Olness and W.–K. Tung, in *Proceedings of the Workshop on From Colliders to Super Colliders*, Madison, Wisconsin, May 11–22, 1987, Int. Journal of Mod. Phys. A, 2, 1413 (1987).

10. Such a calculation is in progress by R. K. Ellis, S. Dawson, and P. Nason.

11. A. H. Mueller and P. Nason, Nucl. Phys. B266, 265 (1986).

12. This large factor arises from color factors and interference effects among the Feynman diagrams contributing to each process.

13. C. Albajar *et al.*, Phys. Lett. 186B, 237 (1987).

14. M. Aguilar–Benitez *et al.*, CERN preprint CERN–EP–87/45, 1987.

15. E. Berger, Argonne preprint ANL–HEP–PR–87–53, 1987 (unpublished).

16. F. Paige and S. Protopopescu, SNOWMASS 86, p. 320.

17. H.–U. Bengtsson and T. Sjostrand, Lund preprint LU–TP–87–3, 1987 and SNOWMASS 86, p. 311.

18. The formalism discussed here closely follows I. I. Y. Bigi and A. I. Sanda, Nucl. Phys. B281, 41 (1987). We use the convention that the B_d meson has quark content $d\bar{b}$, while there is some switching from this definition in the above reference.

19. Other references to this and related topics include A. Pais and S. B. Treiman, Phys. Rev. D12, 2744 (1975); L. B. Okun *et al.*, Nuovo Cim. Lett. 13, 218 (1975); S. Barshay and J. Geris, Phys. Lett. 84B, 319 (1979); M. Bander *et al.*, Phys. Rev. Lett. 43, 242 (1979); A. B. Carter and A. I. Sanda, Phys. Rev. Lett. 45, 952 (1980) and Phys. Rev. D23, 1567 (1981); J. S. Hagelin, Nucl. Phys. B193, 123 (1981); J. Bernabeu and C. Jarlskog, Z. Phys. C8, 233 (1981); H. Y. Cheng, Phys. Rev. D26, 143 (1982); E. A. Paschos and U. Turke, Nucl. Phys. B243, 29 (1984); A. J. Buras, W. Slominski, and H. Steger, Nucl. Phys. B245, 369 (1984); L. Wolfenstein, Nucl. Phys. B246, 45 (1984); I. I. Y. Bigi and A. I. Sanda, Nucl. Phys. B193, 85 (1985); L. L. Chau and H. Y. Cheng, Phys. Lett. 165B, 429 (1985); I. Dunietz and J. L. Rosner, Phys. Rev. D34, 1404 (1986).

20. A. Pais and S. B. Treiman, ref. 19 and L. B. Okun *et al.*, ref. 19.

21. L. Wolfenstein, Phys. Rev. Lett. 51, 1945 (1983). Note that ρ in $\mathrm{Im}\left(\frac{p}{q}\rho\right)$ is a ratio of amplitudes for B decay, and is not to be confused with the parameter (a pure, dimensionless number) in the Wolfenstein parametrization.

22. J. Bernabeu and C. Jarlskog, ref. 19.

23. L. L. Chau and H. Y. Cheng, ref. 19.

24. M. Bander *et al.*, ref. 19.

25. S. Barshay and J. Geris, ref. 19.

26. The utilization of the time dependence has been particularly emphasized by I. Dunietz and J. L. Rosner, ref. 19.

27. I. Wingerter, invited talk at the Topical Conference of the SLAC Summer Institute on Particle Physics, August 10–21, 1987 (unpublished).

28. J. Ellis, J. S. Hagelin, and S. Rudaz, Phys. Lett. 192B, 201 (1987); I. I. Y. Bigi and A. I. Sanda, Phys. Lett. 194B, 307 (1987); F. J. Gilman, invited talk at the International Symposium on the Fourth Family of Quarks and Leptons, Santa Monica, February 26 - 28, 1987 and SLAC-PUB-4315, 1987 (unpublished); G. Altarelli and P. J. Franzini, CERN preprint CERN-TH-4745/87, 1987 (unpublished); H. Harari and Y. Nir, SLAC preprint SLAC-PUB-4341, 1987 (unpublished); V. A. Khose and N. G. Uraltsev, Leningrad preprint, 1987 (unpublished); L. L. Chau and W. Y. Keung, UC Davis preprint UCD-87-02, 1987 (unpublished). J. F. Donoghue *et al.*, SIN preprint SIN-PR-87-05, 1987 (unpublished); A. Ali, DESY preprint DESY-87/083, 1987 (unpublished); J. R. Cudell *et al.*, University of Wisconsin preprint MAD/PH/353, 1987 (unpublished); A. Datta, E. A. Paschos, and U. Turke, Dortmund preprint DO-TH-87/9, 1987 (unpublished); D. Du and Z. Zhao, Phys. Rev. Lett. 59, 1072 (1987).
29. K. Lane and J. Rohlf, SNOWMASS 84, p. 737.
30. E. W. N. Glover and T. D. Gottschalk, SNOWMASS 86, p. 77.
31. G. Herten, these Proceedings.
32. F. Takasaki, invited talk at the 1987 International Symposium on Lepton and Photon Interactions at High Energies, Hamburg, July 27–31, 1987 (unpublished).
33. F. J. Gilman and R. Kauffman, unpublished.
34. I. I. Y. Bigi and H. Krasemann, Z. Phys. C7, 127 (1981).
35. F. Porter, invited talk at the International Symposium on the Fourth Family of Quarks and Leptons, Santa Monica, February 26–28, 1987 and Caltech preprint CALT–68–1434, 1987 (unpublished).
36. F. J. Gilman and H. F.–W. Sadrozinski, Mark II/SLC Working Group Note # 4–03, 1986 (unpublished).

Intermediate Mass Higgs Boson(s)

D.M. Atwood, J.E. Brau, J.F. Gunion, G.L. Kane,
R. Madaras, D.H. Miller, L.E. Price, A.L. Spadafora

INTRODUCTION

Finding and understanding the spectrum of scalar bosons is the central problem of particle physics today. Considerable work has been done to learn how to study Standard Model heavy and obese Higgs bosons; simulations including the problems induced by standard model backgrounds are underway, and some results are reported elsewhere in these proceedings (see the reports of E. Wang et al.).

The mass region $M_H < M_Z/2$ will be covered at SLC and LEP. LEPII will be able to extend this range to about 85 GeV. Above $M_H > 2M_Z$ the search is easy for a standard model H^o at the SSC, though not so simple for the neutral scalars of a supersymmetric theory. The intermediate region, $M_Z/2 \lesssim M_H \lesssim 2M_Z$ is one of the most difficult mass regions to study, and it is the subject of this report. We concentrate on a neutral Standard Model scalar to be specific. The lightest scalar of a supersymmetric theory behaves very much like a Standard Model scalar for most ranges of parameters, so the results generally apply to that case as well, and for any form the scalar spectrum might take our results indicate how the analysis might go. Ultimately, to fully understand spontaneous symmetry breaking and the origin of mass, it will be necessary to find any intermediate mass scalar <u>and</u> to know in what mass ranges no scalars exist.

Our analysis is only "in progress", and our results reported here must be regarded as tentative. We hope to finish the study over the next few months.

The report takes the following form. First the production cross sections are presented, and then the branching ratios. A number of branching ratios must be considered including rather rare ones since the dominant modes are often obscured by large backgrounds. The main modes are $\gamma\gamma, \tau^+\tau^-, b\bar{b}, ZZ^*$. Z^* means a Z forced by kinematics to be below its physical mass. All results must be presented as functions of both M_H and M_t since there is a strong dependence on the presently unknown value of M_t. Each mode is useful in a limited region of the $M_t - M_H$ plane. Since the backgrounds and problems of each mode are different a separate section is given for each, and then the results are combined.

Two recent papers that deal with these questions are available; previous literature can be traced from them. Reference 1 gives a general treatment of the intermediate mass region, with emphasis on $\gamma\gamma$ and ZZ^* modes. A number of other rare modes are discussed. Reference 2 gives a calculation of the $\tau^+\tau^-$ mode opposite a recoil gluon jet, plus some discussion of the $\gamma\gamma$ mode.

PRODUCTION CROSS SECTIONS

The most complete calculation of the H^0 production cross sections is from reference 3, and is shown in Figure 1. The main thing to notice is that in the intermediate region there are typically over 10^6 H^0 produced in an SSC standard year of integrated luminosity of $10^{40} cm^{-2}$.

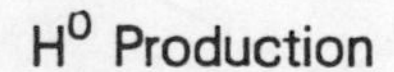

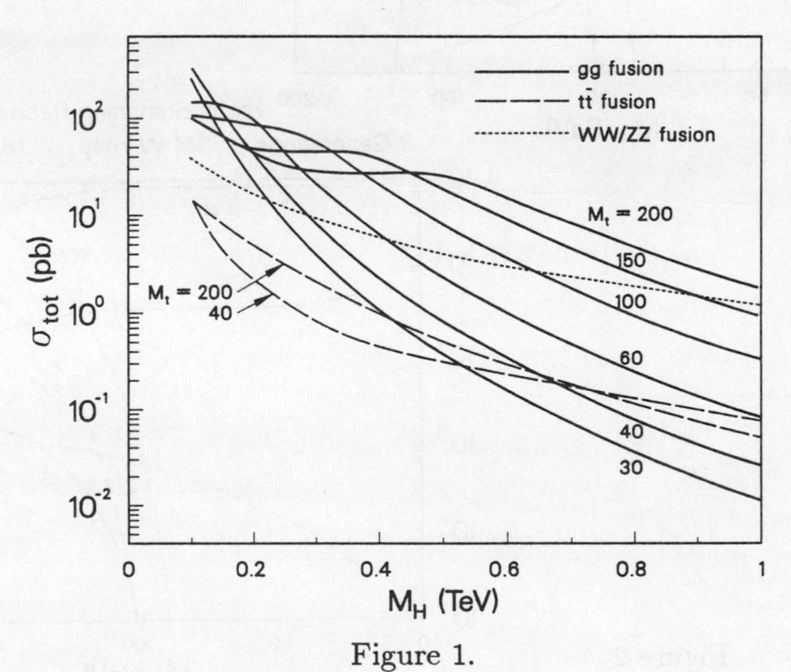

Figure 1.

BRANCHING RATIOS

The H^0 decays to $b\bar{b}$ and $\tau^+\tau^-$ are from the basic tree level vertices. The $Z\,Z^*$ mode has to be calculated [1] with some care, since virtual $\gamma^* + Z^*$ have to be included. The $\gamma\,\gamma$ mode occurs at one loop. The $W^\pm$ contribution to the loop is dominant, but there is a destructive interference with the t-quark contribution that grows as M_t^2. A number of details about the branching ratio calculations are given in reference 1, and references to the original literature are there. Some results are shown in Figure 2, for two values of M_t.

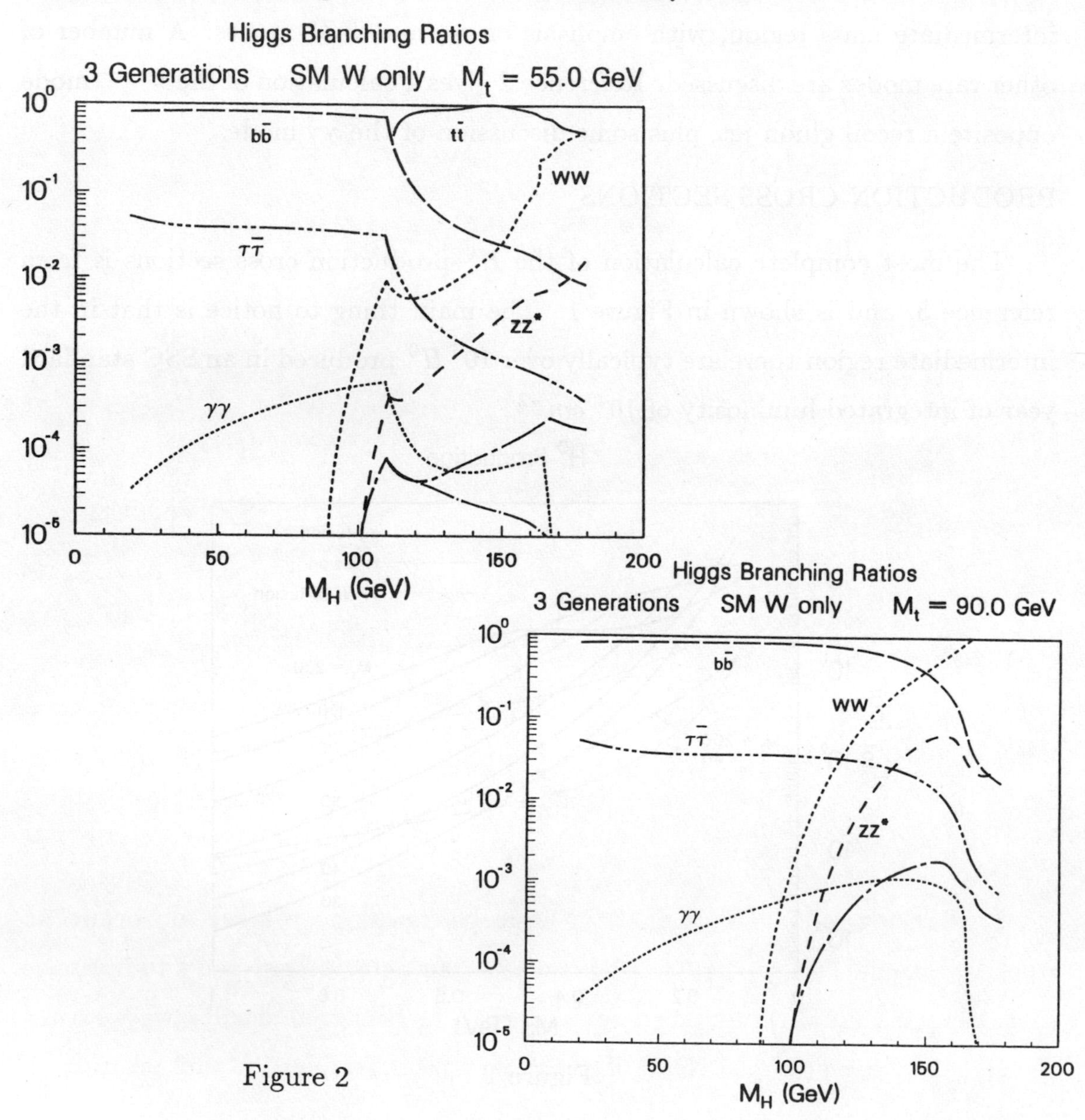

Figure 2

SPECIFIC MODES

A. $\gamma\gamma$

1. Signal. The signal is the product of production cross section and branching ratio. The H^o width is so small in the intermediate mass region that it is always negligible compared to the resolution in $\gamma\gamma$ mass. The number of signal events is shown in Figure 3, for two M_t values, from reference 1. It is clear that in the absence of backgrounds it would be easy to see a signal from $H^o \to \gamma\gamma$ over the entire intermediate mass region.

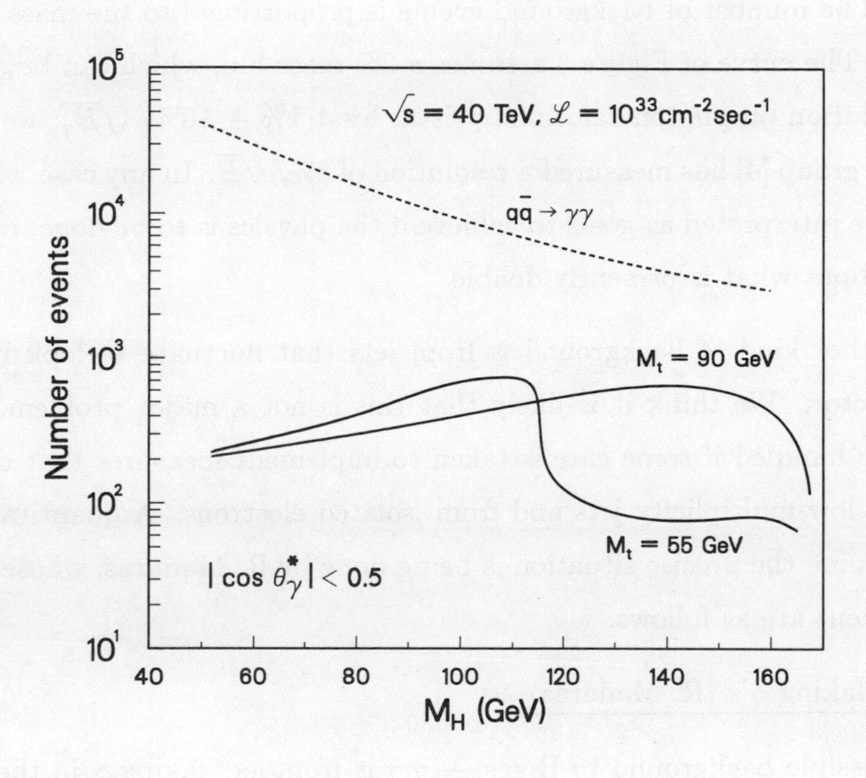

Figure 3.

2. Backgrounds. Unfortunately, large background problems do occur, so whether a signal can be seen depends on whether detectors are built to optimize those aspects that affect finding a signal in $\gamma\gamma$. The background will always exceed the signal by a factor $\gtrsim 5$. One background is real Standard Model production

of $\gamma\,\gamma$, from $q\bar{q} \rightarrow \gamma\,\gamma$ and from $gg \rightarrow \gamma\,\gamma$, the second being mediated by quark loops. The former of these is straightforwardly calculated. The latter has not yet been fully calculated; estimates suggest the two sources are comparable, so the background curve shown in Figure 3 is $q\bar{q} \rightarrow \gamma\,\gamma$ multiplied by two to take account of $gg \rightarrow \gamma\,\gamma$. These backgrounds can be reduced by two effects:

(i) An angular cut can be made, since the γ's from H decay are isotropic while those from $q\bar{q} \rightarrow \gamma\gamma$ are peaked forward and backward; events with $|cos\theta_\gamma| > 0.5$ in the $\gamma\,\gamma$ center-of-mass are not included.

(ii) The number of background events is proportional to the mass resolution in $M_{\gamma\,\gamma}$. The curve of Figure 3 assumes a 2% mass bin, which can be achieved if the resolution on photon calorimetry is at least $1\% + 10\%/\sqrt{E_\gamma}$; we note that the SLD group [4] has measured a resolution of $8\%/\sqrt{E}$. In any case, all numbers should be interpreted as goals to achieve if the physics is to be done, rather than claims about what is presently doable.

Another kind of background is from jets that fluctuate to look like a γ to the detector. We think it is likely that this is not a major problem, and that it can be handled if some care is taken to implement measures that distinguish γ's from low-multiplicity jets and from isolated electrons. A quantitative study to determine the precise situation is being done by R. Madaras, whose results at the moment are as follows.

Jets faking γ's (R. Madaras).

A possible background to Higgs $\rightarrow \gamma\,\gamma$ is from jets faking γ in the detector. This can arise from jet-jet events in which both jets look like isolated γ's, or jet-γ events in which the jet looks like an isolated γ.

It has been estimated [1] that a 2% mass bin in m(jet-jet) [near m(jet-jet) = 100 Gev] contains $< 10^8$ times as many jet-jet events as $\gamma\,\gamma$. Thus one would like the probability of a jet to look like an isolated γ to be less than 10^{-4}, and these estimates have to be checked.

I have used the Pythia monte carlo to calculate this probability, by generating 20K jet-jet events and counting the number of "isolated γ's" arising from the jets. By "γ's" I mean photons, pi-zeros or etas, since I am not assuming that it is possible to resolve both decay photons from the pi-zero or eta. By "isolated γ" I mean a "γ" with:

a) rapidity < 3.0

b) energy $> E_{\min}$

c) D$>$ $D_{\min}$, where D is the distance to the nearest particle (with energy greater than 5 GeV), calculated as $D = \left[\Delta\varphi^2 + \Delta y^2\right]^{1/2}$. The probability for a jet to look like a γ depends on $E_{\min}$ and $D_{\min}$.

The results for 20K jet-jet events (i.e., 40K jets) are:

$D_{\min}$	$E_{\min}$	Number of "isolated γ's"	Probability $j \to$"γ"
0.6	20 Gev	171	$42.8{\times}10^{-4}$
0.6	40 Gev	22	$5.5{\times}10^{-4}$
0.6	50 Gev	5	1.3×10^{-4}
1.0	20 Gev	70	17.5×10^{-4}
1.0	40 Gev	14	3.5×10^{-4}
1.0	50 Gev	2	0.5×10^{-4}

Table 1

It appears that one can obtain low enough probabilities, but only with tight cuts on the "γ" energy and isolation.

Some further work that needs to be done includes:

1) increasing the statistics on the above results a little bit;

2) verifying the ratio of 10^8 stated earlier for N(jet-jet)/N ($\gamma\,\gamma$) with 2% mass bins;

3) calculating similar results for jet-γ events;

4) save the "isolated γ's" found by the monte carlo and pair them in an appropriate way to obtain the m("$\gamma\,\gamma$") spectrum.

An important point to note about detecting H^o in the $\gamma\gamma$ mode is that this is the only accessible mode that does not get tree level contributions from Z decay. Because of this, even at $e^+\,e^-$ colliders the region $M_H \sim M_Z$ may be difficult to do, while it does appear to be tractable at the SSC if sufficient luminosity and resolution are available.

A study is also underway by A. Spadafora to examine the general topology of the signal and background events, in order to see how they look to a detector, to see if our prejudices about the photon isolation and event characteristics are confirmed, and to see if any additional cuts that help reduce background can be found. Since the general situation with $\gamma\,\gamma$ is marginal though promising, even small background reductions could be very helpful.

The region in the $M_t - M_H$ plane where we presently believe [1] $H^o \to \gamma\,\gamma$ is detectable is shown in Figure 6.

B. ZZ^*

1. Signal. Assuming both Z and Z^* can be detected in e^+e^- or $\mu^+\mu^-$ decays, the signal is obtained from Figures 1 and 2 for a Standard Model coupling of $H^o \to ZZ^*$.

2. Background. Backgrounds come from $q\bar{q} \to ZZ^*$, $q\bar{q} \to Z\gamma^*$, $gg \to Z\gamma^*, ZZ^*$. In this case, $q\bar{q}$ provides the serious background and it peaks at small M_{Z^*}. A cut on M_{Z^*} essentially eliminates the background, so the use of this mode is simply rate limited.

The region where ZZ^* can be used is shown in Figure 6, from reference 1.

C. $b\bar{b}$

If $M_H < 2M_t$, essentially all H decays are to $b\bar{b}$. Even so, the QCD background is too large to study H in $pp \to H+$ anything. However, if H is produced in association with a W, used to trigger, then there is a possibility of detecting $H \to b\bar{b}$ for some range of masses. An encouraging study was done at Snowmass 86 [5].

A further study is underway and will be continued. So far its results are as follows.

$H \to b\bar{b}$ (J. Brau)

The following set of cuts have been routinely applied (using PYTHIA 4.8) to sets of real events of the type

$$\text{quark} + \text{antiquark} \to W + \text{Higgs}$$

and background events representing a mix of

$$\text{qluon} + \text{quark} \to W + \text{quark}$$

and

$$\text{quark} + \text{antiquark} \to W + \text{gluon} :$$

1) a minimum transverse momentum in the parton-parton hard scattering process of 50 GeV was required;
2) the two jets not from the W which had the largest transverse energy component and $|y| < 2.5$ were selected;
3) it was required that one of the jets have a lepton with a transverse momentum relative to its reconstructed jet axis of at least 1 GeV;
4) both jets were required to have a value of 0.2 or less for the Gottshalk R parameter. This parameter is a measure of the tightness of the jet;
5) the identified high p_t lepton was required to have an impact distance with the primary vertex of at least 50 microns in the transverse plane;

6) at least one particle from the second jet was required to have an impact distance to the primary vertex in the transverse plane of 50 microns.

After all of these cuts were imposed, the following cross sections (in cm^2) remained (events in the range of M(Higgs)-20GeV$\rightarrow$ M(Higgs) were selected):

M(Higgs)	Signal	Background	Signal/Background
60	1.2×10^{-37}	2.7×10^{-36}	.046
80	6.1×10^{-38}	1.1×10^{-36}	.053
100	5.8×10^{-38}	1.1×10^{-36}	.051
120	3.7×10^{-38}	1.1×10^{-36}	.032

Note that triggering requirements have not yet been imposed, so these number will be reduced by a factor of order 2/9.

These signal/background ratios are considerably smaller than those of ref. 5, which we assume is related to the low statistics of last years study. In all cases the top mass has been assumed to be 150 GeV. In the final sample of background events (drawn from a total of 75,400 generated events) twenty-five events are from gluon + quark and one is from quark + antiquark. The gluon + quark cross section is fifteen times larger before cuts.

It is important to note that these 26 surviving background events are dominated by b-quark jets. That is 21 of the events are two b-quark jets, two are two charm quark jets, and three are a top quark and a bottom quark. Therefore, the cuts are very successfully selecting the types of events that have to mimic successfully the Higgs $\rightarrow b\bar{b}$.

Since the background is largely coming from QCD generated b-quark jets, a further requirement is needed to improve S/N. It turns out that the two b-quark jets from the background do not balance transverse momentum as well as those from Higgs decay. We therefore require that the vectors for the reconstructed Higgs and the reconstructed W be back to back to within 23 degrees in the transverse plane has been added. That leads to the following numbers of events at integrated luminosity of $1.0 \times 10^{40} cm^{-2}$:

M(Higgs)→	60	80	100	120
sig events→	760	410	377	260
bkgnd→	2100	2100	2100	2100
bkgnd flct→	46	46	46	46
sig/bkgnd→	0.36	0.19	0.18	0.12
sig/fluct→	16.6	8.9	8.2	5.7

On the basis of these results it appears likely that some region of the $M_t - M_H$ plane will be measurable in this mode. Backgrounds with $W + Z$ and $g + Z$ followed by $Z \to b\bar{b}$ have not yet been included, so it is not clear whether the region with $M_H \simeq M_Z$ can be studied this way. Since the cross sections are similar while $BR(Z \to b\bar{b}) << BR(H \to b\bar{b})$, the Z background may not be serious. It seems likely the region $M_Z / 2 \lesssim M_H \lesssim 80$ GeV can be searched, and perhaps also the region 100 GeV $\lesssim M_H \lesssim 2M_t$ or $2M_W$.

D. $\tau^+\tau^-$.

This is the final mode that could be large enough to allow detection of H^o. The problem is to obtain good mass resolution on M_H in spite of the missing neutrinos from τ decay. Two methods have been proposed to accomplish this. The first is to produce H opposite a recoil gluon jet or opposite a Z^o. Since the neutrinos from τ decay go in the same direction as the original τ, because its mass is small, if the p_t of the H is assumed to be known because the p_t of the jet or Z is measured, there are enough constraints to determine M_H. The necessary calculations for the gluon jet case have recently been done [2]. Unfortunately, the Monte Carlo calculations done by D. Atwood during this workshop appear to show that the mass resolution obtained is too poor for the method to work. The resulting mass plots do not stand out above background, and have a shape similar to that of the background, so it would be very hard to recognize a signal.

The second method, whose promise was emphasized by F. Paige, appears able to work if sufficient luminosity is available. The resolution is obtained by only using τ decays to three or more pions, so the neutrino carries off little energy. About 22% of τ decays are satisfactory for this, assuming full recognition of charged and neutral pions. The full inclusive rate can be used, so before detection cuts of order $(2 \times 10^6) \times (0.03) \times (0.22)^2 \simeq 3000$ events qualify. The analysis, by D. Atwood, is as follows.

$\tau\bar{\tau}$ Mode - (D. Atwood)

$pp \to$ Higgs $\to \tau^+\tau^-$ is considered as a method of detecting the Higgs boson if it has intermediate mass. It is found that if events are considered where both τ's decay to at least 3 π's it may be feasible to detect the Higgs in some cases. The method is probably useful only if $M_H < 2M_t$.

From ISAJET we found that the Higgs had an rms average transverse momentum depending on its mass of

$$P_{th} \approx 6.27 + 0.318 M_H(GeV).$$

In the parton Monte Carlo we assumed that the distribution of the transverse momenta of the Higgs was a 2-dimensional gaussian with the appropriate average.

DECAY OF $\tau^{\pm}$

We assume the branching ratios given in (Particle Data Group "Review of Particle Properties" Phys. Lett. 170B 1986). In particular we note that the following modes make up 22.4% of the τ decays: $\pi^-\pi^o\pi^o\nu(6\%)$, $\pi^-\pi^o\pi^o\pi^o\nu(3\%)$, $\pi^-\pi^-\pi^+\nu(8.1\%)$, $\pi^-\pi^-\pi^+ \geq 1\pi^o\nu(5.3\%)$ (we will assume $1\pi^o$). We will call these modes multi-π modes. We will consider events where both τ's from a given Higgs undergo multi-π decays which happens in about 5% of Higgs $\to \tau^+\tau^-$ decays.

We also assume that the matrix elements for the various decays are constant in phase space.

BACKGROUND

The background which we consider is from Drell Yan production of τ pairs through the sub process $q\bar{q} \to \gamma^*, Z^* \to \tau^+\tau^-$. We assume the total transverse momentum of the τ pair is as above.

FAKE MISSING MOMENTUM

As suggested by M. Barnett we assume that there is an average missing transverse momentum due to mismeasurement of about 25GeV. We assume the distribution of this missing momentum is a 2-dimensional Gaussian with this rms average.

MASS RECONSTRUCTION

Suppose a relativistic τ with momentum p_τ undergoes a multi-π decay. Let $p_{\rm obs}$ be the momentum of all the observable particles and p_x be the momentum of the neutrino. If the energy of the τ is large, all these momenta will be colinear. In particular we define $p_{\rm obs} = xp_\tau$.

The distribution in x of multi-π decays as determined by Monte Carlo is given in Figure 4. Let us define this function as $f_\tau(x)$.

Given a two τ event we can see only $p_{\rm obs1}$, the observed four-momentum of the τ^+, $p_{\rm obs2}$, the observed four-momentum of the τ^- and $p_{tx{\rm obs}}$, the total observed missing transverse momentum including fake missing momentum due to mismeasurement.

If the four-momenta of the τ's are $p_{\tau 1}$ and $p_{\tau 2}$, assuming the τ's are ultra-relativistic,

$$\begin{aligned} p_{\tau 1} &= p_{\rm obs1} / x_1 \\ p_{\tau 2} &= p_{\rm obs2} / x_2. \end{aligned}$$

Furthermore, the true missing transverse momenta (now using $p_{\rm obs}$ for the transverse momenta) is

$$p_{tx} = p_{\rm obs1} (1 - x_1) / x_1 + p_{\rm obs2} (1 - x_2) / x_2.$$

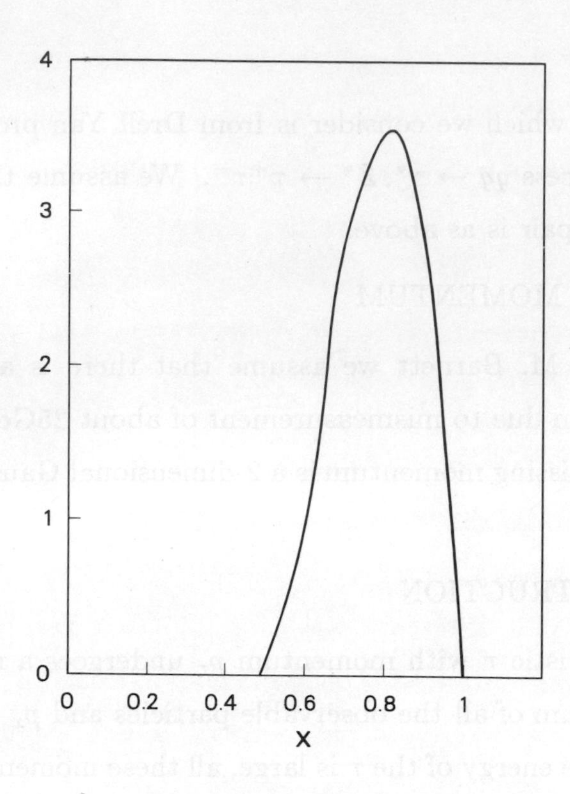

Figure 4

The transverse momentum of the Higgs, p_{tH}, is

$$p_{tH} \;=\; p_{\text{obs1}} \,/\, x_1 \;+\; p_{\text{obs2}} \,/\, x_2,$$

and the fake missing transverse momentum is $p_{t\text{fake}} \;=\; p_{tx\text{obs}} \;-\; p_{tx}$.

Using the assumed statistical properties of these quantities we define a log-liklihood function as follows:

$$\begin{aligned} L(x_1, x_2) = & \log(f_\tau(x_1)) + \log(f_\tau(x_2)) \\ & - p_{t\text{fake}}^2 \,/\, 25^2 + \log(p_{t\text{fake}}^2/25^2) \\ & - \log(x_1^2\, x_2^2) \\ & - p_{tH}^2 \,/\, 50^2 + \log(p_{tH}^2/50^2) \\ & - \log(x_1^2\, x_2^2) \end{aligned}$$

(the 25 in the second line is the average fake missing momentum and the 50 in the second line is an estimate of the average transverse momentum of the Higgs in the mass range we are considering).

This is intended to represent the logarithm of $dlikelihood/dx_1\, dx_2$. Of course we do not know a priori what the values of x_1 and x_2 are. Our reconstruction method will be to select the "most likely" values of x_1, x_2 which maximize $L(x_1, x_2)$. The reconstructed mass of the τ pair, $M_{\rm recon}$, is thus (four-momentum scalar product)

$$M^2_{\rm recon} = 2\, p_{\rm obs2} \cdot p_{\rm obs1} \,/\, x_1 x_2.$$

RESULTS

Figure 5 shows a plot of $d\sigma/dM_{\rm recon}$ for the Drell Yan background and for Higgs production given various $(M_H,\, M_t)$ combinations. Note that the mass which comes out of this reconstruction method tends to be larger than the real mass of the τ pair. This presents no problem since it may be compensated for in the final determination of the Higgs mass. The important thing which is accomplished by this method is to separate the Higgs peak from the strong Drell-Yan peak at the Z° resonance which might otherwise give large backgrounds.

We use the following criteria for the Higgs being detectable:

1) Signal to background must be better than 10% in three consecutive 5 GeV bins in the $d\sigma \,/\, d\, M_{recon}$ plot.
2) Those bins must have 3-σ signal each giving a 5-σ signal for those three bins, assuming an integrated luminosity of $10^4\, pb^{-1}$.

The region of the $M_H - M_t$ plane where this is satisfied is shown in Figure 6.

REMARKS

Because signal to background is not very good, the background must be calibrated with as much precision as possible. The number of Drell Yan events will, of course be equal to the number of Drell Yan $\mu^+\, \mu^-$ events. The main uncertainty therefore arises from the detector efficiency of detecting the τ events.

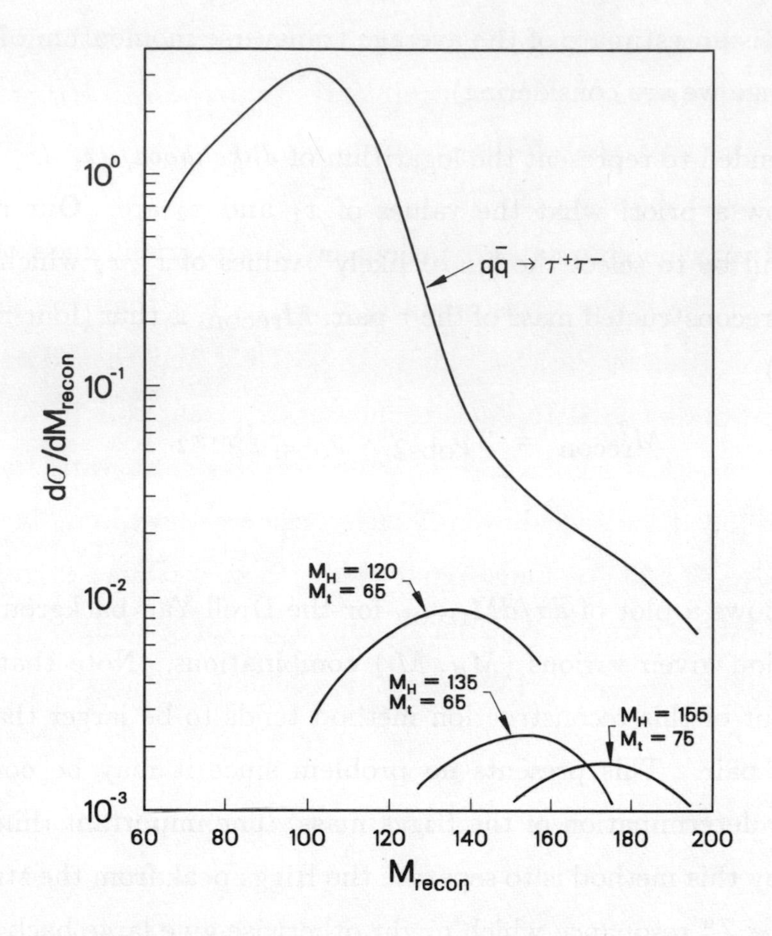

Figure 5.

One approach to resolving this with a specific detector might be to take a large sample of $Z^o \to \tau^+\tau^-$ events where one of the τ's decays to some particular easy to identify mode and then study the statistical properties of detecting the other τ.

THE RECOIL METHOD

As mentioned above, it had been hoped [6,7,2] that producing the Higgs in association with a gluon or Z^o would improve the resolution on M_H sufficiently to allow a signal to be seen. This does not appear to work. The main obstacle seems to be that the fake missing momentum due to mismeasurement and to mismeasurement of the total jet momentum causes the resolution to be too poor.

For example, if we take the fake missing momentum to be 25 GeV on average and the error in determining the momentum of the jet to be 5%, $M_H = 135$, $M_t = 65$, and require the jet transverse momentum be > 100 GeV, we find signal to background ratios on the order of 1:70. Since a lot of the background comes from the Z° peak, it may be eliminated by a cut on the invariant mass of observed decay products of the τ since this quantity must be less than the mass of the τ pair. If one requires an invariant mass > 95 GeV and then does a plot of $d_\sigma/dM_{\text{recon}}$ one finds signal to background ratios of about 1:10 in the above example. The statistical significance is, however, less than with the inclusive $pp \rightarrow$Higgs method discussed above: if one considers the three "best" 5 GeV bins $120 < M_{\text{recon}} < 135$ one finds signal events to background events is 18:168 which is little better than a one sigma significance. Integrating over more bins we find that if $100 < M_{\text{recon}} < 135$ the ratio is 30:330, still less than $2\,\sigma$.

Another problem of more general importance is being able to identify τ's with a given detector.

Examination of the problem of triggering with and identifying τ's has been begun by D. Miller, with initial comments mainly for the case of $\tau^+\,\tau^-$ against a recoiling jet.

τ Identification (D. Miller)

For the mass of the Higgs M_H in the range $100 < M_H < 200$ GeV a decay mode which might allow the observation of the Higgs is $H \rightarrow \tau^+\tau^-$. This branching ratio for this mode is several percent providing that $H \rightarrow t\bar{t}$ is not allowed. Cross sections and production characteristics have been considered by a number of authors [1,2]. In this paper, we will use the calculations of Ellis et al. [2]

$H \rightarrow \tau^+\tau^-$

Gluon-Gluon fusion is the dominant production process and has the following properties for $M_H = 140$ Gev

$$\sigma\text{HIGGS} = 200pb = 2 \times 10^6 \text{ evts/yr}$$

$$\sigma(H \to \tau^+ \tau^-) = 4pb = 4000 \text{ evts/yr}$$

$$\sigma(H \to \tau^+ \tau^-)(p_t > 100GeV) = .2 \; pb = 2000 \text{ evts/yr}$$

These cross sections are for the whole rapidity range.

SSC EVENT RATES

$$\sigma_{\text{TOT}} \sim 100 \; mb \int Ldt = 10^{40} = 10^{15} \text{ evts/yr}$$

$$\text{Event rate} = 10^8 \; Hz.$$

If we assume an event rate to tape of $2Hz = 2 \times 10^7$ events/year this is 2×10^{-8} of σ_{TOT}.

TAU DECAYS

A typical τ decay has missing energy due to one or more neutrinos and will give both electromagnetic (EM) and hadronic energy deposition in the calorimeter system. The decay products of the τ will be in a narrow $< 1^o$ cone.

Main Modes	AVERAGE $E_{vis}(E.M.)/E$	AVERAGE E_{vis} (Hadronic)$/E$
$e\nu\nu$ 10%	1/3	-
$\mu\nu\nu$ 10%	-	-
$\rho\nu$ 22%	1/4	1/4
$\pi\nu$ 11%	-	1/2
ν A1$\to \pi^+\pi^+\pi^-\nu$ 7%	-	3/4
ν A1$\to \pi^+\pi^o\pi^o\nu$ 7%	1/2	
$(3\pi + n\pi^o)$ 6%	3/8	3/8
K,π +neutrals 10%	1/3	1/3

CHARACTERISTICS OF HIGGS

For $M_H = 140$ GeV the p^* of the τ is ~ 70 GeV and $\beta_\tau = .99967$. For a Higgs momentum equal to its rest mass of 140 GeV the most probable lab opening angle is 90°, but one τ can go backwards with respect to the Higgs line of flight $\sim 30\%$ of the time. Even for $p = 500$ GeV $\theta \sim 30^o$ and one τ can still go backwards. This means that in general the τ's are distributed over the whole detector.

RECONSTRUCTION

I consider the case where a recoiling jet is used to help determine the Higgs line of flight and solve for the missing energy of the neutrinos. One must also assume that the trigger will utilize all the characteristics of the event i.e., a jet with an additional two isolated tracks.

Assume trigger requires > 60 GeV of calorimeter energy and $E_T > 20$ GeV.

E.M.	Hadronic
12% $(e\nu\nu)$	-
7% $(\rho\nu)$	7% $(\rho\nu)$
-	8% $(\pi\nu)$
-	5% (3π)
4%$(\pi^+\pi^o\pi^o)$	1% $(\pi^+\pi^o\pi^o)$
1% $(3\pi^+ n\pi^o)$	3% $(3\pi^+ n\pi^o)$
4% $(K, \pi + \text{neutrals})$	2% $(K\pi + \text{neutrals})$
28%	26%

These are estimates of the percentages of τ's accepted for each mode. That is 28% of τ's will give an E.M. trigger, and 26% of τ's will give a hadronic trigger with the isolation criteria.

EVENT RATES

If we use the region $P_T > 100$ GeV, we start with 2000 events/year. Imposing reductions of 0.5 (rapidity range), 0.54×0.54 (efficiency of detecting both τ's), and 0.5 (additional solid angle cuts or loss of the isolation of one τ) gives 146 events/year.

To improve the signal to background it may be necessary to choose events in which most of the energy of the τ's is observed, i.e., choose a region excluding the decay $Z \to \tau^+\tau^-$. This can be done by using 1-1 τ decays with a high fraction of the τ energy visible or perhaps 1-3 τ decays.

events/year with $E_{vis} > M_Z = 30$

1-3 events/year = 30

BACKGROUNDS

There are a large number of processes which could provide the same trigger i.e. jet + 2 isolated

$$pp \to + \text{ anything } (p_t > 100) \sim \text{ few } mb$$

$$pp \to W + \text{anything} \sim 60\ nb\ (y \pm 3)$$

$$pp \to W + \text{jet } (p_t > 100)\ 2 \times 10^7 \text{ events/year}$$
$$\to e\nu, \mu\nu, \tau$$

$$pp \to Z + \text{jet} \quad p_t > 100\ 2 \times 10^6 \text{ events/year}$$
$$\to ee, \mu\mu, \tau\tau$$

$$pp \to q\bar{q}, gg, qg\ p_t > 100\ 2 \times 10^8 \text{ events/year}$$
$$+ W \text{ radiated}$$
$$\to e\,\nu, \mu\nu, \tau\nu$$

$$pp \to \gamma^*, Z^* \to 1^+1^- + \text{anything} \to 10^6/\text{year}.$$

Snowmass 86 P111 claims for W pair triggering an overall rate of 1 Hz with a similar trigger. It is clear that the isolation criteria needs to have a rejection of 10^{-3} to 10^{-4} to have an acceptable trigger rate.

OTHER PROBLEMS

There is probably an intrinsic P_T to the initial collision ($\sim 30 GeV$). This will cause a feed down from lower P_T and give a higher trigger rate e.g., for $W+$ jet. It will also affect the "missing" E_T.

There is also a potential loss of E_T in the forward direction. Example: A 10 TeV particle at 10 mr has a P_T of 100 GeV which would not be observed.

If one takes a scaling factor of 10 (pure guess) from UA1 to SSC it is equivalent to UA1 using P_T jet > 10 and P_T lepton > 2.

CONCLUSIONS $H^o \rightarrow \tau + \bar{\tau}, P_T > 100$ GeV

1) Need to trigger on all characteristics of the event, i.e., Jet + 2 isolated τ's.
2) Guess is trigger rate will be too high to get all the cross section, so one year will not yield a Higgs signal.
3) Detectors should have the capability to trigger on isolated particles, e.g., the τ.

Both the shape analysis of Atwood and the rate analysis of Miller are pessimistic concerning using the jet $+\tau^+\tau^-$ method for finding H^o. The inclusive H^o approach using τ's decaying to ≥ 3 pions seems more hopeful, but has not yet been subjected to a full rate and triggering analysis. The region where the method could work is shown in Figure 6.

COMMENTS

We hope to pursue the analyses discussed here and obtain more definitive results which will be presented elsewhere. We would be happy to have additional collaborators. In part our approach is not to make claims about what can be done, but to say what must be accomplished by detectors in order to find or

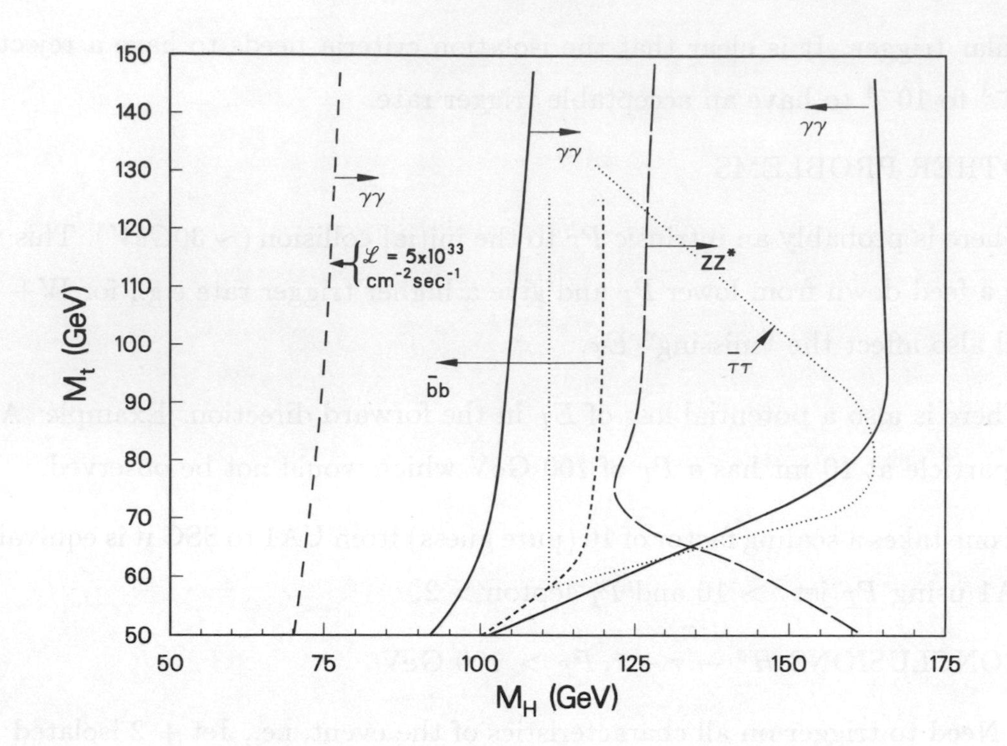

Figure 6. This shows regions where an intermediate mass, Standard Model, Higgs boson appears to detectable at the SSC. The $\gamma\,\gamma$ mode is detectable inside the region marked by the solid line, the ZZ^* mode to the right of the long-dashed line, the $\tau\bar{\tau}$ mode inside the region marked by the dotted line, and the associated production $b\bar{b}$ mode to the left of the short-dashed line. See text for assumptions and the present status of analyses.

exclude an intermediate mass H^o. That is the appropriate attitude since the physics of scalar bosons is so important. Figure 6 shows the regions where the various methods appear workable (generally defined as at least a 4 σ effect).

REFERENCES

1. J.F. Gunion, G.L. Kane, and J. Wudka, Davis preprint UCD-87-28.
2. R.K. Ellis, I. Hinchliffe, M. Soldate, and J.J. van der Bij, preprint FERMILAB-Pub-87/100-T.
3. J.F. Gunion, H.E. Haber, F. Paige, W.-K. Tung, and S. Willenbrock, Davis preprint.
4. C. Baltay, private communication.

5. F. Gilman and L. Price, Proceedings of the Study on the Design and Utilization of the SSC, Snowmass, 1986, ed. R. Donaldson and J. Marx.

6. J.F. Gunion et al., Phys. Rev. Lett 54 (1985) 1226.

7. J.F. Gunion et al., Phys. Rev. D34 (1986) 101.

RARE W DECAYS AT THE SSC?

Frank E. Paige
Physics Department
Brookhaven National Laboratory
Upton, NY 11973

ABSTRACT

The feasibility of studying rare W decays at the SSC is considered. It seems unlikely that hadronic decays can be observed with greater sensitivity than that obtainable at LEP 200. Interesting purely leptonic decays occur in some models and might be observable. The sensitivity will be limited by backgrounds, so excellent lepton identification, jet recognition, and missing p_T resolution are needed.

The SSC will produce approximately 10^9 each of $W^\pm$ and Z^0 per year at a luminosity of $10^{33}\ \mathrm{cm}^{-2}\ \mathrm{sec}^{-1}$, compared to about 10^7 Z^0 at LEP or SLC and 10^4 W^+W^- at LEP 200. This suggests the possibility of using the SSC to search for rare $W^\pm$ and Z^0 decays, either to probe new physics or to measure parameters of the standard model. For the Z^0 the rate advantage of the SSC over the e^+e^- machines is not enormous, while the signal to background ratio is very much worse. The SSC might be competitive only in channels which both can be precisely reconstructed and have negligible background. The only such channel seems to be $Z^0 \to e^\pm\mu^\mp$. However, such a decay with a branching ratio greater than 10^{-9} is ruled out by the existing limit [1]

$$B\left(\mu^\pm \to e^\pm e^+ e^-\right) < 2.4 \times 10^{-12}$$

except possible in very contrived models. Rare Z^0 decays will therefore not be considered further.

This does not imply that the conventional decays $Z^0 \to e^+e^-$, $\mu^+\mu^-$ are of no interest. Indeed, such decays are of great importance as a test of perturbative QCD calculations at SSC energies. Most SSC experiments will involve mass scales M small compared to $\sqrt{s}$, and perturbative QCD is not well tested in this region. Furthermore, the calculations depend on parton distribution functions at small x, and these are not measured by existing experiments. The usual assumption, which is equivalent to assuming a constant hadronic total cross section, is to take at low Q^2

$$f(x) \sim 1/x.$$

and to use this to extrapolate the existing data. Collins [2] has recently argued that the numerical stability of the Altarelli-Parisi equations requires a more singular behavior, roughly

$$f(x) \sim 1/x^p, \qquad p \approx 1.3 - 1.5$$

Certainly the solution of the equations gives an increasingly singular behavior at larger Q^2. This should be tested in the near future by measuring the cross section for low-mass

$\mu^+\mu^-$ pairs at the Tevatron, but additional information at higher mass and energy scales will be useful. On a more practical level, the Z^0 leptonic decays will provide a very useful calibration for the detector; "yesterday's sensation is today's calibration is tommorrow's background."

The SSC will also produce of order 10^9 $W^{\pm}$ events per standard year, compared to only 10^4 W^+W^- pairs per year at LEP 200. Thus the SSC has potentially an enormous advantage, being sensitive in principle to branching ratios $B_W \gtrsim 10^{-8}$ compared to $B_W \gtrsim 10^{-3}$ at LEP 200. The question is whether this potential advantage can be realized in the face of an even more enormous disadvantage in signal to background. High statistics do not require a very large rapidity coverage. While the $W^{\pm}$ rapidity distribution is flat for $|y_W| \lesssim 4$ for the conventional structure functions, and somewhat wider if the structure functions are more singular as $x \to 0$, about half of the events have $|y_W| \lesssim 2$. Thus measurement of the $W^{\pm}$ decay products for $|y| \lesssim 3$ is sufficient.

It seems likely that the most interesting decays would be those into hadronic channels, either into $q\bar{q}$ or into new strongly interacting particles, giving two jets. Unfortunately, these are just the channels in which the QCD background is very large. The raw signal to background ratio for jets at the $W^{\pm}$ mass is

$$\frac{\sigma\left(W^{\pm} \to q\bar{q}\right)}{\sigma\,(\text{jet-jet})} \approx \frac{1}{500}$$

not including any smearing from detector resolution or jet reconstruction. Including these effects would worsen the signal to background ratio by a factor of a few. The interesting level for hadronic decays not observable at LEP 200 is three orders of magnitude smaller, or 1/500000. It of course cannot be proven that no strange hadronic decay could be observed at this level. It is, however, instructive to compare the background rejection required for rare $W^{\pm}$ decays with what has been found in studies of the separation of quark and gluon jets. The separation obtainable between light quark and gluon jets is at best a factor of a few. Hadrons containing c and b quarks can be tagged with a microvertex detector, in principle with small background. But tagging a hadron containing a c or b is not the same thing as tagging a c or b jet, since such hadrons are also produced in gluon jets by $g \to c\bar{c}$, $b\bar{b}$. At $p_T \sim m_W/2$ the probability of a c quark is measured [3] to be of order 10%, in agreement with QCD calculations, and at higher p_T the multiplicity of b hadrons in gluon jets becomes significant. The potential background rejection is set by this unless the momentum of the heavy hadron can be measured and a stringent cut on its z is made, greatly reducing the efficiency.

The most distinctive signature for a heavy quark jet is $t \to b\ell\nu$, giving an isolated lepton, missing p_T, and a b jet. The raw signal to background ratio is

$$\frac{\sigma\,(t\bar{t})}{\sigma\,(\text{jet-jet})} \approx \frac{1}{200}$$

The region $p_T \sim m_t$ is relevant for the UA1 top quark search; limits on m_t have been reported [4] with backgrounds less than but comparable to the signal. For higher p_T Lane and Rohlf [5] found that the obtainable background rejection after requiring an isolated lepton was less than 1/100. This is useful but a far cry from the 1/500000 rejection needed to compete in rare $W^{\pm}$ decays with LEP 200. The ν will give a missing transverse momentum of order 15 GeV, which is very difficult to observe at 40 TeV.

While it is probably possible to observe $W^{\pm} \to t\bar{b}$, unless the signature is truly extraordinary it seems unlikely to observe a $W^{\pm}$ decay into jets at even the 10^{-3} level, let alone the 10^{-7} potential sensitivity of the SSC.

In addition to jet decays one might consider looking for exclusive decays involving hadrons. The branching ratios for $W^{\pm} \to P^{\pm}\gamma$, where $P^{\pm}$ is a pseudoscalar meson, are reliably calculable using PCAC and electromagnetic gauge invariance. The result is [6]

$$\frac{\Gamma\,(W^{\pm} \to \pi^{\pm}\gamma)}{\Gamma\,(W^{\pm} \to \mu^{\pm}\nu)} = 3 \times 10^{-8}$$

implying 2 events per year with

$$\frac{\sigma\,(W^{\pm} \to \pi^{\pm}\gamma)}{\sigma\,(\text{jet} - \text{jet})} = 6 \times 10^{-11}$$

This seems somewhat less than very promising. The largest such branching ratio is into the light pseudo-Goldstone fermions $P_{TC}^{\pm}$ in technicolor theories, where [6]

$$\frac{\Gamma\,(W^{\pm} \to P_{TC}^{\pm}\gamma)}{\Gamma\,(W^{\pm} \to \mu^{\pm}\nu)} = 2 \times 10^{-5}$$

implying 1700 events per year. This may be observable for $P_{TC}^{\pm} \to \ell^{\pm}\nu$, but it is certainly not the way to discover technicolor; the $P^{\pm}$ would easily be found at LEP 200 if not before. Nor would one learn much new, since the calculation depends only on PCAC and electromagnetic gauge invariance. Branching ratios for $W^{\pm} \to \pi^{\pm}\pi^{0}$ are even smaller.

If hadronic decays are rather discouraging, leptonic decays are at least detectable, and there are approximately 10^{8} $W^{\pm} \to \ell^{\pm}\nu$ and 10^{7} $Z^{0} \to \ell^{+}\ell^{-}$ decays per year. Unfortunately, these are not very interesting in the standard model. One could use them to test e-μ-τ universality, but the experimental detection of the three leptons is so different that the systematic errors would necessarily be large, limiting the attainable precision.

In nonstandard electroweak models there can be interesting purely leptonic decays. For example, in a left-right symmetric $SU\,(2) \times SU\,(2) \times U\,(1)$ model, the leptons are placed in two doublets:

$$\begin{pmatrix} \nu_e \\ e^- \end{pmatrix}_L \quad \begin{pmatrix} N_e \\ e^- \end{pmatrix}_R$$

The heavy neutral lepton N_e interacts only through the right-handed gauge bosons and the Higgs sector. To agree with standard model results, $W_R^{\pm}$ and $W_L^{\pm}$ must be almost unmixed, so that N_e is almost decoupled from $W_L^{\pm}$, and indeed from all known particles. If the N_e is lighter than the $W_L^{\pm}$, then the mixing can be measured from the branching ratio for decay chains like

$$\begin{array}{cccc} W_L^{\pm} & \to & e^{\pm} \; + & N_e \\ & & & \downarrow \\ & & & e^{\mp}\mu^{\pm}\nu_{\mu} \end{array}$$

The N_e could even be a Majorana particle, in which case one would also have lepton-number-violating decays like

$$W_L^\pm \rightarrow e^\pm + \underset{\displaystyle e^\pm \mu^\mp \nu_\mu}{\underset{\downarrow}{N_e}}$$

with two like-sign electrons. This is certainly a dramatic — albeit perhaps an unlikely — signature.

The experimental problem is to reject misidentified electrons and muons and real ones coming from heavy quark decays. The degree to which this can be done would require detailed study, but the general requirements on the detector are obvious:

- Excellent electron and muon identification over a reasonably large range in rapidity. Having both not only increases the rate but provides an important systematic check.
- Excellent hadron calorimetry at rather low p_T, since $b\bar{b}$ jets can give similar leptons except for the isolation requirement.
- Missing p_T would provide an additional handle, but the required resolution is of order 10 GeV, which is probably impossible.

Again, this is not the most likely way to discover left-right symmetry. One would probably first find $e^+e^- \rightarrow N_e N_e$ (although its production depends on the mixing of Z_L^0 and Z_R^0) or directly observe the $W_R^\pm$ at the SSC. But observation of the $W_L^\pm \rightarrow e^\pm N_e$ decays might provide valuable information about mixing and other properties of the model. The achievable sensitivity will surely be set by the background rejection and not by the available sample.

In gauge theories the tri-linear couplings of the $W^\pm$ and Z^0 are precisely defined and indeed are required for the renormalizability of the theory. Measuring them experimentally is an important test of the standard model and is often cited as a major justification for LEP 200. The methods of doing this have recently been analysed in detail for $e^+e^- \rightarrow W^+W^-$ by Hagiwara, Peccei, Zeppenfeld, and Hikasa [7]. The most general CP conserving effective Lagrangian $\mathcal{L}$ excluding higher derivative terms is $(V = \gamma, Z^0)$

$$\begin{aligned}
\frac{1}{g_{WWV}}\mathcal{L} = {} & ig_1\left(W_{\mu\nu}^\dagger W^\mu V^\nu - W_\mu^\dagger V_\nu W^{\mu\nu}\right) \\
& + i\kappa W_\mu^\dagger W_\nu V^{\mu\nu} \\
& + \frac{i\lambda}{m_W^2} W_{\lambda\mu} W_\nu^\mu V^{\nu\lambda} \\
& + g_5\, \epsilon^{\mu\nu\rho\sigma} W_\mu^\dagger \overleftrightarrow{\partial} W_\nu V_\sigma \\
W_{\mu\nu} = {} & \partial_\mu W_\nu - \partial_\nu W_\mu
\end{aligned}$$

where κ is conventionally called the "anomalous magnetic moment." In gauge theories $g_1 = \kappa = 1$ and the rest of the couplings are $O(\alpha)$ or smaller. The various couplings can be determined by measuring particular angular distributions for the $W^\pm$ decays using the class of fully reconstructable events

$$\begin{array}{ccccccc} e^+ & + & e^- & \to & W^\pm & + & W^\mp \\ & & & & \downarrow & & \downarrow \\ & & & & \ell^\pm \nu_\ell & & q\bar{q} \end{array}$$

Given the expected statistics at LEP 200, a deviation of $\Delta\kappa = .5$ from the standard model value is clearly visible; a 1σ limit of $\Delta\kappa = .1$ is claimed in Ref. 7. This is a reasonable test of the gauge nature of the interaction but far from the level of radiative corrections.

There is of course every reason to expect that the gauge nature of the $W^\pm$ and Z^0 will be confirmed experimentally, at least to the level that can be probed at LEP 200. If anomalous couplings were found, then the $W^\pm$ and Z^0 cannot be fundamental gauge bosons. Presumably they would be composites of some sort, leading to dramatic new interactions on the 1 TeV scale. Searching for such effects would be the natural way to probe such physics at the SSC.

One might also try to measure the effective $W^+W^-\gamma$ interaction from a precise measurement of $W^\pm\gamma$ production at relatively low $p_{T,\gamma}$. This process does not involve the $W^+W^-Z^0$ vertex and so could provide independent information from that obtainable in e^+e^-. In the standard model the interference between the s-channel and t-channel graphs for $d\bar{u} \to W^-\gamma$ produces a remarkable zero in the cross section at [8]

$$\cos\theta^* = e_d = -1/3;$$

this is absent for $\kappa \neq 1$. The statistics available to make the measurement are large: for $p_{T,\gamma} > 40\text{GeV}$

$$B_W\left(\ell^\pm\nu\right)\sigma\left(WW^\pm\gamma\right) = 2 \times 10^{-36}\text{cm}^2 \quad \Longrightarrow \quad 20000 \text{ events/yr.}$$

But the potential background is also large: for $p_{T,jet} > 40\text{GeV}$

$$B_W\left(\ell^\pm\nu\right)\sigma\left(W^\pm\text{jet}\right) = 1 \times 10^{-32}\text{cm}^2$$

The signal-to-background ratio is even worse at the interesting dip in the cross section.

Triggering on such events is a difficult problem; the ISAJET [9] predictions shown in Fig. 1 for the transverse momenta of the e, the γ, and the ν are all of order 50 GeV, as expected. The rapidity distributions of the e and the γ, which both extend to $|y| \approx 5$, are shown in Fig. 2. But the experiment will be limited by background, not by statistics, so it is sufficient to cover a substantially smaller rapidity interval. To reconstruct $\cos\theta^*$ for the $W^\pm\gamma$ scattering, it is also necessary to measure the missing p_T to have a 0C fit for the $W^\pm$ momentum. Calculating the true missing p_T resolution requires a detailed simulation of the detector, but a rough estimate can be made by putting events into a highly idealized calorimeter having cells with $\Delta y = .1$, $\Delta\phi = 5^\circ$ with no shower spreading and energy resolutions

$$\left(\frac{\Delta E}{E}\right)_{\text{e.m.}} = \frac{.15}{\sqrt{E}} \qquad \left(\frac{\Delta E}{E}\right)_{\text{had.}} = \frac{.50}{\sqrt{E}}$$

This produces the resolution curves shown in Fig. 3 for three different rapidity coverages. Evidently calorimeter coverage for $|y| < 5$ is required and $|y| < 6$ is significantly better.

Since the $W^{\pm}$jet background is so large, a detailed analysis including detector effects is required to determine the sensitivity. This has never been carried out for SSC energies, but it was done [10] for ISABELLE at $\sqrt{s} = 800\,\mathrm{GeV}$. Some simulated distributions from this analysis are shown in Fig. 4, and from these it was concluded that the sensitivity was of order $\Delta\kappa = 1$, less than that quoted for LEP 200. Improvements in event simulation since this analysis was done have made the events more complicated and difficult to reconstruct. Compared to $\sqrt{s} = 800\,\mathrm{GeV}$, the SSC has higher statistics even for a higher cut on $p_{T,\gamma}$, but it has a worse signal to background ratio, worse missing p_T resolution, and poorer separation between $q\bar{q}$ and $\bar{q}q$ production. It seems unlikely, therefore, that the measurement of $W^{\pm}\gamma$ at the SSC at low p_T could compete with LEP 200 in sensitivity, although if LEP were to find an effect, it might provide significant new information.

This work has been supported in part by the United States Department of Energy under contract number DE-AC02-76CH00016.

REFERENCES

1. Particle Data Group, Phys. Letters 170B (1986).
2. J.C. Collins, In *Supercollider Physics*, ed. by D.E. Soper, (World Scientific, 1986), p. 62.
3. UA1 Collaboration, Phys. Letters 147B, 222 (1984).
4. UA1 Collaboration, presented at the 1987 Europhysics Conference and in preparation.
5. K. Lane and J. Rohlf, in *Design and Utilization of the SSC*, ed. by R. Donaldson and J.G. Morfin (Snowmass, 1984), p. 737.
6. L. Arnellos, W.J. Marciano, and Z. Parsa, Nucl. Phys. B196, 378 (1982).
7. K. Hagiwara, R.D. Peccei, D. Zeppenfeld, and K. Hikasa, DESY 86-058.
8. R.W. Brown, D. Sahdev, and K.O. Mikaelian, Phys. Rev. D20, 1164 (1979).
9. F.E. Paige and S.D. Protopopescu, BNL-38774 (1986).
10. S. Kahn, T.J. Killian, M.J. Murtagh, and F.E. Paige, BNL-34020 (unpublished).

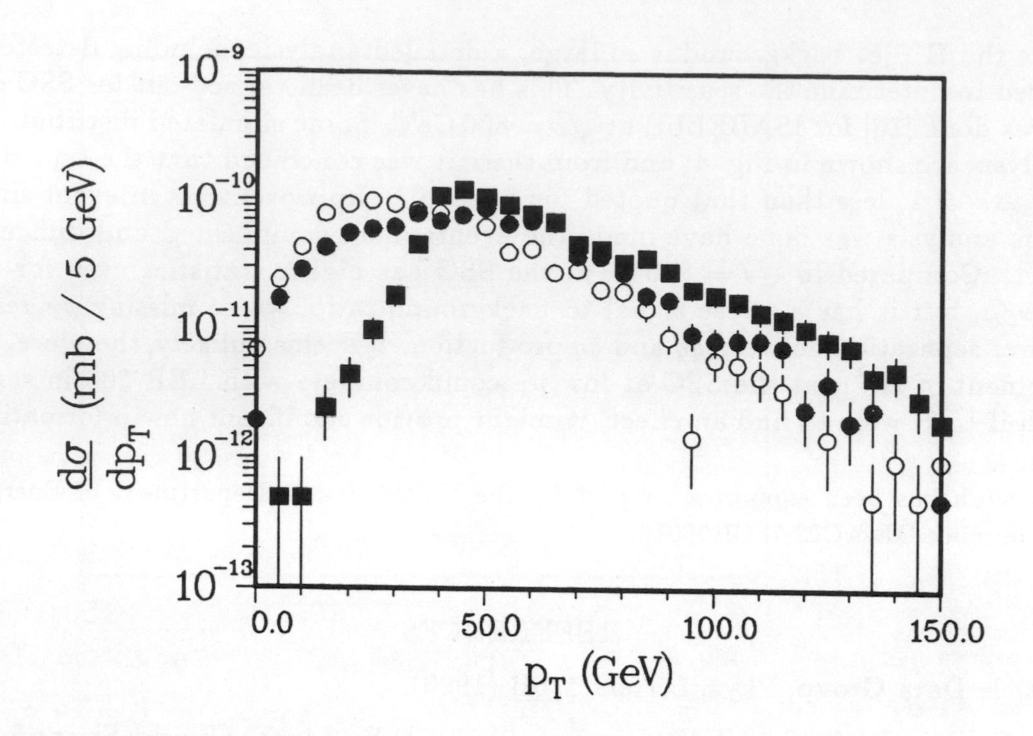

Fig. 1: p_T distributions for $e^{\pm}$ (solid circles), γ (solid squares), and ν (open circles) in $W^{\pm}\gamma$ events at SSC.

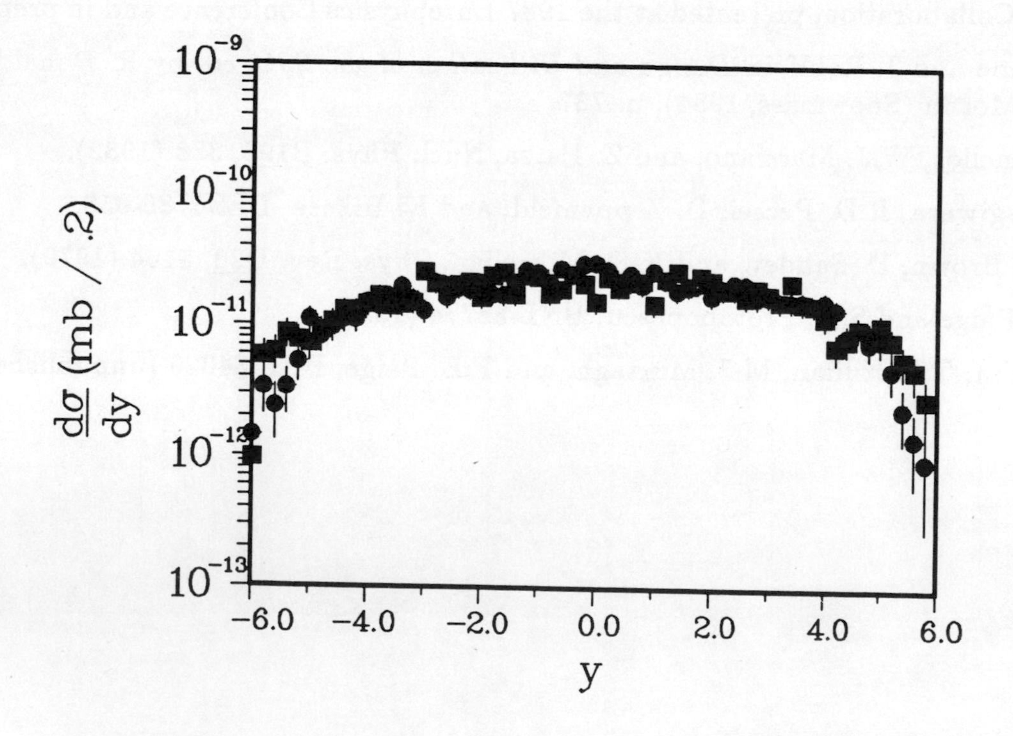

Fig. 2: y distributions for $e^{\pm}$ (solid circles) and γ (solid squares) in $W^{\pm}\gamma$ events at SSC.

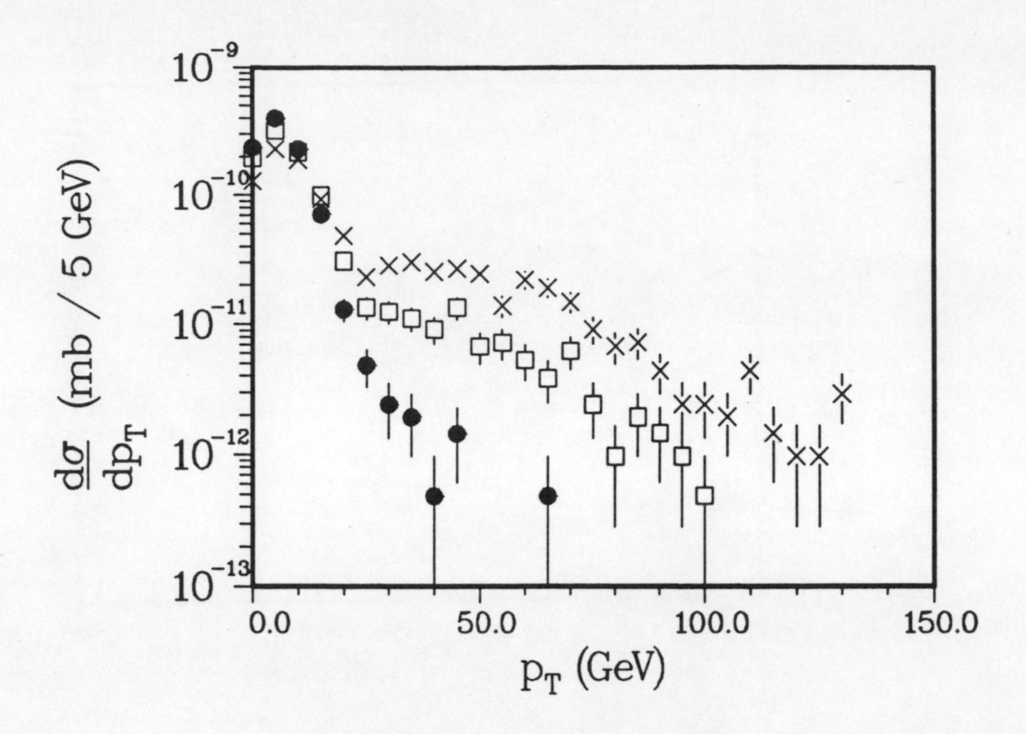

Fig. 3: $|p_{T,\nu} - p_{T,miss}|$ distributions for an idealized calorimeter with $|y| < 4$ (crosses), $|y| < 5$ (open squares), and $|y| < 6$ (solid circles).

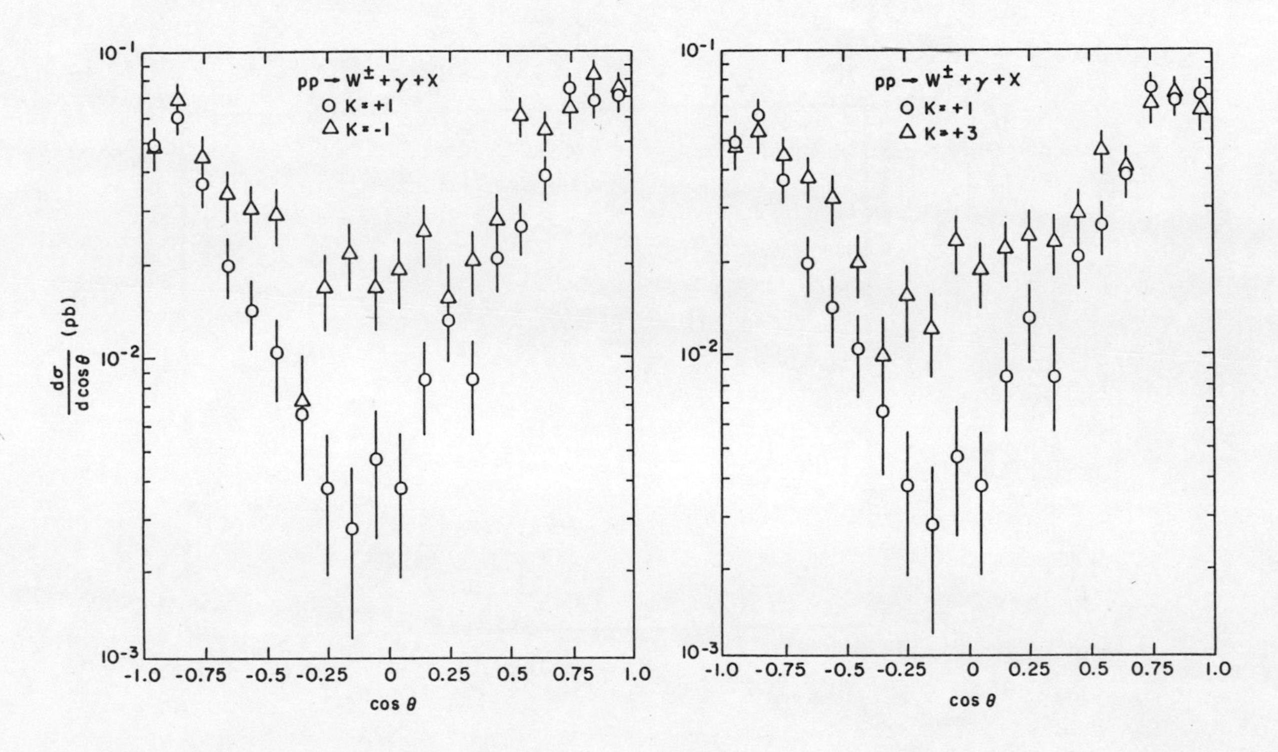

Fig 4: Simulated $\cos\theta^*$ distributions for $W^{\pm}\gamma$ events at $\sqrt{s} = 800\,\text{GeV}$ for different values of κ.

Intermediate p_T Detector Configurations

REPORT OF THE INTERMEDIATE-$p_\perp$ DETECTOR GROUP: A BEAUTY SPECTROMETER FOR THE SSC

K. J. Foley
Brookhaven National Laboratory, Upton, Long Island NY 11973
C.D. Buchanan
University of California, Los Angeles CA 90024
R. J. Morrison, S. W. McHugh, and M. S. Witherell
University of California, Santa Barbara CA 93106
M. Atac, B. Cox, M. V. Purohit, R. Stefanski, and D. E. Wagoner
Fermilab, Batavia IL 60510
K. Schubert
Heidelberg University, Heidelberg, FRG
L. W. Jones
University of Michigan, Ann Arbor MI 48109
I. Leedom
Northeastern University, Boston MA 02115
D. A. Buchholz
Northwestern University, Evanston IL 60201
N. W. Reay
Ohio State University, Columbus OH 43210
N. Lockyer
University of Pennsylvania, Philadelphia PA 19104
S. N. White
Rockefeller University, New York NY 10021
V. G. Lüth and S. L. Shapiro
Stanford Linear Accelerator Center, Stanford CA 94305
D. E. Groom
Superconducting Super Collider Central Design Group, Berkeley CA 94720
A. Jawahery
Syracuse University, Syracuse NY 13244
P. Karchin, M. P. Schmidt, and J. Slaughter
Yale University, New Haven CT 06511

A "Beauty Spectrometer" has been designed for studies of B physics at the SSC. The ultimate goal is a definitive measurement of CP violation in the B system. The spectrometer consists of two stages and occupies one side of an intermediate-luminosity interaction region. An upstream, or intermediate, stage extends from the interaction point to 14 m and covers the angular region from 57 mrad (3.3°) to 350 mrad (20°). The forward stage extends to 77 m and to angles down to 5.7 mrad. The design includes silicon microstrip detectors, conventional tracking, momentum analysis, and hadron and lepton identification. While no fundamental problems have been found, the detector must deal with unprecedented particle fluxes, trigger rates, and data rates.

1. Introduction

The great majority of the intermediate-$p_\perp$ group worked as one subgroup on one problem—a spectrometer for bottom quark physics. The members of this effort set out to make a trial design of a forward beauty spectrometer capable of decisively observing CP violation at the level expected in the standard model. The theoretical groundwork was provided by the B physics group, whose report is contained in this volume [1]. Although there has been a great deal of interest in various techniques for obtaining large samples of B-meson decays, it is doubtful that any of them will be sufficient for studying CP violation before the SSC is operating. Even though there are still many experimental questions that require further study, it was possible to produce a rough design that is not obviously inadequate. It should serve as a good starting point for more detailed studies in the future, which can concentrate on the critical problems pointed out in this study.

Since the forward beauty spectrometer subgroup included most of the members of the intermediate $p_\perp$ group, its description makes up the bulk of this report. Two smaller smaller groups worked on the problem of electron and muon triggers for the B-physics experiment. The description of the electron trigger [2] is included in this volume as a separate contribution. It is treated separately in part to emphasize the fact that the lepton triggers are crucial for the success of the experiment and by themselves present difficult problems that need more attention.

Other members of the intermediate-$p_\perp$ group worked on three spectrometers, which are described in a separate paper [3]. These designs address experimental goals not well matched to the general purpose detectors, and which are discussed in detail in the report. In addition, L. W. Jones considered instrumentation of the forward cone inside that covered by the beauty spectrometer; his design is also presented separately [4].

2. B Physics at the SSC

There is a great deal of interest in the bottom quark system as a source of information on CP violation. In contrast to the charm case, bottom can only decay outside the top-bottom family, greatly enhancing the sensitivity to CP violation. The recent observation of mixing for B_d^0 has generated still more interest in the possibility for observing CP violating effects, since the size of these effects is now expected to be substantially larger.

In the bottom system CP violating effects are expected to appear in many decay modes, but as a general pattern the largest asymmetries occur for modes with extremely small branching ratios. As a result, an experiment must be exposed to a very large number of B's in order to guarantee making a clear statement about CP violation.

The cleanest predictions of CP violating effects occur for decays into CP eigenstates. In these cases, CP violation is observed by comparing the decay time distributions of the CP eigenstates, using the other B to tag whether a particle or antiparticle is decaying. For example, the final state for B (or $\overline{\mathrm{B}}$) $\rightarrow \psi \mathrm{K}_s \mathrm{X}$ is tagged as CP = +1 by the fast 2π decay of the K_s, while the leptonic decay of the other B identifies the parent as a B or $\overline{\mathrm{B}}$. The asymmetry $(\mathrm{N(B)} - \mathrm{N}(\overline{\mathrm{B}}))/(\mathrm{N(B)} + \mathrm{N}(\overline{\mathrm{B}}))$ is expected to be between 0.05 and 0.3 if the violation arises through the phase angle δ in the Kobayashi-Maskawa matrix, large enough to be measured with relative freedom from systematic errors if enough events are available. With the expected branching ratios, tagging efficiency, and other penalties, the asymmetry can be measured at the 3σ level with less than 10^9 B's.

The total cross section for $\mathrm{B}\overline{\mathrm{B}}$ production at the SSC is expected to lie in the range 200 μb to 700 μb.* The spectrometer we envisage makes use of an intermediate-length interaction region, where the luminosity is 10^{32} cm^{-2}s^{-1}. The $\mathrm{B}\overline{\mathrm{B}}$ production rate is then (2 to 7) $\times$ 10^4 s^{-1}, for a total of (2 to 7) $\times$ 10^{11} in a nominal year of running. The rapidity distribution is expected to be fairly flat out to 5 or 6.†

Given the prodigious rates and rich physics, $\mathrm{B}\overline{\mathrm{B}}$ studies promise to be an important part of the SSC program. They are also very difficult. The $\mathrm{B}\overline{\mathrm{B}}$ rate is already formidable, but the total p-p interaction rate is about 500 times greater, with roughly the same rapidity distribution. The spectrometer must operate at unprecedentedly high rates, there is an extreme demand on triggering and readout electronics, and there must be an enormous capability for data transfer and data reduction. To investigate the possibilities and problems of these rates,

---

* We take 200 μb as the "standard" $\mathrm{B}\overline{\mathrm{B}}$ production cross section for purposes of this report. This cross section is to be compared with the 1.1 μb cross section measured at the S$\bar{\mathrm{p}}$pS. The SSC luminosity contributes several additional orders of magnitude to the rate.

† The rapidity y is defined as

$$y = \tfrac{1}{2}\ln\frac{E+p_{\parallel}}{E-p_{\parallel}}$$
$$= \tfrac{1}{2}\ln\frac{\cos^2(\theta/2)+m^2/4p^2+\cdots}{\sin^2(\theta/2)+m^2/4p^2+\cdots}\ .$$

If terms of order $(m/p)^2$ and higher can be neglected, the function depends only upon angle and is called the *pseudorapidity* η. The approximation $y \approx \eta$ breaks down for low energy particles and for angles smaller than $m/p \approx 1/\gamma$. Because of its greater relevance to experimental design, pseudorapidity is used for the remainder of this report.

we have made a first pass at the design of a specialized detector for B physics at the SSC. Our goal was to make a detector which would be as sensitive as possible to CP violation in as many modes as possible, and, in addition, to be sensitive in general to rare B decays. It followed immediately that high precision vertex finding and both hadron and lepton particle identification are required.

Triggering is an extremely difficult problem. In addition to the J/ψ trigger which has been previously discussed [5], we attempted to make a spectrometer which includes electron and muon triggers. The trigger lepton provides a particle-antiparticle tag if the associated D is reconstructed, allowing for a wide variety of detectable decay modes for the other B.

3. Design Considerations

i. Acceptance

If the distribution in rapidity (actually pseudorapidity) $dN/d\eta$ is taken as a constant (C), then $|dN/d\theta| = C/\sin\theta$ and $dN/d\Omega = C/(2\pi\sin^2\theta)$. Therefore (a) the distribution of B's (or secondary particles of any kind) is peaked strongly in the forward direction, falling only at angles smaller than 0.7 mrad,* (b) the counting rate in a small counter at a transverse distance R_T from the beam is independent of distance from the interaction point, down to $\theta \approx 5$ mrad ($\eta \approx 6$), and (c) at a fixed distance from the interaction point the counting rate scales as R_T^{-2}.

These considerations led us to a spectrometer occupying the forward region. The length of the detector is well matched to the available free space in the intermediate-luminosity interaction regions planed for the SSC, and the luminosity in these regions is well matched to rate requirements for CP violation studies and other B studies. The forward region is also well matched to the geometry of silicon microstrip detectors, whose resolution is of fundamental importance for heavy flavor identification.

Since both the B and the $\overline{\mathrm{B}}$ must be accepted, there is a high premium on providing a large acceptance. The expected production characteristics of quarks have been described by B. Cox and D. E. Wagoner in 1986 [5], and by Foley *et al.* elsewhere in this volume [1]. A broad distribution of B mesons extends to about ± 6 units of rapidity. The b and $\bar{b}$ quarks tend to be produced in the same direction, so the directions of the B and $\overline{\mathrm{B}}$ mesons are highly correlated within about ± 1.5 units of rapidity. The resulting angular distribution of B quarks is strongly peaked in the forward-backward direction with respect to the beams,

---

* Since $dN/d\eta$ decreases for $\eta \gtrsim 6$.

with both B quarks on the same side.*

To investigate the acceptance of a generic detector, a study was performed simulating the decays of B's into two and three charged particles. The B and $\overline{\text{B}}$ mesons were generated with a $p_\perp$ greater than two GeV/c and in the rapidity interval $-6 < \eta < 6$. The results of this study are shown in Fig. 1. Here increasing the acceptance interval of the "TASTER", ($1° < \theta < 20°$, or $1.74 < \eta < 4.74$) [6], was studied. All of the decay products from both B's fall in the 1°–20° acceptance for 8.8% of the events. Keeping the 20° ($\eta = 1.74$) intermediate acceptance and lowering the forward angle to 5 milliradians ($\eta = 6$) increases this fraction to 14%. Keeping the forward acceptance limited to 1° and increasing the outer acceptance to 40° increases the fraction to 15%.

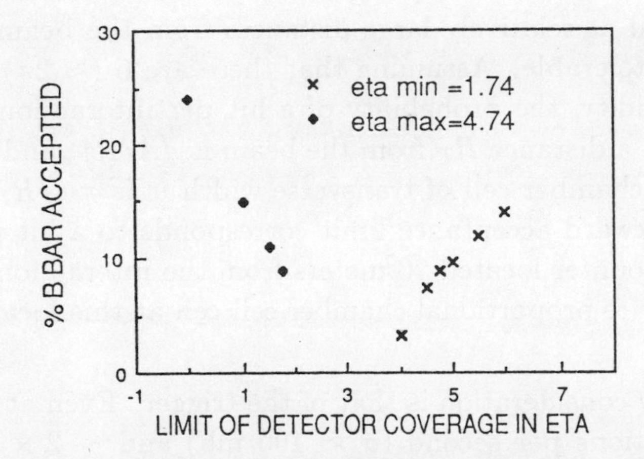

Figure 1. $\text{B}\overline{\text{B}}$ acceptance of the 1986 "TASTER" for both B's into two and three body charged modes as the inner and outer acceptance angles are changed. The dots indicate the fraction of events accepted when the outer acceptance angle is varied and the small angle acceptance is fixed at 1°. The crosses correspond to keeping the outer acceptance fixed at 20° and varying the small angle acceptance. The outer acceptance angle of the spectrometer described in this report is the same as that of the "TASTER," but the inner angle (5.7 mrad) corresponds to $\eta \approx 6$.

Clearly the acceptance should be increased both forward and outward as much as possible. In fact, however, it is only the part of the acceptance which has good vertex finding, charge particle reconstructibility, and particle identification. We have accordingly adopted a minimum angle of 5.7 mrad while leaving the outer angle at 20° (350 mrad).

* Because of this, there is little to be gained in covering both directions. The replication of cost and effort would only double the rate.

ii. General considerations

We have mentioned that for a number of reasons an intermediate luminosity of $10^{32}\,\mathrm{cm}^{-2}\mathrm{s}^{-1}$ is appropriate for B physics at the SSC. If we assume current technology (silicon microstrip vertex detectors) for the vertex finding, then the multiple Coulomb scattering limit on the vertex resolution is $\Delta x = 0.014(R_T/p_\perp)\sqrt{X_0}$. In this expression R_T is the transverse distance from the beams, $p_\perp$ is the transverse momentum of the particle in GeV/c, and X_0 is the thickness of the detector in radiation lengths. This limit requires that the vertex detectors be placed within about 1 cm of the beams. With a luminosity of $10^{32}\,\mathrm{cm}^{-2}\mathrm{s}^{-1}$, current detectors could just survive the dose from a year of running, estimated as 3 megarads at 1 cm from the beam and scaling as the inverse square of this distance.

Another reason is that the approximately ± 100 meters of clear space associated with intermediate luminosity regions allows the small angle regions of detectors to be placed at relatively large distances from the beam, where rates and occupancies are tolerable. Assuming that there are 6 ($\approx 2\pi$) charged particles per unit of rapidity, the probability of a hit per interaction in a counter element of area dA at a distance R_T from the beam is dA/R_T^2, and the probability of a hit in a wire chamber cell of transverse width w is $\pi w/R_T$. The 0.0057 rad chosen for the forward acceptance limit corresponds to a hit probability of $5 \times 10^{-4}\,\mathrm{cm}^{-2}$ for a counter located 70 meters from the interaction region. Similarly, a 2 mm multiwire proportional chamber cell cell at this location has a hit probability of 1.5%.

A final luminosity consideration is that of the trigger. Even at $10^{32}\,\mathrm{cm}^{-2}\mathrm{s}^{-1}$ there are 10^7 interactions per second ($\sigma \approx 100$ mb) and $\sim 2 \times 10^4$ $\mathrm{B\overline{B}}$ pairs produced each second ($\sigma \approx 200\ \mu$b), of which 3000 s^{-1} are "in aperture." It is already a severe challenge to design a trigger which is both open for $\mathrm{B\overline{B}}$ events and which efficiently rejects the 4000 times larger rate of uninteresting events.

Given that the luminosity is limited to $10^{32}\,\mathrm{cm}^{-2}\mathrm{s}^{-1}$, then what determines the small angle acceptance limit? Here a number of factors come in at $\lesssim 10$ mrad. At 0.0057 radians a 0.3 mm thick cylindrical beryllium beam pipe presents about 10% of a radiation length of material. There are probably better shapes than a simple cylinder for the beam pipe, but in any case the beam pipe becomes a serious problem at very small angles. While occupancies and momentum resolution can be improved with still more space and extra magnets, it appears that the ultra-small angular region is not worth the effort. We chose 0.0057 radians as a practical small-angle limit.

A number of experimental problems develop as the large-angle acceptance limit increases. The vertex detector geometry becomes increasingly complicated due to the ± 7 cm length of the collision region. Tracks from different parts of this region necessarily have large angles with respect to the normal to the silicon detector plane. Since the silicon is much thicker than the strip spacing (~50 μm), these large-angle tracks produce partial signals from a number of neighboring strips. These small signals may not be distinguished reliably from noise. If they could be, there would still be a worsened position resolution and a larger probability of overlapping hits.

For this design we decided upon particle tracking and vertex finding before the first magnet, using straight tracks (a condition not completely satisfied for very small angle tracks). It is geometrically very difficult to find an arrangement of magnets and tracking devices which finds reliable pre-magnet track segments and also extends to angles larger than about 20 degrees (350 mrad). Finally, particle and especially electron identification becomes more problematic for these rather low $p_\perp$ tracks at the larger angles. While there is clearly a significant gain to be achieved in extending the angular coverage to larger angles, we decided upon a limit of 350 mrad for purposes of this study.

iii. Minimum size of the beam pipe

The minimum allowable size for the beam pipe can be set by a simple criterion: the limiting aperture of the machine should be somewhere else during all phases of the collider operation. It is sufficient to consider the limiting cases of injection and collision. Both the beam size and the finite crossing angle determine the result.

The r.m.s. width of the beam in the x (or y) direction is given by

$$\sigma_{x(y)} = \sqrt{\epsilon_N \beta_{x(y)}/\gamma}\ ,$$

where ϵ_N is the normalized emittance (nominally 1.0×10^{-6} m-rad), β is the transverse or betatron amplitude function, and γ is the usual Lorentz factor. The optics in the arcs is always the same, with $\beta = 388$ m, and the beam pipe radius is 1.65 cm. During injection β in the interaction region (IR) varies from 60 m at the interaction point (IP) to 320 m at the quadrupole entrance 120 m away. Scaling the arc beam pipe radius by $\sqrt{\beta}$, we find a limiting *inside* radius of 0.7 cm at the IP and 1.4 cm at the quadrupole entrance. Halfway between (60 m) β is about 12 m, so the limiting *inside* radius is 1.0 cm.

After acceleration, the beams are smaller by a factor of $\sqrt{20}$ because of the change in the Lorentz factor. β reaches a very large maximum (2600 m) in the IR quadrupoles. A limiting aperture here would be intolerable for background reasons, so scrapers are moved into place elsewhere in the machine. As a result, the injection optics discussed above still determine the limiting beam pipe dimensions, even though β at the quadrupole entrance is actually larger in the collision optics configuration.

The crossing angle (α) of the beams is 70 μrad in both the high- and intermediate-luminosity IRs, and for a variety of reasons it cannot be decreased much further. Thus $\alpha/2 \times 120$ m $= 0.42$ cm must be added to the vertical clearance at the quadrupole entrance. The addition scales linearly with distance from the IP.

These results are summarized in Table 1.

Table 1

Minimum inside dimensions of the beam pipe in an intermediate-luminosity interaction region. Alignment errors are assumed to be zero.

Distance from IP	r_x	r_y
0 m	0.7 cm	0.7 cm
60 m	1.0 cm	1.2 cm
120 m	1.4 cm	1.8 cm

iv. Forward and intermediate stages

Given that the occupancy of a detector element depends only upon R_T and is independent of the z position, and given the enormous range of particle momenta, the detector naturally needs to be composed of two, and perhaps even three stages, each covering an optimal rapidity range. Two stages were chosen to avoid the extra edge. Given two stages, the hole into the forward stage, should be made as large as possible. For $dN/d\eta \approx 6$ the occupancy of the edge is $2\pi\Delta R_T/R_T$, where ΔR_T is the edge thickness as defined by electromagnetic and hadronic transverse shower thicknesses.

A large hole is also advantageous for detecting K_s's. This is particularly important since ψK_s is one of the prime CP violation modes. We assume that to be observed a K_s must decay before or within a magnet. This leads to a reduced K_s efficiency for the rather fast K_s's which are just outside of the maximum forward spectrometer acceptance angle. Particles analyzed by the forward stage have much better tracking and particle identification and have a much better momentum resolution.

The cost of the downstream detector limits the size of the hole. The 0.057 radian choice for the hole very roughly bisects the rapidity ranged covered by the detector, with an average of 11 charged particles going into the intermediate detectors and 14 through the hole into the forward stage.

v. Trigger

As discussed in the 1986 Summer Study [5], the B$\rightarrow \psi X \rightarrow \ell^+\ell^- X$ is a clean trigger which is fairly easy to implement. However, this branching ratio with ψ going into either $\mu^+\mu^-$ or e^+e^- is only 7×10^{-4}. The semileptonic B branching ratio into a muon or into an electron is about 10%, more than 100 times larger than for the B $\rightarrow \psi \rightarrow \ell^+\ell^-$ case. In addition, the lepton plus an associated and reconstructed D clearly tags the particle-antiparticle character of the B meson (CP violation is then determined by a study of the time evolution of the "other" B meson). It is thus strongly desirable to trigger on a single electron or muon.

To be effective, such a trigger must include reconstruction of the lepton track to the vertex and must require impact parameters of $\gtrsim 200\ \mu$m with respect either to the beam interaction volume or with respect to another found vertex. There will be a large direct charm background for such a trigger, which can be reduced by imposing a $p_\perp \gtrsim 1.5$ GeV/c requirement on the lepton track. Clearly, the spectrometer must be designed to facilitate the trigger. For this study we have focused on the electron trigger but clearly both electron and muon triggers should be implemented (if practical).

4. The Spectrometer

The 1987 B spectrometer for the SSC shown in Figs. 2 and 3. The gross features are an intermediate (wide angle) spectrometer covering the rapidity range from 1.74 to 3.6, and a forward (downstream) spectrometer with rapidity coverage from 3.6 to 5.9. The intermediate spectrometer contains a 1.5 tesla-meter dipole, starting 1 meter downstream from the center of the interaction volume, and supplying a $p_\perp$ kick of 0.45 GeV/c to charged particles. All such charged particles will have traversed 6 silicon microstrip planes with perpendicular anode and cathode readouts, leading to 12 high-resolution position measurements before the particle enters the magnetic field. The microstrip detectors in combination with the two downstream drift chamber tracking stations will result in a momentum resolution of

$$\Delta p/p \sim 3 \times 10^{-3} + 10^{-4} p$$

for the intermediate spectrometer. The corresponding mass resolution $\Delta m/m$ will be ~0.6% for typical 30 GeV particles.

Particles produced at angles smaller than 0.057 radians go into the forward spectrometer, passing through a 4.5 tesla-meter dipole and three drift chamber tracking stations. These particles receive a total $p_\perp$ kick of 1.8 GeV/c. The

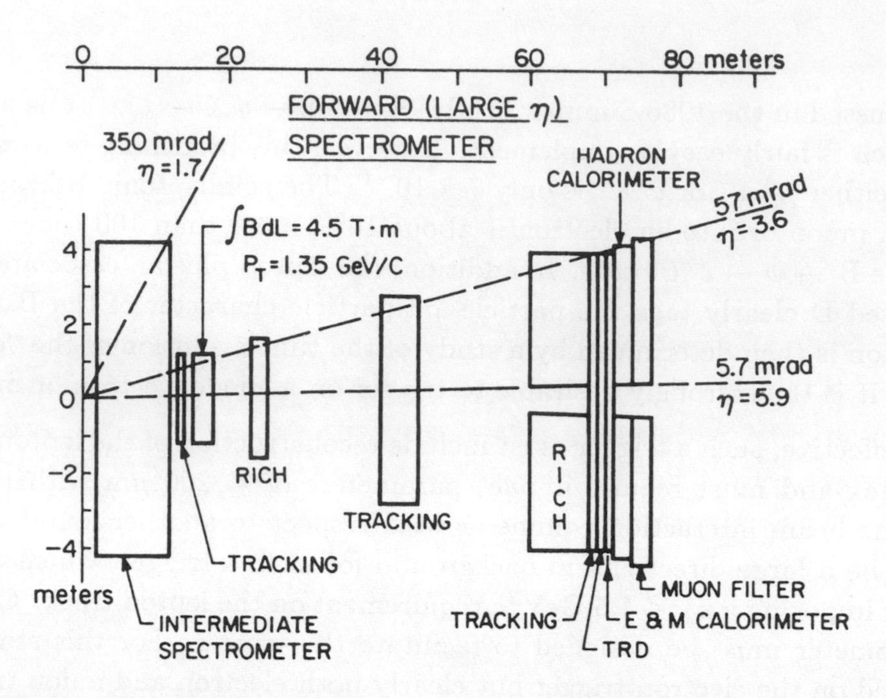

Figure 2. The 1987 B spectrometer. To set the scale, the electromagnetic calorimeter is located 70 meters from the beam crossing point and has transverse dimensions of ± 4 meters. The upstream (intermediate) spectrometer is shown in Fig. 3.

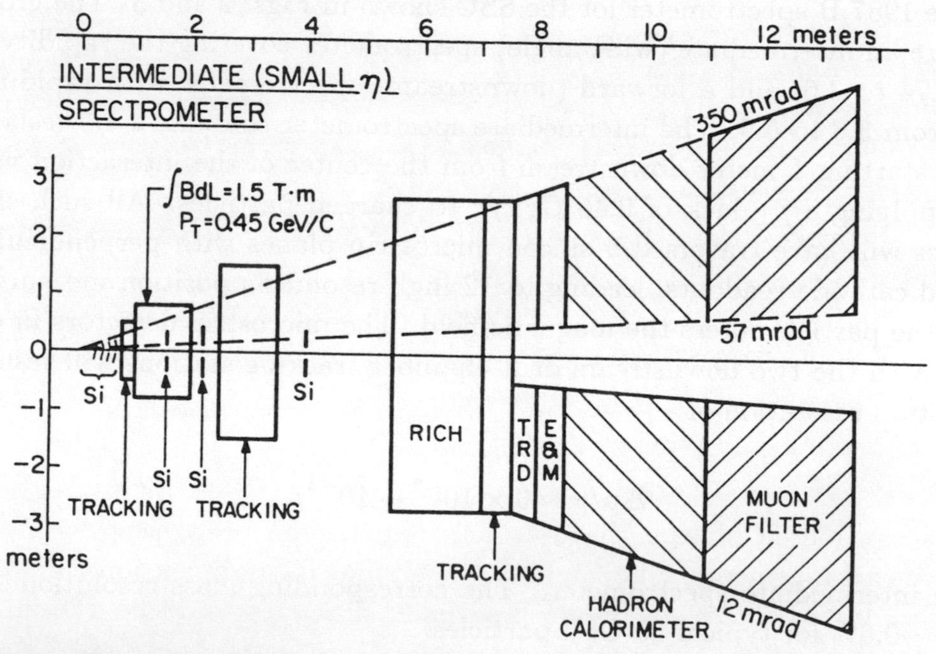

Figure 3. The intermediate beauty spectrometer. To set the scale, the electromagnetic calorimeter is 8 meters from the crossing point and has transverse dimensions of ± 3 meters.

momentum resolution,

$$\Delta p/p \sim 1.5 \times 10^{-3} + 1.3 \times 10^{-5} p \ ,$$

corresponds to a mass resolution of ~0.3% for particles of typically 100 GeV.

Vertex finding, tracking, hadronic and leptonic particle identification, and other issues are discussed in the following sections.

i. Vertex finding and the beam pipe

Precise vertex resolution is absolutely essential for doing B physics at the SSC. In the absence of multiple Coulomb scattering the spatial resolution of a track at a vertex would be

$$\Delta x \simeq \sigma_o \sqrt{n} \Delta S / \Delta L \ ,$$

where σ_o is the intrinsic spatial resolution of a vertex tracking device plane, n is the number of planes per view, ΔS is the distance from the vertex to the first plane and ΔL is the distance from the first to the last plane. Unfortunately, the contribution from multiple Coulomb scattering may dominate. This contribution is

$$\Delta x^{ms} = 0.014 \frac{\Delta S}{p} \sqrt{X/X_0} \ ,$$

where p is the momentum in the GeV/c and X/X_0 is the thickness of the relevant material in radiation lengths. Since the transverse distance from the beam (R_T) is $\Delta S \tan\theta$ and $p_\perp = p \sin\theta$ ($\approx p \tan\theta$), the expression can be conveniently rewritten as

$$\Delta x^{ms}(\mu\text{m}) = \frac{7.6}{p_\perp(\text{GeV})} R_T(\text{cm})$$

for X/X_0 for a 300 μm silicon plane. To do B physics it is essential that Δx^{ms} be much less than $c\tau$ for beauty (~300 μm). Since the beauty decay particles have $p_\perp \sim$ 1–2 GeV/c, the first planes must be very close to the beams and multiple scattering in the beam pipe must be small.

Due to the practical difficulties in operating vertex detectors inside the beam pipe we considered placing the detectors outside the beam pipe with various beam pipe arrangements. The minimum beam pipe inner dimensions summarized in Table 1 indicate that the beam pipe inner diameter could be 1.0 cm to 1.2 cm for the first 60 m. (These requirements allow for an aperture-filling beam during filling, and guarantee that any beam scraping will occur elsewhere in the machine.) A 1 cm Be beam pipe could be 0.3 mm thick ($0.9 \times 10^{-4} X_0$) with a structural safety factor of three.

For such a straight cylindrical beam pipe the spatial resolution at the vertex just due to multiple scattering in the pipe is

$$\Delta x^{\text{beam pipe}}(\mu\text{m}) \sim \frac{4\,R_T(\text{cm})}{p_\perp(\text{GeV})}\frac{1}{\sqrt{\theta}}\,.$$

This is probably acceptable for the angles of the intermediate spectrometer such that $1/\sqrt{\theta} \lesssim 4$. For more forward angles a complicated beam pipe shape will be necessary. A number of beam pipe styles were proposed (see Fig. 4), but the optimal shape remains unclear.

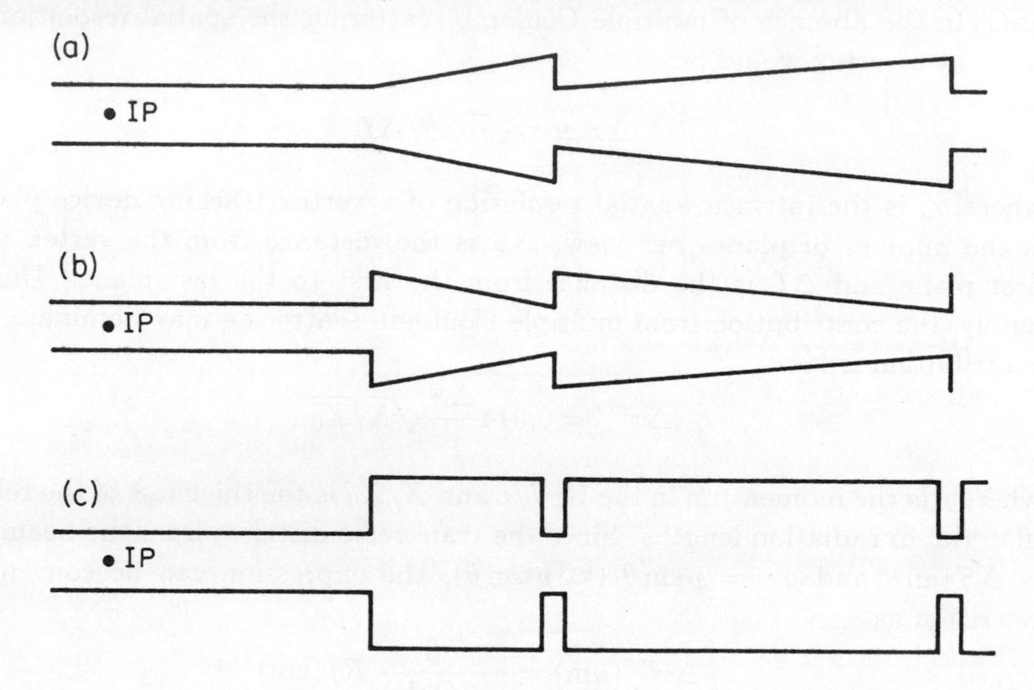

Figure 4. Three classes of ideas for beam pipe designs.

Independently of the beam pipe shape the organization of the vertex detector will be something like that shown in Fig. 5. The design is based upon currently available silicon microstrip type devices. Pixel devices are under active development [7] and could provide significant advantages. Figure 5a shows the approximate distribution of 15 of the 16 silicon microstrip stations along the beam line (the 16th is at about 4 meters from the interaction point). Each station is assumed to provide measurements in each of 4 views if silicon microstrips are used, or to provide a single high resolution pixel measurement if pixel devices are used. The distribution, designed so that each track registers in 3 stations, is approximately exponential.

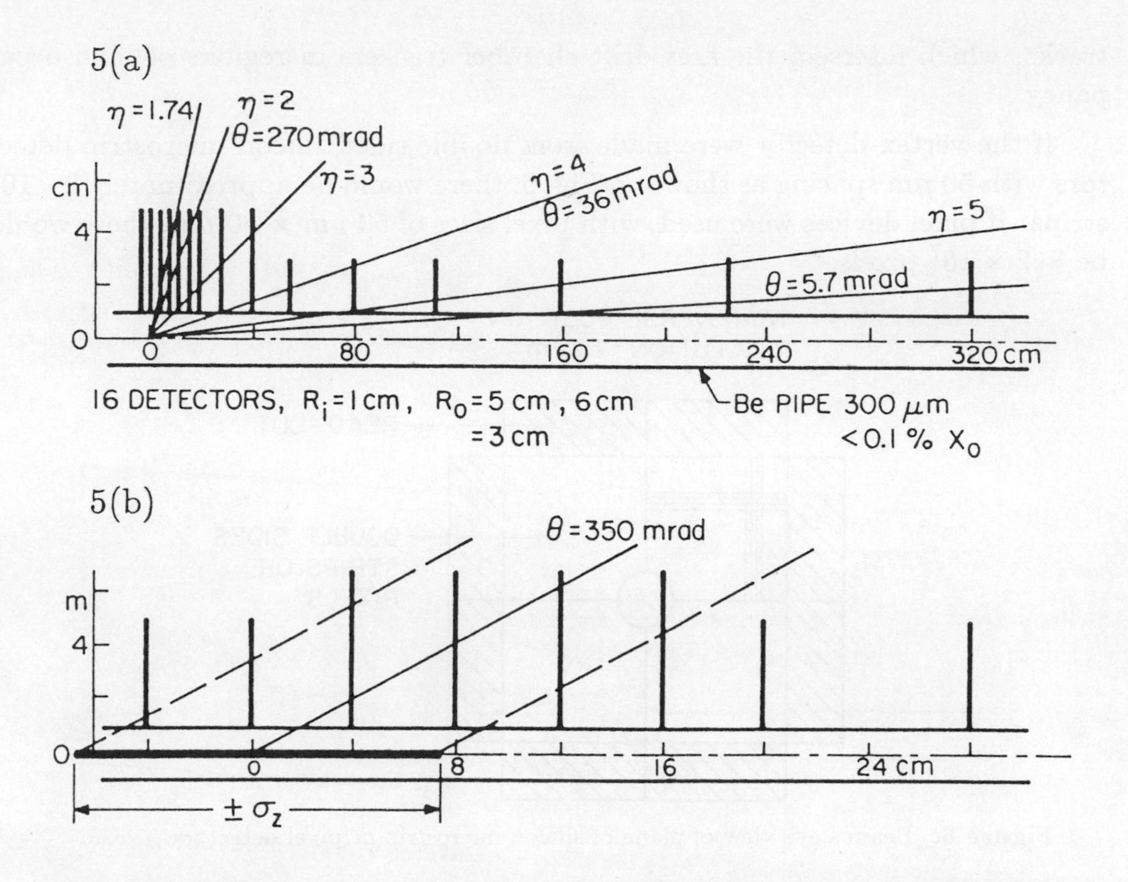

Figure 5. Schematic illustration of vertex detector arrangement. Fig. 5a shows 15 of the 16 measuring stations which surround the beam pipe. Each measuring station contains either 4 silicon microstrip views, or one pixel device. The detectors extend from an inner radius at the beam pipe of 1 cm to outer radii of 3, 5, and 6 cm as shown. Fig. 5b shows the area near the ± 7 cm long beam interaction volume. The detector are actually tilted to be oriented as nearly perpendicular to the tracks as possible.

The problems associated with the long ($\sigma_z = \pm 7$ cm) collision region are evident from Fig. 5b, which shows the vertex detector arrangement near this region. While the detectors are tipped so that the average track is normal to the detector surface, there is still a sizeable spread in angles for these tracks, resulting in partial signals from a number of strips (or pixels).

For angles greater than about 20 milliradians the vertex tracking is accomplished with straight tracks in a nearly field free region. For particles at smaller angles, with momenta of typically 100 GeV/c, there will be some track curvature in the vertex detector.

Finally, we note that the vertex tracking system forms an essential part of the overall track finding and fitting. This is particularly true for the small angle tracks, which intersect the first drift chamber trackers in regions of high occupancy.

If the vertex detector were made from double sided silicon microstrip detectors with 50 μm spacing as shown in Fig. 6, there would be approximately 2×10^5 strips. If pixel devices were used, with pixel sizes of 50 μm $\times$ 50 μm, there would be $\sim 2 \times 10^7$ pixels.

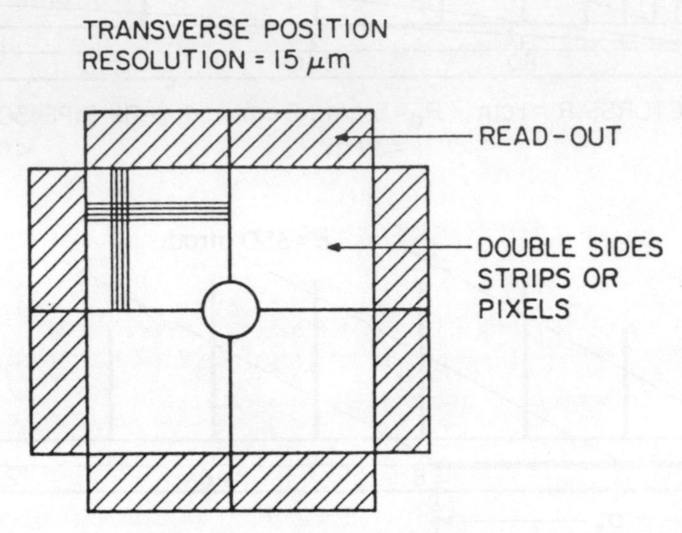

Figure 6. Beam's eye view of plane of silicon microstrip or pixel detectors.

ii. Magnets

The first magnet is a 1.5 tesla-meter dipole and has a ± 0.35 m entrance aperture in both the bend and non-bend directions. This aperture, which is located 1 meter from the center of the interaction volume, probably contains a mirror plate to terminate the fringe field. The downstream aperture must accept particles produced in a cone with 350 milliradian half angle. If the magnet is one meter long the exit aperture is ± 0.7 meters in size.

The second dipole magnet accepts tracks with angles < 57 mrad at a distance of 15 meter from the interaction point, and has a total field integral of 4.5 tesla-meters. The aperture of this magnet is ± 0.9 meters in both dimensions and the magnet is three to four meters long.

iii. Charged particle tracking

The tracking detectors must be highly redundant to achieve a high reconstruction efficiency and to be capable of handling the 10^7 s^{-1} event rate. The readout must be designed to provide fast triggering information. To facilitate track reconstruction in the trigger logic, the detectors have bend-plane and non-bend-plane views.

The design of the spectrometer is partially determined by occupancy considerations. With a charged particle multiplicity of 6 per unit of rapidity, a long chamber cell of width w a transverse distance R_T from the beam has a hit probability per event of

$$P = \pi w / R_T .$$

This gives a cell occupancy

$$O \simeq \frac{\pi w}{R_T} \left(1 + \frac{\Delta t}{\tau} \right) ,$$

where Δt is the relevant time resolution of the device and τ is the average time between events (~100 ns). For example, a 3 mm diameter cylindrical ("soda straw") cell with a drift speed of 50 μm ns^{-1} and with single hit electronics has a time resolution of 30 ns. Such a cell located 20 cm from the beam would have an occupancy of 6%. With larger cells and multiple hit electronics it is less clear how to define the relevant occupancy. If we define the effective width of such a cell as the double hit separation (~1 mm) and the time resolution as 1 mm/(50 μm ns^{-1}), then the occupancy of such a cell is again about 6%. It is clear that even with a luminosity of 10^{32}cm^{-2}s^{-1}, chambers cannot be effective at distances of less than about 15 cm from the beam. On the other hand, an infinitely long 50 μm wide silicon strip at $R_T = 1$ cm has an occupancy of ~2% if the readout time resolution is 50 ns.

The tracking philosophy is evident from Figs. 2 and 3. Straight track segments are found by the silicon vertex detectors in the pre-magnet region where each track will have ~12 high resolution points if silicon strips are used or 3 space points if pixel devices are used. In addition, the wider angle tracks will register in a pre-magnet chamber with very small cells. The main tracker for the wider angle tracks of the intermediate spectrometer is located about 3 meters downstream where these tracks are usually more than 20 cm from the beam. The main tracker might consist of a jet cell system, as shown in Fig. 7. There will be cells with x, y and 45° views, each composed of 16 sense wires with a half cell width of 6 mm. The sense wire planes are oriented at about 10° with respect to the forward track direction. Since a track always passes very close to at least one sense wire, there is a prompt signal which gives the track coordinate and the drift times from neighboring sense wires give the track angle. It might be necessary to reduce the cell spacing at small R_T due to the rates. This chamber contains about 11,000 sense wires. The final tracking device of the intermediate spectrometer is located just upstream of the TRD. This chamber helps in the tracking and improves momentum resolution. It is also required in the electron

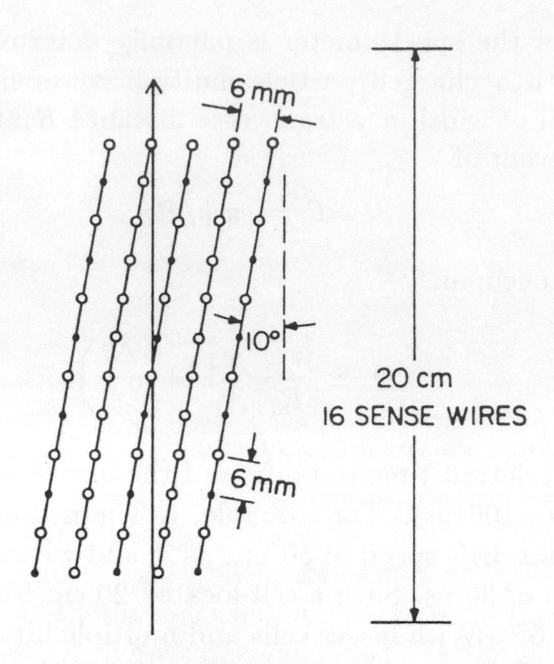

Figure 7. Cell structure of main drift chamber tracker.

trigger to reject photon conversion occurring in the TRD radiator. This tracker has ~6000 sense wires.

At less than 57 mrad particles go through the hole into the downstream spectrometer. A chamber at 13 meters, just before the big magnet, helps in linking tracks found in the downstream tracker with the pre-magnet segments found in the silicon. This chamber also helps with soft tracks which have inadvertently gone through the hole, and it helps identify "shine" off the edge of the hole. This chamber contains about 6000 small cells. At this location, the resolution obtainable with a high pressure straw chamber [10] would significantly improve the momentum resolution of high momentum tracks. The main downstream tracker is at about 40 meters. It is similar to that of the main intermediate tracker and contains about 20,000 sense wires. The occupancy of this device should be satisfactory at 20 cm from the beam, which corresponds to the 5.7 mrad acceptance limit. There is also a far-downstream chamber, just in front of the TRD, which helps with track finding, momentum resolution for high momentum tracks, and is important for the trigger. This device contains about 8000 wires. The estimated total number of sense wires in the charged particle tracking is about 55,000. Portions of wires closer to the beam than 10 cm are deadened.

It should be pointed out that this proposal for tracking is clearly one example of many possibilities. The main problems occur close to the beam where occupancies are 5–10%. Linking the small angle tracks found in the downstream spectrometer to those found in the silicon might present problems. In this case somewhat more silicon could be installed at about 4–8 meters from the interaction point.

iv. Hadron particle identification

For B physics the identification of hadrons, and especially kaons, will be very important, probably essential. The two stage spectrometer greatly facilitates particle identification. Figs. 2 and 3 show three stages of ring-imaging Čerenkov (RICH) detectors, each of which has 32 photon detector planes. Table 2 indicates the basic physics properties of each detector. In the forward spectrometer $K\pi$ separation is achieved from 5–200 GeV/c and K/p separation is obtained from 6 to 370 GeV/c. For the intermediate spectrometer the corresponding momenta are 3–40 GeV/c and 10–70 GeV/c, respectively. The particle identification for the intermediate spectrometer can be extended to very low momentum via time of flight counters located near the electromagnetic calorimeter, 8 meters from the interaction point. At this distance 3 GeV/c K's and π's have a flight time difference of 0.36 ns, nearly identical to the time difference of 5 GeV K's and protons. These values are about 4 times larger than reported time of flight resolution (under 0.1 ns) [8].

Table 2

Properties of the three ring imaging Čerenkov counters. The table gives the difference of the index of refraction n from unity, the length of each counter in meters, the relevant momentum region for each in GeV/c, and the expected number of photoelectrons.

	1	2	3
$n-1$	1.7×10^{-3}	6×10^{-4}	6×10^{-5}
Length (m)	0.8	2.0	20
e/π	< 12	< 20	< 60
π/K	3 – 40	5 – 70	15 – 200
K/p	10 – 70	6 – 120	50 – 370
# p.e.	25	25	25

A typical RICH cell contains an average of $\frac{1}{2}$ of a ring per interaction, i.e., an overall rate of ~10 MHz. The detector planes can be placed (via mirrors) at rather large distances from the beam to reduce radiation problems and the passage of minimum ionizing tracks. The electron rate at a sense wire is still very high, however. The planes can be read out in the same manner as in Fermilab experiment E605 [9], with a multistep proportional chamber. Finely segmented cathode pads may be necessary. Due to the concerns about TMAE related chamber degradation it may be necessary to develop a new readout scheme (perhaps segmented phototubes) or to use conventional threshold Čerenkov counters.

v. Transition radiation detectors

It is clear that a single lepton trigger will be essential for high rate B physics, and here we consider the case of an electron. In addition to greatly helping with electron identification offline, transition radiation detectors (TRDs) are crucial in making fast electron trigger decisions.

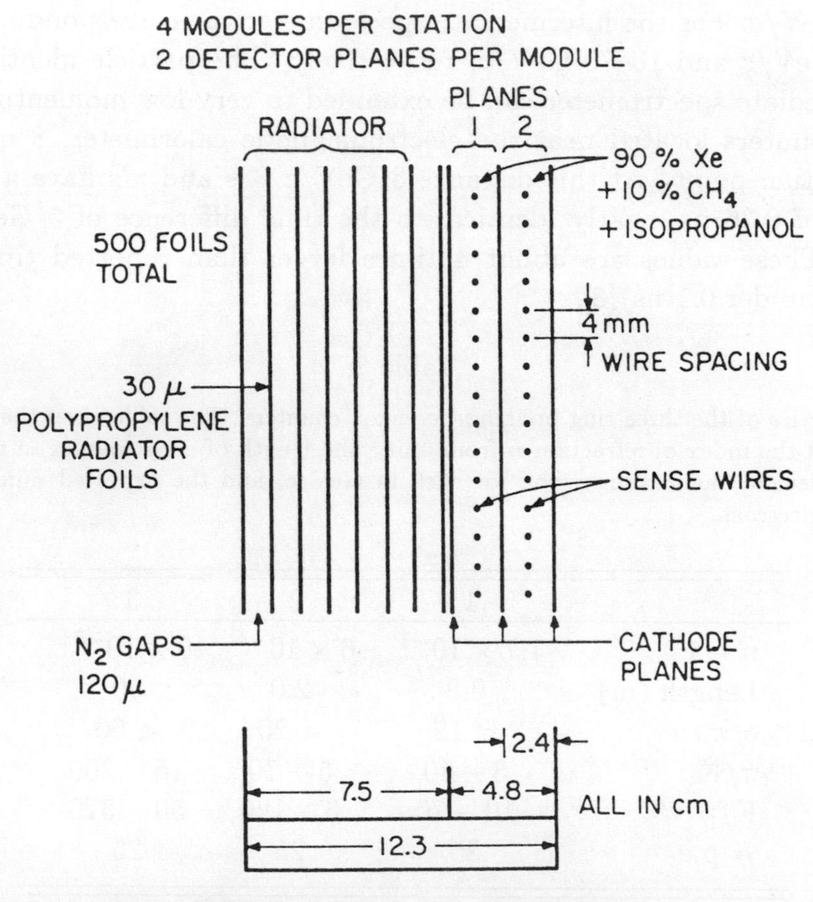

Figure 8. Schematic design of one transition radiation detector. There are four modules per TRD detector.

The TRDs are located just upstream of the electromagnetic calorimeter. Each is composed of 4 modules. As shown in Fig. 8, each module contains 500 thirty-micron polypropylene foils separated by 120 μm nitrogen gas gaps. The average of two TRD photons produced per module are detected in the two 90% Xe +10% CH_4 gas detectors. Two detectors are chosen, each with an absorption probability of 0.5 for each module, to reduce the total drift collection time to 600 ns. The electronics are just sensitive to large ionization clusters and OR the two detectors. Demanding that 3 of the 4 modules has a cluster, results in a 90% efficiency for counting electrons with a rejection of better than 1/50 for pions of momenta less than 100 GeV. Figs. 9a and 9b illustrate a scheme for arranging a mosaic of smaller detectors to cover the large required areas. The wires are ganged to minimize the number of readout channels while maintaining a reasonable occupancy. The distribution of channels is given in Table 3. An estimated 23,000 anode channels are needed.

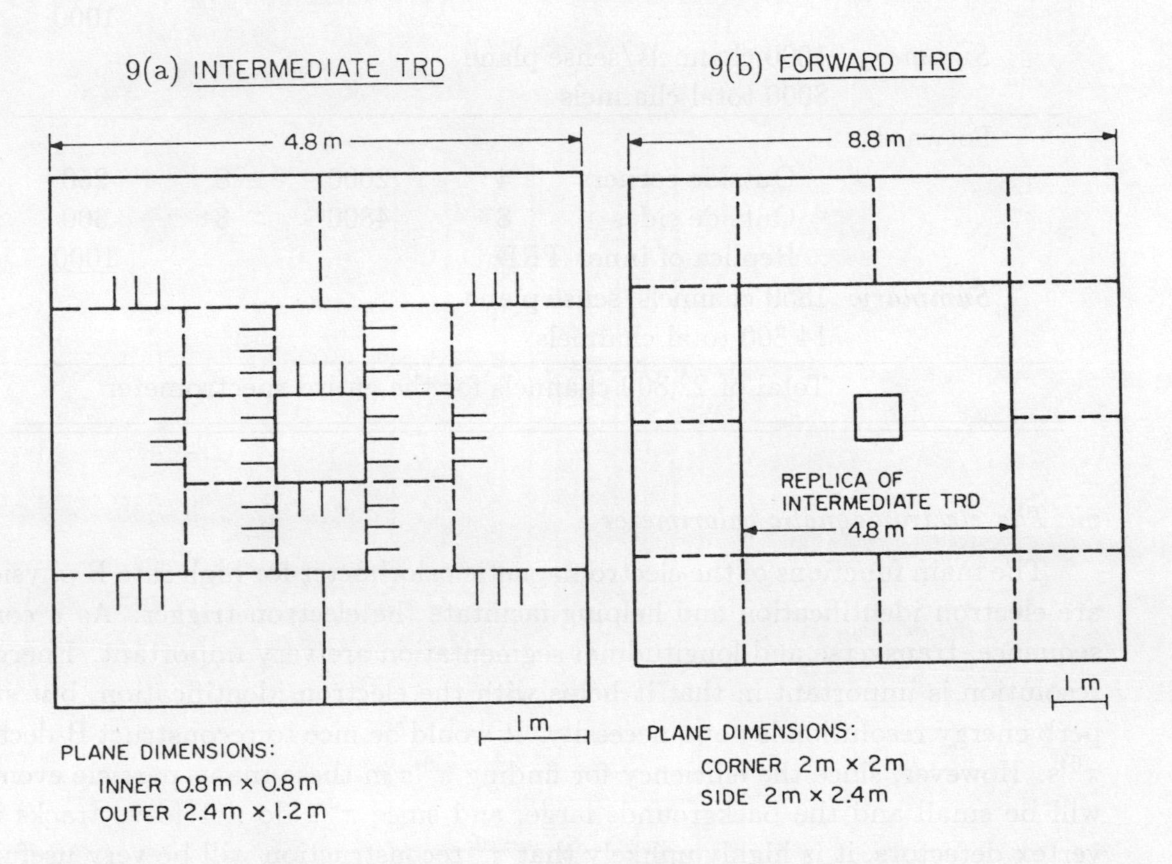

Figure 9. Schematic of the mosaic arrangement of TRD detectors to cover the large required areas of the intermediate (Fig. 9a) and forward (Fig. 9b) detector.

It may also be advantageous to read out cathode pads which are geometrically matched to the towers of the electromagnetic calorimeter. Finally, we note that since the readout is supposed to be sensitive only to the signal clusters associated with the large local ionization from the absorbed x-rays, the occupancy of the TRD should in general be rather low. This will reduce problems caused by the long drift time in xenon.

Table 3

Distribution of TRD channels.

TRD	Chamber type	Chambers per plane	Wires per plane	Ganging factor	Channels per plane
Intermediate					
	Inner	8	1600	4	400
	Outer	6	3600	6	600
					1000
Summary:	1000 channels/sense plane				
	8000 total channels				
Forward					
	Outside corners	4	2000	8	250
	Outside sides	8	4800	8	600
	Replica of inner TRD				1000
Summary:	1850 channels/sense plane				
	14,800 total channels				
	Total of 22,800 channels for the entire spectrometer				

vi. The electromagnetic calorimeter

The main functions of the electromagnetic calorimeter for high rate B physics are electron identification and helping facilitate the electron trigger. As a consequence, transverse and longitudinal segmentation are very important. Energy resolution is important in that it helps with the electron identification, but superb energy resolution is not a necessity. It would be nice to reconstruct B-decay π^0's. However, since the efficiency for finding π^0's in these many-particle events will be small and the backgrounds large, and since π^0's do not leave tracks in vertex detectors, it is highly unlikely that π^0 reconstruction will be very useful.

Two considerations determine the transverse segmentation of the detector. The first is occupancy. The probability per event that a detector element of area dA at a transverse distance R_T from the beam is hit is $dA/R_T^2 = \Delta\eta\Delta\phi$. The occupancy for charged particles or photons then is $(dA/R_T^2)(1+\Delta t/\tau)$, where Δt is the time resolution of the detector and τ is the average time between collisions (~100 ns). To fix ideas, we consider a lead-liquid argon device with narrow gaps and assume that $\Delta t \sim 200$ ns has been achieved through the addition of methane. If we wish to restrict the occupancy to $\alpha = 0.002$, then the size of a transverse cell should be $dA = \alpha R_T^2/(1+\Delta t/\tau)$. The number of cells in an annulus of radius R_T and width ΔR_T is $\Delta N = (2\pi/\alpha)(1+\Delta t/\tau)\,\Delta R_T/R_T$. For a detector with inner radius R_T^{min} and outer radius R_T^{max} we then need a total number of cells

$$N = (2\pi/\alpha)(1+\Delta t/\tau)\,\ln(R_T^{max}/R_T^{min})\ .$$

For the downstream detector, $R_T^{min} \sim 30$ cm and $R_T^{max} \sim 4$ meters, so $N \simeq 24,000$ cells. Faster time resolution (e.g. ~50 ns) could reduce the number of cells by a factor of two. A less stringent occupancy criterion could also reduce the total.

A second criteria is that of shower size. For a lead-liquid argon calorimeter the 90% containment radius is ~2 cm. There is not much reason to make a cell smaller than about 1.4 cm × 1.4 cm, since this is the intrinsic shower size. On the other hand, much of the pion rejection is obtained by observing a narrow shower and by comparing the shower location with that of the incident charged particle. The shower position can be obtained to ~1 mm precision in a fine-grained calorimeter, and this leads to a powerful rejection of pion-photon overlaps. To obtain both the width information and the shower position, shower segmentation should be comparable to the shower width.

The two criteria agree at a radius such that

$$\begin{aligned} R_T &= \sqrt{(1.4\ \text{cm})^2\ (1+\Delta t/\tau)/\alpha} \\ &= 53\ \text{cm}\ . \end{aligned}$$

For smaller radii the occupancy is larger than our 0.002. At $R_T = 30$ cm it is 0.006, which is still acceptable. For larger radii the shower width criterion is the more stringent.

To satisfy both criteria while keeping the number of segments (and therefore the cost) down, a hybridized scheme is suggested for use at larger R_T. Towers will be designed according to an occupancy rule, interleaved in alternating layers with x and y strips whose width is determined by the transverse shower size. In addition, each tower is segmented into 3 longitudinal segments to further aid in pion rejection.

A quadrant of the forward electromagnetic calorimeter is shown in Fig. 10. Within the region where x and y are less than 1 meter from the beam there are a total of 20,408 towers, each 1.4 cm square, and each segmented longitudinally into three readouts. For x and y between 1 m and 2 m there are 7890 towers, each 3.9 cm square and segmented longitudinally for three readouts. In this region tower layers are interleaved with x and y strip layers 1.3 cm $\times$ 11.7 cm. The outer region contains larger towers and strips, as indicated in Table 4. The total thickness of the detector is about 35 radiation lengths and each layer is about 0.5 radiation lengths thick. The total weight of the detector is about 140 tons.

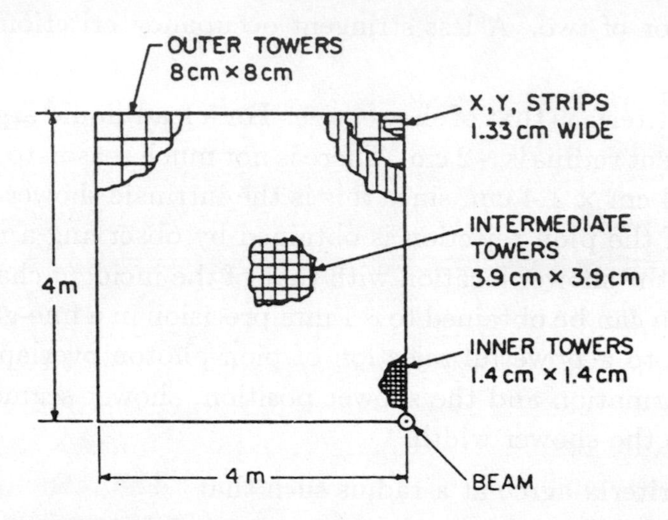

Figure 10. **A schematic view of one quadrant of the forward electromagnetic calorimeter. For larger distances from the beam a hybrid cell structure is chosen with projective towers, x strips and y strips in alternate layers.**

The electromagnetic calorimeter for the intermediate detectors is very similar but smaller. The weight is about 80 tons and there are 116,300 readouts.

vii. Hadron calorimeter and muon detector

The primary purpose of the hadron calorimeter is to help with electron and muon identification. As a consequence a simple iron calorimeter is adequate. This calorimeter is followed by an iron muon filter.

The first two interaction lengths mainly serve to help with the electron identification and are arranged in towers, with two samples per interaction length. Each tower matches four of the electromagnetic towers and is used in the electron trigger to veto hadrons which otherwise appear to be electrons. There are about 9000 and 8000 towers, respectively, for the forward and intermediate calorimeters. This section of the calorimeter might also contain liquid argon and would then share a cryostat with the electromagnetic calorimeter.

Table 4

Distribution of electromagnetic calorimeter readouts for the forward spectrometer. The intermediate spectrometer has somewhat fewer channels.

Type	Size	Minimum x or y	Maximum x or y	Longitudinal Segments	Number of Readouts
Towers	1.4 cm × 1.4 cm	0	1 m	3	61,224
Towers	3.9 cm × 3.9 cm	1 m	2 m	3	23,668
x strips	1.3 cm × 11.7 cm	1 m	2 m	1	7,890
y strips	1.3 cm × 11.7 cm	1 m	2 m	1	7,890
Towers	8 cm × 8 cm	2 m	4 m	3	22,500
	1.3 cm × 48 cm	2 m	4 m	1	7,500
	1.3 cm × 48 cm	2 m	4 m	1	7,500
					138,172

The remainder of the hadron calorimeter, composed of 10 interaction lengths of iron, is designed to help with muon identification and with muon tracking for a single or dimuon trigger. It is read out in 20 sampling layers, using a combination of plastic scintillators for fast timing and gas chambers with anode wire and cathode pad readout for position measurement. The readout configuration has yet to be completely defined. The iron weighs 900 tons for the forward hadron calorimeter and 600 tons for the intermediate calorimeter.

The iron muon filters are each 1.8 meters thick, with a total of four layers of drift chamber readout for muon tracking within and behind the iron. The weight of this iron is about 900 tons for the forward and 700 tons for the intermediate stage. There is clearly space for a longer filter for the forward spectrometer, should it be required.

viii. Acceptance for K_s's

Here we make a very rough estimate acceptance for K_s decay particles. We assume that a K_s must decay before or within a magnetic field to be identified. The typical decay distance L is $\gamma\beta c\tau$. With a $p_\perp$ of a K_s from B decay of about 1.5 GeV/c and with $\gamma\beta \sim p_\perp/\theta M$ we have $L\theta \sim 9$ cm for a K_s decay. A K_s produced at an angle greater than 0.057 radius must decay before about 1.5 meters in order for the decay mesons to be magnetically analyzed. The worst case for these particles is at the edge where the K_s's are fastest. The mean decay distance at 0.057 radians is $\sim$9 cm/0.057 = 1.5 m. At this worst angle about half of the K_s's are accepted.

Similarly, at the inner acceptance of the forward spectrometer the mean decay distance is 15 meters, which is again comparable to the distance to the second magnet. We conclude that the acceptance for K_s's exceeds 75%.

ix. Shine from the hole

In a typical event, the probability that a particle hits the intermediate calorimeters within a distance ΔR_T of the inner hole is about $2\pi\Delta R_T/R_T$ for both photons and charged particles. For photons, we expect some downstream shine when ΔR_T is less than the transverse electromagnetic shower width. The same is true for hadrons when ΔR_T is less than the corresponding width for hadronic showers.

To minimize both effects the inner edge regions of the intermediate calorimeters should be made as compact as possible with tungsten absorber and perhaps silicon readout materials. In tungsten the 90% containment radius is about 1 cm for electromagnetic shower and about 4 cm for hadronic cascades. We then expect that about 13% and 50% of the events will have a photon and a charged hadron, respectively, within the 90% containment radius. (The hadronic fraction should be increased somewhat to account for neutral hadrons).

For events at the 90% containment radius, about 3% of the energy escapes into the hole and produces shine. Typical energies for the edge particles are about 70 GeV, so we might expect about 2 GeV in this shine. This shine will be more or less isotropic and will generate noise in nearby chambers. When the particle strikes nearer to the edge the shine will be more directional and energetic. The biggest problem will occur in the chambers just upstream of the second magnet. These chambers subtend a solid angle which is about $4\pi/50$.

The problem of shine from these interactions deserves a much more careful study. It appears from this very quick look that shine will be a nuisance, but not a major problem.

5. Rates

The crucial question is whether the experiment would produce enough clean, tagged B decays to observe CP violation in a few separate channels. We have chosen those channels which have large asymmetries and have clean experimental signatures. Unfortunately, they also have small branching ratios, but it is unrealistic to think that it will be easier to see smaller CP-violating-effects in more common modes. The efficiencies can only be estimated at present, but such estimates are useful to reach a preliminary number for the sensitivity.

The goal is to collect events of the type $B^0 \to \psi K_s$ (or $\pi^+\pi^-$) with $\overline{B} \to De\nu$ ($D^*e\nu$). The geometric acceptance for all particles, including the requirement

$p_\perp \geq 0.3$ GeV/c, is about 0.05. The branching ratio for one of the semileptonic decays is about 20%, and the fractions of D's decaying into modes with two more charged particles and no neutrals is about 20%. The branching ratio product for $B^0 \to \psi K_s$, $\psi \to e^+e^-$ and $K_s \to \pi^+\pi^-$ is about 5×10^{-5}, and the branching ratio for $B^0 \to \pi^+\pi^-$ may not be much different. The product of all these factors is about 10^{-7}. Efficiencies for vertex cuts, reconstruction, and particle identification are similar to existing experiments, and might be about 0.2, which gives a total factor for acceptance, efficiency, and branching ratios of 2×10^{-8}.

A cross section of 200 μb for $B\overline{B}$ and a luminosity of 10^{32} cm^{-2}s^{-1} cm^{-2}s^{-1} means a total of about 2×10^{11} $B\overline{B}$ pairs in a running year of 10^7 seconds. This would give about 4000 B events in each of the rare modes in which a large asymmetry is expected, with a well-identified $\overline{B}$. The studies of the B-physics subgroup show that about 2000 events of this type are sufficient to see significant CP violation, assuming an asymmetry of about 0.1. The other factor that needs to be taken into account is that part of the trigger efficiency not included in the requirement of a high $p_\perp$ electron within the spectrometer acceptance. There must be more detailed work on the triggers before such effects can be estimated. For the particular case of the ψK_s mode, a dilepton trigger can probably be used without further loss. However, it is important to look at modes which require single-lepton triggers.

Observing 4000 B events per channel for 2×10^{11} $B\overline{B}$ pairs is somewhat more pessimistic than the conclusions of many studies of the CP problem in B decays. This is partly because special modes are used which have ideal properties for CP violation, but small branching ratios. Since the fraction of events with $B\overline{B}$ production is 2×10^{-3}, the series of cuts used may be more restrictive than needed. It is quite possible that it is not necessary to see a decay mode of the D with no neutrals to be able to clearly identify the semileptonic B decay. It may also be possible to sum over particular modes which have the same sign for their asymmetries.

At present we choose to take the most conservative approach, which demands a high rate. It is worth noting that even at an e^+e^- machine with vertex detection, where a trigger is not needed, a similarly conservative calculation shows that about 10^9 produced $B\overline{B}$ pairs are needed.

6. Cost Estimates

The cost estimates for most of the components of the beauty spectrometer have been made using the standardized costs from the Report of the Detector Cost Evaluation Panel [11], and are given in Table 5 below.

Table 5
Estimated cost of the beauty spectrometer.

Item	Cost in K$	
1. Electromagnetic Calorimeter		
A. Forward		
140 tons at $9	1,260	
Electronics 138 K channels at $120	16,560	
		17,820
B. Intermediate		
80 tons at $9	720	
Electronics 116 K channels at $120	13,920	
		14,640
2. Tracking		
55 K wires at $100	5,500	
Electronics 55 K channels at $120	6,600	
		12,100
3. Hadron Calorimeter and Muon Filter		
A. Forward		
L.A. section 300 tons at $9	2,700	
Other section 600 tons at $3.5	2,100	
Electronics 18 K channels at $120	2,160	
Muon Filter iron 900 tons at $3.5	3,150	
4 K wires at $100	400	
4 K channels at $120	480	
		10,990
B. Intermediate		
L.A. section 200 tons at $9	1,800	
Other section 400 tons at $3.5	1,400	
Electronics 16 K channels at $120	1,920	
Muon Filter iron 700 tons at $3.5	2,450	
4 K wires at $100	400	
4 K channels at $120	480	
		8,450
4. Cryogenics for liquid Argon		2,000
5. RICH		
Mechanical	6,000	
Electronics	7,000	
		13,000

6. TRDs			
	23 K channels at \$120	2,760	
	23 K ganged wires at \$100	2,300	
			5,060
7. Vertex Detector			
	2×10^5 strips at \$20 each		4,000
8. Magnets			2,000
9. On-Line Computing			2,500
10. EDIA			25,000
	Total Cost		117,560

7. Conclusion

The purpose of this study was to examine the feasibility of studying B physics at very high rates, with a definitive CP violation experiment as the ultimate goal. We found no problems of a fundamental nature in achieving this goal.

On the other hand, this experiment demands detectors which work at rates well beyond those commonly used in high energy physics. These are a number of questions about radiation hardness that need more work. There are also more subtle problems, such as effects of high rates on tracking efficiency, electron identification, etc. There are also possible solutions involving finer segmentation, faster elements (as suggested in the electron trigger study) and in some cases more clever design. There must be more detailed study of these problems to see in which cases the additional expense of a more elaborate detector is justified.

Major design choices included the acceptance angles for the two spectrometers. This problem deserves further attention in future workshops. The detectors need to be larger in the bend-plane dimension to accept the charged particles with low momentum. Already the transverse sizes of the calorimeters are quite large. The size of the forward spectrometer can be reduced by either making the hole in the intermediate spectrometer smaller or by moving the forward spectrometer closer to the interaction point. In the first case, the bad effects at the edge of the hole become more serious and the intermediate detector must operate closer to the beam pipe. Moving the forward detector closer to the interaction region reduces the acceptance of the spectrometer at small angles. A possible alternative that might be less expensive and easier is a three-stage spectrometer, but then there is a second edge to contend with. Finally, it would help if larger angles could be accepted by the intermediate spectrometer. All of these are important issues that need detailed study and should be addressed in the future. For this study, we assumed vertex detectors outside the beam pipe because they are easier to build and maintain. This does have some negative consequences on the vertex resolution, on pair backgrounds for the electron trigger, and for the generation of secondary hadrons. It appears that these effects are tolerable except perhaps at the smallest angles accepted. Although some effort was spent on producing better beam pipe-vertex detector designs, it became apparent that this would also take more study.

In order to do the definitive CP experiment, it is necessary to measure CP violation in many modes. This requires single lepton triggers which do not make any restriction on the decay mode being studied. (Only particles from the opposite-side $\overline{\mathrm{B}}$ are used in the trigger.) This requires some information from the vertex detectors as well as clean lepton identifications. It is also necessary to select 100 to 1000 events s^{-1} from total event rate of 10^7 s^{-1}. Such a requirement makes demands on the readout electronics, triggers, input/output devices, and data analysis system which are truly unprecedented. The most important outstanding question is the extent to which an ambitious but realistic trigger and data acquisition system will limit the physics results.

To summarize our results, it is reasonably safe to conclude that the SSC will be a good place for a high-rate B experiment. The fraction of events which contain a $\mathrm{B}\overline{\mathrm{B}}$ pair is larger than the fraction of events in fixed target hadron production which contain charm. Recent experience with fixed target charm production experiments indicates that there is no serious question about the availability of clean bottom meson samples at the SSC. With 2.5×10^{11} $\mathrm{B}\overline{\mathrm{B}}$ events produced each year, the size of clean B signals should be beyond that available at any other presently proposed machine, even after accounting for trigger losses.

The second major conclusion is that a forward spectrometer is needed to obtain the desired acceptance. One should take the broadest view of forward, however. It is important to accept the largest rapidity slice possible, keeping in mind that this means within the prime performance region of the detector.

The spectrometer elements are relatively conventional, and nothing outside our present technology is required. There are a series of critical experimental issues, enumerated above, that need further study. The single leptonic triggers are possibly the most challenging problems.

In summary, the prospects that the definitive experiment on CP violation in the $\mathrm{B}\overline{\mathrm{B}}$ system will be built at the SSC are quite good. Given the central nature of this problem for our understanding of the standard model, it is worth pursuing the design of such an experiment. A fresh look at the problems in the next workshop should produce a design which more closely approaches a realistic experiment.

References

1. K. J. Foley *et al.*, "Bottom and Top Physics," these proceedings.
2. R. J. Morrison, S. McHugh, and M. V. Purohit, "A Semileptonic Electron Trigger for Recording $B^0\overline{B^0}$ Events," these proceedings.
3. L. J. Gutay, D. Koltick, J. Hauptmann, D. Stork, and G. Theodosiou, "Intermediate $p_\perp$ Jet Spectrometers, these proceedings.
4. L. W. Jones, "A Small-Angle Spectrometer Addition to the Intermediate-$p_\perp$ B Spectrometer," these proceedings.
5. B. Cox and D. E. Wagoner, pp. 83–91, *Physics of the Superconducting Supercollider*, ed. by R. Donaldson and J. Marx (Snowmass CO, 1986).
6. B. Cox, F. Gilman, T. D. Gottschalk, pp. 33-44, *Physics of the Superconducting Supercollider*, ed. by R. Donaldson and J. Marx (Snowmass CO, 1986).
7. Talks by David Nygren, Sherwood Parker and Steve Shapiro at this Workshop.
8. W. Reay, private communication (1987).
9. M. Adams *et al.*, Nucl. Instr. Meth. **217**, 237 (1983), and H. Glass *et al.*, IEEE Trans. Nucl. Sci. **NS-32**, 692 (1985).
10. R. DeSalvo, "A proposal for an SSC Central tracking detector", Cornell University Laboratory of Nuclear Studies report CLNS 87/52 (1987).
11. "Cost Estimate of Initial SSC Equipment", SSC Central Design Group Report SSC-SR-1023 (1986).

Intermediate P_T Jet Spectrometers

L.J. Gutay and D. Koltick
Physics Department, Purdue University
Lafayette, IN 47907

J. Hauptman
Physics Department and Ames Laboratory
Iowa State University, Ames IA 50011

D. Stork
Physics Department, UCLA, Los Angeles CA 90024

G. Theodosiou
Nuclear Research Center, Demokritos, Athens, Greece

Abstract

A design is presented for a limited solid angle, high resolution double arm spectrometer at 90^0 to the beam, with a vertex detector and particle identification in both arms. The "jet arm" is designed to accept a complete jet, and identify its substructure of sub-jets, hadrons, and leptons. The "particle arm" would measure e,π,K,p ratios for $P_T < 400$ GeV/c. The vertex detector for both arms would measure the associated multiplicity. The dimensions of the experimental hall required are 50 m (transverse to beam) and 20 m (along beam), with drop hatch, and both external and overhead crane.

We also describe a second double arm spectometer at 0^0 to the beam for the purpose of tagging Higgs production by boson fusion, [1] gauge boson (WW, ZZ, and WZ) scattering [2], and other processes involving the interactions of virtual gauge bosons.

Introduction

Most of the interest in this Workshop has rightly concentrated on the design of large 4π detectors capable of the most stringent tests of the standard model. We have instead addressed physics problems which require only limited solid angular coverage, but very specialized performance, by jet spectrometers. The physics problems are three:

- A particle spectrometer to perform a comprehensive particle survey, starting even at very low machine luminosities, of e,π,K,p ratios, event multiplicities and transverse momenta up to 0.4 TeV/c.

- A jet spectrometer to study whole jets, jets within jets, and to identify hadron, photon, muon, and electron components up to 2 TeV/c as stringent tests of QCD.

- A forward specrometer to tag WW, ZZ, and WZ interactions.

The first two spectrometers are highly compatible and complementary, and would naturally fill the same interaction region, as shown in Figure 1. The third spectrometer ought to be matched with a more general central detector.

[1] R. Cahn, S. Ellis, R. Kleiss, and W. Stirling, Phys. Rev. **D35**,1626 (1987)
[2] S. Willenbrock, WW Physics, XXIII Int'l. Conf. on High Energy Physics, 1986, p. 1276.

I. The Particle Spectrometer

The experimental goals are (a) to carry out a general survey of lepton yields and particle flavors as functions of P_T and E_T in both spectrometer arms up to 2. TeV/c, and (b) to measure event multiplicities from 20 to 200 within the spectrometer solid angle as a function of beam energy. The former may provide stringent tests of particle production at high Q^2, provide useful information to the large simulation codes to be used by the larger and more general 4π detectors, while the latter allows a test of quark-gluon plasma formation. Although the original models of particle production were statistical and thermodynamic [3-4], the main emphasis in particle physics has been the study of particles and resonances of definite quantum numbers which could be classified into symmetry groups [5]. Together with the discovery of Bjorken scaling [6] this led to the hypothesis that hadrons are composed of quarks and gluons. The absence of any evidence for the existence of free quarks has led to the concept of quark confinement. From the concept of asymptotic freedom [7] QCD evolved [8] as the theory of stong interactions in agreement with numerous hard scattering experiments. Both thermodynamical bootstrap and lattice gauge calculations of QCD predict a phase transition from hadrons to a quark-gluon plasma at sufficiently high temperature and/or quark densities. [9] There is general theoretical concensus that temperatures of T $\sim$ 215 MeV and/or $q\bar{q}$ densities of $n(q) + n(\bar{q}) \sim 2$ to 4 per cubic Fermi are required to produce a quark-gluon plasma. [10]

A possible signature of the particle-plasma transition is a discontinuity in T as a function of S, as in the phase diagram of H_2 near its boiling point. Since

$$S \sim \frac{dn}{dy} \qquad and \qquad < P_T > \sim T, \tag{.1}$$

we expect discontinuities both in the mean transverse momentum and in the particle ratios as a function of event multiplicity, and a discontinuity in the yields of direct photons and lepton pairs. Theoretical calculations [11] predict a deconfinement temperature in the central region of $T_D \sim 215$ MeV, as shown in Figure 2(a), and an energy density $\epsilon \sim 2$ GeV/F^3.

Both our FNAL data [12] and UA1 data, shown in Figure 2(b), indicate that the deconfinement transition starts at $(dn/dy)_D \sim 12$, and from cosmic ray data the transition is finished at 26. For a pseudo-rapidity range $-3 < \eta < +3$, the detector must reconstruct about $n_C \sim 150$ tracks.

The physics constraints which determine the geometry of the detector are

- The central drift chamber must reconstruct with high efficiency 80 charged tracks inside $\eta < 1.5$.
- The end cap drift chambers must reconstruct 40 tracks inside $1.5 < \eta < 3.0$.

---

[3] E. Fermi, Prog. Theor. Phys. **5**, 570(1950)

[4] L.D. Landau, Izv. Akad. Nauk. S.S.S.R. **17**, 51(1953); P. Carruthers and Minh-Duong Van, Phys. Rev. **D8**, 857(1973)

[5] M.Gell-Mann,Synchrotron Laboratory Report, CTSL-20 (1961) and Phys. Lett. **8**, 214 (1964); G. Zweig, CERN Report Nos. TH 401, 402 (1964).

[6] J.D. Bjorken, Phys. Rev. **179**, 1547 (1969); J.D. Bjorken and E.A. Paschos, Phys. Rev. **185**, 1975; S.J. Drell, D.J. Levy, and T.M. Yan, Phys. Rev. Lett. **22**, 744 (1969)

[7] D. Gross and F. Wilczek, Phys. Rev. Lett. **30**, 1343 (1973); H.D. Politzer, Phys. Rev. Lett. **30**, 1346 (1973)

[8] H. Georgi and H. Politzer, Phys. Rev. **D14**,1829 (1976)

[9] R. Hagedorn and J. Rafelski, Phys. Lett. **97B**, 136 (1980); L.D. McLerran and B. Svetitsky, Phys. Lett. **98B**, 199 (1981); J. Kuti, J.Polonyi, and K. Szlachanyi, Phys. Lett. **98B** (1981)

[10] L. Van Hove, Phys. Lett. **118B**, 138 (1982); J.D. Bjorken, Fermilab Pub. THY-82/44.

[11] J. Kogut, et al. Phys. Rev. Lett. **51**, 869 (1983); J. Engels, et al., Nucl. Phys. **B205**, 545 (1982)

[12] E735 data, Fermilab

- There must be particle identification for e, π, K, p for $P_T < 2$ GeV/c.
- Hanbury-Twiss analysis requires a P_T resolution of 20 MeV/c.
- The error in track counting after reconstruction must be $< 10\%$.

These requirements can be met by a central tracking chamber covering 50 cm $< R <$ 100 cm of the De Salvo design [13], and end caps of the Atac design. [14] Particle identification requirements demand a side arm magnetic spectrometer with both time-of-flight and RICH counters. Triggering requires multi-element hodoscopes both upstream and downstream of the detector providing both a T0 and an interaction trigger, and finely segmented (about 1^0) hodoscopes in both the barrel and end cap regions. A schematic drawing of the apparatus is shown as the left spectrometer arm in Figure 1.

To avoid showering and cascading within the detector, the chamber frame must be built of very low-mass materials (e.g. epoxy impregnated carbon fiber-rohacell composite), and the beam pipe and flanges from beryllium and aluminum. A rough cost estimate is:

Chamber Construction:	
(a) Central Chamber	300 K
(b) Pre and Post Magnet Chambers	100 K
(c) z Chambers	100 K
(d) Straw Chambers	150 K
(e) Radial Drift Chambers	100 K
(f) Magnet	1000 K
Scintillators.	
(a) Trigger Hodoscope	100 K
(b) Time-of-flight	100 K
(c) Halo Counters	50 K
Electronics.	
(a) Flash ADCs	1000 K
(b) Computers	500 K
(c) Terminals, power supplies	500 K
Total:	4000 K

II. The Jet Spectrometer

Jet spectrometers have been designed and studied at earlier Workshops [15], and here the justification is the same: trade solid angle for segmentation and detailed reconstruction and identification capabilities, so that measurements inaccessible to the large 4π detectors can be made. The spectrometer designed by Theodosiou is particularly good, and with supplements and replacements for newer methods and technologies, provides a good benchmark. This spectrometer, shown schematically as the right spectrometer arm in Figure 1, would perform: [16]

[13] R. De Salvo, A Proposal for a SSC Central Tracking Detector, Cornell preprint, 1987

[14] M. Atac, Description of an Operating End Cap Radial Drift Chamber Array, these Proceedings

[15] M. Witherell, Report of the Specialized Detector Group, in Proc. Summer Study on the Design and Utilization of the SSC, Snowmass, 23 July 1984; and G. Theodosiou, SSC Jet Spectrometer: A Multipurpose Detector Scheme for the Central Region, Proc. Workshop on Triggering, Data Acquisition and Computing, Fermilab, 11 November 1985.

[16] Theodosiou, *ibid.*

• Intra-jet studies. [17] This will provide stringent tests of QCD at very high Q^2 through flavor tagging, lepton identification, and detailed reconstruction.

• Compositeness tests. High precision cross section measurement for point cross section deviation measurements.

• A search from heavy stable objects.

• A search for an intermediate mass Higgs.

• τ and B physics studies. At 90^0 the signal to background is much more favorable than at 0^0.

Detailed discussions of triggering capabilities, physics reach, and costs have been previously made (Ref. 15). Here we emphasize that, roughly speaking, a specialized jet spectrometer with a modest solid angle of $.05 \times 4\pi$ can provide fine segmentation, triggering, and reconstruction capabilities well in excess of those possible with a general 4π detector. Furthermore, such a spectrometer can become useful initially at low luminosities and possibly extend its usefulness to luminosities of 10^{34} cm^{-2} s^{-1}.

III. The Forward Spectrometers

Many studies [18] have been of several schemes to search for the existence of the Higgs boson. All of these have considered only the final states of the Higgs. Recently, a proposal [19] has been made to tag the initial state as well. Since the production of a heavy Higgs is dominated by WW boson fusion, the quarks which emit the W's receive sustantial transverse momenta of order M_W, which in general are not equal. These quark jets can be measured in an intermediate P_T forward jet spectrometer with rapidities $2 < \eta < 7$, and energies of about 1 TeV.

We have performed a simple simulation experiment in which we have generated event samples of the Higgs signal and two backgrounds, *viz.*,

• the production of a Higgs by WW fusion, and its subsequent decay to WW, with one W decaying leptonically and the other hadronically:

$$qq \rightarrow qqWW \rightarrow qqH \rightarrow jet + jet + (W \rightarrow \ell\nu) + (W \rightarrow jet + jet) \qquad (.2)$$

• a simple background process, W+jet production with a subsequent leptonic W decay:

$$qq \rightarrow qW \rightarrow jet + (W \rightarrow \ell\nu) \qquad (.3)$$

• and a trivial, but very high-rate background, single parton elastic scattering with a single high energy track satisfying a level 1 single lepton trigger:

$$qq \rightarrow qq \rightarrow jet + jet + (fake\ lepton\ trigger). \qquad (.4)$$

---

[17] Steve Ellis, these Proceedings.
[18] Proc. of the Summer Study on the SSC, Snowmass, July 1984; Snowmass, 1986.
[19] Cahn,Ellis, Kleiss,Stirling, *ibid.*

Events were simulated with the Pythia monte carlo [20], and all stable interacting particles are measured in a typical tower calorimeter with hadronic and electromagnetic energy resolutions at 1 GeV of 0.50 and 0.20, respectively. The forward jet spectrometers (labelled k=+1 for the +z end cap, and k=-1 for the -z end cap) subtend a rapidity range of $2.5 < \eta < 6.4$, and the central calorimeters (k=0) the remainder, and apart from the beam holes beyond $\eta > 6.4$, the system is perfectly hermetic. Trigger quantities are formed directly from calorimeter towers, and are:

- Level 1 Trigger. LEPTON = isolated e or μ, P_T >20 GeV/c.
- Level 2 Trigger.

$$ETSUM = \sum \left|\vec{E}_T(k)\right| \quad for\ k = 0\ \ (Cylindrical\ calorimeter)$$

$$EZSUM = \sum E_z(k) \quad for\ k = -1, +1 (ForwardCalorimeters)$$

- Level 3 Trigger.

$$\Delta ET = \left|\sum \vec{E}_T(k = -1) - \sum \vec{E}_T(k = +1)\right|$$

$$ET0 = \left|\sum \vec{E}_T(k = 0)\right|$$

ΔET is the transverse momentum mismatch of the two scattered quark jets in the forward spectrometers, and ET0 is the vector sum of the tower energies in the central calorimeter.

We take LEPTON to reduce the rate [21] of triggers by 10^{-3}. A trigger also requires $ETSUM > 600$ GeV, $EZSUM > 2000$ GeV, 20 GeV $< \Delta ET <$ 2000 GeV, and ET0 to satisfy a somewhat complicated Level 3 requirement:

$$[ET0 < 1.67 \cdot \Delta ET - 100.GeV] \quad or \quad [ET0 > 0.42 \cdot \Delta ET + 50.GeV]$$

The trigger probabilities, i.e. the numbers of events of each reaction which satisfy LEPTON · ETSUM · EZSUM ·ΔET · ET0, are:

Reaction	triggers/events	Trigger Prob.	Trigger Rate(Hz)
qqH	950 / 5000	0.190	0.0004
qqqq	245 / 20000	0.0122 $\times 10^{-3}$	.012
qW	195 / 10000	0.0195	1.0

These events are subsequently analyzed with the following simple algorithm. For the $W \to \ell\nu$ decay hypothesis, the highest energy e or μ is assigned as ℓ; if the (background) event has no lepton, the highest energy hadron is assigned as ℓ. The missing transverse momentum of all three calorimeters is assigned to the ν, the quadratic equation in Pz(ν) is solved, and the smaller solution for Pz($\ell\nu$) is assigned to W$\to$ $\ell\nu$. The momentum vector for the W$\to$ $q\bar{q}$ hypothesis is taken to be the total momentum vector of the central calorimeter less the lepton momentum vector. This analysis is quite crude.

---

[20] Pythia, vers. 4.7, H-U. Bengtsson. We used subprocesses 25+26 for Higgs production, 1-6 for parton elastic scattering, and 15+18 for W+jet production.

[21] Electronics, Triggering, and Data Acquisition for the SSC, T. Devlin, A. Langford, and H. Williams, Physics of the SSC, Snowmass, July 1986.

The hadronic W decay hypothesis is further subjected to an analysis [22] with passes 10% of the W decays and reduces the QCD jet sample by a factor of 500, i.e. a rejection relative to W's of 50. This assumption is a factor of 2 more pessimistic than ref. 22. In the resulting sample, the WW invariant mass is calculated, and displayed in Fig. 3(a) for "qqH", 3(b) for "qW", and 3(c) for "qq" event samples. A clear Higgs signal of about 8 events/40 GeV is seen in Figure 3(a), however the signal to background ratio is not very favorable relative to the single W production background of about 30 events/40 GeV in 800 GeV/c^2 mass region. Even the "qq" event sample yields a rate of about 5 events/40 GeV below 1 TeV. The trigger rate for this single W channel is too high by at least a factor of 10, and since a successful search will require testing for a wide structure over an extended mass range from, say, 0.3 TeV/c^2 to 1.2 TeV/c^2, the overall acceptance in this channel must be reduced by at least a factor of 10. Clearly, the definition of the background mass distributions is limited by the CPU required to generate the each event and simulate the calorimeter response. We generated 20K "qq" events and 10K "qW" events.

These simple forward spectrometers, and the accompanying simple triggering requirements and analysis, have yielded encouraging but not yet useful results. The large 4π detectors could make good use of sophisticated forward jet spectrometers for triggering and analysis purposes.

---

[22] Kinematic Isolation and Reconstruction of Hadronic W and Z decays at a Super Collider, J.Hauptman and A.Savoy-Navarro, these Proceedings

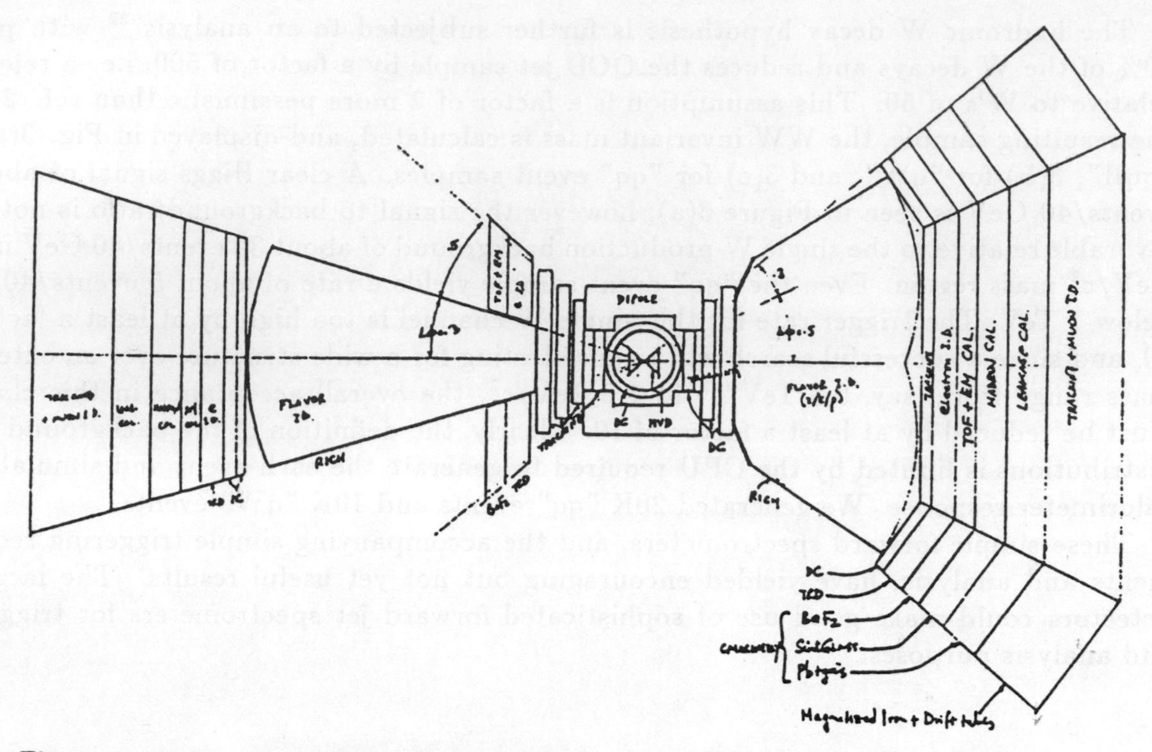

Figure 1. The combined 'jet' and 'particle' spectrometers. The 'jet' spectrometer is on the right, the 'particle' spectrometer on the left.

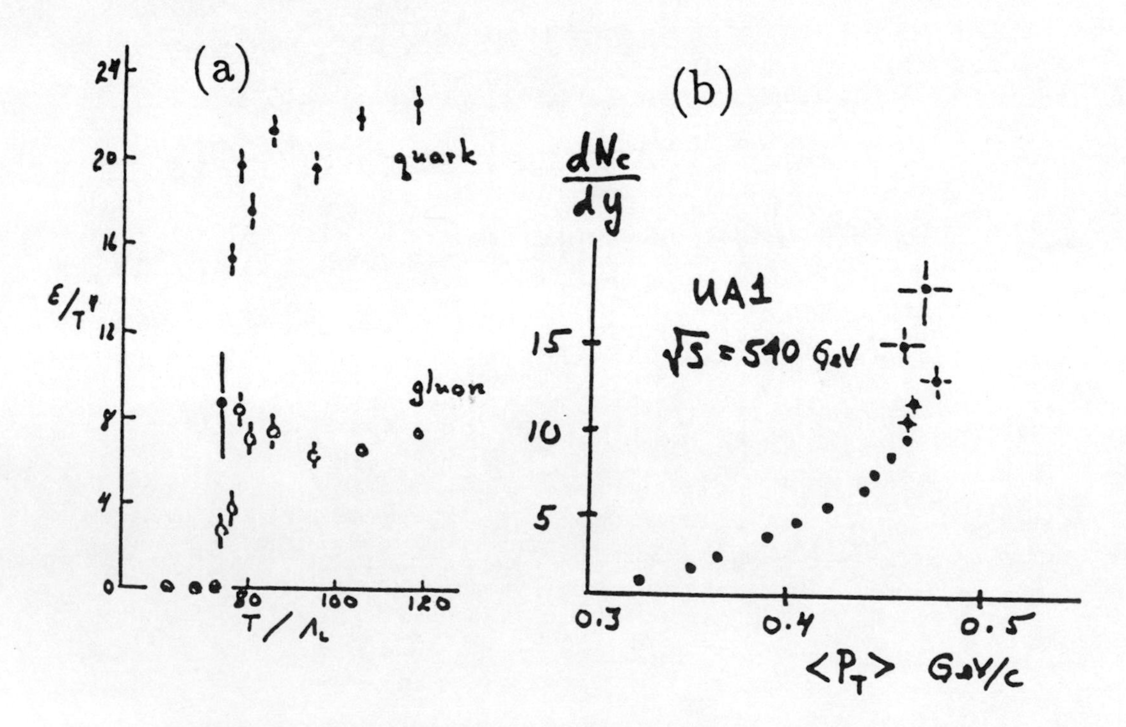

Figure 2. Particle-plasma transition. (a) QCD calculation of Ref. 11, (b) UA1 data.

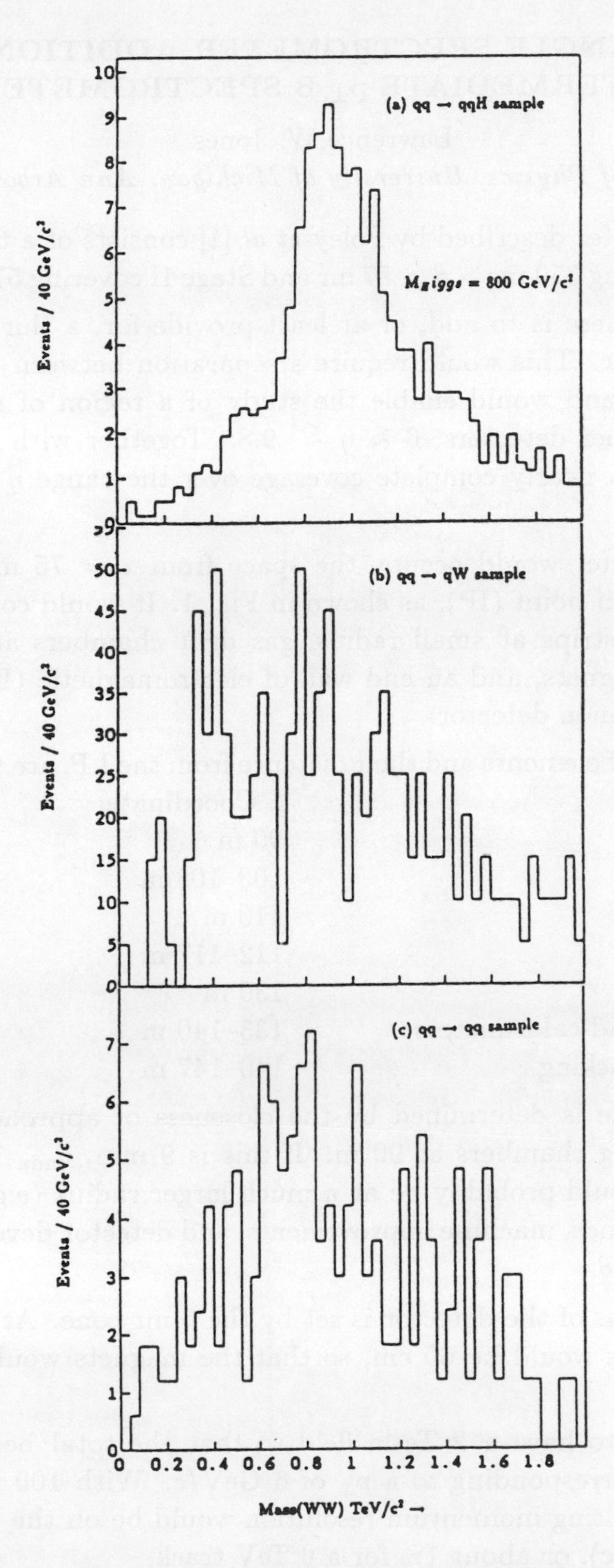

Figure 3. Forward spectrometer. WW mass distributions expected from 1 year of running $L = 10^{40}$ cm^{-2}. (a) Higgs signal, (b) W+anything background, (c) QCD elastic scattering background.

A SMALL-ANGLE SPECTROMETER ADDITION TO THE INTERMEDIATE $p_\perp$ B SPECTROMETER

Lawrence W. Jones
Department of Physics, University of Michigan, Ann Arbor MI 48109

The spectrometer described by Foley *et al.* [1] consists of a two-stage system with Stage I covering 350 mr $> \theta >$ 57 mr and Stage II covering 57 mr $> \theta >$ 5 mr.

The proposal here is to add, or at least provide for, a third stage to cover 5 mr $> \theta >$ 0.1 mr. This would require a separation between the low-β quads of $L^* = \pm 150$ m and would enable the study of a region of rapidity "space" not covered by other detectors, $6 \lesssim \eta \lesssim 9.8$. Together with the Stage I and II spectrometers, a nearly complete coverage over the range $\eta \leq 9.8$ would be realized.

The spectrometer would occupy the space from z = 75 m to z = 150 m from the interaction point (IP), as shown in Fig. 1. It would consist of tracking chambers (silicon strips at small radius, gas drift chambers at larger radius), dipole bending magnets, and an end wall of electromagnetic (EM) and hadron calorimeters and muon detector.

The sequence of elements and their distance from the I.P. are tabulated below:

Element:	z Coordinate:
Tracking chambers	90 m
5 m dipole magnet	103–108 m
Tracking chambers	110 m
5 m dipole magnet	112–117 m
Tracking chambers	130 m
EM and Hadron and calorimeters	133–140 m
Muon filter and tracking	140–147 m

The inner angle is determined by the closeness of approach to the beam pipe of the tracking chambers at 90 m. If this is 9 mm, $\theta_{\min}$ is 10^{-4} radians. Early operation would probably be at a much larger radius (e.g. 3 cm, $\theta_{\min} \sim 3.3 \times 10^{-4}$); experience, machine improvements, and detector development should permit the smaller θ.

The outer radius of the detector is set by the 5 mr cone. At the back of the dipole magnets this would be 60 cm, so that the magnets would require a 120 cm aperture (4 ft).

It is proposed to have a 2 Tesla field so that the total bending would be 20 Tesla-meters, corresponding to a $p_\perp$ of 6 GeV/c. With 100 micron tracking resolution, the resulting momentum resolution would be on the order of $\delta p/p = 1.6 \times 10^{-6}$ p (GeV/c), or about 1% for a 6 TeV track.

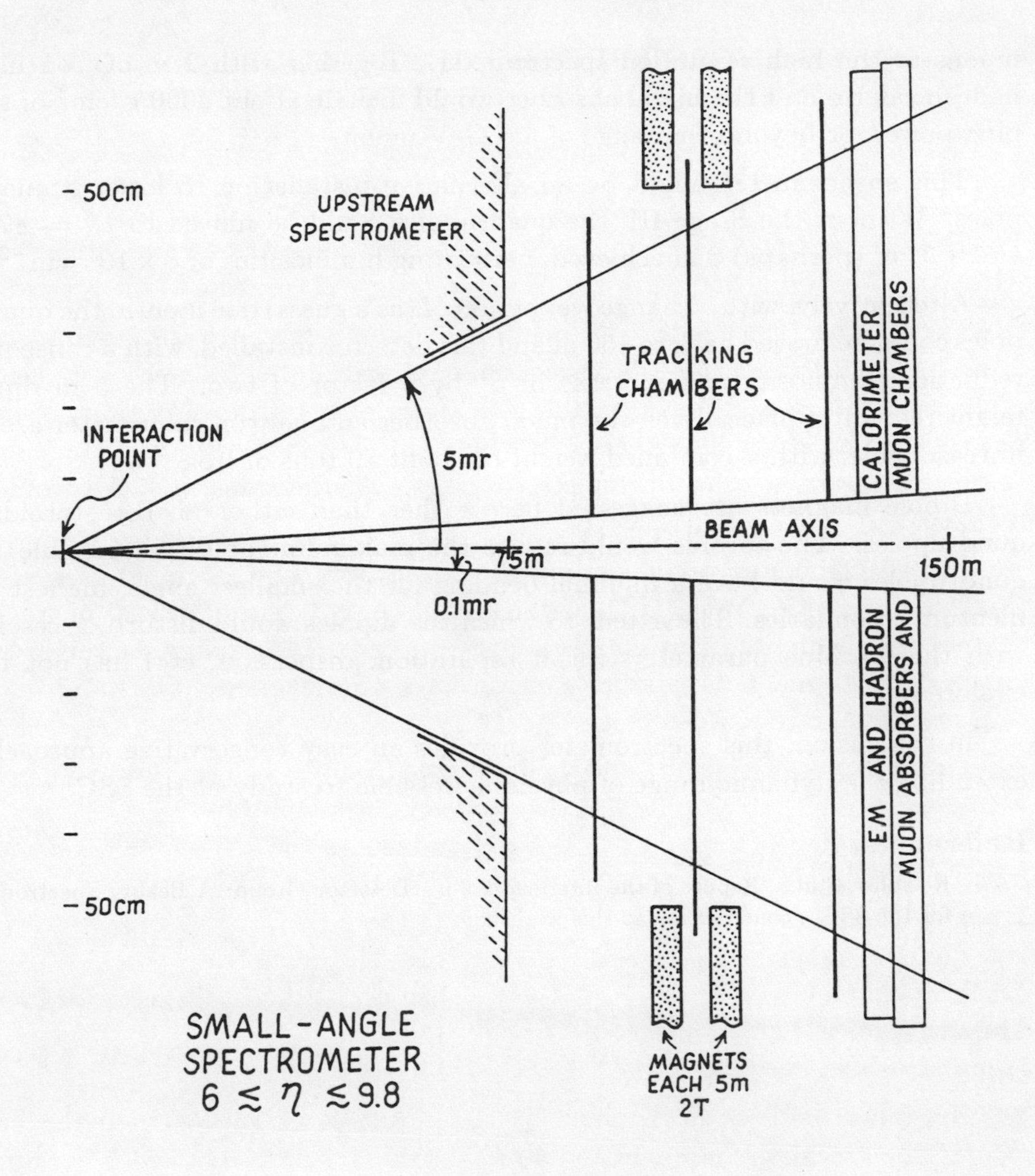

Figure 1. Schematic layout of the small-angle spectrometer.

The calorimetry need not be elegant; iron-scintillator or iron-gas proportional-wire chambers seem practical. The limitation would be the rates and radiation damage at the inner radius. As most tracks will be of high energy ($\gtrsim$ 1 Tev) the calorimeter resolution will be dominated by the constant (systematic uncertainty) term, so that compensation, etc. seem not important. Thus relatively coarse sampling should be satisfactory.

The muon absorber should be at least 3 m of iron. Some tracking might be included within the absorber as well as behind it in order to track cleanly muons to the high resolution spectrometer. Together with 2 m of iron in the hadron calorimeter the muon absorber would include about 4000 g/cm^2 of stopping power, or beyond the range of a 6 GeV muon.

This spectrometer might be an alternative installation to lower-β quadrupoles. Without the Stage III, the quadrupoles could be moved to $L^* = \pm 75$ m and a β^* of (perhaps) 6 m achieved, permitting luminosities of 8×10^{31} cm^{-2}s^{-1}.

Alternatively, with a changeover period of (as a guess) one month, the quadrupoles could be moved back to 150 m and this detector installed, with a consequent reduction luminosity and increase in β^* by a factor of two. The 5 m dipoles might remain in place. Even the muon absorber and hadron calorimeter are not unreasonable, with a combined weight of about 80 tons of iron.

Dipole magnets are suggested here rather than either air-core toroids or quadrupoles. The toroids would reduce the maximum rapidity accessible, and quadrupoles would have a minimal bending for the smallest angle, highest momentum secondaries. The extent to which the dipoles would disturb or resonate with the machine parameters (orbit separation, dispersion, etc) has not been explored.

In conclusion, this spectrometer provides an easy conservative approach to extending the dynamic range of physics accessible to study at the SSC.

References:

1. K. Foley *et al.*, "Report of the Intermediate-$p_\perp$ Detector Group: A Beauty Spectrometer for the SSC", contribution to this workshop.

A SEMILEPTONIC ELECTRON TRIGGER FOR RECORDING $B^0\overline{B^0}$ EVENTS

R. J. Morrison, and S. McHugh
University of California at Santa Barbara
Santa Barbara CA 93106
M. V. Purohit
Fermilab, Batavia IL 60510

The problems associated with a single electron trigger for the beauty spectrometer are studied. Backgrounds from Dalitz pairs and various types of pair conversions are analyzed and the effect of the beam pipe is determined. Other backgrounds studied are those from $K^{\pm} \rightarrow \eta^0 e^{\pm} \nu$ and from imperfect pion rejection. A multilevel trigger architecture is discussed.

1. Introduction

In this contribution we study the problem of triggering the beauty spectrometer [1]. The true $B^0\overline{B^0}$ event rate at an interaction region at the SSC with a luminosity of $10^{32}\mathrm{cm}^{-2}\mathrm{s}^{-1}$ is estimated to be $\sim 2 \times 10^4$ s^{-1}. In order to make a comprehensive study of CP violation in the beauty system it is necessary to be sensitive to a substantial part of this rate.

The full hadronic interaction rate is about 10^7 events s^{-1}. It is clearly essential to substantially reduce this, with a fast signature, in order to have a rate which can conceivably be considered for more sophisticated processing.

A semileptonic lepton trigger is relatively efficient since there is a 40% chance that one of the B's decays to a lepton. Using high precision vertex detectors, it is often possible, to reconstruct the D associated with this semileptonic decay, thus tagging the particle or antiparticle nature of the semileptonic decay. While this will not be a 100% pure tag for the case of the neutral B's due to mixing and other problems, it should be quite adequate for CP violation studies.

There are advantages and disadvantages to choosing muons or electrons. We have chosen to study an electron trigger since the electron shower energy is quickly determined and it's position is well localized. This facilitates fast track reconstruction for the electron track. The ultimate goal is to quickly find that the electron track emerges from a weak decay vertex. To be more specific about the true $B\overline{B}$ rates we take the B branching ratio to electrons to be 0.1. Then with an interaction rate of 10^7 s^{-1} we will trigger on $2 \times 0.1 \times 10^7 \times \epsilon \times a \times g$ $B\overline{B}$ events s^{-1}, where g is the fraction of the SSC cross section which is $B\overline{B}$, a is the acceptance of the spectrometer for B^0 decay electrons, including the effect of a $p_\perp$ cut, and ϵ is the electron trigger detection efficiency. With $a = 0.25$, $\epsilon = 0.5$ and $g = 0.002$ we have a rate of 500 real $B\overline{B}$ triggers s^{-1}. The requirement of an impact parameter of $\sim$ 300 microns or a displaced vertex, will reduce this by about a factor of $\sim$ two to about 250 s^{-1}. A less optimistic estimate of the cross section and/or ϵ could reduce this by a factor of up to ~ 4.

So even with no non-beauty background this leads to trigger rates $\gtrsim 100\ \mathrm{s}^{-1}$. The choice is then either to attempt to write an enormous amount of data to a recording medium (if possible), or to have a very high level on line filter which selects events based on desireable event characteristics such as reconstructability or evidence for particular decay modes.

The basic notion of the trigger is that it starts with a first level trigger which uses information from the electromagnetic and hadron calorimeters, the TRD's, and hits from chambers near the calorimetry and TRD's to select electrons above a $p_\perp$ threshold of ~ 1.5 GeV/c. There are probably many parallel processors to do this job. At a higher level, the electron track is reconstructed, eliminating many downstream conversions and electrons from K decays. At the highest level, the electron impact parameter is computed, using the vertex detector. For large angle electrons this can be computed with respect to the ± 20 micron beam interaction volume. For small angle electrons, it will have to be computed with respect to reconstructed vertices. Another possibility is to compute the distance of closest approach of the electron track to any well measured stiff track.

In the following sections we estimate backgrounds and are a bit more specific about how the trigger might be implemented. We go into some detail on the backgrounds to elucidate how the spectrometer design affects the viability of the trigger.

2. Backgrounds

Backgrounds for this trigger come from many sources. The most serious source is from Dalitz pairs and photon conversions. Charged K's decay to $\pi^0 e\nu$ and charged hadrons will be misidentified as electrons. Charm production cross sections are large, so these semileptonic decays are ~ 10 times as prevalent as those from B's.

1. *Electron Pairs*

The problems in removing electron pair backgrounds from the single electron trigger are different for different types of pairs.

Type A pairs are Dalitz pairs or pairs which convert in material before the first silicon plane (or in the first silicon plane). These can be removed only by finding some evidence of the pair mate. The $p_\perp$ involved in the first level trigger is the $p_\perp$ of the single electron. Since the average $p_\perp$ of the electrons is small, this rate is significantly reduced by the $p_\perp$ cut. An important issue here is the effect of putting the silicon detectors outside the beam pipe.

Type B pairs are conversions downstream of the first silicon plane but upstream of a magnet. These can be removed by the non-existence of tracks upstream of the conversion point, and by finding the pair mate. The $p_\perp$ in the first level trigger is again the $p_\perp$ of a single electron.

Type C pairs are pair conversions downstream of the last magnet. These can be removed by the non-existence of upstream tracks and by the fact that downstream track segments point back to the interaction point with no apparent bend. Since the two electrons go into the same calorimeter cell and the same TRD section, the first level $p_\perp$ observed is the $p_\perp$ of the parent photon of the two electrons. This causes the level one rate to be high, but these triggers can be eliminated cleanly in higher levels of the trigger.

a. Type A pairs

We consider first type A pairs. Here we assume that the vertex detectors are outside the beam pipe. Let f be the fraction of pair electrons with $p_\perp > p_\perp^{cut}$ and assume 3 π^0's/unit rapidity. The number of background electrons per event is

$$N^{el} = 12f\epsilon \int_{\theta^{min}}^{\theta^{max}} \frac{d\theta}{\theta}\left(0.012 + \frac{7\times 10^{-4}}{\theta}\right)$$

where the first term is from Dalitz pairs and the second is from conversion in a 0.3 mm thick Be beam pipe of 1 cm radius. The number from Dalitz pairs for $\theta^{min} = 0.0057$ and $\theta^{max} = 0.35$ is $N^{el}_{Dalitz} = 0.6f\epsilon$. The number due to the beam pipe is $0.12f\epsilon$ and $1.4f\epsilon$ for the intermediate and forward spectrometers, respectively. The cylindrical beam pipe has a serious effect at small angles as expected.

We can also attempt to find the pair mate and reject mated events. If the mate electron is above $E^{min} \sim 1$ GeV there is a good chance that it will appear in the E & M calorimeters and in the TRD's in the same bend plane as the primary trigger electron. Alternatively, if there is no magnetic field before the first silicon plane, the registration of double pulse heights in the first silicon plane or correlated neighbor strip hits can be used as a veto. For Dalitz pairs this method is limited by the natural pair angular spread, m_e/E, so that E^{min} for this method is $\sim 0.5\times 10^{-3}/\theta\Delta x$ where Δx is the strip spacing in cm. For beam pipe pairs this method depends crucially on the shape of the beam pipe.

Assuming the first method, we can estimate the number of background pairs remaining after a veto on a pair mate of energy greater than E^{min}. In the Appendix we show that the non-vetoable fraction is $\sim \left(1 - e^{-2E^{min}\theta/\langle p_\perp\rangle}\right)$ where $\langle p_\perp\rangle$ is the mean $p_\perp$ of the parent π^0 distribution. The number of electrons not vetoed is then

$$N^{el}_{N\cdot veto} = 12f\epsilon \int_{\theta^{min}}^{\theta^{max}} \frac{d\theta}{\theta}\left(0.012 + \frac{7\times 10^{-4}}{\theta}\right)\left(1 - e^{-2E^{min}\theta/\langle p_\perp\rangle}\right).$$

For the Dalitz pairs this is

$$N^{Dalitz}_{N\cdot veto} \sim 0.14f\epsilon\left[0.25 - \frac{2E^{min}\theta^{min}}{\langle p_\perp\rangle} + 0.5\,\ln\left(\frac{2E^{min}\theta^{max}}{0.25p_\perp}\right)\right] .$$

This leads to $N^{Dalitz}_{N\cdot veto} \sim 0.18f\epsilon$, assuming $\langle p_\perp\rangle \sim 0.35$. The non-vetoable pairs from the beam pipe is

$$\sim 8.4 \times 10^{-3}\ f\epsilon \left[\frac{2E^{min}}{\langle p_\perp\rangle} \ln\left(\frac{0.25\,\langle p_\perp\rangle}{2E^{min}\theta^{min}}\right) + 0.5\left(\frac{2E^{min}}{0.25\,\langle p_\perp\rangle} - \frac{1}{\theta^{max}}\right)\right].$$

This is also about 0.18 $f\epsilon$. The total type A pair background rate is therefore about 0.4 $f\epsilon$ per event. The pair veto has reduced the Dalitz contribution by a factor of three. The most significant effect is the removal of the small angle high energy beam pipe pairs. The damage from the beam pipe is greatly reduced by the pair veto.

Monte Carlo studies using the LUND model indicate that f is about 0.01. This is clearly very dependent upon the value of $\langle p_\perp\rangle$, the mean $p_\perp$ of the parent π^0 distribution. For this study the π^0 $\langle p_\perp\rangle$ was 0.40 GeV/c. If we then assume that ϵ is 0.5 we have a non vetoable type A pair rate of 2×10^4 s^{-1}.

b. Type B pairs

The background due to type B pairs is due to the amount of conversion material between the beam pipe and the first magnet, for the intermediate spectrometer, and before the second magnet for the forward spectrometer. Including 6 silicon planes, air, mylar and argon for the chambers and He gas for the forward spectrometer we estimate conversion probabilities χ of 2% and 5% for the intermediate and forward spectrometer, respectively. The number of such pairs per event is

$$N = 12\, f\epsilon\chi\, \ln\left(\frac{\theta^{max}}{\theta^{min}}\right).$$

This gives $0.4f\epsilon$ and $1.4f\epsilon$ for the intermediate and forward spectrometer, respectively. The number which survives the pair veto is

$$N^{N\cdot veto} = 12f\epsilon\chi \int_{\theta^{min}}^{\theta^{max}} \frac{d\theta}{\theta}\left(1 - e^{-2E^{min}\theta/\langle p_\perp\rangle}\right).$$

This is approximately

$$N^{N\cdot veto} \sim 12f\epsilon\chi\left[0.25 - \frac{2E^{min}\theta^{min}}{\langle p_\perp\rangle} + 5\,\ln\left(\frac{2E^{min}\theta^{max}}{0.25\,\langle p_\perp\rangle}\right)\right].$$

For the 2% of conversions before the first magnet this is $\sim 0.3f\epsilon$. For the forward spectrometer conversions between the magnets are vetoed with an E^{min} of $\sim$ 10 GeV due to the higher $p_\perp$ kick of the second magnet and the less favorable geometry. Few of these cases are vetoed so the number of these pairs is $0.8f\epsilon$. The total nonvetoable rate of type B pairs is $1.1f\epsilon$. With the same assumptions for ϵ and f as for type A pairs we get a rate of 5×10^4 s^{-1}. These pairs are eliminated at the higher trigger level where complete electron tracks are required. The veto is of little value for these pairs.

c. Type C pairs

Type C pairs are conversions downstream of the last magnet. The pair electrons appear as a single signal in the electromagnetic calorimeter and give a calorimeter $p_\perp$ which is that of the parent photon. Since this $p_\perp$ distribution is less steep than that of the electrons, there will be a rather high rate at the first level of the trigger.

The number of these triggers/event is $N^c = 6\,\chi f'\epsilon\,\ln(\theta^{max}/\theta^{min})$, where f' is the fraction of photons with $p_\perp > p_\perp^{cut}$ and χ is the fraction of a conversion length of the relevant material. For the intermediate spectrometer χ is about 1%, and we have,

$$N^c_{int} = 0.11 f'\epsilon.$$

For the forward spectrometer χ is about 2% giving $N^c_{for} = 0.3 f'\epsilon$. The total rate of type C pairs is $\sim 0.4 f'\epsilon$. With the same asssumptions as for type A pairs, and assuming that f' is 10 times larger than f, we get a rate of $2.5\times10^5\ \mathrm{s}^{-1}$. These pairs are easily eliminated because the downstream track segment points directly to the interaction point with no bend, and the track can not be linked to an upstream track segment.

it d. Summary of pair background

To summarize, the number of pair triggers which survive a $p_\perp$ cut of about 1.5 GeV/c is about $10^7\times\epsilon\times(3.9f+0.4f')$ or $\sim 4\times10^5\ \mathrm{s}^{-1}$. The breakdown of the various components of this rate is given in Table 1. The number which survive a mate veto and have sensible downstream tracks is $10^7\times 1.5 f\epsilon$, which is about $8\times10^4\ \mathrm{s}^{-1}$. The number which have good tracks in the full tracking system is $0.4 f\epsilon$, which is about $2\times10^4\ \mathrm{s}^{-1}$.

2. Ke₃ Decays

Charged K's can produce an electron trigger through the 4.8% branching to the $\pi^0 e\nu$ mode. To generate an electron of $p_\perp \gtrsim p_\perp^{cut} \sim 1.5$ GeV the parent K must have a $p_\perp \sim \alpha p_\perp^{cut}$ where α is ~ 3. The decay length for these K's is

$$L = \frac{p}{m}\,c\tau \sim \frac{\alpha\, p_\perp^{cut}\times 3.7\ \mathrm{m}}{m\theta}$$

or $L \approx 33.3\ \mathrm{m}/\theta$.

Assume that there are ξ charged K's per unit of rapidity and that $0.048 f''$ of these decay to an electron of $p_\perp > p_\perp^{cut}$. Then in the intermediate spectrometer of length 8 meters the number of triggers/interaction is

$$N_K^{int} = 0.048\ f''\xi\epsilon \int_{\theta=0.057}^{\theta=0.350} \frac{d\theta}{\theta}\left(\frac{8\times\theta}{33.3}\right),$$

or $N_K^{int} = 3.5\times10^{-3}\ f''\xi\epsilon$.

Table 1

Estimates of the components of the electron pair background. In the first column a $p_\perp$ cut greater than 1.5 GeV/c from the electron is assumed. If only calorimetry information is used then only the component of $p_\perp$ perpendicular to the bend plane is available. In this case the numbers correspond to an effective real $p_\perp$ cut which is larger than 1.5 GeV/c with the attendant loss of efficiency. If pad MWPC information is available (as described in the text) the true $p_\perp$ can be determined. In this case most of the term proportional to f' is eliminated. The rates in the text have assumed values of 0.5, 0.01, and 0.1 for ϵ, f and f', respectively, and a multiplicitive total interaction rate of 10^7 s^{-1}.

		Information used		
		Calorimetry + TRD	Pair veto + Downstream tracks	Tracks found + to first vertex plane
Type A	Dalitz	0.6 $f\epsilon$	0.18 $f\epsilon$	0.18 $f\epsilon$
	Beam Pipe	Int 0.12 $f\epsilon$		
		For 1.4 $f\epsilon$	0.18 $f\epsilon$	0.18 $f\epsilon$
		2.1 $f\epsilon$	0.4 $f\epsilon$	0.4 $f\epsilon$
Type B		Int 0.4 $f\epsilon$	0.3 $f\epsilon$	0
		For 1.4 $f\epsilon$	0.8 $f\epsilon$	
		1.8 $f\epsilon$	1.1 $f\epsilon$	
Type C		Int 0.11 $f'\epsilon$	0	0
		For 0.3 $f'\epsilon$		
		0.4 $f'\epsilon$		
		$\epsilon \times (3.9\ f + 0.4 f')$	1.5 $f\epsilon$	0.4 $f\epsilon$

For the forward spectrometer this is

$$N_K^F = 0.048\ f''\xi\epsilon \int_{\theta=0.0057}^{0.057} \frac{d\theta}{\theta}\left(\frac{70\times\theta}{33.3}\right),$$

giving $N_K^F = 5.2\times10^{-3}\ f''\xi\epsilon$. With $\xi = 1$, $\epsilon = 0.5$ and $f'' = 0.02$ this is a total rate of ~ 900 s^{-1}. Most of these decay downstream of a magnet and will not give a good E/p match even if good tracks are found. We expect that only a few percent of these survive to the high level trigger.

3. Hadron Misidentification

There will be approximately 24 hadrons produced/event into the spectrometer. To generate a trigger these hadrons must

a. deposit a $p_\perp > p_\perp^{cut}$ in the electromagnetic calorimeter.

b. not have a large signal in the corresponding region of the hadron calorimeter.

c. have a narrow electromagnetic type transverse shower profile in the electromagnetic calorimeter.

d. have a TRD signal at the x, y coordinate of the electromagnetic shower.

e. e. have a longitudinal electromagnetic type profile in the electromagnetic calorimeter.

f. have a drift chamber track projection consistent with the electromagnetic calorimeter position to within $\pm \sim$ 2mm.

g. have a momentum which is $\sim$ equal to the shower energy.

The rejection power of the calorimetry alone is $\sim 10^{-3}$ and with the addition of the TRD's the rejection should be $\sim 2 \times 10^{-5}$ for charged hadrons of energy up to $\sim$ 200 GeV. This corresponds to a rate of about 5×10^3 s^{-1}. Above 200 GeV the effectiveness of the TRD's diminishe. According to the Monte Carlo 12 % of the events have one or more pions above 200 GeV into the spectrometer acceptance. This gives a rate of about 10^3 s^{-1}. The energy momentum comparison should reduce these rates by a factor of 2 to 4, so we estimate a total rate of about 2,500 s^{-1}.

4. Direct Charm Background

We expect direct charm to be produced at about 10 times the bottom rate, and with a mean $p_\perp \sim$ 1.5 GeV. We expect direct D* production to be about three times that of direct D production with the net effect that we will have D^0's about twice as often as D^+'s. The semilepton branching ratio, the lifetime of the D^0, and especially the $p_\perp$ cut should reduce direct charm background rate to about that of the beauty signal.

3. Possible Trigger Scheme

We assume that all signals from the experiment are stored in analog pipeline type delays until a time τ_0 required to generate an nth level trigger. We assume that there are a number of specialized asynchronous processors available which are triggered by an electromagnetic calorimeter segment signal corresponding to a non-bend plane component of $p_\perp > p_\perp^0$. These trigger processors have available the following additional information which is local to the region of the electromagnetic calorimetry hit:

a. Electromagnetic and hadronic calorimetric energy deposits for neighboring transverse and longitudinal cells.

b. TRD information from corresponding pads or wires.

c. Drift chamber information from corresponding wires of the chamber just upstream of the TRD.

The processor does not do any sophisticated tracking but does determine whether the transverse and longitudinal shower profiles are electromagnetic. It finds the particle's energy and x and y coordinates at the calorimeters. It checks the relevant TRD pads and/or wires and also looks for a track in the closest chamber. Any negative result makes the processor available to analyze another shower.

It would be very nice if the available track information is sufficient to determine the charge of the particle. It would be very useful, for this trigger and also for the general track reconstruction, to have an MWPC, with wires measuring the non-bend plane coordinates and anode pads of size similar to those of the electromagnetic towers, placed near each of the two main tracking drift chambers. With about 10,000 pads per plane and a time resolution of 50 nsec, each pad would have an occupancy of about 0.002. From the known energy and position of the shower at the electromagnetic calorimeters two locations at the pad chambers are determined corresponding to the two possible charges. The pad with the hit determines the charge and therefore the true $p_\perp$ of the particle, and this also eliminates most of the class C pairs. We expect an average of one to two processor calls per event, from high $p_\perp$ photons. With an anticipated average time to rejection of 2 micro seconds there would need to be $\gtrsim 20$ - 40 processors of this type for the experiment.

The rate of triggers from this level is $\sim 4 \times 10^5$ s^{-1}. This can be reduced to about 8×10^4 by elimination of the class C pairs and by the pair mate veto. Most of the class C pairs can be removed by requiring a hit in the appropriate bend plane of the main tracker, or the pad chamber (as discussed above). The mate veto requires a search, again in the appropriate bend plane, for electromagnetic showers and/or TRD hits. In neither case is a full track reconstruction required. However, sensitivity to ~ 1 GeV showers is probably required. At this point it may be appropriate to initiate digitization of the event. The time to complete this calculation defines τ_0, the analog pipeline storage time for all primary signals. Up to this point the system can be made nearly deadtimeless.

The next level of the trigger includes tracking back to the Vertex detectors. This involves more sophistication since the focusing of the magnet must be taken into account in the non bend plane and since the track projection must be done to a precision which is small compared with the average two track separation at the vertex detectors. The result of this trigger level is a rate of $\sim 2 \times 10^4$ s^{-1}. The number of parallel processors of this type needed is $\sim$ (computation time)/10 μsec. If this computation could be done in a millisecond then ~ 100 processers would be required.

The next level trigger requires evidence that the track makes an impact parameter of ~ 300 microns either with respect to the beam interaction volume or with respect to other stiff tracks. In principal this reduces the pair backgrounds to a negligible level. A detailed Monte Carlo of the vertex reconstruction confusion will be required to estimate the actual background. We assume an event rate to $\sim 10^3$ s^{-1} of non heavy quark background plus about 250 events s^{-1} of direct charm and about 250 events s^{-1} of real semileptonic beauty signal. The numbers of processors of this type will be $\sim$ (computation time)/50 μsec.

The next level selects those events which are most likely to be useful. This includes the dielectrons with the mass of the ψ. Other possibilities might be partial reconstruction of the associated D in the B semileption decay, or evidence for another decay vertex. This very high level trigger might be done via a very large farm of 1994 style ACP's. It has to handle an event rate of $\sim$ 1500/sec.

4. Conclusions

We have attempted to analyze the possibilities and problems associated with a semileptonic electron trigger for the beauty spectrometer. This has been done in some detail to understand the impact of various design parameters (e.g. beam pipe) on the viability of the trigger. Under the assumptions made, the main background is from Dalitz pairs. It is assumed that a final background reduction factor of about 20 can be achieved with a fast vertexing algorithm, which computes the distance of closest approach of the electron with respect to another good track or with respect to the interaction volume. From the basic physics point of view this trigger appears difficult but possible.

The goal of the SSC beauty spectrometer is to be sensitive to a very high rate of produced $B\overline{B}$ pairs. This means that not only the background rates but even $B\overline{B}$ pair rates are very large. As a consequence a very large rate of information must be transferred and processed. Perhaps the most important question remaining is whether this is feasible with the resources and technology which are likely to be available.

We would like to thank Bill Reay and Chuck Buchanan for helpful discussions.

APPENDIX

Estimates of Pair Backgrounds

The spectrum of photons from the decay of a π^0 of energy E_π is flat with

$$dn_\gamma = \frac{2}{E_\pi} dE_\gamma,$$

up to the energy E_π. Here the photons have very nearly the same direction as the π^0. In a fraction of a conversion length χ, the electron and positron spectrum from a photon of energy E_γ is very crudely $dn_e = 2\chi\, dE_e/E_\gamma$.

Starting with a primordial π^0 distribution $f(\theta, E_\pi)$, the photon distribution is

$$g(\theta, E_\gamma) = 2 \int_{E_\gamma}^{\infty} \frac{dE_\pi}{E_\pi} f(\theta, E_\pi)$$

and the electron distribution is

$$h(\theta, E_e) = 2\chi \int_{E_e}^{\infty} \frac{dE_\gamma}{E_\gamma} g(\theta, E_\gamma) = 4\chi \int_{E_e}^{\infty} \frac{dE_\gamma}{E_\gamma} \int_{E_\gamma}^{\infty} \frac{f(\theta, E_\pi) dE_\pi}{E_\pi}$$

Since the electron $p_\perp$ is $P_T^e \simeq \theta E$, and we are interested in $P_T^e \gtrsim 1.5$ GeV, f is a steeply falling function of E_π, and g is a still steeper function of E_γ. As a consequence, electrons over the p_T^e cut correspond to photons of not much higher energy and these photons correspond to π^0's of energy not much higher than that of the photons. An important consequence of this is that the mate electron is nearly always the lower energy one, making it harder to find the mate and to reject the pair.

Due to the 0.45 GeV/c p_T kick of the first magnet, a mate electron will not register in an electromagnetic calorimeter if its energy is below $\sim E^{min} \sim 1$ GeV/c.

Considering for the moment pre-magnet conversions, then the number triggering is,

$$N^{trig} = \int_{P_{T/\theta}^{cut}}^{\infty} dE_e h(\theta, E_e) = 4\chi \int_{P_{T/\theta}^{cut}}^{\infty} dE_e \int_{E_e}^{\infty} \frac{dE_\gamma}{E_\gamma} \int_{E_\gamma}^{\infty} \frac{f(\theta, E_\pi) dE_\pi}{E_\pi}$$

The number which trigger but can be rejected because the mate hits a calorimeter is,

$$N^{Trig-rej} = 4\chi \int_{p_{T/\theta}^{cut}}^{\infty} dE_e \int_{E_e+E^{min}}^{\infty} \frac{dE_\gamma}{E_\gamma} \int_{E_\gamma}^{\infty} \frac{f(\theta, E_\pi) dE_\pi}{E_\pi}$$

The fraction which can not be rejected is,

$$fr^{N-Rej} = \frac{\int_{P_{T/\theta}^{cut}}^{\infty} dE_e \int_{E_e}^{E_e+E^{min}} \frac{dE_\gamma}{E_\gamma} \int_{E_\gamma}^{\infty} \frac{f(\theta, E_\pi) dE_\pi}{E_\pi}}{\int_{P_{T/\theta}^{cut}}^{\infty} dE \int_{E_e}^{\infty} \frac{dE_\gamma}{E_\gamma} \int_{E_\gamma}^{\infty} \frac{f(\theta, E_\pi) dE_\pi}{E_\pi}}$$

To make some crude estimates we take:

$$\frac{d\sigma}{dP_T^2} = \frac{2}{\langle p_\perp \rangle^2} e^{-\frac{2p_\perp}{\langle p_\perp \rangle}}$$

where $\langle p_\perp \rangle$ is an average π^o $p_\perp$ and the expresson is normalized to unity. Then,

$$f(\theta, E_\pi) = \frac{4\theta^2 E_\pi}{\langle p_\perp \rangle^2} e^{-\frac{2\theta E_\pi}{\langle p_\perp \rangle}}.$$

Assuming that the energy dependence of $f(\theta, E_\pi)$ is much steeper than $\frac{1}{E_\pi}$ for large $p_\perp$ we get:

$$h(\theta, E_e) = \frac{4\chi}{E_e} e^{-\frac{2\theta E_e}{\langle p_\perp \rangle}} = \frac{4\chi}{P_T^e} e^{-\frac{2p_\perp^e}{\langle p_\perp \rangle}}.$$

The number above a $p_\perp^{cut}$ for the electron is

$$N^{cut} \sim \frac{2\langle p_\perp \rangle}{p_\perp^{cut}} e^{-\frac{2p_\perp^{cut}}{\langle p_\perp \rangle}}.$$

The fraction of electrons where the mate is not found is

$$fr^{N-Rej} \sim \left(1 - (1 - \frac{E^{min}\theta}{p_\perp^{cut}})e^{-\frac{2E^{min}\theta}{\langle p_\perp \rangle}}\right)$$

While the expression from N^{cut} is not reliable, the estimate of the non rejectable fraction is probably not bad. With $\theta < 0.35$ and $E^{min} \sim 1$GeV, $f_r^{N-Rej} \sim (1 - e^{-\frac{E^{min}\theta}{\langle p_\perp \rangle}})$.

For small θ the rejection is quite good since the $p_\perp^{cut}$ corresponds to a large energy and a consequent large probability that the mate is > 1 GeV.

References

1. K. J. Foley *et al.*, "Report of the Intermediate-$p_\perp$ Detector Group: A Beauty Spectrometer for the SSC," contribution to this workshop.

Low p_T Physics and Detectors

LARGE CROSS-SECTION PHYSICS

P.V. Landshoff

Department of Applied Mathematics and Theoretical Physics
University of Cambridge, England

ABSTRACT

Existing data for total cross-sections, elastic scattering, and diffraction dissocation together indicate that the pomeron, whose exchange generates the long-distance part of the strong interaction at high energy, behaves very much like an isoscalar C = +1 photon. This will be explored further in hard-diffraction experiments, where one examines events in which there is a hard interaction and also one of the initial protons changes its momentum only a very little. In this way, we may hope to gain a better understanding of just what the pomeron actually is.

At SSC energy, minijets will be a prominent feature of typical events. The cross-section for producing these will measure the gluon content of the proton at very small x, which is interesting for understanding the interface between perturbative QCD and nonperturbative pomeron exchange. It seems also that minijet production has a very important effect on particle multiplicity distributions, and will be a contributing factor in a rather high average total multiplicity at the SSC.

1. Introduction

All experimentalists at the SSC must have some appreciation of large-cross-section events, if only to understand possible backgrounds to rarer events. But my brief in this talk is to explain why large-cross-section events are interesting in their own right. My discussion will include the following topics, in varying levels of detail:

total cross-sections
elastic scattering
diffraction dissociation
hard diffraction
minijets
multiplicity distributions

Other important topics in large-cross-section physics which the SSC will surely explore, but which I do not cover here, are

charm and bottom production
pomeron-pomeron collisions

I also do not discuss

Centauro events
quark plasma formation

which may possibly occur with large cross-sections.

Large-cross-section physics is sometimes regarded as dirty physics. However, some of its features are actually rather clean, if looked at in the right way. One should give these particular emphasis, because they have the best chance of being interpreted and understood. The aim must always be to try and understand *dynamics*, rather than just making multi-parameter fits to data.

The dynamics in question is that of the long-range part of the strong interaction at high energy. So far, our understanding of this at the fundamental level is not at all good. However, a rather good phenomenological description has already emerged, and the object of further experiments is to help to turn this into a more theoretical analysis. There is little doubt that the long-range force is described by QCD, but the only available calculational tool is perturbation theory, which cannot be used here because the QCD coupling at long distance is not small. Guided very much by the phenomenology, we do have the beginnings of a more basic understanding [1], but a great deal more work will be needed to turn this into a theory.

2. The Total Cross-Section

One of the triumphant successes of particle physics is Regge theory [2], and this is well illustrated by the total-cross-section data. The difference between the $\bar{p}p$ and pp cross-sections is well accounted for by the exchange of the dominant C = -1 trajectories, namely ρ and ω, and the curves in figure 1 satisfy

$$\sigma(\bar{p}p) - \sigma(pp) = 70\ s^{-0.56} \tag{1}$$

with s measured in $(\mathrm{GeV})^2$. The f and A_2 trajectories are to a good approximation degenerate with the ρ and ω and so should also give a contribution varying as $s^{-0.56}$. They are surely responsible for the average of the $\bar{p}p$ and pp cross-sections initially falling with increasing energy, and the average of the two curves in figure 1 includes a contribution 105 $s^{-0.56}$. The remaining part of the average of the cross-sections rises steadily throughout the energy range of the data and to a first approximation it may be parametrised as [3,4]

$$22.7\ s^{0.08} \tag{2}$$

The virtue of the curves in figure 1 is that they are dynamically motivated and their parametrisation is fairly simple. Other authors are often less concerned with dynamics and allow themselves more parameters, and so of course they have even better fits to the data. They commonly use powers of logarithms instead of the fractional rising power of s in (2): see in particular the work of Block and Cahn [5].

The use of powers of logarithms is motivated by the Froissart-Lukaszuk-Martin bound [6]: at large s any total cross-section must obey

$$\sigma^{Tot} < (\pi/m_\pi^2)\ \log^2 s/s_0 \tag{3}$$

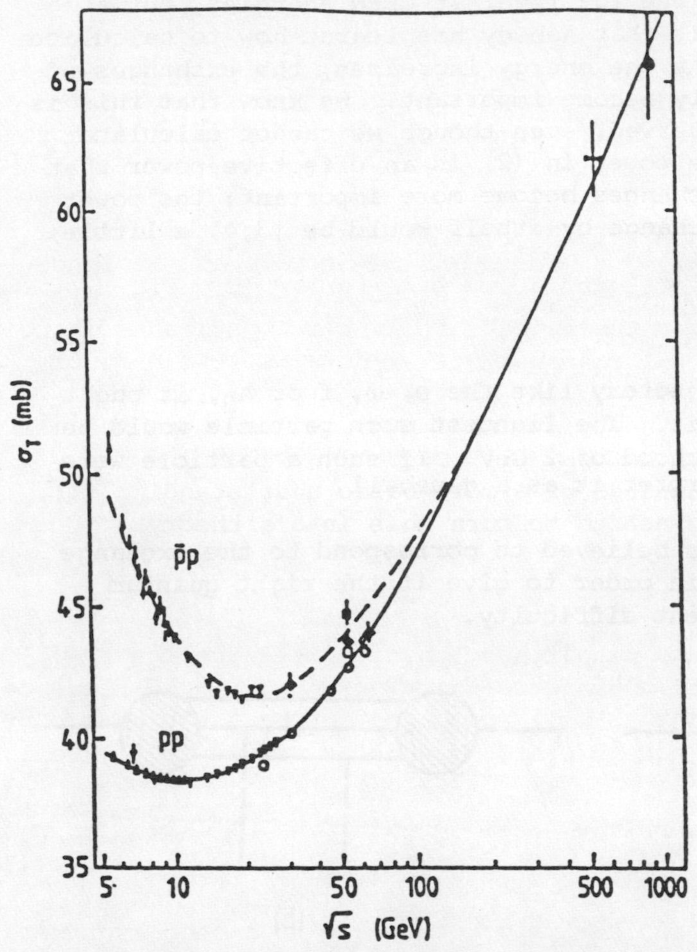

Figure 1 : Data for $\bar{p}p$ and pp total cross sections, with Regge-theory fits from reference 3

For reasonable choices of the unknown parameter s_o, this bound is about 30 *barns* at SSC energy, so it is not really a severe constraint on the physics. The bound implies that the effective power in (2) must decrease at higher energies. The decrease can be very slow, but this introduces an uncertainty in the extrapolation of the curves to SSC energies. Most predictions in the literature fall in the range

$$110 \pm 30 \text{ mb} \tag{4}$$

It has recently been emphasised [7] that when $\sigma(\bar{p}p)$ and $\sigma(pp)$ rise with energy, their difference need not go to zero at high energy. So it may be that at SSC energy the two curves have separated again by a few mb. While this is not excluded by what we know so far, it is certainly not required, but it is worth keeping in mind.

The rising component (2) of $\sigma(\bar{p}p)$ and $\sigma(pp)$ is interpreted as resulting from the exchange of a Regge trajectory known as the pomeron, and the key question of large-cross-section physics is: *what is the pomeron*?

At CERN collider energies and below, the exchange of a single pomeron dominates in (2), but the simultaneous exchange of two or more pomerons is not completely negligible. We have estimated [3] that two-pomeron exchange gives a (negative) contribution at the 10% level at CERN energies, but a severe limitation on Regge theory is that nobody has learnt how to calculate multiple exchanges properly [4]. As the energy increases, the exchanges of more and more pomerons progressively become important. We know that this is how the Froissart bound (3) is preserved, even though we cannot calculate these multiple exchanges. Thus the power in (2) is an effective power that slowly decreases as the further exchanges become more important; the power corresponding to single-pomeron exchange by itself would be [3,4] a little greater than 0.08.

3. Properties of the pomeron

If the pomeron is a Regge trajectory like the ρ, ω, f or A_2, it ought to have particles associated with it. The lightest such particle would be a 2^{++}, with a mass in the neighbourhood of 2 GeV. If such a particle were found, it would be natural to interpret it as a glueball.

This is because the pomeron is believed to correspond to the exchange of a number of gluons, at least 2 in order to give it the right quantum numbers. But this raises an apparent difficulty.

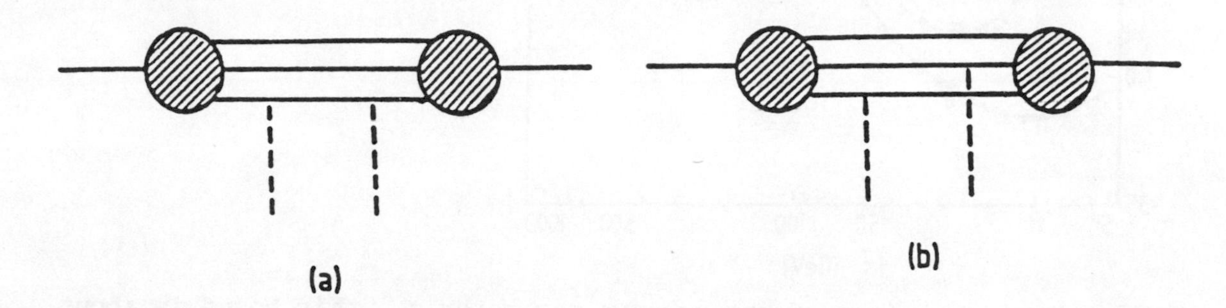

Figure 2 : Couplings of 2 gluons to the quarks in a nucleon

If the pomeron is 2 or more gluons, one would think that the gluons need not couple to the same quark in a hadron: if figure 2a is allowed, why not also figure 2b? But experiment indicates that only one quark seems to be involved. For example, a good rule for relating different hadron-hadron total cross-sections is the additive quark rule, which says that a hadron-hadron cross-section is just the sum of the cross-sections for the possible separate scatterings of the valence quarks and antiquarks of the hadrons involved. The rule is found to be good to within 10%, and is probably even better if it is applied only to the pomeron-exchange parts of total cross-sections (though we cannot extract these accurately from the data).

So figure 2a is wanted by the data, but not figure 2b. This would be hard to understand if the gluons were described by perturbation theory, but perturbation theory cannot be used because we are dealing with the long-range force. There is some understanding [1] of how nonperturbative effects can resolve the problem, but much more work is needed here, helped by input from further experiments such as I discuss in §4.

But experiment has already told us that the properties of the pomeron are surprisingly simple. From total-cross-section data, elastic scattering at small t, inelastic diffraction dissociation, and the small x behaviour of νW_2, we have deduced the following properties [3,8]:

1. The pomeron couples to single valence quarks and antiquarks of hadrons
2. The coupling is γ^{μ} (times a C = +1 signature factor)
3. The pomeron is almost pointlike, with an effective radius of about $\frac{1}{6}$ fm.

That is, the pomeron resembles an isoscalar C = +1 photon.

For example, the contribution from pomeron exchange to pp elastic scattering at small t is

$$\frac{d\sigma}{dt} = \frac{[3\beta_o F_1(t)]^4}{4\pi} (\alpha' s)^{2\alpha(t)-2} \tag{5}$$

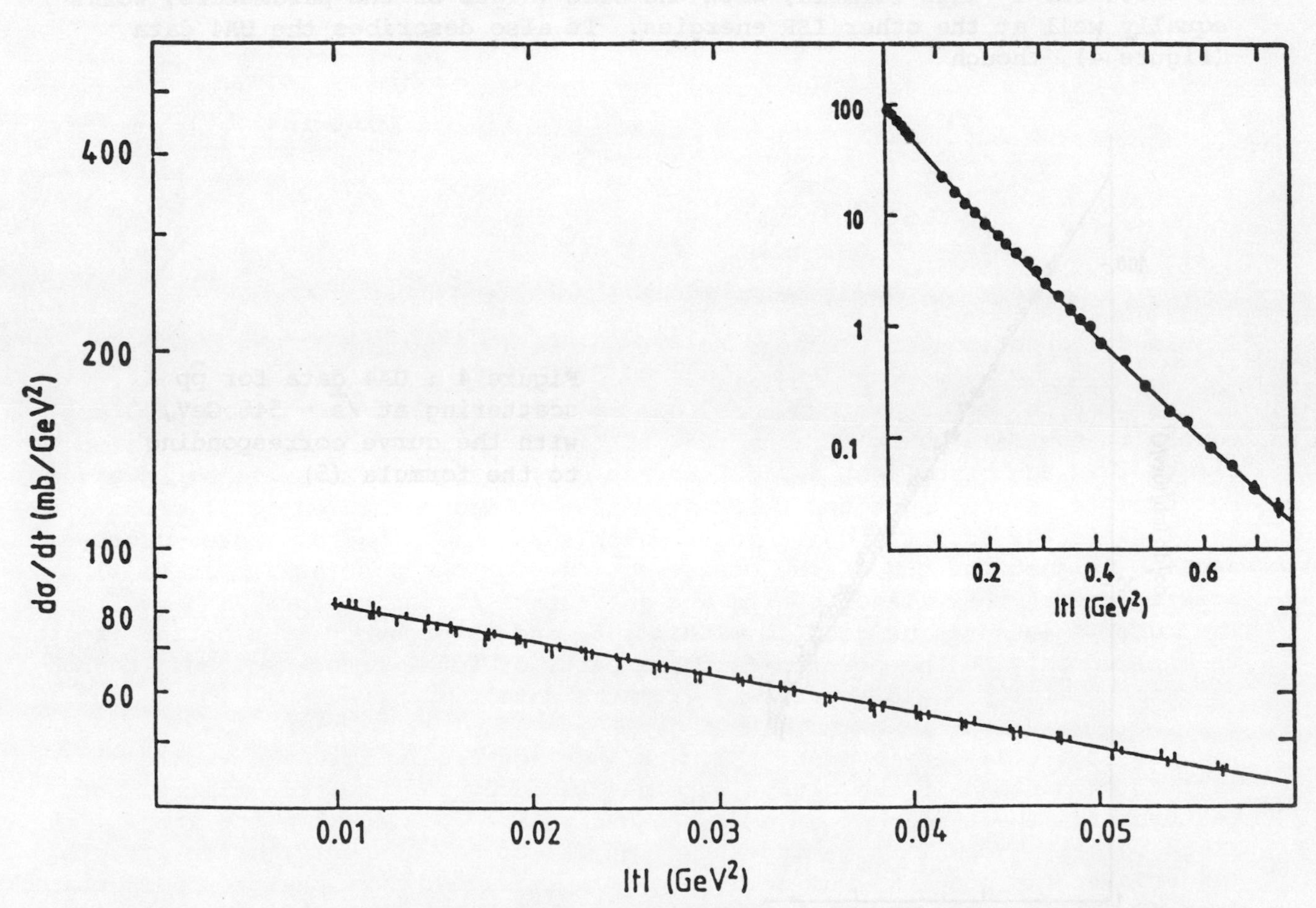

Figure 3 : ISR data for pp elastic scattering at $\sqrt{s}$ = 53 GeV. The curves correspond to the formula (5).

Here β_o is the strength of the coupling of the pomeron to a quark; in photon exchange it would be the charge. The factor 3 arises because each hadron has 3 valence quarks. The function $F_1(t)$ is the isoscalar Dirac elastic form factor; it takes account of the proton wave function. If, as experiment suggests, the electric form factor $F_1^{en}(t)$ is 0, $F_1(t)$ is just the electric form factor $F_1^{ep}(t)$ of the proton, which is measured and is well represented by taking a simple dipole form for $G_M(t)$ and $G_E(t)$. It is natural to assume that the pomeron trajectory $\alpha(t)$ is linear, like that of the exchange-degenerate (ρ, ω, f, A_2):

$$\alpha(t) = 1 + \epsilon + \alpha' t \tag{6}$$

The total cross-section data determine $\epsilon \simeq 0.08$ and $\beta_o^2 \simeq 3\ \text{GeV}^{-2}$. So the only free parameter in (5) is the trajectory slope α'. We fix this from very-small-t ISR data at some energy, which gives

$$\alpha' = 0.25\ \text{GeV}^{-2} \tag{7}$$

We find that the formula (5) then fits the data well right out to $-t = 0.7\ \text{GeV}^2$: see figure 3. Notice that the formula naturally includes what experimentalists commonly refer to as a break in exponential slope at $-t \simeq 0.1$ GeV . This formula, with the same values of the parameters, works equally well at the other ISR energies. It also describes the UA4 data (figure 4), though

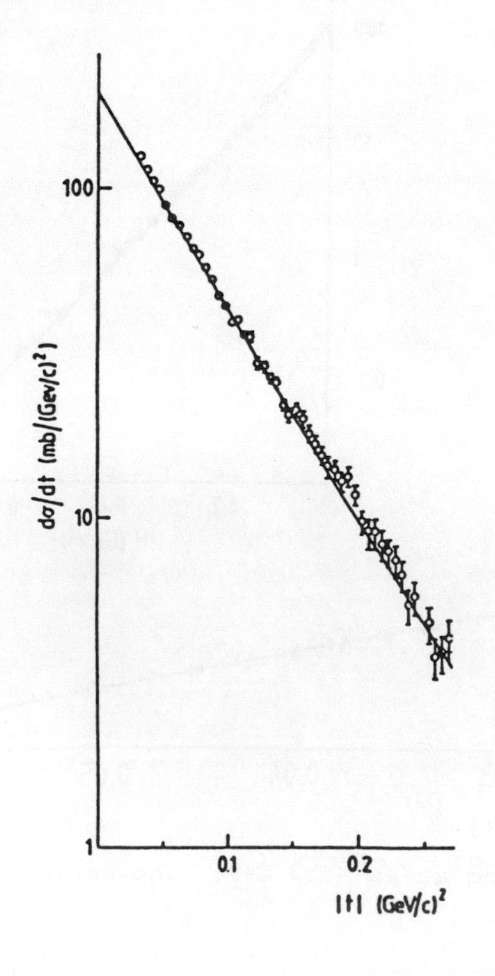

Figure 4 : UA4 data for $\bar{p}p$ scattering at $\sqrt{s}$ = 546 GeV, with the curve corresponding to the formula (5)

now only out to about $-t = 0.2$ GeV^2. This is well understood [3]: as the energy increases, double pomeron exchange has an important modifying effect at progressively smaller t. For this reason, predictions for SSC energy are subject to considerable uncertainty. A sample prediction is shown in figure 5; this shows a dip for pp scattering which has moved in to rather a smaller t value than that found at the ISR. It also shows pp and $\bar{p}p$ scattering significantly different in the dip region, as was found [9] at

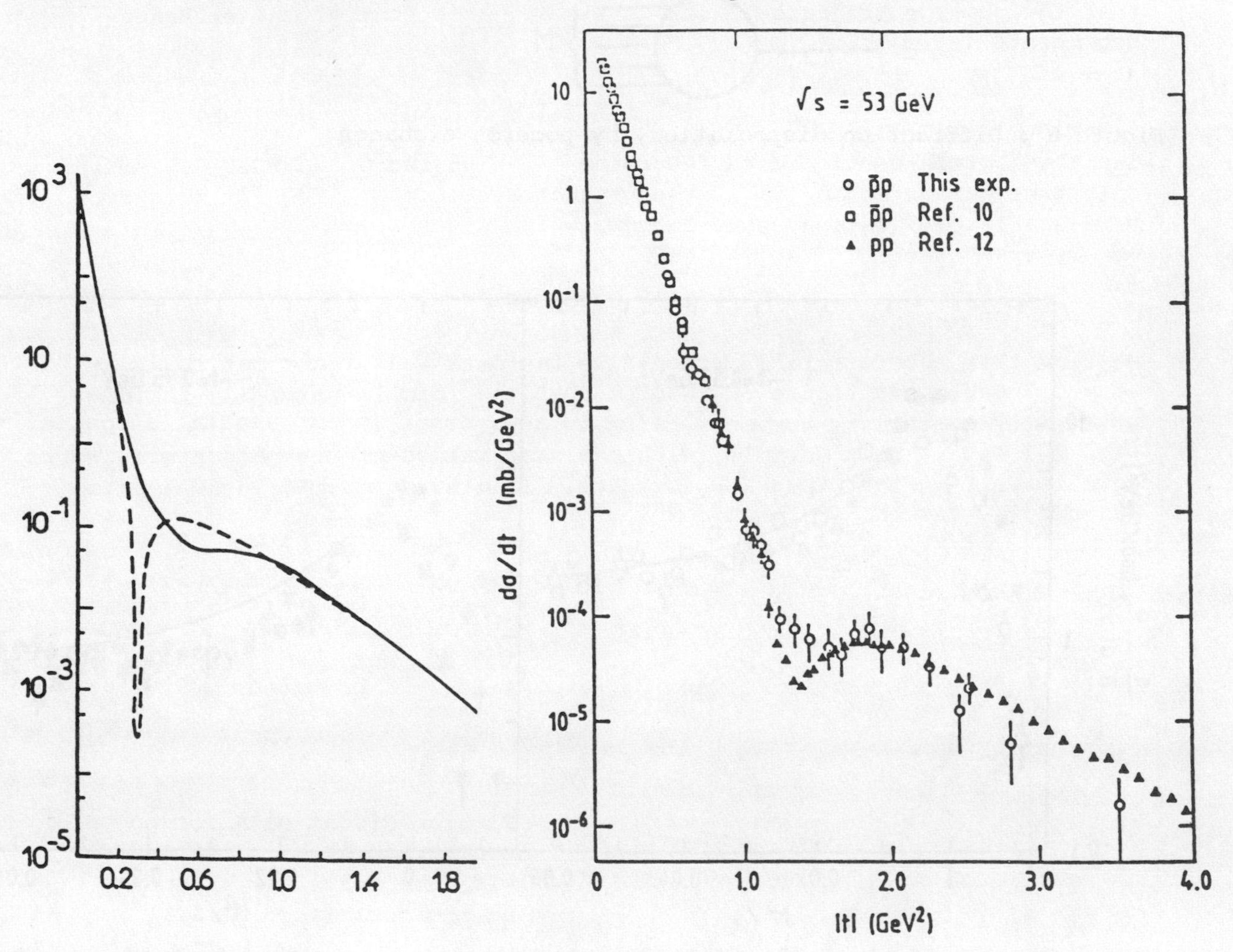

Figure 5 : (a) predictions [7] for pp and $\bar{p}p$ scattering at SSC energy; (b) ISR data [9].

the ISR. The dip is a complicated interference effect and so is particularly difficult to predict accurately, so that it will be interesting to have the results of the measurements in this region of t.

Pomeron exchange may also be studied in inelastic processes, for example single diffraction dissociation (figure 6). If indeed the pomeron behaves like a photon, this may be calculated [3] in terms of the νW_2 measurements in muon and neutrino scattering, with no free parameters at all. The result is compared with data in figure 7. At the SSC, the peak at small M^2/s will be much more prominent, given reasonable resolution.

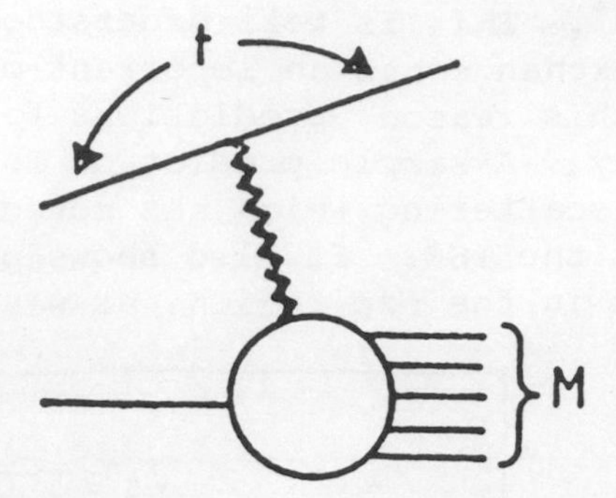

Figure 6 : Diffraction dissociation, by pomeron exchange

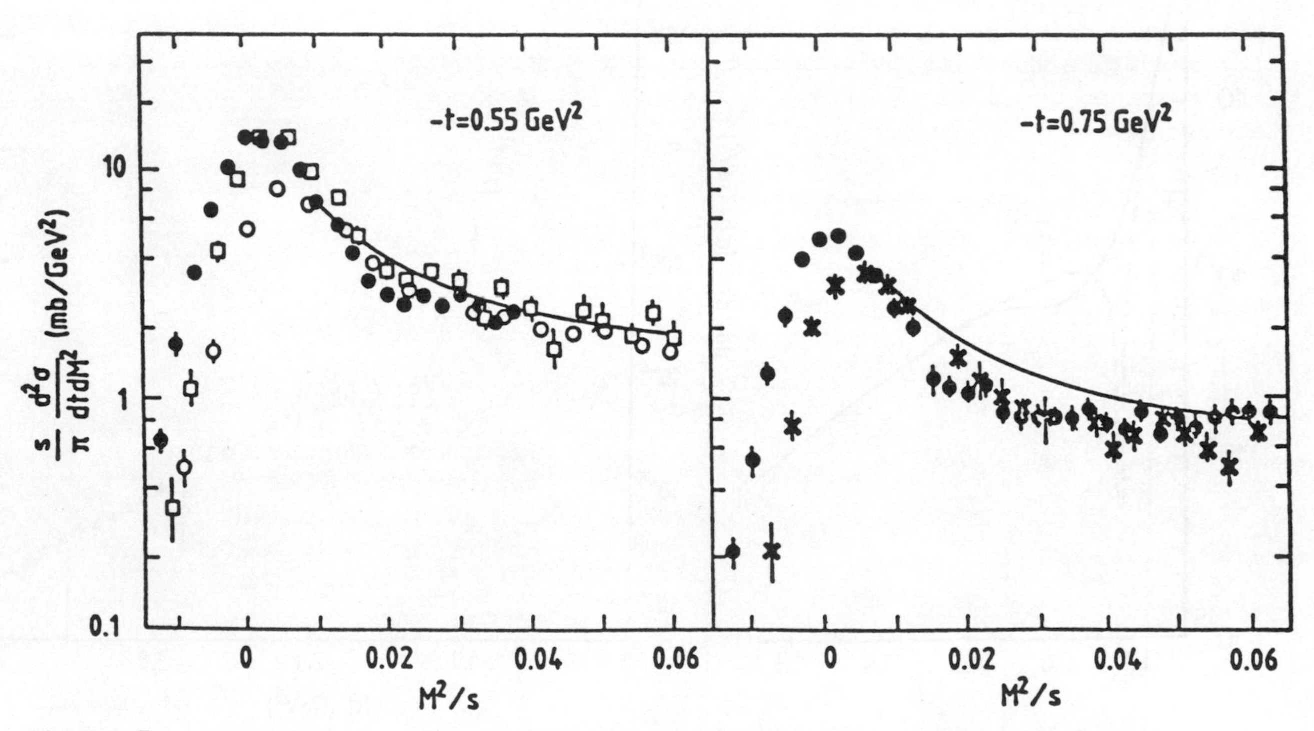

Figure 7 : Calculations [3] of diffraction dissociation with data from the ISR (CHLM collaboration) and UA4

4. Hard Diffraction

One may regard the lower part of figure 6 as the amplitude for a pomeron-proton collision, just as if the pomeron were a hadron. It has been suggested [10] that, as in a hadron-hadron collision, the final state M produced in the collision might occasionally contain a pair of high-p_T jets. This would be an example of a hard collision between the pomeron and the proton, and so is called a hard diffractive event. Other examples of hard diffractive events are the Drell-Yan production of lepton pairs or of W or Z [8] or heavy flavour production [11]. In each case the event is identified as hard diffractive by the presence in the final state of a proton with 90% or more of the initial momentum, so that an in-beam-pipe forward proton tagger is needed.

One can express the cross-section for a given hard diffractive process in terms of the structure functions of the proton and of the pomeron that collides with it. So the object of hard diffractive experiments is to determine the structure function of the pomeron. Like that of an ordinary hadron, the pomeron's structure function presumably has both a gluon component and a quark component. Different types of hard scattering experiments will be needed to separate the two.

Two rather different predictions have been made about the pomeron's structure function. Ingelman and Schlein [10] assume that the pomeron couples mainly to gluons and that its structure function is as large as it could possibly be. For the reasons that I have described in §3, we [8] believe that the pomeron has a significant coupling to quarks; we also believe that its structure function is somewhat smaller. Predictions of the two approaches are compared in figure 8, for the case of jet production at CERN collider energy.

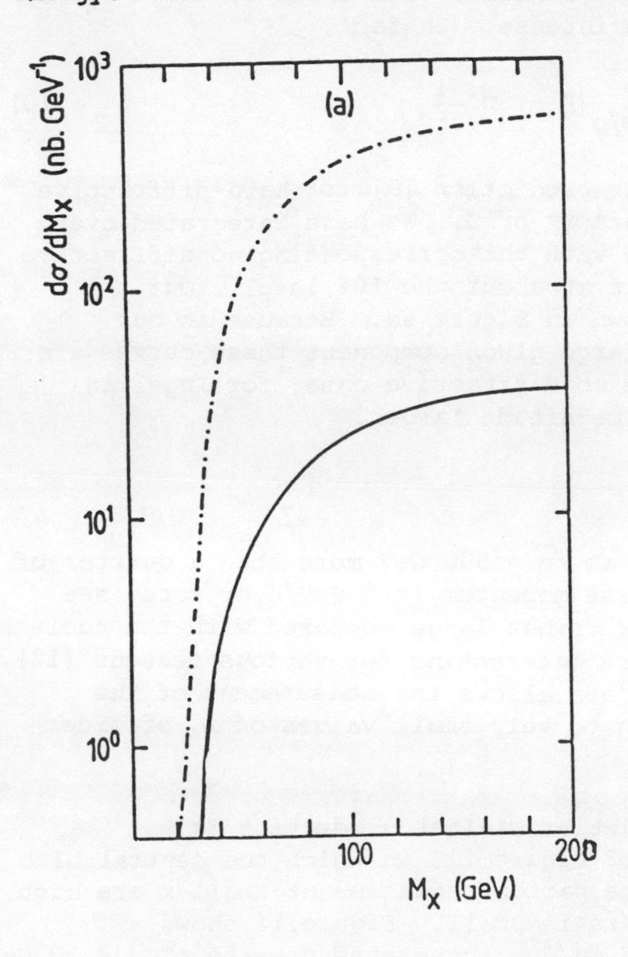

Figure 8 : Predictions from references 8 (lower curve) and 10 (upper curve) for the production of a pair of jets with $p_T > 8$ GeV/c at $\sqrt{s} = 630$ GeV.

In a diffractive collision one of the protons has its initial momentum p changed very little. Suppose that its longitudinal momentum after the interaction is $(1 - \xi)p$ and that the momentum transfer is t. That is, the proton radiates a pomeron of momentum ξp and squared mass t. Then a factorisation property holds:

$$\frac{d^2}{dtd\xi}\sigma_{pp}(s) = F_{P/p}(\xi,t)\,\sigma_{Pp}(\xi s) \tag{8}$$

Here, σ_{Pp} is the cross-section for producing the final state of interest in the pomeron-proton collision, and $F_{P/p}$ measures the intensity of the beam of pomerons that is radiated by the other proton. In terms of the same quantities that appear in the elastic differential cross-section (5),

$$F_{P/p}(\xi,t) = \frac{9\beta_o^2}{4\pi^2}[F_1(t)]^2\,\xi^{1-2\alpha(t)} \tag{9}$$

The diagram 6 and the formula (8) are generally supposed to be applicable for ξ up to about 0.1. But the virtue of the SSC is that a process that requires a given value of $M^2 = \xi s$ can be achieved with small ξ, where (9) predicts that the pomeron beam will be intense. In fact,

$$\int dt\, F_{P/p}(\xi,t) \approx \frac{1}{5\xi} \tag{10}$$

As an example, figure 9a shows our prediction [8] for hard-diffractive muon-pair production via an intermediate γ^* or Z. We have integrated over all t, and over ξ up to 0.1. Compared with the corresponding nondiffractive process, the predicted cross-section is at about the 10% level. Our predictions for jet production are shown in figure 9b. Because in our approach the pomeron does not have a large gluon component these curves are only at the 1% level compared with the nondiffractive case; for Ingelman and Schlein [10] they are an order of magnitude larger.

5. Minijets

The UA1 collaboration finds that at $\sqrt{s}$ = 900 GeV more than a quarter of the events produce jets whose transverse momentum is 5 GeV/c or more: see figure 10. Such jets, for which $p_T \ll \sqrt{s}$ but large compared with the nucleon mass, are called *minijets*, and they are interesting for various reasons [12]. In particular, the production of minijets allows the measurement of the proton's gluon structure function down to very small values of x, of order $p_T/\sqrt{s}$.

The inclusive cross-section for jet or minijet production is calculated from the standard diagram of figure 11, in which the central blob represents a hard interaction. Because parton densities at small x are high, the resulting $d\sigma/dp_T$ is large when $p_T/\sqrt{s}$ is small. Figure 11 shows a "sample" calculation by Halzen [13] of $d\sigma/dp_T$ integrated down to p_T^{min} = 10 GeV, for $|\eta| < 2.5$; according to this, most SSC events will contain such a minijet.

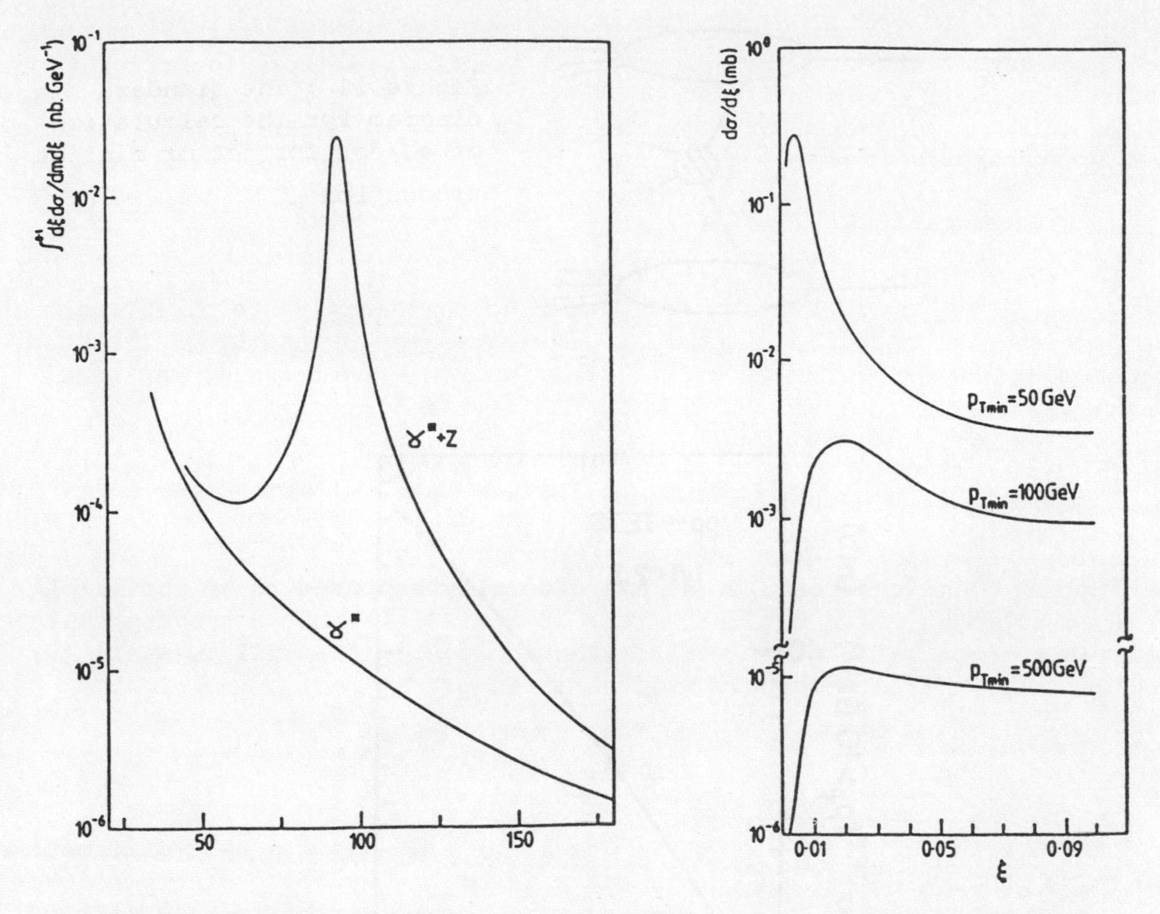

Figure 9 : Predictions [8] for hard-diffractive reactions at $\sqrt{s}$ = 40 TeV (a) muon pair production (b) high-p_T jet production.

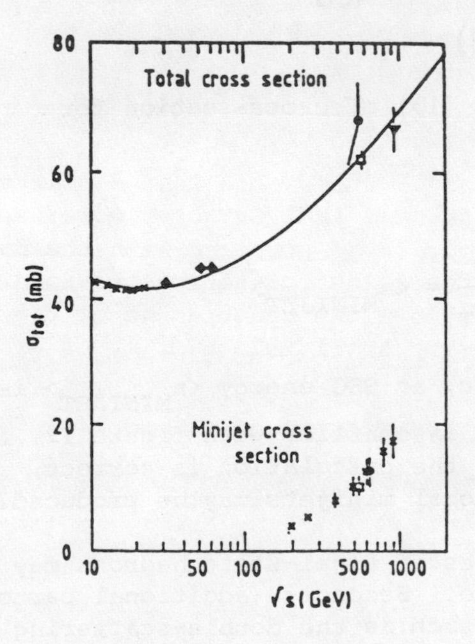

Figure 10 : UA1 cross-section for events having minijets with p_T > 5 GeV/c and $|\eta|$ < 1.5, together with measurements of σ^{Tot}.

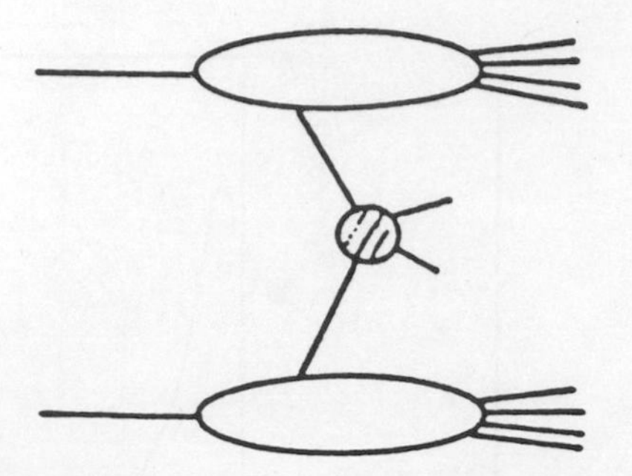

Figure 11 : The standard diagram for the calculation of $d\sigma/dp_T$ for jet or minijet production

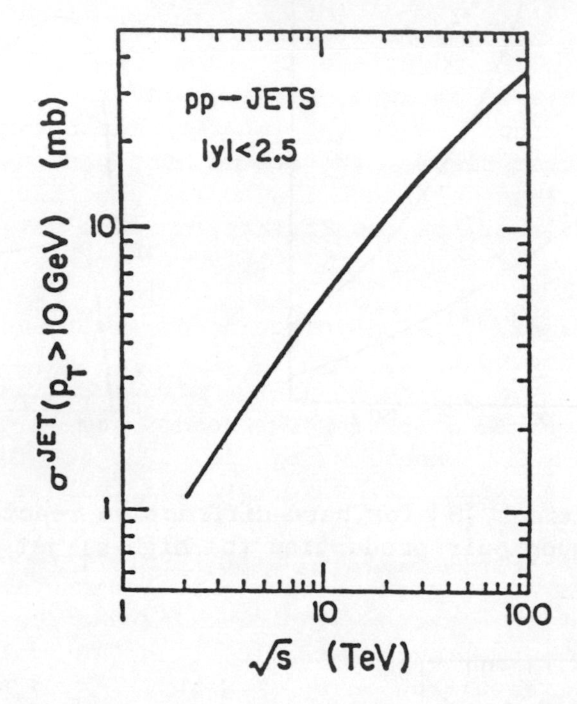

Figure 12 : "Sample" calculation by Halzen [13] of cross-section for minijet production

Recall that

$$\frac{1}{\sigma^{Tot}} \int_{p_T > p_T^{min}} dp_T \frac{d\sigma}{dp_T} = \langle n_{MINIJET} \rangle \tag{11}$$

The prediction is that, with p_T^{min} = 5 GeV/c, at SSC energy $\langle n_{MINIJET} \rangle$ is rather greater than 2. This is apparently in conflict with figure 12, from which the prediction arises, Nevertheless, the calculation is correct, because there are two ways in which additional minijets may be produced.

First [12], the upper and lower bunches of final-state hadrons may well themselves contain minijets when s is large. Secondly, additional parton-parton scatterings may generate minijets, such as the double-scattering term of figure 13a. However, there is an important theorem [14] which says that

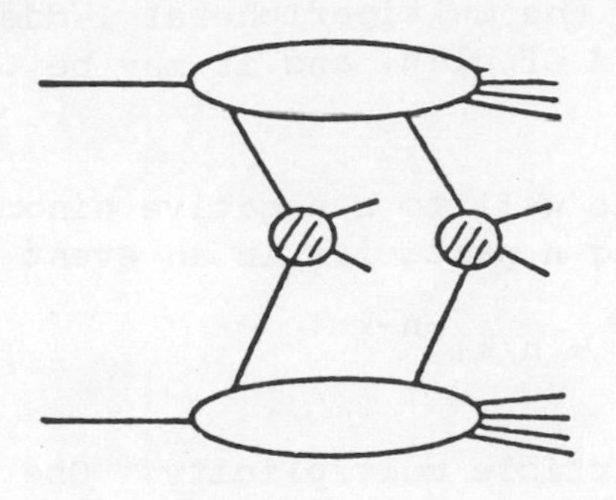

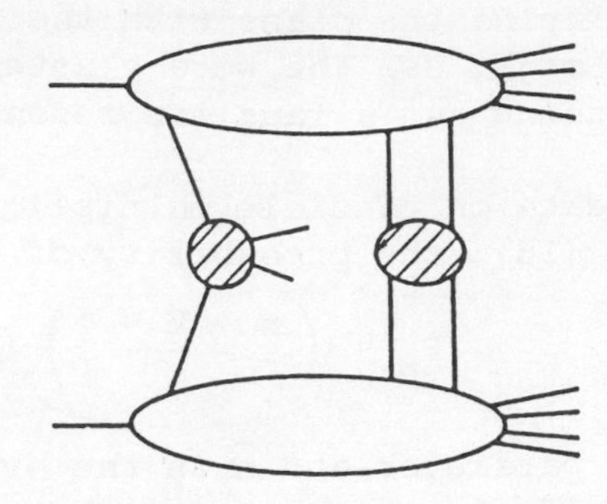

Figure 13 : Modifications to figure 12 : (a) a second real parton-parton scattering (b) a virtual parton-parton scattering

if we only want to calculate $d\sigma/dp_T$ we should not include additional parton-parton scatterings. This is because, for example, the contribution to $d\sigma/dp_T$ from figure 13a cancels with the interference between figure 12 and figure 13b. In the latter figure, the second hard parton-parton scattering is a virtual process. Thus, although figure 13a does matter if we are interested in the details of the final state, it does not enter in the calculation of $d\sigma/dp_T$.

The result that figure 12 alone gives $d\sigma/dp_T$, without any absorption or eikonalisation needing to be applied to it, implies that minijet production does indeed directly measure the proton structure function at small x. This is interesting, because at present we do not have a good theoretical understanding of the small x behaviour of the structure function. A popular assumption is that it is proportional to x^{-1}, but this is surely wrong.

There are two theoretical approaches to the small x behaviour, perturbative and nonperturbative. A perturbative calculation [15], based on the Altarelli-Parisi evolution equation, indicates a behaviour something like $x^{-1.5}$. But nonperturbative arguments [16] link the small x behaviour with pomeron exchange, resulting in $x^{-1-\epsilon}$ with $\epsilon \simeq 0.08$ as in (6). We do not understand how the perturbative and nonperturbative arguments fit together, and so the results of experiment will be useful to help with the theory.

6. Particle Multiplicities

I was asked to include a review of the theory of particle multiplicities. However, so far as I have been able to discover we do not really have a theory, only fits to experimental results.

There are various such fits in the literature. One example is that of Giovannini and Van Hove [17], who assume that an interaction produces a number of primitive objects which they call clans. They assume that the number of clans varies from event to event with a Poisson distribution; I assume that this is because they have in mind some sort of multiperipheral model, identifying the clans with the rungs of the multiperipheral ladder. Other people might use the word cluster instead of clan, and it may be that a clan is nothing but a familiar resonance.

The UA5 data on particle multiplicities fit well to a negative binomial distribution [18]: the probability of producing n particles in an event is

$$P_n = \binom{n + k - 1}{k} (\bar{n}/k)^n \ (1 + \bar{n}/k)^{-n-k} \qquad (12)$$

where k is a parameter and $\bar{n}$ is the average particle multiplicity. One obtains this if [17] one assumes that the average number of clans is $\bar{N} = k \log (1 + \bar{n}/k)$, with the clans decaying into particles with a logarithmic distribution.

What the clan picture does not explain is why the data have the energy variation that is measured. For example, UA5 find that their data at the various energies of the CERN collider in the ramped mode, together with other data down to $\sqrt{s}$ = 10 GeV, fit to an average charged multiplicity given by

$$\bar{n}_{ch} = 7.6 \ s^{0.125} - 7.4 \qquad (13)$$

Extrapolated to SSC energy, this gives about 100 charged particles on average - though it is worth emphasising the uncertainty in the extrapolation. UA5 find also that k^{-1} varies linearly with log s; if they extrapolate this and assume that the negative binomial distribution (12) remains valid they obtain the predicted multiplicity distribution shown in figure 14. If all the measurements at the various energies were to fall on one curve, one would have what is known as KNO scaling.

There has been some suggestion [19] that the breaking of KNO scaling is associated with the emergence of more and more minijets as the energy increases. The presence of minijets may also be correlated with other unusual features. For example, the average p_T per particle is rather constant up to ISR energies, but rises significantly over the CERN collider energy range, where the minijets first become prominent. Also, $\langle p_T \rangle$ rises with the particle multiplicity, which again seems to be associated with the emergence of minijets [20]; the indication of this may be seen in figure 15.

7. Summary

(a) The phenomenology of the pomeron is simple: it couples to single quarks like an I = 0 C = 0 photon

(b) We want to understand this in terms of nonperturbative QCD

(c) It will be useful to have information from hard diffraction experiments about the structure function of the pomeron. Such experiments include hard diffractive deep inelastic lepton scattering, and the production of jets, heavy flavour, W,Z or dimuons in pp collisions. In each case the final state includes a proton whose momentum has changed very little in the collision.

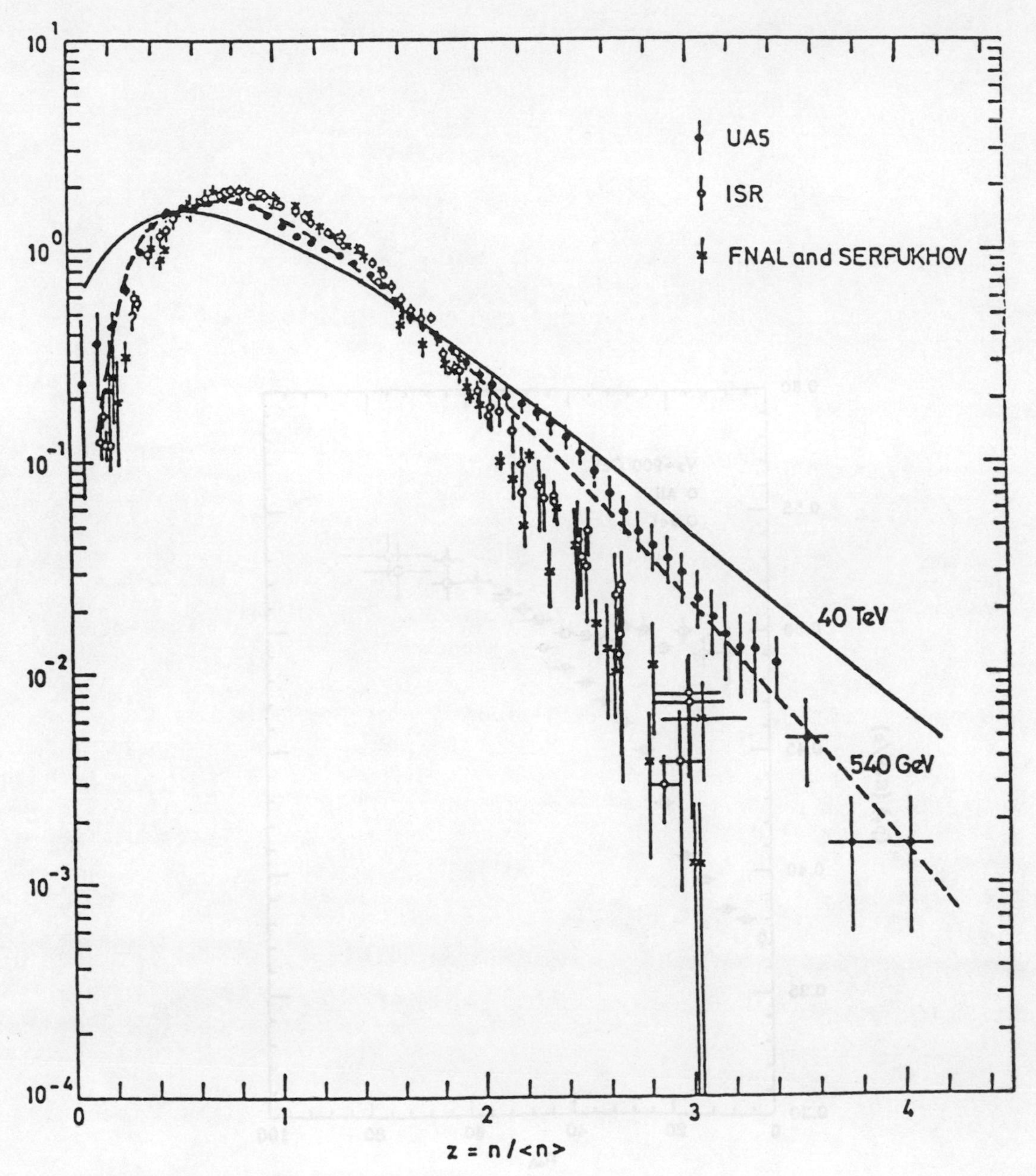

Figure 14 : UA5 prediction for particle multiplicity distribution at SSC energy, with data from lower energies

(d) We expect a large number of minijets. These will enable us to determine the small x behaviour of the proton's structure function, which is not well understood theoretically.

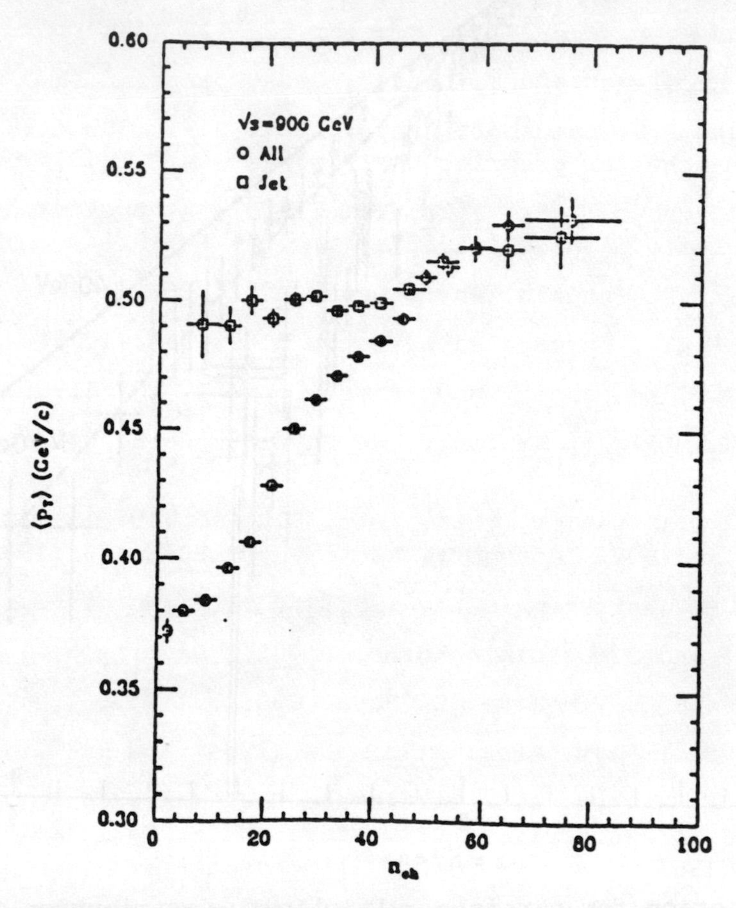

Figure 15 : UA1 data for average p_T against charged particle multiplicity. The upper points are for events that include minijets

References

1. P.V. Landshoff and O. Nachtmann, Z. Phys. C (in press).
2. P.D.B. Collins: An introduction to Regge theory (Cambridge University Press, 1977).
3. A. Donnachie and P.V. Landshoff, Nuclear Physics B244, 322 (1984) and B267. 690 (1986).
4. P.D.B. Collins and F. Gault, Phys. Lett. B112, 255 (1982).
5. M. Block and R.N. Cahn, Rev. Mod. Phys. 57, 563 (1985).
6. L. Lukaszuk and A. Martin, Nuovo Cimento 52, 122 (1967).
7. P. Gauron, B. Nicolescu and E. Leader, Orsay preprint IPNO/TH 86-55.
8. A. Donnachie and P.V. Landshoff, Phys, Lett. B191, 309 (1987) and Cambridge/Manchester preprint DAMTP 87/16 - M/C TH 87/05.
9. S. Erhan et al., Phys. Lett. 152B, 131 (1985); A. Breakstone et al., Phys. Rev. Lett. 54, 2180 (1985).
10. G. Ingelmann and P. Schlein, Phys. Lett. B152, 256 (1985).
11. H. Fritzsch and K.H. Streng, Phys. Lett. B169, 391 (1985).
12. M. Jacob and P.V. Landshoff, Mod. Phys. Lett. 1, 657 (1986).
13. F. Halzen, Review talk at Rencontre de Physique de La Vallée d'Aoste (1987).
14. J.L. Cardy and G.A. Winbow, Phys. Lett. 52B, 95 (1975); C.E. DeTar, S.D. Ellis and P.V. Landshoff, Nucl. Phys. B87, 176 (1975).
15. J.C. Collins, SSC Workshop, UCLA, 1986 and talk at this Workshop.
16. P.V. Landshoff and J.C. Polkinghorne, Physics Reports 5, 1 (1972).
17. A. Giovannini and L. Van Hove, Z. Phys. C30, 391 (1986).
18. G.J. Alner et al., Phys. Lett. B138, 304 (1984) and B160, 193 (1985).
19. T.K. Gaisser, F. Halzen, A.D. Martin and C.J. Maxwell, Phys. Lett. 166B, 219 (1986); G. Pancheri, Y. Srivastava and M. Pallotta, Phys. Lett. 151B, 453 (1985).
20. G. Pancheri and C. Rubbia, Nucl. Phys. A418, 1170 (1984); A.D. Martin and C. Maxwell, Proc. VI International Conf. on Proton-Antiproton Physics, Aachen (1986).

REPORT OF WORKING GROUP 3

LOW P_t PHYSICS AT THE SSC

K. Goulianos

Rockefeller University, New York, NY 10021

R. DeSalvo and J. Orear

Cornell University, Ithaca, NY 14853

P. V. Landshoff

University of Cambridge, England

M. Block

Northwestern University, Evanston, IL 60202

A. Garren and D. Groom

SSC Central Design Group, LBL 90-4040, Berkeley, CA 94720

A. Breakstone

SLAC, Stanford, CA 94305

T. Yasida

Fermilab, Batavia, IL 60510

ABSTRACT

Experimental systems and detectors are presented which can measure (1) small angle elastic scattering into the coulomb region, (2) large angle elastic scattering, (3) single diffractive dissociation, and (4) other small angle processes which would be missed by a typical 4π detector. The small angle elastic scattering is used via the optical theorem to obtain σ_t and the ρ-value. (See Part A.) Requirements on accelerator design are discussed. Beta at the interaction point must be ~4 km in order to do the small angle elastic scattering. High beta is not needed at the detector positions. The intermediate beta region of the conceptual design is well suited for the large angle elastic scattering and the single diffractive scattering. (See Part B.)

PART A

I. Elastic Scattering

The physics goal is to find out just how fast σ_t is rising and what is the precise shape of its energy dependence. Also, is elastic scattering which is expected to be pure imaginary at high energy becoming real instead; i.e., is the ρ-value still increasing? (UA4 claims it has reached 0.25 at $\sqrt{s}$ = 0.546 TeV. With such a high ρ-value, dispersion relations predict an even more rapid rise in σ_t at energies above the SPS - especially for pp as compared to $p\bar{p}$.) Just how fast is the mean squared radius (as measured by the slope parameter B) increasing? Is the proton becoming more black? In order to get the shape of these energy dependencies, short runs at several energies such as $\sqrt{s}$ = 2, 5, 10, 20, 40 TeV would be useful. Fig. 1 shows that the expected pp total cross section could be as large as 140 mb at $\sqrt{s}$ = 40 TeV.[1]

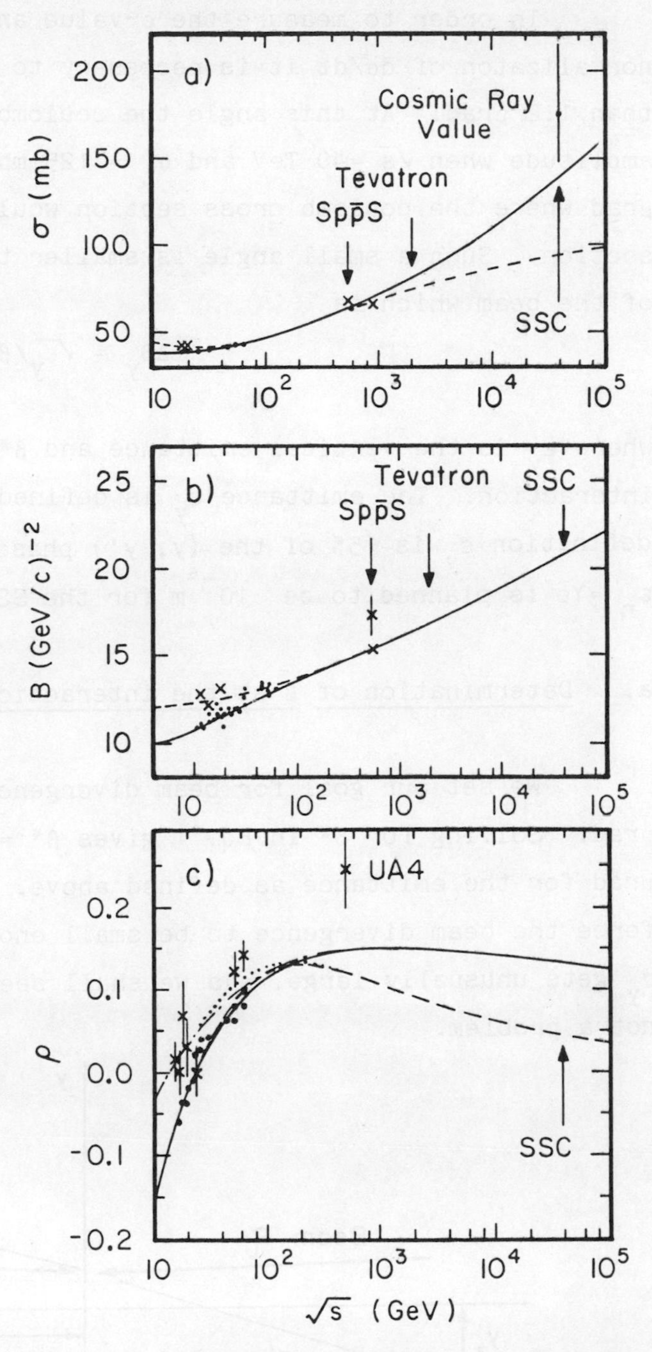

Fig. 1 Extrapolations of σ_t, B, and ρ to higher energies based on fits to Fermilab, ISR, and SPS data. See Reference 1. The UA4 ρ-value of 0.24 ± .04 has been plotted, but not used in the fit.

In order to measure the ρ-value and to use the coulomb amplitude for normalizaton of dσ/dt it is necessary to get to scattering angles smaller than 1.2 μrad. At this angle the coulomb amplitude equals the nuclear amplitude when $\sqrt{s}$ =40 TeV and σ_t = 125 mb. We set our as goal θ_{min} = 0.8 μrad where the coulomb cross section would be 5 times the nuclear cross section. Such a small angle is smaller than the typical angular divergence of the beam which is

$$\Delta\theta_y = \sqrt{\varepsilon_y}/\beta^* \qquad (1)$$

where ε_y is the vertical emittance and β* is the beta function at the interaction. The emittance ε_y is defined as $\varepsilon_y = \sigma_y^2/\beta$. Using this definition ε_y is 15% of the (y, y') phase space. The normalized emittance $\varepsilon_n = \gamma\varepsilon$ is planned to be 10^{-6} m for the SSC.

a. <u>Determination of β at the interaction point</u>

We set our goal for beam divergence $\Delta\theta_y = 0.14\ \theta_{min}$ which is 0.11 μrad. Solving for β* in Eq. 1 gives $\beta^* = \varepsilon/\Delta\theta_y^2$ = 4000m using ε = 1/γ mm-μrad for the emittance as defined above. By using such a high β one can force the beam divergence to be small enough, but then the source spot size σ_y gets unusually large. As we shall see in the next paragraph, this is not a problem.

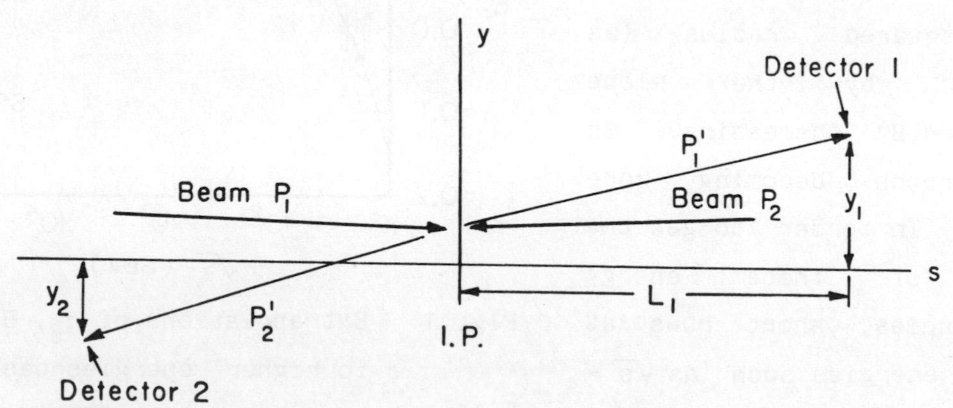

Fig. 2 Schematic of an elastic scattering. The measured vertical displacements form the beam are y_1 and y_2. The measured scattering angle for P_1 is $(y_1')_m = y_1/L_1$ where $L_1 = \sqrt{\beta^*\beta_1}\sin\psi_1$ is the effective length.

b. Beta at the detectors

The usual detection scheme is to use a high resolution particle detector for each of the scattered protons. (See Fig. 2). The detectors are pushed close to the beam after it is in a clean, stored condition. The size of the spot at the IP (interaction point) cancels out if one can achieve a parallel to point focussing condition. This is achievable at those detector positions where

$$\cos \psi_1 = -\alpha^* \qquad (2)$$

At those positions where Eq. 2 is true the bracketed quantity in the following relation will be zero and y will be independent of the spot size y_o.

$$y = \sqrt{\beta/\beta^*}\,[\cos \psi + \alpha^* \sin \psi]y_o + \sqrt{\beta\beta^*}\, \sin \psi\, y_o' \qquad (3)$$

We shall see in Section II that $\alpha^* = 0$, hence the condition for eliminating the effects of spot size is to locate the detectors at regions of phase advance which are odd multiples of $\pi/2$.

We define effective length $L = y/y'$ where y' is the vertical scattering angle and y is the vertical distance from the beam at the detector.

We see from Eq.3 that the effective length is $L = \sqrt{\beta\beta^*} \sin \psi$ where ψ is the phase advance from the IP to the detector. The lattice solution described in Sec. II has β_1 = 331 m and $\psi = 7\pi/2$. Then the effective length is -1150 m and the closest distance to the beam will be y_{min} = 0.92 mm. The beam spot size at the detector position is

$$\sigma_y = \sqrt{\beta_1 \varepsilon} = 127\ \mu m;$$

i.e., the detector is 7 standard deviations from the beam. It is important that this 0.92 mm space from the beam not contain any significant material such as a detector wall. As shown in Sec. IV, this space can be kept free of material. We shall see in the next paragraph that there is no need to have a measuring resolution greater than the spot size; i.e., a detector resolution ~ 100 μm is fine. So as long as the detector can meet these conditions there is no need for high beta at the detector position.

c. Measurement resolution

Now we shall calculate the intrinsic accuracy of the angle measurement and show how it is independent of β at the detector. We shall call $y_m' = y_1/L$ the measured vertical scattering angle. Dividing both sides of Eq. 3 by L gives

$$y'_m = y'_o + 1/\beta^* (\cot \psi_1 + \alpha^*)\, y_o \qquad (4)$$

where y_o is y-position of the scattering and y_o' is its angle. If the detector position meets the condition of Eq. 2, $y'_m = y_o'$ independent of y_o. Then $(y_m')_{rms} = (y_o')_{rms}$ which is the beam angle spread at the IP. This equals $\sqrt{(\varepsilon/\beta^*)} = 0.1$ μrad for $\beta = 4$ km. This can be reduced by the factor $\sqrt{2}$ by averaging measured angles on both sides of the IP.

The detector measuring resolution need be no better than $L\,(y_m')_{rms} = \sqrt{(\beta\varepsilon)}$ which is the same as the spot size at the detector position. With infinitely accurate detectors, the scattering angle can be determined to an accuracy of .07 μrad which is about 10% of the minimum angle to be measured.

We conclude that no matter how high energy is the accelerator, and no matter how poor is the emittance, it is theoretically possible to measure in the coulomb region as long as one has high enough β at the interaction region. The intrinsic angular resolution at the detector position is independent of the local β and is best if the phase advance is near an odd multiple of 90° The main disadvantage of small β at the detector is that the distances to be measured become small. Certainly the distances should be made significantly larger than the wall thickness and the measuring accuracy of the detector.

d. Normalization

If one is to determine σ_t to ~ 1% accuracy, one must know the luminosity to a similar or better accuracy. There are two methods where such accuracy can be achieved: (1) the Coulomb Method, and (2) the Combined Method.

(1) Since the coulomb amplitude is precisely known and becomes much larger than the nuclear amplitude at very small angles, measurements of dN/dt in the coulomb region thus determine the integrated luminosity. The very same data in the larger angle region then give the nuclear amplitude, which when extrapolated to zero gives the product $(1+\rho^2)\sigma_t^2$ via the optical theorem:

$$\left.\frac{d\sigma_n}{dt}\right|_o = \frac{1+\rho^2}{16\pi\hbar^2}\sigma_t^2 \qquad (5)$$

It should be easier to reach the coulomb region when running at lower energies (for the same value of y at the detector, the t-value goes as E^2 where E is the beam energy). For this and other reasons it is important to run at several different energies.

An accurate luminosity monitor could be established by mounting two counters on opposite sides of the IP. the coincidence rate would be proportional to the luminosity (which is obtained from the number of deep coulomb events). The same configuration of 2 counters could be used in any other interaction region for continuous readout of luminosity.

(2) In case one is not able to get deep into the coulomb region, the Combined Method offers a fall-back position. To use this method, one must monitor the total number of interactions at the same time one is measuring dN_{el}/dt. Eq. 5 shows that the latter is proportional to luminosity times $(1+\rho^2)\sigma_t^2$. The total number of interactions $(N_{el}+N_{inel})$ is proportional to luminosity times σ_t. In taking the ratio the luminosity cancels out leaving the product $(1+\rho^2)\sigma_t$. This method is best done in a medium β IR by using a series of "concentric" ring counters on both sides of the IP and extrapolating to zero. dN_{el}/dt would be well measured at $t \sim 3 \times 10^{-3}$ GeV^2 as explained in Sect. III. It could be extrapolated to zero by using the high β results. Then one would have both N_{inel} and $(dN_{el}/dt)_{t=o}$ for the medium β IR. The ratio

$$\frac{(dN_{el}/dt)_o}{N_{el}+N_{inel}} = \frac{(1+\rho^2)\sigma_t}{16\pi\hbar^2} \qquad (6)$$

Another reason for measuring N_{inel} at medium β rather than high β is that $N_{inel} = (N_{double} + 2N_{single})$ where N_{double} is the number of "left-right" coincidences, whereas N_{single} is the number of single arm events. We estimate that for luminosity > 10^{31} $cm^{-2}s^{-1}$ the beam-gas background will be less than the single arm event rate.

It should be noted that the slope parameter B and the ρ-value can be determined without any knowledge of the luminosity. Fig. 3 shows a set of dN_{el}/dt curves all normalized to one at θ = 10 μrad. They all have the same slope of B = 20 GeV^{-2}, but differing values of ρ. A run with ~ 1000 events in the region 2 < θ < 3 μrad could determine ρ even if the data did not reach angles smaller than 2 μrad.

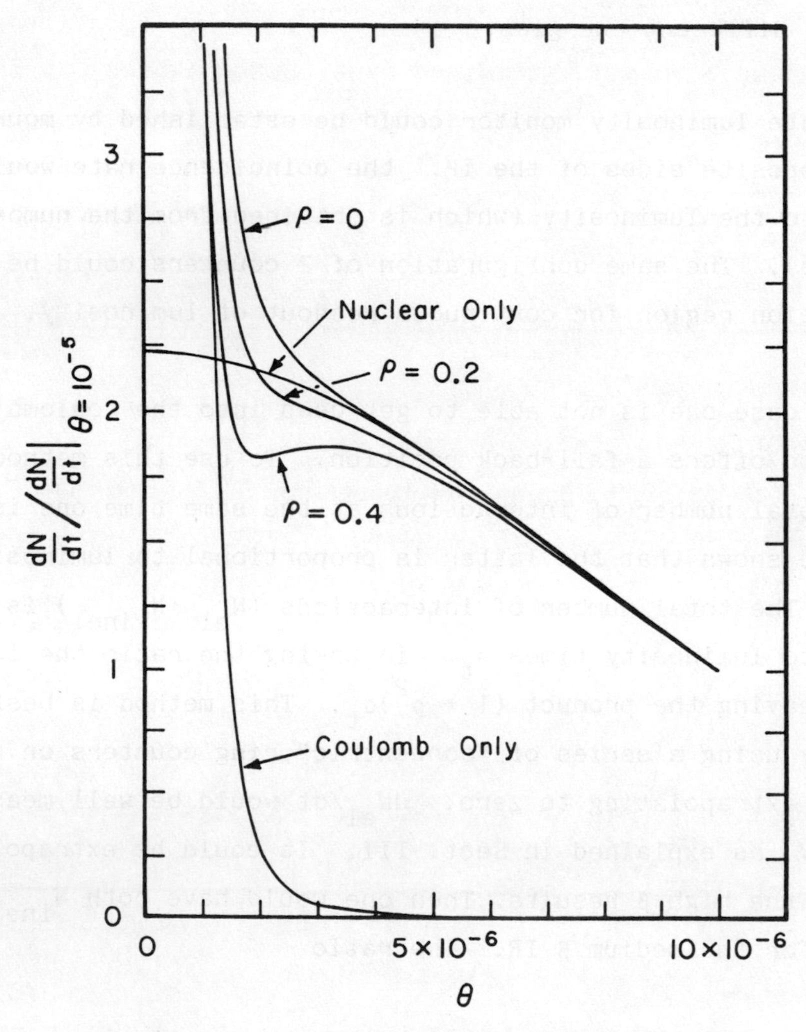

Fig. 3 Plot of dN_{el}/dt vs. θ for different values of ρ. dN/dt has been normalized to one at θ = 10^{-5} radians.

e. Event rate

Let R be the number of elastic events detected per second. Then $R = L\cdot(d\sigma/dt)\Delta t\Delta\phi/2\pi$.

Assume L is 1% of its design value; i.e. $L = 10^{31}\ cm^{-2}s^{-1}$ for $\beta^* = .5m$. Then $L = 1.25 \times 10^{27}$ for $\beta^* = 4$ km. Using $d\sigma/dt = 10^{-24}\ cm^2/GeV^2$, $\Delta t = .04\ GeV^2$, and $\Delta\phi/2\pi = 0.5$ gives R = 25 events per second not including the coulomb part of the cross section. This is 9×10^4 per hour. We see that at very low luminosities the running time to get good statistics is almost insignificant.

II. How to obtain high β

Fig. 4 shows a solution obtained by A. Garren which has $\beta^* = 4$ km and $\beta_1 = \beta_2 = 331$ m with ~3.5π phase difference from IP to detector. It is a modification of one of the "intermediate" β interaction regions and could fit either there or at one of the "future" interaction regions. This

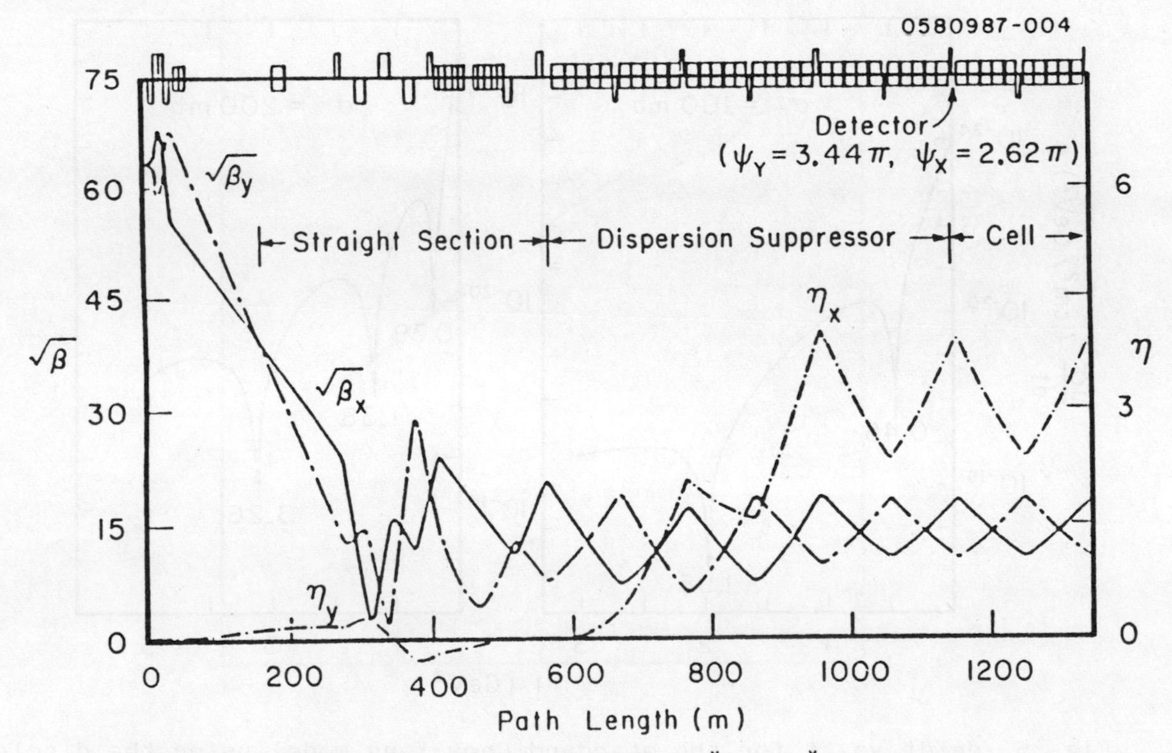

Fig. 4 Magnet configuration to achieve $\beta_x^* = \beta_y^* = 4000$ m and $\alpha_x^* = \alpha_y^* = 0$. At detector position $\sin\psi_y = -0.99$ and $\sin\psi_x = 0.92$.

insert has no adverse effect on operation of the accelerator. The provision for doing this kind of physics should be designed into the machine from the very beginning. It would be wise to dedicate one of the four IR's to this physics from the beginning since early operation is necessarily at low luminosity and this physics does not require high luminosity.

III. Large angle elastic scattering

It is of interest to obtain the shape of the "diffraction" pattern. At Fermilab and ISR energies, there is the prominent t = 1.4 dip, but no sign of a second dip. At $\sqrt{s}$ = 540 GeV this dip has become a sharp kink at t = .9 GeV^2. At the SSC it should move down to ~0.4 GeV^2, and if the proton is becoming blacker as expected, the second dip should show up at ~0.8 GeV. Fig. 5 shows the third dip as well. Assuming 10 mm high detectors 0.8 mm from the beam and L_{eff} = 1000 m, the t-range is .00025 to

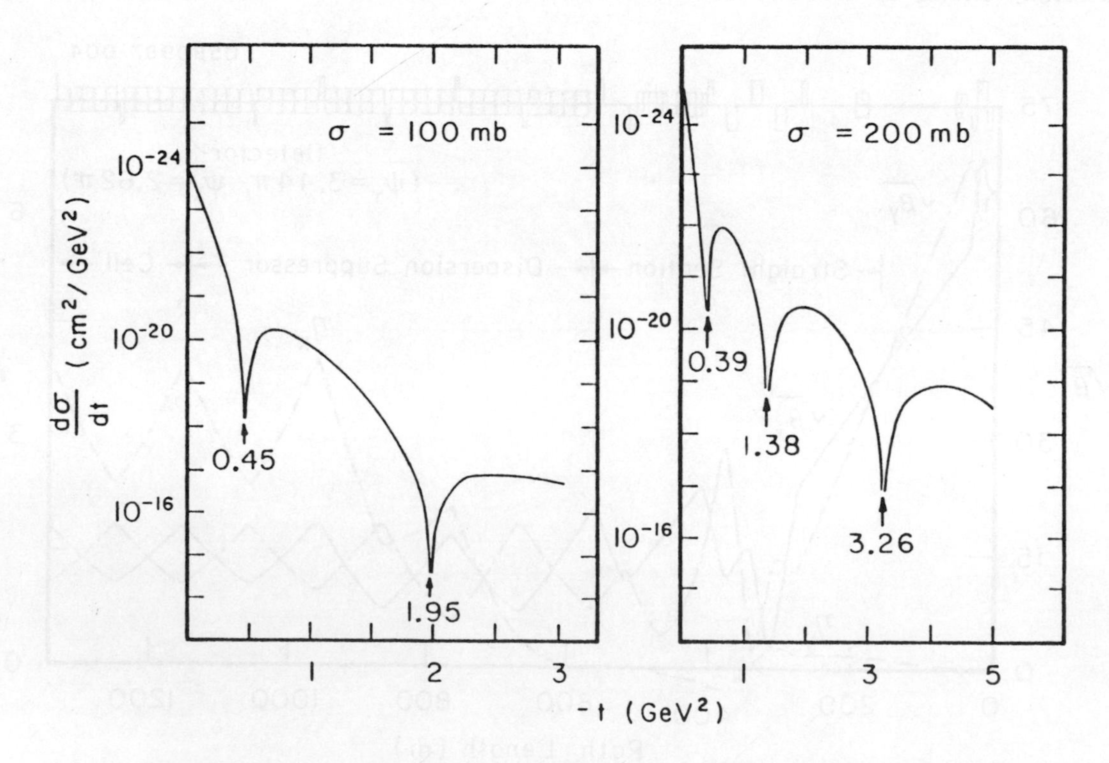

Fig. 5 dσ/dt vs. t for the standard Chow-Yang model using the dipole form factor for the proton. The only free parameter is σ_t which in this case is set to 100 mb and 200 mb.

.04 GeV^2. The same apparatus can be used to measure large angle scattering if the colliding region can be reconfigured from high beta to lower beta. If β^* can be reduced from 4000m to 100 m, the t-range would be from .01 to 1.6 GeV^2. This should give good coverage of the first two dips. As is discussed in Part B, single diffractive and other small angle processes are best studied in one of the medium β interaction regions which are to be tuneable from β = 10 to 60 m. For β^* = 60 m the region covered is $.013 < -t < 3.5$ GeV^2. This would straddle all 3 dips.

IV. Detector design

We have considered drift tube detectors.[2] They have the advantage of low sensitivity to radiation damage. But there is the disadvantage of having to shield from the induced voltage produced by the successive bunches of beam every 15 ns. Even though the tube walls are several skin depths thick, additional shilding between the tubes and beam would probably be needed.

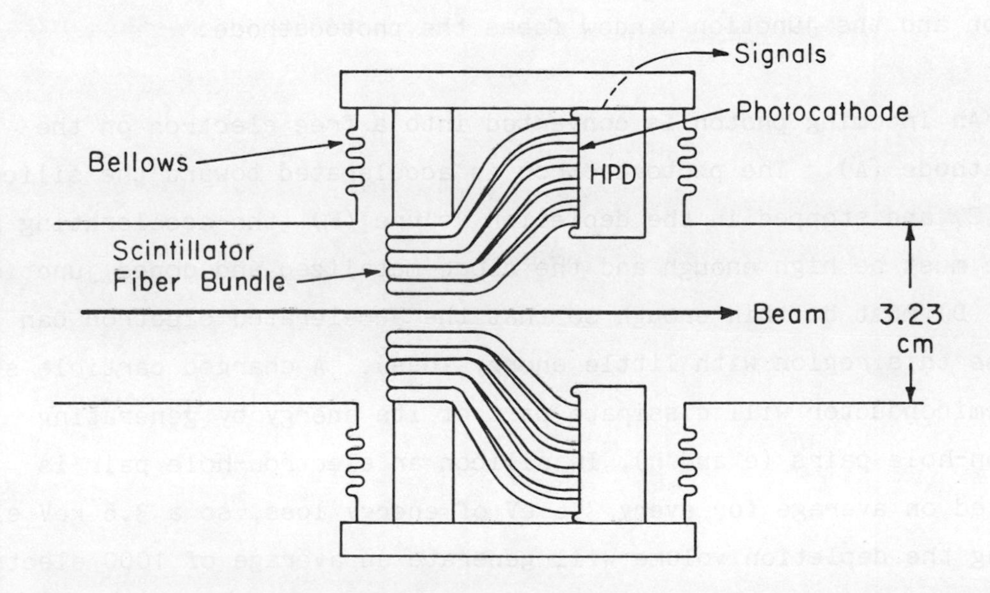

Fig. 6 Schematic of scintillator fiber detector. In this view both a top and bottom detector have been pushed in to about 1 mm from the beam. For 100 μ fibers there would be 100 fibers in both x and y directions. They would spread out over the photocathode of a hybrid photodiode tube (HPD).

At this workshop we gave some thought to a detector which would be unaffected by the beam and would not require any shielding or material between the sensitive volume and the beam. The alternative we propose is a bundle of scintillating fibers oriented parallel to the beam. The fibers would be of 100 μm diameter and ~2 cm in the beam direction. Each fiber would localize x and y with a standard deviation measuring accurary of 25 μm. The fibers would connect to the light collection face of a hybrid photodiode (HPD). They could be spread out over the face as shown in Fig. 6. The main ideas of the HPD were developed by R. Desalvo at this workshop.[3]

The HPD tube (see Fig. 7) is made of an ordinary photomultiplier photocathode coupled with an anode made out of a planar junction, fully depletable, semiconductor diode (a silicon photodiode is a good example of a suitable anode for the HPD). The photocathode and the diode at the anode are separated by a vacuum gap and held at an electric potential difference of a few kilovolts. The diode is reverse biassed to full depletion of the junction and the junction window faces the photocathode.

An incoming photon is converted into a free electron on the photocathode (A). The photoelectron is accelerated toward the silicon diode (C) and stopped in the depletion volume (F) (the accelerating high voltage must be high enough and the diode metalized and doped junction window (D) must be thin enough so that the accelerated electron can traverse this region with little energy loss). A charged particle stopped in a semiconductor will dissipate part of its energy by generating electron-hole pairs (e and h). In silicon an electron-hole pair is generated on average for every 3.6 eV of energy loss, so a 3.6 keV electron entering the depletion volume will generate an average of 1000 electron-hole pairs. The electron-hole pairs are then collected on the diode electrodes (D and G) with the help of the diode bias electric field and read out on a normal amplifier. The distance between photocathode and semiconductor diode need not be more than few millimeters so that the detector case (B) is quite compact.

The charge is deposited in a depth of a few microns of silicon (the average energy loss for a minimum ionizing particle is .34 KeV/μm) and since the typical electron drift speed in silicon is 50 μm/ns, so the charge is collected on the diode electrodes with a temporal spread of a small fraction of a nanosecond. It follows that the HPD speed is limited by the semiconductor diode risetime (suitable planar diodes with less than a nanosecond risetime are easily available).

The HPD is radiation hard. Silicon ionizing particle detectors have been shown to be radiation resistant to several megarads and the metallic photocathode is radiation insensitive. The most radiation sensitive part of the HPD is then the readout amplifier.

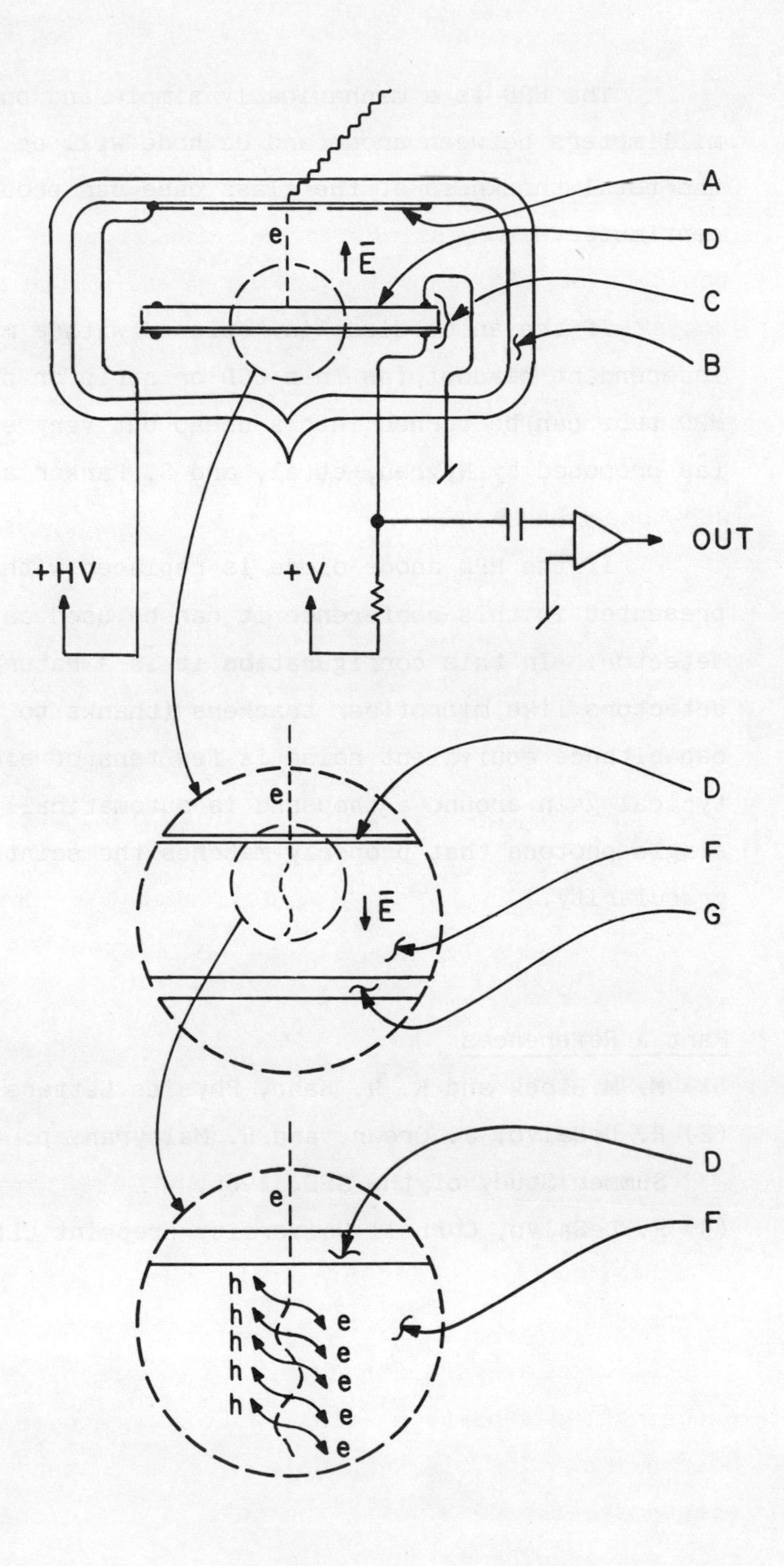

Fig. 7 Schematic of the HPD showing a single photon

The HPD is a mechanically simple and compact device, a gap of few millimeters between anode and cathode will be sufficient in most cases and the total thickness of the glass case can probably be kept around a centimeter.

If the anode diode is replaced with a matrix of photodiodes with independent readout (as in a CCD or strip or pixel silicon detectors) the HPD tube can be turned into a cheap but very efficient imaging detector (as proposed by Nygren, et al, and S. Parker at this workshop).

If the HPD anode diode is replaced with one of these pixel devices presented in this conference it can be used as a very fast imaging detector. In this configuration it is a natural readout device for detectors like microfiber trackers (thanks to the pixel device's low capacitance equivalent noise is few tens of electrons and the HPD with its typical gain around a thousand is automatically a detector sensitive to single photons that properly matches the scintillating fiber's high granularity.)

Part A References

(1) M. M.Block and R. N. Kahn, Physics Letters B188, 1987, p. 143.
(2) R. DeSalvo, J. Orear, and R. Maleyran, p. 464, Proceedings of the 1986 Summer Study of the SSC, 1986.
(3) R. DeSalvo, Cornell University preprint CLNS87-92, 1987.

PART B

DIFFRACTION DISSOCIATION

ABSTRACT

We propose to build "diffractive spectrometers" to measure the momentum of recoil protons from single diffraction dissociation (SD) and double Pomeron exchange (DPE), utilizing the machine magnets and detectors in "Roman Pots". The momentum acceptance will be in the range $4x10^{-4} < 1-x < 0.09$, which corresponds to diffractive masses of 0.8 to 12 TeV and DPE masses of 16 GeV to 3.6 TeV. The spectrometers will be used for measuring differential cross sections and for tagging diffractive masses for further study by general purpose detectors.

1. Introduction.

The study of leading proton diffractive processes, single diffraction dissociation (SD) and double Pomeron exchange (DPE), requires accurate measurements of the angle and momentum of the "recoil" proton(s). These protons, having $\bar{p}_T \sim$ 270 MeV and momentum close to that of the beam, are part of the beam halo and therefore they may be conveniently analyzed by spectrometers employing the magnets of the machine itself. We propose to build such "diffractive spectrometers", consisting of detectors in "Roman Pots" stationed in key positions along the main ring on both sides of an interaction point (IP). The Pots will be movable, so that they may be placed near the coasting-beams after injection, making it possible to measure momenta very close to that of the beam. Two types of spectrometers are proposed, designed to cover two regions of "recoil" momentum. Their combined momentum acceptance is in the region $4x10^{-4} < | p-p_{beam}| / p_{beam} <$ 0.09, which corresponds to SD masses of 0.8-12 TeV and DPE masses of 16 GeV to 3.6 TeV.

In the sections that follow, we first discuss SD, providing some physics motivation for its study and outlining the kinematics relevant to the design of the spectrometers and the detection of the dissociation products. We then proceed to describe the "diffractive spectrometers", comment briefly on their application to the study of DPE and finish with a discussion of event rates, background and triggering.

2. Single Diffraction Dissociation.

A large fraction of the proto-proton cross section at the SSC is expected to be due to the process of single diffraction dissociation,

$$p + p \rightarrow p + X \qquad (1)$$

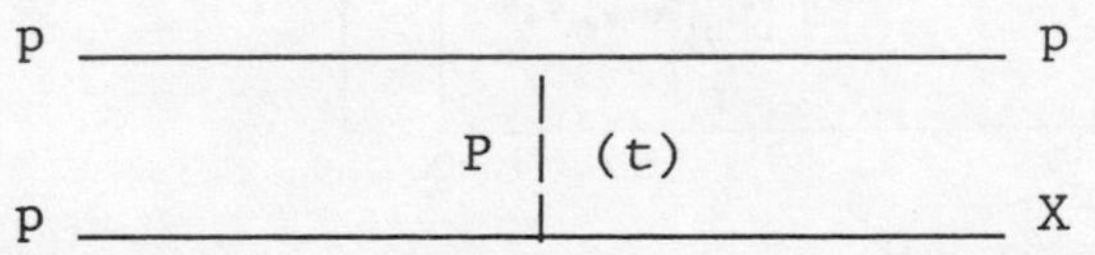

```
p ____________________________ p
              |
            P | (t)
p ____________|_______________ X
```

FIG. 1

in which one of the protons is excited coherently into a high mass state X in a low momentum transfer collision. In terms of the Feynman x-variable, $x = 2 p^*_{11} / \sqrt{s}$, the minimum t-value required for excitation into a mass M is given by

$$|t|^{1/2}_{min} / m_p = 1-x \approx M^2 / s \tag{2}$$

As $|t|^{1/2}_{min}$ becomes smaller than the mass of the pion, the "recoil" proton is likely to remain intact, leading to coherent or diffractive-like phenomena. The excited proton state, upon its creation, dissociates immediately into hadrons, hence the name diffraction "dissociation". The "coherence condition" is given by

$$1 - x = \frac{M^2}{s} \leq \frac{m_\pi}{m_p} \approx 0.15 \tag{3}$$

The differential cross section $d^2\sigma/dtdx$ grows as the process becomes more and more coherent, i.e. as $|t|$ or 1-x decrease. The growth at small $|t|$ follows the well known exponential behavior of classical diffraction, which is also observed in elastic scattering. The increase of the cross section at small 1-x is a purely quantum-mechanical phenomenon involving momentum transfer in the forward rather than in the transverse direction, which kinematically must be associated with a change in mass. Fig. 2 shows the cross section versus 1-x for pp → pX at s = 500 (GeV)^2. The enormous peak in the region 1-x ≤ 0.15 is attributed to diffraction dissociation [1].

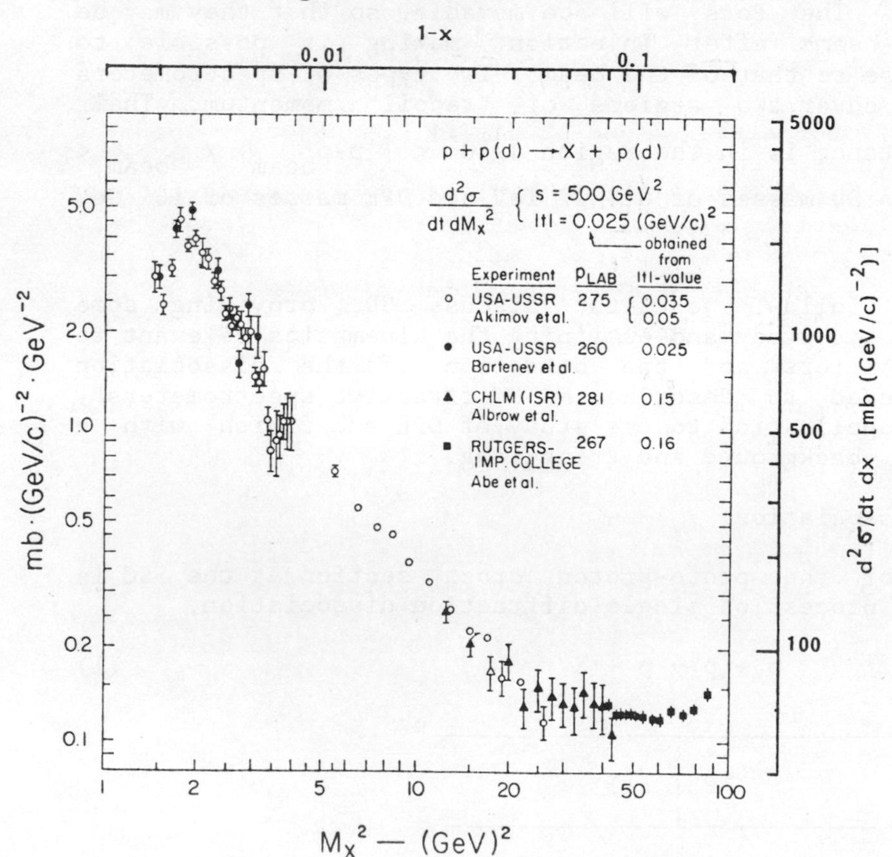

FIG.2:The differential cross section $d^2\sigma/dtdM^2$ versus M^2 for pp→Xp at t=-0.025 (GeV/c)^2 and s = 500 GeV^2 [1]. The axes on top and right give $d^2\sigma/dtdx$ versus 1-x.

Within the diffractive peak, the cross section $d^2\sigma/dtdM^2$ varies as $1/M^2$ and obeys factorization and the finite mass sum rule. This behavior is consistent with the hypothesis that diffraction dissociation is dominated by Pomeron exchange. In the mass region $5\ GeV^2 < M^2 < 0.15s$ and for $|t| < 0.1\ (GeV/c)^2$, a useful parametrization of the diffractive cross section is given by [1]

$$\frac{d^2\sigma}{dt\, dM^2} = \frac{A}{M^2}\ (b\ e^{bt}) \qquad (4)$$

where $A \approx 0.7 \pm 0.05$ mb and $b / b_{e\ell} \approx 1/2$ to $2/3$. At the SSC, the elastic slope is expected to be [2] around $24\ (GeV/c)^2$ and therefore $b \approx 14\ (GeV/c)^2$. The average $|t|$ of the diffractive states is thus $|t|_{ave} = 1/b \approx 0.07\ (GeV/c)^2$, corresponding to an average production angle of ~13 μrad. The total single diffraction dissociation cross section, obtained by integrating Eq. 4 over t and M^2 from threshold (~ 1 GeV^2) to 0.15s, is given by $\sigma_{SD} = 2\ A\ \ell n(\ 0.15s\ /\ GeV^2\ )$, where the factor of 2 accounts for the fact that both protons may dissociate. At $\sqrt{s} = 40$ TeV, $\sigma_{SD} \sim 27$ mb, which is roughly 20 % of the expected [2] total cross section of ~120 mb.

The Pomeron plays an important role not only in SD but also in all other diffractive processes, such as elastic scattering, double diffraction dissociation (DD), DPE and in the total cross section, which is related to elastic scattering through the optical theorem. The QCD description of these "soft" processes will invariably involve a Pomeron-like theoretical structure. In order to formulate such a sructure, one needs to know the cross sections of the above processes, commonly referred to as "ℓns physics", and the nature and properties of the dissociation products, the physics known as "hard diffraction". There is an abundance of theoretical models predicting large heavy flavor or gluon-jet content in hard diffraction, although the theoretical community is not unanimous in its opinion. In view of the small momentum transfer in diffractive processes, it has even been conjectured [3] that the "Pomeron energy" transferred in high mass SD may heat the proton receiving it to very high temperatures, leading to a QCD phase transition at relatively low diffractive masses (~100 GeV), a process that might explain the Centauro-type events observed in cosmic rays. Be as it may, the study of hard diffraction is certainly interesting!

At the SSC, masses as high as 15 TeV fall within the region of the coherence condition, Eq. 3. Although, as pointed out above, the production angles of the masses themselves are small, the angles of their dissociation products are generally quite large. The kinematics of SD are summarized in Fig.s 3, 4 & 5.

Fig. 3 shows the expected shape of the pseudorapidity distribution of a diffractive cluster of mass M. The average η is $\bar{\eta}_M = \ln(\sqrt{s}/M)$. The average laboratory angle, obtained by using the relation $\eta = -\ln\tan(\theta/2)$, is given by

$$\bar{\theta}_M \approx 2\ M/\sqrt{s} = 2\ \sqrt{1-x} \tag{5}$$

Thus, masses in the range 0.8-12 TeV, tagged by the proposed spectrometers, dissociate into clusters of average angle 40-600 mrad. Such large angles are easily covered by general purpose detectors that have been proposed for the SSC.

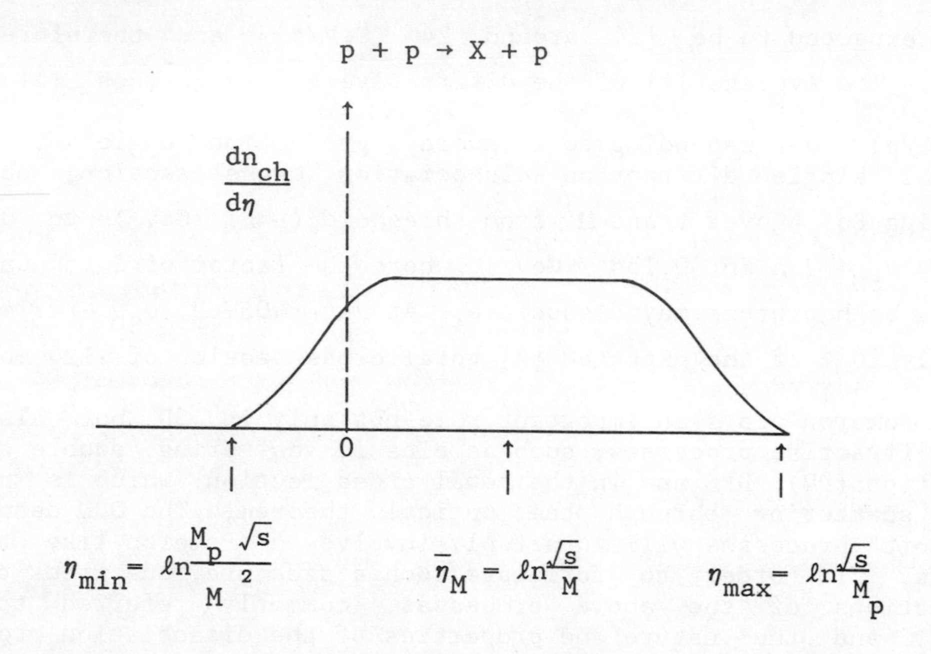

FIG. 3. Schematic drawing of the charged particle pseudorapidity distribution of a diffractive cluster of mass M created in a pp→Xp collision at $\sqrt{s}$.

Fig. 4 shows the cross section $d\sigma/d\bar{\eta}_M$ as a function of $\bar{\eta}_M$. From Eq. 4 and $\bar{\eta}_M = \ln(\sqrt{s}/M)$, it follows that $d\sigma/d\bar{\eta}_M = 2A$, i.e. the cross section is flat and has the value 2A = 1.4 mb. The shaded area in the figure represents the region that can be tagged by the diffractive spectrometers. Clusters from beam-gas non-diffractive collisions will be centered at $\bar{\eta} = \eta_{max}/2 = 5.3$. However, particles from such collisions will spread over the entire hemisphere, from $\eta \approx 0$ to $\eta \approx 10.6$.

Fig. 5 shows the η-distributions expected for diffractive clusters of different masses (see also Fig. 3). The beam-gas clusters are also shown. Beam-gas events will be vetoed in the trigger stage by requiring coincidence between the recoil proton and the diffractive cluster.

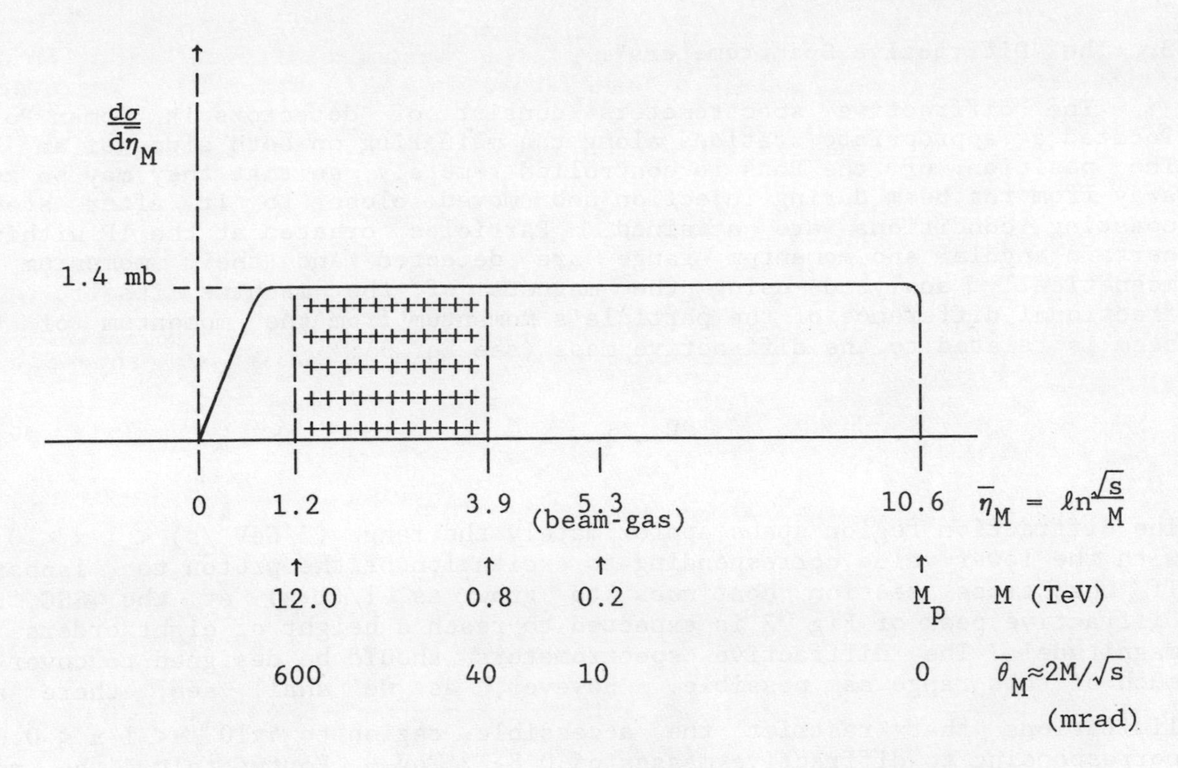

FIG. 4: Single diffraction dissociation csoss section $d\sigma/d\bar{\eta}_M$ versus $\bar{\eta}_M$, where $\bar{\eta}_M$ is the average pseudorapidity of a cluster of mass M. The crossed area corresponds to the acceptance of the "diffractive spectrometers".

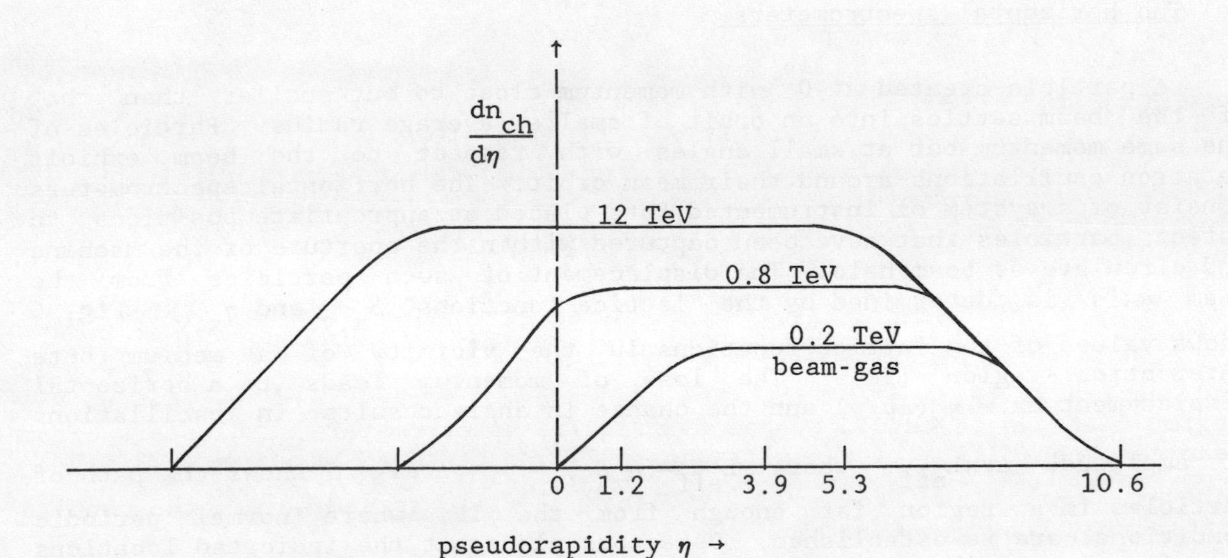

FIG. 5: Charged particle pseudorapidity distribution for various mass values.

3. The "Diffractive Spectrometers".

The diffractive spectrometers consist of detectors in "Roman Pots" located at appropriate stations along the main ring on both sides of an IP. The position of the Pots is controlled remotely, so that they may be kept away from the beam during injection and moved close to it after stable coasting conditions are attained. Particles created at the IP within a certain angular and momentum range are detected and their momentum is magnetically analyzed using the magnets of the machine itself. The fractional difference of the particle's momentum from the momentum of the beam is related to the diffractive mass (see Eq. 3):

$$\frac{\Delta p}{p} = 1\text{-}x = \frac{M^2}{s} \tag{6}$$

The diffraction region spans approximately the range $(2\ \mathrm{GeV}^2/s) < 1\text{-}x < 0.15$ with the lower value corresponding to excitation of the proton to isobars. If the cross section continues to grow as $1/(1\text{-}x)$ at the SSC, the diffractive peak of Fig. 2 is expected to reach a height of eight orders of magnitude! The diffractive spectrometers should be designed to cover as much of this range as possible. However, as we shall see, there are limitations that restrict the accessible region to $4\times10^{-4} < 1\text{-}x < 0.09$, corresponding to diffractive masses of 0.8-12 TeV. Fortunately, the mass region below 0.8 TeV can be covered by the Fermilab Tevatron ($\sqrt{s}$ = 2 TeV).

We have designed two types of spectrometers, a horizontal and a vertical one, covering two different regions of 1-x. Below, we discuss the design characteristics of each type.

(a) The horizontal spectrometers.

A particle created at 0^0 with momentum close to but smaller than that of the beam settles into an orbit of smaller average radius. Particles of the same momentum but at small angles with respect to the beam exhibit betatron oscillations around their mean orbit. The horizontal spectrometers consist of a system of instrumented Pots placed at appropriate positions to detect particles that have been captured within the aperture of the machine and circulate as beam halo. The displacement of such particles from the beam axis is determined by the "lattice functions" $\beta_{x,y}$ and $\eta_{x,y}$. Fig. 6 shows values of the lattice functions in the vicinity of a medium beta interaction region (IR). The loss of momentum leads to a horizontal displacement $\Delta x = \eta_x(\Delta p/p)$ and the change in angle results in oscillations of amplitude $\Delta\theta \times L_{eff}$, where $L_{eff}=(\beta_x\beta^*)^{1/2}$. Fig. 7 shows the path of particles in a region far enough from the IP, where normal periodic conditions are re-established. Detectors placed at the indicated locations comprise the spectrometer. Two of the locations are also shown in Fig. 6. The detectors on the outer side of the beam are used for measuring elastic scattering in conjuction with the mirror image of this arrangement on the other side of the IP, which is also nesssesary for the study of DPE.

The acceptance of the spectrometer is limited on the high momentum side by the size of the beam, which determines the minimum approach of the detectors to the beam axis (~10σ is considered "safe") and on the low side

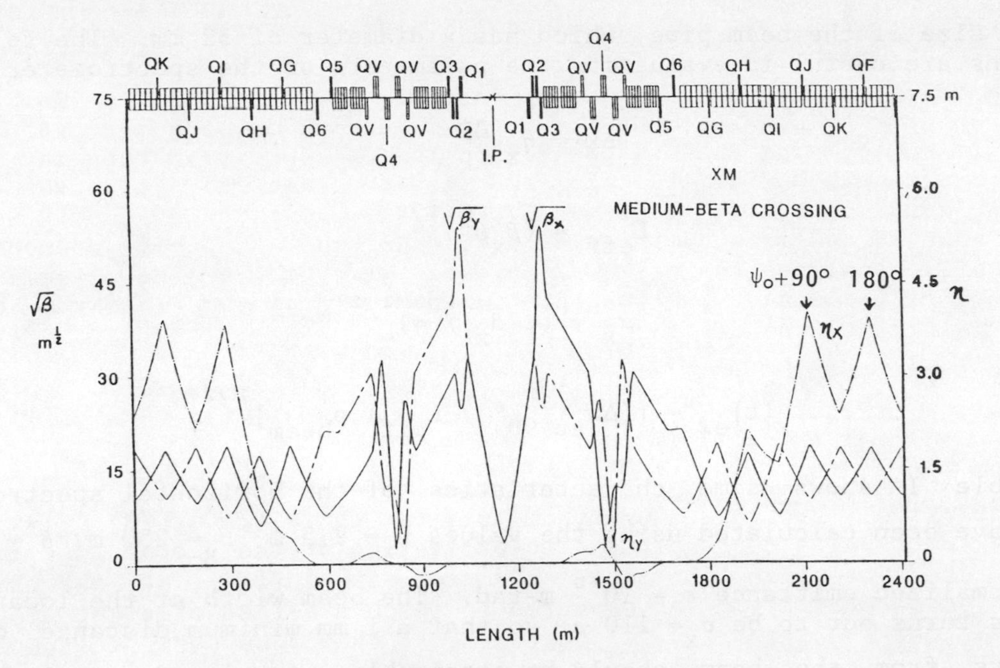

FIG. 6: Lattice functions in the vicinity of a medium-beta crossing IR.

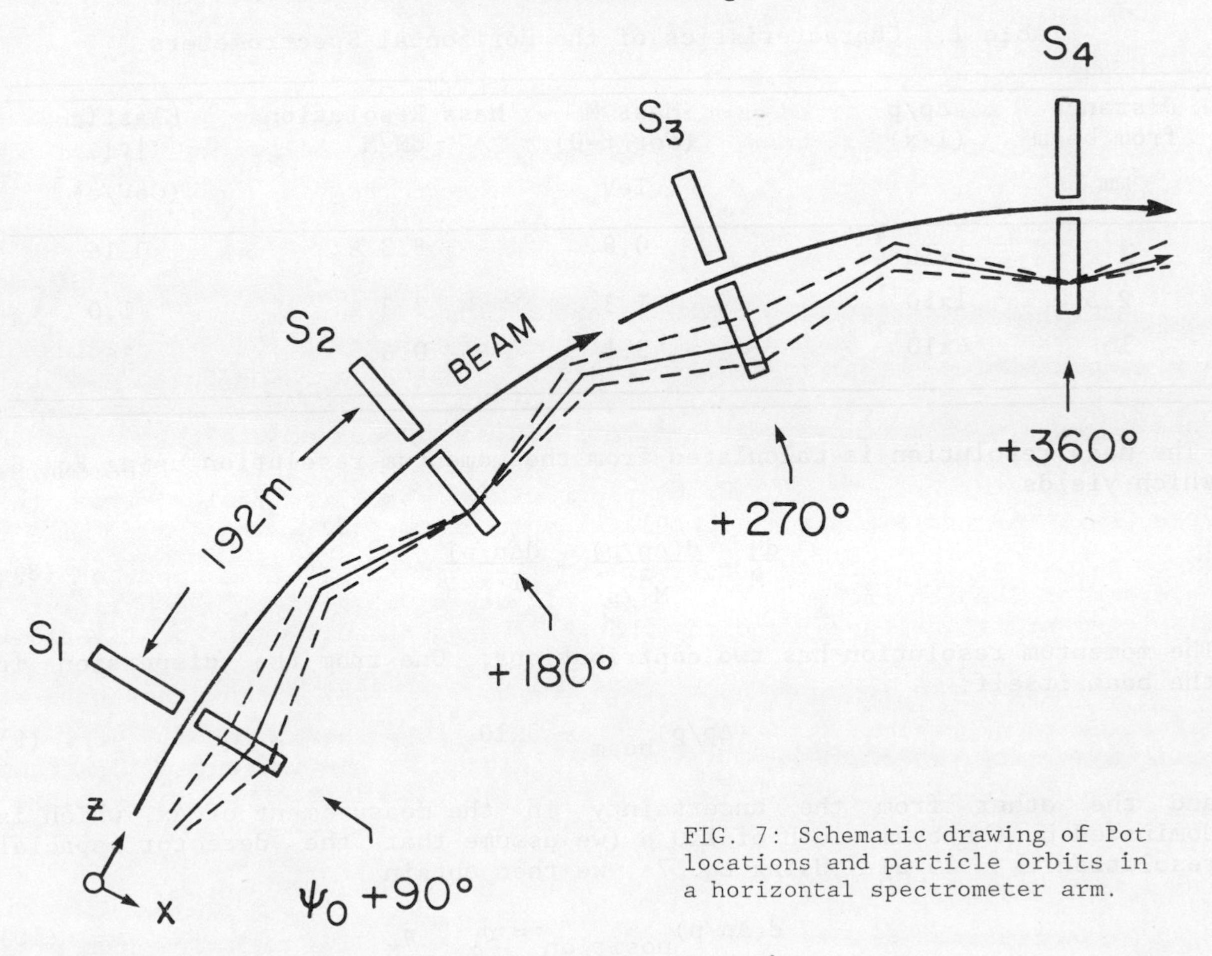

FIG. 7: Schematic drawing of Pot locations and particle orbits in a horizontal spectrometer arm.

by the size of the beam pipe, which has a diameter of 32 mm. The following equations are useful in evaluating the parameters of the spectrometer:

$$\Delta x = \eta_x \frac{\Delta p}{p} \tag{7a}$$

$$L_{eff} = (\beta_x \beta^*)^{1/2} \tag{7b}$$

$$\sigma_x = (\epsilon\ \beta_x\ /\ \gamma)^{1/2} \tag{7c}$$

$$|t|_{e\ell} = [(\Delta x_{at\ 90^0}\ /\ L_{eff})\ p_{beam}]^{1/2} \tag{7d}$$

Table I shows some characteristics of the horizontal spectrometer, which have been calculated using the values $\eta_x = 2.5$ m, $\beta_x = 250$ m, $\beta^* = 10$ m and normalized emittance $\epsilon = 10^{-6}$ m-rad. The beam width at the location of the Pots turns out to be $\sigma_x = 110\ \mu$, so that a 1 mm minimum distance of the detectors from the beam should be attainable. As already mentioned, the maximum distance of 15 mm is imposed by the beam pipe.

Table I. Characteristics of the Horizontal Spectrometers.

Distance from beam mm	Δp/p (1-x)	Mass M (for t=0) TeV	Mass Resolution dM/M	Elastic |t| $(GeV/c)^2$
1	$4x10^{-4}$	0.8	8.3 %	0.16
2.5	$1x10^{-3}$	1.3	3.3 %	1.0
15	$6x10^{-3}$	3.1	0.6	36

The mass resolution is calculated from the momentum resolution using Eq. 6, which yields

$$\frac{dM}{M} = \frac{d(\Delta p/p)}{M^2/s} = \frac{d\Delta p/p)}{1-x} \tag{8}$$

The momentum resolution has two contributions: One from the dispersion in the beam itself,

$$(\Delta p/p)_{beam} = 5x10^{-5} \tag{9}$$

and the other from the uncertainty in the measurement of Δx, which is dominated by the beam width of 110 μ (we assume that the detector spacial resolution is ~ 25 μ). Using Eq. 7a, we then obtain

$$d(\Delta p/p)_{position} \approx \sigma_x\ /\ \eta_x \tag{10}$$

Combining (9) and (10) in quadrature results in a total $d(\Delta p/p) = 6.7\times10^{-5}$, which substituted into (8) gives the values listed in Table I.

The determination of the absolute value of the mass requires knowledge of the absolute position of the Pots relative to the beam. This is obtained by employing two Pots at every station, one on each side of the beam, whose separation is known. Placing the Pots at equal distance from the beam, which can be done by demanding symmetry in the observed elastic scattering events, gives the distance of each one from the beam as one half of their separation.

A final point to be made is that there are four detector stations in each spectrometer arm, as shown in Fig. 7. At two of the stations, those at phase angles $\psi_0 + 180^0$ and $\psi_0 + 360^0$, particles of the same momentum (same mass) but created at different angles (different t-values) register at the same x-position. These detectors may therefore be thought of as "diffractive <u>mass spectrometers</u>". The other two detectors are useful in determining the t-value. The system of four detectors provides the necessary redundancy.

(b) <u>The Vertical Spectrometers.</u>

Protons created with momentum such that $\Delta p/p > 6\times10^{-3}$ do not register in the horizontal spectrometers. In order to extend the acceptance to larger values of $\Delta p/p$ (hence larger masses), we propose the "vertical spectrometers", which take advantage of the vertical crossing magnet arrangement at the IP (Fig. 8). A schematic drawing of the Pot locations is shown in Fig. 9. In Table II we list the spectrometer characteristics, obtained from Eq.s 7 with y substituted for x, using the values $\eta_y = 0.175$ m, $\beta_y = 200$ m and $\beta^* = 10$ m, which are appropriate for these locations. The beam width is calculated to be $\sigma_y = 100\ \mu$. The mass resolution is obtained from Eq. 8, with $d(\Delta p/p)$ dominated now by the position resolution, which has the value $\sim \sigma_y / \eta_y \approx 5.7\times10^{-4}$.

Table II. Characteristics of the Vertical Spectrometers.

Distance from beam mm	$\Delta p/p$ (1-x)	Mass M (for t=0) TeV	Mass Resolution dM/M	Elastic $\lvert t\rvert$ $(GeV/c)^2$
1	5.7×10^{-3}	3.0	5 %	0.2
2.25	1.3×10^{-2}	4.5	2.5 %	1
5	2.9×10^{-2}	6.8	1.1 %	5
15	0.09	12	0.4	58

As with the horizontal spectrometers, two Pots are used at each location, one approaching the beam from the top and the other from the

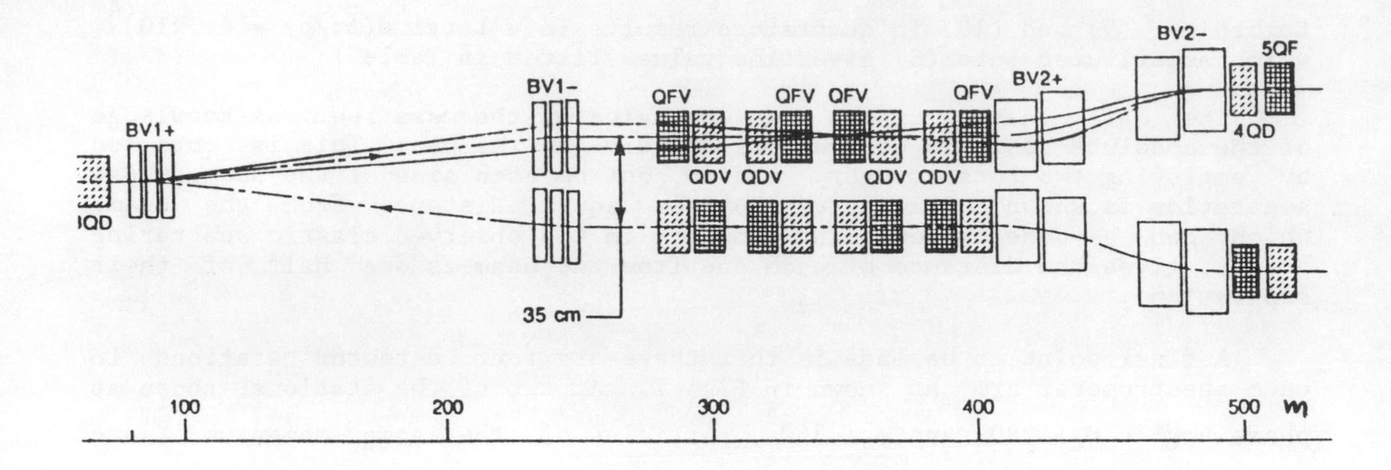

Fig. 8: Magnet arrangement at a low-beta IR vertical crossing (90^0 lattice). The arrangement at the medium-beta IR's is similar.

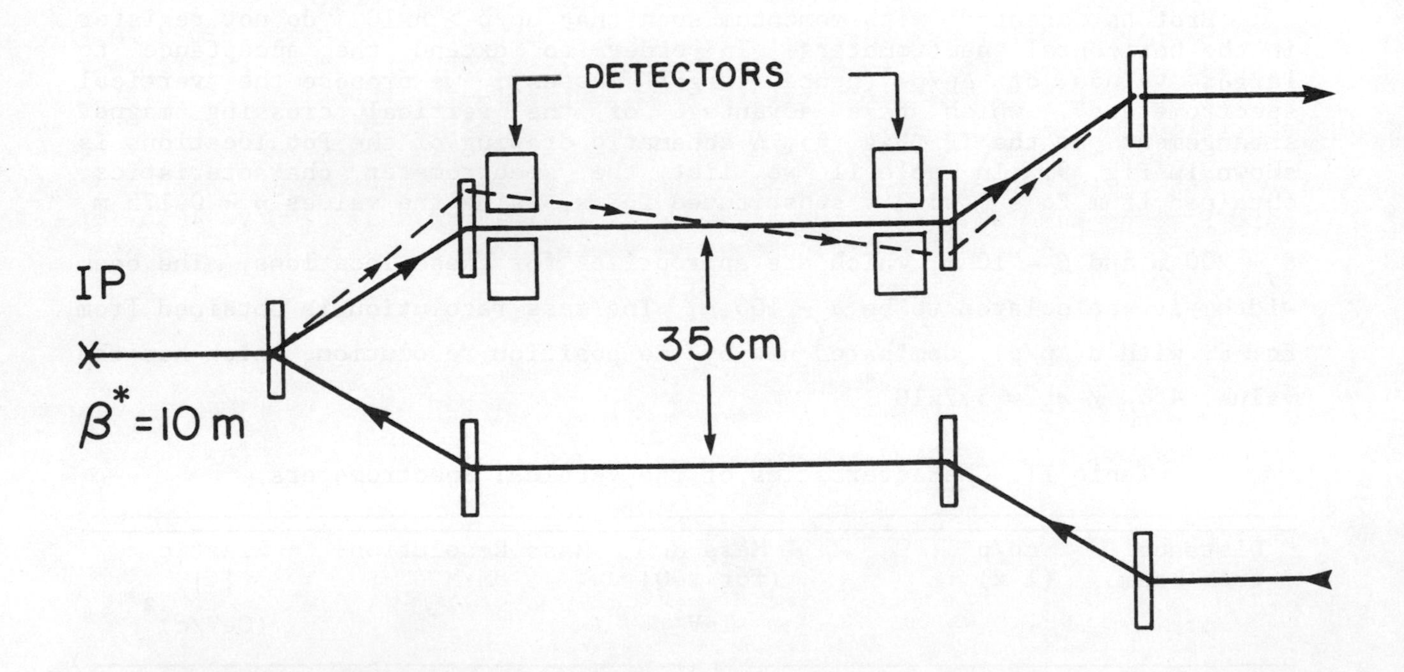

FIG. 9: Schematic drawing of Pot location in a vertical spectrometer arm. The broken line represents the path of a t ≈ 0 "leading proton" created at the IP.

bottom. Eastic events will be used again to position the Pots symmetrically with respect to the beam. The employment of two stations in each arm of the spectrometer is necessary for the simultaneous determination of the momentum and t-value. A mirror image arrangement on the other side of the IP is required for the study of elastic scattering and double Pomeron exchange.

4. Double Pomeron Exchange (DPE).

Double Pomeron exchange is characterized by two leading protons, each having the properties (t and x distribution) of the recoil proton in single diffraction dissociation:

$$p + p \rightarrow p + p + X \qquad (11)$$

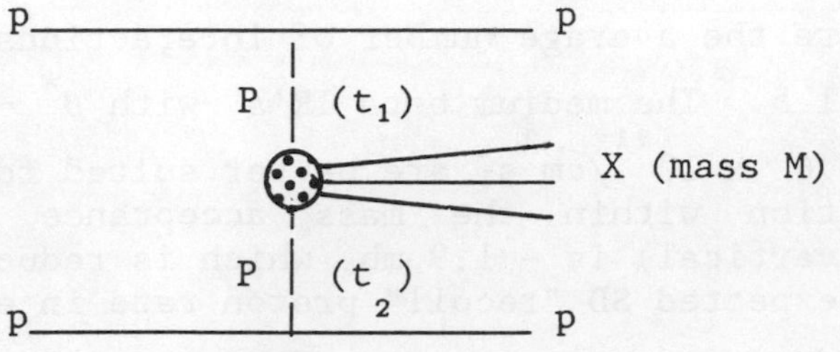

p p
P (t_1)
X (mass M)
P (t_2)
p p

FIG.10

The proposed diffractive spectrometers may therefore be used to study this process, provided they are installed on both sides of an IP. As already mentioned, such an arrangement is also necessary for the study of elastic scattering.

The main interest in DPE is in probing the structure of the Pomeron in pure Pomeron-Pomeron collisions. The differential cross section is also of interest as it provides a strict test of factorization.

The kinematics and cross sections of DPE are summarized below:

$$\frac{M^2}{s} = (1-x_1)(1-x_2) \qquad y_M = \frac{1}{2}\,\ell n\frac{1-x_1}{1-x_2} \quad \text{(rapidity of mass M)}$$

$$\frac{d^4\sigma}{dt_1 dx_1 dt_2 dx_2} = \frac{1}{\sigma_T}\,\frac{d^2\sigma_1}{dt_1 dx_1}\,\frac{d^2\sigma_2}{dt_2 dx_2}$$

$$\text{where} \quad \frac{d^2\sigma}{dt\,dx} = \frac{A}{1-x}\,(b\,e^{dt}) \quad ; \; A = 0.7\text{ mb}$$

$$\frac{d^2\sigma}{d(M^2/s)dy} = \frac{1}{\sigma_T}\,\frac{M^2}{s} \qquad \frac{d\sigma}{dM^2} = \frac{1}{\sigma_T}\,\frac{A^2}{M^2}\,\ell n(M_{max}/M)^2$$

$$\sigma = 2\,\frac{A^2}{\sigma_T}\,\ell n^2\,\frac{M_{max}}{M_{min}}$$

The momentum acceptance of our diffractive spectrometers, $4x10^{-4} < 1-x < 0.09$, corresponds to DPE masses in the range

$$16 \text{ GeV} < M < 3.6 \text{ TeV}$$

The accepted cross section is 0.24 mb, which is about 40 % of the total DPE cross section. The mass resolution varies from ~12 % at low masses to ~0.6 % at 3.6 TeV. Clearly, the proposed double arm diffractive spectrometers are very well suited to study the very promissing DPE process.

5. Event rates, background and triggering.

Studies of "hard diffraction" require high luminosity. However, since events must be tagged, such studies cannot be performed in a high beta IR ($\beta^* = 0.5$), where the average number of interactions per bunch crossing is expected to be ~ 1.5. The medium beta IR's, with $\beta^* = 10$ m and a design luminosity of $5.6 \times 10^{31}/cm^2 s$, are better suited for diffractive tagging. The SD cross section within the mass acceptance of each spectrometer (horizontal or vertical) is ~ 1.9 mb, which is reduced to ~ 1 mb by the t-acceptance. The expected SD "recoil" proton rate in each spectrometer arm is then ~ 6×10^4/sec.

The detectors will, of course, also be counting particles from beam-gas interactions. As an estimate of beam-gas rates, we note that at a pressure of 10^{-9} Torr of hydrogen, the number of interactions expected to occur in a 100 m pipe lengh is ~10^4/sec. Such rates are relativelly small and can be handled by requiring coincidence between the spectrometers and the diffractive clusters.

A more serious problem is posed by the beam halo created at the low beta, high luminosity IR's. As much as 10 % of the pp cross section results in leading protons being captured in the machine, creating a halo at a rate of ~ 10^7/sec. It is imperative that these protons be removed by the use of beam scrapers. Assuming that the halo is "killed" within a few revolutions, the equilibrium rate is expected to be ~ 10^4/sec, which is again tolerable.

References.

1. K. Goulianos, Physics Reports Vol.101, No.3, December 1983.

2. K. Goulianos, "Diffractive and Rising Cross Sections", Proceedings of Workshop on Physics Simulations at High Energy, University of Wisconsin-Madison, 5-16 May 1986, World Scientific (Edited by V. Barger, T. Gottschalk and F. Halzen) pp. 127-140. Also: To be published in "Comments on Nuclear and Particle Physics, Gordon and Breach, Science Publishers, Inc.

2. K. Goulianos, "A Model for Centauro Production", Proceedings of Workshop on Physics Simulations at High Energy, University of Wisconsin-Madison, 5-16 May 1986, World Scientific(Edited by V. Barger, T. Gottschalk and F. Halzen) pp. 299-312. Also: To be published in "Comments on Nuclear and Particle Physics, Gordon and Breach, Science Publishers, Inc.

Exotics

EXOTICS GROUP SUMMARY REPORT

A. M. LITKE
Santa Cruz Institute for Particle Physics
University of California, Santa Cruz, CA 95064
A. C. MELISSINOS
Department of Physics
University of Rochester, Rochester, NY 14627
P. B. PRICE
Department of Physics
University of California, Berkeley, CA 94720

C. Y. Chang, *University of Maryland*
R. Darling, *Rutgers University*
H. Kasha, *Yale University*
K. Kinoshita, *Harvard University*
S. Miyashita, *Fermilab*
S. Olsen, *University of Rochester*
S. Parker, *University of Hawaii*
L. Sarycheva, *Indiana University*
W. Vernon, *Univ. of California, San Diego*
W. Wenzel, *Lawrence Berkeley Lab.*
A. White, *University of Florida*

ABSTRACT

The SSC will open up a new energy region where new and unexpected physics is possible. Therefore, the search for exotic particles, including magnetic monopoles, massive stable particles, and free quarks, will be an important part of the SSC physics program. We propose that the search for exotics be carried out in a dedicated, but modest, facility in a high luminosity interaction region. The exotics search detector would consist of (1) a set of plastic and glass track-etch detectors to search for monopoles (or other highly ionizing particles), (2) a double-arm spectrometer, with the capability of measuring momentum, velocity, and calorimetric energy, to search for massive stable particles (both charged and neutral), and (3) two telescopes of silicon pixel detectors for the measurement of ionization energy loss, to search for quarks (or other fractionally charged or multiply charged particles). We also discuss two types of beam dump experiments to search for particles which are highly penetrating. For detector R&D, we recommend a systematic study of glass track-etch detectors, and the rapid development of silicon pixel detectors, including the associated radiation-hardened electronics.

1. Introduction

According to the nearest dictionary, exotic refers to something which is "strikingly or excitingly different or unusual" [1]. Indeed, the real motivation for the construction of the Superconducting Super Collider is the hope that in the debris of 40 TeV proton-proton collisions we will find something with just these characteristics. In this report, we will summarize the work of the Exotics Group at the 1987 Summer Workshop on Experiments, Detectors, and Experimental Areas for the Supercollider. This group explored some of the experimental possibilities connected with the search for exotic particles at the SSC, particularly the search for magnetic monopoles, massive stable particles, free quarks, and highly penetrating particles.

Our approach was guided by two basic premises. First, the new energy scale that will be opened up by the SSC – with the possibility of new physics and big surprises – implies that a search for exotics at the SSC *must* take place. Second, given the very speculative nature of these searches, the effort expended should be *modest*.

What is a modest effort? Three approaches come to mind:

(1) construct a detector which has a large solid angle but which is inexpensive (and therefore is most likely to be passive);

(2) construct a well-instrumented detector which has, however, a limited solid angle; or

(3) start with a general purpose 4π detector (which will be built in any case) and enhance its capabilities for exotic searches.

In what follows, there are examples of approaches (1) and (2). Our work on option (3) indicated that this was not a competitive approach. However, new ideas and new technology could change this conclusion.

One consequence of our premise of modesty is that we did not pursue the detector design of the 1984 Snowmass Exotics Working Group [2]. This design, with the goal of searching for massive stable particles over the full solid angle, has a radius for the central tracking chamber of 6 m and an estimated price tag of $395M.

There is little discussion of the theoretical background for exotic searches in this report. However, a detailed exposition of these matters can be found in reference 2.

In section 2, we will describe the search for magnetic monopoles (or other highly ionizing particles). The experimental technique is to surround an interaction region with layers of plastic and glass track-etch detectors. Section 3 will describe a double arm magnetic spectrometer, with a 6 m flight path but limited solid angle, for the search for massive stable particles. The search for free quarks (or other fractionally charged or multiply charged particles) is discussed in section 4. We propose the use of silicon pixel detectors to look for quarks inside the dense particle jets we expect at the SSC. Section 5 will discuss the search for highly penetrating particles in "beam dump" experiments. One approach is to collide the beam with an internal gas jet target and look for particles which have penetrated 1 km of earth downstream from the collision. A second approach is to use the pp collisions, and look for penetrating particles ≈ 1.5 mr off-axis after 5 km of earth. Finally, in section 6, we will reach some conclusions and make some recommendations for future R&D efforts. We will stress the desirability of reserving a high luminosity interaction region for specialized detectors and experiments, such as the search for exotic particles.

2. The Search for Magnetic Monopoles – or other Highly Ionizing Particles

A more complete discussion of this topic can be found in the report of Price and references therein [3]. Here we will summarize the main points.

2.1. The Signature – High Ionization Energy Loss

The most important characteristic we will use to identify these particles is the very high rate of ionization energy loss (dE/dx). For electrically charged particles we have the approximate relation:

$$dE/dx \approx (Q/\beta)^2 (dE/dx)_{MIP}$$

where Q is the particle's electric charge (in units of e), β is the particle velocity (in units of c), and $(dE/dx)_{MIP}$ is the ionization energy loss for a minimum ionizing particle with $Q = 1$.

For magnetic monopoles with magnetic charge g, the charge quantization condition

$$ge = (n/2)\hbar c$$

implies that

$$g/e = n/2\alpha \simeq (137/2)n$$

with $n = \pm 1, \pm 2, \ldots$ for a Dirac monopole. A monopole with velocity $\beta \simeq 1$ will have an ionization energy loss given approximately by:

$$(dE/dx)_{MM} \approx (g/e)^2 (dE/dx)_{MIP}$$

where $(dE/dx)_{MIP}$ still refers to an *electrically* charged particle with $Q = 1$. Since $g/e \approx 68.5n$ this is an enormous energy loss, corresponding to about 9.4 GeV/cm in plastic when $n = 1$.

As the monopole slows down the $(dE/dx)_{MM}$ *decreases*, unlike the case for an electrically charged particle. For monopole velocities in the range $3 \times 10^{-4} < \beta < 0.2$ the ionization energy loss is roughly linear with velocity

$$(dE/dx)_{MM} \approx K\beta \, .$$

This can still be a very large energy loss rate. For example, with $n = 1$ and $\beta = 0.1$, $(dE/dx)_{MM} = 3$ GeV/cm in plastic.

Figure 1 shows a plot of the ionization energy loss of a monopole as a function of the monopole velocity. The energy loss is plotted in terms of an equivalent value of Q/β, that is the value of Q/β for an electrically charged particle with the same dE/dx as the monopole. Hence in this plot

$$Q/\beta \equiv \left[\frac{(dE/dx)_{MM}}{(dE/dx)_{MIP}}\right]^{1/2} .$$

Also shown in this plot are detection threshold values for some plastic and glass track-etch detectors, as will be described later.

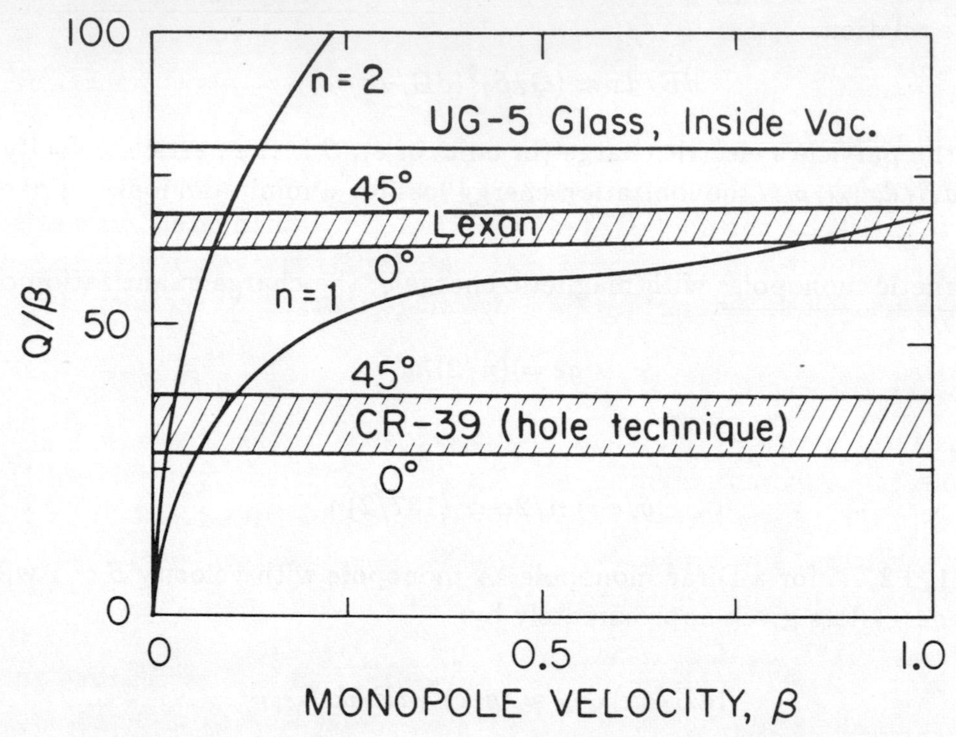

Fig. 1. The ionization energy loss for a magnetic monopole as a function of the monopole velocity. The energy loss is plotted in terms of the equivalent value of Q/β (see text). Also shown are the detection threshold values for some plastic and glass track-etch detectors.

2.2. The Ideal Monopole Detector

The experimental challenge in the search for magnetic monopoles is to find a signal for highly ionizing particles in the face of an enormous background of minimum ionizing particles produced in the p-p collisions. For example, with an integrated luminosity of $\int Ldt = 10^{40}$ cm^{-2}, and a rapidity interval of $\Delta y = \pm 2$, there will be 2×10^{16} MIP's passing through the detector. Nonetheless, we would still like to be able to pick out the signal even if only a few monopoles are produced.

Keeping this signal-to-noise ratio of 10^{-16} in mind, let us list some of the properties we would like to see in an ideal monopole detector for the SSC:

(1) it should be sensitive to monopoles with $n \geq 1$ and with a broad range of velocities;
(2) it should be completely insensitive to normally-ionizing particles (remember, at the 10^{-16} level!);
(3) it should easily cover the full solid angle;
(4) it should be easy to trigger;
(5) it should be capable of tracking monopoles and measuring their properties;
(6) it should be radiation resistant;
(7) it should be possible to place detector elements inside the beampipe in order to detect monopoles with very short range; and
(8) it should be cheap.

Although it might seem an impossible task to come up with a practical detector which comes close to satisfying the above-listed properties, there fortunately does exist a simple and inexpensive solution: place layers of plastic and glass track-etch detectors around an interaction region. These detectors will be described in the next section.

2.3. Track-etch Detectors

Thin sheets of plastic or glass can be used to record the tracks of highly ionizing particles. Substances commonly used for this purpose include materials with the trade names Lexan (a plastic used in airplane windows and Cuisinarts), CR-39 (a plastic used for lenses in eyeglasses and sunglasses), and UG-5 (a cobalt-rich phosphate glass used in blue filters). The technique is illustrated in Figure 2.

In Fig. 2(a), a highly ionizing particle passes through a plastic sheet and leaves a microscopic trail along its trajectory. This trail is susceptible to chemical etching at an accelerated rate.

The latent track is made visible by etching the sheet in a concentrated solution of sodium hydroxide. This is shown in Fig. 2(b). Let V_G be the general etch rate for exposed surfaces, and let V_T be the accelerated rate for etching along the trajectory. As will be shown below, the ratio V_T/V_G is a rapidly increasing function of the ionization energy loss of the particle. Due to this differential etch rate, in analogy to the formation of a conical wavefront in Cherenkov light emission, conical etch pits are formed in the plastic. The half-angle θ of the cone is given by $\sin\theta = V_G/V_T$.

As shown in Fig. 3(a), if the etching process is continued for a sufficient length of time, a hole will be formed in the plastic sheet. These holes can be detected with the technique illustrated in Fig. 3(b). The plastic sheet is placed on top of blueprint paper and the two are sealed with tape around the edges. This package is then exposed to ammonia vapor. Each hole in the plastic is then revealed by a blue spot on the blueprint paper.

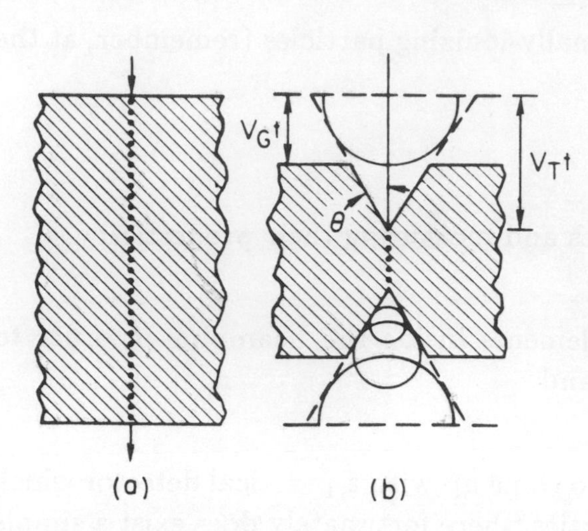

Fig. 2. (a) A highly ionizing particle passes through a track-etch detector. (b) The track-etch detector after etching for a time *t*.

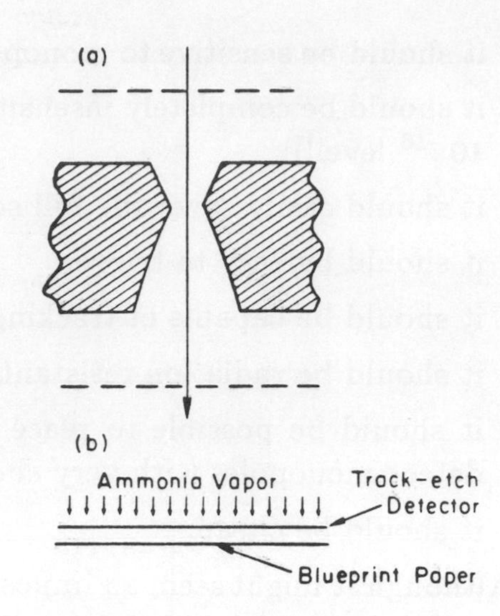

Fig. 3. (a) A hole is etched in the track-etch detector. (b) The technique for locating the etched holes.

Let us now return to the eight properties of an ideal monopole detector listed in section 2.2, and see how well track-etch detectors meet the necessary requirements:

(1) *Sensitivity to Monopoles:* as shown in Fig. 1, monopoles with $n = 1$ will be detectable in CR-39 provided that $\beta \gtrsim 0.1$ and in Lexan for $\beta \gtrsim 0.85$. UG-5 is only sensitive to monopoles with $n \geq 2$.

(2) *Insensitivity to MIP's:* the sensitivity of track-etch detectors to ionizing particles is shown in Figure 4. The value of $V_T/V_G - 1$ (which determines the cone angle of the etched pit) is plotted as a function of the ionization energy loss, expressed in terms of the equivalent value of Q/β. As the ionization energy loss increases, the etch rate along the track increases sharply relative to the general etch rate, and the etched pit becomes correspondingly steeper. We observe that CR-39, Lexan, and UG-5 all have ionization thresholds for hole detection well above the minimum ionizing value of $Q/\beta \approx 1$. So these detectors are insensitive to normally-ionizing particles;

(3) *Solid Angle Coverage:* it is quite easy to cover the full solid angle with stacks of these sheet detectors;

(4) *Trigger:* as these detectors are always sensitive to monopoles and yet insensitive to the normal hadronic events, no trigger is necessary (nor is one possible with these passive devices);

(5) *Monopole Properties:* measurement of the detailed shape of the conical etch pit produced on each side of the plastic or glass sheet gives information on the particle trajectory as well as the equivalent value of Q/β. In a multi-sheet detector, the

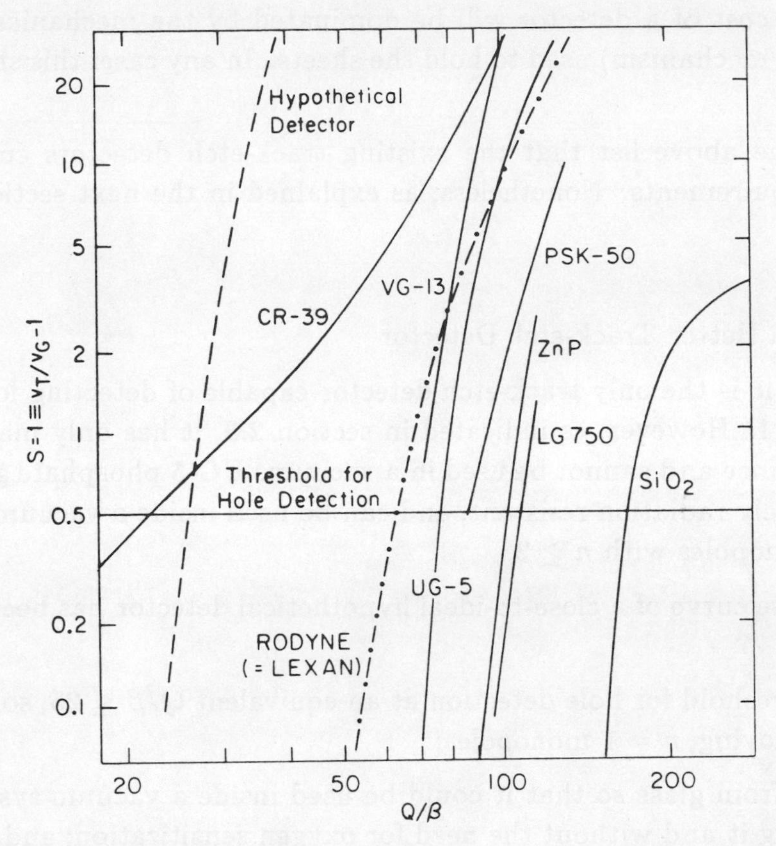

Fig. 4. **$V_T/V_G - 1$ plotted as a function of the equivalent value of Q/β for a number of plastics and glasses. The curve for an ideal hypothetical detector, and the threshold for hole detection, are also shown.**

position and direction information from the individual pits can be combined to provide a very precise trajectory, and the Q/β values can be used to determine the change in ionization energy loss as the particle loses energy passing through the detector (dE/dx should *increase* for an electrically charged particle but *decrease* for a monopole). Showing that the candidate track comes from the interaction region and has consistent dE/dx values will be important checks;

(6) *Radiation Resistance:* the radiation resistance of track-etch detectors is examined in reference 3. At a radius of 10 cm, and taking an SSC year to have an integrated luminosity of 10^{40} cm^{-2}, CR-39, Lexan, and UG-5 will withstand, respectively, ≈ 1 month, ≈ 2.5 years, and ≈ 50 years of operation. Moreover, as the material is inexpensive, it can be replaced when necessary;

(7) *Placement inside the Beampipe:* as explained in reference 3, plastic detectors require oxygen fixation to retain their sensitivity, but glass detectors do not. Therefore, UG-5 can be used inside a vacuum pipe;

(8) *Cost:* the cost of a detector will be dominated by the mechanical structure (with retraction mechanism) used to hold the sheets. In any case, this should be less than $200k.

We see from the above list that the existing track-etch detectors come very close to meeting our requirements. Nonetheless, as explained in the next section, we would like to do better.

2.4. Towards a Better Track-etch Detector

CR-39 plastic is the only track-etch detector capable of detecting low velocity monopoles with $n = 1$. However, as indicated in section 2.3, it has only marginally sufficient radiation resistance and cannot be used in a vacuum. UG-5 phosphate glass, on the other hand, is extremely radiation resistant, and can be used inside a vacuum pipe, but is only sensitive to monopoles with $n \geq 2$.

The response curve of a close-to-ideal hypothetical detector has been shown in Fig. 4. It would:

(1) have a threshold for hole detection at an equivalent $Q/\beta \leq 25$, so as to be sensitive to slow-moving, $n = 1$ monopoles;

(2) be made from glass so that it could be used inside a vacuum system without contaminating it and without the need for oxygen sensitization; and

(3) have a steep response curve (expected for glass detectors). Then it would be less easily degraded by background at ionization rates below that able to produce etchable tracks, and therefore would be more radiation resistant.

In addition to the response curves for CR-39, Lexan, and UG-5, Fig. 4 shows measurements done on a number of commercial glasses in an unsystematic study [4,5,6]. A wide range of sensitivities were found. Clearly, a systematic study of glasses with different compositions should be made; their response curves can be measured with relativistic heavy ions (*e.g.*, at the LBL Bevalac).

2.5 An SSC Monopole Detector

Here, then, are our recommendations for a monopole detector at the SSC:

(1) Use stacks of interleaved CR-39 (high sensitivity) and Lexan (more radiation resistant) in a large solid angle cylindrical array. As discussed in section 2.4, if a glass detector with a sufficiently low ionization rate threshold is found, it could supplement or replace these materials;

(2) The detector should be retractable to avoid radiation damage during periods of machine tune-up or abnormal operation;

(3) UG-5 (or an improved glass detector) can be deployed *inside* the beam pipe to detect monopoles with a very short range. It is very radiation resistant and retains sensitivity in a vacuum; and

(4) The sensitivity of the detector should be verified by irradiating the detector sheets with relativistic heavy ions *before* the run, and demonstrating that these tracks give detectable holes *after* the run.

Finally, Fig. 5 shows limits on the production cross section for $n = 1$ monopoles as a function of monopole mass in accelerator search experiments. Present results, as well as expected future results (assuming no monopoles are found), are shown. The curve for the SSC assumes an integrated luminosity of 10^{39} cm^{-2}.

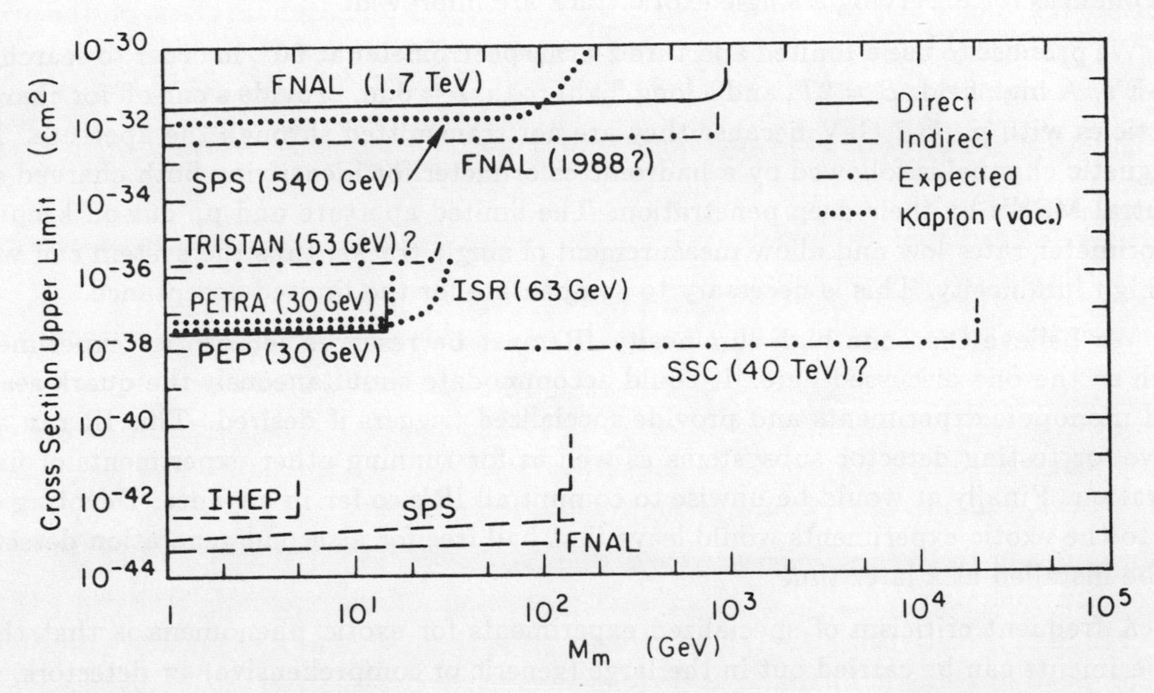

Fig. 5. Cross section upper limits (90% C.L.) for the production of $n = 1$ magnetic monopoles as a function of monopole mass.

3. The Search for Massive Stable Particles

3.1. Introduction

The theoretical motivation for the existence of some sort of massive long-lived particles at the SSC energy scale has been discussed thoroughly in previous SSC workshops [2]. This is based on the fact that most theories beyond the standard model (*i.e.*, SUSY, Technicolor, skyrmions, left-right symmetric models, etc.) generate a spectrum of particles with masses of 100 GeV or more. The ground state in that spectrum may be prohibited from decaying rapidly to the familiar particles of the standard model. We will call such particles with

$$M \gtrsim 100 \text{ GeV}$$
$$\tau \gtrsim 10^{-7} \text{ sec}$$

massive stable exotics (MSE).

It is generally assumed that MSE's are produced by gluon fusion and the standard ISAJET program is used to generate distributions and production cross-sections. Detailed results can be found in ref. 2 and are in agreement with recent runs of the code [7]. As expected, massive particles are produced in a narrow rapidity interval around $y = 0$. In the central region, jet production is suppressed as compared to the forward angles and the chances for observing a single exotic track are improved.

We propose to use a limited aperture 2-arm spectrometer at 90°, in order to search for MSE's. A high field, $B = 2T$, and a long flight path, $\ell = 6$ m, provide a cut-off for charged particles with $p_T \lesssim 7$ GeV because they are not transmitted through the aperture. The magnetic channel is followed by a hadronic calorimeter for identifying both charged and neutral MSE's by their deep penetration. The limited aperture and p_T cut-off keep the calorimeter rates low and allow measurement of single tracks; thus the system can work at high luminosity. This is necessary to compensate for the limited acceptance.

We believe that one high luminosity IR must be reserved for exotics experiments such as the one discussed here. It could accommodate simultaneously the quark search and monopole experiments and provide specialized triggers if desired. This IR can also serve for testing detector subsystems as well as for running other experiments of finite duration. Finally, it would be unwise to commit all IR's so far in advance. Devoting one IR to the exotic experiments would leave that hall free for a second generation detector, to be installed at a later time.

A frequent criticism of specialized experiments for exotic phenomena is that these experiments can be carried out in the large (generic or comprehensive) 4π detectors. We do not believe that this statement is correct. Often the generic detector is not optimized for the exotic search; the press of other research results in delaying and limiting the running for exotics; the complete angular coverage while adding to the event rate is not essential to the search. Since the search for exotics is highly speculative, one wishes to design a detector of relatively low cost; this in turn implies limited acceptance but we have maintained the highest possible resolution. The philosophy behind these choices is that it is imperative to find exotics if they exist; if they are not found, the level at which limits are being set is much less important.

3.2. The Spectrometer

The spectrometer is shown in Fig. 6, each arm consisting of iron core dipole magnets with total effective length $\ell = 5$ m and a field of order $B = 2$ T. The two arms are operated with opposing fields so as to minimize the perturbation at the I.P. Each arm subtends an aperture

$$\Delta\theta = \pm 10^\circ \qquad \Delta\phi = \pm 10^\circ$$

The spectrometer entrance plane is located $z = 1$ m from the I.P. and the exit plane at $z = 6$ m. To reduce the cost, the aperture increases proportionally to z from 0.35×0.35 m^2 at the entrance, to 2.1×2.1 m^2 at the exit. The estimated weight is 570 tons per arm. For construction purposes it may be easier to fabricate three dipoles of progressively larger square aperture for each arm as shown below.

Aperture	$z_2 - z_1$ (m)	Weight (tons)
1.0×1.0	$3.0 - 1.0 = 2.0$	120
1.6×1.6	$4.5 - 3.0 = 1.5$	250
2.1×2.1	$6.0 - 4.5 = 1.5$	450

As a cost estimate for the machined steel we used a figure of $2,000/ton. We note that these dimensions are typical of the spectrometer built for experiment 605 at Fermilab.

The magnetic volume contains 4 sets of tracking chambers as shown in the figure: one at the entrance and one at the exit plane, the other two being placed at $z = 3$ m and $z = 4.5$ m. Each set should contain an X–Y–U–X plane of drift chambers with adequate cell spacing. X is the bending plane where a resolution $\delta x = 200\mu$m should be attained for each set. Three time–of–flight (tof) planes are used, placed at $z = 3$ m, 4.5 m and 6.0 m. They consist of a plastic scintillator picket fence 1" thick (in the z-direction) and 2" wide (in the y-direction). The scintillators are viewed by two phototubes each, so as to measure $(t_1 + t_2)$ and $(t_1 - t_2)$. The tubes are placed outside the magnetic region and we expect a resolution averaged over the three planes of $\delta t = 100$ to 150 psec.

The magnetic volume is followed by an iron-liquid scintillator calorimeter with a 3×3 m^2 active cross section and 4 m depth along the z-direction. This represents 19 interaction lengths and should contain a 10 TeV hadronic shower. The depth of Fe is

$$\ell(Fe) = 3.2 \text{ m}$$

and we propose to sample every 10 cm using 2.5 cm of liquid scintillator. The liquid scintillator modules are 15 cm wide; thus the total number of modules is

$$N = 20 \times 32 = 640$$

and they must be viewed from both ends.

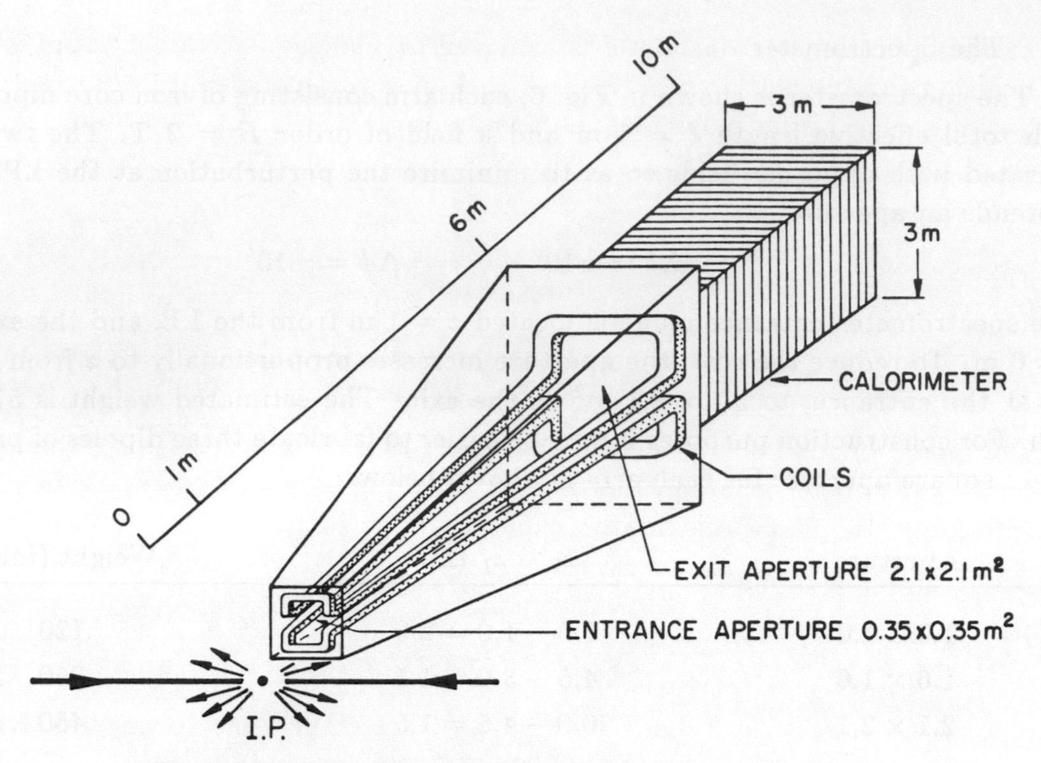

Fig. 6a. **Isometric view of one arm of the spectrometer**

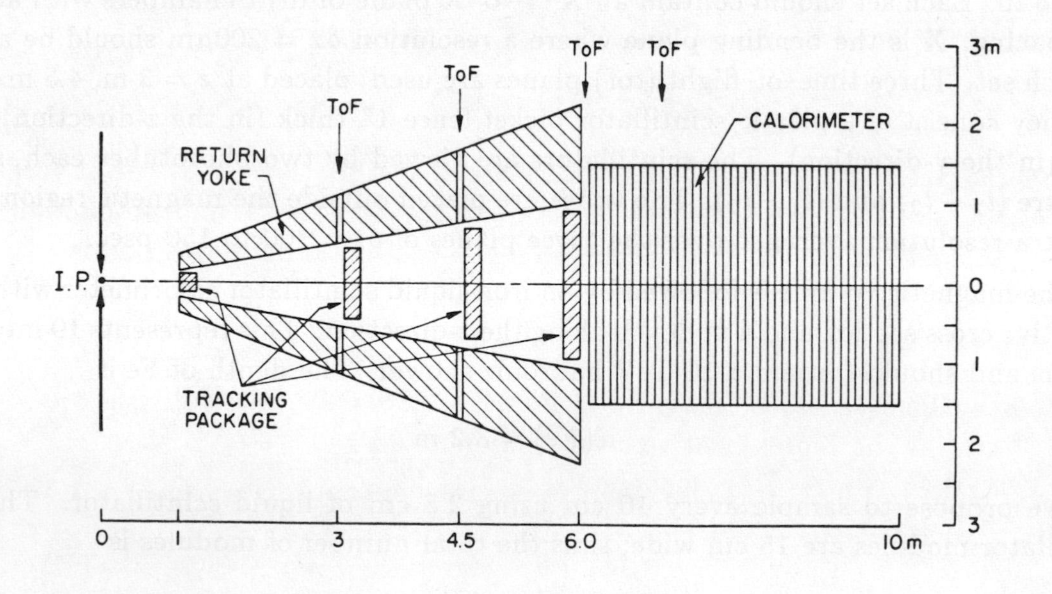

Fig. 6b. **Plan view of the spectrometer showing position of detectors. The magnetic field is along the vertical.**

Tracking inside the calorimeter is achieved by planes of streamer tubes for which we can expect a time resolution of ± 150 nsec and space resolution of ± 1 cm. This is adequate for tracking penetrating particles such as muons or MSE's.

The expected resolution for the calorimeter is

$$\frac{\sigma}{E} = \frac{0.03}{\sqrt{E(\mathrm{TeV})}} + 0.02$$

and the ratio $e/h \sim 1.3$. For a slightly larger cost one could use Pb plates instead of Fe if it is desired to operate in the mode where $e/h = 1$.

One plane of high resolution tof consisting of plastic scintillator elements is placed inside the calorimeter at a depth of ~ 1 m. This plane is used to supplement the timing information of the liquid scintillator, especially for neutral MSE.

The transverse momentum kick of the spectrometer is

$$p_T = 0.3\, B\ell = 3.0 \text{ GeV} .$$

Thus a 10 GeV particle is bent by $\theta = 18°$, and this leads to an effective cut-off in transmission through the magnetic channel of

$$p_T > 7 \text{ GeV} .$$

Since the spectrometer is near 90°, the requirement that transmitted charged particles have $p_T > 7$ GeV will greatly reduce the background from forward jets. Even for jets produced at 90° some of the soft particles will be swept away.

3.3. Acceptance and Rates

The characteristic properties of MSE's are that they are produced at large angle, are massive, and have low velocity. To estimate the acceptance and production rate we consider particles of M = 1 TeV. From ref. 2 the production cross section is

$$\sigma_T(M = 1\text{TeV}) \sim 2 - 200 \text{ pb}$$

and rises by four orders of magnitude if $M \rightarrow 250$ GeV. We will use $\sigma_T = 100$ pb $= 10^{-34}$ cm^2 as the representative cross-section for M = 1 TeV.

For such M = 1 TeV particles, the average momentum is $\langle p \rangle \sim \langle p_T \rangle \sim 1$ TeV so that

$$\langle \beta\gamma \rangle = \langle p \rangle / M \sim 1 .$$

The width of the rapidity distribution is approximately one unit, $\Delta y \sim 1$. Near 90°, $\Delta y \sim \tan \Delta\theta \sim 0.35$ and the acceptance in azimuth is $\Delta\phi/2\pi \sim 0.05$ for each arm. A calculation [7] using ISAJET shows that the acceptance, in each arm, for a 1 TeV particle is

$$\text{Acceptance} = N_{\text{acc}}/N_T = 0.004\ .$$

Thus, using the luminosity $L = 10^{33}$ cm^{-2} sec^{-1} and a detection efficiency $\epsilon = 0.6$, the rate for observed MSE's in the two arms is

$$R(M = 1\text{TeV}) \sim 5 \times 10^{-4}\ \text{Hz}$$

or a total of $\sim$ 5,000 events/year; this is a respectable yield.

To estimate background rates we use

$$\left.\frac{d^2\sigma}{dy\, dp_T}\right|_{y=0} = \frac{100}{p_T^{3.5}}\ \text{mb/GeV}$$

integrate for all $p_T > 7$ GeV, and assume $\Delta y = 0.4$, $\Delta\phi/2\pi = 0.05$ to obtain an effective cross section

$$\sigma_{\text{eff}} = 6\ \mu b\ .$$

To estimate the muon flux we note that for 7 GeV π's the decay probability in $\ell = 6$ m is 1.5×10^{-2} and we expect a contribution of the same order from K-decays, for a total μ/π ratio $\sim 2.5 \times 10^{-2}$.

The table below gives the expected rates at various points in one arm of the detector for $L = 10^{33}$ cm^{-2} sec^{-1}. The results are in agreement with the more detailed Monte Carlo calculations [8].

Location	Rate	Rate/cm^2
Entrance plane $z = 1$ m	50 MHz	100 KHz/cm^2
Exit plane $z = 6$ m	6 KHz	0.1 Hz/cm^2
μ's entering calorimeter	150 Hz	—
Trigger	0.1–1 Hz	—
Signal (M $\sim$ 1 TeV)	2.5×10^{-4} Hz	—

It is more difficult to estimate the rate from albedo background such as neutrons. The rate from devices which intercept forward collisions has been placed at a flux of 10^5 neutrons/cm^2-sec. Assuming little mass in the forward direction and a conversion efficiency of 0.1, one could expect a counting rate of 100 Hz/cm^2 or $\sim$ 0.5 MHz in each liquid scintillator module. Such a rate would be unacceptably high but because of the limited geometry it will be possible to shield the detector against albedo, should this be necessary.

3.4. Resolution and Event Signature

MSE's are selected and identified by their low velocity and penetration through the calorimeter. Coupled with a momentum measurement this yields their mass. The tof system uses the beam crossing as its zero time. While the absolute time for the collision may have a jitter of several 100 psec, the exact zero time of the collision can be determined to a fraction of the crossing jitter by timing the burst of relativistic particles. Thus we will assume that the flight time over $\ell = 6$ m is known to a resolution $\delta t = 100$ psec.

The time difference between the relativistic particles and the MSE's is

$$\Delta t = \frac{\ell}{c}\left(\frac{1}{\beta} - 1\right) = 20 \text{ nsec} \left[\frac{\sqrt{1+\gamma^2\beta^2}}{\gamma\beta} - 1\right] .$$

For $\langle\gamma\beta\rangle \sim 1$ we find $\Delta t = 8$ nsec. Note that the spacing between collisions is $\Delta t = 16$ nsec. In Fig. 7 we show Δt *vs.* $\gamma\beta$ and note that even for $\gamma\beta = 3$, $\Delta t = 1$ nsec which is probably the limit (10σ) at which one can distinguish particles out of the tail of the relativistic muons.

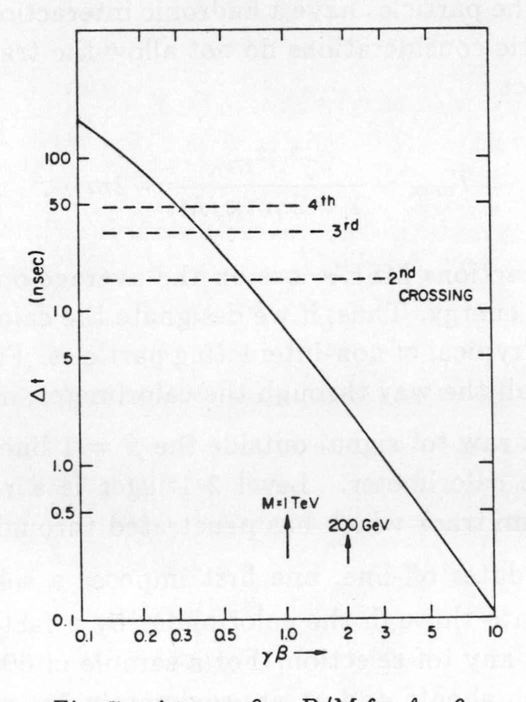

Fig. 7. Δt *vs.* $\gamma\beta = \mathrm{P/M}$ for $\ell = 6$ m

For the momentum resolution we assume a lever arm of $\ell = 5$ m and a position resolution $\delta x = 200\mu$m. Then

$$\frac{\delta p}{p} \sim \frac{\delta\theta}{\theta} \sim \frac{2\delta x/(\ell/2)}{1/2(p_T/p)} \sim 0.1 \text{ [p(TeV)]} .$$

Multiple scattering is not important at these momenta (p $\sim$ 1 TeV) because the path is through air.

The mass resolution from combining the tof and momentum measurements

$$\frac{\delta M}{M} = \sqrt{\left(\frac{\delta p}{p}\right)^2 + \left[\frac{\delta(\gamma\beta)}{\gamma\beta}\right]^2} \sim \frac{\delta p}{p}$$

is dominated by the momentum measurement because

$$\frac{\delta(\gamma\beta)}{\gamma\beta} \sim \frac{\delta t}{\ell/c}(\gamma\beta)^2 \sim 0.02 \text{ at } \gamma\beta = 2 \; .$$

The obvious direction for improving $\delta M/M$ is to increase the magnetic field which has many additional advantages. This can be achieved by using a superconducting coil to excite the magnet.

An important component of the event signature is the penetration of the MSE through the calorimeter. Even if the particles have a hadronic interaction cross section, because of their large mass, kinematic considerations do not allow the transfer of significant energy to a single nucleon. In fact

$$T_{\max} \sim \frac{2\gamma^2\beta^2 m_N}{1 + 2\gamma m_N/M} \sim 2m_N \; .$$

Thus, even after 20 interactions MSE's lose on the average only 20 GeV; this is only a few percent of their total energy. Thus, if we designate the calorimetric energy by E, the ratio of p/E for MSE's is typical of non-interacting particles. Furthermore, one should be able to track the MSE's all the way through the calorimeter, in close analogy to muons.

The basic trigger is a raw tof signal outside the $\beta = 1$ bucket, and penetration (not energy loss) through the calorimeter. Level 2 trigger is a refined tof signal, and an associated high momentum track which has penetrated through the calorimeter.

To select MSE candidates off-line, one first imposes a selection on $p_T > 50$ GeV. This reduces the muon rate through the calorimeter by a factor of 10^3. Thus the ratio of μ/MSE $\sim$ 600 without any tof selection. For a sample of 600 relativistic particles the tail of the tof distribution should end at approximately 5σ, namely at $\Delta t = 500$ psec. We are looking for signals around $\Delta t = 8$ nsec and so should be clearly resolved. Further selection is achieved by rejecting tracks with $p/E \sim 1$ since they must be hadron or electron showers. To exclude confusion from secondary particles (wall scatterings, decays) we demand target pointing and compare the dE/dx in the scintillator with the β from tof.

For neutral MSE's we have fewer handles and expect to see an interaction in the calorimeter followed often, but not always, by a penetrating track. This is of course akin to a neutrino interaction and unless the interaction cross section is hadronic, there is no observable rate. We distinguish neutral MSE's from neutrinos by their long flight time and by the small energy deposition in the calorimeter.

Finally, we note that in most models MSE's are produced in pairs. The pair is fairly colinear in ϕ but the two particles are much less correlated in rapidity. From the ISAJET calculation we find that the probability of detecting both 1 TeV particles in the two arms is

$$N_{\text{pairs}} = 0.02 N_{\text{singles}} \sim 100/\text{yr} \; .$$

This would of course provide a powerful constraint and argues for the use of a double arm system.

Of particular interest is another possible class of exotic particles such as reported now and again in cosmic ray experiments. These are weakly interacting particles $\sigma_{\text{int}} \sim 10^{-38}$ cm^2, but have a moderate lifetime. In view of their weak interaction the probability of interacting in the calorimeter is of order $p_{\text{int}} \sim 10^{-10}$. Thus if they are produced with cross- sections typical of MSE's they would never be detected. However if their lifetime is in the range $10^{-9} < \tau < 10^{-6}$ sec a significant number of decays would take place in the magnetic channel and thus would provide a signature for the detection of the parent, whether charged or neutral.

3.5. Location, Cost and Conclusions

As already stated, the experiment requires a high luminosity IR which will be dedicated to exotics experiments. The clearance transverse to the beam is 12 m on either side but along the beam the requirements are modest, roughly 10 m. This would permit the construction of a hall with a small span even though building small halls around any IR may be a false economy. In fact it may be desirable to share the IR with other experiments especially those designed in the forward/backward direction.

The same IR is ideally suited for mounting the planes of silicon pixel detectors for the fractional charge search. Those detectors can provide important tracking and dE/dx information for the MSE experiment and conversely the spectrometer can study final states in conjunction with triggers from the silicon detector system. The restricted aperture of the spectrometer does not affect the silicon experiment which could cover a larger solid angle if the economics permit it.

We make a rough estimate for the cost of one arm as follows (in millions of $):

Magnet Iron	1.2
Aluminum Coil	0.4
Calorimeter Fe	0.6
Calorimeter Scintillator	0.4
tof and Tracking	0.8
Totals	3.4 M$

For a small incremental cost one could use a superconducting coil to excite part of the magnet. Similarly the calorimeter could be upgraded by using Pb instead of Fe plates. Therefore the overall cost for the two arms could be set in round figures in the vicinity of 10 M$.

A similar spectrometer was built at Fermilab and designed to handle 1 TeV incident protons. The experience from that detector in terms of background, maximum operating rates, resolution, cost etc would be very useful in the design of the proposed spectrometer. In fact one could consider using the major parts of that system for one of the two arms.

We conclude by summarizing the capabilities and special features of the double arm spectrometer that we propose.

(*a*) In one year of operation a limit on MSE's ($M > 100$ GeV, $\tau > 10^{-7}$ sec) can be set at

$$\sigma_T < 50 \text{ fb} .$$

This is about 2×10^6 times below the theoretical prediction for $M = 250$ GeV and 2000 times at $M = 1$ TeV.

(*b*) Unstable massive particles $10^{-9} < \tau < 10^{-6}$ sec are detected with high efficiency.

(*c*) The detector is perfectly matched to the quark search experiment and can provide a good trigger for that experiment or for other exotic searches.

(*d*) The detector offers simplicity, low cost and an alternative to the closed geometry detectors.

(*e*) The detector can (probably) be shielded against neutrons, it can take the full luminosity and it can be later upgraded for higher p_T cut-off. It can be adapted to lower momentum measurements if this is desirable for special experiments (such as search for a quark-gluon plasma).

4. The Search for Free Quarks (or other Fractionally Charged Particles)

In this section we will discuss the search for quarks at the SSC. An extensive review of quark search experiments can be found in the report of Lyons [9].

4.1. Examples of Fractionally Charged Particles

Fractionally charged particles may arise in different ways and have, therefore, different production and interaction properties. Some possibilities are:

(1) *Free Quarks.* Perhaps with a sufficiently energetic, high p_T $gg \to q\bar{q}$, $q\bar{q} \to q\bar{q}$, or $qq \to qq$ collision, quark confinement can be broken and free quarks liberated. As there is no realistic model for this process, the production properties are uncertain, but one might be tempted to look at the particles inside the jets produced in high p_T two-jet events. If free quarks are produced in $qq \to qq$ collisions, they may leave behind fractionally charged spectators in the foward direction. The interaction properties of free quarks are likewise very uncertain, but the suggestion has been made that they may be highly interacting [10].

(2) *Fractionally Charged Hadrons.* These new hadrons could be formed as $q_I qq$ baryons or $q_I \bar{q}$ mesons from new quarks q_I with integer charge, and standard, fractionally charged quarks q. In this case, quark confinement would be preserved. Production could be via gluon fusion, with production properties similar to those discussed for massive stable particles. We would expect the interaction cross section to be hadronic. If the fractionally charged hadron is very massive, then only a small energy transfer to a nucleon is kinematically possible in a single collision, as discussed in section 3.4.

(3) *Fractionally Charged Leptons.* If fractionally charged leptons exist, they would be produced dominantly via the Drell-Yan mechanism (low masses, $M \lesssim 200$ GeV) or gluon fusion (high masses). They would interact electromagnetically in the detector, mainly through ionization energy loss.

In the next sections we will discuss both passive and active searches for fractional charge. Most of our discussion about active quark searches applies as well to the search for multiply charged particles.

4.2. Passive Search Experiments

In a passive type of quark search experiment, one tries to capture quarks emerging from the interaction region in bulk matter placed around the IR (including some inside the beampipe), and then identify the captured quark with stable matter fractional charge search techniques. This approach may be particularly useful for quarks which are very highly interacting, and, therefore, have a reasonable chance to interact and stop in a limited amount of material. In order to reduce the amount of material that must be processed, one may try to concentrate the quarks with chemical or other techniques. Some specific examples of passive experiments that have been carried out at accelerators are given below.

(1) Capture the quarks in tanks of mercury. Concentrate the quarks by distillation of the mercury. The quarks should be left in the residue – a quarked atom will

be attracted to its image charge and will not evaporate when heated. Find the quarks by passing the residue through an automated version of Millikan's oil-drop experiment [11].

(2) Capture the quarks in tanks of insulating liquid. Each tank contains two gold-plated glass fibers with an electrical potential between them. The quarks will concentrate by migrating to one of these charged electrodes. Dissolve the gold-plating in mercury and pass this material through the automated Millikan apparatus [11].

(3) Capture the quarks in small iron balls (these could be inside the beampipe). Search for fractional charge on these balls with a magnetic levitation apparatus [12].

In this type of experiment, if a quark is found, great. If no quark is found, then the interpretation in terms of an upper limit on the production cross section is difficult because the capture probability is unknown.

4.3. Active Search Experiments

The primary handle we will consider in the active quark search experiments is the ionization energy loss rate of the particle. To reduce the smearing and momentum dependent effects of the Landau tail, we will use the *most probable* energy loss rate $(dE/dx)_{MP}$, and not the *mean* energy loss rate.

The most probable energy loss is given approximately by the following expression:

$$(dE/dx)_{MP} \approx (Q/\beta)^2 F(Q/\beta, \beta, t)$$

where F is a slowly varying function of Q/β, β, and the thickness t of the detector. $(dE/dx)_{MP}$ in 1 mm thick silicon is plotted as a function of $\beta\gamma = p/m$ in Fig. 8. In practice, the most probable energy loss is often approximated by the *truncated mean* energy loss, $(dE/dx)_{TM}$. This is calculated by measuring dE/dx N times along the track, and then finding the average value of the lowest fN measurements, where f is typically in the range of 0.6 to 0.7.

For relativistic quarks with charge $Q = 1/3$ or $2/3$, the measurement of $(dE/dx)_{MP}$ alone, with values $\approx 1/9$ or $4/9$ that of a minimum ionizing particle, should be sufficient to identify the fractional charge. However, for particles with $Q > 1$ or $\beta < 1$, $(dE/dx)_{MP}$ can be in the allowed region for normally charged particles. Then additional information is needed for proper identification. For example, since

$$Q \sim \beta\sqrt{(dE/dx)_{MP}}$$

(neglecting the small correction due to F), a measurement of the velocity, along with $(dE/dx)_{MP}$, will give directly the particle charge.

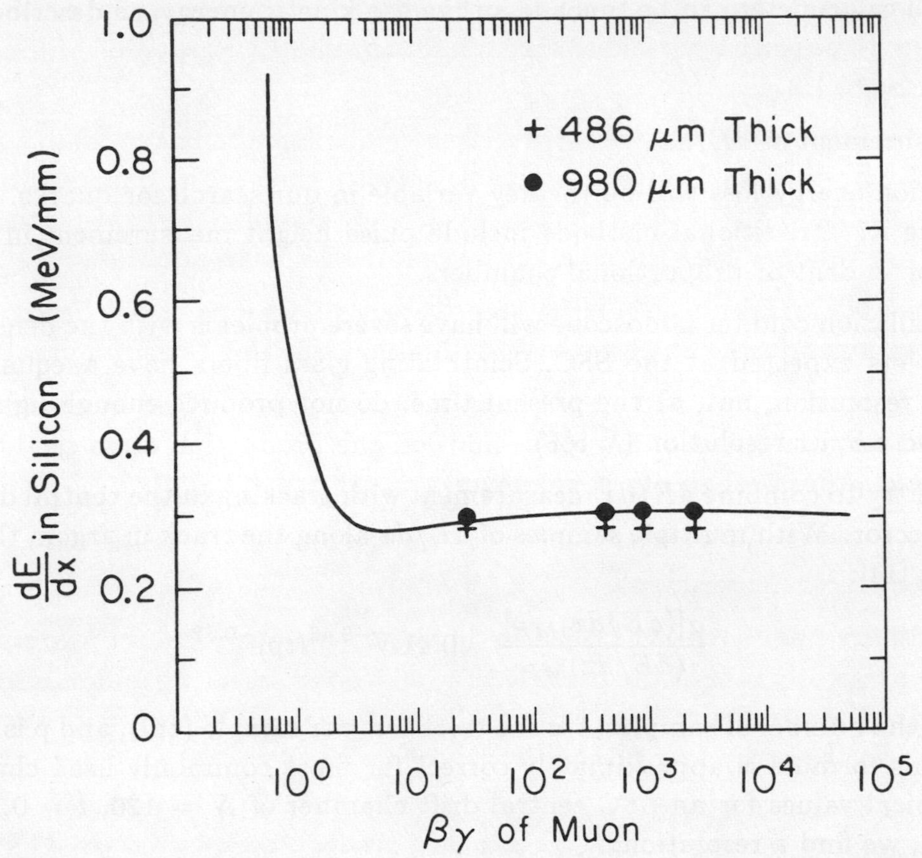

Fig. 8. **The most probable energy loss in 1 mm thick silicon as a function of $\beta\gamma$. Experimental data taken with $980\mu m$ and $486\mu m$ thick silicon detectors are also shown [13].**

Another measurement of interest is the apparent momentum p_A. This is the momentum of the particle as derived from the track curvature in a magnetic field, assuming that $Q = 1$. It is related to the true particle momentum p_{true} by

$$p_A = p_{true}/Q.$$

To search for quarks, we look for values of $(p_A, (dE/dx)_{MP})$ which are inconsistent with known particles. Section 4.5 will demonstrate this approach in more detail. By the way, this technique will not work when $Q/\beta \approx 1$ as there will be overlap with the normally charged particles.

Some other handles that might be used to help establish the existence of something new:

- Measure the mass of the particle, as discussed in section 3.4.
- Look for an inconsistency between the particle momentum and the measured calorimetric energy of the particle. Since $p_{true} = Qp_A$, the measured momentum can differ substantially from the true momentum. For massive particles, the energy deposited

in the calorimeter can be much less than the kinetic energy, as described in section 3.4.

4.4. Measurement of dE/dx

Ionization energy loss will be the key variable in our search for quarks. How should we measure it? Traditional methods include pulse height measurement in scintillation counters, or in drift or proportional chambers.

A scintillation counter hodoscope will have severe problems with the density of tracks inside the jets expected at the SSC. Scintillating glass fibers have adequate two-track separation resolution, but, at the present time, do not produce enough light to provide the required dE/dx resolution ($\lesssim 5\%$).

We can try to combine dE/dx measurement with tracking in the central drift chamber of a 4π detector. With multiple samples of dE/dx along the track in argon, the resolution is given by [14]:

$$\frac{\sigma[(dE/dx)_{MP}]}{(dE/dx)_{MP}} = 0.41N^{-0.46}(tp)^{-0.32}$$

where N is the number of samples, t is the thickness per sample (cm), and p is the pressure (atm.). This formula is approximately correct for most commonly used chamber gases. Taking typical values for an SSC central drift chamber of $N = 120$, $t = 0.725$ cm, and $p = 1$ atm., we find a resolution:

$$\frac{\sigma[(dE/dx)_{MP}]}{(dE/dx)_{MP}} = 5.0\%.$$

This should be adequate for the quark search. However, this approach requires dE/dx measurement capability with low noise, low pulse height electronics on the drift chamber wires. In addition, it is not yet clear that a practical drift chamber system can be constructed with segmentation fine enough to allow the tracking and dE/dx measurement for the particles inside the high energy jets.

We propose a different scheme to search for quarks inside jets that should easily solve the segmentation problem while providing superior dE/dx resolution. The method is to use a telescope of silicon pixel detectors, as will be described in the next section.

4.5. Silicon Pixel Detectors

Silicon pixel detectors have been extensively discussed at this workshop by Shapiro [15], Nygren [16], and Parker [17]. Briefly, a silicon pixel detector consists of a two-dimensional array of separate detector elements (pixels) fabricated on a high purity silicon wafer. Each pixel consists of a pn diode, depleted by application of an external, reverse-bias potential.

When a charged particle passes through the pixel, electron-hole pairs are produced at the rate of one pair per 3.6 ev of deposited energy. The electrons and holes are separated by the internal electric field, and the resulting current pulse can be detected by the read-out electronics. To give some specific dimensions, the detectors being fabricated for Shapiro *et al.* are 256×256 arrays of 30×30 μm^2 pixels on 300 μm thick silicon [15]. In the future, larger detectors may be possible. In any case, it will be possible to cover the area we have in mind for our quark detector ($\approx$ 70 cm^2 per layer) with a mosaic of smaller detectors.

The read-out of the detector can be done with a custom-designed integrated circuit consisting of a two-dimensional array of amplifiers and storage elements that matches the pixel array. Each pixel can be connected to the corresponding amplifier (even on a 30 μm grid!) by bonding the detector to the read-out chip with indium bump bonding. This technology is well developed, and complete detector/read-out assemblies have been produced [18]. Shapiro *et al.* are planning to use an existing read-out chip which provides random access to each pixel [15]. Nygren *et al.* are developing a chip with an architecture suitable for read-out at the SSC [16].

In this architecture, when a pixel is hit (1) the analog signal is stored in the corresponding amplifier cell; (2) a bit corresponding to the pixel row and the beam crossing number is set in a shift register; and (3) likewise for the pixel column. When a valid trigger (due to beam crossing N) arrives, the cells at the intersections of all rows and columns which had hits during beam crossing N, are interrogated. If a cell corresponding to a hit row and a hit column has a signal, then it is read out and digitized. A cell at a "ghost" combination will not have a signal, except for the rare case when a particle at a later beam crossing, but before the trigger arrival, coincidentally passed through this pixel.

A different technological approach is being taken by Parker [17]. He plans to integrate the read-out electronics with the pixel detectors on the same silicon wafer.

4.6. Silicon Pixel Detectors as Quark Detectors

Now we will begin to examine the utility of silicon pixel detectors in the search for quarks at the SSC. First, we will look at the ability of these devices to track individual particles inside high energy jets. We will consider jets with $p_T = 1$ TeV impinging at normal incidence on a pixel detector located at a perpendicular distance of 10 cm from the beamline. Fig. 9 shows the fraction of events in which all tracks are resolved versus the size of the square pixels [16]. With pixel dimensions 100×100 μm^2, each and every track can be resolved in 90% of the events. This should be quite adequate for the quark search experiment.

Next, we will consider the signal, noise, and resolution for the dE/dx measurement in 300 μm thick silicon detectors. The most probable ionization energy loss for $Q = 1$,

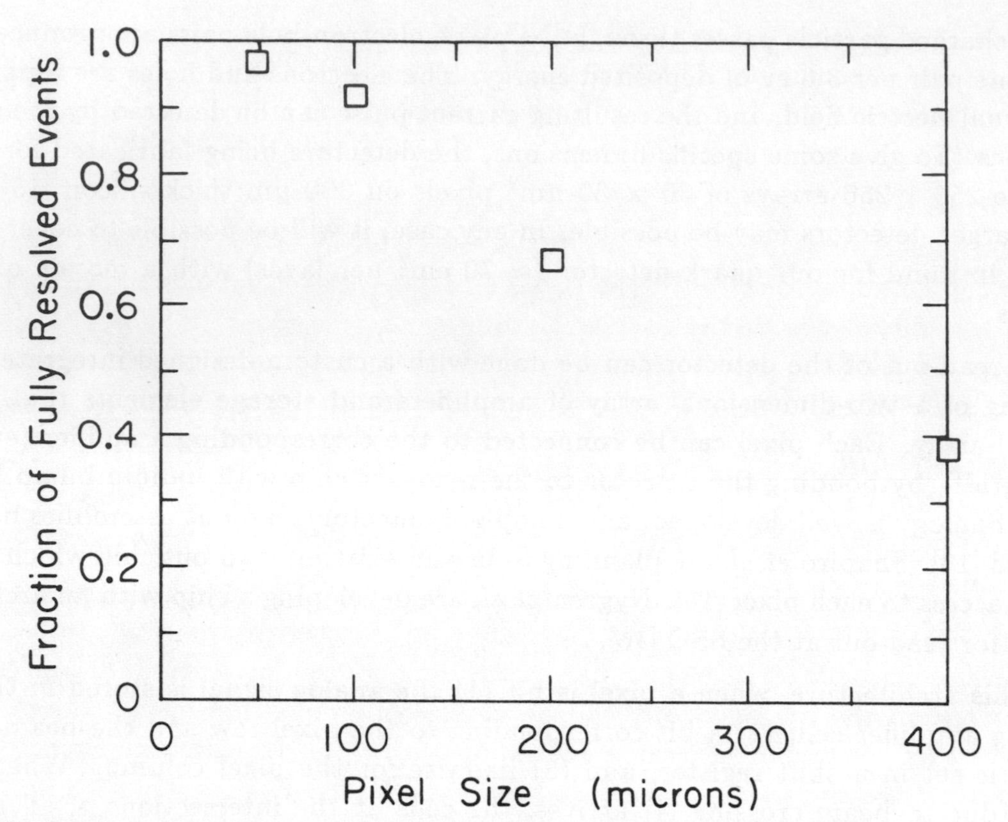

Fig. 9. **The fraction of jets with all tracks resolved versus pixel size. The jets have $p_T = 1$ TeV and are normally incident on a pixel detector 10 cm from the beamline [16].**

$\beta \simeq 1$ particles at this thickness is about 285 kev/mm. Since it takes on average 3.6 ev to produce an electron-hole pair, this implies a total signal of about 24000 electrons (or holes).

The noise on this signal comes from three major, unavoidable sources: statistical fluctuations on the number of electron-hole pairs produced, electronic noise in the read-out amplifier, and Landau fluctuations. If the electron-hole pairs were created in a statistically independent way, then Poisson statistics would apply and the corresponding resolution would be $1/\sqrt{24000}$ or 0.6%. In fact, these processes are *not* statistically independent and the measured statistical fluctuations in silicon detectors are *smaller* than Poisson. The resolution is given by $\sqrt{F_a/N}$ where F_a is the Fano factor, and $F_a \approx 0.1$ for silicon. In any case, the contribution to the noise from these fluctuations are negligible compared to the Landau fluctuations, as we shall see shortly. This is true even for particles with $Q = 1/3$, which will have a resolution $3\times$ bigger.

Due to the small size of the pixel, and therefore small capacitance, the electronic noise should be very small. The read-out chip used by Shapiro *et al.*, bonded to 120×120 μm^2 pixels, has a measured equivalent noise charge (ENC) of 50 electrons rms at a temperature of 10° K. This extrapolates to about 300 electrons rms at room temperature, for a resolution of 1.3% for $Q = 1$ particles, or 11% for $Q = 1/3$. The chip being designed by Nygren *et al.* is expected to have an ENC of 65 electrons rms when bonded to 100×100 μm^2 pixels for a resolution of 0.3% or 2.4% for $Q = 1$ or $Q = 1/3$, respectively. Again, these resolutions are small (first chip) or negligible (second chip) compared to the Landau fluctuations, which we will examine next.

The expected ionization energy loss distribution for $Q = 1$, $\beta \simeq 1$ particles traversing at normal incidence through 300 μm of silicon is shown in Fig. 10. It is calculated by convoluting a Landau distribution with a Gaussian. The Gaussian smearing takes into account the effects due to the atomic binding of the electrons. The σ of the Gaussian is taken to be 5.7 keV, as determined experimentally [19]. The Landau distribution by itself has a FWHM of 26%. The convoluted distribution has a FWHM of 32%.

When Q and/or β are not equal to one, we scale σ as $\sigma = (Q/\beta) \times 5.7$ keV, and correct also the Landau distribution. For fractionally charged particles with $\beta = 1$ the convoluted distributions have FWHM of 65%, 39%, and 29% for Q =1/3, 2/3, and 4/3, respectively.

As mentioned in section 4.3, a convenient way to measure the most probable energy loss is to use the truncated mean. The quark detector we have in mind (and which will be described in the next section) will be a telescope consisting of 20 planes of silicon pixel detectors. Thus the ionization energy loss for a particle will be measured 20 times along its trajectory. The truncated mean energy loss (ΔE_{TM}) is calculated by taking the average of the lowest 70% of the 20 measurements. The distribution of ΔE_{TM}, derived from the distribution of Fig. 10, is shown in Fig. 11. The Gaussian fit drawn has a σ given by $\sigma(\Delta E_{TM})/\Delta E_{TM} = 4.2\%$. It is this 4.2% that we will take to be the resolution of our dE/dx measurement for $Q = 1$, $\beta \simeq 1$ particles. For fractionally charged particles with $\beta \simeq 1$, we have $\sigma(\Delta E_{TM})/\Delta E_{TM}$ =8.1%, 4.8%, and 3.8%, for Q =1/3, 2/3, and 4/3, respectively.

Now that we know our dE/dx resolution, we can begin to evaluate the expected separation between fractionally charged particles and the enormous background of normally charged particles. This separation is illustrated in Fig. 12. In (a) we plot the most probable energy loss ΔE_{MP} in 300 μm thick silicon versus apparent momentum p_A for the $Q = 1$ particles, as well as for particles with $Q = 1/3$, 2/3, and 4/3, and masses of 100 GeV and 1000 GeV. The band represents a $\pm 10\sigma$ region about ΔE_{MP} for the normally charged particles, with σ taken to correspond to the resolution of 4.2%.

To add some dynamics to the curves in 12(a), we used ISAJET to generate events containing heavy quarks of 1000 GeV mass, produced via gluon fusion in 40 TeV pp collisions. The angular region was restricted to be $\pm 10°$ around $\theta = 90°$. In 12(b), (c), and (d), we plot $(p_A, \Delta E_{MP})$ for the emerging heavy hadrons, assuming $Q = 1/3$, 2/3, and 4/3, respectively. The fractions of events lying outside the background band are, respectively, 0.97, 0.51, and 1.0. To separate out the events within the background band, velocity measurement is required.

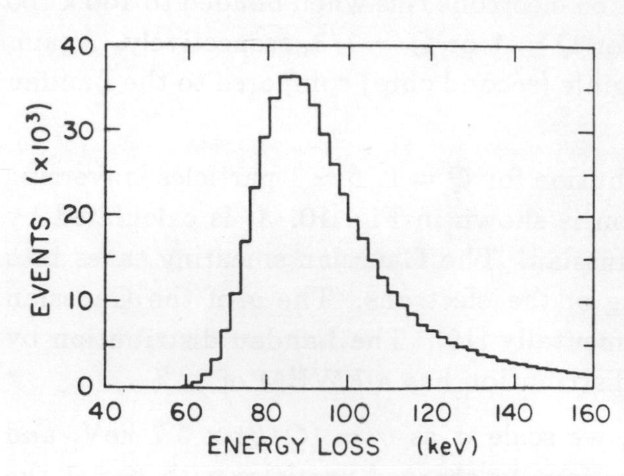

Fig. 10. **The ionization energy loss distribution for $Q = 1$, $\beta \simeq 1$ particles in 300 μm thick silicon. The FWHM of the distribution is 32%.**

Fig. 11. **The truncated mean energy loss distribution for $Q = 1$, $\beta \simeq 1$ particles in 20 layers of 300 μm thick silicon. The Gaussian fit has $\sigma/E = 4.2\%$.**

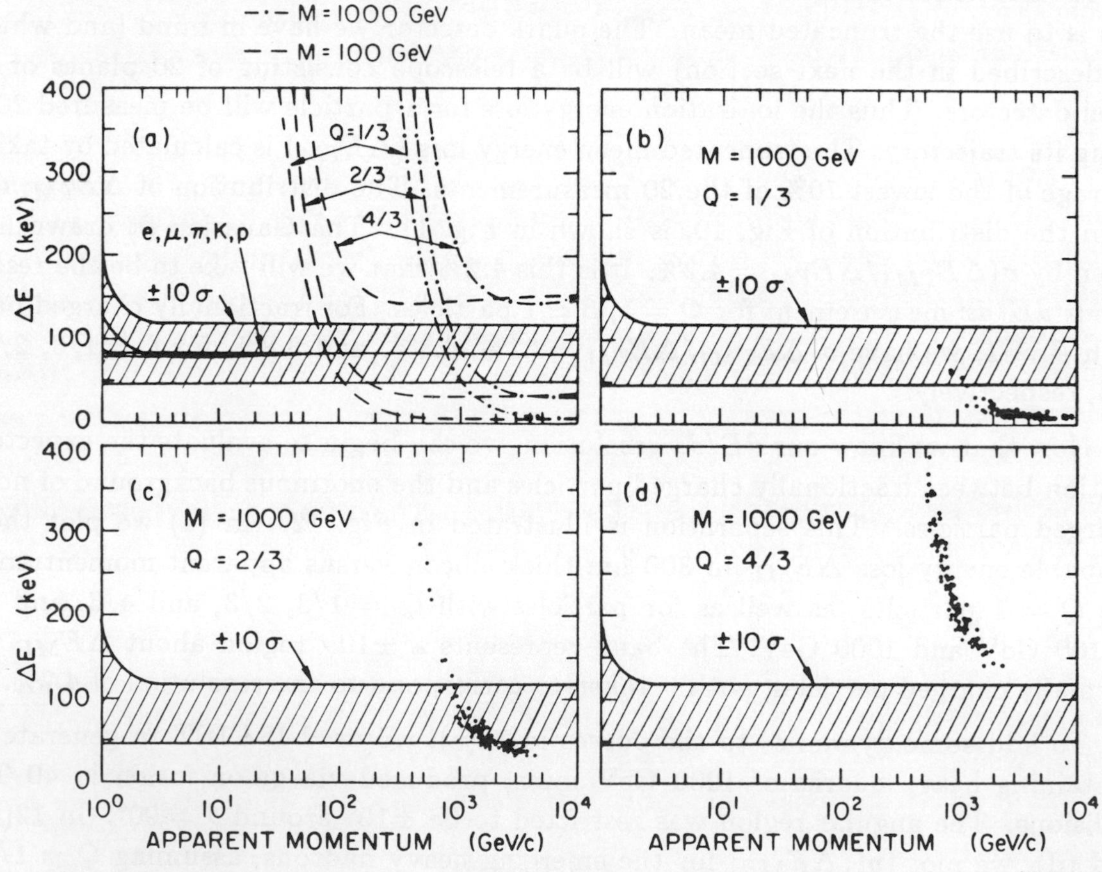

Fig. 12. The most probable energy loss in 300 μm thick silicon versus apparent momentum; (b), (c), and (d) show events generated by ISAJET. Resolution smearing for both ΔE_{MP} and p_A are included, with $\sigma(p_A)/p_A = 0.1p_A$ (p_A in TeV). The band is a $\pm 10\sigma$ region about ΔE_{MP} for the $Q = 1$ stable particles, with $\sigma = 4.2\%$.

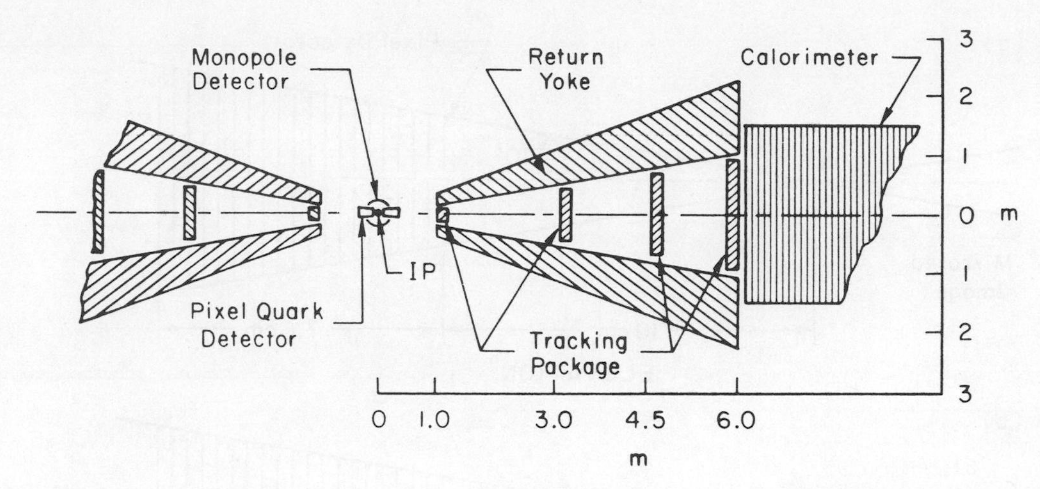

Fig. 13. **A detector for magnetic monopoles, free quarks, and massive stable exotics.**

In a world of Gaussian statistics, the $\pm 10\sigma$ band would contain all the $Q = 1$ particles. In the real world, systematic effects, non-Gaussian tails, zoo-ons, etc. will generate the background [20]. The best weapon against these backgrounds is to have as much information, as much redundancy, and as many consistency checks as possible.

4.7. A Quark Detector for the SSC

The proposed quark detector will have two identical telescopes. Each telescope will consist of twenty planes of silicon pixel detectors, and will be placed in front of one arm of the massive stable exotics spectrometer (see Fig. 13). By combining together the quark telescope with the MSE spectrometer, each will benefit from the additional information and redundancy. The MSE spectrometer will gain tracking data in the 20 pixel planes. In addition, the dE/dx measurement in the quark telescope will provide an important consistency check on the velocity measurement in the spectrometer: since $dE/dx \propto 1/\beta^2$, the 4.2% measurement of dE/dx corresponds to a 2.1% measurement of β.

The spectrometer will be used in the quark search to measure particle velocity, momentum, and energy deposition in the calorimeter. It will also provide triggers based on low velocity or a high p_T jet in the calorimeter.

The quark telescope is shown in more detail in Fig. 14. Each of the 20 planes will consist of a 300 μm thick silicon pixel detector with square pixels 100 μm on a side. The read-out electronics will be indium bump bonded to the detector, or integrated on the detector wafer. The architecture of the read-out will be similar to that of Nygren *et al.* [16], but with additional logic for the generation of a fast quark trigger (see below).

The telescope will be placed 10 cm from the beamline. This is far enough away so that electronics hardened to ≥ 1 Mrad (which appears feasible) will survive the radiation dosage, but close enough so that the total area of silicon is not unreasonable. The solid

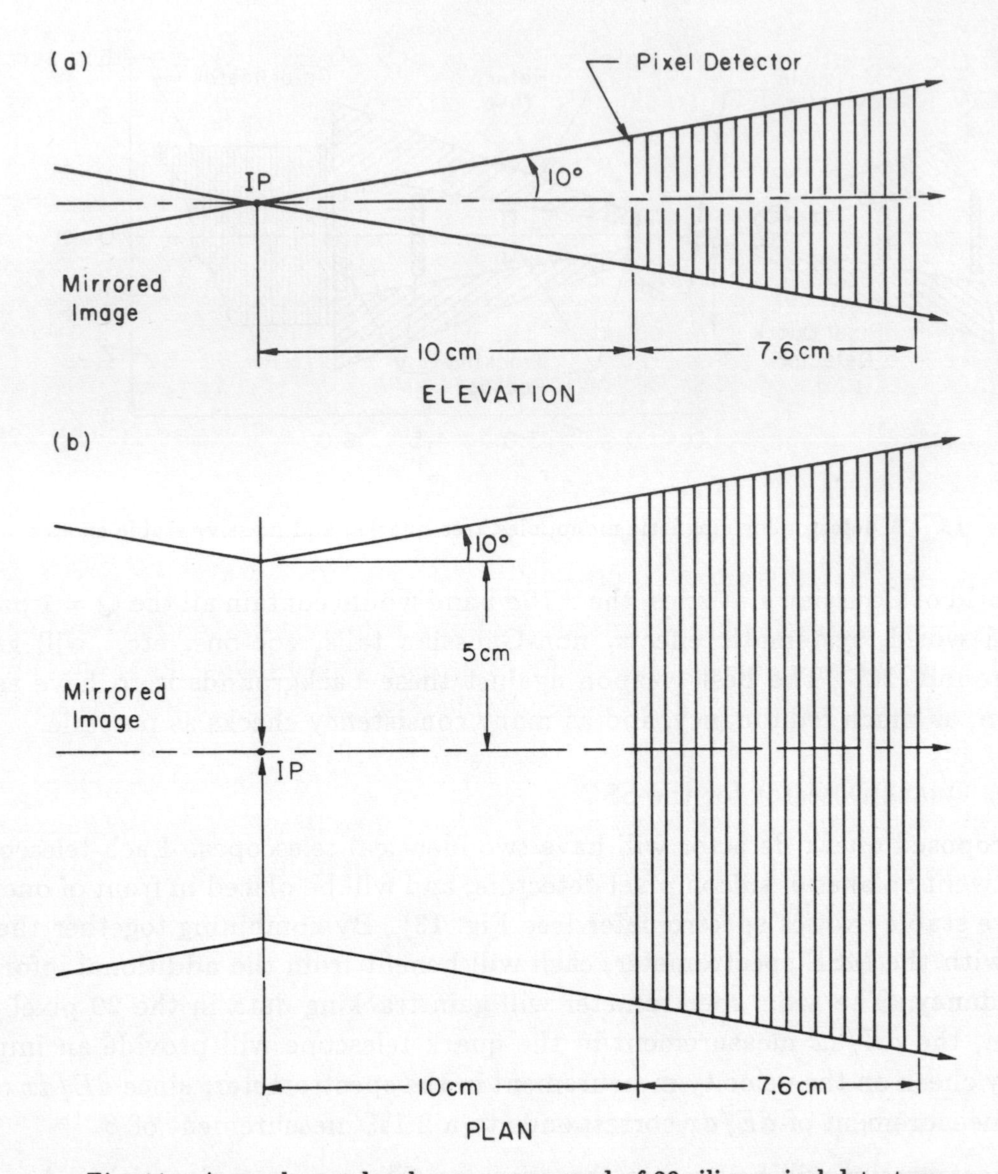

Fig. 14. **A quark search telescope composed of 20 silicon pixel detectors.**

angle covered will be the same as that of the MSE spectrometer, that is:

$$80^\circ < \theta < 100^\circ \quad \text{and} \quad -10^\circ < \phi < +10^\circ .$$

The telescope is lengthened along the beam direction to accommodate the ±5 cm length of the interaction region.

Tracking:

It should be stressed that, in addition to providing a measurement of dE/dx for each particle, the quark telescope will provide excellent tracking. Each track will have 20 space points, and the resolution of each point will be $\approx 30 \times 30\ \mu m^2$. As noted in section 4.6, 90% of 1 TeV jets will have all tracks resolved.

Trigger:

The trigger will be based on data from the MSE spectrometer as well as from the quark telescope. Here are some trigger conditions that should be ORed to form the master trigger:

(1) a low velocity particle is detected by the MSE spectrometer;

(2) a high p_T jet is detected in the MSE calorimeter;

(3) an abnormally low (or high) value for particle dE/dx is measured by the quark telescope.

The dE/dx trigger (3) will require logic that picks out a pixel with abnormal pulse height, but which insures that this signal is not just a spill-over from a near-by $Q = 1$ particle.

The first part of trigger (3) will need to be implemented on the read-out chip. The pixel signal must be in an acceptable range. The sum of the pixel signal and the signals from its 8 neighbors must also be in an acceptable range. If these conditions are satisfied, then a row bit and a column bit corresponding to this pixel are set. An OR of the row bits on a given pixel detector defines a plane bit. External to the read-out chips, majority logic on the 20 plane bits provides the level 1 trigger. The level 2 trigger requires that the row bits in the 20 planes form at least one good track pointing to the interaction region, AND that the same is true for the column bits.

The above trigger (3) only illustrates one possibility. The optimal form of this trigger, and its implementation in silicon, needs further study.

Material, Area, Power, and Cost:

300 μm thick silicon has a radiation length of 0.32%. With the detector and read-out chip of this thickness, and assuming any mechanical support, traces for bus lines, etc., would also be of this order, then the total material for 20 layers is about 19% of a radiation length (the indium for the bump bonding is completely negligible). If the read-out is integrated with the detector, this would be reduced to about 13%.

The total area of silicon is 1460 cm^2 per telescope. At 0.12 W/cm^2 (the estimate of Nygren *et al.*), the total power dissipation would be 175 W per telescope. This estimate should be increased somewhat to take account of the additional trigger logic circuitry, but in any case, with lots of open space, this should be easy to cool. The equipment cost will be dominated by the cost of the silicon detectors. Taking \$130/cm^2 (the present cost for silicon microstrip detectors), this gives \$190k for each telescope. A very rough estimate of the total cost is 1 M\$.

Background Flux and Signal Cross Section Limit:

In one SSC year, with an integrated luminosity of 10^{40} cm^{-2}, about 10^{14} particles will pass through each quark telescope. With a minimum p_T cut of 7 GeV, which can be imposed by demanding that the particle survives the magnetic channel of the MSE spectrometer and reaches the front part of the calorimeter, then this number is reduced to 6×10^{10}. This should be a manageable background flux.

Taking the ISAJET acceptance (given in section 3.3) of 0.004 per telescope for a particle with 1 TeV mass, a detection efficiency of 50%, an integrated luminosity of 10^{40} cm^{-2}, and assuming there are no quark candidates found in either telescope, then the cross section limit can be set at 60 fb.

5. Beam Dump Experiments and the Search for Highly Penetrating Particles

5.1. Introduction

Rare weakly interacting particles can be better identified if the majority of the hadrons produced in the collision are absorbed. Furthermore copious yield of secondary penetrating particles can signal the production of short-lived parents. In such "beam dump" experiments the primary beam is directed onto a dense absorber of suitable length so that only prompt penetrating particles emerge at the other end. Beam dump experiments are in principle simple and can be made sensitive to very small production cross-sections.

At the SSC, dumping the beam at full energy produces a c.m. energy $\sqrt{s} = 200$ GeV which, of course, is much smaller than the available collision energy of 40 TeV. Assuming single turn extraction as in the abort system, the beam could be dumped in 300 μsec. This is much too short to allow particle counting and the detector would be swamped by neutrino interactions. Since no slow extraction is foreseen for the SSC, the alternative is to use a gas jet target which would diffract a small portion of the internal beam into an extraction channel. This scheme has been advocated for establishing low intensity test beams in the 10–20 TeV range. The gas jet, however, can also be considered as the first element of the beam dump. By adjusting the jet density one can control the rate of penetrating particles at the detector.

One particularly attractive feature of a beam dump experiment at the SSC, is the high degree of collimation of the secondaries. Thus the distance between target and detector can be made long without forcing the transverse dimensions of the detector to unrealistic size. For instance, 1 TeV secondaries with a mean transverse momentum $\langle p_T \rangle \sim 0.5$ GeV are contained in a cone of angle $\theta \sim 0.5$ mrad. Thus at 1 km a 1×1 m^2 detector would contain half of the secondaries.

Finally a gas jet target offers the obvious advantage of making possible the study of p-Nucleus collisions at 20 TeV laboratory energy. In terms of a heavy ion collider this is in the same energy range as for RHIC.

5.2. The Targeting System

The target consists of a high pressure gas jet using hydrogen gas or heavy elements in gaseous form. Such jet targets have been operated extensively in the Fermilab internal beam [21] and more recently in the SPS collider. The jet density falls off with distance from the axis and the cross section can be assumed circular; the jet diverges as the distance from the nozzle increases. Typical parameters are: waist of 0.5 cm diameter and density of $\sim 2 \times 10^{-7}$ g/cm^2. For 10^{14} circulating protons this leads to a luminosity $L \simeq 2 \times 10^{34}$ cm^{-2} sec^{-1}.

The technical problem with jet targets is to prevent the gas from diffusing into the beam pipe. This is achieved by differential pumping and by cryopumping. The Fermilab jet operated in a warm machine and with a duty cycle of 10–20%; continuous operation should be possible without serious difficulties. Another consideration is the radiation load on the downstream elements of the lattice. The load is about an order of magnitude higher than at the IR's and the particle spray is more collimated. Thus some radiation hardening may be necessary.

The jet should be placed in a "dog-leg" as shown in Fig. 15. The targeting angle is $\theta = 0.5$ mrad with respect to the beam direction. Neutrals produced in the forward direction intercept the magnet yoke at about 10 m downstream; this flight distance could be reduced. Positive particles are swept away and are clearly separated from the forward direction.

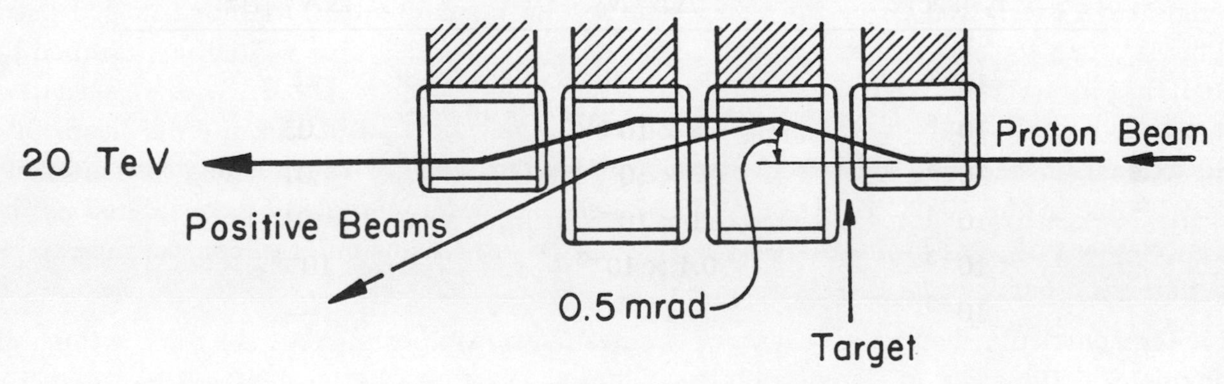

Fig. 15. A possible "dog-leg" for inserting a gas jet target in the SSC beam.

5.3. The Detector

The detector can be placed at a distance of 1 to 3 km from the target, at the machine depth. As a detector we consider a standard calorimeter, 3×3 m^2 in cross section and 10 m deep, for a total mass of 1 kton. The neutrino cross-section at 1 TeV is $\sigma_{\nu N} \sim 2 \times 10^{-36}$ cm^2 so the interaction probability in the calorimeter is $\sim 10^{-8}$. If we assume that there are 0.1 prompt neutrinos produced in the forward direction per pp interaction, and $\sigma_T \sim 10^{-25}$ cm^2, then the rate of neutrino interactions in the detector is ~ 2 Hz. This shows that even if the target to absorber distance is kept short still there is significant neutrino background.

We are primarily interested in neutral particles x which are sufficiently long-lived to reach the detector. We assume a production cross-section $\sigma(x) \simeq 1$ nb and a detector acceptance of 0.5, so that the rate through the detector is ~ 10 Hz. If the x-particles interact weakly (at the $\sigma_{\nu N}$ level) then the interaction rate in the detector is 10^{-7} Hz, which is hopeless. This is why such particles would have escaped detection at the Tevatron and SPS collider even if produced with a 1 nb cross section.

Instead of identifying x-particles by their interactions one could use their decay as their signature. This is possible only for a limited range of lifetimes. The distance from target to detector is ℓ; a decay path is established in front of the detector and it has a length Δ. The decay length of the x-particle is $\lambda = c\beta\gamma\tau_x$. The number of decays occuring in Δ is

$$\Delta N = N_0 \frac{\Delta}{\lambda} e^{-\ell/\lambda}$$

where N_0 is the number of particles produced at the target. For $\Delta = 30$ m, $\ell = 10^3$ m and $\beta\gamma \sim 1$ we find the following rate for observable decays. We have assumed that the production rate multiplied by detector acceptance is $R_0 \sim 10$ Hz.

τ_x (sec)	$\Delta N/N_0$	$R(\Delta N)$ [Hz]
10^{-7}	—	—
10^{-6}	5×10^{-3}	.05
10^{-5}	7×10^{-3}	.07
10^{-4}	1×10^{-3}	.01
10^{-3}	0.1×10^{-3}	10^{-3}
10^{-2}	10^{-5}	—

Still, the x-particles must be identified in the background of neutrino interactions in the detector, and of secondary muons. [Note that in 1 km of earth, 0.5 TeV primary muons have ranged out.] A possible handle is the flight time which for massive x-particles should be in between the r.f. buckets.

5.4. Collider Beam Dump

We now consider the equivalent of a "beam dump" using one of the high luminosity IR's. The nearest absorber for the forward interaction products is the yoke of the quadrupole at 20 m; secondaries emitted at $\theta > 1$ mrad are intercepted. The detector

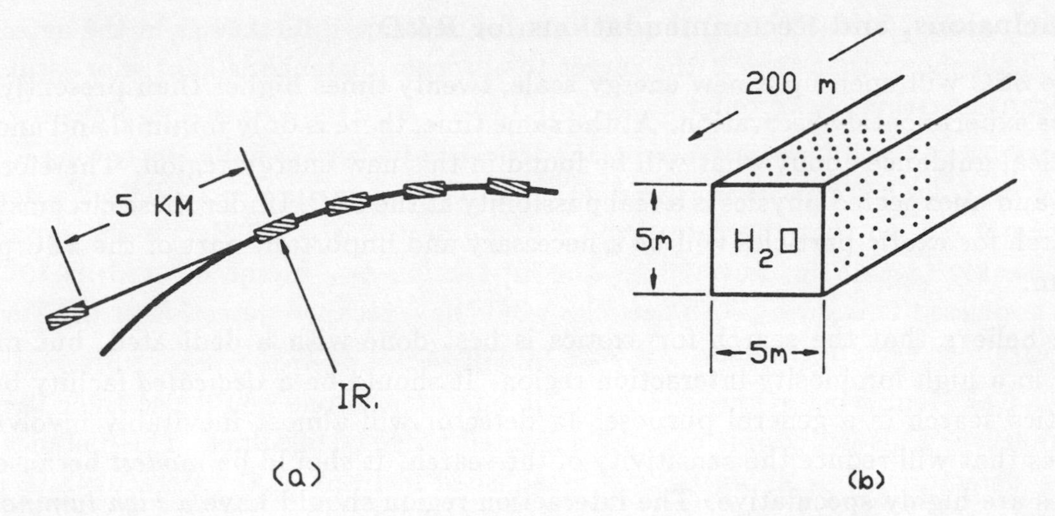

Fig. 16. A collider "beam dump": (a) location of the detector, (b) schematic of the detector.

is centered at $\theta_0 = 1.5$ mr at a distance of 5 km from the IR and at the same depth as the ring. This gives a large offset of approximately 500 m from the ring as shown in Fig. 16(a).

The detector is a water Cherenkov counter 5×5 m^2 in area and 200 m long. 7-inch tubes are placed on a 1×1 m^2 lattice for a total of 4,000 tubes; this gives approximately 3% coverage and ample light collection efficiency. Because of the collimation of the particles of interest, triggering and pattern recognition are relatively simple. The detector covers approximately 1/10 of 2π in azimuth and half a unit of rapidity (for particles with m = 1 GeV); thus an overall acceptance of order $\sim 5 \times 10^{-3}$.

As before we are interested in penetrating particles that survive the 5 km path in the absorber, yet have a finite probability of decay in the detector volume. Since the decay volume is not evacuated, neutrinos will interact and the total material is of the same order as for the calorimeter of the previous section. One will have to distinguish the decaying particles from neutrino interactions on the basis of differing signatures. These are: (a) Timing, since prompt neutrinos will have to be within the r.f. bucket, (b) Most neutrinos will originate from decay of heavy mesons and thus will be of low energy (c) Transverse momentum of the secondaries since the penetrating particles are expected to be massive and (d) Topological differences in the decay *vs.* interaction pattern. Cosmic ray muons do not pose a problem.

This experiment is limited to $10^{-7} \leq \tau \leq 10^{-2}$ sec as in the case of the gas jet target beam dump but it probes much higher c.m. energy, $\sqrt{s} = 40$ TeV. It also provides an excellent detector for measuring inclusive neutrino production since one will be able to reconstruct the neutrino energy from the c.c. interactions. For the length of detector considered here, the water Cherenkov technique appears to be the only realistic approach.

6. Conclusions, and Recommendations for R&D

The SSC will open up a new energy scale, twenty times higher than presently available, for experimental observation. At the same time, there is only minimal and uncertain theoretical guidance about what will be found in this new energy region. Therefore, surprising and unexpected physics is a real possibility at the SSC. Under these circumstances, the search for exotic particles will be a necessary and important part of the SSC physics program.

We believe that the search for exotics is best done with a dedicated, but modest, facility in a high luminosity interaction region. It should be a *dedicated* facility because an exotics search in a general purpose, 4π detector will almost inevitably involve compromises that will reduce the sensitivity of the search. It should be *modest* because these searches are highly speculative. The interaction region should have a *high luminosity* to compensate for the modest solid angle of at least some of the exotic searches.

A specific example of an exotics search detector is shown in Fig. 13. A double arm magnetic spectrometer at 90° measures momentum, velocity, and energy deposition in a hadronic calorimeter to search for massive stable particles. An array of plastic and glass track-etch detectors is used in the search for magnetic monopoles, or other highly ionizing particles. Two telescopes composed of silicon pixel detectors measure ionization energy loss. This measurement of dE/dx, combined with the measurements of velocity and momentum in the spectrometers, is used to search for free quarks, or other fractionally charged particles. This system can also be used to search for multiply charged particles.

The exotics search detector provides flexibility and an open geometry. It can be adapted easily to new physics ideas and/or new detector ideas. It is complementary, in this sense, to the 4π detectors which will be more powerful, but relatively inflexible and resistant to change. As the design of the general purpose detectors will need to be frozen many years ahead of their turn-on date, it would be wise to have an interaction region available which can more easily accommodate new possibilities in physics and detector technology.

The exotics detector can be operated alongside other experiments, such as a forward/backward spectrometer. The interaction region can be used to test new detector prototypes under realistic, high luminosity conditions. After a few years of running, if no exotic particles have been found, it could be replaced with other specialized experiments, or a second generation 4π detector.

In order to have the technology in hand for the best possible exotics detector, we recommend two research and development efforts. First, there should be a systematic search for glass track-etch detectors with low threshold and high radiation resistance. Second, we should move full speed ahead on silicon pixel detectors and the associated radiation-hardened electronics.

We note that our conclusion concerning the non-competitiveness of the general purpose detectors to search for exotics could be modified by future developments. At a future SSC Workshop, with hopefully new and better ideas and improved detector technology, and when the dust has settled a little more on the 4π designs, it may be worthwhile to come back and look again at the feasibility of adapting the general purpose detectors for exotic searches.

We have also discussed in this report two types of "beam dump" experiments to search for highly penetrating particles. In the first type, an internal gas jet target becomes the first element of the beam dump; the center of mass energy for the collisions is 200 GeV. Decaying particles are searched for 1 km downstream from the dump. The gas jet target may have other interesting applications, including the establishment of 10-20 TeV test beams, and the study of p-nucleus collisions at 20 TeV incident momentum. The second type of beam dump experiment looks for decaying particles 5 km downstream and 1.5 mr off-axis from the pp interaction region.

In summary, our study indicates that a variety of interesting and modest searches for exotic particles are both feasible and desirable at the SSC.

Acknowledgements

We would like to thank Chris Adolphsen and Alan Steiner for their help with the dE/dx and ISAJET programs. The ideas on the jet target were developed in conjunction with T. Toohig and others who examined the possibilities for high energy test beams. We very much appreciate all the help we received from Nora Rogers in the preparation of this paper.

REFERENCES

1. Webster's Seventh New Collegiate Dictionary (G. and C. Merriam Co., Springfield, MA, 1967), p. 292.
2. S. Errede *et al.*, "Stable/Exotic Particle Production and Detection at the SSC", 1984 Snowmass SSC Summer Study, p. 175.
3. P. B. Price, "Magnetic Monopoles and Other Highly Ionizing Particles at the SSC", these proceedings.
4. P. B. Price, H.-S. Park, G. Gerbier, J. Drach and M. H. Salamon, *Nucl. Instr. Meth.* **B21**, 60 (1987).
5. P. B. Price, L. M. Cook and A. Marker, *Nature* **325**, 137 (1987).
6. P. B. Price, G. Gerbier, H. S. Park and M. H. Salamon, "Systematics of Annealing of Tracks of Relativistic Nuclei in Phosphate Glass Detectors", *Nucl. Instr. Meth.*, in press (1987).
7. The results reported here were obtained by A. White using ISAJET.

8. See for instance, 1986 SSC Summer Study, p. 450.
9. L. Lyons, *Phys. Rep.* **129**, 225 (1985).
10. A. De Rújula, R. C. Giles and R. L. Jaffe, *Phys. Rev.* **D17**, 285 (1978).
11. H.S. Matis *et al.*, contribution to the *23rd International Conference on High Energy Physics*, Berkeley, CA, July, 1986, and Lawrence Berkeley Lab. Report LBL-21670, July, 1986.
12. L. Lyons *et al.*, contribution to the *23rd International Conference on High Energy Physics*, Berkeley, CA, July, 1986, and Oxford preprint Print-86-1011.
13. E. H. M. Heijne, "Muon Flux Measurement with Silicon Detectors in the CERN Neutrino Beams", CERN report CERN 83-06 (July, 1983), p. 25.
14. W.W.M. Allison and J.H. Cobb, *Ann. Rev. Nucl. Part. Sci.* **30**, 253 (1980).
15. Talk by S. Shapiro at this workshop. The work reported on is being done in collaboration with W. Dunwoodie, J. Arens, J.G. Jernigan, and S. Gaalema.
16. Talk at this workshop by D. Nygren. The work reported on is being done in collaboration with L. Bosisio, S. Kleinfelder, and H. Spieler.
17. Talk by S. Parker at this workshop.
18. S. Gaalema, *IEEE Trans. Nucl. Sci.* **NS-32**, 417 (1985).
19. S. Hancock *et al., Phys. Rev.* **A28**, 615 (1983).
20. One systematic effect that the experimenter should be aware of is *channeling.* In crystalline materials, such as silicon detectors, the dE/dx of a particle will be effected when the particle travels in precise directions relative to the crystal axes and planes. At the SSC, the critical angle for this effect typically will be very small, about 50 microradians at $p = 200$ GeV, and scaling $\propto 1/\sqrt{p}$.
21. A. C. Melissinos and S. L. Olsen, *Phys. Rep.* **17C**, 77 (1975) and references therein.

Magnetic Monopoles and Other Highly Ionizing Particles at the SSC

P. B. Price
University of California, Physics Department
Berkeley, CA 94720

Introduction

Detectors designed to detect magnetic monopoles in flight via their high ionization rate can also be used to search for hypothetical highly ionizing electrically charged particles (e.g., isospin balls[1]) created at SSC. The main background consists of charged, highly ionizing spallation recoils produced by interactions in the beam pipe and nearby apparatus with a roughly thermal spectrum (T ~ 15 MeV). These particles can be discriminated against on the basis of range. Responses of track-etch detectors to electric and magnetic particles are compared in [2], which also discusses a number of other experimental issues. Here we will discuss only magnetic monopoles.

General Remarks about Monopoles

Since the epochal work of 't Hooft and of Polyakov in 1974, it has been customary to divide magnetic monopoles into two classes -- the "classical" point monopoles proposed fifty years ago by Dirac and the monopoles that are demanded in non-Abelian gauge theories characterized by a simple gauge group that spontaneously breaks down into subgroups one of which is U_1. The gauge monopoles have a smooth internal structure without a singularity and are so massive that they could only have been created in the early universe. They are the center of current attention by theorists and experimentalists alike. Their theory is reviewed by Preskill [3]; current searches in nature are reviewed by Groom [4]. Predictions of their mass range from $\sim 10^4$ to $\gtrsim 10^{19}$ GeV. Drukier and Nussinov [5] point out that a gigantic form factor suppression $(\sim 10^{-50})$ associated with even the lightest of these gauge monopoles would prevent them from being produced at an accelerator or in cosmic ray collisions. We can thus restrict discussion to classical Dirac monopoles.

Some properties of Dirac monopoles:

i) quantized magnetic charge: $g\,e = n\,\hbar\,c/2$ or $g = ng_D \approx 68.5ne$, where $n = 3$ if free quarks exist, $n = 1$ if not.
ii) magnetic coupling constant: $g^2/\hbar c = n^2\alpha(g/e)^2 = 34.25\,n^2$
iii) energy gain in a field B: $dE/dx = g\,B = (0.2\,n^2\ \mathrm{GeV/cm})\,B(\mathrm{Tesla})$
iv) ionization rate:

$$-(dE/dx)_m \approx (g\beta/e)^2\,(dE/dx)_e \quad \text{if} \quad \beta >> \alpha$$
$$\approx (n^2/4)\,(dE/dx)_e = K\,\beta \quad \text{if} \quad 3 \times 10^{-4} < \beta < 0.2$$

where $K \approx 33\,n^2$ GeV/cm for plastic, $\approx 124\,n^2$ GeV/cm for Fe
vi) canonical Drell-Yan cross section for point monopole pair production: $\sigma_0 \approx (g/e)^2\,4\pi\alpha^2/3s = \pi n^2\hbar c/3s$

(which disregards the nonperturbative nature of the monopole coupling).

Survey of Searches for Classical Monopoles at Accelerators

I will ignore the many ingenious but indirect searches for classical monopoles in nature [6-8], on the grounds that some assumption about a property of the monopole may have been unjustified. [For example, all searches prior to 1979 had at least one fatal flaw that prevented them from being able to detect a supermassive monopole.] We should be cautious about limits obtained in indirect searches for monopoles at accelerators, which made assumptions about trapping or some other property. For two reasons I am uneasy about drawing conclusions from the searches for monopoles at ISR, Petra, and SPS that utilized kapton plastic detectors inside a vacuum system [9-11]. (1) No calibration data showing the response of kapton to ion beams have ever been published. (2) For every plastic studied [12], when a plastic detector is kept for several days in a vacuum system, it loses from ~20% to ~90% of its sensitivity, a decrease that may render it incapable of detecting a monopole with n = 1. To function well as a detector, a plastic film must contain oxygen absorbed in its structure. The oxygen atoms stabilize the radiation damage by reacting with free radicals created along the trajectory. In contrast, glass detectors [13] do not require oxygen and can be used inside a vacuum system.

Figure 1 shows a selection of data on cross section limits for direct and indirect detection of Dirac monopoles (n = 1), obtained since 1972 at various accelerators. The dotted curves are for kapton detectors. Dot-dash curves indicate expected future limits, assuming negative results. The Fermilab pp collider data from the run in spring, 1987 [14], have just been analyzed, giving a negative result; the dot-dash curve displaced a factor 10 downward assumes that a second run, next fall, will achieve a 10-fold higher luminosity and will obtain a negative result. The dot-dash curve for Tristan assumes that the run now in progress [15] will give an integrated luminosity of at least 10 pb^{-1} and will yield a negative result. The dot-dash curve for SSC assumes an integrated luminosity of 1000 pb^{-1} and a negative result.

Figure 2 attempts to account for the more efficient conversion of beam energy into new particles at electron-positron colliders than at hadron colliders. The dimensionless ratio R is the cross section normalized to the canonical Drell-Yan cross section for production of a monopole pair. For e+e- annihilation this is just $\sigma_0 = (g/e)^2 \, (4\pi\alpha^2/3s)$, where the coupling at the monopole vertex is taken to be $(g/e)^2$. For pp collisions we use the differential cross section data for massive virtual photon production, which exhibits scaling and falls off exponentially with $2M_m/\sqrt{s}$. To determine a total cross section for production of a monopole pair with total mass > $2M_m$, we integrate over mass and over the central one unit of rapidity, and multiply by the monopole coupling ratio $(g/e)^2$. This estimate is conservative: higher order diagrams with more than one virtual photon in the intermediate state will also contribute; and production via gluon-gluon fusion may be more likely still.

One concludes from Fig. 2 that a one-year continuous monopole search at high luminosity at the SSC should provide a test for monopoles with mass up to a few TeV.

Plastic and Glass Track-Etch Detectors

Figure 3 shows side views of a track in a plastic track-etch detector. In (a) we see the trajectory and delta rays; in (b) we see etchpits after an etching time t_1, for the case in which the particle is slowing significantly within one sheet thickness, giving rise to a larger etchpit at the bottom than at the top; in (c), after an etch time t_2 the pits have connected to form a hole that can be detected by passing ammonia gas through the sheet, producing a dark spot on a piece of chemically treated paper in contact with the sheet. The requirement of coincident holes in several successive sheets establishes the existence of a penetrating, highly ionizing particle.

Figure 4 shows the response, $s \equiv \csc \varphi = v_T/v_G$, as a function of Z/β, for CR-39, Rodyne polycarbonate (equivalent to Lexan), and UG-5 phosphate glass. CR-39 is the only track-etch detector sensitive enough to detect low-velocity monopoles with unit magnetic charge, but it has a lower tolerance to background ionization than the others. The very steep response of UG-5 glass means that it is very insensitive to background particles with Z/β below its threshold. Because tracks are formed in glass by a mechanism that does not require oxygen fixation, this type of detector can be left indefinitely inside a vacuum system. UG-5 is being used both inside Tristan and inside the Fermilab collider. Polycarbonate and presently known types of glass detectors are sensitive only to monopoles with $n \gtrsim 2$. The following table is a simplified version of data from [11].

Table 1. Response of Track Detectors in Vacuum

$$S \equiv (s-1)_{vac}/(s-1)_{air}$$

treatment	CR-39	Rodyne (=Lexan)	Cronar	UG-5 glass
2 d in vac.	0.36 ± 0.02	0.57 ± 0.04	0.89 ± 0.05	1.02 ± 0.02
25 d in vac.	0.09 ± 0.01	0.42 ± 0.02	0.78 ± 0.03	1.00 ± 0.02

Note: Response of kapton is unknown, but is probably reduced in sensitivity when in vacuum.

Table 2 compares the resistance of several track-etch detectors to radiation backgrounds. Column 2 gives the radiation dose at which the general etch rate, v_G, increases by ~50%. Column 4 gives the expected density of short etchpits after a year at 10 cm from beams with an integrated luminosity of 10^{40} cm^2. Column 5 gives the minimum detectable value of Z/β for automated detection of a monopole with n = 1 at a 30^0 angle to the sheet, and column 6 gives the minimum detectable monopole velocity. This result is obtained by using Fig. 5, which gives the equivalent Z/β as a function of β for a monopole with n = 1.

Table 2. Radiation Backgrounds and Detector Sensitivity

Detector	Max. Dose	Pits/hadron	Pits cm^{-2} y^{-1}	$(Z/\beta)_{min}$	minimum detectable β (n = 1)
CR-39	2 Mrad	~ 10^{-4}	9×10^8	~27	~0.1
Rodyne	200 Mrad	~4×10^{-6}	4×10^7	~65	~0.85
UG-5	1000 Mrad	1.6×10^{-7}	1×10^6	~80	--

Notes: 1. A penetrating event can probably be recognized as long as the background pit density ≲ 10^8/cm^2.

2. Background etchpits due to spallation recoils (T ≈15 MeV) have a very steeply falling range distribution.

3. At the same ionization dose, hadrons cause a far more serious problem than photons or leptons, because of the spallation recoils they produce.

Design Considerations for a Monopole Detector at SSC

Experience at PEP and at the Fermilab collider has shown that, during the first year of operation, the rate of production of background etchpits will probably be far higher than can be attributed to hadrons created in the colliding beams alone. The most likely source is collisions of the poorly aligned and focussed beams with the beam pipe and with matter far from the collision region, which leads to secondary sources of hadrons and especially to a "gas" of neutrons in the vicinity of the detectors. To combat this problem, one should mount the detectors in a clamshell structure such as the one being used in the highly ionizing particles search at Tristan, which can be automatically retracted except when the beams are colliding at high luminosity.

In my view the detector should consist of stacks of interleaved CR-39 (because of its high sensitivity) and Rodyne (for use when high backgrounds are expected in the early stages of operation), in a cylindrical configuration that can be opened up automatically for retraction. If a glass detector can be developed that is sensitive enough to detect monopoles with equivalent Z/β as low as 40, this glass could advantageously replace the Rodyne. It would probably be resistant enough to background radiation to be left in place for an entire year. Because monopoles with $n \geq 3$ would probably not penetrate a steel beam pipe, one should deploy strips of UG-5 glass inside the vacuum system at the collision region. This glass could remain at a radial distance of 3 cm for a year's run at maximum luminosity.

Because production rates are expected to be very small (if not zero), a monopole detector must have large solid angle and must collect as high an integrated luminosity as possible. The gluon-gluon fusion model indicates that pairs with large masses produced at the SSC will be nearly isotropically distributed, and could be efficiently detected in a cylindrical array which avoids the forward-backward direction with its attendant high radiation background. For several masses Table 3 gives modal values of p_T and β, standard deviation of rapidity, and maximum and minimum values of β obtained from the graphs in ref. 16. Using the curve for equivalent Z/β of a monopole (n=1) as a function of β shown in Fig. 5, in column 6 we give the value of Z/β corresponding to the modal value of β for a monopole with n = 1. Monopoles with these values of Z/β would easily be detectable with an ammonia technique in CR-39. Column 7 gives the range in a steel beam pipe of a monopole with n = 1 at the velocity β_{min} from column 5. Note that a monopole with n = 3 would have only 1/9 of the range in column 7, or about 0.08 cm, and would not penetrate the pipe. The majority of monopoles would have $\beta > \beta_{min}$ and would penetrate the pipe but would have a short range in a detector. With thin plastic or glass sheets this would not be a problem.

Table 3. Features of Monopole Pairs Produced in Gluon-Gluon Fusion

Mass (TeV)	$\overline{p_T}$ (TeV/c)	σ_y	$\overline{\beta}$	β_{min}, β_{max}	$\overline{(Z/\beta)}_{n=1}$	R_{min} in Fe
0.5	0.6	1.2	0.98	0.2, 1	68	0.77 cm
1	0.7	1.0	0.95	0.09, 1	67	0.7 cm
2	1	0.8	0.83	0.04, 0.99	64	0.62 cm
5	1.7	0.5	0.57	0.02, 0.95	58	0.77 cm

To demonstrate that a monopole track could have been detected despite the high etchpit background, we propose irradiating certain areas of the CR-39 detectors with relativistic iron ions ($Z/\beta \approx 30$) and irradiating the Rodyne and UG-5 with gold ions ($Z/\beta \approx 90$) at the LBL Bevalac before the run. Only if these fake monopole tracks were to lead to holes detectable with ammonia after etching could one claim that the detectors were working properly.

Multigamma-ray Bursts as a Possible Monopole-Antimonopole Signature

Because of their superstrong coupling, it is conceivable that a monopole-antimonopole pair could be created at the SSC and collapse, the only indication of its transitory existence being the emission of a burst of tens of energetic gamma rays. The cross section limits set in previous searches[17,18] have not been stringent. I hope that at least one of the all-purpose detectors being considered will be sensitive to a spectacular burst of high-energy gamma rays.

References

1. A. P. Balachandran, B. Rai, G. Sparano and A. M. Srivastava, Syracuse University preprint SU-4428-360 (1987).
2. K. Kinoshita, Ph.D. Thesis, University of California Berkeley,1981.
3. J. Preskill, Ann. Rev. Nucl. Part. Sci. 34, 461 (1984).
4. D.E. Groom, Phys. Reports 140, 323 (1986).
5. A.K. Drukier and S. Nussinov, Phys. Rev. Lett. 49, 102 (1982).
6. See review by R.R. Ross in New Pathways in High-Energy Physics, ed. A. Perlmutter (Plenum Press, New York, 1976), p. 151, and refs. 7 and 8 below.
7. P. B. Price, in Magnetic Monopoles, ed. R.A. Carrigan and W.P. Trower (Plenum Press, New York, 1983), p. 307.
8. G. Giacomelli, Riv. Nuovo Cim. 7, No. 12, 1 (1984).
9. P. Musset, M. Price and E. Lohrmann, Phys. Lett. 128B, 333 (1983).
10. B. Aubert, P. Musset, M. Price and J.P. Vialle, Phys. Lett. 120B, 465 (1983).
11. B. Aubert, P. Musset, M. Price and J.P. Vialle, unpublished results at ISR collider.
12. J. Drach, M. Solarz, Ren Guoxiao and P.B. Price, accepted for publication in Nucl. Instr. Meth. (1987).
13. P.B. Price, L.M. Cook and A. Marker, Nature 325, 137 (1987).
14. P.B. Price, Ren Guoxiao and K. Kinoshita, to be published.
15. By a collaboration consisting of K. Kinoshita (spokesperson), M. Fujii, K. Nakajima, T. Kobayashi, P.B. Price, and S. Tasaka.

16. S. Errede and S.-H. H. Tye, in Proceedings of Snowmass Summer Study on the Design and Utilization of the Superconducting Super Collider (1984), p. 175.

17. D. L. Burke, H. R. Gustafson, L. W. Jones and M. J. Longo, Phys. Lett. 60B, 113 (1975).

18. G. F. Dell et al., Nucl. Phys. B209, 45 (1982).

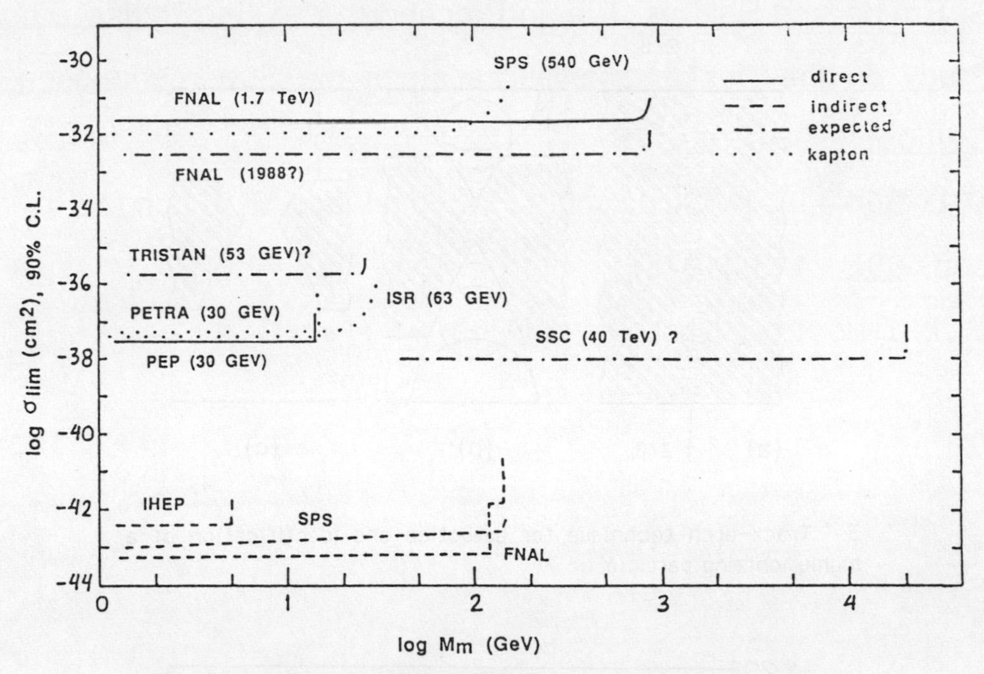

1. Cross section limits (90% C.L.) for production of magnetic monopoles with g = 68.5 e obtained since 1972. For the indirect searches, assumptions had to be made about capture and trapping of monopoles.

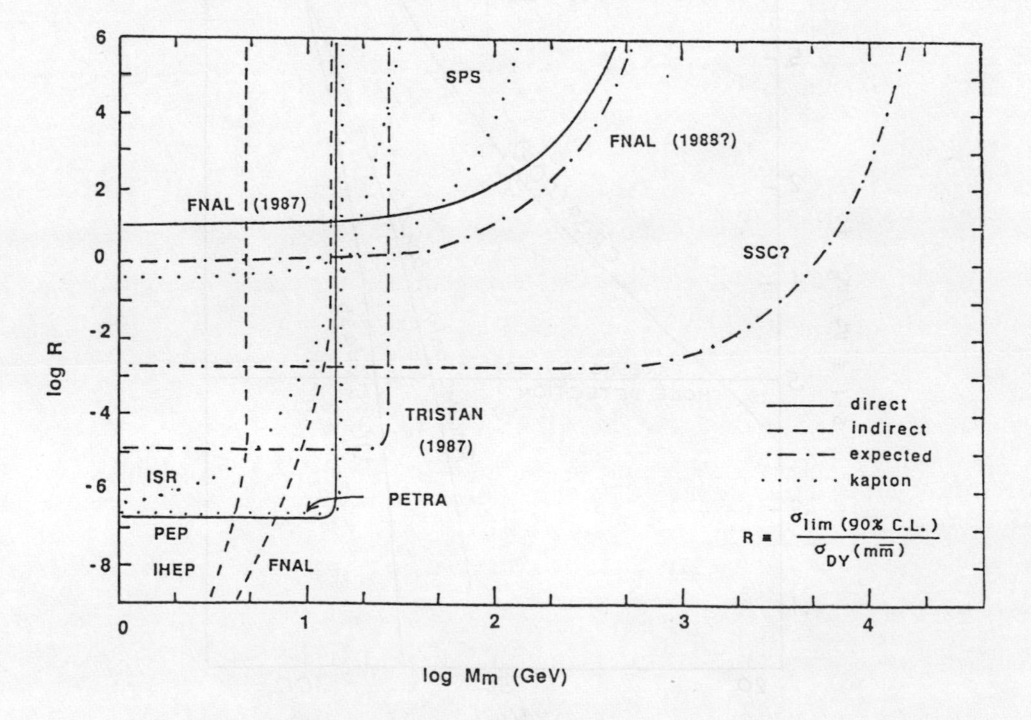

2. Data from Fig. 1, normalized to cross section for monopole production by the Drell-Yan mechanism.

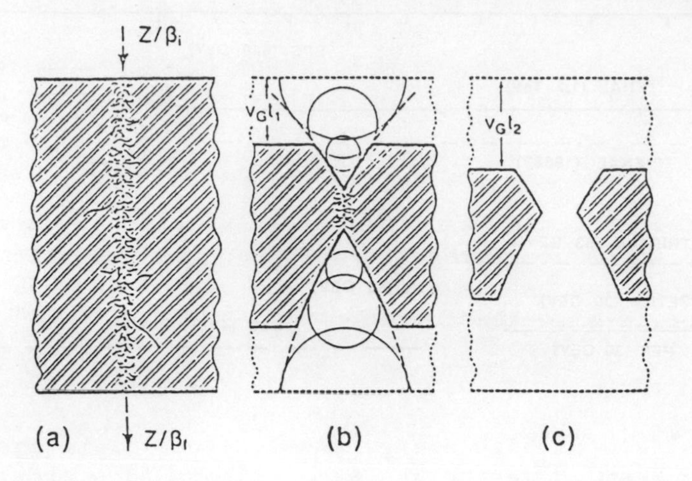

3. Track-etch technique for detection and identification of a highly ionizing particle.

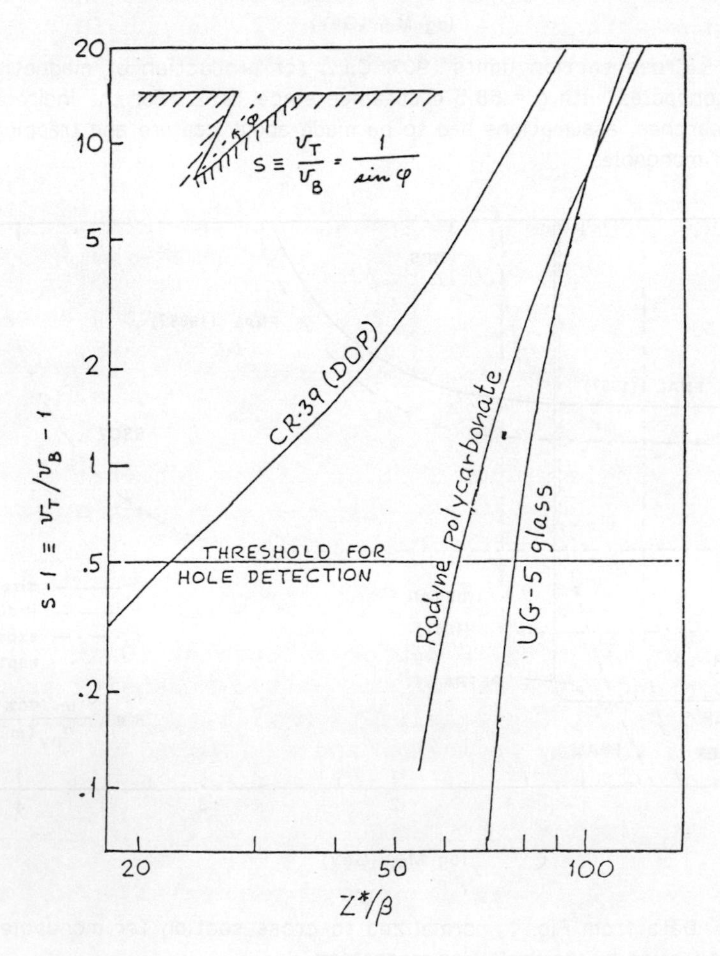

4. Response of several track-recording solids as a function of the ratio Z/β.

GRAVITATION AND LONG-RANGE FORCES AT THE SSC

A. C. Melissinos
Dept. of Physics, University of Rochester, Rochester, NY 14627

ABSTRACT

We discuss the expected magnitude of gravitational effects at the SSC and the level at which long-range forces that manifest themselves at high relative velocities can be probed. The use of the SSC as a gravitational wave detector is also considered.

1. Introduction

The principal aim of high energy accelerators is to probe phenomena at very short distances. Nevertheless accelerators also provide an opportunity to examine the behavior of long range forces at relativistic velocities and/or to search for unknown long or medium range forces that manifest themselves at high relative velocities.

Of particular interest is the gravitational force whose energy dependence is a measure of the spin of the graviton. [1] We consider a stationary detector at a distance b from the circulating beam as shown in Fig. 1.

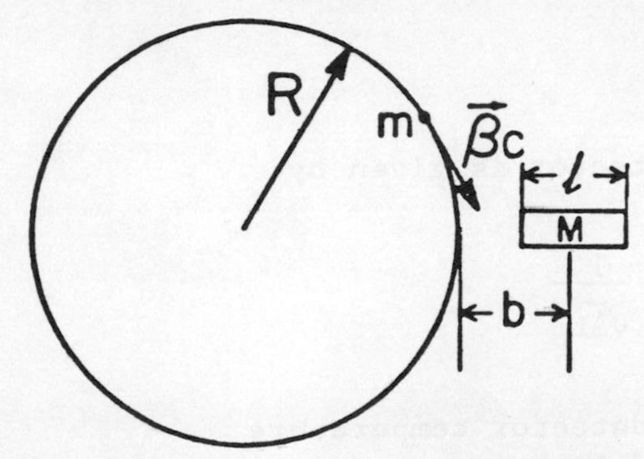

Figure 1
Arrangement for detecting a long range force due to the circulating beam.

The traversal of a single proton imparts to the detector an attractive impulse which corresponds to an average acceleration

$$\langle a \rangle = \frac{F}{M} = \frac{4G}{bL} m\gamma \tag{1}$$

with L the length of the ring, G Newton's constant ($G/c^2 \sim 0.75$ cm/g), m the mass of the proton and $\gamma = E/m$. If the particle density around the ring is distributed sinusoidally, the acceleration will be harmonic at the fundamental revolution frequency $\omega_0 = 2\pi\nu_0$ and

$$a(\omega_0) = \frac{1}{2}\frac{4G}{b}\frac{N}{L} m\gamma \tag{1}$$

where N/L is the average linear density of particles around the ring.

At the SSC the circulating current is expected to be I = 70 mA and therefore

$$\frac{N}{L} = \frac{I}{ec} = 1.4 \times 10^9 /\text{m}$$

Thus if we choose b = 10 cm, we obtain

$$a_G(\omega_0) = 0.6 \times 10^{-20} \ \text{cm/sec}^2$$

$$\omega_0 = 2\pi\nu_0 = 2.3 \times 10^4 \ \text{rad/sec} \tag{2}$$

2. Detectors

Various types of resonant detectors can be used. At the SSC the frequency is low, $\nu_0 = 3.7$ kHz, so that mechanical detectors consisting of a sapphire bar are suitable. In this case the gradient of the force excites the longitudinal mode of the bar. Parametric detectors based on superconducting cavities were used in a similar experiment at Fermilab [2] but at higher frequencies. A free suspension with optical or capacitive readout is another possibility. An example of such a detector is shown in Fig. 2. In general one can expect a mechanical Q-factor $Q_0 = 10^6$.

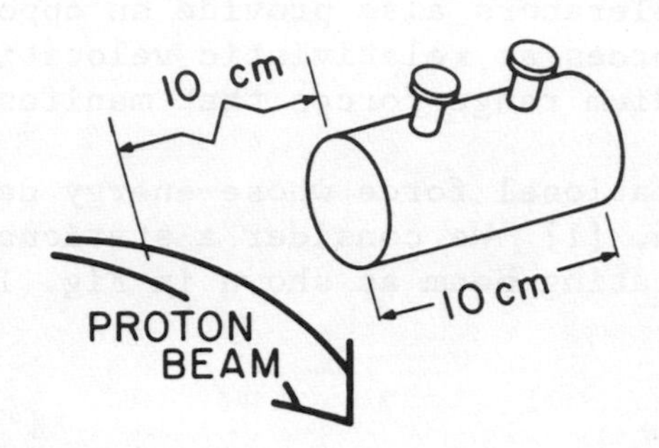

Figure 2
Outline of the parametric converter detector and its placement relative to the beam.

The thermal noise limit for any detector is given by

$$a_B(\omega_0) = \sqrt{\frac{4kT\omega_0}{MQ_0}} \ \frac{1}{\sqrt{\Delta t}} \tag{3}$$

where we choose

$T = 20$ mK	for detector temperature
$M = 1$ Kg	for detector mass
$\Delta t = 10^5$ sec	for the integration time

and $\omega_0 = 2.3 \times 10^4$, $Q_0 = 10^6$, $k = 1.4 \times 10^{-23}$ J/°K. Thus

$$a_B(\omega_0) = 5 \times 10^{-14} \ \text{cm/sec}^2 \tag{4}$$

The ratio

$$a_B/a_G \simeq 10^7 \tag{5}$$

sets the limit on the coupling strength of a force that can be measured as compared to the gravitational coupling. This is shown schematically in Fig. 3 which also indicates the limits from the measurement at Fermilab [2]. In terms of actual detector displacement, the thermal noise corresponds to

$$\delta x_B = a_B \frac{Q_0}{\omega_0^2} = 16^{-16} \ \text{cm} \tag{6}$$

which, for harmonic motion, can be easily measured. The relaxation time of the detector $\tau^* = Q_0/\omega_0 = 40$ sec which is adequately short. The naive quantum limit is given by

$$\delta x = \sqrt{\frac{\hbar Q_0}{2M\omega_0}} \; \frac{1}{\sqrt{\Delta t}} \simeq 10^{-19} \text{ cm} \tag{7}$$

for the parameters used to obtain the result of eq. (4). This is three orders of magnitude below eq. (6) but four orders of magnitude above the displacement generated by the acceleration given in eq. (2).

We conclude that one can search for a long range force with coupling $G_x \sim 10^7$ G, provided the e.m. force can be shielded to that level.

3. Electromagnetic Background

The principal difficulty of this experiment is due to the e.m. effects arising from the passage of the charged beam. These can induce direct signals - by radiative effects - on the readout electronics as well as exert a force because of the appearance of image charges on the detector. These effects can be reduced by shielding with several layers of conducting (and superconducting) material.

At the Fermilab experiment a displacement $\delta x = 10^{-16}$ cm was observed at $\nu = 1.2$ MHz, b = 23 cm, $Q_m = 10^3$ and N/L(effective) = 10^8 m^{-1}. This corresponds to an acceleration $a_{EM} \sim 6\text{X}10^{-6}$ cm/sec^2, and to a ratio between the observed effect and that of the electromagnetic force due to image charges of $F_s/F_{EM} \sim 10^{-14}$. Since image forces are proportional to the square of the electric field, we conclude that the attenuation was of order $|\vec{E}(\text{shielded})|/|\vec{E}(\text{free})| \sim 10^{-7}$. To reach the limit of eq. (4) at the SSC one needs a further suppression of the em force by a factor of 10^{10}. This corresponds to an overall attenuation of the electric field by a factor of 10^{12}. [The observed average field at the detector must be of order

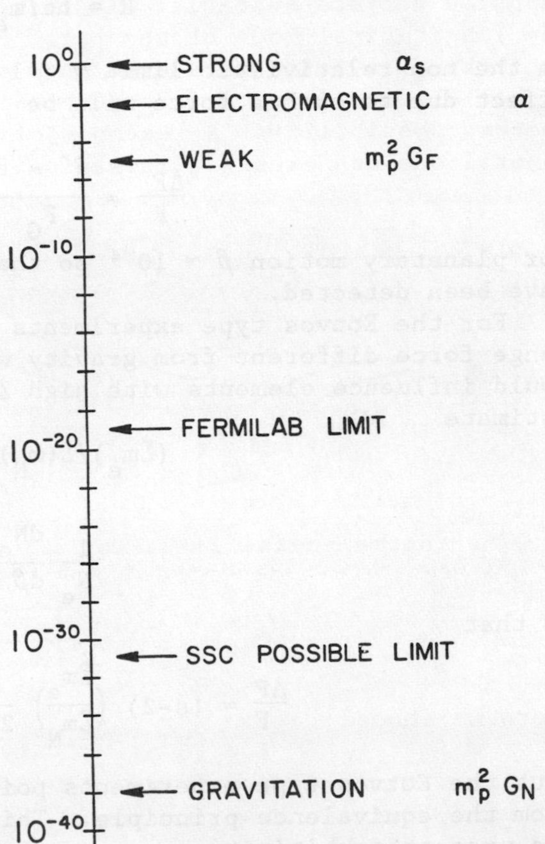

Figure 3
Coupling strength of the fundamental forces and the limits on other long range forces from accelerator experiments.

$$|\vec{E}_{\text{shielded}}| \sim 2\text{X}10^{-11} \text{ V/m.}] \tag{8}$$

To differentiate between the em background and a new long range force one can take advantage of their different energy dependence. The em force is mediated by a vector particle and thus the average impulse is energy independent. The average impulse for a tensor force would grow as, $\gamma = E/m$ i.e. as the energy of the beam. In general if s is the spin of the exchanged quantum the average impulse has an energy dependence

$$a(\gamma) \propto \gamma^{s-1} \quad .$$

4. Existing Evidence on γ-dependent Forces

Recent theoretical models predict the existence of pseudogoldstones which are particles of light mass. As an example the DFS axion [3] is predicted to have a mass $m_a \sim 10^{-5}$ eV. Such a particle would give rise to a force with coupling near G and of range

$$R = \hbar c/m_a = 20 \text{ cm} .$$

In the non-relativistic limit $\gamma \sim 1 + \beta^2/2$ with $\beta = v/c$ and the fractional effect due to such a force will be

$$\frac{\Delta F}{F} = \frac{F' - F_G}{F_G} \simeq (s-2) \frac{\beta^2}{2} \tag{9}$$

For planetary motion $\beta \sim 10^{-4}$ so that $\Delta F/F \sim 10^{-8}$ and such effects would not have been detected.

For the Eötvos type experiments [4] we can take $\Delta F/F \sim 10^{-10}$. If a long range force different from gravity were present, the relativistic electrons would influence elements with high Z differently from those with low Z. We estimate

$$(\Sigma m_e)/\Sigma(m_N) \sim 2.5\times10^{-4}$$

$$\frac{1}{N_e} \frac{dN_e}{d\beta} \bigg|_{\beta > 0.1} \sim 10^{-3}$$

So that

$$\frac{\Delta F}{F} = (s-2) \left(\frac{\Sigma m_e}{\Sigma m_N}\right) \frac{1}{2N_e} \int \frac{dN_e}{d\beta} \beta^2 \, d\beta \sim 10^{-9}$$

Thus the Eötvos type experiments point to a γ-dependence with s=2 as expected from the equivalence principle. This conclusion is however at the limit of the present sensitivity.

5. Gravitational Radiation

The previous considerations dealt with the longitudinal part of the gravitational field due to the circulating beam. We can also calculate the transverse part, that is the gravitational radiation emitted by the beam. This turns out to be too weak to be of practical interest. We do the calculation in two ways:

(a) Evaluate the gravitational potential of the moving protons [1]

$$h(t) = \frac{4(G/c^2)\, m\gamma^2\beta^2}{(b^2 + \gamma^2\beta^2 c^2 t^2)^{1/2}}$$

So that

$$\langle h \rangle_b = \frac{Nc}{L} \int_{-L/2}^{+L/2} h(t)dt \simeq 4(G/c^2)\left(\frac{N}{L} m\gamma\right)\ln\left(\frac{L}{b}\right) \tag{10}$$

and for b = 10 cm

$$\langle h \rangle_b = 1.8 \times 10^{-39} \quad .$$

This result varies weakly with b, and in fact an exact calculation shows that for $b = R = L/2\pi$ the logarithmic term should be replaced by 2π. This led to a suggestion to search for radiation at the center of the ring where electromagnetic radiation would be strongly shielded.

(b) We use the non-relativistic quadrupole radiation formula [5] with $r=R=L/2\pi$.

$$\langle h \rangle_R \simeq 8\pi \left(\frac{G}{c^2}\right) \frac{N}{L} m\gamma \left(\frac{R\omega}{c}\right)^2 \tag{10'}$$

we can set $R\omega/c = \beta \to 1$ in which case eq. (10´) reproduces eq. (10) evaluated at b=R. According to the quadrupole formula, the total radiated energy is

$$\frac{dE}{dt} = \frac{32GI^2\omega^6}{5c^5} = \frac{16\pi}{5}\left(\frac{G}{c^2}\right)\left(m\gamma\frac{N}{L}\right)\omega(Nm\gamma c^2)$$

$$= \frac{2}{5}\langle h \rangle_R \, \omega E_0 \tag{11}$$

where in the second step we used $\omega R = c$, $L = 2\pi/r$. E_0 is the total energy in the beam, which for the SSC (E = 20 TeV, $N = 10^{14}$), is $E_0 = 3\times10^8$ J. Thus

$$\frac{dE}{dt} = \frac{2}{5}\left(2.3\times10^4\right)\left(9\times10^{-40}\right)\left(3\times10^8\right) = 2.5\times10^{-27}\ \text{W} \tag{11'}$$

6. The SSC as a Detector of Gravitational Waves

Consider a circular accelerator with a freely suspended quadrupole placed every 90° as shown in Fig. 4. Let a gravitational wave of amplitude h propagate along the normal to the accelerator's plane and assume that the wave is polarized along the axes defined by the free quads. The quadrupoles will be displaced from their equilibrium position by $\delta\ell$, where

$$\frac{\delta\ell}{\ell/2} = h \tag{12}$$

and ℓ is the distance between the quads, in this case, the ring diameter $2R=\ell$.

We work in the frame of reference where the lattice is fixed and so is the circulating beam. If the quadrupole gradient is G, then the change in magnetic field along the orbit is

$$\delta B = G\delta\ell = GRh \tag{13}$$

The quadrupole length is L and there are 4 quads so that the magnetic kick results in an angular displacement

$$\theta = \frac{0.3(4\delta B \bullet L)}{p} \tag{14}$$

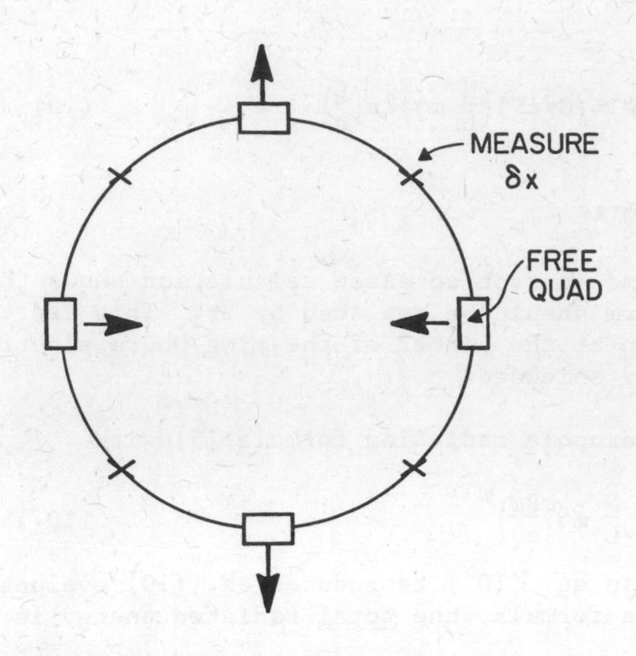

Figure 4
Freely suspended quadrupoles in a storage ring lattice responding to a gravitational wave at normal incidence.

and therefore to a position kick

$$K = \beta\theta \tag{15}$$

where β is the value of the beta function at the quads.

The quadrupole suspension resonates at a frequency ν_Q, typically a few kHz and has a good mechanical Q, $Q \simeq 10^5$. The frequency ν_Q is chosen so as to be in resonance with the machine tune, $\nu_Q = m\nu_0 + n_x$ where m is an integer. In response to a gravitational burst ($\Delta t \sim 10^{-3}$ msec) the quadrupoles will oscillate coherently for atime $\tau_* = Q/2\pi\nu_Q$ which corresponds to N turns of the beam, where

$$N \simeq Q/2\pi m \sim 10^3 - 10^4 \tag{16}$$

Thus the expected displacement of the beam is of order

$$\delta x = KN \tag{17}$$

Below we evaluate the displacement δx for Fermilab and SSC for $h = 10^{-17}$.

	FERMILAB	SSC
h	10^{-17}	10^{-17}
$\delta\ell$	10^{-14} m	1.2×10^{-13} m
G	50 T/m	250 T/m
p	800GeV	20 TeV
L	4 m	10 m
θ	0.3×10^{-14}	1.8×10^{-14}
β	50 m	500 m
N	10^4	10^4
δx	1.5×10^{-9} m	0.9×10^{-7} m

To compare these results with reality we note that R. Siemann [6] has recently measured at Fermilab displacements of the beam centroid of order

$$\langle \delta x \rangle = 10^{-7} \text{ m} \quad .$$

At the SSC the beam emittance is improved over that at Fermilab by a factor $\sqrt{20/0.8} = 5$. Thus one should be able to localize the beam to

$$\delta x \sim 2 \times 10^{-8} \text{ m}$$

making possible the detection of gravitational bursts at the level $h \sim 10^{-17}$. The experiment would run parasitically by using a continuous record of the beam position at a few points around the accelerator.

If one is searching for a CW wave, the suspension is designed to be antiseismic and isolated from the rest of the lattice so that the quadrupoles approach a truly free body. The machine tune will have to be chosen so as to preserve the phase relation between the displacement of the quadrupoles and the betatron amplitude

$$\nu_x \nu_0 = n\nu_f$$

where ν_x is the betatron number, ν_0 is the circulation frequency, ν_f is the frequency of CW gravitational wave and in general $\nu_f \ll \nu_0$; n is an arbitrary integer. As a result of the resonance condition, the betatron amplitude should grow linearly with time and give rise to observable sidebands. This perturbation can be modeled and also measured experimentally by driving one of the suspended quadrupoles. In principle it can reach very large amplitude.

References

1. A. C. Melissinos, Nuovo Cimento 62B, 190 (1981).
2. P. Reiner et al., Phys. Letters B176, 233 (1986).
3. See in particular J. E. Moody and F. Wilczek, Phys. Rev. D30, 130 (1984).
4. See for instance G. Feinberg and J. Sucher, Phys. Rev. D20, 1717 (1979).
5. See for instance J. Foster and J. D. Nightingale, "A short course on general relativity", Longman, London 1979.
6. I thank R. Johnson and R. Siemann for informing me about these results. See Fermilab note "Summary of Measurements of Betatron Line Amplitudes, RF Phase Noise and Emittance Growth" by R. Siemann, May 1987.

Workshop Participants and Author Index

WORKSHOP PARTICIPANTS

Shahriar Abachi
Purdue University

Don Abraham
South Dakota University

Torsten Akesson
CERN

Makoto Asai
Hiroshima Institute of Technology

Muzaffer Atac
Fermi National Accelerator Laboratory

David M. Atwood
McGill University

Robert E. Avery
Lawrence Berkeley Laboratory

Howard Arthur Baer
Argonne National Laboratory

Virgil Barnes
Purdue University

Bruce A. Barnett
Johns Hopkins University

R. Michael Barnett
Lawrence Berkeley Laboratory

James R. Bensinger
Brandeis University

Ikaros Bigi
Stanford Linear Accelerator Center

Harry H. Bingham
Univ. of California, Berkeley

David Bintinger
SSC Central Design Group

James D. Bjorken
Fermi National Accelerator Laboratory

Robert E. Blair
Argonne National Laboratory

Martin M. Block
Northwestern University

Elliott Bloom
Stanford Linear Accelerator Center

Barry J. Blumenfeld
Johns Hopkins University

James E. Brau
University of Tennessee

Alan Breakstone
Stanford Linear Accelerator Center

Charles Buchanan
Univ. of California, Los Angeles

David A. Buchholz
Northwestern University

Robert N. Cahn
Lawrence Berkeley Laboratory

Duncan Carlsmith
University of Wisconsin

Roger Cashmore
Fermi National Accelerator Laboratory

Chung Yun Chang
University of Maryland

Michael Chanowitz
Lawrence Berkeley Laboratory

Min Chen
Massachusetts Inst. of Technology

Masami Chiba
Tokyo Metropolitan University

Chih-Yung Chien
Johns Hopkins University

Sekhar Chivukula
Boston University

John C. Collins
Illinois Institute of Technology

Sergio Conetti
McGill University

Roger Coombs
SSC Central Design Group

Bradley B. Cox
Fermi National Accelerator Laboratory

Per Dahl
SSC Central Design Group

Robert Darling
Rutgers University

Sally Dawson
Brookhaven National Laboratory

Pierre Delpierre
College de France

Riccardo DeSalvo
Cornell University

Nilendra G. Deshpande
University of Oregon

Thomas J. Devlin
Rutgers University

Daryl DiBitonto
Texas A&M University

Robert E. Diebold
US Department of Energy

Marcella Diemoz
INFN, Rome

Jonathan M. Dorfan
Stanford Linear Accelerator Center

Manuel Drees
University of Wisconsin

Gilbert Drouet
SSC Central Design Group

Edmond C. Dukes
CERN

James J. Eastman
Lawrence Berkeley Laboratory

Kenneth W. Edwards
Carleton Univ.

Stephen D. Ellis
University of Washington

Steve Errede
University of Illinois, Urbana

Penny G. Estabrooks
Carleton Univ.

Chris Fabjan
CERN

Gary J. Feldman
Stanford Linear Accelerator Center

Kenneth J. Foley
Brookhaven National Laboratory

Melissa Franklin
University of Illinois, Urbana

Paolo Franzini
Columbia University

James E. Freeman
Fermi National Accelerator Laboratory

Klaus K. Freudenreich
CERN

Raymond E. Frey
Stanford Linear Accelerator Center

Alfred Fridman
CERN

Masaki Fukushima
CERN

Angela Galtieri
Lawrence Berkeley Laboratory

Al Garren
SSC Central Design Group

Gil Gilchriese
SSC Central Design Group

Frederick J. Gilman
Stanford Linear Accelerator Center

Stephen Godfrey
Brookhaven National Laboratory

Mitchell E. Golden
Lawrence Berkeley Laboratory

Ned Goldwasser
SSC Central Design Group

Konstantin Goulianos
Rockefeller University

Giorgio Gratta
Univ. of California, Santa Cruz

Josep Antoni Grifols
Univ. Autonoma de Barcelona

Donald E. Groom
SSC Central Design Group

Paul Grosse-Wiesmann
Stanford Linear Accelerator Center

Jack Gunion
Univ. of California, Davis

Richard Gustafson
University of Michigan

Laszlo J. Gutay
Purdue University

Howard E. Haber
Univ. of California, Santa Cruz

John S. Haggerty
Brookhaven National Laboratory

Kaoru Hagiwara
National Lab. for HEP, Tsukuba

Gail G. Hanson
Stanford Linear Accelerator Center

John M. Hauptman
Iowa State University

David R. Hedin
Northern Illinois Univ.

Steven F. Heppelmann
University of Minnesota

L. Gregor Herten
Massachusetts Inst. Tech.

Clemens Heusch
Univ. of California, Santa Cruz

Ian Hinchliffe
Lawrence Berkeley Laboratory

Christopher L. Hodges
San Francisco State Univ.

Catherine Newman Holmes
Fermi National Accelerator Laboratory

Joey W. Huston
Michigan State University

J. D. Jackson
SSC Central Design Group

Abolhassan Jawahery
Syracuse University

David E. Johnson
SSC Central Design Group

Larry Jones
University of Michigan

Kalpana J. Kallianpur
University of Wisconsin

Gordon L. Kane
University of Michigan

Paul Karchin
Yale University

Henry Kasha
Yale University

Russel Kauffman
Stanford Linear Accelerator Center

V. Paul Kenney
US Department of Energy

Werner F. Kienzle
CERN

Kay Kinoshita
Harvard University

Thomas BW Kirk
Fermi National Accelerator Laboratory

Jasper Kirkby
CERN

Robert Klanner
DESY

Christopher B. Klopfenstein
Lawrence Berkeley Laboratory

Winston Ko
Univ. of California, Davis

David S. Koltick
Purdue University

Sachio Komamiya
Stanford Linear Accelerator Center

Kunitaka Kondo
University of Tsukuba

Takahiko Kondo
National Lab. for HEP, Tsukuba

Richard Van Kooten
Stanford Linear Accelerator Center

Zoltan Kunszt
Fermi National Accelerator Laboratory

Richard L. Lander
Univ. of California, Davis

Peter V. Landshoff
University of Cambridge

Paul G. Langacker
University of Pennsylvania

Andrew J. Lankford
Stanford Linear Accelerator Center

Pierre LeComte
CERN

Ian D. Leedom
Northeastern University

Rudolf Leiste
Akad. der Wissenschaften der DDR

Branko Leskovar
Lawrence Berkeley Laboratory

Alan Litke
Univ. of California, Santa Cruz

Nigel S. Lockyer
University of Pennsylvania

Stewart C. Loken
Lawrence Berkeley Laboratory

Klaus Luebelsmeyer
CERN

Vera G. Luth
Stanford Linear Accelerator Center

Ronald J. Madaras
Lawrence Berkeley Laboratory

Daniel R. Marlow
Princeton University

Jay N. Marx
Lawrence Berkeley Laboratory

Michael D. Marx
State Univ. of New York

Satoshi Matsuda
Kyoto University

Takeshi Matsuda
National Lab. for HEP, Tsukuba

John A. J. Matthews
Johns Hopkins University

Sean F. McHugh
Univ. of California, Santa Barbara

Peter M. McIntyre
Texas A&M University

Adrian C. Melissinos
University of Rochester

Antoni Mendez
Univ. Autonoma de Barcelona

David H. Miller
Purdue University

Geoffrey B. Mills
CERN

Shige Miyashita
University of Tsukuba

Stefano F. Moretti
Ansaldo

Jorge G. Morfin
Fermi National Accelerator Laboratory

Shikegi Mori
University of Tsukuba

Rollin J. Morrison
Univ. of California, Santa Barbara

Uriel Nauenberg
University of Colorado

Harvey B. Newman
California Institute of Technology

David R. Nygren
Lawrence Berkeley Laboratory

Piermaria Oddone
Lawrence Berkeley Laboratory

Takashi Ohsugi
Hiroshima University

Fredrick I. Olness
Illinois Institute of Technology

Stephen L. Olsen
University of Rochester

Jay Orear
Cornell University

Satoshi Ozaki
National Lab. for HEP, Tsukuba

Frank Paige
Brookhaven National Laboratory

Robert S. Panvini
Vanderbilt University

Sherwood Parker
Lawrence Berkeley Laboratory

Hans P. Parr
NIKHEF, Amsterdam

Richard A. Partridge
Brown University

Felicitas Pauss
CERN

Steve Peggs
SSC Central Design Group

David E. Pellett
Univ. of California, Davis

Prisca C. Petersen
Rockefeller University

Jack Peterson
SSC Central Design Group

Mina G. Petradza
Stanford Linear Accelerator Center

Pierre A. Piroue
Princeton University

Lee G. Pondrom
University of Wisconsin

Frank C. Porter
California Institute of Technology

Buford Price
Lawrence Berkeley Laboratory

Lawrence E. Price
Argonne National Laboratory

Serban D. Protopopescu
Brookhaven National Laboratory

Milind V. Purohit
Fermi National Accelerator Laboratory

Chris Quigg
SSC Central Design Group

Rajendran Raja
Fermi National Accelerator Laboratory

Pie-Giorgio Rancoita
INFN Milano

Lisa Randall
Univ. of California, Berkeley

Bill Reay
Ohio State University

Don D. Reeder
University of Wisconsin

Terry Reeves
Vanderbilt University

Jack L. Ritchie
Stanford Linear Accelerator Center

James Rohlf
Harvard University

Michael T. Ronan
Lawrence Berkeley Laboratory

Peter C. Rowson
Columbia University

Roger W. Rusack
Rockefeller University

James S. Russ
Carnegie-Mellon University

John P. Rutherfoord
University of Washington

Hartmut Sadrozinski
Univ. of California, Santa Cruz

Richard Sah
SSC Central Design Group

Anthony I. Sanda
Rockefeller University

Jim Sanford
SSC Central Design Group

Felix R. Sannes
Rutgers University

Ludmila Sarycheva
Moscow State University

Aurore Savoy-Navarro
CEN de Saclay

Walter Scandale
CERN

Michael P. Schmidt
Yale University

Klaus R. Schubert
Universitaet Heidelberg

Frank Sciulli
Columbia University

Abraham Seiden
Univ. of California, Santa Cruz

Paul G. Seiler
Eidgenossische Tech. Hochsch., Zurich

Peter Seyboth
Max Planck Inst. fur Phys.

Mike Shaevitz
Columbia University

Steve Shapiro
Stanford Linear Accelerator Center

Paul D. Sheldon
University of Illinois

A. Jean Slaughter
Yale University

John R. Smith
Univ. of California, Davis

Mark Soldate
Fermi National Accelerator Laboratory

Amarjit S. Soni
University of California, Los Angeles

Davison E. Soper
University of Oregon

Anthony L. Spadafora
Lawrence Berkeley Laboratory

Piero Spillantini
INFN, Florence

Raymond J. Stefanski
Fermi National Accelerator Laboratory

M. Lynn Stevenson
Lawrence Berkeley Laboratory

James L. Stone
Boston University

Donald H. Stork
University of California, Los Angeles

Mark W. Strovink
Lawrence Berkeley Laboratory

Ryszard A. Stroynowski
California Institute of Technology

Lawrence R. Sulak
Boston University

Yasuza Takahiro
Fermi National Accelerator Laboratory

Norio Tamura
Kyoto University

Xerxes R. Tata
University of Wisconsin

Toshiaki Tauchi
National Lab. for HEP, Tsukuba

Kyuzo Teshima
University of Oregon

Jon J. Thaler
University of Illinois

George E. Theodosiou
N.R.C. Demokritos

Dennis Theriot
Fermi National Accelerator Laboratory

Rudolf P. Thun
University of Michigan

Maury Tigner
SSC Central Design Group

Samuel C. C. Ting
CERN

John Tompkins
SSC Central Design Group

Tim Toohig
SSC Central Design Group

George H. Trilling
Lawrence Berkeley Laboratory

Wu-ki Tung
Illinois Institute of Technology

Mike Tuts
Columbia University

George S. Tzanakos
Columbia University

Jurgen Herbert Ulbricht
CERN

Giovanni Valenti
CERN

Richard VanKooten
Stanford Linear Accelerator Center

Wayne Vernon
Univ. of California, San Diego

David E. Wagoner
Fermi National Accelerator Laboratory

W. Wallraff
RWTH, Aachen

Edward Wang
Lawrence Berkeley Laboratory

Alan J. Weinstein
Univ. of California, Santa Cruz

William A. Wenzel
Lawrence Berkeley Laboratory

J. Scott Whitaker
Boston University

Andrew P. White
University of Florida

Sebastian N. White
Rockefeller University

Richard Wigmans
NIKHEF, Amsterdam

Hugh H. Williams
University of Pennsylvania

Robert J. Wilson
Boston University

David R. Winn
Schlumberger-Doll Research

Michael S. Witherell
Univ. of California, Santa Barbara

Stan Wojcicki
SSC Central Design Group

Hiroaki Yamamoto
Lawrence Berkeley Laboratory

Richard K. Yamamoto
Massachusetts Inst. of Technology

N. Yamdagni
University of Stockholm

Takahiro Yasuda
Northeastern University

Chien-Peng Yuan
Univ. of Michigan

Ren-Yuan Zhu
California Institute of Technology

George Zobernig
CERN

Fabio Zwirner
Lawrence Berkeley Laboratory

AUTHOR INDEX